MELATONIN
Therapeutic Value and Neuroprotection

Venkatramanujan Srinivasan
Sri Sathya Sai Medical Educational and Research Foundation
An International Medical Sciences Research Study Center
Tamilnadu, India
NHS Department of Mental Health, Hospital "G. Mazzini", ASL Teramo Italy
Department of Neurosciences; University "G.D Annunzio, Chiety, Italy

Gabriella Gobbi
Department of Psychiatry
McGill University, and McGill University Health Center
Montreal, Quebec, Canada

Samuel D. Shillcutt
Department of Psychiatry and Behavioural Sciences
Mercer University School of Medicine
Macon, Georgia, USA

Sibel Suzen
Departmental of Pharmaceutical Chemistry
Faculty of Pharmacy, Ankara University
Ankara, Turkey

CRC Press
Taylor & Francis Group
Boca Raton London New York

CRC Press is an imprint of the
Taylor & Francis Group, an **informa** business

CRC Press
Taylor & Francis Group
6000 Broken Sound Parkway NW, Suite 300
Boca Raton, FL 33487-2742

First issued in paperback 2019

ISBN-13: 978-1-4822-2009-4 (hbk)
ISBN-13: 978-0-367-37828-8 (pbk)

Library of Congress Cataloging-in-Publication Data

Melatonin (Srinivasan)
Melatonin : therapeutic value and neuroprotection / editors, Venkatramanujan Srinivasan, Gabriella Gobbi, Samuel D. Shillcutt, Sibel Suzen.
p. ; cm.
Includes bibliographical references and index.
Summary: "There is a growing interest in the field of melatonin research concerning its neurobiological mechanisms as well as its repercussions in clinical practice. A decrease in melatonin production and receptors results in numerous dysfunctions in the body. This text explores melatonin's neuroprotective effects and discusses the therapeutic potential of melatonin and melatonin agonists in treating neurodegenerative diseases. In addition to addressing the neuroprotective role of melatonin, the book also explores its protective roles in other tissues such as the brain, heart, and liver"--Provided by publisher.
ISBN 978-1-4822-2009-4 (hardcover: alk. paper)
I. Srinivasan, V. (Venkataramanujan), editor. II. Gobbi, Gabriella, editor. III. Shillcutt, Samuel D., editor. IV. Suzen, Sibel, editor. V. Title.
[DNLM: 1. Melatonin--therapeutic use. 2. Neuroprotective Agents--pharmacology. 3. Neurodegenerative Diseases--drug therapy. WK 350]

RM297.M44
615.3'6--dc23 2014022994

Visit the Taylor & Francis Web site at
http://www.taylorandfrancis.com

and the CRC Press Web site at
http://www.crcpress.com

This book is dedicated to the memory of Professor Venkatramanujam Srinivasan, MSc, PhD, MAMS, whose untimely death has saddened his colleagues and the scientific community. Professor Srinivasan was an eminent neuroscientist, psychopharmacologist, and professor of physiology at multiple prestigious academic institutions. He founded and chaired the Sri Sathya Sai Medical Education and Research Foundation, where he donated his time and effort in the pursuit of knowledge. He devoted his life researching the physiology of sleep, chronobiology, psychoimmunology, and endocrinology. He was an internationally recognized teacher of physiology and biological psychiatry, a noted investigator of melatonin's physiological function, and a leader in the development of melatonin drugs for the treatment of medical conditions. A prolific writer, he published a plethora of his research findings in highly respected journals and books. Professor Srinivasan spent most of his career in the effort to deeply unveil the secrets of melatonin and its functions, with the desire to bring more knowledge and novel treatments to the world. He provided guidance, patience, and stimulus in the completion of this, his final work, allowing an encyclopedic review of melatonin's therapeutic uses and neuroprotective qualities. He passed just before the publication of this book, which celebrates his outstanding contribution to science and humanity.

Contents

Preface

Melatonin was first isolated from the bovine pineal gland in 1958 by the dermatologist Aaron B. Lerner. Since then, thousands of papers have been published on melatonin function, but its action remains to be clarified. However, the nature of melatonin and its diverse physiological actions, ranging from sleep initiation to control of mood and behavior, has greatly increased during the last two decades, primarily due to the impetus that has been given to the identification and cloning of melatonin receptors and the studies related to altered functional receptorial status seen in neurodegenerative diseases like Alzheimer's disease, Parkinson's disease, and cancer. In addition, the discovery of several selective and nonselective melatonergic agonists, displaying a better pharmacokinetic profile compared to melatonin itself, has also allowed us to expand our knowledge of melatonin functions.

A number of clinical trials conducted on the use of these drugs in neurological and psychiatric disorders have testified to the successful therapeutic application of some of these melatonergic agonists in treating brain disorders. Altogether, these preclinical and clinical studies have opened up new vistas in melatonin research, especially in relation to its nature, receptor binding function, and its therapeutic potential.

The aim of this book is to bring together a comprehensive synopsis of the functional importance of melatonin and its receptors with regard to human health as well as the possible mechanisms by which melatonin and its agonists are able to exert their therapeutic effects in diseases like cancer, Alzheimer's disease, Parkinson's disease, epilepsy, stroke, and cardiovascular disorders. The chapters focusing on these themes have been written by experts in specific melatonin fields and/or clinical investigators working with the therapeutic use of either melatonin or melatonergic derivates such as the hypnotic ramelteon or the antidepressant agomelatine.

This book will be useful not only for graduates and research scientists pursuing research studies on melatonin but also for clinical investigators, neurologists, psychiatrists, and physicians with an interest in the field. More importantly, this book will motivate researchers and clinicians to further investigate the therapeutic potential of melatonin.

Acknowledgments

It is our pleasure to acknowledge Lance Wobus, senior editor, Taylor & Francis Group, who showed keen interest in this project and supported it throughout the production process. But for his timely help, this book could not have been published. We are grateful to Prof. Stephan C. Bondy, Prof. Aberto-Dominguez-Rodriguez, Prof. Cesar V. Borlongan, Doz Dr. Alexander M. Mathes, Prof. Gerado B. Ramirez-Rodriguez, Prof. Shuichi-Ueda, Prof. Domenico De Berardis, Prof. Francisco López-Muñoz, Prof. Patrick B. Bradshaw, Prof. Suleyman Kaplan, Dr. Gerardo B. Ramirez Rodriguez, and Prof. Timo Partonen for their unfailing support and encouragement. Our thanks also to all authors for their valuable contributions. Finally, we acknowledge Rachael Panthier, project editor, Taylor & Francis Group, who coordinated this project successfully from beginning to end.

Editors

Professor Venkataramanujam Srinivasan, PhD, MAMS, started his professional career as an assistant professor at the Institute of Physiology, Madurai Medical College, Madurai, Tamil Nadu, India, where he served for 24 years. During his tenure there, he was engaged in both teaching and research studies.

Besides teaching physiology to undergraduate medical students, he undertook various collaborative research studies with the Department of Psychiatry and published a number of papers on lithium, catecholamine metabolism, the pineal gland, and melatonin in both national and international medical and life sciences journals. He has also served as one of the advisory council members for the International Society of Psychoneuroendocrinology (ISPNE) for 10 years. Professor Srinivasan then undertook teaching assignments as associate professor of physiology at the School of Medical Sciences, University Sains Malaysia (USM), Kubang Kerian, Kelantan, Malaysia, after which he continued with his teaching and research activities. He collaborated on research publications with a number of professors and research scientists working in different universities in Argentina, Canada, the United States, the Netherlands, and Spain, which were published and indexed in PubMed. After leaving USM, he continued to work on his collaborative research projects on melatonin with professors working in different universities in Spain, the United States, Italy, Finland, Japan, Germany, Malaysia, and Singapore as well as from Sri Sathya Sai Medical Educational and Research Foundation, an International Medical Sciences Research Study Center located in KOVAI Coimbatore, Tamil Nadu, India. He has also published papers in international medical journals, which can be viewed in PubMed.

Professor Srinivasan has served as guest lecturer and has presented papers in national and international conferences/congresses organized by the Association of Physiologists and Pharmacologists of India (APPI), Federation of the Asian and Oceanian Physiological Societies (FAOPS), International Union of Physiological Sciences (IUPS), British Lithium Congress, Malaysian Society of Pharmacology and Physiology (MSPP), and International Society of Psychoneuroendocrinology (ISPNE) Biotechnology Conference (Tokyo) in addition to numerous symposia and workshops organized in India, Malaysia, Japan, and Singapore. He has also contributed chapters in books published by Springer and Informa Health Care and NOVA Publishers, among others. Recently, he has coedited the book entitled *Melatonin and Melatonergic Drugs in Clinical Practice* along with Prof. A Brzezinski (Israel), Prof. Sukru Oter (Turkey), and Prof. Samuel D. Shillcutt (USA), which was published by Springer in January 2014. Professor Srinivasan is also a life member of the Association of Physiologists and Pharmacologists of India (APPI) and the National Academy of Medical Sciences, India.

Dr. Gabriella Gobbi is an associate professor in the Department of Psychiatry, McGill University; staff psychiatrist at the McGill University Health Center (MUHC), Montreal, Canada; and coleader of Mental Illness and Addiction Axis of the Research Institute of the MUHC.

Dr. Gobbi earned her MD in 1991 and her specialty in psychiatry and psychotherapy in 1995 from the Catholic University of Rome, Italy. She also earned a PhD in neuroscience at the University of Cagliari, Italy. She then moved to McGill University, Montreal, Canada, in 1998, where she initially worked as a postdoc with Drs. Blier, De Montigny, and Debonnel. She then joined the University of Montreal as an assistant professor in 2002 and became associate professor at McGill University in 2008. She leads a laboratory of basic science and works as a psychiatrist at the mood disorder clinic of the MUHC.

Dr. Gobbi's research interests include understanding the pathophysiology of major depression and related disorders and discovering new treatments for them.

In particular, her laboratory conducts research on the short- and long-term effects of cannabis use in mood and anxiety and the potential beneficial effects of drugs acting on the endocannabinoid system (endogenous cannabis) to cure mental diseases. Her lab also conducts research on the effect of melatonin in alleviating mood disorders and anxiety, and sleep regulation in an effort to understand how novel selective ligands for melatonin receptors (called MT1 and MT2 receptors) can be used to treat seasonal depression, major depression, sleep disorders, and other neuropsychiatric conditions. The laboratory research ranges from bench to bedside, bridging the gaps between fundamental and clinical research. The techniques employed in the lab include in vivo electrophysiology, behavioral pharmacology, and neurochemistry.

Dr. Gobbi is the author of more than 50 highly cited manuscripts in international journals, including *Proceedings of the National Academy of Sciences of the United States of America (PNAS), Nature Neuroscience*, and the *Journal of Neuroscience* and holds two international patents in psychopharmacology. She has received several fellowships, awards, and grants from the Canadian Institutes of Health Research (CIHR), Fonds de la Recherche en Santé du Québec (FRSQ, Junior 1, Junior 2, Senior), the Canadian Psychiatric Research Foundation (CPRF), and the Canadian Foundation for Innovation (CFI), as well as from multinational pharmaceutical companies. She won the Canadian College of Neuropsychopharmacology (CCNP) Young Investigator Award in 2012.

Dr. Gobbi has served as a reviewer/editor for many journals and international grant agencies and has been invited to speak at conferences worldwide.

Dr. Samuel D. Shillcutt is a tenured professor and director of neuropsychopharmacology research in the Department of Psychiatry, Mercer University School of Medicine, and a clinical pharmacology consultant for the Medical Center of Central Georgia. He is also the director of the Medical Center of Central Georgia/Mercer University School of Medicine, Psychiatry Research Fellowship Program.

Dr. Shillcutt earned degrees in chemistry (BA), pharmacy (BS), clinical pharmacy (PharmD), and neuropharmacology (PhD) and completed postdoctoral training in clinical neuropsychopharmacology. He has held positions at the University of Nebraska Medical Center, Creighton University School of Medicine, the Veterans Administration Medical Center (Omaha, Nebraska), Southern Illinois University (SIU) School of Medicine, and Mercer University School of Medicine (MUSM). He was the director of the Schizophrenia Research Program with SIU and MacFarland Psychiatric Hospital, Springfield, Illinois, and director of the Center for Clinical Studies at Central State Hospital, Milledgeville, Georgia, and the Medical Center of Central Georgia, Macon, Georgia. For 20 years at MUSM, Dr. Shillcutt was the director of the Neuro-Psychopharmacology Research Laboratory, conducting cerebrovascular, receptor, and neurochemical analysis research, and established collaboration with Harvard University and the University of Arkansas brain banks for pathophysiological investigation of schizophrenia, Alzheimer's disease, and Huntington's disease.

Dr. Shillcutt's 38 years of research have focused on translational studies connecting interpretation of basic science and clinical trials data to lead to effective treatment modalities. His interest in understanding the pathology and treatment of neuropsychiatric disorders has led him to establishing the efficacy and safety of atypical antipsychotics, antidepressants, and medications used for Alzheimer's disease. His research techniques involve using clinical investigation of psychopharmaceuticals, neurochemistry, and electrophysiology to determine human responses and using human tissue in cell culture to establish models for basic science investigation of potential novel compounds. His most current studies investigated the effects of melatonin-like substances in regulation of mood and sleep disorders in dementias.

Dr. Shillcutt has numerous national and international peer-reviewed publications, books, book chapters, and abstracts. He is a member of ten national and international professional societies. He is also a reviewer/editor for several journals, books, and granting agencies and has been an invited speaker at many national and international colloquia and conferences.

Dr. Sibel Süzen currently serves as a professor of pharmaceutical chemistry in the Faculty of Pharmacy, Ankara University, Turkey, and institutional coordinator of the European Union (EU) student exchange programme (Erasmus Programme).

Dr. Süzen pursued her studies at Ankara University and earned her pharmacy degree in 1985 and master's degree in pharmaceutical chemistry in 1989 from the same university. She earned a PhD (1997) from the University of Swansea, Department of Chemistry (United Kingdom), where she worked on the synthesis and hydrazinolysis of dehydroalanine and glycoproteins. She has been involved in several medicinal chemistry projects supported by the government and Ankara University. These projects are related to the synthesis and investigation of biological activity of aldose reductase inhibitory compounds, antioxidant compounds, indole derivatives, in vitro drug metabolisms by electroanalytical, and, most importantly, the synthesis and antioxidant investigation of melatonin derivatives.

Dr. Süzen is the author of more than 50 highly cited manuscripts in international journals. She has organized and participated in a number of international pharmaceutical chemistry meetings. She is on the editorial board and has served as a reviewer for many pharmaceutical and medicinal chemistry journals and has been invited to speak at conferences. Her research interest focuses on the synthesis and biological evaluation of melatonin- and peptide-related antioxidant compounds as well as aldose reductase inhibitors and electrochemical investigations.

Contributors

Pedro Abreu-Gonzalez
Department of Physiology
Universidad de La Laguna
Tenerife, Spain

Niyazi Acer
Department of Anatomy
School of Medicine
Erciyes University
Kayseri, Turkey

Asma Hayati Ahmad
Department of Physiology
School of Medical Sciences
Universiti Sains Malaysia
Kelantan, Malaysia

Hanaa H. Ahmed
Department of Hormones
National Research Centre
Cairo, Egypt

Cecilio Álamo
Faculty of Medicine and Health Sciences
Department of Biomedical Sciences
(Pharmacology Area)
University of Alcalá
Madrid, Spain

Hanan Awad Alkozi
Department of Biochemistry and
Molecular Biology IV
Faculty of Optics and Optometry
University Complutense de Madrid
Madrid, Spain

Daniel Alonso-Alconada
Department of Cell Biology and Histology
School of Medicine and Dentistry
University of the Basque Country
Bizkaia, Spain

Berrin Zühal Altunkaynak
Department of Histology and Embryology
School of Medicine
Ondokuz Mayıs University
Samsun, Turkey

Antonia Alvarez
Department of Cell Biology and Histology
School of Medicine and Dentistry
University of the Basque Country
Bizkaia, Spain

Marcos Aranda
Department of Human Biochemistry
School of Medicine/CEFyBO
University of Buenos Aires/CONICET
Buenos Aires, Argentina

Teoman Aydın
Department of Physical Therapy and
Rehabilitation
School of Medicine
Bezmialem Vakıf University
İstanbul, Turkey

Mustafa Ayyıldız
Department of Physiology
School of Medicine
Ondokuz Mayıs University
Samsun, Turkey

Meral Baka
Department of Histology and Embryology
School of Medicine
Ege University
Izmir, Turkey

Annalida Bedini
Department of Biomolecular Sciences
University of Urbino "Carlo Bo"
Urbino, Italy

Nicolás Belforte
Department of Human Biochemistry
School of Medicine/CEFyBO
University of Buenos Aires/CONICET
Buenos Aires, Argentina

Giovanni Blandino
Translational Oncogenomics Unit
Molecular Medicine Area
Regina Elena National Cancer Institute
Rome, Italy

Stephen C. Bondy
Department of Medicine
Center for Occupational and Environmental Health
University of California, Irvine
Irvine, California

Cesar V. Borlongan
Department of Neurosurgery and Brain Repair
Center of Excellence for Aging and Brain Repair
Morsani College of Medicine
University of South Florida
Tampa, Florida

Mia C. Borlongan
Department of Neurosurgery and Brain Repair
Center of Excellence for Aging and Brain Repair
Morsani College of Medicine
University of South Florida
Tampa, Florida

Patrick C. Bradshaw
Department of Cell Biology, Microbiology, and Molecular Biology
University of South Florida
Tampa, Florida

Gregory M. Brown
Department of Psychiatry
University of Toronto
and
Centre for Addiction and Mental Health
Toronto, Ontario, Canada

Amnon Brzezinski
Department of Obstetrics and Gynecology
Hadassah Medical Center
The Hebrew University
Jerusalem, Israel

Alexandre Budu
Departamento de Fisiologia
Instituto de Biociências
Universidade de São Paulo
São Paulo, Brazil

Daniela Campanella
Department of Mental Health
Psychiatric Service of Diagnosis and Treatment
Hospital "G. Mazzini"
Teramo, Italy

Marilde Cavuto
Istituto Abruzzese di Storia Musicale
L'Aquila, Italy

Emel Öykü Çetin
Department of Biopharmaceutics and Pharmacokinetics
and
Department of Pharmaceutical Technology
Ege University
İzmir, Turkey

Banthit Chetsawang
Research Center for Neuroscience
Institute of Molecular Biosciences
Mahidol University
Nakhonpathom, Thailand

Stefano Comai
Department of Psychiatry
McGill University
and
McGill University Health Centre
Montreal, Québec, Canada

Domenico De Berardis
Department of Mental Health
Psychiatric Service of Diagnosis and Treatment
Hospital "G.Mazzini"
Teramo, Italy

and

Department of Neurosciences and Imaging
University "G.D" Annunzio
Chieti, Italy

Vedad Delic
Department of Cell Biology, Microbiology, and Molecular Biology
University of South Florida
Tampa, Florida

Wuguo Deng
Institute of Cancer Stem Cell
Dalian Medical University Cancer Center
Dalian, Liaodong, People's Republic of China

and

State Key Laboratory of Oncology in South China
Sun Yat-Sen University Cancer Center
Guangzhou, Guangdong, People's Republic of China

Ömür Gülsüm Deniz
Department of Histology and Embryology
Medical School
Ondokuz Mayıs University
Samsun, Turkey

Massimo Di Giannantonio
Department of Neuroscience and Imaging
University "G. D'Annunzio"
Chieti, Italy

Alberto Dominguez-Rodriguez
Department of Cardiology
Hospital Universitario de Canarias
Tenerife, Spain

Ayuka Ehara
Department of Histology and Neurobiology
School of Medicine
Dokkyo Medical University
Tochigi, Japan

Ebru Elibol
Department of Histology and Embryology
School of Medicine
Ondokuz Mayıs University
Samsun, Turkey

Javier Espino
Department of Physiology
University of Extremadura
Badajoz, Spain

Maria Ferraiuolo
Molecular Chemoprevention Group
Molecular Medicine Area
Regina Elena National Cancer Institute
Rome, Italy

María F. González Fleitas
Department of Human Biochemistry
School of Medicine/CEFyBO
University of Buenos Aires/CONICET
Buenos Aires, Argentina

Michele Fornaro
Department of "Scienze della Formazione"
University of Catania
Catania, Italy

Max Franzblau
Department of Neurosurgery and Brain Repair
Center of Excellence for Aging and Brain Repair
Morsani College of Medicine
University of South Florida
Tampa, Florida

Nick Franzese
Department of Neurosurgery and Brain Repair
Center of Excellence for Aging and Brain Repair
Morsani College of Medicine
University of South Florida
Tampa, Florida

Wenyu Fu
Department of Histology and Embryology
Weifang Medical University
Weifang, Shandong, People's Republic of China

Manisha Gautam
Department of Physiology
Maulana Azad Medical College
New Delhi, India

Célia R. S. Garcia
Departamento de Fisiologia
Instituto de Biociências
Universidade de São Paulo
São Paulo, Brazil

Pilar García-García
Faculty of Medicine and Health Sciences
Department of Biomedical Sciences (Pharmacology Area)
University of Alcalá
Madrid, Spain

Somenath Ghosh
Department of Zoology
Banaras Hindu University
Uttar Pradesh, India

Gabriella Gobbi
Department of Psychiatry
McGill University
and
McGill University Health Center
Montreal, Québec, Canada

Chiara Gonzales-Portillo
Department of Neurosurgery and Brain Repair
Center of Excellence for Aging and Brain Repair
Morsani College of Medicine
University of South Florida
Tampa, Florida

Gabriel S. Gonzales-Portillo
Department of Neurosurgery and Brain Repair
Center of Excellence for Aging and Brain Repair
Morsani College of Medicine
University of South Florida
Tampa, Florida

Yingjun Guan
Department of Histology and Embryology
Weifang Medical University
Weifang, Shandong, People's Republic of China

Wei Guo
Institute of Cancer Stem Cell
Dalian Medical University Cancer Center
Dalian, Liaoning, People's Republic of China

Reshu Gupta
Department of Physiology
SMS Medical College
Jaipur, Rajasthan, India

Chandana Haldar
Department of Zoology
Banaras Hindu University
Uttar Pradesh, India

Rüdiger Hardeland
Johann Friedrich Blumenbach Institute of Zoology and Anthropology
University of Göttingen
Göttingen, Germany

Enrique Hilario
Department of Cell Biology and Histology
School of Medicine and Dentistry
University of the Basque Country
Bizkaia, Spain

Felice Iasevoli
Department of Neuroscience
University School of Medicine "Federico II"
Naples, Italy

Kenichi Inoue
School of Medicine
Dokkyo Medical University
Tochigi, Japan

Saime İrkören
Department of Plastic and Reconstructive Surgery
School of Medicine
Adnan Menderes University
Aydın, Turkey

Jae-Kyo Jeong
Bio Safety Research Institute
College of Veterinary Medicine
Chonbuk National University
Jeonju, Republic of Korea

Yuji Kaneko
Department of Neurosurgery and Brain Repair
Center of Excellence for Aging and Brain Repair
Morsani College of Medicine
University of South Florida
Tampa, Florida

Süleyman Kaplan
Department of Histology and Embryology
Medical School
Ondokuz Mayıs University
Samsun, Turkey

Süleyman Karademir
Department of Histology and Embryology
School of Medicine
Adnan Menderes University
Aydın, Turkey

Charanjit Kaur
Department of Anatomy
Yong Loo Lin School of Medicine
National University of Singapore
Singapore, Singapore

Bal Krishana
Department of Physiology
Maulana Azad Medical College
New Delhi, India

Eda Kucuktulu
Department of Radiation Oncology
Trabzon Kanuni Training and Research Hospital
Trabzon, Turkey

Uzer Kucuktulu
Department of General Surgery
Trabzon Kanuni Training and Research Hospital
Trabzon, Turkey

Edward C. Lauterbach
Department of Psychiatry and Behavioural Sciences
and
Department of Internal Medicine (Neurology Section)
School of Medicine
Mercer University
Macon, Georgia

Eng-Ang Ling
Department of Anatomy
Yong Loo Lin School of Medicine
National University of Singapore
Singapore, Singapore

Francisco López-Muñoz
Faculty of Health Sciences
Camilo José Cela University
and
Faculty of Medicine and Health Sciences
Department of Biomedical Sciences (Pharmacology Area)
University of Alcalá
and
Hospital 12 de Octubre Research Institute
Madrid, Spain

Juan A. Madrid
Faculty of Biology
Chronobiology Laboratory
Department of Physiology
University of Murcia
Murcia, Spain

Stefano Marini
Department of Mental Health
Psychiatric Service of Diagnosis and Treatment
Hospital "G. Mazzini"
Teramo, Italy

and

Department of Neuroscience and Imaging
University "G. D'Annunzio"
Chieti, Italy

Giovanni Martinotti
Department of Neuroscience and Imaging
University "G. D'Annunzio"
Chieti, Italy

Tomoyuki Masuda
Department of Neurobiology
School of Medicine
University of Tsukuba
Tsukuba, Japan

Alexander M. Mathes
Department of Anesthesiology
Düsseldorf University Hospital
Düsseldorf, Germany

Monica Mazza
Department of Life, Health and Environmental Sciences
University of L'Aquila
L'Aquila, Italy

Maria D. Mediavilla
Department of Physiology and Pharmacology
School of Medicine and Research Institute
"Marqués de Valdecilla" (IDIVAL)
University of Cantabria
Santander, Spain

Mahaneem Mohamed
Department of Physiology
School of Medical Sciences
Universiti Sains Malaysia
Kelantan, Malaysia

Marco Mor
Department of Pharmacy
University of Parma
Parma, Italy

María C. Moreno
Department of Human Biochemistry
School of Medicine/CEFyBO
University of Buenos Aires/CONICET
Buenos Aires, Argentina

Paola Muti
Department of Oncology
Juravinski Cancer Center
McMaster University
Hamilton, Ontario, Canada

Yoshiji Ohta
School of Medicine
Fujida Health University
Toyoake, Japan

Luigi Olivieri
Department of Mental Health
Psychiatric Service of Diagnosis and Treatment
Hospital "G. Mazzini"
Teramo, Italy

Mehmet Emin Önger
Department of Histology and Embryology
Medical School
Ondokuz Mayıs University
Samsun, Turkey

Leonardo Ortiz-López
Division of Clinical Investigations
National Institute of Psychiatry "Ramón de la Fuente Muñiz"
México, México

Beatriz B. Otalora
Department of Physiology
University of Murcia
Murcia, Spain

Zahiruddin Othman
Department of Psychiatry
School of Medical Sciences
Universiti Sains Malaysia
Kelantan, Malaysia

Heval Selman Özkan
Department of Plastic and Reconstructive Surgery
School of Medicine
Adnan Menderes University
Aydın, Turkey

Meliha Gündağ Papaker
Department of Neurosurgery
School of Medicine
Bezmialem Vakıf University
İstanbul, Turkey

Sang-Youel Park
Bio Safety Research Institute
College of Veterinary Medicine
Chonbuk National University
Jeonju, Republic of Korea

Timo Partonen
Department of Mental Health and Substance Abuse Services
National Institute for Health and Welfare
Helsinki, Finland

Giampaolo Perna
Department of Clinical Neurosciences
Villa San Benedetto Menni
Hermanas Hospitalarias, FoRiPsi
Como, Italy

and

Department of Psychiatry and Behavioral Sciences
Leonard Miller School of Medicine
University of Miami
Miami, Florida

and

Department of Psychiatry and Neuropsychology
University of Maastricht
Maastricht, the Netherlands

Jesús Pintor
Department of Biochemistry and Molecular Biology IV
Faculty of Optics and Optometry
University Complutense de Madrid
Madrid, Spain

Ida Potena
Department of Mental Health
Center of Mental Health
Teramo, Italy

Zheng-Hong Qin
Department of Pharmacology
School of Pharmaceutical Science
Soochow University
Suzhou, Jiangsu, People's Republic of China

Maria Antonia Quera-Salva
AP-HP Sleep Unit
Department of Physiology
Raymond Poincaré Hospital
Garches, France

Gerardo B. Ramírez-Rodríguez
Laboratory of Neurogenesis
Division of Clinical Investigations
National Institute of Psychiatry "Ramón de la Fuente Muñiz"
México, México

Gabriella Rapini
Department of Mental Health
Psychiatric Service of Diagnosis and Treatment
Hospital "G. Mazzini"
Teramo, Italy

Gurugirijha Rathnasamy
Department of Anatomy
Yong Loo Lin School of Medicine
National University of Singapore
Singapore, Singapore

Russel J. Reiter
Department of Cellular and Structural Biology
University of Texas Health Science Center
San Antonio, Texas

Emiliano Ricardo Vasconcelos Rios
Laboratory of Neuropharmacology
Federal University of Ceará
Ceará, Brazil

Silvia Rivara
Department of Pharmacy
University of Parma
Parma, Italy

M. Angeles Rol
Department of Physiology
University of Murcia
Murcia, Spain

Ruth E. Rosenstein
Department of Human Biochemistry
School of Medicine/CEFyBO
University of Buenos Aires/CONICET
Buenos Aires, Argentina

Shin-ichi Sakakibara
Laboratory of Molecular Neurobiology
Institute of Applied Brain Sciences
Waseda University
Saitama, Japan

Emilio J. Sanchez-Barcelo
Department of Physiology and Pharmacology
School of Medicine and Research Institute
"Marqués de Valdecilla" (IDIVAL)
University of Cantabria
Santander, Spain

Pablo Sande
Department of Human Biochemistry
School of Medicine/CEFyBO
University of Buenos Aires/CONICET
Buenos Aires, Argentina

Raffaela Santoro
Molecular Chemoprevention Group
Molecular Medicine Area
Regina Elena National Cancer Institute
Rome, Italy

Ravinder Kumar Saran
Department of Pathology
G B Pant Hospital
New Delhi, India

Nicola Serroni
Department of Mental Health
Psychiatric Service of Diagnosis and Treatment
Hospital "G. Mazzini"
Teramo, Italy

M. Hakan Seyithanoğlu
Department of Neurosurgery
School of Medicine
Bezmialem Vakıf University
İstanbul, Turkey

Aziza B. Shalby
Department of Hormones
National Research Centre
Cairo, Egypt

Edward H. Sharman
Department of Neurology
University of California, Irvine
Irvine, California

Samuel D. Shillcutt
Department of Psychiatry and Behavioral Sciences
School of Medicine
Mercer University
Macon, Georgia

Amaresh Kumar Singh
Department of Zoology
Banaras Hindu University
Uttar Pradesh, India

Gilberto Spadoni
Department of Biomolecular Sciences
University of Urbino "Carlo Bo"
Urbino, Italy

U.S. Srinivasan
Fortis Malar Hospital
Chennai, Tamil Nadu, India

Venkataramanujam Srinivasan
International Medical Sciences Research Study Center
Sri Sathya Sai Medical Educational and Research Foundation
Coimbatore, Tamil Nadu, India

and

Department of Mental Health
Psychiatric Service of Diagnosis and Treatment
Hospital "G. Mazzini"
National Health Service
Teramo, Italy

and

Department of Neurosciences and Imaging
University "G. d'Annunzio"
Chieti, Italy

Meaghan Staples
Department of Neurosurgery and Brain Repair
Center of Excellence for Aging and Brain Repair
Morsani College of Medicine
University of South Florida
Tampa, Florida

Sabrina Strano
Molecular Chemoprevention Group
Molecular Medicine Area
Regina Elena National Cancer Institute
Rome, Italy

and

Department of Oncology
Juravinski Cancer Center
McMaster University
Hamilton, Ontario, Canada

Sibel Suzen
Department of Pharmaceutical Chemistry
Faculty of Pharmacy
Ankara University
Ankara, Turkey

Naoki Tajiri
Department of Neurosurgery and Brain Repair
Center of Excellence for Aging and Brain Repair
Morsani College of Medicine
University of South Florida
Tampa, Florida

Özgür Taşpınar
Department of Physical Therapy and Rehabilitation
School of Medicine
Bezmialem Vakıf University
İstanbul, Turkey

Engin Taştaban
Department of Physical Therapy and Rehabilitation
School of Medicine
Adnan Menderes University
Aydın, Turkey

Mehmet Turgut
Department of Neurosurgery
School of Medicine
Adnan Menderes University
Aydin, Turkey

Saffet Tüzgen
Department of Neurosurgery
School of Medicine
Bezmialem Vakıf University
İstanbul, Turkey

Shuichi Ueda
Department of Histology and Neurobiology
School of Medicine
Dokkyo Medical University
Tochigi, Japan

Yiğit Uyanıkgil
Department of Histology and Embryology
School of Medicine
Ege University
Izmir, Turkey

Alessandro Valchera
Hermanas Hospitalarias, FoRiPsi
Villa S. Giuseppe Hospital
Ascoli Piceno, Italy

Jingshu Wang
State Key Laboratory of Oncology in South China
Sun Yat-Sen University Cancer Center
Guangzhou, Guangdong, People's Republic of China

Xin Wang
Department of Neurosurgery
Brigham and Women's Hospital
Harvard Medical School
Boston, Massachusetts

Özlem Yalçınkaya Yavuz
Department of Physiology
School of Medicine
Trakya University
Edirne, Turkey

Selçuk Yavuz
Department of Physical Therapy and Rehabilitation
Edirne State Hospital
Edirne, Turkey

Kanji Yoshimoto
Department of Food Sciences and Biotechnology
Hiroshima Institute of Technology
Hiroshima, Japan

Rahimah Zakaria
Department of Physiology
School of Medical Sciences
Universiti Sains Malaysia
Kelantan, Malaysia

Shuanhu Zhou
Department of Orthopedic Surgery
Brigham and Women's Hospital
Harvard Medical School
Boston, Massachusetts

1 Melatonin Production and Bioavailability

Emel Öykü Çetin, Yiğit Uyanıkgil, Mehmet Turgut, and Meral Baka

CONTENTS

1.1 INTRODUCTION

The chemical formula of melatonin (Mel) is *N*-acetyl-5-methoxytryptamine, and it is essentially the secretion of the pineal gland. It is controlled by the circadian rhythm (Jung and Ahmad 2006, Lukaszyk and Reiter 1975, Weaver et al. 1993) and is produced during the dark phase of the day–night cycle by the pineal gland (Fourtillan et al. 2001). Conversely, pineal gland is a conical organ, measuring about 5–8 mm in length and 3–5 mm in width in humans (Junqueira et al. 2006). The location of the pineal gland is at the posterior part of the third ventricle, and it is connected to the diencephalon by a short stalk (Junqueira et al. 2006, Krabbe 1916, Weiss 1988). Pineal organ is an organ that has been discovered early in mammals, and Galen called this organ "conarium" (Rodin and Overall 1967). Today, it is called "glandula pinealis," "corpus pineal," "epiphysis cerebri," or "pineal body" (Reitel 1981, Rodin and Overall 1967, Weiss 1988). Pineal hormone Mel, $C_{13}H_{16}N_2O_2$ (molecular weight: 232), is an indoleamine and is a derivative of the essential amino acid tryptophan (Jemima et al. 2011, Lukaszyk and Reiter 1975).

In this chapter, the general histologic structure of the pineal gland, the morphological properties of the pinealocyte cell, and the bioavailability of the Mel hormone that is secreted by the pinealocyte cells will be discussed.

1.1.1 Histological Appearance of the Pineal Gland

The pineal gland, which is formed by the cells with a neurosecretory function, is an endocrine organ and is bound to the brain by a stalk, but there is no direct nerve connection between the pineal gland and the brain. Instead, the pineal gland is supplied by postganglionic sympathetic nerve fibers derived from the superior cervical ganglia (SCG) (Teclemariam-Mesbah et al. 1999). Anatomically, preganglionic fibers to the SCG are derived from the lateral column of the spinal cord (Klein and Moore 1979, Teclemariam-Mesbah et al. 1999).

The function of the pineal gland is regulated by sympathetic nerves. The penetration of the nerve fibers to the pineal gland causes loss of their myelin sheaths; the unmyelinated axons end among pinealocytes, and some of them form synapses (Bowers et al. 1984). A great number of small vesicles that contain norepinephrine are observed in these nerve endings. Serotonin is also present in both pinealocytes and sympathetic nerve terminals (Bowers et al. 1984, Bowers and Zigmond 1982, Deguchi 1982, Lukaszyk and Reiter 1975).

Histologically, the major cells of the pineal gland are pinealocytes (Junqueira et al. 2006, Ross and Pawlina 2006). These cells are arranged in clumps or cords resting on a basal lamina that are located within the lobules formed by the connective tissue septa; the septa extend into the gland from the pia mater, which covers its surface (Figures 1.1a,b and 1.2c,d). Blood vessels lie within the connective tissue surrounding the pinealocytes, and they are lined by fenestrated endothelial cells and nerves (Møller et al. 1978) (Figure 1.1d). A pinealocyte has a large deeply infolded nucleus with one or more prominent nucleoli and contains lipid droplets within its cytoplasm (Junqueira et al. 2006). The pinealocytes have two or more cell process ending in bulbous expansions; one of them ends near capillaries. The cytoplasm contains abundant randomly distributed mitochondria (Ross and Pawlina 2006).

Morphologically, the pinealocyte is composed of a cell body with 7–12 μm in diameter; there are three to five processes emerging from the cell body (Møller and Baeres 2002). In the electron microscopy, it has been seen that dense-core granules are present in the cell body, and the granules are much denser in the club-shaped terminals of the cellular processes. The specific marker for the pinealocyte in the pineal gland is the so-called synaptic ribbon (Vollrath 1981). Though this organelle is common in the pinealocytes of several mammalian species such as rat and guinea pig, in many other species the number of synaptic ribbons is low or absent. The ultrastructural features of the mammalian pinealocyte have been described in some studies (Møller and Baeres 2002, Vollrath 1981, 1984).

1.1.2 Cell Types of Pineal Gland

A total of five pineal cell types have been described in all mammals (Møller and Baeres 2002, Møller et al. 1978):

1. Pinealocyte (hormone-producing): When it is examined by transmission electron microscopy, pinealocytes show typical cytoplasmic organelles along with numerous dense-core, elongated cytoplasmic processes, membrane-bounded vesicles in their elaborate (Cieciura and Krakowski 1991, Møller and Baeres 2002, Redins et al. 2001) (Figure 1.1e,f). The processes also contain numerous parallel bundles of microtubules. The expanded club-like endings of the processes are associated with the blood capillaries, strongly suggesting neuroendocrine activity (Cieciura and Krakowski 1991, Møller and Baeres 2002, Redins et al. 2001) (Figures 1.1b and 1.2c,d).
2. Interstitial cells: These cells are found among pinealocytes. With their glial-like characteristics, interstitial cells provide stromal support to the functional pinealocytes along with the connective tissue. The astrocytes of the pineal gland are observed between the cords of pinealocytes and in the perivascular areas and are characterized by elongated nuclei that stain more heavily than those of parenchymal cells (Møller and Baeres 2002, Nakazato et al. 2002). These cells have long cytoplasmic processes. They are located in the perivascular areas and have long cytoplasmic processes containing a large number of intermediate filaments of 10 nm diameter (Møller and Baeres 2002, Nakazato et al. 2002).

   Histologically, the interstitial cell is smaller than the pinealocyte cell, with a darker triangular nucleus and cytoplasm in comparison to the pinealocyte cell observed in the light and electron microscopes (Møller et al. 1978). The cell is star-shaped with several long and slender processes. The interstitial cell includes a high number of filaments in most species (Møller et al. 1978). A few of the interstitial cells are immunoreactive to the glial fibrillary acid protein (Møller et al. 1978) (Figure 1.2a,b).

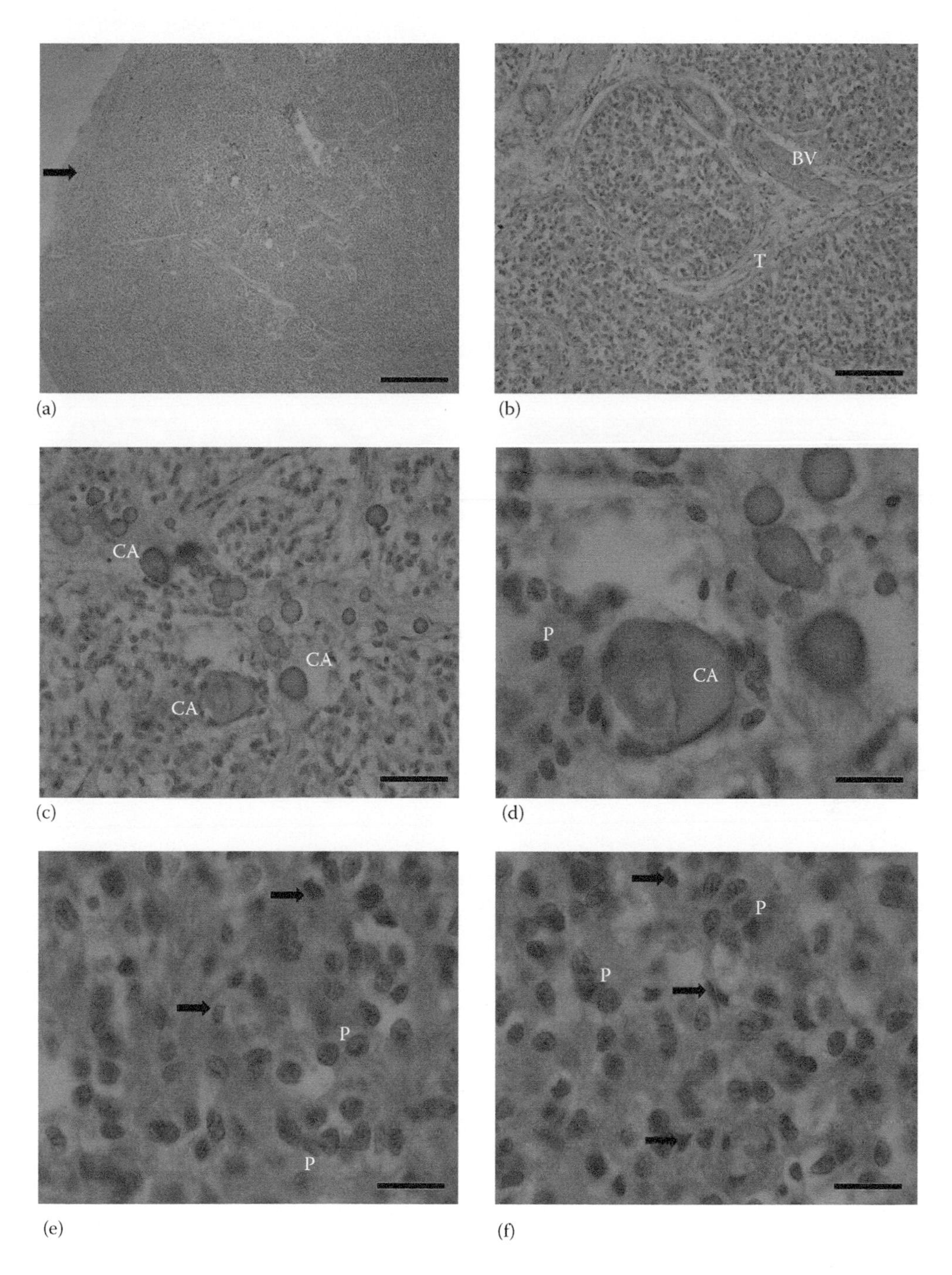

FIGURE 1.1 (See color insert.) (a) Panoramic view of human pineal gland. ×4 magnification. Black arrow, capsule-irregular connective tissue. Hematoxylin & Eosin staining. Scale bar = 1250 μm. (b) Trabecular connective tissue separating cells. T, trabecula; BV, blood vessel. ×20 magnification. Hematoxylin & Eosin staining. Scale bar = 250 μm. (c) Panoramic view of human pineal gland. ×40 magnification. CA, corpora arenacea. Hematoxylin & Eosin staining. Scale bar = 125 μm. (d) Panoramic view of human pineal gland. ×40 magnification. CA, corpora arenacea; P, pinealocyte. Hematoxylin & Eosin staining. Scale bar = 50 μm. (e, f) Human pineal gland. ×100 magnification. P, pinealocyte; E, endothelium; Black arrow, interstitial cell. Scale bar = 50 μm.

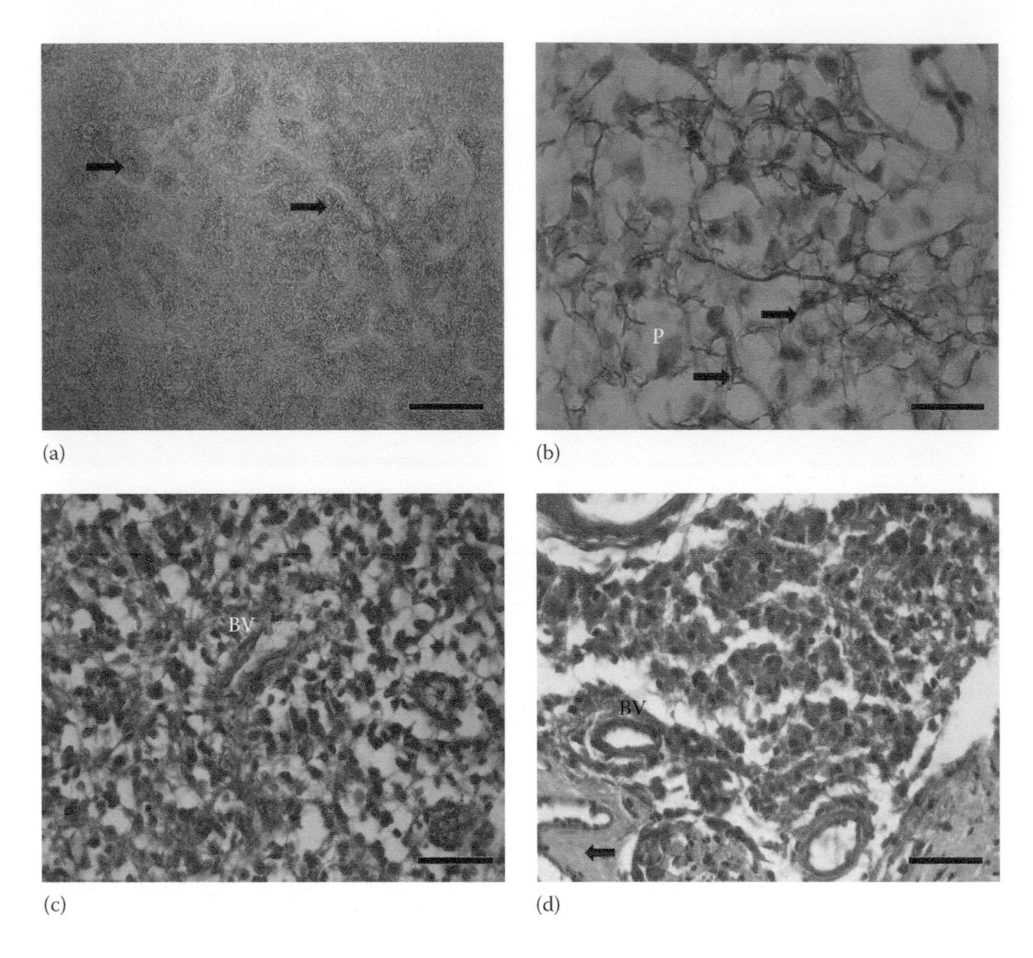

FIGURE 1.2 **(See color insert.)** (a) General distribution of glial fibrillary acid protein (GFAP) positive interstitial cells in human pineal gland (dark brown areas). ×4 magnification. Anti-GFAP staining. Arrow shows dark brown areas. Scale bar = 1250 μm. (b) Interstitial cells and cell processes. Arrow shows cell process. ×100 magnification. Anti-GFAP staining. Scale bar = 50 μm. P, pinealocyte; black arrow, interstitial cell. (c) Parenchyma of the pineal gland. BV, blood vessel (erythrocytes are red in color), ×40 magnification. Gomori trichrome staining. Scale bar = 125 μm. (d) Parenchyma of the pineal gland, ×40 magnification. Gomori trichrome staining. Scale bar = 125 μm. Trabecular connective tissue separating cells. Arrow shows the green-stained connective septa. BV, blood vessel.

3. Perivascular phagocytes: The phagocytic cells are mostly restricted to the perivascular spaces. These cells indicate a high uptake of exogenously applied tracer, for instance, horseradish peroxidase, and might be perivascular microglial-like cells, such as the ones in other parts of the central nervous system (Pedersen et al. 1993, Sato et al. 1996). Recently, immunocytochemical studies have shown that such perivascular cells often include surface proteins that are the markers for macrophages and microglial cells. The phagocytes cause an immune reaction for the class II major histocompatibility system, suggesting that they are also antigen-presenting cells (Pedersen et al. 1993, Sato et al. 1996).
4. Neurons: Neurons of pineal gland are classic neurons characterized by Nissl substance, which are found in various mammalian species, for instance, humans, monkeys, rabbits, ferrets, and cotton rats; they take an input via synapse-like contacts (Matsushima et al. 1994). The neurons form real ganglia within the pineal gland in some species. It is accepted that most of the parasympathetic neurons are innervated by a peripheral ganglion inside

the pineal gland, while intrapineal ganglion receives a synaptic input from the neurons that are located in the epithalamus (David and Herbert 1973).

5. Peptidergic neuron-like cells: These cells have been displayed in various rodent and non-rodent pineals. It is accepted that neuron-like cells that are immunoreactive to enkephalin are present in European hamsters and humans (Coto-Montes et al. 1994). Though the existence and functions of these cells have not been identified yet, their existence in the pineal gland of a few species suggests a paracrine function of the pinealocyte (Møller 1997).

In addition to the five cell types given earlier, the human pineal gland is defined by the presence of calcified concretions known as "corpora arenacea" or "brain sand" (Figure 1.1c,d) (Junqueira et al. 2006, Ross and Pawlina 2006). These concretions may result from the precipitation of calcium carbonates and phosphates upon proteins that are released into the cytoplasm of these cells. In other words, pinealocytes secrete an extracellular matrix (ECM) in which calcium phosphate crystals deposit. The concretions are recognizable in childhood, and their number increases with age. Calcification starts early in childhood and becomes evident in the second decade of life studies (Ivanov 2007, Junqueira et al. 2006, Laure-Kamionowska et al. 2003). On the function of the pineal gland, calcification has no known effect (Junqueira et al. 2006); on the other hand, the calcified concretions are placed in the midline of the brain, and as they are opaque to x-rays, they serve as convenient markers of the midline of the brain in radiographic and computed tomography (CT) studies (Ivanov 2007, Junqueira et al. 2006, Laure-Kamionowska et al. 2003). Like the anterior hypophysis, the pineal gland lacks a blood–brain barrier (BBB). It has been reported that the capillaries in the pineal gland are endowed with continuous endothelium in the human fetus and sheep (Møller et al. 1978).

1.2 BIOAVAILABILITY OF MELATONIN

Mel is an indoleamine, derived from the essential amino acid tryptophan (Jemima et al. 2011). It is synthesized in the pineal gland with acetylation and then methylation of serotonin (Mulchaheya et al. 2004). It is a white powder having a melting point of 117°C. It is an amphiphilic molecule and has a high lipid and water solubility, which simplifies its movement across the cell membrane and various body fluids. As a result, this makes Mel a prominent molecule in almost all the tissues of mammals. The indole ring of Mel acts as a chromophore, showing maximum absorbance at 223 nm, and the functional groups contribute to its fluorescence property. Its functional groups confer it the receptor specificity for its hormonal functions and aid in its oxidation chemistry (Jemima et al. 2011). It is also synthesized in plenty of tissues and cells including platelets, the gastrointestinal tract, bone marrow cells, Harderian gland, skin, retina, and lymphocytes (Figure 1.3) (de Almeida et al. 2011, Fourtillan et al. 2001, Zawilska et al. 2009).

Mel is analyzed by various techniques, which include chemiluminescence method, cyclic voltammetry, and immunological techniques such as enzyme immunoassay and radioimmunoassay (RIA), and chromatographic techniques such as gas chromatography, liquid chromatography, high pressure liquid chromatography (HPLC), and gas chromatography–mass spectrometry coupled with

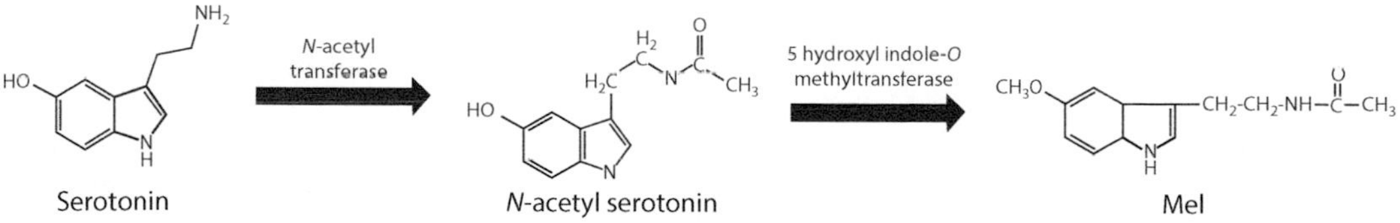

FIGURE 1.3 Schematic formulas showing biochemical pathways regarding biosynthesis of melatonin in mammals.

fluorescence or refractive index detector to cite a few examples (Eriksson et al. 2003, Harumi and Matsushima 2000, Jemima et al. 2011, Lu et al. 2002, Radi and Bekhiet 1998, Reiter et al. 2005, Simonin et al. 1999, Van-Tassel and O'Neill 2001, Vitale et al. 1996, Wang et al. 2011).

In mammals, Mel is recognized as the hormone of darkness, and it is excreted through the hours of darkness between 02:00 and 04:00 a.m. UTC and falls with light. It enters the bloodstream with passive diffusion. In humans, Mel is an endogenous neurohormone, and the secretion of Mel from the pineal gland is controlled by the circadian rhythm, but Mel also effects the regulation of circadian system (Brown et al. 2009, Brzezinski 1997, Johns et al. 2012). It has an important role in the body's internal time system, which controls the sleep–wake cycle along with seasonal rhythms and other circadian rhythms (Hafner et al. 2009).

Mel is also a potent free radical scavenger and a natural antioxidant in almost all the cell functions in the body. Reduction of Mel during advanced ages has been correlated with deterioration of health, disturbance of sleep, and chronic diseases that are related to oxidative damage, including cancer, regulation of seasonal reproduction, blood pressure, immune function, and osteoblast–osteoclast cells of the bone (Altun and Ugur-Altun 2007, Johns et al. 2012). Administration of Mel may provide treatment to jet lag; this disorder is observed after transatlantic flights (Laure-Kamionowska et al. 2003).

In the past, Mel was used in humans at physiological or pharmacological amounts, and there is a general agreement at present that it is a nontoxic molecule. It has been reported that oral administration of Mel in doses of 1–300 mg or 1 g of Mel daily for 1 month has no negative side effects in human volunteers (Altun and Ugur-Altun 2007).

Serum Mel concentrations vary significantly according to the age. During the day, the serum endogenous Mel concentration is low (10–20 pg/mL); at night, it reaches peak values (80–150 pg/mL); and it remains high during sleep. In the blood, most of the Mel is bound reversibly to the glycoproteins and albumin (Bonnefont-Rousselot and Collin 2010, Brzezinski 1997, Li et al. 2013, López-Gamboa et al. 2010). Infants who are younger than 3 months of age produce very little Mel hormone, and its secretion increases and becomes circadian in older infants. The peak nocturnal concentrations are highest (average, 325 pg/mL [1400 pmol/L]) at the age of 1–3 years, after which they decline gradually. In normal young adults, the average daytime and peak nighttime values are 10 and 60 pg/mL (40 and 260 pmol/L), respectively (Brzezinski 1997). Mel levels in plasma show a great intersubject heterogeneity and vary throughout the day and also with age, gender, temperature, and season. RIAs have indicated the production of Mel in healthy adults to range from 28.8 μg/day to 39.2 μg/night. The rate of secretion at night as estimated by GC-MS is about 4.6 μg/h in males and 2.8 μg/h in females (Johns et al. 2012). The oral bioavailability of immediate release Mel is about 3%–76%. The apparent volume of distribution reported in several studies has discrepancy. Fourtillan et al. reported the volume of distribution at a steady state of 0.99 L/kg. In a single passage, up to 90% of blood Mel is cleared by the liver (Fourtillan et al. 2001). This makes its half-life very short (30–60 min). Mel has an elimination half-life ($t_{1/2}$) of 30–45 min (Schemmer et al. 2008). It is metabolized to 6-hydroxymelatonin in the liver by hydroxylation, and over 80% is excreted in the urine following conjugation with sulfuric or glucuronic acid. Except its sulfate metabolites, Mel readily crosses the BBB (Bartoli et al. 2012, Zisapel 2010).

The influence of transdermal drug delivery, oral controlled release, and oral transmucosal systems on plasma concentrations of Mel and its principal metabolite in human subjects using a crossover, single dose design was estimated in a study (Bénès et al. 1997). A total of 12 adult male volunteers participated in this study, and plasma concentrations of Mel and its metabolite, 6-sulfatoxymelatonin, were measured by RIA (Bénès et al. 1997). Administration of the oral controlled-release formulation of Mel resulted in prolonged plasma concentrations of Mel and its metabolite (Bénès et al. 1997). Thus, plasma concentrations of Mel are different after oral controlled release and transdermal drug delivery systems. The ratio of plasma 6-sulfatoxymelatonin area under curve (AUC) to plasma Mel AUC was much greater than the physiological level, when Mel was given orally (Bénès et al. 1997). Transdermal drug delivery systems resulted in a significant delay, and administration of Mel did not mimic the nocturnal plasma level profile of Mel for transdermal drug delivery systems, while transmucosal administration of Mel provided fast systemic drug levels with reduced variability than oral controlled release or transdermal drug delivery. The findings of this study revealed that transmucosal delivery was able to simulate the physiological blood profiles of both Mel and its principal metabolite (Bénès et al. 1997). Mao et al. evaluated Mel microspheres for intranasal intake, and the particle sizes of microspheres were in the range of 30–60 µm (Mao et al. 2004). The absolute bioavailability of Mel microspheres following intranasal intake was found to be 84.1% (Mao et al. 2004).

In another study, the comparison of the new food grade liquid emulsion with a standard Mel formulation in oral tablets in a single dose was done in a total of eight subjects who randomly received 5 mg of oral spray or oral Mel as tablet (Bartoli et al. 2012). Plasma samples were collected after 6 h, and samples were determined by liquid chromatography tandem mass spectrometry method to investigate the main pharmacokinetic parameters. It was obtained that there was a significant difference between the C_{max} values and the AUC. T_{max} of the two Mel formulations did not show statistically significant differences (Bartoli et al. 2012). The amount of Mel reaching the systemic circulation following oral spray intake was found to be higher in comparison to the oral tablet intake (Bartoli et al. 2012).

1.3 CONCLUSION

In summary, Mel, excreted by the pineal gland, is important for synchronizing subjects suffering from blindness, jet lag, old age, shift work, and night work. It has several pharmacological effects, such as sedative, antidepressant, regenerative, anxiolytic, antioxidant, antitoxic, anticonvulsant, and analgesic. Therapeutically, it may be used for the treatment of jet lag, the resynchronization of circadian rhythms, insomnia, cancer therapy, in Alzheimer's disease, and other neurodegenerative disorders. Mel has a short half-life, low and variable bioavailability, when it is administered orally.

REFERENCES

Altun, A. and B. Ugur-Altun. 2007. Melatonin: Therapeutic and clinical utilization. *Int J Clin* 61:835–845.

Bartoli, A.N., S. De Gregori, M. Molinaro, M. Broglia, C. Tinelli, and R. Imberti. 2012. Bioavailability of a new oral spray melatonin emulsion compared with a standard oral formulation in healthy volunteers. *J Bioequiv Availab* 4:96–99.

Bénès, L., B. Claustrat, F. Horrière, M. Geoffriau, J. Konsil, K.A. Parrott, G. DeGrande, R.L. McQuinn, and J.W. Ayres. 1997. Transmucosal, oral controlled-release, and transdermal drug administration in human subjects: A crossover study with melatonin. *J Pharm Sci* 86:1115–1119.

Bonnefont-Rousselot, D. and F. Collin. 2010. Melatonin: Action as antioxidant and potential applications in human disease and aging. *Toxicology* 278:55–67.

Bowers, C.W., C. Baldwin, and R.E. Zigmond. 1984. Sympathetic reinnervation of the pineal gland after postganglionic nerve lesion does not restore normal pineal function. *J Neurosci* 4:2010–2015.

Bowers, C.W. and R.E. Zigmond. 1982. The influence of the frequency and pattern of sympathetic nerve activity on serotonin *N*-acetyltransferase in the rat pineal gland. *J Physiol* 330:279–296.

Brown, G.M., S.R. Pandi-Perumal, I. Trakht, and D.P. Cardinali. 2009. Melatonin and its relevance to jet lag. *Travel Med Infect Dis* 7:69–81.

Brzezinski, A. 1997. Melatonin in humans. *N Engl J Med* 336:186–195.

Cieciura, L. and G. Krakowski. 1991. Junctional systems in the pineal gland of the Wistar rat (*Rattus rattus*). A freeze-fracture and thin section study. *J Submicrosc Cytol Pathol* 23:327–330.

Coto Montes, A., M. Masson-Pévet, P. Pévet, and M. Møller. 1994. The presence of opioidergic pinealocytes in the pineal gland of the European hamster (*Cricetus cricetus*, L.). An immunocytochemical study. *Cell Tissue Res* 278:483–491.

David, G.F.X. and J. Herbert. 1973. Experimental evidence for a synaptic connection between habenula and pineal ganglion in the ferret. *Brain Res* 64:327–343.

de Almeida, E.A., P. Di Mascio, T. Harumi, D.W. Spence, A. Moscovitch, R. Hardeland, D.P. Cardinali, G. M. Brown, and S. R. Pandi-Perumal. 2011. Measurement of melatonin in body fluids: Standards, protocols and procedures. *Childs Nerv Syst* 27:879–889.

Deguchi, T. 1982. Sympathetic regulation of circadian rhythm of serotonin *N*-acetyltransferase activity in pineal gland of infant rat. *J Neurochem* 38:797–802.

Eriksson, K., A. Ostin, and J.O. Levin. 2003. Quantification of melatonin in human saliva by liquid chromatography–tandem mass spectrometry using stable isotope dilution. *J Chromatogr* 794:115–123.

Fourtillan, J.B., A.M. Brisson, M. Fourtillan, I. Ingrand, J.B. Decourt, and J. Girault. 2001. Melatonin secretion occurs at a constant rate in both young and older men and women. *Am J Physiol Endocrinol Metab* 280:E11–E22.

Hafner, A., J. Lovrić, D. Voinovich, and J. Filipović-Grčić. 2009. Melatonin-loaded lecithin/chitosan nanoparticles: Physicochemical characterization and permeability through Caco-2 cell monolayers. *Int J Pharm* 381:205–213.

Harumi, T. and S. Matsushima. 2000. Separation and assay methods for melatonin and its precursors. *J Chromatogr* 747:95–110.

Ivanov, S.V. 2007. Age-dependent morphology of human pineal gland: Supravital study. *Adv Gerontol* 20:60–65.

Jemima, J., P. Bhattacharjee, and R.S. Singhal. 2011. Melatonin: A review on the lesser known potential nutraceutical. *IJPSR* 2:1975–1987.

Johns, J.R., C. Chenboonthai, N.P. Johns, A. Saengkrasat, R. Kuketpitakwong, and S. Porasupatana. 2012. An intravenous injection of melatonin: Formulation, stability, pharmacokinetics and pharmacodynamics. *JAASP* 1:32–43.

Jung, B. and N. Ahmad. 2006. Melatonin in cancer management: Progress and promise. *Cancer Res* 66:9789–9793.

Junqueira, L.C., J. Carneiro, and R.O. Kelley. 2006. *Basic Histology*. Lange Medical Book, New York, pp. 405–406.

Klein, D.C. and R.Y. Moore. 1979. Pineal *N*-acetyltransferase and hydroxyindole-*O*-methyltransferase: Control by the retinohypothalamic tract and the suprachiasmatic nucleus. *Brain Res* 174:245–262.

Kostoglou-Athanassiou, I. 2013. Therapeutic applications of melatonin. *Ther Adv Endocrinol Metab* 4:13–24.

Krabbe, K.H. 1916. Histologische und embryologische Untersuchungenuber die Zirbeldruse des Menschen. *Anatomischer Anzeiger* 54:187–319.

Laure-Kamionowska, M., D. Maślińska, K. Deregowski, E. Czichos, and B. Raczkowska. 2003. Morphology of pineal glands in human foetuses and infants with brain lesions. *Folia Neuropathol* 41:209–215.

Li, C., G. Li, D.-X. Tan, F. Li, and X. Ma. 2013. A novel enzyme-dependent melatonin metabolite in humans. *J Pineal Res* 54:100–106.

López-Gamboa, M., J.S. Canales-Gómez, T. de Jesús Castro Sandoval, E.N. Tovar, M.A. Mejía, M. de los Ángeles, M. Baltazar, and J.A. Palma-Aguirre. 2010. Bioavailability of long acting capsules of melatonin in Mexican healthy volunteers. *J Bioequiv Availab* 2:116–119.

Lu, J., C. Lau, M.K. Lee, and M. Kai. 2002. Simple and convenient chemiluminescence method for the determination of melatonin. *Anal Chim Acta* 455:193–198.

Lukaszyk, A. and R.J. Reiter. 1975. Histophysiological evidence for the secretion of polypeptides by the pineal gland. *Anal Chim Acta* 143:451–464.

Mao, S., J. Chen, Z. Wei, H. Liu, and D. Bi. 2004. Intranasal administration of melatonin starch microspheres. *Int J Pharm* 272:37–43.

Matsushima, S., Y. Sakai, Y. Hira, Y. Oomuri, and S. Daikoku. 1994. Immunohistochemical studies on sympathetic and nonsympathetic nerve fibers and neuronal cell bodies in the pineal gland of cotton rats, *Sigmodon hispidus*. *Arch Histol Cytol* 57:47–58.

Mulchaheya, J.J., D.R. Goldwaterb, and F.P. Zemlan. 2004. A single blind, placebo controlled, across groups' dose escalation study of the safety, tolerability, pharmacokinetics and pharmacodynamics of the melatonin analog h-methyl-6-chloromelatonin. *Life Sci* 75:1843–1856.

Møller, M. 1997. Peptidergic cells in the mammalian pineal gland. Morphological indications for a paracrine regulation of the pinealocyte. *Biol Cell* 89:561–567.

Møller, M. and F.M. Baeres. 2002. The anatomy and innervation of the mammalian pineal gland. *Cell Tissue Res* 309:139–150.

Møller, M., B. van Deurs, and E. Westergaard. 1978. Vascular permeability to proteins and peptides in the mouse pineal gland. *Cell Tissue Res* 195:1–15.

Nakazato, Y., J. Hirato, A. Sasaki, H. Yokoo, H. Arai, Y. Yamane, and S. Jyunki. 2002. Differential labeling of the pinealocytes and pineal interstitial cells by a series of monoclonal antibodies to human pineal body. *Neuropathology* 22:26–33.

Pedersen, E.B., L.M. Fox, A.J. Castro, and J.A. McNulty. 1993. Immunocytochemical and electron-microscopic characterization of macrophage/microglia cells and expression of class II major histocompatibility complex in the pineal gland of the rat. *Cell Tissue Res* 272:257–265.

Radi, A. and G.E. Bekhiet. 1998. Voltammetry of melatonin at carbon electrodes and determination in capsules. *Bioelectrochem Bioenerg* 45:275–279.
Redins, G.M., C.A. Redins, and J.C. Novaes. 2001. The effect of treatment with melatonin upon the ultrastructure of the mouse pineal gland: A quantitative study. *Braz J Biol* 61:679–684.
Reitel, R.J. 1981. The mammalian pineal gland: Structure and function. *Am J Anat* 162:287–313.
Reiter, R.J., L.C. Manchester, and D.X. Tan. 2005. Melatonin in walnuts: Influence on levels of melatonin and total antioxidant capacity of blood. *Nutrition* 21:920–924.
Rodin, A.E. and J. Overall. 1967. Statistical relationship of weight of the human pineal to age and malignancy. *Cancer* 20:1203–1214.
Ross, M.H. and W. Pawlina. 2006. *Histology a Text and Atlas*, 5th edn. Williams &Wilkins, Baltimore, MD, pp. 698–700.
Sato, T., M. Kaneko, A. Hama, T. Kusakari, and H. Fujieda. 1996. Expression of class II MHC molecules in the rat pineal gland during development and effects of treatment with carbon tetrachloride. *Cell Tissue Res* 284:65–76.
Schemmer, P., A. Nickkholgh, H. Schneider, M. Sobirey, M. Weigand, M. Koch, J. Weitz, and M.W. Büchler. 2008. PORTAL: Pilot study on the safety and tolerance of preoperative melatonin application in patients undergoing major liver resection: A double-blind randomized placebo-controlled trial. *BMC Surg* 8:2–6.
Simonin, G., L. Bru, E. Lelievre, J.P. Jeanniot, N. Bromet, B. Walther, and C. Boursier-Neyret. 1999. Determination of melatonin in biological fluids in the presence of the melatonin agonist S20098: Comparison of immunological techniques and GC–MS methods. *J Pharmaceut Biomed* 21:591–601.
Teclemariam-Mesbah, R., G.J. Ter Horst, F. Postema, J. Wortel, and R.M. Buijs. 1999. Anatomical demonstration of the suprachiasmatic nucleus-pineal pathway. *J Comp Neurol* 406:171–182.
Van-Tassel, D.L. and S.D. O'Neill. 2001. Putative regulatory molecules in plants: Evaluating melatonin. *J Pineal Res* 31:1–7.
Vitale, A.A., C.C. Ferrari, H. Aldana, and J.M Affanni. 1996. Highly sensitive method for the determination of melatonin by normal-phase high-performance liquid chromatography with fluorometric detection. *J Chromatogr* 681:381–384.
Vollrath, L. 1981. The pineal organ. In: Oksche, A. and Vollrath, L. (eds.), *Handbuch der Mikroskopischen Anatomie des Menschen*, vol. VI/7. Springer, New York, pp. 1–665.
Vollrath, L. 1984. Functional anatomy of the human pineal gland. In: Reiter, R.J. (ed.), *The Pineal Gland*. Raven Press, New York, pp. 285–322.
Wang, A.Q., B.P. Wei, Y. Zhang, Y.J. Wang, L. Xu, and K. Lan. 2011. An ultra-high sensitive bioanalytical method for plasma melatonin by liquid chromatography–tandem mass spectrometry using water as calibration matrix. *J Chromatogr* 879:2259–2264.
Weaver, D.R., J.H. Stehle, E.G. Stopa, and S.M. Reppert. 1993. Melatonin receptors in human hypothalamus and pituitary: Implications for circadian and reproductive responses to melatonin. *J Clin Endocrinol Metab* 76:295–301.
Weiss, L. 1988. Pineal structure. In: *Cell and Tissue Biology: A Textbook of Histology*, 6th edn. Urban and Schwarzenberg, Baltimore, MA, pp. 997–1004.
Zawilska, J.B., D.J. Skene, and J. Arendt. 2009. Physiology and pharmacology of melatonin in relation to biological rhythms. *Pharmacol Rep* 61:383–410.
Zisapel, N. 2010. Melatonin and sleep. *Open Neuroendocrinol J* 3:85–95.

2 Deviations of Melatonin Levels and Signaling in Aging and Diseases

Options and Limits of Treatment

Rüdiger Hardeland

CONTENTS

2.1 INTRODUCTION

Melatonin is produced in numerous mammalian organs and cells, but the circulating hormone is mostly secreted by the pineal gland (Hardeland et al. 2011), with a prominent maximum at night. The nocturnal peak undergoes age-dependent changes (Waldhauser et al. 1993). A normal decline observed during youth is explained by a dilution effect due to the growth of the body in the absence of substantial changes in pineal secretion. However, decreases at advanced age reflect deficits in melatonin formation, which can be related to a dysfunction of the circadian master clock, the suprachiasmatic nucleus (SCN), which steers pineal activity or signal transmission from the retina via SCN to the pineal gland (Hardeland et al. 2012). However, the aging-related changes have turned out to be interindividually highly variable (Hardeland 2012a). Persons over 60 years may display almost normal nocturnal peaks or, in other cases, may have lost a robust melatonin rhythm. Moreover, the melatonin maximum is frequently phase-advanced in elderly subjects relative to younger ones (Skene and Swaab 2003). However, melatonin levels may be even more decreased in a number of diseases and disorders (summarized by Hardeland et al. 2011; Hardeland 2012a,b). As will be discussed next, these include metabolic diseases, pain and stressful conditions, and, especially, but with exceptions, neurodegenerative diseases. Sleep disturbances can be an indicator of a disturbed melatonin rhythm, which may be corrected by melatonergic treatment (Srinivasan et al. 2009). The changes caused by diseases and disorders may be transient or permanent. Permanently reduced levels of circulating melatonin appear to be stronger predictors of a diseased state than of aging per se. This chapter will analyze the conditions under which melatonin is decreased and discuss the options of treatments as well as their limits.

2.2 REDUCED LEVELS OF MELATONIN

As mentioned, the nocturnal melatonin peak is often decreased to a variable extent in elderly subjects. In some aged individuals, nighttime and daytime values are almost indistinguishable, whereas, in other persons, a well-pronounced rhythm is maintained with only moderate nocturnal reductions. Sometimes, daytime values are also decreased. As soon as melatonin formation is impaired, this can be detected in plasma concentrations, pineal glands (Skene et al. 1990), saliva (Kripke et al. 2005), cerebrospinal fluid (Brown et al. 1979; Liu et al. 1999), and in the urinary amounts of the main metabolite, 6-sulfatoxymelatonin (Youngstedt et al. 2001; Kripke et al. 2005; Mahlberg et al. 2006). The high interindividual variability in pineal melatonin synthesis is also reflected by the urinary 6-sulfatoxymelatonin levels, which can vary in apparently healthy subjects by a factor of 20 (Mahlberg et al. 2006).

Permanent decreases of melatonin formation, frequently in combination with a decomposed rhythm of the pineal hormone (Mishima et al. 1999; Wu et al. 2006; Wu and Swaab 2007), are observed in several neurodegenerative disorders, especially Alzheimer's disease (AD) and other types of senile dementia (Skene et al. 1990; Liu et al. 1999; Skene and Swaab 2003; Srinivasan et al. 2005; Hardeland 2012a,b). These changes are usually more strongly pronounced than in age-matched controls, and, sometimes, the rhythm is more or less abolished. Degeneration of the SCN seems to be the major cause of reduced melatonin and a decomposed rhythm. In cases of macular degeneration, the impaired rhythmic retinal input to the SCN may explain similar reductions in melatonin, as detected by low levels of urinary 6-sulfatoxymelatonin (Rosen et al. 2009).

However, reduced melatonin concentrations are, surprisingly, also observed in various other diseases and disorders that are not related to tissue destruction in the SCN or the light-input pathway. They include various neurological and stressful conditions, but are not always observed in every patient. These changes, which are only present in subpopulations or occur sporadically, have been described for the cases of autism spectrum disorders, schizophrenia, multiple sclerosis, and Menière's disease (summarized in: Hardeland 2012b). In some patients with autism spectrum disorders, decreased melatonin levels can be explained by mutations in the ASMT (acetylserotonin methyltransferase alias hydroxyindole *O*-methyltransferase) gene, which is responsible for the last step of melatonin biosynthesis (Toma et al. 2007; Melke et al. 2008; Jonsson et al. 2010; Goubran Botros et al. 2013). Some of the enzyme variants exhibit only 4%–8% of wild-type activity. These low-activity variants have been found not only in individuals with attention-deficit hyperactivity disorder (ADHD), bipolar disorder (BP), and intellectual disability, but also in the general population (Goubran Botros et al. 2013). Therefore, the low melatonin values cannot provide a monocausal explanation for these disorders. However, the reduced melatonin levels in the subpopulations of the affected subjects may appear plausible on this basis, and the changes should be classified as permanent. Transient changes are seen in other diseases and disorders. Severe epilepsy is a particularly variable condition, which may result in either decreases or increases of melatonin levels (Bazil et al. 2000; Molina-Carballo et al. 2007; Uberos et al. 2011). High serum concentrations are observed in some affected individuals during or after seizures.

In a number of other diseases and disorders, melatonin reductions of the transient type are observed, from which patients can return to normal after therapy and recovery. These include forms of pain such as migraine, neuralgia, and fibromyalgia (Claustrat et al. 1989, 1997; Rohr and Herold 2002), stressful conditions like bulimia (Rohr and Herold 2002), critical illness (Perras et al. 2006, 2007), especially sepsis (Srinivasan et al. 2010), and postoperative stress (Shigeta et al. 2001). In the latter case, these reductions have been only observed in patients devoid of complications, whereas others showed strong increases when entering the phase of delirium. Decreases of melatonin levels are sometimes reported in cancer patients (Grin and Grünberger 1998; Hu et al. 2009) and are frequently found in cardiovascular diseases such as coronary heart disease, cardiac syndrome X, and myocardial infarction (summarized in: Hardeland 2012b), endocrine and metabolic disorders, in particular diabetes type 2 (Peschke et al. 2007), and acute intermittent porphyria (Puy et al. 1993; Bylesjö et al. 2000).

2.3 ALTERED EXPRESSION OF MELATONIN RECEPTORS AND PRESENCE OF MELATONIN RECEPTOR GENE VARIANTS

With regard to melatonin signaling, two types of deviations are relevant to the topic of this chapter: changes in receptor density and variants of receptor genes. This has been mainly studied for the membrane receptors, MT_1 and MT_2. Gene polymorphisms have been occasionally reported for the nuclear transcription factors *Rorα* in macular degeneration and several forms of depression and *Rorβ* (alias "*Rzrβ*"), both discussed as nuclear melatonin receptors, as well as for the melatonin-binding enzyme quinone reductase 2 (*Nqo2*). However, in these cases, the involvement of melatonin has remained uncertain (summarized in: Hardeland et al. 2012). Similar uncertainties exist for gene variants of a melatonin receptor ortholog without an affinity to melatonin, GPR50, which can interact not only with MT_1, but also with other regulatory proteins not related to melatonin (Hardeland 2012b).

Losses of melatonin receptors have been mainly described in neurodegenerative disorders. Although melatonin is not typically reduced in the earlier stages of Parkinson's disease, the expression of MT_1 and MT_2 was found to decline in the substantia nigra and amygdala (Adi et al. 2010). Changes are much more pronounced in AD, in which densities of MT_1 and MT_2 are progressively decreased in the cortex and pineal gland (Brunner et al. 2006), of MT_1 in the cerebrovascular system (Savaskan et al. 2001) and, most importantly, in the SCN (Wu and Swaab 2007; Wu et al. 2007), and of MT_2 in the hippocampus (Savaskan et al. 2005) and retina (Savaskan et al. 2007).

Melatonin receptor polymorphisms have been found to be associated with several diseases (summarized in: Hardeland 2012b; Hardeland et al. 2012). Two major problems exist for a consistent interpretation of these findings. Even if the associations with diseases or disorders are statistically demonstrable, many of the variants are still too rare to be of explanatory value for the populations of affected patients. Moreover, many cases are known in which the variants are present in healthy subjects, without any sign of present or future pathologies. Therefore, the respective polymorphisms may be rather seen as possible contributing or risk factors. Several variants of MT_1 and MT_2 receptors have been identified, which do not only show reduced melatonin binding, but, additionally, changes in the balance of the two major signaling pathways, that is, $G\alpha_{i2/3}$-dependent decrease in cAMP and ERK (extracellular signal-regulated kinase) activation (Chaste et al. 2010). In the extreme case of an MT_1 I212T mutation, melatonin was not bound and, thus, did not display cAMP reductions, but still activated ERK to a certain extent, a finding to be interpreted as a constitutive activation. A truncated MT_1 stop mutant, Y170X, was to date exclusively found in individuals with ADHD (Chaste et al. 2011).

Contrary to the rare or very rare mutant alleles of melatonin receptors associated with diseases, the polymorphism of the *MTNR1B* gene, which encodes the MT_2 receptor, has provided substantial information on the associations of a human risk variant, the G-allele, with prediabetic changes, insulin resistance and the development of diabetes type 2 (extensive literature summarized in Hardeland 2012b; Hardeland et al. 2012). These findings, which have been repeatedly confirmed, certainly reflect deviations in melatonergic signaling, but do not seem to concern a loss but rather a gain of function, as recently discussed (Nagorny and Lyssenko 2012). However, this surprising interpretation contrasts with other data, indicating that rare loss-of-function mutants of the *MTNR1B* gene are also associated with diabetes type 2 (Bonnefond et al. 2012).

2.4 PROPERTIES OF SYNTHETIC MELATONERGIC AGONISTS, PROBLEMS, AND LIMITS

Since melatonin has a very short half-life in the range of mostly 20–30 min in the circulation, longer acting synthetic agonists have been developed. Among the approved drugs, only TIK-301 (β-methyl-6-chloromelatonin) is an indolic compound, which has, however, only received an orphan drug status by the FDA (Food and Drug Administration) for the treatment of circadian rhythm disorders in blind subjects who lack light perception, and for individuals with tardive dyskinesia. Among the other nonindolic compounds, ramelteon {(*S*)-*N*-[2-(1,6,7,8-tetrahydro-2*H*-indeno[5,4-b]furan-8-yl) ethyl]propionamide} was approved by the FDA for the treatment of insomnia in the United States, but the application to EMEA (European Medicines Agency) in Europe was withdrawn. A naphthalenic agonist, agomelatine {*N*-[2-(7-methoxynaphth-1-yl)ethyl]acetamide}, was approved by EMEA, but only for the treatment of major depressive episodes (MDE) in adults. As an alternative to synthetic drugs, slow/controlled release formulations of melatonin have been developed, among which Circadin® was approved by EMEA, but only for the treatment of insomnia in patients aged 55 years and over. Comparisons of these drugs have been published elsewhere (Hardeland 2009; Srinivasan et al. 2009, 2011a), also with regard to pharmacology, pharmacokinetics, and metabolism (Hardeland and Poeggeler 2012). Numerous other agonists of different chemical structures have been developed, which are to date still in the status of investigational drugs (Hardeland 2010).

Some relevant differences of the approved drugs concern receptor affinities, half-life in the circulation, and binding to other receptors (cf. reviews mentioned earlier). TIK-301 binds to the MT_1 receptor with an approximately same affinity as melatonin, but has an affinity almost one order of magnitude higher to MT_2. Moreover, it acts as an antagonist at two serotonergic receptors, $5\text{-}HT_{2B}$ and $5\text{-}HT_{2C}$. Its half-life in the circulation is in the range of 1 h. Ramelteon has a higher affinity to MT_1, by somewhat less than one order of magnitude, and also a higher affinity to MT_2, but not as high as TIK-301. Its half-life is in the range of 1–2 h with some interindividual variability, but one of its metabolites, usually referred to as M-II, which retains about one-tenth of the receptor affinity, has a considerably extended half-life and can amount to concentrations 30 or even 100 times higher than the parent compound. Agomelatine has moderately higher affinities to both MT_1 and MT_2, but acts additionally as a $5\text{-}HT_{2C}$ antagonist. Its antidepressant properties have been interpreted on the basis of the $5\text{-}HT_{2C}$ inhibition or as an interplay between melatonergic and antiserotonergic actions. In this regard, similar properties may be assumed for TIK-301. A major concern with agomelatine is related to possible hepatotoxicity, which has led to the release of warnings and a monitoring guidance to prescribers concerning liver function. No such problems are to date known for ramelteon, which has already been tested in long-term studies. Data on TIK-301 are not yet sufficient for a reliable toxicological comparison. In short-terms studies, all the drugs mentioned are well tolerated, and side effects are not different from placebo. Precautions concerning C→cytochrome P_{450} (CYP)-modulating drugs, hepatic and renal impairment, alcohol, puberty, and pregnancy have been published in detail for ramelteon and may similarly apply to the other melatonergic agonists including melatonin (cf. Hardeland 2009).

2.5 RATIONALE OF MELATONERGIC TREATMENT

A melatonergic replacement therapy has, of course, first to consider eventual contraindications. Apart from diseases, reproductive conditions, and drug interactions mentioned earlier, some other problems may exist, in particular, of immunological nature. Since melatonin can exert both proinflammatory and antiinflammatory effects, depending on the grade and state of the inflammation, melatonergic actions may be undesired (Hardeland 2013). This has been debated especially with regard to autoimmune diseases. Even though this matter may not be definitely settled, it should be taken as a caveat. It is still uncertain whether the immunological actions of synthetic melatonergic agonists are identical with those of the natural hormone, because various effects have been related to RORα subforms which do not interact with nonindolic compounds, but a cautious treatment should avoid a risky treatment. Another debate concerns the usefulness of melatonin or melatonergic agonists in Parkinson's disease, in which a controversy exists with regard to the possible detrimental effects (Srinivasan et al. 2011b).

In patients or elderly subjects in whom a replacement therapy is desired, the first decision to be made is that of the necessity of the duration of action. Very frequently, a treatment is requested because of sleep difficulties (for extensive literature, see the work of Hardeland 2012a). In brief, only short actions are required in patients with problems of falling asleep. In these cases, immediate-release melatonin formulations are fully sufficient, because it reliably reduces sleep onset latency, already at low doses, sometimes below 1 mg. In this regard, synthetic agonists, which likewise exhibit this effect, may not be of advantage, also because the recommended doses are considerably higher (ramelteon: 4 or 8 mg; agomelatine: 25 mg), despite longer half-lives and higher receptor affinities.

Moreover, short actions are sufficient in cases in which the chronobiotic, that is, phase-shifting properties of melatonin are decisive. Resetting of circadian oscillators is required in cases of rhythm perturbations, including circadian rhythm sleep disorders. A short-acting chronobiotic such as melatonin is capable of inducing phase adjustments, because circadian oscillators are entrained by so-called nonparametric resetting, that is, by stimuli in which the relative change is decisive rather than the absolute level of the synchronizer. Perturbations may be caused either externally by light at night or transmeridian flights, by clocks poorly coupled to the environmental cycle, for example, in patients with impaired light input or deviating spontaneous circadian periods, and in dysphased or desynchronized rhythms within the circadian multioscillator system (Hardeland 2012b; Hardeland et al. 2012). Insufficient coupling may result from flattened oscillations, especially under conditions of reduced melatonin secretion because of advanced age or disease. Melatonin has been shown to enhance rhythm amplitudes in some studies and is assumed to do so in other cases too. Impaired coupling and internal desynchronization of rhythms seem to be involved in disturbances of physical and mental fitness, including bipolar and seasonal affective disorders.

As far as circadian malfunctioning is implicated in these latter types of mood disorders, melatonin can be effective in readjusting rhythms and, thereby, improving symptoms. However, it is important not to confuse such actions with direct antidepressant effects, as exerted by agomelatine and TIK-301, which additionally act as 5-HT_{2C} antagonists. Otherwise, treatments with synthetic melatonergic drugs can be expected to be beneficial on a circadian basis, but neither a higher receptor affinity nor a longer half-life would explain a superior efficacy when compared to melatonin. Therefore, the synthetic compounds can only be superior over melatonin if more than phase resetting or promotion of sleep onset is desired.

As far as phase resetting is intended, the treatment has to be based on fundamental chronobiological rules. Resetting signals act according to the specific phase response curve (PRC) for the respective agent. Melatonin has to be administered according to the human PRC (Lewy et al. 1992). Readjustment of rhythms by melatonin will be achieved only if it is given in an appropriate, sufficiently sensitive phase within the circadian cycle. If the rhythm is dysphased because of poor coupling to synchronizers or a deviating period length, it may take several days more until the oscillation has attained the desired phase. Otherwise, a disregard of the chronobiological fundamentals can lead to false conclusions on inefficacy.

Short-term melatonergic actions are different from a replacement therapy, which would be desired in aged individuals and patients suffering from diseases associated with decreases in melatonin levels. Because of melatonin's short half-life, immediate-release formulations of melatonin cannot afford a satisfactory substitution. Synthetic agonists or a controlled-release melatonin formulation such as Circadin should be assumed to be superior. With regard to melatonin's exceptionally good tolerability, the pineal hormone may be tested first. Among the synthetic drugs, ramelteon would be a choice in the United States. Agomelatine, which is licensed in Europe, might give comparably good results, but its approval is restricted to antidepressant treatment. Circadin and ramelteon are licensed only for treating insomnia, the former only for individuals of 55 years and above, and TIK-301 for the use in blind people. The full spectrum of possible applications is, thus, not covered by the approvals.

All licensed melatonergic agonists are effective in promoting sleep onset. Improvements of sleep maintenance or sleep quality have also been reported. However, their extent has remained relatively moderate, although statistical measures have frequently reached significance. In elderly patients with primary chronic insomnia, the efficacy of ramelteon on sleep maintenance was recently found to be highly variable (Pandi-Perumal et al. 2011). To date, no complete restoration of persistent sleep throughout the night has been achieved by any melatonergic treatment, despite statistically demonstrable increases.

2.6 CONCLUSION

Although a number of pathological conditions, including age-related changes, are known to be associated with decreased nocturnal melatonin levels, or with genetic alterations of melatonin signaling, a convincing replacement therapy is not yet possible, neither with the natural hormone nor with synthetic melatonergic agonists. Nevertheless, melatonergic treatment is not without value, because melatonin and its approved analogs can reliably reduce sleep latency and efficiently readjust disturbed or dysphased circadian rhythms.

REFERENCES

Adi, N., D. C. Mash, Y. Ali, C. Singer, L. Shehadeh, and S. Papapetropoulos. 2010. Melatonin MT1 and MT2 receptor expression in Parkinson's disease. *Med. Sci. Monit.* **16**:BR61–BR67.

Bazil, C. W., D. Short, D. Crispin, and W. Zheng. 2000. Patients with intractable epilepsy have low melatonin, which increases following seizures. *Neurology* **55**:1746–1748.

Bonnefond, A., N. Clément, K. Fawcett et al. 2012. Rare *MTNR1B* variants impairing melatonin receptor 1B function contribute to type 2 diabetes. *Nat. Genet.* **44**:297–301.

Brown, G. M., S. N. Young, S. Gauthier, H. Tsui, and L. J. Grota. 1979. Melatonin in human cerebrospinal fluid in daytime; its origin and variation with age. *Life Sci.* **25**:929–936.

Brunner, P., N. Sözer-Topcular, R. Jockers et al. 2006. Pineal and cortical melatonin receptors MT_1 and MT_2 are decreased in Alzheimer's disease. *Eur. J. Histochem.* **50**:311–316.

Bylesjö, I., L. Forsgren, and L. Wetterberg. 2000. Melatonin and epileptic seizures in patients with acute intermittent porphyria. *Epileptic Disord.* **2**:203–208.

Chaste, P., N. Clement, H. Goubran Botros et al. 2011. Genetic variations of the melatonin pathway in patients with attention-deficit and hyperactivity disorders. *J. Pineal Res.* **51**:394–399.

Chaste, P., N. Clement, P. Mercati et al. 2010. Identification of pathway-biased and deleterious melatonin receptor mutants in autism spectrum disorders and in the general population. *PloS One* **5**:e11495. doi: 10.1371/journal.pone.0011495.

Claustrat, B., J. Brun, M. Geoffriau, R. Zaidan, C. Mallo, and G. Chazot. 1997. Nocturnal plasma melatonin profile and melatonin kinetics during infusion in status migrainosus. *Cephalalgia* **17**:511–517.

Claustrat, B., C. Loisy, J. Brun, S. Beorchia, J. L. Arnaud, and G. Chazot. 1989. Nocturnal plasma melatonin levels in migraine: A preliminary report. *Headache* **29**:242–245.

Goubran Botros, H., P. Legrand, C. Pagan et al. 2013. Crystal structure and functional mapping of human ASMT, the last enzyme of the melatonin synthesis pathway. *J. Pineal Res.* **54**:46–57.

Grin, W. and W. Grünberger. 1998. A significant correlation between melatonin deficiency and endometrial cancer. *Gynecol. Obstet. Invest.* **45**:62–65.

Hardeland, R. 2009. New approaches in the management of insomnia: Weighing the advantages of prolonged release melatonin and synthetic melatoninergic agonists. *Neuropsychiatr. Dis. Treat.* **5**:341–354.

Hardeland, R. 2010. Investigational melatonin receptor agonists. *Expert Opin. Investig. Drugs* **19**:747–764.

Hardeland, R. 2012a. Neurobiology, pathophysiology, and treatment of melatonin deficiency and dysfunction. *ScientificWorldJournal* **2012**:640389. doi: 10.1100/2012/640389.

Hardeland, R. 2012b. Melatonin in aging and disease—Multiple consequences of reduced secretion, options and limits of treatment. *Aging Dis.* **3**:194–225.

Hardeland, R. 2013. Melatonin and the theories of aging: a critical appraisal of melatonin's role in antiaging mechanisms. *J. Pineal Res.* **55**:325–356.

Hardeland, R., D. P. Cardinali, V. Srinivasan, D. W. Spence, G. M. Brown, and S. R. Pandi-Perumal. 2011. Melatonin—A pleiotropic, orchestrating regulator molecule. *Prog. Neurobiol.* **93**:350–384.

Hardeland, R., J. A. Madrid, D.-X. Tan, and R. J. Reiter. 2012. Melatonin, the circadian multioscillator system and health: The need for detailed analyses of peripheral melatonin signaling. *J. Pineal Res.* **52**:139–166.

Hardeland, R. and B. Poeggeler. 2012. Melatonin and synthetic melatonergic agonists: Actions and metabolism in the central nervous system. *Cent. Nerv. Syst. Agents Med. Chem.* **12**:189–216.

Hu, S., G. Shen, S. Yin, W. Xu, and B. Hu. 2009. Melatonin and tryptophan circadian profiles in patients with advanced non-small cell lung cancer. *Adv. Ther.* **26**:886–892.

Jonsson, L., E. Ljunggren, A. Bremer et al. 2010. Mutation screening of melatonin-related genes in patients with autism spectrum disorders. *BMC Med. Genomics* **3**:10. doi: 10.1186/1755-8794-3-10.

Kripke, D. F., S. D. Youngstedt, J. A. Elliott et al. 2005. Circadian phase in adults of contrasting ages. *Chronobiol. Int.* **22**:695–709.

Lewy, A. J., S. Ahmed, J. M. Jackson, and R. L. Sack. 1992. Melatonin shifts human circadian rhythms according to a phase-response curve. *Chronobiol. Int.* **9**:380–392.

Liu, R. Y., J. N. Zhou, J. van Heerikhuize, M. A. Hofman, and D. F. Swaab. 1999. Decreased melatonin levels in postmortem cerebrospinal fluid in relation to aging, Alzheimer's disease, and apolipoprotein E-ε4/4 genotype. *J. Clin. Endocrinol. Metab.* **84**:323–327.

Mahlberg, R., A. Tilmann, L. Salewski, and D. Kunz. 2006. Normative data on the daily profile of urinary 6-sulfatoxymelatonin in healthy subjects between the ages of 20 and 84. *Psychoneuroendocrinology* **31**:634–641.

Melke, J., H. Goubran Botros, P. Chaste et al. 2008. Abnormal melatonin synthesis in autism spectrum disorders. *Mol. Psychiatry* **13**:90–98.

Mishima, K., T. Tozawa, K. Satoh, Y. Matsumoto, Y. Hishikawa, and M. Okawa. 1999. Melatonin secretion rhythm disorders in patients with senile dementia of Alzheimer's type with disturbed sleep-waking. *Biol. Psychiatry* **45**:417–421.

Molina-Carballo, A., A. Muñoz-Hoyos, M. Sánchez-Forte, J. Uberos-Fernández, F. Moreno-Madrid, and D. Acuña-Castroviejo. 2007. Melatonin increases following convulsive seizures may be related to its anticonvulsant properties at physiological concentrations. *Neuropediatrics* **38**:122–125.

Nagorny, C. and V. Lyssenko. 2012. Tired of diabetes genetics? Circadian rhythms and diabetes: The *MTNR1B* story? *Curr. Diab. Rep.* **12**:667–672.

Pandi-Perumal, S. R., D. W. Spence, J. C. Verster et al. 2011. Pharmacotherapy of insomnia with ramelteon: Safety, efficacy and clinical applications. *J. Cent. Nerv. Syst. Dis.* **3**:51–65.

Perras, B., V. Kurowski, and C. Dodt. 2006. Nocturnal melatonin concentration is correlated with illness severity in patients with septic disease. *Intensive Care Med.* **32**:624–625.

Perras, B., M. Meier, and C. Dodt. 2007. Light and darkness fail to regulate melatonin release in critically ill humans. *Intensive Care Med.* **33**:1954–1958.

Peschke, E., I. Stumpf, I. Bazwinsky, L. Litvak, H. Dralle, and E. Mühlbauer. 2007. Melatonin and type 2 diabetes — a possible link? *J. Pineal Res.* **42**:350–358.

Puy, H., J. C. Deybach, P. Baudry, J. Callebert, Y. Touitou, and Y. Nordmann. 1993. Decreased nocturnal plasma melatonin levels in patients with recurrent acute intermittent porphyria attacks. *Life Sci.* **53**:621–627.

Rohr, U. D. and J. Herold. 2002. Melatonin deficiencies in women. *Maturitas* **41** (Suppl. 1):S85–S104.

Rosen, R., D. N. Hu, V. Perez et al. 2009. Urinary 6-sulfatoxymelatonin level in age-related macular degeneration patients. *Mol. Vis.* **15**:1673–1679.

Savaskan, E., M. A. Ayoub, R. Ravid et al. 2005. Reduced hippocampal MT_2 melatonin receptor expression in Alzheimer's disease. *J. Pineal Res.* **38**:10–16.

Savaskan, E., R. Jockers, M. Ayoub et al. 2007. The MT_2 melatonin receptor subtype is present in human retina and decreases in Alzheimer's disease. *Curr. Alzheimer Res.* **4**:47–51.

Savaskan, E., G. Olivieri, L. Brydon et al. 2001. Cerebrovascular melatonin MT_1-receptor alterations in patients with Alzheimer's disease. *Neurosci. Lett.* **308**:9–12.

Shigeta, H., A. Yasui, Y. Nimura et al. 2001. Postoperative delirium and melatonin levels in elderly patients. *Am. J. Surg.* **182**:449–454.

Skene, D. J. and D. F. Swaab. 2003. Melatonin rhythmicity: effect of age and Alzheimer's disease. *Exp. Gerontol.* **38**:199–206.

Skene, D. J., B. Vivien-Roels, D. L. Sparks et al. 1990. Daily variation in the concentration of melatonin and 5-methoxytryptophol in the human pineal gland: effect of age and Alzheimer's disease. *Brain Res.* **528**:170–174.

Srinivasan, V., A. Brzezinski, S. R. Pandi-Perumal, D. W. Spence, D. P. Cardinali, and G. M. Brown. 2011a. Melatonin agonists in primary insomnia and depression-associated insomnia: Are they superior to sedative-hypnotics? *Prog. Neuropsychopharmacol. Biol. Psychiatry* **35**:913–923.

Srinivasan, V., D. P. Cardinali, U. S. Srinivasan et al. 2011b. Therapeutic potential of melatonin and its analogs in Parkinson's disease: Focus on sleep and neuroprotection. *Ther. Adv. Neurol. Disord.* **4**:297–317.

Srinivasan, V., S. R. Pandi-Perumal, G. J. M. Maestroni, A. I. Esquifino, R. Hardeland, and D. P. Cardinali. 2005. Role of melatonin in neurodegenerative diseases. *Neurotox. Res.* **7**:293–318.

Srinivasan, V., S. R. Pandi-Perumal, D. W. Spence, H. Kato, and D. P. Cardinali. 2010. Melatonin in septic shock: Some recent concepts. *J. Crit. Care* **25**:656.e1–656.e6.

Srinivasan, V., S. R. Pandi-Perumal, I. Trahkt et al. 2009. Melatonin and melatonergic drugs on sleep: Possible mechanisms of action. *Int. J. Neurosci.* **119**:821–846.

Toma, C., M. Rossi, I. Sousa et al. 2007. Is ASMT a susceptibility gene for autism spectrum disorders? A replication study in European populations. *Mol. Psychiatry* **12**:977–999.

Uberos, J., M. C. Augustin-Morales, A. Molina Carballo, J. Florido, E. Narbona, and A. Muñoz-Hoyos. 2011. Normalization of the sleep-wake pattern and melatonin and 6-sulphatoxy-melatonin levels after a therapeutic trial with melatonin in children with severe epilepsy. *J. Pineal Res.* **50**:192–196.

Waldhauser, F., B. Ehrhart, and E. Förster. 1993. Clinical aspects of the melatonin action: Impact of development, aging, and puberty, involvement of melatonin in psychiatric disease and importance of neuroimmunoendocrine interactions. *Experientia* **49**:671–681.

Wu, Y. H., D. F. Fischer, A. Kalsbeek et al. 2006. Pineal clock gene oscillation is disturbed in Alzheimer's disease, due to functional disconnection from the "master clock." *FASEB J.* **20**:1874–1876.

Wu, Y. H. and D. F. Swaab. 2007. Disturbance and strategies for reactivation of the circadian rhythm system in aging and Alzheimer's disease. *Sleep Med.* **8**:623–636.

Wu, Y. H., J. N. Zhou, J. Van Heerikhuize, R. Jockers, and D. F. Swaab. 2007. Decreased MT_1 melatonin receptor expression in the suprachiasmatic nucleus in aging and Alzheimer's disease. *Neurobiol. Aging* **28**:1239–1247.

Youngstedt, S. D., D. F. Kripke, J. A. Elliott, and M. R. Klauber. 2001. Circadian abnormalities in older adults. *J. Pineal Res.* **31**:264–272.

3 Pineal Gland Volume and Melatonin

Niyazi Acer, Süleyman Karademir, and Mehmet Turgut

CONTENTS

3.1 INTRODUCTION

Circumventricular organs are located in or near the wall of the third ventricle, the cerebral aqueduct of Sylvius, and the fourth ventricle, may be of functional importance with regard to the cerebrospinal fluid (CSF) composition, hormone secretion into the ventricles (Duvernoy and Risold 2007, Horsburgh and Massoud 2013, McKinley et al. 1990). The pineal gland is one of the circumventricular organs in humans, having no blood–brain barrier (BBR) and being attached to the roof of the posterior aspect of the third ventricle. It is located beneath the splenium of the corpus callosum (Figure 3.1). There is no direct connection between the central nervous system (CNS) and the pineal gland. The pineal gland likely affects CNS functions as a result of the release of these hormones into the general circulation and into the brain through BBR (Acer et al. 2011).

Aaron Lerner identified melatonin (Mel) 50 years ago, and it was measured in peripheral body fluids 30 years ago (Arendt 2006). It has been 25 years since bright white light was shown to suppress Mel totally and around 20 years since Mel-binding sites in the suprachiasmatic nuclei (SCN) and pars tuberalis were discovered (Arendt 2006). All of them have enhanced our knowledge concerning the role and effects of Mel in humans (Arendt 2006).

The pineal gland produces Mel, which helps regulate day–night cycles known as the body's circadian rhythm (Acer et al. 2011). Mel secretion rises in the dark and occurs at a low level in daylight (Arendt 2006, Standring 2008). Thus, its secretion fluctuates seasonally with changes in day length. Mel may suppress gonadotropin secretion; removal of the pineal from animals causes premature sexual maturation (Arendt 2006, Standring 2008). The pineal gland is an endocrine gland of major regulatory importance. It modifies the activity of the adenohypophysis, neurohypophysis, endocrine pancreas, parathyroids, adrenal cortex, adrenal medulla, and gonads. Its effect is largely inhibitory (Arendt 2006, Standring 2008).

From animal studies, the fundamental role of Mel has been described as a photoneuroendocrine transducer of information during day length. (Arendt 1995). Interestingly, it has been suggested that Mel is a universal biological signal indicating darkness, possibly due to changing duration of hormone secretion at different day lengths (Arendt 1995, Goldman 2001).

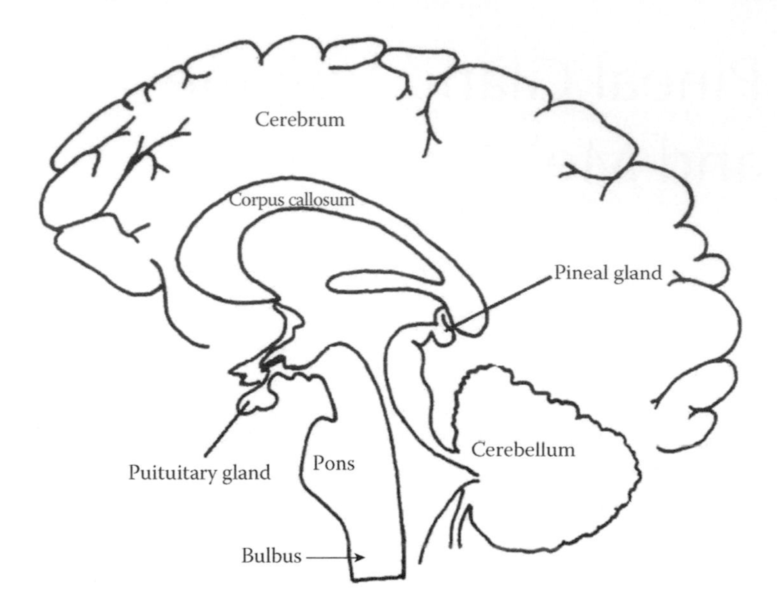

FIGURE 3.1 Schematic drawing of intracranial structures including the pineal gland.

3.2 ANATOMY OF THE HUMAN PINEAL GLAND

The pineal gland is a round or a pine cone-shaped structure like a pine cone in humans. The pineal gland is a reddish-gray organ, about 8 mm long and 5 mm wide, occupying a depression between the superior colliculi (Acer et al. 2011, Duvernoy 1999, Duvernoy et al. 2000). Anatomically, it is composed of three main parts: "pineal body," "pineal apex," and "pineal stalk." The pineal gland is attached by a short broad stalk to the diencephalon, and the stalk lines the pineal recess whose superior lip links the pineal gland to the habenular commissure and habenular nuclei, and inferior lip to the posterior commissure (Acer et al. 2011, Duvernoy 1999, Duvernoy et al. 2000). The stalk of the human pineal gland is invaginated by the pineal recess (Jinkins 2000, Langman 1975).

The pineal gland is located inferior to the splenium of the corpus callosum, rostral to the posterior commissure, and posterior to habenular trigone (Acer et al. 2011, Standring 2008). The human pineal gland is related to the thalamic pulvinar laterally and the posterior commissure and superior colliculi inferiorly (Acer et al. 2011, Duvernoy 1999, Duvernoy et al. 2000). Its base is attached by a peduncle, which divides into inferior and superior laminae by the pineal recess of the third ventricle (Figure 3.1). Importantly, it is well known that the pineal gland is encapsulated within the pial sheath and is bathed in the CSF (Jinkins 2000).

The gland has a rich blood supply, with a flow rate of around 4 mL/min/g (Touitou and Haus 1994). The pineal arteries are branches of the medial posterior choroidal arteries, which are the branches of the posterior cerebral artery (Acer et al. 2011). Within the gland, the branches of the arteries supply fenestrated capillaries whose endothelial cells rest on a tenuous and sometimes incomplete basal lamina. The capillaries drain into numerous pineal veins, which open into the internal cerebral veins and/or into the great cerebral vein (Acer et al. 2011, Standring 2008).

Pineal gland regresses rapidly after the age of 7 and is no more than a tiny shrunken mass of fibrous tissue in the adult (Standring 2008). The pineal gland usually calcifies after the age of 16. From the second decade, calcareous deposits accumulate in the pineal extracellular matrix, where they are deposited concentrically as corpora arenacea or "brain sand" or involution (Standring 2008). Calcification is often detectable in skull radiographs, such as x-rays, enabling radiologists to determine the position of the gland and when it can provide a useful indicator of a space-occupying lesion if the gland is significantly displaced from the midline (Standring 2008). Spiral CT makes

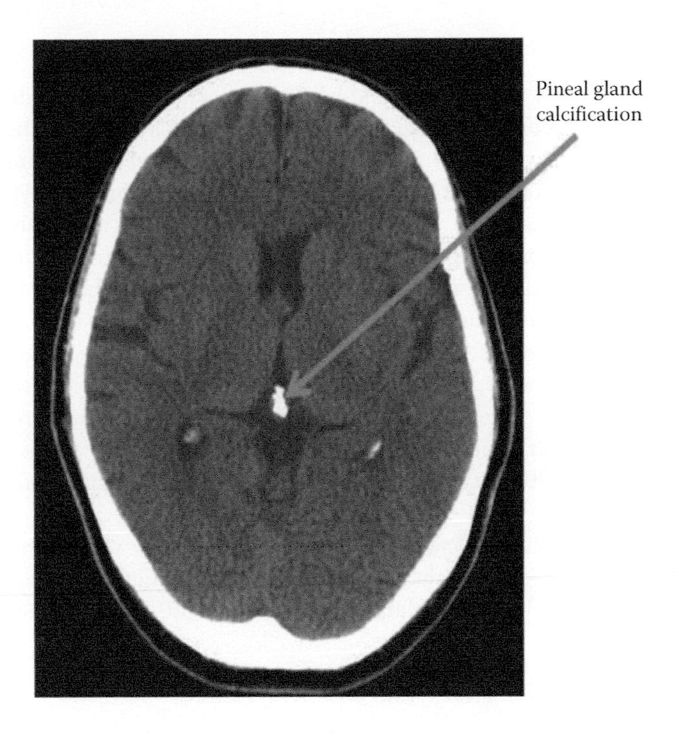

FIGURE 3.2 Axial CT section of the brain demonstrating pineal calcification.

easy the quantification of pineal calcifications (Figure 3.2). Pineal volumetry is ideally carried out using MRI (magnetic resonance imaging) (Nolte et al. 2009).

Although it is well known that the structure of the brain sand results from calcium and magnesium salts, its functional importance is not clear (Vígh et al. 1998). Nevertheless, some investigators suggest that the calcareous concretions are signs of the secretory functions of the gland (Welsh 1985). In general, it is accepted that these concretions are consequences of progressive degeneration of pinealocytes, whose remnants remain visible in the lobule periphery (Kunz et al. 1999). On the other hand, it is interesting to note that the calcifications are not always associated with the pinealocytes; they can be found at the external side of the gland, for instance, in the choroid plexuses (Taraszewska et al. 2008).

Pineal cysts are very common, and the effects of pineal cysts upon pineal function are not known (Nolte et al. 2009, 2010). Pineal cysts are frequently encountered from the fetal period to old age. In autoptic series, the prevalence of pineal cysts has been reported to range between 33% and 40% (Hasegawa et al. 1987), whereas a strong discrepancy was observed in MRI studies that report significantly lower prevalence of 0.1%–4.0% at standard field strengths (1.5 T) (Lum et al. 1987). Hasegawa et al. (1987) studied 168 autopsy cases and detected cysts in 39.7%. Nolte et al. (2010) used true FISP (true-Fast-Imaging-with-Steady-State-Precession) imaging with a voxel size of 800 μm × 800 μm × 400 μm, and they found an overall prevalence of 35.1% (Nolte et al. 2010). They found that pineal cysts were detected frequently in the standard T1–T2 and FLAIR sequences (Nolte et al. 2010). The pineal cyst is usually diagnosed by MRI (Bosnjak et al. 2009).

Pineal cysts are defined when the lesion is 8 mm long or greater in a single plane according to MRI (Engel et al. 2000).

Bumb et al. (2012) found pineal cyst prevalence in children to be 57% using 3D volumetric true FISP sequences with 1.5 T. Whitehead et al. (2013) found that the imaging prevalence of pineal cysts was equally distributed across all pediatric age ranges. Pineal cysts were present in 57% of children. They found that there was a slight trend toward increasing pineal gland volume with age (Whitehead et al. 2013).

3.3 INNERVATION OF THE PINEAL GLAND

The sympathetic innervation of the pineal gland is very important for the production of Mel. The pineal gland receives neural inputs from the sympathetic nervous system via the superior cervical ganglia (Acer et al. 2011). It is well known that the production of Mel in the human pineal gland is received by retinal photoreceptors and transmitted to the SCN of the anterior hypothalamus via the optic nerves (Møller and Baeres 2002). The SCN is a circadian pattern of activity for Mel secretion by the pineal gland and a projection from the SCN to the paraventricular nucleus, a neuronal nucleus in the hypothalamus (Kalsbeek et al. 2000, Vrang et al. 1995). Retinal input to the SCN of the hypothalamus is transported to the reticular formation, and activation of the reticular formation causes excitatory effects on the sympathetic neurons of the intermediolateral nucleus of the upper two thoracic spinal segments in the spinal cord via the reticulospinal tracts (Aslan 2001). Sympathetic neurons carry the secretomotor impulses from the reticular formation to the pineal gland through the nervi conarii (sympathetic postganglionic fibers) (Kalsbeek et al. 2010, Vrang et al. 1995). Fibers of the nervus conarii enter the apex of the pineal gland and extend toward the parenchyma (Kenny 1965).

Anatomically, the parasympathetic pathway consists of cell bodies of the neurons located in the superior salivatory nucleus, the sphenopalatine and otic ganglia of the brain, the fibers of these neurons in the nervus conarii, and the pineal gland (Kalsbeek et al. 2010, Vrang et al. 1995).

3.4 RELATION TO AGING AND MELATONIN LEVEL

Mel secretion primarily reaches the highest level at night (about 10 times increased level) in contrast to the level in the daytime (Touitou 2001). It is well known that the pineal gland produces Mel during the night, with maximum levels at midnight (02:00–03:00 AM) (Arendt and Skene 2005). The highest Mel levels are found in children younger than 4 years; Mel levels begin to decline with age (Pandi-Perumal et al. 2005). Synthesis of melatonin occurs not only in the pineal gland but also in the extrapineal Mel sources, including the retina, skin, and gut (Pandi-Perumal et al. 2005). However, it is accepted that circulating Mel is almost totally derived from the pineal gland, as indicated by its disappearance after pineal removal (Pandi-Perumal et al. 2005).

In all living organisms, including humans, it is well known that the levels of Mel show a circadian rhythm, with low values during the daytime and high values at night (Arendt 1995, Karasek 1999). Interestingly, the rhythm is provided by the circadian pacemaker localized in the SCN of the hypothalamus in vertebrates (Arendt 1995, Karasek 1999). However, it is known that the levels of Mel in the blood diminish gradually after puberty, and the day–night differences in Mel secretion are almost absent in many elderly individuals (Figure 3.3) (Arendt 1995, Karasek 1999).

There are some theories linking Mel to aging. Recent studies suggest that diminished levels of Mel in elderly individuals prevent antioxidant protection against the damage caused by free radicals (Karasek 2004, Reiter et al. 2002).

3.5 PINEAL GLAND VOLUME STUDIES

Several authors have reported the volume of the pineal gland using MRIs (Acer et al. 2012, Bumb et al. 2012, Nolte et al. 2009, Sun et al. 2009). They used indirect methods for determining the pineal gland volume, using 2D images and various formulas (e.g., volume = 1/2 × length × width × height) and manual and stereological measurements (Acer et al. 2012, Bumb et al. 2012, Nolte et al. 2009, Sun et al. 2009).

Hasegawa et al. (1987) reported that the weight of the adult human pineal gland is about 50–150 mg. However, it is well known that the weight and volume of the pineal gland vary with the time of the day, age of the person, and other factors (Sumida et al. 1996). Anatomically, it has been reported that the average size of the pineal gland is 7.4, 6.9, and 2.5 mm in length, width, and height,

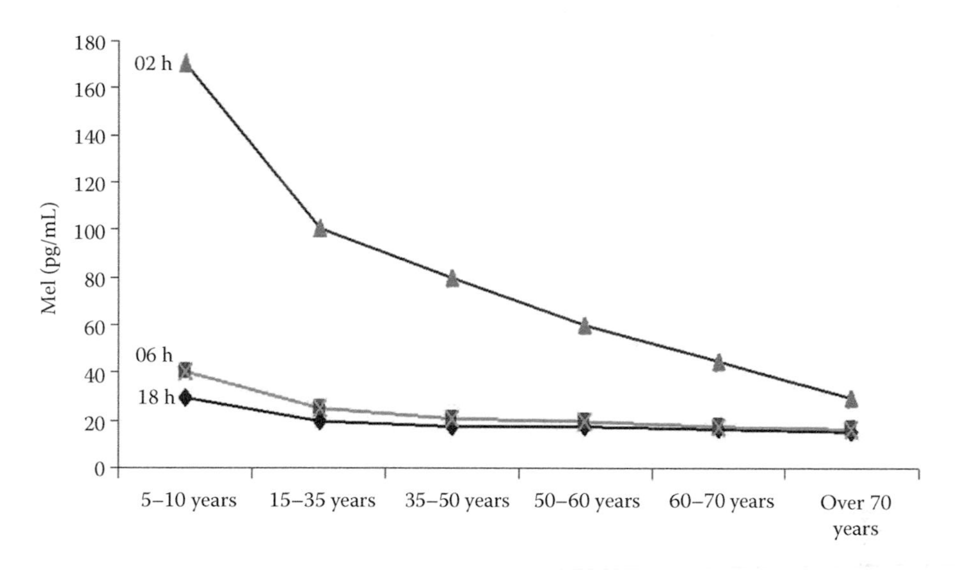

FIGURE 3.3 Circadian profiles of serum Mel concentrations at various years of age and different hours during the day. (Modified from Karasek, M., *Neuroendocrinol. Lett.*, 20(3–4), 179, 1999.)

respectively (Yamamoto and Kageyama 1980). Interestingly, In humans, the pineal gland grows from birth until 2 years of age, and then it remains constant until puberty (Hasegawa et al. 1987), but then there is a gradual increase in the size of the pineal gland from puberty to old age (Tapp and Huxley 1971, 1972). Unfortunately, there are only a few volumetric data of pineal glands that have been reported in the literature to date (Bumb et al. 2012, Sun et al. 2009).

Sun et al. (2009) obtained the accurate reference range of normal pineal volumes in 112 individuals aged 20–30 years, included randomly from a healthy person. Axial and sagittal 3.0 T MRI data were obtained using 3D T1-weighted sequence and true pineal volumes were measured from T1-weighted images, while estimated volumes were calculated using pineal length, width, and height (Sun et al. 2009). (Sun et al. 2009) found that the prevalence of pineal cysts is 25.00% (17 females and 11 males in 112 healthy volunteers). After eliminating the pineal glands with big cysts, the mean pineal volumes of males and females were 95.25 ± 30.70 and 92.70 ± 52.05 mm^3, respectively (Sun et al. 2009). Also, they found that there were no significant correlations between pineal volume and body height, body weight, and head circumference (Sun et al. 2009).

Acer et al. (2012) stated that the pineal gland volumes were calculated in a total of 62 subjects (36 females and 26 males) who were free of any pineal lesions or tumors. They used three methods for pineal gland volume estimation: point-counting, planimetry, and ROI (region of interest); and they found that the mean ± SD pineal gland volumes of the groups were 99.55 ± 51.34, 102.69 ± 40.39, and 104.33 ± 40.45 mm^3, respectively (Acer et al. 2012). Sumida et al. (1996) retrospectively studied 249 consecutive patients (129 males and 120 females) aged 2 weeks to 20 years using MRIs (Sumida et al. 1996). They measured the pineal gland of each patient from the hard copy of the film, with the use of calipers (Sumida et al. 1996). They used the pineal gland of maximum length (L) and height (H) on the T1-weighted sagittal images, and the widths (W) were measured on the T1 weighted coronal or axial images Sumida et al. (1996). Sumida et al. (1996) calculated the pineal gland volume according to Lundin and Pedersen (1992) and Acer et al. (2012). In a previous study, Sumida et al. (1996) found that in individuals younger than 2 years old, the pineal gland volume was 26.9 ± 6 12.4 mm^3; between 2 and 20 years old, the size of the gland was larger; and after attaining the size of 56.6 ± 6 27.6 mm^3, it remained stable. They stated that the gland was significantly smaller in humans younger than 2 years old than in 2–20 years old; also, there was no difference in size noted between males and females (Sumida et al. 1996).

There are major differences among the results for the pineal gland volume calculation. Rajarethinam et al. (1995) used manual tracing of the gland on T1-weighted MRIs, using locally developed software. Nölte et al. (2010) estimated the pineal gland volume on MRI, using the volume analysis program. Bersani et al. (2002) used elliptic formula for the pineal gland volume estimation on MRI sections. Rajarethinam et al. (1995) and Sun et al. (2009) used ROI analysis for pineal gland volume estimation on MRI sections. Acer et al. (2012) used ROI analysis and stereologic method for pineal gland volume estimation on MRI sections. The difference in the estimated pineal gland volume may be attributable to the different methodological techniques used in these studies.

3.6 MELATONIN AND PINEAL GLAND VOLUME

Mel is produced in the pineal gland, a small conical-shaped endocrine gland of about 100 mm^3 (Sun et al. 2009). It has been suggested that the volume of pineal gland varies, possibly due to the reproductive cycle in females, resulting in fluctuations in Mel secretion (Bersani et al. 2002). Some authors demonstrated that there is no correlation between the pineal volume and age (Golan et al. 2002, Sun et al. 2009).

It has been reported that various clinical syndromes related with Mel deficit have no correlation with the blood levels of Mel, because there are lots of factors that determine the dimensions of the pineal gland and the amount of Mel secretion (Gillette and Tischkau 1999, Touitou and Haus 1994). There was a significant correlation between the pineal gland volume and the blood Mel concentration in humans and animals (Hallam et al. 2006, Nolte et al. 2009).

Some studies reported lower levels of plasma Mel in schizophrenia and bipolar disorder (Rajarethinam et al. 1995). Hannsen and Watterberg (1983) reported a higher level of nocturnal Mel in schizophrenia, whereas Beckmann et al. (1984) found no difference in Mel levels between normal controls and patients with schizophrenia (Beckmann et al. 1984). Rajarethinam et al. (1995) used 131 subjects: 86 normal controls and 45 patients with schizophrenia or schizophreniform psychosis. They used the locally developed software BRAINS and T1 MRIs (Rajarethinam et al. 1995). The volume of the pineal gland in the patients was 0.208 mL, and for the controls, it was 0.213 mL. There was no significant difference in the pineal gland volume between the controls and the patients (Rajarethinam et al. 1995).

Using T1 MRIs, Sarrazin et al. (2011) studied 40 participants (20 patients with bipolar disorder and 20 controls) to determine the pineal gland volume. The mean total pineal volume as determined by the 3D method was 115.3 mm^3 (SD = 54.3) in patients with bipolar disorder and 110.4 mm^3 (SD = 40.5) in healthy controls (Sarrazin et al. 2011). They found that there was no significant difference between the two groups (Sarrazin et al. 2011). Bersani et al. (2002) found a significant difference in the total pineal volume in a smaller sample of schizophrenic patients.

Nolte et al. (2009) used 15 healthy male subjects (20–27 years), determined the pineal volume using high-resolution MRI, and determined Mel plasma concentrations every 2 h for 24 h. They found that the mean pineal volume was 125 ± 54 mm^3 (Nolte et al. 2009). The pineal volume correlated linearly to maximum MC and to 24 h Mel (Nolte et al. 2009).

3.7 CONCLUSION

The relation between the pineal gland volume and the Mel level may be useful as a guide for the radiologists and the clinicians regarding these findings. The best diagnostic technique is CT scan for the investigation of calcifications, and T1-weighted MRIs is the best method for pineal volume calculation. Combining MRI volumetry with Mel levels may provide further information in some disorders with disturbed circadian Mel rhythm, such as sleep disorders, Parkinson's disease, aging, schizophrenia, and cancer (Kloeden et al. 1990, Mayo et al. 2005, Pandi-Perumal et al. 2007, Vigano et al. 2001). At present, it seems promising to determine these Mel-related parameters per volume of pineal tissue, as Mel levels might be normal, whereas correction for volume could be a more sensitive marker of pineal dysfunction. Based on these findings, we suggest that the volume of pineal gland and pineal cyst and Mel level may have a role in the etiology of sleep disorder, schizophrenia, headache, and bipolar disorder, suggesting the need for further studies.

REFERENCES

Acer, N., Ilica, A.T., Turgut, A.T., Ozçelik, O., Yıldırım, B., and Turgut, M. 2012. Comparison of three methods for the estimation of pineal gland volume using magnetic resonance imaging. *Scientific World Journal* 2012:123412.

Acer, N., Turgut, M., Yalcın, S.S., and Duvernoy, H.M. 2011. Anatomy of the human pineal gland. In: Turgut, M. and Kumar, R. (Eds.), *Pineal Gland and Melatonin: Recent Advances in Development, Imaging, Disease and Treatment*. Nova Science, New York, pp. 41–54.

Arendt, J. 1995. *Melatonin and the Mammalian Pineal Gland*. Chapman Hall, London, U.K.

Arendt, J. 2006. Melatonin and human rhythms. *Chronobiol Int* 23(1–2):21–37.

Arendt, J. and Skene, D.J. 2005. Melatonin as a chronobiotic. *Sleep Med Rev* 9(1):25–39.

Aslan, O. 2001. *Neuroanatomical Basis of Clinical Neurology*. The Parthenon Publishing Group, New York, pp. 105–107.

Beckmann, H., Wetterberg, L., and Gattaz, W.F. 1984. Melatonin immunoreactivity in cerebrospinal fluid of schizophrenic patients and healthy controls. *Psychiatry Res* 11(2):107–110.

Bersani, G., Garavini, A., Iannitelli, A., Quartini, A., Nordio, M., DiBiasi, C., and Pancheri, P. 2002. Reduced pineal volume in male patients with schizophrenia: No relationship to clinical features of the illness. *Neurosci Lett* 329(2):246–248.

Bosnjak, J., Budisić, M., Azman, D., Strineka, M., Crnjaković, M., and Demarin, V. 2009. Pineal gland cysts—An overview. *Acta Clin Croat* 48(3):355–358.

Bumb, J.M., Brockmann, M.A., Groden, C., Al-Zghloul, M., and Nölte, I. 2012. True FISP of the pediatric pineal gland: Volumetric and microstructural analysis. *Clin Neuroradiol* 22(1):69–77.

Duvernoy, H.M. 1999. *Human Brainstem Vessels*. Springer Verlag, New York, p. 261.

Duvernoy, H.M., Parratte, B., Tatou, L., and Vuillier, F. 2000. The human pineal gland. Relationships with surrounding structures and blood supply. *Neurol Res* 22(8):747–790.

Duvernoy, H.M. and Risold, P.Y. 2007. The circumventricular organs: An atlas of comparative anatomy and vascularization. *Brain Res Rev* 56(1):119–147.

Engel, U., Gottschalk, S., Niehaus, L., Lehmann, R., May, C., Vogel, S., and Janisch, W. 2000. Cystic lesions of the pineal region—MRI and pathology. *Neuroradiology* 42:399–402.

Gillette, M.U. and Tischkau, S.A. 1999. Suprachiasmatic nucleus: The brain's circadian clock. *Recent Prog Horm Res* 54:33–58.

Golan, J., Torres, K., Staskiewicz, G.J., Opielak, G., and Maciejewski, R. 2002. Morphometric parameters of the human pineal gland in relation to age, body weight and height. *Folia Morphol* (Warsz.) 61(2):111–113.

Goldman, B.D. 2001. Mammalian photoperiodic system: Formal properties and neuroendocrine mechanisms of photoperiodic time measurement. *J Biol Rhythms* 16(4):283–301.

Hallam, K.T., Olver, J.S., Chambers, V., Begg, D.P., McGrath, C., and Norman, T.R. 2006. The heritability of melatonin secretion and sensitivity to bright nocturnal light in twins. *Psychoneuroendocrinology* 31(7):867–875.

Hannsen, T. and Watterberg, L. 1983. Serum levels of melatonin in schizophrenic patients: A possible indicator of beta adrenergic function and of responsiveness to beta adrenergic blocking drugs (propranolol). *Neuroendocrinol Lett* 5:143.

Hasegawa, A., Ohtsubo, K., and More, W. 1987. Pineal gland in old age; quantitative and qualitative morphological study of 168 human autopsy cases. *Brain Res* 409(2):343–349.

Horsburgh, A. and Massoud, T.F. 2013. The circumventricular organs of the brain: Conspicuity on clinical 3 T MRI and are view of functional anatomy. *Surg Radiol Anat* 35(4):343–349.

Jinkins, R.J. 2000. *Atlas of Neuroradiologic Embryology, Anatomy and Variants*. Lippincott Williams & Wilkins, Philadelphia, PA, p. 239.

Kalsbeek, A., Garidou, M.L., Palm, I.F., VanDerVliet, J., Simonneaux, V., Pévet, P., and Buijs, R.M. 2000. Melatonin sees the light: Blocking GABA-ergic transmission in the paraventricular nucleus induces daytime secretion of melatonin. *Eur J Neurosci* 12(9):3146–3254.

Karasek, M. 1999. Melatonin in humans: Where we are 40 years after its discovery. *Neuroendocrinol Lett* 20(3–4):179–188.
Karasek, M. 2004. Melatonin, human aging, and age-related diseases. *Exp Gerontol* 39(11–12):1723–1729.
Kenny, G.C. 1965. The innervation of the mammalian pineal body (a comparative study). *Proc Aust Assoc Neurol* 3:133–140.
Kloeden, P.E., Rossler, R., and Rossler, O.E. 1990. Does a centralized clock for ageing exist? *Gerontology* 36(5–6):314–322.
Kunz, D., Schmitz, S., Mahlberg, R., Mohr, A., Stöter, C., Wolf, K.J., and Herrmann, W.M. 1999. A new concept for melatonin deficit: On pineal calcification and melatonin excretion. *Neuropsychopharmacology* 21(6):765–772.
Langman, J. 1975. *Medical Embryology*, 3rd ed. Williams & Wilkins, Baltimore, MD, pp. 175–178, 318–364.
Lum, G.B., Williams, J.P., Machen, B.C., and Akkaraju, V. 1987. Benign cystic pineal lesions by magnetic resonance imaging. *J Comput Tomogr* 11(4):228–235.
Lundin, P. and Pedersen, F. 1992. Volume of pituitary macroadenomas: Assessment by MRI. *J Comput Assist Tomogr* 16(4):519–528.
Mayo, J.C., Sainz, R.M., Tan, D.X., Antolin, I., Rodriguez, C., and Reiter, R.J. 2005. Melatonin and Parkinson's disease. *Endocrine* 27(2):169–178.
McKinley, M.J., McAllen, R.M., Mendelsohn, F.A.O., Allen, A.M., Chai, S.Y., and Oldfield, B.J. 1990. Circumventricular organs: Neuroendocrine interfaces between the brain and the hemal milieu. *Front Neuroendocrinol* 11:91–127.
Møller, M. and Baeres, F.M. 2002. The anatomy and innervation of the mammalian pineal gland. *Cell Tissue Res* 309(1):139–150.
Nolte, I., Brockmann, M.A., Gerigk, L., Christoph, G., and Scharf, J. 2010. TrueFISP imaging of the pineal gland: More cysts and more abnormalities. *Clin Neurol Neurosurg* 112(3):204–208.
Nolte, I., Lutkhoff, A.T., Stuck, B., Lemmer, B., Schredl, M., Findeisen, P., and Groden, C. 2009. Pineal volume and circadian melatonin profile in healthy volunteers: An interdisciplinary approach. *J Magn Reson Imaging* 30(3):499–505.
Pandi-Perumal, S.R., Srinivasan, V., Spence, D.W., and Cardinali, D.P. 2007. Role of the melatonin system in the control of sleep: Therapeutic implications. *CNS Drugs* 21(12):995–1018.
Pandi-Perumal, S.R., Zisapel, N., Srinivasan, V., and Cardinali, D.P. 2005. Melatonin and sleep in aging population. *Exp Gerontol* 40(12):911–925.
Rajarethinam, R., Gupta, S., and Andreasen, N.C. 1995. Volume of the pineal gland in schizophrenia: An MRI study. *Schizophr Res* 14(3):253–255.
Reiter, R.J., Tan, D.X., and Allegra, M. 2002. Melatonin: Reducing molecular pathology and dysfunction die to free radicals and associated reactants. *Neuroendocrinol Lett* 23(1):3–8.
Reiter, R.J., Tan, D.X., and Korkmaz, A. 2007. Light at night, chronodisruption, melatonin suppression, and cancer risk: A review. *Crit Rev Oncogen* 13(4):303–328.
Sarrazin, S., Etain, B., Vederine, F.E., d'Albis, M.A., Hamdani, N., Daban, C., Delavest, M. et al. 2011. MRI exploration of pineal volume in bipolar disorder. *J Affect Disord* 135(1–3):377–379.
Standring, S. 2008. *Grays Anatomy*. Churchill Livingstone, London, U.K., pp. 882–883.
Sumida, M., Barkovich, A.J., and Newton, T.H. 1996. Development of the pineal gland: Measurement with MR. *AJNR Am J Neuroradiol* 17(2):233–236.
Sun, B., Wang, D., Tang, Y., Fan, L., Lin, X., Yu, T., Qi, H., Li, Z., and Liu, S. 2009. The pineal volume: A three-dimensional volumetric study in healthy young adults using 3.0 T MR data. *Int J Dev Neurosci* 27(7):655–660.
Tapp, E. and Huxley, M. 1971. The weight and degree of calcification of the pineal gland. *J Pathol* 105:31–39.
Tapp, E. and Huxley, M. 1972. The histological appearance of the human pineal gland from puberty to old age. *J Pathol* 108:137–144.
Taraszewska, A., Matyja, E., Koszewski, W., Zaczyński, A., Bardadin, K., and Czernicki, Z. 2008. Asymptomatic and symptomatic glial cysts of the pineal gland. *Folia Neuropathol* 46(3):186–195.
Touitou, Y. 2001. Human aging and melatonin. Clinical relevance. *Exp Gerontol* 36(7):1083–1100.
Touitou, Y. and Haus, E. 1994. Aging of the human endocrine and neuroendocrine time structure. *Ann N Y Acad Sci* 719:378–397.
Vigano, D., Lissoni, P., and Rovelli, F. 2001. A study of light/dark rhythm of melatonin in relation to cortisol and prolactin secretion in schizophrenia. *Neuroendocrinol Lett* 22(2):137–141.

Vígh, B., Szél, A., Debreceni, K., Fejér, Z., Manzanoe Silva, M.J., and Vígh-Teichmann, I. 1998. Comparative histology of pineal calcification. *Histol Histopathol* 13(3):851–870.
Vrang, N., Larsen, P.J., Møller, M., and Mikkelsen, J.D. 1995. Topographical organization of the rat suprachiasmatic-paraventricular projection. *J Comp Neurol* 353(4):585–603.
Welsh, M.G. 1985. Pineal calcification: Structural and functional aspects. *Pineal Res Rev* 3:41–68.
Whitehead, M.T., Oh, C.C., and Choudhri, A.F. 2013. Incidental pineal cysts in children who undergo 3-T MRI. *Pediatr Radiol* 43(12):1577–1583.
Yamamoto, I. and Kageyama, N. 1980. Microsurgical anatomy of the pineal region. *J Neurosurg* 53(2):205–221.

4 Melatonin's Beneficial Effects in Metabolic Syndrome with Therapeutic Applications

Venkataramanujam Srinivasan, Yoshiji Ohta, Javier Espino, Rahimah Zakaria, and Mahaneem Mohamed

CONTENTS

4.1 INTRODUCTION

Metabolic syndrome (MetS) is characterized by the symptoms of central obesity, insulin resistance, atherogenic dyslipidemia, and hypertension (Deedwania and Gupta 2006). Raised triglyceride (TG) levels, decreased high-density lipoprotein (HDL), elevated blood pressure, and diabetes mellitus constitute the important criteria for MetS (Cornier et al. 2008; Alberti et al. 2009). The prevalence of MetS in general population ranges from 17% to 25% (Grundy 2008; Al Saraj et al. 2009), but the prevalence is around 59%–61% in diabetes mellitus (Saraj et al. 2009). A high prevalence of MetS is noted with an increase in age (Ford et al. 2002). With regard to the influence of gender on MetS, some studies have shown that the incidence of MetS is higher in men than in women (Fezeu et al. 2007; Ahonen et al. 2009), whereas another study on Chinese has demonstrated that the prevalence of MetS is higher in women than in men (He et al. 2006). In a study conducted on Nigerians with type 2 diabetes mellitus (T2DM), the occurrence of MetS was found to be similar in both sexes, although the prevalence increased in aged subjects (70–79 years, 89%) when compared to young subjects (20–29 years, 11%; Ogbera 2010). Asians are more prone to MetS, and a recent study has shown that the prevalence of MetS in Indians ranges from 35.8% to 45.3% (Ravikiran et al. 2010). The percentage of MetS in Chinese is known to range from 30.5% to 31.5% (Zuo et al. 2009).

The prevalence of MetS is assessed based on different definitions, such as International Diabetes Federation (IDF), World Health Organization (WHO), and National Cholesterol Education Program Adult Treatment Program III (NCEP ATP III). The rates of MetS prevalence assessed by IDF, NCEP ATP III, and WHO definitions in Malaysia were 22.9%, 16.5%, and 6.4%, respectively (Tan et al. 2008). A recent study on Malaysian population has revealed that the rates of MetS prevalence assessed by three definitions ranged from 12.4% to 32.2% (Ruwaida et al. 2011). In another study, the nationwide survey of MetS, the rates of MetS prevalence were found to be 32.1% (WHO), 34.3% (NCEP ATP III), 37.1% (IDF), and 42.5% ("harmonized definition"; Mohamud et al. 2011). According to Chennai Urban Rural Epidemiology Study (CURES-34), the rate of MetS prevalence was 25.8% (IDF), 23.2% (WHO), and 18.3% (NCEP ATP III), with the highest percentage in IDF criteria (Deepa et al. 2007). In another study on MetS prevalence in Malaysia, the prevalence percentage was variable as follows: 32.2% (IDF), 28.5% (NCEP ATP III), and 12.4% (WHO modified). This study also showed the highest rate of MetS prevalence in IDF criteria (Zainuddin et al. 2011). Central obesity plays an important role in diagnosing MetS. In the study conducted in Obesity Clinic at the Hospital Universiti Sains Malaysia (HUSM), the rate of MetS prevalence in obese patients was 40.2%, with the prevalence higher in females (43.7%) than in males (32.3%; Termizy and Mafauzy 2009).

In Malaysia, three ethnic groups exist, namely, Malay, Chinese, and Indian. Studies conducted on the prevalence of MetS among these ethnic groups have shown that the highest prevalence occurs in the Indian subgroup, because Indians have elevated fasting blood sugar levels, low plasma concentrations of HDL cholesterol (HDL-C), and central obesity (Tan et al. 2011). The order of rate of MetS prevalence in Indians (36.4%), Chinese (33.8%), and Malays (27.4%) has been supported in another study conducted on obese Malaysian adolescents (Narayanan et al. 2011). In this study, MetS was diagnosed by NCE ATP III, based on the presence of three of the five risk factors, that is, hypertriglyceridemia, hyperglycemia, hypertension, increased HDL-C, and abnormal waist circumference. MetS was found to be more prevalent in obese boys (40.2%) than in obese girls (17%; Narayanan et al. 2011).

4.2 METABOLIC SYNDROME: ITS PATHOPHYSIOLOGY

The pathophysiological mechanisms involved in MetS are complex in nature and involve dysregulation of many biochemical and physiological regulatory systems in the body. Many clinicians have recognized the involvement of various factors like hypertension, hyperglycemia, and hyperuricemia in the development of MetS (first recognized by Kylin in 1923) and have indicated it as Syndrome X or the insulin resistance syndrome (La Guardia et al. 2012). At present, three definitions of MetS, which were established between 1998 and 2005, are recognized as primary, and have since been widely used (La Guardia et al. 2012). These three definitions based on biochemical parameters are listed in Table 4.1. With regard to the three definitions listed, the primary components of the previously described syndrome include (1) central obesity, (2) dyslipidemia, (3) hyperglycemia, and (4) hypertension (La Guardia et al. 2012). Accordingly, persons who manifest these symptoms are at high risk of developing diabetes and cardiovascular disease, and are prone to have a high mortality rate (Laaksonen et al. 2002; Lakka et al. 2002; Ford and Giles 2003; La Guardia et al. 2012). As to these symptoms, NCEP ATP III definition is regarded as being more powerful in predicting cardiovascular disease and diabetes than the IDF definition (Choi et al. 2007; Nilsson et al. 2007; Tong et al. 2007).

TABLE 4.1
Definitions of Metabolic Syndrome (MetS)

Parameter	WHO (1998)	NCEP ATP 3 (2001)	IDF (2005)
Glucose	>6.1 mmol/L or 110 mg/dL	>5.6 mmol/L or 100 mg/dL	>5.6 mmol/L or 100 mg/dL
HDL-C	<0.9 mmol/L or 35 mg/dL (men)	<1.0 mmol/L or 40 mg/dL (men)	<1.0 mmol/L (men)
	<1.0 mmol/L or 40 mg/dL (women)	<1.3 mmol/L or 50 mg/dL (women)	<1.3 mmol/L (women)
TG	>1.7 mmol/L or 150 mg/L	>1.7 mmol/L or 150 mg/L	>1.7 mmol/L or 150 mg/L
Obesity (waist/hip ratio)	>0.9 (men)	>102 cm (men)	>94 cm (men)
	>0.85 (women)	>88 cm (women)	>80 cm (women)
Hypertension	>140/90 mmHg	>130/85 mmHg	>130/85 mmHg
Number of abnormalities for MetS	>2	>3	>2

4.3 INSULIN RESISTANCE AS THE CAUSE OF METABOLIC SYNDROME

Insulin resistance is an important factor involved in the pathophysiology of MetS. It is characterized by high plasma insulin concentrations that occur due to the failure of insulin function to suppress an increase in plasma glucose concentrations. Specifically, insulin resistance is defined as the inability of insulin to suppress hepatic glucose uptake and to stimulate glucose uptake in muscle and adipose tissue (Leroith 2012). These events lead to glucose intolerance and T2DM in predisposed individuals. When insulin binds to its receptor, specific tyrosine residues are phosphorylated, and then the signaling cascade is initiated. The two major insulin signaling pathways are phosphatidylinositol-3 kinase (PI3K) and mitogen-activated protein kinase (MAPK) pathways (Nystrom and Quon 1999). The PI3K-dependent metabolic actions of insulin directly promote glucose uptake in skeletal muscle by stimulating translocation of insulin-responsive glucose transporters such as glucose transporter-4 (GLUT-4). In diabetes mellitus and MetS, activation of PI3K pathway is blocked, and signaling activation by insulin occurs only through another pathway, namely, MAPK pathway. Activation of MAPK pathway through MEK1 activates extracellular signal-regulated kinases ERK1 and ERK2 that mediate the mitogenic growth and proinflammatory responses of insulin (Nystrom and Quon 1999). Hence, activation of this signaling pathway, that is, MAPK pathway, causes smooth muscle cell growth and proliferation even in insulin-resistant conditions, thereby inducing the development of atherogenesis (Nystrom and Quon 1999).

Under normal healthy conditions, insulin activates PI3K pathway and thereby stimulates its vascular actions and increases the blood flow and capillary recruitment causing a rapid disposal of glucose (Muniyappa et al. 2007). MAPK-dependent pathway regulates the secretion of vasoconstrictor endothelin-1 (ET-1) from endothelium (Nystrom and Quon 1999). ET-1, a potent vasoconstrictor, affects skeletal muscle glucose disposal and induces peripheral insulin resistance. Under normal healthy conditions, however, the effects of insulin's metabolic actions through ET-1 stimulation are antagonized by insulin-stimulated production of nitric oxide (NO) (Muniyappa et al. 2008). In insulin-resistant states, the selective impairment of PI3K signaling pathway in both metabolic and vascular tissues and the diminished sensitivity to the actions of insulin in vascular endothelium result in biochemical and physiological abnormalities of MetS (Muniyappa et al. 2007). As insulin-mediated ET-1 secretion is augmented in MetS, blockade of ET-1 receptors has been shown to improve insulin sensitivity with increased peripheral glucose uptake (Ahlborg et al. 2007). What causes insulin resistance due to obesity and MetS is still not clear. Both genetics and the factors that interfere with the normal insulin actions on metabolic tissues are suggested as the contributory factors (Leroith 2012).

Insulin receptor signaling pathway is blocked by accumulation of diacylglycerol, an intermediate of triacylglycerol (TG) synthesis, and ceramide, or by inflammatory cytokines, such as tumor necrosis factor-α (TNF-α) and interleukin-6 (IL-6), resulting in a vicious cycle that worsens the insulin resistance in MetS (Paz et al. 1997).

4.4 ROLE OF ADIPOSE TISSUE AND ADIPOKINES IN METABOLIC SYNDROME

A variety of pathophysiological mechanisms are linked to central adiposity and MetS. Numerous studies have demonstrated that adipose tissue is an important factor in the pathophysiology of MetS. Adipose tissue secretes into the circulation a number of proteins and nonprotein factors known as adipokines, which regulate glucose and lipid metabolism in the body. Of these adipokines, adiponectin (ADN), leptin, adipsin, and visfatin are exclusively synthesized by adipocytes. ADN is a 30-kDa protein and has insulin-sensitizing, antiatherogenicity, and anti-inflammatory properties, and is present as oligomers in the blood (Arita et al. 1999; Ouchi et al. 1999). Reduction of ADN has been reported in obese patients who exhibit insulin resistance and are prone to develop diabetes mellitus (Hotta et al. 2000; Nawrocki and Scherer 2004; Shimda et al. 2004). ADN administration resulted in the reductions of oxidative stress, inflammation, insulin resistance, and vascular damage (Shimda et al. 2004). ADN deficiency in mice fed a high-fat diet caused greater fat accumulation, impaired glucose tolerance, and hyperlipidemia (Lee et al. 2011). ADN, by its antioxidant, anti-inflammatory, and antithrombotic properties and by its direct antiatherosclerotic properties, exerts counterregulatory role in atherogenesis (Mangge et al. 2010). Plasma ADN concentrations are low in patients with obesity and T2DM, although plasma ADN concentrations are high in patients with chronic kidney disease (Cui et al. 2011).

Since the discovery of the release of leptin (adipose-derived satiety factor) from adipose tissue, this tissue has been regarded as an important endocrine organ (Kershaw and Flier 2004). Moreover, adipose tissues provide vital information about the whole body's nutritional status to other insulin-sensitive organs and tissues by releasing a number of adipokines that are synthesized within the tissue (Galic et al. 2010). Obesity, insulin resistance, and T2DM are all associated with a number of changes in adipose tissues with significant infiltration of macrophages, increases in the size and number of adipocytes, rarefaction of blood vessels, and increased adipocyte turnover and apoptosis (Wellen and Hotamisligil 2003). There are also enhanced release and increased circulating levels of inflammatory cytokines and adipokines, which all interfere with the insulin signaling pathways, resulting in the aggravation of insulin resistance (Lee et al. 2009; Tesauro and Cardillo 2011; Tesauro et al. 2011). Adipokines also have been suggested to contribute to chronic inflammation and MetS-related cardiovascular disease. Furthermore, novel adipokines such as visfatin, apelin, and vaspin act as potential mediators of the interplay between MetS and atherosclerosis (Palios et al. 2012). Accumulated evidence points out that leptin has proinflammatory and proatherogenic properties (Smith and Yellon 2011). In a recent study conducted on patients with chronic heart failure, systemic elevations of both leptin and resistin levels correlated significantly with dilated cardiac myopathy, which is attributed to their effects to cause increased redox stress in cardiac cells (Bobbert et al. 2012). Hence, the exact beneficial or adverse role of adipokines in MetS is still not conclusively known.

4.5 ADIPOCYTES, OBESITY, AND VASCULAR DYSFUNCTION

Adipocytes from obese patients are involved in the pathogenesis of vascular dysfunction and insulin resistance by producing large amounts of nonesterified fatty acids (NEFA), which inhibit carbohydrate metabolism through substrate competition and impaired insulin signaling mechanisms (Shulman 2000). Macrophage recruitment and production of cytokines such as TNF-α, besides contributing to insulin resistance, also result in vascular dysfunction (Hotamisligil et al. 1996). In fact, proinflammatory mediators from adipocytes exert various effects on endothelial dysfunctions like plaque initiation, plaque progression, and plaque rupture, which promote atherogenetic processes (Lau et al. 2005). Initiation and progression of the atherosclerotic process is triggered by endothelial dysfunction (Tesauro and Cardillo 2011). The basis for endothelial dysfunction is the reduction of endothelium-derived NO within the blood vessels. Although NO is essential for vascular relaxation (Bredt and Snyder 1994), it also suppresses atherosclerosis by reducing endothelial cell activation, smooth muscle proliferation, leucocyte activation, leucocyte endothelial interaction, and platelet adhesion and aggregation (Huang 2005; Tesauro et al. 2011). Hence, reduction in the availability of NO is considered as the crucial factor involved in the development of atherosclerotic state (Tesauro et al. 2011). Normally, a balance exists between vasodilator and antiatherosclerotic substances such as NO and other substances such as ET-1, which induces vasoconstrictor effects, as already mentioned earlier.

It is now suggested that increased activity of the ET-1 system of the vascular endothelium seen in obese persons is responsible for derangement of vascular homeostasis (Mather et al. 2004). Indeed, an increased ET-1-dependent vasoconstrictor activity was noted in the forearm circulation of overweight and obese individuals, but not in lean individuals (Tesauro et al. 2011). Stimulation of ET-1 production occurs following exposure of endothelial cells to insulin, and this involves MAPK signaling pathway that induces activation of ERK1/2 (Eringa et al. 2004). The view that obesity-related vascular dysfunction is caused by the imbalance of NO and ET-1 system is supported by another study in which the upregulation of ET-1 system impairs NO-mediated vasodilatation in the blood vessels of insulin-resistant patients with obesity or patients with T2DM, and NO reactivity has been restored through the blockade of ET-1 receptors (Lteif et al. 2007).

4.6 VASCULAR DYSFUNCTION IN METABOLIC SYNDROME

TNF-α, one of the circulating proinflammatory cytokines, also impairs vascular function by altering the balance between endothelial-derived vasodilator and vasoconstrictor substances, since it downregulates expression of endothelial NO synthase (eNOS; Yoshizumi et al. 1993) and upregulates ET-1 production in vascular endothelium (Mohamed et al. 1995). TNF-α also affects vascular endothelial and vascular smooth muscle functions by increasing the production of reactive oxygen species (ROS) from these sites of the blood vessels (De Keulenaer et al. 1998). Evidence for the role of TNF-α in causing abnormal vascular response has been obtained from a number of studies. A recent study has shown that TNF-α blockade improves insulin's stimulating effect both on endothelium-dependent and endothelium-independent vasodilator activities (Tesauro et al. 2008). Moreover, increased oxidative stress has been suggested as the probable cause for TNF-α-related vasculopathy. The role of TNF-α in vascular dysfunction associated with MetS was assessed by the use of the TNF-α-neutralizing antibody infliximab (Tesauro et al. 2008).

Abnormal insulin-stimulated vascular reactivity in patients with central obesity and MetS was assessed by comparing the effects of insulin on vascular responses by use of different vasodilators in healthy subjects and in patients with MetS (Schinzari et al. 2010). Hyperinsulinemia causes enhanced vasodilator responses to acetylcholine (a substance used for endothelial release of NO), sodium nitroprusside (an endogenous NO donor), and verapamil (a calcium channel blocking agent) in normal healthy subjects, but none of these substances caused vasodilator responses following hyperinsulinemia in patients with MetS, and this interesting observation was made by Tesauro and his colleagues (2011). From this study, it was concluded that abnormalities of insulin signaling pathways in resistant vessels affect both vascular endothelium and smooth muscles of blood vessels. The action of insulin to increase contraction of the medial smooth muscle of blood vessels provides a potential link between insulin resistance and obesity-associated hypertension (Tesauro and Cardillo 2011). These observations support the idea that obesity-related vascular disorders caused by hyperinsulinemia and insulin resistance are responsible for the constellation of characteristic abnormalities of MetS (Kahn and Flier 2000). The metabolic abnormalities of MetS are shown in Figure 4.1.

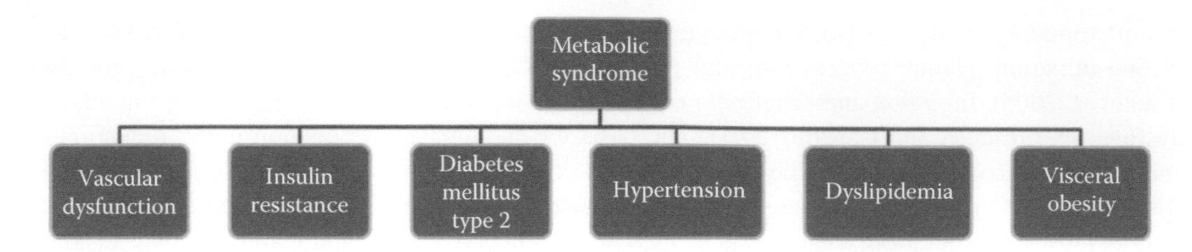

FIGURE 4.1 **(See color insert.)** The metabolic abnormalities of metabolic syndrome.

4.7 DYSLIPIDEMIA

Dyslipidemia is characterized by elevated TG due to increased very low density lipoproteins (VLDL), increased low-density lipoprotein (LDL) particles, and decreased HDL-C levels. The increased concentrations of VLDL and LDL also increase total apolipoprotein-B, thus triggering atherogenic dyslipidemia (Krauss 1995). The small LDL particles penetrate more easily into the arterial walls, resulting in atherogenic modification, and a low HDL level is a risk factor for the atherogenic process (Krauss 1998). The possible mechanism of increased TG production is due to insulin resistance that causes an increase in adipocyte lipolysis, causing an elevated release of NEFA that are taken up by the liver. Stimulation of hepatic lipogenesis by insulin leads to increased hepatic TG production. Cholesterol esters are transferred from HDL to VLDL by lipid transfer protein in plasma. HDL is taken up by the liver, and VLDL is produced by the tissue. All these reactions decrease HDL levels and elevate plasma TG levels seen in MetS (Leroith 2012). Dyslipidemia is prevalent in patients with T2DM, where low plasma HDL-C levels and elevated plasma TG concentrations predominate. It has also been found that as patients progress from insulin-resistant to diabetes mellitus, their plasma LDL cholesterol (LDL-C) levels increase by 15%, while LDL particle size increases by more than 33% (Garvey et al. 2003). This increase of small dense LDL particles in patients with T2DM is highly atherogenic and causes the progression of coronary heart disease (Vakkilainen et al. 2003). Elevated plasma TG levels decrease the activity of lipoprotein lipase, reduce the production of HDL-C, and make synthesized HDL small dense, and all these significantly decrease their antiatherogenic activity (Ginsberg 2002). Thus, elevated plasma TG levels, increased plasma LDL-C levels, and low plasma HDL-C levels that are associated with reduced antiatherogenic activity contribute to dyslipidemia seen in T2DM and MetS (Spratt 2009).

4.8 MELATONIN

Melatonin (*N*-acetyl-5-methoxytryptamine) has been identified in most of the living organisms, ranging from unicellular eukaryotes to invertebrate and vertebrate species. Melatonin is a remarkable molecule with diverse physiological actions, signaling not only the time of the day but also the year. It has been suggested to be the first biological signal that appeared on the earth (Hardeland and Fuhrberg 1996; Claustrat et al. 2005). In vertebrates, melatonin is primarily synthesized by the pineal gland, although many other tissues and organs such as retina, skin, gut, bone marrow, lymphocytes, and thymus also synthesize melatonin (Hardeland et al. 2011). Melatonin secretion is synchronized to the light–dark cycle, the maximum nocturnal and the minimum diurnal plasma levels being ~200 pg/mL and ~10 pg/mL, respectively (Pandi-Perumal et al. 2006). Melatonin is synthesized from L-tryptophan.

Among the various extra pineal sites of melatonin production, gastrointestinal tract is of particular importance as it contains the amount of melatonin exceeding by several hundred folds as those found in the pineal gland (Bubenik 2002). Although melatonin is synthesized by a number of tissues in the body, circulating melatonin is largely derived from the pineal gland. Once formed, melatonin is not stored in the pineal gland but diffuses out into the capillary blood and cerebrospinal fluid (Macchi and Bruce 2004). Circulating melatonin is metabolized mainly in the liver by hydroxylation in the C6 position by cytochrome P_{450} mono-oxygenases (isoenzymes, i.e., CYPIA1, CYPIA2, and CYPIB1) and then conjugated with sulfate to be excreted as 6-sulfatoxymelatonin (aMT6s; Claustrat et al. 2005). Melatonin is also metabolized nonenzymatically by free radicals and other oxidants in all cells. Melatonin is converted to 3-hydroxymelatonin when it scavenges two hydroxyl radicals (Tan et al. 1998). In the central nervous system, melatonin is also metabolized to kynuramine derivatives, namely, N^1-acetyl-N^2-formyl-5-methoxy-kynuramine (AFMK; Hirata et al. 1974) and N^1-acetyl-5-methoxy-kynuramine (AMK; Kelly et al. 1984).

The circadian rhythm of the pineal melatonin biosynthesis is regulated by the suprachiasmatic nucleus (SCN) of the hypothalamus and is synchronized to light–dark cycle through the retinohypothalamic tract (Moore and Klein 1974). Melatonin is involved in the regulation of various important functions in the body like sleep and circadian rhythms, reproduction, immune modulation, anti-inflammation, antitumor actions, antioxidant, and energy metabolism (Pandi-Perumal et al. 2006; Hardeland et al. 2011). The role of melatonin in energy expenditure and body mass regulation has also been explored (Bartness et al. 2002).

4.9 MELATONIN RECEPTORS

Melatonin exerts many physiological actions by acting through membrane-bound MT_1 and MT_2 melatonin receptors, although its free-radical scavenging actions do not require any mediation of melatonin receptors. MT_1/MT_2 melatonin receptors belong to the superfamily of G–protein coupled receptors containing the typical seven transmembrane domains. Some of the actions mediated by melatonin receptors are as follows: MT_1 melatonin receptors mediate vasoconstriction, whereas MT_2 melatonin receptors mediate melatonin's vasodilator effects. Melatonin receptors have been identified in various peripheral tissues concerned with regulation of energy and metabolism and cardiovascular system. The following occurrence and distribution of MT_1 and MT_2 melatonin receptors in various tissues have been studied by different investigators: MT_1 and MT_2 melatonin receptors present in the β-cells of the endocrine pancreas (Muhlbauer and Peschke 2007; Ramracheya et al. 2008; Mulder et al. 2009), aorta, coronary, cerebral, and other peripheral blood vessels (Ekmekcioglu et al. 2001, 2003; Savaskan et al. 2001), as well as in the brown and white adipose tissues (Brydon et al. 2001; Ekmekcioglu 2006).

4.10 ROLE OF MELATONIN IN METABOLIC SYNDROME

Melatonin has been shown to be effective in improving MetS through its antihyperlipidemic action, anti-inflammatory action, modulatory action on insulin's synthesis and release, and its antioxidant actions. These findings have been described in the earlier works published by some coauthors of this chapter (Espino et al. 2011; Kitagawa et al. 2012), as well as by others (Achike et al. 2011; Cardinali et al. 2011; Kozirog et al. 2011; Alemany 2012; Farooqui et al. 2012; Mäntele et al. 2012; Nduhirabandi et al. 2012).

4.11 MELATONIN STUDIES IN OBESITY AND METABOLIC SYNDROME

Obesity is directly or indirectly linked to a number of health disorders like T2DM, cardiovascular diseases, and the associated MetS (Report of a WHO Consultation 2000; Tan et al. 2010; Espino et al. 2011; Vazzana et al. 2011). The normal proportion of adipose tissues in males is around 8%–18%, while it is 14%–28% in females. In obese individuals, however, adipose tissues increase up to 60%–70% of the body weight (Cinti 2006). The effects of melatonin on adiposity and body weight regulation have been studied by a number of investigators, and these effects have been described in an excellent work of Tan and his associates (Vazzana et al. 2011). Daily oral supplementation of melatonin for 12 weeks (0.4–4 μg/mL) in middle-aged rats reduced visceral adiposity and body weight (Rasmussen et al. 1999; Wolden-Hanson et al. 2000; Bojkova et al. 2006; Kassayova et al. 2006). Similar to these studies, melatonin treatment at a dose of 1 mg/kg for 19 weeks in young rats reduced body weight by 25% and visceral fat by 50% (Tan et al. 2010). The effects of melatonin on obesity have been studied using animal models of diet-induced obesity. Melatonin administration (4–10 mg/kg) for 8–12 weeks significantly reduced body weight and plasma levels of glucose, TG, total cholesterol, and leptin in rats with high-fat diet-induced obesity (Prunet-Marcassus et al. 2003; She et al. 2009; Rios-Lugo et al. 2010; Nduhirabandi et al. 2012). In a study conducted on rats with high-calorie diet-induced obesity, the protective effects of melatonin were observed not only on increased body weight and obesity but also on myocardial ischemia and reperfusion damage of the heart. Long-term oral consumption of melatonin (4 mg/kg/day) for 16 weeks, starting before the establishment of obesity, attenuated weight gain and prevented the development of obesity-induced metabolic alterations like elevated visceral fat and significantly increased serum TG, insulin, and leptin levels, as well as plasma HDL-C levels (Nduhirabandi et al. 2012). Melatonin reduced postischemic myocardial infarct size and improved myocardial functional recovery via activation of reperfusion injury salvage protein kinases (PKB/Akt and ERK1/2; Nduhirabandi et al. 2012). This was the first report on the protective effects of melatonin on cardiovascular complications of obesity-induced metabolic syndrome in animal models. The effects of melatonin on metabolic profiles and oxidative stress were studied by some investigators using animal models of diet-induced obesity. Subcutaneous administration of melatonin (1 mg/kg, for 4 weeks) to obese rats reduced heart rate, blood pressure, and sympathetic activities and also ameliorated obesity-related morphological pathologies like fatty changes in the liver, kidney, and blood vessels (Hussain et al. 2007). In addition to these changes, melatonin also caused a significant reduction of oxidative stress in these animals, as indicated by a reduction in plasma malondialdehyde (MDA) levels and an increase in plasma superoxide dismutase (SOD) levels (Hussain et al. 2007). Similar to this study design, the effects of melatonin (4 mg/kg, i.p. for 8 weeks) on both metabolic profile and oxidative stress were studied in diet-induced obese rats. Melatonin reduced body weight gain and plasma glucose, TG, cholesterol, and LDL-C levels, and increased plasma HDL-C levels and the plasma level of glutathione peroxidase (GSH-Px), an antioxidative enzyme (She et al. 2009). These studies clearly pointed at oxidative stress as one of the factors involved in MetS and that the attenuating effect of melatonin on metabolic changes associated with obesity is related to its antioxidant effects. The role of oxidative stress in the etiology of MetS in human subjects was studied by a number of other investigators (Whaley-Connell et al. 2011; Whaley-Connell and Sowers 2012).

4.12 MELATONIN AS AN ANTIOXIDANT

Melatonin has been demonstrated as an efficient antioxidant, both in vivo and in vitro. It scavenges free radicals like O_2^-, OH^-, H_2O_2, and peroxyl radicals (Tan et al. 1993). It participates in the antioxidative defense system of the body, both as a free-radical scavenger and as a hormone that stimulates the synthesis of antioxidative enzymes (Reiter 1997). Melatonin, through its direct genomic action, induces the expression of γ-glutamyl cysteine synthetase, the rate-limiting enzyme in glutathione synthesis (Urata et al. 1999). Melatonin, at physiological concentrations, has been shown to protect cells and tissues from oxidotoxicity (Tan et al. 1994). It has also been shown to reduce the mitochondrial formation of ROS and reactive nitrogen species and to protect against oxidative, nitrosative, or nitrative damage of electron transport chain (ETC) proteins as well as lipid peroxidation in the mitochondrial membranes (Acuña-Castroviejo et al. 2003, 2007). The role of melatonin as an antioxidant and its possible therapeutic use in neurodegenerative conditions has been brought out clearly in the earlier works by the author (Srinivasan 2002; Srinivasan et al. 2005).

4.13 MELATONIN AND METABOLIC SYNDROME: STUDIES ON PATIENTS WITH METABOLIC SYNDROME

The association of oxidative stress with MetS and the beneficial actions of melatonin in human subjects have also been observed. Administration of melatonin (5 mg/day) to patients with MetS reduced body mass index, systolic blood pressure (SBP), plasma fibrinogen levels, and plasma levels of thiobarbituric acid-reactive substances (TBARS), an index of lipid peroxidation, as well. Continuation of melatonin therapy for 2 months (5 mg/day) in the patients with MetS increased plasma catalase (CAT) activity and reduced plasma LDL-C levels (Kozirog et al. 2011). This study was carried out on 30 patients with MetS (12 males:18 females) who did not respond to 3-month lifestyle modification and 33 healthy volunteers as the control group. The effects of melatonin (5 mg/day) on the physiological, biochemical, and oxidative stress parameters in patients with MetS were as follows: in comparison to baseline levels and levels obtained after melatonin treatment for parameters studied, SBP: 132.8 ± 9.8 mmHg versus 120.5 ± 11.0 mmHg ($P < 0.001$); LDL-C: 149.7 ± 26.4 mg/dL versus 139.9 ± 30.2 mg/dL; TBARS: 0.5 ± 0.2 μM/g Hb versus 0.4 ± 0.1 μM/g Hb ($P < 0.01$); CAT: 245.9 ± 46.9 U/g Hb versus 276.8 ± 39.4 U/g Hb ($P < 0.01$). This study points to the possibility that melatonin may be of therapeutic value in treating patients with MetS with hypertension (Kozirog et al. 2011). Reduction of systolic hypertension by melatonin treatment in human subjects has been reported (Scheer et al. 2004). The role of melatonin receptors in mediating antiobese actions of melatonin was also studied. When melatonin agonist NEU-P11 (10 mg/kg/day, i.p. for 8 weeks; She et al. 2009) or MT_1/MT_2 agonist ramelteon (8 mg/kg/day, oral for 8 weeks) was administered (Oxenkrug and Summergrad 2010), both agonists exerted similar effects as melatonin, decreasing body weight and blood pressure. Since MT_1 receptors are present in the adipose tissue (Brydon et al. 2001), it is presumed that melatonin exerts body weight regulation through MT_1 melatonin receptors. However, the exact mechanism of action of melatonin in body weight regulation is a complex phenomenon, and the matter as to whether the action of melatonin depends only on its direct action or through both direct and indirect actions needs further research (Nduhirabandi et al. 2012).

4.14 MELATONIN, INSULIN RESISTANCE, AND METABOLIC SYNDROME

A number of observations indicate that there exists a functional interrelationship between melatonin and insulin, and suggest that a reduction of melatonin may be involved in the genesis of diabetes mellitus (Peschke et al. 2007). There is a report showing that patients with T2DM have reduced circulating melatonin levels and elevated insulin levels and that there is a significant negative correlation between the plasma levels of the two substances in the diabetic patients (Peschke et al. 2006). Moreover, recent molecular and immunocytochemical investigations have established the presence of the melatonin membrane receptors MT_1 and MT_2 in human pancreatic tissues and also in the islets of Langerhans (Kemp et al. 2002). Upregulation of the expression of melatonin receptors in patients with T2DM was also observed through immunocytochemical studies (Peschke et al. 2007). In general, it was found that melatonin receptors on pancreatic β-cells are coupled to three parallel signaling pathways, with different influences on insulin secretion. In terms of insulin release, the insulin-inhibiting action of melatonin is transmitted by the dominantly expressed MT_1 receptor through attenuation of G_i-coupled adenylate cyclase activity, thereby negatively modulating incretin-induced rises in 3′,5′-cyclic adenosine monophosphate (cAMP). Likewise, it was also detected that melatonin inhibits the 3′,5′-cyclic guanosine monophosphate (cGMP) signaling pathway and, consequently, insulin secretion, possibly in an MT_2 receptor-mediated fashion. Meanwhile, melatonin-dependent inositol triphosphate (IP_3) release may play a role in the short-term support of other IP_3-releasing agents, like acetylcholine, or may be related to the long-term regulation of pancreatic cell functions with enhancing effects on insulin secretion (Espino et al. 2011). All these studies support the idea that melatonin plays an important role in the regulation of insulin secretion and glucose/lipid metabolism.

Long-term melatonin consumption (2.5 mg/kg/day, for 9 weeks) increased plasma melatonin levels accompanied by a reduction of plasma insulin levels (Peschke et al. 2010). Numerous studies have shown the beneficial effects of melatonin in increasing insulin sensitivity and glucose tolerance in animals fed on high-fat/high-sucrose diet (Nishida et al. 2002; Kawasaki et al. 2009; Sartori et al. 2009; She et al. 2009; Sheih et al. 2009). Oral administration of melatonin (100 mg/kg/day) to high-fat diet fed mice for 8 weeks improved insulin sensitivity and glucose tolerance (Sartori et al. 2009). A similar melatonin administration (10 mg/kg/day, i.p.) for 2 weeks in high-fat diet-induced diabetic mice caused an increase in hepatic glycogen levels with an improvement of liver steatosis (Sheih et al. 2009). Insulin sensitivity was increased by administration of not only melatonin but also a melatonin agonist NEU-P11 for 8 weeks (She et al. 2009). Reductions of hyperinsulinemia, hyperlipidemia, and hyperleptinemia were seen after chronic melatonin administration (1.1 mg/kg/day for 30 weeks, subcutaneous implantation of melatonin pellets) in rats with T2DM (Nishida et al. 2002). The improving effect of melatonin on MetS in rats fed a high fructose diet (HFD) was evaluated by one of the coauthor of this chapter (Kitagawa et al. 2012). In this study, 4- to 6-week HFD feeding caused a significant increase in relative intraabdominal fat weight of rats, indicating that rats fed HFD for 4 or 6 weeks developed visceral obesity, that is, one of the symptoms of MetS. When melatonin (1 or 10 mg/kg) was administered to HFD-fed rats every early morning for 2 weeks, starting at 4-week of HFD feeding, the higher dose of melatonin attenuated the increased relative intraabdominal fat weight at 6-week of HFD feeding. When rats fed the control diet were administered with melatonin (1 or 10 mg/kg) in the same manner, no change in the relative intraabdominal fat weight was observed. The results indicate that melatonin has an improving effect on MetS induced by HFD (Kitagawa et al. 2012). Melatonin administration also ameliorated high fructose-induced liver hypertrophy. In this study, rats fed HFD for 4 or 6 weeks showed a higher serum insulin response curve when compared with the corresponding rats fed the control diet. Hyperinsulinemia, one of the symptoms of MetS, was noted in rats fed HFD for 4 or 6 weeks. Nevertheless, daily administration of melatonin (1 or 10 mg/kg), starting at 4-week of HFD feeding, improved the abnormal serum insulin response curve in oral glucose tolerance test at 6-week HFD feeding more effectively at its higher dose than at its lower dose. Similarly, rats fed HFD for 4 or 6 weeks showed insulin resistance, another symptom of MetS, as judged by the scores of HOMA-IR and QUICK1 indices of insulin resistance. Once again, daily administration of melatonin (1 or 10 mg/kg) attenuated the abnormal scores of HOMA-IR and QUICK1 at 6-week HFD feeding more effectively at its higher dose than at its lower dose. These results indicate that daily melatonin administration improves continuous high fructose-induced hyperinsulinemia and insulin resistance (Kitagawa et al. 2012). In this study, increases in serum TG, NEFA, leptin, and lipid peroxide concentrations as well as hepatic TG and cholesterol concentrations were noted after HFD feeding for 4 or 6 weeks. The 6-week HFD feeding also increased serum TNF-α and reduced hepatic glutathione (GSH) concentration. Daily administration of melatonin (1 or 10 mg/kg, i.p.), starting at 4-week HFD feeding, attenuated all these changes at 6-week HFD feeding, which indicates that melatonin improves high fructose-induced MetS through its anti-inflammatory, antioxidant, and antilipidemic actions (Kitagawa et al. 2012). Melatonin's beneficial actions in ameliorating the signs and symptoms of MetS are summarized in Figure 4.2. The effects of melatonin in promoting glycogen synthesis through increased phosphorylation of glycogen synthase kinase-3β (GSK-3β) in hepatic cells have also been demonstrated (Sheih et al. 2009). An increase in liver weight at 6-week HFD feeding with characteristic features of nonalcoholic hepatic steatosis has been proven in an earlier study (Kawasaki et al. 2009). The glycogen synthetic actions of melatonin in the hepatic cells are blocked either by a nonselective MT_1 and MT_2 melatonin receptor antagonist luzindole, or an MT_2-selective antagonist 4-phenyl-2-propio-amido-tetraline (4P-PDOT), suggesting the involvement of melatonin receptors (Sheih et al. 2009). At this respect, removal of the MT_1 melatonin receptor significantly impaired the ability of the mice to metabolize glucose and induced insulin resistance in these animals (Contreras-Alcantara et al. 2010).

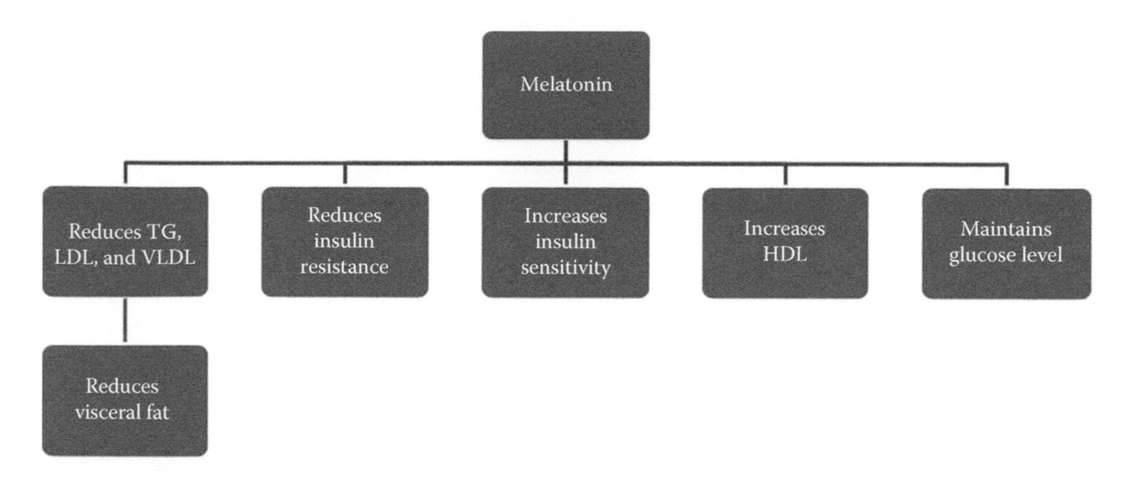

FIGURE 4.2 **(See color insert.)** Melatonin's beneficial actions in ameliorating the signs and symptoms of metabolic syndrome.

With regard to the actions of melatonin on the expression of GLUT-4, different effects have been observed. A decrease in GLUT-4 gene expression was reported after melatonin (1 μM) treatment for 14 days in human brown adipose cell lines (Ramracheya et al. 2008). However, melatonin (50 μg 100 g^{-1} day^{-1}, i.p.) replacement in pinealectomized rats increased the expression of plasma membrane GLUT-4 in white adipose tissues up to the values seen in control rats (Zanquetta et al. 2003). In oxidative stress-induced situations, melatonin also restored the expression of GLUT-4 gene by protecting against oxidative damage (Ghosh et al. 2007).

4.15 MELATONIN'S AMELIORATION OF IMPAIRED ANTIOXIDANT STATUS DURING DIABETIC COMPLICATIONS

Pancreatic β-cells are more susceptible to oxidative stress, as they exhibit a low antioxidative capacity (Tiedge et al. 1997). The relationship between melatonin and impaired oxidative stress in diabetes mellitus is a topic of great interest, and this has been studied by using various experimental models of type 1 diabetes mellitus (Anwar and Meki 2003). Both alloxan and streptozotocin are widely used to induce diabetes in animals. Both compounds cause a selective destruction of pancreatic β-cells, where they induce radical-generating reactions. Alloxan, once taken up inside the cell, generates ROS, especially superoxide anions and hydrogen peroxide, and also reduces intracellular GSH levels (Bromme et al. 1999). Melatonin administration effectively scavenged alloxan-induced production of hydroxyl radicals and inhibited hydroxyl radical-triggered lipid peroxidation in liposomes (Bromme et al. 2000). Melatonin also restored the reduced levels of NO, GSH-Px, and SOD to normalcy in alloxan-induced diabetes (Sailaja Devi and Suresh 2000). Alloxan induced morphological damage in pancreatic β-cells, but alloxan-mediated leakage of insulin from pancreatic β-cells also was effectively prevented by melatonin (Bromme et al. 2000; Ebelt et al. 2000). These studies support the putative use of melatonin in preventing atherosclerosis and other complications of diabetes (Espino et al. 2011). Streptozotocin-induced diabetes increases oxidative stress through generation of free radicals (Takasu et al. 1991). This has been documented in recent studies where streptozotocin-induced diabetes caused increased lipid peroxidation and protein glycosylation (Montilla et al. 1998; Aksoy et al. 2002). Melatonin administration reduced the degree of lipid peroxidation and protein glycosylation (Montilla et al. 1998), and decreased the plasma levels of cholesterol, TG, LDL, and glucose in streptozotocin-induced diabetic rats (Abdel-Wahab and Abd-Allah 2000; Baydas et al. 2002). The protective effects of melatonin against β-cell damage caused by streptozotocin was attributed to its interference with DNA damage and poly(ADP-ribose) polymerase (PARP) activation (Andersson and Sandler 2001). However, it is said that the protective effect of melatonin against β-cell damage caused by streptozotocin is related to its preventive action against the increase of blood NO levels, which is caused during streptozotocin-induced diabetes (Sudnikovich et al. 2007). The most prevalent and incapacitating complication of diabetes is diabetic neuropathy, which is associated with clinically significant morbidities (Bloomgarden 2008). It is speculated that prolonged hyperglycemia, through overproduction of ROS, is likely to damage mitochondrial DNA in dorsal root ganglia, thus contributing to long-term nerve dysfunction (Schmeichel et al. 2003). Melatonin, being a powerful antioxidant, has been shown to exert its beneficial effects against oxidative stress that involves mitochondrial dysfunction (Acuña-Castroviejo et al. 2011; Srinivasan et al. 2011). Antioxidants, in general, have been shown to be beneficial in experimental diabetic neuropathy, and melatonin has been proven to be neuroprotective in diabetic neuropathy in streptozotocin-treated rats (Negi et al. 2010).

4.16 MELATONIN, CIRCADIAN RHYTHMS, AND METABOLIC SYNDROME

The intimate relationship between circadian rhythms and metabolism is getting recognized, and a link between the disruption of circadian rhythms and metabolic perturbations is suggested as one of the major causes of MetS (Karlsson et al. 2001; Green et al. 2008). Circadian rhythms are endogenous rhythms that are driven by the master circadian clock, namely, the SCN of the hypothalamus, whose activity is entrained to the external light–dark cycle through the retinohypothalamic tract that acts to maintain the correct phase position of various rhythms of the body with each other (Srinivasan 1997). The system is hierarchical, the SCN of the hypothalamus being the "master clock" with additional activities of clocks in numerous peripheral tissues.

Circadian rhythms are also sensitive to signals from the metabolism (Rutter et al. 2002). In fact, circadian rhythm of insulin secretion from pancreatic β-cells has been documented (Peschke and Peschke 1998). Disruptions of the phase position of various circadian rhythms give rise to various types of circadian–rhythm-related sleep disorders, and, indeed, the MetS seen in shift workers is attributed to circadian disruption only (Karlsson et al. 2001). Shift workers exhibit disturbances of the circadian rhythms, and sleep deprivation has been associated with both obesity and T2DM (Knutson et al. 2006).

4.17 CONCLUSION

It has been demonstrated that melatonin exerts a beneficial effect in various experimental models of obesity, T2DM, and hypertension by its action on glucose homeostasis (Srivastava and Krishna 2010), by reducing body weight, visceral fat, hyperinsulinemia, plasma levels of leptin, TG, VLDL, NEFA, and C-reactive protein, endothelial dysfunction, insulin resistance, and fasting blood glucose, and by increasing plasma levels of HDL-C and adinopectin, as well as hepatic and muscular glycogen contents. Besides all these metabolic actions that are exerted through MT_1 and MT_2 melatonin receptors, melatonin being an efficient antioxidant is helpful in reducing the oxidative stress involved in the pathophysiology of MetS. Hence, melatonin, its agonist ramelteon, or the melatonergic antidepressant agomelatine, has a great potential therapeutic value in treating patients with MetS. As the effects of melatonin on some of the parameters of MetS are blocked by luzindole, melatonin agonist ramelteon, with greater affinity toward melatonin receptors, will be more beneficial than melatonin itself in ameliorating the signs and symptoms of MetS. The positive effects of melatonin on biochemical, physiological, and hormonal parameters of MetS are summarized in Table 4.2.

TABLE 4.2
Melatonin's Beneficial Effects in Metabolic Syndrome

Effects on	Observed Effect after Melatonin Administration	References
Blood fasting glucose	Decreased	Sheih et al. (2009), Peschke et al. (2010)
Hyperinsulinemia	Decreased	Nishida et al. (2002), Sartori et al. (2009), Sheih et al. (2009), Peschke et al. (2010)
Insulin sensitivity in the muscle	Increased	Sartori et al. (2009), Srivastava and Krishna (2010)
Insulin resistance	Decreased	Kitagawa et al. (2012)
TG	Decreased	Nishida et al. (2002), Hussain et al. (2007), Kozirog et al. (2011)
LDL-C	Decreased	Hussain et al. (2007), Kozirog et al. (2011)
HDL-C	Increased	Hussain et al. (2007), She et al. (2009)
Body weight	Decreased	Wolden-Hanson et al. (2000); Prunet-Marcassus et al. (2003)
Intra-abdominal obesity and Plasma leptin levels	Decreased	Wolden-Hanson et al. (2000)
Visceral fat	Decreased	Rasmussen et al. (1999); Puchalski et al. (2003); She et al. (2009)
TG and VLDL released from the liver	Decreased	Hussain et al. (2007)

REFERENCES

Abdel-Wahab, M. H. and A. R. Abd-Allah. 2000. Possible protective effect of melatonin and/or desferrioxamine against streptozotocin-induced hyperglycaemia in mice. *Pharmacol Res* 41:533–537.

Achike, F. I., N. H. To, H. Wang, and C. Y. Kwan. 2011. Obesity, metabolic syndrome, adipocytes, and vascular function: A holistic viewpoint. *Clin Exp Pharmacol Physiol* 38(1):1–10.

Acuña-Castroviejo, D., G. Escames, J. Leon, A. Carazo, and H. Khaldy. 2003. Mitochondrial regulation by melatonin and its metabolites. *Adv Exp Med Biol* 527:549–557.

Acuña-Castroviejo, D., G. Escames, M. I. Rodriguez, and L. C. Lopez. 2007. Melatonin role in mitochondrial function. *Front Biosci* 12:947–963.

Acuña-Castroviejo, D., L. C. López, G. Escames, A. López, J. A. García, and R. J. Reiter. 2011. Melatonin-mitochondria interplay in health and disease. *Curr Topic Med Chem* 11(2):222–240.

Ahlborg, G., A. Shemyakin, F. Bohm, A. Gonon, and J. Pernow. 2007. Dual endothelial receptor blockade acutely improves insulin sensitivity in obese patients with insulin resistance and coronary heart disease. *Diabetes Care* 30(3):591–596.

Ahonen, T., J. Saltevo, M. Laakso, H. Kautiainen, E. Kumpusalo, and M. Vanhala. 2009. Gender differences relating to metabolic syndrome and proinflammation in Finnish subjects with elevated blood pressure. *Mediators Inflamm* 2009:959281. doi: 10.1155/2009/959281.

Aksoy, N., H. Vural, T. Sabuncu, and S. Aksoy. 2002. Effects of melatonin on oxidative-anti-oxidative status of tissues in streptozotocin induced diabetic rats. *Vascul Pharmacol* 38:127–130.

Alberti, K. G., R. H. Eckel, S. M. Grundy, P. Z. Zimmet, J. I. Cleeman, K. A. Donato, J. C. Fruchart et al. 2009. Harmonizing the metabolic syndrome: A joint interim statement of the International Diabetes Federation Task Force on Epidemiology and Prevention; National Heart, Lung, and Blood Institute; American Heart Association; World Heart Federation; International Atherosclerosis Society; and International Association for the Study of Obesity. *Circulation* 120:1640–1645.

Alemany, M. 2012. Do the interactions between glucocorticoids and sex hormones regulate the development of the metabolic syndrome? *Front Endocrinol (Lausanne)* 3:27.

Al Saraj, F., J. H. McDermott, T. Cawood, S. McAteer, M. Ali, W. Tormey, B. N. Cockburn, and S. Sreenan. 2009. Prevalence of the metabolic syndrome in patients of diabetes mellitus. *Ir J Med Sci* 178(3):309–313.

Andersson, A. K. and S. Sandler. 2001. Melatonin protects against streptozotocin, but not interleukin-I beta-induced damage of rodent pancreatic beta-cells. *J Pineal Res* 30:157–165.

Anwar, M. M. and A. R. Meki. 2003. Oxidative stress in streptozotocin induced diabetic rats: Effects of garlic oil and melatonin. *Comp Biochem Physiol A Mol Integr Physiol* 135:539–547.

Arita, Y., S. Kihara, N. Ouchi, M. Takahashi, K. Maeda, J. Miyagawa, K. Hotta et al. 1999. Paradoxical decrease of an adipose specific protein, adiponectin, in obesity. *Biochem Biophys Res Commun* 257(1):79–83.

Bartness, T. J., G. E. Demas, and C. K. Song. 2002. Seasonal changes in adiposity: The roles of the photoperiod, melatonin, and other hormones, and sympathetic nervous system. *Exp Biol Med (Maywood)* 227:363–376.

Baydas, G., H. Cantan, and A. Turkogllu. 2002. Comparative analysis of the protective effects of melatonin and vitamin E on streptozotocin-induced diabetes mellitus. *J Pineal Res* 32:225–230.

Bloomgarden, Z. T. 2008. Diabetic neuropathy. *Diabetes Care* 31:616–621.

Bobbert, P., A. Jenke, T. Bobbert, U. Kuhl, D. Lassner, C. Scheibenbogen, W. Poller, H. P. Schultheiss, and C. Skurk. 2012. High leptin and resisten expression in chronic heart failure: Adverse outcome in patients with dilated and inflammatory cardiomyopathy. *Eur J Heart Fail* 14(11):1265–1275.

Bojkova, B., M. Markova, E. Ahlersova, L. Ahlers, E. Adamekowa, P. Kubatka, and M. Kassayová. 2006. Metabolic effects of prolonged melatonin administration and short-term fasting in laboratory rats. *Acta Vet Brno* 75:7521–7532.

Bredt, D. S. and S. H. Snyder. 1994. Nitric oxide: A physiologic messenger molecule. *Ann Rev Biochem* 63:175–195.

Bromme, H. J., H. Ebelt, D. Peschke, and E. Peschke. 1999. Alloxan acts as a pro-oxidant only under reducing conditions. Influence of melatonin. *Cell Mol Life Sci* 55:487–493.

Bromme, H. J., W. Morke, D. Peschke, H. Ebelt, and D. Peschke. 2000. Scavenging effect of melatonin on hydroxyl radicals generated by alloxan. *J Pineal Res* 29:201–208.

Brydon, L., L. Petit, P. Delagrange, A. D. Strosberg, and R. Jockers. 2001. Functional expression of MT_2 (Mel1b) melatonin receptors in human PAZ6 adipocytes. *Endocrinology* 142:4264–4271.

Bubenik, G. A. 2002. Gastrointestinal melatonin: Localization, function, and clinical relevance. *Dig Dis Sci* 47:2336–2348.

Cardinali, D. P., P. Cano, J. Ortega, and A. I. Esquifino. 2011. Melatonin and the metabolic syndrome: Physiopathological and therapeutical implications. *Neuroendocrinology* 93(3):133–142.

Choi, K. M., S. M. Kim, Y. E. Kim, D. S. Choi, S. H. Baik, and J. Lee. 2007. International diabetes federation: Prevalence and cardiovascular disease risk of the metabolic syndrome using National Cholesterol Educational Program and International Diabetes Federation definitions in the Korean population. *Metabolism* 56(4):552–558.

Cinti, S. 2006. The role of brown adipose tissue in human obesity. *Nutr Metab Cardiovas Dis* 16:569–574.

Claustrat, B., J. Brun, and G. Chazot. 2005. The basic physiology and pathophysiology of melatonin. *Sleep Med* 9:11–24.

Contreras-Alcantara, S., K. Baba, and G. Tosini. 2010. Removal of melatonin receptor type-1 induces insulin resistance in the mouse. *Obesity (Silver Spring)* 18:1861–1863.

Cornier, M. A., D. Dabelea, T. L. Hernandez, R. C. Lindstrom, A. J. Steig, N. R. Stob, R. E. VanPelt, H. Wang, and R. H. Eckel. 2008. The metabolic syndrome. *Endocr Rev* 29:777–822.

Cui, J., S. Panse, and B. Falkner. 2011. The role of adiponectin in metabolic and vascular disease: A review. *Clin Nephrol* 75:26–33.

De Keulenaer G. W., R. W. Alexander, M. Ushio-Fukai, N. Ishizaka, and K. K. Griendling. 1998. Tumor necrosis factor alpha activates a p22phox-based NADH oxidase in vascular smooth muscle. *Biochem J* 329:653–657.

Deedwania, P. C. and R. Gupta. 2006. Management issues in the metabolic syndrome. *J Assoc Physicians India* 54:797–810.

Deepa, M., S. Farooq, M. Datta, R. Deepa, and V. Mohan. 2007. Prevalence of metabolic syndrome using WHO, ATP III, and IDF definitions in Asian Indians: The Chennai Urban Rural Epidemiology Study (CURES-34). *Diabetes Metab Res Rev* 23(2):127–134.

Ebelt, H., D. Peschke, H. J. Bromme, and D. Peschke. 2000. Influence of melatonin on free-radical induced changes in rat pancreatic beta-cells in vitro. *J Pineal Res* 28:65–72.

Ekmekcioglu, C. 2006. Melatonin receptors in humans, biological role and clinical relevance. *Biomed Pharmacother* 60:97–108.

Ekmekcioglu, C., P. Haslmayer, C. Phillip, M. R. Mehrabi, H. D. Glogar, M. Grimm, T. Thalhammer, and W. Marktl. 2001. 24-h variations in the expression of the mt1 melatonin receptor subtype in coronary heart disease. *Chronobiol Int* 18:973–985.

Ekmekcioglu, C., T. Thahammer, S. Humpeler, M. R. Mehrabi, H. D. Glogar, T. Holzenbein, O. Markovic, V. J. Leibetseder, G. Strauss-Blasche, and W. Marktl. 2003. The melatonin receptor subtype MT2 is present in the cardio-vascular system. *J Pineal Res* 35:40–44.

Eringa, E. C., C. D. Stehouwer, G. P. van Niew Amerongen, L. Ouwhand, N. Westerhof, and P. Siokema. 2004. Vasoconstrictor effects of insulin in skeletal muscle arterioles are mediated by ERK1/2 activation in endothelium. *Am J Physiol Heart Care Circ Physiol* 287:H2043–H2048.

Espino, J., J. A. Pariente, and A. B. Rodriguez. 2011. Role of melatonin on diabetes-related metabolic disorders. *World J Diabetes* 2(6):82–91.

Farooqui, A. A., T. Farooqui, F. Panza, and V. Frisardi. 2012. Metabolic syndrome as a risk factor for neurological disorders. *Cell Mol Life Sci* 69(5):741–762.

Fezeu, L., B. Balkau, A. P. Kengne, E. Sobyngwi, and J. C. Mbanya. 2007. Metabolic syndrome in a sub-Saharan African setting: Central obesity may be the key determinant. *Atherosclerosis* 193(1):70–76.

Ford, E. S. and W. H. Giles. 2003. A comparison of the prevalence of the metabolic syndrome using two proposed definitions. *Diabetes Care* 26(3):575–581.

Ford, E. S., W. H. Giles, and W. H. Dietz. 2002. Prevalence of the metabolic syndrome among US adults: Findings from the third National Health and Nutrition Examination survey. *JAMA* 287(3):356–359.

Galic, S., J. S. Oakhill, and G. R. Steinberg. 2010. Adipose tissue as an endocrine organ. *Mol Cell Endocrinol* 316:129–139.

Garvey, W. T., S. Kwon, D. Zheng, S. Shaughnessy, P. Wallace, A. Hutto, K. Pugh, A. J. Jenkins, R. L. Klein, and Y. Liao. 2003. Effect of insulin resistance and type 2 diabetes on lipoprotein subclass particle size and concentration determined by nuclear magnetic resonance. *Diabetes* 52:453–462.

Ghosh, G., K. De, S. Maity, D. Bandyopadhyay, S. Bhattacharya, R. J. Reiter, and A. Bandyopadhyay. 2007. Melatonin protects against oxidative damage and restores expression of GLUT-4 gene in the hyperthyroid rat heart. *J Pineal Res* 42:71–82.

Ginsberg, H. N. 2002. New perspectives on atherogenesis: Role of abnormal triglyceride rich lipoprotein metabolism. *Circulation* 106:2137–2142.

Green, B., J. S. Takahashi, and J. Bass. 2008. The meter of metabolism. *Cell* 134:728–742.

Grundy, S. M. 2008. Metabolic syndrome pandemic. *Arterioscler Thromb Vas Biol* 28:629–636.

Hardeland, R., D. P. Cardinali, V. Srinivasan, D. W. Spence, G. M. Brown, and S. R. Pandi-Perumal. 2011. Melatonin—A pleiotropic, orchestering regulator molecule. *Prog Neurobiol* 93:350–384.

Hardeland, R. and B. Fuhrberg. 1996. Ubiquitous melatonin. Presence and effects in unicells, plants, and animals. *Trends Comp Biochem Physiol* 2:25–45.

He, Y., B. Jiang, and J. Wang. 2006. Prevalence of the metabolic syndrome and its relation to cardiovascular disease in an elderly Chinese population. *J Am Coll Cardiol* 47:1588–1594.

Hirata, Y., O. Hayaishi, T. Tokuyama, and S. Seno. 1974. In vitro and in vivo formation of two new metabolites of melatonin. *J Biol Chem* 249:1311–1313.

Hotamisligil, G. S., P. Peraldi, A. Budavari, R. Ellis, M. F. White, and B. M. Spiegelman. 1996. IRS-1 mediated inhibition of insulin receptor tyrosine kinase activity in TNF-alpha and obesity-induced insulin resistance. *Science* 271:665–668.

Hotta, K., T. Funahashi, Y. Arita, M. Takahashi, M. Matsuda, Y. Okamoto, H. Iwahashi et al. 2000. Plasma concentrations of a novel adipose-specific protein adiponectin in type 2 diabetic patients. *Arterioscler Thomb Vasc Biol* 20(6):1595–1599.

Hoyos, M., J. M. Guerrero, R. Perez-Cano, J. Olivan, F. Fabiani, A. Garcia-Pergañeda, and C. Osuna. 2000. Serum cholesterol and lipid peroxidation are decreased by melatonin in diet-induced hypercholesterolemic rats. *J Pineal Res* 28(3):150–155.

Huang, P. L. 2005. Unraveling the links between diabetes, obesity, and cardiovascular disease. *Cir Res* 96:1129–1131.

Hussain, M. R., O. G. Ahmed, A. F. Hassan, and M. A. Ahmed. 2007. Intake of melatonin is associated with amelioration of physiological changes, both metabolic and morphological pathologies associated with obesity: An animal model. *Int J Exp Pathol* 88:19–29.

Kahn, B. B. and J. S. Flier. 2000. Obesity and insulin resistance. *J Clin Invest* 106:473–481.

Karlsson, B., A. Knutsson, and B. Lindahl. 2001. Is there an association between shift-work and having a metabolic syndrome? Results from a population based study of 27,485 people. *Occup Environ Med* 58:747–752.

Kassayova, M., M. Markova, B. Bojkova, E. Adamekova, P. Kubartka, E. Ahlersova, and I. Ahlers. 2006. Influence of long-term melatonin administration on basic physiological and metabolic variables of young Wistar Han rats. *Biologia* 61:313–320.

Kawasaki, T., K. Igarashi, T. Koeda, K. Sugimoto, K. Nakagawa, S. Hayashi, R. Yamaji, H. Inui, T. Fukusato, and T. Yamanouchi. 2009. Rats fed fructose enriched diets have characteristics of non-alcoholic hepatic steatosis. *J Nutr* 139:2067–2071.

Kelly, R. W., F. Amato, and R. F. Seamark. 1984. *N*-acetyl-5-methoxy-kynurenamine, a brain metabolite of melatonin is a potent inhibitor of prostaglandin biosynthesis. *Biochem Biophys Res Commun* 121:372–379.

Kemp, D. M., M. Ubeda, and J. F. Habener. 2002. Identification and functional characterization of melatonin Mel1b receptors in pancreatic β-cells: Potential role in incretin-mediated cell function by sensitization of cAMP signaling. *Mol Cell Endocrinol* 191:157–166.

Kershaw, E. E. and J. S. Flier. 2004. Adipose tissue as an endocrine organ. *J Clin Endocrinol Metab* 89(6):2548–2556.

Kitagawa, A., Y. Ohta, and K. Ohashi. 2012. Melatonin improves metabolic syndrome induced by high fructose intake in rats. *J Pineal Res* 52(4):403–413.

Knutson, K. L., A. M. Ryden, B. A. Mander, and E. van Cauter. 2006. Role of sleep duration and quality in the risk and severity of type 2 diabetes mellitus. *Arch Intern Med* 166:1768–1774.

Kozirog, M., A. R. Poliwczak, P. Duchnowicz, M. Koter-Michalak, J. Sikora, and M. Broncel. 2011. Melatonin treatment improves blood pressure, lipid profile and parameters of oxidative stress in patients with metabolic syndrome. *J Pineal Res* 50:261–266.

Krauss, R. M. 1995. Dense low density lipoproteins and coronary heart disease. *Am J Cardiol* 75(6):53B–57B.

Krauss, R. M. 1998. Atherogenicity of triglyceride-rich lipoproteins. *Am J Cardiol* 81(4):13B–17B.

Laaksonen, D. E., H. M. Lakka, L. K. Niskanen, G. A. Kaplan, J. T. Salonen, and T. A. Lakka. 2002. Metabolic syndrome and development of diabetes mellitus: Application and validation of recently suggested definitions of the metabolic syndrome in a prospective cohort study. *Am J Epidemiol* 156(11):1070–1077.

La Guardia, H. A., L. L. Hamm, and J. Chen. 2012. The metabolic syndrome and risk of chronic kidney disease: Pathophysiology and intervention strategies. *J Nutr Metab* 2012:652608. doi: 10.1155/2012/652608.

Lakka, H. M., D. E. Laaksonen, T. A. Lakka, L. K. Niskanen, E. Kumpusalo, J. Tuomilehto, and J. T. Salonen. 2002. The metabolic syndrome and total cardiovascular disease mortality in middle aged men. *JAMA* 288(21):2709–2716.

Lau, D. C., B. Dhillon, H. Yan, P. E. Szmitko, and S. Verma. 2005. Adipokine: Molecular links between obesity and atherosclerosis. *Am J Physiol Heart Care* 288:H2031–H2041.

Lee, D. E., S. Kehlenbrink, H. Lee, M. Hawkins, and J. S. Yudkin. 2009. Getting the message across mechanisms of physiological cross-talk by adipose tissue. *Am J Physiol Endocrinol Metab* 296:E1210–E1229.

Lee, E. B., G. Warmann, R. Dhir, and R. S. Ahima. 2011. Metabolic dysfunction associated with adiponectin deficiency enhances kainic acid-induced seizure severity. *J Neurosci* 31:14361–14362.

Leroith, D. 2012. Pathophysiology of metabolic syndrome: Implications for the cardiometric risks associated with type 2 diabetes. *Am J Med Sci* 343(1):13–16.

Lteif, A., P. Vaishnava, A. D. Baron, and K. J. Mather. 2007. Endothelin limits insulin action in obese/insulin-resistant humans. *Diabetes* 56:728–734.

Macchi, M. M. and J. N. Bruce. 2004. Human pineal physiology and functional significance of melatonin. *Front Neuroendocrinol* 25:177–195.

Mangge, H., G. Almer, M. Truschnig-Wilders, A. Schmidt, R. Gasser, and D. Fuchs. 2010. Inflammation, adiponectin, obesity, and cardiovascular risk. *Curr Med Chem* 17:4511–4520.

Mäntele, S., D. T. Otway, B. Middleton, S. Bretschneider, J. Wright, M. D. Robertson, D. J. Skene, and J. D. Johnston. 2012. Daily rhythms of plasma melatonin, but not plasma leptin or leptin mRNA, vary between lean, obese and type 2 diabetic men. *PLoS One* 7(5):e37123.

Mather, K. J., A. Lteif, H. O. Steinberg, and A. D. Baron. 2004. Interactions between endothelin and nitric oxide in the regulation of vascular tone in obesity and diabetes. *Diabetes* 53:2060–2066.

Mohamed, F., J. C. Monge, A. Gordon, P. Cernacek, D. Blais, and D. J. Stewart. 1995. Lack of role for nitric oxide (NO) in the selective destabilization of endothelial NO synthase mRNA by tumor necrosis factor-alpha. *Arterioscler Thromb Vasc Biol* 15:52–57.

Mohamud, W. N., A. A. Ismail, A. Sharifuddin, I. S. Ismail, K. I. Musa, K. A. Kadir, N. A. Kamaruddin et al. 2011. Prevalence of the metabolic syndrome and its risk factors in adult Malaysians: Results of a nationwide survey. *Diabetes Res Clin Pract* 91(2):239–245.

Montilla, P. L., J. F. Vargas, I. F. Túnez, M. C. Muñoz de Agueda, M. E. Valdelvira, and E. S. Cabrera. 1998. Oxidative stress in diabetic rats induced by streptozotocin: Protective effects of melatonin. *J Pineal Res* 25:94–100.

Moore, R. Y. and D. C. Klein. 1974. Visual pathways and the central neural control of a circadian rhythm in pineal serotonin *N*-acetyltransferase activity. *Brain Res* 71:17–33.

Muhlbauer, E. and E. Peschke. 2007. Evidence for the expression of both MT1- and in addition, the MT2-melatonin receptor, in the rat pancreas, islet, and β-cell. *J Pineal Res* 42:105–106.
Mulder, H., C. L. Nagrony, V. Lyssenko, and L. Groop. 2009. Melatonin receptors in pancreatic islets: Good morning to novel type 2 diabetes gene. *Diabetologia* 52:1240–1249.
Muniyappa, R., M. Iantorno, and M. J. Quon. 2008. An integrated view of insulin resistance and endothelial dysfunction. *Endocr Metab Clin North Am* 37(3):685–711.
Muniyappa, R., M. Montagnani, K. Koh, and M. J. Quon. 2007. Cardiovascular actions of insulin. *Endocr Rev* 28(5):463–491.
Narayanan, P., O. L. Meng, and O. Mahanim. 2011. Do the prevalence and compounds of syndrome differ among different ethnic groups? A cross-sectional study among obese Malaysian adolescents. *Metab Syndr Relat Disord* 9(5):389–395.
Nawrocki, A. R. and P. E. Scherer. 2004. The delicate balance between fat and muscle: Adipokines in metabolic disease and musculoskeletal inflammation. *Curr Opin Pharmacol* 4(3):281–289.
Nduhirabandi, F., E. F. du Toit, and A. Lochner. 2012. Melatonin and the metabolic syndrome: A tool for effective therapy in obesity-associated abnormalities? *Acta Physiol (Oxf)* 205(2):209–223.
Negi, G., A. Kumar, R. K. Kaundal, A. Gulati, and S. S. Sharma. 2010. Functional and biochemical evidence indicating beneficial effect of melatonin and nicotinamide alone and in combination in experimental diabetic neuropathy. *Neuropharmacology* 58(3):585–592.
Nilsson, P. M., G. Engstrom, and B. Hedblad. 2007. The metabolic syndrome and the incidence of cardio vascular disease in non-diabetic subjects—A population-based study comparing three different definitions. *Diabetic Med* 24(5):464–472.
Nishida, S., T. Segawa, I. Murai, and S. Nakagawa. 2002. Long-term melatonin administration reduces hyperinsulinemia and improves the altered fatty acid compositions in type-2 diabetic rats via the restoration of delta-5-desaturase activity. *J Pineal Res* 32:26–33.
Nystrom, F. H. and M. Quon. 1999. Insulin signalling: Metabolic pathways and mechanisms for specificity. *Cell Signal* 11(8):563–574.
Ogbera, A. O. 2010. Prevalence and gender distribution of the metabolic syndrome. *Diabetol Metab Syndr* 2:1.
Ouchi, N., S. Kihara, Y. Arita, K. Maeda, H. Kuriyama, Y. Okamoto, K. Hotta et al. 1999. Novel modulator for endothelial adhesion molecules: Adipocyte-derived plasma adiponectin. *Circulation* 100(25):2473–2476.
Oxenkrug, G. F. and P. Summergrad. 2010. Ramelteon attenuates age-associated hypertension and weight gain in spontaneously hypertensive rats. *Ann N Y Acad Sci* 1199:114–120.
Palios, J., N. P. E. Kadoglou, and S. Lampropoulos. 2012. The pathophysiology of HIV/HAART-related metabolic syndrome leading to cardiovascular disorders: The emerging role of adipokines. *Exp Diabetes Res* 2012: 103063. doi: 10:1155/2012/103063.
Pandi-Perumal, S. R., V. Srinivasan, G. J. M. Maestroni, D. P. Cardinali, B. Poeggeler, and R. Hardeland. 2006. Melatonin: Nature's most versatile signal? *FEBS J* 273:2813–2838.
Paz, K., R. Hemi, R. LeRoith, A. Karasik, E. Elhanany, H. Kanety, and Y. Zick. 1997. A molecular basis for insulin resistance. Elevated serine/threonine phosphorylation of IRS-1 and IRS-2 inhibits their binding to the juxta-glomerulus region of the insulin receptor and impairs their ability to undergo insulin-induced tyrosine phosphorylation. *J Biol Chem* 72:29911–29918.
Peschke, E., T. Frese, E. Chankiewitz, D. Peschke, U. Preiss, U. Schneyer, R. Spessert, and E. Mühlbauer. 2006. Diabetic Goto Kakizaki rats as well as type 2 diabetic patients show decreased diurnal serum melatonin level and an increased pancreatic melatonin-receptor status. *J Pineal Res* 40:135–143.
Peschke, E. and D. Peschke. 1998. Evidence for a circadian rhythm of insulin release from perfused rat pancreatic islet cells. *Diabetologia* 41:1085–1092.
Peschke, E., H. Schucht, and E. Muhlbauer. 2010. Long-term enteral administration of melatonin reduces expression of pineal insulin receptors in both Wistar and type 2 diabetic Goto-Kakizaki rats. *J Pineal Res* 49:373–381.
Peschke, E., L. Stumpf, L. Bazwinsky, L. Litvak, H. Dralle, and E. Muhlbauer. 2007. Melatonin and type 2 diabetes—A possible link? *J Pineal Res* 42:350–358.
Prunet-Marcassus, B., M. Desbazeille, A. Bros, K. Louche, P. Delagrange, P. Renard, L. Casteilla, and L. Pénicaud. 2003. Melatonin reduces body weight in Sprague–Dawley rats with diet induced obesity. *Endocrinology* 144:5347–5352.
Puchalski, S. S., J. N. Green, and D. D. Rasmussen. 2003. Melatonin effect on rat body weight regulation in response to high-fat diet at middle age. *Endocrine* 21(2):163–167.
Ramracheya, R. D., D. S. Muller, P. E. Squires, H. Brereton, D. Sugden, G. C. Huang., S. A. Amiel, P. M. Jones, and S. J. Persaud. 2008. Function and expression of melatonin receptors on human pancreatic islets. *J Pineal Res* 44:273–279.

Rasmussen, D. D., B. M. Boldt, C. W. Wilkinson, S. M. Yellon, and A. M. Matsumoto. 1999. Daily melatonin administration at middle age suppresses male rat visceral fat, plasma leptin, and plasma insulin to youthful levels. *Endocrinology* 140:1009–1012.

Ravikiran, M., A. Bhansali, P. Ravikumar, S. Bhansali, P. Dutta, J. S. Thakur, N. Sachdeva, S. Bhadada, and R. Walia. 2010. Prevalence and risk factors of metabolic syndrome among Asian Indians: A community survey. *Diabetes Res Clin Pract* 89(2):181–189.

Reiter, R. J. 1997. Anti oxidant actions of melatonin. *Adv Pharmacol* 38:103–117.

Report of a WHO Consultation. 2000. Obesity: Preventing and managing the global epidemic. *World Health Organ Tech Rep Ser* 894:i–xii, 1–253.

Rios-Lugo, M. J., P. Cano, V. Jiménez-Ortega, M. P. Fernández-Mateos, P. A. Scacchi, D. P. Cardinalli, and A. I. Esquifino. 2010. Melatonin effects on adinopectin leptin, insulin, glucose, triglycerides, and cholesterol in normal and high fed rats. *J Pineal Res* 49:342–348.

Rutter, J., M. Reick, and S. L. McKnight. 2002. Metabolism and control of circadian rhythms. *Annu Rev Biochem* 71:307–331.

Ruwaida, L., M. Zainuddin, M. Isa, W. M. Wan Muda, and H. J. Mohamed. 2011. The prevalence of metabolic syndrome according to various definitions and hypertriglyceridemic waist in Malaysian adults. *Int J Prev Med* 2(4):229–237.

Sailaja Devi, M. M. and Y. Suresh. 2000. Preservation of the antioxidant status in chemically induced diabetes mellitus by melatonin. *J Pineal Res* 29:108–115.

Sartori, C., P. Dessen, C. Mathieu, A. Monney, J. Bloch, P. Nicod, U. Scherrer, and H. Duplain. 2009. Melatonin improves glucose homeostasis and endothelial vascular function in high-fat fed insulin-resistant mice. *Endocrinology* 150:5311–5347.

Savaskan, E., G. Olivieri, L. Brydon, R. Jockers, K. Krauchi, A. Wirz-Justice, and F. Müller-Spahn. 2001. Cerebrovascular melatonin MT1 receptor alterations in patients with Alzheimer's disease. *Neurosci Lett* 308:9–12.

Scheer, F. A., G. A. Van Montfrans, E. J. Van Someren, G. Mairuhu, and R. M. Buijs. 2004. Daily night time melatonin decreases nocturnal blood pressure in male patients with essential hypertension. *Hypertension* 46:192–197.

Schinzari, F., M. Tesauro, V. Rovella, A. Galli, N. Mores, O. Porzio, D. Lauro, and C. Cardillo. 2010. Generalized impairment of vasodilator reactivity during hyperinsulinemia in patients with obesity-related metabolic syndrome. *Am J Physiol Endocrinol Metab* 299:947–952.

Schmeichel, A. M., J. D. Schmeizer, and P. A. Low. 2003. Oxidative injury and apoptosis of dorsal root ganglion neurons in chronic experimental diabetic neuropathy. *Diabetes* 52:165–171.

She, M., X. Deng, Z. Guo, M. Laudon, Z. Hu, D. Liao, X. Hu et al. 2009. NEU-P11, a novel melatonin agonist, inhibits weight gain and improves insulin sensitivity in high-fat/high-sucrose-fed rats. *Pharmacol Res* 59:248–253.

Sheih, J. M., H. T. Wu, K. C. Cheng, and J. T. Cheng. 2009. Melatonin ameliorates high fat diet induced diabetes and stimulates glycogen synthesis via PKCzeta-Akt-GSK-3beta pathway in hepatic cells. *J Pineal Res* 47(4):339–344.

Shimda, K., T. Miyazaki, and H. Daida. 2004. Adiponectin, and atherosclerotic disease. *Clin Chem Acta* 344(1–2):1–12.

Shulman, G. I. 2000. Cellular mechanisms of insulin resistance. *J Clin Invest* 106:171–176.

Smith, C. C. and D. M. Yellon. 2011. Adipocytokines, cardiovascular pathophysiology and myocardial protection. *Pharmacol Ther* 129:206–219.

Spratt, K. A. 2009. Managing diabetic dyslipidemia: Aggressive approach. *J Am Osteopath Assoc* 109(5):52–57.

Srinivasan, V. 1997. Melatonin, biological rhythm disorders and phototherapy. *Ind J Physiol Pharmacol* 41:309–328.

Srinivasan, V. 2002. Melatonin, oxidative stress and neurodegenerative diseases. *Ind J Exp Biol* 40:668–679.

Srinivasan, V., S. R. Pandi-Perumal, G. J. Maestroni, A. J. Esquifino, R. Hardeland, and D. P. Cardinali. 2005. Role of melatonin in neurodegenerative diseases. *Neurotox Res* 7:293–318.

Srinivasan, V., D. W. Spence, S. R. Pandi-Perumal, G. M. Brown, and D. P. Cardinali. 2011. Melatonin in mitochondrial function and related disorders. *Int J Alzheimers Dis* 2011:326320. doi: 10.4061/2011/326320.

Srivastava, R. K. and A. Krishna. 2010. Melatonin modulates glucose homeostasis during winter dormancy in a vespertilionid bat, *Scotophilus heathi*. *Comp Biochem Physiol A Mol Integr Physiol* 155:392–400.

Sudnikovich, E. J., Y. Z. Maksimchik, S. V. Zabrodskaya, V. L. Kubyshin, E. A. Lapshina, M. Bryszewska, R. J. Reiter, and I. B. Zavodnik. 2007. Melatonin attenuates metabolic disorders due to streptozotocin-induced diabetes in rats. *Eur J Pharmacol* 569:180–187.

Takasu, N., I. Komiya, T. Asawa, Y. Nagasawa, and T. Yamada. 1991. Streptozocin- and alloxan-induced H_2O_2 generation and DNA fragmentation in pancreatic islets. H_2O_2 as mediator for DNA fragmentation. *Diabetes* 40:1141–1145.

Tan, A. K., R. A. Dunn, and S. T. Yen. 2011. Ethnic disparities in metabolic syndrome in Malaysia: An analysis by risk factors. *Metab Syndr Relat Disord* 9(6):441–451.

Tan, B. Y., H. K. Kantilal, and R. Singh. 2008. Prevalence of the metabolic syndrome among Malaysians using International Diabetes Federation, National Cholesterol Education Program and Modified World Health Organization Definitions. *Malays J Nutr* 14(1):65–77.

Tan, D. X., L. D. Chen, B. Poeggeler, L. C. Manchester, and R. J. Reiter. 1993. Melatonin: A potent, endogenous hydroxyl radical scavenger. *Endocr J* 1:57–60.

Tan, D. X., L. C. Manchester, L. Fuentes-Broto, S. D. Paredes, and R. J. Reiter. 2010. Significance and application of melatonin in the regulation of brown adipose tissue metabolism: Relation to human obesity. *Obesity* 12:167–188.

Tan, D. X., L. C. Manchester, R. J. Reiter, B. F. Plummer, L. J. Hardies, S. T. Weintraub, Vijayalaxmi, and A. M. Shepherd. 1998. A novel melatonin metabolite, cyclic 3-hydroxyl-melatonin: A biomarker of in vivo hydroxyl radical generation. *Biochem Biophys Res Commun* 253:614–620.

Tan, D. X., R. J. Reiter, L.-D. Chen, B. Poeggeler, L. C. Manchester, and L. R. Barlow-Walden. 1994. Both physiological and pharmacological levels of melatonin reduce DNA adduct formation induced by the carcinogen saffrole. *Carcinogenesis* 15:215–218.

Termizy, H. M. and M. Mafauzy. 2009. Metabolic syndrome and its characteristics among obese patients attending obesity clinic. *Singapore Med J* 50(4):390–394.

Tesauro, M., M. P. Canale, G. Rodia, N. Di Daniele, D. Lauro, A. Scuteri, and C. Cardillo. 2011. Metabolic syndrome, chronic kidney, and cardiovascular diseases: Role of adipokines. *Cardiol Res Pract* 2011:653182. doi: 10.4061/2011/653182.

Tesauro, M. and C. Cardillo. 2011. Obesity, blood vessels and metabolic syndrome. *Acta Physiol* 203:279–286.

Tesauro, M., F. Schinzari, V. Rovella, D. Melina, N. Mores, A. Barini, M. Mettimano et al. 2008. Tumor necrosis factor-alpha antagonism improves vasodilation during hyperinsulinemia in metabolic syndrome. *Diabetes Care* 31:1439–1441.

Tiedge, M., S. Lortz, J. Drinkgern, and S. Lenzen. 1997. Relation between antioxidant enzyme gene expression and anti-oxidative defense status of insulin-producing cells. *Diabetes* 46:1733–1742.

Tong, P. C. Y., A. P. Kong, W. Y. So, X. Yong, C. S. Ho, R. C. Ma, R. Ozaki et al. 2007. The usefulness of the International Diabetes Federation and the National Cholesterol Education Program's Adult Treatment Panel III definitions of the metabolic syndrome in predicting coronary heart disease in subjects with type 2 diabetes. *Diabetes Care* 30(5):1206–1211.

Urata, Y., S. Honma, S. Goto, S. Todorski, T. Lida, S. Cho, K Honma, and T. Kondo. 1999. Melatonin induces γ-glutamylcysteinesynthetase mediated by activator protein-1 in human vascular endothelial cells. *Free Radic Biol Med* 27:838–847.

Vakkilainen, J., G. Steiner, J. C. Anaquer, F. Aubin, S. Rattier, C. Foucher, A. Hamsten, M. R. Taskinen, and DAIS Group. 2003. Relationship between low-density lipoprotein particle size plasma lipoproteins and progression of coronary artery disease: The Diabetes Atherosclerosis Intervention Study. *Circulation* 107:1733–1737.

Vazzana, N., F. Santilli, S. Sestilli, C. Cuccurullo, and G. Davis. 2011. Determinants of increased cardiovascular disease in obesity and metabolic syndrome. *Curr Med Chem* 18(34):5267–5280.

Wellen, K. E. and G. S. Hotamisligil. 2003. Obesity induced inflammatory changes in adipose tissue. *J Clin Invest* 112(12):1785–1788.

Whaley-Connell, A., P. A. McCullough, and J. R. Sowers. 2011. The role of oxidative stress in the metabolic syndrome. *Rev Cardiovasc Med* 12(1):21–29.

Whaley-Connell, A. and J. A. Sowers. 2012. Oxidative stress in the cardiorenal metabolic syndrome. *Curr Hyperten Rep* 14(4):360–365. doi: 10.1007/s11906-012-0279-2.

Wolden-Hanson, T., D. R. Mitton, R. I. McCants, S. M. Yellon, C. W. Wilkinson, A. M. Matsumoto, and D. D. Rasmussen. 2000. Daily melatonin administration to middle aged male rats suppresses body weight, intra abdominal adiposity, and plasma leptin, and insulin independent of food intake and total body fat. *Endocrinology* 141:487–497.

Yoshizumi, M., M. A. Perrella, J. C. Burnett Jr., and M. E. Lee. 1993. Tumor necrosis factor down regulates an endothelial nitric oxide synthase mRNA by shortening its half life. *Circ Res* 73:205–209.

Zainuddin, L. R., N. Isa, W. M. Muda, and H. J. Mohamed. 2011. The prevalence of metabolic syndrome according to various definitions and hypertriceridemic-waist in Malaysian adults. *Int J Prev Med* 2(4):229–237.

Zanquetta, M. M., P. M. Seraphim, D. H. Sumida, J. Cipolla-Neto, and U. F. Machado. 2003. Calorie restriction reduces pinealectomy induced insulin resistance by improving GLUT-4 gene expression and its translocation to the plasma membrane. *J Pineal Res* 35:141–148.

Zuo, H., Z. Shi, X. Hu, M. Wu, Z. Guo, and A. Hussain. 2009. Prevalence of the metabolic syndrome and factors associated with its components in Chinese adults. *Metabolism* 58(8):1102–1108.

5 Therapeutic Applications of Melatonin in Pediatrics

Emilio J. Sanchez-Barcelo, Maria D. Mediavilla, and Russel J. Reiter

CONTENTS

5.1 INTRODUCTION

Melatonin actions described in experimental studies include the regulation of circadian rhythms (i.e., sleep/alertness or body temperature) (Arendt and Skene, 2005; Cagnacci et al., 1992), effects on reproductive physiology (Reiter et al., 2009), and antioxidant properties (Allegra et al., 2003) among other properties (Macchi and Bruce, 2004). This variety of actions has encouraged the study of its possible clinical applications in different pathologies. A review of the clinical trials carried out to assess the possible usefulness of melatonin as a therapeutic drug can be found in the article of Sanchez-Barcelo et al. (2010). Pediatrics is also listed among the medical areas investigating possible applications of melatonin. Gitto et al. (2011) and Sanchez-Barcelo et al. (2011) recently published review articles analyzing the clinical uses of melatonin in pediatrics. Since this book focuses on the neuroprotective effects of melatonin, we are now going to review the clinical uses of melatonin in pediatric pathologies involving the central neural system, including epilepsy and sleep disorders, as well as in clinical practices such as anesthesia.

5.2 MELATONIN USES IN ANESTHESIA, SEDATION, AND ANALGESIA

Melatonin has sedative, anxiolytic, analgesic, and hypnotic properties (Ebadi et al., 1998; Jan et al., 2011; Wilhelmsen et al., 2011). For these reasons, this indoleamine has been assayed in anesthetic procedures, including premedication and induction as well as postsurgical analgesia (Naguib et al., 2007). Focusing on the question of pediatric anesthesia, several clinical trials have been carried out to compare the effects of melatonin with those of other drugs commonly used in anesthetic procedures. Sometimes, melatonin has been used in association with other conventional anesthetic drugs. Melatonin amplifies the effects of benzodiazepines and barbiturates on the central nervous system (Muñoz-Hoyos et al., 2002); for this reason, it has been assayed as a preoperative drug and its effects compared with those of midazolam (Dormicum), a drug widely used for premedication. The conclusion is that melatonin and midazolam are equally effective in reducing the anxiety that precedes surgery (Özcengiz et al., 2011; Samarkandi et al., 2005), though melatonin is more effective in decreasing postoperative events such as excitement or sleep disturbances (Samarkandi et al., 2005). However, other studies have reported that melatonin is less effective than midazolam (Kain et al., 2009) or even ineffective (Isik et al., 2008). Oral melatonin (0.1 mg/kg) before anesthesia induction reduced the incidence of emergence agitation in children after sevoflurane anesthesia (Özcengiz et al., 2011). Regarding the stage of induction of the anesthesia, melatonin was effective for steal induction in children, although less effective than clonidine (Almenrader et al., 2013).

Melatonin is also useful as an analgesic. From a clinical trial involving a group of 60 preterm infants (30 treated with common sedation and analgesia and 30 with the same treatment plus melatonin), Gitto et al. (2012) demonstrate the utility of melatonin as an adjunct analgesic therapy during procedural pain, especially when an inflammatory component is involved.

When children are to be subjected to examinations that require a certain degree of immobility and relaxation, melatonin could be useful because of its sedative effects. Thus, in a study of 86 children undergoing a sleep EEG after sleep deprivation, melatonin appeared as an alternative treatment to pharmacological sedation (Wassmer et al., 2001). Similarly, in uncooperative children, melatonin (10 mg), especially when associated with sleep deprivation, allows for successful magnetic resonance imaging (MRI) examinations (Johnson et al., 2002). Other authors (Sury and Fairweather, 2006) concluded, from a randomized double-blind study in 98 children undergoing MRI examinations, that melatonin (3–6 mg) did not contribute to sedation of children. It should be noted that the doses of melatonin used in this last study (3–6 mg) were significantly lower than those used in the previously cited work (10 mg). Some investigations in pediatric audiology, such as brainstem-evoked response audiometry (BERA), require sedation. In a clinical trial involving 50 children who underwent BERA, melatonin-induced sleep was a good alternative to conventional sedation (Schmidt et al., 2004).

5.3 MELATONIN USES IN EPILEPSY AND FEBRILE SEIZURES

The fact that suppression of the pineal function by either pinealectomy (associated with parathyroidectomy) (Reiter et al., 1973) or intracerebroventricular administration of melatonin antibodies induced convulsions in rats (Fariello et al., 1977) suggests a possible antiepileptic role for melatonin. Further studies demonstrate that melatonin exerts a neuroprotective role, preventing the damage produced by kainic acid–induced seizures in rats (Giusti et al., 1996). The mechanisms involved in the antiepileptic actions of melatonin could be a reduction of neuronal excitability dependent of its antioxidant and free radicals scavenging properties (Reiter et al., 2009); their actions on both GABA and glutamate receptors (Acuña Castroviejo et al., 1986); and the inhibition of calcium influx into the neurons, reducing NO production, with the subsequent reduction of the excitatory effects of N-methyl-D-aspartate (Muñoz-Hoyos et al., 1998). Although from basic experiments the use of melatonin as an antiepileptic drug seems well founded (Banach et al., 2011), its clinical value is still controversial.

Several authors (Bazil et al., 2000; Molina-Carballo et al., 2007; Muñoz-Hoyos et al., 1998) found that melatonin levels were reduced in patients with epilepsy at baseline, compared with controls, and increased threefold, following seizures. Recent studies carried out in children with refractory epilepsy or febrile seizures (Ardura et al., 2010; Paprocka et al., 2010) confirmed a lowered level of melatonin in these children in comparison to those without seizures. However, whether a lowered level of melatonin is the cause or a consequence of the epilepsy remains unclear. A recent survey in 51 children with epilepsy showed that these patients preserved their rhythm of melatonin secretion as well as other circadian rhythms, and no association between melatonin secretion and seizure characteristics was observed (Praninskiene et al., 2012).

Melatonin alone, at a single evening dose of 5–10 mg, has been reported to reduce the frequency of epileptic attacks in children (Fauteck et al., 1999). However, the association of melatonin with other antiepileptic drugs such as vigabatrin, valproate, and phenobarbital is more commonly used.

Thus, the association of melatonin with phenobarbital was effective for the treatment of a young girl diagnosed with severe myoclonic epilepsy (Molina-Carballo et al., 1997). Two more clinical trials (Peled et al., 2001; Saracz and Rosdy, 2004) on children with severe intractable seizures treated with oral melatonin (3 mg/day) associated with conventional antiepileptic drugs reported significant clinical reduction in seizure activity during treatment, particularly in the night. However, in a trial with a sample of six children treated with melatonin, although treatment improved the sleep quality of the patients, four of six children had elevated seizure activity after treatment (Sheldon, 1998). Two serious weaknesses of this study are the small number of cases and the disparate range of ages of the children enrolled in the trial (from 9 months to 18 years old).

The oxidative status of epileptic children is important since neuronal damage during convulsions depends, at least in part, on the level of oxidative stress. In this regard, the antioxidant properties of melatonin could play a protective role against brain damage linked to epilepsy. When melatonin was associated with carbamazepine or valproate, the changes observed in the antioxidant enzymes, glutathione peroxidase and glutathione reductase, indicated a significant reduction of the oxidative status of the patients when compared with those receiving only conventional antiepileptic drugs (Gupta et al., 2004a,b).

Epilepsy is frequently linked to sleep problems. In epileptic patients, treatment with melatonin has been reported to improve their wake–sleep disorders (Coppola et al., 2004; Elkhayat et al., 2010) with a reduction in the severity of seizures (Elkhayat et al., 2010) and the seizure frequency (Coppola et al., 2004).

From two recent bibliographic reviews that searched for information from the databases of Cochrane Library and Medline in order to assess melatonin effects on epilepsy, the conclusion reached was that, at present, it is not possible to draw any conclusion about the role of melatonin in reducing seizure frequency or improving the quality of life in epileptic patients, and more well-designed clinical trials are necessary to establish the precise role of melatonin in epilepsy treatment (Brigo and Del Felice, 2012; Jain and Besag, 2013).

5.4 MELATONIN IN THE TREATMENT OF PEDIATRIC SLEEP DISORDERS

Numerous lines of evidence indicate that melatonin plays an important role in the regulation of human sleep (Morris et al., 2012). There is a relationship between melatonin nocturnal peak in serum and sleep onset; melatonin increases 10–15 fold in nocturnal blood, 1–2 h before bedtime, and may be one of the triggers for inducing sleep (Lavie, 1997; Tzischinsky et al., 1993). Furthermore, melatonin, at physiologic doses (0.1–0.3 mg), promotes sleep onset and maintenance, decreases sleep latency, increases sleep efficiency, and increases total sleep time (Brzezinski, 1997; Zhdanova and Wurtman, 1997). Finally, melatonin's hypnotic effects might be secondary to its ability to induce a reduction in the core body temperature (Cagnacci et al., 1992). The suppression of nocturnal melatonin, either by treatment with beta-blockers or as a consequence of thoracic spinal cord injury, increases the total wake time during sleep (Morris et al., 2012). Jan et al. (2011) suggested that the promotion of sleep by melatonin or its hypnotic actions occurs within the neurons and independently of its membrane receptors.

The use of melatonin as sleep medicine is particularly efficient in the treatment of sleep disorders involving disturbances of the circadian system, that is to say, those sleep disorders defined by the American Academy of Sleep Medicine as follows: "due primarily to alterations in the circadian timekeeping system or a misalignment between the endogenous circadian rhythm and exogenous factors that affect the timing or duration of sleep."

The most prevalent sleep complaints among school children are bedtime resistance (27%), morning wake-up problems (17%) and morning somnolence and fatigue (17%), delayed sleep onset (11.3%), and night waking (6.5%) (Blader et al., 1997). Melatonin is currently used in the treatment of insomnia in children, and some clinical trials (Pelayo and Yuen, 2012; van Geijlswijk et al., 2010a,b; Zee, 2010) support its efficacy when properly used, that is to say, when the time of administration is 1–2 h before the dim light melatonin onset (DLMO), which is roughly 9–10 h after the patient has awakened (van Geijlswijk et al., 2010a,b; Zee, 2010).

A review of circadian sleep disorders sponsored by the American Academy of Sleep concluded that delayed sleep-phase syndrome (DSPS) is a kind of dyssomnia of circadian etiology, and is successfully treatable with melatonin (Sack et al., 2007). Children with DSPS have affected (delayed) timing of the major sleep episode, compared to the general population and relative to societal requirements. They fall asleep some hours after midnight and have difficulty waking up in the morning. The time of peak melatonin secretion is also significantly delayed in DSPS patients in relation to healthy control subjects (Rahman et al., 2009). Treatment of DSPS with melatonin has been reported to advance the sleep onset and to increase sleep duration; consequently, it reduces the school-learning problems of these children, frequently associated with their sleep deficits (Szeinberg et al., 2006; van Geijlswijk et al., 2010b). As noted earlier, these effects of melatonin are greatest when administered at least 1–2 h before DLMO. Since the determination of DLMO is not included among the ordinary clinical procedures, an alternative practical approach is to administer melatonin just before the desired bedtime (van Geijlswijk et al., 2010b).

Melatonin is also indicated for the treatment of sleep disorders associated with neurological impairment, including mental retardation, intellectual disability, problems of learning, autistic spectrum disorders, Angelman and Rett syndromes, tuberous sclerosis, etc. (Braam et al., 2009; Jan et al., 2000; Pillar et al., 2000; Sanchez-Barcelo et al., 2011; Zhdanova et al., 1999). In this field, several pathologies seem particularly amenable to the treatment of their attendant sleep problems with melatonin. This is the case with the autism spectrum disorders (ASDs). Insomnia is a common concern in ASD children. Interestingly, these children show serum concentrations of melatonin significantly lower than their healthy counterparts over the entire 24 h cycle, and the nocturnal urinary excretion of 6-sulfatoxymelatonin (a metabolite of melatonin, which reflects its circulating levels) was negatively correlated with autism severity in the overall level of verbal language, imitative social play, and repetitive use of objects (Tordjman et al., 2012). It has been suggested that ASD children may have mutations in some of the genes coding for the enzymes involved in melatonin synthesis; however, these are preliminary results and as such are inconclusive (Chaste et al., 2010; Jonsson et al., 2010). Numerous clinical trials have assayed melatonin treatments for sleep disorders associated with ASD. These studies have reported improvements in sleep parameters with exogenous melatonin supplementation in ASD, including longer sleep duration, less nighttime awakenings, quicker sleep onset, and better daytime behaviors, while the reported adverse effects of melatonin were minimal to none (Doyen et al., 2011; Guénolé et al., 2011; Rossignol and Frye, 2013; Sanchez-Barcelo et al., 2011; Wright et al., 2011).

The sequencing of all the genes involved in the melatonin synthesis pathway revealed some mutations in patients with Attention-Deficit Hyperactivity Disorder (ADHD) (Chaste et al., 2011). From the analysis of a total of 139 original articles on sleep and childhood ADHD, including 22 on the treatment of sleep disturbances, it was concluded that, among pharmacological treatments, randomized controlled trials support the use of melatonin to reduce the sleep-onset delay, whereas there is more limited evidence for other medications (Cortese et al., 2013).

Children with Smith Magenis Syndrome (SMS), a multiple congenital anomaly, commonly due to an interstitial deletion of the chromosome 17 band p11.2, present, among other symptoms, sleep disturbances that include early sleep onset, difficulty falling asleep, difficulty staying asleep, frequent awakening, early waking, reduced REM sleep, and decreased sleep time (De Leersnyder et al., 2003). These sleep problems are, in part, responsible for some of the behavioral and cognitive problems of these children. Patients with SMS have a characteristic inversion of the circadian rhythm of melatonin, with lower plasma concentrations at night and higher concentrations during the day (Potocki et al., 2000). This anomalous secretory pattern of melatonin could be, at least in part, responsible for the sleep disorders of children with SMS. On this basis, a treatment regimen consisting of the administration of a β1-adrenergic antagonist in the morning, to abolish the daytime secretion of melatonin, combined with the administration of melatonin in the evening, to restore the normal day/night rhythm of melatonin, has been assessed in these patients, resulting in a dramatic improvement in their sleep quality, reduced irritability, low evening drowsiness, and improved learning capability (Carpizo et al., 2006; De Leersnyder et al., 2003, 2006).

The association between visual impairment/blindness and sleep disorders has been widely confirmed (Gordo et al., 2001; Skene and Arendt, 2007). For this reason, it is extremely surprising to note the scarce randomized controlled studies focusing on the treatment of these sleep problems with melatonin. Only two randomized controlled trials (Coppola et al., 2004; Dodge and Wilson, 2001) have reported improved effects on sleep latencies. Thus, a study by Khan et al. (2011) concludes that currently no high-quality data that support or refute the use of melatonin for sleep disorders in visually impaired children, despite the fact that no significant adverse effects of melatonin were reported in any case.

The role of melatonin as a sleep-promoting compound is limited by its short half-life in the circulation. This fact has led to the development of controlled-release formulations and of various synthetic melatonin receptor agonists, such as Ramelteon (Rozerem™) and Agomelatine (Valdoxan™) (Sanchez-Barcelo et al., 2007). However, according to the limited studies, none of the designed drugs has substantial advantages over melatonin itself (Sanchez-Barcelo et al., 2007).

5.5 UTILITY OF MELATONIN IN NEONATAL CARE

The Neonatal Intensive Care Unit of the University of Messina (Italy) has been developing an interesting research line using melatonin in the management of the oxidative risk of newborns. These children, particularly when they have been delivered preterm, are prone to suffer oxidative damage due to the immaturity of their endogenous antioxidant mechanisms. This research group has carried out different clinical trials in newborns with different pathologies, including perinatal asphyxia, sepsis, respiratory distress syndrome, or those subjected to surgical treatments. Their results show that treatment with melatonin reduced oxidative stress and improved clinical outcomes (Aversa et al., 2012; Gitto et al., 2009, 2011; Marseglia et al., 2013).

5.6 SIDE EFFECTS OF MELATONIN IN PEDIATRIC TREATMENTS

Symptoms related to hypermelatoninemia have been reported in individuals supplemented with melatonin; however, the lack of descriptions of significant side effects after long-term treatment of children with exogenous melatonin is highly remarkable (Carr et al., 2007; Duman et al., 2010). The authors, at this point, comment on a curious case of spontaneous endogenous hypermelatoninemia reported in a 6-year-old girl diagnosed with Shapiro's syndrome (Duman et al., 2010). She had more than 1000 pg/mL of melatonin in serum (normal range 0–150 pg/mL) and showed daily episodes of syncope, hypothermia, and profuse sweating. These symptoms decreased with phototherapy or treatment with propranolol. This clinical case is a possible illustration of the hypothetical (never observed) side effects of high doses of melatonin and its treatment.

5.7 CONCLUSIONS

The multiple properties of melatonin afford perspectives of its beneficial effects in several pediatric therapies, particularly in sleep disorders of circadian etiology and also in epilepsy, anesthetic procedures, and oxidative stress in newborns. In this context, it is both surprising and regrettable to note the current paucity of controlled clinical trials focusing on the clinical usefulness of melatonin in pediatrics. Problems related to dosage, formulations (slow or fast release), and the duration of treatment are awaiting solutions.

REFERENCES

Acuña Castroviejo D, Rosenstein RE, Romeo HE, Cardinali DP. Changes in gamma-minobutyric acid high affinity binding to cerebral cortex membranes after pinealectomy or melatonin administration to rats. *Neuroendocrinology*. 1986; 43(1):24–31.

Allegra M, Reiter RJ, Tan DX, Gentile C, Tesoriere L, Livrea MA. The chemistry of melatoni's interaction with reactive species. *J Pineal Res*. 2003; 34(1):1–10.

Almenrader N, Haiberger R, Passariello M. Steal induction in preschool children: Is melatonin as good of clonidine? A prospective, randomized study. *Paediatr Anaesth*. 2013; 23(4):328–333.

Ardura J, Andres J, Garmendia JR, Ardura F. Melatonin in epilepsy and febrile seizures. *J Child Neurol*. 2010; 25(7):888–891.

Arendt J, Skene D. Melatonin as a chronobiotic. *Sleep Med Rev*. 2005; 9(1):25–39.

Aversa S, Pellegrino S, Barberi I, Reiter RJ, Gitto E. Potential utility of melatonin as an antioxidant during pregnancy and in the perinatal period. *J Matern Fetal Neonatal Med*. 2012; 25(3):207–221.

Banach M, Gurdziel E, Jędrych M, Borowicz KK. Melatonin in experimental seizures and epilepsy. *Pharmacol Rep*. 2011; 63(1):1–11.

Bazil CW, Short D, Crispin D, Zheng W. Patients with intractable epilepsy have low melatonin, which increases following seizures. *Neurology*. 2000; 55(11):1746–1748.

Blader JC, Koplewicz HS, Abikoff H, Foley C. Sleep problems of elementary school children. A community survey. *Arch Pediatr Adolesc Med*. 1997; 151(5):473–480.

Braam W, Smits MG, Didden R, Korzilius H, Van Geijlswijk IM, Curfs LM. Exogenous melatonin for sleep problems in individuals with intellectual disability: A meta-analysis. *Dev Med Child Neurol*. 2009; 51(5):340–349.

Brigo F, Del Felice A. Melatonin as add-on treatment for epilepsy. *Cochrane Database Syst Rev*. 2012; 6:CD006967. doi:10.1002/14651858.CD006967.pub2.

Brzezinski, A. Melatonin in humans. *N Engl J Med*. 1997; 336:186–195.

Cagnacci A, Elliott JA, Yen SS. Melatonin: A major regulator of the circadian rhythm of core temperature in humans. *J Clin Endocrinol Metab*. 1992; 75(2):447–452.

Carpizo R, Martínez A, Mediavilla D, González M, Abad A, Sánchez-Barceló EJ. Smith-Magenis syndrome: A case report of improved sleep after treatment with beta1-adrenergic antagonists and melatonin. *J Pediatr*. 2006; 149(3):409–411.

Carr R, Wasdell MB, Hamilton D, Weiss MD, Freeman RD, Tai J et al. Long-term effectiveness outcome of melatonin therapy in children with treatment-resistant circadian rhythm sleep disorders. *J Pineal Res*. 2007; 43(4):351–359.

Chaste P, Clement N, Botros HG, Guillaume JL, Konyukh M, Pagan C et al. Genetic variations of the melatonin pathway in patients with attention-deficit and hyperactivity disorders. *J Pineal Res*. 2011; 51(4):394–399.

Chaste P, Clement N, Mercati O, Guillaume JL, Delorme R, Botros HG et al. Identification of pathway-biased and deleterious melatonin receptor mutants in autism spectrum disorders and in the general population. *PLoS One*. 2010; 5(7):e11495.

Coppola G, Iervolino G, Mastrosimone M, La Torre G, Ruiu F, Pascotto A. Melatonin in wake-sleep disorders in children, adolescents and young adults with mental retardation with or without epilepsy: A double-blind, cross-over, placebo-controlled trial. *Brain Dev*. 2004; 26(6):373.

Cortese S, Brown TE, Corkum P, Gruber R, O'Brien LM, Stein M et al. Assessment and management of sleep problems in youths with attention-deficit/hyperactivity disorder. *J Am Acad Child Adolesc Psychiatry*. 2013; 52(8):784–796.

De Leersnyder H. Inverted rhythm of melatonin secretion in Smith-Magenis syndrome: From symptoms to treatment. *Trends Endocrinol Metab*. 2006; 17(7):291–298.

De Leersnyder H, de Blois MC, Bresson JL, Sidi D, Claustrat B, Munnich A. Inversion of the circadian melatonin rhythm in Smith-Magenis syndrome. *Rev Neurol* (Paris). 2003; 159(11 Suppl):6S21–6S26.

Dodge NN, Wilson GA. Melatonin for treatment of sleep disorders in children with developmental disabilities. *J Child Neurol*. 2001; 16(8):581–584.

Doyen C, Mighiu D, Kaye K, Colineaux C, Beaumanoir C, Mouraeff Y et al. Melatonin in children with autistic spectrum disorders: Recent and practical data. *Eur Child Adolesc Psychiatry*. 2011; 20(5):231–239.

Duman O, Durmaz E, Akcurin S, Serteser M, Haspolat S. Spontaneous endogenous hypermelatoninemia: A new disease? *Horm Res Paediatr*. 2010; 74(6):444–448.

Ebadi M, Govitrapong P, Phansuwan-Pujito P, Nelson F, Reiter RJ. Pineal opioid receptors and analgesic action of melatonin. *J Pineal Res*. 1998; 24(4):193–200.

Elkhayat HA, Hassanein SM, Tomoum HY, Abd-Elhamid IA, Asaad T, Elwakkad AS. Melatonin and sleep-related problems in children with intractable epilepsy. *Pediatr Neurol.* 2010; 42(4):249–254.

Fariello RG, Bubenik GA, Brown GM, Grota LJ. Epileptogenic action of intraventricularly injected antimelatonin antibody. *Neurology.* 1977; 27(6):567–570.

Fauteck J, Schmidt H, Lerchl A, Kurlemann G, Wittkowski W. Melatonin in epilepsy: First results of replacement therapy and first clinical results. *Biol Signals Recept.* 1999; 8(1–2):105–110.

Gitto E, Aversa S, Reiter RJ, Barberi I, Pellegrino S. Update on the use of melatonin in pediatrics. *J Pineal Res.* 2011; 50:21–28.

Gitto E, Aversa S, Salpietro CD, Barberi I, Arrigo T, Trimarchi G, Reiter RJ, Pellegrino S. Pain in neonatal intensive care: Role of melatonin as an analgesic antioxidant. *J Pineal Res.* 2012; 52(3):291–295.

Gitto E, Pellegrino S, Gitto P, Barberi I, Reiter RJ. Oxidative stress of the newborn in the pre- and postnatal period and the clinical utility of melatonin. *J Pineal Res.* 2009; 46(2):128–139.

Giusti P, Lipartiti M, Franceschini D, Schiavo N, Floreani M, Manev H. Neuroprotection by melatonin from kainate-induced excitotoxicity in rats. *FASEB J.* 1996; 10(8):891–896.

Gordo MA, Recio J, Sánchez-Barceló EJ. Decreased sleep quality in patients suffering from retinitis pigmentosa. *J Sleep Res.* 2001; 10(2):159–164.

Guénolé F, Godbout R, Nicolas A, Franco P, Claustrat B, Baleyte JM. Melatonin for disordered sleep in individuals with autism spectrum disorders: Systematic review and discussion. *Sleep Med Rev.* 2011; 15(6):379–387.

Gupta M, Gupta YK, Agarwal S, Aneja S, Kalaivani M, Kohli K. Effects of add-on melatonin administration on antioxidant enzymes in children with epilepsy taking carbamazepine monotherapy: A randomized, double-blind, placebo-controlled trial. *Epilepsia.* 2004a; 45(12):1636–1639.

Gupta M, Gupta YK, Agarwal S, Aneja S, Kohli K. A randomized, double-blind, placebo controlled trial of melatonin add-on therapy in epileptic children on valproate monotherapy: Effect on glutathione peroxidase and glutathione reductase enzymes. *Br J Clin Pharmacol.* 2004b; 58(5):542–547.

Isik B, Baygin O, Bodur H. Premedication with melatonin vs midazolam in anxious children. *Paediatr Anaesth.* 2008; 18(7):635–641.

Jain S, Besag FM. Does melatonin affect epileptic seizures? *Drug Saf.* 2013; 36(4):207–215.

Jan JE, Hamilton D, Seward N, Fast DK, Freeman RD, Laudon M. Clinical trials of controlled-release melatonin in children with sleep-wake cycle disorders. *J Pineal Res.* 2000; 29(1):34–39.

Jan JE, Reiter RJ, Wong PK, Bax MC, Ribary U, Wasdell MB. Melatonin has membrane receptor-independent hypnotic action on neurons: An hypothesis. *J Pineal Res.* 2011; 50(3):233–240.

Johnson K, Page A, Williams H, Wassemer E, Whitehouse W. The use of melatonin as an alternative to sedation in uncooperative children undergoing an MRI examination. *Clin Radiol.* 2002; 57(6):502–506.

Jonsson L, Ljunggren E, Bremer A, Pedersen C, Landén M, Thuresson K et al. Mutation screening of melatonin-related genes in patients with autism spectrum disorders. *BMC Med Genomics.* 2010; 3:10. doi:10.1186/1755-8794-3-10.

Kain ZN, MacLaren JE, Herrmann L, Mayes L, Rosenbaum A, Hata J, Lerman J. Preoperative melatonin and its effects on induction and emergence in children undergoing anesthesia and surgery. *Anesthesiology.* 2009; 111(1):44–49.

Khan S, Heussler H, McGuire T, Dakin C, Pache D, Cooper D et al. Melatonin for non-respiratory sleep disorders in visually impaired children. *Cochrane Database Syst Rev.* 2011; (11):CD008473.

Lavie P. Melatonin: Role in gating nocturnal rise in sleep propensity. *J Biol Rhythms.* 1997; 12:657–665.

Macchi MM, Bruce JN. Human physiology and functional significance of melatonin. *Front Neuroendocrinol.* 2004; 25:177–195.

Marseglia L, Aversa S, Barberi I, Salpietro CD, Cusumano E, Speciale A et al. High endogenous melatonin levels in critically ill children: A pilot study. *J Pediatr.* 2013; 162(2):357–360.

Molina-Carballo A, Muñoz-Hoyos A, Reiter RJ, Sánchez-Forte M, Moreno-Madrid F, Rufo-Campos M et al. Utility of high doses of melatonin as adjunctive anticonvulsant therapy in a child with severe myoclonic epilepsy: Two years experience. *J Pineal Res.* 1997; 23(2):97–105.

Molina-Carballo A, Muñoz-Hoyos A, Sánchez-Forte M, Uberos-Fernández J, Moreno-Madrid F, Acuña-Castroviejo D. Melatonin increases following convulsive seizures may be related to its anticonvulsant properties at physiological concentrations. *Neuropediatrics.* 2007; 38(3):122–125.

Morris CJ, Aeschbach D, Scheer FA. Circadian system, sleep and endocrinology. *Mol Endocrinol.* 2012; 349:91–104.

Muñoz-Hoyos A, Heredia F, Moreno F, García JJ, Molina-Carballo A, Escames G, Acuña-Castroviejo D. Evaluation of plasma levels of melatonin after midazolam or sodium thiopental anesthesia in children. *J Pineal Res.* 2002; 32:253–256.

Muñoz-Hoyos A, Sánchez-Forte M, Molina-Carballo A, Escames G, Martin-Medina E, Reiter RJ et al. Melatonin's role as an anticonvulsant and neuronal protector: Experimental and clinical evidence. *J Child Neurol.* 1998; 13(10):501–509.
Naguib M, Gottumukkala V, Goldstein PA. Melatonin and anesthesia: A clinical perspective. *J Pineal Res.* 2007; 42(1):12–21.
Özcengiz D, Gunes Y, Ozmete O. Oral melatonin, dexmedetomidine, and midazolam for prevention of postoperative agitation in children. *J Anesth.* 2011; 25(2):184–188.
Paprocka J, Dec R, Jamroz E, Marszał E. Melatonin and childhood refractory epilepsy: A pilot study. *Med Sci Monit.* 2010; 16(9):389–396.
Pelayo R, Yuen K. Pediatric sleep pharmacology. *Child Adolesc Psychiatr Clin N Am.* 2012; 21(4):861–883.
Peled N, Shorer Z, Peled E, Pillar G. Melatonin effect on seizures in children with severe neurologic deficit disorders. *Epilepsia.* 2001; 42(9):1208–1210.
Pillar G, Shahar E, Peled N, Ravid S, Lavie P, Etzioni A. Melatonin improves sleep-wake patterns in psychomotor retarded children. *Pediatr Neurol.* 2000; 23(3):225–228.
Potocki L, Glaze D, Tan DX, Park SS, Kashork CD, Shaffer LG et al. Circadian rhythm abnormalities of melatonin in Smith-Magenis syndrome. *J Med Genet.* 2000; 37(6):428–433.
Praninskiene R, Dumalakiene I, Kemezys R, Mauricas M, Jucaite A. Melatonin secretion in children with epilepsy. *Epilepsy Behav.* 2012; 25(3):315–322.
Rahman SA, Kayumov L, Tchmoutina EA, Shapiro CM. Clinical efficacy of dim light melatonin onset testing in diagnosing delayed sleep phase syndrome. *Sleep Med.* 2009; 10(5):549–555.
Reiter RJ, Blask DE, Talbot JA, Barnett MP. Nature and the time course of seizures associated with surgical removal of the pineal gland from parathyroidectomized rats. *Exp Neurol.* 1973; 38(3):386–397.
Reiter RJ, Tan DX, Manchester L, Paredes SD, Mayo JC, Sainz RM. Melatonin and reproduction revisited. *Biol Reprod.* 2009; 81:445–445.
Rossignol DA, Frye RE. Melatonin in autism spectrum disorders. *Curr Clin Pharmacol.* 2013 [Epub ahead of print]. PMID:24050742.
Sack RL, Auckley D, Auger RR, Carskadon MA, Wright KP, Vitiello MV et al. Circadian rhythm sleep disorders: Part II, advanced sleep phase disorder, delayed sleep phase disorder, free-running disorder, and irregular sleep-wake rhythm. An American Academy of Sleep Medicine review. *Sleep.* 2007; 30(11):1484–1501.
Samarkandi A, Naguib M, Riad W, Thalaj A, Alotibi W, Aldammas F, Albassam A. Melatonin vs. midazolam premedication in children: A double-blind, placebo-controlled study. *Eur J Anaesthesiol.* 2005; 22(3):189–196.
Sanchez-Barcelo EJ, Martínez-Campa, C, Mediavilla MD, Gonzalez A, Alonso-Gonzalez C, Cos S. Melatonin and melatoninergic drugs as therapeutic agents: Ramelteon and Agomelatine, the two most promising melatonin receptor agonists. *Recent Pat Endocr Metab Immune Drug Discov.* 2007; 1:142–151.
Sanchez-Barcelo EJ, Mediavilla MD, Reiter RJ. Clinical uses of melatonin in pediatrics. *Int J Pediatr.* 2011; 2011:892624.
Sanchez-Barcelo EJ, Mediavilla MD, Tan DX, Reiter RJ. Clinical uses of melatonin: Evaluation of human trials. *Curr Med Chem.* 2010; 17(19):2070–2095.
Saracz J, Rosdy B. Effect of melatonin on intractable epilepsies. *Orv Hetil.* 2004; 145(51):2583–2587.
Schmidt CM, Bohlender JE, Deuster D, Knief A, Matulat P, Dinnesen AG. The use of melatonin as an alternative to sedation in children undergoing brainstem audiometry. *Laryngorhinootologie.* 2004; 83(8):523–528.
Sheldon SH. Pro-convulsivant effects of oral melatonin in neurologically disabled children. *Lancet.* 1998; 351(9111):1254.
Skene DJ, Arendt J. Circadian rhythm sleep disorders in the blind and their treatment with melatonin. *Sleep Med.* 2007; 8(6):651–655.
Sury MR, Fairweather K. The effect of melatonin on sedation of children undergoing magnetic resonance imaging. *Br J Anaesth.* 2006; 97(2):220–225.
Szeinberg A, Borodkin K, Dagan Y. Melatonin treatment in adolescents with delayed sleep phase syndrome. *Clin Pediatr (Phila).* 2006; 45(9):809–818.
Tordjman S, Anderson GM, Bellissant E, Botbol M, Charbuy H, Camus F et al. Day and nighttime excretion of 6-sulphatoxymelatonin in adolescents and young adults with autistic disorder. *Psychoneuroendocrinology.* 2012; 37(12):1990–1997.
Tzischinsky O., Shlitner A., Lavie, P. The association between the nocturnal sleep gate and nocturnal onset of urinary 6-sulphatoxymelatonin. *J Biol Rhythms.* 1993; 8:199–209.
van Geijlswijk IM, Korzilius HP, Smits MG. The use of exogenous melatonin in delayed sleep phase disorder: A meta-analysis. *Sleep.* 2010a; 33:1605–1614.

van Geijlswijk IM, van der Heijden KB, Egberts AC, Korzilius HP, Smits MG. Dose finding of melatonin for chronic idiopathic childhood sleep onset insomnia: An RCT. *Psychopharmacology* 2010b; 212:379–391.

Wassmer E, Quinn E, Whitehouse W, Seri S. Melatonin as a sleep inductor for electroencephalogram recordings in children. *Clin Neurophysiol.* 2001; 112(4):683–685.

Wilhelmsen M, Amirian I, Reiter RJ, Rosenberg J, Gögenur I. Analgesic effects of melatonin: A review of current evidence from experimental and clinical studies. *J Pineal Res.* 2011; 51(3):270–277.

Wright B, Sims D, Smart S, Alwazeer A, Alderson-Day B, Allgar V et al. Melatonin versus placebo in children with autism spectrum conditions and severe sleep problems not amenable to behaviour management strategies: A randomised controlled crossover trial. *J Autism Dev Disord.* 2011; 41(2):175–184.

Zee PC. Shedding light on the effectiveness of melatonin for circadian rhythm sleep disorders. *Sleep.* 2010; 33:1581–1582.

Zhdanova IV, Wurtman, RJ Efficacy of melatonin as a sleep-promoting agent. *J Biol Rhythms.* 1997; 112:644–650.

Zhdanova IV, Wurtman RJ, Wagstaff J. Effects of a low dose of melatonin on sleep in children with Angelman syndrome. *J Pediatr Endocrinol Metab.* 1999; 12(1):57–67.

6 Neuroprotective Effect of Melatonin on Perinatal Hypoxia–Ischemia

Daniel Alonso-Alconada, Antonia Alvarez, and Enrique Hilario

CONTENTS

6.1 PERINATAL HYPOXIA–ISCHEMIA

The brain lesion during neonatal life has been mainly related to an asphyctic event. This relationship implies an anomalous gas exchange with an oxygen deficit and a carbon dioxide accumulation, subsequently producing arterial acidosis. If asphyxia persists, systemic hypotension and brain ischemia will develop. Similarly, this cascade of events could take place in utero, in which a reduced or compromised maternal perfusion could affect the fetal gas exchange and blood flow, resulting in a different degree of asphyxia. The term hypoxic–ischemic encephalopathy has been widely used in clinical practice, due to the uncertain role played either by hypoxia or ischemia or by the combination of both, as the main pathogenic determinants of the neurological damage caused by hypoxic–ischemic injury.

Perinatal hypoxia–ischemia is one of the major causes of mortality and neurological morbidity, in both premature and term newborn infants. With an incidence of 2–6/1000 term births (de Haan et al. 2006), it remains the single most important cause of brain injury in the newborn, leading to death or lifelong disability (du Plessis and Volpe 2002; Hamrick and Ferriero 2003). Despite the improvements in perinatal care, the increase in premature born infants and multiple gestations, and survival of these high-risk infants increase the prevalence of the hypoxic–ischemic cerebral lesions. Moreover, in spite of fetal monitoring improvements, the cerebral palsy rate has remained stable for the last decades (Edwards and Azzopardi 2000). Today, 80%–85% of all very-low-birth-weight infants with a birth weight below 1500 g survive, with cerebral palsy developing in 5%–15% of them. Significant neurological sequelae can occur in as many as 50%–75% of these asphyctic children, which may suffer from long-term neurological consequences, ranging from attention deficits and hyperactivity in children and adolescents (Maneru et al. 2001; Yager et al. 2009) to cerebral palsy, mental retardation, and epilepsy (Volpe 2001; Low 2004; Vannucci and Hagberg 2004).

6.2 THERAPEUTIC STRATEGIES

6.2.1 Hypothermia

Today, the most common therapeutic strategy against cerebral brain injury in the neonate is hypothermia, as a therapy for neuronal rescue (Nedelcu et al. 2000). Studies in newborns have shown different levels of neuroprotection from hypothermia, either through inhibiting apoptotic cellular death and DNA fragmentation following hypoxia–ischemia (Esteve et al. 1999; Adachi et al. 2001) or through inducing a reduction in glutamate concentration at the synaptic space, a delay in the accumulation of intracellular calcium, and a reduction in the production of nitric oxide (Hashimoto et al. 2003; Zhu et al. 2004). Although therapeutic hypothermia is a significant advance in the developed world and improves outcome (Edwards et al. 2010), it offers just 11% reduction in risk of death or disability, from 58% to 47%. In this sense, there is an urgent need to develop additional treatment strategies to augment hypothermic purpose and provide safe and effective neuroprotection for neonatal encephalopathy.

6.2.2 Other Strategies

Many experimental studies have been carried out to identify new pharmacological compounds (either natural or synthetic) that could be able to diminish brain damage (Gonzalez and Ferriero 2008; Cilio and Ferriero 2010; Kelen and Robertson 2010). Some therapies have been centered on diminishing brain damage arising from free radicals, thus employing antioxidant molecules such as allopurinol, which inhibits xanthine oxidase (Palmer et al. 1993; Van Bel et al. 1998), *N*-acetylcysteine, which increases the intracellular level of glutathione and reduces apoptosis and inflammation (Ferrari et al. 1995; Yan et al. 1995; Jatana et al. 2006; Lee et al. 2008), or erythropoietin, which has shown antiapoptotic and angiogenic properties (Sola et al. 2005) and also provides neuroprotection and neurogenesis in neonatal rats (Chang et al. 2005; Gonzalez et al. 2007). Anti-inflammatory compounds, such as statins, and the second generation of tetracyclines have also been tested, which were able to reduce the expression of some proinflammatory cytokines and adhesion molecules (Carloni et al. 2006, 2009) and to block microglial activation (Arvin et al. 2002; Jantzie et al. 2005), respectively. During the last decade, interest has grown in the neuroprotective possibilities of cannabinoids after perinatal asphyxia. It has previously been shown that the administration of the synthetic cannabinoid agonist WIN 55,212-2 after a hypoxic–ischemic insult to preterm lambs decreases brain injury, reducing both delayed cell death and glial damage (Alonso-Alconada et al. 2010), an effect developed, at least in part, through the maintenance of mitochondrial integrity and functionality (Alonso-Alconada et al. 2012a).

6.2.3 Melatonin

A large number of potential treatments exist, but optimizing therapy for neonatal brain injury will require capitalizing on multiple pathways which prevent cell death. Since there is still no specific treatment for perinatal brain lesions due to the complexity of neonatal hypoxic–ischemic pathophysiology, interest has grown in the neuroprotective potentials of melatonin after perinatal asphyxia. Robertson et al. (2012) evaluated the potential of several drugs used in preclinical animal models for translation to postnatal therapy of neonatal encephalopathy. Drugs were ranked on (1) ease of administration; (2) knowledge of starting dose; (3) side effects; (4) teratological or toxic effects; and (5) benefit and efficacy. The five most highly ranked drugs in order were: melatonin with the highest score (90%), Epo (88%), *N*-acetylcysteine (84%), Epo-mimetics (82%), and allopurinol (70%). Melatonin is considered the most promising medicine for neuroprotection in terms of efficacy, safety, and practicality in scale-up, as it easily crosses both the placental and the blood–brain barrier, reaching subcellular compartments with a low toxicity and high efficacy (Vitte et al. 1988; Menendez-Pelaez and Reiter 1993; Gupta et al. 2003), and even at high supraphysiological concentrations, there appears to be no adverse side effects (Rees et al. 2011), making it a relatively safe therapy that could be administered to babies.

6.3 OXIDATIVE STRESS

6.3.1 Oxidative Stress in the Newborn

The newborn brain is especially vulnerable to oxidative imbalance, due to its increased fatty acid content, higher concentrations of free iron, high rates of oxygen consumption, low concentrations of antioxidants, an imbalance of antioxidant enzymes, as for example, mitochondrial superoxide dismutase-2 and glutathione peroxidase, and oxygen-induced vasoconstriction leading to reduced brain perfusion (McLean and Ferriero 2004; McQuillen and Ferriero 2004; Sheldon et al. 2004). Neonatal asphyxia brings about the overproduction of free radicals leading to oxidative stress, as the antioxidant capacity of immature neurons is easily overwhelmed by hypoxia-induced reactive oxygen species, thus affecting numerous cellular structures. Lipid peroxidation constitutes a self-feeding factor affecting the cellular membranes; thiol-containing proteins are oxidized and mitochondria damaged, producing a depletion of mitochondrial NADPH and an increase in free cytosolic calcium, which is itself a hazardous intracellular mechanism (Buonocore et al. 2001; Kumar et al. 2008). Moreover, the observed damage during hypoxic–ischemic injury worsens during the reperfusion phase, which induces an overproduction of radical oxygen species in mitochondria, becoming an important cause of cell death after perinatal asphyxia.

6.3.2 Natural Antioxidant

In living organisms, one of the most important functions of melatonin is to protect them from oxidative stress (Tan et al. 2007, 2010). This role is carried out by means of its ability to directly scavenge a wide variety of free radicals (singlet oxygen, superoxide anion radical, hydroperoxide, hydroxyl radical, and the lipid peroxide radical) (Rosen et al. 2006; Tan et al. 2007) and/or by its indirect antioxidant properties: through the strengthening of the antioxidant effect of glutathione and vitamins C and E (Reiter et al. 2000a), the activation of superoxide dismutase, catalase, glucose-6-phosphate dehydrogenase, glutathione reductase, and glutathione peroxidase (Tomas-Zapico and Coto-Montes 2005), or improving the mitochondrial efficiency (Acuna-Castroviejo et al. 2001).

6.3.3 Melatonin against Neonatal Oxidative Stress

Several studies have evaluated the antioxidant effect of melatonin after an asphyctic event, ranging from preclinical analyses using small rodents to clinical evaluation of infants suffering from asphyxia. Using late-gestation fetal sheep submitted to umbilical cord occlusion (an animal model closely related with the clinical practice), melatonin has been able to abolish lipid peroxidation (Miller et al. 2005). Indeed, melatonin confirmed its antioxidant capacity through the evaluation of diverse quantitative biomarkers of oxidative damage. It may prevent protein oxidation caused by hypoxia–ischemia (Eskiocak et al. 2007), as terminal products of protein exposure to free radicals are considered reliable markers of the degree of protein damage in oxidative stress (Witko-Sarsat et al. 1996). It also reduced malondialdehyde levels, an aldehyde-oxygenated compound resulted from the attack of membrane lipoproteins and polyunsaturated fatty acids, an effect not only observed in preclinical studies (Tutunculer et al. 2005), but also in asphyxiated human newborns (Fulia et al. 2001). Isoprostanes, neuroprostanes, and neurofurans are also quantitative biomarkers of oxidative damage (Kohen and Nyska 2002; Arneson and Roberts 2007; Song et al. 2008), and after melatonin administration, their levels were significantly lower than those in hypoxic–ischemic rats (Signorini et al. 2009; Balduini et al. 2012), an effect confirmed in fetal sheep after umbilical cord occlusion, where the production of 8-isoprostane was attenuated (Welin et al. 2007).

6.4 CELL RESCUE

As hypoxic–ischemic brain injury is often unpredictable, the primary approach is to develop postinsult therapies to ameliorate ongoing or secondary injury. In this regard, melatonin administration after neonatal hypoxia–ischemia has been shown to reduce the infarct volume, both administered before or after the injury (Carloni et al. 2008; Cetinkaya et al. 2011; Ozyener et al. 2012). The asphyctic event determines that some cells die, while others remain in a state of ischemic penumbra, a distinction based on the injury severity, the maturational stage, the brain region, and the cell type affected (Puka-Sundvall et al. 2000; Sugawara et al. 2004).

6.4.1 Neurons

Thus, while virtually every cell is affected by asphyxia, they do not respond in the same way during hypoxia–ischemia, being neurons, followed by oligodendrocytes, the most sensitive cells to the lack of oxygen and showing a selective vulnerability (Mattson et al. 1989; Johnston 1998; Northington et al. 2001). After an asphyctic event, melatonin-treated animals maintained the number of morphologically well-preserved neurons (Hamada et al. 2010; Alonso-Alconada et al. 2012b; Watanabe et al. 2012), as represented in Figure 6.1. In this set of microphotographs, we show that both hippocampus (a–c) and cortex (d–f), the cerebral regions normally affected by hypoxia–ischemia when using the commonly known Rice–Vannucci animal model, show a fewer number of well-preserved neurons (b and e). On the contrary, the neurons' number and appearance from the animals treated with melatonin (c and f) are closely similar to those observed in the control group (a and d).

6.4.2 Oligodendrocytes

As referred earlier, oligodendrocytes are particularly vulnerable to asphyxia, affecting myelination that gives rise to white matter lesions and damaging gray matter oligodendrocyte progenitors (Rothstein and Levison 2005). Melatonin may be of therapeutic value in ameliorating hypoxic–ischemic damage to the developing white matter through normalization of the myelination process (Olivier et al. 2009; Kaur et al. 2010; Villapol et al. 2011; Alonso-Alconada et al. 2012b). In this sense, the authors and others have evaluated the expression pattern of myelin basic protein (Figure 6.2a through c), as a decrease in its expression can lead to myelination deficit, which is considered a hallmark of inflammation-associated diffuse white matter damage (Inder et al. 2003; Wang et al. 2007).

6.4.3 Astrocytes

Astrocytes can modulate the extension and degree of severity of the damage (Takuma et al. 2004; Panickar and Norenberg 2005), either conferring neuroprotection or leading to deficiencies in the myelination processes and to neuronal signaling impairment. In Figure 6.2, we show that melatonin administration (Figure 6.2f) reduced reactive gliosis after neonatal hypoxia–ischemia.

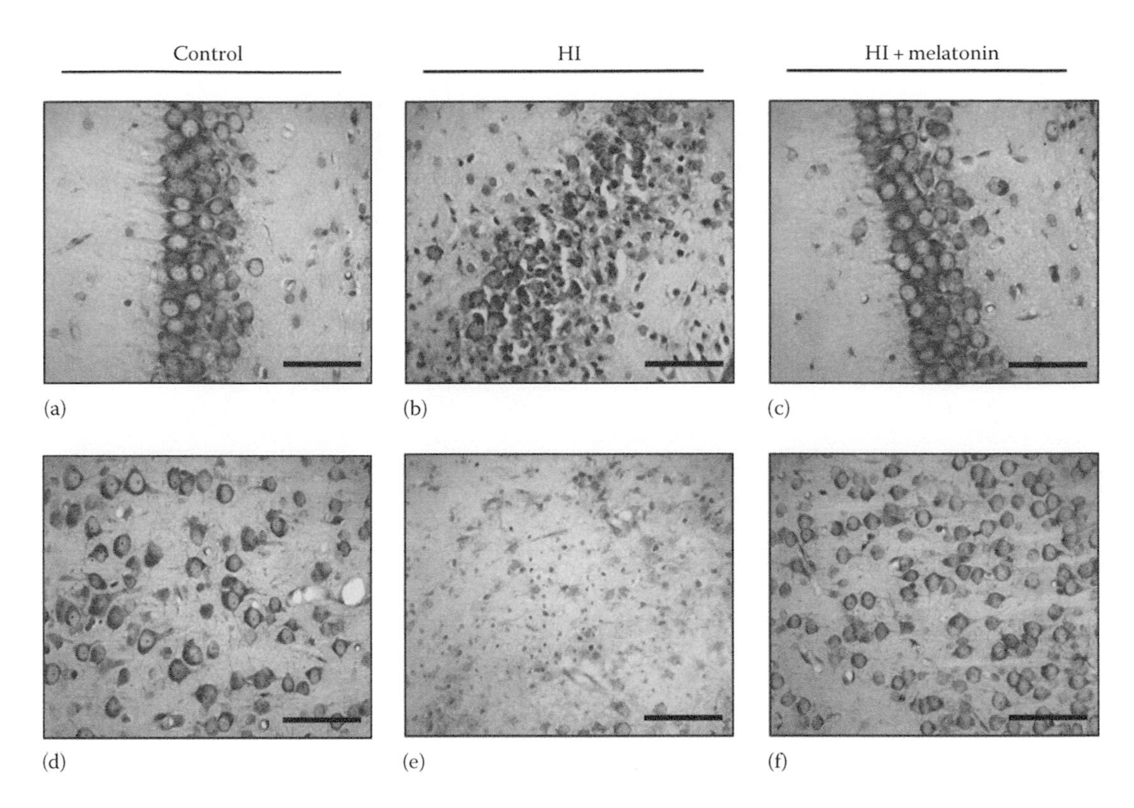

FIGURE 6.1 **(See color insert.)** Nissl-stained brain sections corresponding to the surrounding areas of the CA1 region of the hippocampus (a–c) and cortex (d–f) from neonatal rats showing cell loss after hypoxia-ischemia (b and e) and recovery after melatonin administration (c and f). Bar: 100 μm.

These brain sections correspond to the surrounding areas of the CA1 region of the hippocampus, where the expression of the glial fibrillary acidic protein was increased after the injury (Figure 6.2e). Its accumulation is not only related with the creation of new astrocytic processes, but also with reactive gliosis (Panickar and Norenberg 2005).

6.4.4 Apoptosis

Cell death begins immediately and continues over a period of days or weeks, whose phenotype undergoes a change ranging from an early necrotic morphology to an apoptotic one. This evolution is known as the necrosis–apoptosis continuum. Data suggest that apoptosis plays a prominent role in the neonatal brain, being more important than necrosis after the injury (Hu et al. 2000). Many of these cells involved in a process referred to as delayed cell death will also die if the therapeutic strategies are not effective, in spite of the recovery of cerebral blood flow (Walton et al. 1999; Ferriero 2004). DNA fragmentation and apoptotic figures were evaluated after melatonin administration, showing a reduction in their number, both in neonatal sheep (Welin et al. 2007) and rats (Cetinkaya et al. 2011; Alonso-Alconada et al. 2012b; Ozyener et al. 2012). Indeed, caspase-3 activation and fractin were analyzed using a model of birth asphyxia in the spiny mouse, showing lower levels after its administration (Hutton et al. 2009).

6.4.5 Mitochondria

One of the most important key regulators of apoptotic cell death is mitochondrial impairment, as the disruption of its membrane integrity and loss of membrane potential can determinate cell

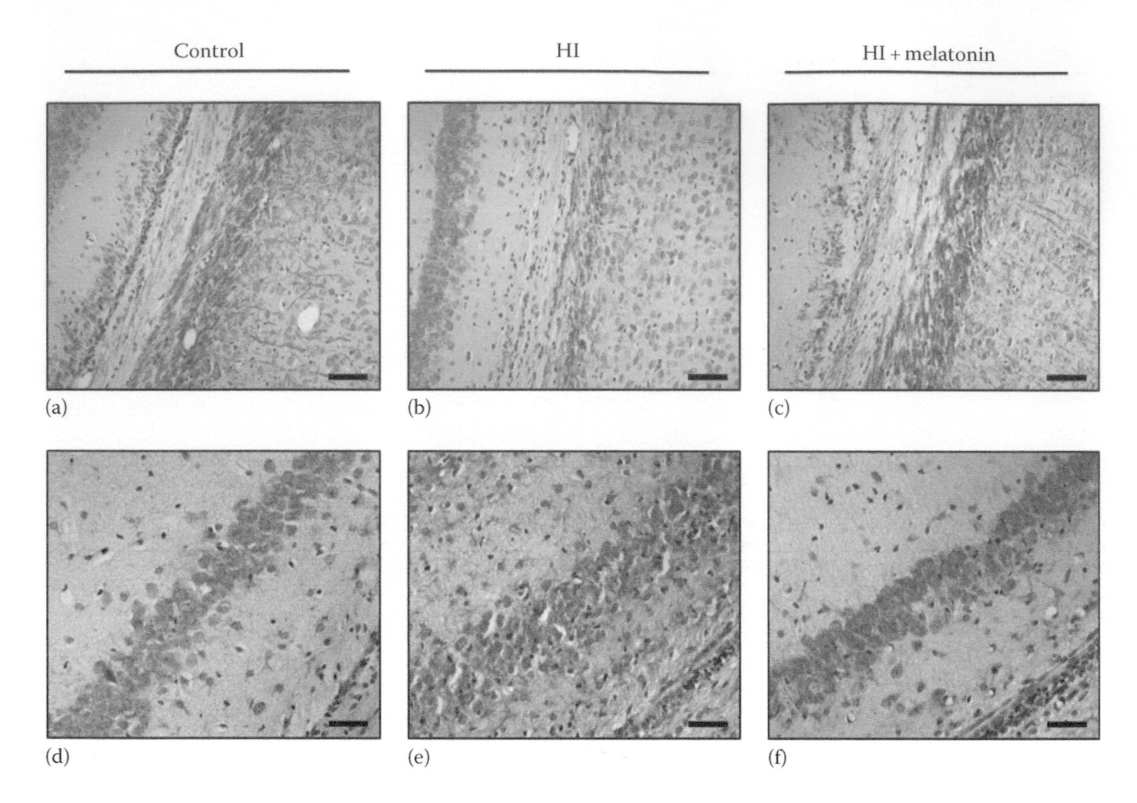

FIGURE 6.2 **(See color insert.)** Myelin basic protein (a–c) and glial fibrillary acidic protein (d–f) immunolabeled brain sections corresponding to the surrounding areas of the CA1 region of the hippocampus (a–c) and the external capsule (d–f) from neonatal rats showing myelination deficit (b) and reactive gliosis (e) after hypoxia-ischemia and recovery after melatonin administration (c and f). Bar: 100 μm.

survival by overproduction of reactive oxygen species, abnormal calcium homeostasis, and release of apoptotic proteins. Given to pregnant rats, melatonin prevented oxidative mitochondria damage after ischemia–reperfusion in premature fetal rat brain (Watanabe et al. 2004) by means of the maintenance of the number of intact mitochondria and the respiratory control index, as well as the reduction in thiobarbituric acid-reactive substances concentration (Hamada et al. 2010; Watanabe et al. 2012).

6.5 INFLAMMATION

6.5.1 Inflammatory Response in the Asphyctic Newborn

Perinatal hypoxia–ischemia triggers an inflammatory response started out by the stimulation of ischemic cells by reactive oxygen species, resulting in the local production of cytokines and chemokines that can be the cause, among other things, of an overexpression of adhesion molecules within the cerebral vascular network and the recruitment of peripheral leukocytes (Maslinska et al. 2002; Chiesa et al. 2003). Once activated, the inflammatory cells can secrete a wide variety of cytotoxic agents, including more cytokines, matrix metalloproteases, nitric oxide, and more reactive oxygen species. These substances are the cause of a greater cellular damage as well as the dismantling of the blood–brain barrier and the extracellular matrix (Wang et al. 1994; Danton and Dietrich 2003; Emsley et al. 2003; Chew et al. 2006; Leonardo and Pennypacker 2009). The damage produced in the blood–brain barrier potentiates the damage in the brain tissue and contributes to secondary damage, allowing the blood and soluble elements to penetrate the brain

(Rosenberg 1999). The secondary damage is developed as a consequence of cerebral edema and vasomotor and hemodynamic deficit, culminating in hypoperfusion and postischemic inflammation, which finally activate the microglia and the infiltration of peripheral inflammatory cells into the brain tissue (Dirnagl et al. 1999; Hailer 2008). This migration of peripheral leukocytes into the brain can result in an increase of the inflammatory cascade, which causes even more damage.

6.5.2 Anti-Inflammatory Effect of Melatonin

Melatonin may be beneficial, as it reduces nitric oxide production, vascular endothelial growth factor concentration, and, hence, vascular permeability, which is normally increased after hypoxic exposure (Kaur et al. 2008). Prophylactic maternal treatment with melatonin has also demonstrated a reduction in central nervous system inflammation, by limiting macrophage infiltration and glial cell activation in a model of birth asphyxia in the spiny mouse (Hutton et al. 2009). Indeed, a reduced number of ED1 positive cells, a marker of activated microglia–macrophages, was found in neonatal rats treated with melatonin when compared with pups without treatment (Balduini et al. 2012). But melatonin and its metabolites are not only protective in neonatal models of brain injury, they also have been able to reverse the inflammatory response and edema after stroke, suppressing the production of inflammatory cytokines (Pei and Cheung 2004; Mayo et al. 2005; Lee et al. 2007), reducing nitric oxide synthase (Koh 2008), preventing the translocation of NF-κB to the nucleus (Mohan et al. 1995; Reiter et al. 2000b) and decreasing cyclooxygenase-2 gene expression (Hardeland 2005), molecular changes correlated with a reduction in the size of brain infarcts. Thus, the use of melatonin can be considered as another meaningful tool against inflammatory response in an effort to improve the clinical course of illnesses with an inflammatory etiology.

REFERENCES

Acuna-Castroviejo, D., M. Martin, M. Macias et al. 2001. Melatonin, mitochondria, and cellular bioenergetics. *Journal of Pineal Research* 30(2): 65–74.

Adachi, M., O. Sohma, S. Tsuneishi, S. Takada, and H. Nakamura. 2001. Combination effect of systemic hypothermia and caspase inhibitor administration against hypoxic-ischemic brain damage in neonatal rats. *Pediatric Research* 50(5): 590–595.

Alonso-Alconada, D., A. Alvarez, F. J. Alvarez, J. A. Martinez-Orgado, and E. Hilario. 2012a. The cannabinoid WIN 55212-2 mitigates apoptosis and mitochondrial dysfunction after hypoxia ischemia. *Neurochemical Research* 37(1): 161–170.

Alonso-Alconada, D., A. Alvarez, J. Lacalle, and E. Hilario. 2012b. Histological study of the protective effect of melatonin on neural cells after neonatal hypoxia-ischemia. *Histology and Histopathology* 27(6): 771–783.

Alonso-Alconada, D., F. J. Alvarez, A. Alvarez et al. 2010. The cannabinoid receptor agonist WIN 55,212-2 reduces the initial cerebral damage after hypoxic-ischemic injury in fetal lambs. *Brain Research* 1362: 150–159.

Arneson, K. O. and L. J. Roberts 2nd. 2007. Measurement of products of docosahexaenoic acid peroxidation, neuroprostanes, and neurofurans. *Methods in Enzymology* 433: 127–143.

Arvin, K. L., B. H. Han, Y. Du, S. Z. Lin, S. M. Paul, and D. M. Holtzman. 2002. Minocycline markedly protects the neonatal brain against hypoxic-ischemic injury. *Annals of Neurology* 52(1): 54–61.

Balduini, W., S. Carloni, S. Perrone et al. 2012. The use of melatonin in hypoxic-ischemic brain damage: An experimental study. *The Journal of Maternal-Fetal & Neonatal Medicine* 25(Suppl 1): 119–124.

Buonocore, G., S. Perrone, and R. Bracci. 2001. Free radicals and brain damage in the newborn. *Biology of the Neonate* 79(3–4): 180–186.

Carloni, S., S. Girelli, G. Buonocore, M. Longini, and W. Balduini. 2009. Simvastatin acutely reduces ischemic brain damage in the immature rat via akt and CREB activation. *Experimental Neurology* 220(1): 82–89.

Carloni, S., E. Mazzoni, M. Cimino et al. 2006. Simvastatin reduces caspase-3 activation and inflammatory markers induced by hypoxia-ischemia in the newborn rat. *Neurobiology of Disease* 21(1): 119–126.

Carloni, S., S. Perrone, G. Buonocore, M. Longini, F. Proietti, and W. Balduini. 2008. Melatonin protects from the long-term consequences of a neonatal hypoxic-ischemic brain injury in rats. *Journal of Pineal Research* 44(2): 157–164.

Cetinkaya, M., T. Alkan, F. Ozyener, I. M. Kafa, M. A. Kurt, and N. Koksal. 2011. Possible neuroprotective effects of magnesium sulfate and melatonin as both pre- and post-treatment in a neonatal hypoxic-ischemic rat model. *Neonatology* 99(4): 302–310.
Chang, Y. S., D. Mu, M. Wendland et al. 2005. Erythropoietin improves functional and histological outcome in neonatal stroke. *Pediatric Research* 58: 106–111.
Chew, L. J., A. Takanohashi, and M. Bell. 2006. Microglia and inflammation: Impact on developmental brain injuries. *Mental Retardation and Developmental Disabilities Research Reviews* 12(2): 105–112.
Chiesa, C., G. Pellegrini, A. Panero et al. 2003. Umbilical cord interleukin-6 levels are elevated in term neonates with perinatal asphyxia. *European Journal of Clinical Investigation* 33(4): 352–358.
Cilio, M. R. and D. M. Ferriero. 2010. Synergistic neuroprotective therapies with hypothermia. *Seminars in Fetal & Neonatal Medicine* 15(5): 293–298.
Danton, G. H. and W. D. Dietrich. 2003. Inflammatory mechanisms after ischemia and stroke. *Journal of Neuropathology and Experimental Neurology* 62(2): 127–136.
de Haan, M., J. S. Wyatt, S. Roth, F. Vargha-Khadem, D. Gadian, and M. Mishkin. 2006. Brain and cognitive-behavioural development after asphyxia at term birth. *Developmental Science* 9(4): 350–358.
Dirnagl, U., C. Iadecola, and M. A. Moskowitz. 1999. Pathobiology of ischaemic stroke: An integrated view. *Trends in Neurosciences* 22(9): 391–397.
du Plessis, A. J. and J. J. Volpe. 2002. Perinatal brain injury in the preterm and term newborn. *Current Opinion in Neurology* 15(2): 151–157.
Edwards, A. D. and D. V. Azzopardi. 2000. Perinatal hypoxia-ischemia and brain injury. *Pediatric Research* 47(4 Pt 1): 431–432.
Edwards, A. D., P. Brocklehurst, A. J. Gunn et al. 2010. Neurological outcomes at 18 months of age after moderate hypothermia for perinatal hypoxic ischaemic encephalopathy: Synthesis and meta-analysis of trial data. *BMJ (Clinical Research Ed.)* 340: c363.
Emsley, H. C., C. J. Smith, C. M. Gavin et al. 2003. An early and sustained peripheral inflammatory response in acute ischaemic stroke: Relationships with infection and atherosclerosis. *Journal of Neuroimmunology* 139(1–2): 93–101.
Eskiocak, S., F. Tutunculer, U. N. Basaran, A. Taskiran, and E. Cakir. 2007. The effect of melatonin on protein oxidation and nitric oxide in the brain tissue of hypoxic neonatal rats. *Brain and Development* 29(1): 19–24.
Esteve, J. M., J. Mompo, J. Garcia de la Asuncion et al. 1999. Oxidative damage to mitochondrial DNA and glutathione oxidation in apoptosis: Studies in vivo and in vitro. *FASEB Journal* 13(9): 1055–1064.
Ferrari, G., C. Y. Yan, and L. A. Greene. 1995. *N*-acetylcysteine (D- and L-stereoisomers) prevents apoptotic death of neuronal cells. *The Journal of Neuroscience* 15(4): 2857–2866.
Ferriero, D. M. 2004. Neonatal brain injury. *The New England Journal of Medicine* 351(19): 1985–1995.
Fulia, F., E. Gitto, S. Cuzzocrea et al. 2001. Increased levels of malondialdehyde and nitrite/nitrate in the blood of asphyxiated newborns: Reduction by melatonin. *Journal of Pineal Research* 31(4): 343–349.
Gonzalez, F. F. and D. M. Ferriero. 2008. Therapeutics for neonatal brain injury. *Pharmacology & Therapeutics* 120(1): 43–53.
Gonzalez, F. F., P. McQuillen, D. Mu et al. 2007. Erythropoietin enhances long-term neuroprotection and neurogenesis in neonatal stroke. *Developmental Neuroscience* 29(4–5): 321–330.
Gupta, Y. K., M. Gupta, and K. Kohli. 2003. Neuroprotective role of melatonin in oxidative stress vulnerable brain. *Indian Journal of Physiology and Pharmacology* 47(4): 373–386.
Hailer, N. P. 2008. Immunosuppression after traumatic or ischemic CNS damage: It is neuroprotective and illuminates the role of microglial cells. *Progress in Neurobiology* 84(3): 211–233.
Hamada, F., K. Watanabe, A. Wakatsuki et al. 2010. Therapeutic effects of maternal melatonin administration on ischemia/reperfusion-induced oxidative cerebral damage in neonatal rats. *Neonatology* 98(1): 33–40.
Hamrick, S. E. and D. M. Ferriero. 2003. The injury response in the term newborn brain: Can we neuroprotect? *Current Opinion in Neurology* 16(2): 147–154.
Hardeland, R. 2005. Antioxidative protection by melatonin: Multiplicity of mechanisms from radical detoxification to radical avoidance. *Endocrine* 27(2): 119–130.
Hashimoto, T., M. Yonetani, and H. Nakamura. 2003. Selective brain hypothermia protects against hypoxic-ischemic injury in newborn rats by reducing hydroxyl radical production. *The Kobe Journal of Medical Sciences* 49(3–4): 83–91.
Hu, B. R., C. L. Liu, Y. Ouyang, K. Blomgren, and B. K. Siesjo. 2000. Involvement of caspase-3 in cell death after hypoxia-ischemia declines during brain maturation. *Journal of Cerebral Blood Flow and Metabolism* 20(9): 1294–1300.

Hutton, L. C., M. Abbass, H. Dickinson, Z. Ireland, and D. W. Walker. 2009. Neuroprotective properties of melatonin in a model of birth asphyxia in the spiny mouse (*Acomys cahirinus*). *Developmental Neuroscience* 31(5): 437–451.

Inder, T. E., S. J. Wells, N. B. Mogridge, C. Spencer, and J. J. Volpe. 2003. Defining the nature of the cerebral abnormalities in the premature infant: A qualitative magnetic resonance imaging study. *The Journal of Pediatrics* 143(2): 171–179.

Jantzie, L. L., P. Y. Cheung, and K. G. Todd. 2005. Doxycycline reduces cleaved caspase-3 and microglial activation in an animal model of neonatal hypoxia-ischemia. *Journal of Cerebral Blood Flow and Metabolism* 25(3): 314–324.

Jatana, M., I. Singh, A. K. Singh, and D. Jenkins. 2006. Combination of systemic hypothermia and *N*-acetylcysteine attenuates hypoxic-ischemic brain injury in neonatal rats. *Pediatric Research* 59(5): 684–689.

Johnston, M. V. 1998. Selective vulnerability in the neonatal brain. *Annals of Neurology* 44(2): 155–156.

Kaur, C., V. Sivakumar, and E. A. Ling. 2010. Melatonin protects periventricular white matter from damage due to hypoxia. *Journal of Pineal Research* 48(3): 185–193.

Kaur, C., V. Sivakumar, J. Lu, F. R. Tang, and E. A. Ling. 2008. Melatonin attenuates hypoxia-induced ultrastructural changes and increased vascular permeability in the developing hippocampus. *Brain Pathology (Zurich, Switzerland)* 18(4): 533–547.

Kelen, D. and N. J. Robertson. 2010. Experimental treatments for hypoxic ischaemic encephalopathy. *Early Human Development* 86(6): 369–377.

Koh, P. O. 2008. Melatonin regulates nitric oxide synthase expression in ischemic brain injury. *The Journal of Veterinary Medical Science/the Japanese Society of Veterinary Science* 70(7): 747–750.

Kohen, R. and A. Nyska. 2002. Oxidation of biological systems: Oxidative stress phenomena, antioxidants, redox reactions, and methods for their quantification. *Toxicologic Pathology* 30(6): 620–650.

Kumar, A., R. Mittal, H. D. Khanna, and S. Basu. 2008. Free radical injury and blood-brain barrier permeability in hypoxic-ischemic encephalopathy. *Pediatrics* 122(3): e722–e727.

Lee, M. Y., Y. H. Kuan, H. Y. Chen et al. 2007. Intravenous administration of melatonin reduces the intracerebral cellular inflammatory response following transient focal cerebral ischemia in rats. *Journal of Pineal Research* 42(3): 297–309.

Lee, T. F., L. L. Jantzie, K. G. Todd, and P. Y. Cheung. 2008. Postresuscitation *N*-acetylcysteine treatment reduces cerebral hydrogen peroxide in the hypoxic piglet brain. *Intensive Care Medicine* 34(1): 190–197.

Leonardo, C. C. and K. R. Pennypacker. 2009. Neuroinflammation and MMPs: Potential therapeutic targets in neonatal hypoxic-ischemic injury. *Journal of Neuroinflammation* 6: 13.

Low, J. A. 2004. Determining the contribution of asphyxia to brain damage in the neonate. *The Journal of Obstetrics and Gynaecology Research* 30(4): 276–286.

Maneru, C., C. Junque, F. Botet, M. Tallada, and J. Guardia. 2001. Neuropsychological long-term sequelae of perinatal asphyxia. *Brain Injury* 15(12): 1029–1039.

Maslinska, D., M. Laure-Kamionowska, A. Kaliszek, and D. Makarewicz. 2002. Proinflammatory cytokines in injured rat brain following perinatal asphyxia. *Folia Neuropathologica* 40(4): 177–182.

Mattson, M. P., P. B. Guthrie, and S. B. Kater. 1989. Intrinsic factors in the selective vulnerability of hippocampal pyramidal neurons. *Progress in Clinical and Biological Research* 317: 333–351.

Mayo, J. C., R. M. Sainz, D. X. Tan et al. 2005. Anti-inflammatory actions of melatonin and its metabolites, *N*1-acetyl-*N*2-formyl-5-methoxykynuramine (AFMK) and *N*1-acetyl-5-methoxykynuramine (AMK), in macrophages. *Journal of Neuroimmunology* 165(1–2): 139–149.

McLean, C. and D. Ferriero. 2004. Mechanisms of hypoxic-ischemic injury in the term infant. *Seminars in Perinatology* 28(6): 425–432.

McQuillen, P. S. and D. M. Ferriero. 2004. Selective vulnerability in the developing central nervous system. *Pediatric Neurology* 30(4): 227–235.

Menendez-Pelaez, A. and R. J. Reiter. 1993. Distribution of melatonin in mammalian tissues: The relative importance of nuclear versus cytosolic localization. *Journal of Pineal Research* 15(2): 59–69.

Miller, S. L., E. B. Yan, M. Castillo-Melendez, G. Jenkin, and D. W. Walker. 2005. Melatonin provides neuroprotection in the late-gestation fetal sheep brain in response to umbilical cord occlusion. *Developmental Neuroscience* 27(2–4): 200–210.

Mohan, N., K. Sadeghi, R. J. Reiter, and M. L. Meltz. 1995. The neurohormone melatonin inhibits cytokine, mitogen and ionizing radiation induced NF-kappa B. *Biochemistry and Molecular Biology International* 37(6): 1063–1070.

Nedelcu, J., M. A. Klein, A. Aguzzi, and E. Martin. 2000. Resuscitative hypothermia protects the neonatal rat brain from hypoxic-ischemic injury. *Brain Pathology (Zurich, Switzerland)* 10(1): 61–71.

Northington, F. J., D. M. Ferriero, E. M. Graham, R. J. Traystman, and L. J. Martin. 2001. Early neurodegeneration after hypoxia-ischemia in neonatal rat is necrosis while delayed neuronal death is apoptosis. *Neurobiology of Disease* 8(2): 207–219.

Olivier, P., R. H. Fontaine, G. Loron et al. 2009. Melatonin promotes oligodendroglial maturation of injured white matter in neonatal rats. *PLoS One* 4(9): e7128.

Ozyener, F., M. Cetinkaya, T. Alkan et al. 2012. Neuroprotective effects of melatonin administered alone or in combination with topiramate in neonatal hypoxic-ischemic rat model. *Restorative Neurology and Neuroscience* 30(5): 435–444.

Palmer, C., J. Towfighi, R. L. Roberts, and D. F. Heitjan. 1993. Allopurinol administered after inducing hypoxia-ischemia reduces brain injury in 7-day-old rats. *Pediatric Research* 33(4 Pt 1): 405–411.

Panickar, K. S. and M. D. Norenberg. 2005. Astrocytes in cerebral ischemic injury: Morphological and general considerations. *Glia* 50(4): 287–298.

Pei, Z. and R. T. Cheung. 2004. Pretreatment with melatonin exerts anti-inflammatory effects against ischemia/reperfusion injury in a rat middle cerebral artery occlusion stroke model. *Journal of Pineal Research* 37(2): 85–91.

Puka-Sundvall, M., C. Wallin, E. Gilland et al. 2000. Impairment of mitochondrial respiration after cerebral hypoxia-ischemia in immature rats: Relationship to activation of caspase-3 and neuronal injury. *Brain Research: Developmental Brain Research* 125(1–2): 43–50.

Rees, S., R. Harding, and D. Walker. 2011. The biological basis of injury and neuroprotection in the fetal and neonatal brain. *International Journal of Developmental Neuroscience* 29(6): 551–563.

Reiter, R. J., J. R. Calvo, M. Karbownik, W. Qi, and D. X. Tan. 2000b. Melatonin and its relation to the immune system and inflammation. *Annals of the New York Academy of Sciences* 917: 376–386.

Reiter, R. J., D. X. Tan, C. Osuna, and E. Gitto. 2000a. Actions of melatonin in the reduction of oxidative stress. A review. *Journal of Biomedical Science* 7(6): 444–458.

Robertson, N. J., S. Tan, F. Groenendaal et al. 2012. Which neuroprotective agents are ready for bench to bedside translation in the newborn infant? *The Journal of Pediatrics* 160(4): 544–552.e4.

Rosen, J., N. N. Than, D. Koch, B. Poeggeler, H. Laatsch, and R. Hardeland. 2006. Interactions of melatonin and its metabolites with the ABTS cation radical: Extension of the radical scavenger cascade and formation of a novel class of oxidation products, C2-substituted 3-indolinones. *Journal of Pineal Research* 41(4): 374–381.

Rosenberg, G. A. 1999. Ischemic brain edema. *Progress in Cardiovascular Diseases* 42(3): 209–216.

Rothstein, R. P. and S. W. Levison. 2005. Gray matter oligodendrocyte progenitors and neurons die caspase-3 mediated deaths subsequent to mild perinatal hypoxic/ischemic insults. *Developmental Neuroscience* 27(2–4): 149–159.

Sheldon, R. A., X. Jiang, C. Francisco et al. 2004. Manipulation of antioxidant pathways in neonatal murine brain. *Pediatric Research* 56(4): 656–662.

Signorini, C., L. Ciccoli, S. Leoncini et al. 2009. Free iron, total F-isoprostanes and total F-neuroprostanes in a model of neonatal hypoxic-ischemic encephalopathy: Neuroprotective effect of melatonin. *Journal of Pineal Research* 46(2): 148–154.

Sola, A., T. C. Wen, S. E. Hamrick, and D. M. Ferriero. 2005. Potential for protection and repair following injury to the developing brain: A role for erythropoietin? *Pediatric Research* 57(5 Pt 2): 110R–117R.

Song, W. L., J. A. Lawson, D. Reilly et al. 2008. Neurofurans, novel indices of oxidant stress derived from docosahexaenoic acid. *The Journal of Biological Chemistry* 283(1): 6–16.

Sugawara, T., M. Fujimura, N. Noshita et al. 2004. Neuronal death/survival signaling pathways in cerebral ischemia. *NeuroRx* 1(1): 17–25.

Takuma, K., A. Baba, and T. Matsuda. 2004. Astrocyte apoptosis: Implications for neuroprotection. *Progress in Neurobiology* 72(2): 111–127.

Tan, D. X., R. Hardeland, L. C. Manchester et al. 2010. The changing biological roles of melatonin during evolution: From an antioxidant to signals of darkness, sexual selection and fitness. *Biological Reviews of the Cambridge Philosophical Society* 85(3): 607–623.

Tan, D. X., L. C. Manchester, M. P. Terron, L. J. Flores, and R. J. Reiter. 2007. One molecule, many derivatives: A never-ending interaction of melatonin with reactive oxygen and nitrogen species? *Journal of Pineal Research* 42(1): 28–42.

Tomas-Zapico, C. and A. Coto-Montes. 2005. A proposed mechanism to explain the stimulatory effect of melatonin on antioxidative enzymes. *Journal of Pineal Research* 39(2): 99–104.

Tutunculer, F., S. Eskiocak, U. N. Basaran, G. Ekuklu, S. Ayvaz, and U. Vatansever. 2005. The protective role of melatonin in experimental hypoxic brain damage. *Pediatrics International* 47(4): 434–439.

Van Bel, F., M. Shadid, R. M. Moison et al. 1998. Effect of allopurinol on postasphyxial free radical formation, cerebral hemodynamics, and electrical brain activity. *Pediatrics* 101(2): 185–193.

Vannucci, S. J. and H. Hagberg. 2004. Hypoxia-ischemia in the immature brain. *The Journal of Experimental Biology* 207(Pt 18): 3149–3154.

Villapol, S., S. Fau, S. Renolleau, V. Biran, C. Charriaut-Marlangue, and O. Baud. 2011. Melatonin promotes myelination by decreasing white matter inflammation after neonatal stroke. *Pediatric Research* 69(1): 51–55.

Vitte, P. A., C. Harthe, P. Lestage, B. Claustrat, and P. Bobillier. 1988. Plasma, cerebrospinal fluid, and brain distribution of 14C-melatonin in rat: A biochemical and autoradiographic study. *Journal of Pineal Research* 5(5): 437–453.

Volpe, J. J. 2001. Perinatal brain injury: From pathogenesis to neuroprotection. *Mental Retardation and Developmental Disabilities Research Reviews* 7(1): 56–64.

Walton, M., B. Connor, P. Lawlor et al. 1999. Neuronal death and survival in two models of hypoxic-ischemic brain damage. *Brain Research. Brain Research Reviews* 29(2–3): 137–168.

Wang, X., H. Hagberg, C. Zhu, B. Jacobsson, and C. Mallard. 2007. Effects of intrauterine inflammation on the developing mouse brain. *Brain Research* 1144: 180–185.

Wang, X., T. L. Yue, F. C. Barone, R. F. White, R. C. Gagnon, and G. Z. Feuerstein. 1994. Concomitant cortical expression of TNF-alpha and IL-1 beta mRNAs follows early response gene expression in transient focal ischemia. *Molecular and Chemical Neuropathology* 23(2–3): 103–114.

Watanabe, K., F. Hamada, A. Wakatsuki et al. 2012. Prophylactic administration of melatonin to the mother throughout pregnancy can protect against oxidative cerebral damage in neonatal rats. *The Journal of Maternal-Fetal & Neonatal Medicine* 25(8): 1254–1259.

Watanabe, K., A. Wakatsuki, K. Shinohara, N. Ikenoue, K. Yokota, and T. Fukaya. 2004. Maternally administered melatonin protects against ischemia and reperfusion-induced oxidative mitochondrial damage in premature fetal rat brain. *Journal of Pineal Research* 37(4): 276–280.

Welin, A. K., P. Svedin, R. Lapatto et al. 2007. Melatonin reduces inflammation and cell death in white matter in the mid-gestation fetal sheep following umbilical cord occlusion. *Pediatric Research* 61(2): 153–158.

Witko-Sarsat, V., M. Friedlander, C. Capeillere-Blandin et al. 1996. Advanced oxidation protein products as a novel marker of oxidative stress in uremia. *Kidney International* 49(5): 1304–1313.

Yager, J. Y., E. A. Armstrong, and A. M. Black. 2009. Treatment of the term newborn with brain injury: Simplicity as the mother of invention. *Pediatric Neurology* 40(3): 237–243.

Yan, C. Y., G. Ferrari, and L. A. Greene. 1995. *N*-acetylcysteine-promoted survival of PC12 cells is glutathione-independent but transcription-dependent. *The Journal of Biological Chemistry* 270(45): 26827–26832.

Zhu, C., X. Wang, X. Cheng, L. Qiu, F. Xu, G. Simbruner, and K. Blomgren. 2004. Post-ischemic hypothermia-induced tissue protection and diminished apoptosis after neonatal cerebral hypoxia-ischemia. *Brain Research* 996(1): 67–75.

7 Melatonin as a Therapeutic Agent for Sepsis

Edward H. Sharman and Stephen C. Bondy

CONTENTS

7.1 INCIDENCE AND CURRENT TREATMENT OF SEPSIS

With an estimated incidence exceeding 1.1 million cases per year in the United States (Moore and Moore, 2012), sepsis is the 10th leading cause of death overall in this country (Yang et al., 2010). As of 2005, "mortality of severe sepsis exceeds other high-profile diseases such as AIDS, venous thromboembolism, and both lung and colon cancers" (Jones, 2006). Moreover, septic shock–related mortality is greater than 40%, and sepsis itself is the leading cause of death in noncardiac intensive care units (Moore and Moore, 2012). Little progress in reducing sepsis is being made: to the contrary, between 2003 and 2007, hospitalizations for severe sepsis in the United States increased by 71% and in 2007 incurred a cost of $24.3 billion (Lagu et al., 2012). Mortality rates have decreased little in the past 30 years (Astiz and Rackow, 1998). As of 2005, multidrug-resistant bacteria and fungi were noted to cause about 25% of cases (Annane et al., 2005), increasing the importance for identifying adjuvant treatments capable of strengthening the immune system of patients.

With deficiencies in key innate immune responses, preterm infants and neonates are particularly susceptible to life-threatening infections (Strunk et al., 2011), including sepsis. Thereafter, the incidence and mortality of sepsis increase dramatically with age: incidence in patients over 85 years of age is 100 times that in children, with a mortality of 38.4% in this elderly age group (Angus et al., 2001). Thus, sepsis is primarily a disease of both the very young and the aged, with increased incidence and mortality occurring in the aged (De Gaudio et al., 2009; McConnell et al., 2011). This may be because the inflammatory response in the aged is exaggerated as compared to the young adult (Leong et al., 2010). This is in part due to the fact that inflammatory genes

TABLE 7.1
Definitions of Sepsis

Systemic inflammatory response syndrome (SIRS)	Two or more of the following conditions: Temperature >38.5°C or <35°C Heart rate >90 beats min^{-1} Ventilatory frequency >20 bpm or Pa_{CO_2}<32 mm Hg or need for mechanical ventilation White blood cell count >12,000 or <4,000 mm^{-3} or >10% immature (band) forms
Sepsis	SIRS and documented infection (culture or Gram-stain of blood, sputum, urine, or normally sterile body fluid, positive for pathogenic organisms; or focus for infection identified by visual inspection, e.g., ruptured bowel with free air or bowel contents found in the abdomen at surgery or a wound with purulent discharge)
Severe sepsis	Sepsis and at least one of the signs of organ hypoperfusion or organ dysfunction: Areas of mottled skin Capillary refill ≥3 s Urinary output of <0.5 mL kg^{-1} for at least 1 h or renal replacement therapy Lactate >2 mmol L^{-1} Abrupt change in mental status or abnormal EEG findings Platelet count <100,000 mL^{-1} or DIC Acute lung injury/acute respiratory distress syndrome Cardiac dysfunction (echocardiography)
Septic shock	Severe sepsis and one of the following conditions: Mean arterial pressure <60 mm Hg (<80 mm Hg if previous hypertension) after 20–30 mL kg^{-1} starch or 40–60 mL kg^{-1} saline solution, or pulmonary capillary wedge pressure between 12 and 20 mm Hg Need for dopamine >5 μg kg^{-1} min^{-1} or norepinephrine or epinephrine of <0.25 μg kg^{-1} min^{-1} to maintain mean arterial pressure >60 mm Hg (80 mm Hg if previous hypertension)
Refractory septic shock	Need for dopamine at >15 μg kg^{-1} min^{-1}, or norepinephrine or epinephrine at >0.25 μg kg^{-1} min^{-1} to maintain a mean arterial pressure >60 mm Hg (80 mm Hg if previously hypertensive)

Source: Hunter, J.D. and Doddi, M., Sepsis and the heart, *Br. J. Anaesth.*, 104, 3–11, 2010 by permission of Oxford University Press.

are expressed at high levels in aged humans and animals even in the absence of a provocative stimulus (Sharman et al., 2007; Böhler et al., 2009). This high resting level is further intensified during sepsis (Wang et al., 2010). The conditions of sepsis, severe sepsis, and septic shock constitute a continuum of increasingly severe clinical response to infection (Hunter and Doddi, 2010, Table 7.1). Patients with sepsis display manifestations of infection and inflammation, consisting of two or more of the following: increased or decreased temperature, increased or decreased leukocyte count, tachycardia, and rapid breathing (Annane et al., 2005). Those progressing to severe sepsis additionally develop hypoperfusion with organ dysfunction. Sepsis often progresses to septic shock, characterized by persistent vascular hypotension refractory to fluid administration and dysfunction of multiple organs, particularly heart, kidney, lung, and liver. Sepsis may best be characterized as a multifactorial condition affecting multiple tissues and organs by multiple mechanisms.

Infections that induce sepsis occur most frequently in the lungs, abdomen, and urinary tract. Gram-negative bacteria are often the causative infectious agents, but increasingly, sepsis also results from infection by Gram-positive bacteria or fungi (van der Poll and Opal, 2008). Multiple infectious organisms may be involved as well. Fungal infections are of particular concern, because they can result in substantially higher mortality (Opal et al., 2003).

The immunological response to septic shock is biphasic: the initial response to the infection is overwhelming inflammation; this is later followed by a period of immune depression that may persist long after the patient has otherwise recovered from the infection.

A critical barrier to progress is that standard therapy has tended to use a select few agents, each with a single mechanism of action, for treating the most prominent aspects of septic shock, such as antibiotics, vasopressors, and anti-inflammatory drugs. While antibiotics may eliminate the causative agents, they do nothing to dampen the body's excessive inflammatory response that precipitates septic shock, and their creation of large quantities of pathogen fragments may actually exacerbate it. Glucocorticoids may seem attractive agents for suppressing such inflammation. Although many clinical trials using a variety of glucocorticoids have been conducted, none has conclusively proven these agents to be beneficial in treating septic shock, and in some trials, glucocorticoids actually increased mortality (Sessler, 2003). The inconclusive or detrimental nature of these outcomes is consonant with the well-known immune-suppressing properties of glucocorticoids (Löwenberg et al., 2007). Although the pathological hypotension associated with sepsis is induced by an increased production of nitric oxide, treatment with inhibitors of nitric oxide synthase either actually increases mortality of septic shock patients (López et al., 2004) or fails to improve their survival (Bakker et al., 2004). These examples illustrate the shortcomings of single-target treatments for a complex multifactorial disorder.

The fact that infections are increasingly resistant to a broad range of antibiotics (Nordmann et al., 2011) further accentuates the urgency of finding novel approaches to sepsis that utilize the intrinsic defense mechanisms of the host.

7.2 OVERVIEW OF SEPSIS PATHOLOGY

The severity of organ dysfunction that accompanies sepsis and septic shock can be measured by evaluating the *sequential organ failure assessment* (SOFA) score. Changes in SOFA scores early in disease progression are meaningful: rapid worsening of this score in the first 48 h led to over 50% mortality, whereas improved organ function in the first 24 h was associated with increased survival (Vincent, 2007).

Hemostasis is a normal and beneficial host response to bacterial infection; bacteria in turn have developed a number of mechanisms for evading, subverting, and dysregulating this process (Fourrier, 2012). A disorder frequently accompanying severe sepsis is disseminated intravascular coagulation (DIC). In this condition, myriad microclots are formed at locales widely distributed throughout small and midsized vessels. This condition seems to result in exhaustion and dysregulation of the clotting system, so that patients become overly sensitive to bleeding, particularly after major surgery. Tissue factor is the main initiator of coagulation in sepsis and is constitutively expressed in the extravascular compartment, thus initiating clotting if blood leaves the confines of the vasculature. During severe sepsis, activated monocytes and endothelial cells, along with circulating microvesicles, become sources of tissue factor. Inhibitors of the factor VIIa–tissue factor pathway in experimental studies in human beings and primates nullify the activation of this coagulation pathway (van der Poll and Opal, 2008). An important characteristic of DIC is an insufficiency of tissue factor pathway inhibitor (TFPI) (Franchini et al., 2006). Melatonin dose-dependently increases levels of TFPI protein in human coronary artery endothelial cells in vitro (Kostovski et al., 2011), suggesting that its use in treating DIC in sepsis may prove useful.

There is substantial consensus that the healthy resolution of the inflammatory state—such as that occurs when the body can overcome sepsis without going into shock—is not a passive process, but is under active regulatory control (Buckley et al., 2013).

7.3 SYNTHESIS, ABSORPTION, AND METABOLISM OF MELATONIN

Although the pineal gland is probably the most well-known site of melatonin synthesis, this hormone is produced in many other mammalian organs, tissues, and cells as well. These include the retina, the gastrointestinal tract, epithelial cells, and—of particular significance for immune function—bone marrow, thymocytes, and a variety of leukocytes, most notably monocytes, eosinophils, mast cells, T lymphocytes, and NK cells (Hardeland et al., 2011). That melatonin synthesized outside of the pineal gland seems not to enter the general circulation implies that its action in extrapineal tissues is likely to be of a local, autocrine, or paracrine nature (Hardeland et al., 2011).

The metabolism, pharmacokinetics, and bioavailability of melatonin in humans have been reviewed (Brzezinski, 1997). Briefly, catabolism of melatonin occurs rapidly by hydroxylation, mainly in the liver; the product, 6-hydroxymelatonin, is conjugated with sulfate or glucuronic acid, and excreted in the urine. Concentrations of urinary 6-sulfatoxymelatonin (the chief metabolite of melatonin) and serum melatonin closely parallel one another (Selmaoui et al., 1996). Melatonin administered i.v. is rapidly distributed (serum $t_{½}$ = 1.35 min) and eliminated (serum decay $t_{½}$ = 28 min) (Mallo et al., 1990). The bioavailability of orally administered melatonin varies widely, but is roughly proportional to the dose. In normal subjects given 80 mg of melatonin, serum melatonin concentrations were 350–10,000 times higher than typical nighttime peak values 1–2½ h later, and remained stable over a 1½ h period (Waldhauser et al., 1984). Lower oral doses of melatonin (1–5 mg) result in roughly proportionately reduced serum concentrations that are 10–100 times higher than the typical nighttime peak within 1 h after ingestion, followed by a decline to baseline values in 4–8 h. Very low oral doses (0.1–0.3 mg) given during the day result in peak serum concentrations that are within the normal nighttime range (Dollins et al., 1994).

A certain proportion of melatonin administered in the diet is able to enter the bloodstream and thence to other organs, including the brain, in an intact and unconjugated form (Lahiri et al., 2004).

7.4 MELATONIN AND AGING

Preterm human infants produce little if any melatonin and require some 15 h to clear it after exogenous administration (Merchant et al., 2013). Human newborns are able to synthesize melatonin, but require ~9–12 weeks to begin secreting normal circadian nighttime pulses (Kennaway et al., 1992). Thus, optimal dosage amounts and regimens that might be contemplated for treating conditions such as sepsis in preterm and newborn infants are likely to differ from those in older patients. After adulthood, melatonin levels decline severely with age in humans and in mice (Waldhauser et al., 1988; Lahiri et al., 2004). While the full range of physiological consequences of this is unknown, it is likely that this may contribute to the impairment of appropriate immune and inflammatory responses in the elderly. There is a large literature on the ability of melatonin to attenuate some of the less desirable consequences of aging (Pierpaoli and Regelson, 1994; Bondy et al., 2004; Akbulut et al., 2008; Sharman et al., 2011). These effects of melatonin have been related to reversal of the age-associated deterioration of the immune system (immunosenescence), including functional decline of granulocytes, macrophages, and T and B cells (Espino et al., 2012). Beneficial effects of melatonin have also been reported for age-associated declines in ovarian biology (Fernández et al., 2013), brain functioning (Ramírez-Rodríguez et al., 2012), liver metabolism (Eşrefoğlu et al., 2012), colonic function (Pascua et al., 2012), cardiac effectiveness (Forman et al., 2011), and increased pancreatic insulin resistance (Cuesta et al., 2013). Many of these effects have been attributed to the improvement of immune function and regulation of inflammatory responses.

Melatonin treatment of aged animals can largely reverse many of the changes in gene expression that characterize aging. This is especially true of genes related to the immune system and to inflammation, which are generally elevated with age and hyperresponsive to inflammatory stimuli (Sharman et al., 2007, 2008). These changes, both in basal mRNA levels and in their reactivity, may be the cause underlying many reported antiaging properties of melatonin. The relevance of these findings to the treatment of sepsis is that the inflammatory response found with sepsis is markedly heightened with aging (Turnbull et al., 2009).

7.5 MELATONIN, THE IMMUNE SYSTEM, AND INFLAMMATORY RESPONSES

Melatonin is synthesized in many types of immune cells; moreover, two plasma membrane melatonin receptors are found to reside in immune cells, and melatonin has been determined to influence many cellular as well as whole animal immune functions (Carrillo-Vico et al., 2013).

Cytokine and chemokine signaling networks are altered in elderly patients, and tend to favor a type-2 cytokine response (largely humoral) over a type-1 (largely cellular) response (Opal et al., 2005). It is the former that can lead to generalized inflammation, while the latter represents a more selective and localized immune response. Thus, this imbalance can lead to excessive production of proinflammatory factors (Opal et al., 2005). A failure of the type-1 arm of the immune system and an overactive type-2 arm are implicated in a wide variety of chronic illnesses. It has been proposed that development of new drugs that specifically regulate the balance of these two kinds of response activity may pave the way for novel therapeutic interventions in sepsis (Matsukawa et al., 2001). Melatonin may be such an agent (Petrovsky and Harrison, 1997), since it can simultaneously upregulate a type-1 immune response while downregulating the type-2 response which involves interleukin (IL)-4, IL-10 production, and splenocyte proliferation (Santello et al., 2008).

Lipopolysaccharide (LPS) treatment is often used as a means of provoking acute inflammatory responses in experimental animals (endotoxemia). However, low levels of LPS have also been used to model sepsis in vitro (Lowes et al., 2011). Responses evoked in this manner have frequently been found to be attenuated by prior treatment with melatonin, both in whole animals (Zhong et al., 2009; Fagundes et al., 2010) and in isolated cell systems (Lowes et al., 2011). This suppression by melatonin may be related to reduced expression of several inflammatory genes, since melatonin is able to reverse the heightened expression of both the basal and LPS-induced levels of expression of inflammatory cytokines found in aging brain (Perreau et al., 2007; Sharman et al., 2007; Yavuz et al., 2007). That the protective effect of melatonin is systemic is well illustrated by the finding that it can markedly reduce the lethality of LPS (Requintina and Oxenkrug, 2003). Following intraperitoneal injection of LPS into rats, there was an increase in the TBARS levels, and in apoptotic cell death. Concurrent administration of melatonin prevented these changes and led to increased activities of the antioxidant enzymes superoxide dismutase and glutathione peroxidase (Ozdemir et al., 2007).

7.6 MELATONIN AND SEPSIS

Melatonin may be a means by which the severity of sepsis can be mitigated, and there are several mechanisms by which this might occur. Melatonin can act to regulate inflammatory processes. The timing of these effects suggests that melatonin can promote the early phases of an immune response and attenuate the later stages of inflammation. Thus, an effective immune defense is initially fostered by melatonin, and this is followed by prevention of chronic and harmful inflammatory events (Radogna et al., 2010). The later stages of melatonin action involve reducing induction of IL-6, cyclooxygenase-2, tumor necrosis factor (TNF)-α, and inducible nitric oxide synthase (iNOS) (Deng et al., 2006; Yavuz et al., 2007). Thus, melatonin is capable of inducing sophisticated regulatory changes rather than acting merely as a broad anti-inflammatory or antioxidant agent. In addition, many oxidant responses associated with sepsis, such as depletion of reduced glutathione levels and elevated lipid peroxidative activity, are reversed by melatonin treatment (Şener et al., 2005). In view of the very low levels of free melatonin encountered within cells, such changes are likely to be mediated by melatonin, either acting through its receptors or effecting alterations in gene expression (Perreau et al., 2007; Sharman et al., 2007).

7.6.1 Animal Models of Sepsis

A wealth of studies in animals suggests the value of melatonin in treating sepsis. The survival rates found in several different animal models of sepsis benefit from melatonin treatment (Wichmann et al., 1996; Reynolds et al., 2003; Zhang et al., 2013). Mice subjected to a combination of hemorrhagic shock followed by septic challenge exhibited reduced mortality when treated for a short time with melatonin (Wichmann et al., 1996). Some aspects of sepsis in specific organs are described, together with an outline of the benefits of melatonin treatment to organ systems impacted by experimental sepsis.

Gastrointestinal tract: The leakage of digestive enzymes from the intestinal lumen into the surrounding tissue is commonly associated with both septic and hemorrhagic shock. According to the autodigestive hypothesis for the initiation of the systemic inflammatory response, errant digestive enzymes not only attack and damage intestinal tissue, but are also responsible for damage to tissues in other organs such as heart and lung, thus contributing to multiorgan failure (Schmid-Schönbein, 2009). Inhibiting these enzymes in the intestinal lumen can substantially increase survival of animals subjected to shock (Delano et al., 2013).

Although the pineal may be more widely known as a source of melatonin, the gastrointestinal tract is in fact the major site of melatonin synthesis in the body (Bubenik, 2002). In the rat cecal ligation and puncture (CLP) model of sepsis, the levels of myeloperoxidase and a measure of lipid peroxide were increased; at the same time, the levels of the antioxidant glutathione were reduced; melatonin treatment reversed all these values (Paskaloğlu et al., 2004). Decreased gastrointestinal motility is characteristic of sepsis (Königsrainer et al., 2011), and melatonin is able to completely reverse the arrest of gastrointestinal motility that is effected by LPS treatment (Paskaloğlu et al., 2004; De Filippis et al., 2008).

Liver: The liver is one of the first organs impacted by sepsis. Sepsis alters hepatic transcription of genes for adaptive acute phase proteins, represses transcription of genes for proteins involved in phase I and II metabolism and transport, and increases levels of transaminases and alkaline phosphatase (Bauer et al., 2013). Intracellularly, i.v. endotoxin injection in rats increased liver iNOS activity and NO levels, accompanied by increases in tissue levels of alkaline phosphatase and transaminase alanine aminotransferase (ALT); i.p. injection of melatonin reversed these effects (Escames et al., 2006). In a rat model of acute sepsis, melatonin ameliorated the liver damage indicated by elevated levels of ALT, and decreased mitochondrial complex IV activity and the ATP:oxygen ratio (Lowes et al., 2013).

Hepatic ischemia/reperfusion (I/R) induces many of the same indicators of inflammatory damage as does sepsis. In rats, during hepatic I/R, levels of endothelin-1 (ET-1) and its receptor, ET(B) mRNA, were elevated but attenuated by melatonin. mRNA levels of endothelial nitric oxide synthase (eNOS), iNOS, heme oxygenase-1 (HO-1), and TNF-α were all elevated after I/R. Melatonin augmented the increased expression of eNOS mRNA, whereas it reduced the increase in iNOS mRNA and TNF-α (Park et al., 2007). In a later study, melatonin was found to lower the expression of numerous proinflammatory cytokines and proapoptotic genes that were raised by I/R, and to improve liver function (Kireev et al., 2012). These studies suggested an overall anti-inflammatory effect.

Septic shock and hemorrhagic shock also inflict rather similar types of damage to the liver. Damage by both is mediated by a reduction in Akt phosphorylation, by increases in iNOS and HO-1, and accompanied by an increase in plasma myeloperoxidase. In the case of hemorrhagic shock in rats, melatonin can normalize liver Akt phosphorylation, plasma myeloperoxidase, and reduce liver damage (Hsu et al., 2012).

Heart: Cardiac dysfunction occurs frequently in septic patients. It is reported that the left ventricular ejection fraction (LVEF) in hemodynamically unstable septic shock patients was depressed, and their stroke volumes were severely reduced; one in six had an LVEF of less than 30% (Rudiger and Singer, 2007). Moreover, occurrence of myocardial depression in septic patients tends to be associated with increased mortality (Court et al., 2002).

Repeated administration of melatonin to rats following CLP improved heart mitochondrial function and increased survival at 48 h post CLP (Zhang et al., 2013). Melatonin (30 mg kg^{-1}, i.p.) was administered at 3 h post CLP and then injected subcutaneously every 3 h thereafter up to the 24 h point. After 48 h, melatonin restored the substantial reductions in base excess and lactate induced by CLP, and increased the proportion of animals surviving from 32% to 57%. Melatonin also restored the more modest reductions in CLP-induced LVEF and cytochrome-c oxidase (COX) activity. In contrast to the many animal studies that begin melatonin treatment prior to injury, in this study, melatonin was administered starting 3 h after CLP and after the onset of symptoms. Consequently, this design is more relevant to clinical applications and provides stronger evidence that melatonin treatment would be of real benefit in the ICU. In a study in mice generally similar to the earlier one, melatonin was also found to improve heart mitochondrial function following CLP: CLP-induced reductions in COX levels were restored, elevated iNOS levels were normalized, as were the indices of lipid peroxidation and protein oxidation (Ortiz et al., 2014). These effects of melatonin were independent of nNOS gene knockout.

Lung: Acute lung injury commonly develops during sepsis, with nearly half of the patients in one observational study developing this condition (Iscimen et al., 2008). In an animal model of acute septic lung injury induced by LPS injection, administration of melatonin was shown to have beneficial effects in rats (Shang et al., 2009). In this study, melatonin reduced the LPS-induced pulmonary leukocyte infiltration, elevated levels of tissue malondialdehyde, myeloperoxidase, and TNF-α, while it augmented levels of the typically anti-inflammatory cytokine IL-10. In an earlier study, i.v. endotoxin injection in rats increased lung mitochondrial NO levels, mitochondrial NOS activity, and decreased activities of complexes I and IV; i.p. injection of melatonin reversed all these effects (Escames et al., 2003). Melatonin administration (i.p.) also has produced beneficial effects in the lungs of rats in which sepsis had been induced by CLP; melatonin reduced the sepsis-elevated levels of malondialdehyde and improved the sepsis-induced pulmonary structural degeneration, vasocongestion, and edema observed (Şener et al., 2005).

Vasculature: Melatonin has been reported to prevent endotoxin-induced circulatory failure in rats (Wu et al., 2001). Such circulatory failure was attributed to the inhibition of (1) the release of TNF-α in plasma, (2) the expression of NOS II in liver, and (3) the production of superoxide in the aorta. Another study suggests that melatonin may improve the outcomes of sepsis due to fungal infections (Yavuz et al., 2007). Rats were first immunosuppressed with cyclophosphamide and then received an i.v. injection of *Candida albicans*. Subsequent daily i.p. injections of 200 μg kg^{-1} melatonin reduced the serum levels of the proinflammatory cytokine IL-6 after 15 days, and reduced the time required for *Candida* to be cleared from the bloodstream by 15%.

Nerve and muscle: Critical illness polyneuropathy and myopathy commonly occur after the onset of severe sepsis (Latronico and Bolton, 2011). Melatonin, particularly in combination with oxytocin, is effective in reducing the electromyographical, inflammatory, and oxidative dysregulation, associated with critical illness polyneuropathy occurring after the induction of sepsis in the CLP rat model (Erbaş et al., 2013).

Central nervous system: The brain plays a key role in sepsis, since it acts as both a mediator of the immune response and a target for the pathologic process (Zampieri et al., 2011; Gofton and Young, 2012). Septic encephalopathy is characterized by alteration of consciousness, occurrence of seizures or focal neurological signs, and involves an ischemic process, secondary to impairment of cerebral blood flow and neuroinflammatory events, including endothelial activation, alteration of the blood–brain barrier, and passage of neurotoxic mediators (Adam et al., 2013). The mortality rate of septic patients with altered mental status was 49% compared with a rate of 26% in septic patients with no neurological symptoms (Sprung et al., 1990). While no clinical study has focused specifically on melatonin and septic encephalopathy, work on septic shock in mice has demonstrated melatonin's protective effect to extend to several organs, including the brain (Carrillo-Vico et al., 2005). A study demonstrating the beneficial effects of melatonin upon an animal model for neonatal encephalopathy (Robertson et al., 2013) is also pertinent.

A vast majority of studies on the effects of melatonin on sepsis have used murine models, undertaken with the implicit assumption that responses in septic humans and rodents are largely similar. Recently, a comparison of 4918 human genes, including those responsive to an i.v. LPS injection, demonstrated that their changes are uncorrelated with changes in the corresponding orthologs in the mice treated similarly (Seok et al., 2013). On the other hand, substantial similarity of gene expression patterns (and some useful disparity) was found between the mice injected i.p. with human feces and severely septic pediatric patients (Lambeck et al., 2012). Thus, while the latter study gives confidence that rodents and humans respond to live bacterial infection in substantially and usefully similar ways, the results of the former (produced under more artificial conditions) imply that caution may be required in applying results of animal studies that utilize LPS to the treatment of sepsis in humans.

7.7 HUMAN CLINICAL TRIALS

Melatonin status is altered under conditions of sepsis. Urinary melatonin excretion of septic patients in a state of shock was higher than that of septic patients not in septic shock and of those of nonseptic patients (Bagci et al., 2011). To date, there has only been one human study relating to the utility of melatonin in the treatment of sepsis—and that in neonates. Treatment with melatonin may be of particular relevance in this group of patients, since human infants do not begin producing melatonin normally until 2–4 months of age. In a clinical trial of melatonin, the drug was found effective in reducing the mortality of neonatal sepsis and at 48 h was associated with statistically significant improvements in the counts of total white blood cells, absolute neutrophils and platelets, and levels of C-reactive protein (Gitto et al., 2001). Numerous laboratory studies provide a rationale for these results: in addition to the earlier-mentioned parameters, a wide variety of inflammatory factors are increased in neonatal sepsis (Sugitharini et al., 2013), and under conditions in which these factors are similarly increased, melatonin has been reported to normalize them in most cases (Table 7.2). Moreover, other clinical trials suggest antibacterial properties of melatonin. These include acceleration of healing of ulcers associated with *Helicobacter pylori* (Celinski et al., 2011) and mitigation of ulcerative colitis associated with several proinflammatory bacterial species (Wang et al., 2007).

Since sepsis involves an excessive inflammatory response leading to massive outpouring of cytokines, other clinical trials reporting anti-inflammatory benefits of melatonin are also relevant to sepsis (Cichoz-Lach et al., 2010; Ochoa et al., 2011). Therefore, it is appropriate to consider reports of melatonin treatment of related conditions. In ulcerative colitis patients, melatonin treatment maintained normal levels of C-reactive protein (between 3.0 and 4.2 mg dL^{-1}), compared to the elevated levels of 13.1 mg dL^{-1} in the untreated patients (Chojnacki et al., 2011). This suggests the possibility that melatonin may be capable of reducing the elevated C-reactive protein levels associated with sepsis.

Burn injuries can often lead to compromise of protective epidermal barriers and generalized sepsis. A relatively low dose of melatonin (3 mg day^{-1}) is sufficient to reduce mortality, shorten healing time, and lower the incidence and distribution of invading bacterial species in patients suffering from severe burns (Sahib et al., 2010).

Serum troponin levels are elevated in critically ill septic patients, and increased levels are associated with lower cardiac stroke ejection fractions and increased mortality (Rudiger, 2007). While perioperative administration of 60 mg of melatonin i.v. to patients undergoing major abdominal aortic surgery failed to reduce the elevated troponin levels associated with this procedure (Kücükakin et al., 2008), similarly elevated troponin levels in a rat model of cardiac ischemia–reperfusion were substantially reduced by melatonin administered 10 mg kg^{-1} i.p., 30 min prior to injury (Acikel et al., 2003).

7.7.1 Sepsis and the Circadian Cycle

The circadian rhythm of melatonin secretion is altered in the early stages of sepsis in ICU patients, with the acrophase of melatonin secretion being shifted from 2 a.m. to 6 p.m. in the nonseptic and

TABLE 7.2
List of Inflammatory Factors Modulated in Neonatal Septic Patients, and Parallel Experimental Conditions under Which Melatonin Normalizes Similarly Modulated Levels of These Factors

	Sepsis-Induced	Melatonin-Related		
Factor	Change	Change	Condition	References
C-reactive protein	⇧	⇩	Human ulcerative colitis	Chojnacki et al. (2011)
		⇩	Rat burn trauma DIC	Bekyarova et al. (2010)
		⇩	Rabbit viral hepatitis	Laliena et al. (2012)
Procalcitonin	⇧		—	
Myeloperoxidase	⇩	⇩	Endotoxemic rat lung	Shang et al. (2009)
		⇩	Rat heat stroke serum	Lin et al. (2011)
Neutrophil Elastase	⇧	⇩	Sl. inhibition in activated human granulocytes	Fjaerli et al. (1999)
TNF-α	⇧	⇩	Mouse CLP	Erbaş et al. (2013)
		⇩	Rat heart, Chagas disease	Oliveira et al. (2013)
		⇩	Endotoxemic rat lung	Shang et al. (2009)
IL-1β	⇧	⇩	Rabbit viral hepatitis	Laliena et al. (2012)
		⇩	Rat heat stroke serum and lung	Lin et al. (2011), Wu et al. (2012)
IL-6	⇧	⇩	Rat heat stroke serum and lung	Lin et al. (2011), Wu et al. (2012)
		⇩	Plasma of rats with *Candida* sepsis	Yavuz et al. (2007)
IL-8	⇧	⇩	Distressed surgical neonate serum	Gitto et al. (2004)
MCP-1	⇧	⇩	Ischemic rat liver mRNA	Kireev et al. (2012)
		⇩	Rat colitis tissue	Li et al. (2008)
IL-10	⇧	⇧	Rat heart, Chagas disease	Oliveira et al. (2013)
		⇧	Endotoxemic rat lung	Shang et al. (2009)
IL-12/IL-23p40	⇧	n.d.		
IL-21	⇧	n.d.		
IL-23	⇧	n.d.		

Source: Adapted from Sugitharini, V. et al., *Inflamm. Res.*, 62, 1025, 2013.

septic patients, respectively (Li et al., 2013). In these patients, peak levels of plasma TNF-α and IL-6 occurred in concordance with peak melatonin secretion, while mRNA levels of the circadian clock genes *cry-1* and *per-2* were suppressed. It is also noteworthy that sleep deprivation after septic insult increases mortality (Friese et al., 2009). Sleep deprivation experienced in the ICU setting during sepsis may thus be deleterious. Thus, another potential beneficial aspect of melatonin is its ability to regulate and restore normal sleep patterns.

7.7.2 Hypotension Associated with Sepsis

Septic shock is characterized by hypotension and vascular hyporeactivity to contractile agents. Melatonin can restore endothelium-derived constricting factor signaling and consequent regulation of the inner diameter in the rat femoral artery after inhibition of nitric oxide–based vasodilation (Paulis et al., 2010). It can also prevent lipopolysaccharide-induced vascular hyporeactivity in rats (d'Emmanuele et al., 2004). By this means, melatonin can increase vascular perfusion and reverse hypotension. Melatonin can also reverse the refractory hypotension associated with multiple organ dysfunction syndrome of septic shock (Wu et al., 2008). Remarkably, melatonin is also able to reduce blood pressure in essential hypertension (Cagnacci et al., 2005). Thus, this moiety is neither intrinsically hypotensive nor hypertensive but appears to be able to regulate blood pressure in a bidirectional manner.

7.8 MECHANISMS UNDERLYING MELATONIN'S ACTIONS

Although a broad range of protective effects of melatonin have been described, it is likely that there is a much more limited number of key mechanisms that underlie these effects. Whether these are mediated by specific receptors or can be attributed to direct effects of melatonin remains uncertain (Reiter et al., 2007). Studies in isolated systems need to take account of physiological concentrations of free melatonin, which are very low; this is the case, especially in tissues other than serum (Lahiri et al., 2004, Table 7.3) and gastrointestinal tract (Bubenik, 2001), in which melatonin concentrations can be substantial. Melatonin is reported to possess direct antimicrobial activity, but at concentrations much higher than those found in intact animals (Tekbas et al., 2008). This intrinsic property against multidrug-resistant Gram-positive and Gram-negative bacteria would be very relevant to sepsis (Srinivasan et al., 2012), but it is more likely that antibiotic effects are mediated by melatonin's promotion of immune targeting.

7.8.1 Receptor-Mediated Mechanisms

Three plasma membrane receptors for melatonin have been identified—the G-protein-coupled receptors MT1, MT2, and the quinone dehydrogenase enzyme NQO2.

The anti-inflammatory properties of melatonin are blocked by luzindole, a nonspecific MT1 and MT2 receptor antagonist (Cevík et al., 2005), suggesting an involvement of receptors in this protective effect. In particular, both the MT1 receptor and its G-protein target G_{16} are found in hematopoietic cells, implying that melatonin may modulate hematopoietic growth and immune function; cytokine production and STAT3 phosphorylation resulting from MT1 activation by melatonin in Jurkat T cells support this notion (Chan and Wong, 2013). It has been proposed that MT1 receptors are key in facilitating some protective roles of melatonin (Renzi et al., 2011). Involvement of the MT2 receptor in immune function has been shown more directly in mice. Melatonin was found to enhance splenocyte proliferation and IgG antibody response in these animals; while these effects

TABLE 7.3
Concentration of Melatonin in Tissues of 6-Month-Old Male B6C3F1 Mice

Values Are pM ± SEM	
Serum	343 ± 41
Liver	8.1 ± 2.4
Kidney	3.3 ± 0.6
Cerebral cortex	3.9 ± 0.8
Heart	2.8 ± 0.4

Source: Adapted from Lahiri, D.K. et al., *J. Pineal. Res.*, 36, 217, 2004.

were unchanged by the knockout of the MT1 receptor gene, they were attenuated by the administration of luzindole (Drazen and Nelson, 2001). However, in view of the incomplete characterization of these receptor types, much remains unresolved.

The third of melatonin's plasma membrane receptors, MT3, is the enzyme NRH:quinone oxidoreductase, NQO2. Knockout of the NQO2 gene in mice lowered the peripheral blood B cell count, altered the homing behavior of those B cells that were present, decreased the germinal center response, and impaired the antibody responses (Iskander et al., 2006). In addition, these mice exhibit decreased expression of NF-κB, suppression of its activation, and altered chemokines and chemokine receptors. These changes were suggested to lead to the deficiency in B cell numbers. Alterations in B cell homing behavior and impaired humoral immune response also were observed in this study (Iskander et al., 2006).

Because of the very low levels of tissue melatonin, it is more likely that, rather than acting directly, its ability to regulate immune function and act as an antioxidant is based on a cascade of magnification. As indicated earlier, this is likely achieved by way of activation of specific receptors, leading to inhibition or stimulation of transcription factors, and thence to altered expression of crucial genes relevant to immune function and antioxidant enzymes. There is considerable evidence for such a postulated trajectory. This comprises demonstration of inhibition by melatonin of activation of proinflammatory transcription factors such as nuclear factor kappa B (Lowes et al., 2011) and inflammatory kinases such as JNK (De Filippis et al., 2008) in LPS modeling of sepsis. This pathway can lead to downregulation of genes associated with inflammation and upregulation of genes for antioxidant enzymes (Sharman et al., 2007, 2008; García et al., 2010; Laothong et al., 2010).

7.8.2 Antioxidant and Anti-Inflammatory Mechanisms

The protective mechanisms against sepsis appear to involve both antioxidant (Li Volti et al., 2012) and anti-inflammatory properties of melatonin (Erbaş et al., 2013). Other potential mechanisms may include the ability of melatonin to increase plasma albumin levels (Oz and Ilhan, 2006; El-Missiry et al., 2007). Low albumin is a mortality risk factor for elderly septic patients. Melatonin may also slow the translocation of bacteria between various organ systems, retarding the development of generalized infection (Akcan et al., 2008).

Bacterially generated LPS induces iNOS synthesis in the host species, resulting in the production of large amounts of NO that are toxic not only to the invading pathogens, but also to the host cells by the inactivation of enzymes, leading to cell death (McCann et al., 1998). Melatonin significantly attenuates the LPS-induced upregulation of both cyclooxygenase and iNOS in RAW264.7 macrophages (Xia et al., 2012).

Melatonin treatment blunts the induction of mitochondrial iNOS isoforms after sepsis and thus protects against the ensuing impaired mitochondrial function. Since heart mitochondria from $iNOS^{-/-}$ mice are unaffected during sepsis, the induction of mitochondrial iNOS is associated with sepsis-related mitochondrial dysfunction (Escames et al., 2007). Similarly, in muscle mitochondria from $iNOS^{-/-}$ mice, ATP production was unaffected by CLP sepsis, and melatonin restored the reduced production of ATP induced by sepsis in $iNOS^{+/+}$ animals (López et al., 2006).

Myeloperoxidase is a key participant in the microbicidal *oxidative burst* generated by neutrophils upon encountering invading pathogenic organisms. Its generation of reactive oxidative species is an important contributor to the inflammatory response to infection, but this response must soon be suppressed in order to avoid causing the excessive oxidative damage to the host associated with severe sepsis and septic shock. The importance of this suppression may be inferred from the increased survival of myeloperoxidase-null mice, following sepsis-induced lung injury (Brovkovych et al., 2008). Melatonin, at physiologically and pharmacologically meaningful concentrations, inhibits myeloperoxidase (Galijasevic et al., 2008); hence, its inhibitory action may contribute to the process of resolving inflammation following sepsis.

7.9 CONCLUSIONS

A distinctive feature of melatonin is the subtlety of its actions. It is not a broad-spectrum anti-inflammatory and immune-quenching agent. Neither is it a potent nonspecific antioxidant. Its low content within cells suggests actions through receptor-linked transcription factors, followed by selective modulation of gene expression. This property allows a more refined approach to the treatment of sepsis.

The mortality associated with sepsis is unacceptably high, and despite substantial effort, survival has improved little over the past few decades. During this time, numerous studies have shown melatonin to have beneficial effects in sepsis and in sepsis-related conditions. These studies have been conducted utilizing a wide variety of systems: in vitro cell cultures, tissues, and whole animals exposed to endotoxin or to live bacteria. These studies reported many kinds of measurements: indicators of immune function, inflammation, and oxidative damage, changes in cell and tissue morphology, changes in animal gene expression patterns, and in animal survival. This large body of evidence, derived both from human and animal studies for over a decade, strongly suggests that melatonin may have significant utility in the treatment of sepsis. No significant toxicity due to melatonin—even at high pharmaceutical dosages—has accompanied the uniformly positive effects reported; moreover, its cost is minimal (Bondy and Sharman, 2010). The results of these many laboratory studies have been confirmed by the results of the single limited clinical trial undertaken to date, in human newborns. It is indeed surprising that only a few human clinical trials have been conducted.

Taken as a whole, results of these studies suggest that, while in no sense should melatonin be considered as a cure or primary treatment for sepsis, incorporation of its administration in standard treatment protocols offers the possibility, both of lowering the mortality due to sepsis and of reducing the long-term damage often inflicted on the survivors' immune systems.

The intent of this chapter is to draw attention to this potential in the hope that more appropriate human trials will be performed. The potential benefits of melatonin therapy are balanced by minimal risks, and it would be tragic if its utility were overlooked for insufficient transmission of information.

In view of the considerable body of evidence supporting the potential value of melatonin, it may be asked why the approach has not received more attention. Some of the answers to this may involve the inexpensiveness and ready availability of melatonin, resulting in a lack of opportunity for commercial development and proprietary marketing by the pharmaceutical industry. However, this industry is currently developing melatonin analogs, which activate specific melatonin receptors. Nevertheless, despite the need for better understanding of the mechanisms underlying this hormone's effects during sepsis, melatonin is deserving of more detailed examination in a clinical situation than it has hitherto received.

REFERENCES

Acikel M, Buyukokuroglu ME, Aksoy H, Erdogan F, Erol MK. Protective effects of melatonin against myocardial injury induced by isoproterenol in rats. *J Pineal Res*. 2003;35:75–79.

Adam N, Kandelman S, Mantz J, Chrétien F, Sharshar T. Sepsis-induced brain dysfunction. *Expert Rev Anti Infect Ther*. 2013;11:211–221.

Akbulut KG, Gonül B, Akbulut H. Exogenous melatonin decreases age-induced lipid peroxidation in the brain. *Brain Res*. 2008;1238:31–35.

Akcan A, Kucuk C, Sozuer E, Esel D, Akyildiz H, Akgun H, Muhtaroglu S, Aritas Y. Melatonin reduces bacterial translocation and apoptosis in trinitrobenzenesulphonic acid-induced colitis of rats. *World J Gastroenterol*. 2008;14:918–924.

Angus DC, Linde-Zwirble WT, Lidicker J, Clermont G, Carcillo J, Pinsky MR. Epidemiology of severe sepsis in the United States: Analysis of incidence, outcome, and associated costs of care. *Crit Care Med*. 2001;29:1303–1310.

Annane D, Bellissant E, Cavaillon JM. Septic shock. *Lancet*. 2005;365:63–78.

Astiz ME, Rackow EC. Septic shock. *Lancet*. 1998;351:1501–1505.

Bagci S, Yildizdas D, Horoz OO, Reinsberg J, Bartmann P, Mueller A. Use of nocturnal melatonin concentration and urinary 6-sulfatoxymelatonin excretion to evaluate melatonin status in children with severe sepsis. *J Pediatr Endocrinol Metab*. 2011;24:1025–1030.

Bakker J, Grover R, McLuckie A, Holzapfel L, Andersson J, Lodato R, Watson D, Grossman S, Donaldson J, Takala J. Administration of the nitric oxide synthase inhibitor NG-methyl-L-arginine hydrochloride (546C88) by intravenous infusion for up to 72 hours can promote the resolution of shock in patients with severe sepsis: Results of a randomized, double-blind, placebo-controlled multicenter study (study no. 144-002). *Crit Care Med*. 2004;32:1–12.

Bauer M, Press AT, Trauner M. The liver in sepsis: Patterns of response and injury. *Curr Opin Crit Care*. 2013;19:123–127.

Bekyarova G, Tancheva S, Hristova M. The effects of melatonin on burn-induced inflammatory responses and coagulation disorders in rats. *Methods Find Exp Clin Pharmacol*. 2010;32:299–303.

Böhler T, Canivet C, Nguyen PN, Galvani S, Thomsen M, Durand D, Salvayre R, Negre-Salvayre A, Rostaing L, Kamar N. Cytokines correlate with age in healthy volunteers, dialysis patients and kidney-transplant patients. *Cytokine*. 2009;45:169–173.

Bondy SC, Lahiri DK, Perreau VM, Sharman KZ, Campbell A, Zhou J, Sharman EH. Retardation of brain aging by chronic treatment with melatonin. *Ann NY Acad Sci*. 2004;1035:197–215.

Bondy SC, Sharman EH. 2010. Melatonin, oxidative stress and the aging brain. In *Oxidative Stress in Basic Research and Clinical Practice: Aging and Age-Related Disorders*, SC Bondy and K Maiese, eds., pp. 339–357. Totowa, NJ: Humana Press.

Brovkovych V, Gao XP, Ong E, Brovkovych S, Brennan ML, Su X, Hazen SL, Malik AB, Skidgel RA. Augmented inducible nitric oxide synthase expression and increased NO production reduce sepsis-induced lung injury and mortality in myeloperoxidase-null mice. *Am J Physiol Lung Cell Mol Physiol*. 2008;295:L96–L103.

Brzezinski A. Melatonin in humans. *N Engl J Med*. 1997;336:186–195.

Bubenik GA. Localization, physiological significance and possible clinical implication of gastrointestinal melatonin. *Biol Signals Recept*. 2001;10:350–366.

Bubenik GA. Gastrointestinal melatonin: Localization, function, and clinical relevance. *Dig Dis Sci*. 2002;47:2336–2348.

Buckley CD, Gilroy DW, Serhan CN, Stockinger B, Tak PP. The resolution of inflammation. *Nat Rev Immunol*. 2013;13:59–66.

Cagnacci A, Cannoletta M, Renzi A, Baldassari F, Arangino S, Volpe A. Prolonged melatonin administration decreases nocturnal blood pressure in women. *Am J Hypertens*. 2005;18:1614–1618.

Carrillo-Vico A, Lardone PJ, Alvarez-Sánchez N, Rodríguez-Rodríguez A, Guerrero JM. Melatonin: Buffering the immune system. *Int J Mol Sci*. 2013;14:8638–8683.

Carrillo-Vico A, Lardone PJ, Naji L, Fernández-Santos JM, Martin-Lacave I, Guerrero JM, Calvo JR. Beneficial pleiotropic actions of melatonin in an experimental model of septic shock in mice: Regulation of pro-/anti-inflammatory cytokine network, protection against oxidative damage and anti-apoptotic effects. *J Pineal Res*. 2005;39:400–408.

Celinski K, Konturek PC, Konturek SJ, Slomka M, Cichoz-Lach H, Brzozowski T, Bielanski W. Effects of melatonin and tryptophan on healing of gastric and duodenal ulcers with *Helicobacter pylori* infection in humans. *J Physiol Pharmacol*. 2011;62:521–526.

Cevík H, Erkanli G, Ercan F, Işman CA, Yeğen BC. Exposure to continuous darkness ameliorates gastric and colonic inflammation in the rat: Both receptor and non-receptor-mediated processes. *J Gastroenterol Hepatol*. 2005;20:294–303.

Chan KH, Wong YH. A molecular and chemical perspective in defining melatonin receptor subtype selectivity. *Int J Mol Sci*. 2013;14:18385–18406.

Chojnacki C, Wisniewska-Jarosinska M, Walecka-Kapica E, Klupinska G, Jaworek J, Chojnacki J. Evaluation of melatonin effectiveness in the adjuvant treatment of ulcerative colitis. *J Physiol Pharmacol*. 2011;62:327–334.

Cichoz-Lach H, Celinski K, Konturek PC, Konturek SJ, Slomka M. The effects of L-tryptophan and melatonin on selected biochemical parameters in patients with steatohepatitis. *J Physiol Pharmacol*. 2010;61:577–580.

Court O, Kumar A, Parrillo JE, Kumar A. Clinical review: Myocardial depression in sepsis and septic shock. *Crit Care*. 2002;6:500–508.

Cuesta S, Kireev R, García C, Rancan L, Vara E, Tresguerres JA. Melatonin can improve insulin resistance and aging-induced pancreas alterations in senescence-accelerated prone male mice (SAMP8). *Age (Dordr)*. 2013;35:659–671.

De Filippis D, Iuvone T, Esposito G, Steardo L, Arnold GH, Paul AP, De Man Joris G, De Winter Benedicte Y. Melatonin reverses lipopolysaccharide-induced gastro-intestinal motility disturbances through the inhibition of oxidative stress. *J Pineal Res*. 2008;44:45–51.

De Gaudio AR, Rinaldi S, Chelazzi C, Borracci T. Pathophysiology of sepsis in the elderly: Clinical impact and therapeutic considerations. *Curr Drug Targets*. 2009;10:60–70.

d'Emmanuele di Villa Bianca R, Marzocco S, Di Paola R, Autore G, Pinto A, Cuzzocrea S, Sorrentino R. Melatonin prevents lipopolysaccharide-induced hyporeactivity in rat. *J Pineal Res*. 2004;36:146–154.

Delano FA, Hoyt DB, Schmid-Schönbein GW. Pancreatic digestive enzyme blockade in the intestine increases survival after experimental shock. *Sci Transl Med*. 2013;5:169ra11.

Deng WG, Tang ST, Tseng HP, Wu KK. Melatonin suppresses macrophage cyclooxygenase-2 and inducible nitric oxide synthase expression by inhibiting p52 acetylation and binding. *Blood*. 2006;108:518–524.

Dollins AB, Zhdanova IV, Wurtman RJ, Lynch HJ, Deng MH. Effect of inducing nocturnal serum melatonin concentrations in daytime on sleep, mood, body temperature, and performance. *Proc Natl Acad Sci US A*. 1994;91:1824–1828.

Drazen DL, Nelson RJ. Melatonin receptor subtype MT2 (Mel 1b) and not mt1 (Mel 1a) is associated with melatonin-induced enhancement of cell-mediated and humoral immunity. *Neuroendocrinology*. 2001;74:178–184.

El-Missiry MA, Fayed TA, El-Sawy MR, El-Sayed AA. Ameliorative effect of melatonin against gamma-irradiation-induced oxidative stress and tissue injury. *Ecotoxicol Environ Saf*. 2007;66:278–286.

Erbaş O, Ergenoglu AM, Akdemir A, Yeniel AO, Taskiran D. Comparison of melatonin and oxytocin in the prevention of critical illness polyneuropathy in rats with experimentally induced sepsis. *J Surg Res*. 2013;183:313–320.

Escames G, Leon J, Macias M, Khaldy H, Acuña-Castroviejo D. Melatonin counteracts lipopolysaccharide-induced expression and activity of mitochondrial nitric oxide synthase in rats. *FASEB J*. 2003;17:932–934.

Escames G, López LC, Ortiz F, López A, García JA, Ros E, Acuña-Castroviejo D. Attenuation of cardiac mitochondrial dysfunction by melatonin in septic mice. *FEBS J*. 2007;274:2135–2147.

Escames G, López LC, Ortiz F, Ros E, Acuña-Castroviejo D. Age-dependent lipopolysaccharide-induced iNOS expression and multiorgan failure in rats: Effects of melatonin treatment. *Exp Gerontol*. 2006;41:1165–1173.

Espino J, Pariente JA, Rodríguez AB. Oxidative stress and immunosenescence: Therapeutic effects of melatonin. *Oxid Med Cell Longev*. 2012;2012:670294.

Eşrefoğlu M, Iraz M, Ates B, Gul M. Melatonin and CAPE are able to prevent the liver from oxidative damage in rats: An ultrastructural and biochemical study. *Ultrastruct Pathol*. 2012;36:171–178.

Fagundes DS, Gonzalo S, Arruebo MP, Plaza MA, Murillo MD. Melatonin and trolox ameliorate duodenal LPS-induced disturbances and oxidative stress. *Dig Liver Dis*. 2010;42:40–44.

Fernández BE, Díaz E, Fernández C, Núñez P, Díaz B. Ovarian aging: Melatonin regulation of the cytometric and endocrine evolutive pattern. *Curr Aging Sci*. 2013;6:1–7.

Fjaerli O, Lund T, Osterud B. The effect of melatonin on cellular activation processes in human blood. *J Pineal Res*. 1999;26:50–55.

Forman K, Vara E, García C, Kireev R, Cuesta S, Escames G, Tresguerres JA. Effect of a combined treatment with growth hormone and melatonin in the cardiological aging on male SAMP8 mice. *J Gerontol A Biol Sci Med Sci*. 2011;66:823–834.

Fourrier F. Severe sepsis, coagulation, and fibrinolysis: Dead end or one way? *Crit Care Med*. 2012;40:2704–2708.

Franchini M, Lippi G, Manzato F. Recent acquisitions in the pathophysiology, diagnosis and treatment of disseminated intravascular coagulation. *Thromb J*. 2006;4:4.

Friese RS, Bruns B, Sinton CM. Sleep deprivation after septic insult increases mortality independent of age. *J Trauma*. 2009;66:50–54.

Galijasevic S, Abdulhamid I, Abu-Soud HM. Melatonin is a potent inhibitor for myeloperoxidase. *Biochemistry*. 2008;47:2668–2677.

García T, Esparza JL, Giralt M, Romeu M, Domingo JL, Gómez M. Protective role of melatonin on oxidative stress status and RNA expression in cerebral cortex and cerebellum of AbetaPP transgenic mice after chronic exposure to aluminum. *Biol Trace Elem Res*. 2010;135:220–232.

Gitto E, Karbownik M, Reiter RJ, Tan DX, Cuzzocrea S, Chiurazzi P, Cordaro S, Corona G, Trimarchi G, Barberi I. Effects of melatonin treatment in septic newborns. *Pediatr Res*. 2001;50:756–760.

Gitto E, Reiter RJ, Amodio A, Romeo C, Cuzzocrea E, Sabatino G, Buonocore G, Cordaro V, Trimarchi G, Barberi I. Early indicators of chronic lung disease in preterm infants with respiratory distress syndrome and their inhibition by melatonin. *J Pineal Res*. 2004;36:250–255.

Gofton TE, Young GB. Sepsis-associated encephalopathy. *Nat Rev Neurol*. 2012;8:557–566.

Hardeland R, Cardinali DP, Srinivasan V, Spence DW, Brown GM, Pandi-Perumal SR. Melatonin—A pleiotropic, orchestrating regulator molecule. *Prog Neurobiol*. 2011;93:350–384.

Hsu JT, Kuo CJ, Chen TH, Wang F, Lin CJ, Yeh TS, Hwang TL, Jan YY. Melatonin prevents hemorrhagic shock-induced liver injury in rats through an Akt-dependent HO-1 pathway. *J Pineal Res*. 2012;53:410–416.

Hunter JD, Doddi M. Sepsis and the heart. *Br J Anaesth*. 2010;104:3–11.

Iscimen R, Cartin-Ceba R, Yilmaz M, Khan H, Hubmayr RD, Afessa B, Gajic O. Risk factors for the development of acute lung injury in patients with septic shock: An observational cohort study. *Crit Care Med*. 2008;36:1518–1522.

Iskander K, Li J, Han S, Zheng B, Jaiswal AK. NQO1 and NQO2 regulation of humoral immunity and autoimmunity. *J Biol Chem*. 2006;281:30917–30924.

Jones AE. Evidence-based therapies for sepsis care in the emergency department: Striking a balance between feasibility and necessity. *Acad Emerg Med*. 2006;13:82–83.

Kennaway DJ, Stamp GE, Goble FC. Development of melatonin production in infants and the impact of prematurity. *J Clin Endocrinol Metab*. 1992;75:367–369.

Kireev RA, Cuesta S, Ibarrola C, Bela T, Moreno Gonzalez E, Vara E, Tresguerres JA. Age-related differences in hepatic ischemia/reperfusion: Gene activation, liver injury, and protective effect of melatonin. *J Surg Res*. 2012;178:922–934.

Königsrainer I, Türck MH, Eisner F, Meile T, Hoffmann J, Küper M, Zieker D, Glatzle J. The gut is not only the target but a source of inflammatory mediators inhibiting gastrointestinal motility during sepsis. *Cell Physiol Biochem*. 2011;28:753–760.

Kostovski E, Dahm AE, Iversen N, Hjeltnes N, Østerud B, Sandset PM, Iversen PO. Melatonin stimulates release of tissue factor pathway inhibitor from the vascular endothelium. *Blood Coagul Fibrinolysis*. 2011;22(4):254–259.

Kücükakin B, Lykkesfeldt J, Nielsen HJ, Reiter RJ, Rosenberg J, Gögenur I. Utility of melatonin to treat surgical stress after major vascular surgery—A safety study. *J Pineal Res*. 2008;44:426–431.

Lagu T, Rothberg MB, Shieh MS, Pekow PS, Steingrub JS, Lindenauer PK. Hospitalizations, costs, and outcomes of severe sepsis in the United States 2003 to 2007. *Crit Care Med*. 2012;40:754–761.

Lahiri DK, Ge Y-W, Sharman EH, Bondy SC. Age-related changes in serum melatonin in mice, higher levels of combined melatonin and melatonin sulfate in the brain cortex than serum, heart, liver and kidney tissues. *J Pineal Res*. 2004;36:217–223.

Laliena A, San Miguel B, Crespo I, Alvarez M, González-Gallego J, Tuñón MJ. Melatonin attenuates inflammation and promotes regeneration in rabbits with fulminant hepatitis of viral origin. *J Pineal Res*. 2012;53:270–278.

Lambeck S, Weber M, Gonnert FA, Mrowka R, Bauer M. Comparison of sepsis-induced transcriptomic changes in a murine model to clinical blood samples identifies common response patterns. *Front Microbiol*. 2012;3:284.

Laothong U, Pinlaor P, Hiraku Y, Boonsiri P, Prakobwong S, Khoontawad J, Pinlaor S. Protective effect of melatonin against *Opisthorchis viverrini*-induced oxidative and nitrosative DNA damage and liver injury in hamsters. *J Pineal Res*. 2010;49:271–282.

Latronico N, Bolton CF. Critical illness polyneuropathy and myopathy: A major cause of muscle weakness and paralysis. *Lancet Neurol*. 2011;10:931–941.

Leong J, Zhou M, Jacob A, Wang P. Aging-related hyperinflammation in endotoxemia is mediated by the alpha2A-adrenoceptor and CD14/TLR4 pathways. *Life Sci*. 2010;86:740–746.

Li CX, Liang DD, Xie GH, Cheng BL, Chen QX, Wu SJ, Wang JL, Cho W, Fang XM. Altered melatonin secretion and circadian gene expression with increased proinflammatory cytokine expression in early-stage sepsis patients. *Mol Med Rep*. 2013;7:1117–1122.

Li JH, Zhou W, Liu K, Li HX, Wang L. Melatonin reduces the expression of chemokines in rat with trinitrobenzene sulfonic acid-induced colitis. *Saudi Med J*. 2008;29:1088–1094.

Li Volti G, Musumeci T, Pignatello R, Murabito P, Barbagallo I, Carbone C, Gullo A, Puglisi G. Antioxidant potential of different melatonin-loaded nanomedicines in an experimental model of sepsis. *Exp Biol Med (Maywood)*. 2012;237:670–677.

Lin XJ, Mei GP, Liu J, Li YL, Zuo D, Liu SJ, Zhao TB, Lin MT. Therapeutic effects of melatonin on heatstroke-induced multiple organ dysfunction syndrome in rats. *J Pineal Res*. 2011;50:436–444.

López A, Lorente JA, Steingrub J, Bakker J, McLuckie A, Willatts S, Brockway M et al. Multiple-center, randomized, placebo-controlled, double-blind study of the nitric oxide synthase inhibitor 546C88: Effect on survival in patients with septic shock. *Crit Care Med.* 2004;32:21–30.

López LC, Escames G, Ortiz F, Ros E, Acuña-Castroviejo D. Melatonin restores the mitochondrial production of ATP in septic mice. *Neuroendocrinol Lett.* 2006;27:623–630.

Löwenberg M, Verhaar AP, van den Brink GR, Hommes DW. Glucocorticoid signaling: A nongenomic mechanism for T-cell immunosuppression. *Trends Mol Med.* 2007;13:158–163.

Lowes DA, Almawash AM, Webster NR, Reid VL, Galley HF. Melatonin and structurally similar compounds have differing effects on inflammation and mitochondrial function in endothelial cells under conditions mimicking sepsis. *Br J Anaesth.* 2011;107:193–201.

Lowes DA, Webster NR, Murphy MP, Galley HF. Antioxidants that protect mitochondria reduce interleukin-6 and oxidative stress, improve mitochondrial function, and reduce biochemical markers of organ dysfunction in a rat model of acute sepsis. *Br J Anaesth.* 2013;110:472–480.

Mallo C, Zaïdan R, Galy G, Vermeulen E, Brun J, Chazot G, Claustrat B. Pharmacokinetics of melatonin in man after intravenous infusion and bolus injection. *Eur J Clin Pharmacol.* 1990;38:297–301.

Matsukawa A, Kaplan MH, Hogaboam CM, Lukacs NW, Kunkel SL. Pivotal role of signal transducer and activator of transcription (Stat)4 and Stat6 in the innate immune response during sepsis. *Exp Med.* 2001;193:679–688.

McCann SM, Licinio J, Wong ML, Yu WH, Karanth S, Rettorri V. The nitric oxide hypothesis of aging. *Exp Geront.* 1998;33:813–826.

McConnell KW, Fox AC, Clark AT, Chang NY, Dominguez JA, Farris AB, Buchman TG, Hunt CR, Coopersmith CM. The role of heat shock protein 70 in mediating age-dependent mortality in sepsis. *J Immunol.* 2011;186:3718–3725.

Merchant NM, Azzopardi DV, Hawwa AF, McElnay JC, Middleton B, Arendt J, Arichi T, Gressens P, Edwards AD. Pharmacokinetics of melatonin in preterm infants. *Br J Clin Pharmacol.* 2013;76:725–733.

Moore LJ, Moore FA. Epidemiology of sepsis in surgical patients. *Surg Clin North Am.* 2012;92:1425–1443.

Nordmann P, Poirel L, Toleman MA, Walsh TR. Does broad-spectrum beta-lactam resistance due to NDM-1 herald the end of the antibiotic era for treatment of infections caused by Gram-negative bacteria? *J Antimicrob Chemother.* 2011;66:689–692.

Ochoa JJ, Díaz-Castro J, Kajarabille N, García C, Guisado IM, De Teresa C, Guisado R. Melatonin supplementation ameliorates oxidative stress and inflammatory signaling induced by strenuous exercise in adult human males. *J Pineal Res.* 2011;51:373–380.

Oliveira LG, Kuehn CC, Santos CD, Miranda MA, da Costa CM, Mendonça VJ, do Prado JC Jr. Protective actions of melatonin against heart damage during chronic Chagas disease. *Acta Trop.* 2013;128:652–658.

Opal SM, Garber GE, LaRosa SP, Maki DG, Freebairn RC, Kinasewitz GT, Dhainaut JF et al. Systemic host responses in severe sepsis analyzed by causative microorganism and treatment effects of drotrecogin alfa (activated). *Clin Infect Dis.* 2003;37:50–58.

Opal SM, Girard TD, Ely EW. The immunopathogenesis of sepsis in elderly patients. *Clin Infect Dis.* 2005;41(Suppl 7):S504–S512.

Ortiz F, García JA, Acuña-Castroviejo D, Doerrier C, López A, Venegas C, Volt H, Luna-Sánchez M, López LC, Escames G. The beneficial effects of melatonin against heart mitochondrial impairment during sepsis: Inhibition of iNOS and preservation of nNOS. *J Pineal Res.* 2014;56(1):71–81. doi: 10.1111/jpi.12099.

Oz E, Ilhan MN. Effects of melatonin in reducing the toxic effects of doxorubicin. *Mol Cell Biochem.* 2006;286:11–15.

Ozdemir D, Uysal N, Tugyan K, Gonenc S, Acikgoz O, Aksu I, Ozkan H. The effect of melatonin on endotoxemia-induced intestinal apoptosis and oxidative stress in infant rats. *Intensive Care Med.* 2007; 33:511–516.

Park SW, Choi SM, Lee SM. Effect of melatonin on altered expression of vasoregulatory genes during hepatic ischemia/reperfusion. *Arch Pharm Res.* 2007;30:1619–1624.

Pascua P, Camello-Almaraz C, Pozo MJ, Martin-Cano FE, Vara E, Fernández-Tresguerres JA, Camello PJ. Aging-induced alterations in female rat colon smooth muscle: The protective effects of hormonal therapy. *J Physiol Biochem.* 2012;68:255–262.

Paskaloğlu K, Sener G, Kapucu C, Ayanoğlu-Dülger G. Melatonin treatment protects against sepsis-induced functional and biochemical changes in rat ileum and urinary bladder. *Life Sci.* 2004;74:1093–1104.

Paulis L, Pechanova O, Zicha J, Liskova S, Celec P, Mullerova M, Kollar J et al. Melatonin improves the restoration of endothelium-derived constricting factor signalling and inner diameter in the rat femoral artery after cessation of L-NAME treatment. *J Hypertens.* 2010;28(Suppl 1):S19–S24.

Perreau VM, Cotman CW, Sharman KG, Bondy SC, Sharman EH. Melatonin treatment in old mice enables a more youthful response to LPS in the brain. *J Neuroimmunol*. 2007;182:22–31.

Petrovsky N, Harrison LC. Diurnal rhythmicity of human cytokine production: A dynamic disequilibrium in T helper cell type 1/T helper cell type 2 balance? *J Immunol*. 1997;158:5163–5168.

Pierpaoli W, Regelson W. Pineal control of aging: Effect of melatonin and pineal grafting on aging mice. *Proc Natl Acad Sci USA*. 1994;91:787–791.

Radogna F, Diederich M, Ghibelli L. Melatonin: A pleiotropic molecule regulating inflammation. *Biochem Pharmacol*. 2010;80:1844–1852.

Ramírez-Rodríguez G, Vega-Rivera NM, Benítez-King G, Castro-García M, Ortíz-López L. Melatonin supplementation delays the decline of adult hippocampal neurogenesis during normal aging of mice. *Neurosci Lett*. 2012;530:53–58.

Reiter RJ, Tan DX, Manchester LC, Pilar Terron M, Flores LJ, Koppisepi S. Medical implications of melatonin: Receptor-mediated and receptor-independent actions. *Adv Med Sci*. 2007;52:11–28.

Renzi A, Glaser S, Demorrow S, Mancinelli R, Meng F, Franchitto A, Venter J et al. Melatonin inhibits cholangiocyte hyperplasia in cholestatic rats by interaction with MT1 but not MT2 melatonin receptors. *Am J Physiol Gastrointest Liver Physiol*. 2011;301:G634–G643.

Requintina PJ, Oxenkrug GF. Differential effects of lipopolysaccharide on lipid peroxidation in F344N, SHR rats and BALB/c mice, and protection of melatonin and NAS against its toxicity. *Ann NY Acad Sci*. 2003;993:325–333.

Reynolds FD, Dauchy R, Blask D, Dietz PA, Lynch D, Zuckerman R. The pineal gland hormone melatonin improves survival in a rat model of sepsis/shock induced by zymosan A. *Surgery*. 2003;134:474–479.

Robertson NJ, Faulkner S, Fleiss B, Bainbridge A, Andorka C, Price D, Powell E et al. Melatonin augments hypothermic neuroprotection in a perinatal asphyxia model. *Brain*. 2013;136:90–105.

Rudiger A, Singer M. Mechanisms of sepsis-induced cardiac dysfunction. *Crit Care Med*. 2007;35:1599–1608.

Sahib AS, Al-Jawad FH, Alkaisy AA. Effect of antioxidants on the incidence of wound infection in burn patients. *Ann Burns Fire Disasters*. 2010;23:199–205.

Santello FH, Frare EO, dos Santos CD, Caetano LC, Alonso Toldo MP, do Prado JC Jr. Suppressive action of melatonin on the TH-2 immune response in rats infected with *Trypanosoma cruzi*. *J Pineal Res*. 2008;45:291–296.

Schmid-Schönbein GW. Landis Award lecture. Inflammation and the autodigestion hypothesis. *Microcirculation*. 2009;16:289–306.

Selmaoui B, Lambrozo J, Touitou Y. Magnetic fields and pineal function in humans: Evaluation of nocturnal acute exposure to extremely low frequency magnetic fields on serum melatonin and urinary 6-sulfatoxymelatonin circadian rhythms. *Life Sci*. 1996;58:1539–1549.

Şener G, Toklu H, Kapucu C, Ercan F, Erkanli G, Kaçmaz A, Tilki M, Yeğen BC. Melatonin protects against oxidative organ injury in a rat model of sepsis. *Surg Today*. 2005;35:52–59.

Seok J, Warren HS, Cuenca AG et al. Genomic responses in mouse models poorly mimic human inflammatory diseases. *Proc Natl Acad Sci USA*. 2013;110:3507–3512.

Sessler CN. Steroids for septic shock: Back from the dead? (Con). *Chest*. 2003;123(5 Suppl):482S–489S.

Shang Y, Xu SP, Wu Y, Jiang YX, Wu ZY, Yuan SY, Yao SL. Melatonin reduces acute lung injury in endotoxemic rats. *Chin Med J*. 2009;122:1388–1393.

Sharman EH, Bondy SC, Sharman KZ., Lahiri D, Cotman CW, Perreau VM. Effects of melatonin and age on gene expression in mouse CNS using microarray analysis. *Neurochem Int*. 2007;50:336–344.

Sharman EH, Sharman KG, Bondy SC. Melatonin causes gene expression in aged animals to respond to inflammatory stimuli in a manner differing from that of young animals. *Curr Aging Sci*. 2008;1:152–158.

Sharman EH, Sharman KG, Bondy SC. Extended exposure to dietary melatonin reduces tumor number and size in aged male mice. *Exp Gerontol*. 2011;46:18–22.

Sprung CL, Peduzzi PN, Shatney CH, Schein RM, Wilson MF, Sheagren JN, Hinshaw LB. Impact of encephalopathy on mortality in the sepsis syndrome. The Veterans Administration Systemic Sepsis Cooperative Study Group. *Crit Care Med*. 1990;18:801–806.

Srinivasan V, Mohamed M, Kato H. Melatonin in bacterial and viral infections with focus on sepsis: A review. *Recent Pat Endocr Metab Immune Drug Discov*. 2012;6:30–39.

Strunk T, Currie A, Richmond P, Simmer K, Burgner D. Innate immunity in human newborn infants: Prematurity means more than immaturity. *J Matern Fetal Neonatal Med*. 2011;24:25–31.

Sugitharini V, Prema A, Thangam EB. Inflammatory mediators of systemic inflammation in neonatal sepsis. *Inflamm Res*. 2013;62:1025–1034.

Tekbas OF, Ogur R, Korkmaz A, Kilic A, Reiter RJ. Melatonin as an antibiotic: New insights into the actions of this ubiquitous molecule. *J Pineal Res*. 2008;44:222–226.

Turnbull IR, Clark AT, Stromberg PE, Dixon DJ, Woolsey CA, Davis CG, Hotchkiss RS, Buchman TG, Coopersmith CM. Effects of aging on the immunopathologic response to sepsis. *Crit Care Med*. 2009;37:1018–1023.

van der Poll T, Opal SM. Host-pathogen interactions in sepsis. *Lancet Infect Dis*. 2008;8:32–43.

Vincent JL. 2007. Setting the scene. In *Mechanisms of Sepsis-Induced Organ Dysfunction and Recovery*, E Abraham and M Singer, eds., p. 1. Berlin, Germany: Springer-Verlag.

Waldhauser F, Waldhauser M, Lieberman HR, Deng MH, Lynch HJ, Wurtman RJ. Bioavailability of oral melatonin in humans. *Neuroendocrinology*. 1984;39:307–313.

Waldhauser F, Weiszenbacher G, Tatzer E, Gisinger B, Waldhauser M, Schemper M, Frisch H. Alterations in nocturnal serum melatonin levels in humans with growth and aging. *J Clin Endocrinol Metab*. 1988;66:648–652.

Wang L, Quan J, Johnston WE, Maass DL, Horton JW, Thomas JA, Tao W. Age-dependent differences of interleukin-6 activity in cardiac function after burn complicated by sepsis. *Burns*. 2010;36:232–238.

Wang M, Molin G, Ahrné S, Adawi D, Jeppsson B. High proportions of proinflammatory bacteria on the colonic mucosa in a young patient with ulcerative colitis as revealed by cloning and sequencing of 16S rRNA genes. *Dig Dis Sci*. 2007;52:620–627.

Wichmann MW, Haisken JM, Ayala A, Chaudry IH. Melatonin administration following hemorrhagic shock decreases mortality from subsequent septic challenge. *J Surg Res*. 1996;65:109–114.

Wu CC, Chiao CW, Hsiao G, Chen A, Yen MH. Melatonin prevents endotoxin-induced circulatory failure in rats. *J Pineal Res*. 2001;30:147–156.

Wu JY, Tsou MY, Chen TH, Chen SJ, Tsao CM, Wu CC. Therapeutic effects of melatonin on peritonitis-induced septic shock with multiple organ dysfunction syndrome in rats. *J Pineal Res*. 2008;45:106–116.

Wu WS, Chou MT, Chao CM, Chang CK, Lin MT, Chang CP. Melatonin reduces acute lung inflammation, edema, and hemorrhage in heatstroke rats. *Acta Pharmacol Sin*. 2012;33:775–782.

Xia MZ, Liang YL, Wang H, Chen X, Huang YY, Zhang ZH, Chen YH et al. Melatonin modulates TLR4-mediated inflammatory genes through MyD88- and TRIF-dependent signaling pathways in lipopolysaccharide-stimulated RAW264.7 cells. *J Pineal Res*. 2012;53:325–334.

Yang Y, Yang KS, Hsann YM, Lim V, Ong BC. The effect of comorbidity and age on hospital mortality and length of stay in patients with sepsis. *J Crit Care*. 2010;25:398–405.

Yavuz T, Kaya D, Behçet M, Ozturk E, Yavuz O. Effects of melatonin on *Candida* sepsis in an experimental rat model. *Adv Ther*. 2007;24:91–100.

Zampieri, FG, Park M, Machado FS, Azevedo LC. Sepsis-associated encephalopathy: Not just delirium. *Clinics*. 2011;66:1825–1831.

Zhang H, Liu D, Wang X, Chen X, Long Y, Chai W, Zhou X et al. Melatonin improved rat cardiac mitochondria and survival rate in septic heart injury. *J Pineal Res*. 2013;55:1–6.

Zhong LY, Yang ZH, Li XR, Wang H, Li L. Protective effects of melatonin against the damages of neuroendocrine-immune induced by lipopolysaccharide in diabetic rats. *Exp Clin Endocrinol Diabetes*. 2009;117:463–469.

8 Melatonin's Cardioprotective Role

Alberto Dominguez-Rodriguez and Pedro Abreu-Gonzalez

CONTENTS

8.1 INTRODUCTION

The clinical importance of circadian biological rhythms has been strengthened by a number of studies showing a circadian distribution of cardiovascular events such as myocardial infarction, complex arrhythmia, or sudden cardiac death (Dominguez-Rodriguez et al., 2009a). Acute coronary occlusion is the leading cause of morbidity and mortality in the Western world, and according to the World Health Organization, it will be the major cause of death in the world by the year 2020 (Lopez and Murray, 1998).

The recognition that thrombotic occlusion of a coronary artery results in a wave front of irreversible myocardial cell injury, extending from the subendocardium to the subepicardium in a time-dependent fashion, led to the introduction of reperfusion therapy for acute myocardial infarction (Reimer et al., 1997). Modalities for reperfusion include thrombolysis and percutaneous coronary intervention. Ischemia/reperfusion injury has been observed in each one of these situations (Moens et al., 2005).

Melatonin has a diverse functional repertoire with actions in essentially all organs, including the heart and other portions of the cardiovascular system (Dominguez-Rodriguez et al., 2010b, 2012a; Dominguez-Rodriguez, 2012). Melatonin, with respect to radical oxygen species (ROS) and radical nitrogen species (RNS), has cardioprotective properties via its direct free radical scavenging and its indirect antioxidant activities. Melatonin efficiently interacts with various reactive oxygen and reactive nitrogen species, and it also upregulates antioxidant enzymes and downregulates pro-oxidant enzymes. In addition, melatonin demonstrated blood pressure lowering, normalization of lipid profile, and anti-inflammatory properties. The lack of these cardioprotective effects due to insufficient melatonin levels might be associated with several cardiovascular pathologies including ischemic heart disease (Tengattini et al., 2008; Dominguez-Rodriguez, 2009b; Reiter et al., 2010a).

Melatonin may influence cardiovascular pathophysiology via both receptor-mediated and receptor-independent mechanisms (Dubocovich and Markowska, 2005; Tengattini et al., 2008). The classic melatonin membrane receptors (MT1 and MT2) are present in the heart and throughout the vascular system. Moreover, nuclear binding sites for melatonin exist. The receptor-independent actions of melatonin relate to its ability, and that of its metabolites, to function as antioxidants (Tan et al., 2007; Peyrot and Ducrocq, 2008).

8.2 MELATONIN AND CARDIAC ISCHEMIA/REPERFUSION IN ANIMAL STUDIES

Investigations have been published, confirming the beneficial effects of melatonin on the physiology and morphology of the hypoxic/reoxygenated heart (Tengattini et al., 2008). Kaneko et al. (2000) showed that, in isolated rat hearts subjected to 30 min ischemia and 30 min reperfusion, the infusion of melatonin (100 μM) significantly reduced the duration of ventricular tachycardia and ventricular fibrillation and restored ventricular function. Simultaneously, Lagneux et al. (2000) confirmed that melatonin's free radical scavenging activity reduced both the abnormal cardiac physiology and infarct volume after ischemia/reperfusion and restored cardiac function. Kacmaz et al. (2005) also confirmed that melatonin has a protective effect on ischemia-/reperfusion-induced oxidative cardiac damage.

Sahna et al. (2002, 2003) were the first to examine whether endogenous, physiologic concentrations of melatonin would change the outcome of such studies. In this case, rats were surgically pinealectomized to reduce the endogenous levels of melatonin and, 2 months later, they, along with the pineal-intact controls, were used in the studies of cardiac ischemia/reperfusion injury. When the left coronary artery was occluded for 7 min followed by 7 min of reperfusion, the degree of cardiac arrhythmia was significantly greater in the pinealectomized rats compared with the controls. Even more importantly, the incidence of mortality was 63% in rats lacking their pineal gland compared with only 25% in the pineal-intact rats after ischemia/reperfusion induction (Shana et al., 2002, 2003). These findings suggest that endogenous melatonin levels are protective to the heart during the episodes of hypoxia and reoxygenation. Moreover, Shana et al. (2005) assumed that some of melatonin's antioxidant actions are probably derived from its stimulatory effect on superoxide dismutase, glutathione peroxidase, glutathione reductase, and glucose-6-phosphate dehydrogenase, and its inhibitory action on inducible nitric oxide synthase.

A recent report by Grossini et al. (2011) documented that the administration of melatonin in the pig primarily increased the coronary blood flow and cardiac function through MT1/MT2 receptors and a β-adrenergic-mediated nitric oxide release. These findings add new information about the mechanisms through which melatonin physiologically modulates cardiovascular function and exerts cardioprotective effects.

8.3 MELATONIN AND CORONARY ARTERY DISEASE IN HUMAN STUDIES

Several studies show that humans with cardiovascular disease have noticeably lower circulating melatonin levels than age-matched subjects without significant cardiovascular deterioration (Dominguez-Rodriguez et al., 2012b)

The first clinical study in humans that demonstrates a relationship between melatonin and coronary artery disease was published by Brugger et al. (1995). They reported reduced levels of plasma melatonin measured at 02:00 h in coronary artery disease patients. Moreover, other investigators studied nocturnal urinary excretion of 6-sulfatoxymelatonin, the major melatonin metabolite, in patients with coronary artery disease, and they demonstrated a low melatonin production rate (Sakotnik et al., 1999; Girotti et al., 2000; Vijayasarathy et al., 2010). Moreover, Yaprak et al. (2003) demonstrated in patients with angiographically documented coronary artery disease, a decreased nocturnal melatonin synthesis and release. Similarly, patients suffering cardiac syndrome X have an attenuated nocturnal rise in serum melatonin levels related to that of age-matched individuals with no cardiac pathology (Altun et al., 2002). Likewise, Dominguez-Rodriguez et al. (2002) analyzed serum levels of melatonin and parameters of oxidative stress in patients with acute myocardial infarction and subjects with no evidence of coronary artery disease as controls. They demonstrated that acute myocardial infarction is associated with a nocturnal serum melatonin deficit as well as increased oxidative stress.

Of particular interest is the emerging prognostic role of melatonin in patients with coronary artery disease. In this respect, Dominguez-Rodriguez et al. (2006) demonstrated in patients with acute myocardial infarction that low nocturnal melatonin levels predict cardiac adverse events during 6-month follow-up. Zaslavskaya et al. (2004) studied melatonin effects on contractile myocardial function in patients with postmyocardial infarction and heart failure, assessed as stage II–III by New York Heart Association. They found melatonin associated with antianginal and anti-ischemic effects, indicating improvement of contractile function. Moreover, endogenous melatonin has been shown to play an important role in predicting left ventricular remodeling during the chronic phase postmyocardial infarction (Dominguez-Rodriguez et al., 2012c).

Importantly, patients with acute myocardial infarction undergoing primary percutaneous coronary intervention were found to have a relationship between intraplatelet melatonin and vascular damage in the coronary artery responsible for the infarct (Dominguez-Rodriguez et al., 2010a). Interestingly, a recent case–control study, carried out by Samimi-Fard et al. (2011), showed a significant association between single nucleotide polymorphisms (rs28383653) of MT1 and coronary artery disease.

Recent investigations in patients with acute myocardial infarction undergoing primary percutaneous coronary intervention confirmed a relationship between melatonin concentrations and ischemia-modified albumin, a biomarker of myocardial ischemia. These data suggest that melatonin can act as a potent antioxidant agent, reducing myocardial damage induced by ischemia/reperfusion (Dominguez-Rodriguez et al., 2008). The available scientific evidence has led Dominguez-Rodriguez and collaborators to initiate a phase II clinical trial (ClinicalTrials.gov no. NCT00640094). They are attempting to demonstrate an inhibition of ischemia/reperfusion damage after administration of intravenous melatonin in patients with acute myocardial infarction, immediately before primary percutaneous coronary intervention (Dominguez-Rodriguez et al., 2007). Moreover, other authors are testing whether intracoronary injection of melatonin can limit ischemia-/reperfusion-related myocardial damage (ClinicalTrials.gov no. NCT01172171).

The importance of these studies is emphasized by the fact that melatonin is quickly distributed throughout the organism via exogenous administration (oral, intravenous, or subcutaneous). Melatonin, a lipophilic molecule, crosses physiological membranes and enters cardiac cells with great ease. Highest intracellular concentrations of melatonin are found at the mitochondrial level. This is especially important, as the mitochondria is a major site of free radical generation and oxidative stress (Tengattini et al., 2008; Dominguez-Rodriguez et al., 2010a).

8.4 MELATONIN AND HYPERTENSIVE CARDIOMYOPATHY

Hypertensive heart disease is defined by the presence of pathologic left ventricular hypertrophy in the absence of a cause other than arterial hypertension. It is characterized by complex changes in myocardial structure, including enhanced cardiomyocyte growth and noncardiomyocyte alterations that induce the remodeling of the myocardium, and ultimately deteriorate the left ventricular function and facilitate the development of heart failure. It is now accepted that a number of pathological processes mediated by mechanical, neurohormonal, and cytokine routes acting on the cardiomyocyte and the noncardiomyocyte compartments are responsible for myocardial remodeling in the context of arterial hypertension (Gonzalez et al., 2012).

The evidence is convincing that melatonin is involved, at least in part, in the regulation of blood pressure (Paulis and Simko, 2007; Simko and Pechanova, 2009a,b). In particular, the nocturnal rise in serum endogenous melatonin levels is related to, or possibly responsible for, the nighttime decline in blood pressure (Enjuanes-Grau et al., 2012). Individuals who experience a melatonin-mediated drop in the nighttime systolic and diastolic pressures are referred to as "dippers," whereas those individuals who maintain daytime blood pressure levels in the night are referred to as "nondippers" (O'Brien et al., 1988; Scheer et al., 2004).

Reduced levels of melatonin have been found in the nocturnal serum of spontaneously hypertensive rats, and the administration of melatonin reduced blood pressure to normal range in these animals (Kawashima et al., 1984; Kawashima et al., 1987). In spontaneously hypertensive rats, blood pressure decreased after 6 weeks of melatonin treatment (10 mg/kg), which was associated with a reduction in interstitial renal tissue inflammation, decreased oxidative stress, and attenuation of proinflammatory transcription factors in the kidney (Nava et al., 2003).

Reduced levels of melatonin have also been found in subjects suffering from nondipper hypertension (Jonas et al., 2003). When 3 mg of melatonin is given to hypertensive patients 1 h before going to bed, improvements were apparent with the day–night rhythm of blood pressure, particularly in women with a blunted nocturnal decline (Cagnacci et al., 2005). Similarly, daily intake of 2.5 mg of melatonin in the night reduced blood pressure to normal range in male subjects with essential hypertension (Scheer et al., 2004).

Although melatonin did not prevent the development of left ventricular hypertrophy, this indolamine reduced hydroxyproline content and concentration in the left ventricle. This antifibrotic effect of melatonin was associated with a reduction of the oxidative load. Therefore, melatonin, presumably because of its antioxidant actions, is able to attenuate fibrosis in the hypertrophied left ventricle, which may be functionally desirable (Reiter et al., 2010b).

8.5 MELATONIN AND CARDIOVASCULAR ATHEROSCLEROSIS

Atherosclerosis is a chronic vascular disease in which inflammation and oxidative stress are commonly implicated as major causative factors. Early stages of plaque development involve endothelial activation induced by inflammatory cytokines, oxidized low-density lipoprotein, and/or changes in endothelial shear stress (Dominguez-Rodriguez et al., 2009c).

Several studies have investigated the antioxidant effect of melatonin on low-density lipoprotein oxidation. Melatonin also has been shown to depress plasma levels of total cholesterol and very low-density lipoprotein cholesterol as well as the low-density lipoprotein cholesterol subfraction in hypercholesterolemic rats (Tengattini et al., 2008). Melatonin may exert these effects by increasing endogenous cholesterol clearance. Because of its lipophilic nature, melatonin readily enters the lipid phase of the low-density lipoprotein particles and prevents lipid peroxidation. Dominguez-Rodriguez et al. (2005) showed an association between nocturnal elevated serum levels of oxidized low-density lipoprotein and reduced circulating melatonin levels in patients with acute myocardial infarction. These findings generally support the notion that melatonin may lower total cholesterol and stimulate high-density lipoprotein levels while reducing the oxidation of low-density lipoprotein, changes that would generally be protective against cardiovascular diseases (Tengattini et al., 2008).

8.6 CONCLUSIONS

Melatonin, an indole produced in several organs but most notably in the pineal gland, has a variety of effects that influence cardiac pathophysiology. The protective actions of melatonin at the level of the heart probably involve membrane melatonin receptors that exist on cardiomyocytes, in addition to the functions of melatonin as an antioxidant that are not receptor-mediated.

Melatonin obviously has a variety of beneficial effects with reference to cardiovascular pathophysiology, including in the treatment of hypertension, ischemia/reperfusion injury, and cardiac hypertrophy (Figure 8.1). The experimental data obtained from both animals and human studies suggest that melatonin may have therapeutic utility in the treatment of several cardiovascular conditions.

Melatonin has been administered in both physiological and pharmacological amounts to humans (Seabra et al., 2000) and animals, and there is widespread agreement that it is a nontoxic molecule. Given the severity of these conditions and the uncommonly low toxicity of melatonin, clinical trials using this indole are highly justified. Unless the findings in animal investigations are totally misleading, it seems likely that melatonin will have similar protective effects at the level of the human heart.

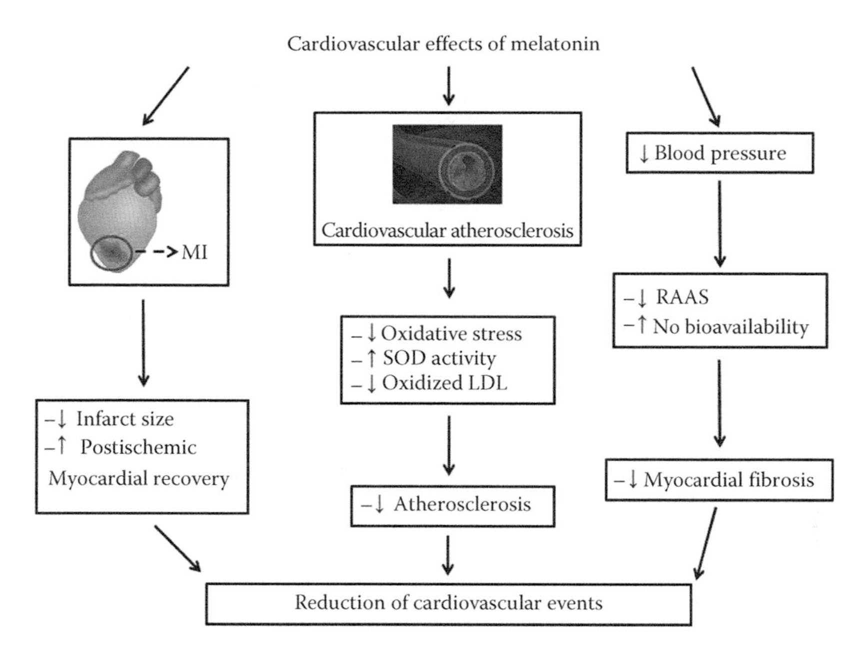

FIGURE 8.1 Cardiovascular effects of melatonin on the cardiovascular system. *Note:* MI, myocardial infarction; SOD, superoxide dismutase; LDL, low-density lipoproteins; RAAS, renin–angiotensin–aldosterone system.

REFERENCES

Altun A, Yaprak M, Aktoz M, Vardar A, Betul UA, and Ozbay G. 2002. Impaired nocturnal synthesis of melatonin in patients with cardiac syndrome X. *Neurosci Lett* 327:143–145.

Brugger P, Marktl W, and Herold M. 1995. Impaired nocturnal secretion of melatonin in coronary heart disease. *Lancet* 345:1408.

Cagnacci A, Cannoletta M, Renzi A, Baldassari F, Arangino S, and Volpe A. 2005. Prolonged melatonin administration decreases nocturnal blood pressure in women. *Am J Hypertens* 18:1614–1618.

Dominguez-Rodriguez A. 2012. Melatonin in cardiovascular disease. *Expert Opin Investig Drugs* 21:1593–1596.

Dominguez-Rodriguez A, Abreu-Gonzalez P, Arroyo-Ucar E, and Reiter RJ. 2012c. Decreased level of melatonin in serum predicts left ventricular remodelling after acute myocardial infarction. *J Pineal Res* 53:319–323.

Dominguez-Rodriguez A, Abreu-Gonzalez P, and Avanzas P. 2012b. The role of melatonin in acute myocardial infarction. *Front Biosci (Landmark Ed)* 17:2433–2441.

Domínguez-Rodríguez A, Abreu-González P, García MJ, Sanchez J, Marrero F, and de Armas-Trujillo D. 2002. Decreased nocturnal melatonin levels during acute myocardial infarction. *J Pineal Res* 33:248–252.

Dominguez-Rodriguez A, Abreu-Gonzalez P, Garcia-Gonzalez M, Ferrer-Hita J, Vargas M, and Reiter RJ. 2005. Elevated levels of oxidized low-density lipoprotein and impaired nocturnal synthesis of melatonin in patients with myocardial infarction. *Atherosclerosis* 180:101–105.

Dominguez-Rodriguez A, Abreu-Gonzalez P, Garcia-Gonzalez M, and Reiter RJ. 2006. Prognostic value of nocturnal melatonin levels as a novel marker in patients with ST-segment elevation myocardial infarction. *Am J Cardiol* 97:1162–1164.

Dominguez-Rodriguez A, Abreu-Gonzalez P, Garcia-Gonzalez MJ, Kaski JC, Reiter RJ, and Jimenez-Sosa A. 2007. A unicenter, randomized, double-blind, parallel-group, placebo-controlled study of Melatonin as an Adjunct in patients with acute myocaRdial Infarction undergoing primary Angioplasty The Melatonin Adjunct in the acute myocaRdial Infarction treated with Angioplasty (MARIA) trial: Study design and rationale. *Contemp Clin Trials* 28:532–539.

Dominguez-Rodriguez A, Abreu-Gonzalez P, Garcia-Gonzalez MJ, Samimi-Fard S, Reiter RJ, and Kaski JC. 2008. Association of ischemia-modified albumin and melatonin in patients with ST-elevation myocardial infarction. *Atherosclerosis* 199:73–78.

Dominguez-Rodriguez A, Abreu-Gonzalez P, Jimenez-Sosa A, Avanzas P, Bosa-Ojeda F, and Kaski JC. 2010a. Usefulness of intraplatelet melatonin levels to predict angiographic no-reflow after primary percutaneous coronary intervention in patients with ST-segment elevation myocardial infarction. *Am J Cardiol* 106:1540–1544.

Dominguez-Rodriguez A, Abreu-Gonzalez P, and Kaski JC. 2009a. Disruption of normal circadian rhythms and cardiovascular events. *Heart Metab* 44:11–15.

Dominguez-Rodriguez A, Abreu-Gonzalez P, and Kaski JC. 2009c. Inflammatory systemic biomarkers in setting acute coronary syndromes—Effects of the diurnal variation. *Curr Drug Targets* 10:1001–1008.

Dominguez-Rodriguez A, Abreu-Gonzalez P, and Reiter RJ. 2009b. Clinical aspects of melatonin in the acute coronary syndrome. *Curr Vasc Pharmacol* 7:367–373.

Dominguez-Rodriguez A, Abreu-Gonzalez P, Sanchez-Sanchez JJ, Kaski JC, and Reiter RJ. 2010b. Melatonin and circadian biology in human cardiovascular disease. *J Pineal Res* 49:14–22.

Dominguez-Rodriguez A, Abreu-Gonzalez P, and Reiter RJ. 2012a. Melatonin and cardiovascular disease: Myth or reality? *Rev Esp Cardiol (Engl Ed)* 65:215–218.

Dubocovich ML and Markowska M. 2005. Functional MT1 and MT2 melatonin receptors in mammals. *Endocrine* 27:101–110.

Enjuanes-Grau C, Dominguez-Rodriguez A, Abreu-Gonzalez P, Jimenez-Sosa A, and Avanzas P. 2012. Blood pressure levels and pattern of melatonin secretion in a population of resident physicians on duty. *Rev Esp Cardiol (Engl Ed)* 65:576–577.

Girotti L, Lago M, Ianovsky O, Carbajales J, Elizari MV, Brusco LI, and Cardinali DP. 2000. Low urinary 6-sulphatoxymelatonin levels in patients with coronary artery disease. *J Pineal Res* 29:138–142.

González A, López B, Ravassa S, Beaumont J, Zudaire A, Gallego I, Brugnolaro C, and Díez J. 2012. Cardiotrophin-1 in hypertensive heart disease. *Endocrine* 42:9–17.

Grossini E, Molinari C, Uberti F, Mary DA, Vacca G, and Caimmi PP. 2011. Intracoronary melatonin increases coronary blood flow and cardiac function through β-adrenoreceptors, MT1/MT2 receptors, and nitric oxide in anesthetized pigs. *J Pineal Res* 51:246–257.

Jonas M, Garfinkel D, Zisapel N, Laudon M, and Grossman E. 2003. Impaired nocturnal melatonin secretion in non-dipper hypertensive patients. *Blood Press* 12:19–24.

Kaçmaz A, User EY, Sehirli AO, Tilki M, Ozkan S, and Sener G. 2005. Protective effect of melatonin against ischemia/reperfusion-induced oxidative remote organ injury in the rat. *Surg Today* 35:744–750.

Kaneko S, Okumura K, Numaguchi Y, Matsui H, Murase K, Mokuno S, Morishima I et al. 2000. Melatonin scavenges hydroxyl radical and protects isolated rat hearts from ischemic reperfusion injury. *Life Sci* 67:101–112.

Kawashima K, Miwa Y, Fujimoto K, Oohata H, Nishino H, and Koike H. 1987. Antihypertensive action of melatonin in the spontaneously hypertensive rat. *Clin Exp Hypertens A* 9:1121–1131.

Kawashima K, Nagakura A, Wurzburger RJ, and Spector S. 1984. Melatonin in serum and the pineal of spontaneously hypertensive rats. *Clin Exp Hypertens A* 6:1517–1528.

Lagneux C, Joyeux M, Demenge P, Ribuot C, and Godin-Ribuot D. 2000. Protective effects of melatonin against ischemia-reperfusion injury in the isolated rat heart. *Life Sci* 66:503–509.

Lopez AD and Murray CC. 1998. The global burden of disease, 1990–2020. *Nat Med* 4:1241–1243.

Moens AL, Claeys MJ, Timmermans JP, and Vrints CJ. 2005. Myocardial ischemia/reperfusion-injury, a clinical view on a complex pathophysiological process. *Int J Cardiol* 100:179–190.

Nava M, Quiroz Y, Vaziri N, and Rodriguez-Iturbe B. 2003. Melatonin reduces renal interstitial inflammation and improves hypertension in spontaneously hypertensive rats. *Am J Physiol Renal Physiol* 284:F447–F454.

O'Brien E, Sheridan J, and O'Malley K. 1988. Dippers and non-dippers. *Lancet* 2:397.

Paulis L and Simko F. 2007. Blood pressure modulation and cardiovascular protection by melatonin: Potential mechanisms behind. *Physiol Res* 56:671–684.

Peyrot F and Ducrocq C. 2008. Potential role of tryptophan derivatives in stress responses characterized by the generation of reactive oxygen and nitrogen species. *J Pineal Res* 45:235–246.

Reimer KA, Lowe JE, Rasmussen MM, and Jennings RB. 1997. The wavefront phenomenon of ischemic cell death. 1. Myocardial infarct size vs duration of coronary occlusion in dogs. *Circulation* 56:786–794.

Reiter RJ, Manchester LC, Fuentes-Broto L, and Tan DX. 2010b. Cardiac hypertrophy and remodelling: Pathophysiological consequences and protective effects of melatonin. *J Hypertens* 28:S7–S12.

Reiter RJ, Tan DX, Paredes SD, and Fuentes-Broto L. 2010a. Beneficial effects of melatonin in cardiovascular disease. *Ann Med* 42:276–285.

Sahna E, Olmez E, and Acet A. 2002. Effects of physiological and pharmacological concentrations of melatonin on ischemia-reperfusion arrhythmias in rats: Can the incidence of sudden cardiac death be reduced? *J Pineal Res* 32:194–198.

Sahna E, Parlakpinar H, Ozer MK, Ozturk F, Ozugurlu F, and Acet A. 2003. Melatonin protects against myocardial doxorubicin toxicity in rats: Role of physiological concentrations. *J Pineal Res* 35:257–261.
Sahna E, Parlakpinar H, Turkoz Y, and Acet A. 2005. Protective effects of melatonin on myocardial ischemia/reperfusion induced infarct size and oxidative changes. *Physiol Res* 54:491–495.
Sakotnik A, Liebmann PM, Stoschitzky K, Lercher P, Schauenstein K, Klein W, and Eber B. 1999. Decreased melatonin synthesis in patients with coronary artery disease. *Eur Heart J* 20:1314–1317.
Samimi-Fard S, Abreu-Gonzalez P, Dominguez-Rodriguez A, and Jimenez-Sosa A. 2011. A case-control study of melatonin receptor type 1A polymorphism and acute myocardial infarction in a Spanish population. *J Pineal Res* 51:400–404.
Scheer FA, Van Montfrans GA, van Someren EJ, Mairuhu G, and Buijs RM. 2004. Daily nighttime melatonin reduces blood pressure in male patients with essential hypertension. *Hypertension* 43:192–197.
Seabra ML, Bignotto M, Pinto LR Jr, and Tufik S. 2000. Randomized, double-blind clinical trial, controlled with placebo, of the toxicology of chronic melatonin treatment. *J Pineal Res* 29:193–200.
Simko F and Pechanova O. 2009a. Potential roles of melatonin and chronotherapy among the new trends in hypertension treatment. *J Pineal Res* 47:127–133.
Simko F and Pechanova O. 2009b. Recent trends in hypertension treatment: Perspectives from animal studies. *J Hypertens Suppl.* 27:S1–S4.
Tan DX, Manchester LC, Terron MP, Flores LJ, and Reiter RJ. 2007. One molecule, many derivatives: A never-ending interaction of melatonin with reactive oxygen and nitrogen species. *J Pineal Res* 42:28–42.
Tengattini S, Reiter RJ, Tan DX, Terron MP, Rodella LF, and Rezzani R. 2008. Cardiovascular diseases: Protective effects of melatonin. *J Pineal Res* 44:16–25.
Vijayasarathy K, Shanthi Naidu K, and Sastry BK. 2010. Melatonin metabolite 6-Sulfatoxymelatonin, Cu/Zn superoxide dismutase, oxidized LDL and malondialdehyde in unstable angina. *Int J Cardiol* 144:315–317.
Yaprak M, Altun A, Vardar A, Aktoz M, Ciftci S, and Ozbay G. 2003. Decreased nocturnal synthesis of melatonin in patients with coronary artery disease. *Int J Cardiol* 89:103–107.
Zaslavskaya RM, Lilitsa GV, Dilmagambetova GS, Halberg F, Cornélissen G, Otsuka K, Singh RB et al. 2004. Melatonin, refractory hypertension, myocardial ischemia and other challenges in nightly blood pressure lowering. *Biomed Pharmacother* 58:S129–S134.

9 Melatonin
Therapeutic Value and Neuroprotection in Stroke

Max Franzblau, Nick Franzese, Gabriel S. Gonzales-Portillo, Chiara Gonzales-Portillo, Meaghan Staples, Mia C. Borlongan, Naoki Tajiri, Yuji Kaneko, and Cesar V. Borlongan

CONTENTS

9.1 INTRODUCTION

Neurological disorders are often marked by oxidative stress, an increase in damage by free radicals (Du et al. 1996, Lee et al. 1999, Nakao et al. 1995). This correlation hints at the potential of free radical scavengers (e.g., deprenyl, 7-nitroindazole, iron chelator, vitamin E) for therapeutic treatments, especially considering the capability of free radical scavengers and antioxidants in preventing cell death (Leker et al. 2002, Suzuki et al. 2002). Due to the observed negative impact of oxidative stress on the pathophysiology and behavior of stroke victims, administration of free radical scavengers and antioxidants may serve as an effective treatment to lessen the detrimental effects of ischemic injury. In fact, numerous antioxidants have made it to clinical trials after extensive preclinical research. Melatonin, however, because of its endogenous nature, may prove even more effective than its antioxidative stress properties would suggest alone. The transplantation of melatonin-secreting cells, postischemic injury, has incredible potential as a melatonin-based therapy for stroke. Transplantation of melatonin-secreting pineal glands into rats as well as exogenous melatonin treatment have both been found to stimulate neuroprotection (Borlongan et al. 2000, 2003). Melatonin receptor type 1A (MT1) has been implicated in the neuroprotective mechanism of action for stem cells in in vivo stroke models (Kaneko et al. 2011).

The neuroprotective effects of melatonin have been demonstrated in several animal stroke models. For example, the harmful effects of experimentally induced stroke are intensified in rats that have been pinealectomized (Kilic et al. 1999, Manev et al. 1996). Furthermore, melatonin treatment following ischemic injury shrinks the infarct zone in both white and gray matter (Lee et al. 2005, Pei et al. 2003, Sinha et al. 2001). Among melatonin's many other beneficial properties relevant to stroke neuroprotection are its abilities to reduce blood–brain barrier permeability (Chen et al. 2006), inflammatory response (Lee et al. 2007), and cerebral edema formation (Kondoh et al. 2002). Of note, blood–brain barrier permeability (Chen et al. 2006), inflammatory response (Lee et al. 2007), and cerebral edema formation are all stroke-associated pathological symptoms. Melatonin also affords behavioral effects, including increased functional capabilities such as grip strength and motor coordination, while diminishing hyperactivity and anxiety (Kilic et al. 2008). Again, it is noteworthy that motor dysfunction and anxiety are behavioral symptoms of stroke. Because stroke is primarily an aging disease, finding a treatment for aged patients may be therapeutic for stroke and vice versa. Indeed, the elderly are at an increased risk of more serious injury from stroke, due to the decrease of melatonin secretion with age (Brzezinski 1997). Interestingly, melatonin pretreatment may be an excellent way to combat such melatonin deficiencies in aged patients as evidenced by reduction in inflammation and the volume of infarction in stroke animals (Pei and Cheung 2004, Pei et al. 2002, 2003). A plethora of information on the neuroprotective effects of melatonin and its possible use in stroke treatment make melatonin-based therapy an exciting area of study.

9.2 MELATONIN'S PROTECTION OF GLIAL CELLS IN STROKE

After ischemic injury, glial cells become actively involved in homeostasis and brain repair. Several studies have revealed the necessity of glial cells for neural survival after ischemic stroke (Diamond et al. 1998, LeRoux and Reh 1995, Luscher et al. 1998, Vernadakis 1996). Similarly, astrocytes have been shown to guide proliferation and migration of neurons in the developing nervous system, and maintain homeostasis and synaptic plasticity in the adult nervous system (LeRoux and Reh 1995, Vernadakis 1996). Studies have shown that astrocytes possess receptors (Diamond et al. 1998, Luscher et al. 1998) and signaling molecules that can result in neuronal survival (Blanc et al. 1998) or death (Lin et al. 1998). Because of these findings, mixed astrocyte–neuronal cultures have been used to investigate models of neuronal cell death, as they better resemble conditions in vivo and promote better neuronal survival than neuron cultures alone (Bronstein et al. 1995, Langeveld et al. 1995).

Findings demonstrating the crucial role of glial cells in neuronal survival and brain function and the identification of trophic factors, including glial cell line–derived neurotrophic factor (GDNF) (Lin et al. 1993), have promoted interest in the therapeutic use of glial cells. Glial cells have been shown to be the primary source of transforming growth factor b, and studies have demonstrated that astrocytes release many growth factors under normal conditions as well as in response to brain injury (Lehrmann et al. 1998, Ridet et al. 1997). As a result, the support and trophic factor properties of glial cells have been explored as a treatment option for neurodegenerative diseases such as Parkinson's disease (Hoffer et al. 1994). For instance, a major indicator of successful cell therapy in Parkinson's disease is the localization of surviving donor glial cells at the grafted site (Isacson et al. 1995, Svendsen et al. 1997), allowing axons of passage to reach their host targets (Deacon et al. 1997). Additionally, transplantation of astrocytes, modified to increase synthesis, and secretion of GDNF, or the dopamine precursor L-DOPA, have been shown to improve symptoms of Parkinson's disease (Horellou and Mallet 1997). A similar improvement has been demonstrated as a result of embryonic dopaminergic neuron grafting treatments combined with the infusion of astrocytic growth factor or GDNF. Astrocytes have also been shown to control water balance by siphoning extracellular water and potassium ions into their networks or surrounding blood vessels, and diminish glutamate toxicity by transporting glutamate into the soma and detoxicating glutamate by converting OH_2 into less harmful H_2O_2 (Hansson and Ronnback 1995, White et al. 1992). Such findings indicate the potential of glial cell therapy as a method of increasing neuronal survival through their trophic, siphoning, and detoxifying actions.

Widespread reactive gliosis has been demonstrated to accompany ischemic stroke, but there is still uncertainty as to whether this gliosis occurs in response to cell death or as an early neuroprotective response. Studies of experimental models of ischemic stroke report conflicting conclusions, some documenting that astrocytes are more resistant to ischemic injury than neurons (Goldberg and Choi 1993, Sochocka et al. 1994), while others demonstrate that neurons are more resistant than astrocytes (Pantoni et al. 1996, Petito et al. 1998). It is also widely debated whether dense glial cell accumulations in the ischemic penumbra propagate or limit infarction size (Ridet et al. 1997). Despite these disagreements, cellular treatments targeting the highly glial cell–populated ischemic penumbra show promise of therapeutic benefit (Fisher 1997, Siesjo 1992). Preclinical studies have demonstrated that transplantation of fetal (Aihara et al. 1994, Borlongan et al. 1998) or cultured neurons (Nishino et al. 1994), near or within the ischemic penumbra, induces significant behavioral recovery in animals affected by ischemic injury. Current clinical treatment of the ischemic penumbra employs anticoagulants or thrombolytics to dissolve blood clots, but there is no conclusive evidence that lasting motor and cognitive improvement accompanies treatment with current drugs (Lee et al. 1999). As a result, stroke remains a prominent cause of mortality and morbidity, and development of new efficacious treatments for the ischemic central nervous system remains a major topic of current research.

Notable alterations in glial cells after cerebral ischemia led us to explore the protective benefits of melatonin; we reasoned that if melatonin prompted therapeutic affects in response to cerebral ischemia, then it could also afford protection for glial cells and neurons (Borlongan et al. 2000). Our data showed that the protective action of glial cells afforded by melatonin led to functional recovery from stroke, both in vitro and in vivo (Borlongan et al. 2000). Observations of reduction in glial cell loss and gliosis in melatonin-treated ischemic animals were accompanied by greatly improved motor function in comparison to the control. Results suggest that behavioral deficits caused by ischemic injury were mediated by the functionality of the cortex in the melatonin-treated group of animals. These animals displayed minimal cortical infarction size in comparison to those treated with saline, and while they also exhibited a reduction in total striatal infarction, the lateral aspect of the striatum still presented significant damage, suggesting that the protection of the cortex is integral for the improvement of motor behaviors. Interestingly, the melatonin-treated group did not exhibit behavioral protection during the 1 h occlusion, indicating that the drug did not prevent the functional deficits caused by interruption of cerebral blood flow. It seems that the protective action of melatonin occurs by prevention of secondary cell death processes. These results in vivo were reproduced in vitro by demonstrating increased survival of melatonin-treated astrocytes following serum deprivation or toxin exposure (3-nitropropionic acid and sodium nitroprusside), which was consistent with the cellular response to ischemic injury observed in vivo.

A widely accepted mechanism for the robust neuroprotective effects of melatonin points to direct free radical scavenging from neurons. Melatonin has been implicated as an effective free radical scavenger (Matuszak et al. 1997), and it has been shown to protect against neurotoxicity in rats (Giusti et al. 1995, Hirata et al. 1998, Iacovitti et al. 1997). The protective effects of melatonin against experimental ischemic damage are well documented (Cho et al. 1997, Joo et al. 1998, Kilic et al. 1999, Manev et al. 1996, Reiter 1998), and melatonin deficiency has been suggested in stroke patients (Fiorina et al. 1996). Melatonin demonstrates efficacy as a free radical scavenger and indirect antioxidant (Cho et al. 1997, Cuzzocrea et al. 1998, Giusti et al. 1995, Hirata et al. 1998, Iacovitti et al. 1997, Miller et al. 1996, Reiter 1998, Shinohara et al. 2009, Zhang et al. 1999). For example, melatonin scavenges hydroxyl radicals generated by hydrogen peroxide via the Fenton reaction and peroxynitrite anions (Cuzzocrea et al. 1998), and blocks singlet oxygen toxicity (Cagnoli et al. 1995). Additionally, melatonin reduces lipid peroxidation in the brain as a result of the intoxication of free-radical-generating agents (Giusti et al. 1995, Melchiorri et al. 1995). These findings indicate the protection from free radical toxicity that melatonin affords neural tissue.

Alteration in glial cells following melatonin treatment is a topic that the earlier studies leave unexplored. Our results suggest that enhanced survival of glial cells after melatonin treatment may exert a protective effect on injured neurons (Borlongan et al. 2000). This mechanism of protection may involve the assistance that glial cells provide with the maintenance of homeostasis of the neural membrane through siphoning of excess potassium or through water capacity enhancement. Another explanation may involve prevention of glutamate toxicity that glial cells afford neurons (Diamond et al. 1998, Luscher et al. 1998, Mawatari et al. 1996). Additionally, the secretion of trophic factors such as GDNF by glial cells may provide a neuroprotective effect against cerebral ischemia

(Wang et al. 1997). These combined capabilities demonstrate the robust capability of melatonin-treated glial cells to combat ischemia/reperfusion injury.

9.3 MELATONIN FOR STROKE THERAPY IN EXPERIMENTAL ANIMALS

Preclinical studies with both animal models and clinical trials have established cell replacement therapy as an effective treatment for neurological disorders (Borlongan and Sanberg 2002a,b, Isacson and Deacon 1996, Redmond 2002). Fetal striatal or cortical cells as well as genetically engineered cells and stem cells have proven successful as treatments when transplanted into stroke animal models (Aihara et al. 1994, Grabowski et al. 1996, Sinden et al. 2000). The first ever clinical trial for neural transplantation therapy for stroke began in 1998 using human-derived cells (called NT2N cells), which are similar to neurons (Kondziolka et al. 2000). The NT2N cells were transplanted near the infarct zone with the hope that they would replace the damaged and dead cells affected by stroke while exhibiting neurotropic and anti-inflammatory effects on the surrounding tissue. Intracerebral transplantation therapy is supported as an efficacious option for stroke treatment by positive clinical results (Meltzer et al. 2001, Nelson et al. 2002). In addition, intravenous stem cell transplantation has been investigated more recently for acute stroke patients in clinical trials (Bang et al. 2005, Savitz et al. 2011b).

To determine whether melatonin-secreting pineal gland grafts exhibit neuroprotective effects, they were transplanted into rats with acute stroke (Borlongan et al. 2003). As a result of the pineal graft, the treated rats displayed considerably less motor asymmetry and reduced cerebral infarction compared to animals in the control group. In this study, the demonstrated neuroprotection was attained when the rats' pineal glands were intact. Rats subjected to pinealectomy did not experience the neuroprotective effects of the pineal graft. This indicates that an intact pineal gland is most likely required to make the pineal graft method an effective treatment. CSF melatonin elevation was linked to the neuroprotection in rats that were granted neuroprotection by the pineal grafts. These findings correspond with the observed positive effects of chronic, exogenous melatonin administration in the rat stroke model (Borlongan et al. 2000, Kondoh et al. 2002). These studies also found improvements in motor and histological deficits in animals given melatonin treatment. Melatonin's neuroprotective properties have also been studied in alternative models of stroke and CNS ailments (Bubenik et al. 1998, Gupta et al. 2002, Skaper et al. 1998).

Naturally, the next step is to determine whether pineal gland grafting or exogenous melatonin treatment is more effective for stroke therapy. Because the measured decrease in infarct size appeared in days 2 and 3 after stroke, and not day 1, it seems pineal gland grafting mainly targets secondary cell death (apoptosis). This may mean that the pineal gland graft treatment is not effective in combating the initial damage caused by ischemic injury. Chronic melatonin treatment must be maintained to combat the secondary cell death that occurs following the actual stroke (Du et al. 1996, Kondoh et al. 1995, Lee et al. 1999). While one treatment option may prove more effective, the benefits and risks of each must be carefully weighed. For example, exogenous melatonin treatment puts much less stress on the patient than the invasive surgery required to properly implement a pineal gland graft. However, the pineal gland graft replacement therapy may be more effective than the exogenous treatment due to the characteristic loss of many cells after stroke (Borlongan et al. 1995a,b, Bubenik et al. 1998, Gupta et al. 2002, Skaper et al. 1998). In addition, the different treatments may be better tailored for the distinct stages of stroke. For example, it is possible that exogenous administration better serves acute stroke patients, while pineal gland grafts may prove more capable of battling chronic stroke. Furthermore, combining the two methods may prove to be the most effective approach. Of note, the previously discussed study (Borlongan et al. 2003) focused on the acute phase of posttransplantation, and the rats were not immunosuppressed. However, immunosuppression may be necessary for long-term treatments in order to ensure the survival of the grated tissue. Although the brain is traditionally considered "immune-privileged," it is still possible for graft rejection to occur. Immune suppression is thus another factor that must be considered when evaluating the clinical application of pineal gland transplantation as a stroke therapy.

Coadministration of melatonin antagonists or free radicals during transplantation may be necessary to determine the mechanism of action of the pineal gland graft. Nonmelatonin-secreting tissues may also function as a negative control to help shed light on the relationship between the pineal gland grafts and melatonin. The administration of antibodies that bind growth factors in conjunction with pineal gland grafting may support or detract from the hypothesis that the grafts produce neurotrophic factors (Borlongan and Sanberg 2002a, Borlongan et al. 2001, Dillon-Carter et al. 2002, Johnston et al. 2001). Further studies are still required to fully examine the usefulness of pineal gland grafting in chronic stroke and to better understand the long-term functional outcomes of the therapy.

The inefficiency of pinealectomy to completely block the pineal gland graft-induced neuroprotection points to the possible interaction of the grafted pineal gland with the host tissue. Similar interactions have been observed in numerous transplant studies (Boer and Griffioen 1990, Isacson and Deacon 1996, Lu et al.1991, Tonder et al.1989). If there are interactions between the grafted tissues and host tissues, this neural network may promote neuronal repair of the ischemic damage. That being said, the pineal glands were transplanted in the striatum, which is physically separate from the host pineal gland, making it unlikely that such connections were made between the tissues. Additionally, because graft maturation lasted only 3 days, it is unlikely that any axonal sprouting occurred in either the grafted tissue or the host gland. Therefore, it is most likely due to the increased levels of melatonin produced by both the intact and grafted glands that afforded the neuroprotection observed exclusively in treated rats with existing pineal glands. This is supported by the requirement of high melatonin levels in the brain for neuroprotection. Should sustaining a high level of melatonin be necessary for neuroprotection, a grafted pineal gland may be more effective than exogenous melatonin administration. The pineal graft's advantage is its ability to secrete a steady amount of melatonin in the brain (Grosse and Davis 1998, Reiter and Maestroni 1999), as opposed to the intermittent spurts of high melatonin concentration that result from exogenous delivery. While intracerebral minipump infusion of melatonin may solve the issue of transient melatonin bursts, the procedure to install the minipump is similarly invasive when compared to the pineal graft. In addition, the graft has the advantage of being able to respond to the microenvironmental cues (e.g., free radicals, inflammatory responses, etc.) to which the minipump cannot react. This suggests that secreting the highest level of melatonin may not be the most effective treatment, but rather maintaining a level that varies according to the stroke stage. In fact, there exists a thorough documentation of irregular free radical levels during stroke in both animal models and humans (Cai et al. 2002, Leker et al. 2002, Maier and Chan 2002, Suzuki et al. 2002, Wang et al. 2002). Inflammation, which may have a positive or negative impact on ischemic injury (Kawashima et al. 2003, Maier and Chan 2002, Takeda et al. 2002), may be another factor affecting graft survival. As hypothesized (Borlongan et al. 2000, Poulos and Borlongan 2000, Reiter et al. 2000, Tan et al. 1998, 2002), the free radical scavenging property of the pineal gland is related to its neuroprotective capabilities. The rescue of the host microenvironment is typically the most important result of cell transplantation to achieve neuroprotection (Borlongan and Sanberg 2002a,b, Borlongan et al. 2001). Considering the fluctuation in the levels of stroke-induced free radicals and inflammation in the host brain (Marquardt et al. 2002, Tabuchi et al. 2002, Walder et al. 1997), pineal grafting may be the most effective melatonin treatment because of the gland's ability to respond to the endogenous cues for dynamic secretion.

Pineal gland grafting and pinealectomy have been examined in animal models to illustrate the importance of the pineal gland in circadian rhythm modulation (Huang et al. 2012, Lesnikov and Pierpaoli 1994, Palaoglu et al. 1994, Pierpaoli and Regelson 1994, Pierpaoli et al. 1991, Wu et al. 1991). The use of novel cells for transplantation (Borlongan et al. 1998, 2001, Chiang et al. 1999, 2001,

Dillon-Carter et al. 2002, Johnston et al. 2001), along with the feasibility of pineal gland grafting, together, led to the investigation of the pineal gland's efficacy as a graft source for stroke therapy. The demonstrated usefulness of the pineal gland as a tool for neural transplantation therapy also highlights the importance of free radical scavengers in the injured brain.

9.4 MELATONIN'S MODULATORY EFFECT ON STEM CELLS

Stem cells have long been used in preclinical trials for stroke therapy. However, it is only recently that melatonin receptors have been implicated in the stem cell's mechanism of action (Kaneko et al. 2011). Of course, there are many other mechanisms with which stem cells provide neuroprotection, including trophic factors (Parr et al. 2007). Recently, melatonin receptors have been examined in stem cells, in an attempt to find a link between melatonin and cell therapies. Amniotic epithelial cells (AEC), pluripotent stem cells found in placental tissue and amniotic fluid (Antonucci et al. 2009a,b, Cargnoni et al. 2009, Yu et al. 2009), have been used to study this relationship. In particular, AECs have been utilized for neural function research and to treat intracerebral hemorrhage and ischemia (Cargnoni et al. 2009, Yu et al. 2009). Unfortunately, these studies have failed to elucidate the pathway that regulates AEC differentiation, though sources are aware of this hole in our knowledge (Shinya et al. 2010). Additionally, there are few findings that support the possibility of AEC having therapeutic benefits for a specific neurological disease.

Melatonin's neuroprotective properties against ischemic and hemorrhagic stroke have been thoroughly explored in the laboratory (Borlongan et al. 2000, Koh 2008, Lee et al. 2010, Lekic et al. 2010, Lin and Lee 2009, Reiter et al. 2005, Tocharus et al. 2010). Inhibition of apoptosis (Beni et al. 2004, Ramirez-Rodriguez et al. 2009, Wang et al. 2009) and limitation of oxidative stress (Hardeland et al. 2009, Reiter et al. 2007, 2009, Samantaray et al. 2009, Xu et al. 2010) have been proposed as mechanisms of action for melatonin, following stroke. However, we are most focused on the possibility of melatonin providing neuroprotection by endogenous neurogenesis in conjunction with stem cells (Moriya et al. 2007, Ramirez-Rodriguez et al. 2009, Rennie et al. 2009) due to our extensive experience with stem cell therapy (Borlongan et al. 2003). Melatonin receptor 1 (MT1) and melatonin receptor 2 (MT2) are both expressed by stem cells and are controlled by the melatonin ligand. Despite these discoveries, the connection between melatonin and stem cells has not yet been sufficiently researched (Niles et al. 2004, Sharma et al. 2007).

To assess the potential use of AECs in stroke therapy, we examined the underlying mechanisms of their neural differentiation using an in vitro stroke model (Kaneko et al. 2011). Our study yielded five key observations. We found that AECs are among the stem cells that express only the MT1 receptor. This suggests that it may be possible to target MT1 in an attempt to change the differential fate of AECs. This suggestion is supported by an earlier study, which found neural stem cells expressing MT1, thus signifying melatonin's important role in mammalian neuronal development as a pleiotropic molecule (Niles et al. 2004). Second, we found that the AEC's neuroprotective impact was inhibited when antagonizing MT1. Antagonizing MT2 did not appear to affect the neuroprotection afforded by AECs. Third, a direct relationship was observed between melatonin levels and enhanced AEC proliferation and differentiation. This relationship was especially true for AECs with MT1. The significance of the melatonin receptor–ligand mechanism that regulates neuronal function is emphasized by melatonin's preparation of neural stem cells for differentiation (Moriya et al. 2007) and AEC's expression of MT1. Our study's novel approach of combining melatonin and AECs to harness the therapeutic powers of both has the potential to create a more effective therapy (Borlongan et al. 2003, Kaneko et al. 2011) than the purely exogenous melatonin therapies previously studied (Kong et al. 2008, Lekic et al. 2010, Lin and Lee 2009). Additionally, we have found evidence suggesting that combined AEC and melatonin treatment may be a viable therapy for other diseases related to oxidative stress (Chen et al. 2009, Kilic et al. 2004, 2005, 2008, Lin and Lee 2009, Ramirez-Rodriguez et al. 2009, Wang 2009, Xu et al. 2010). Melatonin may also suppress neurodegeneration due to its antioxidative properties at several stages of cell death, in addition to its impact on cell proliferation and differentiation. Consequently, the AEC and melatonin combined therapy technique seems to yield better results than either therapy on its own. Finally, our findings implicate neurotrophic factors in the neuroprotection resulting from AEC–melatonin treatment. An upregulation of vascular endothelial growth factor (VEGF) was observed as a result of the treatment and further supports the involvement of growth factors in neuroprotection. This finding is supported by the previously documented interaction between melatonin and VEGF in the periphery (Romeu et al. 2011) and brain-derived neurotrophic factor (BDNF) in the cerebral neurons (Imbesi et al. 2008). In addition, our discovery of increased VEGF levels is reinforced by VEGF's correlation with the expression of MT1 (Kaneko et al. 2011), while BDNF is linked with MT2 expression (Imbesi et al. 2008).

Our study is distinct from others that have hinted at the interaction between melatonin receptors and neurons (Borlongan et al. 2003, Chen et al. 2009, Hill et al. 2011, Imbesi et al. 2008, Lee et al. 2010, Mao et al. 2010, Mor et al. 2010, Sharma et al. 2007), because we directly tied MT1 to stem cell fate as a critical receptor. This pharmacological and cellular link between MT1 and stem cell function is strengthened by the discovery that higher melatonin and MT1 expression are linked to the reduced growth of mammary tumors (Hill et al. 2011). Our study opens up many possibilities for future research including in vivo studies of AEC–melatonin therapy for various brain disease models. In short, MT1 stimulation applies a neuroprotective effect. Moreover, the combination of AEC and melatonin treatment created a powerful neuroprotective influence mediated by MT1 stimulation. These findings suggest that melatonin can be used as a kind of switch to activate or deactivate growth, secretion of growth factors, and differentiation, thus furthering our understanding of the capabilities of melatonin receptor technology in stem cell therapy. Melatonin receptor technology might be able to allow for the remote regulation of transplanted or host stem cells.

Separate from our research elucidating the role of melatonin receptors in stem cell transplantation, there exists substantial evidence outlining the therapeutic properties of MT1 and MT2 in neuroprotection not having to do with grafted stem cells (Chern et al. 2012, Hardeland et al. 2011). This is important due to the existing clinical melatonin receptor agonists (e.g., ramelteon) with improved pharmacokinetic properties (i.e., plasma half-life, MT specificity/affinity), when compared to melatonin (Hardeland et al. 2008, Simpson and Curran 2008).

9.5 FUTURE OF MELATONIN THERAPY IN THE CLINIC

The therapeutic potential of exogenous melatonin treatment, transplantation of melatonin-expressing pineal gland, and stimulating MT1 in stem cells for stroke therapy have all been discussed in the preceding sections. Despite the evidence of beneficial effects, it is extremely important that the safety of melatonin-based therapies is studied and demonstrated in order to allow for a move to clinical trials in any of these treatment approaches. In order to establish the safety of these therapies, it is necessary to monitor the treated animals for an extended period of time and study the brain tissue as well as surrounding tissues to rule out any unexpected adverse effects. STAIR (Stroke Treatment Academic Industry Roundtable) (Albers et al. 2011) and STEPS (Stem Cell Therapies as an Emerging Paradigm in Stroke 2009, Savitz et al. 2011a) also outline recommendations for moving these therapies toward clinical use. The previously stated stroke committee guidelines call for testing the treatments on two models/species of stroke, incorporating comorbidity factors (e.g., hypertension, aging, diabetes, etc.), testing in multiple labs, and adhering to the standard of care controls. Should all of these guidelines be met, the outlined therapies will be much closer to clinical application.

ACKNOWLEDGMENTS

This research was supported by the Department of Neurosurgery and Brain Repair funds. CVB is supported by National Institutes of Health, National Institute of Neurological Disorders and

Stroke 1R01NS071956-01, Department of Defense W81XWH-11-1-0634, James and Esther King Foundation for Biomedical Research Program, SanBio, Inc., KMPHC, and NeuralStem, Inc.

REFERENCES

Aihara, N., K. Mizukawa, K. Koide, H. Mabe, and H. Nishino. 1994. Striatal grafts in infarct striatopallidum increase GABA release, reorganize GABAA receptor and improve water-maze learning in the rat. *Brain Res Bull* 33(5):483–488.

Albers, G. W., L. B. Goldstein, D. C. Hess, L. R. Wechsler, K. L. Furie, P. B. Gorelick, P. Hurn et al. 2011. Stroke Treatment Academic Industry Roundtable (STAIR) recommendations for maximizing the use of intravenous thrombolytics and expanding treatment options with intra-arterial and neuroprotective therapies. *Stroke* 42(9):2645–2650. doi: 10.1161/STROKEAHA.111.618850.

Antonucci, I., I. Iezzi, E. Morizio, F. Mastrangelo, A. Pantalone, M. Mattioli-Belmonte, A. Gigante et al. 2009a. Isolation of osteogenic progenitors from human amniotic fluid using a single step culture protocol. *BMC Biotechnol* 9:9. doi: 10.1186/1472-6750-9-9.

Antonucci, I., A. Pantalone, D. De Amicis, S. D'Onofrio, L. Stuppia, G. Palka, and V. Salini. 2009b. Human amniotic fluid stem cells culture onto titanium screws: A new perspective for bone engineering. *J Biol Regul Homeost Agents* 23(4):277–279.

Bang, O. Y., J. S. Lee, P. H. Lee, and G. Lee. 2005. Autologous mesenchymal stem cell transplantation in stroke patients. *Ann Neurol* 57(6):874–882. doi: 10.1002/ana.20501.

Beni, S. M., R. Kohen, R. J. Reiter, D. X. Tan, and E. Shohami. 2004. Melatonin-induced neuroprotection after closed head injury is associated with increased brain antioxidants and attenuated late-phase activation of NF-kappaB and AP-1. *FASEB J* 18(1):149–151. doi: 10.1096/fj.03-0323fje.

Blanc, E. M., A. J. Bruce-Keller, and M. P. Mattson. 1998. Astrocytic gap junctional communication decreases neuronal vulnerability to oxidative stress-induced disruption of Ca^{2+} homeostasis and cell death. *J Neurochem* 70(3):958–970.

Boer, G. J. and H. A. Griffioen. 1990. Developmental and functional aspects of grafting of the suprachiasmatic nucleus in the Brattleboro and the arrhythmic rat. *Eur J Morphol* 28(2–4):330–345.

Borlongan, C. V., D. W. Cahill, and P. R. Sanberg. 1995a. Locomotor and passive avoidance deficits following occlusion of the middle cerebral artery. *Physiol Behav* 58(5):909–917.

Borlongan, C. V., R. Martinez, R. D. Shytle, T. B. Freeman, D. W. Cahill, and P. R. Sanberg. 1995b. Striatal dopamine-mediated motor behavior is altered following occlusion of the middle cerebral artery. *Pharmacol Biochem Behav* 52(1):225–229.

Borlongan, C. V. and P. R. Sanberg. 2002a. Neural transplantation for treatment of Parkinson's disease. *Drug Discov Today* 7(12):674–682.

Borlongan, C. V. and P. R. Sanberg. 2002b. Neural transplantation in the new millenium. *Cell Transplant* 11(6):615–618.

Borlongan, C. V., T. P. Su, and Y. Wang. 2001. Delta opioid peptide augments functional effects and intrastriatal graft survival of rat fetal ventral mesencephalic cells. *Cell Transplant* 10(1):53–58.

Borlongan, C. V., I. Sumaya, D. Moss, M. Kumazaki, T. Sakurai, H. Hida, and H. Nishino. 2003. Melatonin-secreting pineal gland: A novel tissue source for neural transplantation therapy in stroke. *Cell Transplant* 12(3):225–234.

Borlongan, C. V., Y. Tajima, J. Q. Trojanowski, V. M. Lee, and P. R. Sanberg. 1998. Transplantation of cryopreserved human embryonal carcinoma-derived neurons (NT2N cells) promotes functional recovery in ischemic rats. *Exp Neurol* 149(2):310–321. doi: 10.1006/exnr.1997.6730.

Borlongan, C. V., M. Yamamoto, N. Takei, M. Kumazaki, C. Ungsuparkorn, H. Hida, P. R. Sanberg, and H. Nishino. 2000. Glial cell survival is enhanced during melatonin-induced neuroprotection against cerebral ischemia. *FASEB J* 14(10):1307–1317.

Borlongan, C. V., F. C. Zhou, T. Hayashi, T. P. Su, B. J. Hoffer, and Y. Wang. 2001. Involvement of GDNF in neuronal protection against 6-OHDA-induced parkinsonism following intracerebral transplantation of fetal kidney tissues in adult rats. *Neurobiol Dis* 8(4):636–646. doi: 10.1006/nbdi.2001.0410.

Bronstein, D. M., I. Perez-Otano, V. Sun, S. B. Mullis Sawin, J. Chan, G. C. Wu, P. M. Hudson, L. Y. Kong, J. S. Hong, and M. K. McMillian. 1995. Glia-dependent neurotoxicity and neuroprotection in mesencephalic cultures. *Brain Res* 704(1):112–116.

Brzezinski, A. 1997. Melatonin in humans. *N Engl J Med* 336(3):186–195. doi: 10.1056/NEJM199701163360306.

Bubenik, G. A., D. E. Blask, G. M. Brown, G. J. Maestroni, S. F. Pang, R. J. Reiter, M. Viswanathan, and N. Zisapel. 1998. Prospects of the clinical utilization of melatonin. *Biol Signals Recept* 7(4):195–219.

Cagnoli, C. M., C. Atabay, E. Kharlamova, and H. Manev. 1995. Melatonin protects neurons from singlet oxygen-induced apoptosis. *J Pineal Res* 18(4):222–226.

Cai, H., Z. Li, A. Goette, F. Mera, C. Honeycutt, K. Feterik, J. N. Wilcox, S. C. Dudley, Jr., D. G. Harrison, and J. J. Langberg. 2002. Downregulation of endocardial nitric oxide synthase expression and nitric oxide production in atrial fibrillation: Potential mechanisms for atrial thrombosis and stroke. *Circulation* 106(22):2854–2858.

Cargnoni, A., M. Di Marcello, M. Campagnol, C. Nassuato, A. Albertini, and O. Parolini. 2009. Amniotic membrane patching promotes ischemic rat heart repair. *Cell Transplant* 18(10):1147–1159. doi: 10.3727/096368909X12483162196764.

Chen, H. Y., Y. C. Hung, T. Y. Chen, S. Y. Huang, Y. H. Wang, W. T. Lee, T. S. Wu, and E. J. Lee. 2009. Melatonin improves presynaptic protein, SNAP-25, expression and dendritic spine density and enhances functional and electrophysiological recovery following transient focal cerebral ischemia in rats. *J Pineal Res* 47(3):260–270. doi: 10.1111/j.1600-079X.2009.00709.x.

Chen, T. Y., M. Y. Lee, H. Y. Chen, Y. L. Kuo, S. C. Lin, T. S. Wu, and E. J. Lee. 2006. Melatonin attenuates the postischemic increase in blood-brain barrier permeability and decreases hemorrhagic transformation of tissue-plasminogen activator therapy following ischemic stroke in mice. *J Pineal Res* 40(3):242–250. doi: 10.1111/j.1600-079X.2005.00307.x.

Chern, C. M., J. F. Liao, Y. H. Wang, and Y. C. Shen. 2012. Melatonin ameliorates neural function by promoting endogenous neurogenesis through the MT2 melatonin receptor in ischemic-stroke mice. *Free Radic Biol Med* 52(9):1634–1647. doi: 10.1016/j.freeradbiomed.2012.01.030.

Chiang, Y. H., S. Z. Lin, C. V. Borlongan, B. J. Hoffer, M. Morales, and Y. Wang. 1999. Transplantation of fetal kidney tissue reduces cerebral infarction induced by middle cerebral artery ligation. *J Cereb Blood Flow Metab* 19(12):1329–1335. doi: 10.1097/00004647-199912000-00006.

Chiang, Y., M. Morales, F. C. Zhou, C. Borlongan, B. J. Hoffer, and Y. Wang. 2001. Fetal intra-nigral ventral mesencephalon and kidney tissue bridge transplantation restores the nigrostriatal dopamine pathway in hemi-parkinsonian rats. *Brain Res* 889(1–2):200–207.

Cho, S., T. H. Joh, H. H. Baik, C. Dibinis, and B. T. Volpe. 1997. Melatonin administration protects CA1 hippocampal neurons after transient forebrain ischemia in rats. *Brain Res* 755(2):335–338.

Cuzzocrea, S., G. Costantino, and A. P. Caputi. 1998. Protective effect of melatonin on cellular energy depletion mediated by peroxynitrite and poly (ADP-ribose) synthetase activation in a non-septic shock model induced by zymosan in the rat. *J Pineal Res* 25(2):78–85.

Deacon, T., J. Schumacher, J. Dinsmore, C. Thomas, P. Palmer, S. Kott, A. Edge et al. 1997. Histological evidence of fetal pig neural cell survival after transplantation into a patient with Parkinson's disease. *Nat Med* 3(3):350–353.

Diamond, J. S., D. E. Bergles, and C. E. Jahr. 1998. Glutamate release monitored with astrocyte transporter currents during LTP. *Neuron* 21(2):425–433.

Dillon-Carter, O., R. E. Johnston, C. V. Borlongan, M. E. Truckenmiller, M. Coggiano, and W. J. Freed. 2002. T155g-immortalized kidney cells produce growth factors and reduce sequelae of cerebral ischemia. *Cell Transplant* 11(3):251–259.

Du, C., R. Hu, C. A. Csernansky, C. Y. Hsu, and D. W. Choi. 1996. Very delayed infarction after mild focal cerebral ischemia: A role for apoptosis? *J Cereb Blood Flow Metab* 16(2):195–201. doi: 10.1097/00004647-199603000-00003.

Fiorina, P., G. Lattuada, O. Ponari, C. Silvestrini, and P. DallAglio. 1996. Impaired nocturnal melatonin excretion and changes of immunological status in ischaemic stroke patients. *Lancet* 347(9002):692–693.

Fisher, M. 1997. Characterizing the target of acute stroke therapy. *Stroke* 28(4):866–872.

Giusti, P., M. Gusella, M. Lipartiti, D. Milani, W. Zhu, S. Vicini, and H. Manev. 1995. Melatonin protects primary cultures of cerebellar granule neurons from kainate but not from *N*-methyl-D-aspartate excitotoxicity. *Exp Neurol* 131(1):39–46.

Goldberg, M. P. and D. W. Choi. 1993. Combined oxygen and glucose deprivation in cortical cell culture: Calcium-dependent and calcium-independent mechanisms of neuronal injury. *J Neurosci* 13(8):3510–3524.

Grabowski, M., B. B. Johansson, and P. Brundin. 1996. Fetal neocortical grafts placed in brain infarcts do not improve paw-reaching deficits in adult spontaneously hypertensive rats. *Acta Neurochir Suppl* 66:68–72.

Grosse, J. and F. C. Davis. 1998. Melatonin entrains the restored circadian activity rhythms of syrian hamsters bearing fetal suprachiasmatic nucleus grafts. *J Neurosci* 18(19):8032–8037.

Gupta, Y. K., G. Chaudhary, and K. Sinha. 2002. Enhanced protection by melatonin and meloxicam combination in a middle cerebral artery occlusion model of acute ischemic stroke in rat. *Can J Physiol Pharmacol* 80(3):210–217.

Hansson, E. and L. Ronnback. 1995. Astrocytes in glutamate neurotransmission. *FASEB J* 9(5):343–350.

Hardeland, R., D. P. Cardinali, V. Srinivasan, D. W. Spence, G. M. Brown, and S. R. Pandi-Perumal. 2011. Melatonin—A pleiotropic, orchestrating regulator molecule. *Prog Neurobiol* 93(3):350–384. doi: 10.1016/j.pneurobio.2010.12.004.

Hardeland, R., B. Poeggeler, V. Srinivasan, I. Trakht, S. R. Pandi-Perumal, and D. P. Cardinali. 2008. Melatonergic drugs in clinical practice. *Arzneimittelforschung* 58(1):1–10. doi: 10.1055/s-0031-1296459.

Hardeland, R., D. X. Tan, and R. J. Reiter. 2009. Kynuramines, metabolites of melatonin and other indoles: The resurrection of an almost forgotten class of biogenic amines. *J Pineal Res* 47(2):109–126. doi: 10.1111/j.1600-079X.2009.00701.x.

Hill, S. M., C. Cheng, L. Yuan, L. Mao, R. Jockers, B. Dauchy, T. Frasch, and D. E. Blask. 2011. Declining melatonin levels and MT1 receptor expression in aging rats is associated with enhanced mammary tumor growth and decreased sensitivity to melatonin. *Breast Cancer Res Treat* 127(1):91–98. doi: 10.1007/s10549-010-0958-0.

Hirata, H., M. Asanuma, and J. L. Cadet. 1998. Melatonin attenuates methamphetamine-induced toxic effects on dopamine and serotonin terminals in mouse brain. *Synapse* 30(2):150–155. doi: 10.1002/(SICI)1098-2396(199810)30:2<150::AID-SYN4>3.0.CO;2-B.

Hoffer, B. J., A. Hoffman, K. Bowenkamp, P. Huettl, J. Hudson, D. Martin, L. F. Lin, and G. A. Gerhardt. 1994. Glial cell line-derived neurotrophic factor reverses toxin-induced injury to midbrain dopaminergic neurons in vivo. *Neurosci Lett* 182(1):107–111.

Horellou, P. and J. Mallet. 1997. Gene therapy for Parkinson's disease. *Mol Neurobiol* 15(2):241–256. doi: 10.1007/BF02740636.

Huang, H. F., F. Guo, Y. Z. Cao, W. Shi, and Q. Xia. 2012. Neuroprotection by manganese superoxide dismutase (MnSOD) mimics: Antioxidant effect and oxidative stress regulation in acute experimental stroke. *CNS Neurosci Ther* 18(10):811–818. doi: 10.1111/j.1755-5949.2012.00380.x.

Iacovitti, L., N. D. Stull, and K. Johnston. 1997. Melatonin rescues dopamine neurons from cell death in tissue culture models of oxidative stress. *Brain Res* 768(1–2):317–326.

Imbesi, M., T. Uz, and H. Manev. 2008. Role of melatonin receptors in the effects of melatonin on BDNF and neuroprotection in mouse cerebellar neurons. *J Neural Transm* 115(11):1495–1499. doi: 10.1007/s00702-008-0066-z.

Isacson, O. and T. W. Deacon. 1996. Specific axon guidance factors persist in the adult brain as demonstrated by pig neuroblasts transplanted to the rat. *Neuroscience* 75(3):827–837.

Isacson, O., T. W. Deacon, P. Pakzaban, W. R. Galpern, J. Dinsmore, and L. H. Burns. 1995. Transplanted xenogeneic neural cells in neurodegenerative disease models exhibit remarkable axonal target specificity and distinct growth patterns of glial and axonal fibres. *Nat Med* 1(11):1189–1194.

Johnston, R. E., O. Dillon-Carter, W. J. Freed, and C. V. Borlongan. 2001. Trophic factor secreting kidney cell lines: In vitro characterization and functional effects following transplantation in ischemic rats. *Brain Res* 900(2):268–276.

Joo, J. Y., T. Uz, and H. Manev. 1998. Opposite effects of pinealectomy and melatonin administration on brain damage following cerebral focal ischemia in rat. *Restor Neurol Neurosci* 13(3–4):185–191.

Kaneko, Y., T. Hayashi, S. Yu, N. Tajiri, E. C. Bae, M. A. Solomita, S. H. Chheda, N. L. Weinbren, O. Parolini, and C. V. Borlongan. 2011. Human amniotic epithelial cells express melatonin receptor MT1, but not melatonin receptor MT2: A new perspective to neuroprotection. *J Pineal Res* 50(3):272–280. doi: 10.1111/j.1600-079X.2010.00837.x.

Kawashima, S., T. Yamashita, Y. Miwa, M. Ozaki, M. Namiki, T. Hirase, N. Inoue, K. Hirata, and M. Yokoyama. 2003. HMG-CoA reductase inhibitor has protective effects against stroke events in stroke-prone spontaneously hypertensive rats. *Stroke* 34(1):157–163.

Kilic, E., U. Kilic, M. Bacigaluppi, Z. Guo, N. B. Abdallah, D. P. Wolfer, R. J. Reiter, D. M. Hermann, and C. L. Bassetti. 2008. Delayed melatonin administration promotes neuronal survival, neurogenesis and motor recovery, and attenuates hyperactivity and anxiety after mild focal cerebral ischemia in mice. *J Pineal Res* 45(2):142–148. doi: 10.1111/j.1600-079X.2008.00568.x.

Kilic, E., U. Kilic, R. J. Reiter, C. L. Bassetti, and D. M. Hermann. 2004. Prophylactic use of melatonin protects against focal cerebral ischemia in mice: Role of endothelin converting enzyme-1. *J Pineal Res* 37(4):247–251. doi: 10.1111/j.1600-079X.2004.00162.x.

Kilic, E., Y. G. Ozdemir, H. Bolay, H. Kelestimur, and T. Dalkara. 1999. Pinealectomy aggravates and melatonin administration attenuates brain damage in focal ischemia. *J Cereb Blood Flow Metab* 19(5):511–516. doi: 10.1097/00004647-199905000-00005.

Kilic, U., E. Kilic, R. J. Reiter, C. L. Bassetti, and D. M. Hermann. 2005. Signal transduction pathways involved in melatonin-induced neuroprotection after focal cerebral ischemia in mice. *J Pineal Res* 38(1):67–71. doi: 10.1111/j.1600-079X.2004.00178.x.

Koh, P. O. 2008. Melatonin regulates nitric oxide synthase expression in ischemic brain injury. *J Vet Med Sci* 70(7):747–750.

Kondoh, T., S. H. Lee, and W. C. Low. 1995. Alterations in striatal dopamine release and reuptake under conditions of mild, moderate, and severe cerebral ischemia. *Neurosurgery* 37(5):948–954.

Kondoh, T., H. Uneyama, H. Nishino, and K. Torii. 2002. Melatonin reduces cerebral edema formation caused by transient forebrain ischemia in rats. *Life Sci* 72(4–5):583–590.

Kondziolka, D., L. Wechsler, S. Goldstein, C. Meltzer, K. R. Thulborn, J. Gebel, P. Jannetta et al. 2000. Transplantation of cultured human neuronal cells for patients with stroke. *Neurology* 55(4):565–569.

Kong, X., X. Li, Z. Cai, N. Yang, Y. Liu, J. Shu, L. Pan, and P. Zuo. 2008. Melatonin regulates the viability and differentiation of rat midbrain neural stem cells. *Cell Mol Neurobiol* 28(4):569–579. doi: 10.1007/s10571-007-9212-7.

Langeveld, C. H., C. A. Jongenelen, E. Schepens, J. C. Stoof, A. Bast, and B. Drukarch. 1995. Cultured rat striatal and cortical astrocytes protect mesencephalic dopaminergic neurons against hydrogen peroxide toxicity independent of their effect on neuronal development. *Neurosci Lett* 192(1):13–16.

Le Roux, P. D. and T. A. Reh. 1995. Independent regulation of primary dendritic and axonal growth by maturing astrocytes in vitro. *Neurosci Lett* 198(1):5–8.

Lee, C. H., K. Y. Yoo, J. H. Choi, O. K. Park, I. K. Hwang, Y. G. Kwon, Y. M. Kim, and M. H. Won. 2010. Melatonin's protective action against ischemic neuronal damage is associated with up-regulation of the MT2 melatonin receptor. *J Neurosci Res* 88(12):2630–2640. doi: 10.1002/jnr.22430.

Lee, E. J., M. Y. Lee, H. Y. Chen, Y. S. Hsu, T. S. Wu, S. T. Chen, and G. L. Chang. 2005. Melatonin attenuates gray and white matter damage in a mouse model of transient focal cerebral ischemia. *J Pineal Res* 38(1):42–52. doi: 10.1111/j.1600-079X.2004.00173.x.

Lee, J. M., G. J. Zipfel, and D. W. Choi. 1999. The changing landscape of ischaemic brain injury mechanisms. *Nature* 399(6738 Suppl):A7–A14.

Lee, M. Y., Y. H. Kuan, H. Y. Chen, T. Y. Chen, S. T. Chen, C. C. Huang, I. P. Yang, Y. S. Hsu, T. S. Wu, and E. J. Lee. 2007. Intravenous administration of melatonin reduces the intracerebral cellular inflammatory response following transient focal cerebral ischemia in rats. *J Pineal Res* 42(3):297–309. doi: 10.1111/j.1600-079X.2007.00420.x.

Lehrmann, E., R. Kiefer, T. Christensen, K. V. Toyka, J. Zimmer, N. H. Diemer, H. P. Hartung, and B. Finsen. 1998. Microglia and macrophages are major sources of locally produced transforming growth factor-beta1 after transient middle cerebral artery occlusion in rats. *Glia* 24(4):437–448.

Leker, R. R., A. Teichner, G. Lavie, E. Shohami, I. Lamensdorf, and H. Ovadia. 2002. The nitroxide antioxidant tempol is cerebroprotective against focal cerebral ischemia in spontaneously hypertensive rats. *Exp Neurol* 176(2):355–363.

Lekic, T., R. Hartman, H. Rojas, A. Manaenko, W. Chen, R. Ayer, J. Tang, and J. H. Zhang. 2010. Protective effect of melatonin upon neuropathology, striatal function, and memory ability after intracerebral hemorrhage in rats. *J Neurotrauma* 27(3):627–637. doi: 10.1089/neu.2009.1163.

Lesnikov, V. A. and W. Pierpaoli. 1994. Pineal cross-transplantation (old-to-young and vice versa) as evidence for an endogenous "aging clock." *Ann NY Acad Sci* 719:456–460.

Lin, H. W. and E. J. Lee. 2009. Effects of melatonin in experimental stroke models in acute, sub-acute, and chronic stages. *Neuropsychiatr Dis Treat* 5:157–162.

Lin, J. H., H. Weigel, M. L. Cotrina, S. Liu, E. Bueno, A. J. Hansen, T. W. Hansen, S. Goldman, and M. Nedergaard. 1998. Gap-junction-mediated propagation and amplification of cell injury. *Nat Neurosci* 1(6):494–500. doi: 10.1038/2210.

Lin, L. F., D. H. Doherty, J. D. Lile, S. Bektesh, and F. Collins. 1993. GDNF: A glial cell line-derived neurotrophic factor for midbrain dopaminergic neurons. *Science* 260(5111):1130–1132.

Lu, S. Y., M. T. Shipley, A. B. Norman, and P. R. Sanberg. 1991. Striatal, ventral mesencephalic and cortical transplants into the intact rat striatum: A neuroanatomical study. *Exp Neurol* 113(2):109–130.

Luscher, C., R. C. Malenka, and R. A. Nicoll. 1998. Monitoring glutamate release during LTP with glial transporter currents. *Neuron* 21(2):435–441.

Maier, C. M. and P. H. Chan. 2002. Role of superoxide dismutases in oxidative damage and neurodegenerative disorders. *Neuroscientist* 8(4):323–334.

Manev, H., T. Uz, A. Kharlamov, and J. Y. Joo. 1996. Increased brain damage after stroke or excitotoxic seizures in melatonin-deficient rats. *FASEB J* 10(13):1546–1551.

Mao, L., Q. Cheng, B. Guardiola-Lemaitre, C. Schuster-Klein, C. Dong, L. Lai, and S. M. Hill. 2010. In vitro and in vivo antitumor activity of melatonin receptor agonists. *J Pineal Res* 49(3):210–221. doi: 10.1111/j.1600-079X.2010.00781.x.

Marquardt, L., A. Ruf, U. Mansmann, R. Winter, M. Schuler, F. Buggle, H. Mayer, and A. J. Grau. 2002. Course of platelet activation markers after ischemic stroke. *Stroke* 33(11):2570–2574.

Matuszak, Z., K. Reszka, and C. F. Chignell. 1997. Reaction of melatonin and related indoles with hydroxyl radicals: EPR and spin trapping investigations. *Free Radic Biol Med* 23(3):367–372.

Mawatari, K., Y. Yasui, K. Sugitani, T. Takadera, and S. Kato. 1996. Reactive oxygen species involved in the glutamate toxicity of C6 glioma cells via xc antiporter system. *Neuroscience* 73(1):201–208.

Melchiorri, D., R. J. Reiter, E. Sewerynek, L. D. Chen, and G. Nistico. 1995. Melatonin reduces kainate-induced lipid peroxidation in homogenates of different brain regions. *FASEB J* 9(12):1205–1210.

Meltzer, C. C., D. Kondziolka, V. L. Villemagne, L. Wechsler, S. Goldstein, K. R. Thulborn, J. Gebel, E. M. Elder, S. DeCesare, and A. Jacobs. 2001. Serial [18F] fluorodeoxyglucose positron emission tomography after human neuronal implantation for stroke. *Neurosurgery* 49(3):586–591; discussion 591–592.

Miller, J. W., J. Selhub, and J. A. Joseph. 1996. Oxidative damage caused by free radicals produced during catecholamine autoxidation: Protective effects of *O*-methylation and melatonin. *Free Radic Biol Med* 21(2):241–249.

Mor, M., S. Rivara, D. Pala, A. Bedini, G. Spadoni, and G. Tarzia. 2010. Recent advances in the development of melatonin MT(1) and MT(2) receptor agonists. *Expert Opin Ther Pat* 20(8):1059–1077. doi: 10.1517/13543776.2010.496455.

Moriya, T., N. Horie, M. Mitome, and K. Shinohara. 2007. Melatonin influences the proliferative and differentiative activity of neural stem cells. *J Pineal Res* 42(4):411–418. doi: 10.1111/j.1600–079X.2007.00435.x.

Nakao, N., E. M. Frodl, H. Widner, E. Carlson, F. A. Eggerding, C. J. Epstein, and P. Brundin. 1995. Overexpressing Cu/Zn superoxide dismutase enhances survival of transplanted neurons in a rat model of Parkinson's disease. *Nat Med* 1(3):226–231.

Nelson, P. T., D. Kondziolka, L. Wechsler, S. Goldstein, J. Gebel, S. DeCesare, E. M. Elder et al. 2002. Clonal human (hNT) neuron grafts for stroke therapy: Neuropathology in a patient 27 months after implantation. *Am J Pathol* 160(4):1201–1206. doi: 10.1016/S0002-9440(10)62546-1.

Niles, L. P., K. J. Armstrong, L. M. Rincon Castro, C. V. Dao, R. Sharma, C. R. McMillan, L. C. Doering, and D. L. Kirkham. 2004. Neural stem cells express melatonin receptors and neurotrophic factors: Colocalization of the MT1 receptor with neuronal and glial markers. *BMC Neurosci* 5:41. doi: 10.1186/1471-2202-5-41.

Nishino, H., A. Czurko, K. Onizuka, A. Fukuda, H. Hida, C. Ungsuparkorn, M. Kunimatsu, M. Sasaki, Z. Karadi, and L. Lenard. 1994. Neuronal damage following transient cerebral ischemia and its restoration by neural transplant. *Neurobiology (Bp)* 2(3):223–234.

Palaoglu, S., O. Palaoglu, E. S. Akarsu, I. H. Ayhan, T. Ozgen, and A. Erbengi. 1994. Behavioural assessment of pinealectomy and foetal pineal gland transplantation in rats: Part II. *Acta Neurochir (Wien)* 128(1–4):8–12.

Pantoni, L., J. H. Garcia, and J. A. Gutierrez. 1996. Cerebral white matter is highly vulnerable to ischemia. *Stroke* 27(9):1641–1646; discussion 1647.

Parr, A. M., C. H. Tator, and A. Keating. 2007. Bone marrow-derived mesenchymal stromal cells for the repair of central nervous system injury. *Bone Marrow Transplant* 40(7):609–619. doi: 10.1038/sj.bmt.1705757.

Pei, Z. and R. T. Cheung. 2004. Pretreatment with melatonin exerts anti-inflammatory effects against ischemia/reperfusion injury in a rat middle cerebral artery occlusion stroke model. *J Pineal Res* 37(2):85–91. doi: 10.1111/j.1600-079X.2004.00138.x.

Pei, Z., H. T. Ho, and R. T. Cheung. 2002. Pre-treatment with melatonin reduces volume of cerebral infarction in a permanent middle cerebral artery occlusion stroke model in the rat. *Neurosci Lett* 318(3):141–144.

Pei, Z., S. F. Pang, and R. T. Cheung. 2003. Administration of melatonin after onset of ischemia reduces the volume of cerebral infarction in a rat middle cerebral artery occlusion stroke model. *Stroke* 34(3):770–775. doi: 10.1161/01.STR.0000057460.14810.3E.

Petito, C. K., J. P. Olarte, B. Roberts, T. S. Nowak, Jr., and W. A. Pulsinelli. 1998. Selective glial vulnerability following transient global ischemia in rat brain. *J Neuropathol Exp Neurol* 57(3):231–238.

Pierpaoli, W., A. Dall'Ara, E. Pedrinis, and W. Regelson. 1991. The pineal control of aging. The effects of melatonin and pineal grafting on the survival of older mice. *Ann NY Acad Sci* 621:291–313.

Pierpaoli, W. and W. Regelson. 1994. Pineal control of aging: Effect of melatonin and pineal grafting on aging mice. *Proc Natl Acad Sci USA* 91(2):787–791.

Poulos, S. G. and C. V. Borlongan. 2000. Artificial lighting conditions and melatonin alter motor performance in adult rats. *Neurosci Lett* 280(1):33–36.

Ramirez-Rodriguez, G., F. Klempin, H. Babu, G. Benitez-King, and G. Kempermann. 2009. Melatonin modulates cell survival of new neurons in the hippocampus of adult mice. *Neuropsychopharmacology* 34(9):2180–2191. doi: 10.1038/npp.2009.46.

Redmond, D. E., Jr. 2002. Cellular replacement therapy for Parkinson's disease—Where we are today? *Neuroscientist* 8(5):457–488.

Reiter, R. J. 1998. Oxidative damage in the central nervous system: Protection by melatonin. *Prog Neurobiol* 56(3):359–384.

Reiter, R. J. and G. J. Maestroni. 1999. Melatonin in relation to the antioxidative defense and immune systems: Possible implications for cell and organ transplantation. *J Mol Med (Berl)* 77(1):36–39.

Reiter, R. J., S. D. Paredes, L. C. Manchester, and D. X. Tan. 2009. Reducing oxidative/nitrosative stress: A newly-discovered genre for melatonin. *Crit Rev Biochem Mol Biol* 44(4):175–200. doi: 10.1080/10409230903044914.

Reiter, R. J., D. X. Tan, J. Leon, U. Kilic, and E. Kilic. 2005. When melatonin gets on your nerves: Its beneficial actions in experimental models of stroke. *Exp Biol Med (Maywood)* 230(2):104–117.

Reiter, R. J., D. X. Tan, L. C. Manchester, and H. Tamura. 2007. Melatonin defeats neurally-derived free radicals and reduces the associated neuromorphological and neurobehavioral damage. *J Physiol Pharmacol* 58(Suppl 6):5–22.

Reiter, R. J., D. X. Tan, W. Qi, L. C. Manchester, M. Karbownik, and J. R. Calvo. 2000. Pharmacology and physiology of melatonin in the reduction of oxidative stress in vivo. *Biol Signals Recept* 9(3–4):160–171. doi: 10.1159/000014636.

Rennie, K., M. De Butte, and B. A. Pappas. 2009. Melatonin promotes neurogenesis in dentate gyrus in the pinealectomized rat. *J Pineal Res* 47(4):313–317. doi: 10.1111/j.1600-079X.2009.00716.x.

Ridet, J. L., S. K. Malhotra, A. Privat, and F. H. Gage. 1997. Reactive astrocytes: Cellular and molecular cues to biological function. *Trends Neurosci* 20(12):570–577.

Romeu, L. R., E. L. da Motta, C. C. Maganhin, C. T. Oshima, M. C. Fonseca, K. F. Barrueco, R. S. Simoes, R. Pellegrino, E. C. Baracat, and J. M. Soares-Junior. 2011. Effects of melatonin on histomorphology and on the expression of steroid receptors, VEGF, and PCNA in ovaries of pinealectomized female rats. *Fertil Steril* 95(4):1379–1384. doi: 10.1016/j.fertnstert.2010.04.042.

Samantaray, S., A. Das, N. P. Thakore, D. D. Matzelle, R. J. Reiter, S. K. Ray, and N. L. Banik. 2009. Therapeutic potential of melatonin in traumatic central nervous system injury. *J Pineal Res* 47(2):134–142. doi: 10.1111/j.1600-079X.2009.00703.x.

Savitz, S. I., M. Chopp, R. Deans, S. T. Carmichael, D. Phinney, and L. Wechsler. 2011a. Stem cell therapy as an emerging paradigm for stroke (STEPS) II. *Stroke* 42(3):825–829. doi: 10.1161/STROKEAHA.110.601914.

Savitz, S. I., V. Misra, M. Kasam, H. Juneja, C. S. Cox, Jr., S. Alderman, I. Aisiku, S. Kar, A. Gee, and J. C. Grotta. 2011b. Intravenous autologous bone marrow mononuclear cells for ischemic stroke. *Ann Neurol* 70(1):59–69. doi: 10.1002/ana.22458.

Sharma, R., C. R. McMillan, and L. P. Niles. 2007. Neural stem cell transplantation and melatonin treatment in a 6-hydroxydopamine model of Parkinson's disease. *J Pineal Res* 43(3):245–254. doi: 10.1111/j.1600-079X.2007.00469.x.

Shinohara, Y., I. Saito, S. Kobayashi, and S. Uchiyama. 2009. Edaravone (radical scavenger) versus sodium ozagrel (antiplatelet agent) in acute noncardioembolic ischemic stroke (EDO trial). *Cerebrovasc Dis* 27(5):485–492. doi: 10.1159/000210190.

Shinya, M., H. Komuro, R. Saihara, Y. Urita, M. Kaneko, and Y. Liu. 2010. Neural differentiation potential of rat amniotic epithelial cells. *Fetal Pediatr Pathol* 29(3):133–143. doi: 10.3109/15513811003777292.

Siesjo, B. K. 1992. Pathophysiology and treatment of focal cerebral ischemia. Part I: Pathophysiology. *J Neurosurg* 77(2):169–184. doi: 10.3171/jns.1992.77.2.0169.

Simpson, D. and M. P. Curran. 2008. Ramelteon: A review of its use in insomnia. *Drugs* 68(13):1901–1919.

Sinden, J. D., P. Stroemer, G. Grigoryan, S. Patel, S. J. French, and H. Hodges. 2000. Functional repair with neural stem cells. *Novartis Found Symp* 231:270–283; discussion 283–288, 302–306.

Sinha, K., M. N. Degaonkar, N. R. Jagannathan, and Y. K. Gupta. 2001. Effect of melatonin on ischemia reperfusion injury induced by middle cerebral artery occlusion in rats. *Eur J Pharmacol* 428(2):185–192.

Skaper, S. D., B. Ancona, L. Facci, D. Franceschini, and P. Giusti. 1998. Melatonin prevents the delayed death of hippocampal neurons induced by enhanced excitatory neurotransmission and the nitridergic pathway. *FASEB J* 12(9):725–731.

Sochocka, E., B. H. Juurlink, W. E. Code, V. Hertz, L. Peng, and L. Hertz. 1994. Cell death in primary cultures of mouse neurons and astrocytes during exposure to and 'recovery' from hypoxia, substrate deprivation and simulated ischemia. *Brain Res* 638(1–2):21–28.

Stem Cell Therapies as an Emerging Paradigm in Stroke, Participants. 2009. Stem cell therapies as an emerging paradigm in stroke (STEPS): Bridging basic and clinical science for cellular and neurogenic factor therapy in treating stroke. *Stroke* 40(2):510–515. doi: 10.1161/STROKEAHA.108.526863.

Suzuki, M., M. Tabuchi, M. Ikeda, and T. Tomita. 2002. Concurrent formation of peroxynitrite with the expression of inducible nitric oxide synthase in the brain during middle cerebral artery occlusion and reperfusion in rats. *Brain Res* 951(1):113–120.

Svendsen, C. N., M. A. Caldwell, J. Shen, M. G. ter Borg, A. E. Rosser, P. Tyers, S. Karmiol, and S. B. Dunnett. 1997. Long-term survival of human central nervous system progenitor cells transplanted into a rat model of Parkinson's disease. *Exp Neurol* 148(1):135–146. doi: 10.1006/exnr.1997.6634.

Tabuchi, M., K. Umegaki, T. Ito, M. Suzuki, I. Tomita, M. Ikeda, and T. Tomita. 2002. Fluctuation of serum NO(x) concentration at stroke onset in a rat spontaneous stroke model (M-SHRSP). Peroxynitrite formation in brain lesions. *Brain Res* 949(1–2):147–156.

Takeda, H., M. Spatz, C. Ruetzler, R. McCarron, K. Becker, and J. Hallenbeck. 2002. Induction of mucosal tolerance to E-selectin prevents ischemic and hemorrhagic stroke in spontaneously hypertensive genetically stroke-prone rats. *Stroke* 33(9):2156–2163.

Tan, D. X., L. C. Manchester, R. J. Reiter, W. Qi, S. J. Kim, and G. H. El-Sokkary. 1998. Melatonin protects hippocampal neurons in vivo against kainic acid-induced damage in mice. *J Neurosci Res* 54(3):382–389.

Tan, D. X., R. J. Reiter, L. C. Manchester, M. T. Yan, M. El-Sawi, R. M. Sainz, J. C. Mayo, R. Kohen, M. Allegra, and R. Hardeland. 2002. Chemical and physical properties and potential mechanisms: Melatonin as a broad spectrum antioxidant and free radical scavenger. *Curr Top Med Chem* 2(2):181–197.

Tocharus, J., C. Khonthun, S. Chongthammakun, and P. Govitrapong. 2010. Melatonin attenuates methamphetamine-induced overexpression of pro-inflammatory cytokines in microglial cell lines. *J Pineal Res* 48(4):347–352. doi: 10.1111/j.1600-079X.2010.00761.x.

Tonder, N., T. Sorensen, J. Zimmer, M. B. Jorgensen, F. F. Johansen, and N. H. Diemer. 1989. Neural grafting to ischemic lesions of the adult rat hippocampus. *Exp Brain Res* 74(3):512–526.

Vernadakis, A. 1996. Glia-neuron intercommunications and synaptic plasticity. *Prog Neurobiol* 49(3):185–214.

Walder, C. E., S. P. Green, W. C. Darbonne, J. Mathias, J. Rae, M. C. Dinauer, J. T. Curnutte, and G. R. Thomas. 1997. Ischemic stroke injury is reduced in mice lacking a functional NADPH oxidase. *Stroke* 28(11):2252–2258.

Wang, X. 2009. The antiapoptotic activity of melatonin in neurodegenerative diseases. *CNS Neurosci Ther* 15(4):345–357. doi: 10.1111/j.1755-5949.2009.00105.x.

Wang, X., B. E. Figueroa, I. G. Stavrovskaya, Y. Zhang, A. C. Sirianni, S. Zhu, A. L. Day, B. S. Kristal, and R. M. Friedlander. 2009. Methazolamide and melatonin inhibit mitochondrial cytochrome C release and are neuroprotective in experimental models of ischemic injury. *Stroke* 40(5):1877–1885. doi: 10.1161/STROKEAHA.108.540765.

Wang, Y., C. F. Chang, M. Morales, Y. H. Chiang, and J. Hoffer. 2002. Protective effects of glial cell line-derived neurotrophic factor in ischemic brain injury. *Ann NY Acad Sci* 962:423–437.

Wang, Y., S. Z. Lin, A. L. Chiou, L. R. Williams, and B. J. Hoffer. 1997. Glial cell line-derived neurotrophic factor protects against ischemia-induced injury in the cerebral cortex. *J Neurosci* 17(11):4341–4348.

White, H. S., S. Y. Chow, Y. C. Yen-Chow, and D. M. Woodbury. 1992. Effect of elevated potassium on the ion content of mouse astrocytes and neurons. *Can J Physiol Pharmacol* 70 (Suppl):S263–S268.

Wu, W., D. E. Scott, and R. J. Reiter. 1991. No difference in day-night serum melatonin concentration after pineal grafting into the third cerebral ventricle of pinealectomized rats. *J Pineal Res* 11(2):70–74.

Xu, S. C., M. D. He, M. Zhong, Y. W. Zhang, Y. Wang, L. Yang, J. Yang, Z. P. Yu, and Z. Zhou. 2010. Melatonin protects against nickel-induced neurotoxicity in vitro by reducing oxidative stress and maintaining mitochondrial function. *J Pineal Res* 49(1):86–94. doi: 10.1111/j.1600-079X.2010.00770.x.

Yu, S. J., M. Soncini, Y. Kaneko, D. C. Hess, O. Parolini, and C. V. Borlongan. 2009. Amnion: A potent graft source for cell therapy in stroke. *Cell Transplant* 18(2):111–118.

Zhang, H., G. L. Squadrito, R. Uppu, and W. A. Pryor. 1999. Reaction of peroxynitrite with melatonin: A mechanistic study. *Chem Res Toxicol* 12(6):526–534. doi: 10.1021/tx980243t.

10 Melatonin's Antiapoptotic Activity in Neurodegenerative Diseases*

Xin Wang, Yingjun Guan, Wenyu Fu,
Shuanhu Zhou, and Zheng-Hong Qin

CONTENTS

10.1 INTRODUCTION

10.1.1 Melatonin May Be Beneficial in the Treatment of Neurodegenerative Diseases

Melatonin (*N*-acetyl-5-methoxytryptamine) is a natural hormone secreted by the pineal gland and other tissues of mammals, as well as plants (Dubbels et al., 1995; Esposito and Cuzzocrea, 2010). In clinical use for many years, melatonin is safe and well-tolerated even at high doses (Weishaupt et al., 2006) and easily crosses the blood–brain barrier. Besides being used to increase sleep efficiency, treat jet lag, improve the cardiovascular system (Sewerynek, 2002), as an antiaging drug (Gutierrez-Cuesta et al., 2008), and as a dietary supplement and cancer-protective hormone (Ravindra et al., 2006), intensive research in the past roughly 15 years has indicated melatonin's beneficial effects in the experimental models of neurodegenerative diseases. In particular, its broad spectrum of antioxidant activities in many neurodegenerative diseases (Tan et al., 2002) of the central nervous system (CNS) has been well documented and reviewed (Reiter et al., 1999). This small amphiphilic molecule, a powerful antioxidant and free-radical scavenger, directly scavenges hydroxyl, carbonate, and reactive nitrogen species as well as various organic radicals (Pandi-Perumal et al., 2013; Reiter et al., 2010) and enhances the antioxidant potential of the cell by stimulating the activity of superoxide

* *Source:* Wang, X., The antiapoptotic activity of melatonin in neurodegenerative diseases, *CNS Neuroscience & Therapeutics*, 15, 345–357, 2009.

dismutase (SOD), catalase, glutathione peroxidase, and glutathione reductase and by augmenting glutathione levels (Pandi-Perumal et al., 2013). Alongside its anti-inflammatory and mitochondrial protection, there is growing evidence that melatonin's antiapoptotic effects play an important role in neurodegeneration. This review will summarize the antiapoptotic activities of melatonin *via* the inhibition of intrinsic apoptotic pathways, the activation of survival signal pathways, and the alteration of the autophagy pathway in Parkinson's disease (PD), Alzheimer's disease (AD), amyotrophic lateral sclerosis (ALS), and Huntington's disease (HD).

10.1.2 Intrinsic and Extrinsic Apoptotic Pathways in Neurodegenerative Diseases

Two types of cell death occur in neurodegeneration: apoptosis and necrosis. Apoptosis (also called programmed cell death) occurs naturally under normal physiological conditions and in a variety of diseases, while necrosis is caused by external factors, such as infection or toxins. There are two major apoptotic signaling pathways: extrinsic (the death receptor pathway) and intrinsic (the mitochondrial pathway) (Jin and El-Deiry, 2005). This review focuses only on the intrinsic pathway, since there have been no obvious reports of the involvement of extrinsic pathways in the neuroprotection of melatonin (Figure 10.1 and Table 10.1) (Wang, 2009).

Proapoptotic mitochondrial molecules cytochrome *c*, second mitochondrion-derived activator of caspase (Smac)/Diablo, apoptosis-inducing factor (AIF), and endonuclease G (Endo G), when released into the cytoplasm from mitochondria, induce both caspase-dependent and caspase-independent mitochondrial death pathways in neurodegenerative diseases (Figure 10.1) (Friedlander, 2003; Vila and Przedborski, 2003; Wang, 2009; Wang et al., 2003, 2008, 2011; Zhang et al., 2013; Zhu et al., 2002). The release of cytochrome *c* is pivotal in the activation of caspases (Kroemer and Reed, 2000). During the progression of neurodegenerative diseases, once cytochrome *c* is released, it binds to Apaf-1 and dATP, which stimulates the activation of caspase-9, and then in turn cleaves the key effector, caspase-3, and two other effectors, caspase-6 and caspase-7 (Friedlander, 2003; Graham et al., 2006; Guegan et al., 2001; Li et al., 2000; Vila and Przedborski, 2003; Wang, 2009; Zhu et al., 2002). In addition, the DNA-repair enzyme poly (ADP-ribose) polymerase (PARP) is cleaved (Abeti and Duchen, 2012), and transcription factors such as Nuclear-Factor kappa B (NF-κB) (Jang et al., 2005; Jesudason et al., 2007; Kratsovnik et al., 2005; Lezoualc'h et al., 1998), TNFα-induced activator protein-1 (AP-1) (Jang and Surh, 2005), and p53 (Deigner et al., 2000; de la Monte et al., 1998) are activated. Nuclear condensation and DNA fragmentation are induced, as shown by terminal deoxynucleotidyl transferase–mediated DNA nick-end labeling of (TUNEL)-positive cells, Hoechst 33342 stain, propidium iodide (PI) or diamidino-2-phenylindole (DAPI) staining, and DNA ladder. These events ultimately cause neuronal cell death (Feng et al., 2004b; Ling et al., 2007; Ortiz et al., 2001; Wang et al., 2003, 2011). Other mitochondrial factors include mitochondrial permeability transition pores (mtPTP), mitochondrial membrane potential (ΔΨm), and calcium. mtPTP is a multiprotein complex, including inner- and outer-membrane components whose pores regulate the transport of ions and peptides into and out of mitochondria. The activation of the permeability transition and the irreversible opening of mitochondria pores is a major step in the development of neurodegeneration (Du et al., 2000; Jordan et al., 2003; Sas et al., 2007). The dissipation of ΔΨm and concomitant neuronal death have been reported in the experimental models of neurodegeneration (Jordan et al., 2003; Wang et al., 2003, 2011). Ca^{2+} can trigger apoptosis by the induction/activation of proapoptotic proteins, such as Bax, prostate apoptosis response-4 (Par-4), and p53, leading to mitochondrial membrane permeability changes, release of cytochrome *c* and caspase activation, induction of oxidative stress, or direct activation of calpains and caspases, which degrade a variety of substrates (Mattson, 2007).

Calpains may play an important role in the triggering of apoptotic cascades through their activation of caspases (Mattson, 2007). Caspase-1 activation is an early event in neurodegenerative diseases (Chen et al., 2000; Li et al., 2000). The upregulation of caspase-1 activator receptor-interacting protein-2 (Rip2) has already been reported by us and other researchers in AD (Engidawork et al., 2001), ALS (Zhang et al., 2013), and HD (Wang et al., 2005a). Rip2 stimulates

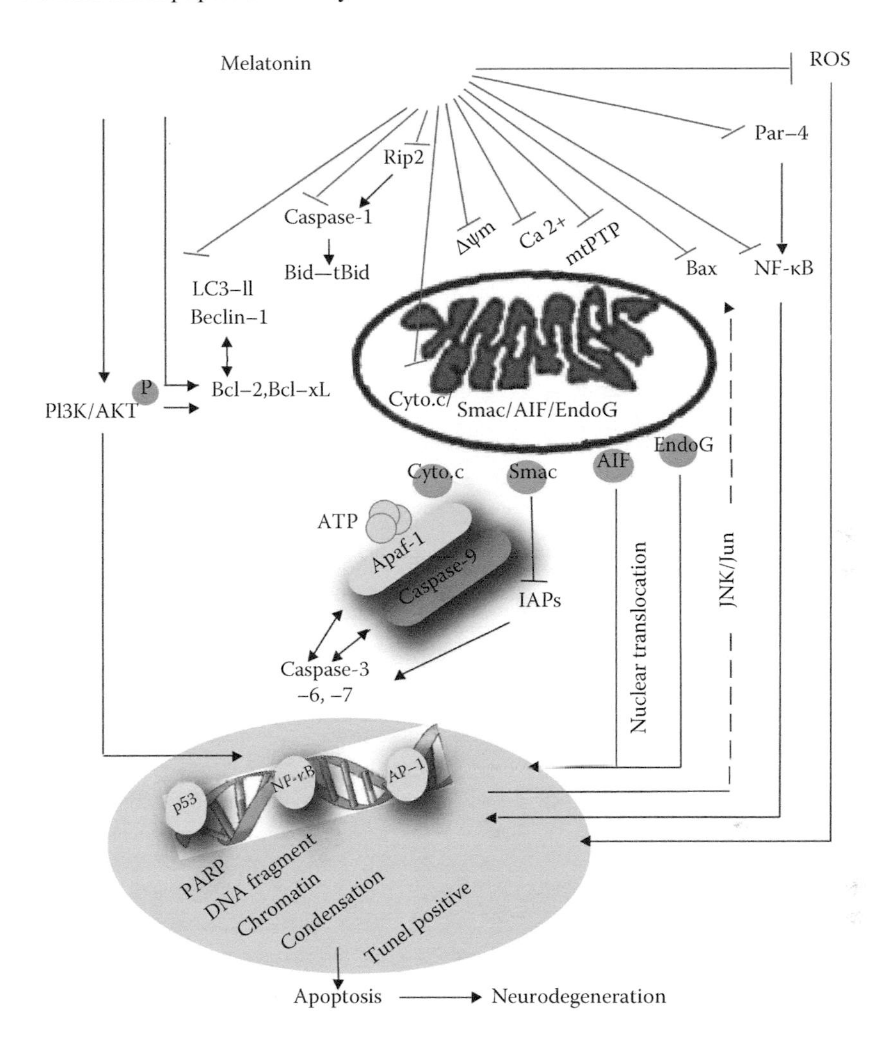

FIGURE 10.1 Scheme of neuroprotection of melatonin. The possible inhibition of the intrinsic cell death pathway and activation of the survival pathway by melatonin is schematized. (From Wang, X., *CNS Neurosci. Ther.*, 15, 346, 2009.)

caspase-1 to activate IL-1β by truncating the proinflammatory cytokine. The release of mature IL-1β indicates caspase-1 activation (Li et al., 1995). The inhibition of pro-IL-1β cleavage and secretion of mature IL-1β are associated with the inhibition of apoptosis in neurodegeneration (Wang et al., 2009; Zhang et al., 2003). Autophagy is an important process for preserving cell homeostasis. Alongside as a cell survival process in neurodegenerative diseases, excessive or hyperreactive autophagy could promote cell death (Sheng et al., 2012), known as "autophagic cell death" in neurodegenerative diseases (Nopparat et al., 2010). Autophagy also suppresses IL-1β signaling (Lee et al., 2012). Beclin-1 (a Bcl-2 homology [BH]–3 domain-only protein) and LC3-II (a hallmark protein of autophagy) are two pacemakers in the autophagic cascade whose expression is increased during autophagic cell death (Chang et al., 2012; Nopparat et al., 2010).

Bcl-2 family members include proapoptotic molecules (Bax, Bak, Bok, Bad, Bid, Bik, Blk, Hrk, BNIP3, and BimL) and antiapoptotic molecules (Bcl-2, Bcl-xL, Bcl-w, Mcl-1, and A1). Bcl-2 family proteins participate in the modulation and execution of cell death (Deigner et al., 2000) and can preserve or disrupt mitochondrial integrity by regulating the release of cytochrome *c*/Smac/AIF/endonuclease G (Danial and Korsmeyer, 2004; Yang et al., 1997). Cytosolic Bax translocates

TABLE 10.1
Summary of the Inhibition of the Antiapoptotic Cell Death Pathway by Melatonin

Inhibits Death Pathway Event	Diseases/ Models	Effects of Melatonin	Species/Cell Line	References
Cyto. *c*	PD	Prevents cyto. *c* release	Astrocyte	Jou et al. (2004)
	ALS	Decreases cyto. *c* release	Mouse	Zhang et al. (2013)
	HD	Decreases cyto. *c* release	Mouse; mutant-htt ST14A cell	Wang et al. (2011)
Smac/Diablo	ALS	Decreases Smac release	Mouse	Zhang et al. (2013)
	HD	Decreases Smac release	Mouse; mutant-htt ST14A cell	Wang et al. (2011)
AIF	HD	Decreases AIF release	Mouse; mutant-htt ST14A cell	Wang et al. (2011)
$\Delta\Psi_m$	PD	Prevents $\Delta\Psi_m$ depolarization	Astrocyte	Jou et al. (2004)
	PD	Inhibits disruption of $\Delta\Psi_m$	PC12 cell	Hausman et al. (2011)
	ALS	Maintains whole-cell $\Delta\Psi_m$	VSC4.1 motoneuron	Das et al. (2010)
	HD	Prevents $\Delta\Psi_m$ depolarization	Mutant-htt ST14A cell	Wang et al. (2011)
Ca^{2+}	PD	Decreases intracellular Ca^{2+} level	Rat astrocytoma cell, C6	Niranjan et al. (2010)
	ALS	Decreases intracellular Ca^{2+} level	VSC4.1 motoneuron	Das et al. (2010)
mtPTP	PD	Prevents mtPTP opening	Astrocyte	Jou et al. (2004)
ROS	PD	Prevents ROS formation	Astrocyte	Jou et al. (2004)
	ALS	Reduces ROS in ALS model	NSC34 motoneuron	Weishaupt et al. (2006)
	ALS	Attenuates production of ROS	VSC4.1 motoneuron	Das et al. (2010)
Bax	PD	Inhibits Bax expression	Mouse; Hela cell	Singhal et al. (2011), Zhou et al. (2012)
	PD	Decreases Bax expression	NB SH-SY5Y cell	Weinreb et al. (2003)
	PD	Inhibits Bax expression	HeLa cell	Zhou et al. (2012)
	AD	Attenuates Aβ25-35-induced apoptosis	Microglial cell	Jang et al. (2005)
	ALS	Inhibits Bax mRNA and protein expression	VSC4.1 motoneuron	Das et al. (2010)
Bad	AD	Prevents nitric oxide–induced apoptosis by increasing the interaction between p-Bad and p-Bad	SK-N-MC cell	Choi et al. (2008)
tBid	ALS	Inhibits tBid cleavage	VSC4.1 motoneuron	Das et al. (2010)
Rip2	ALS	Prevents Rip2 upregulation	Mouse	Zhang et al. (2013)
	HD	Prevents Rip2 upregulation	Mouse; mutant-htt ST14A cell	Wang et al. (2011)
IL-1β	PD	Inhibits the release of IL-1β	Rat astrocytoma cell, C6	Niranjan et al. (2010)
Caspase-1	ALS	Prevents caspase-1 activation	Mouse	Zhang et al. (2013)
Caspase-8	ALS	Inhibits caspase-8 activation	VSC4.1 motoneuron	Das et al. (2010)

Caspase-9	PD	Attenuates caspase-9 expression	Mouse	Singhal et al. (2011)
	ALS	Prevents caspase-9 activation	Mouse	Zhang et al. (2013)
	HD	Prevents caspase-9 activation	Mouse; mutant-htt ST14A cell	Wang et al. (2011)
Caspase-3	PD	Blocks caspase-3 activation	Astrocyte; dopaminergic neuron; CGN	Alvira et al. (2006), Ebadi et al. (2005), Jou et al. (2004)
	PD	Inhibits caspase-3 activation	PC12 cell	Hausman et al. (2011)
	AD	Attenuates Aβ25-35-induced apoptosis	Microglial cell	Jang et al. (2005)
	ALS	Prevents caspase-3 activation	Mouse	Zhang et al. (2013)
	HD	Prevents caspase-3 activation	Mouse; mutant-htt ST14A cell	Wang et al. (2011)
Caspase-7	PD	Inhibits caspase-7 activation	PC12 cell	Hausman et al. (2011)
Caspase-12	PD	Prevents caspase-12 activation	Mouse	Chang et al. (2012)
Calpain	ALS	Decreases calpain activity	VSC4.1 motoneuron	Das et al. (2010)
DNA fragmentation	PD	Prevents DNA fragmentation	SK-N-SH cell; astrocyte; mesencephalic cell; striatal neuron; mouse; PC12 cell	Chetsawang et al. (2007), Jou et al. (2004), Mayo et al. (1998), Ortiz et al. (2001)
	PD	Attenuates DNA fragmentation	Mouse	Chang et al. (2012)
	AD	Attenuates Aβ25-35- or Aβ1-42-induced apoptosis	Astroglioma C6 cell	Feng and Zhang (2004)
	AD	Blocks Aβ25-35-induced DNA fragmentation	Hippocampal neurons	Shen et al. (2002)
TUNEL-positive	AD/OVX	Improves spatial memory performance; reduces apoptosis	Rat	Feng et al. (2004b)
	AD	Protects the wortmannin-induced tau hyperphosphorylation	N2a cells	Deng et al. (2005)
p53	PD	Attenuates the expression of phosphorylated p53 and the change of p53 level	Mouse	Singhal et al. (2011)
JNK	PD	Inhibits cell death	SK-N-SH cell	Chetsawang et al. (2004, 2007)
Par-4	AD	Reduces Par-4 upregulation	Mouse	Feng et al. (2006)
NF-κB	PD	Inhibits MPTP-induced NF-κB translocation	Rat astrocytoma cell (C6)	Niranjan et al. (2010)
	AD	Blocks NF-κB activation	Microglial cell; mouse; rat	Gutierrez-Cuesta et al., (2008), Jang et al. (2005), Jesudason et al. (2007), Jun et al. (2013)
p38 MAPK	PD	Inhibits phosphorylation of p38, MAPK	Rat astrocytoma cell, C6	Niranjan et al. (2010)
	ALS	Attenuates phosphorylation of p38, MAPK	VSC4.1 motoneuron	Das et al. (2010)
LC3-II	PD	Prevents LC3-II activation and formation of aggregates	SK-N-SH dopaminergic cell	Xie et al. (2005)
	PD	Reduces LC3-II levels along with the reduction of α-synuclein aggregation	Mouse	Tan et al. (2005)
	PD	Reduces the formation of LC3-II	HeLa cell	Zhou et al. (2012)
Beclin-1	PD	Decreases the number of Beclin-1-positive cells	SK-N-SH dopaminergic cell	Xie et al. (2005)

OVX, ovariectomized.

to mitochondria upon death stimulus (Gross et al., 1998; Guegan et al., 2001), promoting cytochrome *c* release (Gross et al., 1998). Besides the involvement of the Fas/Caspase-8/Bid cascade, Bid also mediates cytochrome *c* release while binding to both proapoptotic members (e.g., Bax) and antiapoptotic members (e.g., Bcl-2 and Bcl-xL) (Luo et al., 1998); moreover, cleavage of Bid by caspase-8 and caspase-1 mediates mitochondrial damage (Guegan et al., 2002; Li et al., 1998). Bax mediates cell death related to mitochondrial permeability transition (Jin and El-Deiry, 2005). Bcl-2 and Bcl-xL bind to Apaf-1, inhibiting the association of caspase-9 with Apaf-1 (Hu et al., 1998). Additionally, Bax-mediated autophagy has been reported (Zhou et al., 2012).

Prostate apoptosis response-4 (Par-4) induces changes in mitochondrial membrane permeability and promotes mitochondrial dysfunction (Mattson, 2003). Par-4 increases the secretion of Amyloid beta (Aβ) and neuronal degeneration (Guo et al., 2001). Par-4 levels are augmented in AD patients (Xie and Guo, 2005). RNAi knockdown of Par-4 inhibits neurosynaptic degeneration in ALS-linked mice (Xie et al., 2005). Par-4 interacts with Bcl-2, caspase-8, and PKCζ, thus inhibiting NF-κB-dependent survival signaling (Culmsee and Landshamer, 2006).

The MAPK family includes three members: extracellular signal-regulated kinase (ERK), p38 mitogen-activated protein kinase (p38 MAPK), and c-Jun NH(2)-terminal kinase (JNK). The JNK pathway has been characterized in neurodegenerative diseases mostly as activating apoptosis (Chetsawang et al., 2004, 2007) and also as inhibiting cell death (Nopparat et al., 2010). DNA damage causes JNK activation, which contributes to the mitochondrial transduction of Bax (Tournier et al., 2000). The absence of JNK causes a defect in the mitochondrial death signaling pathway, including the failure to release cytochrome *c* (Tournier et al., 2000). Moreover, SP600125 (a JNK inhibitor) enhances the activation of the JNK pathway and attenuation of apoptosis through protection against mitochondrial dysfunction and reduction of caspase-9 activity in PC12 cells (Marques et al., 2003).

10.1.3 Survival Signaling Pathways in Neurodegenerative Diseases

During the progression of neurodegenerative diseases, survival signaling cascades are activated by neuroprotective mechanisms (Mattson et al., 2000), including the phosphoinositol-3 kinase (PI3K)/Akt pathway, the Bcl-2 pathway, the NF-κB pathway, and the MAPK pathway (Figure 10.1 and Table 10.2). AKT (v-akt murine thymoma viral oncogene)/PKB (protein kinase-B) has been identified as an important mediator of neuronal cell survival that helps counteract apoptotic stimuli. PI3K/Akt pathways play essential roles in neuronal cell survival. PI3K is activated, and the membrane

TABLE 10.2
Summary of Activation of Antiapoptotic Survival Signal Pathway by Melatonin

Activates Element of Survival Pathway	Diseases/ Models	Effects of Melatonin	Species/Cell Line	References
PI3K/Akt	AD	Impairs NADPH oxidase *via* PI3K/Akt signaling pathway	Microglia	Zhou et al. (2008)
Bcl-2	PD	Enhances Bcl-2 upregulation dopaminergic cell	SK-N-SH	Nopparat et al. (2010)
	AD	Enhances Bcl-2 upregulation	Microglial cell	Jang et al. (2005)
	ALS	Increases Bcl-2 mRNA and protein expression	VSC4.1 motoneuron	Das et al. (2010)
Bcl-xL	PD	Elevates Bcl-xL expression	NB SH-SY5Y cell	Weinreb et al. (2003)
JNK1	PD	Alters the activation of JNK1	SK-N-SH dopaminergic cell	Nopparat et al. (2010)
	ALS	Attenuates phosphorylation of JNK1	VSC4.1 motoneuron	Das et al. (2010)

phospholipid phosphatidylinositol-3,4,5-trisphosphate is generated, which in turn recruits Akt to the membrane, where it becomes phosphorylated. Once Akt is activated, it phosphorylates survival-mediated targets, including Bcl-2 family members, promoting cell survival and inhibiting apoptosis (Yang et al., 2004). The antiapoptotic Bcl-2 family encodes Bcl-2, Bcl-xL, and BfI-1 (A1) (Lukiw and Bazan, 2006)—all of which repress mitochondrial death pathways (Lukiw and Bazan, 2006). Depletion of endogenous neuroprotective Bcl-2 family signals directly contributes to neuronal loss in neurodegenerative diseases (Lukiw and Bazan, 2006).

The apoptosis regulator Bcl-2 plays a role in negatively regulating autophagy *via* its inhibitory interaction with Beclin-1 in neurodegenerative diseases (Nopparat et al., 2010; Pattingre et al., 2005; Salminen et al., 2013). The Bcl-2/Beclin-1 complex controls the threshold between cell survival and cell death (Pattingre et al., 2005). The JNK pathway is involved in neurodegenerative diseases by inhibiting cell death (Kilic et al., 2005). Bcl-2 is activated by c-Jun N-terminal kinase 1 (JNK 1) (upstream of Bcl-2 phosphorylation) to induce Bcl-2/Beclin 1 dissociation (Nopparat et al., 2010) and dissociation of Beclin 1 and Bcl-2/Bcl-xL, resulting in increased autophagy and antiapoptotic effects (Nixon and Yang, 2012). Conversely, Beclin 1 interacts with Bcl-2 and Bcl-xL, which diminishes its interactions with Beclin 1 complexes and deprives cells of autophagy's antiapoptotic effects. Beclin-1 binding of Bcl-2/Bcl-xL also decreases the PI3K activity associated with Beclin-1 and inhibits autophagic activation (Maiuri et al., 2007).

NF-κB is an inducible transcription factor. The activation of NF-κB signaling not only induces apoptotic signaling (Lezoualc'h et al., 1998) but also activates prosurvival signals in neurodegenerative diseases, depending on the context and cell types in which the signaling is activated (Teng and Tang, 2010). NF-κB activation produces neuroprotective effects in AD brains (Teng and Tang, 2010); in PC12 cells, a cellular model of PD (Lee et al., 2001); and in striatal neurons of mice, a 3-nitropropionic acid model of HD (Yu et al., 2000). Redox-sensitive transcriptional factors NF-κB and p53 regulate basal autophagy with a direct impact on the development of neurodegeneration (Caballero and Coto-Montes, 2012).

10.2 MELATONIN IN NEURODEGENERATIVE DISEASES

10.2.1 Melatonin in Parkinson's Disease

PD is the second most common neurodegenerative disease, affecting ~1.8% of people over 65 years old (de Rijk et al., 2000). PD is characterized by a progressive loss of dopamine (DA)-producing neurons in the substantia nigra (SN), and the resulting severe depletion of DA in the striatum leads to clinical symptoms (Srivastava et al., 2010). Though its detailed etiology and pathogenesis have yet to be specified, PD is considered a complex interaction of normal aging, genetic susceptibilities, and environmental exposure to toxins, with the close involvement of inflammatory and immune factors, while oxidative stress and free radicals from both mitochondrial impairment and DA metabolism are considered to play critical roles in its etiology.

In animal models, toxins and drugs that have been reported to contribute neurotoxic effects include 1-Methyl-4-phenyl-1,2,3,6-tetrahydropyridine (MPTP), 6-hydroxy DA (6-OHDA), rotenone, paraquat and maneb, hydrogen peroxide (H_2O_2), kainic acid, and methamphetamine. Their effects are similar to the symptoms, anatomical changes, and biochemical alterations found in sporadic PD. Next, we will summarize the protective effects of these toxins and drugs on PD.

MPTP has been reported to induce reactive astrogliosis and a cellular manifestation of neuroinflammation and cause parkinsonism *via* its neurotoxic form, 1-methyl-4-phenylpyridinium ion (MPP+), which inhibits mitochondrial complex I of the mitochondrial respiratory chain. MPP+ has been commonly used as an experimental model of PD (Alvira et al., 2006; Chetsawang et al., 2007). Neurodegeneration occurs in PD, at least in part, through the activation of the mitochondria-dependent apoptotic molecular pathway (Vila and Przedborski, 2003). As shown in Table 10.1 and Figure 10.1, melatonin exerts neuroprotective effects against the MPP+-induced mitochondrial apoptotic pathway.

Melatonin promotes mitochondrial homeostasis. It increases the activity of the complex I and complex IV of the electron transport chain, thereby improving mitochondrial respiration and increasing ATP synthesis (Leon et al., 2005). Mitochondria have been identified as a target for melatonin (Andrabi et al., 2004; Leon et al., 2005); in fact, the highest levels of melatonin are found in the mitochondria (Martin et al., 2000). It is known that melatonin scavenges oxygen- and nitrogen-based reactants generated in mitochondria and that mitochondria play a critical role in the neuroprotective function of melatonin in PD.

We screened a library of 1040 FDA-approved drugs assembled by the Neurodegeneration Drug Screening Consortium of the National Institute of Neurological Disorders and Stroke for their ability to inhibit the release of cytochrome *c* from Ca^{2+}-stimulated mitochondria (Wang et al., 2008). Melatonin occupied one of the top positions (14th) (Wang et al., 2008). Furthermore, we and others have demonstrated that melatonin not only is effective in the cell-free purified mitochondrial system but also inhibits cytochrome *c* release in a variety of models of neurodegenerative diseases (Jou et al., 2004; Wang, 2009; Wang et al., 2008, 2009, 2011; Zhang et al., 2013), including MPTP-dependent release of mitochondrial apoptogenic factor cytochrome *c* in rat brain astrocytes, a model of PD (Jou et al., 2004). Alterations in cellular Ca^{2+} homeostasis also contribute to the neurodegenerative process. Melatonin inhibits MPTP-induced neuroinflammation by decreasing the levels of intracellular Ca^{2+} (Niranjan et al., 2010). Melatonin also inhibits MPT-dependent activation of caspase-3 in a rat brain astrocyte model of PD (Jou et al., 2004). This conclusion is further supported by the finding that melatonin suppresses 3-morpholinosydnonimine-induced caspase-3 activation in dopaminergic neurons (Ebadi et al., 2005) and diminishes the activation of caspase-3 enzyme activity in both MPP(+)-treated SK-N-SH cultured cells (Chetsawang et al., 2007) and cerebellar granule neurons (CGNs) (Alvira et al., 2006). In addition, melatonin inhibits the release of proinflammatory cytokines such as IL-1β, IL-1α, TNF-α, and IL-6 in C6 rat astrocytoma cells (Niranjan et al., 2010).

Mounting evidence indicates that melatonin protects dopaminergic cells from the degeneration of MPTP (Ma et al., 2009) and prevents MPTP-induced mouse brain cell DNA fragmentation in vivo (Ortiz et al., 2001) and MPP(+)-mediated cleavage of DNA fragmentation factors in SK-N-SH cultured cells in vitro (Chetsawang et al., 2007, 2009). Other experiments also indicate that melatonin blocks the MPT-dependent apoptotic fragmentation of nuclear DNA in rat brain astrocytes (Jou et al., 2004), rat mesencephalic cultures (Iacovitti et al., 1997), and mouse striatal neurons (Iacovitti et al., 1999).

The JNK pathway is involved in PD by activating apoptosis (Chetsawang et al., 2004, 2007), and transcription factors also play a role, as shown in several experiments demonstrating the action of melatonin to inhibit the JNK signaling cascade (Chetsawang et al., 2004, 2007; Das et al., 2010; Nopparat et al., 2010) and diminish the induction of phosphorylation of c-Jun in MPP(+)-treated SK-N-SH cultured cells (Chetsawang et al., 2007). Melatonin inhibits MPTP-induced neuroinflammation by decreasing nitrative and oxidative stress, lowering the levels of phosphorylated p38 MAPK, inhibiting the translocation of NF-κB, and reducing the expression of inflammatory proteins, including cyclooxygenase-2 (COX-2) and glial fibrillary acidic protein (GFAP) (Niranjan et al., 2010).

6-OHDA is widely used in the animal models of PD in vivo by us and other researchers, because it induces free-radical production, inhibits mitochondrial respiratory chain complexes I and IV, and induces apoptosis (Bove et al., 2005; Fu et al., 2013). In addition, 6-OHDA is widely used in the cellular models of PD in vitro, since it can induce the death of cells, including neuronal PC12 cells and SH-SY5Y cells (Levites et al., 2002), and DNA fragmentation in PC12 cells (Mayo et al., 1998).

Melatonin offers neuroprotection against 6-OHDA-induced PD (Singh et al., 2006). Studies have shown that melatonin plays critical roles in ameliorating symptomatic features of 6-OHDA-induced PD. Melatonin treatment significantly reduced apomorphine-induced motor deficits and prolonged the survival of dopaminergic neurons in SN and tyrosine hydroxylase (TH)-immuoreactive terminals in the striatum (Kim et al., 1998). In addition, melatonin restores lipid peroxidation levels, TH enzyme activity, and DA contents (Ferrer et al., 2001). In addition, melatonin diminishes the induction of phosphorylation of c-Jun in 6-OHDA-induced SK-N-SH cultured cells (Chetsawang et al., 2004).

Melatonin's signaling in PD is related to melatonin receptors (Niles et al., 2004; Sharma et al., 2007). It has been reported that melatonin downregulates the melatonin receptor 1A (MT1) and melatonin receptor 1B (MT2) in the SN and the amygdala in PD patients (Adi et al., 2010).

Like MPTP, rotenone (a botanical pesticide of natural origin) induces selective dopaminergic neurodegeneration of the nigrostriatal pathways, leading to α-synuclein aggregation by inhibiting mitochondrial complex I of the respiratory chain and increasing oxidative stress (Lin et al., 2008). However, unlike MPTP, rotenone uptake is not mediated by the DA transporter (Tapias et al., 2010). Autophagy has been described in the SN neurons of patients with PD and in dopaminergic cells of neurotoxin-induced PD models (Chu, 2006; Larsen and Sulzer, 2002). Additionally, it has been reported that rotenone induces autophagic cell death (Chen et al., 2007; Zhou et al., 2012) through reactive oxygen species (ROS) generation (Chen et al., 2007).

In isolated rat brain mitochondria, Ca^{2+} strongly stimulates the release of ROS in rotenone-treated isolated rat brain mitochondria, and melatonin inhibits rotenone- and calcium ion–induced mitochondrial oxidative stress (Sousa and Castilho, 2005). Melatonin also inhibits ROS release in PC12 cells treated with rotenone plus Ca^{2+} ionophore (Sousa and Castilho, 2005). Although one report does not support the neuroprotection of melatonin against rotenone-induced PD in rats (Dubbels et al., 1995), a number of reports demonstrate that melatonin suppresses cell death induced by rotenone (Zhou et al., 2012) while inhibiting Bax expression and free-radical generation, resisting TH-positive neuronal loss and restoring DA levels in the striatum. Melatonin protects against rotenone-induced cell injury *via* the inhibition of Bax-mediated autophagy and LC3-II formation in Hela cells (Zhou et al., 2012). In addition, melatonin has been reported to reduce symptomatic impairment and loss of dopaminergic neurons in a chronic rotenone-induced *Drosophila* model of PD (Coulom and Birman, 2004).

Paraquat, a potent redox compound, inhibits mitochondrial complex I and generates superoxide anions, leading to oxidative stress and consequently neuronal cell damage and death (Suntres, 2002). Maneb, a fungicide, preferentially inhibits mitochondrial complex III and produces oxidative stress in vitro (Fitsanakis et al., 2002). The combination of paraquat and maneb has been suggested to be a risk factor for the PD phenotype (Costello et al., 2009). Melatonin scavenges hydroxyl free radicals induced by rotenone and restores not only glutathione levels but also altered catalytic activity of superoxide and catalase in the SN (Saravanan et al., 2007). Melatonin prevents the dopaminergic-neuron degeneration caused by paraquat and maneb by inhibiting free radicals, neuroinflammation, and apoptosis. Similar to the protective effect of melatonin against the degeneration of dopaminergic neurons induced by MPTP, melatonin reduces the levels of inducible nitric oxide synthase, nitrites, lipid peroxidation, and the number of degenerating neurons in a maneb- and paraquat-induced PD mouse model (Singhal et al., 2011). Melatonin counteracts the toxicity of paraquat in *Drosophila melanogaster*, preserves nicotinamide adenine dinucleotide (NADH) levels, and improves the efficiency of NADH in both cell-free systems and cultured PC12 cells (Bonilla et al., 2006; Tan et al., 2005). Melatonin inhibits maneb-induced α-synuclein aggregation, mitochondrial dysfunction, and neurodegeneration in PC12 cells (Hausman et al., 2011).

H_2O_2 has been used to induce cell death in the models of PD (Jou et al., 2004). Proper $\Delta\Psi m$ is critical for the maintenance of cellular bioenergetic homeostasis, and dissipation of $\Delta\Psi m$ has been shown to be involved in PD. Studies showed that melatonin effectively inhibits H_2O_2-induced $\Delta\Psi m$ depolarization and mitochondrial calcium overload. These effects reflect the ability of melatonin to ameliorate harmful reductions in $\Delta\Psi m$, which may trigger mtPTP opening and the apoptotic cascade. Melatonin prevents H_2O_2-induced mtPTP opening and ROS formation in the rat brain astrocyte model of PD (Jou et al., 2004). Melatonin significantly suppressed rotenone-induced elevated expression of GFAP and caspase-3 in C6 cells (Swarnkar et al., 2012). In addition, melatonin prevents NF-κB activation by oxidative stress (Camello-Almaraz et al., 2008).

Methamphetamine is a common drug of abuse that induces toxicity and autophagic cell death, both of which are connected to PD (Nopparat et al., 2010). Methamphetamine induces autophagy by inhibiting the dissociation of the Bcl-2/Beclin 1 complex and increasing Beclin-1–positive cell numbers, LC3-II activation, and the formation of aggregates. Melatonin protects SK-N-SH dopaminergic cells from autophagic death by reducing both the numbers of Beclin-1-positive cells and LC3-II expression, while inhibiting the activation of the JNK 1 and Bcl-2 upstream pathway (Nopparat et al., 2010).

The abnormal α-synuclein levels in the brains of Parkinsonian patients may result from the deteriorated ubiquitination and degradation of α-synuclein. Kainic acid induces α-synuclein aggregation and autophagy-lysosomal activation in the hippocampus of mice, while melatonin ameliorates those changes (Chang et al., 2012). In more detail, melatonin reduces kainic acid–induced increases in the levels of LC3-II (a hallmark protein of autophagy), lysosomal-associated membrane protein 2 (a biomarker of lysosomes), and cathepsin B (a lysosomal cysteine protease) (Chang et al., 2012). Additionally, it reduces kainic acid–induced HO-1 levels and caspase-3/12 activation and attenuates DNA fragmentation (Chang et al., 2012).

Melatonin has concentration-dependent neuroprotective effects in NB SH-SY5Y cells, a cellular model of PD. In the range of 1–10 μM, melatonin shows antiapoptotic effects (Weinreb et al., 2003), including increasing cell viability, significantly elevating Bcl-xL expression, and decreasing Bax level (Weinreb et al., 2003). However, it shows proapoptotic activity at high concentrations (50 μM) (Weinreb et al., 2003).

Taken together, evidence to date suggests that melatonin may be a viable treatment to delay the cellular and behavioral alterations in PD.

10.2.2 Melatonin in Alzheimer's Disease

AD is the most common neurodegenerative disease, associated with the progressive loss of memory and deterioration of comprehensive cognition. Its pathology is characterized by the extracellular senile plaques of aggregated β-amyloid (Aβ) and intracellular neurofibrillary tangles that contain hyperphosphorylated tau protein. Aβ and tau therefore represent important therapeutic targets. The early phase of AD is treatable by the inhibitors of β and γ secretase, which degrade amyloid precursor protein (APP) to produce Aβ peptide (Rojas-Fernandez et al., 2002), and the late phase is amenable to treatment that prevents or reverses tau phosphorylation (Iqbal et al., 2002; Kostrzewa and Segura-Aguilar, 2003). Mild cognitive impairment can be a transitional stage between the cognitive decline of normal aging and the more serious problems caused by AD; many people with mild cognitive impairment eventually develop AD. Studies show that melatonin levels are lower in AD patients than in age-matched control subjects (Liu et al., 1999; Mahlberg et al., 2008; Ozcankaya and Delibas, 2002). Moreover, melatonin inhibits Aβ-induced toxicity (Feng et al., 2004a,b, 2006; Jang et al., 2005; Pappolla et al., 1997; Zhou et al., 2008) and attenuates tau hyperphosphorylation (Deng et al., 2005; Kostrzewa and Segura-Aguilar, 2003; Li et al., 2005; Liu and Wang, 2002; Wang et al., 2004, 2005b; Wang and Wang, 2006). In addition to its antioxidant properties, the antiamyloidogenic properties of melatonin for AD have been studied (Pappolla et al., 2000, 2002). Melatonin improved learning and memory deficits in an APP695 in vivo transgenic mouse model of AD (Feng et al., 2004b). In vitro experiments showed that Aβ-treated cultures exhibited characteristic features of apoptosis, and melatonin attenuated Aβ-induced apoptosis in a number of cellular models of AD, including mouse microglial BV2 cells, rat astroglioma C6 cells, and PC12 cells (Feng et al., 2004a,b; Feng and Zhang, 2004; Jang et al., 2005; Pappolla et al., 1997), as well as hippocampal neurons (Shen et al., 2002).

As in PD, melatonin may treat AD by inhibiting mitochondrial cell death pathways and activating survival pathways. We therefore summarize the reports of neuroprotection by melatonin through two pathways (Tables 10.1 and 10.2) in the experimental models of AD (Figure 10.1). As shown in Table 10.1, studies in transgenic AD mice and cultured cells suggest that an administration of melatonin inhibits the Aβ-induced increase in the levels of mitochondria-related Bax (Feng et al., 2006; Jang et al., 2005). In addition, melatonin restores mitochondrial function in APP transgenic mouse and cell models of AD (Dragicevic et al., 2012), reduces the generation of Aβ-induced intracellular ROS in mouse microglial BV2 cells (Jang et al., 2005), and protects SK-N-MC cells from nitric oxide–induced apoptosis by increasing the interaction between p-Bad and 14-3-3β (Choi et al., 2008).

Furthermore, melatonin prevented upregulated expression of Par-4 and suppressed Aβ-induced caspase-3 activity (Feng et al., 2006). Another experiment in mouse microglial BV2 cells in vitro showed that melatonin also decreased caspase-3 activity and inhibited NF-κB activation (Jang et al., 2005). In vivo observations showed that melatonin-treated mice had diminished expression of NF-κB compared with untreated animals in both a mouse (Jesudason et al., 2007) and a rat model of AD (Jun et al., 2013). Other experiments demonstrate that melatonin significantly decreased the number of TUNEL-positive neurons along with improving spatial memory performance in cognitively impaired, ovariectomized adult rats (Feng et al., 2004b) and reduced Alzheimer-like tau hyperphosphorylation in wortmannin-induced N2a cells (Deng et al., 2005).

On the other hand, melatonin may activate survival signal pathways, including the Bcl-2 pathway, which stabilizes mitochondrial function by antiapoptotic Bcl-2 family modulators. As demonstrated in Table 10.2, melatonin both enhanced and Bcl-2 expression and inhibited Aβ-induced cell death (Jang et al., 2005). Another experiment demonstrated that melatonin inhibited the phosphorylation of NADPH oxidase *via* a PI3K/Akt-dependent signaling pathway in microglia exposed to Aβ1-42 (Zhou et al., 2008). Overall, evidence suggests that melatonin may be an effective treatment for AD due to its antiapoptotic effects.

10.2.3 Melatonin in Amyotrophic Lateral Sclerosis

ALS is a fatal chronic neurodegenerative disease characterized by the degeneration of motor neurons and activation of neighboring nonmotor neuron cells, including microglia, astrocytes, and other glial-type cells, as well as targeted muscle cells (Li et al., 2013; Lobsiger and Cleveland, 2007). Various molecular mechanisms, including oxidative stress, apoptosis, inflammation, mitochondrial dysfunction, and excitotoxicity, have been implicated in the pathogenesis of ALS (Pandya et al., 2012, 2013). Riluzole, an antagonist of the glutamate receptor, is the only FDA-approved treatment for ALS. However, it typically prolongs the patient's life by only 3 months. Since the common basis of cellular and extracellular alterations in this disease seems to be oxidative stress, treatment strategies emphasize antioxidant molecules.

Rival et al. report that, like the administration of riluzole to *dEAAT1 RNAi Drosophila*, melatonin significantly enhances performance in a *Drosophila* model with remarkable similarity to some ALS symptoms (Rival et al., 2004). Our current findings and those of other researchers from both animal models in vivo and a cellular model in vitro (Figure 10.1, Tables 10.1 and 10.2) support the results of the *Drosophila* model. In mSOD1^{G93A} transgenic mice, melatonin (30 mg/kg) administered by intraperitoneal injection (Zhang et al., 2013) or high oral doses (Weishaupt et al., 2006) significantly delays disease onset and increases survival, although another report of different doses of melatonin in a small number of mice failed to find any neuroprotection (Dardiotis et al., 2013). However, western blot analysis of spinal cord protein lysates in the latter study found no differences in the total phosphorylation of AKT or ERK ½ in SOD1(G93A)-transgenic mice with melatonin treatment compared with untreated controls (Weishaupt et al., 2006). Another study showed that melatonin alters the expression of SOD1 in the lumbar spinal cord of neonatal rats (Rogerio et al., 2005) and attenuates superoxide-induced cell death and modulates glutamate toxicity in cultured NSC-34 motoneuron cells in vitro (Weishaupt et al., 2006). Evidence indicates that mSOD1-induced cell death among spinal cord motor neurons and cultured motor neuronal cells involves apoptotic machinery (Friedlander, 2003; Przedborski, 2004; Ryu et al., 2005; Sathasivam et al., 2005).

Caspase-1 plays a critical role as an apical activator in the models of ALS (Zhang et al., 2003). Our recent report demonstrates that melatonin administered by intraperitoneal injection inhibits caspase-1 activation (Zhang et al., 2013). In addition, melatonin also prevents caspase-3 activation and preserves MT1 along with inhibiting motor neuron loss (Zhang et al., 2013). Astrogliosis and microgliosis are notable hallmarks of ALS disease and are associated with motor-neuron degeneration. We found that melatonin reduced the expression of GFAP (a marker for astrocytes) and ricinus communis agglutinin-1 (RCA-1) (a marker for microglia/macrophages) in mSOD1^{G93A} ALS mice (Zhang et al., 2013). Interestingly, we demonstrate that ALS disease progression is associated with the loss of both melatonin and the MT1 in the spinal cord of mSOD1^{G93A} ALS mice (Zhang et al., 2013).

Melatonin prevents apoptosis in VSC4.1 motoneurons following exposure to the toxins H_2O_2, glutamate (LGA), or TNF-α (Das et al., 2010). Melatonin attenuates ROS production; the levels of intracellular Ca^{2+}; the phosphorylation of p38, MAPK, and JNK1; and calpain activity. Melatonin also maintains whole-cell membrane potential, decreases Bax expression, and increases Bcl-2 expression. In addition, results suggest that melatonin's neuroprotection of motoneurons is receptor mediated, since melatonin upregulates MT1 and MT2 (Das et al., 2010).

Autophagy markers Beclin-1 and LC3-II are increased in mSOD1^{G93A} mice and in motor neurons under oxidative stress (Gal et al., 2009; Pandya et al., 2013). To date, the neuroprotection afforded by melatonin through the activation of survival pathways remains essentially uninvestigated.

Melatonin also offers protection in human ALS (Jacob et al., 2002; Weishaupt et al., 2006). Importantly, circulating serum protein carbonyls (a surrogate marker for oxidative stress) were elevated in ALS patients but were reported to be normalized to control values by the administration of melatonin in the second clinical trial (Weishaupt et al., 2006). In other words, reduced oxidative damage was reported in an ALS trial of high-dose enteral melatonin (Weishaupt et al., 2006). Chronic high-dose (300 mg/day) (Weishaupt et al., 2006) rectally administered melatonin was well tolerated in patients with sporadic ALS (Jacob et al., 2002; Weishaupt et al., 2006). Because melatonin is neuroprotective in the animal and cellular models of ALS, and in patients with ALS, and relatively nontoxic, it should be considered for further larger clinical trials as a novel pharmacotherapeutic agent to treat ALS.

10.2.4 Melatonin in Huntington's Disease

HD, a hereditary disease, is universally fatal and has no effective treatment. It is characterized by movement disorder (Huntington's chorea), cognitive deterioration, emotional distress, and dementia (Kandel et al., 2000). This degenerative brain disorder is caused by the expansion of cytosine-adenine-guanine (CAG) repeats in exon 1 of the huntingtin gene (The Huntington's Disease Collaborative Research Group, 1993), initially affecting the striatum and then the cortex. Since oxidative stress plays an important role in the etiology of neuronal damage and degeneration in HD (Browne and Beal, 2006), therapeutic strategies focus on antioxidant defense.

3-nitropropionic acid, a mitochondrial complex II inhibitor, closely replicates the neurochemical, histological, and clinical features of HD and hence has been used in an experimental model of HD (Schulz and Beal, 1994; Tunez et al., 2004). Melatonin has been suggested to defer the signs of HD in a 3-nitropropionic acid–induced rat model of HD (Tunez et al., 2004) and to reduce lipid peroxidation induced by quinolinic acid (a causative agent in HD) (Southgate and Daya, 1999). Additionally, we report that melatonin remarkably delays disease onset and increases survival in R6/2 HD transgenic mice (Wang et al., 2011) and is a remarkably potent neuroprotective agent in mutant-huntingtin (mutant-htt) ST14A cells, a cellular model of HD (Rigamonti et al., 2000, 2001; Wang et al., 2008). It protects 76.2% of mutant-htt ST14A cells from temperature shift–induced cell death (Wang et al., 2008). Furthermore, melatonin prevents the cell death of primary cortical neurons that have been challenged with proapoptotic inducer (Wang et al., 2009). Another report in a quinolinic acid rat model of HD further confirms the neuroprotective effect of melatonin, including its antioxidant properties (Antunes Wilhelm et al., 2013).

Melatonin prevents the release of cell death mediator AIF from mitochondria in mutant-htt ST14A striatal cells upon insult and in R6/2 transgenic mice (Andrabi et al., 2004). Thus, melatonin is likely to interfere with both caspase-dependent (cytochrome *c*) and caspase-independent (AIF) mitochondrial cell death pathways.

One of our compelling findings regarding the mechanism of melatonin's action against HD is neuroprotection by the inhibition of mutant-htt-induced cell death through countering caspase-dependent (release of cytochrome *c* and Smac) and caspase-independent (release of AIF) mitochondrial cell death pathways. Melatonin also prevents the activation of caspase-9 (Wang et al., 2011). Additionally, mutant-htt-mediated toxicity in striatal cells, R6/2 transgenic mice, and HD humans is associated with the loss of the expression of MT1, while melatonin preserves MT1 expression (Wang et al., 2011). Furthermore, administration of melatonin significantly inhibits Rip2 upregulation in mutant-htt ST14A cells under insult (Wang et al., 2011) (Figure 10.1). On the other hand, to date, there are no reports of the activation of survival pathways by melatonin in HD. Based on its lack of toxicity, melatonin may be a good candidate for human HD treatment. Further, human trials on the impact of melatonin on HD are needed.

10.3 CONCLUSION AND PERSPECTIVE

Neurodegenerative diseases have become a major health problem worldwide, and vigorous research efforts to date have achieved poor results in identifying effective treatments. The combination of preclinical effectiveness and proven safety of melatonin in humans, animals, and cultured cells recommends melatonin as a particularly interesting candidate neuroprotectant in clinical trials seeking protection against neurodegeneration. Blood concentrations of melatonin are significantly decreased in patients with AD (Ozcankaya and Delibas, 2002), while low levels of melatonin and a prolonged signal of melatonin are found in PD patients (Blazejova et al., 2000), and HD patients demonstrate delayed onset of diurnal melatonin rise (Aziz et al., 2009). The incidence in neurodegenerative diseases increases with age, while the secretion of melatonin in the human body also gradually decreases with age. Thus, it is believed that reduced secretion of melatonin is associated with the development of neurodegenerative diseases (Ozcankaya and Delibas, 2002). Understanding the molecular mechanisms of melatonin's declining potency should tell us about the pathogenesis of related neurodegenerative diseases and will guide the contemplated translation to the clinic. Furthermore, pharmacological strategies to raise the levels of antioxidant melatonin may benefit those suffering from neurodegenerative diseases. Interestingly, regular intake of antioxidants in the elderly has been recommended for the prevention of age-associated neurodegenerative diseases (Hausman et al., 2011). As a preventive and adjunct therapy for neurodegenerative diseases in high-risk populations, melatonin may hold great potential for widespread application.

Biological rhythm disorders in neurodegenerative diseases increase with age. Clinical studies indicate that melatonin can improve sleep and circadian rhythm disruption in AD patients (Cardinali et al., 2013). Melatonin significantly delays the development of the signs of AD and prevents cognitive impairment and ameliorates sundowning in AD patients (Asayama et al., 2003; Brusco et al., 1998, 1999; Cardinali et al., 2002; Olde Rikkert and Rigaud, 2001; Pandi-Perumal et al., 2013). In addition, light therapy or music therapy coordinated with the levels of melatonin may have effects on AD patients (Kumar et al., 1999; Mishima et al., 1994). Furthermore, relatively small-scale human trials suggest that melatonin can improve mild cognitive impairment (Furio et al., 2007; Jean-Louis et al., 1998). However, how melatonin affects disease initiation or the progression of neuropathology and whether antiapoptotic activity drives melatonin's function remain to be seen. Melatonin is useful in treating disturbed sleep in PD, in particular rapid eye movement–associated sleep behavior disorder (Srinivasan et al., 2011).

On the other hand, some researchers report that the impact of melatonin is relatively small in the later stages of AD or that it completely fails to improve sleep or agitation and suggest that there is insufficient evidence to support the effectiveness of melatonin for managing cognitive impairment (Jansen et al., 2006). Therefore, some maintain that melatonin is not an effective soporific agent in patients with AD (Gehrman et al., 2009; Singer et al., 2003). In addition, some clinical trials of melatonin's effect on sleep disturbances in PD show small improvement (Dowling et al., 2005; Medeiros et al., 2007; Pandi-Perumal et al., 2013). Some clinical studies also show evidence that melatonin used for PD either has unremarkable effects or may even make the disease worse (Pandi-Perumal et al., 2013; Willis, 2008).

Melatonin offers protection in human ALS. The first clinical trial of melatonin in three human ALS patients was reported in 2002 (Jacob et al., 2002), and the second human trial in a group of 31 patients with sporadic ALS was reported in 2006 (Weishaupt et al., 2006). In a human study with the addition of tryptophan, though melatonin levels rose significantly in both control and HD patients, there was a larger mean increase among HD patients (Christofides et al., 2006). Moreover, a delayed onset of the diurnal melatonin rise in patients with HD on a small scale has been reported (Aziz et al., 2009). Larger-scale studies of the levels of melatonin in HD patients and further human trials on the impact of melatonin on HD are needed.

Cell death–based therapies are becoming an active area of drug development. For a multidrug regimen to effectively protect neurons from inappropriate apoptosis, several pathways could be coactivated, including antiapoptotic pathways and survival pathways. Besides its traditional role as an antioxidant and free-radical scavenger, melatonin has been shown to target a variety of pathways. Its systemic effect relates to the drug's disruption of the intrinsic mitochondrial cell death pathway, silencing of the Rip2/Caspase-1 pathway, blocking of the autophagic cell death pathways, the activation of survival pathways, and the alteration of melatonin receptors. These actions may be synchronistic and complementary in the models of neurodegenerative diseases.

Melatonin is capable of interfering with mitochondrial cell death pathways and activating survival pathways, both of which would be useful in treating common events in AD, PD, ALS, and HD. Future therapeutic strategies could be directed at identifying and developing drugs from among the analogues of melatonin (Cardinali et al., 2013). Candidate drugs may have more powerful effects than natural melatonin does on inhibiting the mitochondrial cell death pathway, activating the survival pathway, and slowing the progression of neurodegenerative diseases. In addition, effective prevention of neurodegeneration could be achieved by a combination of melatonin and other pharmacological agents and even stem cell therapies (Sharma et al., 2007) that act on different apoptosis targets.

ACKNOWLEDGMENTS

This work was supported by grants from the Muscular Dystrophy Association (To X.W., 254530), the ALS Therapy Alliance (2013D001622 to X.W.), the Bill & Melinda Gates Foundation (BMGF: 01075000191 to X.W.), Brigham and Women's Hospital BRI Fund to Sustain Research Excellence (to S.Z.), and the Shandong Province Taishan Scholar Project.

REFERENCES

Abeti, R. and M. R. Duchen. 2012. Activation of PARP by oxidative stress induced by β-amyloid: Implications for Alzheimer's disease. *Neurochemical Research* 37:2589–2596.

Adi, N., D. C. Mash, Y. Ali, C. Singer, L. Shehadeh, and S. Papapetropoulos. 2010. Melatonin MT1 and MT2 receptor expression in Parkinson's disease. *Medical Science Monitor* 16:BR61–BR67.

Alvira, D., M. Tajes, E. Verdaguer, D. Acuna-Castroviejo, J. Folch, A. Camins, and M. Pallas. 2006. Inhibition of the cdk5/p25 fragment formation may explain the antiapoptotic effects of melatonin in an experimental model of Parkinson's disease. *Journal of Pineal Research* 40:251–258.

Andrabi, S. A., I. Sayeed, D. Siemen, G. Wolf, and T. F. Horn. 2004. Direct inhibition of the mitochondrial permeability transition pore: A possible mechanism responsible for anti-apoptotic effects of melatonin. *FASEB Journal* 18:869–871.

Antunes Wilhelm, E., C. Ricardo Jesse, C. Folharini Bortolatto, and C. Wayne Nogueira. 2013. Correlations between behavioural and oxidative parameters in a rat quinolinic acid model of Huntington's disease: Protective effect of melatonin. *European Journal of Pharmacology* 701:65–72.

Asayama, K., H. Yamadera, T. Ito, H. Suzuki, Y. Kudo, and S. Endo. 2003. Double blind study of melatonin effects on the sleep-wake rhythm, cognitive and non-cognitive functions in Alzheimer type dementia. *Journal of Nippon Medical School* 70:334–341.

Aziz, N. A., H. Pijl, M. Frolich, J. P. Schroder-van der Elst, C. van der Bent, F. Roelfsema, and R. A. Roos. 2009. Delayed onset of the diurnal melatonin rise in patients with Huntington's disease. *Journal of Neurology* 256:1961–1965.

Blazejova, K., S. Nevsimalova, H. Illnerova, I. Hajek, and K. Sonka. 2000. Sleep disorders and the 24-hour profile of melatonin and cortisol. *Sb Lek* 101:347–351.

Bonilla, E., S. Medina-Leendertz, V. Villalobos, L. Molero, and A. Bohorquez. 2006. Paraquat-induced oxidative stress in drosophila melanogaster: Effects of melatonin, glutathione, serotonin, minocycline, lipoic acid and ascorbic acid. *Neurochemical Research* 31:1425–1432.

Bove, J., D. Prou, C. Perier, and S. Przedborski. 2005. Toxin-induced models of Parkinson's disease. *NeuroRx* 2:484–494.

Browne, S. E. and M. F. Beal. 2006. Oxidative damage in Huntington's disease pathogenesis. *Antioxidants & Redox Signaling* 8:2061–2073.

Brusco, L. I., I. Fainstein, M. Marquez, and D. P. Cardinali. 1999. Effect of melatonin in selected populations of sleep-disturbed patients. *Biological Signals and Receptors* 8:126–131.

Brusco, L. I., M. Marquez, and D. P. Cardinali. 1998. Monozygotic twins with Alzheimer's disease treated with melatonin: Case report. *Journal of Pineal Research* 25:260–263.

Caballero, B. and A. Coto-Montes. 2012. An insight into the role of autophagy in cell responses in the aging and neurodegenerative brain. *Histology and Histopathology* 27:263–275.

Camello-Almaraz, C., P. J. Gomez-Pinilla, M. J. Pozo, and P. J. Camello. 2008. Age-related alterations in Ca^{2+} signals and mitochondrial membrane potential in exocrine cells are prevented by melatonin. *Journal of Pineal Research* 45:191–198.

Cardinali, D. P., L. I. Brusco, C. Liberczuk, and A. M. Furio. 2002. The use of melatonin in Alzheimer's disease. *Neuro Endocrinology Letters* 23(Suppl 1):20–23.

Cardinali, D. P., E. S. Pagano, P. A. Scacchi Bernasconi, R. Reynoso, and P. Scacchi. 2013. Melatonin and mitochondrial dysfunction in the central nervous system. *Hormones and Behaviours* 63:322–330.

Chang, C. F., H. J. Huang, H. C. Lee, K. C. Hung, R. T. Wu, and A. M. Lin. 2012. Melatonin attenuates kainic acid-induced neurotoxicity in mouse hippocampus via inhibition of autophagy and alpha-synuclein aggregation. *Journal of Pineal Research* 52:312–321.

Chen, M., V. O. Ona, M. Li, R. J. Ferrante, K. B. Fink, S. Zhu, J. Bian et al. 2000. Minocycline inhibits caspase-1 and caspase-3 expression and delays mortality in a transgenic mouse model of Huntington disease. *Nature Medicine* 6:797–801.

Chen, Y., E. McMillan-Ward, J. Kong, S. J. Israels, and S. B. Gibson. 2007. Mitochondrial electron-transport-chain inhibitors of complexes I and II induce autophagic cell death mediated by reactive oxygen species. *Journal of Cell Science* 120:4155–4166.

Chetsawang, B., J. Chetsawang, and P. Govitrapong. 2009. Protection against cell death and sustained tyrosine hydroxylase phosphorylation in hydrogen peroxide- and MPP-treated human neuroblastoma cells with melatonin. *Journal of Pineal Research* 6:36–42.

Chetsawang, B., P. Govitrapong, and M. Ebadi. 2004. The neuroprotective effect of melatonin against the induction of c-Jun phosphorylation by 6-hydroxydopamine on SK-N-SH cells. *Neuroscience Letters* 371:205–208.

Chetsawang, J., P. Govitrapong, and B. Chetsawang. 2007. Melatonin inhibits MPP+-induced caspase-mediated death pathway and DNA fragmentation factor-45 cleavage in SK-N-SH cultured cells. *Journal of Pineal Research* 43:115–120.

Choi, S. I., S. S. Joo, and Y. M. Yoo. 2008. Melatonin prevents nitric oxide-induced apoptosis by increasing the interaction between 14-3-3beta and p-Bad in SK-N-MC cells. *Journal of Pineal Research* 44:95–100.

Christofides, J., M. Bridel, M. Egerton, G. M. Mackay, C. M. Forrest, N. Stoy, L. G. Darlington, and T. W. Stone. 2006. Blood 5-hydroxytryptamine, 5-hydroxyindoleacetic acid and melatonin levels in patients with either Huntington's disease or chronic brain injury. *Journal of Neurochemistry* 97:1078–1088.

Chu, C. T. 2006. Autophagic stress in neuronal injury and disease. *Journal of Neuropathology and Experimental Neurology* 65:423–432.

Costello, S., M. Cockburn, J. Bronstein, X. Zhang, and B. Ritz. 2009. Parkinson's disease and residential exposure to maneb and paraquat from agricultural applications in the central valley of California. *American Journal of Epidemiology* 169:919–926.

Coulom, H. and S. Birman. 2004. Chronic exposure to rotenone models sporadic Parkinson's disease in *Drosophila melanogaster*. *Journal of Neuroscience* 24:10993–10998.

Culmsee, C. and S. Landshamer. 2006. Molecular insights into mechanisms of the cell death program: Role in the progression of neurodegenerative disorders. *Current Alzheimer Research* 3:269–283.

Danial, N. N. and S. J. Korsmeyer. 2004. Cell death: Critical control points. *Cell* 116:205–219.

Dardiotis, E., E. Panayiotou, M. L. Feldman, A. Hadjisavvas, S. Malas, I. Vonta, G. Hadjigeorgiou, K. Kyriakou, and T. Kyriakides. 2013. Intraperitoneal melatonin is not neuroprotective in the G93ASOD1 transgenic mouse model of familial ALS and may exacerbate neurodegeneration. *Neuroscience Letters* 548:170–175.

Das, A., M. McDowell, M. J. Pava, J. A. Smith, R. J. Reiter, J. J. Woodward, A. K. Varma, S. K. Ray, and N. L. Banik. 2010. The inhibition of apoptosis by melatonin in VSC4.1 motoneurons exposed to oxidative stress, glutamate excitotoxicity, or TNF-α toxicity involves membrane melatonin receptors. *Journal of Pineal Research* 48:157–169.

de la Monte, S. M., Y. K. Sohn, N. Ganju, and J. R. Wands. 1998. P53- and CD95-associated apoptosis in neurodegenerative diseases. *Laboratory Investigation* 78:401–411.

de Rijk, M. C., L. J. Launer, K. Berger, M. M. Breteler, J. F. Dartigues, M. Baldereschi, L. Fratiglioni et al. 2000. Prevalence of Parkinson's disease in Europe: A collaborative study of population-based cohorts. Neurologic Diseases in the Elderly Research Group. *Neurology* 54:S21–S23.

Deigner, H. P., U. Haberkorn, and R. Kinscherf. 2000. Apoptosis modulators in the therapy of neurodegenerative diseases. *Expert Opinion on Investigational Drugs* 9:747–764.

Deng, Y. Q., G. G. Xu, P. Duan, Q. Zhang, and J. Z. Wang. 2005. Effects of melatonin on wortmannin-induced tau hyperphosphorylation. *Acta Pharmacologica Sinica* 26:519–526.

Dowling, G. A., J. Mastick, E. Colling, J. H. Carter, C. M. Singer, and M. J. Aminoff. 2005. Melatonin for sleep disturbances in Parkinson's disease. *Sleep Medicine* 6:459–466.

Dragicevic, N., V. Delic, C. Cao, N. Copes, X. Lin, M. Mamcarz, L. Wang, G. W. Arendash, and P. C. Bradshaw. 2012. Caffeine increases mitochondrial function and blocks melatonin signaling to mitochondria in Alzheimer's mice and cells. *Neuropharmacology* 63:1368–1379.

Du, C., M. Fang, Y. Li, L. Li, and X. Wang. 2000. Smac, a mitochondrial protein that promotes cytochrome c-dependent caspase activation by eliminating IAP inhibition. *Cell* 102:33–42.

Dubbels, R., R. J. Reiter, E. Klenke, A. Goebel, E. Schnakenberg, C. Ehlers, H. W. Schiwara, and W. Schloot. 1995. Melatonin in edible plants identified by radioimmunoassay and by high performance liquid chromatography-mass spectrometry. *Journal of Pineal Research* 18:28–31.

Ebadi, M., S. K. Sharma, P. Ghafourifar, H. Brown-Borg, and H. El Refaey. 2005. Peroxynitrite in the pathogenesis of Parkinson's disease and the neuroprotective role of metallothioneins. *Methods in Enzymology* 396:276–298.

Engidawork, E., T. Gulesserian, B. C. Yoo, N. Cairns, and G. Lubec. 2001. Alteration of caspases and apoptosis-related proteins in brains of patients with Alzheimer's disease. *Biochemical and Biophysical Research Communications* 281:84–93.

Esposito, E. and S. Cuzzocrea. 2010. Antiinflammatory activity of melatonin in central nervous system. *Current Neuropharmacology* 8:228–242.

Feng, Z., Y. Chang, Y. Cheng, B. L. Zhang, Z. W. Qu, C. Qin, and J. T. Zhang. 2004a. Melatonin alleviates behavioral deficits associated with apoptosis and cholinergic system dysfunction in the APP 695 transgenic mouse model of Alzheimer's disease. *Journal of Pineal Research* 37:129–136.

Feng, Z., Y. Cheng, and J. T. Zhang. 2004b. Long-term effects of melatonin or 17 beta-estradiol on improving spatial memory performance in cognitively impaired, ovariectomized adult rats. *Journal of Pineal Research* 37:198–206.

Feng, Z., C. Qin, Y. Chang, and J. T. Zhang. 2006. Early melatonin supplementation alleviates oxidative stress in a transgenic mouse model of Alzheimer's disease. *Free Radical Biology and Medicine* 40:101–109.

Feng, Z. and J. T. Zhang. 2004. Protective effect of melatonin on beta-amyloid-induced apoptosis in rat astroglioma C6 cells and its mechanism. *Free Radical Biology and Medicine* 37:1790–1801.

Ferrer, I., R. Blanco, M. Carmona, and B. Puig. 2001. Phosphorylated mitogen-activated protein kinase (MAPK/ERK-P), protein kinase of 38 kDa (p38-P), stress-activated protein kinase (SAPK/JNK-P), and calcium/calmodulin-dependent kinase II (CaM kinase II) are differentially expressed in tau deposits in neurons and glial cells in tauopathies. *Journal of Neural Transmission* 108:1397–1415.

Fitsanakis, V. A., V. Amarnath, J. T. Moore, K. S. Montine, J. Zhang, and T. J. Montine. 2002. Catalysis of catechol oxidation by metal-dithiocarbamate complexes in pesticides. *Free Radical Biology and Medicine* 33:1714–1723.

Friedlander, R. M. 2003. Apoptosis and caspases in neurodegenerative diseases. *The New England Journal of Medicine* 348:1365–1375.

Fu, W., Z. Zheng, W. Zhuang, D. Chen, X. Wang, and X. Sun. 2013. Neural metabolite changes in corpus striatum after rat multipotent mesenchymal stem cells transplanted in hemiparkinsonian rats by magnetic resonance spectroscopy. *The International Journal of Neuroscience* 123:883–891.

Furio, A. M., L. I. Brusco, and D. P. Cardinali. 2007. Possible therapeutic value of melatonin in mild cognitive impairment: A retrospective study. *Journal of Pineal Research* 43:404–409.

Gal, J., A. L. Strom, D. M. Kwinter, R. Kilty, J. Zhang, P. Shi, W. Fu, M. W. Wooten, and H. Zhu. 2009. Sequestosome 1/p62 links familial ALS mutant SOD1 to LC3 via an ubiquitin-independent mechanism. *Journal of Neurochemistry* 111:1062–1073.

Gehrman, P. R., D. J. Connor, J. L. Martin, T. Shochat, J. Corey-Bloom, and S. Ancoli-Israel. 2009. Melatonin fails to improve sleep or agitation in double-blind randomized placebo-controlled trial of institutionalized patients with Alzheimer disease. *The American Journal of Geriatric Psychiatry* 17:166–169.

Graham, R. K., Y. Deng, E. J. Slow, B. Haigh, N. Bissada, G. Lu, J. Pearson et al. 2006. Cleavage at the caspase-6 site is required for neuronal dysfunction and degeneration due to mutant huntingtin. *Cell* 125:1179–1191.

Gross, A., J. Jockel, M. C. Wei, and S. J. Korsmeyer. 1998. Enforced dimerization of BAX results in its translocation, mitochondrial dysfunction and apoptosis. *EMBO Journal* 17:3878–3885.

Guegan, C., M. Vila, G. Rosoklija, A. P. Hays, and S. Przedborski. 2001. Recruitment of the mitochondrial-dependent apoptotic pathway in amyotrophic lateral sclerosis. *The Journal of Neuroscience* 21:6569–6576.

Guegan, C., M. Vila, P. Teissman, C. Chen, B. Onteniente, M. Li, R. M. Friedlander, and S. Przedborski. 2002. Instrumental activation of bid by caspase-1 in a transgenic mouse model of ALS. *Molecular and Cellular Neuroscience* 20:553–562.

Guo, Q., J. Xie, X. Chang, and H. Du. 2001. Prostate apoptosis response-4 enhances secretion of amyloid beta peptide 1–42 in human neuroblastoma IMR-32 cells by a caspase-dependent pathway. *Journal of Biological Chemistry* 276:16040–16044.

Gutierrez-Cuesta, J., M. Tajes, A. Jimenez, A. Coto-Montes, A. Camins, and M. Pallas. 2008. Evaluation of potential pro-survival pathways regulated by melatonin in a murine senescence model. *Journal of Pineal Research* 45:497–505.

Hausman, D. B., J. G. Fischer, and M. A. Johnson. 2011. Nutrition in centenarians. *Maturitas* 68:203–209.

Hu, Y., M. A. Benedict, D. Wu, N. Inohara, and G. Nunez. 1998. Bcl-XL interacts with Apaf-1 and inhibits Apaf-1-dependent caspase-9 activation. *Proceedings of the National Academy of Sciences of the United States of America* 95:4386–4391.

Huntington's Disease Collaborative Research Group. 1993. A novel gene containing a trinucleotide repeat that is expanded on Huntington's disease chromosomes. *Cell* 72:971–983.

Iacovitti, L., N. D. Stull, and K. Johnston. 1997. Melatonin rescues dopamine neurons from cell death in tissue culture models of oxidative stress. *Brain Research* 768:317–326.

Iacovitti, L., N. D. Stull, and A. Mishizen. 1999. Neurotransmitters, KCl and antioxidants rescue striatal neurons from apoptotic cell death in culture. *Brain Research* 816:276–285.

Iqbal, K., C. Alonso Adel, E. El-Akkad, C. X. Gong, N. Haque, S. Khatoon, I. Tsujio, and I. Grundke-Iqbal. 2002. Pharmacological targets to inhibit Alzheimer neurofibrillary degeneration. *Journal of Neural Transmission Supplement* 62:309–319.

Jacob, S., B. Poeggeler, J. H. Weishaupt, A. L. Siren, R. Hardeland, M. Bahr, and H. Ehrenreich. 2002. Melatonin as a candidate compound for neuroprotection in amyotrophic lateral sclerosis (ALS): High tolerability of daily oral melatonin administration in ALS patients. *Journal of Pineal Research* 33:186–187.

Jang, J. H. and Y. J. Surh. 2005. AP-1 mediates beta-amyloid-induced iNOS expression in PC12 cells via the ERK2 and p38 MAPK signaling pathways. *Biochemical and Biophysical Research Communications* 331:1421–1428.

Jang, M. H., S. B. Jung, M. H. Lee, C. J. Kim, Y. T. Oh, I. Kang, J. Kim, and E. H. Kim. 2005. Melatonin attenuates amyloid beta25–35-induced apoptosis in mouse microglial BV2 cells. *Neuroscience Letters* 380:26–31.

Jansen, S. L., D. A. Forbes, V. Duncan, and D. G. Morgan. 2006. Melatonin for cognitive impairment. *Cochrane Database of Systematic Reviews* 1:CD003802.

Jean-Louis, G., H. von Gizycki, and F. Zizi. 1998. Melatonin effects on sleep, mood, and cognition in elderly with mild cognitive impairment. *Journal of Pineal Research* 25:177–183.

Jesudason, E. P., B. Baben, B. S. Ashok, J. G. Masilamoni, R. Kirubagaran, W. C. Jebaraj, and R. Jayakumar. 2007. Anti-inflammatory effect of melatonin on A beta vaccination in mice. *Molecular and Cellular Biochemistry* 298:69–81.

Jin, Z. and W. S. El-Deiry. 2005. Overview of cell death signaling pathways. *Cancer Biology and Therapy* 4:139–163.

Jordan, J., V. Cena, and J. H. Prehn. 2003. Mitochondrial control of neuron death and its role in neurodegenerative disorders. *Journal of Physiology and Biochemistry* 59:129–141.

Jou, M. J., T. I. Peng, R. J. Reiter, S. B. Jou, H. Y. Wu, and S. T. Wen. 2004. Visualization of the antioxidative effects of melatonin at the mitochondrial level during oxidative stress-induced apoptosis of rat brain astrocytes. *Journal of Pineal Research* 37:55–70.

Jun, Z., Z. Li, W. Fang, Y. Fengzhen, W. Puyuan, L. Wenwen, S. Zhi, and S. C. Bondy. 2013. Melatonin decreases levels of S100β and NFKB, increases levels of synaptophysinina rat model of alzheimer's disease. *Current Aging Science* 6:142–149.

Kandel, E. R., J. H. Schwartz, and T. M. Jessell. 2000. *Principles of Neural Science*, 4th edn. McGraw-Hill, New York.

Kilic, U., E. Kilic, R. J. Reiter, C. L. Bassetti, and D. M. Hermann. 2005. Signal transduction pathways involved in melatonin-induced neuroprotection after focal cerebral ischemia in mice. *Journal of Pineal Research* 38:67–71.

Kim, Y. S., W. S. Joo, B. K. Jin, Y. H. Cho, H. H. Baik, and C. W. Park. 1998. Melatonin protects 6-OHDA-induced neuronal death of nigrostriatal dopaminergic system. *NeuroReport* 9:2387–2390.

Kostrzewa, R. M. and J. Segura-Aguilar. 2003. Novel mechanisms and approaches in the study of neurodegeneration and neuroprotection. a review. *Neurotoxicity Research* 5:375–383.

Kratsovnik, E., Y. Bromberg, O. Sperling, and E. Zoref-Shani. 2005. Oxidative stress activates transcription factor NF-kB-mediated protective signaling in primary rat neuronal cultures. *Journal of Molecular Neuroscience* 26:27–32.

Kroemer, G. and J. C. Reed. 2000. Mitochondrial control of cell death. *Nature Medicine* 6:513–519.

Kumar, A. M., F. Tims, D. G. Cruess, M. J. Mintzer, G. Ironson, D. Loewenstein, R. Cattan, J. B. Fernandez, C. Eisdorfer, and M. Kumar. 1999. Music therapy increases serum melatonin levels in patients with Alzheimer's disease. *Alternative Therapy in Health and Medicine* 5:49–57.

Larsen, K. E. and D. Sulzer. 2002. Autophagy in neurons: A review. *Histology and Histopathology* 17:897–908.

Lee, H. J., S. H. Kim, K. W. Kim, J. H. Um, H. W. Lee, B. S. Chung, and C. D. Kang. 2001. Antiapoptotic role of NF-kappaB in the auto-oxidized dopamine-induced apoptosis of PC12 cells. *Journal of Neurochemistry* 76:602–609.

Lee, J., H. R. Kim, C. Quinley, J. Kim, J. Gonzalez-Navajas, R. Xavier, and E. Raz. 2012. Autophagy suppresses interleukin-1beta (IL-1beta) signaling by activation of p62 degradation via lysosomal and proteasomal pathways. *Journal of Biological Chemistry* 287:4033–4040.

Leon, J., D. Acuna-Castroviejo, G. Escames, D. X. Tan, and R. J. Reiter. 2005. Melatonin mitigates mitochondrial malfunction. *Journal of Pineal Research* 38:1–9.

Levites, Y., M. B. Youdim, G. Maor, and S. Mandel. 2002. Attenuation of 6-hydroxydopamine (6-OHDA)-induced nuclear factor-kappaB (NF-kappaB) activation and cell death by tea extracts in neuronal cultures. *Biochemical Pharmacology* 63:21–29.

Lezoualc'h, F., M. Sparapani, and C. Behl. 1998. *N*-acetyl-serotonin (normelatonin) and melatonin protect neurons against oxidative challenges and suppress the activity of the transcription factor NF-kappaB. *Journal of Pineal Research* 24:168–178.

Li, H., H. Zhu, C. J. Xu, and J. Yuan. 1998. Cleavage of BID by caspase 8 mediates the mitochondrial damage in the Fas pathway of apoptosis. *Cell* 94:491–501.

Li, M., V. O. Ona, C. Guegan, M. Chen, V. Jackson-Lewis, L. J. Andrews, A. J. Olszewski et al. 2000. Functional role of caspase-1 and caspase-3 in an ALS transgenic mouse model. *Science* 288:335–339.

Li, P., H. Allen, S. Banerjee, S. Franklin, L. Herzog, C. Johnston, J. McDowell et al. 1995. Mice deficient in IL-1 beta-converting enzyme are defective in production of mature IL-1 beta and resistant to endotoxic shock. *Cell* 80:401–411.

Li, X., Y. Guan, Y. Chen, C. Zhang, C. Shi, F. Zhou, L. Yu, J. Juan, and X. Wang. 2013. Expression of Wnt5a and its receptor Fzd2 is changed in the spinal cord of adult amyotrophic lateral sclerosis transgenic mice. *International Journal of Clinical and Experimental Pathology* 6:1245–1260.

Li, X. C., Z. F. Wang, J. X. Zhang, Q. Wang, and J. Z. Wang. 2005. Effect of melatonin on calyculin A-induced tau hyperphosphorylation. *European Journal of Pharmacology* 510:25–30.

Lin, C. H., J. Y. Huang, C. H. Ching, and J. I. Chuang. 2008. Melatonin reduces the neuronal loss, downregulation of dopamine transporter, and upregulation of D2 receptor in rotenone-induced parkinsonian rats. *Journal of Pineal Research* 44:205–213.

Ling, F. A., D. Z. Hui, and S. M. Ji. 2007. Protective effect of recombinant human somatotropin on amyloid beta-peptide induced learning and memory deficits in mice. *Growth Hormone & IGF Research* 17:336–341.

Liu, R. Y., J. N. Zhou, J. van Heerikhuize, M. A. Hofman, and D. F. Swaab. 1999. Decreased melatonin levels in postmortem cerebrospinal fluid in relation to aging, Alzheimer's disease, and apolipoprotein E-epsilon4/4 genotype. *Journal of Clinical Endocrinology and Metabolism* 84:323–327.

Liu, S. J. and J. Z. Wang. 2002. Alzheimer-like tau phosphorylation induced by wortmannin in vivo and its attenuation by melatonin. *Acta Pharmacologica Sinica* 23:183–187.

Lobsiger, C. S. and D. W. Cleveland. 2007. Glial cells as intrinsic components of non-cell-autonomous neurodegenerative disease. *Nature Neuroscience* 10:1355–1360.

Lukiw, W. J. and N. G. Bazan. 2006. Survival signalling in Alzheimer's disease. *Biochemical Society Transactions* 34:1277–1282.

Luo, X., I. Budihardjo, H. Zou, C. Slaughter, and X. Wang. 1998. Bid, a Bcl2 interacting protein, mediates cytochrome c release from mitochondria in response to activation of cell surface death receptors. *Cell* 94:481–490.

Ma, J., V. E. Shaw, and J. Mitrofanis. 2009. Does melatonin help save dopaminergic cells in MPTP-treated mice? *Parkinsonism Related and Disorders* 15:307–314.

Mahlberg, R., S. Walther, P. Kalus, G. Bohner, S. Haedel, F. M. Reischies, K. P. Kuhl, R. Hellweg, and D. Kunz. 2008. Pineal calcification in Alzheimer's disease: An in vivo study using computed tomography. *Neurobiology of Aging* 29:203–209.

Maiuri, M. C., G. Le Toumelin, A. Criollo, J. C. Rain, F. Gautier, P. Juin, E. Tasdemir et al. 2007. Functional and physical interaction between Bcl-X(L) and a BH3-like domain in Beclin-1. *EMBO Journal* 26:2527–2539.

Marques, C. A., U. Keil, A. Bonert, B. Steiner, C. Haass, W. E. Muller, and A. Eckert. 2003. Neurotoxic mechanisms caused by the Alzheimer's disease-linked Swedish amyloid precursor protein mutation: Oxidative stress, caspases, and the JNK pathway. *Journal of Biological Chemistry* 278:28294–28302.

Martin, M., M. Macias, G. Escames, J. Leon, and D. Acuna-Castroviejo. 2000. Melatonin but not vitamins C and E maintains glutathione homeostasis in t-butyl hydroperoxide-induced mitochondrial oxidative stress. *FASEB Journal* 14:1677–1679.

Mattson, M. P. 2003. Excitotoxic and excitoprotective mechanisms: Abundant targets for the prevention and treatment of neurodegenerative disorders. *Neuromolecular Medicine* 3:65–94.

Mattson, M. P. 2007. Calcium and neurodegeneration. *Aging Cell* 6:337–350.

Mattson, M. P., C. Culmsee, and Z. F. Yu. 2000. Apoptotic and antiapoptotic mechanisms in stroke. *Cell and Tissue Research* 301:173–187.

Mayo, J. C., R. M. Sainz, H. Uria, I. Antolin, M. M. Esteban, and C. Rodriguez. 1998. Melatonin prevents apoptosis induced by 6-hydroxydopamine in neuronal cells: Implications for Parkinson's disease. *Journal of Pineal Research* 24:179–192.

Medeiros, C. A., P. F. Carvalhedo de Bruin, L. A. Lopes, M. C. Magalhaes, M. de Lourdes Seabra, and V. M. de Bruin. 2007. Effect of exogenous melatonin on sleep and motor dysfunction in Parkinson's disease. A randomized, double blind, placebo-controlled study. *Journal of Neurology* 254:459–464.

Mishima, K., M. Okawa, Y. Hishikawa, S. Hozumi, H. Hori, and K. Takahashi. 1994. Morning bright light therapy for sleep and behavior disorders in elderly patients with dementia. *Acta Psychiatrica Scandinavica* 89:1–7.

Niles, L. P., K. J. Armstrong, L. M. Rincon Castro, C. V. Dao, R. Sharma, C. R. McMillan, L. C. Doering, and D. L. Kirkham. 2004. Neural stem cells express melatonin receptors and neurotrophic factors: Colocalization of the MT1 receptor with neuronal and glial markers. *BMC Neuroscience* 5:41.

Niranjan, R., C. Nath, and R. Shukla. 2010. The mechanism of action of MPTP-induced neuroinflammation and its modulation by melatonin in rat astrocytoma cells, C6. *Free Radical Research* 44:1304–1316.

Nixon, R. A. and D. S. Yang. 2012. Autophagy and neuronal cell death in neurological disorders. *Cold Spring Harbor Perspectives in Biology* 4(10):pii: a008839.

Nopparat, C., J. E. Porter, M. Ebadi, and P. Govitrapong. 2010. The mechanism for the neuroprotective effect of melatonin against methamphetamine-induced autophagy. *Journal of Pineal Research* 49:382–389.

Olde Rikkert, M. G. and A. S. Rigaud. 2001. Melatonin in elderly patients with insomnia. A systematic review. *Gerontology Geriatrics* 34:491–497.

Ortiz, G. G., M. F. Crespo-Lopez, C. Moran-Moguel, J. J. Garcia, R. J. Reiter, and D. Acuna-Castroviejo. 2001. Protective role of melatonin against MPTP-induced mouse brain cell DNA fragmentation and apoptosis in vivo. *Neuro Endocrinology Letters* 22:101–108.

Ozcankaya, R. and N. Delibas. 2002. Malondialdehyde, superoxide dismutase, melatonin, iron, copper, and zinc blood concentrations in patients with Alzheimer disease: Cross-sectional study. *Croatian Medical Journal* 43:28–32.

Pandi-Perumal, S. R., A. S. BaHammam, G. M. Brown, D. W. Spence, V. K. Bharti, C. Kaur, R. Hardeland, and D. P. Cardinali. 2013. Melatonin antioxidative defense: Therapeutical implications for aging and neurodegenerative processes. *Neurotoxicity Research* 23:267–300.

Pandya, R. S., L. J. Mao, E. W. Zhou, R. Bowser, Z. Zhu, Y. Zhu, and X. Wang. 2012. Neuroprotection for amyotrophic lateral sclerosis: Role of stem cells, growth factors, and gene therapy. *Central Nervous System Agents in Medicinal Chemistry* 12:15–27.

Pandya, R. S., H. Zhu, W. Li, R. Bowser, R. M. Friedlander, and X. Wang. 2013. Therapeutic neuroprotective agents for amyotrophic lateral sclerosis. *Cellular and Molecular Life Sciences* 70:4729–4745.

Pappolla, M. A., Y. J. Chyan, B. Poeggeler, B. Frangione, G. Wilson, J. Ghiso, and R. J. Reiter. 2000. An assessment of the antioxidant and the antiamyloidogenic properties of melatonin: Implications for Alzheimer's disease. *Journal of Neural Transmission* 107:203–231.

Pappolla, M. A., M. J. Simovich, T. Bryant-Thomas, Y. J. Chyan, B. Poeggeler, M. Dubocovich, R. Bick, G. Perry, F. Cruz-Sanchez, and M. A. Smith. 2002. The neuroprotective activities of melatonin against the Alzheimer beta-protein are not mediated by melatonin membrane receptors. *Journal of Pineal Research* 32:135–142.

Pappolla, M. A., M. Sos, R. A. Omar, R. J. Bick, D. L. Hickson-Bick, R. J. Reiter, S. Efthimiopoulos, and N. K. Robakis. 1997. Melatonin prevents death of neuroblastoma cells exposed to the Alzheimer amyloid peptide. *Journal of Neuroscience* 17:1683–1690.

Pattingre, S., A. Tassa, X. Qu, R. Garuti, X. H. Liang, N. Mizushima, M. Packer, M. D. Schneider, and B. Levine. 2005. Bcl-2 antiapoptotic proteins inhibit Beclin 1-dependent autophagy. *Cell* 122:927–939.

Przedborski, S. 2004. Programmed cell death in amyotrophic lateral sclerosis: A mechanism of pathogenic and therapeutic importance. *Neurologist* 10:1–7.

Ravindra, T., N. K. Lakshmi, and Y. R. Ahuja. 2006. Melatonin in pathogenesis and therapy of cancer. *Indian Journal of Medical Sciences* 60:523–535.

Reiter, R. J., J. Cabrera, R. M. Sainz, J. C. Mayo, L. C. Manchester, and D. X. Tan. 1999. Melatonin as a pharmacological agent against neuronal loss in experimental models of Huntington's disease, Alzheimer's disease and parkinsonism. *Annals of the New York Academy of Sciences* 890:471–485.

Reiter, R. J., L. C. Manchester, and D. X. Tan. 2010. Neurotoxins: Free radical mechanisms and melatonin protection. *Current Neuropharmacology* 8:194–210.

Rigamonti, D., J. H. Bauer, C. De-Fraja, L. Conti, S. Sipione, C. Sciorati, E. Clementi et al. 2000. Wild-type huntingtin protects from apoptosis upstream of caspase-3. *Journal of Neuroscience* 20:3705–3713.

Rigamonti, D., S. Sipione, D. Goffredo, C. Zuccato, E. Fossale, and E. Cattaneo. 2001. Huntingtin's neuroprotective activity occurs via inhibition of procaspase-9 processing. *Journal of Biological Chemistry* 276:14545–14548.

Rival, T., L. Soustelle, C. Strambi, M. T. Besson, M. Iche, and S. Birman. 2004. Decreasing glutamate buffering capacity triggers oxidative stress and neuropil degeneration in the *Drosophila* brain. *Current Biology* 14:599–605.

Rogerio, F., S. A. Teixeira, A. C. de Rezende, R. C. de Sa, L. de Souza Queiroz, G. De Nucci, M. N. Muscara, and F. Langone. 2005. Superoxide dismutase isoforms 1 and 2 in lumbar spinal cord of neonatal rats after sciatic nerve transection and melatonin treatment. *Brain Research Developmental Brain Research* 154:217–225.

Rojas-Fernandez, C. H., M. Chen, and H. L. Fernandez. 2002. Implications of amyloid precursor protein and subsequent beta-amyloid production to the pharmacotherapy of Alzheimer's disease. *Pharmacotherapy* 22:1547–1563.

Ryu, H., K. Smith, S. I. Camelo, I. Carreras, J. Lee, A. H. Iglesias, F. Dangond et al. 2005. Sodium phenylbutyrate prolongs survival and regulates expression of anti-apoptotic genes in transgenic amyotrophic lateral sclerosis mice. *Journal of Neurochemistry* 93:1087–1098.

Salminen, A., K. Kaarniranta, A. Kauppinen, J. Ojala, A. Haapasalo, H. Soininen, and M. Hiltunen. 2013. Impaired autophagy and APP processing in Alzheimer's disease: The potential role of Beclin 1 interactome. *Progress in Neurobiology* 106–107:33–54.

Saravanan, K. S., K. M. Sindhu, and K. P. Mohanakumar. 2007. Melatonin protects against rotenone-induced oxidative stress in a hemiparkinsonian rat model. *Journal of Pineal Research* 42:247–253.

Sas, K., H. Robotka, J. Toldi, and L. Vecsei. 2007. Mitochondria, metabolic disturbances, oxidative stress and the kynurenine system, with focus on neurodegenerative disorders. *Journal of Neurological Sciences* 257:221–239.

Sathasivam, S., A. J. Grierson, and P. J. Shaw. 2005. Characterization of the caspase cascade in a cell culture model of SOD1-related familial amyotrophic lateral sclerosis: Expression, activation and therapeutic effects of inhibition. *Neuropathology and Applied Neurobiology* 31:467–485.

Schulz, J. B. and M. F. Beal. 1994. Mitochondrial dysfunction in movement disorders. *Current Opinion in Neurology* 7:333–339.

Sewerynek, E. 2002. Melatonin and the cardiovascular system. *Neuro Endocrinology Letters Supplement* 1:79–83.

Sharma, R., C. R. McMillan, and L. P. Niles. 2007. Neural stem cell transplantation and melatonin treatment in a 6-hydroxydopamine model of Parkinson's disease. *Journal of Pineal Research* 43:245–254.

Shen, Y. X., S. Y. Xu, W. Wei, X. L. Wang, H. Wang, and X. Sun. 2002. Melatonin blocks rat hippocampal neuronal apoptosis induced by amyloid beta-peptide 25–35. *Journal of Pineal Research* 32:163–167.

Sheng, R., X. Q. Liu, L. S. Zhang, B. Gao, R. Han, Y. Q. Wu, X. Y. Zhang, and Z. H. Qin. 2012. Autophagy regulates endoplasmic reticulum stress in ischemic preconditioning. *Autophagy* 8:310–325.

Singer, C., R. E. Tractenberg, J. Kaye, K. Schafer, A. Gamst, M. Grundman, R. Thomas, and L. J. Thal. 2003. A multicenter, placebo-controlled trial of melatonin for sleep disturbance in Alzheimer's disease. *Sleep* 26:893–901.

Singh, S., R. Ahmed, R. K. Sagar, and B. Krishana. 2006. Neuroprotection of the nigrostriatal dopaminergic neurons by melatonin in hemiparkinsonium rat. *Indian Journal of Medical Research* 124:419–426.

Singhal, N. K., G. Srivastava, D. K. Patel, S. K. Jain, and M. P. Singh. 2011. Melatonin or silymarin reduces maneb- and paraquat-induced Parkinson's disease phenotype in the mouse. *Journal of Pineal Research* 50:97–109.

Sousa, S. C. and R. F. Castilho. 2005. Protective effect of melatonin on rotenone plus Ca^{2+}-induced mitochondrial oxidative stress and PC12 cell death. *Antioxidants & Redox Signaling* 7:1110–1116.

Southgate, G. and S. Daya. 1999. Melatonin reduces quinolinic acid-induced lipid peroxidation in rat brain homogenate. *Metabolic Brain Disease* 14:165–171.

Srinivasan, V., D. P. Cardinali, U. S. Srinivasan, C. Kaur, G. M. Brown, D. W. Spence, R. Hardeland, and S. R. Pandi-Perumal. 2011. Therapeutic potential of melatonin and its analogs in Parkinson's disease: Focus on sleep and neuroprotection. *Therapeutic Advances in Neurological Disorders* 4:297–317.

Srivastava, G., K. Singh, M. N. Tiwari, and M. P. Singh. 2010. Proteomics in Parkinson's disease: Current trends, translational snags and future possibilities. *Expert Review of Proteomics* 7:127–139.

Suntres, Z. E. 2002. Role of antioxidants in paraquat toxicity. *Toxicology* 180:65–77.

Swarnkar, S., S. Singh, P. Goswami, R. Mathur, I. K. Patro, and C. Nath. 2012. Astrocyte activation: A key step in rotenone induced cytotoxicity and DNA damage. *Neurochemical Research* 37:2178–2189.

Tan, D. X., L. C. Manchester, R. M. Sainz, J. C. Mayo, J. Leon, R. Hardeland, B. Poeggeler, and R. J. Reiter. 2005. Interactions between melatonin and nicotinamide nucleotide: NADH preservation in cells and in cell-free systems by melatonin. *Journal of Pineal Research* 39.185–194.

Tan, D. X., R. J. Reiter, L. C. Manchester, M. T. Yan, M. El-Sawi, R. M. Sainz, J. C. Mayo, R. Kohen, M. Allegra, and R. Hardeland. 2002. Chemical and physical properties and potential mechanisms: Melatonin as a broad spectrum antioxidant and free radical scavenger. *Current Topics in Medicinal Chemistry* 2:181–197.

Tapias, V., J. R. Cannon, and J. T. Greenamyre. 2010. Melatonin treatment potentiates neurodegeneration in a rat rotenone Parkinson's disease model. *Journal of Neuroscience Research* 88:420–427.

Teng, F. Y. and B. L. Tang. 2010. NF-kappaB signaling in neurite growth and neuronal survival. *Reviews in the Neurosciences* 21:299–313.

Tournier, C., P. Hess, D. D. Yang, J. Xu, T. K. Turner, A. Nimnual, D. Bar-Sagi, S. N. Jones, R. A. Flavell, and R. J. Davis. 2000. Requirement of JNK for stress-induced activation of the cytochrome c-mediated death pathway. *Science* 288:870–874.

Tunez, I., P. Montilla, M. Del Carmen Munoz, M. Feijoo, and M. Salcedo. 2004. Protective effect of melatonin on 3-nitropropionic acid-induced oxidative stress in synaptosomes in an animal model of Huntington's disease. *Journal of Pineal Research* 37:252–256.

Vila, M. and S. Przedborski. 2003. Targeting programmed cell death in neurodegenerative diseases. *Nature Reviews Neuroscience* 4:365–375.

Wang, D. L., Z. Q. Ling, F. Y. Cao, L. Q. Zhu, and J. Z. Wang. 2004. Melatonin attenuates isoproterenol-induced protein kinase A overactivation and tau hyperphosphorylation in rat brain. *Journal of Pineal Research* 37:11–16.

Wang, J. Z. and Z. F. Wang. 2006. Role of melatonin in Alzheimer-like neurodegeneration. *Acta Pharmacologica Sinica* 27:41–49.

Wang, X. 2009. The antiapoptotic activity of melatonin in neurodegenerative diseases. *CNS Neuroscience & Therapeutics* 15:345–357.

Wang, X., B. E. Figueroa, I. G. Stavrovskaya, Y. Zhang, A. C. Sirianni, S. Zhu, A. L. Day, B. S. Kristal, and R. M. Friedlander. 2009. Methazolamide and melatonin inhibit mitochondrial cytochrome C release and are neuroprotective in experimental models of ischemic injury. *Stroke* 40:1877–1885.

Wang, X., A. Sirianni, Z. Pei, K. Cormier, K. Smith, J. Jiang, S. Zhou et al. 2011. The melatonin MT1 receptor axis modulates mutant huntingtin-mediated toxicity. *Journal of Neuroscience* 31:14496–14507.

Wang, X., H. Wang, B. E. Figueroa, W. H. Zhang, C. Huo, Y. Guan, Y. Zhang, J. M. Bruey, J. C. Reed, and R. M. Friedlander. 2005a. Dysregulation of receptor interacting protein-2 and caspase recruitment domain only protein mediates aberrant caspase-1 activation in Huntington's disease. *Journal of Neuroscience* 25:11645–11654.

Wang, X., S. Zhu, M. Drozda, W. Zhang, I. G. Stavrovskaya, E. Cattaneo, R. J. Ferrante, B. S. Kristal, and R. M. Friedlander. 2003. Minocycline inhibits caspase-independent and -dependent mitochondrial cell death pathways in models of Huntington's disease. *Proceedings of the National Academy of Sciences of the United States of America* 100:10483–10487.

Wang, X., S. Zhu, Z. Pei, M. Drozda, I. G. Stavrovskaya, S. J. Del Signore, K. Cormier et al. 2008. Inhibitors of cytochrome c release with therapeutic potential for Huntington's disease. *Journal of Neuroscience* 28:9473–9485.

Wang, X. C., J. Zhang, X. Yu, L. Han, Z. T. Zhou, Y. Zhang, and J. Z. Wang. 2005b. Prevention of isoproterenol-induced tau hyperphosphorylation by melatonin in the rat. *Sheng Li Xue Bao* 57:7–12.

Weinreb, O., S. Mandel, and M. B. Youdim. 2003. Gene and protein expression profiles of anti- and pro-apoptotic actions of dopamine, R-apomorphine, green tea polyphenol (–)-epigallocatechine-3-gallate, and melatonin. *Annals of the New York Academy of Sciences* 993:351–361; discussion 387–393.

Weishaupt, J. H., C. Bartels, E. Polking, J. Dietrich, G. Rohde, B. Poeggeler, N. Mertens et al. 2006. Reduced oxidative damage in ALS by high-dose enteral melatonin treatment. *Journal of Pineal Research* 41:313–323.

Willis, G. L. 2008. Parkinson's disease as a neuroendocrine disorder of circadian function: Dopamine-melatonin imbalance and the visual system in the genesis and progression of the degenerative process. *Reviews of Neuroscience* 19:245–316.

Xie, J., K. S. Awad, and Q. Guo. 2005. RNAi knockdown of Par-4 inhibits neurosynaptic degeneration in ALS-linked mice. *Journal of Neurochemistry* 92:59–71.

Xie, J. and Q. Guo. 2005. PAR-4 is involved in regulation of beta-secretase cleavage of the Alzheimer amyloid precursor protein. *Journal of Biological Chemistry* 280:13824–13832.

Yang, J., X. Liu, K. Bhalla, C. N. Kim, A. M. Ibrado, J. Cai, T. I. Peng, D. P. Jones, and X. Wang. 1997. Prevention of apoptosis by Bcl-2: Release of cytochrome c from mitochondria blocked. *Science* 275:1129–1132.

Yang, Z. Z., O. Tschopp, A. Baudry, B. Dummler, D. Hynx, and B. A. Hemmings. 2004. Physiological functions of protein kinase B/Akt. *Biochemical Society Transactions* 32:350–354.

Yu, Z., D. Zhou, G. Cheng, and M. P. Mattson. 2000. Neuroprotective role for the p50 subunit of NF-kappaB in an experimental model of Huntington's disease. *Journal of Molecular Neuroscience* 15:31–44.

Zhang, W. H., X. Wang, M. Narayanan, Y. Zhang, C. Huo, J. C. Reed, and R. M. Friedlander. 2003. Fundamental role of the Rip2/caspase-1 pathway in hypoxia and ischemia-induced neuronal cell death. *Proceedings of the National Academy of Sciences of the United States of America* 100:16012–16017.

Zhang, Y., A. Cook, J. Kim, S. V. Baranov, J. Jiang, K. Smith, K. Cormier et al. 2013. Melatonin inhibits the caspase-1/cytochrome c/caspase-3 cell death pathway, inhibits MT1 receptor loss and delays disease progression in a mouse model of amyotrophic lateral sclerosis. *Neurobiology of Disease* 55:26–35.

Zhou, H., J. Chen, X. Lu, C. Shen, J. Zeng, L. Chen, and Z. Pei. 2012. Melatonin protects against rotenone-induced cell injury via inhibition of Omi and Bax-mediated autophagy in Hela cells. *Journal of Pineal Research* 52:120–127.

Zhou, J., S. Zhang, X. Zhao, and T. Wei. 2008. Melatonin impairs NADPH oxidase assembly and decreases superoxide anion production in microglia exposed to amyloid-beta1–42. *Journal of Pineal Research* 45:157–165.

Zhu, S., I. G. Stavrovskaya, M. Drozda, B. Y. Kim, V. Ona, M. Li, S. Sarang et al. 2002. Minocycline inhibits cytochrome c release and delays progression of amyotrophic lateral sclerosis in mice. *Nature* 417:74–78.

11 MT_2 Melatonin Receptors *Their Role in Sleep and Neuropsychiatric Disorders*

Stefano Comai and Gabriella Gobbi

CONTENTS

11.1 INTRODUCTION

Melatonin (MLT) is a neurohormone involved in numerous physiological processes, including circadian rhythms, mood regulation, anxiety, sleep, appetite, immune responses, and cardiac functions (Arendt, 1988; Reiter, 2003). Most of the effects of MLT in the brain result from the activation of two high-affinity G-protein coupled receptors (GPCRs), MT_1 and MT_2 (Dubocovich et al., 2010). A great number of ligands for the MLT receptors have been developed, but these advances have been met with difficulty obtaining compounds with selectivity toward only one of the two MLT receptor subtypes, especially the MT_1 receptor (Rivara et al., 2012). Since structure–activity relationships for the binding at the MT_2 receptor are quite consolidated (Rivara et al., 2007), several MT_2 receptor ligands belonging to different chemical classes have been synthesized (Mor et al., 2010). Despite this large number of selective MT_2 receptor ligands, only four have been tested and employed in neurobiological/neuropharmacological studies.

In this chapter, we thus focus our attention on the recent experimental evidence regarding the role of MT_2 receptors in sleep, anxiety, and other neuropsychiatric disorders.

11.2 MT_2 RECEPTORS REGULATION

The MT_2 receptor, also known as MEL_{1B}, Mel_{1b}, or MTNR1B, belongs to the superfamily of GPCRs, and in recombinant systems, it inhibits adenylyl cyclase activity (Reppert et al., 1995) and 3′-5′-monophosphate (cGMP) levels via the soluble guanylyl cyclase pathway (Petit et al., 1999).

MT_2 receptors have also been demonstrated to modulate the activity of the protein kinase C (Hunt et al., 2001), the production of phosphoinositide (MacKenzie et al., 2002), and voltage-gated K^+ channels (Yang et al., 2011).

However, the signaling pathways engaged following the MT_2 receptor activation, as well as the mechanisms responsible for the regulation of MT_2 receptor, are, for many aspects, still unknown. In vitro experiments carried out in CHO or NIH3T3 cells expressing MT_2 receptors showed that the expression of the receptors varies during the day due to physiological circadian oscillations in circulating MLT levels, which produces desensitization of these receptors (Witt-Enderby et al., 2003). But whether this effect also occurs in vivo and at the level of the brain needs to be demonstrated.

11.3 MT_2 RECEPTORS IN THE BRAIN

Study of the distribution of both MLT receptors within the central nervous system is a matter of ongoing research and has been mainly performed in animals. MT_2 receptor mRNA expression has been observed in the retina, suprachiasmatic nucleus (SCN), thalamus, hippocampus, vestibular nuclei, and cerebral and cerebellar cortex (Sallinen et al., 2005). Using polyclonal antibodies, we found that MT_2 receptors are present in the reticular thalamus, substantia nigra (pars reticulata), supraoptic nucleus, red nucleus, and CA_2 and CA_3 areas of hippocampus (Ochoa-Sanchez et al., 2011). A recent study in human postmortem brains has highlighted the presence of MT_2 receptors at the level of the SCN, the supraoptic nucleus, and the paraventricular nucleus (Wu et al., 2013).

Very recently, our laboratory completed the mapping of MT_1 and MT_2 receptors in the rat brain using polyclonal anti-MT_1 and anti-MT_2 antibodies (Angeloni et al., 2000). We found that MT_2 receptors are located in distinct areas of the brain compared to MT_1 receptors, suggesting that each of them plays a specific brain function (unpublished results). Very importantly, both MT_2 and MT_1 receptors were exclusively somatodendritic, which is in agreement with earlier electron microscopy observations demonstrating the predominance of GPCRs at the neuronal plasma membrane (Lacoste et al., 2009).

11.4 TOOLS USED TO ASSESS THE ROLE OF MT_2 RECEPTORS IN BRAIN FUNCTION: MT_2 RECEPTOR SELECTIVE LIGANDS, MT_2 RECEPTOR KNOCKOUT MICE, AND GENETIC STUDIES OF THE GENE ENCODING THE MT_2 RECEPTOR (*MTNR1B*)

Several selective MT_2 receptor ligands (Mor et al., 2010; Zlotos, 2012) belonging to different chemical classes have been developed during the last decade, but only a very limited number of them have been tested in preclinical psychopharmacology tests and neurobiological studies aimed at dissecting the role of MT_2 receptors in brain function. To date, only four selective MT_2 receptor ligands have been tested: the MT_2 receptor partial agonist UCM765 (Rivara et al., 2007), the MT_2 receptor full agonist IIK7 (Faust et al., 2000), and the MT_2 receptor antagonists 4-phenyl-2-propionamidotetralin (4P-PDOT) (Dubocovich et al., 1997) and K-185 (Faust et al., 2000). It is important to consider that while UCM765 is an MT_2 receptor partial agonist (alpha = 0.6) that possesses 100-fold higher affinity for MT_2 compared with MT_1 receptors (Rivara et al., 2007), IIK7 is a full MT_2 receptor agonist with 90-fold higher affinity for MT_2 than MT_1 receptors (Fisher and Sugden, 2009).

Another important instrument that has contributed to our understanding of the role of MT_2 receptors in brain function is the MT_2 receptor knockout mouse (MT_2KO) (Jin et al., 2003). In-vivo and in-vitro experiments using MT_2KO mice have demonstrated and supported an important role of MT_2 receptors in sleep (Comai et al., 2013), the control of SCN activity (Jin et al., 2003), anxiety (Ochoa-Sanchez et al., 2012), and learning and/or memory process (Larson et al., 2006).

In the last decade, researchers have also started looking at the gene encoding the MT_2 receptor (*MTNR1B*) in order to understand whether genetic or hereditary factors related to the MT_2 receptor could contribute to the etiology of several diseases. A large body of research has demonstrated an association between several variants within the *MTNR1B* gene locus and impaired fasting glucose and increased risk of type-2 diabetes (Karamitri et al., 2013). However, only one study to date has examined possible variations within the *MTNR1B* gene locus associated with mental diseases. Galecka et al. (2011) found that the *MTNR1B* gene may play a role in determining the risk of recurrent depressive disorders.

11.5 MELATONIN, MT_2 RECEPTORS, AND SLEEP

Many preclinical and clinical studies (Zhdanova, 2005) have highlighted the hypnotic effects of MLT, but several others have failed to find significant effects at both levels (James et al., 1987). Consequently, its clinical efficacy is still a matter of debate. Moreover, it is still unknown whether MLT acts directly on sleep regulation or on circadian rhythms associated with sleep.

Until now, it was hypothesized that MT_1 receptors control sleep whereas MT_2 receptors are involved in the time shift mainly by controlling the neural activity of the SCN (Dubocovich, 2007). Pharmacological and genetic studies using selective MT_2 receptor ligands and MT_2KO mice, respectively, have allowed researchers to better investigate and verify this hypothesis. UCM765, at the dose of 40 mg/kg (s.c.), significantly reduced the latency to the onset of non-rapid eye movement sleep (NREMS) while increasing the amount of NREMS during the 24 h. This increase in NREMS amount was mainly evident during the light/inactive phase of the 24 h light/dark cycle. No effects were observed for rapid eye movement sleep (REMS) amount or latency, whereas the amount of wakefulness was significantly decreased (Ochoa-Sanchez et al., 2011) (Figure 11.1). These findings have been replicated with UCM924 (Rivara et al., 2007), a class congener of UCM765 (Ochoa-Sanchez et al., 2014).

Similarly, Fisher and Sugden (2009) found that IIK7 (10 mg/kg, i.p.) decreased the latency to NREMS and increased the amount of NREMS during the first hour after injection. No effects on REMS amount and latency were reported.

It is noteworthy that the MT_2 receptor antagonist 4P-PDOT blocked the effects of UCM765 on NREMS and wakefulness, thus suggesting an MT_2 receptor–dependent mechanism (Ochoa-Sanchez et al., 2011). Looking more in detail at how UCM765 modulates the vigilance states, we found that this compound increased the number of sleep spindles that are considered the epitome of electroencephalographic (EEG) synchronization at sleep onset (Figure 11.1). In addition, UCM765 significantly enhanced the NREMS EEG power of the delta band. On the contrary, IIK7 did not affect NREMS EEG power spectra.

The same 24 h EEG/EMG experiment carried out with the non-selective MT_1–MT_2 receptor ligand UCM793, and MLT failed to produce similar results (unpublished data). Consequently, MT_2 receptor ligands possess more potent hypnotic properties compared with the nonselective MT_1–MT_2 ligands or MLT itself.

The reticular thalamus is an important component of the neural network controlling sleep; it is implicated in the generation of delta waves during deep slow wave sleep (SWS) (Bal and McCormick, 1993) and also of spindle waves during light SWS and is thus referred to as the spindle *pacemaker* (Steriade et al., 1985). Since we found that all neurons in the reticular thalamus (almost all GABAergic [Houser et al., 1980]) bear MT_2 receptors (Ochoa-Sanchez et al., 2011), we tested whether the effects we observed on NREMS were mediated by the reticular thalamus. Using in-vivo electrophysiology, we observed that UCM765 (20 mg/kg, i.v.) significantly increased the firing and burst activities of reticular thalamic neurons. Moreover, microinfusion of UCM765 directly into the reticular thalamus decreased the latency to NREMS, increased the amount of NREMS, and had no effect on REMS. These effects were similar to those obtained with systemic injection of the drug (see the earlier text). When UCM765 was instead microinfused in a brain area bearing MT_2 receptors but not involved in sleep regulation, such as the substantia nigra pars reticulata, no effects on NREMS and REMS were observed.

We have recently analyzed the 24 h sleep–wake pattern of MT_2KO mice (Comai et al., 2013). If, in agreement with the literature, MT_2 receptors are only involved in the time shift of SCN neural

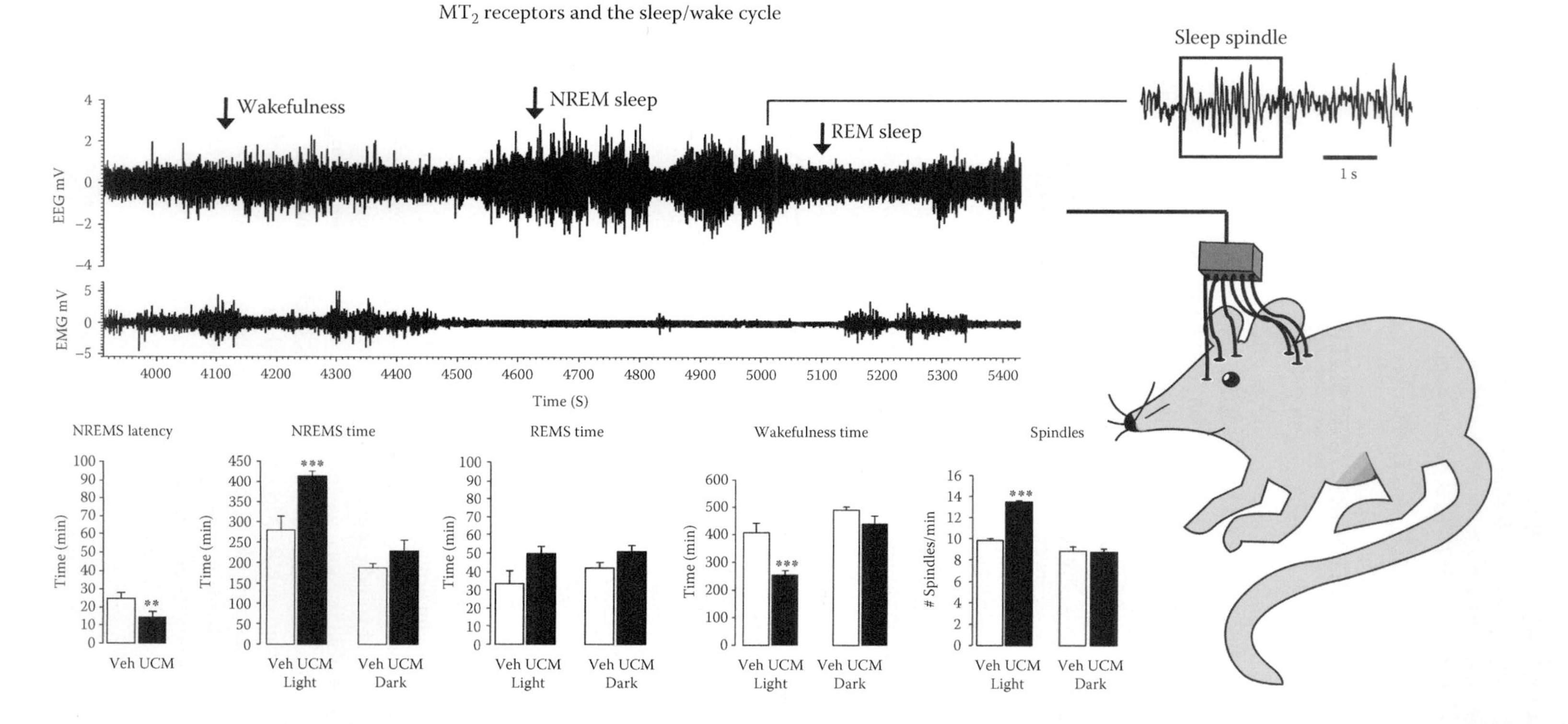

FIGURE 11.1 The MT_2 receptor partial agonist UCM765 promotes NREMS. Upper part, right: Schematic depiction of a rat with electrodes implanted for EEG and EMG recordings. Upper part: Example of an EEG/EMG recording highlighting a period of wakefulness, NREMS and REMS, and amplification of a sleep spindle. Lower part: Bar graphs indicate the effect of UCM765 (40 mg/kg, s.c.) on NREMS latency, NREMS time, REMS time, wakefulness time, and the number of sleep spindles during the light and dark phases of the 12:12 h light:dark cycle. $^{**}P < 0.01$ and $^{***}P < 0.001$ vs. Vehicle, two-way mixed-design ANOVA plus Student-Newman-Keuls test for post hoc comparison. (From S. Comai and G. Gobbi, *J. Psychiatry Neurosci.*, 39(1), 6, 2014. With permission.)

activity, the physiological sleep–wake cycle of MT_2KO mice should mirror that of wild-type control (WT) mice. Interestingly, MT_2KO mice displayed a significant decrease in the amount of NREMS during the 24 h, mainly due to changes occurring during the light/inactive phase of the 12 h/12 h light/dark cycle. No variation in the amount of REMS was found. The decline in the amount of NREMS was likely attributed to the decrease of NREMS delta and sigma powers. In the same study, we found that mice with the genetic inactivation of MT_1 receptors also have an altered sleep–wake cycle, but they displayed perturbations at the level of REMS rather than NREMS.

In conclusion, pharmacological and genetic studies targeting MT_2 receptors have revealed a new important function of this MLT receptor subtype: the selective modulation of NREMS. Consequently, MT_2 receptors may be considered a novel target for hypnotic agents.

11.6 MELATONIN, MT_2 RECEPTORS, AND ANXIETY

MLT and agomelatine (a nonselective MT_1/MT_2 receptor agonist and 5-HT_{2c} receptor antagonist) have displayed anxiolytic-like properties in classical animal paradigms of anxiety (Golombek et al., 1993; Golus and King, 1981; Ochoa-Sanchez et al., 2012; Papp et al., 2006). However, clinical evidence for using MLT as an anxiolytic drug is limited.

Little is known about the neurobiological mechanism mediating anxiolytic properties of MLT, but through the use of selective MT_2 receptor agonists and MT_2KO mice, we were able to investigate whether MT_2 receptors are implicated in the regulation of anxiety.

UCM765 and MLT were tested in three well-established animal paradigms used to study anxiety-like behavior: the elevated plus maze test (EPMT), the open field test (OFT), and the novelty suppressed feeding test (NSFT) (Ochoa-Sanchez et al., 2012). Both UCM765 (10 mg/kg) and MLT (20 mg/kg) displayed anxiolytic properties by increasing the time spent in the open arms of the EPMT and by decreasing the latency to eat in a new environment in the NSFT. In the attempt to elucidate whether both MT_1 and MT_2 receptors were implicated in the anxiolytic properties of MLT, we found that when UCM765 and MLT were injected in rats pretreated with the nonselective MT_1/MT_2 receptor antagonist luzindole or the selective MT_2 receptor antagonist 4P-PDOT, the anxiolytic effects of the two compounds were blocked in both behavioral tests. Altogether, these experiments suggest that the anxiolytic properties of UCM765 and MLT are MT_2 receptor mediated. In the OFT, UCM765 and MLT did not alter the measures of anxiety behavior such as the number of entries and time spent in the center of the arena. However, unlike diazepam, UCM765 and MLT did not reduce locomotor activity (Ochoa-Sanchez et al., 2012) or increase the number of falls in the rotorod test (unpublished results), two measures related to the sedative and motor impairments side effects of a putative psychotropic drug.

The phenotype displayed by MT_2KO mice in the EPMT, NSFT, and OFT is somewhat complex depending on the animal paradigm used and is thus a matter of ongoing research (unpublished data). For instance, MT_2KO mice spend more time in the central platform of the EPMT and display more entries and time spent in the central quadrant of the OFT (indicative of an anxiolytic response) but conversely show a longer latency to eat in a new environment in the NSFT (indicative of an anxiogenic response) when compared with controls.

Altogether, pharmacological and genetic studies targeting MT_2 receptors have demonstrated that this receptor subtype modulates anxiety-like responding in preclinical paradigms without producing the sedative side effects of conventional anxiolytic compounds. Consequently, the MT_2 receptor deserves consideration as a novel putative target for the treatment of anxiety disorders.

11.7 MT_2 RECEPTORS, MELATONIN, AND DEPRESSION

Preclinical evidence suggests that MLT has antidepressant-like effects due to its modulation of the serotonergic (Micale et al., 2006) and dopaminergic (Binfare et al., 2010) systems, although the effectiveness of MLT as an antidepressant in clinical studies has not been supported.

However, when administered in combination with other antidepressants such as buspirone, MLT has been shown to increase the efficacy of the pharmacological treatment (Ahn et al., 2012). On the contrary, MLT-derived ligands appear to be a superior antidepressant strategy. Indeed, agomelatine has been approved by the European Medicines Agency for the treatment of major depression in Europe (de Bodinat et al., 2010). However, the current knowledge on whether and how MT_1 and MT_2 receptors are involved in the antidepressant activity of MLT and melatonergic ligands is still limited.

Sumaya et al. (2005) found that luzindole displayed antidepressant-like activity in mice, and importantly, this pharmacological effect was MT_2 receptor mediated since the compound was not active as an antidepressant in MT_2KO mice.

In agreement with a possible role of MT_2 receptors in depression, the presence of the rs4753426 C allele within the *MTNR1B* gene locus increases the risk of recurrent depressive disorders (Galecka et al., 2011).

11.8 MELATONIN, MT_2 RECEPTORS, AND MEMORY FUNCTION

MLT regulates memory formation acting directly on hippocampal neurons (El-Sherif et al., 2003; Musshoff et al., 2002), which are involved in the processes of memory acquisition and consolidation (Deng et al., 2010). Wang et al. (2005) demonstrated that MLT, through the activation of MT_2 and not MT_1 receptors, produces a concentration-dependent inhibition of long-term potentiation (LTP) in mice hippocampal slices. LTP is considered the mechanism by which memory traces are encoded and stored in the central nervous system. In the hippocampus, MLT was also able to modulate evoked potentials, and in particular, the attenuation of synaptic transmission observed during the first phase of a biphasic ligand-induced effect was MT_2 receptor mediated (El-Sherif et al., 2004). In accordance with a role of MT_2 receptors in memory, Larson et al. (2006) showed that memory and LTP maintenance are impaired in MT_2KO mice. In addition, we observed that MT_2KO mice spend more time in the central platform of the EPMT compared with WT mice (unpublished data), a measure that could be a symptom of impaired decision making and cognitive flexibility.

In patients diagnosed with Alzheimer's disease, a progressive neurodegenerative disease characterized by a loss of memory and cognitive function and the emergence of dementia, the intensity of MT_2 receptor immunoreactivity was significantly reduced in the hippocampus, retina, pineal gland, and pyramidal and nonpyramidal cells of cortical layers II–V (Brunner et al., 2006; Savaskan et al., 2005).

The neurobiological mechanisms underlying the process of learning and memory formation is still a matter of ongoing research, and the studies described earlier suggest that MLT is involved, likely via the activation of MT_2 receptors. However, this topic deserves further investigation with selective MT_2 receptor ligands in the preclinical models of Alzheimer's disease and behavioral tasks designed to examine learning and memory processes.

11.9 MT_2 RECEPTORS, MELATONIN, AND PAIN

A large body of research has indicated that MLT possesses analgesic properties and plays a role in pain regulation through several mechanisms such as the activation of MLT and opioid receptors, ion channels (K^+ and Ca^{2+}), etc. (Ambriz-Tututi et al., 2009; Wilhelmsen et al., 2011). MLT displayed antinociceptive properties in preclinical tests such as the formalin test, the hot water tail-flick test, tactile allodynia induced by L5/L6 spinal nerve ligation, flinching behavior in the 0.5% formalin test, and tactile allodynia in diabetic rats (Ambriz-Tututi and Granados-Soto, 2007; Arreola-Espino et al., 2007; Yoon et al., 2008; Yu et al., 2000). In these experiments, the analgesic effects of MLT were blocked by luzindole and by the selective MT_2 receptor antagonist 4P-PDOT or K-185. Consequently, MT_2 receptors seem to mediate the analgesic properties of MLT, but no research has been conducted using selective MT_2 receptor agonists in these behavioral paradigms to investigate the neurobiological mechanism through which MT_2 receptor activation induces analgesia.

11.10 CONCLUSION

MLT is a pleiotropic compound that mediates a wide range of physiological functions, and consequently, alterations in the endogenous MLT system are implicated in the etiopathogenesis of several diseases. This neurohormone acts at cellular and intracellular levels through receptor-dependent (but also receptor-independent) mechanisms. Mainly due to the lack of selective MT_1 and MT_2 receptor agonists and antagonists, it is still unknown whether several of the physiological functions mediated by MLT are dependent on the activation of only one or both MLT receptor subtypes. In this chapter, we have examined the current knowledge on the role of MT_2 receptors in brain function. We need to remember that selective MT_2 receptor agonists/antagonists, MT_2KO mice, and selective MT_2 receptor antibodies have been developed mainly during the last 10–15 years and, consequently, the scientific evidence highlighting a role of MT_2 receptors in sleep and neuropsychiatric diseases is still limited. However, the results obtained to date are straightforward and promising and justify all further economical and scientific resources that will be spent on this topic in the future.

Sleep studies in rats and mice have corroborated a pivotal role of MT_2 receptors in sleep; in particular, the activation of MT_2 receptors promotes NREMS without altering REMS. Importantly, unlike currently used hypnotics such as benzodiazepines and nonbenzodiazepine compounds whose target is the GABAergic system, MT_2 receptor agonists appear to have an important advantage in that they do not yield sedation. On the other hand, similar to benzodiazepines, MT_2 receptor agonists evidently modulate brain function in a dose-dependent manner, producing anxiolytic effects at lower doses (10 mg/kg) and hypnotic effects at higher doses (40 mg/kg). This similar pharmacological profile may derive from the fact that in some brain areas such as the reticular thalamus, MT_2 receptors are densely expressed on GABAergic neurons. MLT modulates pain, memory, and mood, and even though still limited in number, the experiments undertaken until now have shown that MT_2 receptors are very likely implicated.

A link between MT_2 receptors and diabetes has been recently highlighted (Bonnefond et al., 2012), and patients suffering from diabetes have a higher-than-average risk of having stroke and myocardial infarction. These pathological conditions exhibit a high grade of comorbidity with neurological/psychiatric diseases. For example, sleep disorders are a risk factor for obesity and type-2 diabetes (Taylor et al., 2007). Consequently, we predict that MT_2 receptors may be a common etiological determinant, but how MT_2 receptors may be implicated in these comorbid diseases needs to be further investigated. However, if this link can be demonstrated, it is very likely that MT_2 receptor ligands may become a novel synergic pharmacological strategy for these debilitating conditions whose prevalence is increasing, especially in industrialized countries.

Our recent data regarding the localization of the MT_2 receptor in the rat brain will inevitably offer additional novel insights into the underlying physiological mechanism subserving MT_2 receptor–mediated effects of MLT. For example, MLT controls uterine contractions and parturition (Olcese et al., 2012). Since we observed MT_2 receptor labeling in the paraventricular nucleus and supraoptic nuclei (Ochoa-Sanchez et al., 2011), brain areas directly and indirectly implicated in the secretion of oxytocin and vasopressin, we expect that new research regarding how MT_2 receptors modulate uterine contraction and parturition will soon follow.

In conclusion, several novel studies have shown that MT_2 receptors are implicated in sleep, anxiety, memory, pain, and mood. Moreover, MT_2 receptor agonist/partial agonists have shown promising hypnotic and anxiolytic effects. Since we are still in need of better therapeutics for the treatment of many neuropsychiatric conditions, in light of the recent experimental findings, MT_2 receptors may represent a novel target in neuropsychopharmacology.

REFERENCES

Ahn S.K., Khalmuratova R., Hah Y.S. et al. 2012. Immunohistochemical and biomolecular identification of melatonin 1a and 1b receptors in rat vestibular nuclei. *Auris Nasus Larynx* 39: 479–483.

Ambriz-Tututi M. and Granados-Soto V. 2007. Oral and spinal melatonin reduces tactile allodynia in rats via activation of MT2 and opioid receptors. *Pain* 132: 273–280.

Ambriz-Tututi M., Rocha-Gonzalez H.I., Cruz S.L., and Granados-Soto V. 2009. Melatonin: A hormone that modulates pain. *Life Sci* 84: 489–498.

Angeloni D., Longhi R., and Fraschini F. 2000. Production and characterization of antibodies directed against the human melatonin receptors Mel-1a (mt1) and Mel-1b (MT2). *Eur J Histochem* 44: 199–204.

Arendt J. 1988. Melatonin. *Clin Endocrinol* 29: 205–229.

Arreola-Espino R., Urquiza-Marin H., Ambriz-Tututi M. et al. 2007. Melatonin reduces formalin-induced nociception and tactile allodynia in diabetic rats. *Eur J Pharmacol* 577: 203–210.

Bal T. and McCormick D.A. 1993. Mechanisms of oscillatory activity in guinea-pig nucleus reticularis thalami in vitro: A mammalian pacemaker. *J Physiol* 468: 669–691.

Binfare R.W., Mantovani M., Budni J., Santos A.R., and Rodrigues A.L. 2010. Involvement of dopamine receptors in the antidepressant-like effect of melatonin in the tail suspension test. *Eur J Pharmacol* 638: 78–83.

Bonnefond A., Clement N., Fawcett K. et al. 2012. Rare MTNR1B variants impairing melatonin receptor 1B function contribute to type 2 diabetes. *Nat Genet* 44: 297–301.

Brunner P., Sozer-Topcular N., Jockers R. et al. 2006. Pineal and cortical melatonin receptors MT1 and MT2 are decreased in Alzheimer's disease. *Eur J Histochem* 50: 311–316.

Comai S and Gobbi G. 2014. Unveiling the role of melatonin MT2 receptors in sleep, anxiety and other neuropsychiatric diseases: A novel target in psychopharmacology. *J Psychiatry Neurosci*. 39: 6–21.

Comai S., Ochoa-Sanchez R., and Gobbi G. 2013. Sleep–wake characterization of double MT1/MT2 receptor knockout mice and comparison with MT1 and MT2 receptor knockout mice. *Behav Brain Res* 243: 231–238.

de Bodinat C., Guardiola-Lemaitre B., Mocaer E. et al. 2010. Agomelatine, the first melatonergic antidepressant: Discovery, characterization and development. *Nat Rev Drug Discov* 9: 743.

Deng W., Aimone J.B., and Gage F.H. 2010. New neurons and new memories: How does adult hippocampal neurogenesis affect learning and memory? *Nat Rev Neurosci* 11: 339–350.

Dubocovich M.L. 2007. Melatonin receptors: Role on sleep and circadian rhythm regulation. *Sleep Med* 8: 34–42.

Dubocovich M.L., Delagrange P., Krause D.N. et al. 2010. International union of basic and clinical pharmacology. LXXV. Nomenclature, classification, and pharmacology of G protein-coupled melatonin receptors. *Pharmacol Rev* 62: 343–380.

Dubocovich M.L., Masana M.I., Iacob S., and Sauri D.M. 1997. Melatonin receptor antagonists that differentiate between the human Mel(1a), and Mel(1b) recombinant subtypes are used to assess the pharmacological profile of the rabbit retina ML(1) presynaptic heteroreceptor. *Naunyn Schmiedebergs Arch Pharmacol* 355: 365–375.

El-Sherif Y., Tesoriero J., Hogan M.V., and Wieraszko A. 2003. Melatonin regulates neuronal plasticity in the hippocampus. *J Neurosci Res* 72: 454–460.

El-Sherif Y., Witt-Enderby P., Li P.K. et al. 2004. The actions of a charged melatonin receptor ligand, TMEPI, and an irreversible MT2 receptor agonist, BMNEP, on mouse hippocampal evoked potentials in vitro. *Life Sci* 75: 3147–3156.

Faust R., Garratt P.J., Jones R. et al. 2000. Mapping the melatonin receptor. 6. Melatonin agonists and antagonists derived from 6H-isoindolo[2,1-a]indoles, 5,6-dihydroindolo[2,1-a]isoquinolines, and 6,7-dihydro-5H-benzo[c]azepino[2,1-a]indoles. *J Med Chem* 43: 1050–1061.

Fisher, S.P. and Sugden D. 2009. Sleep-promoting action of IIK7, a selective MT2 melatonin receptor agonist in the rat. *Neurosci Lett* 457: 93–96.

Galecka E., Szemraj J., Florkowski A. et al. 2011. Single nucleotide polymorphisms and mRNA expression for melatonin MT(2) receptor in depression. *Psychiatry Res* 189: 472–474.

Golombek D.A., Martini M., and Cardinali D.P. 1993. Melatonin as an anxiolytic in rats: Time dependence and interaction with the central GABAergic system. *Eur J Pharmacol* 237: 231–236.

Golus P. and King M.G. 1981. The effects of melatonin on open field behavior. *Pharmacol Biochem Behav* 15: 883–885.

Houser C.R., Vaughn J.E., Barber R.P., and Roberts E. 1980. GABA neurons are the major cell type of the nucleus reticularis thalami. *Brain Res* 200: 341–354.

Hunt A.E., Al-Ghoul W.M., Gillette M.U., and Dubocovich M.L. 2001. Activation of MT2 melatonin receptors in rat suprachiasmatic nucleus phase advances the circadian clock. *Am J Physiol Cell Physiol* 280: C110–C118.

James S.P., Mendelson W.B., Sack D.A., Rosenthal N.E., and Wehr T.A. 1987. The effect of melatonin on normal sleep. *Neuropsychopharmacology* 1: 41–44.

Jin X., von G.C., Pieschl R.L. et al. 2003. Targeted disruption of the mouse Mel(1b) melatonin receptor. *Mol Cell Biol* 23: 1054–1060.

Karamitri A., Renault N., Clement N., Guillaume J.L., and Jockers R. 2013. Minireview: Toward the establishment of a link between melatonin and glucose homeostasis: Association of melatonin MT2 receptor variants with type 2 diabetes. *Mol Endocrinol* 27: 1217–1233.

Lacoste B., Riad M., Ratte M.O. et al. 2009. Trafficking of neurokinin-1 receptors in serotonin neurons is controlled by substance P within the rat dorsal raphe nucleus. *Eur J Neurosci* 29: 2303–2314.

Larson J., Jessen R.E., Uz T. et al. 2006. Impaired hippocampal long-term potentiation in melatonin MT2 receptor-deficient mice. *Neurosci Lett* 393: 23–26.

MacKenzie R.S., Melan M.A., Passey D.K., and Witt-Enderby P.A. 2002. Dual coupling of MT(1) and MT(2) melatonin receptors to cyclic AMP and phosphoinositide signal transduction cascades and their regulation following melatonin exposure. *Biochem Pharmacol* 63: 587–595.

Micale V., Arezzi A., Rampello L., and Drago F. 2006. Melatonin affects the immobility time of rats in the forced swim test: The role of serotonin neurotransmission. *Eur Neuropsychopharmacol* 16: 538–545.

Mor M., Rivara S., Pala D. et al. 2010. Recent advances in the development of melatonin MT(1) and MT(2) receptor agonists. *Expert Opin Ther Pat* 20: 1059–1077.

Musshoff U., Riewenherm D., Berger E., Fauteck J.D., and Speckmann E.J. 2002. Melatonin receptors in rat hippocampus: Molecular and functional investigations. *Hippocampus* 12: 165–173.

Ochoa-Sanchez R., Comai S., Spadoni G. et al. 2014. Melatonin, selective and non-selective MT1/MT2 receptors agonists: Differential effects on the 24-hr vigilance states. *Neurosci Lett* 561: 156–161.

Ochoa-Sanchez R., Comai S., Lacoste B. et al. 2011. Promotion of non-rapid eye movement sleep and activation of reticular thalamic neurons by a novel MT2 melatonin receptor ligand. *J Neurosci* 31: 18439–18452.

Ochoa-Sanchez R., Rainer Q., Comai S. et al. 2012. Anxiolytic effects of the melatonin MT(2) receptor partial agonist UCM765: Comparison with melatonin and diazepam. *Prog Neuropsychopharmacol Biol Psychiatry* 39: 318–325.

Olcese J., Lozier S., and Paradise C. 2012. Melatonin and the circadian timing of human parturition. *Reprod Sci* 20: 168–174.

Papp M., Litwa E., Gruca P., and Mocaer E. 2006. Anxiolytic-like activity of agomelatine and melatonin in three animal models of anxiety. *Behav Pharmacol* 17: 9–18.

Petit L., Lacroix I., de Coppet P., Strosberg A.D., and Jockers R. 1999. Differential signaling of human Mel1a and Mel1b melatonin receptors through the cyclic guanosine 3′-5′-monophosphate pathway. *Biochem Pharmacol* 58: 633–639.

Reiter R.J. 2003. Melatonin: Clinical relevance. *Best Pract Res Clin Endocrinol Metab* 17: 273–285.

Reppert S.M., Godson C., Mahle C.D. et al. 1995. Molecular characterization of a second melatonin receptor expressed in human retina and brain: The Mel1b melatonin receptor. *Proc Natl Acad Sci USA* 92: 8734–8738.

Rivara S., Lodola A., Mor M. et al. 2007. *N*-(substituted-anilinoethyl)amides: Design, synthesis, and pharmacological characterization of a new class of melatonin receptor ligands. *J Med Chem* 50: 6618–6626.

Rivara S., Pala D., Lodola A. et al. 2012. MT1-selective melatonin receptor ligands: Synthesis, pharmacological evaluation, and molecular dynamics investigation of *N*-{[(3-*O*-substituted)anilino]alkyl}amides. *ChemMedChem* 7: 1954–1964.

Sallinen P., Saarela S., Ilves M., Vakkuri O., and Leppaluoto J. 2005. The expression of MT1 and MT2 melatonin receptor mRNA in several rat tissues. *Life Sci* 76: 1123–1134.

Savaskan E., Ayoub M.A., Ravid R. et al. 2005. Reduced hippocampal MT2 melatonin receptor expression in Alzheimer's disease. *J Pineal Res* 38: 10–16.

Steriade M., Deschenes M., Domich L., and Mulle C. 1985. Abolition of spindle oscillations in thalamic neurons disconnected from nucleus reticularis thalami. *J Neurophysiol* 54: 1473–1497.

Sumaya I.C., Masana M.I., and Dubocovich M.L. 2005. The antidepressant-like effect of the melatonin receptor ligand luzindole in mice during forced swimming requires expression of MT2 but not MT1 melatonin receptors. *J Pineal Res* 39: 170–177.

Taylor D.J., Mallory L.J., Lichstein K.L. et al. 2007. Comorbidity of chronic insomnia with medical problems. *Sleep* 30: 213–218.

Wang L.M., Suthana N.A., Chaudhury D., Weaver D.R., and Colwell C.S. 2005. Melatonin inhibits hippocampal long-term potentiation. *Eur J Neurosci* 22: 2231–2237.

Wilhelmsen M., Amirian I., Reiter R.J., Rosenberg J., and Gogenur I. 2011. Analgesic effects of melatonin: A review of current evidence from experimental and clinical studies. *J Pineal Res* 51: 270–277.

Witt-Enderby P.A., Bennett J., Jarzynka M.J., Firestine S., and Melan M.A. 2003. Melatonin receptors and their regulation: Biochemical and structural mechanisms. *Life Sci* 72: 2183–2198.

Wu Y.H., Ursinus J., Zhou J.N. et al. 2013. Alterations of melatonin receptors MT1 and MT2 in the hypothalamic suprachiasmatic nucleus during depression. *J Affect Disord* 148: 357–367.
Yang X.F., Miao Y., Ping Y. et al. 2011. Melatonin inhibits tetraethylammonium-sensitive potassium channels of rod ON type bipolar cells via MT2 receptors in rat retina. *Neuroscience* 173: 19–29.
Yoon M.H., Park H.C., Kim W.M. et al. 2008. Evaluation for the interaction between intrathecal melatonin and clonidine or neostigmine on formalin-induced nociception. *Life Sci* 83: 845–850.
Yu C.X., Zhu C.B., Xu S.F., Cao X.D., and Wu G.C. 2000. Selective MT(2) melatonin receptor antagonist blocks melatonin-induced antinociception in rats. *Neurosci Lett* 282: 161–164.
Zhdanova I.V. 2005. Melatonin as a hypnotic: Pro. *Sleep Med Rev* 9: 51–65.
Zlotos D.P. 2012. Recent progress in the development of agonists and antagonists for melatonin receptors. *Curr Med Chem* 19: 3532–3549.

12 Melatonin
Its Hepatoprotective Actions and the Role of Melatonin Receptors

Alexander M. Mathes

CONTENTS

12.1 MELATONIN: ITS HEPATOPROTECTIVE ACTIONS

Administration of exogenous melatonin has been established as a powerful hepatoprotective strategy in over 120 experimental in vivo studies, using various animal species, applying over 40 different models of stress [1–125] (Table 12.1). The data presented here are certainly incomplete: there are more studies investigating hepatoprotective effects of melatonin, not included in this overview. For this chapter, only a selection of the most important in vivo studies with acute onset models of stress or toxicities were chosen (Pubmed™ literature research from 1966 until today; inclusion criteria: melatonin and liver; hand-by-hand search for in vivo studies with a hepatoprotective aim of melatonin therapy). Other studies, especially the ones on chronic diseases (cirrhosis, pancreatitis, and diabetes), tumor or cancer development/metastases, aging, dietary changes, exercise-induced stress, remote organ injuries, and studies coadministering melatonin with other substances were excluded, even though they may be showing similar and very convincing results. The available literature from this research indicates that melatonin may protect from hepatic injury after different types of shock, ischemia/reperfusion, trauma, cholestasis, radiation, application of over 30 types of toxic agents, and damages inflicted by different microorganisms, in rats, mice, chicks, and hamsters [1–125]. Many models were investigated by more than one researcher, and all studies—independent of the type of hepatic injury—show similar results with respect to melatonin's effects on hepatic oxidative stress, interleukin signaling, hepatocellular integrity, and survival.

TABLE 12.1
Hepatoprotective Effects of Melatonin in Different Models of Stress

Model	Induction/Type	Melatonin Treatment	Hepatoprotective Effects of Melatonin	Species	References
Septic shock	CLP/LPS/LPS + BCG	0.25–60 mg/kg ip/iv/po 1–10×	hLPO ↓, AST/ALT/GGT/ALP/BIL ↓, hGSH/hGPx/hSOD/hCAT ↑, hNEC ↓, hPMN infiltration ↓, hTNF-α/hIL-1/hNO ↓, mitochondrial respiration ↑, 72 h survival rate ↑	Rats, mice	[1–11]
Hemorrhagic shock	90 min (MAP 35)/40%	10 mg/kg iv 1 dose	AST/ALT/LDH ↓, liver function PDR-ICG ↑, hepatic perfusion ↑, hNEC ↓, hHO-1 ↑, hCAS ↓	Rats	[12–15]
Ischemia/reperfusion	40–60 min ischemia/ ischemia + resection	10–20 mg/kg ip/im 1–5×	hLPO ↓, AST/ALT/LDH ↓, hGSH ↑, hNEC ↓, hMPO ↓, hPMN infiltration ↓, hTNF-α/hCAS/hAPO/hiNOS ↓, DLP-1 ↑, hTLR-3/4 ↓, 7-day survival rate ↑	Rats	[16–27]
Surgical trauma	70% hepatectomy	10 mg/kg/d ip for 7d	HLPO ↓, hGSH ↑, histological alterations ↓	Rats	[28]
Toxic liver injury	δ-Aminolevulinic acid	10 mg/kg/d ip 7–14 d	hLPO ↓, hepatic DNA damage ↓	Rats	[29,30]
	Acetaminophen	10–100 mg/kg ip/po/sc 1×	hLPO ↓, AST/ALT ↓, hGSH ↑, hMPO ↓, hNEC ↓, 72 h survival rate ↑	Mice	[31–33]
	Adriamycin	2–6 mg/kg ip/sc 1–7×	hLPO ↓, hGSH/hGPx/hCAT ↑, hHSP 40/60/70 ↓	Rats, mice	[34–36]
	Aflatoxins	5–40 mg/kg/d ig/ip for 3–8 wks	hLPO ↓, hGSH/hGPx ↑, hCAS/hNO ↓, hHSP-70 ↓, hNEC ↓	Rats, chicks	[37–41]
	Allyl alcohol	100 mg/kg ip 1×	hLPO ↓, AST/ALT/LDH ↓, hGSH ↑, hNEC ↓	Rats	[42]
	Arsenic	10 mg/kg ip for 5d	hLPO ↓, hGSH/hSOD/hCAT ↑	Rats	[43]
	Cadmium	10–12 mg/kg/d ip/po for 3–15 d	hLPO ↓, hGSH/hGPx ↑, hNEC ↓	Rats, mice	[44–47]
	Carbon tetrachloride	10–100 mg/kg ip/sc 1–30×	hLPO ↓, AST/ALT/ALP/LDH/BIL ↓, hGSH/hSOD/hCAT ↑, hXO ↓, hNO ↓, hTNF-α/hIL-1b/hNF-kB ↓, hNEC ↓	Rats, mice	[48–59]
	Cyclophosphamid	100 μg/kg/d po for 15 d	hLPO ↓, hGSH ↑	Mice	[60]
	Cyclosporin A	715 μg/kg/d ip for 14 d	hLPO ↓, AST/ALT/GGT ↓, hNEC ↓	Rats	[61–63]
	Diazepam	5 mg/kg/d sc for 30 d	hLPO ↓, hSOD/hGSH ↑	Rats	[64]
	Dimethylnitrosamine	50–100 mg/kg/d ip for 14 d	hLPO ↓, AST/ALT/ALP/BIL ↓, hSOD/hGSH/hGPx/hHO-1 ↑, hTNF-α/hIL-1b/hIL-6/hNF-kB ↓	Rats	[65,66]
	Diquat	20 mg/kg ip 1×	ALT ↓, hepatic content of F2-isoprostane ↓, 24 h survival rate ↑	Rats, Mice	[67,68]
	Doxorubicin	10 mg/kg sc for 7 d	hLPO ↓, GGT/LDH ↓	Rats	[69]
	Endosulfan	10 mg/kg ip for 5 d	hLPO ↓, AST/ALT/LDH ↓, hGSH ↑, hMPO ↓, hTNF-α/IL-1b ↓	Rats	[70]
	Iodine	1 mg/kg/d ip for 14 d	Hepatic content of Schiff's bases ↓	Rats	[71]
	Kainic acid	4–10 mg/kg ip 1×	Hepatic DNA damage ↓	Rats	[72]
	Lead	10–30 mg/kg/d ig for 7–30 d	hLPO ↓, hGSH/hGPx/hSOD ↑, hNEC ↓	Rats	[73,74]
	Letrozole	0.5 mg/kg/d sc	AST/LDH/ALP/BIL ↓	Rats	[75]

	Methanol	10 mg/kg ip 2×	hLPO ↓, hGSH/hGPx/hSOD/hCAT ↑, hMPO/hNO ↓, histological alterations ↓	Rats	[76,77]
	Methotrexate	10 mg/kg/d ip for 5 d	hLPO ↓, hGSH ↑, hNEC ↓	Rats	[78]
	Mercury-(II)	10 mg/kg ip 2×	hLPO ↓, hGSH ↑, hMPO ↓	Rats	[79]
	α-Naphthylisothiocyanate	10–100 mg/kg ip/po 1–4×	hLPO ↓, AST/ALT/LDH/GGT/ALP/BIL ↓, hSOD/hCAT ↑, hMPO ↓	Rats	[80–83]
	Nicotine	10 mg/kg ip 7× over 21 d	Histological alterations ↓	Mice	[84]
	Nodularin	5–15 mg/kg/d ip for 7 d	hGPx/hSOD/hCAT ↑	Mice	[85]
	Ochratoxin A	5–20 mg/kg ig/po 1–28×	hLPO ↓, GGT/ALP ↓, hGSH/hGPx/hSOD/hCAT ↑, hNEC ↓	Rats	[86–90]
	Paraquat	1–10 mg/kg ip 5–6×	hLPO ↓, hGSH ↑, LD_{50} of paraquat ↑	Rats	[91,92]
	Phosphine	10 mg/kg ip 1×	hLPO ↓, hGSH ↑	Rats	[93]
	Safrole	0.1–0.2 mg/kg sc 2×	Hepatic DNA damage ↓	Rats	[94]
	Thioacetamid	3 mg/kg ip 3–5×	hLPO ↓, AST/ALT/LDH/ammonia ↓.hGSH/hCAT ↑, hiNOS/hNEC ↓	Rats	[95–97]
	Toluene inhalation	10 mg/kg ip for 28 d	hMPO ↓, AST/ALT ↓, histological alterations ↓	Rats	[98,99]
	Zymosan	5–50 mg/kg ip 1–7×	hLPO/hMPO ↓	Rats	[100,101]
Cholestasis	Bile duct ligation	0.5–100 mg/kg/d ip/po for 7–13 d	hLPO ↓, AST/ALT/GGT/ALP/BIL ↓. hGSH/hGPx/hSOD/hCAT ↑, hMPO ↓, hNO ↓, hNEC ↓, iron disturbances ↓, ADMA ↓	Rats	[102–113]
Ionizing radiation	Full body; 0.8–6.0 Gray	5–50 mg/kg ip 1–5×	hLPO ↓, AST/ALT/GGT ↓, hGSH/hSOD/hGPx ↑, hMPO/hNO ↓, hepatic DNA damage ↓	Rats	[114–119]
Malaria	*Schistosoma mansoni*	10 mg/kg/d ip for 30 d	hLPO ↓, AST/ALT ↓ hGSH/hSOD ↑, 56-day survival rate ↑	Mice	[120]
Liver fluke	*Opisthorchis viverrini*	5–10 mg/kg/d po for 30 d	ALT ↓, hHO-1 ↓, bile duct proliferation ↑	Hamsters	[121]
Viral hepatic failure	Rabbit hemorrhagic disease virus	10–20 mg/kg ip 3×	Apoptosis ↓, hCAS ↓, hIL-1β/hIL-6 ↓, CRP ↓, hTLR-4 ↓	Rabbits	[122–124]
Amoebiasis	*Entamoeba histolytica*	15 mg/kg sc 6×	HSP-70 ↓, hNEC ↓. histological alterations ↓	Hamsters, Rats	[125]

Source: Modified from Mathes, A.M., *World J. Gastroenterol.*, 16, 6087, 2010. With permission.

Notes: Not all investigators showed all effects. For details, please refer to the text. *Legend*: ↑, upregulation/increase/improvement; ↓, downregulation/decrease/deterioration; ADMA, asymmetric dimethylarginine; ALT, alanine transaminase; ALP, alkaline phosphatase; AST, aspartate transaminase; BCG, Bacillus Calmette-Guérin; BIL, bilirubin; CLP, cecal ligation and puncture; d, day; DLP-1, dynamic-like protein-1; DNA, deoxyribonucleic acid; GGT, gamma glutamyl transferase; hAPO, hepatic apoptosis; hCAT, hepatic catalase; hCAS, hepatic caspase; hGPx, hepatic glutathione peroxidase; hGSH, hepatic glutathione; hHSP, hepatic heat shock protein; hHO-1, hepatic heme oxygenase 1; hIL, hepatic interleukin; hiNOS, hepatic inducible nitric oxide synthase; hLPO, hepatic lipid peroxidation; hMPO, hepatic myeloperoxidase; hNEC, hepatocellular necrosis; hNF-kB, nuclear factor kappa-light-chain-enhancer of activated B cells; hNO, hepatic nitric oxide; hPMN, hepatic polymorphonuclear granulocytes; hSOD, hepatic superoxide dismutase; hTLR, hepatic toll-like receptor; hTNF-α, hepatic tumor necrosis factor alpha; hXO, hepatic xanthine oxidase; ig, intragastrically; im, intramuscularly; ip, intraperitoneally; iv, intravenously; kg, kilograms; LD, lethal dose; LDH, lactate dehydrogenase; LPS, lipopolysaccharide; MAP, mean arterial pressure; mg, milligrams; min, minutes; μg, micrograms; PDR-ICG, plasma disappearance rate of indocyanine green; po, per os; Ref, references; sc, subcutaneously; wks, weeks.

12.1.1 Antioxidant Effects

Melatonin appears to reduce hepatic oxidative stress, measured by different techniques, after various types of injury. In the majority of studies, hepatic lipid peroxidation is strongly attenuated by melatonin, indicating an intense antioxidant effect [1–7,16–66,69–71,73,74,76–83,86–93,95–97,100–120]. Also, hepatic antioxidant enzymes, like glutathione, glutathione peroxidase, and superoxide dismutase, show either an increased activity or an upregulated expression after treatment with melatonin [1–7,16–28,31–60,64–66,70,76–83,85–93,95–97,102–120]. Similar results are obtained for hepatic catalase after melatonin treatment [5,36,43,48–50,54,76,80,85,86,88,95,104,107,111]. For a detailed overview on melatonin's antioxidant capacity, see Chapter 21.

12.1.2 Interleukin Signaling and Granulocyte Infiltration

Exogenous melatonin seems to be able to suppress the proinflammatory pathway in sepsis and after ischemia/reperfusion, as well as after carbon tetrachloride and dimethylnitrosamine toxicity, by attenuating tumor necrosis factor alpha (TNF-α), interleukin-1 (IL-1), IL-1β, IL-6, and cellular interleukin response protein nuclear factor kappa-light-chain-enhancer of activated B cells (NF-κB) [3,4,16,27,53,66]. Neutrophil granulocyte infiltration, measured by the hepatic levels of myeloperoxidase, was also strongly reduced after treatment with the pineal hormone [2,18,32,76,77,80,106,116].

12.1.3 Hepatocellular Integrity

The rise in serum enzyme levels of aspartate transaminase (AST), alanine transaminase (ALT), lactate dehydrogenase (LDH), alkaline phosphatase (ALP), gamma glutamyl transferase (GGT), and bilirubin, as markers of hepatocellular integrity, is significantly reduced by melatonin after almost all types of injury [1–23,31–33,42,48–57,61–63,65–70,80–84,86–90,95–97,102–120]. Further, melatonin treatment may reduce hepatocellular necrosis and a variety of histological alterations after all types of injury. Sinusoidal flow and integrity are preserved by melatonin; this was also investigated using intravital microscopy [12,13]. Liver function, as measured by plasma disappearance rate of indocyanine green, may also be improved by melatonin after hemorrhagic shock and resuscitation [12,13].

12.1.4 Survival

Although not specifically associated with liver injury, survival was evaluated in a small number of the studies on the hepatoprotective effect of melatonin. Whenever examined, melatonin was able to improve survival rate or mean survival time after sepsis, ischemia/reperfusion, acetaminophen and diquat toxicity, and malaria [2–4,10,16,23,33,68,120]. As a strong indicator for protective effects, this shows that melatonin may exert many of its beneficial actions on a broad level, improving the viability of the whole organism.

12.1.5 Hepatoprotective Mechanisms of Melatonin

With respect to melatonin's effects on hepatic gene expression, the focus of most investigators was on heat shock protein (HSP) gene regulation. After adriamycin or aflatoxin challenges, and after amoebiasis, melatonin was able to attenuate the rise in hepatic HSP-70 significantly [34,38,125]. Further, microarray experiments after hemorrhagic shock in rat indicate a significant influence of melatonin on hepatic HSP expression, as well as on the regulation of various pathways of G-protein coupled receptors [126]. Other potential hepatoprotective mechanisms of melatonin may involve attenuation of inducible nitric oxide synthase in the liver [37–41,76,77,95–97], modifications of hepatic heme oxygenase-1 expression [14,15,65,66], attenuation of asymmetric dimethylarginine formation [113], upregulation of dynamic-like protein-1 [26], and reduction of hepatic DNA damage [29,30,72,94,119]. However, the precise signaling cascade of melatonin, leading to hepatoprotective effects, remains to be determined.

12.1.6 Effective Dose and Route of Melatonin Administration

Most investigators did not measure melatonin plasma levels after administration, and data on dose–response relationships are sparse; no prospective, randomized, double-blind, placebo-controlled studies were performed for dose-finding purposes. Effective treatment doses in all animal species range from 100 μg/kg [60,94] to 100 mg/kg [42] melatonin, given as a single dose, or repetitively over days to weeks. Different routes of administration were chosen (orally, intraperitoneally, subcutaneously, intramuscular, and intravenously), while intravenous treatment was performed only after hemorrhagic and septic shock [12,13]. Many researchers used melatonin as a pretreatment or for preconditioning, and not as a therapy after the damage was inflicted; this limits interpretation for a potential clinical use. Yet, dose–response data from survival experiments after sepsis indicate that the most effective dose of intravenous melatonin therapy may be in the lower range of the applied amounts in rats, between 100 μg/kg and 1.0 mg/kg bodyweight [127]; higher doses do not appear to be associated with a better outcome.

12.1.7 Limitations of Melatonin

Not all investigators were able to demonstrate hepatoprotective effects of melatonin in similar experiments. While 12 studies support the idea of melatonin being able to protect from carbon tetrachloride–induced liver injury [48–59], one researcher was unable to reproduce all of these results in the same setting [128]. Also, melatonin had no effect on 2-nitropropane-induced lipid peroxidation or on hypobaric hypoxia–induced liver injury in rat [129,130]. With respect to ethanol toxicity, treatment with melatonin modified neither hepatic lipid peroxidation nor glutathione or glutathione peroxidase activities in rat in one investigation [131]. Two other studies, however, imply a reduction of ethanol-induced liver injury by melatonin, showing less hepatocellular injury and improved matrix metalloproteinase-9 expression [132,133]. Yet, despite these little controversies, there is generally no doubt that melatonin possesses a highly significant and reproducible hepatoprotective capacity [1–125].

12.2 HEPATIC MELATONIN RECEPTORS

The first melatonin receptor subtype to be discovered is of nonmammalian origin and was called Mel1c, using the former terminology [134]. The two subtypes of mammalian melatonin receptors have previously been called Mel1a and Mel1b but are now officially named MT1 and MT2 and are classified as membrane bound, high-affinity G-protein coupled receptors [135]. MT1 and MT2 are both coupled to heterotrimeric G-proteins, and their signaling involves inhibition of cyclic adenosinmonophosphat (cAMP) formation, protein kinase A activity and phosphorylation of cAMP responsive element binding, and effects on adenyl cyclases, phospholipase A2 and C, and calcium and potassium channels [136–140]. A third receptor, named MT3, was demonstrated to be equivalent to intracellular quinone-reductase-2 [141].

MT1, MT2, and MT3 mRNA and binding sites have been discovered in the liver of different species [142–154] (Table 12.2). Hepatic melatonin receptor mRNA and protein of MT1 and MT2 appear to underlie relevant circadian variations [143,148,150–152]. Concerning melatonin receptor protein, however, it has to be noted that the available antibodies for MT1 and MT2 are not without controversy concerning specificity and reproducibility of results [155]. Therefore, research demonstrating MT1 and MT2 protein has to be viewed with caution, and further data are needed to support these findings. The amount of melatonin receptor mRNA in the liver is very low; this raises

TABLE 12.2
Melatonin Receptors in the Liver of Various Species

Class	Species	MT1	MT2	MT3/QR2	Technique	References
Mammals	CD-1 mouse	n/t	n/t	+	Iodine ligand	[142]
	CH3/He mouse	+	+	n/t	RT-PCR	[143]
	Swiss mouse	+	–	n/t	RT-PCR	[144]
	Sprague-Dawley rat	–	+	n/t	RT-PCR	[145]
	Wistar rat	+	+	n/t	RT-PCR	[146–148]
	Syrian hamster	n/t	n/t	+	Iodine ligand	[142,149]
	Dog	n/t	n/t	+	Iodine ligand	[142]
	Cynomolgus monkey	n/t	n/t	+	Iodine ligand	[142]
Fish	Golden rabbitfish	+	+	n/t	RT-PCR	[150,151]
	European sea bass	–	+	n/t	RT-PCR	[152]
	Senegalese sole	+	–	n/t	RT-PCR	[153]
	Tench	n/s	n/s	n/s	Iodine ligand	[154]

Source: Modified from Mathes, A.M., *World J. Gastroenterol.*, 16, 6087, 2010. With permission.
Note: +, detected; –, not detected; MT1, melatonin receptor type 1; MT2, melatonin receptor type 2; MT3/QR2, melatonin receptor type 3/quinone reductase-2; n/s, not specified; n/t, not tested; RT-PCR, real-time polymerase chain reaction.

the question on how these receptors may be involved in intense alterations of hepatocellular function, as observed for hepatoprotective effects. Further, there are almost no data on the physiological significance of these receptors. Hepatic melatonin receptors could be involved in regulating blood glucose, as two studies suggest [144,156]. Melatonin receptor double knockout mice appear to have an unaltered phenotype [127], and the absence of hepatic melatonin receptors seems to have no disadvantages under physiological conditions.

12.3 ROLE OF MELATONIN RECEPTORS FOR HEPATOPROTECTION

The observation that melatonin may have a strong hepatoprotective potential, combined with the fact that melatonin receptors are expressed in the liver, raises the question whether some or all of the observed effects may be attributable to melatonin receptor activation. Considering the strong direct antioxidant capacity of the pineal hormone (see Chapter 21), it seems unlikely that all beneficial findings of melatonin after liver injury could be receptor-mediated. Yet, some studies have addressed this question in vivo, using melatonin receptor antagonist luzindole [12,13], selective melatonin receptor agonist ramelteon [157], and melatonin receptor knockout mice [127].

After hemorrhagic shock, luzindole antagonized the hepatoprotective effects of melatonin pretreatment or therapy significantly [12,13]. Yet, it needs to be mentioned that not all effects of melatonin were antagonized. Further, luzindole itself was shown to possess a strong direct antioxidant capacity [158], and to reduce lipid peroxidation in vitro [159]. In the same model, therapy with ramelteon improved liver function and hepatic microcirculation in rat [157]. This melatonin receptor agonist was demonstrated not to have any relevant direct radical scavenging property, and again, these effects are antagonized by luzindole [157]. These results are further supported by investigations in other organ systems: luzindole has been reported to antagonize the beneficial effects of melatonin after myocardial ischemia [160], cyclosporine-A cardiotoxicity [161], neonatal brain injury [161], and stress-induced gastric lesions [163]. In an investigation using melatonin receptor double knockout mice, both melatonin and ramelteon were able to improve survival after polymicrobial sepsis in wild-type mice, but not in knockout mice [127]. This finding may serve as a strong indicator that at least some of the protective effects of melatonin are mediated by melatonin receptors. The limited evidence on the relevance of melatonin receptors for organ protection may be summarized as follows:

1. No radical scavenging properties appear to be necessary to provide organ protection via melatonin receptor activation by ramelteon [127,157].
2. Melatonin receptor antagonist luzindole may abolish almost all the protective effects of melatonin [12,13,127,160–163].
3. Melatonin and ramelteon may improve survival in wild-type, but not in melatonin receptor double knockout mice [127].

As a consequence, it may be postulated that melatonin receptors are at least partially involved in the mediation of organ protection, especially regarding hepatoprotective effects. Yet, the signaling pathways following melatonin receptor activation for hepatoprotection are presently unknown. Preliminary data from microarray experiments indicate that the receptor agonist ramelteon may induce different modifications of hepatocellular transcription, including HSPs, as well as intense regulations of other membrane-bound receptors and signal transduction factors, in a rat model of hemorrhagic shock [126]. These data allow the speculation that melatonin therapy may regulate hepatic gene transcription in the environment of oxidative stress. However, whether these influences on hepatic gene expression are mediated by melatonin receptors remains to be determined.

12.4 POTENTIAL CLINICAL APPLICATIONS

The evidence on the hepatoprotective effects of melatonin from animal studies is convincing; it is therefore tempting to challenge these experiences for clinical practice. Unfortunately, there is very limited information concerning melatonin-induced hepatoprotection in human patients. There are currently no prospective, randomized, double-blind, placebo-controlled studies that evaluate potential doses of melatonin for human applications. Yet, after transcatheter arterial chemoembolization (TACE) and treatment with melatonin in patients with inoperable advanced hepatocellular carcinoma, one study reports attenuated immunological activity, reduced liver damage, and improvements in survival, compared with patients that underwent TACE without melatonin administration [164]. Oral treatment with a single preoperative bolus of melatonin 50 mg/kg was well tolerated by another group of patients undergoing liver resection, but this did not influence outcome parameters, with a nonsignificant trend toward lower liver enzymes and shorter hospital stay [165]. In yet another investigation, a 12-week treatment period with 5 mg melatonin orally twice daily resulted in lower plasma liver enzyme levels in patients with nonalcoholic steatohepatitis, compared to placebo [166].

These human studies, in combination with the results from animal experiments, may be regarded a promising beginning to translate our knowledge on melatonin as a hepatoprotective agent from bench to bedside. Future studies are needed to demonstrate whether melatonin can meet our high expectations not only in the laboratory but also for our patients. Nonetheless, the available literature allows us to believe that melatonin will successfully continue its way as a powerful hepatoprotective agent in clinical practice.

REFERENCES

1. Crespo E, Macías M, Pozo D, Escames G, Martín M, Vives F, Guerrero JM, Acuña-Castroviejo D. 1999. Melatonin inhibits expression of the inducible NO synthase II in liver and lung and prevents endotoxemia in lipopolysaccharide-induced multiple organ dysfunction syndrome in rats. *FASEB J* 13:1537–1546.
2. Sener G, Toklu H, Kapucu C, Ercan F, Erkanli G, Kaçmaz A, Tilki M, Yeğen BC. 2005. Melatonin protects against oxidative organ injury in a rat model of sepsis. *Surg Today* 35:52–59.

3. Carrillo-Vico A, Lardone PJ, Naji L, Fernández-Santos JM, Martín-Lacave I, Guerrero JM, Calvo JR. 2005. Beneficial pleiotropic actions of melatonin in an experimental model of septic shock in mice: Regulation of pro-/anti-inflammatory cytokine network, protection against oxidative damage and anti-apoptotic effects. *J Pineal Res* 39:400–408.
4. Wu CC, Chiao CW, Hsiao G, Chen A, Yen MH. 2001. Melatonin prevents endotoxin-induced circulatory failure in rats. *J Pineal Res* 30:147–156.
5. Xu DX, Wei W, Sun MF, Wei LZ, Wang JP. 2005. Melatonin attenuates lipopolysaccharide-induced down-regulation of pregnane X receptor and its target gene CYP3A in mouse liver. *J Pineal Res* 38:27–34.
6. Sewerynek E, Abe M, Reiter RJ, Barlow-Walden LR, Chen L, McCabe TJ, Roman LJ, Diaz-Lopez B. 1995. Melatonin administration prevents lipopolysaccharide-induced oxidative damage in phenobarbital-treated animals. *J Cell Biochem* 58:436–444.
7. Sewerynek E, Melchiorri D, Reiter RJ, Ortiz GG, Lewinski A. 1995. Lipopolysaccharide-induced hepatotoxicity is inhibited by the antioxidant melatonin. *Eur J Pharmacol* 293:327–334.
8. Wang H, Wei W, Shen YX, Dong C, Zhang LL, Wang NP, Yue L, Xu SY. 2004. Protective effect of melatonin against liver injury in mice induced by Bacillus Calmette-Guerin plus lipopolysaccharide. *World J Gastroenterol* 10:2690–2696.
9. Wang H, Xu DX, Lv JW, Ning H, Wei W. 2007. Melatonin attenuates lipopolysaccharide (LPS)-induced apoptotic liver damage in D-galactosamine-sensitized mice. *Toxicology* 237:49–57.
10. Wu JY, Tsou MY, Chen TH, Chen SJ, Tsao CM, Wu CC. 2008. Therapeutic effects of melatonin on peritonitis-induced septic shock with multiple organ dysfunction syndrome in rats. *J Pineal Res* 45:106–116.
11. Lowes DA, Webster NR, Murphy MP, Galley HF. 2013. Antioxidants that protect mitochondria reduce interleukin-6 and oxidative stress, improve mitochondrial function, and reduce biochemical markers of organ dysfunction in a rat model of acute sepsis. *Br J Anaesth* 110:472–480.
12. Mathes AM, Kubulus D, Pradarutti S, Bentley A, Weiler J, Wolf B, Ziegeler S, Bauer I, Rensing H. 2008. Melatonin pretreatment improves liver function and hepatic perfusion after hemorrhagic shock. *Shock* 29:112–118.
13. Mathes AM, Kubulus D, Weiler J, Bentley A, Waibel L, Wolf B, Bauer I, Rensing H. 2008. Melatonin receptors mediate improvements of liver function but not of hepatic perfusion and integrity after hemorrhagic shock in rats. *Crit Care Med* 36:24–29.
14. Yang FL, Subeq YM, Lee CJ, Lee RP, Peng TC, Hsu BG. 2011. Melatonin ameliorates hemorrhagic shock-induced organ damage in rats. *J Surg Res* 167:315–321.
15. Hsu JT, Kuo CJ, Chen TH, Wang F, Lin CJ, Yeh TS, Hwang TL, Jan YY. 2012. Melatonin prevents hemorrhagic shock-induced liver injury in rats through an Akt-dependent HO-1 pathway. *J Pineal Res* 53:410–416.
16. Rodríguez-Reynoso S, Leal C, Portilla E, Olivares N, Muñiz J. 2001. Effect of exogenous melatonin on hepatic energetic status during ischemia/reperfusion: Possible role of tumor necrosis factor-alpha and nitric oxide. *J Surg Res* 100:141–149.
17. Bülbüller N, Cetinkaya Z, Akkus MA, Cifter C, Ilhan YS, Dogru O, Aygen E. 2003. The effects of melatonin and prostaglandin E1 analogue on experimental hepatic ischemia reperfusion damage. *Int J Clin Pract* 57:857–860.
18. Sener G, Tosun O, Sehirli AO, Kaçmaz A, Arbak S, Ersoy Y, Ayanoğlu-Dülger G. 2003. Melatonin and *N*-acetylcysteine have beneficial effects during hepatic ischemia and reperfusion. *Life Sci* 72:2707–2718.
19. Zhang WH, Li JY, Zhou Y. 2006. Melatonin abates liver ischemia/reperfusion injury by improving the balance between nitric oxide and endothelin. *Hepatobiliary Pancreat Dis Int* 5:574–579.
20. Park SW, Choi SM, Lee SM. 2007. Effect of melatonin on altered expression of vasoregulatory genes during hepatic ischemia/reperfusion. *Arch Pharm Res* 30:1619–1624.
21. Sewerynek E, Reiter RJ, Melchiorri D, Ortiz GG, Lewinski A. 1996. Oxidative damage in the liver induced by ischemia-reperfusion: Protection by melatonin. *Hepatogastroenterology* 43:898–905.
22. Kim SH, Lee SM. 2008. Cytoprotective effects of melatonin against necrosis and apoptosis induced by ischemia/reperfusion injury in rat liver. *J Pineal Res* 44:165–171.
23. Liang R, Nickkholgh A, Hoffmann K, Kern M, Schneider H, Sobirey M, Zorn M, Büchler MW, Schemmer P. 2009. Melatonin protects from hepatic reperfusion injury through inhibition of IKK and JNK pathways and modification of cell proliferation. *J Pineal Res* 46:8–14.
24. Baykara B, Tekmen I, Pekcetin C, Ulukus C, Tuncel P, Sagol O, Ormen M, Ozogul C. 2009. The protective effects of carnosine and melatonin in ischemia-reperfusion injury in the rat liver. *Acta Histochem* 111:42–51.
25. Li JY, Yin HZ, Gu X, Zhou Y, Zhang WH, Qin YM. 2008. Melatonin protects liver from intestine ischemia reperfusion injury in rats. *World J Gastroenterol* 14:7392–7396.

26. Cho EH, Koh PO. 2010. Proteomic identification of proteins differentially expressed by melatonin in hepatic ischemia-reperfusion injury. *J Pineal Res* 49:349–355.
27. Kang JW, Koh EJ, Lee SM. 2011. Melatonin protects liver against ischemia and reperfusion injury through inhibition of toll-like receptor signaling pathway. *J Pineal Res* 50:403–411.
28. Kirimlioglu H, Ecevit A, Yilmaz S, Kirimlioglu V, Karabulut AB. 2008. Effect of resveratrol and melatonin on oxidative stress enzymes, regeneration, and hepatocyte ultrastructure in rats subjected to 70% partial hepatectomy. *Transplant Proc* 40:285–289.
29. Carneiro RC, Reiter RJ. 1998. Delta-aminolevulinic acid-induced lipid peroxidation in rat kidney and liver is attenuated by melatonin: An in vitro and in vivo study. *J Pineal Res* 24:131–136.
30. Karbownik M, Reiter RJ, Garcia JJ, Tan DX, Qi W, Manchester LC. 2000. Melatonin reduces rat hepatic macromolecular damage due to oxidative stress caused by delta-aminolevulinic acid. *Biochim Biophys Acta* 1523:140–146.
31. Matsura T, Nishida T, Togawa A, Horie S, Kusumoto C, Ohata S, Nakada J, Ishibe Y, Yamada K, Ohta Y. 2006. Mechanisms of protection by melatonin against acetaminophen-induced liver injury in mice. *J Pineal Res* 41:211–219.
32. Sener G, Sehirli AO, Ayanoğlu-Dülger G. 2003. Protective effects of melatonin, vitamin E and *N*-acetylcysteine against acetaminophen toxicity in mice: A comparative study. *J Pineal Res* 35:61–68.
33. Kanno S, Tomizawa A, Hiura T, Osanai Y, Kakuta M, Kitajima Y, Koiwai K, Ohtake T, Ujibe M, Ishikawa M. 2006. Melatonin protects on toxicity by acetaminophen but not on pharmacological effects in mice. *Biol Pharm Bull* 29:472–476.
34. Catalá A, Zvara A, Puskás LG, Kitajka K. 2007. Melatonin-induced gene expression changes and its preventive effects on adriamycin-induced lipid peroxidation in rat liver. *J Pineal Res* 42:43–49.
35. Rapozzi V, Comelli M, Mavelli I, Sentjurc M, Schara M, Perissin L, Giraldi T. 1999. Melatonin and oxidative damage in mice liver induced by the prooxidant antitumor drug, adriamycin. *In Vivo* 13:45–50.
36. Othman AI, El-Missiry MA, Amer MA, Arafa M. 2008. Melatonin controls oxidative stress and modulates iron, ferritin, and transferrin levels in adriamycin treated rats. *Life Sci* 83:563–568.
37. Meki AR, Abdel-Ghaffar SK, El-Gibaly I. 2001. Aflatoxin B1 induces apoptosis in rat liver: Protective effect of melatonin. *Neuro Endocrinol Lett* 22:417–426.
38. Meki AR, Esmail Eel-D, Hussein AA, Hassanein HM. 2004. Caspase-3 and heat shock protein-70 in rat liver treated with aflatoxin B1: Effect of melatonin. *Toxicon* 43:93–100.
39. Gesing A, Karbownik-Lewinska M. 2008. Protective effects of melatonin and N-acetylserotonin on aflatoxin B1-induced lipid peroxidation in rats. *Cell Biochem Funct* 26:314–319.
40. Ozen H, Karaman M, Ciğremiş Y, Tuzcu M, Ozcan K, Erdağ D. 2009. Effectiveness of melatonin on aflatoxicosis in chicks. *Res Vet Sci* 86:485–489.
41. Sirajudeen M, Gopi K, Tyagi JS, Moudgal RP, Mohan J, Singh R. 2011. Protective effects of melatonin in reduction of oxidative damage and immunosuppression induced by aflatoxin B1-contaminated diets in young chicks. *Environ Toxicol* 26:153–160.
42. Sigala F, Theocharis S, Sigalas K, Markantonis-Kyroudis S, Papalabros E, Triantafyllou A, Kostopanagiotou G, Andreadou I. 2006. Therapeutic value of melatonin in an experimental model of liver injury and regeneration. *J Pineal Res* 40:270–279.
43. Pal S, Chatterjee AK. 2006. Possible beneficial effects of melatonin supplementation on arsenic-induced oxidative stress in Wistar rats. *Drug Chem Toxicol* 29:423–433.
44. Kim CY, Lee MJ, Lee SM, Lee WC, Kim JS. 1998. Effect of melatonin on cadmium-induced hepatotoxicity in male Sprague-Dawley rats. *Tohoku J Exp Med* 186:205–213.
45. Eybl V, Kotyzova D, Koutensky J. 2006. Comparative study of natural antioxidants—curcumin, resveratrol and melatonin—in cadmium-induced oxidative damage in mice. *Toxicology* 225:150–156.
46. Kara H, Cevik A, Konar V, Dayangac A, Servi K. 2008. Effects of selenium with vitamin E and melatonin on cadmium-induced oxidative damage in rat liver and kidneys. *Biol Trace Elem Res* 125:236–244.
47. El-Sokkary GH, Nafady AA, Shabash EH. 2010. Melatonin administration ameliorates cadmium-induced oxidative stress and morphological changes in the liver of rat. *Ecotoxicol Environ Saf* 73:456–463.
48. Ohta Y, Kongo M, Sasaki E, Nishida K, Ishiguro I. 2000. Therapeutic effect of melatonin on carbon tetrachloride-induced acute liver injury in rats. *J Pineal Res* 28:119–126.
49. Ohta Y, Kongo-Nishimura M, Matsura T, Yamada K, Kitagawa A, Kishikawa T. 2004. Melatonin prevents disruption of hepatic reactive oxygen species metabolism in rats treated with carbon tetrachloride. *J Pineal Res* 36:10–17.
50. Ohta Y, Kongo M, Sasaki E, Nishida K, Ishiguro I. 1999. Preventive effect of melatonin on the progression of carbon tetrachloride-induced acute liver injury in rats. *Adv Exp Med Biol* 467:327–332.

51. Kus I, Ogeturk M, Oner H, Sahin S, Yekeler H, Sarsilmaz M. 2005. Protective effects of melatonin against carbon tetrachloride-induced hepatotoxicity in rats: A light microscopic and biochemical study. *Cell Biochem Funct* 23:169–174.
52. Zavodnik LB, Zavodnik IB, Lapshina EA, Belonovskaya EB, Martinchik DI, Kravchuk RI, Bryszewska M, Reiter RJ. 2005. Protective effects of melatonin against carbon tetrachloride hepatotoxicity in rats. *Cell Biochem Funct* 23:353–359.
53. Wang H, Wei W, Wang NP, Gui SY, Wu L, Sun WY, Xu SY. 2005. Melatonin ameliorates carbon tetrachloride-induced hepatic fibrogenesis in rats via inhibition of oxidative stress. *Life Sci* 77:1902–1915.
54. Noyan T, Kömüroğlu U, Bayram I, Sekeroğlu MR. 2006. Comparison of the effects of melatonin and pentoxifylline on carbon tetrachloride-induced liver toxicity in mice. *Cell Biol Toxicol* 22:381–391.
55. Ogeturk M, Kus I, Pekmez H, Yekeler H, Sahin S, Sarsilmaz M. 2008. Inhibition of carbon tetrachloride-mediated apoptosis and oxidative stress by melatonin in experimental liver fibrosis. *Toxicol Ind Health* 24:201–208.
56. Hong RT, Xu JM, Mei Q. 2009. Melatonin ameliorates experimental hepatic fibrosis induced by carbon tetrachloride in rats. *World J Gastroenterol* 15:1452–1458.
57. Shaker ME, Houssen ME, Abo-Hashem EM, Ibrahim TM. 2009. Comparison of vitamin E, L-carnitine and melatonin in ameliorating carbon tetrachloride and diabetes induced hepatic oxidative stress. *J Physiol Biochem* 65:225–233.
58. Cheshchevik VT, Lapshina EA, Dremza IK, Zabrodskaya SV, Reiter RJ, Prokopchik NI, Zavodnik IB. 2012. Rat liver mitochondrial damage under acute or chronic carbon tetrachloride-induced intoxication: Protection by melatonin and cranberry flavonoids. *Toxicol Appl Pharmacol* 261:271–279.
59. Ebaid H, Bashandy SA, Alhazza IM, Rady A, El-Shehry S. 2013. Folic acid and melatonin ameliorate carbon tetrachloride-induced hepatic injury, oxidative stress and inflammation in rats. *Nutr Metab (Lond)* 10:20.
60. Manda K, Bhatia AL. 2003. Prophylactic action of melatonin against cyclophosphamide-induced oxidative stress in mice. *Cell Biol Toxicol* 19:367–372.
61. Kwak CS, Mun KC. 2000. The beneficial effect of melatonin for cyclosporine hepatotoxicity in rats. *Transplant Proc* 32:2009–2010.
62. Rezzani R, Buffoli B, Rodella L, Stacchiotti A, Bianchi R. 2005. Protective role of melatonin in cyclosporine A-induced oxidative stress in rat liver. *Int Immunopharmacol* 5:1397–1405.
63. Kurus M, Esrefoglu M, Sogutlu G, Atasever A. 2009. Melatonin prevents cyclosporine-induced hepatotoxicity in rats. *Med Princ Pract* 18:407–410.
64. El-Sokkary GH. 2008. Melatonin and vitamin C administration ameliorate diazepam-induced oxidative stress and cell proliferation in the liver of rats. *Cell Prolif* 41:168–176.
65. Tahan V, Ozaras R, Canbakan B, Uzun H, Aydin S, Yildirim B, Aytekin H, Ozbay G, Mert A, Senturk H. 2004. Melatonin reduces dimethylnitrosamine-induced liver fibrosis in rats. *J Pineal Res* 37:78–84.
66. Jung KH, Hong SW, Zheng HM, Lee DH, Hong SS. 2009. Melatonin downregulates nuclear erythroid 2-related factor 2 and nuclear factor-kappaB during prevention of oxidative liver injury in a dimethylnitrosamine model. *J Pineal Res* 47:173–183.
67. Zhang L, Wei W, Xu J, Min F, Wang L, Wang X, Cao S, Tan DX, Qi W, Reiter RJ. 2006. Inhibitory effect of melatonin on diquat-induced lipid peroxidation in vivo as assessed by the measurement of F2-isoprostanes. *J Pineal Res* 40:326–331.
68. Xu J, Sun S, Wei W, Fu J, Qi W, Manchester LC, Tan DX, Reiter RJ. 2007. Melatonin reduces mortality and oxidatively mediated hepatic and renal damage due to diquat treatment. *J Pineal Res* 42:166–171.
69. Oz E, Ilhan MN. 2006. Effects of melatonin in reducing the toxic effects of doxorubicin. *Mol Cell Biochem* 286:11–15.
70. Omurtag GZ, Tozan A, Sehirli AO, Sener G. 2008. Melatonin protects against endosulfan-induced oxidative tissue damage in rats. *J Pineal Res* 44:432–438.
71. Swierczynska-Machura D, Lewinski A, Sewerynek E. 2004. Melatonin effects on Schiff's base levels induced by iodide administration in rats. *Neuro Endocrinol Lett* 25:70–74.
72. Tang L, Reiter RJ, Li ZR, Ortiz GG, Yu BP, Garcia JJ. 1998. Melatonin reduces the increase in 8-hydroxy-deoxyguanosine levels in the brain and liver of kainic acid-treated rats. *Mol Cell Biochem* 178:299–303.
73. El-Missiry MA. 2000. Prophylactic effect of melatonin on lead-induced inhibition of heme biosynthesis and deterioration of antioxidant systems in male rats. *J Biochem Mol Toxicol* 14:57–62.
74. El-Sokkary GH, Abdel-Rahman GH, Kamel ES. 2005. Melatonin protects against lead-induced hepatic and renal toxicity in male rats. *Toxicology* 213:25–33.

75. Aydin M, Oktar S, Ozkan OV, Alçin E, Oztürk OH, Nacar A. 2011. Letrozole induces hepatotoxicity without causing oxidative stress: The protective effect of melatonin. *Gynecol Endocrinol* 27:209–215.
76. Kurcer Z, Oğuz E, Iraz M, Fadillioglu E, Baba F, Koksal M, Olmez E. 2007. Melatonin improves methanol intoxication-induced oxidative liver injury in rats. *J Pineal Res* 43:42–49.
77. Koksal M, Kurcer Z, Erdogan D, Iraz M, Tas M, Eren MA, Aydogan T, Ulas T. 2012. Effect of melatonin and n-acetylcysteine on hepatic injury in rat induced by methanol intoxication: A comparative study. *Eur Rev Med Pharmacol Sci* 16:437–444.
78. Jahovic N, Cevik H, Sehirli AO, Yeğen BC, Sener G. 2003. Melatonin prevents methotrexate-induced hepatorenal oxidative injury in rats. *J Pineal Res* 34:282–287.
79. Sener G, Sehirli AO, Ayanoglu-Dülger G. 2003. Melatonin protects against mercury(II)-induced oxidative tissue damage in rats. *Pharmacol Toxicol* 93:290–296.
80. Ohta Y, Kongo M, Sasaki E, Ishiguro I, Harada N. 2000. Protective effect of melatonin against alpha-naphthylisothiocyanate-induced liver injury in rats. *J Pineal Res* 29:15–23.
81. Ohta Y, Kongo M, Kishikawa T. 2001. Effect of melatonin on changes in hepatic antioxidant enzyme activities in rats treated with alpha-naphthylisothiocyanate. *J Pineal Res* 31:370–377.
82. Calvo JR, Reiter RJ, García JJ, Ortiz GG, Tan DX, Karbownik M. 2001. Characterization of the protective effects of melatonin and related indoles against alpha-naphthylisothiocyanate-induced liver injury in rats. *J Cell Biochem* 80:461–470.
83. Ohta Y, Kongo M, Kishikawa T. 2003. Preventive effect of melatonin on the progression of alpha-naphthylisothiocyanate-induced acute liver injury in rats. *J Pineal Res* 34:185–193.
84. Mercan S, Eren B. 2013. Protective role of melatonin supplementation against nicotine-induced liver damage in mouse. *Toxicol Ind Health* 29:888–896.
85. Lankoff A, Banasik A, Nowak M. 2002. Protective effect of melatonin against nodularin-induced oxidative stress. *Arch Toxicol* 76:158–165.
86. Meki AR, Hussein AA. 2001. Melatonin reduces oxidative stress induced by ochratoxin A in rat liver and kidney. *Comp Biochem Physiol C Toxicol Pharmacol* 130:305–313.
87. Aydin G, Ozçelik N, Ciçek E, Soyöz M. 2003. Histopathologic changes in liver and renal tissues induced by Ochratoxin A and melatonin in rats. *Hum Exp Toxicol* 22:383–391.
88. Soyöz M, Ozçelik N, Kilinç I, Altuntaş I. 2004. The effects of ochratoxin A on lipid peroxidation and antioxidant enzymes: A protective role of melatonin. *Cell Biol Toxicol* 20:213–219.
89. Abdel-Wahhab MA, Abdel-Galil MM, El-Lithey M. 2005. Melatonin counteracts oxidative stress in rats fed an ochratoxin A contaminated diet. *J Pineal Res* 38:130–135.
90. Sutken E, Aral E, Ozdemir F, Uslu S, Alatas O, Colak O. 2007. Protective role of melatonin and coenzyme Q10 in ochratoxin A toxicity in rat liver and kidney. *Int J Toxicol* 26:81–87.
91. Melchiorri D, Reiter RJ, Attia AM, Hara M, Burgos A, Nistico G. 1995. Potent protective effect of melatonin on in vivo paraquat-induced oxidative damage in rats. *Life Sci* 56:83–89.
92. Melchiorri D, Reiter RJ, Sewerynek E, Hara M, Chen L, Nisticò G. 1996. Paraquat toxicity and oxidative damage. Reduction by melatonin. *Biochem Pharmacol* 51:1095–1099.
93. Hsu C, Han B, Liu M, Yeh C, Casida JE. 2000. Phosphine-induced oxidative damage in rats: Attenuation by melatonin. *Free Radic Biol Med* 28:636–642.
94. Tan DX, Pöeggeler B, Reiter RJ, Chen LD, Chen S, Manchester LC, Barlow-Walden LR. 1993. The pineal hormone melatonin inhibits DNA-adduct formation induced by the chemical carcinogen safrole in vivo. *Cancer Lett* 70:65–71.
95. Bruck R, Aeed H, Avni Y, Shirin H, Matas Z, Shahmurov M, Avinoach I, Zozulya G, Weizman N, Hochman A. 2004. Melatonin inhibits nuclear factor kappa B activation and oxidative stress and protects against thioacetamide induced liver damage in rats. *J Hepatol* 40:86–93.
96. Túnez I, Muñoz MC, Villavicencio MA, Medina FJ, de Prado EP, Espejo I, Barcos M, Salcedo M, Feijóo M, Montilla P. 2005. Hepato- and neurotoxicity induced by thioacetamide: Protective effects of melatonin and dimethylsulfoxide. *Pharmacol Res* 52:223–228.
97. Túnez I, Muñoz MC, Medina FJ, Salcedo M, Feijóo M, Montilla P. 2007. Comparison of melatonin, vitamin E and L-carnitine in the treatment of neuro- and hepatotoxicity induced by thioacetamide. *Cell Biochem Funct* 25:119–127.
98. Tas U, Ogeturk M, Meydan S, Kus I, Kuloglu T, Ilhan N, Kose E, Sarsilmaz M. 2011. Hepatotoxic activity of toluene inhalation and protective role of melatonin. *Toxicol Ind Health* 27:465–473.
99. Tas U, Ogeturk M, Kuloglu T, Sapmaz HI, Kocaman N, Zararsiz I, Sarsilmaz M. 2013. HSP70 immune reactivity and TUNEL positivity in the liver of toluene-inhaled and melatonin-treated rats. *Toxicol Ind Health* 29:514–522.

100. Cuzzocrea S, Zingarelli B, Costantino G, Caputi AP. 1998. Protective effect of melatonin in a non-septic shock model induced by zymosan in the rat. *J Pineal Res* 25:24–33.
101. El-Sokkary GH, Reiter RJ, Cuzzocrea S, Caputi AP, Hassanein AF, Tan DX. 1999. Role of melatonin in reduction of lipid peroxidation and peroxynitrite formation in non-septic shock induced by zymosan. *Shock* 12:402–408.
102. López PM, Fiñana IT, De Agueda MC, Sánchez EC, Muñoz MC, Alvarez JP, De La Torre Lozano EJ. 2000. Protective effect of melatonin against oxidative stress induced by ligature of extra-hepatic biliary duct in rats: Comparison with the effect of *S*-adenosyl-L-methionine. *J Pineal Res* 28:143–149.
103. Montilla P, Cruz A, Padillo FJ, Túnez I, Gascon F, Muñoz MC, Gómez M, Pera C. 2001. Melatonin versus vitamin E as protective treatment against oxidative stress after extra-hepatic bile duct ligation in rats. *J Pineal Res* 31:138–144.
104. Ohta Y, Kongo M, Kishikawa T. 2003. Melatonin exerts a therapeutic effect on cholestatic liver injury in rats with bile duct ligation. *J Pineal Res* 34:119–126.
105. Bülbüller N, Akkuş MA, Cetinkaya Z, Ilhan YS, Ozercan I, Kirkil C, Doğru O. 2002. Effects of melatonin and lactulose on the liver and kidneys in rats with obstructive jaundice. *Pediatr Surg Int* 18:677–680.
106. Ohta Y, Kongo M, Kishikawa T. 2003. Therapeutic effect of melatonin on cholestatic liver injury in rats with bile duct ligation. *Adv Exp Med Biol* 527:559–565.
107. Padillo FJ, Cruz A, Navarrete C, Bujalance I, Briceño J, Gallardo JI, Marchal T et al. 2004. Melatonin prevents oxidative stress and hepatocyte cell death induced by experimental cholestasis. *Free Radic Res* 38:697–704.
108. Esrefoglu M, Gül M, Emre MH, Polat A, Selimoglu MA. 2005. Protective effect of low dose of melatonin against cholestatic oxidative stress after common bile duct ligation in rats. *World J Gastroenterol* 11:1951–1956.
109. Ohta Y, Imai Y, Matsura T, Yamada K, Tokunaga K. 2005. Successively postadministered melatonin prevents disruption of hepatic antioxidant status in rats with bile duct ligation. *J Pineal Res* 39:367–374.
110. Muñoz-Castañeda JR, Túnez I, Herencia C, Ranchal I, González R, Ramírez LM, Arjona A et al. 2008. Melatonin exerts a more potent effect than *S*-adenosyl-L-methionine against iron metabolism disturbances, oxidative stress and tissue injury induced by obstructive jaundice in rats. *Chem Biol Interact* 174:79–87.
111. Emre MH, Polat A, Eşrefoğlu M, Karabulut AB, Gül M. 2008. Effects of melatonin and acetylsalicylic acid against hepatic oxidative stress after bile duct ligation in rat. *Acta Physiol Hung* 95:349–363.
112. Huang LT, Tiao MM, Tain YL, Chen CC, Hsieh CS. 2009. Melatonin ameliorates bile duct ligation-induced systemic oxidative stress and spatial memory deficits in developing rats. *Pediatr Res* 65:176–180.
113. Tain YL, Kao YH, Hsieh CS, Chen CC, Sheen JM, Lin IC, Huang LT. 2010. Melatonin blocks oxidative stress-induced increased asymmetric dimethylarginine. *Free Radic Biol Med* 49:1088–1098.
114. Karbownik M, Reiter RJ, Qi W, Garcia JJ, Tan DX, Manchester LC, Vijayalaxmi. 2000. Protective effects of melatonin against oxidation of guanine bases in DNA and decreased microsomal membrane fluidity in rat liver induced by whole body ionizing radiation. *Mol Cell Biochem* 211:137–144.
115. Taysi S, Koc M, Büyükokuroğlu ME, Altinkaynak K, Sahin YN. 2003. Melatonin reduces lipid peroxidation and nitric oxide during irradiation-induced oxidative injury in the rat liver. *J Pineal Res* 34:173–177.
116. Sener G, Jahovic N, Tosun O, Atasoy BM, Yeğen BC. 2003. Melatonin ameliorates ionizing radiation-induced oxidative organ damage in rats. *Life Sci* 74:563–572.
117. Koc M, Taysi S, Buyukokuroglu ME, Bakan N. 2003. Melatonin protects rat liver against irradiation-induced oxidative injury. *J Radiat Res (Tokyo)* 44:211–215.
118. El-Missiry MA, Fayed TA, El-Sawy MR, El-Sayed AA. 2007. Ameliorative effect of melatonin against gamma-irradiation-induced oxidative stress and tissue injury. *Ecotoxicol Environ Saf* 66:278–286.
119. Shirazi A, Mihandoost E, Ghobadi G, Mohseni M, Ghazi-Khansari M. 2013. Evaluation of radio-protective effect of melatonin on whole body irradiation induced liver tissue damage. *Cell J* 14:292–297.
120. El-Sokkary GH, Omar HM, Hassanein AF, Cuzzocrea S, Reiter RJ. 2002. Melatonin reduces oxidative damage and increases survival of mice infected with *Schistosoma mansoni*. *Free Radic Biol Med* 32:319–332.
121. Laothong U, Pinlaor P, Hiraku Y, Boonsiri P, Prakobwong S, Khoontawad J, Pinlaor S. 2010. Protective effect of melatonin against *Opisthorchis viverrini*-induced oxidative and nitrosative DNA damage and liver injury in hamsters. *J Pineal Res* 49:271–282.
122. Tuñón MJ, San Miguel B, Crespo I, Jorquera F, Santamaría E, Alvarez M, Prieto J, González-Gallego J. 2011. Melatonin attenuates apoptotic liver damage in fulminant hepatic failure induced by the rabbit hemorrhagic disease virus. *J Pineal Res* 50:38–45.

123. Laliena A, San Miguel B, Crespo I, Alvarez M, González-Gallego J, Tuñón MJ. 2012. Melatonin attenuates inflammation and promotes regeneration in rabbits with fulminant hepatitis of viral origin. *J Pineal Res* 53:270–278.
124. Tuñón MJ, San-Miguel B, Crespo I, Laliena A, Vallejo D, Álvarez M, Prieto J, González-Gallego J. 2013. Melatonin treatment reduces endoplasmic reticulum stress and modulates the unfolded protein response in rabbits with lethal fulminant hepatitis of viral origin. *J Pineal Res* 55:221–228.
125. França-Botelho AC, França JL, Oliveira FM, Franca EL, Honório-França AC, Caliari MV, Gomes MA. 2011. Melatonin reduces the severity of experimental amoebiasis. *Parasit Vectors* 4:62.
126. Mathes A, Ruf C, Fink T, Abend M, Rensing H. 2010. Molecular effects of melatonin and ramelteon administration after hemorrhagic shock in rat liver [abstract]. *Eur J Anesthesiol* 27:S47–S42.
127. Fink T, Glas M, Wolf A, Kleber A, Reus E, Wolff M, Kiefer D et al. 2014. Melatonin receptors mediate improvements of survival in a model of polymicrobial sepsis. *Crit Care Med* 42(1):e22–e31, PMID 24145838.
128. Daniels WM, Reiter RJ, Melchiorri D, Sewerynek E, Pablos MI, Ortiz GG. 1995. Melatonin counteracts lipid peroxidation induced by carbon tetrachloride but does not restore glucose-6 phosphatase activity. *J Pineal Res* 19:1–6.
129. Kim SJ, Reiter RJ, Rouvier Garay MV, Qi W, El-Sokkary GH, Tan DX. 1998. 2-Nitropropane-induced lipid peroxidation: Antitoxic effects of melatonin. *Toxicology* 130:183–190.
130. Farías JG, Zepeda AB, Calaf GM. 2012. Melatonin protects the heart, lungs and kidneys from oxidative stress under intermittent hypobaric hypoxia in rats. *Biol Res* 45:81–85.
131. Genç S, Gürdöl F, Oner-Iyidoğan Y, Onaran I. 1998. The effect of melatonin administration on ethanol-induced lipid peroxidation in rats. *Pharmacol Res* 37:37–40.
132. Hu S, Yin S, Jiang X, Huang D, Shen G. 2009. Melatonin protects against alcoholic liver injury by attenuating oxidative stress, inflammatory response, and apoptosis. *Eur J Pharmacol* 616:287–292.
133. Mishra A, Paul S, Swarnakar S. 2011. Downregulation of matrix metalloproteinase-9 by melatonin during prevention of alcohol-induced liver injury in mice. *Biochimie* 93:854–866.
134. Ebisawa T, Karne S, Lerner MR, Reppert SM. 1994. Expression cloning of a high-affinity melatonin receptor from *Xenopus* dermal melanophores. *Proc Natl Acad Sci USA* 91:6133–6137.
135. Dubocovich ML, Rivera-Bermudez MA, Gerdin MJ, Masana MI. 2003. Molecular pharmacology, regulation and function of mammalian melatonin receptors. *Front Biosci* 8:d1093–d1108.
136. Jockers R, Maurice P, Boutin JA, Delagrange P. 2008. Melatonin receptors, heterodimerization, signal transduction and binding sites: What's new? *Br J Pharmacol* 154:1182–1195.
137. von Gall C, Stehle JH, Weaver DR. 2002. Mammalian melatonin receptors: Molecular biology and signal transduction. *Cell Tissue Res* 309:151–162.
138. New DC, Tsim ST, Wong YH. 2003. G protein-linked effector and second messenger systems involved in melatonin signal transduction. *Neurosignals* 12:59–70.
139. Pandi-Perumal SR, Trakht I, Srinivasan V, Spence DW, Maestroni GJ, Zisapel N, Cardinali DP. 2008. Physiological effects of melatonin: Role of melatonin receptors and signal transduction pathways. *Prog Neurobiol* 85:335–353.
140. Barrett P, Morris M, Choi WS, Ross A, Morgan PJ. 1999. Melatonin receptors and signal transduction mechanisms. *Biol Signals Recept* 8:6–14.
141. Nosjean O, Ferro M, Coge F, Beauverger P, Henlin JM, Lefoulon F, Fauchere JL et al. 2000. Identification of the melatonin-binding site MT3 as the quinone reductase 2. *J Biol Chem* 275:31311–31317.
142. Nosjean O, Nicolas JP, Klupsch F, Delagrange P, Canet E, Boutin JA. 2001. Comparative pharmacological studies of melatonin receptors: MT1, MT2 and MT3/QR2. Tissue distribution of MT3/QR2. *Biochem Pharmacol* 61:1369–1379.
143. Naji L, Carrillo-Vico A, Guerrero JM, Calvo JR. 2004. Expression of membrane and nuclear melatonin receptors in mouse peripheral organs. *Life Sci* 74:2227–2236.
144. Mühlbauer E, Gross E, Labucay K, Wolgast S, Peschke E. 2009. Loss of melatonin signalling and its impact on circadian rhythms in mouse organs regulating blood glucose. *Eur J Pharmacol* 606:61–71.
145. Sallinen P, Saarela S, Ilves M, Vakkuri O, Leppäluoto J. 2005. The expression of MT1 and MT2 melatonin receptor mRNA in several rat tissues. *Life Sci* 76:1123–1134.
146. Ishii H, Tanaka N, Kobayashi M, Kato M, Sakuma Y. 2009. Gene structures, biochemical characterization and distribution of rat melatonin receptors. *J Physiol Sci* 59:37–47.
147. Sánchez-Hidalgo M, Guerrero Montávez JM, Carrascosa-Salmoral Mdel P, Naranjo Gutierrez Mdel C, Lardone PJ, de la Lastra Romero CA. 2009. Decreased MT1 and MT2 melatonin receptor expression in extrapineal tissues of the rat during physiological aging. *J Pineal Res* 46:29–35.

148. Venegas C, García JA, Doerrier C, Volt H, Escames G, López LC, Reiter RJ, Acuña-Castroviejo D. 2013. Analysis of the daily changes of melatonin receptors in the rat liver. *J Pineal Res* 54:313–321.
149. Paul P, Lahaye C, Delagrange P, Nicolas JP, Canet E, Boutin JA. 1999. Characterization of 2-[125I] iodomelatonin binding sites in Syrian hamster peripheral organs. *J Pharmacol Exp Ther* 290:334–340.
150. Park YJ, Park JG, Hiyakawa N, Lee YD, Kim SJ, Takemura A. 2007. Diurnal and circadian regulation of a melatonin receptor, MT1, in the golden rabbitfish, *Siganus guttatus*. *Gen Comp Endocrinol* 150:253–262.
151. Park YJ, Park JG, Kim SJ, Lee YD, Saydur Rahman M, Takemura A. 2006. Melatonin receptor of a reef fish with lunar-related rhythmicity: Cloning and daily variations. *J Pineal Res* 41:166–174.
152. Sauzet S, Besseau L, Herrera Perez P, Covès D, Chatain B, Peyric E, Boeuf G, Muñoz-Cueto JA, Falcón J. 2008. Cloning and retinal expression of melatonin receptors in the European sea bass, *Dicentrarchus labrax*. *Gen Comp Endocrinol* 157:186–195.
153. Confente F, Rendón MA, Besseau L, Falcón J, Muñoz-Cueto JA. 2010. Melatonin receptors in a pleuronectiform species, *Solea senegalensis*: Cloning, tissue expression, day-night and seasonal variations. *Gen Comp Endocrinol* 167:202–214.
154. López Patiño MA, Guijarro AI, Alonso-Gómez AL, Delgado MJ. 2012. Characterization of two different melatonin binding sites in peripheral tissues of the teleost *Tinca tinca*. *Gen Comp Endocrinol* 175:180–187.
155. Weaver D, Prof., GRC Pineal Cell Biology, Galveston, TX, 2010. Personal Communication.
156. Poon AM, Choy EH, Pang SF. 2001. Modulation of blood glucose by melatonin: A direct action on melatonin receptors in mouse hepatocytes. *Biol Signals Recept* 10:367–379.
157. Mathes A, Kubulus D, Waibel L, Weiler J, Heymann P, Wolf B, Rensing H. 2008. Selective activation of melatonin receptors with ramelteon improves liver function and hepatic perfusion after hemorrhagic shock in rat. *Crit Care Med* 36:2863–2870.
158. Mathes A, Wolf B, Rensing H. 2008. Melatonin receptor antagonist luzindole is a powerful radical scavenger in vitro. *J Pineal Res* 45:337–338.
159. Requintina PJ, Oxenkrug GF. 2007. Effect of luzindole and other melatonin receptor antagonists on iron- and lipopolysaccharide-induced lipid peroxidation in vitro. *Ann NY Acad Sci* 1122:289–294.
160. Lochner A, Genade S, Davids A, Ytrehus K, Moolman JA. Short- and long-term effects of melatonin on myocardial post-ischemic recovery. *J Pineal Res* 40:56–63.
161. Rezzani R, Rodella LF, Bonomini F, Tengattini S, Bianchi R, Reiter RJ. 2006. Beneficial effects of melatonin in protecting against cyclosporine A-induced cardiotoxicity are receptor mediated. *J Pineal Res* 41:288–295.
162. Husson I, Mesplès B, Bac P, Vamecq J, Evrard P, Gressens P. 2002. Melatoninergic neuroprotection of the murine periventricular white matter against neonatal excitotoxic challenge. *Ann Neurol* 51:82–92.
163. Brzozowski T, Konturek PC, Zwirska-Korczala K, Konturek SJ, Brzozowska I, Drozdowicz D, Sliwowski Z, Pawlik M, Pawlik WW, Hahn EG. 2005. Importance of the pineal gland, endogenous prostaglandins and sensory nerves in the gastroprotective actions of central and peripheral melatonin against stress-induced damage. *J Pineal Res* 39:375–385.
164. Yan JJ, Shen F, Wang K, Wu MC. 2002. Patients with advanced primary hepatocellular carcinoma treated by melatonin and transcatheter arterial chemoembolization: A prospective study. *Hepatobiliary Pancreat Dis Int* 1:183–186.
165. Nickkholgh A, Schneider H, Sobirey M, Venetz WP, Hinz U, Pelzl le H, Gotthardt DN et al. 2011. The use of high-dose melatonin in liver resection is safe: First clinical experience. *J Pineal Res* 50:381–388.
166. Gonciarz M, Gonciarz Z, Bielanski W, Mularczyk A, Konturek PC, Brzozowski T, Konturek SJ. 2012. The effects of long-term melatonin treatment on plasma liver enzymes levels and plasma concentrations of lipids and melatonin in patients with nonalcoholic steatohepatitis: A pilot study. *J Physiol Pharmacol* 63:35–40.
167. Mathes AM. 2010. Hepatoprotective actions of melatonin: Possible mediation by melatonin receptors. *World J Gastroenterol* 16:6087–6097.

13 Melatonin Signaling in the Control of *Plasmodium* Development and Replication

Alexandre Budu and Célia R. S. Garcia

CONTENTS

13.1 MALARIA AND SYNCHRONICITY: THE MELATONIN CONNECTION

Malaria is a disease caused by parasites of the genus *Plasmodium*. The life cycle of the parasite occurs within a mosquito vector, *Anopheles*, and a vertebrate host (reviewed in Bannister and Mitchell 2003). A plethora of malaria species that infect a wide range of vertebrates such as birds (*P. cathemerium*), lizards (around 100 species), murines (*P. berghei*, *P. chabaudi*, *P. yoelii*), monkeys (*P. brasilianum*, *P. cynomolgi*), and humans (*P. falciparum*, *P. vivax*) exist. In humans, the burden of malaria is devastating with an estimate of 655,000 deaths in 2010, according to the *World Health Organization*.

In the bloodstream of the vertebrate, the parasite invades erythrocytes and undergoes development in the ring, trophozoite, and schizont stages. During the late stages of intraerythrocytic development, division occurs (schizogony), originating merozoites that burst the erythrocytes and invade new red blood cells. Some of the intraerythrocytic parasites experience maturation into the sexual forms (male and female gametocytes), which are taken up by the *Anopheles* mosquito during a blood meal. The mosquito hosts the sexual reproduction of the parasite (Bannister and Mitchell 2003).

A common feature of some *Plasmodium* species is synchronicity of development inside erythrocytes (i.e., the majority of parasites are at the same stage at a given time). This is the case for *P. falciparum* and *P. malariae* (Golgi 1889), *P. brasilianum* (Taliaferro and Taliaferro 1934), and *P. knowlesi* and *P. cynomolgi* (Hawking et al. 1968). Moreover, the duration of the intraerythrocytic cycle of many species of *Plasmodium* is a multiple of 24 h (reviewed in Mideo et al. 2013). Thus, it is tempting to speculate that circadian rhythms regulate the parasite life cycle.

The mechanism that dictates parasite synchronization remained elusive for a long time. Boyd (1929) began to shed some light on the problem when he decided to verify whether the periodicity of schizogony of the parasite was determined by the inherent length of its cell cycle or whether it was a result of the interaction of the parasite with the host. If the latter assumption were correct, the alteration of the environmental conditions of the host could affect the parasite cycle. Indeed, when the light cycle of *P. cathemerium*-infected birds was inverted, schizogony was shifted too, occurring at 6 a.m. rather than 6 p.m. (Boyd 1929). A similar result was obtained with *P. brasilianum*-infected monkeys, exposed to an inverted light regime (Taliaferro and Taliaferro 1934). Another line of evidence suggests the importance of the host environment to elicit *Plasmodium* synchronization: the growth of *Plasmodium falciparum* in continuous culture (Trager and Jensen 1976). This was accomplished by the culture of infected blood from an *Aotustrivigatus* monkey. At first, the infected red blood cells displayed only ring-stage parasites, but after 50 days in culture, the authors observed ring, trophozoite, and schizont stages in the blood smears. The fact that intraerythrocytic synchronic development of the *Plasmodium falciparum* population is lost in vitro suggests that a host factor may be responsible for its maintenance (Trager and Jensen 1976).

A question then arises: Which is the circadian host cue responsible for the synchronization of the *Plasmodium* intraerythrocytic cycle? The hormone melatonin seemed a likely candidate, as (1) its synthesis is regulated by the light/dark circadian rhythm of the host; (2) it displays a circadian rhythm of release in the blood, peaking during the night; and (3) it is hydrophobic, thus having the potential of traversing the erythrocyte and PVMs to reach the parasite. For a thorough review on the modulation of melatonin synthesis and physiological effects in humans, the reader is referred to Reiter et al. (2010).

13.2 HOST MELATONIN AND TRYPTOPHAN DERIVATIVES AS MODULATORS OF *PLASMODIUM* INTRAERYTHROCYTIC DEVELOPMENT AND SYNCHRONICITY

Hotta et al. (2000) showed that when melatonin is added to in vitro cultures of *P. chabaudi*, a rise in parasitemia relative to the control is observed after an 18 h incubation period. This result could indicate that (1) melatonin increases the ability of parasites to infect erythrocytes or (2) it has a stimulatory effect on parasite growth. To address the hypotheses, the authors performed the same experiment with *P. falciparum*, which has a longer intraerythrocytic cycle (48 h) compared to *P. chabaudi* (24 h). After a 24 h incubation of desynchronized cultures of *P. falciparum*-infected erythrocytes with melatonin, the authors noted that the hormone induced a rise in the frequency of the late stage (schizont) while diminishing the frequency of earlier stages (rings and trophozoites, Hotta et al. 2000). This result indicated that melatonin is able to stimulate parasite growth and to promote a synchronization of parasite stages (Hotta et al. 2000). In vivo experiments with *P. chabaudi* revealed that when the melatonin source, the pineal gland of the host mice, was removed, intraerythrocytic synchronous development was reduced. This effect could be reverted by exogenous injection of melatonin (Hotta et al. 2000). Desynchronization of the intraerythrocytic development could also be achieved pharmacologically, via the administration of luzindole, an antagonist of melatonin receptors (Hotta et al. 2000). Luzindole precluded the melatonin-induced parasite synchronicity in vivo, suggesting that melatonin receptors are involved in the process (Hotta et al. 2000).

Some *Plasmodium* species do not display synchronicity in their intraerythrocytic development. This is the case of *P. berghei* and *P. yoelii*, two murine parasites that have a 22 and 18 h intraerythrocytic cycle, respectively (reviewed in Mideo et al. 2013). Melatonin does not have an effect on the synchrony of any of these parasites (Bagnaresi et al. 2009). Taken together with the data on *P. falciparum* and *P. chabaudi*, the data on *P. berghei* and *P. yoelii* strengthen the link between melatonin and synchronization.

Interestingly, tryptophan, *N*-acetylserotonin, and serotonin, intermediates of the melatonin synthesis pathway (Beraldo and Garcia 2005, Schuck et al. 2011), and a melatonin metabolite, *N*-acetyl-*N*-formyl-5-metoxykinuramine (Budu et al. 2007), promote *P. falciparum* intraerythrocytic development synchronization. This raises questions about the molecular identity and properties of the receptor for melatonin in *Plasmodium*. Although the *Plasmodium* genome does not possess any annotated genes that resemble mammalian receptors for extracellular signals, Madeira et al. (2008), using stringent in silico methods, identified four strong candidates for seven transmembrane receptors in the parasite genome. The functional characterization of these receptors is the subject of current research.

13.3 MELATONIN-INDUCED SIGNAL TRANSDUCTION IN *PLASMODIUM*

13.3.1 Second Messengers and Sensors

The intracellular signaling events triggered by melatonin in *Plasmodium* are intricate. In nanomolar concentrations, melatonin triggered a cytosolic calcium concentration ($[Ca^{2+}]_{cyt}$) rise, as measured by a fluorescent chemical Ca^{2+} indicator loaded into the isolated parasite (Hotta et al. 2000). Importantly, this result indicates that the PVM and the erythrocyte membrane are dispensable for the perception of melatonin. Moreover, the $[Ca^{2+}]_{cyt}$ rise evoked by melatonin persisted in the parasite even if when it was not isolated from erythrocytes, which suggests that melatonin is able to traverse the erythrocyte membrane and the PVM (Hotta et al. 2003). The melatonin-elicited $[Ca^{2+}]_{cyt}$ rise was not abrogated by chelating extracellular Ca^{2+} with EGTA (ethylene glycol tetraacetic acid), indicating that the second messenger is released from intracellular stores (Hotta et al. 2000). Accordingly, when a phospholipase C (PLC) inhibitor was used, the melatonin-induced $[Ca^{2+}]_{cyt}$ rise was abrogated (Hotta et al. 2000).

PLC classically leads to the production of the second messengers inositol 1,4,5-trisphosphate (IP_3) and diacylglycerol (DAG) by cleaving phosphatidylinositol 4,5-bisphosphate (PIP_2). Pharmacological data suggested that IP_3 could be produced upon melatonin stimulation of *Plasmodium* (Hotta et al. 2000). To show that this indeed was the case, radioactive myo-inositol was provided to *Plasmodium falciparum*, and the parasite was challenged with melatonin. After column purification, a melatonin dose–dependent rise in IP_3 was observed relative to the control (Alves et al. 2011). By simultaneously loading parasites with caged IP_3, which can be released by a UV pulse, and the fluorescent calcium dye Fluo-4AM, it was also demonstrated that IP_3 is capable of generating a $[Ca^{2+}]_{cyt}$ rise in the parasite (Alves et al. 2011). Importantly, the Ca^{2+} rise evoked by IP_3 in the cytosol originated from intracellular stores, as when the store was depleted, the $[Ca^{2+}]_{cyt}$ evoked by a UV pulse was abrogated (Alves et al. 2011). Moreover, melatonin and IP_3 converge to the same mechanism of $[Ca^{2+}]_{cyt}$ increase (Alves et al. 2011, Figure 13.1).

The classic calcium pathway is not the only player in the melatonin-triggered signal transduction in *Plasmodium*. Indeed, a membrane-permeable analogue of another second messenger, cyclic adenosine monophosphate (6-benzyl cyclic AMP, 6-Bz-cAMP), could mimic the melatonin-induced synchronization of *Plasmodium*, and this effect was abrogated by protein kinase A (PKA) inhibitors (Beraldo et al. 2005). The fact that cAMP could bypass the need of melatonin to trigger downstream effects leading to synchronization raised the question whether cAMP played a role in the signal transduction elicited by melatonin, and, if this were the case, how were the Ca^{2+} and cAMP pathways related? In fact, a complex cross-talk between Ca^{2+} and cAMP was found in *Plasmodium* (Beraldo et al. 2005, Gazarini et al. 2011). Melatonin triggered a rise in cAMP levels and enhanced the activity of a cAMP sensor, the PKA, while BAPTA-AM, an intracellular Ca^{2+} chelator, and U73122, a PLC inhibitor, blocked the melatonin-elicited cAMP rise (Beraldo et al. 2005). These results indicated that the Ca^{2+} and cAMP pathways intersected in the parasite and that the $[Ca^{2+}]_{cyt}$ rise was upstream of the cAMP rise. When the membrane-permeable cAMP analogue was added to the parasite, a $[Ca^{2+}]_{cyt}$ rise was observed. This $[Ca^{2+}]_{cyt}$ rise was blocked by a PKA inhibitor but not by U73122, indicating that this $[Ca^{2+}]_{cyt}$ rise was downstream of PLC and that PKA was a key player in provoking the $[Ca^{2+}]_{cyt}$ increase. To determine which the cAMP-released Ca^{2+} source was, 6-Bz-cAMP was added prior to or after thapsigargin (THG), a sarcoplasmic reticulum calcium ATPase (SERCA) inhibitor, which promotes leakage of Ca^{2+} from the endoplasmic reticulum. THG was not able to promote a $[Ca^{2+}]_{cyt}$ rise when it was added after 6-Bz-cAMP,

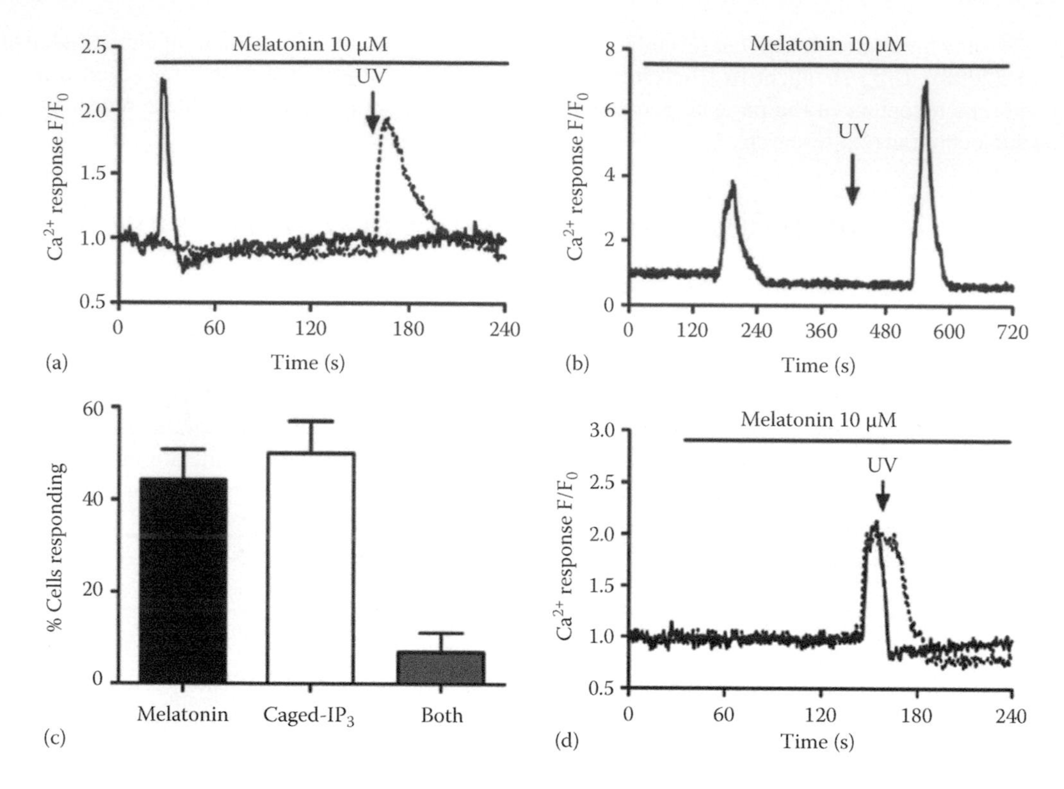

FIGURE 13.1 $[Ca^{2+}]_{cyt}$ measurements in caged-IP_3 and Fluo-4AM-loaded intraerythrocytic parasites (*Plasmodium falciparum*). Melatonin and IP_3 were added sequentially to the parasite preparations. Representative fluorescence traces of parasites that display $[Ca^{2+}]_{cyt}$ increase elicited by (a) melatonin or by uncaging of IP_3 with and UV pulse; (b) melatonin and by uncaging of IP_3 with a UV pulse; (c) percentage of cells responding to melatonin, UV pulse, or both stimuli; (d) representative traces showing that parasites were not able to display a further rise in $[Ca^{2+}]_{cyt}$ by a UV pulse while displaying a $[Ca^{2+}]_{cyt}$ rise evoked by melatonin. Data in (c) and (d) indicate that melatonin and IP_3 increase $[Ca^{2+}]_{cyt}$ from a same pool of Ca^{2+}. (This research figure was originally published in Alves, E. et al., *J. Biol. Chem.*, 286, 5905, 2011. Copyright 2011 The American Society for Biochemistry and Molecular Biology.)

which led to the conclusion that THG and 6-Bz-cAMP converged to the same mechanism of Ca^{2+} release (i.e., the endoplasmic reticulum, Beraldo et al. 2005). Moreover, the experiments provided evidence that 6-Bz-cAMP mobilizes Ca^{2+} from other sources besides the endoplasmic reticulum, as the compound continued to promote $[Ca^{2+}]_{cyt}$ increase when added after THG. Indeed, in the presence of the calcium chelator EGTA, the 6-Bz-cAMP-evoked Ca^{2+} was abrogated. This result indicated that 6-Bz-cAMP is able to promote the opening of plasma membrane Ca^{2+} channels (Beraldo et al. 2005).

Melatonin signaling in *Plasmodium* can also, in all probability, trigger capacitative Ca^{2+} entry (CCE) in the parasite (Beraldo et al. 2007). CCE is triggered by the reduction of the $[Ca^{2+}]$ in the endoplasmic reticulum, which provokes the opening of plasma membrane Ca^{2+} channels. When 8-Br-cAMP, a membrane-permeable cAMP analogue, was added to the parasite in nominally free calcium medium a rise in $[Ca^{2+}]_{cyt}$ which originated from intracellular stores was observed (Beraldo et al. 2007). When Ca^{2+} was added to the extracellular buffer after 8-Br-cAMP, a further rise in $[Ca^{2+}]_{cyt}$ was observed, which originated from influx through plasma membrane Ca^{2+} channels (Beraldo et al. 2007).

Besides the endoplasmic reticulum, another intracellular Ca^{2+} store participates in the melatonin signaling: the mitochondrion. Using double labeling with Fluo-3 AM, a fluorescent cytosolic Ca^{2+} sensor, and Rhod-2 AM, a fluorescent Ca^{2+} sensor, with affinity to mitochondria, it was

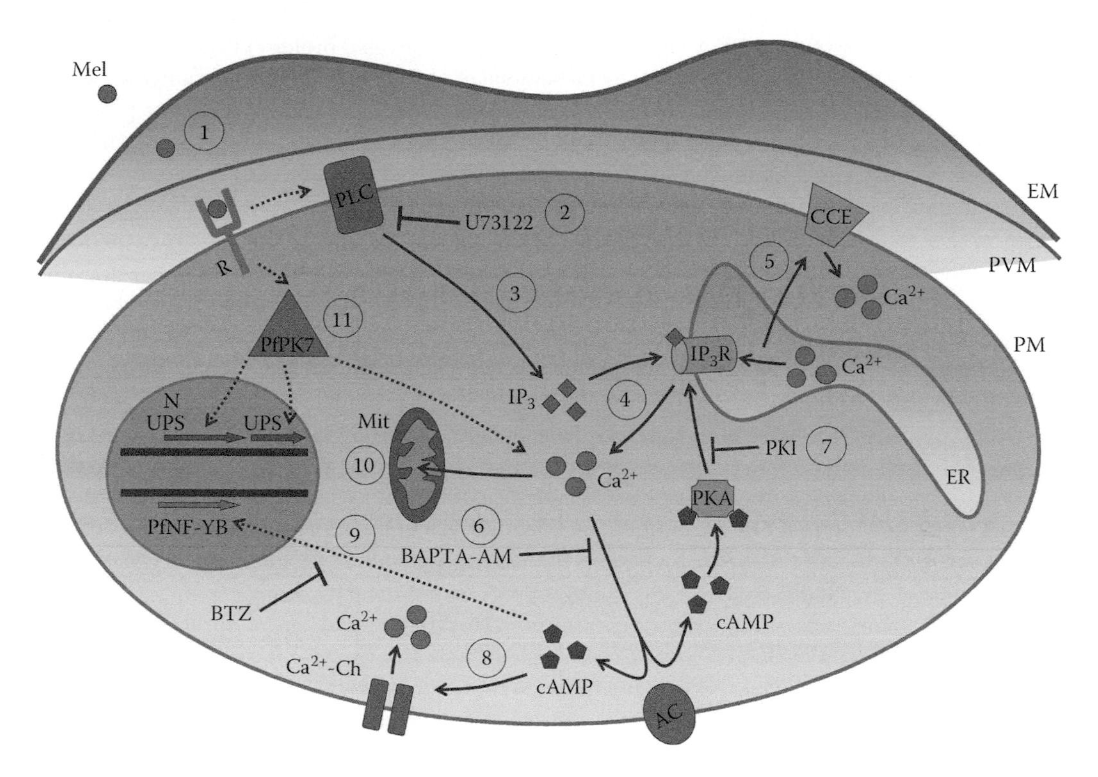

FIGURE 13.2 Schematic model of melatonin-triggered signal transduction in *Plasmodium*. (1) Melatonin traverses the erythrocyte membrane (Hotta et al. 2003). (2) U73122 inhibits $[Ca^{2+}]_{cyt}$ rise (Hotta et al. 2000) and cAMP formation (Beraldo et al. 2005). (3) Melatonin leads to IP_3 formation (Alves et al. 2011). (4) IP_3 leads to $[Ca^{2+}]_{cyt}$ rise from the endoplasmic reticulum (Alves et al. 2011). (5) $[Ca^{2+}]$ reduction in intracellular stores leads to capacitative calcium entry (Beraldo et al. 2007). (6) BAPTA-AM, an intracellular calcium chelator, inhibits cAMP formation (Beraldo et al. 2005). (7) PKI inhibits Ca^{2+} release from the ER (Beraldo et al. 2005). (8) cAMP leads to the opening of plasma membrane calcium channels (Beraldo et al. 2005). (9) cAMP promotes transcription of PfNF-YB at throphozoite stage; bortezomib, a proteasome inhibitor, inhibits transcription of PfNF-YB (Lima et al. 2013). (10) Ca^{2+} in the cytosol is taken up by the *Plasmodium* mitochondrion (Gazarini and Garcia 2004). (11) Melatonin induces transcription of UPS genes via PfPK7, and PfPK7 controls melatonin-induced $[Ca^{2+}]_{cyt}$ rise (Koyama et al. 2012). AC, adenylyl cyclase; BTZ, bortezomib; Ca^{2+}-Ch, membrane Ca^{2+} channel opened via cAMP; cAMP, cyclic adenosine monophosphate; CCE, capacitative calcium entry channel; EM, erythrocyte plasma membrane; ER, endoplasmic reticulum; IP_3, Inositol 1,4,5 trisphosphate; IP_3R, putative IP_3 receptor; Mel, melatonin; Mit, mitochondria; N, nucleus; PfNF-YB, transcription factor NF-YB from *P. falciparum*; PfPK7, protein kinase 7 from *P. falciparum*; PKA, protein kinase A; PKI, protein kinase inhibitor peptide; PLC, Phospholipase C; PM, parasite plasma membrane; PVM, parasitophorous vacuole membrane; R, Putative receptor; UPS, ubiquitin-proteasome system.

demonstrated that the mitochondrion of *Plasmodium* senses $[Ca^{2+}]_{cyt}$ fluctuations evoked by melatonin (Gazarini and Garcia 2004). The authors provide strong evidence for a transient $[Ca^{2+}]$ rise in the cytosol, followed by a transient $[Ca^{2+}]$ rise in the organelle. This latter $[Ca^{2+}]$ rise potentially activates metabolism-related enzymes (Gazarini and Garcia 2004). Taken together, the data unveil a complex scenario for melatonin-triggered signal transduction in *Plasmodium* (Figure 13.2).

13.3.2 Effectors of Melatonin-Induced Signal Transduction in *Plasmodium*

In order to understand how synchronicity is achieved in *Plasmodium*, it is vital to unveil which effector proteins are modulated by melatonin signaling. The atypical kinase PfPK7 is a likely effector of melatonin, as its knockout provokes a reduced growth rate of intraerythrocytic stages and as it has regions resembling both fungal PKA and MAPKK (mitogen-activated protein kinase kinase), which are signal transduction–related enzymes (Dorin-Semblat et al. 2008). Indeed, using a PfPK7 knock-out parasite strain (PfPK7⁻), Koyama et al. (2012) showed that the kinase is a key player in promoting melatonin-induced synchronicity in *Plasmodium*, as melatonin was not able to promote synchronization in PfPK7⁻ parasites. Moreover, PfPK7 could be involved in the control of melatonin-induced $[Ca^{2+}]_{cyt}$ increase as only a minor calcium increase was observed in PfPK7⁻ parasites challenged with melatonin (Koyama et al. 2012). The MAPK (mitogen-activated protein kinase) pathway classically modulates the activity of proteins involved in cell cycle control, which could be the case for PfPK7, as PfPK7⁻ parasites produce a lower number of merozoites (Dorin-Semblat et al. 2008). However, the mechanism by which PfPK7 triggers these events remains to be determined, as the classical three-component cascade of MAPK is absent in *Plasmodium* (Dorin-Semblat et al. 2008).

Melatonin also promotes the increase in the number of transcripts of genes of the ubiquitin-proteasome system (UPS, Koyama et al. 2012). This included upregulation of E1-activating enzymes, E3 ubiquitin ligases, and proteasome subunits (Koyama et al. 2012, reviewed in Lima et al. 2013). Ubiquitination is involved with various processes in the eukaryotic cell, such as signal transduction, cell cycle regulation, control of transcription, and degradation of proteins (reviewed in Hershko and Ciechanover 1998). Accordingly, the melatonin-induced upregulation of UPS genes in *Plasmodium* could account for the modulation of many cellular processes of the parasite.

One link between melatonin, ubiquitination gene transcription, and *Plasmodium falciparum* is the PfNF-YB transcription factor, which is known to bind the CCAAT box of gene promoters (reviewed in Lima et al. 2013). PfNF-YB is expressed throughout the intraerythrocytic stages of the parasites, with partial localization with the nucleus (Lima et al. 2013, Figure 13.3).

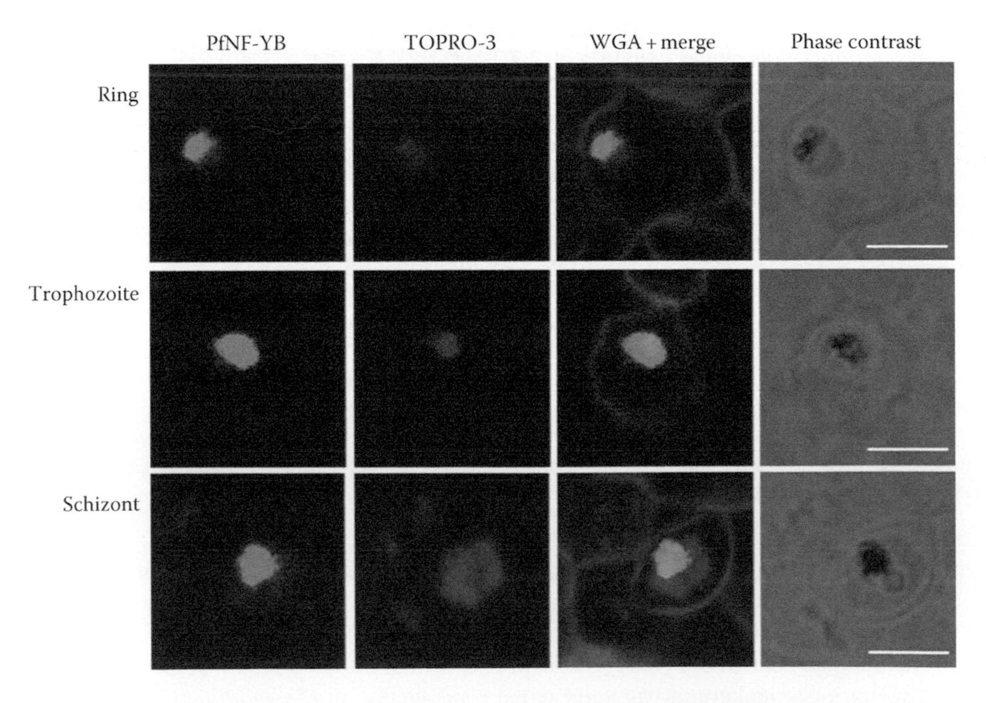

FIGURE 13.3 PfNF-YB localization in intraerythrocytic stages of *P. falciparum* (ring, trophozoite, and schizont). TOPRO-3: DNA marker, WGA (wheat germ agglutinin): plasma membrane marker staining erythrocyte membrane. DNA staining and PfNF-YB staining partially overlap. (This research figure was originally published in Lima, W.R., Moraes, M., Alves, E., Azevedo, M.F., Passos, D.O., and Garcia, C.R.S.: The PfNF-YB transcription factor is a downstream target of melatonin and cAMP signalling in the human malaria parasite *Plasmodium falciparum*. *J. Pineal. Res.*, 2013, 54, 145–153. Copyright Wiley-VCH Verlag GmbH & Co. KGaA. Reproduced with permission.)

In trophozoites, melatonin and 6-Bz-cAMP caused an increase in the expression of PfNF-YB, as assessed by western blot and real-time PCR. Moreover, melatonin was able to cause an increase in the ubiquitination state of PfNF-YB in trophozoites, assessed by immunoprecipitation (Lima et al. 2013; Figure 13.4).

In order to explore the implications of PfNF-YB ubiquitination for its stability, bortezomib, an inhibitor of the proteasome, was used. Surprisingly, the transcription and expression of PfNF-YB was diminished in rings, trophozoites, and schizonts, suggesting that PfNF-YB transcription is under tight control of ubiquitination (Lima et al. 2013).

13.4 SYNCHRONICITY AS AN ADAPTIVE FEATURE IN *PLASMODIUM*: IMPLICATION FOR THERAPIES

There are some hypotheses to explain why intraerythrocytic synchronicity has evolved in the malaria parasite. An interesting review on the question suggests that synchronicity may come as a result of parasites bursting at the same time or as a consequence of parasites bursting at a particular time of the day (Mideo et al. 2013). Depending on the selective pressure, synchronicity may have risen due to the burst of parasites at a particular time of the day, for example, avoiding the circadian release of parasite killing immune responses. On the other hand, the strategy of the parasite could be to find *safety in numbers* from immune attacks (i.e., achieved by the release of many merozoites at the same time, which overwhelms the capacity of the immune system to deal with the parasite). In the latter case, it would be important for parasites to burst at the same time, but not necessarily at a particular time (Mideo et al. 2013).

The strategies *Plasmodium* could use for a synchronous intraerythrocytic development could be *quorum* sensing, allowing parasites to communicate directly with one another, which implies only in bursting of erythrocytes at the same time in one host (but the time of egress could be different in another host); or it could also be by perceiving circadian cues from its environment. In the latter case, parasites would burst erythrocytes at a particular time (Mideo et al. 2013). This hypothesis is in agreement with the host melatonin-synchronizing effect on intraerythrocytic stages of malaria parasites (Hotta et al. 2000).

Importantly, there is experimental evidence supporting that circadian rhythms play an important role in the evolution of host-*Plasmodium* interactions (O'Donnell et al. 2011). The experimental design that supports this hypothesis is elegant: two groups of mice were infected with synchronous *P. chabaudi*, in rooms with standard regime of light/dark (12 h/12 h) or inverted regime of light/dark. The infected blood from mice in the standard light regime was used to infect mice in the same room (matched infection) and mice in the inverted light regime (unmatched infection), whereas the infected blood from mice in the inverted light regime was used to infect mice in the same room (matched infection) and mice in the standard light regime (unmatched infection). After 24 h, the authors observed that parasitemia was higher in matched infections compared with unmatched infections, thus demonstrating that there is a fitness cost for the parasite if the circadian rhythm of the host is disrupted (O'Donnell et al. 2011). These results are also in agreement with the existence of a host circadian factor that can be perceived by the parasite.

If one takes into account the importance of the circadian rhythm of the host for the replication of *Plasmodium* (O'Donnell et al. 2011) and melatonin as the molecule that links host circadian rhythm to *Plasmodium* synchronicity (Hotta et al. 2000) blocking the melatonin synchronizing effect on the parasite might have therapeutic implications. Indeed, a study suggests that there are therapeutic implications for disrupting synchronicity in *Plasmodium* development: by treating *P. chabaudi*-infected mice with suboptimal doses of chloroquine, a classical antimalarial drug, and with the concurrent administration of luzindole, which desynchronizes the intraerythrocytic development of *Plasmodium*, it was demonstrated that there is an enhancement in the survival of the mice when compared with the chloroquine-only-treated control (Bagnaresi et al. 2008). Another study also demonstrates that high concentrations of melatonin, which act as antioxidant, ameliorate the liver

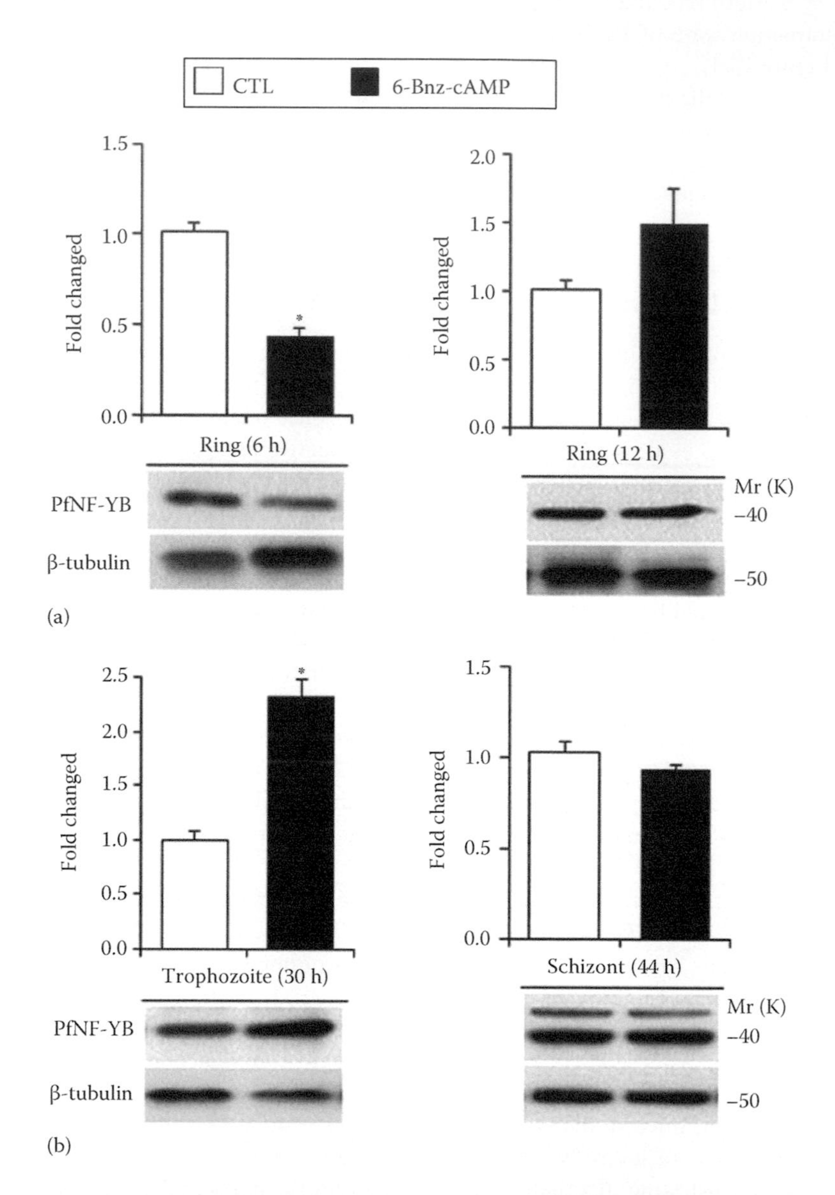

FIGURE 13.4 Effect of 6-Bz-cAMP (6-Bzn-cAMP) on the transcription of PfNF-YB mRNA, assessed by real-time PCR (bar graphics) and on the protein level of PfNF-YB (blots). (a) Rings (6 h postinvasion), rings (12 h postinvasion); (b) trophozoites (30 h postinvasion), schizonts (44 h postinvasion). Transcription and translation data concur on the rise of expression of PfNF-YB induced by 6-Bz-cAMP in trophozoites. In rings 6 h postinvasion, 6-Bz-cAMP causes a decrease in PfNF-YB transcription and translation, while on rings 12 h postinvasion and schizonts 44 h postinvasion, no statistically significant effect on transcription is observed. (This research figure was originally published in Lima, W.R., Moraes, M., Alves, E., Azevedo, M.F., Passos, D.O., and Garcia, C.R.S.: The PfNF-YB transcription factor is a downstream target of melatonin and cAMP signalling in the human malaria parasite *Plasmodium falciparum*. *J. Pineal. Res.*, 2013, 54, 145–153. Copyright Wiley-VCH Verlag GmbH & Co. KGaA. Reproduced with permission.)

damage provoked by the parasite (Guha et al. 2007). Thus, by combining high doses of melatonin with melatonin receptor inhibitors, an efficient treatment could be achieved (Srinivasan et al. 2010).

13.5 CONCLUSIONS AND PERSPECTIVES

Synchronicity is a common feature of many *Plasmodium* species and may confer important adaptive advantages to the parasite (reviewed in Mideo et al. 2013). Intraerythrocytic cell cycle synchronization can be achieved by the parasite via the perception of the host hormone melatonin (Hotta et al. 2000) which is released in a circadian manner in the bloodstream. Although detailed information about melatonin-evoked signal transduction in the parasite exists, unanswered questions that could have therapeutic implications still remain: (1) What is the molecular identity and pharmacological properties of the melatonin receptor? (2) What are the genes regulated by the transcription factor PfNF-YB? (3) Does *Plasmodium* possess clock-like genes? (4) Do asynchronous malaria parasites use different strategies to evade host defenses?

As a large proportion of genes in *Plasmodium* have unassigned functions, the answer for these questions is challenging. Thus, a better understanding of the parasite biology is paramount to successfully address them.

REFERENCES

Alves, E., P. J. Bartlett, C. R. Garcia, and A. P. Thomas. 2011. Melatonin and IP_3-induced Ca^{2+} release from intracellular stores in the malaria parasite *Plasmodium falciparum* within infected red blood cells. *J Biol Chem* 286(7):5905–5912.

Bagnaresi, P., E. Alves, H. B. da Silva, S. Epiphanio, M. M. Mota, and C. R. Garcia. 2009. Unlike the synchronous *Plasmodium falciparum* and *P. chabaudi* infection, the *P. berghei* and *P. yoelii* asynchronous infections are not affected by melatonin. *Int J Gen Med* 2:47–55.

Bagnaresi, P., R. P. Markus, C. T. Hotta, T. Pozzan, and C. R. S. Garcia. 2008. Desynchronizing *Plasmodium* cell cycle increases chloroquine protection at suboptimal doses. *Open Parasitol J* 2:55–58.

Bannister, L. and G. Mitchell. 2003. The ins, outs and roundabouts of malaria. *Trends Parasitol* 19(5):209–213.

Beraldo, F. H., F. M. Almeida, A. M. da Silva, and C. R. Garcia. 2005. Cyclic AMP and calcium interplay as second messengers in melatonin-dependent regulation of *Plasmodium falciparum* cell cycle. *J Cell Biol* 170(4):551–557.

Beraldo, F. H. and C. R. Garcia. 2005. Products of tryptophan catabolism induce Ca^{2+} release and modulate the cell cycle of *Plasmodium falciparum* malaria parasites. *J Pineal Res* 39(3):224–230.

Beraldo, F. H., K. Mikoshiba, and C. R. Garcia. 2007. Human malarial parasite, *Plasmodium falciparum*, displays capacitative calcium entry: 2-aminoethyl diphenylborinate blocks the signal transduction pathway of melatonin action on the *P. falciparum* cell cycle. *J Pineal Res* 43(4):360–364.

Boyd, G. H. 1929. Induced variations in the asexual cycle of *Plasmodium cathemerium*. *Am J Epidemiol* 9 (1):181–187.

Budu, A., R. Peres, V. B. Bueno, L. H. Catalani, and C. R. Garcia. 2007. N1-acetyl-N2-formyl-5-methoxykynuramine modulates the cell cycle of malaria parasites. *J Pineal Res* 42(3):261–266.

Dorin-Semblat, D., A. Sicard, C. Doerig, L. Ranford-Cartwright, and C. Doerig. 2008. Disruption of the PfPK7 gene impairs schizogony and sporogony in the human malaria parasite *Plasmodium falciparum*. *Eukaryot Cell* 7(2):279–285.

Gazarini, M. L., F. H. Beraldo, F. M. Almeida, M. Bootman, A. M. Da Silva, and C. R. Garcia. 2011. Melatonin triggers PKA activation in the rodent malaria parasite *Plasmodium chabaudi*. *J Pineal Res* 50(1):64–70.

Gazarini, M. L. and C. R. Garcia. 2004. The malaria parasite mitochondrion senses cytosolic Ca^{2+} fluctuations. *Biochem Biophys Res Commun* 321(1):138–144.

Golgi, C. 1889. Sul ciclo evolutivo dei parassiti malarici nella febbre terzana: Diagnosi differenziale tra i parassiti endoglobulari malarici della terzana e quelli della quartana. *Arch Sci Med (Torino)* 13:173–193.

Guha, M., P. Maity, V. Choubey, K. Mitra, R. J. Reiter, and U. Bandyopadhyay. 2007. Melatonin inhibits free radical-mediated mitochondrial-dependent hepatocyte apoptosis and liver damage induced during malarial infection. *J Pineal Res* 43(4):372–381.

Hawking, F., M. J. Worms, and K. Gammage. 1968. 24- and 48-hour cycles of malaria parasites in the blood; their purpose, production and control. *Trans R Soc Trop Med Hyg* 62(6):731–765.

Hershko, A. and A. Ciechanover. 1998. The ubiquitin system. *Annu Rev Biochem* 67:425–479.

Hotta, C. T., M. L. Gazarini, F. H. Beraldo, F. P. Varotti, C. Lopes, R. P. Markus, T. Pozzan, and C. R. Garcia. 2000. Calcium-dependent modulation by melatonin of the circadian rhythm in malarial parasites. *Nat Cell Biol* 2(7):466–468.

Hotta, C. T., R. P. Markus, and C. R. Garcia. 2003. Melatonin and *N*-acetyl-serotonin cross the red blood cell membrane and evoke calcium mobilization in malarial parasites. *Braz J Med Biol Res* 36(11):1583–1587.

Koyama, F. C., R. Y. Ribeiro, J. L. Garcia, M. F. Azevedo, D. Chakrabarti, and C. R. Garcia. 2012. Ubiquitin proteasome system and the atypical kinase PfPK7 are involved in melatonin signaling in *Plasmodium falciparum*. *J Pineal Res* 53(2):147–153.

Lima, W. R., A. A. Holder, and C. R. Garcia. 2013. Melatonin signaling and its modulation of PfNF-YB transcription factor expression in *Plasmodium falciparum*. *Int J Mol Sci* 14(7):13704–13718.

Lima, W. R., M. Moraes, E. Alves, M. F. Azevedo, D. O. Passos, and C. R. Garcia. 2013. The PfNF-YB transcription factor is a downstream target of melatonin and cAMP signalling in the human malaria parasite *Plasmodium falciparum*. *J Pineal Res* 54(2):145–153.

Madeira, L., P. A. Galante, A. Budu, M. F. Azevedo, B. Malnic, and C. R. Garcia. 2008. Genome-wide detection of serpentine receptor-like proteins in malaria parasites. *PLoS One* 3(3):e1889.

Mideo, N., S. E. Reece, A. L. Smith, and C. J. Metcalf. 2013. The Cinderella syndrome: Why do malaria-infected cells burst at midnight? *Trends Parasitol* 29(1):10–16.

O'Donnell, A. J., P. Schneider, H. G. McWatters, and S. E. Reece. 2011. Fitness costs of disrupting circadian rhythms in malaria parasites. *Proc Biol Sci* 278(1717):2429–2436.

Reiter, R. J., D. X. Tan, and L. Fuentes-Broto. 2010. Melatonin: A multitasking molecule. *Prog Brain Res* 181:127–151.

Schuck, D. C., R. Y. Ribeiro, A. A. Nery, H. Ulrich, and C. R. Garcia. 2011. Flow cytometry as a tool for analyzing changes in *Plasmodium falciparum* cell cycle following treatment with indol compounds. *Cytometry A* 79(11):959–964.

Srinivasan, V., D. W. Spence, A. Moscovitch, S. R. Pandi-Perumal, I. Trakht, G. M. Brown, and D. P. Cardinali. 2010. Malaria: Therapeutic implications of melatonin. *J Pineal Res* 48(1):1–8.

Taliaferro, W.H. and L.G. Taliaferro. 1934. Alteration in the time of sporulation of *Plasmodium brasilianum* in monkeys by reversal of light and dark. *Am J Epidemiol* 20:10.

Trager, W. and J. B. Jensen. 1976. Human malaria parasites in continuous culture. *Science* 193(4254):673–675.

14 Melatonin
Regulator of Adult Hippocampal Neurogenesis

Gerardo B. Ramírez-Rodríguez and Leonardo Ortiz-López

CONTENTS

14.1 INTRODUCTION

Hippocampal neurogenesis consists in the generation of new neurons in the dentate gyrus of the adult brain (Kempermann et al., 2004a). Initial reports about the neurogenic process in the hippocampus derived from studies of Joseph Altman in the 1960s (Altman and Das, 1966). Further studies supported the existence of new neurons in the hippocampus during adulthood (Cameron and McKay, 1999; Gage et al., 1998; Gould et al., 1997; Gould and Tanapat, 1999; Horner et al., 2000; Kempermann et al., 1997; Kuhn et al., 1996; van Praag et al., 1999). Adult hippocampal neurogenesis originates from neural stem cells (NSCs), with the characteristics of radial glial cell (NS-RGC) (Kempermann et al., 2004a). The NS-RGC goes through asymmetric divisions to give rise to subpopulations of rapid amplifying cells and neuroblasts, cells that will form immature neurons, which at their latest stage form new neurons that incorporate into the existent brain circuits (Ge et al., 2006, 2007; Kempermann et al., 2004a,b; Nollet et al., 2012; Zhao et al., 2006b, 2008). This process occurs in the hippocampus due to the presence of a permissive microenvironment in which several populations of immature and mature cells coexist to promote neurogenesis (Palmer et al., 2000). Thus, the generation of new neurons becomes a highly regulated process in which the involvement of hormones such as melatonin has been reported (Ramirez-Rodriguez et al., 2009).

14.2 ADULT NEUROGENESIS

Adult neurogenesis takes place in two canonical regions of the central nervous system (CNS) during adulthood: the olfactory bulb (OB) and the hippocampus. New neurons derive from neural stem cells, which reside in the subventricular zone (SVZ) of the lateral ventricles or in the subgranular zone (SGZ) of the dentate gyrus in the hippocampus (Alvarez-Buylla and Garcia-Verdugo, 2002; Kempermann et al., 2004a).

In the SVZ-OB system, NS-RGC may proliferate and generate the rapid amplifying progenitor and neuroblast populations that will migrate and differentiate in neurons to finally be integrated into the neuronal network of the OB (Alvarez-Buylla and Garcia-Verdugo, 2002).

In a similar manner in the hippocampus, NS-RGCs reside in the SGZ, in which they will divide to give rise to the rapid amplifying progenitors and neuroblasts. These cells will migrate into the granular cell layer (GCL) to fully differentiate and integrate in the neuronal circuitry (Gage et al., 1998; Ge et al., 2007; Kempermann et al., 2004a).

14.3 HIPPOCAMPAL NEURONAL DEVELOPMENT

In the hippocampus, adult neurogenesis starts from the slow division of NSCs that reside in the SGZ (Kempermann et al., 2004a). The NS-RGCs show radial glial characteristics, expressing markers as the glial fibrilar acidic protein (GFAP) and the undifferentiated cell marker nestin; both proteins belong to the family of intermediate filaments, the component of the cytoskeleton (Kempermann et al., 2004a). The NS-RGCs, also known as *Type 1* cells, through asymmetric divisions give rise to the rapid amplifying population that corresponds to the neural precursor cells (NPCs) (Kempermann et al., 2004a).

Considering the temporal expression of proteins used as specific markers for each cellular population involved in the neuronal hippocampal development, it has been possible to study different events of neurogenesis (Kempermann et al., 2004a). Thus, the phenotypic analysis has allowed cell classification in three types according to their protein expression marker: *Type 2a*, *2b*, and *3*. The *Type 2a* and *Type 2b* cells exhibit short neurite processes parallel to the GCL, but *Type 2a* cells are characterized by the expression of nestin but not GFAP, while *Type 2b* cells express nestin and start to express doublecortin (DCX). Interestingly, considering the role of DCX in migration, *Type 3* cells expressed DCX but not nestin and show vertical dendrites integrated into the GCL. In the next event of this process in the hippocampus, immature neurons show longer dendritic processes crossing the GCL. Once immature neurons become postmitotic, DCX is coexpressed with calcium-binding proteins as calretinin (CR) followed by both calbindin and for the nuclear protein (NeuN) (Brandt et al., 2003; Kempermann et al., 2004a; Plumpe et al., 2006; Rao and Shetty, 2004). At this final stage, the cells are fully differentiated, and their electrophysiological properties are similar to the old neurons (Zhao et al., 2006a, 2008), Figure 14.1.

14.4 REGULATION OF ADULT HIPPOCAMPAL NEUROGENESIS

Hippocampal neurogenesis is a widely regulated process (Kempermann et al., 2004a; Zhao et al., 2008). The major regulation of the process occurs at the survival step while the rest happens at the cell proliferation level (Kempermann et al., 2006). The regulation of hippocampal neurogenesis depends on different factors in which the niche or microenvironment plays an important role (Palmer et al., 2000). The hippocampal niche is composed by NS-RGC, astrocytes, microglia, neurons, endothelial cells, and the secreted soluble factors released by all of these cells (Palmer et al., 2000). In a very dynamic form, all of these cells work in synchrony not only to promote neurogenesis but also to maintain their own cellular populations (Palmer et al., 2000).

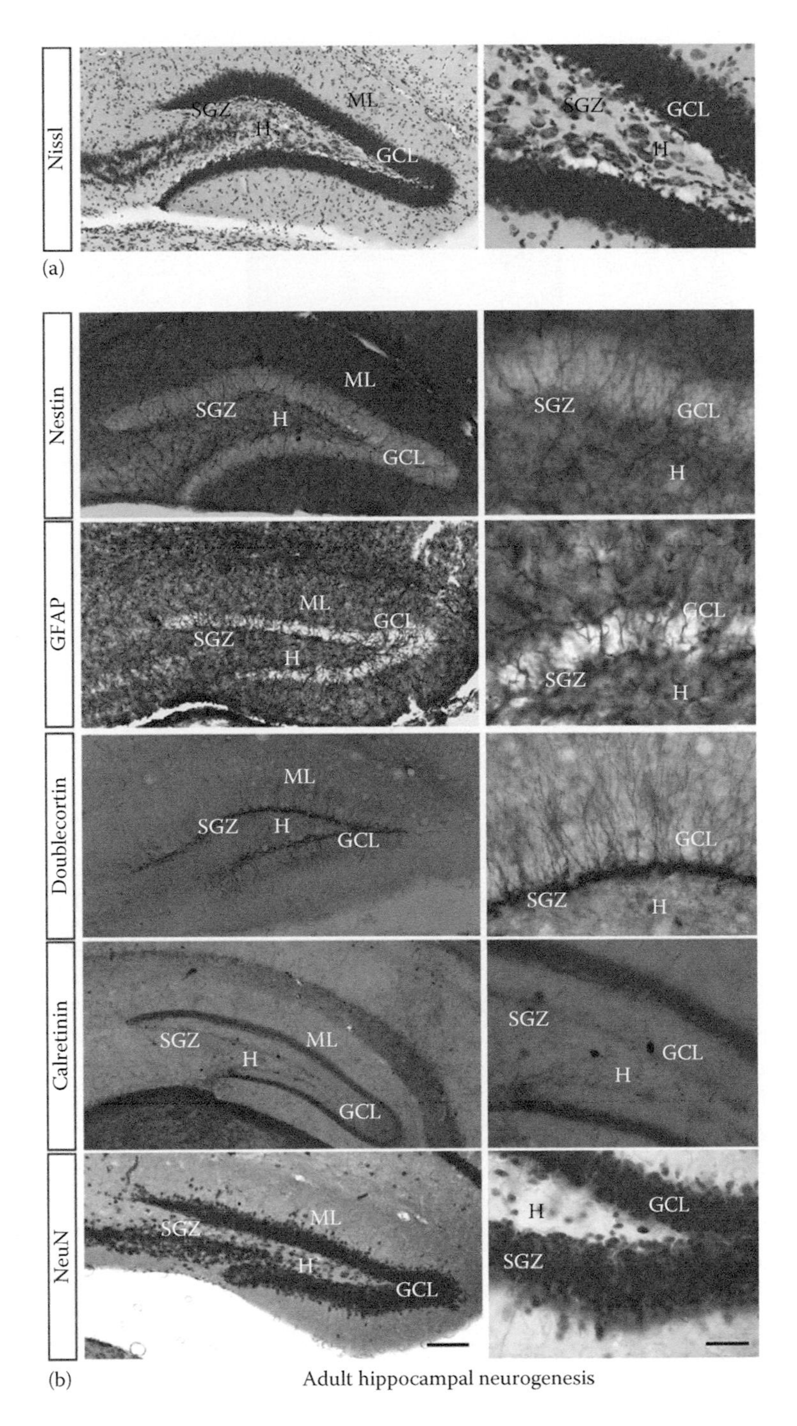

FIGURE 14.1 Adult hippocampal neurogenesis. Panel (a) shows a coronal section within the hippocampus. The box indicates fragments of the SGZ and GCL, regions of the hippocampus where neurogenesis occurs. The picture also shows the molecular layer (Cleveland et al., 1977) and the hilar zone (H). Panel (b) shows representative images of Type 1 cells expressing the GFAP or nestin. Neuroblasts identified by DCX expression or immature neurons with DCX-elaborated dendritic trees are also shown. Also, neurons identified by the expression of calretinin (CR) or NeuN are presented in panel b. Pictures show SGZ, GCL, ML, and H. Scale bar in panoramic microphotographs = 120 μm; and scale bar of amplifications of panoramic microphotographs = 50 μm. (For more details, see Kempermann, G. et al., *Trends Neurosci.*, 27, 447, 2004a.)

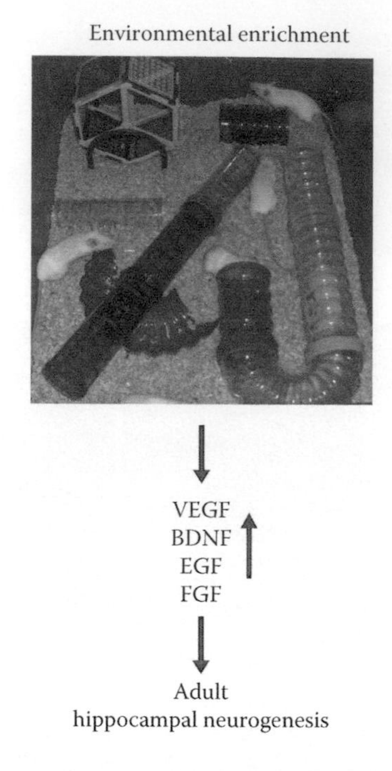

FIGURE 14.2 Paradigms with proneurogenic effects in the hippocampus. Mice housed under environmental enrichment (ENR) conditions are shown. The housing in ENR or in the presence of a running wheel promotes cellular and neurochemical changes, increasing neurogenesis and the levels of positive factors for neuronal development. (For more details, see Nithianantharajah, J. and Hannan, A.J., *Nat. Rev. Neurosci.*, 7, 697, 2006.)

Additional positive factors for the hippocampal neurogenic process are neurotransmitters (i.e., GABA, glutamate, and serotonin) (Bolteus and Bordey, 2004; Ge et al., 2007; Klempin et al., 2010; Petrus et al., 2009; Platel et al., 2007); growth factors (i.e., fibroblastic growth factor, FGF; epidermal growth factor, EGF; vascular endothelial growth factor, VEGF) (Bick-Sander et al., 2006; Fabel et al., 2003; Kuhn et al., 1997); neurotrophins (i.e., brain-derived neurotrophic factor, BDNF; neurotrophin-3, NT-3) (Babu et al., 2009; Barnabe-Heider and Miller, 2003; Taliaz et al., 2010); and also hormones (i.e., prolactin, growth hormone, and melatonin) (Bridges and Grattan, 2003; Crupi et al., 2011; Ramirez-Rodriguez et al., 2009, 2011; Rennie et al., 2009; Shingo et al., 2003). Besides these factors, the physical activity and the exposure to an environmental enrichment also promote neurogenesis concomitant to neurochemical changes by increasing the levels of growth factors and neurotrophins (Bick-Sander et al., 2006; Brown et al., 2003; Fabel et al., 2009; Kempermann et al., 2010; Nithianantharajah and Hannan, 2006; van Praag et al., 1999), Figure 14.2.

Although the importance of the earlier-mentioned regulators is widely reported, negative regulators that affect the generation of new neurons in the hippocampus also exist. Some of the negative factors are as follows: stress (Bessa et al., 2009; Gould and Tanapat, 1999; Mirescu and Gould, 2006; Mirescu et al., 2004; Schoenfeld and Gould, 2012), abuse drugs such as toluene and cocaine (Bowman and Kuhn, 1996; Paez-Martinez et al., 2013), or during aging (Couillard-Despres et al., 2009; Fabel and Kempermann, 2008; Hattiangady and Shetty, 2008; Heine et al., 2004; Klempin and Kempermann, 2007; Kuhn et al., 1996; Ramirez-Rodriguez et al., 2012).

In the context of adult hippocampal neurogenesis, melatonin, a conserved endogenous molecule, has gained attention for its role in modulating several cellular processes (AlAhmed and

Herbert, 2010; Bellon et al., 2007; Benitez-King, 2000, 2006; Caballero et al., 2009; Ortiz-Lopez et al., 2009; Reiter, 1998a,b).

14.5 MELATONIN

Melatonin, the main product of the pineal gland, is cyclically synthesized in synchrony with the dark phase of the circadian cycle (Reiter, 1991). Melatonin has pleiotropic neurobiological roles mediated through cell membrane receptors (Dubocovich, 1991, 1995) and by the activation of intracellular signaling cascades (Bellon et al., 2007; Benitez-King, 2000; Ortiz-Lopez et al., 2009). Additionally, melatonin might scavenge free oxygen radicals, acting as a neuroprotector (Reiter, 1998a) and also as a modulator of the rearrangement of the cytoskeleton components such as microtubules and microfilaments, and also intermediate filaments (Bellon et al., 2007; Benitez-King, 2000, 2006; Huerto-Delgadillo et al., 1994). Interestingly, melatonin levels fall markedly during aging and in some neuropsychiatric disorders, which show a disrupted circadian rhythm that affects sleep patterns (Brusco et al., 2000). Also, preclinical studies have suggested an antidepressant-like effect of melatonin (Crupi et al., 2010; Ramirez-Rodriguez et al., 2009).

Considering that some of the aspects mentioned earlier affect hippocampal neuronal development, they have pointed in the direction that melatonin is an interesting molecule in the context of adult neurogenesis in the hippocampus.

14.6 MELATONIN AND NEUROGENESIS

14.6.1 Initial Findings

Initial evidence about the role of the indole in adult hippocampal neurogenesis, specifically in cell proliferation, derived from postnatal rats (Kim et al., 2004). Further studies performed in vitro revealed that melatonin promotes cell proliferation and differentiation of embryonic neural stem cells and in cells derived from rat midbrain (Kong et al., 2008; Moriya et al., 2007). However, the role of melatonin in hippocampal neurogenesis during the adulthood may be differently regulated by environmental cues (Romer et al., 2011).

14.6.2 Physiological Conditions

Several works have pointed out a specific role of melatonin in neuronal development. Melatonin administered to young adult C57Bl6 mice positively modulates cell survival of newborn cells without affecting cell proliferation. Additionally, the hormone produces positive regulation of immature neurons identified by DCX expression and DCX/CR coexpression in the dentate gyrus of the hippocampus. In the same line, melatonin promotes adult hippocampal derived precursor cell differentiation in vitro in a mechanism that partially involves the membrane receptors of the hormone (Ramirez-Rodriguez et al., 2009) (Figures 14.3 and 14.4). Interestingly, the combination of melatonin plus running, a paradigm that has widely demonstrated its proneurogenic effect, promotes hippocampal neurogenesis in C3H/HeN mice (Liu et al., 2013). Contrary to C57Bl6 mice, melatonin alone was not able to alter neurogenesis in C3H/HeN mice (Liu et al., 2013). However, BalbC mice supplemented with melatonin show positive effects of the hormone on cell proliferation and survival (Ramirez-Rodriguez et al., 2012).

Melatonin also increases the population of DCX cells in the hippocampus of rodents (Ramirez-Rodriguez et al., 2011). This effect is accompanied with the increase in dendrite complexity of new immature neurons (Ramirez-Rodriguez et al., 2011). This effect is interesting because melatonin also modulates the rearrangement of microtubules and DCX is a microtubule binding protein that, in addition to its expression in neuroblast, the protein is also expressed in immature neurons (Hattiangady et al., 2008; Huerto-Delgadillo et al., 1994; Rao and Shetty, 2004). Thus, melatonin

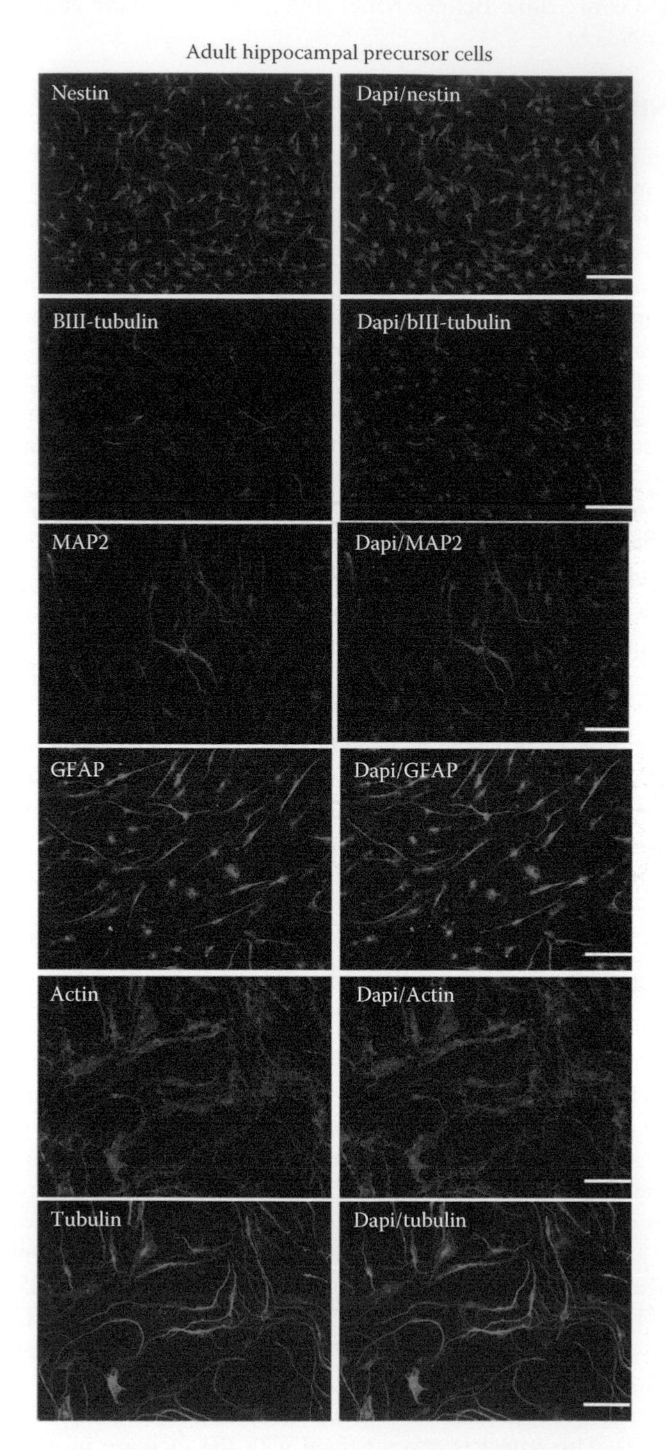

FIGURE 14.3 Adult hippocampal neurogenesis from isolated precursor cells. Pictures show adult hippocampal precursor cells (AHPC) cultured in vitro. Precursor cells express nestin and once they go to differentiation express βIII-tubulin, microtubule associated tubulin protein-2 (MAP2) or the GFAP. Scale bars = 100 μm. Representative pictures of microfilaments and microtubules are also shown. Scale bars = 70 μm.

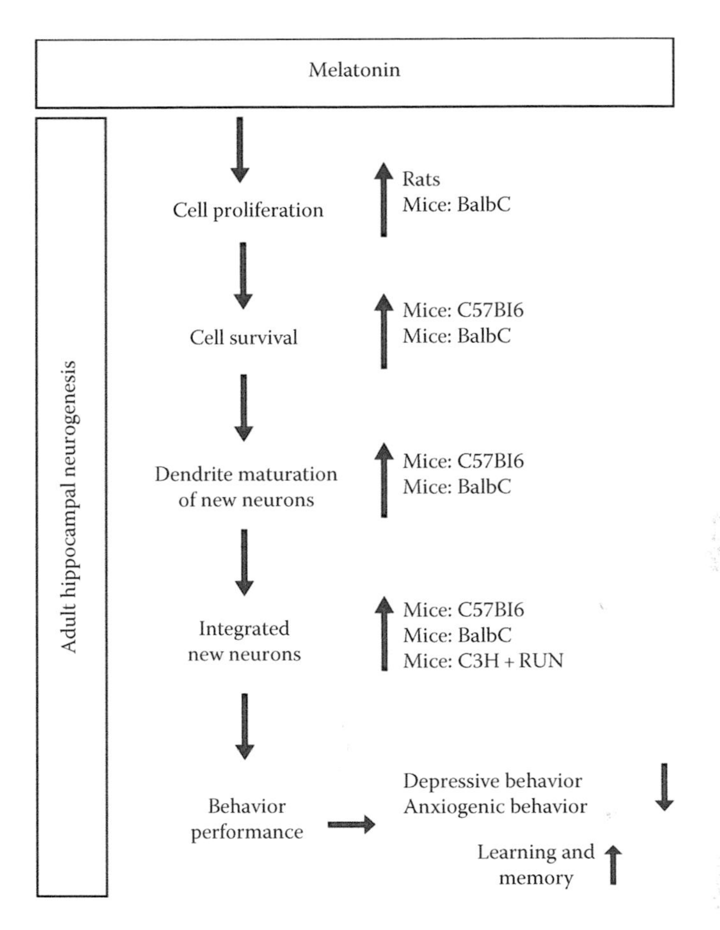

FIGURE 14.4 Melatonin acts as a regulator of adult hippocampal neurogenesis. Scheme summarizes the effects of melatonin under physiological conditions to promote cell proliferation, survival, and dendritic and neuronal maturation. Also, the strain of mice in which melatonin regulates specific events of the neurogenic process in the hippocampus is shown.

could modulate dendrite complexity of new neurons in the hippocampus through the modulation of cytoskeletal components.

14.6.3 Nonphysiological Conditions

14.6.3.1 Stress

Hippocampal neurogenesis is affected by stress, a key precipitating factor for depression (Bessa et al., 2009; Caspi et al., 2003; Tanti et al., 2013). The cellular populations involved in the hippocampal neurogenic process express glucocorticoid receptors (Garcia et al., 2004), making neurogenesis susceptible to stress (Bessa et al., 2009; Gould and Tanapat, 1999; Schoenfeld and Gould, 2012). Melatonin levels also decrease in neuropsychiatric disorders, and preclinical studies show the antidepressant-like action of the hormone tested in several behavioral test (Brusco et al., 2000; Crupi et al., 2010; Ramirez-Rodriguez et al., 2009). Thus, the correlation of increasing neurogenesis with less depressive behavior caused by melatonin have been shown in rodents pretreated with the hormone or in stressed mice further treated with melatonin. Interestingly, the combination of melatonin with other drugs, as buspirone to treat depression, also promotes neurogenesis and behavioral benefits in rodents and in humans (Fava et al., 2012), Figure 14.4.

14.6.3.2 Ischemia

Hippocampal neurogenesis is affected by ischemia, and the administration of melatonin shows that in mice, the hormone is able to improve the survival of mature neurons and increase neurogenesis within one-day delayed administration of melatonin (Kilic et al., 2008; Rennie et al., 2008). This cellular change occurred in parallel to the improvement of locomotor behavior (Kilic et al., 2008). However, in a gerbil's brain submitted to global ischemia, the major benefit of melatonin on immature DCX neurons is observed in the Cornus Amonios 1 (CA1), the region of the hippocampus, but without producing changes in the dentate gyrus (Rennie et al., 2008), Figure 14.4.

14.6.3.3 Irradiation

Some evidence has supported that not only melatonin can attenuate the negative effects of irradiation, but also its metabolite, the compound *N*(1)-acetyl-*N*(2)-formyl-5-methoxykyrunamine (AFMK), is capable of ameliorating the negative effects of irradiation in rodents (Manda et al., 2008). Interestingly, melatonin attenuates the inhibition of hippocampal neurogenesis caused by irradiation; specifically, the pretreatment with the hormone significantly decreases the negative effects of irradiation on immature neurons identified by DCX expression and on Ki67 proliferative cells (Manda and Reiter, 2010; Manda et al., 2009). In a similar manner, the melatonin metabolite, AFMK, ameliorate the effects of irradiation on neurogenesis (Manda et al., 2008). The beneficial effects of melatonin are also accompanied with the improvement of cognitive functions, Figure 14.4.

14.6.3.4 Aging

Hippocampal neurogenesis and melatonin levels decrease during aging (Brusco et al., 2000; Kuhn et al., 1996). Thus, the decline of hippocampal neurogenesis during aging might be related to changes in the content of brain milieu (Villeda et al., 2011). Melatonin supplementation exerts positive effects on hippocampal neurogenesis (Ramirez-Rodriguez et al., 2012; Rennie et al., 2009). In rats, supplementation of melatonin promoted neurogenesis, specifically maintaining the population of DCX cells (Rennie et al., 2009). Additional effects of supplementation of melatonin during normal aging reflect positive regulation on cell proliferation and survival and on immature DCX neurons in the dentate gyrus of BalbC mice (Ramirez-Rodriguez et al., 2012). In fact, at least in BalbC mice, melatonin exhibited a time window effect to maintain or to delay the decay of hippocampal neurogenesis determined by DCX expression. This is interesting because in addition to melatonin, other positive factors for the neurogenic process also fall during aging (Brusco et al., 2000; Villeda et al., 2011), Figure 14.4.

14.7 CONCLUSION

Adult hippocampal neurogenesis is a widely regulated process in which the role of melatonin, as one of its regulators, has gained importance. In this regard, the evidence reviewed in this chapter supports the involvement of melatonin in the neuronal development of the hippocampus by acting on cellular proliferation, survival, and maturation of new neurons under physiological and non-physiological conditions. These events are differentially regulated by melatonin due to the genetic variability of individuals (Kempermann et al., 2006).

In addition to the regulation at the cellular level caused by melatonin, the benefits of the hormone to improve cognition and to decrease depressive behavior are an interesting area that deserves further exploration to support the use of melatonin as an adjuvant for the treatment of neuropsychiatric disorders.

Also, further investigations will help to elucidate in depth the mechanisms by which melatonin modulates hippocampal neurogenesis, especially if we consider that the hormone acts through several mechanisms involving the reorganization of the different components of the cytoskeleton that drives the morphological changes occurring during hippocampal neuronal development, which are important for the function of new neurons in the hippocampus.

REFERENCES

AlAhmed, S. and Herbert, J. (2010). Effect of agomelatine and its interaction with the daily corticosterone rhythm on progenitor cell proliferation in the dentate gyrus of the adult rat. *Neuropharmacology* 59, 375–379.

Altman, J. and Das, G.D. (1966). Autoradiographic and histological studies of postnatal neurogenesis. I. A longitudinal investigation of the kinetics, migration and transformation of cells incorporating tritiated thymidine in neonate rats, with special reference to postnatal neurogenesis in some brain regions. *J Comp Neurol* 126, 337–389.

Alvarez-Buylla, A. and Garcia-Verdugo, J.M. (2002). Neurogenesis in adult subventricular zone. *J Neurosci* 22, 629–634.

Babu, H., Ramirez-Rodriguez, G., Fabel, K., Bischofberger, J., and Kempermann, G. (2009). Synaptic network activity induces neuronal differentiation of adult hippocampal precursor cells through BDNF signaling. *Front Neurosci* 3, 49.

Barnabe-Heider, F. and Miller, F.D. (2003). Endogenously produced neurotrophins regulate survival and differentiation of cortical progenitors via distinct signaling pathways. *J Neurosci* 23, 5149–5160.

Bellon, A., Ortiz-Lopez, L., Ramirez-Rodriguez, G., Anton-Tay, F., and Benitez-King, G. (2007). Melatonin induces neuritogenesis at early stages in N1E-115 cells through actin rearrangements via activation of protein kinase C and Rho-associated kinase. *J Pineal Res* 42, 214–221.

Benitez-King, G. (2000). PKC activation by melatonin modulates vimentin intermediate filament organization in N1E-115 cells. *J Pineal Res* 29, 8–14.

Benitez-King, G. (2006). Melatonin as a cytoskeletal modulator: Implications for cell physiology and disease. *J Pineal Res* 40, 1–9.

Bessa, J.M., Ferreira, D., Melo, I., Marques, F., Cerqueira, J.J., Palha, J.A., Almeida, O.F., and Sousa, N. (2009). The mood-improving actions of antidepressants do not depend on neurogenesis but are associated with neuronal remodeling. *Mol Psychiatry* 14, 764–773, 739.

Bick-Sander, A., Steiner, B., Wolf, S.A., Babu, H., and Kempermann, G. (2006). Running in pregnancy transiently increases postnatal hippocampal neurogenesis in the offspring. *Proc Natl Acad Sci USA* 103, 3852–3857.

Bolteus, A.J. and Bordey, A. (2004). GABA release and uptake regulate neuronal precursor migration in the postnatal subventricular zone. *J Neurosci* 24, 7623–7631.

Bowman, B.P. and Kuhn, C.M. (1996). Age-related differences in the chronic and acute response to cocaine in the rat. *Dev Psychobiol* 29, 597–611.

Brandt, M.D., Jessberger, S., Steiner, B., Kronenberg, G., Reuter, K., Bick-Sander, A., von der Behrens, W., and Kempermann, G. (2003). Transient calretinin expression defines early postmitotic step of neuronal differentiation in adult hippocampal neurogenesis of mice. *Mol Cell Neurosci* 24, 603–613.

Bridges, R.S. and Grattan, D.R. (2003). Prolactin-induced neurogenesis in the maternal brain. *Trends Endocrinol Metab* 14, 199–201.

Brown, J., Cooper-Kuhn, C.M., Kempermann, G., Van Praag, H., Winkler, J., Gage, F.H., and Kuhn, H.G. (2003). Enriched environment and physical activity stimulate hippocampal but not olfactory bulb neurogenesis. *Eur J Neurosci* 17, 2042–2046.

Brusco, L.I., Marquez, M., and Cardinali, D.P. (2000). Melatonin treatment stabilizes chronobiologic and cognitive symptoms in Alzheimer's disease. *Neuro Endocrinol Lett* 21, 39–42.

Caballero, B., Vega-Naredo, I., Sierra, V., Huidobro-Fernandez, C., Soria-Valles, C., De Gonzalo-Calvo, D., Tolivia, D. et al. (2009). Melatonin alters cell death processes in response to age-related oxidative stress in the brain of senescence-accelerated mice. *J Pineal Res* 46, 106–114.

Cameron, H.A. and McKay, R.D. (1999). Restoring production of hippocampal neurons in old age. *Nat Neurosci* 2, 894–897.

Caspi, A., Sugden, K., Moffitt, T.E., Taylor, A., Craig, I.W., Harrington, H., McClay, J. et al. (2003). Influence of life stress on depression: Moderation by a polymorphism in the 5-HTT gene. *Science* 301, 386–389.

Cleveland, D.W., Fischer, S.G., Kirschner, M.W., and Laemmli, U.K. (1977). Peptide mapping by limited proteolysis in sodium dodecyl sulfate and analysis by gel electrophoresis. *J Biol Chem* 252, 1102–1106.

Couillard-Despres, S., Wuertinger, C., Kandasamy, M., Caioni, M., Stadler, K., Aigner, R., Bogdahn, U., and Aigner, L. (2009). Ageing abolishes the effects of fluoxetine on neurogenesis. *Mol Psychiatry* 14, 856–864.

Crupi, R., Mazzon, E., Marino, A., La Spada, G., Bramanti, P., Cuzzocrea, S., and Spina, E. (2010). Melatonin treatment mimics the antidepressant action in chronic corticosterone-treated mice. *J Pineal Res* 49, 123–129.

Crupi, R., Mazzon, E., Marino, A., La Spada, G., Bramanti, P., Spina, E., and Cuzzocrea, S. (2011). Melatonin's stimulatory effect on adult hippocampal neurogenesis in mice persists after ovariectomy. *J Pineal Res* 51, 353–360.

Dubocovich, M.L. (1991). Melatonin receptors in the central nervous system. *Adv Exp Med Biol* 294, 255–265.

Dubocovich, M.L. (1995). Melatonin receptors: Are there multiple subtypes? *Trends Pharmacol Sci* 16, 50–56.

Fabel, K. and Kempermann, G. (2008). Physical activity and the regulation of neurogenesis in the adult and aging brain. *Neuromolecular Med* 10, 59–66.

Fabel, K., Tam, B., Kaufer, D., Baiker, A., Simmons, N., Kuo, C.J., and Palmer, T.D. (2003). VEGF is necessary for exercise-induced adult hippocampal neurogenesis. *Eur J Neurosci* 18, 2803–2812.

Fabel, K., Wolf, S.A., Ehninger, D., Babu, H., Leal-Galicia, P., and Kempermann, G. (2009). Additive effects of physical exercise and environmental enrichment on adult hippocampal neurogenesis in mice. *Front Neurosci* 3, 50.

Fava, M., Targum, S.D., Nierenberg, A.A., Bleicher, L.S., Carter, T.A., Wedel, P.C., Hen, R., Gage, F.H., and Barlow, C. (2012). An exploratory study of combination buspirone and melatonin SR in major depressive disorder (MDD): A possible role for neurogenesis in drug discovery. *J Psychiatr Res* 46, 1553–1563.

Gage, F.H., Kempermann, G., Palmer, T.D., Peterson, D.A., and Ray, J. (1998). Multipotent progenitor cells in the adult dentate gyrus. *J Neurobiol* 36, 249–266.

Garcia, A., Steiner, B., Kronenberg, G., Bick-Sander, A., and Kempermann, G. (2004). Age-dependent expression of glucocorticoid- and mineralocorticoid receptors on neural precursor cell populations in the adult murine hippocampus. *Aging Cell* 3, 363–371.

Ge, S., Pradhan, D.A., Ming, G.L., and Song, H. (2007). GABA sets the tempo for activity-dependent adult neurogenesis. *Trends Neurosci* 30, 1–8.

Ge, W., He, F., Kim, K.J., Blanchi, B., Coskun, V., Nguyen, L., Wu, X. et al. (2006). Coupling of cell migration with neurogenesis by proneural bHLH factors. *Proc Natl Acad Sci USA* 103, 1319–1324.

Gould, E., McEwen, B.S., Tanapat, P., Galea, L.A., and Fuchs, E. (1997). Neurogenesis in the dentate gyrus of the adult tree shrew is regulated by psychosocial stress and NMDA receptor activation. *J Neurosci* 17, 2492–2498.

Gould, E. and Tanapat, P. (1999). Stress and hippocampal neurogenesis. *Biol Psychiatry* 46, 1472–1479.

Hattiangady, B., Rao, M.S., and Shetty, A.K. (2008). Plasticity of hippocampal stem/progenitor cells to enhance neurogenesis in response to kainate-induced injury is lost by middle age. *Aging Cell* 7, 207–224.

Hattiangady, B. and Shetty, A.K. (2008). Aging does not alter the number or phenotype of putative stem/progenitor cells in the neurogenic region of the hippocampus. *Neurobiol Aging* 29, 129–147.

Heine, V.M., Maslam, S., Joels, M., and Lucassen, P.J. (2004). Prominent decline of newborn cell proliferation, differentiation, and apoptosis in the aging dentate gyrus, in absence of an age-related hypothalamus-pituitary-adrenal axis activation. *Neurobiol Aging* 25, 361–375.

Horner, P.J., Power, A.E., Kempermann, G., Kuhn, H.G., Palmer, T.D., Winkler, J., Thal, L.J., and Gage, F.H. (2000). Proliferation and differentiation of progenitor cells throughout the intact adult rat spinal cord. *J Neurosci* 20, 2218–2228.

Huerto-Delgadillo, L., Anton-Tay, F., and Benitez-King, G. (1994). Effects of melatonin on microtubule assembly depend on hormone concentration: Role of melatonin as a calmodulin antagonist. *J Pineal Res* 17, 55–62.

Kempermann, G., Chesler, E.J., Lu, L., Williams, R.W., and Gage, F.H. (2006). Natural variation and genetic covariance in adult hippocampal neurogenesis. *Proc Natl Acad Sci USA* 103, 780–785.

Kempermann, G., Fabel, K., Ehninger, D., Babu, H., Leal-Galicia, P., Garthe, A., and Wolf, S.A. (2010). Why and how physical activity promotes experience-induced brain plasticity. *Front Neurosci* 4, 189.

Kempermann, G., Jessberger, S., Steiner, B., and Kronenberg, G. (2004a). Milestones of neuronal development in the adult hippocampus. *Trends Neurosci* 27, 447–452.

Kempermann, G., Kuhn, H.G., and Gage, F.H. (1997). Genetic influence on neurogenesis in the dentate gyrus of adult mice. *Proc Natl Acad Sci USA* 94, 10409–10414.

Kempermann, G., Wiskott, L., and Gage, F.H. (2004b). Functional significance of adult neurogenesis. *Curr Opin Neurobiol* 14, 186–191.

Kilic, E., Kilic, U., Bacigaluppi, M., Guo, Z., Abdallah, N.B., Wolfer, D.P., Reiter, R.J., Hermann, D.M., and Bassetti, C.L. (2008). Delayed melatonin administration promotes neuronal survival, neurogenesis and motor recovery, and attenuates hyperactivity and anxiety after mild focal cerebral ischemia in mice. *J Pineal Res* 45, 142–148.

Kim, M.J., Kim, H.K., Kim, B.S., and Yim, S.V. (2004). Melatonin increases cell proliferation in the dentate gyrus of maternally separated rats. *J Pineal Res* 37, 193–197.

Klempin, F., Babu, H., De Pietri Tonelli, D., Alarcon, E., Fabel, K., and Kempermann, G. (2010). Oppositional effects of serotonin receptors 5-HT1a, 2, and 2c in the regulation of adult hippocampal neurogenesis. *Front Mol Neurosci* 3, pii: 14.

Klempin, F. and Kempermann, G. (2007). Adult hippocampal neurogenesis and aging. *Eur Arch Psychiatry Clin Neurosci* 257, 271–280.

Kong, X., Li, X., Cai, Z., Yang, N., Liu, Y., Shu, J., Pan, L., and Zuo, P. (2008). Melatonin regulates the viability and differentiation of rat midbrain neural stem cells. *Cell Mol Neurobiol* 28, 569–579.

Kuhn, H.G., Dickinson-Anson, H., and Gage, F.H. (1996). Neurogenesis in the dentate gyrus of the adult rat: Age-related decrease of neuronal progenitor proliferation. *J Neurosci* 16, 2027–2033.

Kuhn, H.G., Winkler, J., Kempermann, G., Thal, L.J., and Gage, F.H. (1997). Epidermal growth factor and fibroblast growth factor-2 have different effects on neural progenitors in the adult rat brain. *J Neurosci* 17, 5820–5829.

Liu, J., Somera-Molina, K.C., Hudson, R.L., and Dubocovich, M.L. (2013). Melatonin potentiates running wheel-induced neurogenesis in the dentate gyrus of adult C3H/HeN mice hippocampus. *J Pineal Res* 54, 222–231.

Manda, K. and Reiter, R.J. (2010). Melatonin maintains adult hippocampal neurogenesis and cognitive functions after irradiation. *Prog Neurobiol* 90, 60–68.

Manda, K., Ueno, M., and Anzai, K. (2008). Space radiation-induced inhibition of neurogenesis in the hippocampal dentate gyrus and memory impairment in mice: Ameliorative potential of the melatonin metabolite, AFMK. *J Pineal Res* 45, 430–438.

Manda, K., Ueno, M., and Anzai, K. (2009). Cranial irradiation-induced inhibition of neurogenesis in hippocampal dentate gyrus of adult mice: Attenuation by melatonin pretreatment. *J Pineal Res* 46, 71–78.

Mirescu, C. and Gould, E. (2006). Stress and adult neurogenesis. *Hippocampus* 16, 233–238.

Mirescu, C., Peters, J.D., and Gould, E. (2004). Early life experience alters response of adult neurogenesis to stress. *Nat Neurosci* 7, 841–846.

Moriya, T., Horie, N., Mitome, M., and Shinohara, K. (2007). Melatonin influences the proliferative and differentiative activity of neural stem cells. *J Pineal Res* 42, 411–418.

Nithianantharajah, J. and Hannan, A.J. (2006). Enriched environments, experience-dependent plasticity and disorders of the nervous system. *Nat Rev Neurosci* 7, 697–709.

Nollet, M., Gaillard, P., Tanti, A., Girault, V., Belzung, C., and Leman, S. (2012). Neurogenesis-independent antidepressant-like effects on behavior and stress axis response of a dual orexin receptor antagonist in a rodent model of depression. *Neuropsychopharmacology* 37, 2210–2221.

Ortiz-Lopez, L., Morales-Mulia, S., Ramirez-Rodriguez, G., and Benitez-King, G. (2009). ROCK-regulated cytoskeletal dynamics participate in the inhibitory effect of melatonin on cancer cell migration. *J Pineal Res* 46, 15–21.

Paez-Martinez, N., Flores-Serrano, Z., Ortiz-Lopez, L., and Ramirez-Rodriguez, G. (2013). Environmental enrichment increases doublecortin-associated new neurons and decreases neuronal death without modifying anxiety-like behavior in mice chronically exposed to toluene. *Behav Brain Res* 256, 432–440.

Palmer, T.D., Willhoite, A.R., and Gage, F.H. (2000). Vascular niche for adult hippocampal neurogenesis. *J Comp Neurol* 425, 479–494.

Petrus, D.S., Fabel, K., Kronenberg, G., Winter, C., Steiner, B., and Kempermann, G. (2009). NMDA and benzodiazepine receptors have synergistic and antagonistic effects on precursor cells in adult hippocampal neurogenesis. *Eur J Neurosci* 29, 244–252.

Platel, J.C., Lacar, B., and Bordey, A. (2007). GABA and glutamate signaling: Homeostatic control of adult forebrain neurogenesis. *J Mol Histol* 38, 602–610.

Plumpe, T., Ehninger, D., Steiner, B., Klempin, F., Jessberger, S., Brandt, M., Romer, B., Rodriguez, G.R., Kronenberg, G., and Kempermann, G. (2006). Variability of doublecortin-associated dendrite maturation in adult hippocampal neurogenesis is independent of the regulation of precursor cell proliferation. *BMC Neurosci* 7, 77.

Ramirez-Rodriguez, G., Klempin, F., Babu, H., Benitez-King, G., and Kempermann, G. (2009). Melatonin modulates cell survival of new neurons in the hippocampus of adult mice. *Neuropsychopharmacology* 34, 2180–2191.

Ramirez-Rodriguez, G., Ortiz-Lopez, L., Dominguez-Alonso, A., Benitez-King, G.A., and Kempermann, G. (2011). Chronic treatment with melatonin stimulates dendrite maturation and complexity in adult hippocampal neurogenesis of mice. *J Pineal Res* 50, 29–37.

Ramirez-Rodriguez, G., Vega-Rivera, N.M., Benitez-King, G., Castro-Garcia, M., and Ortiz-Lopez, L. (2012). Melatonin supplementation delays the decline of adult hippocampal neurogenesis during normal aging of mice. *Neurosci Lett* 530, 53–58.

Rao, M.S. and Shetty, A.K. (2004). Efficacy of doublecortin as a marker to analyse the absolute number and dendritic growth of newly generated neurons in the adult dentate gyrus. *Eur J Neurosci* 19, 234–246.

Reiter, R.J. (1991). Melatonin: The chemical expression of darkness. *Mol Cell Endocrinol* 79, C153–C158.

Reiter, R.J. (1998a). Melatonin, active oxygen species and neurological damage. *Drug News Perspect* 11, 291–296.

Reiter, R.J. (1998b). Oxidative damage in the central nervous system: Protection by melatonin. *Prog Neurobiol* 56, 359–384.

Rennie, K., de Butte, M., Frechette, M., and Pappas, B.A. (2008). Chronic and acute melatonin effects in gerbil global forebrain ischemia: Long-term neural and behavioral outcome. *J Pineal Res* 44, 149–156.

Rennie, K., de Butte, M., and Pappas, B.A. (2009). Melatonin promotes neurogenesis in dentate gyrus in the pinealectomized rat. *J Pineal Res* 47, 313–317.

Romer, B., Krebs, J., Overall, R.W., Fabel, K., Babu, H., Overstreet-Wadiche, L., Brandt, M.D., Williams, R.W., Jessberger, S., and Kempermann, G. (2011). Adult hippocampal neurogenesis and plasticity in the infrapyramidal bundle of the mossy fiber projection: I. Co-regulation by activity. *Front Neurosci* 5, 107.

Schoenfeld, T.J. and Gould, E. (2012). Stress, stress hormones, and adult neurogenesis. *Exp Neurol* 233, 12–21.

Shingo, T., Gregg, C., Enwere, E., Fujikawa, H., Hassam, R., Geary, C., Cross, J.C., and Weiss, S. (2003). Pregnancy-stimulated neurogenesis in the adult female forebrain mediated by prolactin. *Science* 299, 117–120.

Taliaz, D., Stall, N., Dar, D.E., and Zangen, A. (2010). Knockdown of brain-derived neurotrophic factor in specific brain sites precipitates behaviors associated with depression and reduces neurogenesis. *Mol Psychiatry* 15, 80–92.

Tanti, A., Westphal, W.P., Girault, V., Brizard, B., Devers, S., Leguisquet, A.M., Surget, A., and Belzung, C. (2013). Region-dependent and stage-specific effects of stress, environmental enrichment and antidepressant treatment on hippocampal neurogenesis. *Hippocampus* 23, 797–811.

van Praag, H., Kempermann, G., and Gage, F.H. (1999). Running increases cell proliferation and neurogenesis in the adult mouse dentate gyrus. *Nat Neurosci* 2, 266–270.

Villeda, S.A., Luo, J., Mosher, K.I., Zou, B., Britschgi, M., Bieri, G., Stan, T.M. et al. (2011). The ageing systemic milieu negatively regulates neurogenesis and cognitive function. *Nature* 477, 90–94.

Zhao, C., Deng, W., and Gage, F.H. (2008). Mechanisms and functional implications of adult neurogenesis. *Cell* 132, 645–660.

Zhao, C., Teng, E.M., Summers, R.G., Jr., Ming, G.L., and Gage, F.H. (2006a). Distinct morphological stages of dentate granule neuron maturation in the adult mouse hippocampus. *J Neurosci* 26, 3–11.

Zhao, Z., Sun, P., Chauhan, N., Kaur, J., Hill, M.D., Papadakis, M., and Buchan, A.M. (2006b). Neuroprotection and neurogenesis: Modulation of cornus ammonis 1 neuronal survival after transient forebrain ischemia by prior fimbria-fornix deafferentation. *Neuroscience* 140, 219–226.

15 Comparison of Neuroprotective Effect of Melatonin in the Nigrostriatal and Mesolimbic Dopaminergic Systems of the Zitter Rat

Shuichi Ueda, Ayuka Ehara, Kenichi Inoue, Tomoyuki Masuda, Shin-ichi Sakakibara, and Kanji Yoshimoto

CONTENTS

15.1 INTRODUCTION

The major secretory product of the pineal gland, melatonin (MEL), is known to have strong reactive oxygen species (ROS) scavenger properties, in addition to hormonal functions (Tan et al. 2003). Due to its solubility in both lipids and water, MEL can easily pass the blood–brain barrier and access the neurons and glial cells, which suggests the possibility of its utilization as an effective neuroprotective agent. The efficacy of MEL has been investigated in both animal models for several neurological diseases such as Parkinson's disease (PD) and Alzheimer's disease and in traumatic brain injury and cerebral artery occlusion models (Esposito and Cuzzocrea 2010). Experimental data suggest that there might be a common link in the pathogenesis for ROS. Monoaminergic neuron systems, which include the dopaminergic and serotonergic neuron systems, especially appear to be susceptible to ROS-induced damage.

The zitter rat is an autosomal recessive mutant derived from the Sprague-Dawley (SD) strain. Homozygous zitter (*zi*/*zi*) rats are characterized by a loss of function due to an 8 bp deletion mutation in attractin (Atrn), which is a glycosylated transmembrane protein (Kuramoto et al. 2001).

In addition, several neuropathological changes are found in this strain, including postnatal onset of impaired oligodendrocyte differentiation (Sakakibara et al. 2008) and age-dependent degeneration of both the serotonergic (Ueda et al. 1998) and dopaminergic neuron systems (Nakadate et al. 2006; Ueda et al. 2000). Although the detailed mechanisms underlying the neuropathological changes in the *zi/zi* rats remain unclear, this mutant strain displays an abnormal metabolism of ROS (Gomi et al. 1994; Ueda et al. 2002). As the degeneration of the dopaminergic neurons in the substantia nigra pars compacta (SNc) of *zi/zi* rats has been reported to be attenuated by a chronic intake of vitamin E, which is a direct scavenger of ROS, this evidence strongly supports the concept that the degeneration is caused by oxidative stress (Ueda et al. 2005).

Our recent experiments in these mutant rats demonstrated that chronic MEL treatment attenuated the age-related decline of dopamine (DA) and 3,4-dihydroxyphenylacetic acid (DOPAC) concentrations in the caudate-putamen (CPU). These findings suggest that MEL affects the dopaminergic neuron systems not only by direct scavenger action but also via alteration in the proinflammatory cytokine and antioxidant enzyme (catalase: CAT, superoxide dismutase: SOD, and glutathione peroxidase: GPx) gene expressions in each brain area (Hashimoto et al. 2012). In fact, when the mesolimbic and SNc–CPU dopaminergic systems were compared during normal physiological and 1-methyl-4-phenyl-1, 2, 3-6-tetrahydropyridine (MPTP)-induced oxidative stress conditions, higher levels of CAT, SOD, and GPx activities were noted in the mesolimbic dopaminergic system, which includes the ventral tegmental area (VTA)–nucleus accumbens (NA) tract and the VTA–olfactory tubercle (OT) tract (Hung and Lee 1998; Trepanier et al. 1996). When taken together, these findings suggest the possibility that the actual neuroprotective effect of MEL within the DA systems exists between the nigrostriatal and mesolimbic tract. In fact, previous data from our laboratory have demonstrated that chronic MEL treatment differentially attenuated the degeneration of the serotonergic fibers in the CPU, NA, and OT of *zi/zi* rats. Furthermore, sprouting and hyper-reinnervation were also observed in the OT, which provides additional support for the supposition that these differences do indeed exist (Ueda et al. 2008).

It has been previously proven that specific neurotrophic factors, including brain-derived neurotrophic factor (BDNF) and glia cell line–derived neurotrophic factor (GDNF), provide protection from dopaminergic injury (Hu and Russek 2008; Sharma et al. 2006; Sun et al. 2005) and that BDNF promotes neuronal survival and the sprouting of serotonergic fibers via stimulation of the TrkB receptor (Mattson et al. 2004). Additionally, MEL treatment has been shown to increase receptor-mediated *GDNF* and *BDNF* expression in neurons and glia cells in several brain areas (Sharma et al. 2006; Tang et al. 1998).

The present study attempted to elucidate the effects of chronic high-dose MEL treatment on the mesostriatal DA systems, which included the nigrostriatal DA system, the VTA-NA DA system, and the VTA–OT DA system in *zi/zi* rats. Furthermore, we also used quantitative polymerase chain reaction (qPCR) to examine whether chronic MEL administration changed the expression of *GDNF* mRNA, and *BDNF* and its receptor *TrkB* mRNA.

15.2 MATERIALS AND METHODS

15.2.1 Animals and Experimental Plan

All rats used in the present study were raised and maintained in the Laboratory Animal Research Center at Dokkyo University School of Medicine. All animals were housed in groups of two or three in cages with ad libitum access to food and water. Cages were maintained on a 12 h light/dark cycle. All procedures in the present study were certified by the University's Animal Welfare Committee. To inhibit the oxidation of MEL that occurs via drinking water, we used highly reduced electrolytic water (HRE) (pH 7.0, –500 mV) (Super Red Water®, Miz, Fujisawa, Japan) as the drinking water source (Ueda et al. 2008). Plasma MEL concentrations during the chronic MEL treatment were determined from the total blood of the rats after 2M of drinking water with and without MEL (3-month-old rats). Blood was collected by exsanguinations between 10:00 and 11:00 in the morning, followed by the measurement of the plasma MEL concentrations via the use of our previously reported method (Ueda et al. 2008).

Male *zi/zi* rats were used to investigate the effects of MEL (0.05 mg/mL) in drinking water. At 1M (weaning period), *zi/zi* rats were divided into two groups (with or without MEL) until 10M.

15.2.2 HPLC Procedure, Tissue Preparation, Immunostaining, and Cell Counting

Details of our high-performance liquid chromatography (HPLC) methods have been reported in our previous papers (Nakamura et al. 2010; Ueda et al. 2000, 2008). The 1M, 10M, and 10M MEL *zi/zi* groups (five in each group) were examined under deep anesthesia achieved by using sodium pentobarbital (50 mg/kg body weight). Tissues were weighed and homogenized in 0.1 M perchloric acid containing 0.1 mM of ethylenediaminetetraacetate (EDTA), followed by filtration. After centrifugation of the homogenate, the supernatant was analyzed.

For immunohistochemical analysis rats in the 1M, 10M, and 10M MEL groups (four in each group) were deeply anesthetized with sodium pentobarbital (50 mg/kg), transcardially perfused with physiological saline, followed by perfusion with 800 mL of ice-cold fixative containing 4% paraformaldehyde and 0.2% picric acid in 0.1 M of phosphate-buffered saline (PBS) (pH 7.4). Sectioning and immunohistochemical staining procedures with anti-TH-monoclonal antiserum have been previously reported (Ueda et al. 2000).

Unbiased stereological counting using Stereo Investigator (Micro Bright Field Japan, Tokyo, Japan) was utilized for counting the total number of TH-immunoreactive neurons in the SNc and SN reticulata (SNr) and VTA of the 1M and 10M *zi/zi* rats with and without MEL (Ueda et al. 2005).

15.2.3 qPCR Method

For qPCR analysis, 3M *zi/zi* rats with and without MEL and age-matched SD controls (four in each group) were used. The tissue preparation and the procedures have been reported in our previous paper (Sakakibara et al. 2008). The sequences for the primers and predicted sizes are as follows: *Bdnf*, forward 5′-gcccaacgaagaaaaccata-3′ and reverse 5′-caaaggcacttgactgctga-3′ (73 bp), trkb, forward 5′-ctacctggcatcccaacact-3′ and reverse 5′-tcaccagcaggttctctcct-3′ (77 bp), and *Gdnf*, forward 5′-cggacgggactctaagatga-3′ and reverse 5′-cgcttcgagaagcctcttac-3′ (109 bp). For quantitative real-time PCR, THUNDERBIRD™ SYBR® qPCR Mix (TOYOBO) was used in accordance with the manufacturer's instruction. We used an ABI PRISM 7000 (Applied Biosystems Japan, Tokyo, Japan) for the signal detection and analysis. Each target sequence was amplified using 40 cycles of PCR (denatured at 95°C for 5 s, with annealing/extension at 59°C for 31 s). The cellular origins of BDNF in the dorsal striatum (including CPU) and ventral striatum (including NA and OT) have been shown to be in the SNc and frontal cortex, and in the VTA, respectively (Alter et al. 1997; Guillin et al. 2001). Therefore, we focused on the frontal cortex, striatum (including dorsal and ventral striatum), and midbrain (including SNc and VTA) in our qPCR experiments.

15.2.4 Statistical Analysis

Data were expressed as mean ± standard error of the mean (SEM). For comparing the differences between the groups, analysis of variance (ANOVA) was used. If the ANOVA test showed a significant difference, a further post hoc Bonferroni–Dunn test was applied. Significance was defined as p values less than 0.05. All statistical analyses were performed using Stat View software (Abacus Concepts, Inc., Berkeley, CA, USA).

15.3 RESULTS

Similar to the results of our previous study (Ueda et al. 2008), serum MEL levels after 2M of drinking MEL (3 month of age) (3025.0 ± 988.5 pg/mL, mean ± SEM) were significantly higher than those seen for the age-matched *zi/zi* controls (19.0 ± 5.5 pg/mL) ($p < 0.05$).

15.3.1 Immunohistochemical Analysis

A large number of TH-immunoreactive neurons were present in the SNc of the 1M rats, with apparent dendrites running in a dorsomedial to ventrolateral direction in the SNr. Figure 15.1 shows the TH-immunoreactive neurons in the SNc (Figure 15.1a and b) and VTA (Figure 15.1c and d) of the 10M rats with and without MEL treatment. ANOVA showed that there were significant differences in the number of TH-immunoreactive neurons in the SNc. TH-immunoreactive neurons in the SNc were significantly decreased in the 10M rats. However, this reduction was ameliorated by the MEL treatment (Figure 15.1a and b, Table 15.1). When the experimental groups were examined, no significant differences were found for the number of TH-immunoreactive neurons in the SNr and VTA (Figure 15.1c and d; Table 15.1).

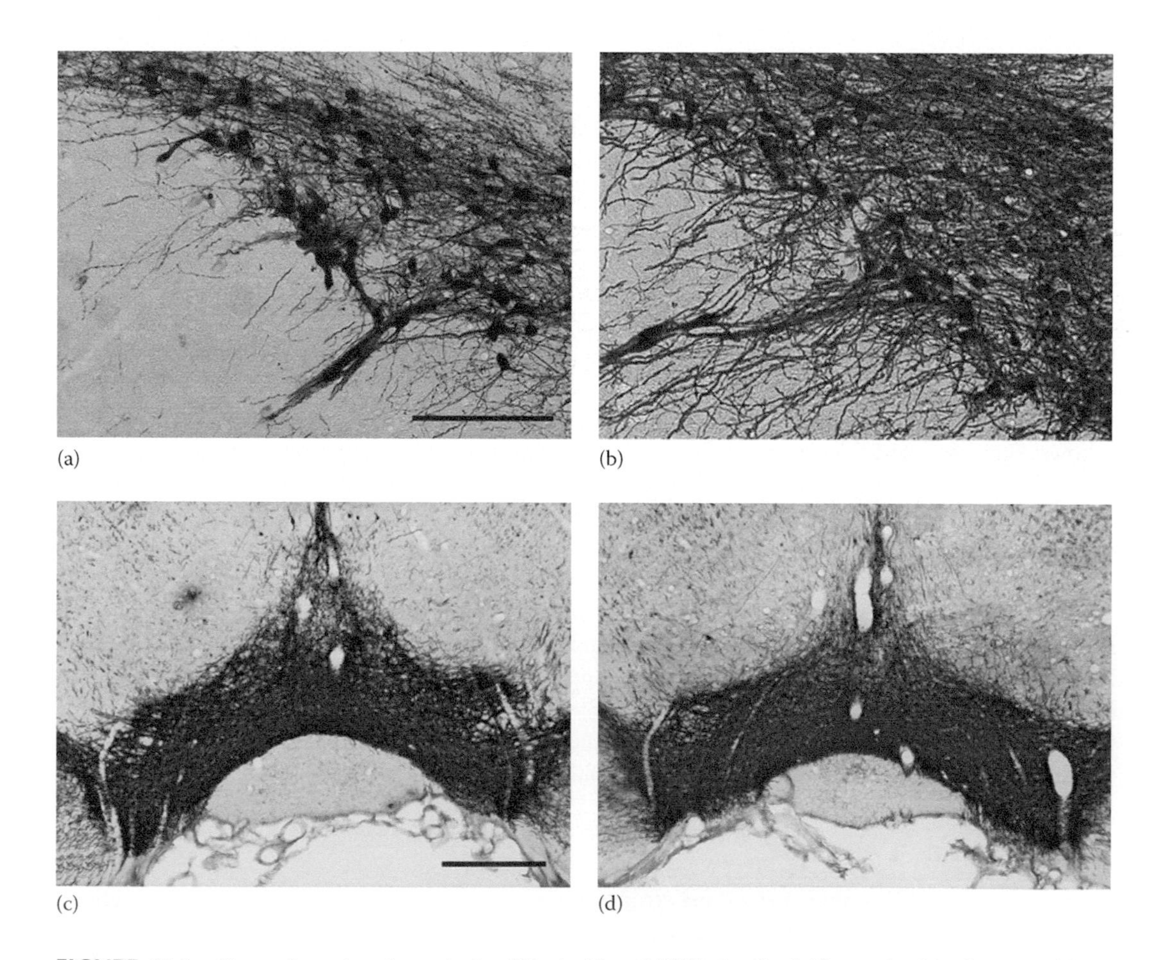

FIGURE 15.1 Coronal section through the SNc (a, b) and VTA (c, d) of 10-month-old *zi/zi* rats without treatment (a, c) and 10-month-old *zi/zi* rats with a 9-month exposure to drinking water that contained melatonin (b, d). All sections were immunohistochemically stained for tyrosine hydroxylase. Scale bar = 200 μm (a, b) and 500 μm (c, d).

TABLE 15.1
Number of TH-Immunoreactive Neurons in the SNc, SN Pars Reticulata (SNr), and VTA of 1-Month-Old (1M), 10-Month-Old *zi/zi* Rat without Treatment (10M), and 10-Month-Old *zi/zi* Rat with MEL Treatment (10M MEL)

Group	SNc	SNr	VTA
1M	11,148 ± 839	463 ± 40	9832 ± 1901
10M	4,450 ± 531[a]	521 ± 65	7415 ± 451
10M MEL	8,255 ± 808[a,b]	533 ± 45	7763 ± 1915

Data are given as mean ± SEM.
[a] $p < 0.05$ versus 1M group.
[b] $p < 0.05$ versus 10M group.

As seen in Figure 15.2, there was TH-immunohistochemical staining in the middle and caudal parts of the CPU, NA, and OT for the 1M (Figure 15.2a and b), 10M (Figure 15.2c and d), and 10M MEL (Figure 15.2e and f) groups. TH-immunoreactive fibers were densely distributed in the CPU, NA, and OT of the 1M rats, which was similar to that seen for the control SD and *zi*/+ rats of the same age. In the 10M rats, a noticeable decrease in the TH-immunoreactive fibers was seen for the lateral and caudal parts of the CPU. However, a dense innervation of TH-immunoreactive fibers was also observed in the fundus striati (ventral part of the caudal CPU) (Figure 15.2d). In the lateral part of the CPU, chronic MEL treatment resulted in a relative attenuation of reductions in the TH-immunoreactive fibers, whereas in the caudal part of the CPU, the staining pattern resembled that seen in the nontreated 10M rats. When observed at a higher magnification, a different morphology of the TH-immunoreactive fibers was seen in the NA when the 10M and 10M MEL rats were compared. In the 10M rats, numerous TH-immunoreactive fibers with abnormal morphology were observed in the CPU and NA. These fibers were characterized by swollen varicosities, irregularly thickened intervaricose segments, with intense TH-immunoreactivity (Figure 15.2g and i).

Although these abnormal fibers were rarely seen in the NA of the 10M MEL rats (Figure 15.2h and j), they were still present in the CPU.

15.3.2 Neurochemical Analysis and qPCR Analysis

Figure 15.3 shows the results of the HPLC analysis. Age-related decreases in the DA concentration were observed in the CPU, NA, and OT of the 10M rats. Significant differences were observed in the DA levels of the CPU, NA, and OT, and in the DOPAC/DA ratio of the CPU between 1M and 10M rats. These reductions in the DA were ameliorated by MEL treatment in the CPU and NA. In terms of DOPAC levels, significant differences were noted in the NA between the 1M and MEL groups. The highest DOPAC/DA ratio was found in the OT of the 10M rats.

These were no statistically significant differences between the groups (SD, *zi/zi*, and *zi/zi* with MEL) for the *GDNF*, *BDNF*, and *TrkB* m RNA expressions in the frontal cortex, striatum (which includes the dorsal and ventral striatum), or midbrain (which includes the SN and the VTA) (Figure 15.4).

15.4 DISCUSSION

The current study evaluated the effects of MEL on the DA systems in zitter mutant rats, which display abnormal metabolism of ROS (Gomi et al. 1994; Ueda et al. 2002, 2005). Similar to previous

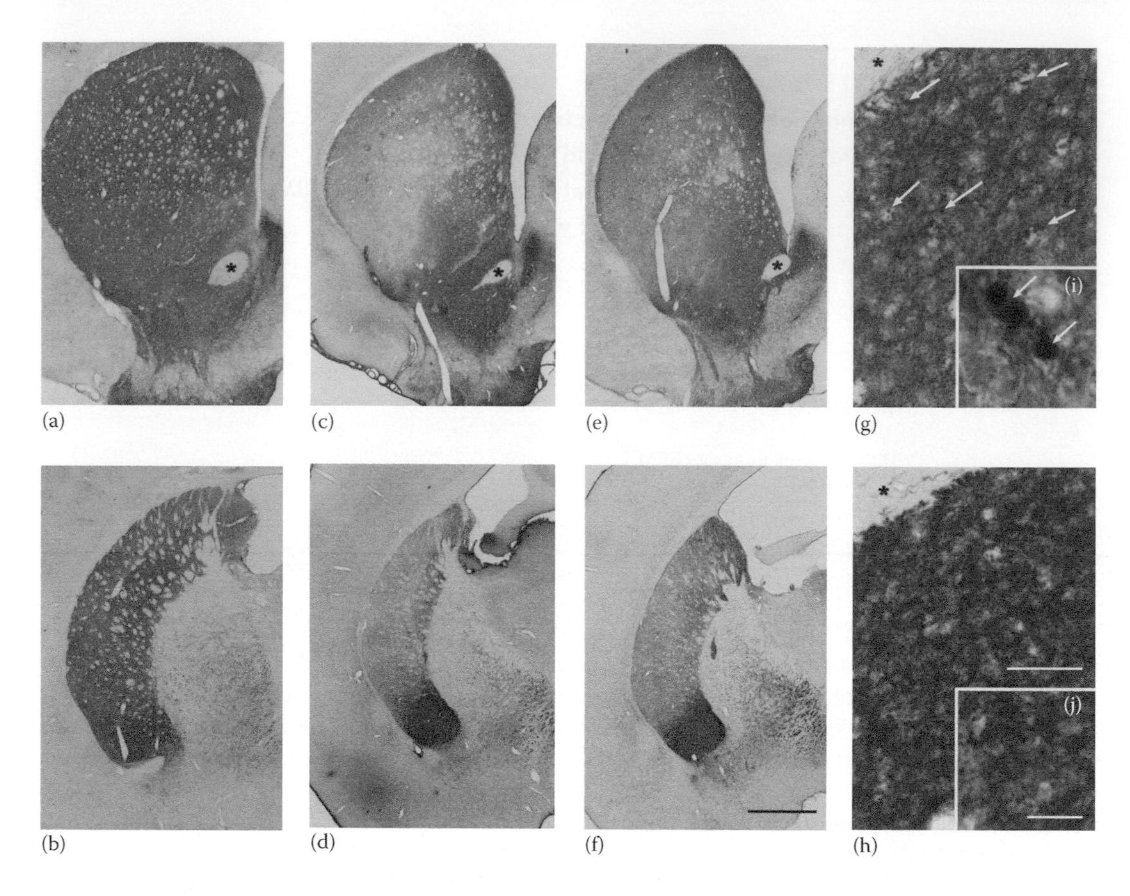

FIGURE 15.2 (**See color insert.**) Low-power photomicrographs through the middle (a, c, e) and caudal (b, d, f) parts of the striatum in 1-month-old (a, b) and 10-month-old *zi/zi* rats without treatment (c, d) and the 10-month-old *zi/zi* rat with a 9-month exposure to drinking water containing melatonin (e, f). Sections were immunostained for TH. Scale bar = 1 mm. (g–j) High-power photomicrographs of TH-immunohistochemically stained sections of the NA of 10-month-old *zi/zi* rats without treatment (g, i) and 10-month-old *zi/zi* rats with a 9-month exposure to drinking water that contained melatonin (h, j). Arrows indicate swollen TH-immunoreactive fibers (g, i). ★ = anterior commissure. Scale bar = 1 mm (a–f), 50 μm (g, h), and 2 μm (i, j).

studies of neurotoxin-induced PD models (Acuna-Castroviejo et al. 1997; Antolin et al. 2002; Sharma et al. 2006), chronic MEL treatment significantly inhibited both neuronal and terminal degenerations in the zitter DA systems. Furthermore, the present evidence showed that the neuroprotective effects of MEL were more effective in the mesolimbic DA system than in the nigrostriatal DA system.

Several lines of previous evidence suggested possible mechanisms by which MEL can exert these neuroprotective effects. It has been reported that MEL can dose-dependently induce the synthesis of the antioxidative enzymes SOD, CAT, and GSH, in addition to directly acting as an ROS scavenger (Galano et al. 2011; Tomas-Zapico and Coto-Montes 2005). In our recent zitter rat study, we confirmed there was an upregulation of these antioxidative defense systems after chronic MEL treatment, and furthermore, we demonstrated that MEL had an anti-inflammatory effect on this mutant rat (Hashimoto et al. 2012). Therefore, neuroinflammation and oxidative stress may synergistically induce neuropathological changes in this mutant rat strain.

It has been previously demonstrated that the levels of SOD, CAT, and GSH are higher in the VTA and NA as compared to the SN and CPU (Hung and Lee 1998; Trepanier et al. 1996). In the present study, we additionally showed that DA content was completely restored in the NA

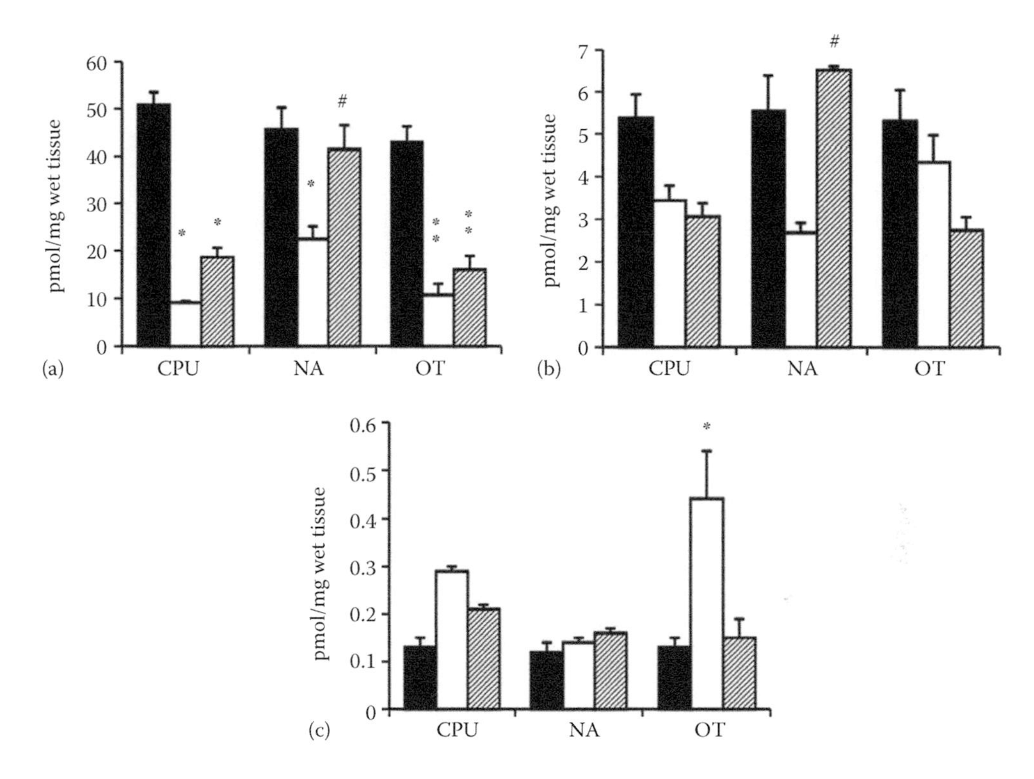

FIGURE 15.3 Effects of aging and melatonin treatment on DA (a) and DOPAC (b) concentration (pmol/mg wet tissue) and on the DOPAC/DA ratio (c) in the CPU, NA, and OT. *Closed column*: 1-month-old *zi/zi* rat without treatment. *Open column*: 10-month-old *zi/zi* rat without treatment. *Hatched column*: 10-month-old *zi/zi* rat with a 9-month exposure to drinking tap water that contained melatonin. Data are given as mean ± SEM. *$p<0.05$ versus the 1M group; **$p<0.01$ 1M group; #$p<0.05$ versus the 10M group.

and partially restored in the CPU of the *zi/zi* rat after a chronic high-dose MEL treatment. The different levels of DA recovery and metabolite levels that were noted between the NA and CPU may reflect the different preventive effects that MEL has on DA fiber degeneration in these DA systems. Since chronic MEL treatments increase the expression of *SOD*, *CAT*, and *GPx* mRNA in the *zi/zi* striatum (which includes ventral and dorsal striata) (Hashimoto et al. 2012), the different activities seen between the nigrostriatal and mesolimbic DA systems for the antioxidative enzymes suggest that MEL might very well contribute different neuroprotective effects within DA neuron systems in these mutant rats.

The present findings also confirm our previous results that showed that the abnormal TH-immunoreactive fibers were characterized by swollen varicosities and irregularly thickened intervaricose segments in the CPU and NA of aged *zi/zi* rats (Ueda et al. 2000). Double-labeling studies of the *zi/zi* rat with TH and Fluoro-Jade C, which is a definitive marker for neuronal degeneration, demonstrated that the swollen fibers that were observed in the striatum of 1M *zi/zi* rats represented degenerative DA fibers (Ehara and Ueda 2009). When taken together with our previous results (Ehara and Ueda 2009; Nakadate et al. 2006; Ueda et al. 2002), these findings suggest that the terminal degeneration of the DA fibers occurs prior to the degeneration of the DA neurons in the *zi/zi* SNc. In addition, the number of swollen DA fibers in the CPU and NA *zi/zi* rats increased with age, while the contrary normal DA fibers decreased within the same areas (Ueda et al. 2000). After a chronic high-dose MEL treatment, there was a reduction in the number of these swollen fibers in the NA, along with a concurrent increase in the density of TH-immunoreactive fibers with normal morphologies. In contrast, the same treatment appeared

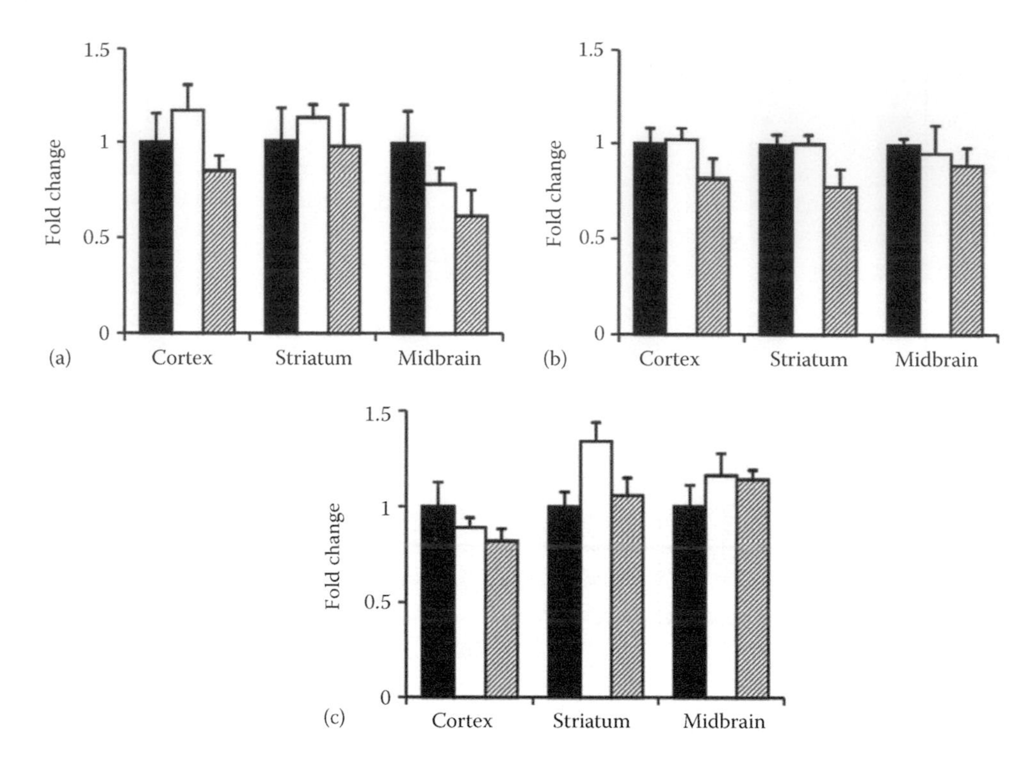

FIGURE 15.4 Effects of *zi/zi* and melatonin treatment on the expression of *GDNF* (a), *BDNF* (b), and *TrkB* (c) mRNA in the cortex, striatum, and midbrain of 3-month-old SD rats (closed column), *zi/zi* rats without treatment (open column), and *zi/zi* rats with 2-month exposure to drinking tap water that contained melatonin (hatched column). The mRNA levels are presented as a fold change (mean ± SEM).

to only partially affect the DA fibers in the CPU, as swollen TH-immunoreactive fibers were still observed in this area. Thus, these noted morphological changes also support the theory that MEL might be responsible for different neuroprotective effects within the mesostriatal DA neuron systems.

Several other studies have shown that BDNF and GDNF attenuate the lesion-induced degeneration of DA neurons in PD models (Hu and Russek 2008; Hung and Lee 1998; Sharma et al. 2006; Sun et al. 2005). Sharma et al. (2006) reported there was an upregulation of *GDNF* mRNA in the striatum after an MEL treatment in the 6-hydroxydopamine-induced PD model. However, unlike this previous study, we did not observe any significant differences in the *GDNF* mRNA levels among any of our experimental groups. Even so, the present study did further confirm our previous in situ hybridization data that showed there were no significant difference for the *BDNF* and *TrkB* mRNA levels in the frontal cortex and striatum of *zi/zi* rats as compared to age-matched *zi/+* rats (Joyce et al. 2004). And finally, the data from the present study also demonstrated there were no significant differences in the expressions of *BDNF* and *TrkB* mRNA in the frontal cortex, striatum, and midbrain among any of our experimental groups. Therefore, our results demonstrate that these neurotrophic factors do not participate in the neuroprotective effects of MEL in this mutant rat strain.

ACKNOWLEDGMENTS

We would like to thank Dr. K. Kato of Nihon University School of Medicine for his advice on MEL measurements and his helpful suggestions for our initial experiments, Mr. B. Sato and Dr. T. Naito

(Miz Co. Ltd., Fujisawa, Japan) for their generous donation of HRE, Ms. S. Nihei for technical assistance, and Ms. F. Terauchi with the preparation of the manuscript.

REFERENCES

Acuna-Castroviejo, D., A. Coto-Montes, M. G. Monti et al. 1997. Melatonin is protective against MPTP-induced striatal and hippocampal lesions. *Life Sci* 60: PL23–PL29.

Alter, C. A., N. Cai, T. Bliven et al. 1997. Anterograde transport of brain-derived neurotrophic factor and its role in the brain. *Nature* 389: 856–860.

Antolin, I., J. C. Mayo, R. M. Sainz et al. 2002. Protective effect of melatonin in a chronic experimental model of Parkinson's disease. *Brain Res* 94: 163–173.

Ehara, A. and S. Ueda. 2009. Application of Fluoro-Jade C in acute and chronic neurodegeneration models: Utilities and staining differences. *Acta Histochem Cytochem* 42: 171–179.

Esposito, E. and S. Cuzzocrea. 2010. Anti-inflammatory activity of melatonin in central nervous system. *Curr Neuropharmacol* 8: 228–242.

Galano, A., D. X. Tan, and R. J. Reiter. 2011. Melatonin as a naturally ally against oxidative stress: A physicochemical examination. *J Pineal Res* 51: 1–16.

Gomi, H., I. Ueno, and K. Yamanouchi. 1994. Antioxidant enzymes in the brain of zitter rats: Abnormal metabolism of oxygen species and its relevance to pathogenic changes in the brain of zitter rats with genetic spongiform encephalopathy. *Brain Res* 653: 66–72.

Guillin, O., J. Diaz, P. Carroll et al. 2001. BDNF controls dopamine D3 receptor expression and triggers behavioural sensitization. *Nature* 411: 86–89.

Hashimoto, K., S. Ueda, A. Ehara et al. 2012. Neuroprotective effects of melatonin on the nigrostriatal dopamine system in the zitter rat. *Neurosci Lett* 506: 79–83.

Hu, Y. and S. J. Russek. 2008. BDNF and diseased nervous system: A delicate balance between adaptive and pathological processes of gene regulation. *J Neurochem* 105: 1–17.

Hung, H. C. and E. H. Y. Lee. 1998. MPTP produces differential oxidative stress and antioxidative responses in the nigrostriatal and mesolimbic dopaminergic pathways. *Free Radic Biol Med* 24: 76–84.

Joyce, J. N., T. C. Der, L. Renish et al. 2004. Loss of D3 receptors in the zitter mutant rat is not reversed by L-DOPA treatment. *Exp Neurol* 187: 178–189.

Kuramoto, T., K. Kitada, T. Inui et al. 2001. Attractin/Mahogany/Zitter plays a critical role in myelination of the central nervous system. *Proc Natl Acad Sci USA* 98: 559–564.

Mattson, M. P., S. Maudsley, and B. Martin. 2004. BDNF and 5-HT: A dynamic duo in age-related neuronal plasticity and neurodegenerative disorders. *Trends Neurosci* 27: 569–594.

Nakadate, K., T. Noda, S. Sakakibara et al. 2006. Progressive dopaminergic neurodegeneration of substantia nigra in the zitter mutant rat. *Acta Neuropathol* 112: 64–73.

Nakamura, A., T. Kadowaki, S. Sakakibara et al. 2010. Regeneration of 5-HT fibers in hippocampal heterotopia of methylazoxymethanol-induced micrencephalic rats after neonatal 5,6-DHT injection. *Anat Sci Int* 85: 38–45.

Sakakibara, S., K. Nakadate, S. Ookawara et al. 2008. Non-cell autonomous impairment of oligodendrocyte differentiation precedes CNS degeneration in the zitter rat: Implications of macrophage/microglial activation in the pathogenesis. *BMC Neurosci* 9: 35.

Sharma, R., C. R. McMillan, C. C. Tenn et al. 2006. Physiological neuroprotection by melatonin in a 6-hydroxydopamine model of Parkinson's disease. *Brain Res* 1068: 230–236.

Sun, M., L. Kong, X. Wang et al. 2005. Comparison of the capability of GDNF, BDNF, or both, to protect nigrostriatal neurons in a rat model of Parkinson's disease. *Brain Res* 1052: 119–129.

Tan, D. X., L. C. Manchester, R. Hardeland et al. 2003. Melatonin: A hormone, a tissue factor, an autocoid, a paracoid, and an oxidant vitamin. *J Pineal Res* 34: 75–78.

Tang, Y. P., Y. L. Ma, C. C. Chao et al. 1998. Enhanced glial cell line-derived neurotrophic factor mRNA expression upon (−)-deprenyl and melatonin treatments. *J Neurosci Res* 53: 593–604.

Tomas-Zapico, C. and A. Coto-Montes. 2005. A proposed mechanism to explain the stimulatory effects of melatonin on antioxidative enzymes. *J Pineal Res* 29: 99–104.

Trepanier, G., D. Furling, J. Puymirat et al. 1996. Immunocytochemical localization of seleno-glutathion peroxidase in the adult mouse brain. *Neuroscience* 75: 231–243.

Ueda, S., M. Aikawa, A. Ishizuya-Oka et al. 1998. Age-related degeneration of the serotoninergic fibers in the zitter rat brain. *Synapse* 30: 62–70.

Ueda, S., M. Aikawa, A. Ishizuya-Oka et al. 2000. Age-related dopamine deficiency in the mesostriatal dopamine system of zitter mutant rats: Regional fiber vulnerability in the striatum and the olfactory tubercle. *Neuroscience* 95: 389–398.

Ueda, S., S. Sakakibara, T. Kadowaki et al. 2008. Chronic treatment with melatonin attenuates serotonergic degeneration in the striatum and olfactory tubercle of zitter mutant rats. *Neurosci Lett* 448: 212–216.

Ueda, S., S. Sakakibara, K. Nakadate et al. 2005. Degeneration of dopaminergic neurons in the substantia nigra of zitter mutant rat and protection by chronic intake of vitamin E. *Neurosci Lett* 380: 252–256.

Ueda, S., S. Sakakibara, E. Watanabe et al. 2002. Vulnerability of monoaminergic neurons in the brainstem of the zitter rat in oxidative stress. *Prog Brain Res* 136: 293–302.

16 Melatonin's Neuroprotective Role in Parkinson's Disease

Venkataramanujam Srinivasan, Timo Partonen, Mahaneem Mohamed, Rahimah Zakaria, Asma Hayati Ahmad, Charanjit Kaur, Domenico De Berardis, Edward C. Lauterbach, Samuel D. Shillcutt, and U.S. Srinivasan

CONTENTS

16.1 INTRODUCTION

Parkinson's disease (PD) is a neurodegenerative disorder with a multifactorial etiology, but it is primarily suggested to be due to a loss of dopaminergic function, which produces multiple neurological and psychiatric symptoms, including motor, cognitive, and emotional dysfunction. Additionally, most patients with PD experience sleep-related symptoms, including difficulty in initiating and maintaining sleep, excessive daytime sleepiness (EDS), and parasomnias such as rapid eye movement (REM) sleep behavior disorder (Kumar et al. 2002; Brotini and Gigli 2004; Postuma et al. 2009). Most studies point out that the prevalence of sleep disturbance in PD is nearly 100% (Lees et al. 1988; Kumar et al. 2002; Poryazova and Zacchariev 2005). These sleep disorders are classified as primary sleep disorders that are intrinsic to PD and secondary sleep disorders due to either medications or motor impairment (Friedman and Chou 2004). Most studies have shown that the use of dopamine (DA) agonists cause somnolence with a reversal of the sleep–wake cycle (Ondo et al. 2001; Sanjiv et al. 2001).

16.2 ETIOLOGY OF PARKINSON'S DISEASE

PD is a neurodegenerative disorder with a multifactorial etiology but mainly due to a loss of dopaminergic nigrostriatal function. Research studies by Braak and colleagues (Braak et al. 2003, 2007) reveal that Lewy body pathology is pronounced not only in substantia nigra (SN) but also in many other central nervous system (CNS) regions, including the lower brainstem and autonomic nervous system, which have been supported by recent studies (Beach et al. 2009). It is suggested that neurological and psychiatric manifestations may precede the traditional motor manifestations of PD (Savica et al. 2010). Insomnia and depression form the integral feature of PD and have been suggested as the primary manifestations in PD (Willis and Armstrong 1999). Case control and cohort studies suggest that depressive and anxiety disorders may be one of the earliest manifestations of PD (Shiba et al. 2000; Weisskopf et al. 2003). Since sleep disturbances seen in PD are associated with cognitive decline and psychiatric symptoms, attention should be focused on the development of targeted interventions in this direction (Naismith et al. 2010). There is also much evidence for the possible involvement of the retino-hypothalamic system in the etiology of PD as circadian rhythm disturbances are common in this disease (Willis 2008). Parkinsonian symptoms themselves undergo circadian fluctuations. Patients with PD often experience worsening of symptoms in the afternoon and evening. Patients with PD experience time-dependent responsiveness to dopaminergic stimulation (Bruguerolle and Simon 2002). Recently, many studies on molecular clock mechanisms regulating circadian physiology and behavior in mammals have been undertaken in the central circadian pacemaker, the suprachiasmatic nucleus (SCN), and in various peripheral tissues and cells (Liu et al. 2007). Approximately 20 canonical circadian genes, known as the key *clock genes*, have been identified. Of these genes, *PER1* and *BMAL1* (*ARNTL* or *Mop3*) are regarded as the best markers of the molecular clock. Disruptions of *PER1* and *BMAL1* in mice have been shown to cause altered circadian behavior and dysregulation of circadian patterns in gene expression (Cermakian et al. 2001; Kondratov et al. 2006). Circadian clock genes *PER1* and *BMAL1* have been located in leukocytes of healthy humans (Boivin et al. 2003; Fukuya et al. 2007), and hence, study of these genes in patients with PD has been undertaken recently. *PER1* and *BMAL1* expression in leukocytes in patients with PD and normal controls was undertaken between 21:00 and 09:00 h. It was noticed that during this dark phase, expression of *BMAL1* but not *PER1* was much reduced in PD, suggesting thereby that a peripheral molecular clock is altered in PD patients. Moreover, expression of *BMAL1* in PD patients correlated with the United Parkinson's Disease Rating Scale score at 06:00 and 09:00 h and with the Pittsburgh Sleep Quality Index Score at 06:00 h (Cai et al. 2010).

Melatonin, the major neurohormone secreted by the pineal gland, is a chronobiotic regulating circadian clock functions (Srinivasan 1989) and exerts its action through MT_1 and MT_2 melatonin receptors expressed in the central circadian clock, namely, the SCN of the anterior hypothalamus. Also, there are melatonin receptors distributed elsewhere in a range of brain regions (Reppert et al. 1994, 1995). Melatonin MT_1 and MT_2 receptors are expressed in the human amygdala and SN. The decreased expression of MT_1 and MT_2 in these regions in patients with PD suggests the possible involvement of melatonin receptors in the pathophysiology of PD (Adi et al. 2010). Various factors that trigger PD and the role of oxidative stress in the pathogenesis of PD are shown in Figures 16.1 and 16.2, respectively.

16.3 REM SLEEP BEHAVIOR DISORDER IN PARKINSON'S DISEASE

REM sleep behavior disorder (RSBD) in PD is poorly understood and may occur as a prodromal feature predating motor symptoms by several years. Its prevalence is suggested to be 60% in patients, and it has been suggested to be predictive of dementia in longitudinal studies (Vendette et al. 2007; Marion et al. 2008). RSBD is characterized by a loss of skeletal muscle atonia with prominent motor activity during dreaming (Olson et al. 2000). Numerous cases of RSBD have been found in clinically diagnosed PD (Silber and Ahlskog 1992; Schenck et al. 1996; Boeve et al. 2004, 2007).

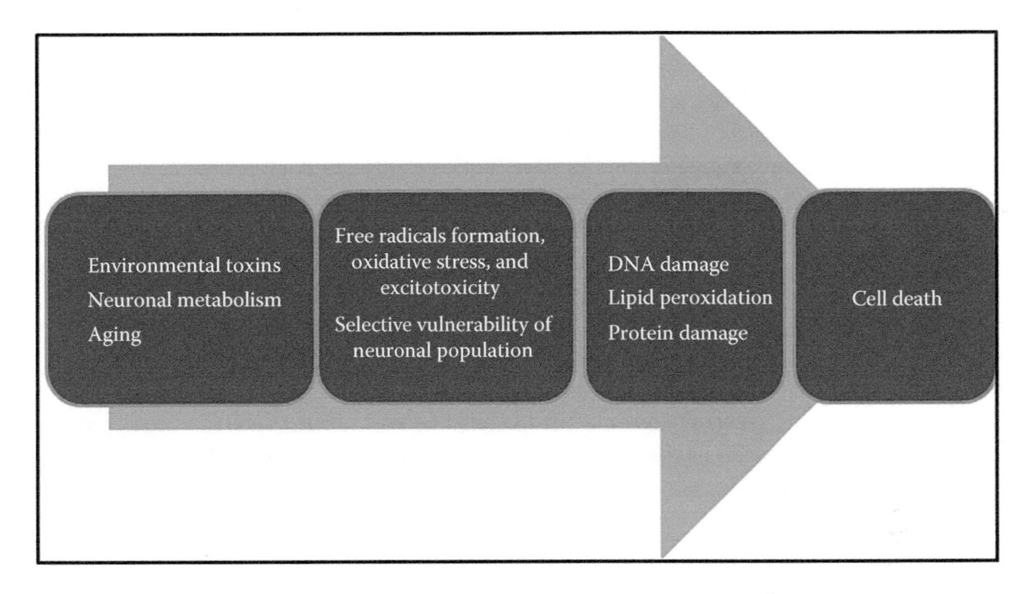

FIGURE 16.1 **(See color insert.)** Various factors that trigger Parkinson's disease.

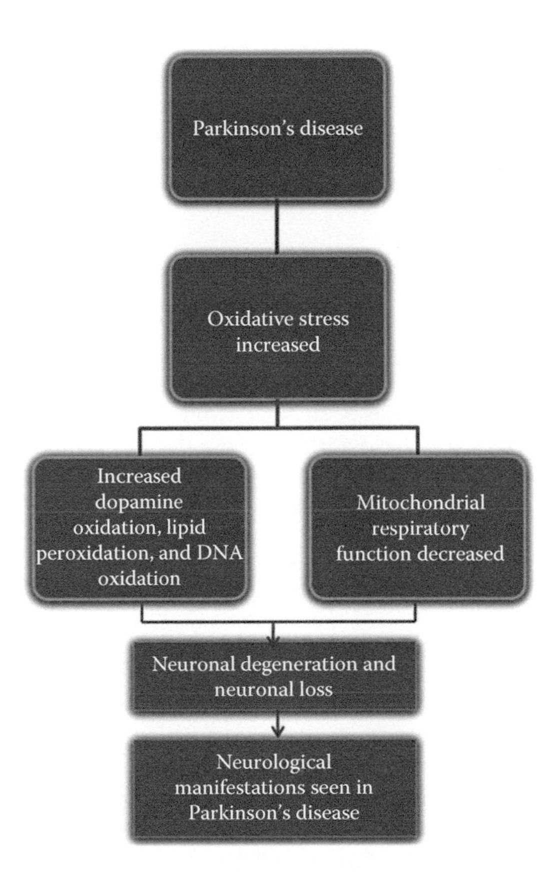

FIGURE 16.2 **(See color insert.)** Role of oxidative stress in the pathogenesis of Parkinson's disease.

Loss of REM sleep atonia and/or an increased locomotor drive is suggested as the likely mechanism for the clinical expression of human RSBD (Mahowald and Schenck 2000). It is a dream disorder similar to REM motor disorder, and there is a tendency for the dream content to involve an aggressive, attacking, or chasing theme. Nightmare behaviors like screaming, kicking, punching, and injuring the bed partner are common (Fantini et al. 2005; Jahan et al. 2009). Nocturnal disturbance and sleep arousals as measured by actigraphy are specific to RSBD seen in PD (Naismith et al. 2010). Loss of hypocretin neurons and cells secreting melanin-concentrating hormone in the hypothalamus of PD patients is said to be responsible for nocturnal insomnia, RSBD, and hallucinations. In a study of the hypothalamus in 11 PD patients and 5 control subjects, a loss of hypothalamic neurons containing hypocretin and melanin-concentrating hormone has been found in PD (Thannickal et al. 2007). However, a recent study undertaken on this line of research found that changes in orexin/hypocretin do not necessarily underlie RSBD sleep disturbances (Compta et al. 2009). It is suggested that probing into the components of the circadian system that mediate the onset and timing of REM sleep, including the pattern and timing of melatonin secretion, combined with clinico pathological studies may prove to be vital for defining the neuroanatomical correlate of RSBD in PD (Naismith et al. 2010).

16.4 EXCESSIVE DAYTIME SLEEPINESS IN PARKINSON'S DISEASE

EDS occurs in PD patients, and the percentage of its occurrence ranges from 15% to 50% (Tandberg et al. 1999; Arnulf et al. 2002). EDS is found in 41% of PD patients compared with 24% of a control population (Brodsky et al. 2003). It is recently reported that daytime sleepiness using the Epworth Sleepiness Scale score in PD is associated with cognitive impairment in PD, especially in the setting of dementia, and attention/working memory, executive function, memory, and visuospatial deficits (Goldman et al. 2013). Although intake of both DA and levodopa are said to be contributory factors for the development of EDS (Mehta et al. 2008), EDS also could be underlying in the pathogenesis of PD (Arnulf et al. 2002). Among the nonmotor symptoms of PD, sleep disturbances, particularly RSBD, are important and may even predict the diagnosis of PD based on motor symptoms (Erro et al. 2010; Salawu et al. 2010).

16.5 SLEEP APNEA IN PARKINSON'S DISEASE

Central sleep apnea is reported to occur in most of the neurodegenerative diseases, including PD. The prevalence of sleep apnea is estimated to be from 20% to 27% and varies from moderate to severe (Arnulf et al. 2002; Maria et al. 2003). A greater incidence of obstructive sleep apnea in PD patients (73%) than would be expected has been noted (Maria et al. 2003). PD patients with obstructive sleep apnea have been shown to have a higher arousal index on the polysomnogram than PD patients without obstructive sleep apnea, although the two groups had similar daytime sleepiness using the Epworth Sleepiness Scale scores (Nomura et al. 2013).

16.6 INSOMNIA IN PARKINSON'S DISEASE

Insomnia is recognized as a common complaint among elderly people (Van Someren 2000; Bellon 2006) and in patients with PD (Factor et al. 1990; Young et al. 2002). It is found that 55% of PD patients with insomnia have impaired mobility in bed. Furthermore, PD patients with insomnia and the self-report of impaired mobility in bed have reduced sleep-related body position changes (i.e., nocturnal hypokinesia) and decreased sleep efficiency (Louter et al. 2013). Insomnia is considered to be either a direct consequence of the disease process (Kales et al. 1971) or due to the effects of medications such as painful dystonia and mood disturbances (Leeman et al. 1987; Jahan et al. 2009). While studying the effects of antiparkinsonian medication on endogenous melatonin levels and circadian rhythms, it was found that the melatonin rhythm was phase advanced in patients receiving levodopa therapy

(Fertl et al. 1993; Bordet et al. 2003). A progressive decrease in amplitude and phase advance of the melatonin rhythm was noted with progression and an increased duration of the disease. Increase in the dosage of levodopa therapy was found to exacerbate these changes (Bordet et al. 2003).

16.7 MELATONIN IN PARKINSON'S DISEASE

Melatonin (*N*-acetyl-5-methoxytryptamine) has diverse physiological functions and is synthesized mainly by the pineal gland of all animals. Synthesis also occurs in various other areas of the body like the retina, gut, lymphocytes, skin, and thymus. In all of these tissues, melatonin has either an autocrine or paracrine role. In the pineal gland, melatonin secretion is regulated by the central circadian pacemaker or SCN and is synchronized to the light/dark cycle, with a nocturnal maximum of 200 pg/mL (in plasma), and levels of less than 10 pg/mL during the day. Melatonin is synthesized from serotonin through two enzymatic steps. The first is *N*-acetylation by serotonin-*N*-acetyltransferase to yield *N*-acetylserotonin. The activity of this enzyme is high at night, suggesting it as a rate regulating enzyme in melatonin biosynthesis. The second step is transfer of a methyl group from *S*-adenosylmethionine to the 5-hydroxy group of *N*-acetylserotonin to yield melatonin, and this reaction is catalyzed by the enzyme hydroxyl-indole-*O*-methyl transferase. Once formed, melatonin is not stored within the pineal gland but diffuses into the capillary blood and cerebrospinal fluid (CSF) (Arendt 2000; Tricoire et al. 2003). In a recent study conducted in human beings, CSF melatonin levels were higher in the third ventricle than the lateral, showing thereby that melatonin enters the CSF through the pineal recess even during daytime (Leston et al. 2010). As melatonin passes with ease through all biological membranes, the brain has much higher concentrations of melatonin than any other tissue in the body (Reiter and Tan 2002).

In all mammals, pineal melatonin biosynthesis is regulated by the SCN, which receives input from the retinohypothalamic tract. Special photoreceptive retinal ganglion cells containing melanopsin as a photopigment are involved in the projection from the retina (Brainard et al. 2001; Berson et al. 2002). Fibers from the SCN pass through a circuitous route involving the paraventricular nucleus, medial forebrain bundle, reticular formation, lateral horn cells of the spinal cord, and superior cervical ganglion and then proceed to innervate the pineal gland as postganglionic sympathetic fibers. Norepinephrine, released from these fibers, regulates melatonin biosynthesis through the cascade of β-adrenergic receptors, adenyl cyclase, and cyclic adenosine monophosphate (Klein et al. 1971). Cyclic adenosine monophosphate stimulates arylalkylamine *N*-acetyltransferase expression and phosphorylation via protein kinases (Schomerus and Korf 2005). Circulating melatonin is metabolized mainly in the liver where it is first hydroxylated in the 6 position by cytochrome P450 monooxygenases (isoenzymes CYP1A2, CYP1A1) and thereafter conjugated with sulfate to be excreted as 6-sulfatoxymelatonin. Melatonin can be metabolized nonenzymatically in all cells of the body. It is converted into 3-hydroxymelatonin when it scavenges two hydroxyl (·OH) radicals (Tan et al. 1998). In the brain, a substantial amount of melatonin is metabolized to kynuramine derivatives (Hirata et al. 1974). These metabolites of melatonin formed in the brain, namely, N^1-acetyl-N^2-formyl-5-methoxykynuramine and N^1-acetyl-5-methoxykynuramine (AMK), share the antioxidant and anti-inflammatory properties of melatonin (Tan et al. 2001).

16.8 MELATONIN RECEPTORS AND PARKINSON'S DISEASE

Melatonin exerts its physiological actions through G-protein MT_1 and MT_2 melatonin receptors expressed both singly and together in various cells and tissues of the body (Dubocovich et al. 2000; Dubocovich and Markowska 2005). Functional melatonin receptors have been localized broadly in the brain, including structures like the SCN (Liu et al. 1997), cerebellum (Al Ghoul et al. 1998), hippocampus (Savaskan et al. 2002), SN, caudate-putamen, ventral tegmental areas, and nucleus accumbens (Uz et al. 2005). In this context, it is significant to note that the expression of both MT_1 and MT_2 receptors are decreased in the SN of PD patients (Adi et al. 2010).

16.9 MELATONIN'S NEUROPROTECTIVE ROLE IN EXPERIMENTAL ANIMAL MODELS OF PARKINSON'S DISEASE

Table 16.1 summarizes the studies of the neuroprotective role of melatonin using experimental animal models of PD. Animal models employing altered brain DA function by injecting 6-hydroxydopamine (6-OHDA) into the nigrostriatal pathway of the rat or by injecting the neurotoxin 1-methyl-4-phenyl-1,2,3,6 tetrahydropyridine (MPTP), which produces behavioral deficits, are commonly employed for studying the efficacy of various therapeutic agents used for the treatment of PD (Chieuh et al. 1985). The loss of DA neurons occurring in these animal models causes severe sensory and motor impairments that give rise to tremor, rigidity, and akinesia similar to those seen in PD patients (Ben Shachar et al. 1986). In a study using the MPTP model of PD, melatonin was able to counteract MPTP-induced lipid peroxidation in the striatal, hippocampal, and midbrain regions (Acuna-Castroviejo et al. 1997). Using the same MPTP model, melatonin's ability to prevent neuronal cell death in the nigrostriatal pathway was again demonstrated (Antolin et al. 2002). MPTP elicits its neurotoxic effects by increasing nitric oxide (NO) radicals derived from inducible NO synthase (iNOS), which act mainly on DA neurons while NO radicals derived from neuronal NO synthase (nNOS) damage striatal dopaminergic fibers and terminals. Hence, it is suggested that optimal therapy for treating PD requires agents that inhibit the degenerative effects of iNOS in the SN pars compacta (Zhang et al. 2000). As melatonin can effectively downregulate iNOS and prevent NO radical formation in the brain (Cuzzocrea et al. 1997), it can be considered as a potential therapeutic agent for treating PD (Srinivasan et al. 2005).

TABLE 16.1
Melatonin's Neuroprotective Role in Experimental Animal Models of Parkinson's Disease

Model Used for PD Phenotype	Effects Seen	Melatonin's Beneficial Neuroprotective Effects against PD	References
6-OHDA injection model	Loss of DA neurons and severe sensory and motor impairments with tremor, rigidity, and akinesia as seen in PD patients	Melatonin counteracted the neurotoxic effects of 6-OHDA injection and prevented the occurrence of sensory and motor impairments	Thomas and Mohankumar (2004), Saravanan et al. (2005)
Ferrous-ascorbate DA generating system (in vitro model)	Increased 6-OHDA generation	Melatonin administration decreased 6-OHDA generation in dose-dependent manner	Borah and Mohankumar (2009)
MPTP injection	Generates peroxynitrate, inhibits ETC and mitochondrial function, and produces consequent neuronal cell death	Both melatonin and AMP counteracted the effects of MPTP, increased complex-I, reduced lipid peroxidation and nitrate, and prevented neuronal cell loss	Tapias et al. (2009)
Maneb- or paraquat-induced nigrostriatal degeneration	Decreased striatal DA and tyrosine hydroxylase activity, increased lipid peroxidation and degenerative changes seen in nigrostriatal pathway, with the loss of locomotor activity	Melatonin administration delayed degenerative changes by reducing oxidative stress and improved locomotor activity	Singhal et al. (2011)

The neuroprotective action of melatonin in experimental animal models has been found in certain other studies (Dabbeni-Sala et al. 2001; Thomas and Mohankumar 2004).

In a recent study exploring melatonin's neuroprotective effect, the effect of rotenone (a specific inhibitor of mitochondrial complex 1) was studied in rats. Rotenone produces behavioral, pathological, and biochemical lesions that resemble PD (Saravanan et al. 2005; Sindhu et al. 2005). Rotenone induces neurodegeneration by releasing ·OH radicals that cause oxidative stress following complex 1 inhibition in the SN (Saravanan et al. 2006). Rotenone has been shown to cause glutathione depletion in the cell body region of SN. Administration of melatonin in various doses of 10, 20, and 30 mg/kg independently attenuated the rotenone-induced glutathione depletion. Moreover, a significant reduction in rotenone-induced ·OH formation in the mitochondria and an increase in the activity of superoxide dismutase and catalase were also observed in the rotenone-damaged SN (Saravanan et al. 2007). The results of this study support the beneficial effects of melatonin treatment in PD. It is suggested that 6-OHDA could be generated per se from excessive DA in the brain and increased production of this neurotoxin could be the reason for selective degeneration of the DA containing SN as seen in PD. Following L-3,4-dihydroxyphenylalanine (L-DOPA) administration, dopaminergic nuclei and serotonergic nuclei in the brain are overloaded with DA, and this increased DA can cause the production of 6-OHDA in the brain (Maharaj et al. 2005). Pursuing this concept, an in vitro study has been undertaken involving the generation of 6-OHDA and melatonin's ability to prevent its formation. Employing a ferrous-ascorbate dopamine ·OH radical generating system, the addition of DA (1 mM) caused 6-OHDA production. The DA-dependent production of 6-OHDA was decreased by melatonin in a dose-dependent manner. Similarly in an in vivo study carried out in mice, daily administration of melatonin (30 mg/kg) along with L-DOPA for 7 days significantly reduced the L-DOPA- or L-DOPA + MPTP-induced generation of striatal 6-OHDA by 21% and 32%, respectively, when compared with those groups without melatonin. This study demonstrates that melatonin has the potential to influence 6-OHDA generation in vitro and in vivo in the brain (Borah and Mohankumar 2009).

16.10 MITOCHONDRIAL DYSFUNCTION IN PD AND THE NEUROPROTECTIVE EFFECTS OF MELATONIN AND *N*(1)-ACETYL-5-METHOXYKYNURAMINE

The MPTP model of PD is a valuable tool for studying not only the participation of various factors, like oxidative/nitrosative stress, excitotoxicity, and inflammation, in the pathogenesis of PD, but also the role of mitochondrial dysfunction in the pathogenesis of PD. MPTP is metabolized into 1-methyl-4-phenyl pyridinium (MPP^+), which is taken up into dopaminergic neurons through the DA transporter and accumulates in the mitochondria of the SN pars compacta (Przedborski et al. 2004). MPP^+ binds with complex 1 of the electron transport chain (ETC) and inhibits it (Greenmyre et al. 2001), thereby causing increased generation of reactive oxygen species (ROS). This results in oxidative damage to the ETC and ATP production, and in nigral cell death (Rego and Oliveira 2003; Przedborski et al. 2004; Tretter et al. 2004). MPP^+, by inducing microglial activation and iNOS expression in the SN, has been shown to produce large amounts of NO and neuronal cell death (Liberatore et al. 1999; Brown and Bal-Price 2003). NO by reacting with O_2^- generates highly toxic peroxynitrite ($ONOO^-$), which impairs mitochondrial function, causing irreversible inhibition of all ETC complexes (Brown and Borutaite 2004) and neuronal cell death (Muravchick and Levy 2006; Zhang et al. 2006). Recently, the participation of inducible mitochondrial nitric oxide synthase (i-mtNOS) in mitochondrial dysfunction and nigrostriatal degeneration in PD was studied by using the MPTP mouse model. In this study, it was found that MPTP administration induced i-mtNOS in the mitochondria of the SN, leading to high production of NO (Tapias et al. 2009). Moreover, complex 1 inhibition, NO production, and lipid peroxide (LPO) levels were significantly higher in the SN than in the striatum after treatment with MPTP.

Treatment with melatonin and AMK counteracted the effects of MPTP in both brain nuclei, increasing complex 1 activity above control values in SN ($P < 0.001$) and striatal ($P < 0.05$) mitochondria. Both melatonin and AMK counteracted the effects of MPTP on LPO levels in the cytosol ($P < 0.05$) and in the mitochondria of the SN ($P < 0.001$). Similarly, both melatonin and AMK reduced nitrites to control values in the cytosol ($P < 0.05$) and mitochondria ($P < 0.001$). An emerging feature of this study is that AMK, the brain metabolite of melatonin, was as efficient as melatonin itself in counteracting i-mtNOS production, oxidative stress, and mitochondrial dysfunction induced by MPTP (Tapias et al. 2009). This study points out that future directions should involve the development of i-mtNOS antagonists, mainly melatonin, AMK, and other melatonin agonists, as therapeutic strategies for treating PD (Tapias et al. 2009).

16.11 MELATONIN AS A THERAPEUTIC AGENT IN THE TREATMENT OF PARKINSON'S DISEASE

Studies undertaken in experimental animals point out that melatonin, as an effective antioxidant, has the potential for offering neuroprotection in patients with PD and in treating the cognitive disorders of PD (Arushanian 2010). Clinical studies in this direction are very few. Melatonin has been used for treating sleep problems, insomnia, and daytime sleepiness. In a study undertaken in 40 patients (11 women, 29 men, mean age 61.7 ± 8.4 years, range 43–76 years), melatonin was administered in doses ranging from 5 mg to 50 mg/day 30 min before bedtime for 2 weeks (Dowling et al. 2005). All subjects were taking stable doses of antiparkinsonian medications during the course of the study. Treatment with 50 mg of melatonin significantly increased nighttime sleep compared to placebo as revealed by actigraphy. Subjective reports of overall sleep disturbance improved significantly on 5 mg of melatonin compared to 50 mg or placebo. It was found that high doses of melatonin 50 mg were well tolerated in this study (Dowling et al. 2005). With the finding of reduced expression of melatonin MT_1 and MT_2 receptors in patients with PD (Adi et al. 2010), there is a possibility that the melatonergic system is involved in the abnormal sleep mechanisms seen in PD and in the pathophysiology of PD as well. Hence, therapeutic strategies should aim at targeting the use of melatonin and its agonists like ramelteon not only in treating the nonmotor symptoms of PD but also in preventing the progression of the disease itself. However, the use of melatonin as an adjunct therapy to either halt the progression or provide symptomatic relief in PD has been questioned (Willis and Armstrong 1999).

16.12 POTENTIAL USE OF MELATONIN AGONISTS IN THE TREATMENT OF PARKINSON'S DISEASE

Available evidence indicates that both sleep induction and maintenance of sleep at appropriate circadian phases are greatly affected in PD patients. Moreover, the onset and timing of REM sleep also is very much impaired in PD patients (Naismith et al. 2010). RSBD seen in PD patients occurs much earlier and is predictive of dementia (Vendette et al. 2007; Marion et al. 2008). Treatment of sleep disturbances seen in PD patients with appropriate drugs may help not only in solving sleep problems but may also help to prevent the progression of PD as well. As the conventional drugs like benzodiazepines used for the treatment of insomnia may worsen the cognitive and memory impairment associated with PD, a hypnotic drug without any of these adverse effects will be beneficial. As melatonin exerts its hypnotic and chronobiotic effects by acting through MT_1 and MT_2 receptors located in the SCN, it can be helpful. Although melatonin significantly improves the quality of sleep, it is seen that sleep abnormalities persist in PD (Medeiros et al. 2006). Since melatonin has a short elimination half-life (shorter than 30 min), a melatonin agonist having a longer duration of action and an enhanced bioavailability might be of greater benefit than melatonin in promoting sleep initiation efficiency (Turek and Gillette 2004).

Recently, the melatonin agonist ramelteon, a chronohypnotic, has been introduced for treating insomniacs (Pandi-Perumal et al. 2009). Ramelteon is a novel melatonin receptor agonist that has been shown to bind to MT_1 and MT_2 receptors and has a longer duration of action than melatonin (Kato et al. 2005). The efficacy and safety of ramelteon in treating insomnia have been proven in a number of clinical studies conducted on elderly insomniacs (Erman et al. 2006; Roth et al. 2007; Zammit et al. 2007, 2009). Ramelteon may have important therapeutic potential in treating sleep problems seen in PD, including RSBD. Apart from treating sleep disturbances, ramelteon can alter the sleep–wake rhythm as well and, hence, can correct REM rhythm abnormality.

PD is often complicated by depression, and drugs like nortryptyline and selective serotonin reuptake inhibitors have been employed to treat depression of PD, although the results have not been promising (Weintraub et al. 2010; Dobkin et al. 2011). In this context, the melatonergic antidepressant agomelatine has been studied in treating major depressive disorder and has been found to be effective in alleviating the symptoms of depression in European clinical trials (Loo et al. 2002; Zupancic and Guilleminault 2006; Pjrek et al. 2007). Agomelatine has a dually phased mechanism of action. At night, its sleep-promoting melatonergic effects prevail over its potentially antihypnotic 5-HT_{2C} antagonism, while during the day, its antidepressant action via 5-HT_{2C} is uncoupled from melatonin's nocturnal actions. This sequential mode of action is a major advantage for agomelatine (Millan 2006). Furthermore, it might turn out to be helpful in the treatment of depressive disorders and sleep disorders associated with PD, but this needs to be tested and verified.

16.13 ANTIPARKINSONIAN DRUGS AND THEIR EFFECTS ON SLEEP IN PD

Imaging studies have shown the role of dopaminergic alterations in the pathogenesis of PD-related sleep disorders (Mehta et al. 2008). In a study conducted in 10 PD patients, polysomnographic parameters were compared with dopaminergic function in the striatum and upper brainstem using fluoro-DOPA (F-DOPA) uptake. It was found that decreased F-DOPA uptake correlated with increased REM sleep duration in early PD (Hilker et al. 2003). Treatment of PD with dopaminergic drugs has been shown to promote sleep when given in low doses, but when administered in higher doses, it prolonged sleep latency and caused fragmentation of sleep (Askenasy and Yahr 1984; Van Hilten et al. 1994). Selegiline caused alerting effects and difficulty in falling asleep when given to PD patients (Lavie et al. 1980). Parkinson's patients, when given pergolide as an add-on drug, experienced increased nocturnal activity and worsened sleep fragmentation as compared to placebo (Comella et al. 2005). Thus, both dopaminergic drugs and nondopaminergic drugs that are commonly used in treating PD affect sleep and worsen the quality of life in these patients (Mehta et al. 2008). Hence, it is important and timely to introduce a clinically effective nontoxic hypnotic drug for treatment of sleep disorders seen in PD that at the same time could also address the etiology of PD itself. In this context, the neurohormone melatonin and particularly its agonist melatonergic drug ramelteon offer promising considerations for treating sleep disorders associated with PD.

16.14 CONCLUSIONS

Symptoms of PD, including sleep disorders, and particularly REM sleep disorder, occur in a majority of PD patients, often even earlier than the manifestation of PD motor symptoms and may even serve as a *preclinical marker*. Treating the sleep disorders of PD may be essential in deterring both the prevention and progression of this disease. Current evidence points to melatonin and melatonin receptors in the pathophysiology of PD. Administration of melatonin in the animal models of PD has been effective in preventing neuronal cell death and has also been effective in ameliorating the symptoms of PD. The involvement of melatonin in the possible etiology of PD is further strengthened by the recent finding of reduced expression of MT_1 and MT_2 melatonin receptors in the SN and amygdala, structures that are affected by PD. Hence, therapeutic strategies for the treatment of PD should consider the application of melatonin or its receptor agonists, like ramelteon, and the melatonergic antidepressant agomelatine since these agents can improve both the sleep problems and the progression of PD by their neuroprotective actions. Clinical trials pursuing this line of research can prove the efficacy of melatonin and melatonin agonists in the treatment and management of PD and the sleep disorders associated with it.

REFERENCES

Acuna-Castroviejo, D., A. Coto-Montes, M. M. Gaia, G. G. Ortiz, and R. J. Reiter. 1997. Melatonin is protective against MPTP-induced striatal and hippocampal lesions. *Life Sci* 60:L23–L29.

Adi, N., D. C. Mash, Y. Ali, C. Singer, L. Shehadeh, and S. Papapetropoulos. 2010. Melatonin MT1 and MT2 receptor expression in Parkinson's disease. *Med Sci Monit* 16(2):BR61–BR67.

Al Ghoul, W. M., M. D. Herman, and M. L. Dubocovich. 1998. Melatonin receptor subtype expression in human cerebellum. *Neuroreport* 9:4063–4068.

Antolin, I., J. C. Mayo, R. M. Sainz, M. L. del Brio, F. Herrera, V. Martin, and C. Rodríguez. 2002. Protective effect of melatonin in chronic experimental model of Parkinson's disease. *Brain Res* 943:163–173.

Arendt, J. 2000. Melatonin, circadian rhythms and sleep. *N Engl J Med* 343:1114–1116.

Arnulf, I., E. Konofal, M. Merino-Andreu, J. L. Houeto, V. Mesnage, M. L. Welter, L. Lacomblez, J. L. Golmard, J. P. Derenne, and Y. Agid. 2002. Parkinson's disease and sleepiness: An integral part of PD. *Neurology* 58(7):1019–1024.

Arushanian, E. B. 2010. A hormonal drug melatonin in the treatment of cognitive function disorders in Parkinsonism. *Eksp Klin Farmacol* 73(3):35–39.

Askenasy, J. J. and M. D. Yahr. 1984. Suppression of REM rebound by pergolide. *J Neural Trans* 59(2):151–159.

Beach, T. G., C. H. Adler, L. Lue, L. I. Sue, J. Bachalakuri, J. Henry-Watson, J. Sasse et al. 2009. Unified staging system for Lewy body disorders: Correlation with nigrosriatal degeneration, cognitive impairment and motor dysfunction. *Acta Neuropathol* 117(6):613–634.

Bellon, A. 2006. Searching for new options for treating insomnia: Are melatonin and Ramelteon beneficial? *J Psychiatr Pract* 12(4):229–242.

Ben Shachar, D., R. Ashkenazi, and M. B. Youdim. 1986. Long term consequence of early iron-efficiency on dopaminergic neurotransmission in rats. *Int J Dev Neurosci* 4:81–88.

Berson, D. M., F. A. Dunn, and M. Takao. 2002. Phototransduction by retinal ganglion cells that set the circadian clock. *Science* 295:1070–1073.

Boeve, B., M. Silber, and T. Ferman. 2004. REM sleep behaviour disorder in Parkinsonn's disease and dementia with Lewy bodies. *J Geriatr Psychiatry Neurol* 17:146–157.

Boeve, B. F., M. H. Silber, C. B. Saper, T. J. Ferman, D. W. Dickson, J. E. Parisi, E. E. Benarroch et al. 2007. Pathophysiology of REM sleep behaviour disorder and relevance to neurodegenerative disease. *Brain* 130:2770–2788.

Boivin, D. B., F. O. James, A. Wu, P. F. Cho-Park, H. Xiong, and Z. S. Sun. 2003. Circadian clock genes oscillate in human peripheral blood mononuclear cells. *Blood* 102:4143–4145.

Borah, A. and K. P. Mohankumar. 2009. Melatonin inhibits 6-hydroxydopamine production in the brain to protect against experimental Parkinsonism in rodents. *J Pineal Res* 47:293–300.

Bordet, R., D. Devos, S. Brique, Y. Touitou, J. D. Guieu, C. Libersa, and A. Destée. 2003. Study of circadian melatonin secretion pattern at different stages of Parkinson's disease. *Clin Neuropsychopharmacol* 26(2):65–72.

Braak, H., K. Del Tredici, U. Rub, R. A. I. de Vas, E. N. H. Jansen Steur, and E. Braak. 2003. Staging of brain pathology related to sporadic Parkinson's disease. *Neurobiol Aging* 24(2):197–211.

Braak, H., M. Sastre, J. R. E. Bohl, R. A. I. de Vos, and K. Del Tradici. 2007. Parkinson's disease: Lesions in dorsal horn layer 1, involvement of parasympathetic and sympathetic pre and post ganglionic neurons. *Acta Neuropathol* 113(4):421–429.

Brainard, G. C., J. P. Hanifin, J. M. Greeson, B. Byrne, G. Glickman, E. Gerner, and M. D. Rollag. 2001. Action spectrum for melatonin regulation in humans: Evidence for a novel circadian photoreceptor. *J Neurosci* 21:6405–6412.

Brodsky, M. A., J. Godbold, T. Roth, and C. W. Olanow. 2003. Sleepiness in Parkinson's disease. A controlled study. *Mov Disord* 18(6):668–672.

Brotini, S. and G. L. Gigli. 2004. Epidemiology and clinical features of sleep disorders in extrapyramidal disease. *Clin Neuropharmacol* 5(2):169–179.

Brown, G. C. and A. Bal-Price. 2003. Inflammatory neurodegeneration mediated by nitric oxide, glutamate and mitochondria. *Mol Neurobiol* 27:325–355.

Brown, G. C. and V. Borutaite. 2004. Inhibition of mitochondrial respiratory complex 1 by nitric-oxide, peroxynitrite and S-nitrosothiols. *Biochim Biophys Acta* 1658:44–49.

Bruguerolle, B. and N. Simon. 2002. Biologic rhythms and Parkinson's disease: A chrono-pharmacologic approach to considering fluctuations in function. *Clin Neuropharmacol* 25:194–201.

Cai, Y., S. Liu, R. B. Sothern, S. Xu, and P. Chan. 2010. Expression of clock genes Per1 and Bmal1 in total leukocytes in health and Parkinson's disease. *Eur J Neurol* 17:550–554.

Cermakian, N., L. Monaco, M. P. Pando, A. Dierich, and P. Sasone-Corsi. 2001. Altered behavioural rhythms and clock gene expression in mice with a targeted mutation in the Period 1 gene. *EMBO J* 20:3967–3974.

Chieuh, C. C., R. S. Burns, D. M. Markey, D. M. Jacobowitz, and I. J. Kopin. 1985. Primate model of parkinsonism: Selective lesion of nigrostrial neurons by 1-methyl-4-phenyl-1,2,3,6 tetrahydropyridine produces an extrapyramidal syndrome in rhesus monkeys. *Life Sci* 36:213–218.

Comella, C. L., M. Momissey, and K. Janko. 2005. Nocturnal activity with night time pergolide in Parkinson's disease: A controlled study using actigraphy. *Neurology* 64(8):1450–1451.

Compta, Y., J. Santamaria, L. Ratti, E. Tolosa, A. Iranzo, E. Muñoz, F. Valldeoriola, R. Casamitjana, J. Ríos, and M. J. Marti. 2009. Cerebrospinal hypocretin, daytime sleepiness and sleep architecture in Parkinson's disease dementia. *Brain* 132:3308–3317.

Cuzzocrea, S., B. Zingarelli, E. Gilad, P. Hake, A. L. Salzman, and C. Szabo. 1997. Protective effect of melatonin in carrageenan-induced models of local inflammation: Relationship to its inhibitory effect on nitric oxide production and its peroxynitrite scavenging activity. *J Pineal Res* 23:106–116.

Dabbeni-Sala, F., S. Di Santo, D. Franceschini, S. D. Skaper, and P. Giusti. 2001. Melatonin protects against 6-OHDA induced neurotoxicity in rats: A role for mitochondrial complex 1 activity. *FASEB J* 15:164–170.

Dobkin, R. D., M. Menza, K. L. Bienfait, M. Gara, H. Marin, M. H. Mark, A. Dicke, and J. Friedman. 2011. Depression in Parkinson's disease: Symptom improvement and residual symptoms after acute pharmacologic management. *Am J Geriatr Psychiatry* 19(3):222–229.

Dowling, G. A., J. Mastick, E. Colling, J. H. Carter, C. M. Singer, and M. J. Aminoff. 2005. Melatonin for sleep disturbances in Parkinson's disease. *Sleep Med* 6(5):459–466.

Dubocovich, M. L., D. P. Cardinali, P. Delagrange, D. N. Krause, D. Strosberg, D. Sugden, and F. D. Yocca. 2000. Melatonin receptors. In *The IUPHAR Compendium of Receptor Characterization and Classification*, 2nd edn., IUPHAR, ed., pp. 271–277. London, U.K.: IUPHAR Media.

Dubocovich, M. L. and M. Markowska. 2005. Functional MT_1 and MT_2 melatonin receptor in mammals. *Endocrine* 27:101–110.

Erman, M., D. Seiden, G. Zammit, S. Sainati, and J. Zhang. 2006. An efficacy, safety, and dose-response study of ramelteon in patients with chronic primary insomnia. *Sleep Med Rev* 7:17–24.

Erro, M. E., M. P. Moreno, and B. Zandio. 2010. Pathophysiological bases of the non-motor symptoms in Parkinson's disease. *Rev Neurol* 50(Suppl 2):S7–S13.

Factor, S. A., T. McAlarney, J. R. Sanchez-Ramos, and W. J. Weiner. 1990. Sleep disorders and sleep effect in Parkinson's disease. *Mov Disord* 5(4):280–285.

Fantini, M. L., A. Corona, S. Clerisi, and L. Ferini-Strambi. 2005. Aggressive dream content without daytime aggressiveness in REM sleep behaviour disorder. *Neurology* 65:1010–1015.

Fertl, E., E. Auff, A. Doppelbauer, and F. Waldhauser. 1993. Circadian secretion pattern of melatonin de novo parkinsonian patients: Evidence for phase shifting properties of L-dopa. *J Neural Transm Park Dis Sect* 5(3):227–234.

Friedman, J. H. and K. L. Chou. 2004. Sleep and fatigue in Parkinson's disease. *Parkinsoniam Relat Disord* 10(Suppl):S27–S35.

Fukuya, H., N. Emoto, H. Nonaka, K. Yagita, H. Okamura, and M. Yokoyama. 2007. Circadian expression of clock genes in human peripheral leukocytes. *Biochem Biophys Res Commun* 354:924–928.

Goldman, J. G., R. A. Ghode, B. Ouyang, B. Bernard, C. G. Goetz, and G. T. Stebbins. 2013. Dissociations among daytime sleepiness, nighttime sleep, and cognitive status in Parkinson's disease. *Parkinsonism Relat Disord* 19(9):806–811. doi: 10.1016/j.parkreldis.2013.05.006.

Greenmyre, J. T., T. B. Sherer, R. Betarbet, and A. V. Panov. 2001. Complex 1 and Parkinson's disease. *IUBMB Life* 52:135–141.

Hilker, R., N. Razai, M. Ghaemi, S. Weisenbach, J. Rudolf, B. Szelies, and W. D. Heiss. 2003. [18F]fluorodopa uptake in the upper brainstem measured with positron emission tomography correlates with decreased REM sleep duration in early Parkinson's disease. *Clinical Neurol Neurosurg* 105(4):262–269.

Hirata, F., O. Hayaishi, T. Tokuyama, and S. Seno. 1974. In vitro and in vivo formation of two new metabolites of melatonin. *J Biol Chem* 249:1311–1313.

Jahan, I., R. A. Hauser, K. L. Sullivan, A. Miller, and T. A. Zesiewicz. 2009. Sleep disorders in Parkinson's disease. *Neuropsychiatr Dis Treat* 5:535–540.

Kales, A., R. D. Ansel, C. H. Markham, M. B. Scharf, and T. L. Tan. 1971. Sleep in patients with Parkinson's disease and normal subjects prior to and following levodopa administration. *Clin Pharmacol Ther* 12:397–407.

Kato, K., K. Hirai, K. Nishiyama, O. Uchikawa, K. Fukatsu, S. Ohkawa, Y. Kawamata, S. Hinuma, and M. Miyamoto. 2005. Neurochemical properties of ramelteon (TAK 375), a selective MT1/MT2 receptor agonist. *Neuropharmacology* 48:301–310.

Klein, D. C., J. L. Weller, and R. Y. Moore. 1971. Melatonin metabolism: Neural regulation of pineal serotonin. Acetyl coenzyme A *N*-acetyltransferase activity. *Proc Natl Acad Sci USA* 68:3107–3110.

Kondratov, R. V., A. A. Kondratova, V. Y. Gorbacheva, O. V. Vykhovanets, and M. P. Antoch. 2006. Early age related pathologies in mice deficient in BMAL1, the core components of the circadian clock. *Genes Dev* 20:1868–1873.

Kumar, S., M. Bhatia, and M. Behari. 2002. Sleep disorders in Parkinson's disease. *Mov Disord* 17(4):775–781.

Lavie, P., J. Wajsbort, and M. B. Youdim. 1980. Deprenyl does not cause insomnia in parkinsonian patients. *Commun Psychopharmacol* 4(4):303–307.

Leeman, A. L., C. J. O' Neill, P. W. Nicholson, A. A. Deshmukh, M. J. Denham, J. P. Royston, R. J. Dobbs, and S. M. Dobbs. 1987. Parkinson's disease in the elderly: Response to and optimal spacing of night time dosing with levodopa. *Br J Clin Pharmacol* 24:637–643.

Lees, A. J., N. A. Blackburn, and V. L. Campbell. 1988. The nighttime problems of Parkinson's disease. *Clin Neuropharmacol* 11(6):512–519.

Leston, J., C. Harthe, J. Brun, C. Mottolese, P. Mertens, M. Sindou, and B. Claustrat. 2010. Melatonin is released in the third ventricle in humans. A study in movement disorders. *Neurosci Lett* 469(3):294–297.

Liberatore, G. T., V. Jackson-Lewis, S. Vukosavic, A. S. Mandir, M. Vila, W. G. McAuliffe, V. L. Dawson, T. M. Dawson, and S. Przedborski. 1999. Inducible nitric oxide synthase stimulates dopaminergic neurodegeneration in the MPTP model of Parkinson's disease. *Nat Med* 5:1403–1409.

Liu, C., D. R. Weaver, X. Jin, L. P. Shearman, R. L. Pieschl, V. K. Gribkoff, and S. M. Reppert. 1997. Molecular dissection two distinct actions of melatonin on the suprachiasmatic circadian clock. *Neuron* 19:91–102.

Liu, S., Y. Cai, R. B. Sothern, Y. Guan, and P. Chan. 2007. Chronobiological analysis of circadian patterns in transcription of seven clock genes in six peripheral tissues in mice. *Chronobiol Int* 24:793–820.

Loo, H., A. Hale, and H. D'haenen. 2002. Determination of the dose of agomelatine, a melatonergic agonist and selective 5-HT_{2C} antagonist in the treatment of major depressive disorder: A placebo controlled dose range study. *Int Clin Psychopharmacol* 17:239–247.

Louter, M., R. J. van Sloun, D. A. Pevernagie, J. B. Arends, P. J. Cluitmans, B. R. Bloem, and S. Overeem. 2013. Subjectively impaired bed mobility in Parkinson disease affects sleep efficiency. *Sleep Med* 14(7):668–674. doi: 10.1016/j.sleep.2013.03.010.

Maharaj, H., D. Sukhdev Maharaj, M. Scheepers, R. Mokokong, and S. Daya. 2005. L-DOPA administration enhances 6-hydroxydopamine generation. *Brain Res* 1063:180–186.

Mahowald, M. and C. Schenck. 2000. REM sleep behaviour disorder. In *Principles and Practice of Sleep Medicine*, M. Kryger, T. Roth, and W. Dement, eds., pp. 724–741. Philadelphia, PA: WB Saunders.

Maria, B., S. Sophia, M. Mihalis, L. Charalampos, P. Andreas, M. E. John, and S. M. Nikolaos. 2003. Sleep breathing disorders in patients with idiopathic Parkinson's disease. *Respir Med* 97(10):1151–1157.

Marion, M. H., M. Qurashi, G. Marshall, and O. Foster. 2008. Is REM sleep behaviour disorder (RBD) a risk factor of dementia in idiopathic Parkinson's disease? *J Neurol* 255:192–196.

Medeiros, C. A. M., P. F. C. de Bruin, L. A. Lopes, M. C. Magalhaes, M. de Lourdes Seabra, and V. M. S. de Bruin. 2006. Effect of exogenous melatonin on sleep and motodysfunction in Parkinson's disease: A randomized, double blind, placebo-controlled study. *J Neurol* 254(4):1–7.

Mehta, S. H., J. C. Morgan, and K. D. Sethi. 2008. Sleep disorders associated with Parkinson's disease: Role of dopamine, epidemiology, and clinical scales of assessment. *CNS Spectr* 13:3(Suppl 4):6–11.

Millan, M. J. 2006. Multi-target strategies for the improvement of depressive states: Conceptual foundations and neuronal substrates, drug discovery and therapeutic application. *Pharmacol Therapeut* 110:135–370.

Muravchick, S. and R. J. Levy. 2006. Clinical implications of mitochondrial dysfunction. *Anesthesiology* 105:819–837.

Naismith, S. L., N. L. Rogers, J. Mackenzie, I. B. Hickie, and S. J. Lewis. 2010. The relationship between actigraphically defined sleep disturbance and REM sleep behaviour disorder in Parkinson's disease. *Clin Neurol Neurosurg* 112(5):420–423.

Nomura, T., Y Inoue, M. Kobayashi, K. Namba, and K. Nakashima. 2013. Characteristics of obstructive sleep apnea in patients with Parkinson's disease. *J Neurol Sci* 327(1–2):22–24.

Olson, E. J., B. F. Boeve, and M. H. Silber. 2000. Rapid eye movement sleep behaviour disorder: Demographic, clinical and laboratory findings in 93 cases. *Brain* 123:331–339.

Ondo, W. G., K. Dat Vuong, H. Khan, F. Atassi, C. Kwak, and J. Jankovic. 2001. Daytime sleepiness and other sleep disorders in Parkinson's disease. *Neurology* 57(8):1392–1396.

Pandi-Perumal, S. R., V. Srinivasan, D. W. Spence, A. Moscovitch, R. Hardeland, G. M. Brown, and D. P. Cardinali. 2009. Ramelteon: A review of its therapeutic potential in sleep disorders. *Adv Ther* 26(6):613–636.

Pjrek, E., D. Winkler, A. Konstantinidis, M. Willeit, N. Praschak-Reider, and S. Kasper. 2007. Agomelatine in the treatment of seasonal affective disorder. *Psychopharmacology (Berlin)* 190:575–579.

Poryazova, R. G. and Z. I. Zacchariev. 2005. REM sleep behaviour disorder in patients of Parkinson's disease. *Folia Med (Plodiv)* 47(1):5–10.

Postuma, R. B., J. F. Gagnon, M. Vendette, M. L. Fantini, J. Massicote-Marquez, and J. Montplaisir. 2009. Quantifying the risk of neurodegenerative disease in idiopathic REM sleep behaviour disorder. *Neurology* 72(15):1296–1300.

Przedborski, S., K. Tieu, C. Perier, and M. Vila. 2004. MPTP as a mitochondrial neurotoxic model of Parkinson's disease. *J Bioenerg Biomembr* 36(4):375–379.

Rego, A. C. and C. R. Oliveira. 2003. Mitochondrial dysfunction and reactive oxygen species in excitotoxicity and apoptosis: Implications for the pathogenesis of neurodegenerative diseases. *Neurochem Res* 28:1563–1574.

Reiter, R. J. and D. X. Tan. 2002. Melatonin: An antioxidant in edible plants. *Ann N Y Acad Sci* 957:341–344.

Reppert, S. M., C. Godson, C. D. Mahle, D. R. Weaver, S. A. Slaugenhaupt, and J. F. Gusella. 1995. Molecular characterization of a second melatonin receptor expressed in human retina and brain: The Melib melatonin receptor. *Proc Natl Acad Sci USA* 92:8734–8738.

Reppert, S. M., D. R. Weaver, and T. Ebisawa. 1994. Cloning and characterization of a mammalian melatonin receptor that mediates reproductive and circadian responses. *Neuron* 13:1177–1185.

Roth, T., D. Seiden, S. Wang-Weigand, and J. Zhang. 2007. A 2-night, 3 period cross over study of ramelteon's efficacy and safety in older adults with chronic insomnia. *Curr Med Res Opin* 23:1005–1014.

Salawu, F. K., A. Danburam, and A. B. Olokoba. 2010. Non-motor symptoms of Parkinson's disease: Diagnosis and management. *Niger J Med* 19(2):126–131.

Sanjiv, C. C., M. J. Schulzer, E. Mak, J. Fleming, W. R. Martin, T. Brown, S. M. Calne et al. 2001. Daytime somnolence in patients with Parkinson's disease. *Parkinsonism Relat Disord* 7(4):283–286.

Saravanan, K. S., K. M. Sindhu, and K. P. Mohankumar. 2005. Acute intranigral infusion of rotenone in rats causes progressive biochemical lesions in the striatum similar to Parkinson's disease. *Brain Res* 1049:147–155.

Saravanan, K. S., K. M. Sindhu, and K. P. Mohankumar. 2007. Melatonin protects against rotenone-induced oxidative stress in a hemiparkinsonian model. *J Pineal Res* 42:247–253.

Saravanan, K. S., K. M. Sindhu, K. S. Senthilkumar, and K. P. Mohanakumar. 2006. L-Deprenyl protects against rotenone-induced oxidative stress mediated dopaminergic neurodegeneration in rats. *Neurochem Int* 49:28–40.

Savaskan, E., G. Olivieri, F. Meier, L. Brydon, R. Jockers, R. Ravid, A. Wirz-Justice, and F. Müller-Spahn. 2002. Increased melatonin 1a-receptor immunoreactivity in the hippocampus of Alzheimer's disease patients. *J Pineal Res* 32:59–62.

Savica, R., W. A. Rocca, and E. Ahlskog. 2010. When does Parkinson disease start? *Arch Neurol* 67(7):798–801.

Schenck, C. H., S. R. Bundlie, and M. W. Mahowald. 1996. Delayed emergence of a parkinsonian disorder in 38% of 29 older women initially diagnosed with idiopathic rapid eye movement sleep behaviour disorder. *Neurology* 46:388–393.

Schomerus, C. and H. W. Korf. 2005. Mechanisms regulating melatonin synthesis in the mammalian pineal organ. *Ann N Y Acad Sci* 1057:372–383.

Shiba, M., J. H. Bower, D. M. Maraganore, S. K. McDonnell, B. J. Peterson, J. E. Ahlskog, D. J. Schaid, and W. A. Rocca. 2000. Anxiety disorders and depression disorders preceding Parkinson's disease: A case controlled study. *Mov Disord* 15(4):669–677.

Silber, M. H. and J. E. Ahlskog. 1992. REM sleep behaviour disorder in parkinsonian syndromes. *Sleep Res (Abstract)* 21:313.

Sindhu, K. M., K. S. Saravanan, and K. P. Mohankumar. 2005. Behavioural differences in a rotenone-induced hemiparkinsonian rat model developed following intranigral or median forebrain bundle infusion. *Brain Res* 1051:25–34.

Singhal, N. K., G. Srivastava, D. K. Patel, S. K. Jain, and M. P. Singh. 2011. Melatonin or silymarin reduces maneb- and paraquat-induced Parkinson's disease phenotype in the mouse. *J Pineal Res* 50:97–109.

Srinivasan, V. 1989. The pineal gland: Its physiological and pharmacological role. *Ind J Physiol Pharmacol* 33(4):263–272.

Srinivasan, V., S. R. Pandi-Perumal, G. J. M. Maestroni, A. I. Esquifino, R. Hardeland, and D. P. Cardinali. 2005. Role of melatonin in neurodegenerative diseases. *Neurotox Res* 7(4):293–318.

Tan, D. X., L. C. Manchester, S. Burkhardt, R. M. Sainz, J. C. Mayo, R. Kohen, E. Shohami, Y. S. Huo, R. Hardeland, and R. J. Reiter. 2001. N^1-acetyl-N^2-formyl-5-methoxy-kynuramine, a biogenic amine and melatonin metabolite, functions as a potent antioxidant. *FASEB J* 15:2294–2296.

Tan, D. X., L. C. Manchester, R. J. Reiter, B. F. Plummer, L. J. Hardies, S. D. Weintraub, Vijayalaxmi, and A. M. Shepherd. 1998. A novel melatonin metabolite, cyclic 3-hydroxyl melatonin: A biomarker of in vivo hydroxyl radical generation. *Biochem Biophys Res Commun* 253:614–620.

Tandberg, E., J. P. Larsen, and N. K. Karlsen. 1999. Excessive daytime sleepiness and sleep benefit in Parkinson's disease: A community based study. *Mov Disord* 14(6):922–927.

Tapias, V., G. Escames, L. C. Lopez, A. Lopez, E. Camacho, M. D. Carrion, A. Entrena, M. A. Gallo, A. Espinosa, and D. Acuña-Castroviejo. 2009. Melatonin and its brain metabolite N^1-acetyl-5-methoxykynuramine prevent mitochondrial nitric oxide synthase induction in Parkinsonian mice. *J Neurosci Res* 87:3002–3010.

Thannickal, T. C., Y. Y. Lai, and J. M. Siegel. 2007. Hypocretin (orexin) cell loss in Parkinson's disease. *Brain* 130(Pt 6):1586–1595.

Thomas, B. and K. P. Mohankumar. 2004. Melatonin protects against oxidative stress caused by 1-methyl-4-phenyl-1,2,3,6-tetra hydropyridine in the mouse nigrostriatum. *J Pineal Res* 36:25–32.

Tretter, L., I. Sipos, and V. Adam-Vizi. 2004. Inhibition of neuronal damage by complex 1 deficiency and oxidative stress in Parkinson's disease. *Neurochem Res* 29:569–577.

Tricoire, H., M. Moller, P. Chemineau, and B. Malpaux. 2003. Origin of cerebrospinal melatonin and possible function in the integration of photoperiod. *Reprod Suppl* 61:311–321.

Turek, F. W. and M. U. Gillette. 2004. Melatonin, sleep, and circadian rhythms: Rationale for development of specific melatonin agonists. *Sleep Med* 5:523–532.

Uz, T., A. D. Arsian, M. Kurtuncu, M. Imbesi, M. Akhisaroglu, Y. Dwivedi, G. N. Pandey, and H. Manev. 2005. The regional and cellular expression profile of the melatonin receptor MT_1 in the central dopaminergic system. *Brain Res Mol Brain Res* 136:45–53.

Van Hilten, B., J. I. Hoff, H. A. Middlekoop, E. A. Vander Velde, G. A. Kerkhof, A. Wauquier, H. A. Kamphuisen, and R. A. Roos. 1994. Sleep disruption in Parkinson's disease. Assessment by continuous activity monitoring. *Arch Neurol* 51(9):922–928.

Van Someren, E. J. W. 2000. Circadian and sleep disturbances in the elderly. *Exp Gerontol* 35:1229–1237.

Vendette, M., J. F. Gagnon, A. Decary, J. Massicotte-Marquez, R. B. Postuma, J. Doyan, M. Panisset, and J. Montplaisir. 2007. REM sleep behaviour disorder predicts cognitive impairment in Parkinson's disease without dementia. *Neurology* 69:1843–1849.

Weintraub, D., S. Mavandadi, E. Mamikonyan, A. D. Siderowf, J. E. Duda, H. I. Hurtig, A. Colcher et al. 2010. Atomoxetine for depression and other neuropsychiatric symptoms in Parkinson's disease. *Neurology* 75(5):448–455.

Weisskopf, M. G., H. Chen, M. A. Schwarzschild, I. Kawachi, and A. Ascherio. 2003. Prospective study of phobic anxiety and risk of Parkinson's disease. *Mov Disord* 18(6):646–651.

Willis, G. L. 2008. Parkinson's disease as a neuroendocrine disorder of circadian function: Dopamine-melatonin imbalance and the visual system in the genesis and progression of the degenerative process. *Rev Neurosci* 19(4–5):245–316.

Willis, G. L. and S. M. Armstrong. 1999. A therapeutic role for melatonin antagonism in experimental models of Parkinson's disease. *Physiol Behav* 66(5):785–795.

Young, A., M. Home, T. Churchward, N. Freezer, P. Holmes, and M. Ho. 2002. Comparison of sleep disturbance in mild versus severe Parkinson's disease. *Sleep* 25:573–577.

Zammit, G., M. Erman, S. Wang-Weigand, S. Sainati, J. Zhang, and T. Roth. 2007. Evaluation of the efficacy and safety of ramelteon in subjects with chronic insomnia. *J Clin Sleep Med* 3:495–504.

Zammit, G., H. Schwartz, T. Roth, S. Wang-Weigand, S. Sainati, and J. Zhang. 2009. The effects of ramelteon in a first night model of transient insomnia. *Sleep Med* 10:55–59.

Zhang, I., V. L. Dawson, and T. M. Dawson. 2006. Role of nitric oxide in Parkinson's disease. *Pharmacol Ther* 109:33–41.

Zhang, Y., V. L. Dawson, and T. M. Dawson. 2000. Oxidative stress and genetics in the pathogenesis of Parkinson's disease. *Neurobiol Dis* 7:240–250.

Zupancic, M. and C. Guilleminault. 2006. Agomelatine: A preliminary review of a new antidepressant. *CNS Drugs* 20:981–992.

17 Neuroprotective Role of Melatonin in Glaucoma

Ruth E. Rosenstein, María C. Moreno, Pablo Sande, Marcos Aranda, María F. González Fleitas, and Nicolás Belforte

CONTENTS

ABBREVIATIONS

BDNF Brain-derived neurotrophic factor
CAT Cationic amino acid transporter
CS Chondroitin sulfate
EAAT-1 Excitatory amino acid transporter type 1
GABA Gamma aminobutyric acid
GAD Glutamic acid decarboxylase
GATs GABA transporters
GPX Glutathione peroxidase
GS Glutamine synthetase
GSH Reduced glutathione
HA Hyaluronic acid
iNOS Inducible NOS or NOS-2
IOP Intraocular pressure
NMDA N-methyl d-aspartate
NOS NO synthase
nNOS Neuronal NOS or NOS-1
ONH Optic nerve head
PLR Pupil light reflex
POAG Primary open-angle glaucoma

mRGCs	Retinal ganglion cells expressing melanopsin
RGC	Retinal ganglion cell
ROS	Reactive oxygen species
SCN	Suprachiasmatic nuclei
SOD	Superoxide dismutase

17.1 GLAUCOMA

More than 60 million people around the world are affected by glaucoma, and it has been estimated that ~8 million suffer from bilateral blindness caused by this disease (Quigley and Broman 2006). Glaucoma is characterized by specific visual field defects due to the loss retinal ganglion cells (RGCs) and damage to the optic nerve head (ONH). Visual loss often starts in the periphery and advances to involve the central vision, with devastating consequences to the patient's quality of life (Almasieh et al. 2012). It is estimated that half of those affected may be not aware of their condition because symptoms may not occur during the early stages of the disease. When vision loss appears, considerable and permanent damage has already occurred. Medications and surgery can help to slow the progression of some forms of the disease, but at present, there is no cure. An increase in intraocular pressure (IOP) definitely plays a causal role in glaucomatous neuropathy. However, although ocular hypertension is common among open-angle glaucoma patients, only a limited subset of individuals with ocular hypertension will develop this disease (Friedman et al. 2004). Moreover, a significant number of patients with glaucoma continue to lose vision despite responding well to therapies that lower eye pressure (Caprioli 1997; Georgopoulos et al. 1997; Harbin et al. 1976; Leske 2003). Thus, the mechanisms that lead to RGC death in glaucoma are still under debate. Since glaucoma is probably a complex and multifactorial disease, it is likely that several molecular pathways converge to induce RGC loss. Signals that promote RGC death in glaucoma might be exacerbated by risks factors, tilting the neuron's fate toward dysfunction and demise. In recent years, there has been considerable progress in our understanding of multiple pathways that lead to RGC degeneration following optic nerve injury. This body of work has notably increased our knowledge of RGC neurobiology (Almasieh et al. 2012). In this vein, several factors such as a glutamate excitotoxicity (Moreno et al. 2005a), decrease in gamma aminobutyric acid (GABA) levels (Moreno et al. 2008), reduced antioxidant defense system activity (Aslan et al. 2008; Tezel 2006), and an increase in the nitridergic pathway activity (Belforte et al. 2007; Neufeld et al. 1999) have been suggested as possible additional causes for early or advanced stages of glaucomatous damage. Although the current management of glaucoma is mainly directed at the control of IOP, a therapy that prevents the death of ganglion cells should be the main goal of treatment.

Unraveling which are the most critical mechanisms involved in glaucoma is unlikely to be achieved in studies that are limited to the clinically observable changes to the retina and ONH that are seen in human glaucoma. Far more detailed and invasive studies are required, preferably in a readily available animal model. An experimental model system of pressure-induced optic nerve damage would greatly facilitate the understanding of the cellular events leading to RGC death, and how they are influenced by IOP and other risk factors associated to glaucoma. Several groups have developed various ways to increase IOP in the rat eye, generally by impeding the outflow of aqueous humor (Morrison et al. 1997; Shareef et al. 1995; Ueda et al. 1998). All of these models have both advantages and disadvantages. We have developed a model of glaucoma in rats through weekly intracameral injections of 1% hyaluronic acid (HA). Weekly injections of HA in the rat anterior chamber significantly increase IOP as compared with vehicle-injected contralateral eye (Benozzi et al. 2002; Moreno et al. 2005b). Although multiple injections of HA may be needed to obtain a sustained hypertension, we have shown that the injection procedure itself does not affect IOP and retinal function and histology. On the contrary, several advantages support our model: (1) a highly consistent hypertension is achieved, (2) it may have a reasonably long course, (3) daily variations in IOP persist in HA-injected eyes, (4) in contrast to other models, in all likelihood, HA does not impede the blood flow out of the eye, and (5) it is easy to perform. Furthermore, we have shown that this model may be useful for pharmacological studies, since the HA-induced hypertension was significantly reduced by the topic and acute application of therapeutically used hypotensive drugs (Benozzi et al. 2002). The chronic administration of HA significantly decreases the scotopic electroretinographic activity and provokes a significant loss of RGCs and optic nerve fibers (Moreno et al. 2005b). Based on both functional and histological evidence, these results indicate that intracameral injections of HA in the rat eye anterior chamber appear to mimic some key features of primary open-angle glaucoma (POAG), and therefore, it may be a useful tool to understand this ocular disease and to develop new therapeutic strategies. In particular, using this model, recent results indicate that melatonin could be a new therapy for glaucoma treatment.

17.2 MELATONIN IN THE RETINA

Numerous studies have firmly established that melatonin synthesis occurs in the retina of vertebrates, including mammals (Tosini and Fukuhara 2003). Although available data indicate that photoreceptors synthesize melatonin independently of the rest of the retina (Cahill and Besharse 1993), we have shown that melatonin could also be synthesized in chick RGCs (Garbarino-Pico et al. 2004). In the vertebrate species studied so far, melatonin synthesis in the retina is elevated at night and reduced during the day in a fashion similar to events in the pineal gland. Melatonin synthesis in the retina is under the control of a circadian oscillator, and circadian rhythms in melatonin synthesis have been described in several species. These rhythms occur in vivo, persist in vitro, and are entrained by the light/dark signal (Iuvone et al. 2005; Tosini and Menaker 1996). Retinal biosynthesis of melatonin has been extensively studied and revised (Tosini and Fukuhara 2003). In contrast, although a mutual inhibitory relationship between melatonin and dopamine (Dubocovich 1983; Jaliffa et al. 2000), as well as the involvement of melatonin in the regulation of photoreceptor disc shedding and phagocytosis (Besharse and Dunis 1983), melanosome aggregation in pigment epithelium, and cone photoreceptor retinomotor movements (Pierce and Besharse 1985), were conclusively demonstrated, the full range of physiological actions of melatonin in the retina is far from being completely known. Retinal melatonin does not contribute to circulating levels, suggesting that it acts locally as a neurohormone and/or neuromodulator. We will discuss evidence supporting the therapeutic effect of melatonin for the treatment of glaucoma.

17.3 OXIDATIVE DAMAGE IN GLAUCOMA AND MELATONIN AS A RETINAL ANTIOXIDANT

There are a considerable variety of free radicals in the organism that are produced as by-products of molecular oxygen and that are able to exert extensive damage, particularly over time. Oxidative stress, characterized by the imbalance between the production of reactive oxygen species (ROS) and their elimination system, plays an essential role in the injury and death of neuron cells, including RGCs. Excessive ROS could cause protein modification and DNA damage, consequently activating cell death signals. The retina is especially susceptible to oxidative stress because of its high oxygen consumption, its high proportion of polyunsaturated fatty acids, and its exposure to light. Among others, glutathione, and antioxidant enzymes such as superoxide dismutase (SOD), catalase, and glutathione peroxidase (GPX), provide a powerful antioxidant defense in the retina (Armstrong et al. 1981; Castorina et al. 1992; Ohta et al. 1996). SOD catalyzes the conversion of superoxide radicals (O_2) to hydrogen peroxide (H_2O_2), which is the first step in the metabolic defense against cellular oxidative stress. Although H_2O_2 is not a free radical, it is highly reactive, membrane permeable, and can be converted to highly reactive metabolites of oxygen such as hydroxyl radical. Under normal conditions, most of the H_2O_2 molecules generated by SOD are further metabolized to water by catalase and GPX. Thus, it is critical for the cellular survival that SOD activity should be coupled with similar GPX and catalase activities to safely detoxify H_2O_2. Despite having high levels of antioxidants, the retina is still susceptible to oxidative stress, which has been observed in several retinal conditions (Rajesh et al. 2003; Tanito et al. 2002; van Reyk et al. 2003; Wu et al. 1997). It has been confirmed that oxidative damage occurs in experimental models of optic nerve injury and in human glaucoma, and insufficiency in ROS-neutralizing mechanisms in RGCs was also discovered in glaucoma (Kanamori et al. 2010; Moreno et al. 2004; Yuki et al. 2011). Moreover, it has been reported that the level of lipid peroxidation products increases more than twofold and that the ocular antioxidant defense mechanism decreases in the anterior chamber of patients with advanced glaucoma (Kurysheva et al. 1996). We have demonstrated that SOD and catalase activities decrease in eyes with ocular hypertension induced by HA (Moreno et al. 2004). Although the mechanism(s) involved in the changes of these retinal enzymatic activities from hypertensive eyes is not yet understood, it is highly probable that a decrease of SOD and catalase activities could provoke an imbalance of the endogenous antioxidant defense system. This hypothesis is strongly supported by a significant increase in retinal lipid peroxidation observed in the retina from hypertensive eyes (Moreno et al. 2004).

The possibility that melatonin could detoxify highly ROS was originally suggested by Ianas et al. (1991). Three years later, Reiter and coworkers (1994), using spin trapping and electron resonance spectroscopy, demonstrated that melatonin has capacity to directly scavenge highly reactive hydroxyl radicals. Since then, several reports have shown that melatonin acts as a free radical scavenger and an efficient antioxidant (Hardeland et al. 1995; Pandi-Perumal et al. 2006; Reiter 1998; Reiter et al. 1997, 2000; Turjanski et al. 1998). Not only melatonin but also several of its metabolites generated during its free radical scavenging action may act as antioxidants (Tan et al. 2007). The kynurenic pathway of melatonin metabolism includes a series of radical scavengers with the possible sequence: Melatonin → cyclic 3-hydroxymelatonin → N^1-acetyl-N^2-formyl-5-methoxykynuramine (AFMK) → N^1-acetyl-5-methoxykynuramine (AMK). In the metabolic step from melatonin to AFMK, up to four free radicals can be consumed (Adler et al. 1997; Guenther et al. 2005; Hardeland 2005; Tan et al. 2007), greatly increasing melatonin's efficacy as an antioxidant. Melatonin has been shown to scavenge free radicals generated in mitochondria, reduce electron leakage from the respiratory complexes, and improve adenosine triphosphate (ATP) synthesis (Acuña-Castroviejo et al. 2003; Leon et al. 2005). Moreover, melatonin preserves mitochondrial glutathione levels, thereby enhancing the antioxidant potential (Leon et al. 2004). It was recently demonstrated that melatonin significantly increases SOD activity and reduced glutathione (GSH) levels whereas it decreases retinal lipid peroxidation in the rat retina (Belforte et al. 2010a). By scavenging free radicals, increasing the antioxidant defense system activity, and improving the electron transport chain at the mitochondrial level, melatonin is able to protect ocular tissues from oxidative damage (Lundmark et al. 2006; Siu et al. 2006). In the experimental model of glaucoma induced by injections of HA, we have demonstrated a significant decrease in retinal melatonin levels (Moreno et al. 2004). Taking into account the conclusive evidence on the role of melatonin as antioxidant, together with the fall in retinal melatonin levels and with the decrease in the antioxidant defense system activity in hypertensive eyes, it is tempting to speculate about a causal relationship between these latter phenomena.

Whether retinal oxidative damage is involved in glaucomatous RGC death is far from being understood. Compared with other retinal cells, neonatal ganglion cells are remarkably resistant to cell death induced by superoxide anion, H_2O_2, or hydroxyl radical, and it was postulated that this resistance may be mediated by the possession of sufficient constitutive levels of one or more peroxidases, probably catalase and/or GPX (Kortuem et al. 2000). Thus, it seems possible that a decrease of some of these enzymatic activities may overcome the capacity of these cells to resist oxidative damage. In summary, these results support the involvement of oxidative stress in glaucomatous damage. Thus, manipulation of intracellular redox status using antioxidants such as melatonin may be a new therapeutic strategy to prevent glaucomatous cell death.

17.4 NITROSATIVE STRESS IN GLAUCOMA AND ANTINITRIDERGIC EFFECTS OF MELATONIN

Nitric oxide (NO) is a ubiquitous signaling molecule that participates in a variety of cellular functions. However, in concert with ROS, NO can be transformed into a highly potent and effective cytotoxic entity of pathophysiological significance. In fact, NO modulates the activity of various proteins that contribute to apoptosis (Melino et al. 1997). Furthermore, it was demonstrated that an extracellular proteolytic pathway in the retina contributes to RGC death via NO-activated metalloproteinase-9 (Manabe et al. 2005). Several lines of evidence support a link between NO and glaucoma. In that vein, an increased presence of neuronal nitric oxide synthase (NOS) (NOS-1 or nNOS) and inducible NOS (NOS-2 or iNOS) was reported in astrocytes of the lamina cribrosa and ONH of patients with POAG (Liu and Neufeld 2000; Neufeld et al. 1997). In rats whose extraocular veins were cauterized to produce chronic ocular hypertension and retinal damage, expression of NOS-2 but not NOS-1 increases in ONH astrocytes (Neufeld and Liu 2003; Shareef et al. 1999). Moreover, elevation of hydrostatic pressure in vitro upregulates the expression of NOS-2 in human astrocytes derived from the ONH (Liu and Neufeld 2001). Most importantly, inhibition of NOS-2 by aminoguanidine or L-*N*[6]-[1-iminoethyl]lysine 5-tetrazole amide protects against RGC loss in the rat cautery model of glaucoma (Neufeld 2004; Neufeld et al. 1999). These data support that activation of NOS, especially NOS-2, may play a significant role in glaucomatous optic neuropathy. However, later on, Pang et al. (2005) showed that chronically elevated IOP in the rat induced by episcleral injection of hypertonic saline does not increase NOS-2 immunoreactivity in the optic nerve, ONH, or ganglion cell layer. Moreover, retinal and ONH NOS-2 mRNA levels did not correlate with either IOP level or severity of optic nerve injury. In addition, there was no difference in NOS-2 immunoreactivity in the optic nerve or ONH between POAG and nonglaucomatous eyes (Pang et al. 2005), and aminoguanidine treatment did not affect the development of pressure-induced optic neuropathy in rats (Pang et al. 2005). A significant activation of the retinal nitridergic pathway was described in the experimental model of glaucoma induced by intracameral injections of HA (Belforte et al. 2007). Despite that other studies (mostly based on Western blotting or immunohistochemical analysis) previously addressed the issue of NO involvement in glaucoma, they did not assess changes in the functional capacity of the retinal nitridergic pathway. Although no changes in the levels of NOS isoforms were observed in HA-treated eyes, a significant increase of the retinal arginine to citrulline conversion was demonstrated in HA-injected eyes (Belforte et al. 2007). The intracellular events triggered by ocular hypertension that could explain the increase in retinal NOS activity remain to be established. However, since glutamate acting through *N*-methyl D-aspartate (NMDA) receptors is one of the most conspicuous activators of NOS-1 activity, the raise in glutamate synaptic levels in HA-treated eyes (as discussed later) could account for it. In this sense, it was shown that RGCs in the nNOS-deficient mouse were relatively resistant to NMDA, while damage in the retina of the endothelial NOS-deficient mouse was not distinguishable from that observed in control animals (Vorwerk et al. 1997). Moreover, it was demonstrated that intravitreal injection of NMDA in rats induces accumulation of nitrite/nitrate (El-Remessy et al. 2003).

A significant increase in the retinal uptake of L-arginine (an NOS substrate) was demonstrated in HA-treated eyes. Purified NOS from different sources has been reported to have a low half-saturating L-arginine concentration (EC_{50}) ~10 μM. Since high levels of intracellular L-arginine ranging from 0.1 to 1 mM have been measured in many systems (Block et al. 1995), it is expected that endogenous L-arginine would support maximal activation of NOS. However, a number of in vivo and in vitro studies indicate that NO production under physiological conditions can be increased by extracellular L-arginine, despite saturating intracellular L-arginine concentrations. This has been termed "the arginine paradox" (Kurz and Harrison 1997). One possible explanation could be that intracellular L-arginine is sequestered in one or more pools that are poorly, if at all, accessible to NOS, whereas extracellular L-arginine transported into the cells is preferentially delivered to NO biosynthesis (Kurz and Harrison 1997). Accordingly, it was demonstrated that

L-arginine availability controls NMDA-induced NO synthesis in the rat central nervous system (Grima et al. 1998). Therefore, it seems likely that to induce the activation of NOS, an obligatory influx of L-arginine is required. The coordination between NOS activity and L-arginine uptake has been demonstrated in several systems such as rat brain (Stevens et al. 1996) and diabetic rat retina (do Carmo et al. 1998). A similar coordination between NO biosynthesis and intracellular L-arginine availability seems to occur in hypertensive eyes. It was demonstrated that activation of NMDA receptors in cultured retinal cells promotes an increase of the intracellular L-arginine pool available for NO synthesis (Cossenza et al. 2006). This way, the increase in both NOS activity and L-arginine influx could be triggered by higher levels of synaptic glutamate levels.

Four amino acid transport systems (denoted by y^+, $b^{o,+}$, $B^{o,+}$, or y^+L) have been defined on the basis of substrate specificity and sodium dependence (Deves and Boyd 1998). Only one of them (y^+) is selective for cationic amino acids and sodium-independent. It was demonstrated that the uptake of L-arginine in retinas from rats and hamsters occurs through a transporter resembling the y^+ system (Carmo et al. 1999; Sáenz et al. 2002a). This transport system encompasses three homologous proteins (named cationic amino acid transporter (CAT)-1, CAT-2, and CAT-3) that have been characterized in several tissues. Real time-reverse transcription polymerase chain reaction (RT-PCR) analysis using primers for the aforementioned isoforms demonstrated an increase of mRNAs for both CAT-1 and CAT-2 in retinas from hypertensive eyes, suggesting that ocular hypertension could induce an upregulation of L-arginine transporters (Belforte et al. 2007; Carmo et al. 1999).

It was demonstrated that melatonin inhibits the nitridergic pathway activity in the golden hamster (Sáenz et al. 2002b) and rat retina (Belforte et al. 2010a). Melatonin significantly decreases retinal NOS activity and L-arginine uptake and inhibits the accumulation of cGMP induced by both L-arginine and an NO donor. The inhibitory effect of melatonin on retinal NOS activity is consistent with the previously described effect of melatonin on this enzyme from other neural structures (Bettahi et al. 1996; Leon et al. 1998; Pozo et al. 1997). However, while the effect of melatonin in those tissues was evident up to 1 nM, a much higher sensitivity to the methoxyindole was evident in the hamster retina, since it is effective even at 1 pM, suggesting that the retinal nitridergic pathway is regulated by physiological concentrations of melatonin (Saenz et al. 2002b). In addition to inhibiting NOS activity, melatonin is able to directly scavenge NO, generating at least one stable product, that is, *N*-nitrosomelatonin (Turjanski et al. 2000). Moreover, melatonin reduces NO-induced lipid peroxidation in rat retinal homogenates and ileum tissue sections (Cuzzocrea et al. 2000; Siu et al. 1999). Taken together, these results indicate that melatonin modulates the nitridergic pathway in an opposite way to that induced by ocular hypertension.

17.5 EXCITOTOXICITY IN GLAUCOMA AND THE EFFECT OF MELATONIN ON GLUTAMATE RETINAL SYNAPTIC LEVELS

Glutamate is the main excitatory neurotransmitter in the retina, but it is toxic when present in excessive amounts. Retinal tissue is, in fact, an established paradigm for glutamate neurotoxicity for several reasons: different insults lead to accumulation of relatively high levels of glutamate in the extracellular fluid (Louzada-Junior et al. 1992), administration of glutamate leads to neuronal cell death (David et al. 1988), and glutamate receptor antagonists can protect against neuronal degeneration (Mosinger et al. 1991). Thus, an appropriate clearance of synaptic glutamate is required for the normal function of retinal excitatory synapses and for the prevention of neurotoxicity. Glial cells, mainly astrocytes and Müller glia, surround glutamatergic synapses and express glutamate transporters and the glutamate-metabolizing enzyme, glutamine synthetase (GS) (Riepe and Norenburg 1977; Sarthy and Lam 1978). Glutamate is transported into glial cells and amidated by GS to the nontoxic aminoacid glutamine. Glutamine is then released by the glial cells and taken up by neurons, where it is hydrolyzed by glutaminase to form glutamate again, completing the retinal glutamate/glutamine cycle (Sarthy and Lam 1978; Thoreson and Witkovsky 1999). In this way, the neurotransmitter pool is replenished, and glutamate neurotoxicity is prevented. Glutamatergic injury has been proposed to contribute to the RGC death in glaucoma. This hypothesis is supported by the demonstration that vitreal glutamate is elevated in glaucomatous dogs (Brooks et al. 1997) and quail with congenital glaucoma (Brooks et al. 1997; Dkhissi et al. 1999). In addition, high glutamine levels have been found in retinal Müller cells of glaucomatous rat eyes (Brooks et al. 1997; Shen et al. 2004). In contrast, other authors showed no significant elevation of glutamate in the vitreous of patients with glaucoma (Honkanen et al. 2003), or in rats (Levkovitch-Verbin et al. 2002), and monkeys with experimental glaucoma (Carter-Dawson et al. 2002; Wamsley et al. 2005). In any case, the assumption that high levels of glutamate in the vitreous are a necessary condition for the involvement of excitotoxicity in glaucomatous neuropathy is limited. The local concentration of glutamate at the membrane receptors of RGCs is the important issue for toxicity. This could be very different from the level in samples of vitreous. Vitreous humor must be removed for experimental measurement by a process that inevitably disturbs its state before removal. These manipulations could themselves alter the measured amount of glutamate. At present, there are no available tools to directly assess retinal glutamate synaptic concentrations in vivo. However, glutamate synaptic concentrations could be estimated by studying the retinal mechanisms that regulate glutamate clearance and recycling. We have demonstrated a significant alteration of the retinal glutamate/glutamine cycle activity in rats exposed to experimentally elevated IOP (Moreno et al. 2005a). Since no enzymes exist extracellularly that degrade glutamate, glutamate transporters are responsible for maintaining low synaptic glutamate concentrations. Retinal glutamate uptake significantly decreases in HA-treated eyes. In agreement, a significant reduction in the amount of the main retinal glutamate transporter (excitatory amino acid transporter type 1 [EAAT-1]) assessed by Western blot analysis in a rat glaucoma model (Martin et al. 2002) and a downregulation of this transporter in retinal Müller cells from glaucoma patients (Naskar et al. 2000) were demonstrated. While these studies did not assess changes in the functional capacity of glutamate transporters, our results demonstrated a removing glutamate disability in retinas from hypertensive eyes. The synaptically released glutamate is taken up into glial cells, where GS converts it into glutamine. Since Müller cells rapidly convert glutamate to glutamine, the driving force for glutamate uptake would be stronger in these cells than in neurons, which have much higher intracellular free glutamate concentrations (Pow and Robinson 1994). In fact, although glutamate uptake is controlled by the expression and posttranslational modifications, physiological measurements suggest that glutamate uptake may also depend on its metabolism (Gegelashvili and Schousboe 1998; Tanaka 2000; Tanaka et al. 2007). Indeed, an increase in internal glutamate concentrations significantly slows down the net transport of glutamate, and it was suggested that instantaneous intracellular glutamate metabolism may be needed for efficient glutamate clearance of the extracellular milieu (Attwell et al. 1993; Otis and Jahr 1998). Thus, a decrease in GS activity could account for a decrease in glutamate uptake. Glutamine is released from Müller cells and could be a precursor for neuronal glutamate synthesis. The increase in the basal release and the uptake of glutamine in HA-treated eyes could provoke a raise in the availability of substratum for glutamate synthesis. Moreover, this increase in glutamate production could be further potentiated by the augment of GS. Decreasing the levels of expression of the EAAT1 increases vitreal glutamate, which is toxic to RGCs (Vorwerk et al. 2000). Thus, the decrease in glutamate influx could provoke an increase in synaptic glutamate levels. In addition, a decrease of GS activity, as well as an increase in glutaminase activity in retinas form hypertensive eyes, could contribute synergically and/or redundantly to an excessive increase in synaptic glutamate levels (Moreno et al. 2008).

Nanomolar concentrations of melatonin significantly modulate the glutamate/glutamine cycle activity in the golden hamster (Saenz et al. 2004) and rat retina (Belforte et al. 2010a). In that sense, it was demonstrated that low concentrations of melatonin significantly increase retinal glutamate uptake and GS activity and decrease glutaminase activity. This way, melatonin may contribute to the conversion of glutamate to glutamine through a possibly redundant mechanism. The physiological consequences of a modulation by melatonin of the retinal glutamate/glutamine cycle are yet to be determined, although this effect could provide new insights into the neuroprotective potential of melatonin. In that respect, it was demonstrated that an increase in GS provides neuroprotection in experimental models of neurodegeneration (Gorovits et al. 1997; Heidinger et al. 1999). Induction of GS in vivo or in vitro by glucocorticoids was clearly demonstrated in different tissues, including the retina (Patel et al. 1983; Sarkar and Chaudhury 1983). Physiological levels of glucocorticoids regulate GS expression by stimulating the gene transcription. This effect of glucocorticoids has been associated to their ability to protect against neuronal degeneration (Gorovits et al. 1997), as shown in animal models of brain injury (Hall 1985), as well as after retinal photic injury (Rosner et al. 1992). However, since the induction of GS expression by glucocorticoids takes about 24 h, there are some potential weaknesses in glucocorticoid treatment. In contrast, since the effect of melatonin is much faster (in the range of minutes), a treatment with the methoxyindole may circumvent this obstacle. Furthermore, this beneficial effect of melatonin may be further improved by its effect on glutamate uptake, and glutaminase activity. In summary, these findings suggest that a treatment with melatonin could be considered as a new approach to handling glutamate-mediated neuronal degeneration, such as that induced by glaucoma.

17.6 GABAERGIC DYSFUNCTION IN GLAUCOMA AND THE EFFECT OF MELATONIN ON THE RETINAL GABAERGIC SYSTEM

The neurochemical organization maintained throughout vertebrate retinas is that glutamate is the neurotransmitter in the photoreceptor cell → bipolar cell → ganglion cell chain, whereas GABA is used by numerous horizontal and amacrine cells in the lateral pathway, modulating neural transmission in both synaptic layers (Kalloniatis and Tomisich 1999; Yang 2004). Retinal output neurons communicate by liberating glutamate, while tonically or phasically active inhibitory neurons (mostly GABAergic) modulate the passage of information, offering resistance against the firing tendencies, which results in variable levels of neural activity. The prevailing view is that the balance between excitatory and inhibitory signaling plays a pivotal role in mechanisms underlying the modulation and maintenance of a variety of retinal functions and sensory information encoding. In fact, the loss of this balance could provoke cell death. Despite the putative involvement of glutamate in glaucomatous cell death previously discussed, and the key role of GABA in retinal function, the GABAergic activity was not extensively examined in experimental models of glaucoma. Recently, a significant dysfunction of the retinal GABAergic system was demonstrated in rats exposed to experimentally elevated IOP (Moreno et al. 2008). These results indicate that retinal GABA steady-state concentrations, GABA turnover rate, glutamic acid decarboxylase (GAD) activity, GABA transporters (GATs), and GABA receptors are susceptible to ocular hypertension.

Retinal GABA release involves two distinct components; one requires extracellular calcium, while the other is calcium-independent and involves reversal of the operating direction of a high affinity GABA carrier (Schwartz 1987). Although differences in retinal GABA release were observed among species, glutamate induces GABA release mostly via a Ca^{2+}-independent and Na^{+}-dependent carrier mechanism, while high K^{+}-induced GABA release is partially calcium-dependent (Andrade da Costa et al. 2000; do Nascimento et al. 1998; Lopez-Costa et al. 1999). Retinal GABA release induced by both stimuli significantly decreases in the retina from HA-injected eyes, supporting that ocular hypertension affects the carrier-mediated component. As significant changes in GABA influx are also observed in HA-treated eyes (Moreno et al. 2008), ocular hypertension may influence the preferred direction of GATs provoking a switch from release to uptake. Indeed, since GAD activity significantly decreases in HA-treated eyes, the driving force for GABA uptake would be stronger in cells with lower intracellular GABA concentrations, favoring its uptake over its release.

GATs in cells surrounding the release site are responsible for terminating GABA signals within the retina (Gadea and Lopez-Colome 2001). GATs were mainly localized in amacrine, displaced amacrine, RGCs, and Müller cells (Honda et al. 1995; Johnson et al. 1996). GABA uptake is significantly higher in retinas from eyes injected with HA, which could contribute synergically and/or redundantly to a significant decrease of synaptic GABA levels in retinas exposed to ocular hypertension.

To our knowledge, there is no other retinal pathological condition in which a GABAergic deficit has been previously described. However, a decrease in GABAergic neurotransmission has been implicated in the pathophysiology of several central nervous system disorders, particularly in epilepsy. In this case, excessive glutamate-mediated neurotransmission and impaired GABA-mediated inhibition, among other mechanisms, may trigger a cascade of events leading to neuronal damage and cell death. Excitotoxicity is a common pathogenic mechanism in neurodegenerative diseases, including probably glaucoma, which may result from the failure of normal compensatory antiexcitatory mechanisms, necessary to maintain cellular homeostasis. An abnormal glutamate outflow may play a crucial role in triggering cellular events leading to excitotoxic neuronal death. Reduced synaptic inhibition as a result of a GABAergic dysfunction could be one of the major causes for this imbalance. In this way, ocular hypertension may greatly shift the balance between retinal excitation and inhibition. As mentioned before, we have demonstrated an increase in glutamate synaptic concentrations in retinas from eyes injected with HA (Moreno et al. 2008). Since an increase in synaptic glutamate or its residence time in the synaptic region is toxic to RGCs (Kawasaki et al. 2000; Nucci et al. 2005), the decrease in GABAergic activity could worse glutamate toxicity. In this way, the deficit in GABAergic activity could contribute to glaucomatous neuropathy.

It was demonstrated that melatonin increases retinal GABA levels, as shown by its effect on GABA turnover rate and GAD activity (Belforte et al. 2010a). The effect of melatonin on the GABAergic activity seems not to be exclusive for the retina, since it was previously demonstrated that melatonin increases GABA turnover rate and GAD activity in rat hypothalamus, cerebellum, and cerebral cortex (Rosenstein et al. 1986, 1989). These results indicate that glaucomatous damage may involve a decrease in GABA synaptic levels, which can be restored by a treatment with melatonin.

17.7 EFFECT OF MELATONIN ON GLAUCOMATOUS DAMAGE

As already discussed in this chapter, previous results indicate that melatonin is able to impair retinal glutamate neurotoxicity, decrease NO levels, increase GABA concentrations, and reduce oxidative stress, and that an increase in glutamate and NO levels, a decrease in the GABAergic activity, and oxidative damage could be involved in glaucomatous neuropathy. Thus, it seems likely that melatonin could have a beneficial effect against glaucoma. This hypothesis was analyzed in the glaucoma model induced by HA injections. For this purpose, a pellet of melatonin was implanted subcutaneously 24 h before the first injection of HA. Melatonin, which did not affect IOP, prevented the effect of ocular hypertension on retinal function (assessed by electroretinography) and diminished the vulnerability of RGCs to the deleterious effects of ocular hypertension (Belforte et al. 2010a). Although melatonin conferred neuroprotection in the experimental model of glaucoma induced by HA, the translational relevance of this result is limited by the fact that melatonin was administered before the induction of ocular hypertension (e.g., 24 h before the first injection of HA).

Human open-angle glaucoma is a progressive optic neuropathy. In agreement, different stages were identified in the experimental model of glaucoma induced by HA, showing the following characteristics:

(i) Three weeks of ocular hypertension: increase in glutamate and NO levels, decrease in GABA levels, and incipient oxidative stress, without changes in the ERG and retinal morphology (e.g., *asymptomatic ocular hypertension*)
(ii) Six weeks of ocular hypertension: increase in NO levels, higher oxidative stress, and decrease in the ERG without histological changes (e.g., *moderated glaucoma*)
(iii) Ten weeks of ocular hypertension: higher oxidative stress, further decrease in the ERG (vs. 6 weeks), and loss of RGCs and optic nerve fibers (e.g., *advanced glaucoma*) (Moreno et al. 2005b).

To analyze the therapeutic effect of melatonin, a pellet of the methoxyindole was implanted at 6 weeks of ocular hypertension, a time point in which functional alterations are already evident. The results indicate that the delayed treatment with melatonin of eyes with ocular hypertension resulted in similar protection when compared with eyes treated from the onset of ocular hypertension. We do not have any clear explanation for these results. We have shown that an increase in synaptic glutamate concentrations (Moreno et al. 2005a) and NO production, as well as a GABAergic dysfunction (Moreno et al. 2005a), occurs mostly prior to 6 weeks of ocular hypertension induced by HA. Based on these results, it seems possible that alterations in glutamate, NO, and GABA may trigger an initial insult responsible for initiation of damage that is followed by a slower secondary degeneration that ultimately results in cell death. In that sense, we showed that oxidative stress is a longer lasting phenomenon, which can be observed even at 10 weeks of ocular hypertension (Moreno et al. 2004). In this scenario, the preventive effect of melatonin (shown by the administration of melatonin before the first injection of HA) could be explained by the decrease in glutamate and NO levels and an increase in GABA concentrations, while its therapeutic effect (shown by the administration of melatonin at 6 weeks of ocular hypertension) can be explained essentially by its antioxidant effect. In this context, the fact that melatonin was similarly effective in the chronically treated animals and the delayed treatment could support the hypothesis that in both cases melatonin is able to reverse oxidative damage, which could be a key factor in glaucomatous dysfunction and cell death.

Neuroprotection in glaucoma implies the use of drugs or chemicals to slow down whatever causes loss of vision (RGC death), without influencing IOP. In order to be effective, a neuroprotectant must reach the ONH and/or RGCs and will therefore probably have to be taken orally (Osborne et al. 1999). Because it will reach other parts of the body, any side effect of an appropriate neuroprotectant must be reduced to a minimum. Melatonin, a very safe compound for human use, is highly lipophilic and readily diffuses into tissues. In fact, subcutaneously administered, melatonin reaches the retina (Sande et al. 2008), increasing the local levels of the methoxyindole. RGCs are induced to die by different triggers in glaucoma, suggesting that neuroprotectants with multiple modes of actions are likely be effective in the therapeutic management of glaucoma (Marcic et al. 2003).

Besides the mechanisms already described, there are other beneficial mechanisms of melatonin that can support its usefulness for glaucoma treatment. Several lines of evidence support that the obstruction of retrograde transport at the ONH results in the deprivation of neurotrophic support to RGCs, leading to apoptotic cell death in glaucoma (Tang et al. 2006). An important corollary to this concept is the implication that appropriate enhancement of neurotrophic support will prolong the survival of injured RGCs. Of particular importance is the fact that brain-derived neurotrophic factor (BDNF) not only promotes RGC survival following damage to the ON but also helps to preserve the structural integrity of the surviving neurons, which in turn results in enhanced visual function (Weber et al. 2008). As for the link between melatonin and neurotrophins, it has been suggested that melatonin may participate in neurodevelopment and in the regulation of neurotrophic factors (Jimenez-Jorge et al. 2007; Niles et al. 2004). In vitro, melatonin promotes the viability and neuronal differentiation of neural stem cells and increases their production of BDNF (Kong et al. 2008). Moreover, ramelteon (a melatonin receptor agonist) is capable of increasing BDNF protein in primary cultures of cerebellar granule cells (Imbesi et al. 2008).

In addition to ocular hypertension, the majority of glaucoma patients show signs of reduced ocular blood flow as well as ischemic signs in the eye, supporting that hemodynamic factors are involved as well in glaucomatous neuropathy. Several animal and human studies have indicated that vascular dysregulation and ischemia play a role in glaucoma pathogenesis (Flammer et al. 2002; Fuchsjager-Mayrl et al. 2004; Galassi et al. 2003; Zink et al. 2003). Retinal ischemia develops when retinal blood flow is insufficient to match the metabolic needs of the retina, one of the highest oxygen-consuming tissues. Ischemia impairs retinal energy metabolism and triggers a reaction cascade that can result in cell death. Oxidative stress, excitotoxicity, calcium influx, and other mechanisms acting in tandem are of considerable importance in retinal ischemic damage (Osborne et al. 2004). Notably, most of these mechanisms are also involved in glaucomatous neuropathy

(Moreno et al. 2004). In this sense, it was shown that melatonin could increase the survival rate and rescue and restore injured RGCs in an experimental model of ischemia/reperfusion in rats (Tang et al. 2006), and it counteracts ischemia-induced apoptosis in human retinal pigment epithelial cells (Osborne et al. 2004). Finally, while the cellular mechanisms involved in RGC loss observed in glaucomatous neuropathy are based on apoptosis phenomenon, melatonin was shown to have antiapoptotic properties acting through several mechanisms, such as reduction of caspases, cytochrome c release, and modulation of Bcl-2 and Bax genes, among others. Taken together, these data indicate that melatonin seems to fulfill all the requirements to be considered a promissory neuroprotectant for glaucoma treatment. Alone or combined with an ocular hypotensive therapy, a treatment with melatonin, a very safe compound for human use, could be a new therapeutic tool helping the challenge faced by ophthalmologists treating glaucoma.

17.8 GLAUCOMA, CIRCADIAN RHYTHMS, AND MELATONIN

Circadian rhythms are controlled by a biological clock that in humans is located in the suprachiasmatic nuclei (SCN) and have a period of approximately 24 h (Moore and Silver 1998). Because SCN neurons contain an internal pacemaker that generates an endogenous rhythmic electrical activity (Buijs and Kalsbeek 2001), changes supervised by the SCN occur even in the absence of external stimuli. Nevertheless, the SCN are influenced by environmental changes, especially by the light–dark cycle (Golombek and Rosenstein 2010). The inherited period of the human circadian clock is not precisely 24 h; however, the light–dark and other cyclic environmental and behavioral clues modify the period to precisely 24 h, thereby supporting the circadian daytime activity–nocturnal sleep routine, among many other biological rhythms (Klerman et al. 1998; Duffy and Czeisler 2009).

Although the eye has been largely recognized as the organ of sight (in man and animals), over the past decade, a second role for the eye has been uncovered: even in the absence of *cognitive* vision, the eye can serve as a sensor for ambient lighting, akin to the light meter in a camera (Van Gelder et al. 2003). A host of light-regulated functions, including entrainment of circadian clocks, suppression of activity by light, photic suppression of pineal melatonin synthesis, and pupillary light responses are retained in animals that are blind as a result of mutations causing complete or near-complete degeneration of the classical photoreceptors, the rods and cones (Freedman et al. 1999; Lucas et al. 1999, 2001). These light-regulated functions are controlled by nonclassical retinal photoreceptors because animals lacking RGCs lose circadian photoresponses, behavioral masking, and pupillary light responses (Wee et al. 2002). The identification of intrinsically photosensitive RGCs has given to the nonvisual phototransduction an anatomical basis (Berson et al. 2002; Hattar et al. 2002; Provencio et al. 2002). A subtype of RGCs (mRGCs) containing a novel photopigment, melanopsin, responds to light independently of the input from rods and cones and project their axons through the retinohypothalamic tract to the SCN. A strong body of evidence supports that mRGCs are involved in modulating circadian rhythms (Mrosovsky and Hattar 2003; Panda et al. 2002; Ruby et al. 2002). In addition, mRGCs send monosynaptic projections to the olivary pretectal nucleus, responsible for the pupil light reflex (PLR) (Hattar et al. 2006). In that sense, although it was originally assumed that the light-evoked neural signals driving the PLR originated exclusively from rods and cones, it was later shown that mRGCs significantly contribute to the pupil constriction (Kardon et al. 2009; Lucas et al. 2003; Nissen et al. 2011).

Most studies on glaucoma have focused on RGCs involved in conscious visual functions and have found that a chronic increase in IOP causes a progressive loss of these cells. However, clinical studies in glaucomatous patients show a high prevalence of sleep-disordered breathing, characterized by snoring, excessive daytime sleepiness, and insomnia accompanied by large swings in blood pressure and repetitive hypoxic periods during sleep (Onen et al. 2000). Moreover, an abnormal light-induced nocturnal melatonin suppression in glaucoma patients was described (Perez-Rico et al. 2010), and afferent pupillary defects during the early stages of the disease were reported

(Kalaboukhova et al. 2007; Kohn et al. 1979). In addition, it was recently shown that the postillumination pupil response is dysfunctional in patients with early, moderate, or severe glaucoma (Feigl et al. 2012).

The involvement of mRGCs in glaucoma has been postulated, and two studies in animal models of glaucoma demonstrate a loss of mRGCs (Drouyer et al. 2008; Wang et al. 2008). These clues imply that mRGCs as other RGCs can be damaged in glaucomatous patients. However, other studies have shown that mRGCs may be spared in some animal models of chronic ocular hypertension (Jakobs et al. 2005; Li et al. 2006). In an experimental model of glaucoma in rats induced by chronic injections of chondroitin sulfate (CS) in the eye anterior chamber (Belforte et al. 2010b), we have recently demonstrated that a similar decrease in the number of RGCs projecting to the superior colliculus (which are involved in the image-forming visual system) and in the number of mRGCs (which participate in the non-image-forming visual system) occurs in glaucomatous eyes. In addition, we have demonstrated that experimental glaucoma induces a significant decrease in the afferent PLR and alterations in the light-induced nocturnal melatonin suppression (de Zavalía et al. 2011). These results indicate that experimental glaucoma can affect the non-image-forming visual system.

Notwithstanding, further studies are warranted to clarify the involvement of mRGCs in the natural history of human glaucoma and to investigate the occurrence of circadian rhythm abnormalities in patients with advanced glaucoma. In this regard, we have recently analyzed the circadian rhythm in locomotor activity and the sleep/wake cycle, estimated by wrist actigraphy, in patients with advanced stages of glaucoma (Lanzani et al. 2012). The results of this pilot study indicate significant changes in the sleep pattern of advanced glaucoma patients, which showed more awakenings at night, and a decrease in the amount of sleep duration and sleep efficiency, as compared with control subjects. Although no significant changes were found for circadian parameters such as period, amplitude, or acrophase of locomotor activity, the general pattern of actimetry, from which sleep parameters were estimated, indicated alterations in its robustness, suggesting a disrupted sleep–wake cycle. In particular, clear differences between control and glaucoma patients were observed in the sleep, but not in the wakefulness state, which supports that circadian misalignment in glaucomatous patients could be attributed to sleep disturbances.

Although several groups have examined the involvement of mRGCs or melanopsin in experimental glaucoma in rodents (Drouyer et al. 2008; Jakobs et al. 2005; Wang et al. 2008), to date, few studies investigated the mRGCs system directly in human glaucoma (Kankipati et al. 2011; Perez-Rico et al. 2010). In that sense, a disturbance of the autonomic nervous system circadian rhythm was shown in patients with glaucoma (Kashiwagi et al. 2000). It was demonstrated that the light-induced suppression of melatonin secretion is substantially unaffected in other neurodegenerative optic neuropathies such as Leber hereditary optic neuropathy and dominant optic atrophy (La Morgia et al. 2011), in which the PLR is also preserved (Bremner et al. 2001; Wakakura and Yokoe 1995). Thus, it is tempting to speculate that mRGCs are particularly susceptible to glaucomatous damage and that glaucoma is a key model that needs to be further investigated in regard to circadian photoreception in optic neuropathies. These results suggest that the effects of glaucoma on the circadian timing system might be twofold: (1) a direct impact through degeneration of RGCs and eventually, also of mRGCs; and (2) an indirect impact through social isolation due to blindness, as is the case for other ophthalmic diseases (Wee et al. 2002).

Although these findings should be replicated in large-scale studies before definitive conclusions can be reached, these results are consistent with sleep alterations in advanced glaucoma, suggesting that attention should be paid to non-image-forming visual functions, such as control of circadian rhythms and its clinical impact in patients with glaucoma, even when most of them retain some residual visual function.

Besides the already discussed beneficial effects of melatonin on retinal glaucomatous damage, a treatment with melatonin could also be useful to treat sleep alterations in patients with glaucoma. Treatment of circadian rhythm disorders, whether precipitated by intrinsic factors (e.g., sleep disorders, blindness, mental disorders, aging) or by extrinsic factors (e.g., shift work, jet lag) has led to the development of a new type of agents called "chronobiotics." The term "chronobiotic" defines a substance displaying the therapeutic activity of shifting the phase or increasing the amplitude of the circadian rhythms. The prototype of this therapeutic group is melatonin, whose administration synchronizes the sleep–wake cycle in blind people and in individuals suffering from circadian rhythm sleep disorders, like delayed sleep phase syndrome, jet lag, or shift work (Cardinali et al. 2011). Melatonin and melatoninergic drugs have hypnotic effects mediated through specific receptors, especially those in the SCN, which acts on the hypothalamic sleep switch. They favor sleep initiation and reset the circadian clock, allowing persistent sleep, a requirement in circadian rhythm sleep alterations. The action of melatonin on sleep is mainly of a chronobiological nature; melatonin acts in a dual way, resetting the circadian clock and suppressing neuronal firing (Cardinali et al. 2011; Carpentieri et al. 2012; Hardeland 2008). In this regard, it was demonstrated that melatonin is effective in advancing sleep–wake rhythm and endogenous melatonin rhythm in delayed sleep phase disorder (van Geijlswijk et al. 2010) and that melatonin administration significantly improves sleep and behavioral disorders in the elderly and facilitates discontinuation of therapy with conventional hypnotic drugs (Garzon et al. 2009).

17.9 CONCLUSIONS

Glaucoma may affect the quality of life in several ways. These include the visual effects of the disease itself (decreased visual field), the psychological effects of diagnosis (specifically fear of blindness), the potential side effects of treatment (either medical or surgical), and financial effects (as the cost of visits and therapy). Recent results suggest another risk to the quality of life of patients with glaucoma, that is, alterations of the circadian physiology. Circadian rhythm disorders may include poor concentration, sleep problems, impaired performance, decrease in cognitive skills, poor psychomotor coordination, and headaches, among many others. Based on all the considerations discussed in this chapter, melatonin appears as a highly attractive resource for glaucoma treatment, because besides its properties as a retinal neuroprotectant, it could also contribute to restore circadian balance and, in this way, significantly help to improve the quality of life of patients with this ocular disease.

REFERENCES

Acuña-Castroviejo, D., Escames, G., León, J., Carazo, A., and H. Khaldy. 2003. Mitochondrial regulation by melatonin and its metabolites. *Adv Exp Med Biol* 527:549–557.

Adler, L.J., Gyulai, F.E., Diehl, D.J., Mintun, M.A., Winter, P.M., and L.L. Firestone. 1997. Regional brain activity changes associated with fentanyl analgesia elucidated by positron emission tomography. *Anesth Analg* 84:120–126.

Almasieh, M., Wilson, A.M., Morquette, B., Cueva Vargas, J.L., and P.A. Di. 2012. The molecular basis of retinal ganglion cell death in glaucoma. *Prog Retin Eye Res* 31:152–181.

Andrade da Costa, B.L., de Mello, F.G., and J.N. Hokoc. 2000. Transporter-mediated GABA release induced by excitatory amino acid agonist is associated with GAD-67 but not GAD-65 immunoreactive cells of the primate retina. *Brain Res* 863:132–142.

Armstrong, D., Santangelo, G., and E. Connole. 1981. The distribution of peroxide regulating enzymes in the canine eye. *Curr Eye Res* 1:225–242.

Aslan, M., Cort, A., and I. Yucel. 2008. Oxidative and nitrative stress markers in glaucoma. *Free Radic Biol Med* 45:367–376.

Attwell, D., Barbour, B., and M. Szatkowski. 1993. Nonvesicular release of neurotransmitter. *Neuron* 11:401–407.

Belforte, N., Moreno, M.C., Cymeryng, C., Bordone, M., Keller Sarmiento, M.I., and R.E. Rosenstein. 2007. Effect of ocular hypertension on retinal nitridergic pathway activity. *Invest Ophthalmol Vis Sci* 48:2127–2133.

Belforte, N.A., Moreno, M.C., de Zavalía, N. et al. 2010a. Melatonin: A novel neuroprotectant for the treatment of glaucoma. *J Pineal Res* 48:353–364.

Belforte, N., Sande, P, de Zavalía, N., Knepper, P.A., and R.E. Rosenstein. 2010b. Effect of chondroitin sulfate on intraocular pressure in rats. *Invest Ophthalmol Vis Sci* 51:5768–5775.

Benozzi, J., Nahum, L.P., Campanelli, J.L., and R.E. Rosenstein. 2002. Effect of hyaluronic acid on intraocular pressure in rats. *Invest Ophthalmol Vis Sci* 43:2196–2200.

Berson, D.M., Dunn, F.A., and M. Takao. 2002. Phototransduction by retinal ganglion cells that set the circadian clock. *Science* 295:1070–1073.

Besharse, J.C. and D.A. Dunis. 1983. Methoxyindoles and photoreceptor metabolism: Activation of rod shedding. *Science* 219:1341–1343.

Bettahi, I., Pozo, D., Osuna, C., Reiter, R.J., Acuña-Castroviejo, D., and J.M. Guerrero. 1996. Melatonin reduces nitric oxide synthase activity in rat hypothalamus. *J Pineal Res* 20:205–210.

Block, E.R., Herrera, H., and M. Couch. 1995. Hypoxia inhibits L-arginine uptake by pulmonary artery endothelial cells. *Am J Physiol* 269:L574–L580.

Bremner, F.D., Tomlin, E.A., Shallo-Hoffmann, J., Votruba, M., and S.E. Smith. 2001. The pupil in dominant optic atrophy. *Invest Ophthalmol Vis Sci* 42:675–678.

Brooks, D.E., Garcia, G.A., Dreyer, E.B., Zurakowski, D., and R.E. Franco-Bourland. 1997. Vitreous body glutamate concentration in dogs with glaucoma. *Am J Vet Res* 58:864–867.

Buijs, R.M. and A. Kalsbeek. 2001. Hypothalamic integration of central and peripheral clocks. *Nat Rev Neurosci* 2:521–526.

Cahill, G.M. and J.C. Besharse. 1993. Circadian clock functions localized in xenopus retinal photoreceptors. *Neuron* 10:573–577.

Caprioli, J. 1997. Neuroprotection of the optic nerve in glaucoma. *Acta Ophthalmol Scand* 75:364–367.

Cardinali, D.P., Furio, A.M., and L.I. Brusco. 2011. The use of chronobiotics in the resynchronization of the sleep/wake cycle. Therapeutical application in the early phases of Alzheimer's disease. *Recent Pat Endocr Metab Immune Drug Discov* 5:80–90.

Carmo, A., Cunha-Vaz, J.G., Carvalho, A.P., and M.C. Lopes. 1999. L-arginine transport in retinas from streptozotocin diabetic rats: Correlation with the level of IL-1 beta and NO synthase activity. *Vision Res* 39:3817–3823.

Carpentieri, A., Diaz de Barboza, G., Areco, V., Peralta López, M., and N. Tolosa de Talamoni. 2012. New perspectives in melatonin uses. *Pharmacol Res* 65:437–444.

Carter-Dawson, L., Crawford, M.L., Harwerth, R.S. et al. 2002. Vitreal glutamate concentration in monkeys with experimental glaucoma. *Invest Ophthalmol Vis Sci* 43:2633–2637.

Castorina, C., Campisi, A., Di, G.C., Sorrenti, V., Russo, A., and A. Vanella. 1992. Lipid peroxidation and antioxidant enzymatic systems in rat retina as a function of age. *Neurochem Res* 17:599–604.

Cossenza, M., Cadilhe, D.V., Coutinho, R.N., and R. Paes-de-Carvalho. 2006. Inhibition of protein synthesis by activation of NMDA receptors in cultured retinal cells: A new mechanism for the regulation of nitric oxide production. *J Neurochem* 97:1481–1493.

Cuzzocrea, S., Costantino, G., Mazzon, E., Micali, A., De Sarro, A., and A.P. Caputi. 2000. Beneficial effects of melatonin in a rat model of splanchnic artery occlusion and reperfusion. *J Pineal Res* 28:52–63.

David, P., Lusky, M., and V.I. Teichberg. 1988. Involvement of excitatory neurotransmitters in the damage produced in chick embryo retinas by anoxia and extracellular high potassium. *Exp Eye Res* 46:657–662.

Deves, R. and C.A. Boyd. 1998. Transporters for cationic amino acids in animal cells: Discovery, structure, and function. *Physiol Rev* 78:487–545.

de Zavalía, N., Plano, S.A., Fernandez, D.C. et al. 2011. Effect of experimental glaucoma on the non-image forming visual system. *J Neurochem* 117:904–914.

Dkhissi, O., Chanut, E., Wasowicz, M. et al. 1999. Retinal TUNEL-positive cells and high glutamate levels in vitreous humor of mutant quail with a glaucoma-like disorder. *Invest Ophthalmol Vis Sci* 40:990–995.

do Carmo, A., Lopes, C., Santos, M., Proença, R., Cunha-Vaz, J., and A.P. Carvalho. 1998. Nitric oxide synthase activity and L-arginine metabolism in the retinas from streptozotocin-induced diabetic rats. *Gen Pharmacol* 30:319–324.

do Nascimento, J.L., Ventura, A.L., and R. Paes de Carvalho. 1998. Veratridine- and glutamate-induced release of [3*H*]-GABA from cultured chick retina cells: Possible involvement of a GAT-1-like subtype of GABA transporter. *Brain Res* 798:217–222.

Drouyer, E., Dkhissi-Benyahya, O., Chiquet, C. et al. 2008. Glaucoma alters the circadian timing system. *PLoS One* 3:e3931.

Dubocovich, M.L. 1983. Melatonin is a potent modulator of dopamine release in the retina 1. *Nature* 306:782–784.

Duffy, J.F. and C.A. Czeisler. 2009. Effect of light on human circadian physiology. *Sleep Med Clin* 4:165–177.

El-Remessy, A.B., Khalil, I.E., Matragoon, S. et al. 2003. Neuroprotective effect of (−)Delta9-tetrahydrocannabinol and cannabidiol in *N*-methyl-D-aspartate-induced retinal neurotoxicity: Involvement of peroxynitrite. *Am J Pathol* 163:1997–2008.

Feigl, B., Zele, A.J., Fader, S.M. et al. 2012. The post-illumination pupil response of melanopsin-expressing intrinsically photosensitive retinal ganglion cells in diabetes. *Acta Ophthalmol* 90:e230–e234.
Flammer, J., Orgul, S., Costa, V.P. et al. 2002. The impact of ocular blood flow in glaucoma. *Prog Retin Eye Res* 21:359–393.
Freedman, M.S., Lucas, R.J., Soni, B. et al. 1999. Regulation of mammalian circadian behavior by non-rod, non-cone, ocular photoreceptors. *Science* 284:502–504.
Friedman, D.S., Wilson, M.R., Liebmann, J.M., Fechtner, R.D., and R.N. Weinreb. 2004. An evidence-based assessment of risk factors for the progression of ocular hypertension and glaucoma. *Am J Ophthalmol* 138:S19–S31.
Fuchsjager-Mayrl, G., Wally, B., Georgopoulos, M. et al. 2004. Ocular blood flow and systemic blood pressure in patients with primary open-angle glaucoma and ocular hypertension. *Invest Ophthalmol Vis Sci* 45:834–839.
Gadea, A. and A.M. Lopez-Colome. 2001. Glial transporters for glutamate, glycine, and GABA: II. GABA transporters. *J Neurosci Res* 63:461–468.
Galassi, F., Sodi, A., Ucci, F., Renieri, G., Pieri, B., and M. Baccini. 2003. Ocular hemodynamics and glaucoma prognosis: A color Doppler imaging study. *Arch Ophthalmol* 121:1711–1715.
Garbarino-Pico, E., Carpentieri, A.R., Contin, M.A. et al. 2004. Retinal ganglion cells are autonomous circadian oscillators synthesizing *N*-acetylserotonin during the day. *J Biol Chem* 279:51172–51181.
Garzon, C., Guerrero, J.M., Aramburu, O., and T. Guzman. 2009. Effect of melatonin administration on sleep, behavioral disorders and hypnotic drug discontinuation in the elderly: A randomized, double-blind, placebo-controlled study. *Aging Clin Exp Res* 21:38–42.
Gegelashvili, G. and A. Schousboe. 1998. Cellular distribution and kinetic properties of high-affinity glutamate transporters. *Brain Res Bull* 45:233–238.
Georgopoulos, G., Andreanos, D., Liokis, N., Papakonstantinou, D., Vergados, J., and G. Theodossiadis. 1997. Risk factors in ocular hypertension. *Eur J Ophthalmol* 7:357–363.
Golombek, D.A. and R.E. Rosenstein. 2010. Physiology of circadian entrainment. *Physiol Rev* 90:1063–1102.
Gorovits, R., Avidan, N., Avisar, N., Shaked, I., and L. Vardimon. 1997. Glutamine synthetase protects against neuronal degeneration in injured retinal tissue. *Proc Natl Acad Sci USA* 94:7024–7029.
Grima, G., Cuenod, M., Pfeiffer, S., Mayer, B., and K.Q. Do. 1998. Arginine availability controls the *N*-methyl-D-aspartate-induced nitric oxide synthesis: Involvement of a glial-neuronal arginine transfer. *J Neurochem* 71:2139–2144.
Guenther, A.L., Schmidt, S.I., Laatsch, H. et al. 2005. Reactions of the melatonin metabolite AMK (*N*1-acetyl-5-methoxykynuramine) with reactive nitrogen species: Formation of novel compounds, 3-acetamidomethyl-6-methoxycinnolinone and 3-nitro-AMK. *J Pineal Res* 39:251–260.
Hall, E.D. 1985. High-dose glucocorticoid treatment improves neurological recovery in head-injured mice. *J Neurosurg* 62:882–887.
Harbin, T.S. Jr., Podos, S.M., Kolker, A.E., and B. Becker. 1976. Visual field progression in open-angle glaucoma patients presenting with monocular field loss. *Trans Sect Ophthalmol Am Acad Ophthalmol Otolaryngol* 81:253–257.
Hardeland, R. 2005. Antioxidative protection by melatonin: Multiplicity of mechanisms from radical detoxification to radical avoidance. *Endocrine* 27:119–130.
Hardeland, R. 2008. Melatonin, hormone of darkness and more: Occurrence, control mechanisms, actions and bioactive metabolites. *Cell Mol Life Sci* 65:2001–2018.
Hardeland, R., Balzer, I., Poeggeler, B. et al. 1995. On the primary functions of melatonin in evolution: Mediation of photoperiodic signals in a unicell, photooxidation, and scavenging of free radicals. *J Pineal Res* 18:104–111.
Hattar, S., Kumar, M., Park, A. et al. 2006. Central projections of melanopsin-expressing retinal ganglion cells in the mouse. *J Comp Neurol* 497:326–349.
Hattar, S., Liao, H.W., Takao, M., Berson, D.M., and K.W. Yau. 2002. Melanopsin-containing retinal ganglion cells: Architecture, projections, and intrinsic photosensitivity. *Science* 295:1065–1070.
Heidinger, V., Hicks, D., Sahel, J., and H. Dreyfus. 1999. Ability of retinal Muller glial cells to protect neurons against excitotoxicity in vitro depends upon maturation and neuron-glial interactions. *Glia* 25:229–239.
Honda, S., Yamamoto, M., and N. Saito. 1995. Immunocytochemical localization of three subtypes of GABA transporter in rat retina. *Brain Res Mol Brain Res* 33:319–325.
Honkanen, R.A., Baruah, S., Zimmerman, M.B. et al. 2003. Vitreous amino acid concentrations in patients with glaucoma undergoing vitrectomy. *Arch Ophthalmol* 121:183–188.
Ianas, O., Olinescu, R., and I. Badescu. 1991. Melatonin involvement in oxidative processes. *Endocrinologie* 29:147–153.

Imbesi, M., Uz, T., and H. Manev. 2008. Role of melatonin receptors in the effects of melatonin on BDNF and neuroprotection in mouse cerebellar neurons. *J Neural Transm* 115:1495–1499.

Iuvone, P.M., Tosini, G., Pozdeyev, N., Haque, R., Klein, D.C., and S.S. Chaurasia. 2005. Circadian clocks, clock networks, arylalkylamine *N*-acetyltransferase, and melatonin in the retina. *Prog Retin Eye Res* 24:433–456.

Jakobs, T.C., Libby, R.T., Ben, Y., John, S.W., and R.H. Masland. 2005. Retinal ganglion cell degeneration is topological but not cell type specific in DBA/2J mice. *J Cell Biol* 171:313–325.

Jaliffa, C.O., Lacoste, F.F., Llomovatte, D.W., Sarmiento, M.I., and R.E. Rosenstein. 2000. Dopamine decreases melatonin content in golden hamster retina. *J Pharmacol Exp Ther* 293:91–95.

Jimenez-Jorge, S., Guerrero, J.M., Jimenez-Caliani, A.J. et al. 2007. Evidence for melatonin synthesis in the rat brain during development. *J Pineal Res* 42:240–246.

Johnson, J., Chen, T.K., Rickman, D.W., Evans, C., and N.C. Brecha. 1996. Multiple gamma-Aminobutyric acid plasma membrane transporters (GAT-1, GAT-2, GAT-3) in the rat retina. *J Comp Neurol* 375:212–224.

Kalaboukhova, L., Fridhammar, V., and B. Lindblom. 2007. Relative afferent pupillary defect in glaucoma: A pupillometric study. *Acta Ophthalmol Scand* 85:519–525.

Kalloniatis, M. and G. Tomisich. 1999. Amino acid neurochemistry of the vertebrate retina. *Prog Retin Eye Res* 18:811–866.

Kanamori, A., Catrinescu, M.M., Kanamori, N., Mears, K.A., Beaubien, R., and L.A. Levin. 2010. Superoxide is an associated signal for apoptosis in axonal injury. *Brain* 133:2612–2625.

Kankipati, L., Girkin, C.A., and P.D. Gamlin. 2011. The post-illumination pupil response is reduced in glaucoma patients. *Invest Ophthalmol Vis Sci* 52:2287–2292.

Kardon, R., Anderson, S.C., Damarjian, T.G., Grace, E.M., Stone, E., and A. Kawasaki. 2009. Chromatic pupil responses: Preferential activation of the melanopsin-mediated versus outer photoreceptor-mediated pupil light reflex. *Ophthalmology* 116:1564–1573.

Kashiwagi, K., Tsumura, T., Ishii, H., Ijiri, H., Tamura, K., and S. Tsukahara. 2000. Circadian rhythm of autonomic nervous function in patients with normal-tension glaucoma compared with normal subjects using ambulatory electrocardiography. *J Glaucoma* 9:239–246.

Kawasaki, A., Otori, Y., and C.J. Barnstable. 2000. Muller cell protection of rat retinal ganglion cells from glutamate and nitric oxide neurotoxicity. *Invest Ophthalmol Vis Sci* 41:3444–3450.

Klerman, E.B., Rimmer, D.W., Dijk, D.J., Kronauer, R.E., Rizzo, J.F. III., and C.A. Czeisler. 1998. Nonphotic entrainment of the human circadian pacemaker. *Am J Physiol* 274:R991–R996.

Kohn, A.N., Moss, A.P., Hargett, N.A., Ritch, R., Smith, H. Jr., and S.M. Podos. 1979. Clinical comparison of dipivalyl epinephrine and epinephrine in the treatment of glaucoma. *Am J Ophthalmol* 87:196–201.

Kong, X., Li, X., Cai, Z. et al. 2008. Melatonin regulates the viability and differentiation of rat midbrain neural stem cells. *Cell Mol Neurobiol* 28:569–579.

Kortuem, K., Geiger, L.K., and L.A. Levin. 2000. Differential susceptibility of retinal ganglion cells to reactive oxygen species. *Invest Ophthalmol Vis Sci* 41:3176–3182.

Kurysheva, N.I., Vinetskaia, M.I., Erichev, V.P., Demchuk, M.L., and S.I. Kuryshev. 1996. Contribution of free-radical reactions of chamber humor to the development of primary open-angle glaucoma. *Vestn Oftalmol* 112:3–5.

Kurz, S. and Harrison, D. 1997. Insulin and the arginine paradox. *J Clin Invest* 99:369–370.

La Morgia, C., Ross-Cisneros, F.N., Hannibal, J., Montagna, P., Sadun, A.A., and V. Carelli. 2011. Melanopsin-expressing retinal ganglion cells: Implications for human diseases. *Vision Res* 51:296–302.

Lanzani, M.F., deZavalía, N., Fontana, H., Sarmiento, M.I., Golombek, D., and R.E. Rosenstein. 2012. Alterations of locomotor activity rhythm and sleep parameters in patients with advanced glaucoma. *Chronobiol Int* 29:911–919.

León, J., Acuña-Castroviejo, D., Escames, G., Tan, D.X., and R.J. Reiter. 2005. Melatonin mitigates mitochondrial malfunction. *J Pineal Res* 38:1–9.

León, J., Acuña-Castroviejo, D., Sainz, R.M., Mayo, J.C., Tan, D.X., and R.J. Reiter. 2004. Melatonin and mitochondrial function. *Life Sci* 75:765–790.

León, J., Vives, F., Crespo, E. et al. 1998. Modification of nitric oxide synthase activity and neuronal response in rat striatum by melatonin and kynurenine derivatives. *J Neuroendocrinol* 10:297–302.

Leske, M.C. 2003. Glaucoma and mortality: A connection? *Ophthalmology* 110:1473–1475.

Levkovitch-Verbin, H., Martin, K.R., Quigley, H.A., Baumrind, L.A., Pease, M.E., and D. Valenta. 2002. Measurement of amino acid levels in the vitreous humor of rats after chronic intraocular pressure elevation or optic nerve transection. *J Glaucoma* 11:396–405.

Li, R.S., Chen, B.Y., Tay, D.K., Chan, H.H., Pu, M.L., and K.F. So. 2006. Melanopsin-expressing retinal ganglion cells are more injury-resistant in a chronic ocular hypertension model. *Invest Ophthalmol Vis Sci* 47:2951–2958.

Liu, B. and A.H. Neufeld. 2000. Expression of nitric oxide synthase-2 (NOS-2) in reactive astrocytes of the human glaucomatous optic nerve head. *Glia* 30:178–186.

Liu, B. and A.H. Neufeld. 2001. Nitric oxide synthase-2 in human optic nerve head astrocytes induced by elevated pressure in vitro. *Arch Ophthalmol* 119:240–245.

Lopez-Costa, J.J., Goldstein, J., Pecci-Saavedra, J. et al. 1999. GABA release mechanism in the golden hamster retina. *Int J Neurosci* 98:13–25.

Louzada-Junior, P., Dias, J.J., Santos, W.F., Lachat, J.J., Bradford, H.F., and J. Coutinho-Netto. 1992. Glutamate release in experimental ischaemia of the retina: An approach using microdialysis. *J Neurochem* 59:358–363.

Lucas, R.J., Douglas, R.H., and R.G. Foster. 2001. Characterization of an ocular photopigment capable of driving pupillary constriction in mice. *Nat Neurosci* 4:621–626.

Lucas, R.J., Freedman, M.S., Muñoz, M., Garcia-Fernández, J.M., and R.G. Foster. 1999. Regulation of the mammalian pineal by non-rod, non-cone, ocular photoreceptors. *Science* 284:505–507.

Lucas, R.J., Hattar, S., Takao, M., Berson, D.M., Foster, R.G., and K.W. Yau. 2003. Diminished pupillary light reflex at high irradiances in melanopsin-knockout mice. *Science* 299:245–247.

Lundmark, P.O., Pandi-Perumal, S.R., Srinivasan, V., and D.P. Cardinali. 2006. Role of melatonin in the eye and ocular dysfunctions. *Vis Neurosci* 23:853–862.

Manabe, S., Gu, Z., and S.A. Lipton. 2005. Activation of matrix metalloproteinase-9 via neuronal nitric oxide synthase contributes to NMDA-induced retinal ganglion cell death. *Invest Ophthalmol Vis Sci* 46:4747–4753.

Marcic, T.S., Belyea, D.A., and B. Katz. 2003. Neuroprotection in glaucoma: A model for neuroprotection in optic neuropathies. *Curr Opin Ophthalmol* 14:353–356.

Martin, K.R., Levkovitch-Verbin, H., Valenta, D., Baumrind, L., Pease, M.E., and H.A. Quigley. 2002. Retinal glutamate transporter changes in experimental glaucoma and after optic nerve transection in the rat. *Invest Ophthalmol Vis Sci* 43:2236–2243.

Melino, G., Bernassola, F., Knight, R.A., Corasaniti, M.T., Nistico, G., and A. Finazzi-Agro. 1997. *S*-nitrosylation regulates apoptosis. *Nature* 388:432–433.

Moore, R.Y. and R. Silver. 1998. Suprachiasmatic nucleus organization. *Chronobiol Int* 15:475–487.

Moreno, M.C., Campanelli, J., Sande, P., Sanez, D.A., Keller Sarmiento, M.I., and R.E. Rosenstein. 2004. Retinal oxidative stress induced by high intraocular pressure. *Free Radic Biol Med* 37:803–812.

Moreno, M.C., de Zavalía, N., Sande, P. et al. 2008. Effect of ocular hypertension on retinal GABAergic activity. *Neurochem Int* 52:675–682.

Moreno, M.C., Marcos, H.J., Oscar, C.J. et al. 2005b. A new experimental model of glaucoma in rats through intracameral injections of hyaluronic acid. *Exp Eye Res* 81:71–80.

Moreno, M.C., Sande, P., Marcos H.A., de Zavalía, N., Keller Sarmiento, M.I., and R.E. Rosenstein. 2005a. Effect of glaucoma on the retinal glutamate/glutamine cycle activity. *FASEB J* 19:1161–1162.

Morrison, J.C., Moore, C.G., Deppmeier, L.M., Gold, B.G., Meshul, C.K., and E.C. Johnson. 1997. A rat model of chronic pressure-induced optic nerve damage. *Exp Eye Res* 64:85–96.

Mosinger, J.L., Price, M.T., Bai, H.Y., Xiao, H., Wozniak, D.F., and J.W. Olney. 1991. Blockade of both NMDA and non-NMDA receptors is required for optimal protection against ischemic neuronal degeneration in the in vivo adult mammalian retina. *Exp Neurol* 113:10–17.

Mrosovsky, N. and S. Hattar. 2003. Impaired masking responses to light in melanopsin-knockout mice. *Chronobiol Int* 20:989–999.

Naskar, R., Vorwerk, C.K., and E.B. Dreyer. 2000. Concurrent downregulation of a glutamate transporter and receptor in glaucoma. *Invest Ophthalmol Vis Sci* 41:1940–1944.

Neufeld, A.H. 2004. Pharmacologic neuroprotection with an inhibitor of nitric oxide synthase for the treatment of glaucoma. *Brain Res Bull* 62:455–459.

Neufeld, A.H., Hernandez, M.R., and M. Gonzalez. 1997. Nitric oxide synthase in the human glaucomatous optic nerve head. *Arch Ophthalmol* 115:497–503.

Neufeld, A.H. and B. Liu. 2003. Glaucomatous optic neuropathy: When glia misbehave. *Neuroscientist* 9:485–495.

Neufeld, A.H., Sawada, A., and B. Becker. 1999. Inhibition of nitric-oxide synthase 2 by aminoguanidine provides neuroprotection of retinal ganglion cells in a rat model of chronic glaucoma. *Proc Natl Acad Sci USA* 96:9944–9948.

Niles, L.P., Armstrong, K.J., Rincon Castro, L.M. et al. 2004. Neural stem cells express melatonin receptors and neurotrophic factors: Colocalization of the MT1 receptor with neuronal and glial markers. *BMC Neurosci* 5:41.

Nissen, C., Sander, B., and H. Lund-Andersen. 2011. The effect of pupil size on stimulation of the melanopsin containing retinal ganglion cells, as evaluated by monochromatic pupillometry. *Front Neurol* 2:92.

Nucci, C., Tartaglione, R., Rombola, L., Morrone, L.A., Fazzi, E., and G. Bagetta. 2005. Neurochemical evidence to implicate elevated glutamate in the mechanisms of high intraocular pressure (IOP)-induced retinal ganglion cell death in rat. *Neurotoxicology* 26:935–941.

Ohta, Y., Yamasaki, T., Niwa, T., Niimi, K., Majima, Y., and I. Ishiguro. 1996. Role of catalase in retinal antioxidant defence system: Its comparative study among rabbits, guinea pigs, and rats. *Ophthalmic Res* 28:336–342.

Onen, S.H., Alloui, A., Eschalier, A., and C. Dubray. 2000. Vocalization thresholds related to noxious paw pressure are decreased by paradoxical sleep deprivation and increased after sleep recovery in rat. *Neurosci Lett* 291:25–28.

Osborne, N.N., Casson, R.J., Wood, J.P., Chidlow, G., Graham, M., and J. Melena. 2004. Retinal ischemia: Mechanisms of damage and potential therapeutic strategies. *Prog Retin Eye Res* 23:91–147.

Osborne, N.N., Chidlow, G., Nash, M.S., and J.P. Wood. 1999. The potential of neuroprotection in glaucoma treatment. *Curr Opin Ophthalmol* 10:82–92.

Otis, T.S. and C.E. Jahr. 1998. Anion currents and predicted glutamate flux through a neuronal glutamate transporter. *J Neurosci* 18:7099–7110.

Panda, S., Sato, T.K., Castrucci, A.M. et al. 2002. Melanopsin (Opn4) requirement for normal light-induced circadian phase shifting. *Science* 298:2213–2216.

Pandi-Perumal, S.R., Srinivasan, V., Maestroni, G.J., Cardinali, D.P., Poeggeler, B., and R. Hardeland. 2006. Melatonin: Nature's most versatile biological signal? *FEBS J* 273:2813–2838.

Pang, I.H., Johnson, E.C., Jia, L. et al. 2005. Evaluation of inducible nitric oxide synthase in glaucomatous optic neuropathy and pressure-induced optic nerve damage. *Invest Ophthalmol Vis Sci* 46:1313–1321.

Patel, A.J., Hunt, A., and C.S. Tahourdin. 1983. Regulation of in vivo glutamine synthetase activity by glucocorticoids in the developing rat brain. *Brain Res* 312:83–91.

Pérez-Rico, C., de la Villa, P., Arribas-Gómez, I., and R. Blanco. 2010. Evaluation of functional integrity of the retinohypothalamic tract in advanced glaucoma using multifocal electroretinography and light-induced melatonin suppression. *Exp Eye Res* 91:578–583.

Pierce, M.E. and J.C. Besharse. 1985. Circadian regulation of retinomotor movements. I. Interaction of melatonin and dopamine in the control of cone length. *J Gen Physiol* 86:671–689.

Pow, D.V. and S.R. Robinson. 1994. Glutamate in some retinal neurons is derived solely from glia. *Neuroscience* 60:355–366.

Pozo, D., Reiter, R.J., Calvo, J.R., and J.M. Guerrero. 1997. Inhibition of cerebellar nitric oxide synthase and cyclic GMP production by melatonin via complex formation with calmodulin. *J Cell Biochem* 65:430–442.

Provencio, I., Rollag, M.D., and A.M. Castrucci. Photoreceptive net in the mammalian retina. This mesh of cells may explain how some blind mice can still tell day from night. *Nature* 2002 415:493.

Quigley, H.A. and A.T. Broman. 2006. The number of people with glaucoma worldwide in 2010 and 2020. *Br J Ophthalmol* 90:262–267.

Rajesh, M., Sulochana, K.N., Punitham, R., Biswas, J., Lakshmi, S., and S. Ramakrishnan. 2003. Involvement of oxidative and nitrosative stress in promoting retinal vasculitis in patients with Eales' disease. *Clin Biochem* 36:377–385.

Reiter, R.J. 1998. Oxidative damage in the central nervous system: Protection by melatonin. *Prog Neurobiol* 56:359–384.

Reiter, R.J., Guerrero, J.M., Escames, G., Pappolla, M.A., and D. Acuña-Castroviejo. 1997. Prophylactic actions of melatonin in oxidative neurotoxicity. *Ann N Y Acad Sci* 825:70–78.

Reiter, R.J., Tan, D.X., Osuna, C., and E. Gitto. 2000. Actions of melatonin in the reduction of oxidative stress. A review. *J Biomed Sci* 7:444–458.

Reiter, R.J., Tan, D.X., Poeggeler, B., Menendez-Pelaez, A., Chen, L.D., and S. Saarela. 1994. Melatonin as a free radical scavenger: Implications for aging and age-related diseases. *Ann N Y Acad Sci* 719:1–12.

Riepe, R.E. and M.D. Norenburg. 1977. Muller cell localisation of glutamine synthetase in rat retina. *Nature* 268:654–655.

Rosenstein, R.E. and D.P. Cardinali. 1986. Melatonin increases in vivo GABA accumulation in rat hypothalamus, cerebellum, cerebral cortex and pineal gland. *Brain Res* 398:403–406.

Rosenstein, R.E., Estevez, A.G., and D.P. Cardinali. 1989. Time-dependent effect of melatonin on glutamic acid decarboxylase activity and CI influx in rat hypothalamus. *J Neuroendocrinol* 1:443–447.

Rosner, M., Lam, T.T., and M.O. Tso. 1992. Therapeutic parameters of methylprednisolone treatment for retinal photic injury in a rat model. *Res Commun Chem Pathol Pharmacol* 77:299–311.

Ruby, N.F., Brennan, T.J., Xie, X. et al. 2002. Role of melanopsin in circadian responses to light. *Science* 298:2211–2213.

Sáenz, D.A., Cymeryng, C.B., De Nichilo, A., Sacca, G.B., Keller Sarmiento, M.I., and R.E. Rosenstein. 2002a. Photic regulation of L-arginine uptake in the golden hamster retina. *J Neurochem* 80:512–519.

Sáenz, D.A., Goldin, A.P., Minces, L., Chianelli, M., Sarmiento, M.I., and R.E. Rosenstein. 2004. Effect of melatonin on the retinal glutamate/glutamine cycle in the golden hamster retina. *FASEB J* 18:1912–1913.

Sáenz, D.A., Turjanski, A.G., Sacca, G.B. et al. 2002b. Physiological concentrations of melatonin inhibit the nitridergic pathway in the Syrian hamster retina. *J Pineal Res* 33:31–36.

Sande, P.H., Fernandez, D.C., Aldana Marcos, H.J. et al. 2008. Therapeutic effect of melatonin in experimental uveitis. *Am J Pathol* 173:1702–1713.

Sarkar, P.K. and S. Chaudhury. 1983. Messenger RNA for glutamine synthetase. Review article. *Mol Cell Biochem* 53–54:233–244.

Sarthy, P.V. and D.M. Lam. 1978. Biochemical studies of isolated glial (Muller) cells from the turtle retina. *J Cell Biol* 78:675–684.

Schwartz, E.A. 1987. Depolarization without calcium can release gamma-aminobutyric acid from a retinal neuron. *Science* 238:350–355.

Shareef, S., Sawada, A., and A.H. Neufeld. 1999. Isoforms of nitric oxide synthase in the optic nerves of rat eyes with chronic moderately elevated intraocular pressure. *Invest Ophthalmol Vis Sci* 40:2884–2891.

Shareef, S.R., Garcia-Valenzuela, E., Salierno, A., Walsh, J., and S.C. Sharma. 1995. Chronic ocular hypertension following episcleral venous occlusion in rats. *Exp Eye Res* 61:379–382.

Shen, F., Chen, B., Danias, J. et al. 2004. Glutamate-induced glutamine synthetase expression in retinal Muller cells after short-term ocular hypertension in the rat. *Invest Ophthalmol Vis Sci* 45:3107–3112.

Siu, A.W., Maldonado, M., Sanchez-Hidalgo, M., Tan, D.X., and R.J. Reiter. 2006. Protective effects of melatonin in experimental free radical-related ocular diseases. *J Pineal Res* 40:101–109.

Siu, A.W., Reiter, R.J., and C.H. To. 1999. Pineal indoleamines and vitamin E reduce nitric oxide-induced lipid peroxidation in rat retinal homogenates. *J Pineal Res* 27:122–128.

Stevens, B.R., Kakuda, D.K., Yu, K., Waters, M., Vo, C.B., and M.K. Raizada. 1996. Induced nitric oxide synthesis is dependent on induced alternatively spliced CAT-2 encoding L-arginine transport in brain astrocytes. *J Biol Chem* 271:24017–24022.

Tan, D.X., Manchester, L.C., Terron, M.P., Flores, L.J., and R.J. Reiter. 2007. One molecule, many derivatives: A never-ending interaction of melatonin with reactive oxygen and nitrogen species? *J Pineal Res* 42:28–42.

Tanaka, K. 2000. Functions of glutamate transporters in the brain. *Neurosci Res* 37:15–19.

Tanaka, D., Furusawa, K., Kameyama, K., Okamoto, H., and M. Doi. 2007. Melatonin signaling regulates locomotion behavior and homeostatic states through distinct receptor pathways in *Caenorhabiditis elegans*. *Neuropharmacology* 53:157–168.

Tang, Q., Hu, Y., and Y. Cao. 2006. Neuroprotective effect of melatonin on retinal ganglion cells in rats. *J Huazhong Univ Sci Technolog Med Sci* 26:235–237, 253.

Tanito, M., Nishiyama, A., Tanaka, T. et al. 2002. Change of redox status and modulation by thiol replenishment in retinal photooxidative damage. *Invest Ophthalmol Vis Sci* 43:2392–2400.

Tezel, G. 2006. Oxidative stress in glaucomatous neurodegeneration: Mechanisms and consequences. *Prog Retin Eye Res* 25:490–513.

Thoreson, W.B. and P. Witkovsky. 1999. Glutamate receptors and circuits in the vertebrate retina. *Prog Retin Eye Res* 18:765–810.

Tosini, G. and C. Fukuhara. 2003. Photic and circadian regulation of retinal melatonin in mammals. *J Neuroendocrinol* 15:364–369.

Tosini, G. and M. Menaker. 1996. Circadian rhythms in cultured mammalian retina. *Science* 272:419–421.

Turjanski, A., Chaia, Z.D., Doctorovich, F., Estrin, D., Rosenstein R., and O.E. Piro. 2000. *N*-nitrosomelatonin. *Acta Crystallogr C* 56:682–683.

Turjanski, A.G., Rosenstein, R.E., and D.A. Estrin. 1998. Reactions of melatonin and related indoles with free radicals: A computational study. *J Med Chem* 41:3684–3689.

Ueda, J., Sawaguchi, S., Hanyu, T. et al. 1998. Experimental glaucoma model in the rat induced by laser trabecular photocoagulation after an intracameral injection of India ink. *Jpn J Ophthalmol* 42:337–344.

van Geijlswijk, I.M., Korzilius, H.P., and M.G. Smits. 2010. The use of *exogenous* melatonin in delayed sleep phase disorder: A meta-analysis. *Sleep* 33:1605–1614.

Van Gelder, R.N. 2003. Making (a) sense of non-visual ocular photoreception. *Trends Neurosci* 26:458–461.

van Reyk, D.M., Gillies, M.C., and M.J. Davies. 2003. The retina: Oxidative stress and diabetes. *Redox Rep* 8:187–192.

Vorwerk, C.K., Hyman, B.T., Miller, J.W. et al. 1997. The role of neuronal and endothelial nitric oxide synthase in retinal excitotoxicity. *Invest Ophthalmol Vis Sci* 38:2038–2044.

Vorwerk, C.K., Naskar, R., Schuettauf, F. et al. 2000. Depression of retinal glutamate transporter function leads to elevated intravitreal glutamate levels and ganglion cell death. *Invest Ophthalmol Vis Sci* 41:3615–3621.

Wakakura, M. and J. Yokoe. 1995. Evidence for preserved direct pupillary light response in Leber's hereditary optic neuropathy. *Br J Ophthalmol* 79:442–446.

Wamsley, S., Gabelt, B.T., Dahl, D.B. et al. 2005. Vitreous glutamate concentration and axon loss in monkeys with experimental glaucoma. *Arch Ophthalmol* 123:64–70.

Wang, H.Z., Lu, Q.J., Wang, N.L., Liu, H., Zhang, L., and G.L. Zhan. 2008. Loss of melanopsin-containing retinal ganglion cells in a rat glaucoma model. *Chin Med J (Engl)* 121:1015–1019.

Weber, A.J., Harman, C.D., and S. Viswanathan. 2008. Effects of optic nerve injury, glaucoma, and neuroprotection on the survival, structure, and function of ganglion cells in the mammalian retina. *J Physiol* 586:4393–4400.

Wee, R., Castrucci, A.M., Provencio, I., Gan, L., and R.N. Van Gelder. 2002. Loss of photic entrainment and altered free-running circadian rhythms in math5–/– mice. *J Neurosci* 22:10427–10433.

Wu, G.S., Zhang, J., and N.A. Rao. 1997. Peroxynitrite and oxidative damage in experimental autoimmune uveitis. *Invest Ophthalmol Vis Sci* 38:1333–1339.

Yang, X.L. 2004. Characterization of receptors for glutamate and GABA in retinal neurons. *Prog Neurobiol* 73:127–150.

Yuki, K., Ozawa, Y., Yoshida, T. et al. 2011. Retinal ganglion cell loss in superoxide dismutase 1 deficiency. *Invest Ophthalmol Vis Sci* 52:4143–4150.

Zink, J.M., Grunwald, J.E., Piltz-Seymour, J., Staii, A., and J. Dupont. 2003. Association between lower optic nerve laser Doppler blood volume measurements and glaucomatous visual field progression. *Br J Ophthalmol* 87:1487–1491.

18 Melatonin Receptors and Their Preventive Role in Carcinogenesis

Raffaela Santoro, Maria Ferraiuolo, Giovanni Blandino, Paola Muti, and Sabrina Strano

CONTENTS

18.1 INTRODUCTION

Melatonin (*N*-acetyl-5-methoxytryptamine) is mainly secreted by the pineal gland at the hypothalamic level, as well as in the gastrointestinal tract. Melatonin synthesis occurs mainly at night and is inhibited by the sunlight (Axelrod et al. 1965, Pandi-Perumal et al. 2008). Epidemiological studies have clearly demonstrated that high levels of melatonin decrease the risk of developing cancer (Alpert et al. 2009, Flynn-Evans et al. 2009, Kliukiene et al. 2001, Kloog et al. 2008, Lipton et al. 2009, Schernhammer and Hankinson 2009, Schernhammer et al. 2008, 2010). In particular, two epidemiological studies have demonstrated an inverse correlation between overnight urinary levels of melatonin and the incidence of breast cancer (Schernhammer and Hankinson 2009, Schernhammer et al. 2008). The first, conducted within the ORDET cohort, tested the concentration of the melatonin metabolite 6-sulfatoxymelatonin (aMT6s) in 178 postmenopausal women with incident breast cancer and 710 matched controls (Schernhammer and Hankinson 2009). The second one is a case-control study conducted within the NHIS cohort on 357 postmenopausal women with incident breast cancer and 533 matched controls, in which the authors found that there was a negative correlation between aMT6s levels and the risk of invasive breast cancer, regardless of ER and Her2 status (Schernhammer and Hankinson 2009). Light at night (LAN), which causes reduced melatonin levels, is believed to increase the incidence of breast cancer: blindness has an inverse correlation with breast cancer because the total absence of LAN in totally blind people leads to inhibition of nocturnal decrease of melatonin production by light (Blask et al. 2011, Reiter et al. 2007); on the contrary, exposure to LAN in night-shift workers reduces melatonin production during night and therefore increases cancer risk (Blask et al. 2011) as well as exposure of animals to long light periods enhances the growth of chemically induced carcinomas in female rats (Cos et al. 2006). Along the same line, when mouse xenografts and rat hepatomas are perfused with blood collected at night from healthy premenopausal women, they show a marked suppression in proliferation as compared with tumors perfused with blood collected during daytime and therefore melatonin-depleted (Blask et al. 2005). The same type of results has been observed in men: LAN exposure increases the risk of developing cancer at several sites (Parent et al. 2012). These studies suggest that melatonin might be able to induce tumor suppressor and/or inhibit oncogenic pathways in vivo. In particular, molecular studies have shown that melatonin is able to modulate the p53 tumor suppressor pathway (Kim and Yoo 2010, Mediavilla et al. 1999, Santoro et al. 2012, 2013), the JNK pathway (Carbajo-Pescador et al. 2011), and the proinflammatory NFκB pathway (Cuesta et al. 2010). Based on this and other evidence, trials on melatonin use as a chemopreventive agent are ongoing (Schernhammer et al. 2012).

Being highly soluble in both lipids and water, melatonin can diffuse across the cellular membrane and act as a radical scavenger. In the same way, it can cross the nuclear membrane and bind to its nuclear receptor RZR/RORα. This is then recruited onto RZR regulatory elements present in the promoters of several genes, such as p21 and 5-lipoxygenase, and modulates their transcription (Missbach et al. 1996, Schrader et al. 1996, Steinhilber et al. 1995). In addition, melatonin regulates signaling pathways by binding to its membrane receptors, MTNR1A and MTNR1B, also known as MT1 and MT2. Melatonin also binds with lower affinity to the MT3 receptor, a quinone reductase (Nosjean et al. 2000).

18.2 MELATONIN RECEPTORS

Melatonin nuclear receptors RZR/RORα belong to the retinoid orphan receptor subfamily. Structurally, they show several conserved domains: a ligand-binding domain, a DNA binding domain, and a transactivation domain (Smirnov 2001). These receptors have an intrinsic transactivating activity, which is enhanced by melatonin binding. They bind to response elements (ROREs) represented by 5′-extended half-sites with the A/GGGTCA motif (Carlberg et al. 1994, Giguere et al. 1994, 1995, Ginestier et al. 2006, Medvedev et al. 1996, Schrader et al. 1996, Winrow et al. 1998) (Figure 18.1a), and they are involved in several physiological processes, such as lipid metabolism (human-5-lipoxygenase, rat and mouse apolipoprotein A-I), cancer development (mouse laminin B1, N-myc, p21), immune response (5-lipoxygenase, IL-2), and central nervous system (CNS) development. Although interesting, there is not much literature on the activities of melatonin nuclear receptors (Missbach et al. 1996, Wiesenberg et al. 1998).

Melatonin membrane receptors MT1 and MT2 are expressed in several areas of the CNS, as well as in the retina, immune cells, gastrointestinal tract, and arteries (Borjigin et al. 1999, Cardinali, Vacas and Boyer 1979, Dubocovich et al. 2010, Klein 1999, Reiter 1991). Both are G protein-coupled membrane receptors (GPCRs), which, upon ligand binding, activate G proteins and inhibit cAMP formation (Dubocovich et al. 2010). The GPCR superfamily is quite complex and includes several classes of receptors, including the rhodopsin family, to which both MT1 and MT2 belong (Dubocovich et al. 2010). They have seven transmembrane domains, which are connected by three extracellular and three intracellular loops (Fredriksson et al. 2003). MT1 and MT2 exist in both homodimeric and heterodimeric state (Maurice et al. 2010) and show high sequence homology. Unlike for nuclear receptors, many studies have focused on the activities of MT1 and MT2 receptors. It has been shown that, upon melatonin binding, both MT1 and MT2 receptors can also activate MAP kinases, such as c-Jun, in *C. aethiops* Cos-7 cells (Chan et al. 2002) and p38 (Santoro et al. 2012, 2013) (Figure 18.1b).

18.3 MELATONIN RECEPTORS AND CANCER

Recent evidence has shown a correlation between melatonin receptors expression and cancer occurrence for several types of cancers.

Nakamura et al. found that MT1 is frequently downregulated in primary oral squamous cell carcinomas (OSCC) as compared with normal oral epithelia, through hypermethylation of the CpG islands in its promoter. They also found that there is significant positive association between hypermethylation of MT1 and T staging, as tumors with higher T stages showed lower expression of MT1 (Nakamura et al. 2008).

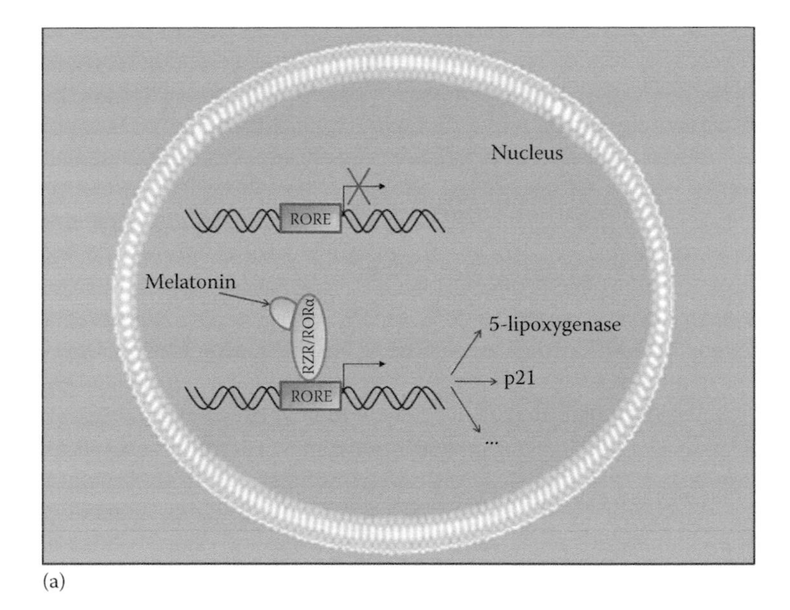

(a)

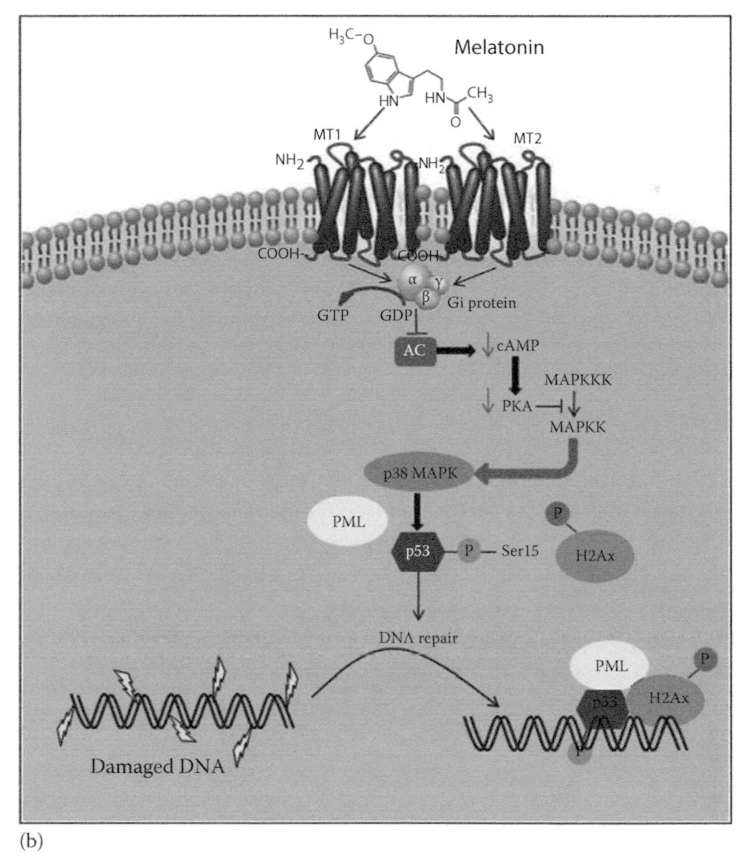

(b)

FIGURE 18.1 (See color insert.) Melatonin receptors. (a) Upon melatonin binding, melatonin nuclear receptor RZR/RORα is recruited onto ROREs present in its target genes and drives their transcription. (b) Melatonin membrane receptors MT1 and MT2 activate G-proteins following melatonin binding. This results in inhibition of adenylate cyclise (AC) and consequent decrease in cAMP intracellular levels. This, in turn, leads to activation of p38MAPK (red arrows) and its downstream targets, such as p53.

Oprea-Ilies et al. investigated the expression of MT1 in triple negative breast cancers (TNBCs) in relation to survival. They found that, among TNBC, MT1 negative tumors showed lower overall survival and progression-free survival as compared with MT1 positive TNBC. This was more evident in African-American women than in Caucasian (Oprea-Ilies et al. 2013).

Jablonska et al. showed an increase in MT1 expression in invasive ductal carcinoma (IDC) as compared with fibrocystic breast disease. In addition, they showed a marked decrease in MT1 expression with higher tumor grade and TNM staging (Jablonska et al. 2013). It is worth noting that no information about mutational status of the receptor in normal and tumoral tissues is provided in this report.

Leon et al. showed a correlation between MT1 and MT2 expression and colorectal staging (Leon, *Molecular Carcinogenesis*, 2012). In particular, MT1 levels were significantly lower in both early- and advanced-stage colon cancers as compared with their matched controls. On the contrary, MT2 levels were significantly lower only in advanced-stage tumors (Leon et al. 2012).

Clinical trials have been conducted on small numbers of patients with colorectal carcinoma. In particular, a group of 30 patients were administered a combination of melatonin (20 mg/day at bedtime) and irinotecan (CPT-11), which proved more effective in terms of progression-free survival than CPT-11 alone (Cerea et al. 2003). This finding suggests that melatonin could enhance the effectiveness of chemotherapeutic drugs. A similar study was conducted on 35 patients with tumors of the gastrointestinal tract (Lissoni et al. 1993), as melatonin has been found at a very high concentration in the liver (Messner et al. 2001). Here, melatonin (50 mg/day) was administered 7 days before the beginning of therapy with interleukin 2 (IL-2), which was administered simultaneously with melatonin for one cycle (4 weeks). The outcome was that 23% of patients showed a reduction of at least 50% in the tumor mass, which made the authors conclude that low doses of IL-2 plus melatonin can be well tolerated and effective for tumors of the gastrointestinal tract.

In cases of high mass liver tumors, transarterial chemoembolization (TACE) is the election treatment, as it allows delivering high doses of chemotherapy in the inner tumor mass and simultaneously destroy the tumor's blood supply (Verslype et al. 2012). In a study by Yan et al. (2002) conducted on 100 patients with advanced primary hepatocellular carcinoma (HCC) and treated with TACE, administration of melatonin significantly increased patients survival rate, and liver functions were protected from TACE-caused damage.

The growth of prostate cancer cells has been effectively blocked by melatonin treatment, as reported by Shiu et al. (2003), both in vitro and in nude mice xenografts.

In addition to these reports, analysis of several casuistries of breast and colon cancers deposited in Oncomine (www.oncomine.org) reveals a downregulation of both MT1 and MT2 mRNA expression in tumors as compared with normal tissues (Desmedt et al. 2007, Farmer et al. 2005, Ginestier et al. 2006, Hong et al. 2010, Kaiser et al. 2007, Skrzypczak et al. 2010). This strongly supports the involvement of melatonin receptors in carcinogenesis.

Besides in vivo studies, many in vitro studies have been performed, which depict the role of both melatonin and melatonin receptors in the carcinogenic process. Many of these have been conducted in the estrogen receptor (ER) positive breast cancer cell line MCF-7 and have aimed at studying the capability of melatonin to induce the p53 pathway (Cos et al. 1991, 1998, Mediavilla et al. 1999, Santoro et al. 2012, 2013). In particular, some authors found that melatonin induction of p53 results in cell cycle arrest and apoptosis (Cos et al. 1991, 1998, Mediavilla et al. 1999), while others have identified a role for melatonin stimulation of the p53 pathway toward DNA damage repair (Santoro et al. 2012, 2013). Overexpression of MT1 in MCF-7 cells caused inhibition of cell proliferation (Yuan et al. 2002) while silencing of either receptors abolished the activity of melatonin toward inhibition of cell proliferation (Santoro et al. 2013). On the other hand, silencing of MT1 and/or MT2 functions caused an impairment of melatonin-induced DNA repair (Santoro et al. 2013). Altogether, these data strengthen the correlation between melatonin receptors' activation and induction of the p53 pathway.

Studies conducted on human ovarian cancer cell lines showed that melatonin can inhibit cell proliferation to various extent, depending on the characteristics of the cell line used (Bartsch et al. 2000, Futagami et al. 2001, Petranka et al. 1999), by acting through the activation of MT1 and MT2. In colorectal carcinoma cell lines, melatonin activates MT1 and MT2, thus inhibiting cell proliferation and DNA repair (Santoro et al. 2013) and potentiating chemotherapy-induced apoptosis (Wenzel et al. 2005). Moreover, activation of MT1 results in inhibition of fatty acid uptake in HCC cell lines, thus resulting in oncostatic action (Blask et al. 1999, Sauer et al. 2001).

The use of MT2 agonists and antagonists in inhibition and enhancement of cell growth of melanoma cells, respectively (Roberts et al. 2000, Souza et al. 2003), again points out at a major role of melatonin receptor signaling in melatonin's oncostatic actions. Also, different doses of melatonin produced antitumor effects, ranging from receptor-dependent oncostatic to oncotoxic actions in diverse melanoma cell lines (Hu et al. 1998, Roberts et al. 2000).

Growth of prostate cancer cell lines can be inhibited by melatonin in both androgen-dependent and androgen-independent manner, through activation of melatonin receptors (Gilad et al. 1999, Kim and Yoo 2010, Rimler et al. 2001, 2002a,b, 2006, 2007, Sampson et al. 2006, Xi et al. 2001).

Taken together, all these in vivo and in vitro studies underline a major role for melatonin receptors in the carcinogenic process, ranging from maintenance of genome integrity to inhibition of cell proliferation and control of cancer metabolism.

18.4 CONCLUSIONS AND PERSPECTIVES

Epidemiological studies have clearly demonstrated that high levels of melatonin decrease the risk of developing cancer, including breast and colon cancer (Alpert et al. 2009, Flynn-Evans et al. 2009, Kliukiene et al. 2001, Kloog et al. 2008, Lipton et al. 2009, Schernhammer and Hankinson 2009, Schernhammer et al. 2008, 2010), and that there is no association between nocturnal melatonin concentration and either ER or HER2 status (Schernhammer and Hankinson 2009). However, only recently molecular studies are dissecting the mechanisms underlying melatonin's chemopreventive actions and are focusing on the role of melatonin membrane receptors MT1 and MT2. Unfortunately, discordant reports have been published, some of them evidencing an inverse correlation between MT1 expression and tumor development (Hill et al. 2011, Nakamura et al. 2008, Nemeth et al. 2011), while others found a positive correlation (Danielczyk and Dziegiel 2009, Dillon et al. 2002, Lai et al. 2009). The inverse correlation between melatonin receptor levels and tumor progression could be explained by the fact that some authors showed that higher MT1 expression corresponds to low melatonin serum levels (Barrett et al. 1996), shading the possibility that enhanced expression of melatonin receptors in cancer could be due to lower levels of circulating melatonin in cancer subjects. Based on these observations, it can be concluded that perturbation of melatonin receptor signaling, either through inhibition of melatonin secretion or deletion/reduction of melatonin receptors expression, is strongly correlated to the carcinogenic process.

To study the capability of melatonin to counteract cancer growth in vivo, several trials have been conducted with melatonin as an adjuvant for chemotherapy in cancer patients (Lissoni et al. 1991, 1993, 1999, Mills et al. 2005), and systematic reviews on these data have been published (Mills et al. 2005, Wang et al. 2012). The main aim of these studies was to understand whether melatonin had an active role in suppressing tumor growth, but little attention was paid to the side effects of chemotherapy. Instead, a detailed systematic review considered and analyzed the data related to the response to chemotherapy (Wang et al. 2012). Meta-analysis showed that the use of melatonin as an adjuvant for chemotherapy significantly improved relapse and survival rate as compared with chemotherapy alone. Moreover, melatonin treatment ameliorated chemotherapy side effects, such as immune response and thrombocytopenia.

Data from Santoro et al. (2013) provide in vitro and in vivo evidence that melatonin anticancer activities, including DNA damage response and inhibition of cell proliferation, are mediated by the activation of the MT1 and MT2 receptors and, together with the earlier-mentioned studies on the levels of MT1 and MT2 in cancer as compared with normal breast and colon samples, represent important information for both the design and the analysis of trials involving melatonin either as an adjuvant for chemotherapy or as a chemopreventive agent. In fact, when analyzing the effects of melatonin on cell growth and relapse-free survival, MT1 and MT2 status and expression levels in tumors under study should be taken into consideration. In fact, it would be expected that patients with tumors expressing low levels of melatonin receptors will not respond to melatonin treatment in terms of decreasing tumor progression and increasing relapse-free survival. However, the same patients could benefit from melatonin's ability to reduce bystander effects, increase immune response, and ameliorate side effects following chemotherapy administration (thrombocytopenia, anemia, asthenia). In addition, further studies should be performed to understand whether higher doses of melatonin or melatonin analogues with higher affinity for melatonin receptors could have an effect on cancer progression or any adverse effects and how they could influence the expression of melatonin receptors themselves.

REFERENCES

Alpert, M., E. Carome, V. Kubulins, and R. Hansler (2009) Nighttime use of special spectacles or light bulbs that block blue light may reduce the risk of cancer. *Med Hypotheses*, 73, 324–325.

Axelrod, J., R. J. Wurtman, and S. H. Snyder (1965) Control of hydroxyindole *O*-methyltransferase activity in the rat pineal gland by environmental lighting. *J Biol Chem*, 240, 949–954.

Barrett, P., A. MacLean, G. Davidson, and P. J. Morgan (1996) Regulation of the Mel 1a melatonin receptor mRNA and protein levels in the ovine pars tuberalis: Evidence for a cyclic adenosine 3′,5′-monophosphate-independent Mel 1a receptor coupling and an autoregulatory mechanism of expression. *Mol Endocrinol*, 10, 892–902.

Bartsch, H., A. Buchberger, H. Franz, C. Bartsch, I. Maidonis, D. Mecke, and E. Bayer (2000) Effect of melatonin and pineal extracts on human ovarian and mammary tumor cells in a chemosensitivity assay. *Life Sci*, 67, 2953–2960.

Blask, D. E., G. C. Brainard, R. T. Dauchy, J. P. Hanifin, L. K. Davidson, J. A. Krause, L. A. Sauer et al. (2005) Melatonin-depleted blood from premenopausal women exposed to light at night stimulates growth of human breast cancer xenografts in nude rats. *Cancer Res*, 65, 11174–11184.

Blask, D. E., S. M. Hill, R. T. Dauchy, S. Xiang, L. Yuan, T. Duplessis, L. Mao, E. Dauchy, and L. A. Sauer (2011) Circadian regulation of molecular, dietary, and metabolic signaling mechanisms of human breast cancer growth by the nocturnal melatonin signal and the consequences of its disruption by light at night. *J Pineal Res*, 51, 259–269.

Blask, D. E., L. A. Sauer, R. T. Dauchy, E. W. Holowachuk, M. S. Ruhoff, and H. S. Kopff (1999) Melatonin inhibition of cancer growth in vivo involves suppression of tumor fatty acid metabolism via melatonin receptor-mediated signal transduction events. *Cancer Res*, 59, 4693–4701.

Borjigin, J., X. Li, and S. H. Snyder (1999) The pineal gland and melatonin: Molecular and pharmacologic regulation. *Annu Rev Pharmacol Toxicol*, 39, 53–65.

Carbajo-Pescador, S., A. Garcia-Palomo, J. Martin-Renedo, M. Piva, J. Gonzalez-Gallego, and J. L. Mauriz (2011) Melatonin modulation of intracellular signaling pathways in hepatocarcinoma HepG2 cell line: Role of the MT1 receptor. *J Pineal Res*, 51, 463–471.

Cardinali, D. P., M. I. Vacas, and E. E. Boyer (1979) Specific binding of melatonin in bovine brain. *Endocrinology*, 105, 437–441.

Carlberg, C., R. Hooft van Huijsduijnen, J. K. Staple, J. F. DeLamarter, and M. Becker-Andre (1994) RZRs, a new family of retinoid-related orphan receptors that function as both monomers and homodimers. *Mol Endocrinol*, 8, 757–770.

Cerea, G., M. Vaghi, A. Ardizzoia, S. Villa, R. Bucovec, S. Mengo, G. Gardani, G. Tancini, and P. Lissoni (2003) Biomodulation of cancer chemotherapy for metastatic colorectal cancer: A randomized study of weekly low-dose irinotecan alone versus irinotecan plus the oncostatic pineal hormone melatonin in metastatic colorectal cancer patients progressing on 5-fluorouracil-containing combinations. *Anticancer Res*, 23, 1951–1954.

Chan, A. S., F. P. Lai, R. K. Lo, T. A. Voyno-Yasenetskaya, E. J. Stanbridge, and Y. H. Wong (2002) Melatonin MT1 and MT2 receptors stimulate c-Jun N-terminal kinase via pertussis toxin-sensitive and -insensitive G proteins. *Cell Signal*, 14, 249–257.

Cos, S., D. E. Blask, A. Lemus-Wilson, and A. B. Hill (1991) Effects of melatonin on the cell cycle kinetics and "estrogen-rescue" of MCF-7 human breast cancer cells in culture. *J Pineal Res*, 10, 36–42.

Cos, S., R. Fernandez, A. Guezmes, and E. J. Sanchez-Barcelo (1998) Influence of melatonin on invasive and metastatic properties of MCF-7 human breast cancer cells. *Cancer Res*, 58, 4383–4390.

Cos, S., D. Mediavilla, C. Martinez-Campa, A. Gonzalez, C. Alonso-Gonzalez, and E. J. Sanchez-Barcelo (2006) Exposure to light-at-night increases the growth of DMBA-induced mammary adenocarcinomas in rats. *Cancer Lett*, 235, 266–271.

Cuesta, S., R. Kireev, K. Forman, C. Garcia, G. Escames, C. Ariznavarreta, E. Vara, and J. A. Tresguerres (2010) Melatonin improves inflammation processes in liver of senescence-accelerated prone male mice (SAMP8). *Exp Gerontol*, 45, 950–956.

Danielczyk, K. and P. Dziegiel (2009) The expression of MT1 melatonin receptor and Ki-67 antigen in melanoma malignum. *Anticancer Res*, 29, 3887–3895.

Desmedt, C., F. Piette, S. Loi, Y. Wang, F. Lallemand, B. Haibe-Kains, G. Viale et al. (2007) Strong time dependence of the 76-gene prognostic signature for node-negative breast cancer patients in the TRANSBIG multicenter independent validation series. *Clin Cancer Res*, 13, 3207–3214.

Dillon, D. C., S. E. Easley, B. B. Asch, R. T. Cheney, L. Brydon, R. Jockers, J. S. Winston, J. S. Brooks, T. Hurd, and H. L. Asch (2002) Differential expression of high-affinity melatonin receptors (MT1) in normal and malignant human breast tissue. *Am J Clin Pathol*, 118, 451–458.

Dubocovich, M. L., P. Delagrange, D. N. Krause, D. Sugden, D. P. Cardinali, and J. Olcese (2010) International Union of Basic and Clinical Pharmacology. LXXV. Nomenclature, classification, and pharmacology of G protein-coupled melatonin receptors. *Pharmacol Rev*, 62, 343–380.

Farmer, P., H. Bonnefoi, V. Becette, M. Tubiana-Hulin, P. Fumoleau, D. Larsimont, G. Macgrogan et al. (2005) Identification of molecular apocrine breast tumours by microarray analysis. *Oncogene*, 24, 4660–4671.

Flynn-Evans, E. E., R. G. Stevens, H. Tabandeh, E. S. Schernhammer, and S. W. Lockley (2009) Total visual blindness is protective against breast cancer. *Cancer Causes Control*, 20, 1753–1756.

Fredriksson, R., M. C. Lagerstrom, L. G. Lundin, and H. B. Schioth (2003) The G-protein-coupled receptors in the human genome form five main families. Phylogenetic analysis, paralogon groups, and fingerprints. *Mol Pharmacol*, 63, 1256–1272.

Futagami, M., S. Sato, T. Sakamoto, Y. Yokoyama, and Y. Saito (2001) Effects of melatonin on the proliferation and cis-diamminedichloroplatinum (CDDP) sensitivity of cultured human ovarian cancer cells. *Gynecol Oncol*, 82, 544–549.

Giguere, V., L. D. McBroom, and G. Flock (1995) Determinants of target gene specificity for ROR alpha 1: Monomeric DNA binding by an orphan nuclear receptor. *Mol Cell Biol*, 15, 2517–2526.

Giguere, V., M. Tini, G. Flock, E. Ong, R. M. Evans, and G. Otulakowski (1994) Isoform-specific amino-terminal domains dictate DNA-binding properties of ROR alpha, a novel family of orphan hormone nuclear receptors. *Genes Dev*, 8, 538–553.

Gilad, E., M. Laufer, H. Matzkin, and N. Zisapel (1999) Melatonin receptors in PC3 human prostate tumor cells. *J Pineal Res*, 26, 211–220.

Ginestier, C., N. Cervera, P. Finetti, S. Esteyries, B. Esterni, J. Adelaide, L. Xerri et al. (2006) Prognosis and gene expression profiling of 20q13-amplified breast cancers. *Clin Cancer Res*, 12, 4533–4544.

Hill, S. M., C. Cheng, L. Yuan, L. Mao, R. Jockers, B. Dauchy, T. Frasch, and D. E. Blask (2011) Declining melatonin levels and MT1 receptor expression in aging rats is associated with enhanced mammary tumor growth and decreased sensitivity to melatonin. *Breast Cancer Res Treat*, 127, 91–98.

Hong, Y., T. Downey, K. W. Eu, P. K. Koh, and P. Y. Cheah (2010) A 'metastasis-prone' signature for early-stage mismatch-repair proficient sporadic colorectal cancer patients and its implications for possible therapeutics. *Clin Exp Metastasis*, 27, 83–90.

Hu, D. N., S. A. McCormick, and J. E. Roberts (1998) Effects of melatonin, its precursors and derivatives on the growth of cultured human uveal melanoma cells. *Melanoma Res*, 8, 205–210.

Jablonska, K., B. Pula, A. Zemla, T. Owczarek, A. Wojnar, J. Rys, A. Ambicka, M. Podhorska-Okolow, M. Ugorski, and P. Dziegiel (2013) Expression of melatonin receptor MT1 in cells of human invasive ductal breast carcinoma. *J Pineal Res*, 54, 334–345.

Kaiser, S., Y. K. Park, J. L. Franklin, R. B. Halberg, M. Yu, W. J. Jessen, J. Freudenberg et al. (2007) Transcriptional recapitulation and subversion of embryonic colon development by mouse colon tumor models and human colon cancer. *Genome Biol*, 8, R131.

Kim, C. H. and Y. M. Yoo (2010) Melatonin induces apoptotic cell death via p53 in LNCaP cells. *Korean J Physiol Pharmacol*, 14, 365–369.

Klein, D. C. (1999) Serotonin *N*-acetyltransferase. A personal historical perspective. *Adv Exp Med Biol*, 460, 5–16.
Kliukiene, J., T. Tynes, and A. Andersen (2001) Risk of breast cancer among Norwegian women with visual impairment. *Br J Cancer*, 84, 397–399.
Kloog, I., A. Haim, R. G. Stevens, M. Barchana, and B. A. Portnov (2008) Light at night co-distributes with incident breast but not lung cancer in the female population of Israel. *Chronobiol Int*, 25, 65–81.
Lai, L., L. Yuan, Q. Cheng, C. Dong, L. Mao, and S. M. Hill (2009) Alteration of the MT1 melatonin receptor gene and its expression in primary human breast tumors and breast cancer cell lines. *Breast Cancer Res Treat*, 118, 293–305.
Leon, J., J. Casado, A. Carazo, L. Sanjuan, A. Mate, P. Munoz de Rueda, P. de la Cueva et al. (2012) Gender-related invasion differences associated with mRNA expression levels of melatonin membrane receptors in colorectal cancer. *Mol Carcinog*, 51, 608–618.
Lipton, J., J. T. Megerian, S. V. Kothare, Y. J. Cho, T. Shanahan, H. Chart, R. Ferber et al. (2009) Melatonin deficiency and disrupted circadian rhythms in pediatric survivors of craniopharyngioma. *Neurology*, 73, 323–325.
Lissoni, P., S. Barni, G. Cattaneo, G. Tancini, G. Esposti, D. Esposti, and F. Fraschini (1991) Clinical results with the pineal hormone melatonin in advanced cancer resistant to standard antitumor therapies. *Oncology*, 48, 448–450.
Lissoni, P., S. Barni, G. Tancini, A. Ardizzoia, F. Rovelli, M. Cazzaniga, F. Brivio et al. (1993) Immunotherapy with subcutaneous low-dose interleukin-2 and the pineal indole melatonin as a new effective therapy in advanced cancers of the digestive tract. *Br J Cancer*, 67, 1404–1407.
Lissoni, P., G. Tancini, F. Paolorossi, M. Mandala, A. Ardizzoia, F. Malugani, L. Giani, and S. Barni (1999) Chemoneuroendocrine therapy of metastatic breast cancer with persistent thrombocytopenia with weekly low-dose epirubicin plus melatonin: A phase II study. *J Pineal Res*, 26, 169–173.
Maurice, P., A. M. Daulat, R. Turecek, K. Ivankova-Susankova, F. Zamponi, M. Kamal, N. Clement et al. (2010) Molecular organization and dynamics of the melatonin MT(1) receptor/RGS20/G(i) protein complex reveal asymmetry of receptor dimers for RGS and G(i) coupling. *EMBO J*, 29, 3646–3659.
Mediavilla, M. D., S. Cos, and E. J. Sanchez-Barcelo (1999) Melatonin increases p53 and p21WAF1 expression in MCF-7 human breast cancer cells in vitro. *Life Sci*, 65, 415–420.
Medvedev, A., Z. H. Yan, T. Hirose, V. Giguere, and A. M. Jetten (1996) Cloning of a cDNA encoding the murine orphan receptor RZR/ROR gamma and characterization of its response element. *Gene*, 181, 199–206.
Messner, M., G. Huether, T. Lorf, G. Ramadori, and H. Schworer (2001) Presence of melatonin in the human hepatobiliary-gastrointestinal tract. *Life Sci*, 69, 543–551.
Mills, E., P. Wu, D. Seely, and G. Guyatt (2005) Melatonin in the treatment of cancer: A systematic review of randomized controlled trials and meta-analysis. *J Pineal Res*, 39, 360–366.
Missbach, M., B. Jagher, I. Sigg, S. Nayeri, C. Carlberg, and I. Wiesenberg (1996) Thiazolidine diones, specific ligands of the nuclear receptor retinoid Z receptor/retinoid acid receptor-related orphan receptor alpha with potent antiarthritic activity. *J Biol Chem*, 271, 13515–13522.
Nakamura, E., K. Kozaki, H. Tsuda, E. Suzuki, A. Pimkhaokham, G. Yamamoto, T. Irie et al. (2008) Frequent silencing of a putative tumor suppressor gene melatonin receptor 1 A (MTNR1A) in oral squamous-cell carcinoma. *Cancer Sci*, 99, 1390–1400.
Nemeth, C., S. Humpeler, E. Kallay, I. Mesteri, M. Svoboda, O. Rogelsperger, N. Klammer, T. Thalhammer, and C. Ekmekcioglu (2011) Decreased expression of the melatonin receptor 1 in human colorectal adenocarcinomas. *J Biol Regul Homeost Agents*, 25, 531–542.
Nosjean, O., M. Ferro, F. Coge, P. Beauverger, J. M. Henlin, F. Lefoulon, J. L. Fauchere, P. Delagrange, E. Canet, and J. A. Boutin (2000) Identification of the melatonin-binding site MT3 as the quinone reductase 2. *J Biol Chem*, 275, 31311–31317.
Oprea-Ilies, G., E. Haus, L. Sackett-Lundeen, Y. Liu, L. McLendon, R. Busch, A. Adams, and C. Cohen (2013) Expression of melatonin receptors in triple negative breast cancer (TNBC) in African American and Caucasian women: Relation to survival. *Breast Cancer Res Treat*, 137, 677–687.
Pandi-Perumal, S. R., I. Trakht, V. Srinivasan, D. W. Spence, G. J. Maestroni, N. Zisapel, and D. P. Cardinali (2008) Physiological effects of melatonin: Role of melatonin receptors and signal transduction pathways. *Prog Neurobiol*, 85, 335–353.
Parent, M. E., M. El-Zein, M. C. Rousseau, J. Pintos, and J. Siemiatycki (2012) Night work and the risk of cancer among men. *Am J Epidemiol*, 176, 751–759.
Petranka, J., W. Baldwin, J. Biermann, S. Jayadev, J. C. Barrett, and E. Murphy (1999) The oncostatic action of melatonin in an ovarian carcinoma cell line. *J Pineal Res*, 26, 129–136.

Reiter, R. J. (1991) Pineal melatonin: Cell biology of its synthesis and of its physiological interactions. *Endocr Rev*, 12, 151–180.

Reiter, R. J., D. X. Tan, A. Korkmaz, T. C. Erren, C. Piekarski, H. Tamura, and L. C. Manchester (2007) Light at night, chronodisruption, melatonin suppression, and cancer risk: A review. *Crit Rev Oncog*, 13, 303–328.

Rimler, A., Z. Culig, G. Levy-Rimler, Z. Lupowitz, H. Klocker, H. Matzkin, G. Bartsch, and N. Zisapel (2001) Melatonin elicits nuclear exclusion of the human androgen receptor and attenuates its activity. *Prostate*, 49, 145–154.

Rimler, A., Z. Culig, Z. Lupowitz, and N. Zisapel (2002a) Nuclear exclusion of the androgen receptor by melatonin. *J Steroid Biochem Mol Biol*, 81, 77–84.

Rimler, A., R. Jockers, Z. Lupowitz, S. R. Sampson, and N. Zisapel (2006) Differential effects of melatonin and its downstream effector PKCalpha on subcellular localization of RGS proteins. *J Pineal Res*, 40, 144–152.

Rimler, A., R. Jockers, Z. Lupowitz, and N. Zisapel (2007) Gi and RGS proteins provide biochemical control of androgen receptor nuclear exclusion. *J Mol Neurosci*, 31, 1–12.

Rimler, A., Z. Lupowitz, and N. Zisapel (2002b) Differential regulation by melatonin of cell growth and androgen receptor binding to the androgen response element in prostate cancer cells. *Neuro Endocrinol Lett*, 23(Suppl 1), 45–49.

Roberts, J. E., D. N. Hu, L. Martinez, and C. F. Chignell (2000) Photophysical studies on melatonin and its receptor agonists. *J Pineal Res*, 29, 94–99.

Sampson, S. R., Z. Lupowitz, L. Braiman, and N. Zisapel (2006) Role of protein kinase Calpha in melatonin signal transduction. *Mol Cell Endocrinol*, 252, 82–87.

Santoro, R., M. Marani, G. Blandino, P. Muti, and S. Strano (2012) Melatonin triggers p53Ser phosphorylation and prevents DNA damage accumulation. *Oncogene*, 31, 2931–2942.

Santoro, R., F. Mori, M. Marani, G. Grasso, M. A. Cambria, G. Blandino, P. Muti, and S. Strano (2013) Blockage of melatonin receptors impairs p53-mediated prevention of DNA damage accumulation. *Carcinogenesis*, 34, 1051–1061.

Sauer, L. A., R. T. Dauchy, and D. E. Blask (2001) Melatonin inhibits fatty acid transport in inguinal fat pads of hepatoma 7288CTC-bearing and normal Buffalo rats via receptor-mediated signal transduction. *Life Sci*, 68, 2835–2844.

Schernhammer, E. S., F. Berrino, V. Krogh, G. Secreto, A. Micheli, E. Venturelli, S. Grioni et al. (2010) Urinary 6-sulphatoxymelatonin levels and risk of breast cancer in premenopausal women: The ORDET cohort. *Cancer Epidemiol Biomarkers Prev*, 19, 729–737.

Schernhammer, E. S., F. Berrino, V. Krogh, G. Secreto, A. Micheli, E. Venturelli, S. Sieri et al. (2008) Urinary 6-sulfatoxymelatonin levels and risk of breast cancer in postmenopausal women. *J Natl Cancer Inst*, 100, 898–905.

Schernhammer, E. S., A. Giobbie-Hurder, K. Gantman, J. Savoie, R. Scheib, L. M. Parker, and W. Y. Chen (2012) A randomized controlled trial of oral melatonin supplementation and breast cancer biomarkers. *Cancer Causes Control*, 23, 609–616.

Schernhammer, E. S. and S. E. Hankinson (2009) Urinary melatonin levels and postmenopausal breast cancer risk in the Nurses' Health Study cohort. *Cancer Epidemiol Biomarkers Prev*, 18, 74–79.

Schrader, M., C. Danielsson, I. Wiesenberg, and C. Carlberg (1996) Identification of natural monomeric response elements of the nuclear receptor RZR/ROR. They also bind COUP-TF homodimers. *J Biol Chem*, 271, 19732–19736.

Shiu, S. Y., I. C. Law, K. W. Lau, P. C. Tam, A. W. Yip, and W. T. Ng (2003) Melatonin slowed the early biochemical progression of hormone-refractory prostate cancer in a patient whose prostate tumor tissue expressed MT1 receptor subtype. *J Pineal Res*, 35, 177–182.

Skrzypczak, M., K. Goryca, T. Rubel, A. Paziewska, M. Mikula, D. Jarosz, J. Pachlewski, J. Oledzki, and J. Ostrowski (2010) Modeling oncogenic signaling in colon tumors by multidirectional analyses of microarray data directed for maximization of analytical reliability. *PLoS One*, 5, e13091.

Smirnov, A. N. (2001) Nuclear melatonin receptors. *Biochemistry (Mosc)*, 66, 19–26.

Souza, A. V., M. A. Visconti, and A. M. Castrucci (2003) Melatonin biological activity and binding sites in human melanoma cells. *J Pineal Res*, 34, 242–248.

Steinhilber, D., M. Brungs, O. Werz, I. Wiesenberg, C. Danielsson, J. P. Kahlen, S. Nayeri, M. Schrader, and C. Carlberg (1995) The nuclear receptor for melatonin represses 5-lipoxygenase gene expression in human B lymphocytes. *J Biol Chem*, 270, 7037–7040.

Verslype, C., O. Rosmorduc, P. Rougier, and ESMO Guidelines Working Group (2012) Hepatocellular carcinoma: ESMO–ESDO Clinical Practice Guidelines for diagnosis, treatment and follow-up. *Ann Oncol*, 23(Suppl 7), vii41–vii48.

Wang, Y. M., B. Z. Jin, F. Ai, C. H. Duan, Y. Z. Lu, T. F. Dong, and Q. L. Fu (2012) The efficacy and safety of melatonin in concurrent chemotherapy or radiotherapy for solid tumors: A meta-analysis of randomized controlled trials. *Cancer Chemother Pharmacol*, 69, 1213–1220.

Wenzel, U., A. Nickel, and H. Daniel (2005) Melatonin potentiates flavone-induced apoptosis in human colon cancer cells by increasing the level of glycolytic end products. *Int J Cancer*, 116, 236–242.

Wiesenberg, I., M. Chiesi, M. Missbach, C. Spanka, W. Pignat, and C. Carlberg (1998) Specific activation of the nuclear receptors PPARgamma and RORA by the antidiabetic thiazolidinedione BRL 49653 and the antiarthritic thiazolidinedione derivative CGP 52608. *Mol Pharmacol*, 53, 1131–1138.

Winrow, C. J., J. P. Capone, and R. A. Rachubinski (1998) Cross-talk between orphan nuclear hormone receptor RZRalpha and peroxisome proliferator-activated receptor alpha in regulation of the peroxisomal hydratase-dehydrogenase gene. *J Biol Chem*, 273, 31442–31448.

Xi, S. C., S. W. Siu, S. W. Fong, and S. Y. Shiu (2001) Inhibition of androgen-sensitive LNCaP prostate cancer growth in vivo by melatonin: Association of antiproliferative action of the pineal hormone with mt1 receptor protein expression. *Prostate*, 46, 52–61.

Yan, J. J., F. Shen, K. Wang, and M. C. Wu (2002) Patients with advanced primary hepatocellular carcinoma treated by melatonin and transcatheter arterial chemoembolization: A prospective study. *Hepatobiliary Pancreat Dis Int*, 1, 183–186.

Yuan, L., A. R. Collins, J. Dai, M. L. Dubocovich, and S. M. Hill (2002) MT(1) melatonin receptor overexpression enhances the growth suppressive effect of melatonin in human breast cancer cells. *Mol Cell Endocrinol*, 192, 147–156.

19 Enhancement of Antitumor Activity of Ursolic Acid by Melatonin in Colon Cancer Cells

Wei Guo, Jingshu Wang, and Wuguo Deng

CONTENTS

19.1 INTRODUCTION

Colon cancer, a malignant tumor arising from the inner wall of the large intestine, has become the global leading cause of death from cancers (André et al., 2004). Although surgery is introduced as the most common treatment strategy, 40%–50% of patients ultimately relapse and die of the metastatic disease. Thus, novel therapy strategies are necessary to reduce the low survival rate of colon cancer patients. Increased knowledge and studies about cancers indicated that the consumption of diets rich in fruits and vegetables could lower cancer incidence. Natural products existing in the fruits and vegetables have been increasingly and widely used to prevent and treat cancer, including colon cancer (Volate et al., 2005).

Ursolic acid (UA), a natural pentacyclic triterpenoid carboxylic acid, largely distributed in medical herbs and edible plants, possesses a wide range of biological activities, such as antibacterial, antiviral, anti-inflammatory, and hepatoprotective (Ikeda et al., 2008; Tsai and Yin, 2008). Antitumor properties have also been attributed to UA. It has been reported to display a variety of roles to influence cancer development process, including tumorigenesis, tumor promotion, invasion, metastasis and angiogenesis, induction of tumor cell differentiation, and DNA synthesis enzyme interruption. Furthermore, UA has been shown to be able to induce apoptosis and inhibit proliferation in colon cancer cells (Shan et al., 2009). However, its anticancer effectiveness is not powerful enough, especially to the variation among individuals of cancer patients and the variation at different stages of carcinogenic progression. Therefore, more information and studies concerning its combination with other antitumor reagents and the action mechanisms of such combination should be required to enhance its therapeutic effect.

Melatonin (MT) is an indoleamine compound produced in the pineal gland and also a plant-derived product (Vitalini et al., 2011). It is a pleiotropic and multitasking molecule by contributing to diverse physiological functions, including antioxidant (Bonnefont-Rousselot et al., 2011; Dominguez-Rodriguez and Breu-Gonzalez 2011; Galano et al., 2011; Mukherjee et al., 2010), immunomodulatory (Couto-Moraes et al., 2009), antiaging (Petrosillo et al., 2010; Reiter et al., 1994), and tumor inhibition (Cabrera et al., 2010; Hill et al., 2009; Joo and Yoo, 2009; Jung-Hynes et al., 2011;

Mao et al., 2010; Margheri et al., 2012; Orendas et al., 2009; Zha et al., 2012). MT has been shown to inhibit growth of different tumors in experimental preclinical models in vitro and in vivo, including colon cancer (Garcia-Navarro et al., 2007; Gonzalez et al., 2011; Jung-Hynes et al., 2010; Motilva et al., 2011; Padillo et al., 2010; Proietti et al., 2011; Srinivasan et al., 2008; Um et al., 2011; Wang et al., 2012). It is capable of modulating the related signaling pathways associated with antiproliferation, proapoptotic effect, and potent antioxidation ability in cancer cells (Sánchez-Hidalgo et al., 2012). MT has been shown to potentiate flavone-induced apoptosis in human cancer cells by increasing the level of glycolytic end products. It also helps decrease angiogenesis in cancer cells, which means that it helps block blood supply to the tumor, resulting in tumor suppression. However, the precise therapeutic effects and underlying mechanisms of the action of MT are still not so clear, and its combination with other antitumor agents, such as UA, to achieve additional potential benefits in anticancer treatment deserves better investigation.

Cyclooxygenase-2 (COX-2) plays a key role in multiple pathophysiological processes, including inflammation and tumorigenesis (Kajita et al., 2005; Smith et al., 2000; Wang and Dubois, 2010). It is overexpressed in a wide range of human cancers (Kang et al., 2008). COX-2 inhibitors have been shown potentially useful as chemotherapeutic agents in colorectal cancer treatment (Arber et al., 2006). COX-2 expression is transcriptionally controlled by the binding of transactivators such as NF-κB and coactivators such as p300 to the corresponding sites of its promoters (Deng et al., 2003, 2004, 2007; Shi et al., 2012; Xiao et al., 2011). However, little information is available about the regulation of COX-2 expression by UA and MT combination in human colon cancer cells.

In this study, we hypothesized that MT might play a role in sensitizing or synergizing colon cancer cellular response to UA treatment. To test this hypothesis, we first analyzed the combined effect of UA and MT on tumor cell proliferation, migration, and apoptosis in colon cancer SW480 and LoVo cells. Next, we analyzed the combined effects of UA and MT on some key proteins involved in cell proliferation, migration, and apoptosis signaling pathways, to uncover the molecular mechanisms of the combination of these two natural products in colon cancer cells. Our study showed that the enhanced effects of UA in combination with MT on antiproliferation and proapoptosis in colon cancer cells was mediated through simultaneous modulation of the multiple signaling pathways, suggesting that such a combinational treatment might potentially become an effective way in colon cancer therapy.

19.2 RESULTS

To determine whether MT could potentiate the UA-induced inhibition of colon cancer cell proliferation, we first quantitatively analyzed the effect of MT or UA alone on cell proliferation in colon cancer SW480 and LoVo cells by an 3-(4,5-Dimethylthiazol-2-yl)-2,5-diphenyltetrazolium bromide (MTT) assay. As shown in Figure 19.1a, treatment with MT alone at 1.0 mM significantly inhibited cell proliferation, resulting in 18% and 27% inhibitions of cell viability in SW480 and LoVo cells, respectively. Treatment of cells with UA alone at the dose of 10–60 μM also considerably suppressed colon cancer cell viability in a dose-dependent manner (Figure 19.1b). We next examined the combined effect of both agents on cell viability in colon cancer cells. Combined treatment with MT (1.0 mM) and UA at the doses of 10–60 μM significantly enhanced the UA-mediated inhibition of cell viability in both SW480 and LoVo cells as compared with those cells treated with UA alone (Figure 19.1b).

We next evaluated the degree of UA-and-MT-combination-mediated enhancement of antiproliferative effect by analyzing the IC_{50} values of UA for cell viability inhibition in SW480 and LoVo cells treated with or without MT. As shown in Figure 19.1c, treatment with MT (1.0 mM) considerably enhanced the sensitivities of SW480 and LoVo cells to UA, resulting in a marked reduction of the IC_{50} value in LoVo cells when compared with the cells treated with UA alone. The LoVo cells are more sensitive to MT than the SW480 cells.

We also detected the changes in cell morphology and spreading in colon cancer SW480 cell cotreated with UA (20 μM) and MT (1.0 mM). As shown in Figure 19.1d, the combined treatment exhibited highly reduced cell-to-cell contact and was mostly individualized as compared with the cells treated with UA or MT alone. The cells treated with UA or MT alone formed a cell layer, and more spread and filopodia were observed. By contrast, the combined treatment with UA and MT markedly reduced cell-to-cell contact and had lower spreading with fewer formation of filopodia. These results demonstrate that UA and MT combination induces changes in cell morphology and spreading in colon cancer cells.

We also investigated the effect of UA in combination with MT on cell migration in colon cancer SW480 cells by employing a scratch assay. Consistent with the data from cell proliferation inhibition, treatment with UA (20 μM) or MT (1.0 mM) alone also inhibited cell migration, but the

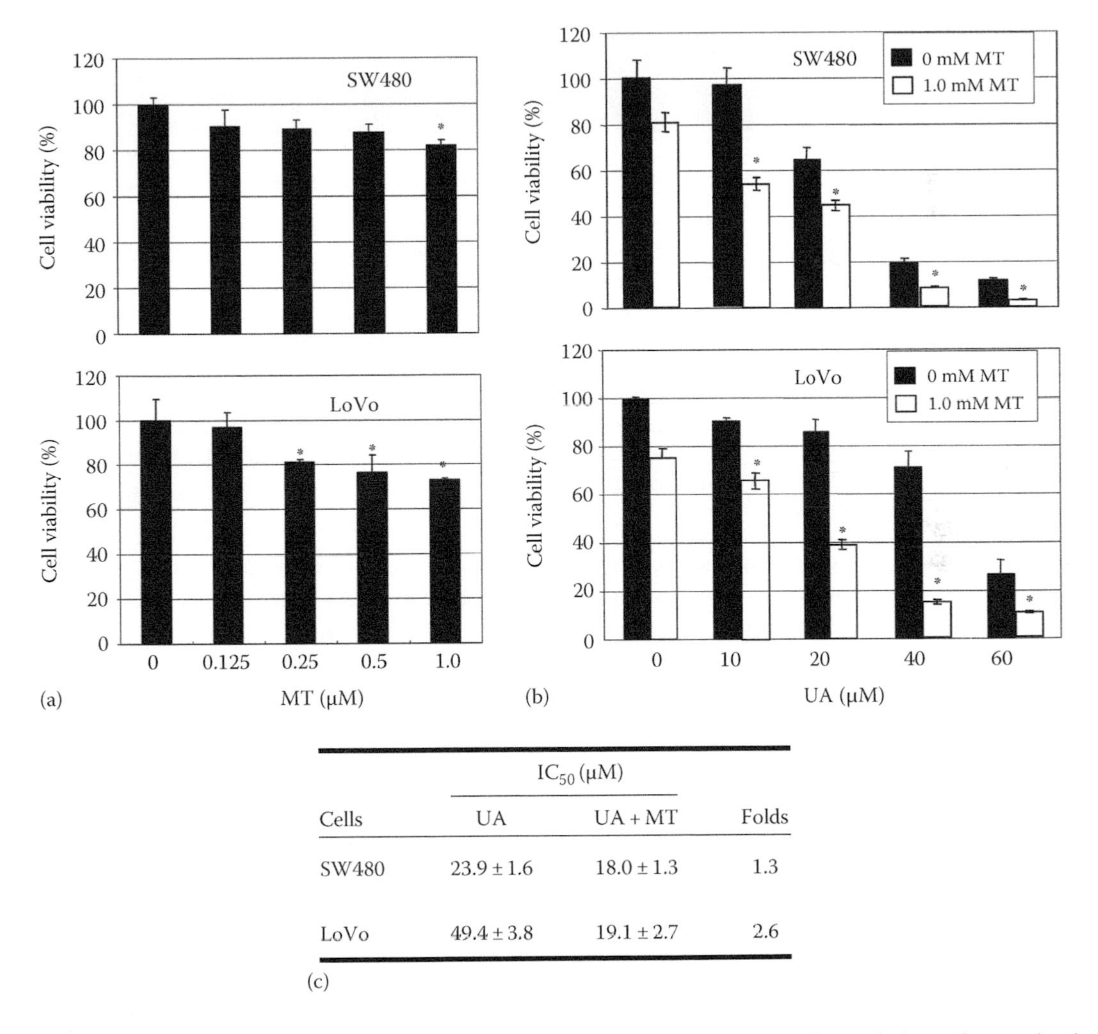

Cells	IC_{50} (μM) UA	UA + MT	Folds
SW480	23.9 ± 1.6	18.0 ± 1.3	1.3
LoVo	49.4 ± 3.8	19.1 ± 2.7	2.6

FIGURE 19.1 MT potentiated the UA-mediated cell proliferation inhibition and morphology change. (a–c) Human SW480 and LoVo cells were treated with MT or UA at the indicated doses. At 48 h after treatment, the cell viability was determined by an MTT assay (a, b), and the IC_{50} values of UA for cell viability inhibition in cells treated with or without MT were determined (c). Cells treated with vehicle control dimethyl sulfoxide (DMSO) were used as the referent group with cell viability set at 100%. The percent cell viability in each treatment group was calculated relative to cells treated with vehicle control. The data are presented as the mean ± SD of three separate experiments. (d) The changes in cell morphology and spreading in SW480 cells treated with UA (20 μM) and MT (1.0 mM) for 48 h were observed, and cells were photographed using a microscope fitted with digital camera. *, $P < 0.05$, significant differences between treatment groups and DMSO control groups (a) or between the UA+MT-treated groups and the UA-treated groups (b). *(Continued)*

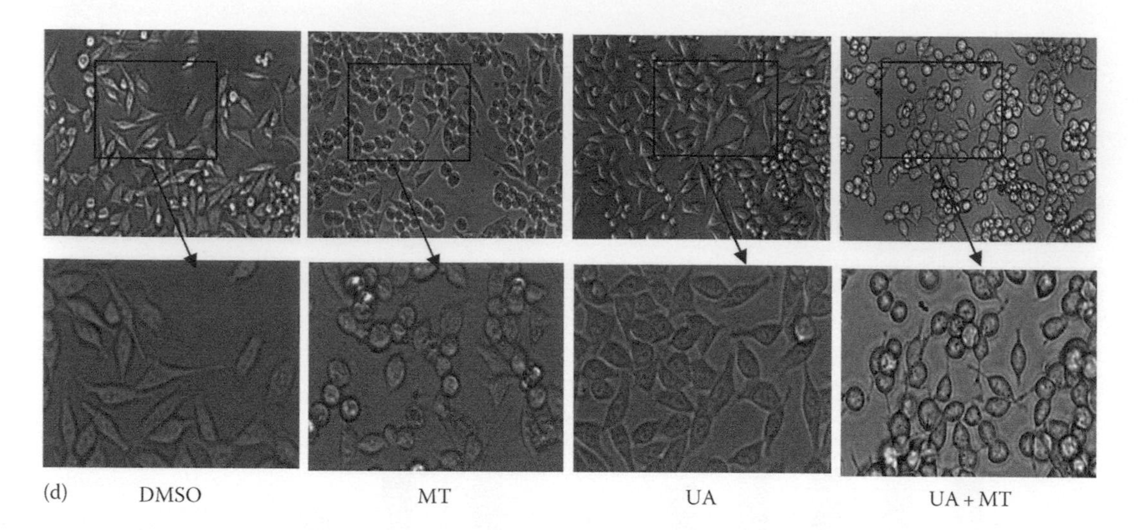

FIGURE 19.1 (*Continued*) MT potentiated the UA-mediated cell proliferation inhibition and morphology change. (a–c) Human SW480 and LoVo cells were treated with MT or UA at the indicated doses. At 48 h after treatment, the cell viability was determined by an MTT assay (a, b), and the IC_{50} values of UA for cell viability inhibition in cells treated with or without MT were determined (c). Cells treated with vehicle control DMSO were used as the referent group with cell viability set at 100%. The percent cell viability in each treatment group was calculated relative to cells treated with vehicle control. The data are presented as the mean ± SD of three separate experiments. (d) The changes in cell morphology and spreading in SW480 cells treated with UA (20 μM) and MT (1.0 mM) for 48 h were observed, and cells were photographed using a microscope fitted with digital camera. *, $P < 0.05$, significant differences between treatment groups and DMSO control groups (a) or between the UA+MT-treated groups and the UA-treated groups (b).

combined treatment with two agents together markedly enhanced the inhibition of cell migration (Figure 19.2a). The part of gap or wounding space between cell layers after making a scratch was occupied completely by the migrating cells after 56 h in the control group. By contrast, the empty space of the cells was not occupied by the migrating cells cotreated with UA and MT (Figure 19.2a). Similar inhibitory effect of UA and MT combination on cell migration was also found in LoVo cells (data not shown). These observations suggest that MT can enhance UA-mediated inhibition of cell migration in colon cancer cells.

Matrix metallopeptidase 9 (MMP9) and cadherin 1 (CDH1) are two key molecules involved in the cell migration signaling pathway. We next tested the combined effects of UA and MT on the expression of these two proteins in SW480 cells by reverse transcription polymerase chain reaction (RT-PCR) analysis. As shown in Figure 19.2b, treatment with UA (20 and 40 μM) or MT (1.0 and 2.0 mM) alone led to a reduction of MMP9; however, the combined treatment with UA (20 μM) and MT (1.0 mM) significantly enhanced the inhibition of MMP9 (Figure 19.2b). The expression of CDH1 was not significantly altered by UA and MT alone or a combination of them (Figure 19.2b).

We also quantitatively analyzed the combined effects of both agents on the expression of MMP9 and CDH1 by a real-time qPCR analysis in SW480 cells. Similar to the inhibition of cell migration, a combined treatment with UA (20 μM) and MT (1.0 mM) significantly enhanced the inhibition of the mRNA level of MMP9 gene, whereas the combination did not change the CDH1 levels as compared with those treatment with UA or MT alone (Figure 19.2c). These results indicate that MT plays an important role in regulating colon cancer cell migration via a modulation of MMP9 signaling.

To determine whether the enhancement of cell proliferation inhibition induced by the combination of UA and MT is associated with the increase of apoptosis, we next analyzed the effect of UA and MT on apoptosis by a FACS analysis. In the cells examined, treatment with UA alone at

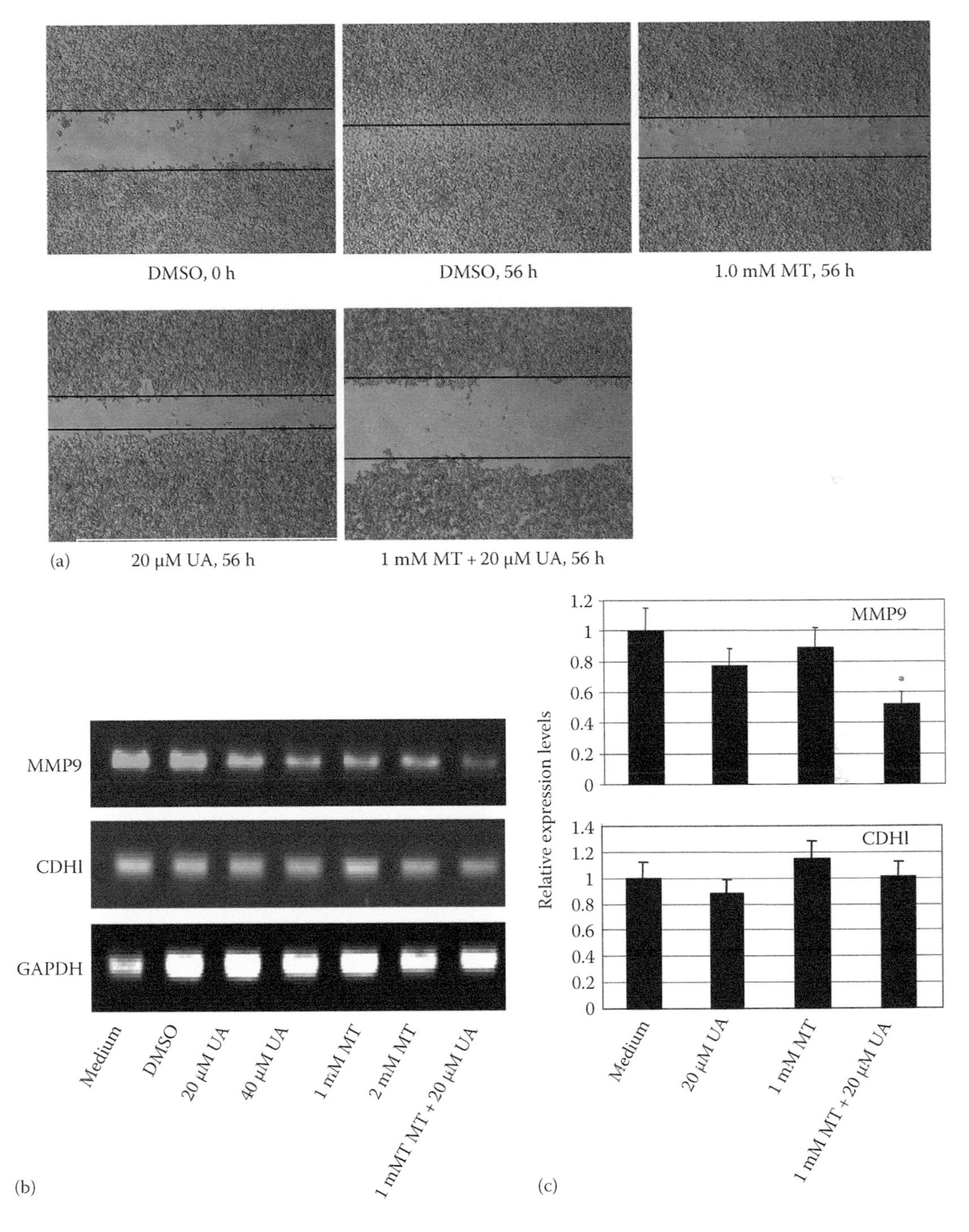

FIGURE 19.2 MT enhanced the UA-mediated cell migration inhibition by regulating MMP9. (a) Cell migration was analyzed by a scratch assay. SW480 cells were grown to full confluency. The cell monolayers were wounded with a sterile pipette tip and washed with medium to remove detached cells from the plates. Cells were left either untreated or treated with the indicated doses of UA or MT. After 56 h, the wound gap was observed, and cells were photographed. (b, c) SW480 cells were treated with UA and MT at the indicated doses. At 56 h after treatment, the expression of MMP9 or CDH1 was detected by RT-PCR (b) or quantitatively analyzed by a real-time qPCR analysis (c). *, $P < 0.05$, significant differences between the UA+MT-treated groups and the UA-treated groups.

the doses of 20 μM induced 9.8% and 15.4% apoptotic cells in SW480 (Figure 19.3a) and LoVo cells (Figure 19.3b), respectively, at 48 h after treatment. However, the addition of MT (1.0 mM) greatly increased the UA-induced apoptosis, resulting in a 21.5%–21.6% induction of apoptotic cells in SW480 (Figure 19.3a) and LoVo cells (Figure 19.3b), respectively. Similarly, the propidium iodide (PI) staining–based Fluorescence-activated cell sorting (FACS) assay also showed that pretreatment with MT (1.0 mM) significantly increased the UA-mediated induction of the apoptotic sub-G1 cell population in both cells (data not shown).

Activation of the caspase signaling is an important event in apoptosis pathway. We next detected the combined effect of UA and MT on the expression of the cleaved proteins of three key apoptosis-related proteins: caspase-3, -9 and protease poly adenosine diphosphate (ADP)-ribose polymerase (PARP) in SW480 cells at 48 h after treatment by Western blot analysis. As shown in Figure 19.3c, cotreatment with MT (1.0 mM) and UA at the doses of 20 and 40 μM resulted in a marked induction of the cleaved caspase-3, -9 and PARP proteins (Figure 19.3c). These results indicate that UA and MT may function as an important and specific mediator to facilitate the activation of multiple caspase cascades.

Cytochrome-c (cyt-c) is the upstream molecule of the caspase-dependent apoptosis pathway. Many apoptotic stimuli induce cyt-c release from the mitochondrial intermembrane space into the cytosol, thereby inducing apoptosis. We next performed immunofluorescence imaging (IFI) analysis to monitor changes in the subcellular localization of cyt-c in UA- and MT-cotreated SW480 cells to determine whether a combined treatment with UA and melatonin could trigger cyt-c release. As shown in Figure 19.3d, UA (20 μM) alone also effectively induced the release of cyt-c from the intermitochondrial space into the cytosol in colon cancer cells. However, the combined treatment with UA (20 μM) and MT (1.0 mM) greatly triggered the release of cyt-c. These results indicate a novel function of MT in coordinating cyt-c release from the intermitochondrial membrane space and facilitating the downstream cyt-c-dependent apoptosome assembly and caspase activation in the cytosol in colon cancer cells.

COX-2 expression has been shown to induce cell proliferation, migration, and angiogenesis in cancer cells. To determine the combined effects of UA and MT on COX-2 signaling in colon cancer cells, we evaluated the expression of COX-2 at protein and mRNA levels by Western blot and RT-PCR in SW480 and LoVo cells. Treatment of the SW480 cells with UA alone at the dose of 20 or 40 μM slightly inhibited COX-2 protein (Figure 19.4a) and mRNA levels (Figure 19.4b); however, a combined treatment with UA (20 μM) and MT (1.0 mM) markedly increased the inhibition of COX-2 expression at protein (Figure 19.4a) and mRNA levels (Figure 19.4b). Similarly, the results from LoVo cells also showed that treatment with UA (20 μM) in combination with MT (1.0 mM) significantly enhanced the inhibition of COX-2 protein expression as compared with those cells treated with UA alone (Figure 19.4c).

To further validate the role of the UA and MT combination in regulating the COX-2 signaling in colon cancer cells, the SW480 cells were pretreated with a COX-2-selective inhibitor celecoxib (CB) (20 μM) for 8 h, followed by the combined treatment with UA (20 μM) and MT (1.0 mM). After 48 h, the effect of the combined treatment on the CB-mediated inhibition of cell proliferation was analyzed by an MTT assay. As shown in Figure 19.4d, pretreatment with CB inhibited cell proliferation, whereas a combination with UA and MT did not significantly alter cell viability inhibition mediated by the COX-2-selective inhibitor. These results indicate that the inhibition of colon cancer cell proliferation by the UA and MT combination might also be partially mediated through the inactivation of the COX-2 signaling.

Transcriptional coactivator p300 has been shown to bind to promoter-bound transactivators such as NF-κB in cell nuclei to regulate COX-2 gene expression. We next performed immunofluorescence assay to confirm the nuclear localization and interaction of p300 and NF-κB in colon cancer LoVo cells. Constitutive translocation of NF-κB p50/p65 and p300 to the cell nuclei (Figure 19.5a and b) and the colocalization of p65 with p50 (Figure 19.5a) and p300 (Figure 19.5b) were detected in LoVo cells. Treatment with UA alone at 20 μM caused translocation of the NF-κB p65 (Figure 19.5a) and p300 (Figure 19.5b) from cell nuclei to cytoplasm. However, the combined

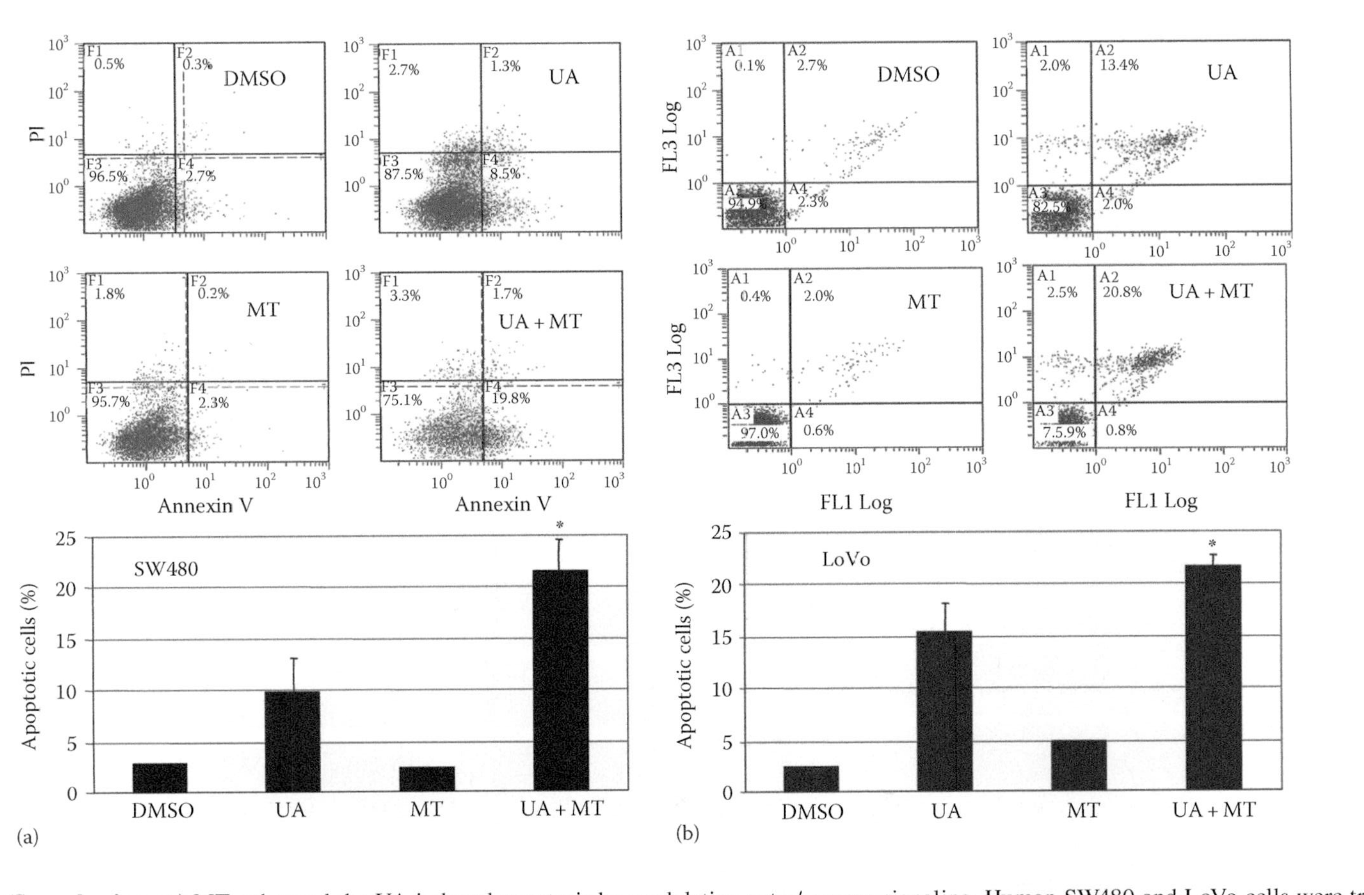

FIGURE 19.3 **(See color insert.)** MT enhanced the UA-induced apoptosis by modulating cyt-c/caspase signaling. Human SW480 and LoVo cells were treated with UA (20 μM) and MT (1.0 mM). At 48 h after treatment, the apoptosis was determined by an FACS analysis (a, b), and the levels of the cleaved caspase-3, -9 and PARP proteins (c) were analyzed by Western blot. The release of cyto-c was determined by IFI analysis to monitor cyto-c release from the intermitochondrial space into the cytosol (d). The apoptoses are represented by the relative percentages of apoptotic cells versus that in DMSO-treated cells. *, $P < 0.05$, significant differences between the UA+MT-treated groups and the UA-treated groups.

(Continued)

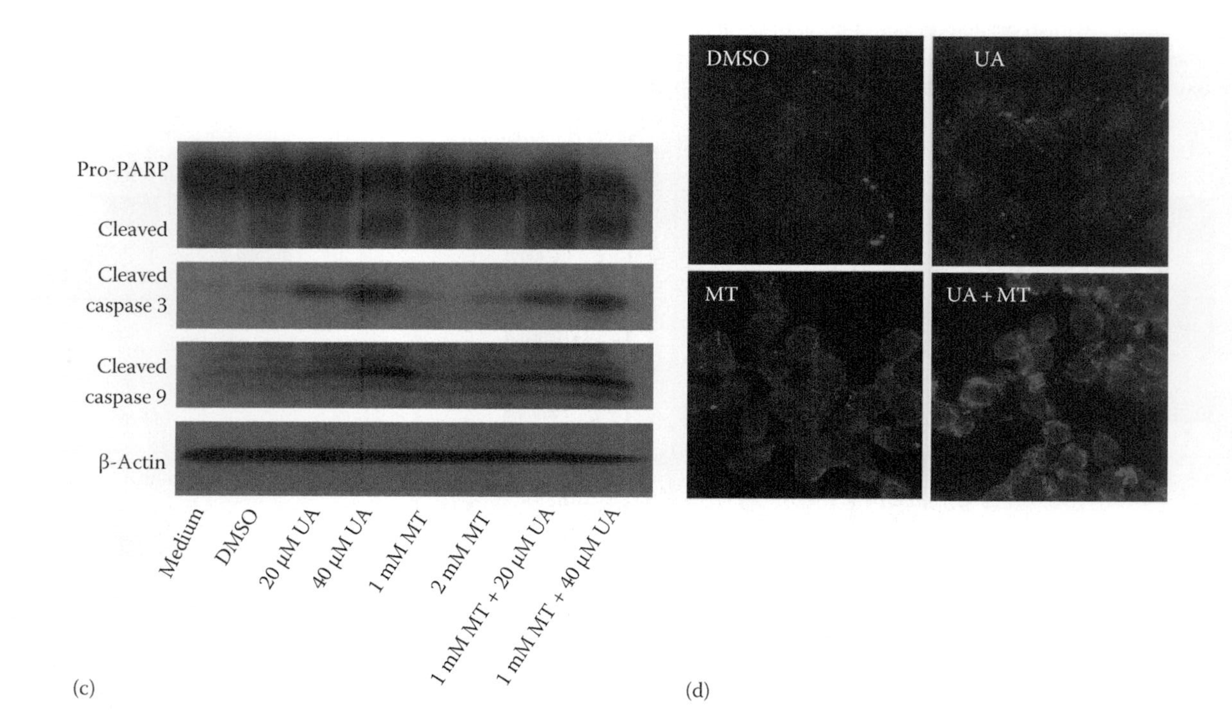

FIGURE 19.3 (*Continued*) **(See color insert.)** MT enhanced the UA-induced apoptosis by modulating cyt-c/caspase signaling. Human SW480 and LoVo cells were treated with UA (20 μM) and MT (1.0 mM). At 48 h after treatment, the apoptosis was determined by an FACS analysis (a, b), and the levels of the cleaved caspase-3, -9 and PARP proteins (c) were analyzed by Western blot. The release of cyto-c was determined by IFI analysis to monitor cyto-c release from the intermitochondrial space into the cytosol (d). The apoptoses are represented by the relative percentages of apoptotic cells versus that in DMSO-treated cells. *, $P < 0.05$, significant differences between the UA+MT-treated groups and the UA-treated groups.

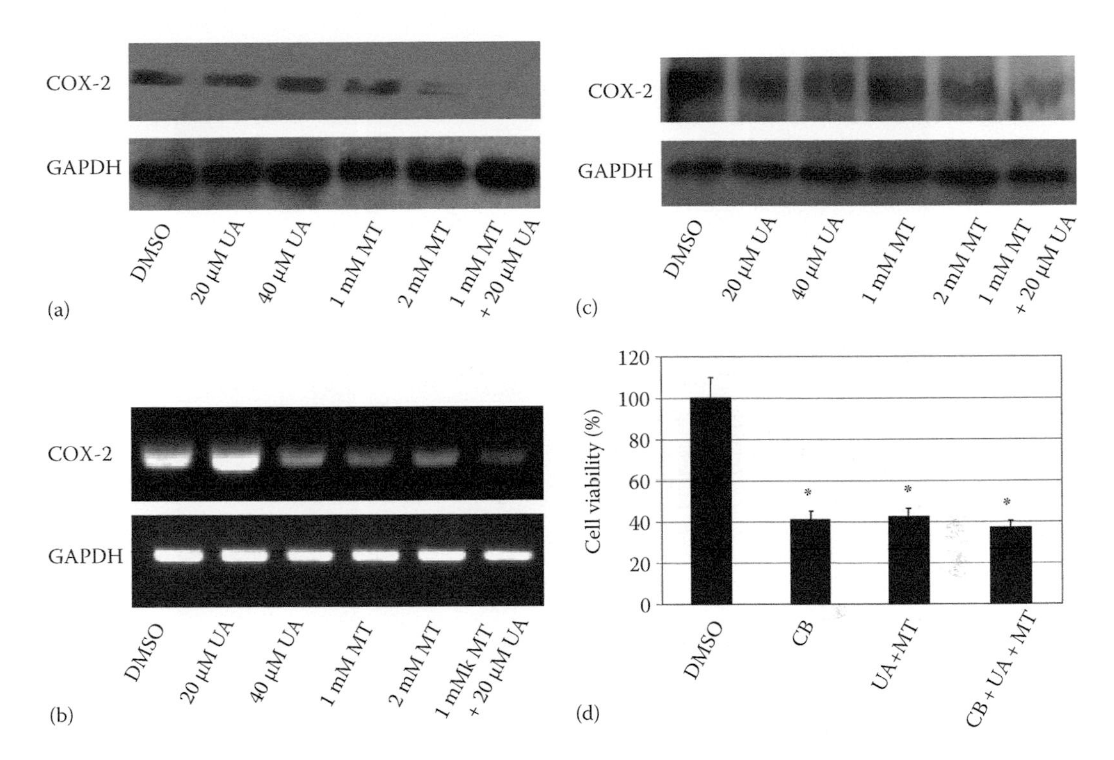

FIGURE 19.4 MT increased the UA-mediated suppression of COX-2 expression. (a–d) Human SW480 and LoVo cells were treated with UA and MT at the indicated doses. At 48 h after treatment, the COX-2 proteins (a, c) and mRNA (b) were analyzed by Western blotting and RT-PCR, respectively. Glyceraldehyde phosphate dehydrogenase (GAPDH) was used as a control for sample loading. (d) SW480 cells were treated with the COX-2 selective inhibitor CB (20 μM) for 24 h and then treated with UA (20 μM) in combination with MT (1.0 mM). At 48 h after treatment, cell viability was determined by MTT analysis. The percent cell viability in each treatment group was calculated relative to cells treated with the vehicle control. The data are presented as the mean ± SD of three separate experiments. *, $P < 0.05$, significant differences between treatment groups and DMSO control groups.

treatment with UA (20 μM) and MT (1.0 mM) markedly enhanced the UA-stimulated translocation of NF-κB p65 (Figure 19.5a) and p300 (Figure 19.5b) proteins from cell nuclei to cytoplasma in comparison with either agent alone. The results indicate that the inhibition of colon cancer cell proliferation by UA and MT combination might be mediated by potentiating NF-κB and p300 translocation from cell nuclei to cytoplasm.

The expression of COX-2 is regulated by the binding activity of NF-κB on COX-2 promoter structure. We next determined whether the UA-and-MT-combination-induced inhibition of cell proliferation and COX-2 expression is mediated by the inhibition of the binding of NF-κB to COX-2 promoter in colon cancer cells by a chromatin immunoprecipitation (ChIP) assay. We used antibodies directed against p50, p65, and p300 to precipitate chromatin in SW480 cells. The cells were treated with UA (20 μM) and MT (1.0 mM) for 48 h, and the COX-2 promoter region in the precipitated chromatin was amplified by PCR. As shown in Figure 19.6a, the combined treatment with UA and MT markedly increased the inhibition of NF-κB p50 and p65 binding to chromatin COX-2 promoter in comparison with those cells treated with UA alone (Figure 19.6a). The quantitative densitometric analysis also showed that the UA in combination with MT significantly enhanced the inhibition of the binding activity of NF-κB p50 and p65 on COX-2 promoter in SW480 cells (Figure 19.6b).

P300 integrates the transcriptional signal by interacting with promoter-bound transcriptional factors such as NF-κB. Since MT inhibited binding of NF-κB to COX-2 promoter, we suspected a

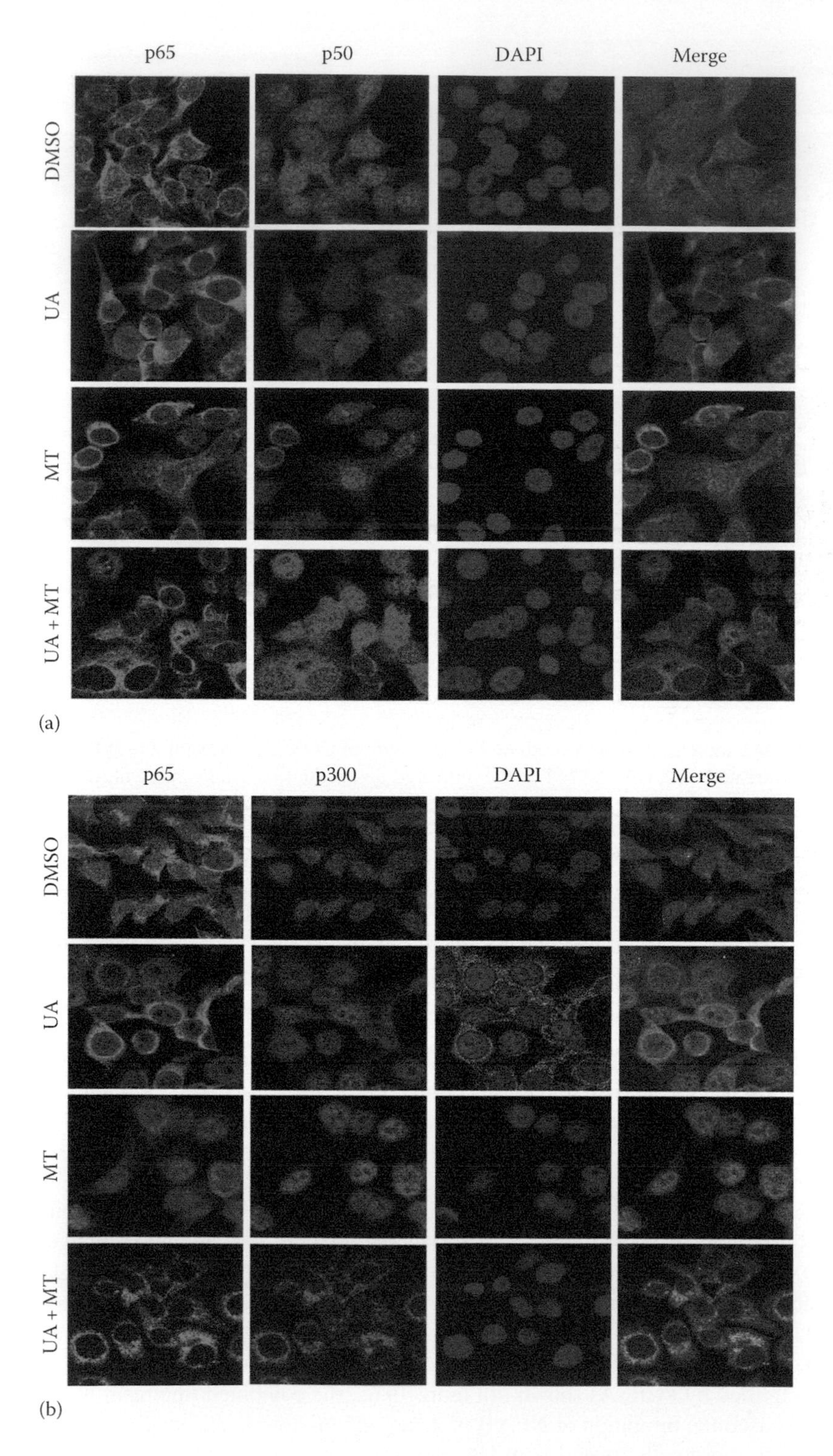

FIGURE 19.5 **(See color insert.)** MT promoted the UA-induced translocation of p300 and NF-κB from nuclei to cytoplasm. Human LoVo cells grown on chamber slides were treated with UA (20 μM) and MT (1.0 mM). At 48 h after treatment, the subcellular localization of p50, p65, and p300 and the colocalization of p65 with p50 (a) or p300 (b) were examined by confocal microscopy analysis with a confocal microscope. More than 100 cells were inspected per experiment, and cells with typical morphology were presented.

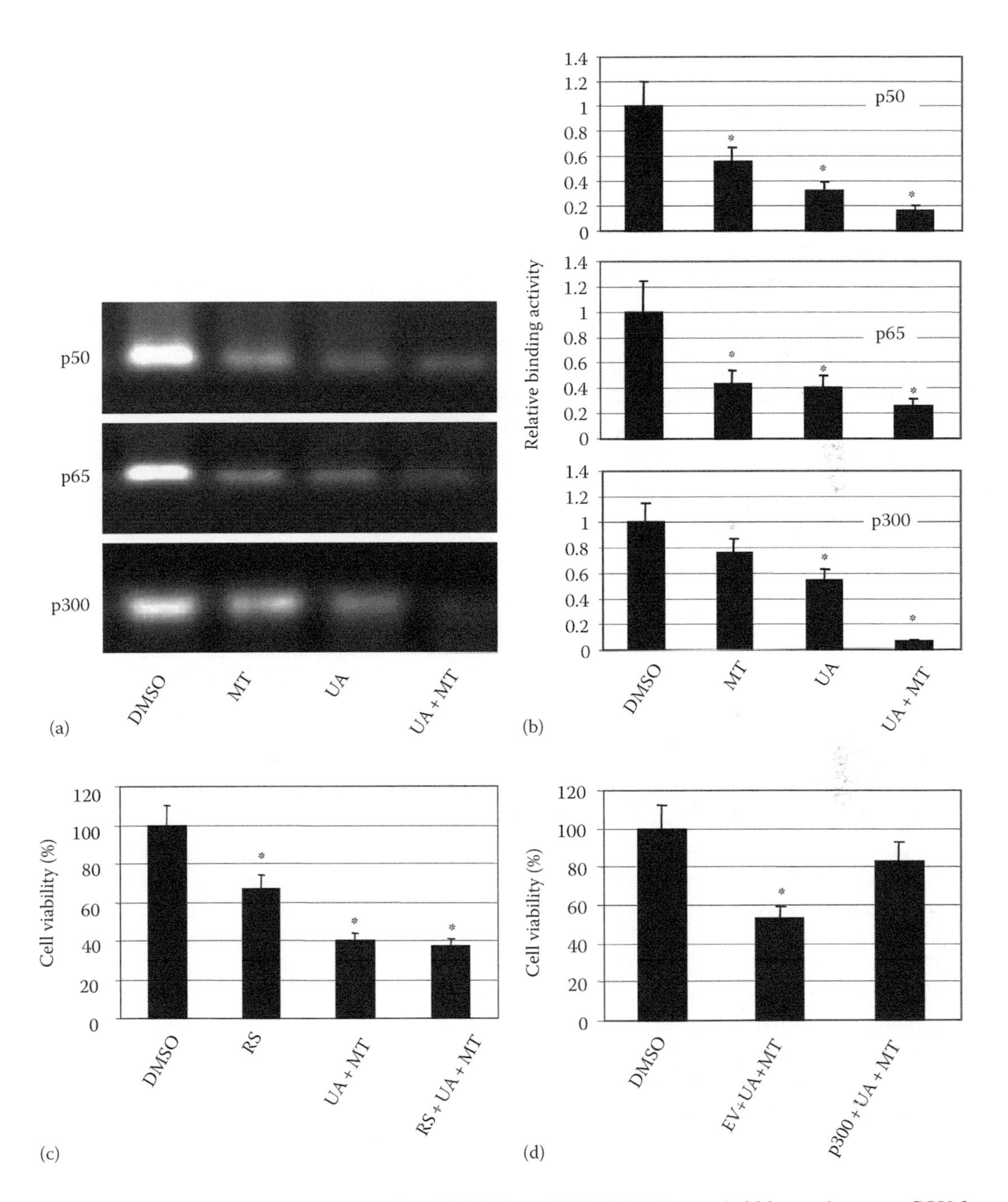

FIGURE 19.6 MT enhanced the UA-mediated inhibition of NF-κB binding and p300 recruitment to COX-2 promoter. (a, b) Human SW480 cells were treated with UA (20 μM) and MT (1.0 mM). Chromatin in the treated cells was immunoprecipitated with antibodies to p50, p65, and p300; and the COX-2 promoter region in the precipitated chromatin was amplified by PCR (a). The relative binding activities of p50, p65, and p300 to COX-2 promoter was analyzed by densitometric analysis (b). (c, d) SW480 cells were pretreated with p300 selective inhibitor RS (20 μM) (c) or transfected with a FLAG-p300 or empty vector (d) for 24 h and then treated with UA (20 μM) and MT (1.0 mM). At 48 h after treatment, the cell viability was determined. Each bar represents the mean ± SD of three experiments. *, $P < 0.05$, significant differences between treatment groups and DMSO control groups.

consequent reduction in the level of p300 in the DNA-transactivator complex. ChIP and densitometric analysis showed that the combined treatment with UA (20 μM) and MT (1.0 mM) also markedly inhibited p300 recruitment to the transactivators-promoter complex in SW480 cells (Figure 19.6a and b). These results indicate that the enhanced inhibition of cell proliferation by UA and MT combination might be partially mediated by the inhibition of NF-κB binding and p300 recruitment on COX-2 promoter.

To further confirm the involvement of p300 signaling in the UA-and-MT-combination-mediated inhibition of cell proliferation, the SW480 cells were pretreated with roscovitine (RS) (20 μM), an inhibitor of p300, and the effect of UA and MT combination on the RS-mediated inhibition of cell proliferation was analyzed. As shown in Figure 19.6c, treatment with RS (20 μM) alone significantly inhibited cell proliferation, whereas pretreatment with RS did not significantly alter the inhibition of cell viability mediated by the combination of UA (20 μM) and MT (1.0 mM), indicating that p300 is an important target for UA and MT.

We also transfected SW480 cells with an expressing vector of constitutively active p300 and analyzed the effects of the overexpressed p300 on the UA-and-MT-combination-mediated proliferation inhibition. As shown in Figure 19.6d, an overexpression of constitutively active p300 effectively reversed the inhibition of cell proliferation by the combination of UA (20 μM) and MT (1.0 mM) as compared with the transfection with the control empty vector (EV). These results further confirm that the UA-and-MT-combination-mediated inhibition is mediated at least in part through p300 signaling pathway in colon cancer cells.

19.3 DISCUSSION

UA has been shown to exhibit a broad range of pharmacological properties such as antitumor, antiangiogenesis, and antimetastasis activities. MT is capable of modulating the signaling pathways associated with antiproliferation and proapoptotic effect in cancer cells to inhibit the growth of different tumors. Both UA and MT have been shown to inhibit cancer cell growth in a huge number of studies, and they have been used in combination with other chemotherapeutic agents in various cancer cells, but they have never been combined altogether as an anticolon cancer treatment. In this study, we hypothesized that MT might play a role in sensitizing or synergizing a colon cancer cellular response to UA treatment and actually analyzed the combined effect of UA and MT on cell proliferation, migration, and apoptosis in colon cancer cells. We found that MT indeed potentiated the effects of UA alone on cell proliferation, migration, inhibition, and apoptosis induction. Furthermore, we found that such combined effects of UA and MT on colon cancer cells are mediated through activating cyt-c/caspase-dependent apoptotic pathway, downregulating the expression of migration marker proteins MMP9, and inhibiting COX-2 signaling through abrogating NF-κB binding and p300 recruitment on COX-2 promoter. To the best of our knowledge, it might be the first time to report the combinational treatment of UA and MT on colon cancer cells and to demonstrate the underlying mechanisms under such a combinational treatment in colon cancer cells. All the results might serve as a basis for guiding the combinational treatment of natural antitumor compounds in improving the therapy efficiency for colon cancer.

COX-2 is an inducible enzyme that converts arachidonic acid to prostaglandins. The increased expression of COX-2 significantly enhances carcinogenesis and inflammatory reactions, and its regulation may be a reasonable target for cancer chemoprevention. COX-2 overexpression commonly appears in a wide range of human cancers. Its expression and the sequential Human Prostaglandin E2 (PGE2) production could upregulate EGFR, PI3K, and ERK1/2 signaling to induce angiogenesis, cell proliferation, invasion, and metastasis of tumor cells (Yang et al., 2009). It has been reported that MT plays its antitumor and anti-inflammation roles partially through inhibiting COX-2 expression (Martínez-Campa et al., 2009; Murakami et al., 2011), which is also proved in our previous studies (Deng et al., 2006). Our results in this study demonstrated that UA could also inhibit COX-2 expression in colon cancer cells, and the pretreatment of MT enhanced the UA-mediated inhibition of COX-2 expression not only at protein level but also at mRNA level, suggesting that MT potentiated the UA-mediated inhibition of colon cancer cell proliferation partially through the suppression of COX-2 signaling.

The mechanism by which COX-2 is commonly overexpressed in most cancer cells is still not clear enough. It has been shown that its expression is transcriptionally controlled by the binding of multiple transactivators and coactivators to the corresponding sites located in its promoter. Among the known several regulatory elements distributing in the core promoter region of COX-2 transcription start site, NF-κB binding site is essential for COX-2 promoter activity (Deng et al., 2003). Also, p300, serving as a transcription coactivator, is necessary for COX-2 expression by exerting a global effect in maintaining COX-2 promoter chromatin structure to enhance the binding of transactivators, such as NF-κB (Deng et al., 2003, 2007). Because UA in combination with MT inhibits COX-2 expression, we were interested in whether UA and MT would modulate NF-κB and p300 signaling in colon cancer cells. In our study, we confirmed the nuclear localization and interaction of NF-κB and p300 in colon cancer cells. We found that the enhanced inhibition of COX-2 expression in colon cancer cells by the cotreatment of these two natural products is partially mediated by stimulating p300 and NF-κB translocation from nuclear to cytosol. We also further demonstrated that the increased inhibitory effects of the combined treatment with UA and MT in colon cancer cells were mediated by inhibiting the binding of NF-κB and p300 to COX-2 promoter, thereby abrogating COX-2 transcriptional activation.

P300 is a transcription coactivator that integrates the transcriptional signal by interacting with promoter-bound transactivators. P300 overexpression augments transcriptional activation of COX-2 (Deng et al., 2003, 2004). P300 HAT plays an essential role in promoter activation of COX-2. It is involved in making functional enhancer elements in the chromatin COX-2 promoter region accessible to transactivators as well as augmenting NF-κB binding and p300 recruitment via acetylation (Deng et al., 2003, 2004). As p300 HAT has been shown to acetylate NF-κB and UA and MT combination exerts an enhanced inhibitory effect on NF-κB binding, it is possible that UA in combination with MT may potentiate the inhibition of p300 HAT activity, thereby altering p300 HAT conformation and catalytic activity. Further studies are needed to elucidate the mechanisms by which UA in combination with MT inhibits p300 HAT activity.

We have also shown that tumor cell killing induced by UA and MT combination is mediated by cyt-c and caspase-dependent apoptosis pathways. Apoptosis has been implicated in an extensive variety of diseases, including cancer. It has been demonstrated to play an important role in human cancer development and response to chemotherapy and radiation therapy (Kondo et al., 2005). The activation of caspase cascade forms the essential basis of apoptosis, while the release of cyt-c from the mitochondrial intermembrane space into the cytosol is the precondition of caspase-dependent apoptosis pathway. The released cyt-c could bind Apaf-1 and adenosine triphosphate (ATP), and then bind pro-caspase-9 to form apoptosome, which cleaves the pro-caspase to caspase 9, and in turn activates the effector caspase-3 (Zou et al., 1999). In this study, we also found that the combinational treatment of UA and MT induced the increase of apoptosis in colon cancer cells by the enhanced activation of caspase and PARP proteins and the promoted release of cyt-c from mitochondria to cytosol. Our results therefore suggest the enhancement of cell proliferation inhibition induced by UA and MT combination in colon cancer cells is associated with the increased activation of the cyt-c and caspase-dependent apoptotic pathway.

In summary, we demonstrated that UA in combination with MT leads to an enhanced antiproliferative, antimigration, and proapoptotic activities and identified the underlying mechanisms of action in colon cancer SW480 and LoVo cells. We found that the combined effects of UA and MT on colon cancer cells are achieved through multiple mechanisms by simultaneously targeting cyt-c/caspase, MMP9/COX-2, and p300/NF-κB signaling. Our findings provide new insights into the molecular mechanisms of UA-and-melatonin-mediated colon cancer cell suppression and suggest that such a combinational treatment might potentially become an effective way in colon cancer therapy.

REFERENCES

André, T., Boni, C., Mounedji-Boudiaf, L. et al. 2004. Oxaliplatin, fluorouracil, and leucovorin as adjuvant treatment for colon cancer. *N Engl J Med* 350:2343–2351.

Arber, N., Eagle, G.C., Spicak, J. et al. 2006. Celecoxib for the prevention of colorectal adenomatous polyps. *N Engl J Med* 355:885–895.

Bonnefont-Rousselot, D., Collin, F., Jore, D. et al. 2011. Reaction mechanism of melatonin oxidation by reactive oxygen species in vitro. *J Pineal Res* 50:328–335.

Cabrera, J., Negrín, G., Estévez, F. et al. 2010. Melatonin decreases cell proliferation and induces melanogenesis in human melanoma SK-MEL-1 cells. *J Pineal Res* 49:45–54.

Couto-Moraes, R., Palermo-Neto, J., Markus, R.P. 2009. The immune-pineal axis: Stress as a modulator of pineal gland function. *Ann NY Acad Sci* 1153:193–202.

Deng, W.G., Montero, A.J., Wu, K.K. 2007. Interferon-gamma suppresses cyclooxygenase-2 promoter activity by inhibiting C-Jun and C/EBP beta binding. *Arterioscler Thromb Vasc Biol* 27:1752–1759.

Deng, W.G., Tang, S.T., Tseng, H.P. et al. 2006. Melatonin suppresses macrophage cyclooxygenase-2 and inducible nitricoxide synthase expression by inhibiting p52 acetylation and binding. *Blood* 108:518–524.

Deng, W.G., Zhu, Y., Wu, K.K. 2003. Up-regulation of p300 binding and p50 acetylation in tumor necrosis factor-alpha-induced cyclooxygenase-2 promoter activation. *J Biol Chem* 278:4770–4777.

Deng, W.G., Zhu, Y., Wu, K.K. 2004. Role of p300 and PCAF in regulating cyclooxygenase-2 promoter activation by inflammatory mediators. *Blood* 103:2135–2142.

Dominguez-Rodriguez, A., Breu-Gonzalez, P. 2011.Melatonin: Still a forgotten antioxidant. *Int J Cardiol* 149:382.

Galano, A., Tan, D.X., Reiter, R.J. 2011. Melatonin as a natural ally against oxidative stress: A physicochemical examination. *J Pineal Res* 51:1–16.

Garcia-Navarro, A., Gonzalez-Puga, C., Escames, G. et al. 2007. Cellular mechanisms involved in the melatonin inhibition of HT-29 human colon cancer cell proliferation in culture. *J Pineal Res* 43:195–205.

Gonzalez, A., del Castillo-Vaquero, A., Miro-Moran, A. et al. 2011. Melatonin reduces pancreatic tumor cell viability by altering mitochondrial physiology. *J Pineal Res* 50:250–260.

Hill, S.M., Frasch, T., Xiang, S. et al. 2009. Molecular mechanisms of melatonin anticancer effects. *Integr Cancer Ther* 8:337–346.

Ikeda, Y., Murakami, A., Ohigashi, H. 2008. Ursolic acid: An anti- and pro-inflammatory triterpenoid. *Mol Nutr Food Res* 52:26–42.

Joo, S.S., Yoo, Y.M. 2009. Melatonin induces apoptotic death in LNCaP cells via p38 and JNK pathways: Therapeutic implications for prostate cancer. *J Pineal Res* 47:8–14.

Jung-Hynes, B., Reiter, R.J., Ahmad, N. 2010. Sirtuins, melatonin and circadian rhythms: Building a bridge between aging and cancer. *J Pineal Res* 48:9–19.

Jung-Hynes, B., Schmit, T.L., Reagan-Shaw, S.R. et al. 2011. Melatonin, a novel Sirt1 inhibitor, imparts antiproliferative effects against prostate cancer in vitro in culture and in vivo in TRAMP model. *J Pineal Res* 50:140–149.

Kajita, S., Ruebel, K.H., Casey, M.B. et al. 2005. Role of COX-2, Thromboxane A2 synthase, and prostaglandin I2 synthase in papillary thyroid carcinoma growth. *Mod Pathol* 18:221–227.

Kang, C.H., Chiang, P.H., Huang, S.C. 2008. Correlation of COX-2 expression in stromal cells with high stage, high grade, and poor prognosis in urothelial carcinoma of upper urinary tracts. *Urology* 72:153–157.

Kondo, Y., Kanzawa, T., Sawaya, R. et al. 2005. The role of autophagy in cancer development and response to therapy. *Nat Rev Cancer* 5:726–734.

Mao, L., Yuan, L., Slakey, L.M. et al. 2010. Inhibition of breast cancer cell invasion by melatonin is mediated through regulation of the p38 mitogen-activated protein kinase signaling pathway. *Breast Cancer Res* 12:R107.

Margheri, M., Pacini, N., Tani, A. et al. 2012. Combined effects of melatonin and all-transretinoic acid and somatostatin on breast cancer cell proliferation and death: Molecular basis for the anticancer effect of these molecules. *Eur J Pharmacol* 681:34–43.

Martínez-Campa, C., González, A., Mediavilla, M.D. et al. 2009. Melatonin inhibits aromatase promoter expression by regulating cyclooxygenases expression and activity in breast cancer cells. *Br J Cancer* 101:1613–1619.

Motilva, V., García-Mauriño, S., Talero, E. et al. 2011. New paradigms in chronic intestinal inflammation and colon cancer: Role of melatonin. *J Pineal Res* 51:44–60.

Mukherjee, D., Roy, S.G., Bandyopadhyay, A. et al. 2010. Melatonin protects against isoproterenol-induced myocardial injury in the rat: Antioxidative mechanisms. *J Pineal Res* 48:251–262.

Murakami, Y., Yuhara, K., Takada, N. et al. 2011. Effect of melatonin on cyclooxygenase-2 expression and nuclear factor-kappa B activation in RAW264.7 macrophage-like cells stimulated with fimbriae of *Porphyromonas gingivalis*. *In Vivo* 25:641–647.

Orendas, P., Kassayova, M., Kajo, K. et al. 2009. Celecoxib and melatonin in prevention of female rat mammary carcinogenesis. *Neoplasma* 56:252–258.

Padillo, F.J., Ruiz-Rabelo, J.F., Cruz, A. et al. 2010. Melatonin and celecoxib improve the outcomes in hamsters with experimental pancreatic cancer. *J Pineal Res* 49:264–270.

Petrosillo, G., Moro, N., Paradies, V. et al. 2010. Increased susceptibility to Ca (2+)-induced permeability transition and to cytochrome c release in rat heart mitochondria with aging: Effect of melatonin. *J Pineal Res* 48:340–346.

Proietti, S., Cucina, A., D'Anselmi, F. et al. 2011. Melatonin and vitamin D3 synergistically down-regulate Akt and MDM2 leading to TGFbeta-1-dependent growth inhibition of breast cancer cells. *J Pineal Res* 50:150–158.

Reiter, R.J., Tan, D.X., Poeggeler, B. et al. 1994. Melatonin as a free radical scavenger: Implications for aging and age-related diseases. *Ann NY Acad Sci* 719:1–12.

Sánchez-Hidalgo, M., Guerrero, J.M., Villegas, I. et al. 2012. Melatonin, a natural programmed cell death inducer in cancer. *Curr Med Chem* 19:3805–3821.

Shan, J.H., Xuan, Y.Y., Zheng, S. et al. 2009. Ursolic acid inhibits proliferation and induces apoptosis of HT-29 colon cancer cells by inhibiting the EGFR/MAPK pathway. *J Zhejiang Univ Sci B* 10:668–674.

Shi, D., Wang, J., Chen, W. et al. 2012. Melatonin suppresses proinflammatory mediators in lipopolysaccharide-stimmulated CRL1999 cells via targeting MAPK, NF-kB, c/EBPβ signaling. *J Pineal Res* 53:154–165.

Smith, W.L., DeWitt, D.L., Garavito, R.M. 2000. Cyclooxygenases: Structural, cellular, and molecular biology. *Annu Rev Biochem* 69:145–182.

Srinivasan, V., Spence, D.W., Pandi-Perumal, S.R. et al. 2008. Therapeutic actions of melatonin in cancer: Possible mechanisms. *Integr Cancer Ther* 7:189–203.

Tsai, S.J., Yin, M.C. 2008. Antioxidative and anti-inflammatory protection of oleanolic acid and ursolic acid in PC12 cells. *J Food Sci* 73:H174–H178.

Um, H.J., Park, J.W., Kwon, T.K. 2011. Melatonin sensitizes Caki renal cancer cells to kahweol-induced apoptosis through CHOP-mediated up-regulation of PUMA. *J Pineal Res* 50:359–366.

Vitalini, S., Gardana, C., Zanzotto, A. et al. 2011. The presence of melatonin in grapevine (*Vitis vinifera* L.) berry tissues. *J Pineal Res* 51:331–337.

Volate, S.R., Davenport, D.W., Muga, S.J. et al. 2005. Modulation of aberrant crypt foci and apoptosis by dietary herbal supplements (quercetin, curcumin, silymarin, ginseng and rutin). *Carcinogenesis* 26:1450–1456.

Wang, D., Dubois, R.N. 2010. The role of COX-2 in intestinal inflammation and colorectal cancer. *Oncogene* 29:781–788.

Wang, J., Xiao, X., Zhang, Y. et al. 2012. Simultaneous modulation of COX-2, p300, Akt, and Apaf-1 signaling by melatonin to inhibit proliferation and induce apoptosis in breast cancer cells. *J Pineal Res* 53:77–90.

Xiao, X., Liu, L., Shi, D. et al. 2011. Quercetin suppresses cyclooxygenase-2 expression and angiogenesis through inactivation of P300 signaling. *PLoS ONE* 6:e22934.

Yang, C.M., Lee, I.T., Lin, C.C. et al. 2009. Cigarette smoke extract induces COX-2 expression via a PKCalpha/c-Src/EGFR, PDGFR/PI3K/Akt/NF-kappaB pathway and p300 in tracheal smooth muscle cells. *Am J Physiol Lung Cell Mol Physiol* 297: L892–L902.

Zha, L., Fan, L., Sun, G. et al. 2012. Melatonin sensitizes human hepatoma cells to endoplasmic reticulum stress-induced apoptosis. *J Pineal Res* 52:322–331.

Zou, H., Li, Y., Liu, X. et al. 1999. An APAF-1.cytochrome c multimeric complex is a functional apoptosome that activates procaspase-9. *J Biol Chem* 274:11549–11556.

20 Synthetic Melatonin Receptor Ligands

Silvia Rivara, Marco Mor, Annalida Bedini, and Gilberto Spadoni

CONTENTS

20.1 NONSELECTIVE MT_1 AND MT_2 MELATONIN RECEPTOR LIGANDS

There is a long-standing interest in the use of melatonin (compound **1**, MLT, Figure 20.1) and its derivatives for the treatment of a number of pathological conditions, mainly sleep disturbances and depression. Neuroprotective properties have also been shown for MLT in different animal models and in clinical studies (Srinivasan et al., 2011; Robertson et al., 2013).

Rational medicinal chemistry approaches have been successfully applied to the discovery of many structurally diverse MLT receptor ligands, two of which, ramelteon (compound **2**) and agomelatine (compound **3**), have reached the market. The structure of MLT has been modified in several ways to identify the pharmacophoric groups and the key structural fragments able to confer potency improvements, metabolic stability, or selectivity between MT_1 and MT_2 receptors. For more detailed information on structure–activity relationships, the reader is referred to review articles (Zlotos, 2005, 2012; Rivara et al., 2008; Mor et al., 2010; Spadoni et al., 2011a). These studies have shown that suitably spaced methoxy and amido groups are critical elements for both MLT receptor affinity and efficacy. Replacement of the 5-methoxy substituent with bromine is tolerated, whereas its substitution with H, OH, or other alkoxy groups resulted in a decrease in receptor binding affinity and intrinsic activity. Interestingly, by replacing the 5-methoxy with a hydrophilic group, such as a 5-hydroxyethoxy, and the simultaneous insertion of bromine in the C-2 indole position gave a high-affinity mixed MT_1 agonist/MT_2 antagonist (compound **4**, Figure 20.1) (Spadoni et al., 2006a). The methoxy group can be incorporated into a five- or six-membered ring, as can be seen in ramelteon and in several other recently developed MLT receptor ligands. This can be regarded as a form of conformational restrain, an approach that has also been extensively applied to the C3-ethanamido side chain to establish the bioactive conformation of MLT. Among conformationally constrained MLT receptor ligands, it is worth mentioning that the tricyclic derivatives **5** (Davies et al., 1998) and **6** (Spadoni et al., 1997) played a key role in the definition of pharmacophore models for MLT receptor agonists. Both compounds show a stereoselective behavior, with the eutomer endowed with more than 100-fold higher binding affinity than the distomer. The alkyl side chain can also be constrained in a *trans*-cyclopropyl fragment as in the clinically advanced tasimelteon (compound **7**) or in recently developed *N*-(arylcyclopropyl)acetamides (Morellato et al., 2013). The naphthalene derivative **8** has MT_1 and MT_2 binding affinities similar to those of MLT. The prominent role of the *N*-acyl chain

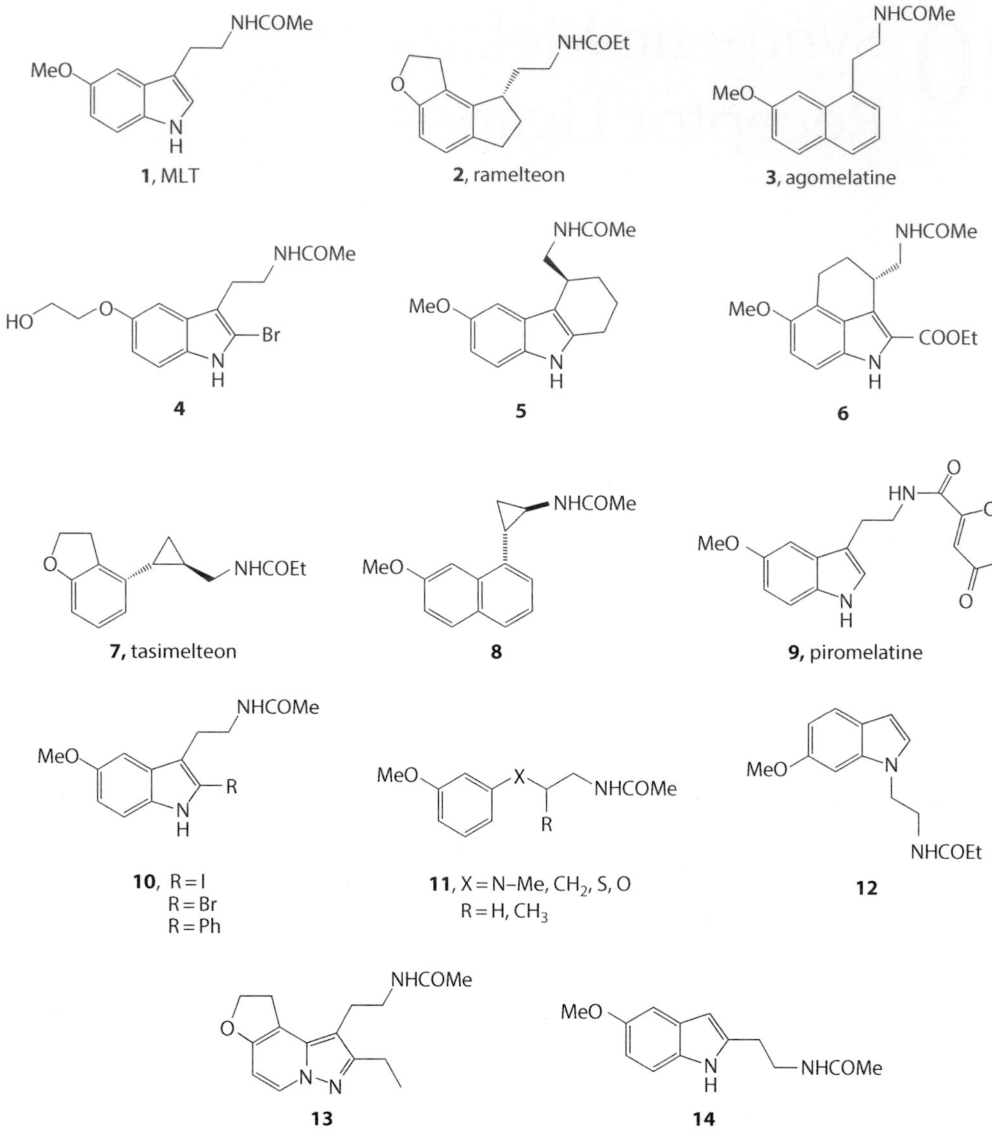

FIGURE 20.1 MLT and representative nonselective MLT receptor ligands.

has also been pointed out. MLT receptor agonists usually bear an acetamide or a propionamide group. Bulkier groups, such as cyclobutanecarboxamide, lead to a decrease of binding affinity and intrinsic activity, and thus to a partial agonist/antagonist behavior. Interestingly, piromelatine (Neu-P11, compound **9**), a combined MT_1/MT_2 and 5-$HT_{1A/1D}$ agonist, which was recently advanced to clinical studies for insomnia treatment, is an indole MLT derivative in which the acetamide group has been replaced with a 4-oxo-4H-pyran-2-carboxamide moiety, prompting further exploration of this region to design novel ligands (Yalkinoglu et al., 2010). The MLT indole ring has been substituted at all the available positions, evidencing an increase in binding affinity with the introduction of a phenyl ring or a halogen atom at position 2 (compounds **10**). 2-Iodomelatonin is characterized by 10-fold higher binding affinity than MLT at both MT_1 and MT_2 receptors, and its radiolabeled derivative, 2-[^{125}I]iodomelatonin, is the reference agonist used in radioligand displacement binding studies. Replacing the indole nucleus with bioisosteric nuclei, such as naphthalene, indane, tetralin, benzothiophene, benzofuran, indene, quinoline, azaindole, and indoline, is another common strategy. For instance, the two marketed drugs ramelteon and agomelatine were developed by bioisosteric replacement of the indole scaffold of MLT with an indane or naphthalene ring, respectively. Further investigations have determined that a bicyclic aromatic core is not an essential requirement for melatonergic activity. In fact, compounds in which the methoxy-phenyl moiety is connected to the alkylamido portion by atoms or groups of different nature, such as CH_2, N-R, O, and S (compounds **11**), were found to retain good activity. The influence of shifting the amide side chain of MLT from C-3 to C-2 and N-1 indole positions was also investigated. In particular, moving the side chain from C-3 to C-2 indole position led to compounds with MLT receptor antagonist and partial agonist properties. On the contrary, potent MLT receptor agonists were obtained by shifting the side chain from C-3 to N-1 and the methoxy group from 5- to 6-position (i.e. compound **12**). Other MLT receptor ligand structures have adopted various heterocyclic scaffolds such as tricyclic and tetracyclic indole–based rings or other aromatic ring system as reviewed in Zlotos, 2005. In this context, it is worth noting the potent tricyclic MT_1/MT_2 agonist **13**. It exhibited good oral absorption, BBB permeability in rats, and a sleep-promoting action in freely moving cats at 0.1 mg/kg (Koike et al., 2011a). Recent observations suggest a possible role of MLT in the prevention and therapy of neurodegenerative pathologies. MLT is a potent free-radical scavenger, and it can also indirectly reduce the damage caused by oxidative stress by enhancing the synthesis of antioxidant enzymes, as well as their activity. Furthermore, altered MT_1 and MT_2 receptor expression levels have been reported in models of Alzheimer's and Parkinson's diseases. However, the involvement of MT_1/MT_2 receptors in providing MLT-mediated neuroprotection remains controversial. A set of indole MLT derivatives, with changes in the 5-methoxy and acylamino groups, the side chain position, and the lipophilic/hydrophilic balance, were selected and tested for their in vitro antioxidant potency and for their cytoprotective activity against kainate excitotoxicity on cerebellar cell cultures (Mor et al., 2004). Poor correlation was observed between antioxidant potency, cytoprotective activity, and MT_1 or MT_2 receptor affinity. Compound **14** is a low-affinity antagonist at MLT membrane receptors, and one of the most potent compounds in the cytoprotection and antioxidant assays (Spadoni et al., 2006b). Similar results were recently obtained by testing the antioxidant potency of nonindole MLT derivatives such as some *N*-(phenoxyalkyl)amides of the general formula **11** (X = O). The low-affinity MT_1/MT_2 compound (*R*)-*N*-[2-(3-methoxyphenoxy)propyl]butanamide displayed potent in vitro antioxidant activity and strong cytoprotective activity (Carocci et al., 2013).

20.2 MT_1-SELECTIVE COMPOUNDS

MT_1- or MT_2-selective synthetic ligands could potentially give advantages over MLT or the two marketed drugs ramelteon and agomelatine by activating some but not all of the targets of the natural compound. Even if the interplays of single MLT receptor subtypes in many pathological conditions still need to be unraveled, relevant roles for receptor-mediated actions of MLT in neuroprotection are emerging from recent studies. For example, loss of the mitochondrial MT_1 receptor enhances neuronal vulnerability and potentially accelerates the neurodegenerative process caused by mutant huntingtin–mediated toxicity in mice. MLT-mediated protection in mutant huntingtin striatal cells is dependent on the presence and activation of the MT_1 receptor (Wang et al., 2011). Thus, MT_1-selective synthetic ligands would be valuable tools to investigate the roles of receptor subtypes in physiological processes and in therapy. Few examples of MLT receptor ligands characterized by selectivity for the MT_1 receptor are reported in the literature.

The first MT_1-selective derivatives were developed following the bivalent ligand approach, linking two agomelatine units through their methoxy substituent by a polymethylene chain (Descamps-Francois et al., 2003). The length of the spacer can vary from 2 to 10, but 3 gives the highest selectivity ratio (compound **15**, 224-fold MT_1 selectivity, Figure 20.2). This product has long been considered a reference MT_1-selective antagonist, even if only threefold MT_1 selectivity has been recently reported on a cell line different from that employed in the original work (Markl et al., 2011).

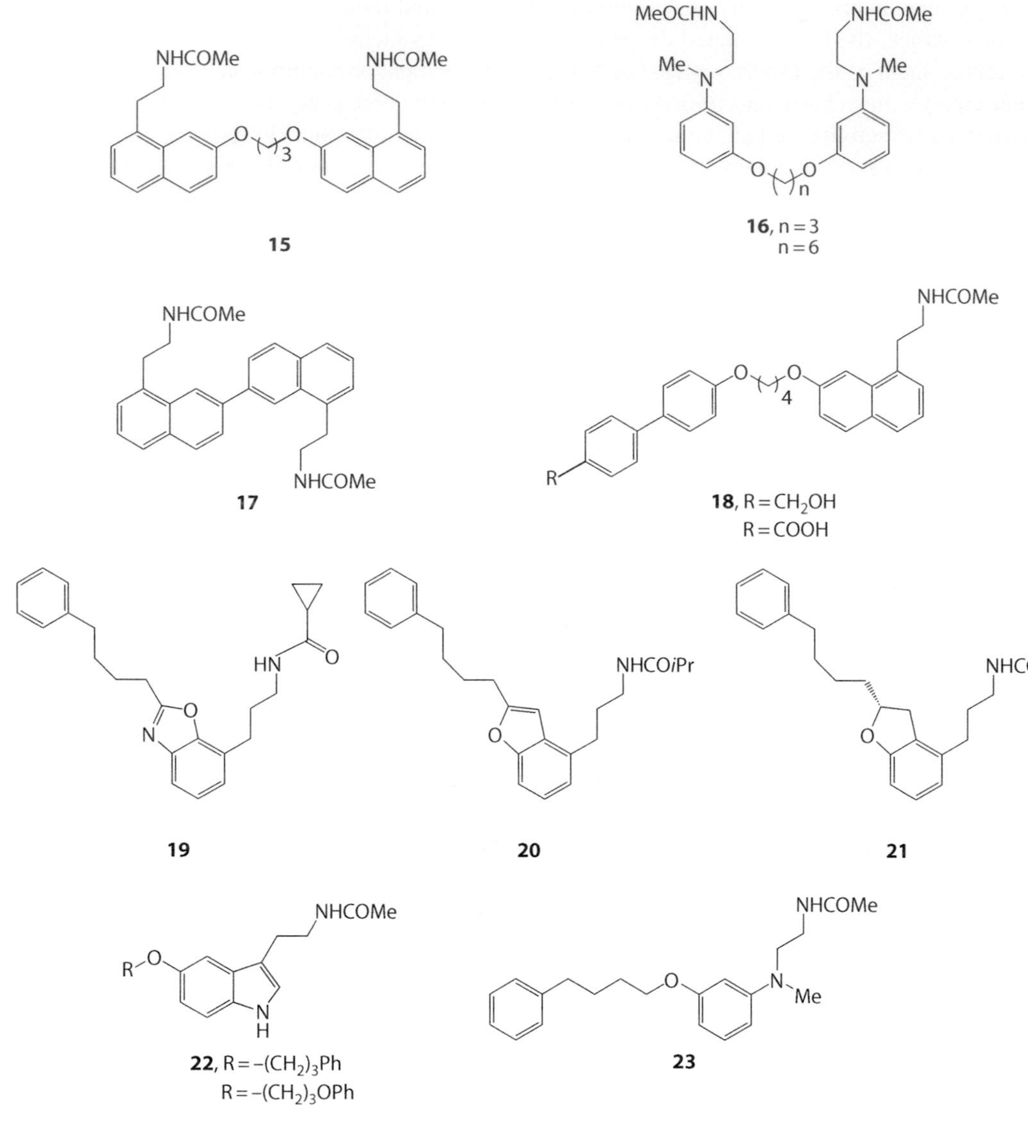

FIGURE 20.2 Representative MT_1-selective ligands.

According to the same strategy, modest MT_1-selective dimers in the nanomolar affinity range ($K_{iMT2/MT1}$ selectivity ratios approximately ranging from 50 to 110) were developed by linking two molecules of the monomer ligand based on the 3-methoxyanilino skeleton (compounds **16**) (Spadoni et al., 2011b). Also in this case a three methylene linker conferred the highest MT_1 selectivity (112-fold), while the C_6-dimer showed the highest MT_1 binding affinity ($pK_i = 8.47$, 54-fold MT_1 selectivity) of the series. Both products behave as MT_1/MT_2 partial agonists in the GTPγS assay.

The binaphtyl derivative **17** is another example of a dimeric MT_1-selective agonist ($K_{iMT2/MT1} = 50–100$, in CHO and HEK cells, respectively) prepared by connecting two desmethoxy agomelatine units via the C-7 position of the aromatic ring.

A set of novel asymmetric heterodimers, formally obtained by replacing one of the agomelatine units of compound **15** with structurally different aryl moieties, was recently reported; some of these compounds (derivatives **18**) were described as MT_1-selective partial agonists ($K_{iMT2/MT1} = 70–90$)

with subnanomolar affinity. These results indicate that a bivalent ligand structure is not necessary to get MT_1 selectivity.

This is further supported by structure–activity relationship studies of simple monomeric ligands. For instance, compounds endowed with good binding affinity and MT_1 moderate selectivity (approximately 20–40 fold) are characterized by a bicyclic scaffold (benzoxazole, benzofuran, or dihydrobenzofuran), mimicking the methoxyphenyl fragment of MLT, carrying a bulky lipophilic moiety. The optimal lipophilic group seems to be a 4-phenylbutyl substituent in position 2 (compounds **19–21**) (Sun et al., 2004a,b, 2005). A lipophilic chain connected to the oxygen atom in position 5 of MLT appears to confer some MT_1 selectivity (ca. 10-fold), as highlighted by the receptor binding affinity data of some *N*-acetyl-5-arylalkoxytryptamine analogs, such as compound **22** (Markl et al., 2011). Another series of MT_1-selective ligands are *N*-(anilinoethyl)amides bearing 3-arylalkyloxy or 3-alkyloxy substituents at the aniline ring. Derivative with a 3-phenylbutoxy substituent (i.e. compound **23**) was shown to be a potent MLT receptor partial agonist, displaying 78-fold selectivity for the MT_1 receptor (Spadoni et al., 2012).

Structure-based and ligand-based information was used to build homology models of the MT_1 receptor in its active state, providing possible hypotheses on the receptor elements responsible for MT_1 subtype selectivity. In particular, the comparison between the amino acid sequences of the extracellular ends of transmembrane helices 3 and 4 in MT_1 and MT_2 receptors showed that the MT_1 receptor is characterized by the presence of some smaller amino acids, which could favor the accommodation of the phenylbutoxy substituent and, in general, of the lipophilic substituent carried by MT_1-selective ligands (Pala et al., 2013).

20.3 MT_2-SELECTIVE COMPOUNDS

In the last decades, a number of MT_2-selective ligands have been reported, in particular compounds acting as antagonists, even if some classes of MT_2-selective agonists are now available (Wan et al., 2013), and the stereochemical requirements for MT_2 selectivity are being defined (Bedini et al., 2011).

Analysis of structure–activity relationships shows that the presence of certain substituents on the indole nucleus of MLT, or in the corresponding positions of bioisosteric rings, confers some selectivity for the MT_2 subtype. This is the case of position 6 of MLT, where insertion of a chlorine atom (compound **24**, Figure 20.3) or a methoxy group leads to compounds having 60-fold selectivity for the MT_2 receptor. Moderate selectivity can also be achieved with substituents in the amide side chain, such as vinyl, allyl, or trifluoromethyl, usually leading to no more than 10-fold selectivity compared with their methyl analogues. A hydroxyethyl substituent, one of the very few examples of hydrophilic groups tolerated by MLT receptors, also confers some preference for the MT_2 receptor. The best characterized and widely exploited element conferring MT_2 selectivity is a lipophilic substituent in a position corresponding to N1 or C2 of the MLT indole ring, not coplanar with the indole. This out-of-plane substituent usually lowers the intrinsic activity of the compound, leading to MT_2-selective partial agonists or antagonists. An explanation for this behavior has been hypothesized, based on the homology models of MLT receptors (Rivara et al., 2005). In the MT_2 receptor model, the out-of-plane substituent occupies a lipophilic region, close to the tryptophan residue of the CWXP motif in transmembrane helix 6 involved in the process of receptor activation. The corresponding region of the MT_1 receptor is characterized by amino acids with bulkier side chains, which hamper the accommodation of the substituent, conferring lower MT_1 binding affinity and MT_2 selectivity. Relevant compounds carrying this type of substituent can be found among indole derivatives or bioisosteric analogs, such as benzofuran, benzothiophene, or tetrahydronaphthalene derivatives. Luzindole (compound **25**), having a benzyl substituent in position 2 and lacking the methoxy group, is an antagonist with 15-fold selectivity for the MT_2 receptor. Despite its limited potency and selectivity, it has been widely used to investigate the role of MLT receptors in in vitro and in vivo experiments. Other antagonists carrying a benzyl substituent are the indole derivative **26**, in which the amide side chain is shifted from position 3 to position 2 and the methoxy group

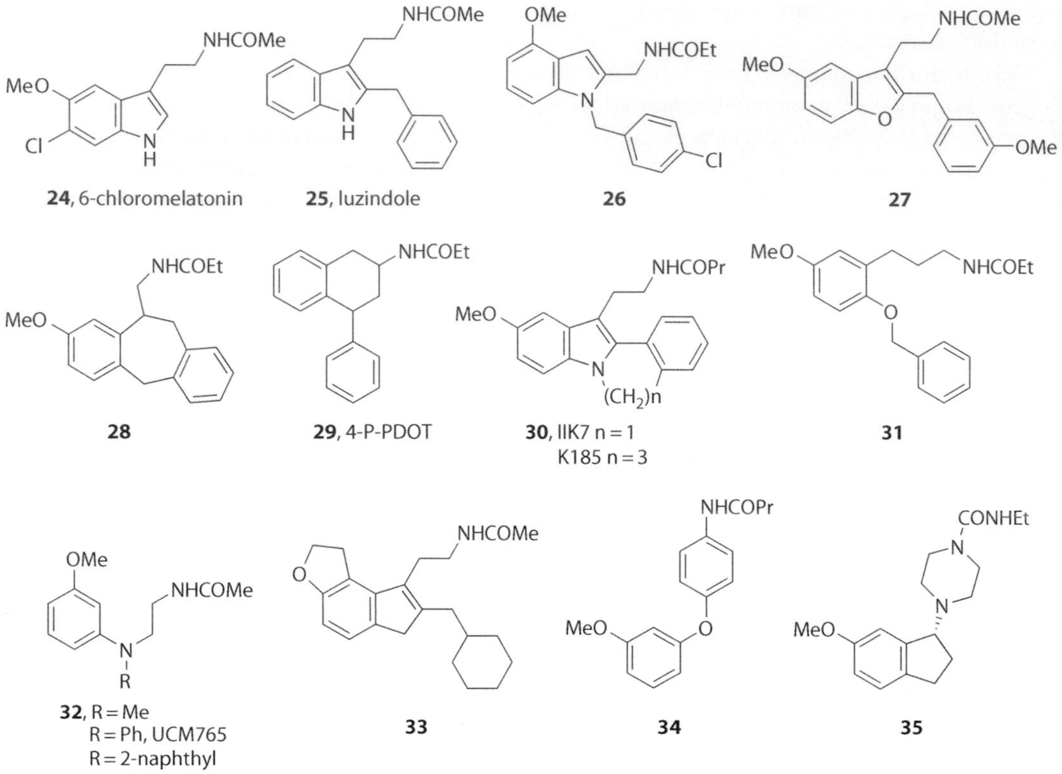

24, 6-chloromelatonin **25**, luzindole **26** **27**

28 **29**, 4-P-PDOT **30**, IIK7 n = 1 K185 n = 3 **31**

32, R = Me R = Ph, UCM765 R = 2-naphthyl **33** **34** **35**

FIGURE 20.3 Representative MT_2-selective ligands.

from position 5 to position 4, and the benzofuran derivative **27**, which has MT_2 binding affinity similar to that of MLT and 123-fold selectivity over the MT_1 subtype. To validate the out-of-plane hypothesis previously described, a series of tricyclic dihydrodibenzocycloheptene derivatives were designed, in which the two aromatic rings are not coplanar due to the skewed conformation of the central ring. Compound **28** is indeed an antagonist with 10-fold selectivity for the MT_2 receptor.

Another relevant class of MT_2-selective compounds is that of 4-phenyl-tetralines. 4-P-PDOT (compound **29**) has 300-fold selectivity for the MT_2 receptor, and, depending on the test performed, it has been described as a partial agonist with low intrinsic activity or an antagonist. The out-of-plane arrangement of the phenyl substituent could be the explanation for its behavior. 4-P-PDOT is another pharmacological tool often used in experiments aimed at dissecting the role of MT_1 and MT_2 receptors. The tetracyclic compounds **30** are characterized by a five- to seven-membered ring connecting the indole nitrogen to the phenyl ring. Interestingly, their intrinsic activity decreases as the size of the bridging ring becomes bigger. IIK7 ($n = 1$) is an agonist with 89-fold selectivity for the MT_2 receptor; K185 ($n = 3$) is an antagonist with 132-fold MT_2 selectivity. Good selectivity for the MT_2 receptor can be found also in structurally simplified compounds, in which the aromatic core of the compound is a benzene ring. Compound **31** is a phenyl-propanamide derivative in which the benzyloxy substituent provides high MT_2 selectivity compared to the nonselective profile of the unsubstituted precursor. It behaves as an MT_2 agonist with 412-fold MT_2 selectivity. A series of anilino-ethylamide derivatives (compounds **32**) has been described in which selectivity and intrinsic activity could be modulated based on the size and shape of the substituent on the nitrogen atom. Indeed, the derivative carrying a methyl substituent behaves as an MT_1/MT_2 nonselective agonist, the phenyl derivative is an MT_2-selective (63-fold) partial agonist, and the bulkier naphthyl analog is an antagonist with more than 1000-fold MT_2 selectivity (Rivara et al., 2007). Remarkable potency and MT_2 selectivity have been recently reported for indene derivatives in which the methoxy group was incorporated into a fused dihydrofuran ring. Insertion of substituents in position 7 confers MT_2 selectivity while intrinsic activity depends on the nature of the substituent. The cyclohexylmethyl derivative **33** displayed picomolar binding affinity at the MT_2 receptor with 800-fold selectivity and a full agonist behavior (Koike et al., 2011b). Two series of MT_2-selective agonists have been reported some years ago, lacking the classical acylaminoethyl side chain of MLT. In the first series, the methoxyphenyl portion has been linked to an anilide fragment by a monoatomic linker. The MT_2 agonist **34** is one of the most selective compounds for the MT_2 receptor (84-fold) in the series. (*R*)-Indanyl-piperazines also displayed selectivity for the MT_2 receptor. Urea derivative **35** has good MT_2 binding affinity, 120-fold MT_2 selectivity, and it behaves as a full agonist. Interestingly, its enantiomer is devoid of any MT_1 and MT_2 binding affinity, and the different behavior of the two enantiomers has been rationalized on the basis of a pharmacophore model for MLT receptor ligands (Rivara et al., 2006).

20.4 SYNTHETIC LIGANDS WITH IMPROVED METABOLIC STABILITY

MLT is characterized by low oral bioavailability, due to various factors, such as limited uptake from the gastrointestinal tract and high first-pass metabolism. Moreover, it has a short oral half-life, usually in the 30–45 min range. While MLT can be used when a short action is sufficient (e.g., reduction of sleep onset latency, phase shifting), its application is prevented when a longer effect is required (e.g., sleep maintenance) (Hardeland and Poeggeler, 2012). In this case, the availability of synthetic analogs with improved pharmacokinetic properties could be a valid alternative. TIK-301 ((*R*)-β-methyl-6-chloromelatonin, compound **36**, Figure 20.4) is a MLT derivative in which a chlorine atom has been inserted in position 6 of the indole ring. The chlorine atom hampers the formation of the major metabolite of MLT, 6-hydroxymelatonin, responsible for its rapid elimination as 6-sulfatoxymelatonin. In humans, the reported half-life of TIK-301 is 1 h, but data on MLT half-life in the same experimental conditions are not available for comparison.

Attempts to improve the metabolic stability have been reported for some investigational compounds. UCM765 (compound **32**, Figure 20.3) is an MT_2-selective partial agonist, which selectively promoted nonrapid eye movement sleep in rodents at the dose of 40 mg/kg (Ochoa-Sanchez et al., 2011). The need for such a high dose, in spite of its subnanomolar binding affinity for the MLT receptors, could be related to its low oral bioavailability (<2% in rats) and is suggestive of

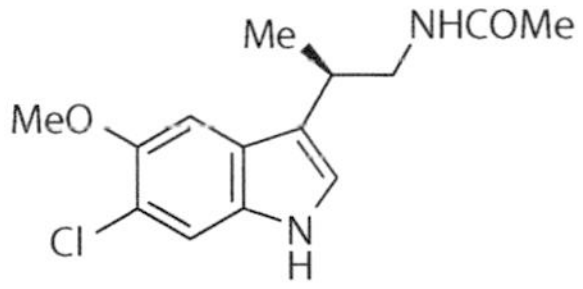

36, (*R*)-β-methyl-6-chloromelatonin

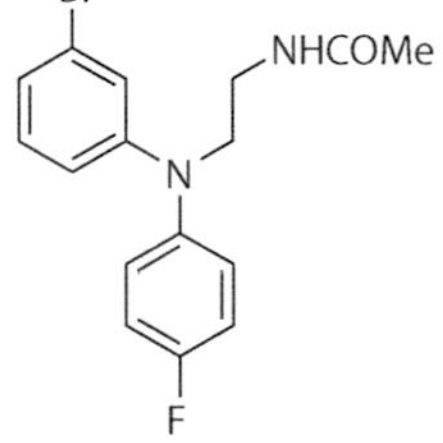

37, UCM924

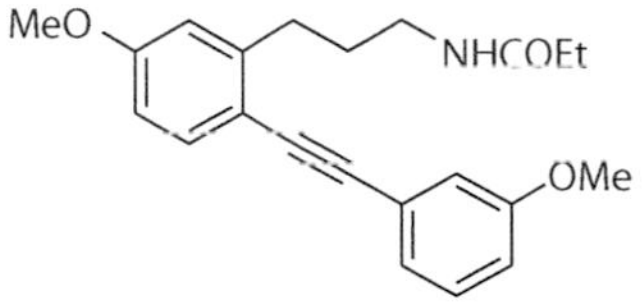

38

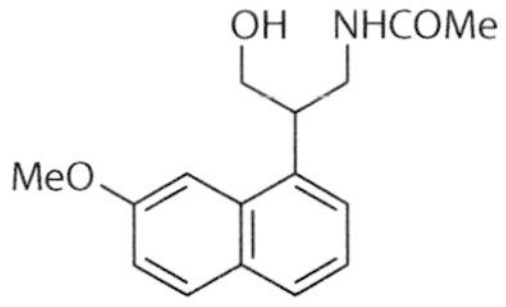

39

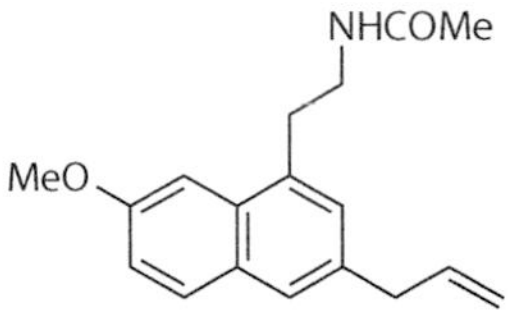

40

FIGURE 20.4 Examples of MLT receptor ligands with improved metabolic stability.

extensive first-pass metabolism. To design metabolically protected derivatives, the stability of UCM765 to oxidative metabolism was evaluated in vitro by assessing its behavior in the presence of liver S9 fraction and microsomal preparations (Rivara et al., 2009). UCM765 showed a half-life of 8.0 min in rat liver S9 fraction and the presence of two major metabolically liable positions: the para-position of the unsubstituted phenyl ring, which is hydroxylated, and the methoxy substituent, which undergoes demethylation. To prevent these metabolic transformations, protection was achieved by the introduction of a fluorine atom in the para-position and replacement of the methoxy substituent with groups known from structure–activity relationships to maintain high binding affinity for MLT receptors. The best derivative, UCM924 (compound **37**, Figure 20.4), is a meta-bromo, para-fluoro derivative of UCM765 exhibiting improved resistance to oxidative metabolism in both rat and human liver preparations. Indeed, 40% of UCM924 remains unaltered after 1 h in rat liver S9 fraction. This compound has the same binding properties as the parent UCM765, behaving as an MT_2-selective partial agonist.

A structural optimization aimed at decreasing the metabolic liability and ultimately improving the pharmacokinetic properties has been reported also for 6-substituted-phenyl-propylamides. Compound **31** (Figure 20.3) is characterized by a short half-life in rat and human liver microsomes (5.5 min in human microsomes), significantly shorter than MLT (73 min). Metabolite profiling evidenced that the methoxy group and the C6-ether linkage are the major metabolic soft spots. Structural modification of the linker connecting the two phenyl rings was therefore performed, replacing the oxy-methylene portion with different connecting elements (Hu et al., 2013). The best result was achieved with the insertion of an alkynyl spacer. Compound **38** is characterized by improved MT_2 binding affinity and selectivity, and it has higher oxidative stability. It displayed a half-life of 18 min in human microsomes, longer than the precursor **31**, even if not yet ideal. Compound **38** is described as a new lead compound for further structural modifications aimed at protecting other metabolically liable positions.

Structural modulation of agomelatine was also performed aimed at maintaining the binding profile at the MLT receptors while improving metabolic stability and binding affinity for 5-HT_{2c} receptors (Ettaoussi et al., 2013). Different substituents were inserted either in position 3 of the naphthalene nucleus or in the beta position of the ethylamide side chain, since these are two major sites of oxidative metabolic transformation for agomelatine. Some derivatives showed slightly improved binding affinity for the 5-HT_{2c} receptor (e.g., compounds **39** and **40**), but no data on their metabolic protection have been reported.

Also considering the interesting pharmacological activities recently described for MT_2-selective ligands in preclinical models (Ochoa-Sanchez et al., 2011, 2012), the development of novel compounds with improved pharmacokinetics could provide new therapeutic opportunities for the treatment of sleep disturbances, anxiety, and other neuropsychiatric diseases (Comai and Gobbi, 2013).

REFERENCES

Bedini, A., Lucarini, S., Spadoni, G., Tarzia, G., Scaglione, F., Dugnani, S., Pannacci, M. et al. 2011. Toward the definition of stereochemical requirements for MT_2-selective antagonists and partial agonists by studying 4-phenyl-2-propionamidotetralin derivatives. *J. Med. Chem.* 54:8362–8372.

Carocci, A., Catalano, A., Bruno, C., Lovece, A., Roselli, M. G., Cavalluzzi, M. M., De Santis, F. et al. 2013. *N*-(Phenoxyalkyl)amides as MT_1 and MT_2 ligands: Antioxidant properties and inhibition of Ca^{2+}/CaM-dependent kinase II. *Bioorg. Med. Chem.* 21:847–851.

Comai, S., Gobbi, G. 2014. Unveiling the role of melatonin MT_2 receptors in sleep, anxiety and other neuropsychiatric diseases: A novel target in psychopharmacology. *J. Psychiatry Neurosci.* 39:6–21.

Davies, D. J., Garratt, P. J., Tocher, D. A., Vonhoff, S., Davies, J., Teh, M.-T., Sugden, D. 1998. Mapping the melatonin receptor. 5. Melatonin agonists and antagonists derived from tetrahydrocyclopent[*b*]indoles, tetrahydrocarbazoles and hexahydrocyclohept[*b*]indoles. *J. Med. Chem.* 41:451–467.

Descamps-Francois, C., Yous, S., Chavatte, P., Audinot, V., Bonnaud, A., Boutin, J. A., Delagrange, P., Bennejean, C., Renard, P., Lesieur, D. 2003. Design and synthesis of naphthalenic dimers as selective MT_1 melatoninergic ligands. *J. Med. Chem.* 46:1127–1129.

Ettaoussi, M., Sabaouni, A., Pérès, B., Landagaray, E., Nosjean, O., Boutin, J. A., Caignard, D. H., Delagrange, P., Berthelot, P., Yous, S. 2013. Synthesis and pharmacological evaluation of a series of the agomelatine analogues as melatonin MT_1/MT_2 agonist and $5\text{-}HT_{2C}$ antagonist. *ChemMedChem* 8:1830–1845.

Hardeland, R., Poeggeler, B. 2012. Melatonin and synthetic melatonergic agonists: Actions and metabolism in the central nervous system. *Cent. Nerv. Syst. Agents Med. Chem.* 12:189–216.

Hu, Y., Zhu, J., Chan, K. H., Wong, Y. H. 2013. Development of substituted *N*-[3-(3-methoxylphenyl)propyl] amides as MT_2-selective melatonin agonists: Improving metabolic stability. *Bioorg. Med. Chem.* 21:547–552.

Koike, T., Hoashi, Y., Takai, T., Nakayama, M., Yukuhiro, N., Ishikawa, T., Hirai, K., Uchikawa, O. 2011b. 1,6-Dihydro-2H-indeno[5,4-b]furan derivatives: Design, synthesis and pharmacological characterization of a novel class of highly potent MT_2-selective agonists. *J. Med. Chem.* 54:3436–3444.

Koike, T., Takai, T., Hoashi, Y., Nakayama, M., Kosugi, Y., Nakashima, M., Yoshikubo, S., Hirai, K., Uchikawa, O. 2011a. Synthesis of a novel series of tricyclic dihydrofuran derivatives: Discovery of 8,9-dihydrofuro[3,2-c]pyrazolo[1,5-a]pyridines as melatonin receptor (MT_1/MT_2) ligands. *J. Med. Chem.* 54:4207–4218.

Markl, C., Clafshenkel, W. P., Attia, M. I., Sethi, S., Witt-Enderby, P. A., Zlotos, D. A. 2011. *N*-Acetyl-5-arylalkoxytryptamine analogs: Probing the melatonin receptors for MT_1-selectivity. *Arch. Pharm. Chem. Life Sci.* 344:666–674.

Mor, M., Rivara, S., Pala, D., Bedini, A., Spadoni, G., Tarzia, G. 2010. Recent advances in the development of melatonin MT_1 and MT_2 receptor agonists. *Expert Opin. Ther. Pat.* 20:1059–1077.

Mor, M., Silva, C., Vacondio, F., Plazzi, P. V., Magnanini, F., Spadoni, G., Diamantini, G. et al. 2004. Indole-based analogs of melatonin: In vitro antioxidant and cytoprotective activities. *J. Pineal Res.* 36:95–102.

Morellato, L., Lefas-Le Gall, M., Langlois, M., Caignard, D. H., Renard, P., Delagrange, P., Mathé-Allainmat, M. 2013. Synthesis of new *N*-(arylcyclopropyl)acetamides and *N*-(arylvinyl)acetamides as conformationally-restricted ligands for melatonin receptors. *Bioorg. Med. Chem. Lett.* 23:430–434.

Ochoa-Sanchez, R., Comai, S., Lacoste, B., Bambico, F. R., Dominguez-Lopez, S., Spadoni, G., Rivara, S. et al. 2011. Promotion of non-rapid eye movement sleep and activation of reticular thalamic neurons by a novel MT_2 melatonin receptor ligand. *J. Neurosci.* 31:18439–18452.

Ochoa-Sanchez, R., Rainer, Q., Comai, S., Spadoni, G., Bedini, A., Rivara, S., Fraschini, F., Mor, M., Tarzia, G., Gobbi, G. 2012. Anxiolytic effects of the melatonin MT(2) receptor partial agonist UCM765: Comparison with melatonin and diazepam. *Prog. Neuropsychopharmacol. Biol. Psychiatry* 39:318–325.

Pala, D., Lodola, A., Bedini, A., Spadoni, G., Rivara, S. 2013. Homology models of melatonin receptors: Challenges and recent advances. *Int. J. Mol. Sci.* 14:8093–8121.

Rivara, S., Diamantini, G., Di Giacomo, B., Lamba, D., Gatti, G., Lucini, V., Pannacci, M., Mor, M., Spadoni, G., Tarzia, G. 2006. Reassessing the melatonin pharmacophore: Enantiomeric resolution, pharmacological activity, structure analysis, and molecular modeling of a constrained chiral melatonin analogue. *Bioorg. Med. Chem.* 14:3383–3391.

Rivara, S., Lodola, A., Mor, M., Bedini, A., Spadoni, G., Lucini, V., Pannacci, M. et al. 2007. *N*-(Substituted-anilinoethyl)amides: Design, synthesis, and pharmacological characterization of a new class of melatonin receptor ligands. *J. Med. Chem.* 50:6618–6626.

Rivara, S., Lorenzi, S., Mor, M., Plazzi, P. V., Spadoni, G., Bedini, A., Tarzia, G. 2005. Analysis of structure-activity relationships for MT_2 selective antagonists by melatonin MT_1 and MT_2 receptor models. *J. Med. Chem.* 48:4049–4060.

Rivara, S., Mor, M., Bedini, A., Spadoni, G., Tarzia, G. 2008. Melatonin receptor agonists: SAR and applications to the treatment of sleep-wake disorders. *Curr. Top. Med. Chem.* 8:954–968.

Rivara, S., Vacondio, F., Fioni, A., Silva, C., Carmi, C., Mor, M., Lucini, V. et al. 2009. *N*-(Anilinoethyl)amides: Design and synthesis of metabolically stable, selective melatonin receptor ligands. *ChemMedChem* 4:1746–1755.

Robertson, N. J., Faulkner, S., Fleiss, B., Bainbridge, A., Andorka, C., Price, D., Powell, E. et al. 2013. Melatonin augments hypothermic neuroprotection in a perinatal asphyxia model. *Brain* 136:90–105.

Spadoni, G., Balsamini, C., Diamantini, G., Di Giacomo, B., Tarzia, G., Mor, M., Plazzi, P. V. et al. 1997. Conformationally restrained melatonin analogs: Synthesis, binding affinity for the melatonin receptor, evaluation of the biological activity, and molecular modeling study. *J. Med. Chem.* 40:1990–2002.

Spadoni, G., Bedini, A., Guidi, T., Tarzia, G., Lucini, V., Pannacci, M., Fraschini, F. 2006a. Towards the development of mixed MT_1-agonist/MT_2-antagonist melatonin receptor ligands. *ChemMedChem.* 1:1099–1105.

Spadoni, G., Bedini, A., Rivara, S., Mor, M. 2011a. Melatonin receptor agonists: New options for insomnia and depression treatment. *CNS Neurosci. Ther.* 17:733–741.

Spadoni, G., Bedini, A., Orlando, P., Lucarini, S., Tarzia, G., Mor, M., Rivara, S., Lucini, V., Pannacci, M., Scaglione, F. 2011b. Bivalent ligand approach on *N*-{2-[(3-methoxyphenyl)methylamino]ethyl} acetamide: Synthesis, binding affinity and intrinsic activity for MT_1 and MT_2 melatonin receptors. *Bioorg. Med. Chem.* 19:4910–4916.

Spadoni, G., Diamantini, G., Bedini, A., Tarzia, G., Vacondio, F., Silva, C., Rivara, M. et al. 2006b. Synthesis, antioxidant activity and structure-activity relationships for a new series of 2-(*N*-acylaminoethyl)indoles with melatonin-like cytoprotective activity. *J. Pineal Res.* 40:259–269.

Spadoni, G., Rivara, S., Pala, D., Lodola, A., Mor, M., Lucini, V., Dugnani, S. et al. 2012. MT_1-selective melatonin receptor ligands: Synthesis, pharmacological evaluation and molecular dynamics investigation of *N*-{[(3-*O*-substituted)anilino]alkyl}amides. *ChemMedChem* 7:1954–1964.

Srinivasan, V., Cardinali, D. P., Srinivasan, U. S., Kaur, C., Brown, G. M., Spence, D. W., Hardeland, R., Pandi-Perumal, S. R. 2011. Therapeutic potential of melatonin and its analogs in Parkinson's disease: Focus on sleep and neuroprotection. *Ther. Adv. Neurol. Disord.* 4:297–317.

Sun, L. Q., Chen, J., Bruce, M., Deskus, J. A., Epperson, J. R., Takaki, K., Johnson, G. et al. 2004a. Synthesis and structure-activity relationship of novel benzoxazole derivatives as melatonin receptor agonists. *Bioorg. Med. Chem. Lett.* 14:3799–3802.

Sun, L. Q., Takaki, K., Chen, J., Bertenshaw, S., Iben, L., Mahle, C. D., Ryan, E., Wu, D., Gao, Q., Xu, C. 2005. (*R*)-2-(4-Phenylbutyl)dihydrobenzofuran derivatives as melatoninergic agents. *Bioorg. Med. Chem. Lett.* 15:1345–1349.

Sun, L. Q., Takaki, K., Chen, J., Iben, L., Knipe, J. O., Pajor, L., Mahle, C. D., Ryan, E., Xu, C. 2004b. *N*-[2-[2-(4-Phenylbutyl)benzofuran-4-yl]cyclopropylmethyl]acetamide: An orally bioavailable melatonin receptor agonist. *Bioorg. Med. Chem. Lett.* 14:5157–5160.

Wan, N., Zhang, F. F., Ju, J., Liu, D. Z., Zhou, S. Y., Zhang, B. L. 2013. Investigational selective melatoninergic ligands for receptor subtype MT_2. *Mini Rev. Med. Chem.* 13:1462–1474.

Wang, X., Sirianni, A., Pei, Z., Cormier, K., Smith, K., Jiang, J., Zhou, S. et al. 2011. The melatonin MT_1 receptor axis modulates mutant Huntingtin-mediated toxicity. *J. Neurosci.* 31:14496–14507.

Yalkinoglu, O., Zisapel, N., Nir, T., Piechatzek, R., Schorr-Neufing, U., Bitterlich, N., Oertel, R., Allgaier, C., Kluge, A., Laudon, M. 2010. Phase-I study of the safety, tolerability, pharmacokinetics and sleep promoting activity of Neu-P11, a novel putative insomnia drug in healthy humans. *Sleep* 33:A220 (Abstract Supplement).

Zlotos, D. P. 2005. Recent advances in melatonin receptor ligands. *Arch. Pharm. Chem. Life Sci.* 338:229–247.

Zlotos, D. P. 2012. Recent progress in the development of agonists and antagonists for melatonin receptors. *Curr. Med. Chem.* 19:3532–3549.

21 Evaluation of Synthetic Melatonin Analog Antioxidant Compounds

Sibel Suzen

CONTENTS

21.1 INTRODUCTION

Antioxidants scavenge and prevent the formation of free radicals, so they are exceedingly important for the treatment and management of these kinds of diseases (Emerit et al. 2004). Therefore, in recent years, there has been an increasing interest in finding new antioxidant compounds. The damages caused by oxidative stress occur due to the imbalance between prooxidants and antioxidants. Then antioxidant barrier represented mainly by enzymes and nonenzymatic antioxidant factors are weakened, which allows the accumulation of cytotoxic compounds that consume antioxidant reserves of the body (Sies 1997; Padurariu et al. 2013). Reactive oxygen species (ROS) can react with all biological macromolecules and cells. The first reaction produces a new radical, and this one can create a chain reaction. Polyunsaturated fatty acids (PUFAs) are the best targets since they have multiple double bonds. Removal of a hydrogen atom from a PUFA initiates the process of lipid peroxidation (Suzen 2013).

Brain is one of the most exposed organs to free radical attacks in the body since its high consumption of O_2. Oxidative damage that has a common link in the pathogenesis and neuropathology of a variety of neurodegenerative disorders, such as Alzheimer's disease (AD), Parkinson's disease, neuropathy, ischemia-reperfusion injury, hyperoxia, can be measured in the brain after toxin exposure (Poeggeler et al. 1993; Emerit et al. 2004).

Antioxidant compounds have an essential part in oxidative stress–related diseases and reduce the risk of many chronic diseases, including cancer. The development of new synthetic potent antioxidant compounds is of paramount importance of antioxidant effectiveness in preventing diseases (Gupta et al. 2012). Antioxidant properties of melatonin (MLT) has been found and evaluated by Reiter et al. (2001). Since it has significant free radical scavenger and antioxidant properties, MLT-related compounds such as MLT metabolites and synthetic analogs are under study to find out which display the highest activity with the lowest side effects (Suzen 2006, 2007; Hardeland 2011; Suzen et al. 2013).

21.2 ANTIOXIDANT ACTION OF MELATONIN

MLT (*N*-acetyl-5-methoxytryptamine), a tryptophan metabolite and synthesized mainly in the pineal gland, has a number of physiological functions (Ates-Alagoz and Suzen 2001b). Main functions include regulating circadian rhythms, reimbursing of free radicals, helping to improve immunity, and generally inhibiting the oxidation of biomolecules directly or indirectly (Suzen et al. 2000). In many studies, it was established that decreased MLT in serum and cerebrospinal fluid are observed in patients who suffer with neuropathological diseases (Zhou et al. 2003; Wu and Swaab 2005; Lin et al. 2013).

One of the most interesting properties of MLT is that its metabolites also have the ability to scavenge ROS and reactive nitrogen species (RNS) (Galano et al. 2013). The continuous protection is performed by MLT and its metabolites, referred to as the free radical scavenging cascade (Tan et al. 2007). MLT and its two metabolites, *N*(1)-acetyl-*N*(2)-formyl-5-methoxykynuramine (AFMK) and *N*-acetyl-5-methoxykynuramine (AMK), are all predicted to be excellent ·OH scavengers. It was found that AFMK is a poorer scavenger than AMK and MLT. Accordingly, the efficiency of MLT for scavenging free radicals is predicted to be reduced when it is metabolized to AFMK (Galano et al. 2013).

MLT is a lipophilic molecule, which can cross all membranes (including blood–brain barrier) easily and be presumably present in all subcellular compartments. However, still there is a significant restriction in its therapeutic use, which is its short half-life as a result of its rapid metabolic inactivation (Yu-Chieh et al. 2013). This disadvantage can be subjected by designing and synthesizing new MLT analogs with longer half-life than MLT. The use of MLT derived from animal pineal tissues may also carry the risk of contamination or transmission of viral material (Suzen et al. 2013). Neither it is approved as a drug by FDA (Food and Drug Administration) nor a pharmaceutical company applied for a regulatory approval to market it as a drug, which is suggested to be, because of its nonpatentable situation (Bhavsar et al. 2009).

21.3 INDOLE-BASED ANALOGS OF MELATONIN AS ANTIOXIDANTS

In the design of new drugs, the development of hybrid molecules through the combination of different pharmacophores in one structure may lead to compounds with interesting pharmacological properties (Gupta et al. 2012). The design of small molecule agents to fight cellular oxidative stress has become an important therapeutic purpose, given the wide-ranging damage to cellular macromolecules caused by reactive oxygen (and reactive nitrogen) radical species (Karaaslan et al. 2013).

A number of studies have been published by Suzen and coworkers (Ates-Alagoz et al. 2005; Suzen 2006; Das-Evcimen et al. 2009; Gurkok et al. 2009; Shirinzadeh et al. 2010; Suzen et al. 2012; Yılmaz et al. 2012) on the discovery of indole-based MLT analogs, mainly designed to improve free radical scavenging antioxidant properties, including 2-phenylindoles (Suzen et al. 2006), that have particular relevance to design, synthesis, and antioxidant activity of a series of substituted 2-(4-aminophenyl)-1*H*-indoles and 2-(methoxyphenyl)-1*H*-indoles (Figure 21.1) (Karaaslan et al. 2013). The new compounds are structurally associated to the antioxidant lead compound MLT and the antitumor 2-(4-aminophenyl)benzothiazole and 2-(3,4-dimethoxyphenyl) benzothiazole series. Compounds in particular showed potent antioxidant activity in the 2,2-diphenyl-1-picrylhydrazyl (DPPH) and superoxide radical scavenging tests (80% and 81% inhibition at 1 mM concentration of 3b, respectively), at a level comparable with the reference standard MLT (98% and 75% at 1 mM).

One of the comparative studies was performed by Poeggeler et al. (2002) on the oxidation chemistry of MLT and its analogs. Redox chemistry of MLT and tryptamine, *N*-acetyltryptamine, serotonin, *N*-acetylserotonin, 5-methoxytryptamine, 6-chloromelatonin, and 2-iodomelatonin was investigated by scavenging of hydroxyl radicals (·OH) using an assay based on ABTS (2,2'-azino-bis(3-ethylbenzothiazoline-6-sulphonic acid)) cation radical ($ABTS^{\cdot+}$) formation. The results

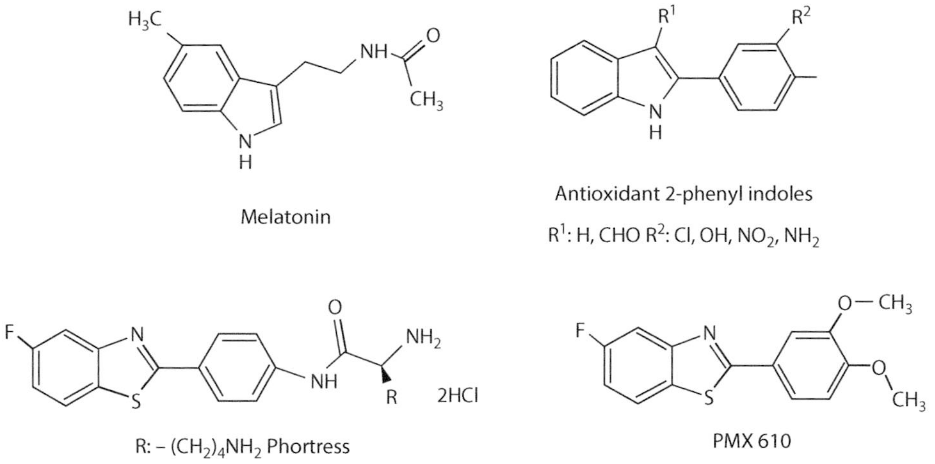

Melatonin

Antioxidant 2-phenyl indoles
R^1: H, CHO R^2: Cl, OH, NO_2, NH_2

R: – $(CH_2)_4NH_2$ Phortress

PMX 610

FIGURE 21.1 Antioxidant and antitumor lead compounds.

indicated that the 5-methoxy group is important for avoiding the formation of O-centered radical intermediates, as occurring in serotonin (Perez-Reyes and Mason 1981). The N-acetyl group plays a part in the oxidation reactions by cyclization reaction to form 3-OHM (Tan et al. 1998). It was significant that tryptamine showed a better ·OH scavenger in the ABTS assay than *N*-acetyltryptamine and 5-methoxytryptamine. The results evidently suggest that radical scavenging by MLT analog indolic compounds does not depend on the existence of an indole moiety, but their functional groups.

Antioxidants that decrease the free radical generation could be used as protective agents against excitotoxic injuries. Considering that hydroxyindoles are efficient free radical scavengers and antioxidants, antioxidant profile of the previously synthesized compounds was studied (Buemi et al. 2013). The designed hybrid derivative 3,4-dihydroxy-*N*-[1-[2-(5-hydroxy-1*H*-indol-3-yl)-2-oxoethyl] piperidin-4-yl]benzamide (Figure 21.2) was the most effective antioxidant agent (>94.1 ± 0.1% of inhibition at 17 μM).

Antioxidant properties of some selected substituted 2-indolyl carbohydrazides, substituted 3-indolyl carbohydrazide, 3-(3-hydrazinylpropyl)-1*H*-indole, and 3-(1*H*-indol-3-yl)propanehydrazide (Figure 21.3), throughout the assessment of their antioxidative potential using different antioxidant assays such as DPPH, lipid peroxidation in the APPH (2,2-azobis(2-amidinopropane) dihydrochloride), or the DMSO (dimethylsulfoxide) method were studied (Hadjipavlou-Litina et al. 2013). It was observed that these compounds are promising for the design of useful drugs to treat AD.

The combination of 2-oxindole derivatives (Figure 21.4) and the chalcones in the same molecule showed interesting challenge for the development of new pharmacologically active antioxidants.

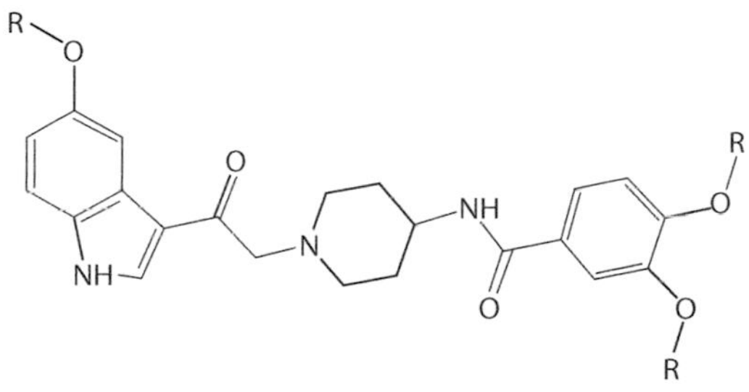

FIGURE 21.2 Designed hybrid derivative.

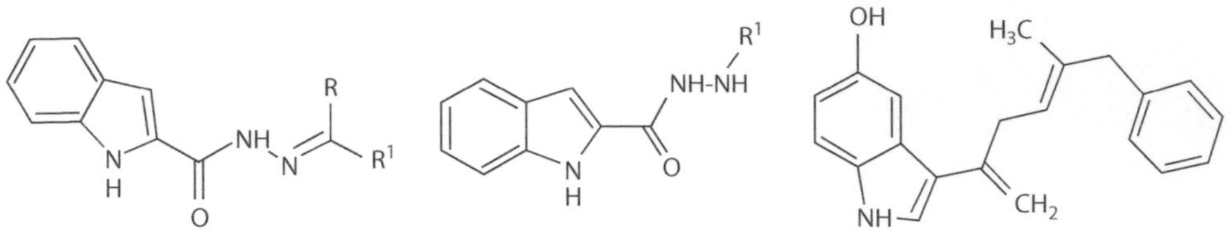

R^1: Me, R^2: Me, Bn, etc. R^1: *n*-Bu, $(CH_2)_3Ph$, etc.

FIGURE 21.3 Substituted 2-indolyl carbohydrazides.

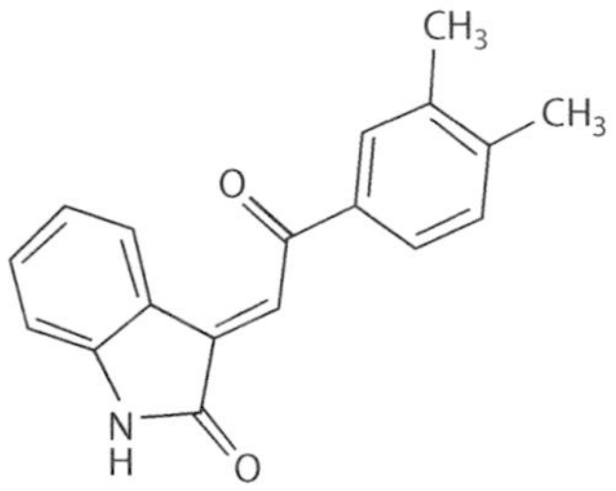

FIGURE 21.4 3-Substituted-2-oxindole derivatives.

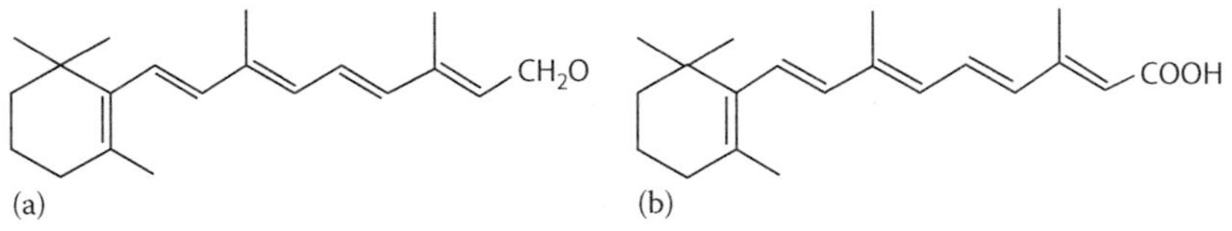

(a) (b)

FIGURE 21.5 Structures of (a) retinol (vitamin A) and (b) *all-trans*-retinoic acid.

In vitro antioxidant activity of a new series of compounds synthesized by Knoevenagel reaction of substituted isatins with various acetophenones using DPPH assay was performed (Gupta et al. 2012). The majority of 3-aroyl methylene indol-2-ones showed good antioxidant activity within a concentration range of 5–100 μg/mL. Findings indicated that heterocyclic systems with oxindole nucleus possess moderate-to-good antioxidant activity at low concentrations. It may be due to keto lactam ring being responsible to initiate the free radical scavenging activity due to its N–H and C=O moieties.

Protection and defense of vital biological molecules such as lipids, carbohydrates, proteins, and DNA from oxidative stress is crucial to prevent many diseases caused by high levels of ROS (Dong et al. 2000). Retinoids (Figure 21.5) have received extensive notice as agents that may be useful for both cancer prevention and treatment due to their cell differentiation, proliferation, and antioxidant effects (Tallman and Wiernik 1992). Additionally, retinoids have been shown to function as efficient antioxidants by inhibiting lipid peroxidation (Samokyszyn and Marnett 1990; Ates-Alagoz 2013).

In a study (Ates-Alagoz et al. 2006), novel MLT retinamide derivatives (containing the MLT moiety integrated with the structure of retinoic acid and Am580), which can be classified into three sets (**IV**, **V**, **VII**), were synthesized (Figure 21.6). Superoxide anion, DPPH free radical scavenging activities, and inhibition of lipid peroxidation assays were used to determine the antioxidative activities. The compounds had a strong effect on the lipid peroxidation and did not show a strong effect on the DPPH radical, whereas they showed moderate scavenger effect on

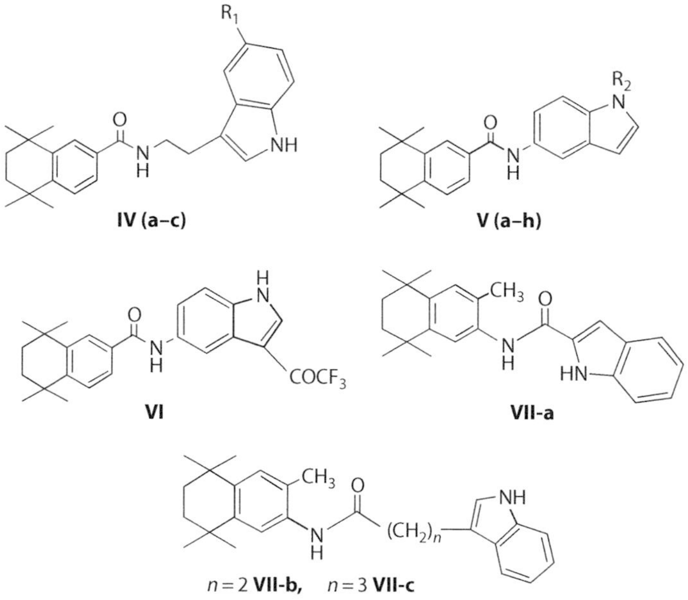

FIGURE 21.6 Structures of synthesized MLT retinamide derivatives: **IV (a–c)**, **V (a–h)**, **VI**, **VII (a–c)**.

superoxide anion when compared with MLT. Compounds **IV (a–c)**, **V (a–b)**, **V (g)**, and **VII (a–c)** showed a strong inhibitory effect on the lipid peroxidation at the 10^{-3} M concentration, and the inhibition rates were in the range of 90%–99%. In addition, these compounds showed even better activity than MLT at 10^{-3} M concentration. Results indicated the importance of the alkyl chain inclusion at the first position on the indole ring since the alkyl substitution of the indole ring obviously decreased the LP (lipid peroxidation) inhibition. From these data, it can be concluded that the structural requirements for optimal LP inhibitory activity are rather strict, in particular that (a) the presence of the linker group (amide) between retinoid head and indole moiety might be required for activity and (b) alkyl residues linked to the indole moiety may be more critical for antioxidant activity than with linker chain.

Suzen and coworkers basically work on the synthesis and antioxidant evaluation by measuring their potential radical scavenging activity and protective effect against oxidative damage of new MLT analog compounds (Figure 21.7). These compounds can be summarized as indole-3-propionamide derivatives (Suzen et al. 2001; Ates-Alagoz et al. 2005), 5-bromoindole derivatives (Gurkok et al. 2009), 2-*p*-florophenylindole derivatives (Suzen et al. 2013), *N*-methylindole derivatives (Shirinzadeh et al. 2010), 2-indole aldehyde derivatives (Suzen et al. 2013), indolyl-2-thiohydantoin derivatives (Suzen et al. 2003), 5-chloroindole derivatives (Yilmaz et al. 2012), indole amino acids (Suzen et al. 2012), and 2-phenylindole derivatives (Bozkaya et al. 2006; Suzen et al. 2006). The studies showed that the majority of the compounds evaluated established significant antioxidant activity, especially in DPPH scavenging and lipid peroxidation assays (Ates-Alagoz et al. 2005; Suzen et al. 2006, 2013; Gurkok et al. 2009; Shirinzadeh et al. 2010; Yilmaz et al. 2012). Anisic acid and nicotinic acid hydrazide derivatives (Gurkok et al. 2009; Yilmaz et al. 2012) were the only exception. Interestingly, they were found inactive as antioxidants in DPPH assay. Furthermore, they establish prooxidant behavior in superoxide radical scavenging activity assay. Introduction of Cl or Br instead of methoxy group in the fifth position of MLT (Gurkok et al. 2009; Yilmaz et al. 2012), which gives different electronic and lipophilic properties, did not result in considerable difference. Even though the most powerful compound was found to be *o*-Cl substituted derivative, in general, the active compounds were recognized as difluoro substituted in LP assay and

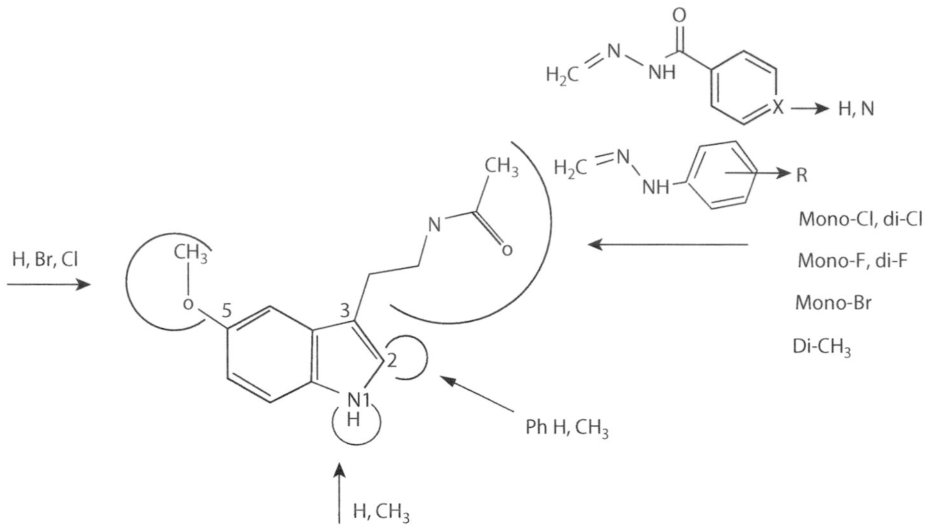

FIGURE 21.7 Modifications made on MLT to synthesize new analog compounds.

difluoro and dichloro substituted in DPPH assay. The derivative that has 3,5-dichloro substitutions on phenyl showed the best antioxidant.

2′,7′-Dichlorofluorescin (DCFH) and its diacetate form (DCFHDA) are widely used to measure oxidative stress in cells due to the high sensitivity of fluorescence-based assays. The assay consists of the oxidation of DCFH (after hydrolysis of the diacetate form) to fluorescein by ferryl-type intermediates and/or oxygen and nitrogen reactive species, whose fluorescence can be measured at 522 nm (Tsuchiya et al. 1994; Bonini et al. 2006). The protective effect of newly synthesized indole-based MLT analogs against DCFH-DA oxidation was assayed in human erythrocytes (Shirinzadeh et al. 2010). N-methylated indole ring decreased the free radical scavenging capacity of indole derivatives comparing the N–H substituted derivatives. Moreover lactate dehydrogenase (LDH) activity, which was used to assess the membrane stabilizing effect of MLT analogs, showed no increases in LDH activity with any of the tested compounds. Interestingly, some of the compounds were found to decrease LDH leakage, which indicates membrane stabilizing effect. This explains a possible cytotoxic effect of the tested indole derivatives. It can be concluded that none of the synthesized MLT analogs are found to have cytotoxic effect.

Suzen et al. (2013) recently synthesized and evaluated the antioxidant activity of new indole-based MLT analogs by evaluating their reducing activity against oxidation in human erythrocytes that were oxidatively challenged by H_2O_2. Almost all compounds exhibited an antioxidant activity, excluding the hydrazide compounds. Among three halogens that were present in the synthesized analogs, F substitution seemed to increase the antioxidant activity where Cl and Br substitution resulted in similar antioxidant activity, which were lower than F. Difluoro compounds were found the most active, compared to the compounds with a dichlorinated aromatic side chain. LDH activity was examined in medium of CHO–K1 cells, which were incubated either with MLT or the analog compounds. None of the compounds were found to induce membrane damage.

21.4 CONCLUSIONS

ROS and RNS can cause vital damages to all cellular macromolecules, including nucleic acids, proteins, carbohydrates, and lipids (Suzen et al. 2012). Antioxidants can interact with reactive species and prevent their chain reactions before vital and essential molecules are damaged (Shirinzadeh et al. 2010).

The mechanism by which the indole ring interacts with free radicals is still not completely recognized (Karaaslan et al. 2013). Even though it is thought that MLT interacts with free radicals by the contribution of an electron to form the melatoninyl cation radical through a radical addition at C3, other possibilities include hydrogen donation from the nitrogen atom or substitution at position C2, C4, and C7 on the indole ring (Rosenstein et al. 1998; Tan et al. 2002). It is possible that the 4-aminophenyl side chain might assist the formation of an indolyl cation radical during scavenging free radicals in the in vitro assays. The outcome of recent research evidently proved that for antioxidant activity of MLT analog indole derivatives, both the indole ring and side chain are almost equally important (Ates-Alagoz and Suzen 2001; Karaaslan et al. 2013). These results suggest a new approach for the in vitro antioxidant activity properties and structure–activity relationships of substituted indole rings.

REFERENCES

Ates-Alagoz, Z. 2013. Antioxidant activities of retinoidal benzimidazole or indole derivatives in vitro model systems. *Curr Med Chem* 20(36):4633–4639.

Ates-Alagoz, Z., Çoban, T., Buyukbingol, B. 2006. Synthesis and antioxidant activity of new tetrahydronaphthalene-indole derivatives as retinoid and melatonin analogs. *Arch Pharm Chem Life Sci* 339:193–200.

Ates-Alagoz, Z., Coban, T., Suzen, S. A. 2005. Comparative study: Evaluation of antioxidant activity of melatonin and some indole derivatives. *Med Chem Res* 14:169–179.

Ates-Alagoz, Z., Suzen, S. 2001a. Structure-activity relationships of melatonin analogues. *J Fac Pharm Ankara Univ* 30(4):41–52.

Ates-Alagoz, Z., Suzen, S. 2001b. Oxidative damage in the central nervous system and protection by melatonin. *J Fac Pharm Ankara Univ* 30(3):47–62.

Bhavsar, B., Farooq, M. U., Bhatt, A. 2009. The therapeutic potential of melatonin in neurological disorders. *Recent Pat Endocr Metab Immune Drug Discov* 3:60–64.

Bonini, M. G., Rota, C., Tomasi, A., Mason, R. P. 2006. The oxidation of 2′,7′-dichlorofluorescin to reactive oxygen species: A self-fulfilling prophesy? *Free Radic Biol Med* 40(6):968–975.

Bozkaya, P., Dogan, B., Suzen, S., Nebioglu, D., Ozkan, S. A. 2006. Determination and investigation of electrochemical behaviour of 2-phenylindole derivatives: Discussion on possible mechanistic pathways. *Can J Anal Sci Spec* 51:125–139.

Buemi, M. R., De Luca, L., Chimirri, A., Ferro, S., Gitto, R., Alvarez-Builla, J., Alajarin, R. 2013. Indole derivatives as dual-effective agents for the treatment of neurodegenerative diseases: Synthesis, biological evaluation, and molecular modeling studies. *Bioorg Med Chem* 21(15):4575–4580.

Das-Evcimen, N., Yildirim, O., Suzen, S. 2009. Relationship between aldose reductase and superoxide dismutase inhibition capacities of indole-based analogs of melatonin derivatives. *Arch Biol Sci* 61(4):675–681.

Dong, Y., Venkatachalam, T. K., Narla, R. K., Trieu, V. N., Sudbeck, E. A., Uckun, F. M. 2000. Antioxidant function of phenethyl-5-bromo-pyridyl thiourea compounds with potent anti-HIV activity. *Bioorg Med Chem Lett* 10:87–90.

Emerit, J., Edeas, M., Bricaire, F. 2004. Neurodegenerative diseases and oxidative stress. *Biomed Pharmacother* 58:39–46.

Galano, A., Tan, D. X., Reiter, R. J. 2013. On the free radical scavenging activities of melatonin's metabolites, AFMK and AMK. *J Pineal Res* 54(3):245–257.

Gupta, A. K., Kalpana, S., Malik, J. K. 2012. Synthesis and in vitro antioxidant activity of new 3-substituted-2-oxindole derivatives. *Indian J Pharm Sci* 74(5):481–486.

Gurkok, G., Coban, T., Suzen, T. 2009. Melatonin analogue new indole hydrazide/hydrazone derivatives with antioxidant behavior: Synthesis and structure-activity relationships. *J Enzyme Inhib Med Chem* 24:506–515.

Hadjipavlou-Litina, D., Samadi, A., Unzeta, M., Marco-Contelles, J. 2013. Analysis of the antioxidant properties of differently substituted 2- and 3-indolyl carbohydrazides and related derivatives. *Eur J Med Chem* 63:670–674.

Hardeland, R. 2011. Melatonin and its metabolites as anti-nitrosating and anti-nitrating agents. *J Exp Integr Med* 1(2):67–81.

Karaaslan, C., Kadri, H., Coban, T., Suzen, S., Westwell, A. D. 2013. Synthesis and antioxidant properties of substituted 2-phenyl-1*H*-indoles. *Bioorg Med Chem Lett* 23(9):2671–2674.

Lin, L., Huang, Q. X., Yang, S. S., Chu, J., Wang, J. Z., Tian, Q. 2013. Melatonin in Alzheimer's disease. *Int J Mol Sci* 14(7):14575–14593.

Padurariu, M., Ciobica, A., Lefter, R., Serban, I. L., Stefanescu, C., Chirita, R. 2013. The oxidative stress hypothesis in Alzheimer's disease. *Psychiatr Danub* 25(4):401–409.

Perez-Reyes, E., Mason, R. P. 1981. Characterization of the structure and reactions of free radicals from serotonin and related indoles. *J Biol Chem* 256:2427–2432.

Poeggeler, B., Reiter, R. J., Tan, D. X., Chen, L. D., Manchester, L. C. 1993. Melatonin, hydroxyl radical-mediated oxidative damage, and aging: A hypothesis. *J. Pineal Res* 14:151–168.

Poeggeler, B., Thuermann, S., Dose, A., Schoenke, M., Burkhardt, S., Hardeland, R. 2002. Melatonin's unique radical scavenging properties—Roles of its functional substituents as revealed by a comparison with its structural analogs. *J Pineal Res* 33(1):20–30.

Reiter, R. J., Tan, D. X., Manchester, L. C., Qi, W. 2001. Biochemical reactivity of melatonin with reactive oxygen and nitrogen species: A review of the evidence. *Cell Biochem Biophys* 34(2):237–256.

Rosenstein, R. E., Estrin, D. A., Turjanski, A. G. 1998. Reactions of melatonin and related indoles with free radicals: A computational study. *J Med Chem* 41:3684–3689.

Samokyszyn, V. M., Marnett, L. J. 1990. Inhibition of microsomal lipid peroxidation by 13-cis-retinoic acid. *Methods Enzymol* 190:281–288.

Shirinzadeh, H., Gumustas, M., Suzen, S., Ozden, S., Ozkan S. A. 2010. Synthesis and investigation of electrochemical behaviour of *N*-methylindole-3-carboxaldehyde izonicotinoyl hydrazone. *Molecules* 15(4):2187–2202.

Sies, H. 1997. Oxidative stress: Oxidants and antioxidants. *Exp Physiol* 82:291–295.

Suzen, S. 2006. Recent developments of melatonin related antioxidant compounds. *Comb Chem High Throughput Screen* 9:409–419.

Suzen, S. 2007. Topics in heterocyclic chemistry, bioactive heterocycles V. In: Khan M. T. H., ed., *Antioxidant Activities of Synthetic Indole Derivatives and Possible Activity Mechanisms*. Berlin, Germany: Springer-Verlag, Vol. 11, pp. 145–178.

Suzen, S. 2013. Melatonin and synthetic analogs as antioxidants. *Curr Drug Deliv* 10(1):71–75.

Suzen, S., Ateş-Alagoz, Z., Demircigil, T., Ozkan, S. A. 2001. Synthesis and analytical evaluation by voltammetric studies of some new indole-3-propionic acid derivatives. *Farmaco* 56:835–840.

Suzen, S., Ates-Alagoz, Z., Puskullu, O. 2000. Antioxidant activities of indole and benzimidazole derivatives. *FABAD J Pharm Sci* 25(3):113–119.

Suzen, S., Bozkaya, P., Coban, T., Nebioglu, D. 2006. Investigation of in vitro antioxidant behaviour of some 2-phenylindole derivatives: Discussion on possible antioxidant mechanisms and comparison with melatonin. *J Enzyme Inhib Med Chem* 21(4):405–411.

Suzen, S., Cihaner, S. S., Coban, T. 2012. Synthesis and comparison of antioxidant properties of indole-based melatonin analogue indole amino acid derivatives. *Chem Biol Drug Des* 79(1):76–83.

Suzen, S., Demircigil, T., Buyukbingol, E., Ozkan, S. A. 2003. Electroanalytical evaluation and determination of 5-(3′-indolyl)-2-thiohydantoin derivatives by voltammetric studies: Possible relevance to in vitro metabolism. *New J Chem* 27:1007–1011.

Suzen, S., Tekiner-Gulbas, B., Shirinzadeh, H., Uslu, D., Gurer-Orhan, H., Gumustas, M., Ozkan, S. A. 2013. Antioxidant activity of indole-based melatonin analogues in erythrocytes and their voltammetric characterization. *J Enzyme Inhib Med Chem* 28(6):1143–1155.

Tallman, M. S., Wiernik, P. H. 1992. Retinoids in cancer treatment. *J Clin Pharmacol* 32:868–888.

Tan, D. X., Manchester, L. C., Reiter, R. J., Plummer, B. F., Hardies, L. J., Weintraub, S. T., Vijayalaxmi, Shepherd, A. M. M. 1998. A novel melatonin metabolite, cyclic 3-hydroxymelatonin: A biomarker of in vivo hydroxyl radical generation. *Biochem Biophys Res Commun* 253:614–620.

Tan, D. X., Manchester, L. C., Terron, M. P., Flores, L. J., Reiter, R. J. 2007. One molecule, many derivatives: A never-ending interaction of melatonin with reactive oxygen and nitrogen species? *J Pineal Res* 42:28–42.

Tan, D. X., Reiter, R. J., Manchester, L. C., Yan, M.-T., El-Sawi, M., Sainz, R. M., Mayo, J. C., Kohen, R., Allegra, M., Hardeland, R. 2002. Chemical and physical properties and potential mechanisms: Melatonin as a broad spectrum antioxidant and free radical scavenger. *Curr Top Med Chem* 2:181–197.

Tsuchiya, M., Suematsu, M., Suzuki, H. 1994. In vivo visualization of oxygen radical-dependent photoemission. *Methods Enzymol* 233:128–140.

Wu, Y. H., Swaab, D. F. 2005. The human pineal gland and melatonin in aging and Alzheimer's disease. *J Pineal Res* 38:145–152.

Yılmaz, A. D., Coban, T., Suzen, S. 2012. Synthesis and antioxidant activity evaluations of melatonin-based analogue indole-hydrazide/hydrazone derivatives. *J Enzyme Inhib Med Chem* 27:428–436.

Yu-Chieh, C., Jiunn-Ming, S., Miao-Meng, T., You-Lin, T., Li-Tung, H. 2013. Roles of melatonin in fetal programming in compromised pregnancies. *Int J Mol Sci* 14:5380–5401.

Zhou, J. N., Liu, R. Y., Kamphorst, W., Hofman, M. A., Swaab, D. F. 2003. Early neuropathological Alzheimer's changes in aged individuals are accompanied by decreased cerebrospinal fluid melatonin levels. *J Pineal Res* 35:125–130.

22 Melatonin's Analogues in Glaucoma

Hanan Awad Alkozi and Jesús Pintor

CONTENTS

22.1 MELATONIN IS NOT ONLY SYNTHESIZED IN THE PINEAL GLAND

Many studies have established where melatonin synthesis occurs. The first location in which melatonin synthesis was described was the pineal gland (Lerner et al. 1959). The pineal gland, which is the main source of this neurohormone, is the source of melatonin distributed via blood vessels to most other organs; nevertheless, there are other tissues where the synthesis of melatonin has been detected, including ocular tissues.

One of the most relevant aspects of melatonin biochemistry is its variable rate of synthesis during the day. Melatonin synthesis is elevated at night and reduced during the day (Tosini and Fukuhara 2003) (Figure 22.1). This circadian rhythm is generated by the suprachiasmatic nucleus (Lincoln et al. 1985) and the superior cervical ganglion, the latter being responsible for adrenergic innervation of the gland (Arendt 1998). Melatonin synthesis is regulated by norepinephrine, which is released from nerves that innervate the gland exclusively at night (Schomerus and Korf 2005). Norepinephrine acts through a dual α-β-adrenergic cAMP/Ca^{2+} system to increase the activity of the serotonin *N*-acetyltransferase also known as arylalkylamine *N*-acetyltransferase (AANAT) (Klein et al. 1997). Thus, the neural input to the gland is norepinephrine and the output is melatonin (Schomerus and Korf 2005).

Aside from the pineal gland, which is the classic location for the production of melatonin, another organ where this substance is synthesized, and therefore can modulate tissue behavior, is the eye. Melatonin is synthesized in the retina of vertebrates, including mammals, where it also shows a daily rhythm (Tosini and Fukuhara 2002; Zawilska et al. 2006). Since the retina is highly vascularized, it is assumed that the melatonin present in the retina comes from the pineal gland via the bloodstream, especially as it is present with a circadian rhythm. Nonetheless, although the

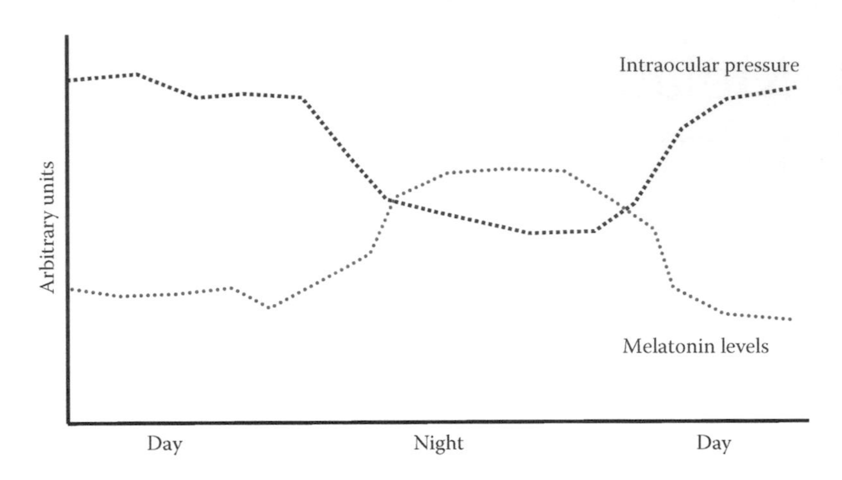

FIGURE 22.1 Relationship between IOP and melatonin levels during night and day. There is a surprising correlation between the reduction of IOP that occurs during the night and the elevated levels of melatonin in blood in this period.

retina has a complicated neuronal architecture formed by five different types of neurons, melatonin synthesis has been identified in a subpopulation of photoreceptors (probably the cones) (Tosini and Fukuhara 2003). The retina itself contains a circadian pacemaker that drives the rhythm of local melatonin synthesis (Tosini 2000). This fact has been demonstrated since retinal production of melatonin has been confirmed after pinealectomy. This process drops the melatonin levels in blood to 40% of their maximum values during the day, but retinal melatonin levels did not fall. Moreover, the local production of melatonin does not seem to contribute significantly to plasma levels (Vaughan and Reiter 1986). This fact indicates that it is synthesized in the retina principally for local purposes (Vanecek 1998). Once produced, melatonin is not stored but freely diffuses out of the cells. The amount of melatonin produced by the retina is small compared to that in the pineal gland and retinal melatonin is thought to act as a local neuromodulator within the eye. However, in a few instances (e.g., quails) retinal melatonin may contribute to the levels of the hormone in the blood (Underwood et al. 1984). Accordingly, melatonin seems to act locally as a neurohormone or neuromodulator (Tosini and Fukuhara 2003). The melatonin concentration in rat retina is 2 pg/mg protein, this concentration being adequate to activate retinal melatonin receptors (Chanut et al. 1998).

The retina is probably the main ocular location of melatonin synthesis, but it is not the only place where this substance is produced. The ciliary body has also been shown to be capable of synthesizing melatonin. All the machinery necessary for its synthesis is present: indolamines and the enzymatic activity of AANAT and hydroxyindole *O*-methyltransferase (HIOMT) necessary for melatonin synthesis have been found in both the aqueous humor and the ciliary body of the human eye. Metabolites of melatonin, such as 5-hydroxy-tryptophan and *N*-acetylserotonin, have been detected in the human ciliary processes at concentrations of 7.2–8.6 ng/mL and 273 ± 25 pmol/mg protein/h, respectively, thus suggesting the presence and metabolism of melatonin in this human tissue (Martin et al. 1992). Indeed, melatonin has been identified and quantified in the human aqueous humor, being present at levels approximating 0.47 ng/mL (Martin et al. 1992). A more recent evaluation of melatonin levels in the human aqueous humor indicates that the concentration of this substance is 6.4 pg/mL (Chiquet et al. 2006). These are extremely important data since melatonin has been reported to modulate aqueous humor dynamics.

The crystalline lens is bathed by the aqueous humor. In this ocular structure, the presence of AANAT and HIOMT activity has also been reported. In the rabbit lens, the production of melatonin shows significant circadian changes. The concentration of this neurohormone in the rabbit lens changes from 0.14 pmol/lens under light conditions to 0.25 pmol/lens in the darkness (Abe et al. 1999). These findings strongly suggest that the rabbit lens may synthesize melatonin from serotonin and it is possible that, as in the retina, it may function as a local regulator of rhythmic activity.

Another ocular structure where melatonin has been identified is the Harderian gland (Djeridane et al. 1998). This sebaceous gland acts as an accessory to the lacrimal gland (Kittner et al. 1978), and it secretes fluid that facilitates movement of the third eyelid in some animals (Chieffi et al. 1996). Melatonin confirmed its presence, synthesis, and release from isolated rodent Harderian glands (Djeridane et al. 1998). The influence of this gland is relevant since it releases melatonin into the tears, as occurs with the lachrymal gland (Mahtre et al. 1988). The concentration of melatonin in tears is 200 ng/mL, which should be enough to stimulate the melatonin receptors present on the ocular surface. Therefore, it is clear that the tears will be the main source of melatonin at the ocular surface, at least in rodents and lagomorphs.

22.2 RECEPTORS FOR MELATONIN AND THEIR PRESENCE IN OCULAR STRUCTURES

Melatonin actions in tissues are performed by means of several receptors. Melatonin receptors are referred to by the letters MT. "MT" represents melatonin receptors with a well-defined functional pharmacology in a native tissue, as well as known molecular structure, followed by a number subscript (Dubocovich et al. 2000). Italicization (*MT*) followed by the corresponding number is reserved for receptor pharmacology characterized in native tissues for which the molecular structure is not known (Dubocovich et al. 2000).

Three mammalian melatonin receptors have been proposed and cloned so far: MT_1 (Reppert et al. 1994; Roca et al. 1996), MT_2 (Reppert et al. 1995), and MT_3 (Nosjean et al. 2000). The two first receptors are classified as unique subtypes based on their molecular structure and chromosomal localization, and the third receptor has been affinity-purified from Syrian hamster kidney (Nosjean et al. 2000) (Figure 22.2).

Melatonin receptors in mammalian tissues types MT_1 and MT_2 belong to the class A group of rhodopsin-like G-protein coupled receptor or GPCRs (www.gpcr.org/7tm) and contain seven hydrophobic transmembrane domains. MT_1 and MT_2 melatonin receptors are formed from 350 and 362 aminoacids, respectively, and their calculated molecular weights are 39–40 kDa (Navajas et al. 1996).

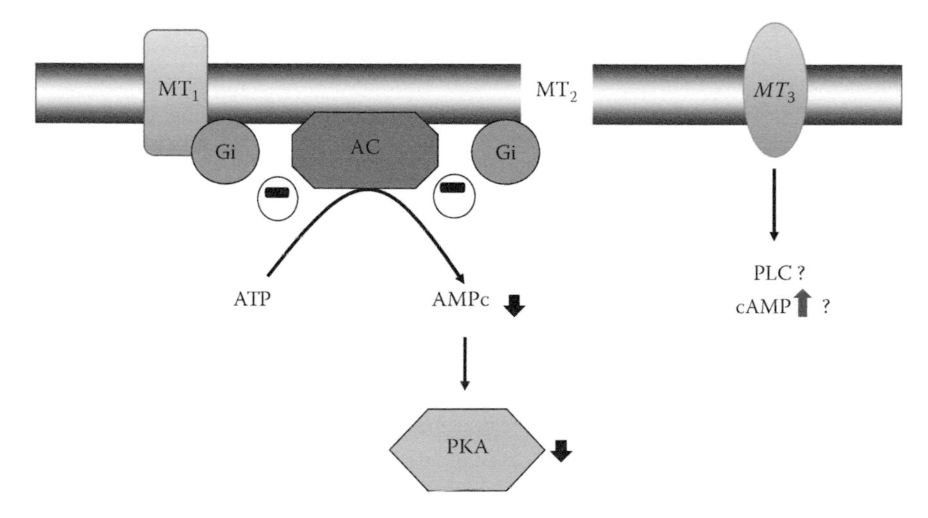

FIGURE 22.2 (See color insert.) Melatonin receptors and second messenger cascades. Three melatonin receptors have been described. MT_1 and MT_2 are negatively coupled to adenylate cyclase, whereas the putative MT_3 could be coupled to PLC and/or positively to adenylate cyclase.

These two melatonin receptors have different chromosomal locations (Reppert et al. 1996). The gene for the MT_1 receptor is located at position 4q35-1 (Slaugenhaupt et al. 1995) and the gene for the MT_2 receptor at 11q21-22 (Reppert et al. 1995) and they share approximately 60% homology with each another (Reppert et al. 1996). The existence of three extracellular loops alternating with three intracellular loops to link the seven transmembrane regions (Wess 1993) suggests the existence of potential sites for glycosylation and phosphorylation. The MT_1 melatonin receptor has two potential glycosylation sites in the N-terminal region (Navajas et al. 1996), and it may exist in more than one glycosylated form (Reppert et al. 1994, 1996), while MT_2 has one potential glycosylation site in the N-terminal region (Navajas et al. 1996; Reppert et al. 1996).

The third mammalian receptor, according to binding and functional studies the MT_3 melatonin receptor, seems to belong to the quinone reductase (QR) family. Uniquely, the MT_3 has been identified as QR2, an enzyme involved in detoxification (Nosjean et al. 2000). QR2 seems to have antioxidant properties, although it is difficult to establish a clear role in detoxification on the basis of the currently available experiments (Vella et al. 2005). Many tissues deprived of QR2 genes lack melatonin binding sites, indicating that in many cases QR2 is indeed the MT_3 melatonin receptor (Mailliet et al. 2004). There are other cases in which it has not been possible to match the presence of QR2 and the putative MT_3 melatonin receptor (Nosjean et al. 2001). Moreover, unexpectedly, the binding of melatonin and derivatives to QR2 does not inhibit the activity of the enzyme when the natural substrate is present (menadione), although it occupies the enzyme's active site (Mailliet et al. 2004). However, binding approaches for melatonin and derivatives on the QR2/MT_3 receptor are not enough to rule out the existence of a real receptor with a clear-cut physiological action. Moreover, in both mammals and nonmammals there is a correlation between the pharmacological profile of the putative MT_3 receptor and the generation of IP3 and diacylglycerol (Eison and Mullins 1993; Mullins et al. 1997).

MT_1 receptors are widespread in several ocular structures as well as in visual central nervous system areas, and they are also expressed in the suprachiasmatic nucleus and *pars tuberalis*, appearing to mediate the circadian and reproductive effects of melatonin (Reppert 1997; Von Gall et al. 2002). MT_2 receptors are expressed in the retina and several brain regions and could mediate the physiological effects of melatonin in the retina (Reppert et al. 1995). The MT_3 receptor was initially shown to be widely distributed in the hamster brain, including those areas related to visual processing, by means of the radioligand [125I]iodomelatonin (Dubocovich 1985).

There is evidence that all three melatonin receptor subtypes are expressed in ocular tissues. The presence of melatonin receptor subtypes in the retina, cornea, ciliary body, lens, choroid, and sclera has been described (Rada and Wiechmann 2006). Moreover, differential distributions of melatonin receptors in ocular tissues mediate distinct cellular functions of melatonin and suggest potential roles for melatonin in modulating rhythms in ocular growth, anterior chamber depth, and aqueous humor production (Rada and Wiechmann 2006).

It is important to emphasize that some of the points raised in the following paragraphs have been described in nonmammalian models. This generates controversy regarding whether or not findings in nonmammalian ocular tissues can be extrapolated to mammals. In some cases, the existence of the corresponding receptor orthologue in mammals invites one to think about their presence in these animals. Nevertheless, receptors such as the Mel1c, without an equivalent orthologue in mammals, give rise to questions about their possible existence in these animals.

The cornea is the most superficial ocular structure. It consists of five main layers: epithelium, Bowman's membrane, stroma, Descemet's membrane, and endothelium. The presence of melatonin receptors in the cornea has been investigated in fish, frogs, birds, and, very recently, rabbits. Several studies have described the presence of MT_1 (Mel1a) and MT_2 (Mel1b) receptors in the cornea of various species. In addition, and as commented previously, the presence of Mel1c has been described in *Xenopus laevis* cornea.

When the presence of melatonin receptors is studied in mammalian corneas, the existence of MT_1 receptors has been demonstrated in human corneal endothelial cells, keratocytes, and the basal corneal epithelial cells (Meyer et al. 2002). The corneal epithelium is the most superficial part of the eye and therefore suffers from accidents such as those caused by the introduction of a foreign body, inadequate contact lens wearing, and even from refractive surgery. Even without the previous examples, it has been shown that the renewal of the corneal epithelium exhibits circadian regulation (Doughty 1990). This day–night difference suggests that melatonin of nonretinal origin might be involved in the circadian mitotic rhythm of the cornea. The melatonin receptor responsible for the enhancement of the corneal wound healing process is an MT_1 receptor (Buffa et al. 1993; Wiechmann and Rada 2003). Corneal wound healing experiments performed in New Zealand white rabbits, in agreement with the previous nonmammalian models, also confirm this point, since low micromolar concentrations of melatonin significantly accelerate the rate of healing, causing the wounds to close faster than in the absence of this neurohormone (Pintor et al. 2005). This effect is antagonized by luzindole, suggesting the presence of MT_1/MT_2 melatonin receptors. In primary rabbit corneal epithelial cells in culture, the same behavior was observed for melatonin and luzindole. It has been possible to demonstrate that the effect of melatonin is to increase the rate of cell migration rather than mitosis. Moreover, immunocytochemical staining of these cells demonstrate the presence of MT_2 melatonin receptors (Pintor et al. 2005). So, both MT_1 and MT_2 receptors are present in the corneal epithelium, but it is still not clear whether both contribute to accelerate the rate of epithelial cell migration or if they combine sequentially: one to facilitate cell migration and the other triggering mitosis to complete corneal re-epithelialization.

Harmonizing the previous ideas are the results presented by Wiechmann. This author indicates that alteration of melatonin receptors in neural and nonneural ocular tissues, in particular on the corneal surface, may lead to impaired wound healing, indicating the relevance of melatonin and its receptor in corneal re-epithelialization (Wiechmann and Rada 2003).

Another possible role for melatonin in the cornea is the inhibition of reactive oxygen species such as superoxide. Superoxide radicals create deleterious changes in the collagen arrays that normally permit corneal transparency (Ciuffi et al. 2003). Very recently, it has been possible to investigate the role of melatonin on tear secretion. Melatonin alone, topically applied to New Zealand rabbits, inhibits tear production. On the other hand, the dinucleotide diadenosine tetraphosphate, Ap_4A, enhances tear production. Surprisingly, when melatonin and Ap_4A are both applied simultaneously, the amount of tear production is enhanced 34%, by means of a mechanism that is sensitive to luzindole (Hoyle et al. 2006). Cross talk between melatonin and purinergic receptors may be the reason for this increase in tear production, although the detailed mechanism is still under study.

The iris and ciliary body are two ocular tissues with different but closely related tasks. The iris regulates the amount of light that arrives at the retina by using a mechanism similar to a camera diaphragm. The ciliary body is responsible for the production of the aqueous humor that controls intraocular pressure (IOP).

The existence of melatonin receptors in the iris ciliary body of animal models, such as the rabbit, has been described (Osborne and Chidlow 1994). Autoradiographic experiments have shown that 125I-iodomelatonin binds specific sites associated with the rabbit iris–ciliary body (Osborne and Chidlow 1994). A melatonin receptor subtype (probably Mel1c) has been located on the basolateral surfaces of nonpigmented epithelial cells of *X. laevis* ciliary body. The existence of this receptor on the nonpigmented epithelial cells may support for the hypothesis that melatonin may decrease the rate of aqueous humor secretion by the nonpigmented ciliary epithelium, resulting in a modulation of IOP (Wiechmann and Wirsig-Wiechmann 2001). Because of their location on the basolateral surface of the nonpigmented ciliary epithelium, the melatonin receptors are situated where they may be reached by melatonin released either from the circulation or from the ciliary body (Wiechmann and Wirsig-Wiechmann 2001). Accordingly, these melatonin receptors may be involved in maintaining IOP (Osborne 1994; Osborne and Chidlow 1994; Pintor et al. 2001, 2003; Wiechmann and Wirsig-Wiechmann 2001).

The lens is the ocular structure that focuses images onto the retina, adapting its shape depending on the distance of the object that is observed. It is a nonvascular, noninnervated structure, so that all nutrients and chemical messengers come from the aqueous humor that surrounds it.

Little is known about the presence of melatonin receptor subtypes in the lens. Only *X. laevis* studies have been published, wherein neither MT_1 (Mel1a) nor Mel1c receptor immunoreactivity has been detected. However, MT_2 (Mel1b) immunoreactive bands appeared in the western blots of the lens fiber cell membranes, and these punctuate immunolabeling throughout the lens (Wiechmann et al. 2004). It is possible that melatonin in the lens, as in the retina, may function as a local mediator of rhythmic activity. Melatonin can activate lens fiber receptors because it is present in the aqueous humor, but it is important to note that the lens also contains the enzymes necessary for its synthesis. Indeed, both melatonin and AANAT activity exhibit significant diurnal changes with the highest levels in the rabbit lens during the dark period of a 12/12 h light/dark cycle (Abe et al. 1999).

The retina is an extension of the brain highly specialized for the detection of light. It is formed at least by five different types of neurons: photoreceptors, horizontal cells, bipolar cells, amacrine cells, and ganglion cells. These cells are perfectly organized scaffolding in which all the connections among the cells are well known. Much of the research in ocular melatonin receptors has been performed in the retina from different species. High-affinity melatonin receptor subtypes have been identified in distinct parts of the retina, suggesting that retinal cells in this structure undergo biochemical changes after being challenged with melatonin. Although the function of melatonin in the human retina is not well known, various actions have been described in other experimental models. For example, melatonin modulates dopamine release from rabbit neurones of the retina (Dubocovich 1983), it is involved in the regulation of retinomotor movements in lower vertebrates (Pierce and Besharse 1985), it participates in regulation of horizontal cell sensitivity to light in salamander retina (Wiechmann et al. 1988), and it mediates circadian disc shedding in rat photoreceptor cells (White and Fisher 1989).

Concerning the MT_1 melatonin receptor, its expression has been localized to the inner segments of rod and cone photoreceptors, in ganglion cells of human retina (Meyer et al. 2002; Scher et al. 2002), and on photoreceptors of *Xenopus* (Wiechmann and Smith 2001). Also, in domestic chicks, MT1 (Mel1a) and Mel1c receptor RNA present similar patterns of expression, primarily in the inner segments of the photoreceptors, the vitreal portion of the inner nuclear layer, and in the ganglion cell layer (Natesan and Cassone 2002). These data suggest a role for melatonin in rod phototransduction. In particular, since MT_1 receptors are negatively coupled to adenylate cyclase, it could be the case that melatonin produces a reduction in the activity of protein kinase A (PKA) and a consequent fall in the phosphorylation of proteins related to the neurosecretory machinery. MT_1 melatonin receptor expression has also been confirmed in the horizontal cells of human and rat retina (Fujieda et al. 1999; Scher et al. 2002), and immunocytochemical data have revealed the expression of the MT_1 (Mel1a) receptor in horizontal cells (Huang et al. 2005). This coincidence between mammal and nonmammalian animal models suggests a possible role for melatonin in the synaptic transmission from cones to horizontal cells in retina.

The MT_1 receptor has also been immunocytochemically localized to both the inner and outer plexiform layers in rodent retina (Fujieda et al. 1999, 2000) and in the inner plexiform layer in human retina (Savaskan et al. 2002). This MT_1 receptor therefore participates in the two main areas of synapses existing in the mammalian retina. Specifically, melatonin is involved in retinal physiology by acting on amacrine cells, via the MT_1 receptor expressed in their synapses. This probably indicates that melatonin regulates the release of neurotransmitters, such as acetylcholine (Fujieda et al. 1999), dopamine (Scher et al. 2002), or even gamma aminobuytric acid (GABA) (Meyer et al. 2002), the most typical amacrine cell chemical messengers. Since amacrine cells, like horizontal cells, connect peripheral photoreceptors with central photoreceptors in the receptive fields, the role of MT_1 receptors seem to be similar to that previously described for horizontal cells. An indication of how complex the distribution of MT1 receptors can be regarding the species under study, MT_1 immunoreactivity has been analyzed in postmortem nonpathological human eyes and in the eyes of the long-tailed macaque, *Macaca fascicularis*. MT_1 appeared in 100% of AII amacrine cells (containing acetylcholine) in the monkey and between 78% and 86% in human retina, again demonstrating differences between two species not so far away phylogenetically (Scher et al. 2003).

Regarding the MT_2 (or Mel1b in the case of nonmammalian models) receptor, expression has been located to the apical membrane of retinal pigmented epithelial cells (the nonneural retina) and immunoreactivity has also been observed in both rod and cone photoreceptor inner segments. The role of melatonin in retinal pigmented epithelial cells may be to modulate the phagocytosis of photoreceptors as occurs with other transmitters that are also coupled to the adenylate cyclase second messenger system, although this has not yet been demonstrated (Gregory et al. 1994). MT_2 (Mel1b) staining appears in horizontal cells but it is not expressed in dopaminergic or GABAergic amacrine cells, in clear contrast with what happens with the MT_1 melatonin receptor (Wiechmann et al. 2004). Also, MT_2 (Mel1b) receptor mRNA is expressed in photoreceptors, throughout the inner nuclear layer and in the ganglion cell layer in domestic chick (Natesan and Cassone 2002). In mammalian models and, in particular, humans, comparative reverse transcription polymerase chain reaction (PCR) shows that MT_2 receptors are expressed in the retina. Therefore, MT_2 receptors may participate in some of the neurobiological effects of melatonin described in mammals (Reppert et al. 1995). In this sense, it has been proposed that inhibition of dopamine release by melatonin is mediated through the MT_2 receptor in rabbit retina (Dubocovich et al. 1997).

It has been suggested that MT_1 and MT_2 melatonin receptors are involved in changes related to aging (Scher et al. 2002). The physiological decrease in melatonin levels in elderly people has been related to the initiation of age-related macular degeneration (Yi et al. 2005), suggesting the participation of these receptors in the protection against macular degeneration (Yi et al. 2005). There is also evidence to suggest that melatonin treatment prevents proliferative vitreoretinopathy (Er et al. 2006).

22.3 GLAUCOMA: A PATHOLOGY RELATED TO INTRAOCULAR PRESSURE

Glaucoma pathology, known as the silent thief of sight, is a common ocular condition of multifactorial etiology characterized by progressive loss of retinal ganglion cells leading to optic neuropathy with irreversible visual field defects (Quigley 2011). It is the second leading cause of blindness worldwide and it is often underdiagnosed. At least 50% of glaucoma patients are left without diagnosis (Dielemans et al. 1994). However, the estimated number of glaucoma patients is 60.5 million worldwide, from which over 8.4 million are bilaterally blind from glaucoma. By 2020, this number will increase to by 11.1 million as the number of glaucoma cases will increase by 20 million cases (Quigley and Broman 2006).

There are several types of glaucoma, the most common type in particular is the primary open-angle glaucoma (POAG), in which the anterior chamber of the eye is deep and there is a reduction in the aqueous humor outflow through the trabecular meshwork, so in turn, causes an increase of IOP (Lee et al. 2001).

Many risk factors have been associated with glaucoma. These risk factors include aging or race (Figure 22.3). In this sense, the chance of developing glaucoma in Africans is six times more than for Caucasians. However, the only risk factor that can be pharmacomodulated, and which has been the most important risk factor, is elevated IOP (Sommer et al. 1991).

Normal IOP is normally between 10 and 21 mmHg. Ocular hypertension is defined as elevated IOP, when it is above 21 mmHg (Leske 1983). High IOP can be produced due to a reduction of aqueous humor outflow through the trabecular meshwork and through the uveoscleral way, or when there is a high production of aqueous humor from the ciliary body (Figure 22.4). Therefore, the main mechanisms of action for antiglaucoma therapeutic agents work through the mentioned ways in order to achieve a normal IOP (Lee and Goldberg 2011; Johnson 2006).

In the early manifest glaucoma treatment trial (EMGT), glaucoma progression decreased by 10% for each millimeter mercury reduction of IOP (Heijl et al. 2002). Even in other types of glaucoma

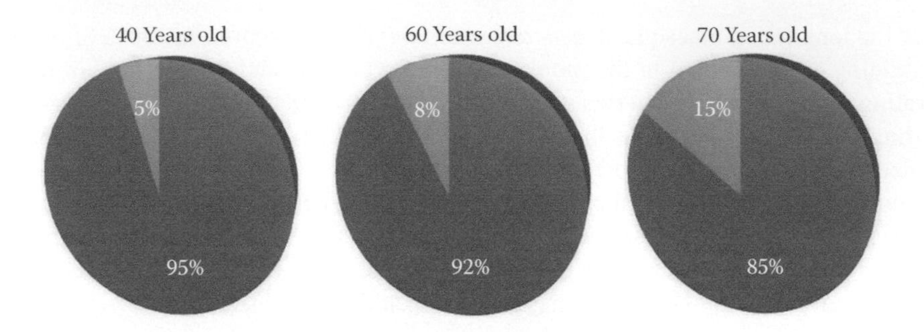

FIGURE 22.3 Prevalence of glaucoma regarding aging. The development of the glaucomatous pathology increases with aging, being 15% when age is 70 years or more.

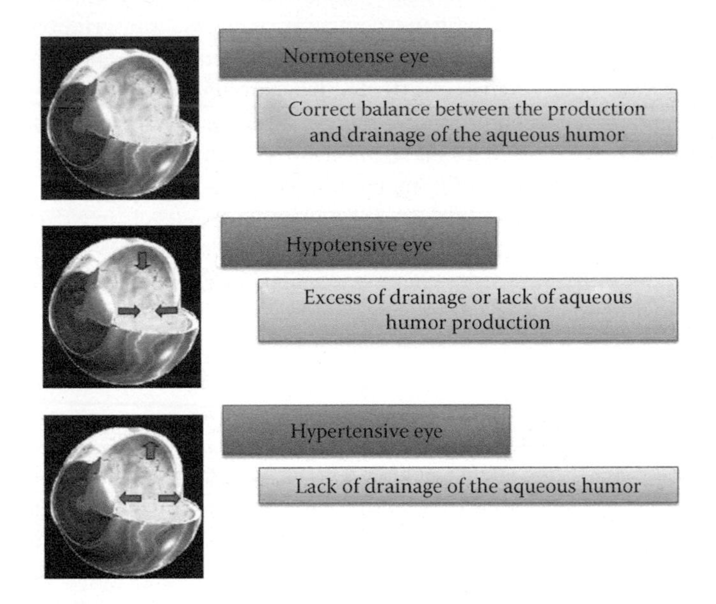

FIGURE 22.4 **(See color insert.)** Ocular conditions related with IOP. Ocular shape and function depends on IOP. Under normal conditions, the production and drainage of the aqueous humor (responsible for IOP) is balanced. The imbalance can produce either an increase in the IOP and therefore a risk of glaucoma or a reduction that may collapse the ocular globe (see the arrows in the figure).

with normal IOP, it has been shown that a 30% reduction in IOP can slow the rate of visual field loss progression. Furthermore, It has been shown that topical ocular hypotensive medication is effective in reducing the incidence of glaucomatous visual field loss and/or optic nerve deterioration in individuals with high IOP, and in a 60-month study, POAG was 4.4% in treated patients and 9.5% in nontreated ones (Kass et al. 2002; Varma et al. 2004).

Current ocular hypotensive agents in use that reduce aqueous inflow are β-adrenergic antagonists, carbonic anhydrase inhibitors, and α2-adrenergic agonists. Alternatively, it is possible to increase the aqueous humor outflow such as prostaglandin analogues and prostamides do (Lee and Goldberg 2011).

However, there is still a need for new studies regarding the treatment of glaucoma by means of lowering IOP due to the side effects of the common medications. For example, the group of adrenergic antagonists such as timolol can cause bradycardia, bronchospasm, hypoglycemia, and depression. Another category of medications are the carbonic anhydrase inhibitors, for example, dorzolamide, which can lead to irreversible corneal edema (Stewart et al. 2004). Furthermore, prostaglandin derivatives, like latanoprost, also have important side effects such as conjunctival hyperemia and iris pigmentation (Parrish et al. 2003; Lee and Goldberg 2011). Under these circumstances, there are a number of new antiglaucoma pharmacological agents that are currently under investigation, one of which is melatonin and its analogues.

22.4 SHORT-TERM EFFECT OF MELATONIN AND ANALOGUES

22.4.1 Melatonin and Analogues Reduce Intraocular Pressure

IOP is higher during daytime than at night and it has been suggested that melatonergic mechanisms in the eye could be responsible for the diurnal rhythm in IOP (Chiou and McLaughlin 1984; Aimoto et al. 1985; Rohde et al. 1985) and that the synthesis and diurnal rhythm of this melatonin are independent of the pineal gland (Rohde et al. 1985). This functional response may be regulated by melatonin MT_3 receptors since there is evidence to support a role of melatonin in the circadian changes of IOP (Pintor et al. 2001). Thus, depending on the experimental model, method of measurement, and experimental conditions, a reduction in melatonin content in the eye implies a reduction in the IOP and vice versa (Chiou and McLaughlin 1984). Contrarily, in other cases it has been shown that melatonin administration lowers IOP (Samples et al. 1988). Moreover, a potent hypotensive effect of 5-MCA-NAT, a melatonin MT_3 receptor agonist, has been observed in rabbit IOP (Pintor et al. 2001, 2003). In concordance with these results, hypotensive effects have also been observed in glaucomatous monkey eyes (Serle et al. 2004). Therefore, we suggest that the MT_3 receptor subtype may be localized in the ciliary processes and trabecular meshwork. The strong hypotensive effect of the MT_3 agonist, 5-MCA-NAT, reducing IOP about 60% and the effect lasting for more than 8 h, suggests that this compound may be a useful agent for treating those pathologies where IOP is abnormally elevated (Pintor et al. 2003; Serle et al. 2004).

Due to the identification of MT_3 as QR2, this receptor appears to be another molecular target to explore the multiple aspects of melatonin action (Nosjean et al. 2000). Thus, QR2 has been suspected to detoxify quinones because of its sequence homology with QR1. If melatonin binds to MT_3 and activates QR2, this could explain the antioxidant role of melatonin (Vella et al. 2005) and consequently the antioxidant activity of the lens.

The decrease of IOP can also occur by the MT_2 melatonin receptor, which is stimulated by the selective agonist IIK7, indicating the involvement of this receptor in the modulation of IOP (Alarma-Estrany et al. 2008). The hypotensive effect of melatonin and IIK7 is concentration-dependent, and it can be blocked by selective MT_2 receptor antagonists, this confirming the activation of such receptors. In fact, the MT_2 melatonin receptor selective antagonists, DH-97 or 4P-PDOT, were much more potent at blocking responses induced by IIK7 than other antagonists, such as prazosin (MT_3) or luzindole (nonselective melatonin antagonist). The study of its location by means of immunohistochemistry and western blot analysis demonstrated the existence of MT_2 melatonin receptors in the ciliary processes of New Zealand rabbit eye. MT_2 has been observed mostly in the region corresponding to basolateral nonpigmented epithelium of the ciliary processes and bordering the stroma probably in the vessels. This matches with the results in nonmammals where the presence of a melatonin receptor subtype has been shown, in this case, probably Mel1c, on the basolateral surfaces of nonpigmented epithelial cells of *X. laevis* ciliary body (Wiechmann et al. 2001).

Interestingly, experimental results have confirmed that the effect on IOP produced by melatonin and IIK7 decreased substantially in sympathetically denervated animals. Therefore, it is suggested that the sympathetic nervous system plays an additional role in the regulation of aqueous humor dynamics by modulating MT_2 melatonin receptors activity. In this sense, chemically sympathectomized animals treated with 6-OHDA affected to the responses triggered by melatonin, 5-MCA-NAT, as well as IIK7 (Alarma-Estrany et al. 2008).

Concerning the relationship between the melatoninergic and the adrenergic nervous systems, some observations can be made. For instance, the α2-adrenergic agonist clonidine did not modify the IIK7 IOP reduction, whereas the β-adrenergic agonists, terbutaline and salbutamol, potentiated the hypotensive effect of IIK7. This indicates that MT_2 melatonin receptors can be modulated by noradrenaline, as this neurotransmitter interacts mostly with β-adrenoceptors in the ciliary processes (Alarma-Estrany et al. 2008).

On the other hand, immunohistochemical studies with antibodies against the vesicular monoamine transporter 2 (VMAT2) and MT_2 melatonin receptors showed no co-localization of MT_2 melatonin receptors with autonomic nervous fibers. Labeling against MT_2 melatonin receptor has also been observed in ciliary processes of sympathectomized rabbits. It suggests that, somehow, MT_2 melatonin receptors are modulated by the sympathetic nervous system, as a neural regulator of this melatonin receptor activity.

22.4.2 Molecular Mechanism of Melatonin Actions Reducing Intraocular Pressure

The molecular mechanism that underlies the reduction in IOP, considering that the place where melatonin acts is the ciliary body, has to be the regulation of the mechanisms for aqueous humor production. In this sense, the main action of melatonin and its analogues is the modulation of intracellular chloride concentrations, which is important since chloride rules water movement, and therefore is the key ion driving the production of the aqueous humor (Civan and Macknight 2004). It has been demonstrated that the chloride efflux from ciliary body cells is one of the most, if not the most, important ion controlling aqueous humor secretion, as previously commented (Jacob and Civan 1996; Forrester 2002). The nature of the proteins ruling the movement of Cl^- is not clear. Some evidence points to CIC-3 as one of the predominant channels in the NPE, as well as the pICln (Paulmichl et al. 1992; Chen et al. 1999; Do et al. 2006). There is, nevertheless, an incomplete model to understand how different neurotransmitters and other relevant substances can regulate chloride secretion and, secondarily, IOP.

Melatonin and 5-MCA-NAT can reduce IOP because they decrease the efflux of chloride from the cytoplasm toward the extracellular space. The effect of both, melatonin and 5-MCA-NAT inhibiting this ion movement is strong during the first 15 min. After this interval of inhibition, the rate of secretion increases, but it is always below normal conditions. This dramatic inhibition of the chloride release and the presumable inhibition of the aqueous humor formation may be responsible for the IOP reduction observed with these compounds (Pintor et al. 2003).

It is a matter of interest to investigate the second messengers produced by melatonin receptors and how these can link receptor activation with chloride efflux. Originally, MT_1 and MT_2 receptors are negatively coupled to adenylate cyclase, while MT_3 is coupled to PLC (Eison and Mullins 1993; Mullins et al. 1997) (Figure 22.2). Interestingly, new works are appearing indicating that new second messenger systems can be coupled to melatonin receptors. For instance, melatonin receptors produce a reduction in Single-nucleotide polymorphism SNP-released nitric oxide and cGMP levels in human nonpigmented ciliary epithelial cells (Dortch-Carnes and Tosini 2013). These authors indicate that, at least in part, the effects of melatonin and analogues can use this pathway in human NPE cells and that this second messenger system may be responsible for the hypotensive effect of melatoninergic compounds on IOP.

In rabbit ciliary body cells, neither melatonin nor 5-MCA-NAT is able to activate the phospholipase C/protein kinase C PLC/PKC pathway. Surprisingly, melatonin and its analogue increase the concentrations of cAMP in a concentration-dependent manner. This is not a common mechanism of signal amplification but it has been described in some models (Raviola 1974; Beraldo and Garcia 2005; Schuster et al. 2005). The relationship between increases of cAMP and decrease of IOP has been widely studied in the eye. In rabbit and monkey, it is known that this decrease in IOP is associated with a significant decrease in aqueous humor secretion. Interestingly, it has been reported that 5-MCA-NAT is able to produce important increases in cAMP in chick retinas by a mechanism that may involve an MT_3 (Mel1c) binding site (Sampaio 2009).

22.4.3 Optimizing the Formulations for Ocular Treatments

One of the main problems regarding melatonin and analogues is their low solubility in water. This implies that it is necessary to prepare these indoles in preparations containing either ethanol or dimethyl sulfoxyde (DMSO). None of these two substances are approved for topical application either by Food and Drug Administration (FDA) or European Medicines Agency (EMEA). Therefore, one of the main challenges is to find out vehicles to dissolve these molecules.

One of the most active hypotensive compounds is 5-MCA-NAT, but one big problem is the lack of data available about its solubility in other solvents apart from DMSO. Melatonin can be taken as a reference, as data are available, regarding its solubility in solvents such as propylene glycol (PG), polyethylene glycol (PEG), and ethanol, and their binary mixtures with phosphate buffered saline (PBS). The results obtained in those works demonstrated that the solubility of melatonin was found to be similar in pure PG and ethanol, being around 100 mg/mL at 32°C. These experiments required the use of high amounts of drug (solutions must be oversaturated with drug) as well as continuous shaking of the samples for several days (Oh et al. 2001), at the risk of losing the drug if it had low stability in dissolution. When the compound is not melatonin but 5-MCA-NAT, solubility PG:PBS mixtures to prepare ophthalmic solutions is possible (Andres-Guerrero et al. 2011). In this sense, studies performed with ratios of PG:PBS 10:90, 25:75, and 50:50 were suitable and brought different results. Altogether, it is necessary to be aware that when comparing 5-MCA-NAT alone and with PG:PBS proportions that several factors are involved, such as those listed in the following:

1. If we look for a similar maximal hypotensive effect of the active compound, only 5-MCA-NAT 500 mM prepared in PG and further diluted in PBS (PG 1.43%) will be able to reach IOP values close to those obtained when the melatonin analogue was prepared in DMSO. The difference is probably due to DMSO acting as a permeant agent since it can break the epithelial barrier, allowing 5-MCA-NAT to enter the aqueous humor more easily.
2. It is necessary to consider the mean time effect of this analogue. When 5-MCA-NAT is dissolved in DMSO, the mean time effect is greater than 7 h, so it would be recommendable to choose a vehicle with the same mean time effect if not a longer one.
3. It is important to reduce the number of toxic/secondary effects of the vehicles. In vivo experiments demonstrated that the formulation constituted with PBS produced virtually no side effects.

Also, it is important to point out that the PBS formulation containing 5-MCA-NAT 500 mM and PG (1.43%) is the most efficient in reducing IOP, when compared with the other formulations. An explanation for this may be the lack of tearing and blepharospasm in the presence of this vehicle, in clear contrast with the other preparations containing higher PG concentrations, as it was previously indicated by Reinhardt and Sprout (1978). PG has been reported to be an eye irritant depending on its concentration. This implies that part of the active principle would be drained with the excess tear production at higher concentrations of PG.

Cytotoxicity assays demonstrated good tolerance for those formulations including 1.43% and 11.66% of PG. Nevertheless, cell viability decreased dramatically in formulations with high percentages of PG (25.02%). Although the use of PG in topical medications is allowed up to 80%, high concentrations of the mentioned vehicle were not well tolerated by corneal cells. Based on the results reported in this work, PG could be employed up to 10%, this being the best preparation for human administration.

22.4.4 Synthesis of New Melatonin Analogues for the Treatment of Glaucoma

The synthesis of new possible candidates for the treatment of ocular hypertension and glaucoma based in melatonin structure indicates that a feature associated with their high efficacy as agonists

TABLE 22.1
Structures and Chemical Names of Newly Synthesized Melatonin Analogues for the Treatment of Ocular Hypertension and Glaucoma

Name	Chemical Name	Chemical Structure
INS 48848	[methyl-1-methylene-2,3,4,9-tetrahydro-1*H*-carbazol-6-ylcarbamate]	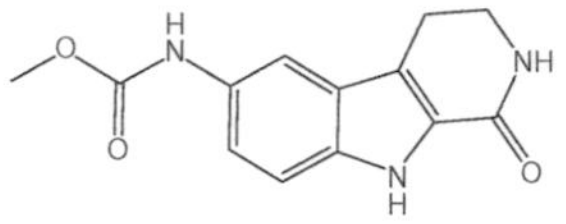
INS 48862	[methyl-2-bromo-3-(2-ethanamidoethyl)-1*H*-indol-5-ylcarbamate]	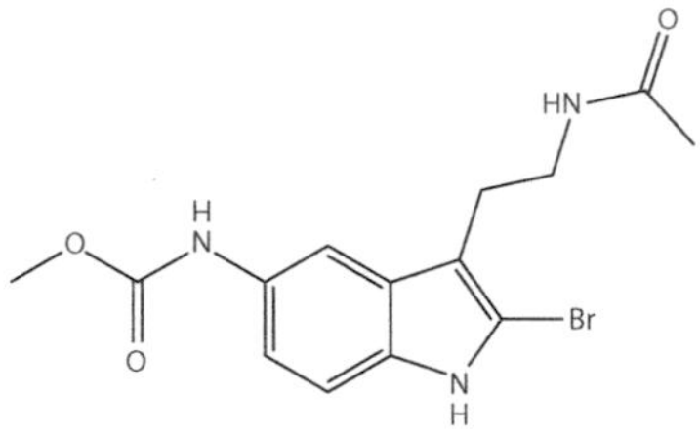
INS 48852	[(*E*)-*N*-(2-(5-methoxy-1*H*-indol-3-yl)ethyl)-3-phenylprop-2-enamide]	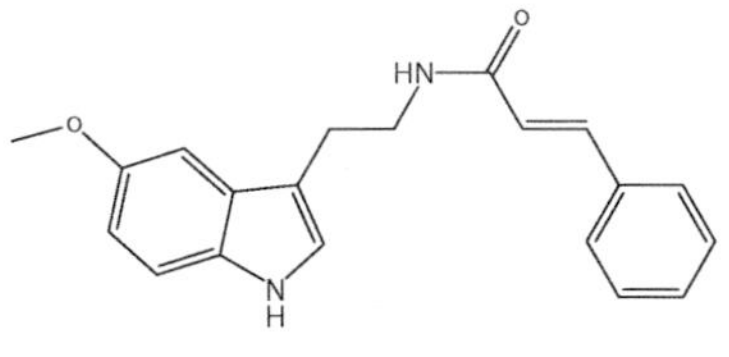

is related to the structural rigidity and increased hydrophobicity on the eastern hemisphere of the molecule. This conclusion can be obtained from the work by Alarma-Estrany et al. (2011), where a series of compounds were synthesized, exemplified by INS48848, INS48852, and INS48862 (Table 22.1). In the case of compounds INS48879 and INS48864, the positive effect of added hydrophobicity may be mitigated by the higher charge density of the phosphoryl and sulfonyl groups compared with an acyl group. Substitution of the N-acetyl group of the three-position side chain with a charged functional group or a highly polarized group almost abolished IOP-lowering activity, as did methylation of the R4-carbamate, suggesting that hydrogen-bonding ability at R4 may be a critical point for activity.

Using a 0.1 mM concentration of INS48848 (total dose, 0.259 μg), INS48852 (total dose, 0.320 μg), and INS48862 (total dose, 0.354 μg) to easily compare with concentrations of applied melatonin, 5-MCA-NAT, and IIK7 in the studies previously described (Alarma-Estrany et al. 2007, 2008, 2009), responses were similar to those obtained using 0.25 mM.

Designing and developing new drugs to treat glaucoma require consideration of many factors, including efficacy, specificity, bioavailability, safety, and toxicity. In this sense, the newly synthesized compounds were well tolerated and did not cause short-term ocular surface irritation.

One interesting aspect is that these new compounds may have two possible targets to stimulate IOP decrease, either MT_2 and/or MT_3 melatonin receptors. To understand the selectivity of these compounds it was necessary to deal with selective melatonin receptor antagonists. In this sense, the effects of INS48848 were completely blocked by prazosin, an antagonist of MT_3 melatonin receptors (Paul et al. 1999; Pintor et al. 2003; Xia et al. 2008), and were potently inhibited by luzindole, a nonselective antagonist of melatonin receptors (Pintor et al. 2003). However, DH97, an MT_2 receptor antagonist (Chen et al. 2005; Alarma-Estrany et al. 2007; Mendoza-Vargas et al. 2009), had little effect against INS48848. In clear contrast, the results obtained for INS48862 and INS48852 were the opposite. Luzindole and prazosin had no significant effects against these two compounds, whereas DH97 blocked them completely. These data strongly suggest that the compound INS48848 could be acting through the MT_3 melatonin receptors and that the compounds INS48862 and INS48852 could be acting preferentially through MT_2 melatonin receptors. Considering that the compounds described here can activate MT_2 and MT_3 receptors, an interesting approach could be to combine them to get stronger reductions in IOP. In this sense, a combination of INS48852 (which produces 30% IOP reduction via MT_2 receptors) plus INS48848 (40% reduction via MT_3) could be an interesting one, although to date it has not been tested.

22.5 LONG-TERM EFFECTS OF MELATONIN AND ANALOGUES

There is evidence pointing toward a long-term effect of melatonin and analogues. This effect can be observed when after a single dose of melatonin, IOP do not return to the initial values. Only after 5–7 days, it is possible to measure IOP values similar to the ones measured before any melatonin or analogue was added. This slow return to normal IOP may suggest that melatonin receptors can modulate the expression of some genes which would be, at the end, the ones responsible for the commented slow return to normal IOP values.

22.5.1 Regulation of Carbonic Anhydrases by Melatonin and Analogues

CA enzymes are involved in a variety of physiological processes including maintaining the acid–base balance and the transport and secretion of fluids. In the eye, these enzymes are involved in aqueous humor production, so CA inhibitors are being used to treat glaucoma (McLaren 2009). Carbonic anhydrase genes seem to be good targets through which 5-MCA-NAT exerts its long-term hypotensive effect. The results indicate that these two substances were able to reduce the expression of mRNA for carbonic anhydrases 2 (CA2) and 12 (CA12).

It is interesting to indicate that the effects of melatonin and 5-MCA-NAT showed different behavior for each treatment after 24 h. A similar and significant reduction in CA2 mRNA was observed in melatonin- and 5-MCA-NAT-treated ciliary body cells while in the case of CA12, a target gene in glaucoma (Liao et al. 2003). The mRNA levels were clearly more reduced with 5-MCA-NAT treatment. These results are in concordance with those obtained from immunocytochemical studies, where a significant reduction in both CA was also observed in each treatment; but in melatonin-treated cells, CAII was more reduced than CAXII. After 48 h, only CA12 mRNA levels were decreased in both treatments, but the effect was more marked in 5-MCA-NAT-treated cells. CA12 transcript levels were still diminished after 72 h, while in melatonin-treated cells, CA12 mRNA levels were increased and CA2 transcripts were significantly reduced. This different behavior may be due to the presence of different melatonin receptors in the rabbit ciliary body as previously indicated (Crooke et al. 2012a).

So, in summary, the long-term effect of melatonin and 5-MCA-NAT, reducing the expression of carbonic anhydrases, mimics the effect depicted by classical carbonic anhydrase inhibitors such as dorzolamide or azetazolamide commonly used for the treatment of glaucoma.

22.5.2 Regulation of Adrenoceptors by Melatonin and Analogues

Carbonic anhydrases are not the only genes in which expression is modified by melatonin and analogues. Other targets closely related to the modulation of IOP are adrenergic receptors.

Melatonin and 5-MCA-NAT modulate the expression of both α- and β-adrenoceptors. This action seems to be a two-step process, in which the first part is the down-regulation of ciliary β2-adrenergic receptors that begins 24 h after the treatment with melatoninergic compounds (principally by melatonin). This effect was quantifiable at the mRNA level. In addition, immunocytochemical studies revealed a qualitative reduction of β2-adrenergic receptor protein levels after melatoninergic treatment, but to a lesser extent than mRNA reduction (Crooke et al. 2013). This reduced correlation between mRNA and protein content may be explained by a different turnover rate for adrenergic receptor mRNA and its cognate protein in ciliary body cells treated with melatoninergic substances. In this sense, it has been previously reported that in vitro or in vivo changes at mRNA level of other adrenergic receptors, such as α1B-adrenergic receptor were not accompanied by similar changes at the protein level (Coon et al. 1997).

The second step of the melatoninergic action is the up-regulation of α2A-adrenergic receptors that begins after 48 h of melatonin and 5-MCA-NAT treatment. In this case, both melatoninergic compounds exert a similar action on mRNA levels as well as protein levels. In addition, there is a high correlation between up-regulation of α2A-adrenergic receptor mRNA and its cognate protein (Crooke et al. 2013).

Interestingly, these findings can be applied to potentiate the ocular hypotensive action of β-adrenergic receptor antagonist or α2-adrenergic receptor agonist drugs, such as timolol maleate or brimonidine, respectively. In this context, the administration of timolol once daily in normotensive rabbits pretreated with a single dose of melatonin or 5-MCA-NAT evokes an additional IOP reduction of 14% or 17% at peak, in comparison with experimental animals treated with timolol alone for 24 h.

Also, and regarding the up-regulation of α2A-adrenergic receptors, the administration of brimonidine, once daily in normotensive rabbits pretreated with melatoninergic compounds, evokes an additional IOP reduction of 29%–39% at peak, in comparison with animals treated with brimonidine alone for 24 h.

These interesting results confirm a remarkable potentiating effect of melatoninergic drugs upon the ocular hypotensive action of brimonidine, and to a lesser extent of timolol, and suggest a possible additional beneficial effect of this melatoninergic treatment combined with the classical glaucoma treatments.

22.6 FUTURE DEVELOPMENTS

It is clear that melatonin regulates IOP both in animal models and in humans. The modulatory role of this neurohormone clearly suggests the possibility of using this compound or any of its analogues for the treatment of the ocular hypertension related to the glaucomatous pathology.

This therapeutic perspective can be taken in two possible ways, either developing new compounds as suggested earlier, or by using compounds already in the market that act via melatonin receptors.

Considering the first possibility, it is necessary not only to develop careful studies about the effect, efficacy, and tolerance of the newly synthesized compounds, but also to fulfill the regulatory items depicted by FDA and EMEA to permit a compound to be in the market. This implies a long preclinical study on toxicity, lethality, etc., in animal models (Crooke et al. 2012b), prior to the classical clinical trials in humans.

Some of the already tested compounds, such as 5-MCA-NAT, have been tried in a glaucomatous model developed in monkeys. In this case, a clear reduction of IOP was measured from the third day of treatment in advance. The reduction in IOP was roughly 20%, and interestingly, it was not possible to detect any apparent side effect (Serle et al. 2004).

An alternative to reviewing the data from clinical trials is to look to the existing melatoninergic compounds in the market. The easiest thing to do is to test oral melatonin in patients with ocular hypertension. This has not been done so far, but it is noteworthy to point out that oral melatonin is being used in cataract surgery to reduce IOP. In these patients, IOP gets reduced 22%, therefore suggesting it might be useful in hypertensive patients (Ismail and Mowafi 2009).

There are two compounds with melatoninergic activity currently marketed and their uses are diverse: ramelteon, which is used for insomnia, and agomelatine, which is used to treat depression. There is no available information for ramelteon regarding its possible ability to reduce IOP. On the contrary, there are data supporting that agomelatine, topically applied in experimental models, can significantly reduce IOP (Martinez-Aguila et al. 2013). The reduction in IOP, as happens for melatonin, is about 20% of reduction. Considering that this compound is already in the pharmaceutical market and therefore preclinical tests have been already completed, we should not be surprised that clinical trials, using agomelatine, are initiated, and it becomes the first melatoninergic compound joining the group of glaucoma treatment substances.

ACKNOWLEDGMENTS

This work was supported by Universidad Complutense de Madrid (Project GR35/10-A-920777), the Ministry of Economy (Project SAF 2010/16024), and the Institute Carlos III (RETICS RD12/0034/0003). HAA is a fellowship holder of Saudi Arabia Government.

REFERENCES

Abe, M., Itoh, M. T., Miyata, M., Ishikawa, S., and Sumi, Y. 1999. Detection of melatonin, its precursors and related enzyme activities in rabbit lens. *Exp. Eye Res.* 68:255–262.

Aimoto, T., Rohde, B. H., Chiou, G. C., and Lauber, J. K. 1985. *N*-acetyltransferase activity and melatonin level in the eyes of glaucomatous chickens. *J. Ocul. Pharmacol.* 1:149–160.

Alarma-Estrany, P., Crooke, A., Mediero, A., Peláez, T., and Pintor, J. 2008. Sympathetic nervous system modulates the ocular hypotensive action of MT2-melatonin receptors in normotensive rabbits. *J. Pineal Res.* 45:468–475.

Alarma-Estrany, P., Crooke, A., Peral, A., and Pintor, J. 2007. Requirement of intact sympathetic transmission for the ocular hypotensive effects of melatonin and 5-MCA-NAT. *Auton. Neurosci.* 137:63–66.

Alarma-Estrany, P., Crooke, A., and Pintor, J. 2009. 5-MCA-NAT does not act through NQO2 to reduce intraocular pressure in New-Zealand white rabbit. *J. Pineal Res.* 47:201–209.

Alarma-Estrany, P., Guzman-Aranguez, A., Huete, F., Peral, A., Plourde Jr., R., Pelaez, T., Yerxa, B., and Pintor, J. 2011. Design of novel melatonin analogs for the reduction of intraocular pressure in normotensive rabbits. *J. Pharmacol. Exp. Ther.* 337:703–709.

Andres-Guerrero, V., Molina-Martinez, I. T., Peral, A., de las Heras, B., Pintor, J., and Herrero-Vanrell, R. 2011. The use of mucoadhesive polymers to enhance the hypotensive effect of a melatonin analogue, 5-MCA-NAT, in rabbit eyes. *Invest. Ophthalmol. Vis. Sci.* 52:1507–1515.

Arendt, J. 1998. Melatonin and the pineal gland: Influence on mammalian seasonal and circadian physiology. *Rev. Reprod.* 3:13–22.

Beraldo, F. H. and Garcia, C. R. 2005. Products of tryptophan catabolism induce Ca^{2+} release and modulate the cell cycle of *Plasmodium falciparum* malaria parasites. *J. Pineal Res.* 39:224–230.

Buffa, A., Rizzi, E., Falconi, M. et al. 1993. Bromodeoxyuridine incorporation in corneal epithelium: An immunocytochemical study in rats. *Boll. Soc. Ital. Biol. Sper.* 69:767–773.

Chanut, E., Nguyen-Legros, J., Versaux-Botteri, C., Trouvin, J. H., and Launay, J. M. 1998. Determination of melatonin in rat pineal, plasma and retina by high-performance liquid chromatography with electrochemical detection. *J. Chromatogr. Biomed. Sci. Appl.* 709:11–18.

Chen, L., Wang, L., and Jacob, T. J. 1999. Association of intrinsic pICln with volume-activated Cl^- current and volume regulation in a native epithelial cell. *Am. J. Physiol.* 276:C182–C192.

Chen, Y., Tjong, Y. W., Ip, S. F., Tipoe, G. L., and Fung, M. L. 2005. Melatonin enhances the hypoxic response of rat carotid body chemoreceptor. *J. Pineal Res.* 38:157–163.

Chieffi, G., Baccari, G. C., Di Matteo, L., d'Istria, M., Minucci, S., and Varriale, B. 1996. Cell biology of the Harderian gland. *Int. Rev. Cytol.* 168:1–80.

Chiou, G. C. and McLaughlin, M. A. 1984. Studies on the involvement of melatonergic mechanism in intraocular pressure regulation. *Ophthalmic Res.* 16:302–306.

Chiquet, C., Claustrat, B., Thuret, G., Brun, J., Cooper, H. M., and Denis, P. 2006. Melatonin concentrations in aqueous humor of glaucoma patients. *Am. J. Ophthalmol.* 142:325–327.

Ciuffi, M., Pisanello, M., Pagliai, G. et al. 2003. Antioxidant protection in cultured corneal cells and whole corneas submitted to UV-B exposure. *J. Photochem. Photobiol.* B71:59–68.

Civan, M. M. and Macknight, A. D. 2004. The ins and outs of aqueous humour secretion. *Exp. Eye Res.* 78:625–631.

Collaborative Normal-Tension Glaucoma Study Group. 1998. Comparison of glaucomatous progression between untreated patients with normal-tension glaucoma and patients with therapeutically reduced intraocular pressures. *Am. J. Ophthalmol.* 126:487–497.

Coon, S. L., McCune, S. K., Sugden, D., and Klein, D. C. 1997. Regulation of pineal alpha1B- adrenergic receptor mRNA: Day/night rhythm and beta-adrenergic receptor/cyclic AMP control. *Mol. Pharmacol.* 51:551–557.

Crooke, A., Huete-Toral, F., Martinez-Aguila, A., Colligris, B., and Pintor, J. 2012b. Ocular disorders and the utility of animal models in the discovery of melatoninergic drugs with therapeutic potential. *Expert Opin. Drug Discov.* 7(10):989–1001.

Crooke, A., Huete-Toral, F., Martínez-Águila, A., Martín-Gil, A., and Pintor, J. 2012a. Involvement of carbonic anhydrases in the ocular hypotensive effect of melatonin analogue 5-MCA-NAT. *J. Pineal Res.* 52:265–270.

Crooke, A., Huete-Toral, F., Martínez-Águila, A., Martín-Gil, A., and Pintor, J. 2013. Melatonin and its analog 5-methoxycarbonylamino-*N*-acetyltryptamine potentiate adrenergic receptor-mediated ocular hypotensive effects in rabbits: Significance for combination therapy in glaucoma. *J. Pharmacol. Exp. Ther.* 346(1):138–145.

Dielemans, I., Vingerling, J. R., Wolfs, R. C. W. et al. 1994. The prevalence of primary open-angle glaucoma in a population-based study in the Netherlands. *Ophthalmology* 11:1851–1855.

Djeridane, Y., Vivien-Roels, B., Simonneaux, V., Miguez, J. M., and Pevet, P. 1998. Evidence for melatonin synthesis in rodent Harderian gland: A dynamic in vitro study. *J. Pineal Res.* 25:54–64.

Do, C. W., Peterson-Yantorno, K., and Civan, M. M. 2006. Swelling-activated Cl^- channels support Cl^- secretion by bovine ciliary epithelium. *Invest. Ophthalmol. Vis. Sci.* 47:2576–2582.

Dortch-Carnes, J. and Tosini, G. 2013. Melatonin receptor agonist-induced reduction of SNP-released nitric oxide and cGMP production in isolated human non-pigmented ciliary epithelial cells. *Exp. Eye Res.* 107:1–10.

Doughty, M. J. 1990. Morphometric analysis of the surface cells of rabbit corneal epithelium by scanning electron microscopy. *Am. J. Anat.* 189:316–328.

Dubocovich, M. L. 1983. Melatonin is a potent modulator of dopamine release in the retina. *Nature* 306:782–784.

Dubocovich, M. L. 1985. Characterization of a retinal melatonin receptor. *J. Pharmacol. Exp. Ther.* 234:395–401.

Dubocovich, M. L., Masana, M., and Benloucif, S. 2000. Molecular pharmacology and function of melatonin receptor subtypes. In *Melatonin after Four Decades*, ed. Olcese, J., pp. 181–188. New York: Kluwer Academic/Plenum Publishers.

Dubocovich, M. L., Masana, M. I., Iacob, S., and Sauri, D. M. 1997. Melatonin receptor antagonists that differentiate between the human Mel1a and Mel1b recombinant subtypes are used to assess the pharmacological profile of the rabbit retina ML1 presynaptic heteroreceptor. *Naunyn Schmiedebergs Arch. Pharmacol.* 355:365–375.

Eison, A. S. and Mullins, U. L. 1993. Melatonin binding sites are functionally coupled to phosphoinositide hydrolysis in Syrian hamster RPMI 1846 melanoma cells. *Life Sci.* 53:393–398.

Er, H., Turkoz, Y., Mizrak, B., and Parlakpinar, H. 2006. Inhibition of experimental proliferative vitreoretinopathy with protein kinase C inhibitor (chelerythrine chloride) and melatonin. *Ophthalmologica* 220:17–22.

Forrester, J. V., ed. 2002. *The Eye. Basic Sciences in Practice*. Philadelphia, PA: Sanders Ltd.

Fujieda, H., Hamadanizadeh, S. A., Wankiewicz, E., Pang, S. F., and Brown, G. M. 1999. Expression of mt1 melatonin receptor in rat retina: Evidence for multiple cell targets for melatonin. *Neuroscience* 93:793–799.

Fujieda, H., Scher, J., Hamadanizadeh, S. A., Wankiewicz, E., Pang, S. F., and Brown, G. M. 2000. Dopaminergic and GABAergic amacrine cells are direct targets of melatonin: Immunocytochemical study of mt1 melatonin receptor in guinea pig retina. *Vis. Neurosci.* 17:63–70.

Gregory, C. Y., Abrams, T. A., and Hall, M. O. 1994. Stimulation of A2 adenosine receptors inhibits the ingestion of photoreceptor outer segments by retinal pigment epithelium. *Invest. Ophthalmol. Vis. Sci.* 35:819–825.

Heijl, A., Leske, M. C., Bengtsson, B., Hyman, L., Bengtsson, B., Hussein, M., Early Manifest Glaucoma Trial Group. 2002. Reduction of intraocular pressure and glaucoma progression: Results from the Early Manifest Glaucoma Trial. *Arch. Ophthalmol.* 120:1268–1279.

Hoyle, C. H. V., Peral, A., and Pintor, J. 2006. Melatonin potentiates tear secretion induced by diadenosine tetraphosphate in the rabbit. *Eur. J. Pharmacol.* 552:159–161.

Huang, H., Lee, S. C., and Yang, X. L. 2005. Modulation by melatonin of glutamatergic synaptic transmission in the carp retina. *J. Physiol.* 569(Pt 3):857–871.

Ismail, S. A. and Mowafi, H. A. 2009. Melatonin provides anxiolysis, enhances analgesia, decreases intraocular pressure, and promotes better operating conditions during cataract surgery under topical anesthesia. *Anesth. Analg.* 108(4):1146–1151.

Jacob, T. J. and Civan, M. M. 1996. Role of ion channels in aqueous humor formation. *Am. J. Physiol.* 271:C703–C720.

Johnson, M. 2006. What controls aqueous outflow resistance? *Exp. Eye Res.* 82:545–557.

Kass, M. A., Heuer, D. K., Higginbotham, E. J. et al. 2002. The Ocular Hypertension Treatment Study: A randomized trial determines that topical ocular hypotensive medication delays or prevents the onset of primary open-angle glaucoma. *Arch. Ophthalmol.* 120:701–713.

Kittner, Z., Olah, I., and Toro, I. 1978. Histology and ultrastructure of the Harderian glands—accessory lacrimal gland—of the chicken. *Acta Biol. Acad. Sci. Hung.* 29:29–41.

Klein, D. C., Coon, S. L., Roseboom, P. H. et al. 1997. The melatonin rhythm-generating enzyme: Molecular regulation of serotonin *N*-acetyltransferase in the pineal gland. *Recent Prog. Horm. Res.* 52:307–357 (discussion 357–358).

Lee, A. J. and Goldberg, I. 2011. Emerging drugs for ocular hypertension. *Expert Opin. Emerg. Drugs* 16(1):137–161.

Lerner, A. B., Case, J. D., and Heinzelman, R. V. 1959. Structure of melatonin. *J. Am. Chem. Soc.* 81:6084–6085.

Leske, M. C. 1983. The epidemiology of open-angle glaucoma. *Am. J. Epidemiol.* 118:166–191.

Liao, S. Y., Ivanov, S., Ivanova, A. et al. 2003. Expression of cell surface transmembrane carbonic anhydrase genes CA9 and CA12 in the human eye: Overexpression of CA12 (CAXII) in glaucoma. *J. Med. Genet.* 40(4):257–261.

Lincoln, G. A., Ebling, F. J., and Almeida, O. F. 1985. Generation of melatonin rhythms. *Ciba Found. Symp.* 117:129–148.

Mahtre, M. C., van Jaarsveld, A. S., and Reiter, R. J. 1988. Melatonin in the lachrymal gland: First demonstration and experimental manipulation. *Biochem. Biophys. Res. Commun.* 153:1186–1192.

Mailliet, F., Ferry, G., Vella, F., Thiam, K., Delagrange, P., and Boutin, J. A. 2004. Organs from mice deleted for NRH:quinone oxidoreductase 2 are deprived of the melatonin binding site MT3. *FEBS Lett.* 578116–578120.

Martin, X. D., Malina, H. Z., Brennan, M. C., Hendrickson, P. H., and Lichter, P. R. 1992. The ciliary body—The third organ found to synthesize indoleamines in humans. *Eur. J. Ophthalmol.* 2:67–72.

Martinez-Aguila, A., Fonseca, B., Bergua, A., and Pintor, J. 2013. Melatonin analogue agomelatine reduces rabbit's intraocular pressure in normotensive and hypertensive conditions. *Eur. J. Pharmacol.* 701(1–3):213–217.

McLaren, J. W. 2009. Measurement of aqueous humor flow. *Exp. Eye Res.* 88:641–647.

Mendoza-Vargas, L., Solís-Chagoyán, H., Benítez-King, G., and Fuentes-Pardo, B. 2009. MT2-like melatonin receptor modulates amplitude receptor potential in visual cells of crayfish during a 24-hour cycle. *Comp. Biochem. Physiol. A Mol. Integr. Physiol.* 154:486–492.

Meyer, P., Pache, M., Loeffler, K. U. et al. 2002. Melatonin MT-1 receptor immunoreactivity in the human eye. *Br. J. Ophthalmol.* 86:1053–1057.

Mullins, U. L., Fernandes, P. B., and Eison, A. S. 1997. Melatonin agonists induce phosphoinositide hydrolysis in *Xenopus laevis* melanophores. *Cell Signal* 9:169–173.

Natesan, A. K. and Cassone, V. M. 2002. Melatonin receptor mRNA localization and rhythmicity in the retina of the domestic chick *Gallus domesticus*. *Vis. Neurosci.* 19:265–274.

Navajas, C., Kokkola, T., Poso, A., Honka, N., Gynther, J., and Laitinen, J. T. 1996. A rhodopsin-based model for melatonin recognition at its G protein-coupled receptor. *Eur. J. Pharmacol.* 304:173–183.

Nosjean, O., Ferro, M., Coge, F. et al. 2000. Identification of the melatonin-binding site MT3 as the quinone reductase 2. *J. Biol. Chem.* 275:31311–31317.

Nosjean, O., Nicolas, J. P., Klupsch, F., Delagrange, P., Canet, E., and Boutin, J. A. 2001. Comparative pharmacological studies of melatonin receptors: MT1, MT2 and MT3/QR2. Tissue distribution of MT3/QR2. *Biochem. Pharmacol.* 61:1369–1379.

Oh, H. J., Oh, Y. K., and Kim, C. K. 2001. Effect of vehicles and enhancers on transdermal delivery of melatonin. *Int. J. Pharm.* 212(1):63–71.

Osborne, N. N. and Chidlow, G. 1994. The presence of functional melatonin receptors in the iris-ciliary processes of the rabbit eye. *Exp. Eye Res.* 59:3–9.

Parrish, R. K., Palmberg, P., and Sheu, W. P. 2003. A comparison of latanoprost, bimatoprost, and travoprost in patients with elevated intraocular pressure: A 12-week, randomized, masked-evaluator multicenter study. *Am. J. Ophthalmol.* 135:688–703.

Paul, P., Lahaye, C., Delagrange, P., Nicolas, J. P., Canet, E., and Boutin, J. A. 1999. Characterization of 2-[125I] iodomelatonin binding sites in Syrian hamster peripheral organs. *J. Pharmacol. Exp. Ther.* 290:334–340.

Paulmichl, M., Li, Y., Wickman, K., Ackerman, M., Peralta, E., and Clapham, D. 1992. New mammalian chloride channel identified by expression cloning. *Nature* 356:238–241.

Pierce, M. E. and Besharse, J. C. 1985. Circadian regulation of retinomotor movements: I. Interaction of melatonin and dopamine in the control of cone length. *J. Gen. Physiol.* 86:671–689.

Pintor, J., Martin, L., Pelaez, T., Hoyle, C. H., and Peral, A. 2001. Involvement of melatonin MT(3) receptors in the regulation of intraocular pressure in rabbits. *Eur. J. Pharmacol.* 416:251–254.

Pintor, J., Pelaez, T., Hoyle, C. H., and Peral, A. 2003. Ocular hypotensive effects of melatonin receptor agonists in the rabbit: Further evidence for an MT3 receptor. *Br. J. Pharmacol.* 138:831–836.

Pintor, J. J., Carracedo, G., Mediero, A. et al. 2005. Melatonin increases the rate of corneal re-epithelialisation in New Zealand white rabbits. *Invest. Ophthalmol. Vis. Sci.* 46:A2152.

Quigley, H. A. 1996. Number of people with glaucoma worldwide. *Br. J. Ophthalmol.* 80:389–393.

Quigley, H. A. 2011. Glaucoma. *Lancet* 377:1367–1377.

Quigley, H. A. and Broman, A. T. 2006. The number of people with glaucoma worldwide in 2010 and 2020. *Br. J. Ophthalmol.* 90:262–267.

Rada, J. A. and Wiechmann, A. F. 2006. Melatonin receptors in chick ocular tissues: Implications for a role of melatonin in ocular growth regulation. *Invest. Ophthalmol. Vis. Sci.* 47:25–33.

Raviola, G. 1974. Effects of paracentesis on the blood-aqueous barrier: An electron microscope study on *Macaca mulatta* using horseradish peroxidase as a tracer. *Invest. Ophthalmol.* 13:828–858.

Reinhardt, C. F. and Sprout, W. L. 1978. Propylene glycol eye wash. *J. Occup. Med.* 20(3):164.

Reppert, S. M., Godson, C., Mahle, C. D., Weaver, D. R., Slaugenhaupt, S. A., and Gusella, J. F. 1995. Molecular characterization of a second melatonin receptor expressed in human retina and brain: The Mel1b melatonin receptor. *Proc. Natl. Acad. Sci. USA.* 92:8734–8738.

Reppert, S. M., Weaver, D. R., and Ebisawa, T. 1994. Cloning and characterization of a mammalian melatonin receptor that mediates reproductive and circadian responses. *Neuron* 13:1177–1185.

Reppert, S. M., Weaver, D. R., and Godson, C. 1996. Melatonin receptors step into the light: Cloning and classification of subtypes. *Trends Pharmacol. Sci.* 17:100–102.

Roca, A. L., Godson, C., Weaver, D. R., and Reppert, S. M. 1996. Structure, characterization, and expression of the gene encoding the mouse Mel1a melatonin receptor. *Endocrinology* 137:3469–3477 (Erratum in: *Endocrinology* 1997;138:2307).

Rohde, B. H., McLaughlin, M. A., and Chiou, L. Y. 1985, Fall. Existence and role of endogenous ocular melatonin. *J. Ocul. Pharmacol.* 1:235–243.

Sampaio, L. de F. 2009. An unexpected effect of 5-MCA-NAT in chick retinal development. *Int. J. Dev. Neurosci.* 27:511–515.

Samples, J. R., Krause, G., and Lewy, A. J. 1988. Effect of melatonin on intraocular pressure. *Curr. Eye Res.* 7:649–653.

Savaskan, E., Wirz-Justice, A., Olivieri, G. et al. 2002. Distribution of melatonin MT1 receptor immunoreactivity in human retina. *J. Histochem. Cytochem.* 50:519–526.

Scher, J., Wankiewicz, E., Brown, G. M., and Fujieda, H. 2002. MT(1) melatonin receptor in the human retina: Expression and localization. *Invest. Ophthalmol. Vis. Sci.* 43:889–897.

Scher, J., Wankiewicz, E., Brown, G. M., and Fujieda, H. 2003. AII amacrine cells express the MT1 melatonin receptor in human and macaque retina. *Exp. Eye Res.* 77:375–382.

Schomerus, C. and Korf, H. W. 2005. Mechanisms regulating melatonin synthesis in the mammalian pineal organ. *Ann. N. Y. Acad. Sci.* 1057:372–383.

Schuster, C., Williams, L. M., Morris, A., Morgan, P. J., and Barrett, P. 2005. The human MT1 melatonin receptor stimulates cAMP production in the human neuroblastoma cell line SH-SY5Y cells via a calcium-calmodulin signal transduction pathway. *J. Neuroendocrinol.* 17:170–178.

Serle, J. B., Wang, R. F., Peterson, W. M., Plourde, R., and Yerxa, B. R. 2004. Effect of 5-MCA-NAT, a putative melatonin MT3 receptor agonist, on intraocular pressure in glaucomatous monkey eyes. *J. Glaucoma* 13:385–388.

Slaugenhaupt, S. A., Roca, A. L., Liebert, C. B., Altherr, M. R., Gusella, J. F., and Reppert, S. M. 1995. Mapping of the gene for the Mel1a-melatonin receptor to human chromosome 4(MTNR1A) and mouse chromosome 8 (Mtnr1a). *Genomics* 27:355–357.

Sommer, A., Tielsch, J. M., Katz, J. et al. 1991. Racial differences in the cause-specific prevalence of blindness in East Baltimore. *N. Engl. J. Med.* 325:1412–1421.

Stewart, W. C., Day, D. G., Stewart, J. A. et al. 2004. Short-term ocular tolerability of dorzolamide 2% and brinzolamide 1% vs placebo in primary open-angle glaucoma and ocular hypertension subjects. *Eye (Lond)* 18:905–910.

Tosini, G. 2000. Melatonin circadian rhythm in the retina of mammals. *Chronobiol. Int.* 17:599–612.

Tosini, G. and Fukuhara, C. 2002. The mammalian retina as a clock. *Cell Tissue Res.* 309(1):119–126.

Tosini, G. and Fukuhara, C. 2003. Photic and circadian regulation of retinal melatonin in mammals. *J. Neuroendocrinol.* 15:364–369.

Underwood, H., Binkley, S., Siopes, T., and Mosher, K. 1984. Melatonin rhythms in the eyes, pineal bodies, and blood of Japanese quail (*Coturnix coturnix japonica*). *Gen. Comp. Endocrinol.* 56:70–81.

Vanecek, J. 1998. Cellular mechanisms of melatonin action. *Physiol. Rev.* 78:687–721.

Varma, R., Ying-Lai, M., Francis, B. A. et al. 2004. Prevalence of open-angle glaucoma and ocular hypertension in Latinos. *Ophthalmology* 111:1439–1448.

Vaughan, G. M. and Reiter, R. J. 1986. Pineal dependence of the Syrian hamster's nocturnal serum melatonin surge. *J. Pineal Res.* 3:9–14.

Vella, F., Ferry, G., Delagrange, P., and Boutin, J. A. 2005. NRH:quinone reductase 2: An enzyme of surprises and mysteries. *Biochem. Pharmacol.* 71:1–12.

Von Gall, C., Stehle, J. H., and Weaver, D. R. 2002. Mammalian melatonin receptors: Molecular biology and signal transduction. *Cell. Tissue Res.* 309:151–162.

Wess, J. 1993. Molecular basis of muscarinic acetylcholine receptor function. *Trends Pharmacol. Sci.* 14:308–313.

White, M. P. and Fisher, L. J. 1989. Effects of exogenous melatonin on circadian disc shedding in the albino rat retina. *Vision Res.* 29:167–179.

Wiechmann, A. F. and Rada, J. A. 2003. Melatonin receptor expression in the cornea and sclera. *Exp. Eye Res.* 77:219–225.

Wiechmann, A. F. and Smith, A. R. 2001. Melatonin receptor RNA is expressed in photoreceptors and displays a diurnal rhythm in *Xenopus retina. Brain Res. Mol. Brain Res.* 91:104–111.

Wiechmann, A. F., Udin, S. B., and Rada, J. A. 2004. Localization of Mel1b melatonin receptor-like immunoreactivity in ocular tissues of *Xenopus laevis*. *Exp. Eye Res.* 79:585–594.

Wiechmann, A. F. and Wirsig-Wiechmann, C. R. 2001. Melatonin receptor mRNA and protein expression in *Xenopus laevis* nonpigmented ciliary epithelial cells. *Exp. Eye Res.* 73:617–623.

Wiechmann, A. F., Yang, X. L., Wu, S. M., and Hollyfield, J. G. 1988. Melatonin enhances horizontal cell sensitivity in salamander retina. *Brain Res.* 453:377–380.

Xia, C. M., Shao, C. H., Xin, L., Wang, Y. R., Ding, C. N., Wang, J., Shen, L. L., Li, L., Cao, Y. X., and Zhu, D. N. 2008. Effects of melatonin on blood pressure in stress-induced hypertension in rats. *Clin. Exp. Pharmacol. Physiol.* 35:1258–1264.

Yi, C., Pan, X., Yan, H., Guo, M., and Pierpaoli, W. 2005. Effects of melatonin in age-related macular degeneration. *Ann. N. Y. Acad. Sci.* 1057:384–392.

Zawilska, J. B., Lorenc, A., Berezinska, M., Vivien-Roels, B., Pevet, P., and Skene, D. J. 2006. Diurnal and circadian rhythms in melatonin synthesis in the turkey pineal gland and retina. *Gen. Comp. Endocrinol.* 145:162–168.

23 Melatonergic Drugs as Therapeutic Agents for Insomnia and Depressive Disorders

Venkataramanujam Srinivasan, Domenico De Berardis, Amnon Brzezinski, Rahimah Zakaria, Zahiruddin Othman, Samuel D. Shillcutt, Edward C. Lauterbach, Gregory M. Brown, and Timo Partonen

CONTENTS

23.1 INTRODUCTION

Sleep disturbances are the most common complaint of the elderly and one of the most prevalent symptoms of mental illness. The co-occurrence of disturbed sleep and disturbed mood also has been reported in a number of clinical studies in patients with major depressive disorder (MDD).

Insomnia is a sleep disorder characterized by poor quality of sleep with symptoms including difficulty in falling asleep, frequent nocturnal awakenings, early morning awakenings, etc., resulting in fatigue, reduced alertness, irritability, and impaired concentration, memory, and performance, with a major negative impact on the quality of life (Cricco et al. 2001; Bastien 2011). It is common among elderly people and is a major cause of impaired physical and mental health in this population (Van Someren 2000).

Nearly 30%–40% of the adult population suffers from mild to severe insomnia. In addition to this, most depressed patients suffering from MDD or bipolar disorder (BD) exhibit profound disturbances in sleep architecture (Srinivasan et al. 2006). As diminished melatonin secretion is seen in elderly individuals suffering from chronic insomnia, as well as in certain categories of patients suffering from MDD, a deficiency of melatonin has been implicated as a major contributory factor for chronic insomnia with or without depressive disorders, particularly MDD (Lieberman 1986). Hence, this chapter focuses attention on the use of melatonin or its agonist ramelteon for treating insomnia, and on the use of the melatonergic drug agomelatine for treating depressive disorders.

23.2 MELATONIN'S ROLE IN SLEEP

The role of melatonin in the control of sleep has been investigated in both diurnal and nocturnal species. Local injection of pharmacological amounts of melatonin (1–50 μg) in the medial preoptic area of the rat hypothalamus during daytime increased total sleep time (TST) in a dose-dependent manner, mainly by increasing nonrapid eye movement (NREM) sleep (Mendelson 2002).

Melatonin has been shown to induce sleep by altering the functions of the gamma-aminobutyric acid $(GABA)_A$–benzodiazepine receptor complex (Golombek et al. 1996). In diurnal species, the suppression of electrical activity in the suprachiasmatic nucleus (SCN) is suggested as a possible mechanism by which melatonin regulates sleep. This effect is absent in MT_1 knockout mice, thereby demonstrating the importance of MT_1 receptors in melatonin's acute inhibitory effects on SCN electrical activity (Liu et al. 1997). The MT_1 and MT_2 melatonin receptor subtypes are complementary in their actions and, to some extent, mutually substitute for each other. The enhancement of neuronal (GABAergic) activity by melatonin is one of the possible mechanisms by which this hormone contributes to the regulation of sleep. Melatonin is a natural hypnotic. It is suitable for long-term use in elderly people due to its low toxicity and limited adverse effect profile. Melatonin increases TST and improves sleep quality and reduces sleep onset latency (SOL) (Dollins et al. 1994; Garfinkel et al. 1995; Zhdanova et al. 1995, 1996; Monti et al. 1999; Brzezinski et al. 2005).

The relationship between sleep disturbances and low nocturnal melatonin production was investigated in a large population of insomniacs (aged 55 years or more). These elderly patients with sleep problems excreted 9.0 ± 8.3 μg of the urinary melatonin metabolite 6-sulfatoxymelatonin per night, whereas age-matched healthy controls excreted 18.1 ± 12.7 μg of 6-sulfatoxymelatonin per night, and younger subjects excreted 24.2 ± 11.9 μg of 6-sulfatoxymelatonin per night. It was also observed that half of the elderly insomniacs excreted less than 8.0 μg of 6-sulfatoxymelatonin per night. Within this subpopulation of 372 subjects, 112 had urinary 6-sulphatoxymelatonin values lower than 3.5 μg/night (Dijk and Cajochen 1997).

Studies carried out using 0.3–1 mg of melatonin that attained "physiological" melatonin blood levels have shown that melatonin reduced sleep latency (SL) and increased sleep efficacy (SE) when administered to healthy human subjects during the evening (Dollins et al. 1994). However, in most studies, higher amounts of melatonin (2–6 mg) were given in order to obtain similar effects (Buscemi et al. 2006). Brain imaging studies on subjects during wakefulness show that melatonin modulates the brain activity pattern to one resembling actual sleep (Gorfine et al. 2006). Despite clinical studies, the general efficacy of melatonin as a sleep-promoting substance has been a subject of debate (Mendelson 1997). A possible explanation for this is that administered melatonin doses are too low. The reported lack of efficacy of melatonin could be related to the extremely short elimination half-life of the fast-release melatonin preparations, and this prompted the development of active slow-release formulations (Dalton et al. 2000). Circadin®, a slow-release preparation of melatonin, 2 mg, developed by Neurim (Tel Aviv, Israel), was approved by the European Medicines Agency (EMEA) as a monotherapy for short-term treatment of primary insomnia in elderly subjects in 2007. Circadin has been shown to improve the quality of sleep and morning alertness, reduce SOL, and ameliorate the quality of life in middle-aged and elderly patients with insomnia (Lemoine et al. 2007; Wade et al. 2011). Moreover, in a trial, 58 adults with subsyndromal seasonal affective disorder (SAD) randomized to either Circadin, 2 mg or placebo tablets administered 1–2 h before a desired bedtime for 3 weeks, melatonin improved the quality of sleep and vitality (Leppämäki et al. 2003). Generally, patients who have low melatonin levels responded better to melatonin replacement therapy compared to other patients with insomnia.

An unknown aspect of melatonin activity in the brain with regard to its hypnotic and chronobiotic activities is the extent to which it desensitizes its membrane MT_1 and MT_2 receptors. Since MT_1 and MT_2 melatonin receptors are G protein–coupled receptors, desensitization is an expected normal phenomenon in these receptors. Gerdin et al. (2004) demonstrated the desensitization of endogenous MT_2 melatonin receptors by physiological melatonin concentrations simulating the nocturnal surge in the rat SCN. This finding questioned the efficacy of using supraphysiological doses of melatonin to treat insomnia. However, in a study conducted on SCN neurons, neither in vivo studies (intraperitoneal or iontophoretic application of melatonin) nor in vitro studies on SCN neuronal cells revealed any desensitization phenomenon (Ying et al. 1998). Moreover, the melatonin receptor concentration has been shown to increase in parallel with the increase of melatonin concentration (Masana et al. 2000), which raises doubt over the theory of receptor desensitization phenomenon after long-term use of either melatonin or its agonists.

23.3 RAMELTEON: THE MELATONERGIC DRUG FOR INSOMNIA

Ramelteon (Rozerem®, Takeda Pharmaceuticals, Japan) is a melatonergic hypnotic medication that has been demonstrated to be clinically effective and safe. It is a tricyclic synthetic analogue of melatonin, chemically designated as (*S*)-*N*-[2-(1, 6,7,8-tetrahydro-2*H*-indeno[5,4-*b*]furan-8-yl)-ethyl] propionamide. In 2005, the Food and Drug Administration (FDA) approved its use for the treatment of insomnia. Ramelteon is a selective MT_1/MT_2 receptor agonist without significant affinity for other receptor sites (Kato et al. 2005; Miyamoto 2009). In vitro binding studies have shown that ramelteon's affinity for MT_1 and MT_2 receptors is 3–16 times higher than that of melatonin. The selectivity of ramelteon for MT_1 has been found to be greater than that of MT_2 receptors. The selectivity of MT_1 receptors by ramelteon suggests that it targets sleep onset more specifically than melatonin itself (Miyamoto 2009). Figure 23.1 is a schematic diagram on mechanisms of actions of ramelteon on sleep.

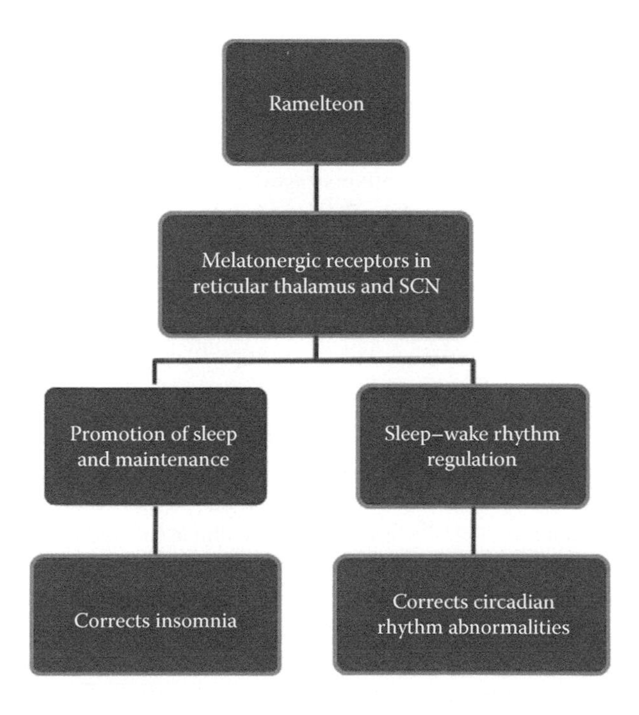

FIGURE 23.1 Mechanisms of action of ramelteon on sleep.

Ramelteon is extensively metabolized into active metabolites primarily via oxidation to hydroxyl and carbonyl species, with secondary metabolism to form glucuronide conjugates. Cytochrome P450 1A2 is the major hepatic enzyme involved in ramelteon metabolism. Four principal metabolites of ramelteon (M-I, M-II, M-III, M-IV) have been identified. Among these, M-II has been found to occur in much higher concentrations with systemic levels 20–100-fold greater than ramelteon itself. Although the activity of M-II is 30-fold lower than that of ramelteon, its exposure exceeds that of ramelteon by a factor of 30. Hence, it is suggested that M-II may contribute significantly to the net clinical effect of ramelteon intake (Miyamoto 2009).

23.4 USE OF RAMELTEON IN INSOMNIA: CLINICAL STUDIES

The first study of the effects of ramelteon on sleep was conducted by Roth and his colleagues (2005). In this study involving 117 patients (aged 16–64 years) drawn from 13 centers in Europe, the efficacy, safety, and dose response of ramelteon were examined. Patients were randomized to a dosing sequence of 4, 8, 16, or 32 mg of ramelteon. All doses of ramelteon produced a statistically significant reduction in latency to persistent sleep (LPS) and increased TST as shown by polysomnography (Roth et al. 2005).

In a follow-up study, the same group of investigators administered ramelteon for a period of 5 weeks to 829 patients (aged 65 years or more). In this double-blind study, ramelteon at doses of either 4 or 8 mg/day brought about a significant reduction in SOL (16%–35%) and increased TST (Roth et al. 2006). In another randomized, multicenter double-blind, placebo-controlled crossover study involving 107 patients, ramelteon was administered in doses of 4–32 mg/day. The treatment decreased LPS and increased TST significantly as measured objectively with polysomnography (Erman et al. 2006).

A short-term evaluation of the efficacy of ramelteon performed in 100 elderly patients by administering 4 and 8 mg doses in a 2-night/3-day period crossover design revealed decreased LPS, and improved TST and SE as compared to placebo (Roth et al. 2007). Likewise, the efficacy of ramelteon in reducing SOL and in increasing TST and SE was evaluated in 371 patients administered 8 or 16 mg of ramelteon for 5 weeks in a double-blind placebo-controlled study. The results were consistent in confirming the efficacy of ramelteon to reduce SOL and to increase SE and TST (Zammit et al. 2007).

A long-term evaluation of the efficacy of ramelteon was conducted in 1213 adult and elderly patients with chronic insomnia (DeMicco et al. 2006). The adult patients (aged 18–64 years) were given ramelteon 16 mg, and the elderly patients (aged 65 years or more) were given ramelteon 8 mg. Five hundred and ninety-seven patients remained in the study at 6 months, and 473 patients completed the full 12 months (Richardson et al. 2006). Both elderly and adult patients experienced improvement at month 1 (34.0% and 35.1%, respectively), month 6 (44.7% and 49.1%), and month 12 (50.3% and 52.1%) (DeMicco et al. 2006). TST estimates also showed steady improvement for elderly and adult patients at month 1 (15.2% and 16.9%), month 6 (21.6% and 22.7%), and month 12 (25.5% and 23.9%).

In a post hoc analysis of a previously published 5-week, randomized, double-blind, placebo-controlled study involving 405 patients with chronic insomnia (aged 18–64 years), the rapid onset of action of ramelteon 8 mg caused significant reductions in SOL within a week (63% for ramelteon vs. 39.7% for placebo, $P<0.001$). This reduction in SOL was sustained throughout the 5 weeks of study (63% and 65.9% ramelteon vs. 41.2% and 48.9% placebo at the end of the third and fifth week, respectively) (Mini et al. 2008). Reduction in LPS and increase in TST after ramelteon were also noted in healthy human subjects in a 6-week-long study using an 8 mg dose (Dobkin et al. 2009). In another 6-month study performed in 451 adults suffering from chronic insomnia drawn from different centers across the globe (mainly from the USA, Europe, Russia, and Australia), ramelteon consistently reduced LPS when compared to placebo (Mayer et al. 2009). The baseline LPS decreased from 70.7 to 32.0 min at week 1 (with ramelteon) and this reduction in LPS was maintained at months 1, 3, 5, and 6. No adverse effects like next-morning residual effects, rebound insomnia, or withdrawal effects were noted (Mayer et al. 2009).

In a double-blind placebo-controlled study involving a large number of Japanese patients with chronic insomnia ($N = 1130$), the efficacy and safety of 4 and 8 mg ramelteon doses were evaluated. At a 4 mg dose of ramelteon, no statistically significant differences were found in subjective SOL when compared with the placebo group, while at 8 mg, a significant increase in TST and decrease in SOL were observed (Uchimura et al. 2011).

The same investigators evaluated the efficacy and safety of ramelteon in 190 Japanese adults with chronic insomnia treated for 24 weeks. TST significantly increased with ramelteon 8 mg/day and this was maintained for 20 weeks. In this study, ramelteon was well tolerated and it did not cause residual effects, rebound insomnia, withdrawal symptoms, or dependence even after 24 weeks of continuous treatment (Uchiyama et al. 2011). In all clinical studies undertaken so far to evaluate the efficacy and safety of ramelteon in various doses ranging from 4 to 32 mg/day in patients with chronic insomnia, the drug reduced SOL and increased sleep duration (Table 23.1).

Besides acting as a sedative–hypnotic drug, ramelteon also exhibited chronobiotic properties. In a study conducted in 75 healthy human subjects, the administration of ramelteon at doses of 1, 2, 4, and 8 mg for 6 days was associated with a significant advancement of dim light melatonin offset (Richardson et al. 2008). As a melatonergic hypnotic and chronobiotic drug, ramelteon has a unique place in the development of novel drugs for treatment of insomnia (Srinivasan et al. 2009a).

Interestingly, a recent randomized, placebo-controlled study suggested that ramelteon also can be beneficial for outpatients with bipolar 1 disorder (BD 1) suffering from manic symptoms and sleep disturbances. Twenty-one outpatients with BD 1 with mild-to-moderate manic symptoms and sleep disturbances were randomized to receive either ramelteon ($N = 10$) or placebo ($N = 11$) in an 8-week, double-blind, fixed-dose (8 mg/day) study. Ramelteon and placebo had similar rates of reduction in ratings of symptoms of insomnia, mania, and global severity of illness. However, ramelteon was associated with improvement in a global rating of depressive symptoms. It was also well tolerated and not associated with serious adverse events (McElroy et al. 2011).

23.5 MECHANISM OF RAMELTEON SEDATIVE–HYPNOTIC ACTION

Although MT_1 and MT_2 receptors are widely distributed in the brain outside of the SCN (Wu et al. 2006), the high density of melatonin receptors in the SCN and their relationship to the circadian pacemaker function and especially to the sleep–wake cycle are highly suggestive of the SCN melatonin receptor role in sleep regulation (Reppert et al. 1988). Ramelteon's specificity for MT_1 and MT_2 melatonin receptors present in SCN suggests that its sleep-related site of action is due to its effects on these receptors.

A "sleep-switch" model to describe the regulation of sleep–wakefulness was originally proposed by Saper and his colleagues (2005). It consists of "flip-flop" reciprocal inhibitions among sleep-associated activities in the ventrolateral preoptic nucleus and wakefulness-associated activities in the locus coeruleus, dorsal raphe, and tuberomammilary nuclei. The SCN has an active role both in promoting wakefulness as well as in promoting sleep and this depends upon a complex neuronal network and also involves a number of neurotransmitters including GABA, histamine, glutamate, arginine vasopressin, somatostatin, etc. (Kalsbeek et al. 2006; Reghunandanan and Reghunandanan 2006).

Ramelteon may accelerate sleep onset by influencing the hypothalamic sleep switch downstream from the SCN in the same way as melatonin (Saper et al. 2005; Fuller et al. 2006). Ramelteon promotes sleep onset through inhibition of SCN electrical activity and the consequent inhibition of the circadian wake signal, thereby activating the specific sleep-circuit pathway (Srinivasan et al. 2012b). Contrary to this earlier concept that MT_1 receptors control sleep whereas MT_2 receptors control sleep–wakefulness timing through their actions on melatonergic receptors present in SCN,

TABLE 23.1
Ramelteon's Beneficial Effects in Sleep Disorders: Primary Insomnia

Dosage (mg/Day)	Duration of Administration	Number of Insomnia Patients	Sleep Onset Latency	Sleep Efficacy and Quality	Total Sleep Time	Reference
4 and 8	5 weeks	829 (mean age: 72.4 years)	Reduced	Enhanced	Increased at the end of first week, third week and fifth week,	Roth et al. (2006)
4	5 weeks	100 elderly patients	Reduced ($P>0.001$)	Increased	Increased	Roth et al. (2007)
8	5 weeks	270 patients with chronic insomnia	63% reduction in week 1 and 3 ($P>0.001$), 65.9% reduction at week 5 ($P>0.05$)	—	—	Mini et al. (2008)
	6 weeks	20 healthy peri- and postmenopausal women	Reduced	Increased	Increased	Dobkin et al. (2009)
8	6 months	451 adults with chronic insomnia	Reduced latency to persistent sleep consistently	—	—	Mayer et al. (2009)
4 and 8	2 nights	65 patients with insomnia	Reduced	Increased sleep quality	Increased	Kohsaka et al. (2011)
4 and 8	2 weeks	1130 adults	Reduced with 8 mg only	Increased in the first week	Increased	Uchimura et al. (2011)
4, 8, and 16	24 weeks	190 adults with chronic insomnia	Reduced	Increased	Increased up to 20 weeks and then it was maintained	Uchiyama et al. (2011)

recent studies by Stefano Comai and colleagues (Comai et al. 2013) reveal that MT_2 melatonin receptors influence non-REM sleep (NREMS), by acting through MT_2 melatonin receptors present in reticular thalamus. The reticular thalamus is an important nucleus in the neural network controlling sleep and is implicated in the generation of delta waves during slow wave sleep (Bal and McCormick 1993). Since MT_2 receptors are highly expressed in the reticular thalamus, UCM 765, a selective MT_2 receptor agonist, was injected into this nucleus by microinfusion through bilateral cannula implanted into this nucleus to demonstrate its effect on sleep. Indeed, microinfusion of UCM 765 increased the amount of NREM sleep and also decreased the latency to NREM sleep. This excellent study modified the conceptual role of MT_2 melatonin receptors in the actual promotion of sleep in addition to their sleep–wake rhythm regulating activity due to the observation of UCM765's action on hypothalamic SCN MT_2 receptors. Hence, ramelteon, being an MT_1 and MT_2 melatonin receptor agonist, can regulate and promote sleep by acting through these melatonergic receptors of SCN.

23.6 DEPRESSIVE DISORDERS

MDD is a heterogeneous syndrome that comprises a variety of physiological, neuroendocrine, behavioral, and psychological symptoms (Nestler et al. 2002). The common sleep disturbances reported by patients with depression include delayed sleep onset, early morning wakefulness, and a reversal of the normal morning peaks in subjective mood, energy, and alertness. Several studies point out that there are profound disturbances in sleep and circadian rhythms in depressive disorders (Germain and Kupfer 2008; Srinivasan et al. 2009b). Although unipolar and bipolar depressive patients can be differentiated on the basis of symptoms that they manifest, the sleep disturbances exhibited by manic patients, unipolar depressed patients, and bipolar depressed patients are almost similar (Hudson et al. 1988, 1992; Riemann et al. 2001; Plante and Winkelman 2008). Polysomnographic studies of sleep disturbances in patients with MDD show reduced REM latency, elevated REM density, decreased slow-wave sleep and increased stage 1 and stage 2 sleep (Steiger and Kimura 2010). REM sleep is regulated by a circadian time-keeping system while NREM sleep is regulated by a homeostatic sleep system (Dijk and Czeisler 1995). The presence of sleep abnormalities is seen not only in depressive patients, but also in their first-degree relatives, suggesting thereby that sleep changes can be viewed as trait-related "markers" of depression (Lustberg and Reynolds 2000). As sleep and circadian rhythm disturbances constitute primary symptoms in affective disorders including MDD, BD, and SAD, the study of a common underlying sleep pathophysiology in these disorders has become a necessity for using appropriate pharmacological agents to normalize the sleep issues in order to achieve complete remission of symptoms. As the key endogenous neurohormone involved in the physiological regulation of circadian rhythms and sleep–wake cycles, melatonin plays an important role in the etiology of mood disorders like MDD, BD, and SAD (Lewy et al. 1998; Mayeda and Nurnberger 1998; Wetterberg 1998).

23.7 PHARMACOTHERAPY OF DEPRESSIVE DISORDERS

There is increasing evidence suggesting that mood disorders are caused by several other hormonal and neurotransmitter systems like brain-derived neurotrophic factor (BDNF), amino acid neurotransmitters, GABA, corticotrophin releasing factor (CRF), substance P, etc. (Rakofsky et al. 2009). As monoaminergic antidepressants were identified primarily through serendipity, there was a need to develop drugs that could act through the circadian system and sleep mechanisms and, thus, melatonergic antidepressants were developed (Srinivasan et al. 2009b). Moreover, most of the antidepressants that are currently used have varying effects on sleep. While some antidepressants like TCAs and 5-HT receptor antagonists promote sleep initiation and maintenance, many other antidepressants like selective serotonin reuptake inhibitors (SSRIs), such as fluoxetine, and serotonin-norepinephrine reuptake inhibitors (SNRIs), such as venlafaxine, exert adverse effects on sleep (Pandi-Perumal et al. 2008). Most pharmacoepidemiologic surveys indicate that at least one-third of patients taking SSRIs receive concomitant sedative–hypnotic medications (Thase 2006). Hence, clinicians consider antidepressant effects on sleep as a potentially important determining factor in selecting the therapeutic option to treat patients with depressive symptoms (Winokur et al. 2001; De Martinis and Winokur 2007). The possibility that insomnia does not simply reflect a consequence or accompanying phenomenon of affective disorders, but rather represents a "major triggering factor for the development of depressive disorders," is a new concept. Hence, there is a need for the development of novel antidepressants that can ameliorate symptoms of insomnia and circadian rhythm dysfunction (Srinivasan et al. 2010; Kennedy et al. 2011).

23.8 AGOMELATINE: CHEMISTRY AND PHARMACODYNAMICS

Agomelatine, a naphthalenic compound chemically designated as *N*-[2-(7-methoxynaphth-1-yl) ethyl] acetamide or S-20098, is a newly developed selective agonist for MT_1 and MT_2 receptors with antagonism of $5\text{-}HT_{2C}$ receptors (Papp et al. 2003). Agomelatine displays overall selectivity (>100-fold) for MT_1 and MT_2 melatonin receptors sites. Its half-life is longer than melatonin (about 2 h) in humans and it is metabolized in the liver by three CYP isoenzymes; $CYP1_{A1}$, $CYP1_{A2}$, and $CYP2_{C9}$. The main metabolites of agomelatine are 3-hydroxy S20098, 3-hydoxy, 7-methoxy S20098, 7-desmethyl S20098, and dihydrodiol S20098. Agomelatine has no significant affinity for histaminergic, muscarinic, dopaminergic, or adrenergic receptors. The mean terminal elimination half-life is 2.3 h (Papp et al. 2003).

23.9 AGOMELATINE'S MECHANISM OF ANTIDEPRESSANT EFFECTS

Agomelatine (Valdoxan®) was developed by Servier and Novartis in the United States. In February 2009, valdoxan was approved by the EU-EMEA (Europe, Middle East, and Africa) for the treatment of MDD in European countries and is currently available in all European countries. Agomelatine acts synergistically on both melatonergic (MT_1/MT_2) and $5\text{-}HT_{2C}$ receptors (San and Arranz 2008). It is an interesting finding that both melatonergic MT_1 and MT_2 receptors and $5\text{-}HT_{2C}$ receptors are expressed in the SCN and other brain areas involved in the pathophysiology of depression, namely cerebral cortex, hippocampus, amygdala, and thalamus (Masana et al. 2000). Both receptors (MT_1/MT_2 and $5\text{-}HT_{2C}$) exhibit circadian fluctuations and are regulated by light and the biological clock (Masana et al. 2000). The combined actions of agomelatine on melatonergic and $5\text{-}HT_{2C}$ receptors help to resynchronize disturbed circadian rhythms and abnormal sleep patterns and will be effective in treating the mood spectrum disorders including MDD, BD, and SAD (Kennedy et al. 2011).

Agomelatine's antidepressant actions are attributed to its sleep promoting and chronobiotic actions mediated by MT_1 and MT_2 melatonergic receptors present in the SCN as well as due to its effects on the blockade of $5\text{-}HT_{2C}$ receptors. Blockade of $5\text{-}HT_{2C}$ receptors causes a release of both norepinephrine (NE) and dopamine (DA) at the frontocortical dopaminergic and noradrenergic pathways. It is well known that dopaminergic and adrenergic mechanisms in the frontal cortex modulate mood and cognitive functions and antidepressants improve mood and cognition by enhancing the release of NE and DA (Millan et al. 2000).

Agomelatine exhibits similar actions to that of melatonin in synchronizing circadian rhythms following brief exposure, and this has been demonstrated in various animal models (Redman et al. 1995; Pitrosky et al. 1999; Van Reeth et al. 2001). Agomelatine influences the daily patterns of locomotor activity, running wheel activity, and body temperature rhythm (Redman et al. 1995). The circadian rhythm regulating effect of agomelatine has been demonstrated in young healthy subjects in a double-blind crossover design study in which administration of 5–100 mg of agomelatine in the early evening induced phase advance of various rhythms like salivary dim-light melatonin onset, core body temperature minimum, and proximal skin temperature (Krauchi et al. 1997). In another double-blind cross-over study, a prolonged administration of agomelatine (50 mg for 50 days) significantly advanced the circadian rhythm of body temperature by an average of 2 h and cortisol by an average of 1.5–2.0 h in healthy older men (aged 51–76 years) (Leproult et al. 2005). These studies demonstrate agomelatine's circadian entrainment and phase-shifting effects in humans.

The antidepressant mechanism of action of agomelatine has been studied in animals in regard to its neurogenic effects on ventral hippocampus. The ventral hippocampus is implicated in mood and anxiety regulation (Bannerman et al. 2004). Agomelatine reversed decreases in neurogenesis in glucocorticoid receptor–impaired mice, an animal model of depression, and in corticosterone-treated mice, demonstrating that the hippocampus is also one of the target areas through which agomelatine exerts its antidepressant effects (Paizanis et al. 2010; Rainer et al. 2012). Figure 23.2 is a schematic diagram of mechanisms contributing to the antidepressant actions of agomelatine. Agomelatine has demonstrated its antidepressant effects in several animal models of depression such as the forced swimming test, learned helplessness model, chronic mild stress, and psychosocial stress, which have been discussed in an earlier review paper (Pandi-Perumal et al. 2006).

In addition, it is possible, but not demonstrated yet, that agomelatine induces the sensitization of adenylate cyclase to support the circadian clock protein PER1 in the morning hours (Wagner et al. 2008), which leads to repression of the actions of the circadian clock protein PER2 and to inhibition of monoamine oxidase A (MAOA) enzyme activity (Hampp et al. 2008), thereby increasing dopamine release specifically from the ventral tegmental area, where neurons have MT_1 and MT_2 receptors as well as 5-HT_{2C} receptors. Here, normalized dopamine release in the prefrontal cortex and striatum produces a resynchronization effect, as dopamine is key to the resynchronization of the circadian rhythms and the sleep–wake cycle (Hirsh et al. 2010).

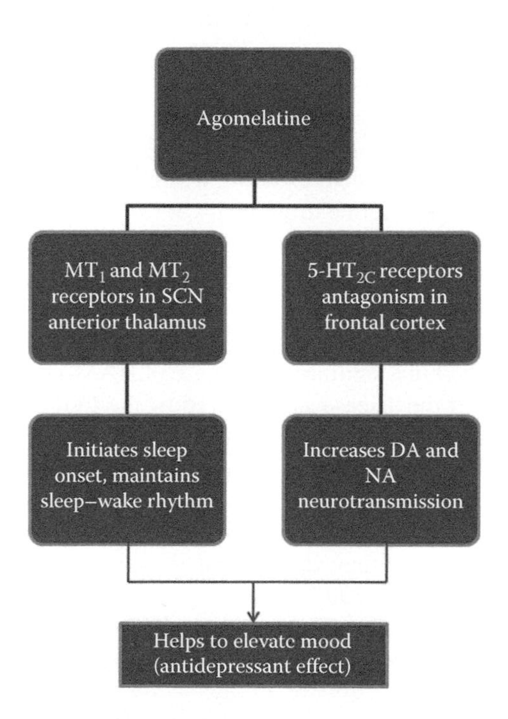

FIGURE 23.2 **(See color insert.)** Mechanism of melatonergic antidepressant's actions in depressive disorders.

23.10 USE OF AGOMELATINE IN PATIENTS WITH MAJOR DEPRESSIVE DISORDER

The assessment of clinical efficacy of agomelatine is based on symptomatic relief as measured by Hamilton Depression Rating Scale, HAM-D (Hamilton 1967). A 50% reduction in the HAM-D baseline score is defined as "response" to antidepressant therapy, whereas the absence of depressive symptoms and return to premorbid level of functioning is considered as "remission" (Eser et al. 2009). Agomelatine's efficacy as an antidepressant has been assessed both in acute phase trials and relapse prevention trials as compared to placebo.

In three acute phase studies conducted with agomelatine, the superiority of agomelatine over placebo was found for 25 mg/day and with an increased dose of 50 mg/day (Loo et al. 2002; Kennedy and Emsley 2006; Olie and Kasper 2007). The primary outcome for efficacy in these studies was the change in the 17-item HAM-D score from baseline. Remission rates were found to be significantly (15%) higher for agomelatine in the study of Loo and colleagues (2002), whereas in another study the remission rate was only 7.5% higher in the total population and 9.1% higher in the subgroup with severe depression (Loo et al. 2002; Kennedy and Emsley 2006). In the randomized, double-blind placebo-control trial study of Loo and colleagues (2002) consisting of 711 patients (MDD $N=698$ and BD $N=13$) diagnosed as per DSM-IV, patients were randomized to receive agomelatine 1, 5, or 25 mg once daily in the evening or placebo during the 8-week study period (Loo et al. 2002). Paroxetine 20 mg/day was also used in the study but as a validator of the study methodology and study population. One-third of these patients had severe depression (HAM-D 17 score > 25). Agomelatine at a dose of 25 mg/day showed a statistically significant efficacy compared to placebo based on the mean final HAM-D score in this study. The time to respond was significantly shorter in the agomelatine 25 mg/day group in comparison to the placebo group and also was shorter with reference to the paroxetine group, which showed significant improvement after 4 weeks of treatment, whereas agomelatine 25 mg/day was significantly superior to placebo after 2 weeks of treatment (Loo et al. 2002).

Similarly, the clinical efficacy of agomelatine in MDD was confirmed in another double-blind, placebo-controlled study (Kennedy and Emsley 2006). In this study ($N=212$), agomelatine was administered in 25 and 50 mg doses. On completion of 6 weeks of treatment, agomelatine was found to be significantly more effective than placebo. Significant improvement in the severity of disease as measured by the Clinical Global Impression of Severity (CGI-S) was noted with agomelatine compared to placebo. Among the agomelatine 50 mg treated group (34% of patients), the HAM-D score decreased from 26.1 ± 2.6 at baseline to 17.5 ± 7.4 at week 6, compared to the placebo group, which showed a decrease from 26.7 ± 2.8 to 20.4 ± 6.0. Also, agomelatine at a dose of 50 mg/day was observed to be effective in patients who failed to show improvement after 2 weeks on a dose of 25 mg/day (Kennedy and Emsley 2006).

In another study of 238 patients with moderate-to-severe major depression, treatment with agomelatine (25–50 mg) resulted in a significant decrease of HAM-D final scores with reference to baseline scores. The severity of the disease as measured by the CGI-S also significantly improved with agomelatine (Olie and Kasper 2007). The clinical efficacy of agomelatine was also proved in another double-blind randomized study of 332 patients treated with either agomelatine 25–50 mg/day or venlafaxine (75–150 mg/day). After 6 weeks of treatment, the antidepressant efficacy of agomelatine was similar to that of venlafaxine as measured by CGI-S. However, agomelatine showed greater efficacy in improving subjective sleep than venlafaxine as measured by the Leeds Sleep Evaluation Questionnaire (LSEQ) (Lemoine et al. 2007).

Agomelatine's efficacy was studied in patients with SAD (Pjrek et al. 2007). Thirty-seven acutely depressed patients with SAD diagnosed as per DSM-IV-TR criteria were selected in an open study and agomelatine was administered at a dose of 25 mg/day for over 14 weeks. Clinical efficacy was assessed by the Structured Interview Guide for the Hamilton Rating Scale (SIGH-SAD) and CGI-S and improvement by the Circa Screen, a self-rated scale for the assessment of sleep and circadian rhythm disorders, and a hypomania scale. Response and remission rates were computed from the SIGH-SAD score. Response was defined as a reduction of SIGH-SAD total score of more than 50% from baseline score. Remission was defined as SIGH-SAD total score of less than 8 points. A significant reduction in the SIGH-SAD total score was noted from the second week onward with agomelatine ($P<0.001$). Treatment effects were progressive and sustained throughout the course of the trial (baseline score 29.8 ± 4.6; final score at 14 weeks 8.4 ± 10.1). CGI-S score was 4.5 ± 1.0 at baseline; at 14 weeks 1.9 ± 1.3. The reduction in the CGI-S score with agomelatine was significant ($P<0.001$). From this study it was noted that 75.7% of patients had responded to agomelatine treatment. This study, being the first of its kind on agomelatine's effect in SAD, showed that agomelatine exerted its antidepressant action from the second week onward. The antidepressant efficacy of agomelatine was sustained in a large number of patients, revealing a sustained remission during the entire period of study (Pjrek et al. 2007).

The clinical efficacy of agomelatine (50 mg) was compared with venlafaxine (target dose of 150 mg) in a total sample of 276 male and female patients. Both treatments resulted in equivalently high rates of remission (agomelatine 73%, venlafaxine XR 67%), but treatment-emergent sexual dysfunction was significantly less prevalent among patients who received agomelatine, whereas venlafaxine XR was associated with significantly greater deterioration on the sex effects scale domain of desire and orgasm. From this study it is evident that agomelatine is endowed with an efficacious antidepressant effect with a superior sexual side effect profile compared with venlafaxine XR (Kennedy et al. 2008).

The antidepressant efficacy of agomelatine was also evaluated in patients with BD 1. Patients with BD on lithium ($N=14$) or valpromide ($N=7$) were given adjunctive open-label agomelatine at 25 mg/day for a minimum of 6 weeks followed by an additional optional extension up to 46 weeks. Marked improvement (HAM-D score>50%) was noted. Among the severely depressed patients (HAM-D score>25.2), 47.6% responded as early as 1 week of initiating treatment. Of the 19 patients who entered the optional extension period for a mean of 211 days, 11 completed the 1 year extension on agomelatine. During the optional extension period, three lithium-treated patients experienced manic or hypomanic episodes, one of which was treatment-related. Otherwise, agomelatine 25 mg/day has been effective for treating patients with BD 1 experiencing a major depressive episode, when co-medicated with lithium or valpromide (Calabrese et al. 2007).

In a 24-week randomized, double-blind treatment study, patients were randomly assigned to receive agomelatine ($N=165$) or placebo ($N=174$) and the time to relapse was evaluated by using the Kaplan–Meier method of survival analysis. During the 6-month evaluation period, the incidence of relapse was significantly lower in patients who continued to receive agomelatine than those switched to placebo ($P=0.0001$). The cumulative relapse rate at 6 months for agomelatine-treated patients was 22%, and for placebo-treated patients, it was 47%. From this long-term study, it was concluded that agomelatine is an effective and safe antidepressant and that the incidence of relapse was significantly lower in patients who continued agomelatine compared to those who switched to placebo (Goodwin et al. 2009).

In an 8-week double-blind trial, the efficacy and safety of two fixed doses of agomelatine was evaluated in patients with moderate-to-severe MDD (Zajecka et al. 2010). The primary efficacy variable was assessed by the HAM-D in 511 MDD patients. Secondary efficacy was assessed by the Clinical Global Impression-Improvement score (CGI-I), sleep disability by LSEQ, and disability by Sheehan Disability Scale. Patients were randomized (1:1:1) to once daily agomelatine 25 mg; agomelatine 50 mg; placebo. Patients who received agomelatine 50 mg showed statistically significant improvement in the HAM-D score from their first baseline visit throughout the 8-week treatment period. At week 8, the statistical significance was $P=0.004$ with agomelatine 50 mg whereas with 25 mg it was $P=0.505$. Also, agomelatine 50 mg was found to be superior to placebo in all other secondary efficacy variables in this study, including the CGI-I ($P=0.012$), CGI-S ($P=0.003$), patient's ability to get to sleep ($P<0.001$), and quality of sleep ($P=0.002$). These results confirm the significant antidepressant efficacy of agomelatine 50 mg/day with positive effects on sleep. Transient amino transferase elevations were noted in 4.5% of the patients in the agomelatine 50 mg group (Zajecka et al. 2010).

The efficacy, safety, and tolerability of fixed doses of agomelatine 25 and 50 mg/day were evaluated during an 8-week, multicenter double-blind parallel group trial in outpatients with moderate-to-severe MDD as compared to placebo (Stahl et al. 2010). In this study, patients were randomly assigned (1:1:1) to receive once daily dose of agomelatine 25 mg, agomelatine 50 mg, or placebo. Agomelatine 50 mg/day caused a statistically significant reduction in HAM-D total score from 2 to 6 weeks but not at week 8 ($P=0.144$). But a higher proportion of patients receiving agomelatine 25 mg/day showed clinical response ($P=0.013$), clinical remission ($P=0.07$), and improvement according to the CGI-I ($P=0.065$) compared to those receiving placebo. However, no statistically significant difference between patients receiving agomelatine 50 mg/day and placebo was noted. From this study, agomelatine 25 mg/day was found effective in the treatment of patients with moderate to severe intensity depression and was safe and well tolerated throughout the 8 week period whereas agomelatine 50 mg/day provided evidence for its antidepressant efficacy until week 6 of treatment (Stahl et al. 2010).

In a group of severely depressed patients ($N=252$) with HAM-D scores of more than 25 and CGI-S scores of 4 and above, an 8-week randomized, double-blind study was carried out with agomelatine and fluoxetine (agomelatine 25–50 mg/day; fluoxetine 20–40 mg/day). The mean decrease in HAM-D score over 8 weeks was significantly greater with agomelatine than with fluoxetine with a group difference of 1.49. The percentage of responders at the last post-baseline assessment was higher with agomelatine on both the HAM-D 17 and CGI. A 50% decrease in total score from baseline was evident in 72% with agomelatine versus 69% with fluoxetine for the HAM-D ($P=0.060$) and 78% with agomelatine versus 69% with fluoxetine for the CGI-improvement score ($P=0.023$). Although both the treatments were safe and well tolerated, agomelatine showed superior antidepressant efficacy over fluoxetine (Hale et al. 2010).

The efficacy of agomelatine over sertraline was evaluated in a randomized, double-blind study carried out for 6 weeks. Outpatients with DSM-IV-TR diagnosis of MDD received either agomelatine 25–50 mg/day ($n=154$) or sertraline 50–100 mg/day ($n=159$) for 6 weeks and efficacy against depressive symptoms was evaluated using the HAM-D 17 scale, CGI-S, and measures of sleep efficiency, SL, and circadian-rest activity. Over the 6-week treatment period, depressive symptoms improved significantly more with agomelatine than with sertraline ($P<0.05$), as did anxiety symptoms ($P<0.05$). A significant improvement in SL ($P<0.001$) and sleep efficiency from week 1 to week 6 was observed with agomelatine as compared to sertraline. The relative amplitude of the circadian rest–activity cycle was also in favor of agomelatine as compared to sertraline. All of these findings indicate that agomelatine demonstrated superior beneficial effects in depressed patients (Kasper et al. 2010).

Di Giannantonio and his coinvestigators carried out an 8-week open-label study to evaluate the efficacy of agomelatine (25–50 mg/day) on depressive symptoms in 32 patients with MDD. Secondary endpoints were the effect of agomelatine on anhedonia assessed by the Hamilton rating scale. Of the 24 patients who completed 8 weeks of treatment, significant improvements were noted at all visits on the HAM-D ($P<0.05$), Hamilton Anxiety Rating Scale, HAM-A ($P<0.01$), Snaith-Hamilton-Pleasure-Scale, SHAPS ($P<0.05$), and LSEQ-sleep scale ($P<0.05$). Five subjects (17%) had remitted at week 1. At the end of the trial period, 18 subjects (60%) had remitted. No serious adverse effects or amino transferase elevations were noted. Not only was evidence for early response to agomelatine treatment and clinical improvements obtained from this study, but this study is also the first of its kind where agomelatine was found as effective in the treatment of anhedonia (Di Giannantonio et al. 2011).

In addition to mood disorders, obsessive–compulsive disorder (OCD) has also been successfully treated with agomelatine (De Berardis et al. 2013). In this single case study, agomelatine was administered at 25 mg/day and symptoms of OCD were evaluated at 2 weeks, 3 weeks, and at the end of 6 weeks. After 2 weeks of treatment, the dose was titrated to 50 mg/day, resulting in a gradual improvement of symptoms, and full clinical remission was maintained after 6 weeks of treatment at this dose. In a case series report of six patients with SSRI-refractory OCD, agomelatine 50 mg/day was initiated and patients were followed up for 12 weeks. Three out of six patients showed a Yale-Brown Obsessive–Compulsive Scale (Y-BOCS) score reduction of ≥35%, indicating a possible role for agomelatine in SSRI-refractory OCD patients (Kennedy and Rizvi 2010). Agomelatine's antidepressant effects are summarized in Table 23.2.

23.11 SAFETY AND TOLERABILITY OF AGOMELATINE

Agomelatine has exhibited good tolerability and safety in all the clinical studies that have been undertaken so far. The frequency of adverse effects reported with both 25 and 50 mg/day, such as headache, anxiety, abdominal pain, diarrhea, etc., are similar to that reported for placebo (Loo et al. 2002). The cardiovascular profile of agomelatine is also the same as that of placebo, with the mean heart rate and blood pressure of patients unchanged (Montejo et al. 2010). The side effect profile of agomelatine has been shown to be benign in individual trials as well as pooled analyses. The specific side effects like increases in body weight or sexual dysfunction, which are common with some other antidepressants, are not seen with agomelatine (Rosen et al. 1999; Montgomery et al. 2004). Agomelatine's ability to improve sleep-related complaints is not associated with daytime sedation (Rosen et al. 1999).

Impaired sexual function that often occurs with other antidepressants is a major cause of noncompliance (Fornaro et al. 2010). Specific trials were conducted regarding the effects of agomelatine on sexual function (Kennedy et al. 2008). Agomelatine's effect on sexual dysfunction was compared with venlafaxine ER by a sexual function questionnaire. Desire, arousal, orgasm scores, and total sexual dysfunction were significantly greater in venlafaxine-treated patients. By analysis of medication-induced side effects and sexual symptoms associated with the intake of antidepressants (either agomelatine or venlafaxine), it was shown that 7.3% of agomelatine-treated patients and 15.7% of venlafaxine-treated patients reported deterioration of sexual function (Kennedy et al. 2008).

With regard to discontinuation symptoms, fewer patients discontinued treatment with agomelatine compared to fluoxetine (12% vs. 19%), venlafaxine (2% vs. 9%), and sertraline (14% vs. 19%) (Kennedy et al. 2008; Hale et al. 2010; Zajecka et al. 2010). A double-blind placebo-controlled study in 192 patients was undertaken to observe for discontinuation symptoms with agomelatine using the Discontinuation Emergent Signs and Symptoms (DESS) checklist (De Berardis et al. 2011). These patients were randomized to receive either agomelatine 25 mg/day or paroxetine 20 mg/day for 12 weeks, followed by an abrupt discontinuation of treatment for 2 weeks, after which they were randomized to placebo or their initial antidepressant. Compared to paroxetine, no discontinuation symptoms were noted in patients who discontinued agomelatine (De Berardis et al. 2011).

In terms of parameters, mild elevations in serum aminotransferases were reported in 1.1% of the patients treated with agomelatine and these increases were isolated, reversible, and occurred without any clinical signs of liver damage (Rosen et al. 1999). In other recent studies, aminotransferase elevations were noted in 2.4% of patients treated with agomelatine 50 mg/day in one study and 4.5% of patients in another (Zajecka et al. 2010). These elevations in aminotransferases seen with 50 mg/day of agomelatine are not accompanied by any clinical signs of liver damage but only reflected the higher prevalence of hepatobiliary disorders at baseline, according to medical history, in the agomelatine 50 mg/day group when compared to agomelatine 25 mg or placebo groups (both at 0.6%). As discontinuation rates for agomelatine were lower than for venlafaxine, sertraline, and fluoxetine, its use is not associated with any discontinuation symptoms, and the incidence of relapse over 6 months was much lower for agomelatine than for any other antidepressants that are in clinical use today, it has a promising role for the treatment of patients with depressive disorders (De Berardis et al. 2013). The mechanism of agomelatine's antidepressant action also

TABLE 23.2
Agomelatine's Antidepressant Effects

Agomelatine Dosage	Type of Patient	Number of Patients	Type of Study	Duration of Study (Weeks)	Antidepressant Response	Reference
1, 5, 25 mg/day	MDD	711	Double-blind, placebo-controlled	8	25 mg/day was more effective than placebo	Loo et al. (2002)
25 and 50 mg/day	MDD	212	Double-blind, placebo-controlled	6	Effective in depressed and severely depressed patients	Kennedy and Emsley (2006)
25 and 50 mg/day	MDD	238	Double-blind, placebo-controlled	6	Improved depressive symptoms and sleep	Olie and Kasper (2007)
25 mg/day	MDD	332 agomelatine ($n=165$); venlafaxine ($n=167$)	Double-blind randomization	6	Agomelatine showed superior antidepressant and sleep quality	Lemoine et al. (2007)
25 mg/day	SAD	37	Open study	14	Remission was sustained	Pjrek et al. (2007)
50 mg/day	MDD	276 agomelatine ($n=137$); venlafaxine XR ($n=139$).	Double-blind	12	Agomelatine showed superior antidepressant efficacy	Kennedy et al. (2008)
25 mg/day	BD I	21 lithium ($n=14$); valpromide ($n=7$)	Open-label	6	Improved depression	Calabrese et al. (2007)
25 and 50 mg/day	MDD	339	Double-blind, placebo-controlled	24	Very effective antidepressant effect	Goodwin et al. (2009)
25 and 50 mg/day	MDD	511	Double-blind, placebo-controlled	8	50 mg/day had significant antidepressant efficacy with positive effects on sleep	Zajecka et al. (2010)
25 and 50 mg/day	MDD	403	Double-blind, placebo-controlled	8	25 mg/day was effective for moderate-to-severe depression	Stahl et al. (2010)
25–50 mg/day	MDD (severe)	515 agomelatine ($n=252$); fluoxetine ($n=263$)	Double-blind randomization	8	Agomelatine showed superior antidepressant efficacy	Hale et al. (2010)
25–50 mg/day	MDD	313 agomelatine ($n=154$); sertraline ($n=159$)	Double-blind randomization	6	Agomelatine showed superior antidepressant efficacy and sleep quality	Kasper et al. (2010)
25–50 mg/day	MDD	30	Open-label study	8	Significant response	Di Giannantonio et al. (2011)

is unique in the sense that it acts on $MT_1$and MT_2 melatonergic receptors present in the SCN and reticular thalamus and acts as an antagonist at 5-HT_{2C} receptors in frontocortical dopaminergic pathways, functioning as both a chronohypnotic and an antidepressant drug, helping to improve sleep efficiency and depressive symptoms more quickly when compared to other antidepressants (Srinivasan et al. 2012a,c).

23.12 CONCLUSION

Ramelteon (Rozerem), a melatonergic drug with a rapid onset and sustained duration of action, has been effective in treating sleep disorders and sleep disturbances associated with depressive disorders. This melatonergic drug, by acting through MT_1 and MT_2 melatonergic receptors in the SCN and also in the reticular thalamus, has demonstrated its superior efficacy in promoting sleep with its rapid onset of action. Moreover, it did not exhibit any adverse side effects that have been associated with the use of benzodiazepine and non benzodiazepine sedative drugs that are commonly used today. Ramelteon exerts a promising effect on sleep by enhancing daytime alertness and sleep quality and by displaying an effective sleep-inducing effect. The long-term use of melatonin or ramelteon has not been associated with dependency, next-day hangover, memory impairment, cognitive dysfunction, or psychomotor retardation.

Agomelatine, the melatonergic antidepressant, is effective in treating patients with MDD and other mood disorders, partly attributable to its property of improving sleep quality and efficiency. Agomelatine's 5-HT_{2C} antagonism also enhances frontocortical dopaminergic and noradrenergic transmission, which is essential for the modulation of mood and producing antidepressant effects (Hudson et al. 1988; Srinivasan et al. 2009b). In addition to its therapeutic efficacy, agomelatine does not manifest any of the adverse effects like sexual dysfunction, sleep disturbances, or discontinuation effects commonly seen with the use of other antidepressants. Its clinical efficacy in MDD, combined with its early onset of action and its good tolerability and safety, has been supported by a number of clinical research studies and recent treatment reviews.

REFERENCES

Bal, T., D. A. McCormick. 1993. Mechanisms of oscillatory activity in guinea-pig nucleus reticularis thalami in vitro: A mammalian pacemaker. *J Physiol* 468;669–691.

Bannerman, D. M., R. M. Deacon, S. Brady, A. Bruce, R Sprengel, P. H. Seeburg, J. N. Rawlins. 2004. A comparison of GluR-A deficient and wild type mice on a test battery assessing sensorimotor, affective and cognitive behaviours. *Behav Neurosci* 118:643–647.

Bastien, C. H. 2011. Insomnia: Neurophysiological and neuropsychological approaches. *Neuropsychol Rev* 21:22–40.

Brzezinski, A., M. G. Vangel, R. J. Wurtman, G. Norrie, I. Zhdanova, A. Ben-Shushan, I. Ford. 2005. Effects of exogenous melatonin on sleep: A meta-analysis. *Sleep Med Rev* 9:41–50.

Buscemi, N., B. Vandermeer, N. Hooton, R. Pandya, L. Tjosvold, L. Hartling, S. Vohra, T. P. Klassen, G. Baker. 2006. Efficacy and safety of exogenous melatonin for secondary sleep disorders and sleep disorders accompanying sleep restriction: Meta-analysis. *BMJ* 332:385–393.

Calabrese, J. R., J. D. Guelfi, C. Perdrizet-Chevallier, Agomelatine Bipolar Study Group. 2007. Agomelatine adjunctive therapy for acute bipolar depression: Preliminary open data. *Bipolar Disord* 6:628–635.

Comai, S., R. Ochoa-Sanchez, G. Gobbi. 2013. Sleep-wake characterization of double MT_1/MT_2 receptor knockout mice and comparison with MT_1/MT_2 receptor knockout mice. *Behav Brain Res* 243:231–238.

Cricco, M., E. M. Simonsick, D. J. Foley. 2001. The impact of insomnia on cognitive functioning in older adults. *J Am Geriatr Soc* 49:1185–1189.

Dalton, E. J., D. Rotondi, R. D. Levitan, S. H. Kennedy, G. M. Brown. 2000. Use of slow-release melatonin in treatment-resistant depression. *J Psychiatry Neurosci* 25:48–52.

De Berardis, D., G. Di Lorio, T. Acciavatti, C. Conti, N. Serroni, L. Olivieri, M. Cavuto et al. 2011. The emerging role of melatonin agonists in the treatment of major depression: Focus on agomelatine. *CNS Neurol Disord Drug Targets* 10(1):119–132.

De Berardis, D., S. Marini, M. Fornaro, V. Srinivasan, F. Iasvoli, C. Tomasetti, A. Valchera et al. 2013. The melatonergic system in mood and anxiety disorders and the role of agomelatine: Implications for clinical practice. *Int J Mol Sci* 14:12458–12483.

De Martinis, N. A., A. Winokur. 2007. Effects of psychiatric medications on sleep and sleep disorders. *CNS Neurol Disord Drug Targets* 6:17–29.

DeMicco, M., S. Wang-Weigand, J. Zhang. 2006. Long-term therapeutic effects of ramelteon treatment in adults with chronic insomnia: A 1 year study. *Sleep* 29(Abstract Suppl):A234.

Di Giannantonio, M., G. Di Iorio, R. Guglielmo, D. DeBerardis, C. M. Conti, T. Acciavatti, M. Cornelio, G. Martinotti. 2011. Major depressive disorder, anhedonia and agomelatine: An open-label study. *J Biol Regul Homeost Agents* 25(1):109–114.

Dijk, D. J., C. Cajochen. 1997. Melatonin and the circadian regulation of sleep initiation, consolidation, structure and sleep EEG. *J Biol Rhythm* 12:627–635.

Dijk, D. J., C. A. Czeisler. 1995. Contribution of the circadian pacemaker and the sleep homeostat to sleep propensity, sleep structure, electroencephalographic slow waves and sleep spindle activity in humans. *J Neurosci* 15:3526–3538.

Dobkin, R. D., M. Menza, K. L. Bienfait, L. A. Allen, H. Marin, M. A. Gara. 2009. Ramelteon for the treatment of insomnia in menopausal women. *Menopause Int* 15:13–18.

Dollins, A. B., I. V. Zhdanova, R. J. Wurtman, H. J. Lynch, M. H. Deng. 1994. Effect of inducing nocturnal serum melatonin concentrations in daytime on sleep, mood, body temperature, and performance. *Proc Natl Acad Sci USA* 91:1824–1828.

Erman, M., D. Seiden, G. Zammit, S. Sainati, J. Zhang. 2006. An efficacy, safety, and dose-response study of ramelteon in patients with chronic primary insomnia. *Sleep Med* 7:17–24.

Eser, D., T. C. Baghai, H. J. Moller. 2009. Agomelatine: The evidence for its place in the treatment of depression. *Core Evid* 3:171–179.

Fornaro, M., S. Prestia Colicchio, G. Perugi. 2010. A systematic, updated review on the antidepressant agomelatine focusing on its melatonergic modulation. *Curr Neuropharmacol* 8:287–304.

Fuller, P. M., J. J. Gooley, C. B. Saper. 2006. Neurobiology of the sleep-wake cycle: Sleep architecture, circadian regulation, and regulatory feedback. *J Biol Rhythms* 21:482–493.

Garfinkel, D., M. Laudon, D. Nof, N. Zisapel. 1995. Improvement of sleep quality in elderly people by controlled-release melatonin. *Lancet* 346:541–544.

Gerdin, M. J., M. I. Masana, M. A. Rivera-Bermudez, R. L. Hudson, D. J. Earnest, M. U. Gillette, M. L. Dubocovich. 2004. Melatonin desensitizes endogenous MT_2 melatonin receptors in the rat suprachiasmatic nucleus: Relevance for defining the periods of sensitivity of the mammalian circadian clock to melatonin. *FASEB J* 18:1646–1656.

Germain, A., D. J. Kupfer. 2008. Circadian rhythm disturbances in depression. *Hum Psychopharmacol* 23:571–585.

Golombek, D. A., P. Pevet, D. P. Cardinali. 1996. Melatonin effect on behavior: Possible mediation by the central GABAergic system. *Neurosci Biobehav Rev* 20:403–412.

Goodwin, G. M., R. Emsley, S. Rembry, F. Rouillon, Agomelatine Study Group. 2009. Agomelatine prevents relapse in patients with major depressive disorder without evidence of a discontinuation syndrome: A 24-week randomized, double-blind, placebo controlled trial. *J Clin Psychiatry* 70(8):1128–1137.

Gorfine, T., Y. Assaf, Y. Goshen-Gottstein, Y. Yeshurun, N. Zisapel. 2006. Sleep-anticipating effects of melatonin in the human brain. *Neuroimage* 31:410–418.

Hale, A., R. M. Corral, C. Mencacci, J. S. Ruiz, C. A. Severo, V. Gentil. 2010. Superior antidepressant efficacy results of agomelatine versus fluoxetine in severe MDD patients: A randomized double-blind study. *Int Clin Psychopharmacol* 25(6):305–314.

Hamilton, M. 1967. Development of a rating scale for primary depressive illness. *Br J Soc Clin Psychol* 6:278–296.

Hampp, G., J. A. Ripperger, T. Houben, I. Schmutz, C. Blex, S. Perreau-Lenz, I. Brunk et al. 2008. Regulation of monoamine oxidase A by circadian-clock components implies clock influence on mood. *Curr Biol* 18:678–683.

Hirsh, J., T. Riemensperger, H. Coulom, M. Iché, J. Coupar, S. Birman. 2010. Roles of dopamine in circadian rhythmicity and extreme light sensitivity of circadian entrainment. *Curr Biol* 20:209–214.

Hudson, J. I., J. F. Lipinski, F. R. Frankenburg, V. J. Grochocinski, D. J. Kupfer. 1988. Electroencephalographic sleep in mania. *Arch Gen Psychiatry* 45:267–273.

Hudson, J. I., J. F. Lipinski, P. E. Keck Jr., H. G. Aizley, S. E. Lukas, A. J. Rothschild, C. M. Waternaux, D. J. Kupfer. 1992. Polysomnographic characteristics of young manic patients. Comparison with unipolar depressed patients and normal control subjects. *Arch Gen Psychiatry* 49(5):378–383.

Kalsbeek, A., S. Perreau-Lenz, R. M. Buijs. 2006. A network of (autonomic) clock outputs. *Chronobiol Int* 23:521–535.

Kasper, S., G. Hajak, K. Wulff, W. J. Hoogendijk, A. L. Montejo, E. Smeraldi, J. K. Rybakowski et al. 2010. Efficacy of the novel antidepressant agomelatine on the circadian rest-activity cycle and depressive and anxiety symptoms in patients with major depressive disorder: A randomized, double blind comparison with sertraline. *J Clin Psychiatry* 71(2):109–120.

Kato, K., K. Hirai, K. Nishiyama, O. Uchikawa, K. Fukatsu, S. Ohkawa, Y. Kawamata, S. Hinuma, M. Miyamoto. 2005. Neurochemical properties of ramelteon (TAK-375), a selective MT1/MT2 receptor agonist. *Neuropharmacology* 48:301–310.

Kennedy, S. H., R. A. Emsley. 2006. Placebo-controlled trial of agomelatine in the treatment of major depressive disorder. *Eur Neuropsychopharmacol* 16(2):93–100.

Kennedy, S. H., S. J. Rizvi. 2010. Agomelatine in the treatment of major depressive disorder: Potential for clinical effectiveness. *CNS Drugs* 24(6):479–499.

Kennedy, S. H., S. Rizvi, K. Fulton, J. Rasmussen. 2008. A double blind comparison of sexual functioning, antidepressant efficacy, and tolerability between agomelatine and venlafaxine XR. *J Clin Psychopharmacol* 28(3):329–333.

Kennedy, S. H., A. H. Young, P. Blier. 2011. Strategies to achieve clinical effectiveness: Refining existent therapies and pursuing emerging targets. *J Affect Disorder* 132(Suppl. 1):S21–S28.

Kohsaka, M., T. Kanemura, M. Taniguchi, H. Kuwahara, A. Mikami, K. Kamikawa, H. Uno, A. Ogawa, M. Murasaki, Y. Sugita. 2011. Efficacy and tolerability of ramelteon in a double-blind, placebo-controlled, crossover study in Japanese patients with chronic primary insomnia. *Expert Rev Neurother* 11(10):1389–1397.

Krauchi, K., C. Cajochen, D. Mori, P. Graw, A. Wirz-Justice. 1997. Early evening melatonin and S-20098 advance circadian phase and nocturnal regulation of core body temperature. *Am J Physiol* 272:R1178–R1188.

Lemoine, P., C. Guilleminault, E. Alvarez. 2007. Improvement in subjective sleep in major depressive disorder with a novel antidepressant, agomelatine: Randomized, double blind comparison with venlafaxine. *J Clin Psychiatry* 68:1723–1732.

Leppämäki, S., T. Partonen, O. Vakkuri, J. Lönnqvist, M. Partinen, M. Laudon. 2003. Effect of controlled-release melatonin on sleep quality, mood, and quality of life in subjects with seasonal or weather-associated changes in mood and behaviour. *Eur Neuropsychopharmacol* 13:137–145.

Leproult, R., A. Van Onderbergen, M. L'Hermite-Baleriaux, E. Van Cauter, G. Copinschi. 2005. Phase-shifts of 24-h rhythms of hormonal release and body temperature following early evening administration of the melatonin agonist agomelatine in healthy older men. *Clin Endocrinol (Oxf)* 63:298–304.

Lewy, A. J., R. L. Sack, N. L. Cutler. 1998. Melatonin in circadian phase sleep and mood disorders. In *Melatonin in Psychiatric and Neoplastic Disorders*, eds. M. Shafii, S. L. Shafii, pp. 81–104. Washington, DC: American Psychiatry Press.

Lieberman, H. R. 1986. Behaviour, sleep and melatonin. *J Neural Transm (Suppl)* 21:233–241.

Liu, C., D. R. Weaver, X. Jin, L. P. Shearman, R. L. Pieschl, V. K. Gribkoff, S. M. Reppert. 1997. Molecular dissection of two distinct actions of melatonin on the suprachiasmatic circadian clock. *Neuron* 19:91–102.

Loo, H., A. Hale, H. D'haenen. 2002. Determination of the dose of agomelatine, a melatonergic agonist and selective 5-HT(2C) antagonist in the treatment of major depressive disorder. *Int Clin Psychopharmacol* 17:239–247.

Lustberg, L., C. E. Reynolds. 2000. Depression and insomnia: Questions of cause and effect. *Sleep Med Rev* 4:253–262.

Masana, M. J., S. Benloucif, M. L. Dubocovich. 2000. Circadian rhythm of MT_1 melatonin receptor expression in the suprachiasmatic nucleus of the C3H/HeN mouse. *J Pineal Res* 28:185–192.

Mayeda, A., J. I. Nurnberger. 1998. Melatonin and circadian rhythms in bipolar mood disorder. In *Melatonin in Psychiatric and Neoplastic Disorders*, eds. M. Shafii, S. L. Shafii, pp. 105–123. Washington, DC: American Psychiatry Press.

Mayer, G., S. Wang-Weigand, B. Roth-Schechter, R. Lehmann, C. Staner, M. Partinen. 2009. Efficacy and safety of 6-month nightly ramelteon administration in adults with chronic primary insomnia. *Sleep* 32:351–360.

McElroy, S. L., E. L. Winstanley, B. Martens, N. C. Patel, N. Mori, D. Moeller, J. McCoy, P. E. Keck Jr. 2011. A randomized, placebo-controlled study of adjunctive ramelteon in ambulatory bipolar I disorder with manic symptoms and sleep disturbance. *Int Clin Psychopharmacol* 26(1):48–53.

Mendelson, W. B. 1997. A critical evaluation of the hypnotic efficacy of melatonin. *Sleep* 20:916–919.

Mendelson, W. B. 2002. Melatonin microinjection into the medial preoptic area increases sleep in the rat. *Life Sci* 71:2067–2070.

Millan, M. J., A. Gobert, J. M. Rivet, A. Adhumeau-Auclair, D. Cussac, A. Newman-Tancredi, A. Dekeyne, J. P. Nicolas, F. Lejeune. 2000. Mirtazapine enhances frontocortical dopaminergic and cortico limbic adrenergic but not serotonergic transmission by blockade of alpha-2 adrenergic and serotonergic 2c receptors; a comparison with citalopram. *Eur J Neurosci* 12:1079–1095.

Mini, L., S. Wang-Weigand, J. Zhang. 2008. Ramelteon 8 mg/d versus placebo in patients with chronic insomnia: Post hoc analysis of a 5-week trial using 50% or greater reduction in latency to persistent sleep as a measure of treatment effect. *Clin Ther* 30:1316–1323.

Miyamoto, M. 2009. Pharmacology of ramelteon, a selective MT1/MT2 receptor agonist: A novel therapeutic drug for sleep disorders. *CNS Neurosci Ther* 15:32–51.

Montejo, A. L., N. Prieto, A. Terleira, J. Matias, S. Alonso, G. Paniagua, S. Naval et al. 2010. Better sexual acceptability of agomelatine (25–50 mg) compared with paroxetine (20 mg) in healthy male volunteers. An 8 week, placebo controlled study using the PRSEXDQ-SALSEX scale. *J Psychopharmacol* 24:111–120.

Montgomery, S. A., S. H. Kennedy, G. D. Burrows, M. Lejoyeux, I. Hindmarch. 2004. Absence of discontinuation symptoms with agomelatine and occurrence of discontinuation with paroxetine. A randomized double-blind placebo controlled discontinuation study. *Int Clin Psychopharmacol* 19:271–280.

Monti, J. M., F. Alvarino, D. Cardinali, I. Savio, A. Pintos. 1999. Polysomnographic study of the effect of melatonin on sleep in elderly patients with chronic primary insomnia. *Arch Gerontol Geriatr* 28:85–98.

Nestler, E. J., M. Barrot, R. J. DeLeone, A. J. Eisch, S. J. Gold, L. M. Monteggia. 2002. Neurobiology of depression. *Neuron* 34:13–25.

Olie, J. P., S. Kasper. 2007. Efficacy of agomelatine, a MT1/MT2 receptor agonist with 5-HT_{2C} antagonistic properties in major depressive disorder. *Int J Neuropsychopharmacol* 10(5):661–673.

Paizanis, E., T. Renoir, V. Lelievre, F. Saurini, M. Melfort, C. Gabriel, N. Barden, E. Mocaër, M. Hamon, L. Lanfumey. 2010. Behavioural and neoplastic effects of the new generation antidepressant agomelatine compared to fluoxetine in glucocorticoid receptor impaired mice. *Int J Neuropsychopharmacol* 13:759–774.

Pandi-Perumal, S. R., V. Srinivasan, D. P. Cardinali, M. J. Monti. 2006. Could agomelatine be the ideal antidepressant? *Exp Rev Neurother* 6(11):1595–1608.

Pandi-Perumal, S. R., I. Trakht, V. Srinivasan, D. W. Spence, B. Poeggeler, R. Hardeland, D. P. Cardinali. 2008. The effect of melatonergic and non-melatonergic antidepressants on sleep: Weighing the alternatives. *World J Biol Psychiatry* 4:1–13.

Papp, M., P. Gruca, P. A. Boyer, E. Mocaër. 2003. Effect of agomelatine in the chronic mild stress model of depression in the rat. *Neuropsychopharmacology* 28:604–703.

Pitrosky, B., R. Kirsch, A. Malan, E. Mocaer, P. Pevet. 1999. Organization of rat circadian rhythms during daily infusions or melatonin or S20098, a melatonin agonist. *Am J Physiol* 277:812–828.

Pjrek, E., D. Winkler, A. Konstantinidis, M. Willeit, N. Praschak-Rieder, S. Kasper. 2007. Agomelatine in the treatment of seasonal affective disorder. *Psychopharmacology* 190:575–579.

Plante, D. T., J. W. Winkelman. 2008. Sleep disturbance in bipolar disorder: Therapeutic implications. *Am J Psychiatry* 165(7):830–843.

Rainer, Q., L. Xia, J. P. Guilloux, C. Gabriel, E. Mocaër, R. Hen, E. Enhamre, A. M. Gardier, D. J. David. 2012. Beneficial behavioural and neurogenic effects of agomelatine in a model of depression/anxiety. *Int J Neuropsychopharmacol* 15:321–335.

Rakofsky, J. J., P. E. Holtzheimer, C. B. Nemeroff. 2009. Emerging targets for antidepressant therapies. *Curr Opin Chem Biol* 13:291–302.

Redman, J. R., B. Guardiola-Lemaitre, M. Brown, P. Delagrange, S. M. Armstrong. 1995. Dose dependent effects of S-20098 a melatonin agonist on direction of re-entrainment of rat circadian activity rhythms. *Psychopharmacology (Berl)* 118:385–390.

Reghunandanan, V., R. Reghunandanan. 2006. Neurotransmitters of the suprachiasmatic nuclei. *J Circadian Rhythms* 4:2.

Reppert, S. M., D. R. Weaver, S. A. Rivkees, E. G. Stopa. 1988. Putative melatonin receptors in a human biological clock. *Science* 242(4875):78–81.

Richardson, G. S., S. Wang-Weigand, J. Zhang, M. DeMicco. 2006. Long-term safety of ramelteon treatment in adults with chronic insomnia: A 1-year study. *Sleep* 29(Abstract Suppl):A233.

Richardson, G. S., P. C. Zee, S. Wang-Weigand, L. Rodriguez, X. Peng. 2008. Circadian phase-shifting effects of repeated ramelteon administration in healthy adults. *J Clin Sleep Med* 4:456–461.

Riemann, D., M. Berger, U. Voderholzer. 2001. Sleep and depression: Results from psychobiological studies. *Biol Psychol* 57:67–103.

Rosen, R. C., R. M. Lane, M. Menza. 1999. Effects of SSRIs on sexual function: A critical review. *J Clin Psychopharmacol* 19:67–85.

Roth, T., D. Seiden, S. Sainati, S. Wang-Weigand, J. Zhang, P. Zee. 2006. Effects of ramelteon on patient-reported sleep latency in older adults with chronic insomnia. *Sleep Med* 7:312–318.

Roth, T., D. Seiden, S. Wang-Weigand, J. Zhang. 2007. A 2-night, 3-period, crossover study of ramelteon's efficacy and safety in older adults with chronic insomnia. *Curr Med Res Opin* 23:1005–1014.

Roth, T., C. Stubbs, J. K. Walsh. 2005. Ramelteon (TAK-375), a selective MT1/MT2-receptor agonist, reduces latency to persistent sleep in a model of transient insomnia related to a novel sleep environment. *Sleep* 28:303–307.

San, L., B. Arranz. 2008. A novel mechanism of antidepressant action involving melatonergic serotonergic system. *Eur Psychiatry* 23:396–402.

Saper, C. B., J. Lu, T. C. Chou, J. Gooley. 2005. The hypothalamic integrator for circadian rhythms. *Trends Neurosci* 28:152–157.

Srinivasan, V., A. Brzezinski, D. W. Spence, S. R. Pandi-Perumal, R. Hardeland, G. M. Brown, D. P. Cardinali. 2010. Sleep, mood disorders and anti depressants the melatonergic antidepressant agomelatine offers a new strategy for treatment. *Psychiatria Fennica* 41:168–180.

Srinivasan, V., D. De Berardis, S. D. Shillcutt, A. Brzezinski. 2012a. Role of melatonin in mood disorders and the antidepressant effects of agomelatine. *Expert Opin Invest Drugs* 21:1503–1522.

Srinivasan, V., S. R. Pandi-Perumal, I. Trakht, D. W. Spence, R. Hardeland, B. Poeggeler, D. P. Cardinali. 2009b. Pathophysiology of depression: Role of sleep and the melatonergic system. *Psychiatr Res* 165(3):201–214.

Srinivasan, V., S. R. Pandi-Perumal, I. Trakht, D. W. Spence, B. Poeggeler, R. Hardeland, D. P. Cardinali. 2009a. Melatonin and melatonergic drugs on sleep: Possible mechanisms of action. *Int J Neurosci* 119:821–846.

Srinivasan, V., Z. Rahimah, O. Zahiruddin, E. C. Lauterbach, D. Acuna-Castroviejo. 2012c. Agomelatine in depressive disorders. Its novel mechanisms of action. *J Neuropsychiatry Clin Neurosci* 24:290–308.

Srinivasan, V., M. Smits, W. Spence, A. D. Lowe, L. Kayumov, S. R. Pandi-Perumal, B. Parry, D. P. Cardinali. 2006. Melatonin in mood disorders. *World J Biol Psychiatry* 7:138–151.

Srinivasan, V., R. Zakaria, Z. Othman, A. Brzezinski, A. Prasad, G. Brown. 2012b. Melatonergic drugs for therapeutic use in insomnia and sleep disturbances of mood disorders. *CNS Neurol Disord* 11:180–189.

Stahl, S. M., M. Fava, M. H. Trivedi, A. Caputo, A. Shah, A. Post. 2010. Agomelatine in the treatment of major depressive disorder: An 8 week, multicenter, randomized, placebo controlled trial. *J Clin Psychiatry* 71(5):616–626.

Steiger, A., M. Kimura. 2010. Wake and sleep EEG provide biomarkers in depression. *J Psychiatr Res* 44(4):242–252.

Thase, M. E. 2006. Pharmacotherapy of bipolar depression: An update. *Curr Psychiatry Rep* 8:478–488.

Uchimura, N., A. Ogawa, M. Hamamura, T. Hashimoto, H. Nagata, M. Uchiyama. 2011. Efficacy and safety of ramelteon in Japanese adults with chronic insomnia: A randomized, double-blind, placebo-controlled study. *Expert Rev Neurother* 11:215–224.

Uchiyama, M., M. Hamamura, T. Kuwano, H. Nagata, T. Hashimoto, A. Ogawa, N. Uchimura. 2011. Long-term safety and efficacy of ramelteon in Japanese patients with chronic insomnia. *Sleep Med* 12:127–133.

Van Reeth, O., L. Weibel, E. Olivares, S. Maccari, E. Mocaer, F. W. Turek. 2001. Melatonin or a melatonin agonist corrects age-related changes in circadian response to environmental stimulus. *Am J Physiol Regul Integr Comp Physiol* 80:1582–1591.

Van Someren, J. 2000. Circadian and sleep disturbances in the elderly. *Exp Gerontol* 35:1229–1237.

Wade, A. G., G. Crawford, I. Ford, A. McConnachie, T. Nir, M. Laudon, N. Zisapel. 2011. Prolonged release melatonin in the treatment of primary insomnia: Evaluation of the age cut-off for short- and long-term response. *Curr Med Res Opin* 27:87–98.

Wagner, G. C., J. D. Johnston, I. J. Clarke, G. A. Lincoln, D. G. Hazlerigg. 2008. Redefining the limits of day length responsiveness in a seasonal mammal. *Endocrinology* 149:32–39.

Wetterberg, L. 1998. Melatonin in adult depression. In *Melatonin in Psychiatric and Neoplastic Disorders*, eds. M. Shafii, S. L. Shafii, pp. 43–79. Washington, DC: American Psychiatry Press.

Winokur, A., K. A. Gary, S. Rodner, C. Rae-Red, A. T. Fernando, M. P. Szuba. 2001. Depression, sleep physiology and antidepressant drugs. *Depress Anxiety* 14:19–28.

Wu, Y. H., J. N. Zhou, R. Balesar, U. Unmehopa, A. Bao, R. Jockers, J. VanHeerikhuize, D. F. Swaab. 2006. Distribution of MT1 melatonin receptor immunoreactivity in the human hypothalamus and pituitary gland: Colocalization of MT1 with vasopressin, oxytocin, and corticotropin-releasing hormone. *J Comp Neurol* 499:897–910.

Ying, S. W., B. Rusak, B. Mocaer. 1998. Chronic exposure to melatonin receptor agonists does not alter their effects on suprachiasmatic nucleus neurons. *Eur J Pharmacol* 342:29–37.

Zajecka, J., A. Schatzberg, S. Stahl, A. Shah, A. Caputo, A. Post. 2010. Efficacy and safety of agomelatine in treatment of major depressive disorder: A multicenter, randomized double-blind, placebo controlled trial. *J Clin Psychopharmacol* 30(2):135–144.

Zammit, G., M. Erman, S. Wang-Weigand, S. Sainati, J. Zhang, T. Roth. 2007. Evaluation of the efficacy and safety of ramelteon in subjects with chronic insomnia. *J Clin Sleep Med* 3:495–504.

Zhdanova, I. V., R. J. Wurtman, H. J. Lynch, J. R. Ives, A. B. Dollins, C. Morabito, J. K. Matheson, D. L. Schomer. 1995. Sleep-inducing effects of low doses of melatonin ingested in the evening. *Clin Pharmacol Ther* 57:552–558.

Zhdanova, I. V., R. J. Wurtman, C. Morabito, V. R. Piotrovska, H. J. Lynch. 1996. Effects of low oral doses of melatonin, given 2–4 hours before habitual bedtime, on sleep in normal young humans. *Sleep* 19:423–431.

24 Agomelatine
A Neuroprotective Agent with Clinical Utility beyond Depression and Anxiety

Cecilio Álamo, Francisco López-Muñoz, and Pilar García-García

CONTENTS

24.1 INTRODUCTION

Agomelatine is a psychopharmacological agent, which acts as a powerful agonist of the MT_1 and MT_2 melatonin receptors ($K_i = 6.15 \times 10^{-11}$ M and 2.68×10^{-10} M, respectively). Their capacity for binding to the receptors is comparable to melatonin ($K_i = 8.52 \times 10^{-11}$ M and 2.63×10^{-10} M, respectively) (Ying et al. 1996; Álamo et al. 2008). On the other hand, agomelatine is an antagonist of the $5\text{-}HT_{2C}$ serotonergic receptors (2.7×10^{-7} M) (Millan et al. 2003). Furthermore, agomelatine lacks affinity for adrenergic, dopaminergic, muscarinic, or histaminergic receptors (Bourin et al. 2004), responsible in part for the adverse effects of other antidepressants. In summary, agomelatine has a different profile to the other antidepressants present in our pharmacological arsenal, both "for what it does," melatonergic agonist and $5\text{-}HT_{2C}$ antagonist, and for "what it does not do"; it does not inhibit the reuptake of monoamines and lacks the capacity to block other receptors (Álamo et al. 2008).

In this chapter, we target some aspects of agomelatine not related with its principal profile. Agomelatine has antidepressant and anxiolytic properties, aspects that have been discussed in other chapters of this book so that we will not go into details in this chapter.

24.2 SWITCHING OF THERAPEUTICS EFFECTS OF AGOMELATINE

For many years, there have been several theories to explain how antidepressants exert their effects beyond the structure of the synapse (López-Muñoz and Álamo 2009). In this regard, it is considered that the initial action of antidepressants on the synapse is similar to a switch that activates complex intracellular postsynaptic and extrasynaptic mechanisms involved in its therapeutic action. Thus, classical antidepressants act as a switch, for example, inhibition of monoamine uptake, for the activation of the "hard drive" of the slow machinery involved in their therapeutic effects.

In this way, agomelatine "switching on" the melatonergic receptors produces a decrease in AMPc and CREB (*cAMP related element binding*), which modifies the expression of early genes such as *c-fos* and *jun-B*. Furthermore, it inhibits the activity of protein kinases activated by various mitogenic factors (MAPkinase, $ERK_{1/2}$, $MEK_{1/2}$) and activates phospholipase C, which stimulates protein kinase C, facilitating the intracellular entry of calcium. Likewise, the stimulation of MT_1 receptors causes the stimulation of hyperpolarizing potassium currents through Kir3 channels coupled to the PG_i. On the other hand, MT_2 receptors inhibit the accumulation of GMPc. The set of all these data suggest the existence of a large number of cellular responses induced by melatonin, hence imitated by agomelatine, after acting on their high-affinity receptors (Guardiola-Lemaitre 2005; Álamo et al. 2008).

On the other hand, agomelatine "switching off" the $5\text{-}HT_{2C}$ receptors, since it acts as antagonist, inhibits the activity of the $G_{q/11}$ and G_{i3} proteins and antagonizes phosphoinositol depletion mediated by the activation of phospholipase C (Millan et al. 2003). This effect is not observed with melatonin, which lacks $5\text{-}HT_{2C}$ blocking capacity (Chagraoui et al. 2003).

Serotonin provides excitatory input to the $5\text{-}HT_{2C}$ receptors and it stimulates the GABA interneurons in prefrontal cortex, which in turn inhibit norepinephrine and dopamine cortical circuits. The antagonist effect of agomelatine on $5\text{-}HT_{2C}$ receptors removes the serotonergic inhibition, facilitating the release of dopamine and norepinephrine in the prefrontal cortex (Stahl 2007). This effect is specific, since it was not produced in the accumbens or in the striatal nucleus and it is due to $5\text{-}HT_{2C}$ antagonists since it is not modified by melatonin or melatonin antagonists. The increase in monoamines in the prefrontal cortex constitutes an important characteristic, which contributes to the antidepressant effect (Millan et al. 2000) and may be related to its anxiolytic profile (Millan 2005; Álamo et al. 2008; Stein et al. 2008).

In summary, we can consider that agomelatine stimulating receptors MT_1 and MT_2 and antagonizing $5\text{-}HT_{2C}$ receptor produces a cascade of biochemical events that may be involved in its therapeutics effects.

24.3 CHRONIC STRESS AND NEUROTOXICITY

Animal models of chronic stress represent a valuable tool to investigate behavioral, endocrinal, and neurobiological changes underlying stress-related psychopathologies (Nestler et al. 2002). The chronic mild stress (CMS) paradigm has been proposed as an experimental model of depression and mimics some of the dysfunctions associated with human depressive disorder, like anhedonia (Bessa et al. 2009). Moreover, CMS was shown to suppress adult hippocampal neurogenesis, suggested to play a role in depressive disorder (Elizalde et al. 2010; Dagyte et al. 2011). In addition, chronic stress exposure also causes atrophy of neurons in rodent prefrontal cortex and hippocampus, effects that could contribute to decreased volume of these regions reported in brain imaging studies of major depression disorder (MDD) patients (Li et al. 2011). It has been demonstrated that neural cells may react to chronic stress by destroying apical dendrites or with spine loss, and these changes are strictly associated to rest and activity daily periods (Pérez-Cruz et al. 2009).

In the physiopathology of depression, especially in those that progress with continued stress, three factors have greater significance. On the one hand, the hyperfunction of the hypothalamic-pituitary-adrenal (HPA) axis, which maintains high corticosteroid levels with an increase in the release of excitatory amino acids, has been considered neurotoxic factors. On the other hand, CMS induces a deficit of neurotrophins, in special BDNF (*brain-derived neurotrophic factor*), which limits this neuroprotector factor. Therefore, CMS can be involved in the physiopathology of depression and probably in other mental and neurologic status, increasing two neurotoxic factors, corticosteroids and excitatory amino acids, would act in a harmful manner on neuronal survival and neuroplasticity. Furthermore, the deficit of neurotrophins would not counteract the aggressive factors cited (Álamo and López-Muñoz 2009).

Sustained corticosteroid exposure is considered as one of the mediators of stress-induced changes in neurogenesis (Fuchs et al. 2001). Some researchers report mild-to-moderate increases in plasma corticosterone concentrations after exposure to CMS (Ushijima et al. 2006). A reduction in hippocampal volume might be due to glial and neuronal atrophy, which is related in part to increases in corticosteroids and excitatory amino acids (Fuchs and Flugge 1998; Banasr et al. 2006). In the dentate gyrus, there is a corticosteroid's important circadian rhythm involved in the regulation of cell proliferation. Increased corticosteroid suppresses mitosis, whereas removal of endogenous corticoid increases it above normal baseline levels (Wong and Herbert 2006; AlAhmed and Herbert 2008). The corticosteroid rhythm drives a corresponding rhythm in mitosis rates (Pinnock et al. 2006) and also seems to be essential for the action of other agents controlling neurogenesis. It appears that the corticosteroid rhythm opens gates to other systems regulating the rate of progenitor proliferation in the adult dentate gyrus (AlAhmed and Herbert 2008).

The second neurotoxic factor increased by chronic stress is glutamate. CMS produces transient glutamate effluxes in the hippocampus that remains constant in length and magnitude. After a subsequent single stress challenge, extracellular glutamate levels remained high in chronically stressed rats compared to naive rats that were subjected to the same acute stressor. In contrast, in the prefrontal cortex (PFC) glutamate levels decrease upon subsequent applications of acute stress. These results suggest a selective adaptation of glutamate release to stress in the PFC (Yamamoto and Reagan 2006; Popoli et al. 2011).

Early studies suggest that chronic stress, at least in the hippocampus, causes prolonged periods of stimulated glutamate release following acute stress exposure. Possibly as a compensatory response to elevated synaptic glutamate activity, there are changes in the surface expression of AMPA receptors (AMPAR) and NMDA receptors (NMDAR) subunits that seem to be associated with decreased transmission efficiency and potentially impaired synaptic plasticity. Last, there is growing evidence from animal studies that chronic stress has effects on glial cell morphology, metabolism, and function in the PFC and possibly also in the hippocampus (Popoli et al. 2011).

Furthermore, corticosteroids secreted during the diurnal rhythm and during stress affect the basal release of glutamate in several limbic and cortical areas, including the hippocampus, amygdale, and PFC. For example, in vivo microdialysis studies have shown that exposure of rats to tail-pinch, forced-swim, or restraint stress induces a marked and transitory increase of extracellular glutamate levels in the PFC (Bagley and Moghaddam 1997; Popoli et al. 2011).

On the other hand, the structure of neurons and the maintenance of their dendrite network and synaptic connectivity require neurotrophic support not only during development but also in adulthood. BDNF is a growth factor enriched in the rodent hippocampus that is released from neurons in an activity-dependent manner and plays a key role in neurogenesis, synaptic plasticity, differentiation of newly generated cells, and neuronal cell survival (Magariños et al. 2011).

CMS was shown to reduce expression of BDNF, and reduced neurotrophins expression has been repeatedly reported in the CMS model. The reduced expression of BDNF can contribute to structural anomalies and functional impairment in the central nervous system (Gronli et al. 2006; Dagyte et al. 2011). In the CA3 region of the hippocampus, the effects of chronic stress on the shrinkage of dendrites are mediated in part by BDNF (Magariños et al. 2011; Nowacka and Obuchowicz 2013).

In summary, chronic stress, excess concentrations of glutamate and corticosteroids, and by contrast, a deficit of BDNF affect the morphology of hippocampal neurons, resulting in a pronounced loss of apical dendrites.

24.4 AGOMELATINE, STRESS NEUROPROTECTION, AND NEUROGENESIS

In animal models of depression, neuroprotection, cell proliferation, and neurogenesis are generally reduced. These negatives effects are reversed by chronic antidepressant treatments (Soumier et al. 2008). There is a great interest in the effects of agomelatine, with a mechanism of action different from the rest of antidepressants (Álamo and López-Muñoz 2010), on neuroprotection, neurogenesis, and neuroplasticity.

Chronic treatment with agomelatine has been reported to increase progenitor cell proliferation and neurogenesis in the ventral (Banasr et al. 2006) and the dorsal part of the dentate gyrus of adults rats (AlAhmed and Herbert 2010). Dagyté et al. (2011) show that CMS affected distinct aspects of the neurogenesis process. CMS did not alter the rate of cell proliferation, but it decreased the newborn cell survival and the expression of doublecortin, a marker of dendrite growth of newly generated neurons, in dentate gyrus. Importantly, treatment with agomelatine interfered with these stress-associated changes in the brain. This antidepressant completely normalized stress-affected cell survival and partly reversed reduced doublecortin expression.

According to the biochemical changes caused by continued stress, it is important to highlight the effects of agomelatine on the excess of corticosteroids and glutamate and in the deficit of BDNF that affects the neuroplasticity of hippocampal neurons.

It is clearly established that stress, which stimulates the HPA axis, leading to high glucocorticoid levels, exerts an inhibitory influence on hippocampal neurogenesis via glucocorticoid receptor (GR) activation. Conversely, chronic administration of antidepressants enhances both GR expression and hippocampal cell proliferation, the latter effect resulting, at least in part, from an enhanced efficacy of the negative feedback control of the HPA axis (Duman et al. 2001).

Experimental mice with a genetic deficit of GR (GR-i) in the hypothalamus, hippocampus, and cortex exhibit a marked deficit in HPA axis feedback regulation similar to that observed in most depressed patients (Froger et al. 2004). In these GR-i mice, cell proliferation is significantly less in the dentate gyrus compared to control mice. Chronic treatment with agomelatine promoted cell proliferation in the hippocampus of GR-i mice. These data strongly suggest that the deficit in hippocampal neurogenesis observed in GR-i mice was, at least in part, related to a downregulation of GR. Agomelatine reversed these alterations and increased cell survival in the ventral hippocampus and counteracted GR downregulation (Païzanis et al. 2010).

In this sense, psychosocial stress in subordinate monkeys produces, in addition to a loss of synchronization in the circadian rhythms, a significant rise in urinary cortisol. The administration of agomelatine in prolonged treatment (4 weeks) normalizes the cortisol to the levels before the stressful situation. In parallel, agomelatine resynchronizes body temperature in these animals (Fuchs et al. 1998, 2006).

Furthermore, the proliferation of progenitor cell of the dorsal dentate gyrus induced by agomelatine requires the presence of an intact diurnal corticosterone rhythm. According to AlAhmed and Herbert (2010), melatonin was unable to alter progenitor mitosis rates, but SB242084, an antagonist 5-HT_{2C}, increased cell mitosis. More importantly, the combined administration of RO600175, a 5-HT_{2C} agonist, effectively prevented the stimulating action of agomelatine. These results indicate that the action of agomelatine on neurogenesis is not through melatonin receptors, but through 5-HT_{2C} receptors (AlAhmed and Herbert 2010).

The second aggressive factor cited would be the neurotoxicity produced by excitatory amino acids. Acute footshock-stress induces a marked increase of depolarization-evoked overflow of glutamate from prefrontal and frontal cortex synaptosomes, via glucocorticoid receptors activation and SNARE complex formed by two synaptic membrane proteins (syntaxin 1 or syntaxin 2 and SNAP25) and a vesicular protein (synaptobrevin 1 or synaptobrevin 2) accumulation in synaptic membranes. Glutamate induces an excitatory effect, in particular due to the action on NMDA receptors. Agomelatine lacks the capacity of blocking these receptors (Gressens et al. 2008), but it could decrease glutamate release up to 30%, which one would translate into less excitotoxic effect

(Bonanno et al. 2005). In fact, agomelatine blocks the increase in glutamate release induced by acute stress in the rat frontal and prefrontal cortex (Musazzi et al. 2010).

In line with these data, Dagyte et al. (2010) have found that agomelatine treatment reverses the reduction in Fos protein expression in the hippocampus of adult rats exposed to daily footshock-stress. As *c-fos* induction is an established marker of neuronal activation, these data demonstrate that agomelatine normalizes the pathological changes in neuronal activity induced by chronic stress.

Recent studies indicate that the repetitive administration of agomelatine (40 mg/kg, 2 weeks) completely inhibits the release of glutamate induced by stress in the frontal and prefrontal cortex of rats (Musazzi et al. 2010). These data suggest that agomelatine modulates glutamatergic transmission in hippocampus. Its action seems to be mediated by molecular mechanisms located on the presynaptic membrane and related with the size of the vesicle pool ready for release (Milanese et al. 2013). This effect that is produced by agomelatine, but not by melatonin or by the selective antagonist at 5-HT_{2C} receptors, S32006, suggests a potential synergy between melatonergic and serotonergic pathways in the action of the melatonergic antidepressant (Tardito et al. 2010). The reduced accumulation of SNARE complexes in presynaptic membranes suggests selected mechanisms in the exocytotic machinery as possible molecular targets of these drugs (Milanese et al. 2013).

Altogether, these studies show a protective action of agomelatine against the harmful effect of stress on the CNS. Agomelatine appears to correct the "allostatic" changes in neuroplasticity that develop to chronic stress, and eventually lead to behavioral abnormalities, mood disorder, and anxiety. The identification of the precise mechanisms by which activation of MT_1/MT_2 receptors and 5-HT_{2C} receptor blockade restores neuronal homeostasis is a challenge for the near future.

On the other hand, the influence of agomelatine (40 mg/kg, 3 weeks) on different cellular and molecular parameters related to neurogenesis has been studied on rats. In these conditions, agomelatine significantly increased BDNF levels in the hippocampus by 20%. Furthermore, agomelatine increased the $ERK_{1/2}$ signaling pathways (protein kinase activated by mitogens), Akt (protein kinase B), and GSK3β (glycogen synthetase kinase-3beta) in 91%, 45%, and 45%, respectively. It should be highlighted that these signal transduction pathways are involved in the control of neuronal proliferation and survival and are also modulated by antidepressant and mood stabilizing drugs. Furthermore, it is important to point out that these molecular changes are parallel to an increase in cell proliferation of 39% in the ventral hippocampus, induced by agomelatine administration. These effects, as occurred in the experimental models of antidepressants, are due to the joint melatonergic and 5-HT_{2C} antagonist properties exhibited by the agomelatine, since the melatonin or 5-HT_{2C} agonists lacked said effects (Soumier et al. 2008).

Calabrese et al. (2011) studied the influence of agomelatine in acute administration on the genetic expression of RNAm of BDNF and the gene related to the activity-regulated cytoskeleton-associated protein (Arc). More important effects were observed in the prefrontal cortex, in which a 46% decrease in the RNAm gene of BDNF was observed in the controls, but not in the animals treated with agomelatine, neither for BDNF gene nor for Arc. The acute upregulation of BDNF mRNA levels appears to be the result of a synergistic effect between the melatonergic properties of agomelatine as MT_1/MT_2 agonist and its serotonergic 5-HT_{2C} antagonism, since either melatonin or the 5-HT_{2C} antagonist S32006 does not mimic the effects of agomelatine (Molteni et al. 2010). These authors evaluate the influence of treatment for 21 days with agomelatine, venlafaxine, or a vehicle to the messenger RNA (mRNA) and protein expression of two neuroplastic markers, fibroblast growth factor (FGF-2) and Arc, in the hippocampus and prefrontal cortex. Agomelatine, but not venlafaxine, produced major transcriptional changes in the hippocampus, where significant upregulations of BDNF and FGF-2 were observed. Both drugs up-regulate the Arc transcription levels. The ability of agomelatine to modulate the expression of these neuroplastic molecules, which follows a circadian rhythm and required the action on melatonergic and 5-HT_{2C} receptors, may contribute to its antidepressant action (Calabrese et al. 2011).

These results seem to reveal that in the neuroprotective activity of agomelatine, its receptorial action as MT_1 and MT_2 receptor agonist and the 5-HT_{2C} receptor antagonist capacity are fundamental. Likewise, it is interesting to highlight that both actions set in motion transynaptic mechanisms responsible for a neuroprotector effect. In this effect, decreasing the negative effects of cortisol and glutamate and enhancing the neuroprotector effects, fundamentally via BDNF, induced by agomelatine are involved.

Furthermore, several studies showed that agomelatine influences neurogenesis, cell proliferation, maturation, and survival, in adult ventral hippocampus. The regulation of cell proliferation is mediated by the 5-HT_{2C} antagonism, whereas control of cell survival depends upon the joint action at MT_1/MT_2 and 5-HT_{2C} receptors. Recently it was reported that agomelatine, in the presence of an intact diurnal corticosterone rhythm, stimulates progenitor cell proliferation also in the neurogenic layer of the dorsal dentate gyrus in adult rats and that, again, this effect is mimicked by a selective5-HT_{2C} antagonist, but not by melatonin (AlAhmed and Herbert 2010).

On the other hand, adult male rats treated with agomelatine (40 mg/kg, i.p.) once a day for 22 days exerted procognitive and antidepressant activity and enhanced microtubule dynamics in the hippocampus and to a higher magnitude in the amygdala. By contrast, in the PFC, a decrease in microtubule dynamics was observed. Spinophilin, a dendritic spines marker, was decreased, and BDNF increased in the hippocampus. Presynaptic synaptophysin and spinophilin were increased in the PFC and amygdala, while PSD-95, a postsynaptic marker, was increased in the amygdala, consistent with the phenomena of synaptic remodeling. The modulation of cytoskeletal microtubule and synaptic markers may play a role in the pharmacological behavioral effects of agomelatine (Ladurelle et al. 2012).

These data provide new information regarding the molecular mechanisms that contribute to the chronic effects of the new antidepressant agomelatine on brain function. The ability of agomelatine to modulate the expression of these neuroplastic molecules, which follows a circadian rhythm, may contribute to its antidepressant action. The increase in monoamines in the PFC constitutes an important characteristic that contributes to the antidepressant effect (Millan et al. 2000) and may be related to its anxiolytic profile (Millan 2005; Álamo et al. 2008; Stein et al. 2008).

24.5 NO ANTIDEPRESSANT OR ANTIANXIETY EFFECTS OF AGOMELATINE

Agomelatine represents an innovative approach to treating depression as it is the first regulatory approved agent to incorporate a nonmonoaminergic mechanism. Extensive clinical trials have established both the short-term and long-term efficacy of agomelatine in major depression in mildly and severely ill patients, with an improvement of sleep quality, preservation of sexual function, absence of weight gain, and good tolerability (Bodinat et al. 2010). These antidepressant properties were demonstrated in several randomized placebo-controlled studies (Loo et al. 2002; Kennedy and Emsley 2006; Olie and Kasper 2007; Goodwin et al. 2009; Zajecka et al. 2010). Furthermore, agomelatine has demonstrated antidepressant properties in comparative studies with sertraline, fluoxetine, and venlafaxine as active controls (Kennedy et al. 2008; Hale et al. 2010; Kasper et al. 2010). On the other hand, in a small open add-on study to lithium or valproic acid, agomelatine shows efficacy for bipolar depression (Calabrese et al. 2007).

Preclinical and clinical data (Lôo et al. 2002; Lemoine et al. 2007; Hale et al. 2010) suggest that agomelatine may have anxiolytic properties in patients with MDD. Moreover, a randomized, double-blind, placebo-controlled trial in generalized anxiety disorder (GAD) suggests that agomelatine is effective in the treatment of anxiety (Stein et al. 2008) and in prevention of relapse in patients with GAD (Stein et al. 2012). Furthermore, some cases report suggest a potential role of agomelatine in some patients with obsessive–compulsive disorder (OCD) refractory to serotonin selective reuptake inhibitors (SSRIs) (Fornaro 2011a) and in a patient resistant to sertraline, clomipramine, risperidone, and aripiprazole (Da Rocha and Correa 2011). Moreover, there are some anecdotal cases about the efficacy of agomelatine in panic disorder (Fornaro 2011b), social anxiety disorder (Crippa et al. 2010), and posttraumatic stress disorder (De Berardis et al. 2012).

Antidepressant and anxiolytic properties of agomelatine are not the objective of this chapter. However, there is a rich potential, as highlighted earlier, for the exploration of the broader use of agomelatine in the treatment of other central nervous system disorders.

24.5.1 Agomelatine and Sleep Disorders

Disturbance of sleep–wake cycles is one of the core symptoms of MDD. Agomelatine improves disturbed sleep–wake cycles in depressed patients and improves both nighttime sleep and daytime function (Kupfer 2006; Quera-Salva et al. 2010).

On the other hand, animal studies have shown that agomelatine can resynchronize disrupted circadian rhythms and induce phase advancement (Zupancic and Guilleminault 2006). The chronobiotic effect of agomelatine is mediated through the suprachiasmatic nucleus and appears to be more prolonged than melatonin (Quera-Salva et al. 2010). These results are in accordance with the findings from studies on healthy human subjects, in whom evening administration of agomelatine induced a phase advance in circadian rhythm (body temperature, cortisol secretion, and growth hormone) (Leproult et al. 2005).

The effects of agomelatine on sleep appear to be linked not only to its action as a melatonergic agonist, but also to simultaneous activity as a 5-HT_{2C} receptor antagonist, both mechanisms being possibly synergic. In this context, alterations in slow wave sleep are thought to be predominantly mediated via the 5-HT_{2C} receptor subtype and agomelatine enhances restorative slow wave sleep (Bodinat et al. 2010).

Regardless of the sleep regulatory role of agomelatine in depressed patients, this agent may act on other sleep disorders not related to depression. Here we review some studies that show that agomelatine may be beneficial in these sleep disorders.

24.5.1.1 Rapid Eye Movement Sleep Behavior Disorder and Agomelatine

Rapid eye movement (REM) sleep behavior disorder is characterized by increased muscle tone during REM sleep, resulting in the patient acting out dreams with possible harmful consequences. It is diagnosed based on history and polysomnography findings, and treated with environmental safety measures and with symptomatic therapy with melatonin or clonazepam (Ramar and Olson 2013). To date no compound with convincing evidence of disease-modifying or neuroprotective efficacy has been identified in REM sleep behavior disorder patients, but these patients are considered ideal candidates for neuroprotective studies (Schenck et al. 2013). Since agomelatine has neuroprotective properties and is considered to improve sleep with a good side effect profile, mainly lack of sedation and morning somnolence (San and Arranz 2008; Álamo and López-Muñoz 2010), can be a candidate for the treatment of REM sleep behavior disorder.

Recently, Bonakis et al. (2012) studied three patients with a clinically and video-polysomnographically confirmed diagnosis of REM sleep behavior disorder whose conditions drastically improved in frequency and severity of episodes and in vivid and violent content of their dreams with agomelatine. The authors postulate that this improvement is due to the action of agomelatine on frontal and limbic structures, which are particularly active during the dream state. Agomelatine has been reported to enhance dopamine and noradrenaline activity in the frontal cortex and that long-term administration decreases depolarization-evoked release of glutamate, thus inhibiting excitatory neurotransmission in the rat hippocampus (Dubovsky and Warren 2009). Moreover, agomelatine's benefit in REM sleep disorders may be attributed also to its melatonergic action, through normalization of the circadian timing of REM sleep, which decreases REM sleep behaviors.

In conclusion, this small case series shows that agomelatine may be a promising treatment for REM sleep behavior disorder, but further controlled studies are needed to assess its efficacy in this parasomnia.

24.5.1.2 Agomelatine in Fatal Familial Insomnia

Fatal familial insomnia (FFI) is a rapidly progressive prion disease with profound disruption of circadian rhythmicity and associated neurocognitive symptoms. It has been suggested that some of the most distressing symptoms in FFI were secondary to chronic sleep deprivation and should therefore improve with a restoration of sleep (Harder et al. 1999).

Recently, Froböse et al. (2012) described a case of a young patient with FFI treated with agomelatine 25 mg to medicate nocturnal insomnia. Under this treatment, sleep efficiency was improved, slow wave sleep was high, and awakenings during sleep period time were far less than before. Clinically, the patient was less restless during nighttime. This case is promising but further studies will have to replicate this finding.

24.5.1.3 Agomelatine, Sleep Disorders, and Parkinson's Disease

Currently, there are no controlled clinical trials proving that agomelatine is effective for treating Parkinson's disease (PD). However, there are a number of arguments in favor of the efficacy of agomelatine in some of the nonmotor symptoms of PD and probably in its evolution.

Anxiety or depression alone, as well as comorbid, are associated with deleterious effects on physical and interpersonal functioning, negatively impacting quality of life and well-being of Parkinson's patients. The prevalence of depressive symptoms was present in 45% residents with PD in nursing home. Also, sleep disorders have been frequently reported in patients with PD. The most common sleep problems were overall poor nighttime sleep quality, daytime sleepiness, and nocturia (Weerkamp et al. 2013). Moreover, some idiopathic REM behavior disorders have been associated with a risk for developing neurodegenerative diseases, including PD (Bruin et al. 2012).

On the other hand, some of the components of the circadian system, including melatonin secretion, can be involved in the physiopathology of PD. Melatonin is a regulator of the sleep–wake cycle and also acts as an effective antioxidant and mitochondrial function protector. A reduction in the expression of melatonin MT_1 and MT_2 receptors has been documented in the substantia nigra of PD patients. The efficacy of melatonin for preventing neuronal cell death and for ameliorating PD symptoms has been demonstrated in animal models of PD employing neurotoxins. A small number of controlled trials (Dowling et al. 2005; Medeiros et al. 2007) indicate that melatonin is useful in treating disturbed sleep in PD, in particular REM behavior disorders (Srinivasan et al. 2011).

Taking all these data into consideration it is not surprising that agomelatine may have therapeutic potential in PD. As discussed earlier, agomelatine has antidepressant and anxiolytic properties, while improving sleep architecture and sleep–wake circadian rhythms. The use of agomelatine in PD may have considerable therapeutic potential because of its dual action for treating both the symptoms of depression and disturbed sleep. Since it also shares melatonin's chronobiotic properties, it may thus represent an ideal drug for treating PD (Srinivasan et al. 2011). It is also possible that the neuroprotective properties of agomelatine can improve the development of neurodegenerative diseases such as PD.

For now, the role of agomelatine in PD is speculative and should be confirmed with the necessary clinical studies. However, there is a preliminary open clinical trial with agomelatine (25 mg/day) in 13 patients with PD and depression that shows a statistically significant improvement over time in the Hamilton depression rating scale ($p<0.0005$), the SCOPA-S daytime sleepiness subscale ($p=0.004$), the Parkinson's disease sleep scale PDSS ($p=0.002$), and the UPDRS motor subscale ($p<0.0005$). If these preliminary results are confirmed, agomelatine ought to be considered a good option for treating depression in Parkinson's (Cardona et al. 2012). However, there are sufficient scientific and clinical arguments that stimulate the study of agomelatine in this disease and not just in motor symptomatology.

24.5.2 Agomelatine Antiaddictive Properties in Patients Treated with Benzodiazepines

There is some evidence from animal trials that concurrent use of agomelatine might reduce the consumption of benzodiazepines (Loiseau et al. 2006). On the other hand, agomelatine did not have reinforcing or discriminative stimulus effects similar to those of methohexital and diazepam, respectively (Wiley et al. 1998).

There are also hints that agomelatine could be clinically effective in the discontinuation of benzodiazepine use in patients (Morera-Fumero and Abreu-González 2010). Furthermore, recently published studies have shown the significantly superior anxiolytic effects of 25–50 mg of agomelatine when compared to placebo trials (Stein et al. 2008, 2012).

Müller et al. (2012) report three cases of benzodiazepine addicts, with histories of unsuccessful withdrawal attempts, who experienced marked reductions in craving and improved relapse prognoses under add-on administration of agomelatine. The extent to which this effect is due to the anticraving effects of agomelatine or its profile of receptor activation should be further investigated in larger clinical and experimental studies.

These cases demonstrate a possible area of use for the antidepressant agomelatine in the treatment of benzodiazepine withdrawal and addiction.

24.5.3 Agomelatine in Attention-Deficit Hyperactivity Disorder

Circadian rhythm problems are very common in children and adults with attention-deficit hyperactivity disorder (ADHD). ADHD patients are more often and more extreme evening chronotypes and more often meet the criteria for delayed sleep phase disorder (DSPD). Furthermore, some antidepressants, like tricyclic antidepressants or SSRIs (Quintana et al. 2007) and specially atomoxetine, have been reported to have some possible benefits in treating ADHD. This suggests that treatment with agomelatine, which regulates circadian rhythm, improved some sleep disorders and its antidepressant and antianxiety properties could decrease ADHD symptoms.

Moreover, many of the symptoms of ADHD are thought to arise from the dysfunction of the PFC and its connections with cortical and subcortical brain regions. Dopaminergic and adrenergic mechanisms in the PFC modulate cognitive–attentional performance, mood, and motor behavior. Current data suggest that the lateral PFC regions regulating attention and behavior are especially sensitive to the influence of noradrenaline and dopamine (Arnsten and Pliszka 2011). These findings may explain why all currently approved medications for ADHD increase or mimic noradrenaline and/or dopamine signaling. The stimulant medications and atomoxetine appear to enhance PFC function by indirectly increasing these catecholamine actions through blockade of noradrenaline and/or dopamine transporters (Koda et al. 2010). In contrast, guanfacine mimics the enhancing effects of noradrenaline at postsynaptic α_{2A}-receptors in the PFC, strengthening network connectivity (Arnsten 2011).

In this sense, agomelatine increases levels of serotonin and noradrenaline in PFC via a different mechanism of psychostimulants or atomoxetine. Serotonin provides excitatory input to the 5-HT_{2C} receptors and it stimulates the GABA interneurons in PFC that in turn inhibit noradrenaline and dopamine cortical circuits. The antagonist effect of agomelatine on 5-HT_{2C} receptors removes the serotonergic inhibition, facilitating the release of dopamine and noradrenaline in the PFC (Millan et al. 2003; Stahl 2007). This catecholamine increase in PFC induced by agomelatine, despite being through a different mechanism, could be the basis of a hypothetical benefit in patients with ADHD.

According to these clinical and experimental data, Niederhofer (2012a) conducted a clinical study versus placebo in 10 patients with ADHD (DSM-III-R). In general, agomelatine's effect was superior to placebo but seems to be less than that of first-line medications such as methylphenidate. The major beneficial effects of agomelatine are in behaviors reflecting high levels of arousal and activity and in cognitive aspects of inhibition and selective attention. In particular, a wide spectrum of ADHD patients with associated oppositional defiant behavior may respond preferentially to agomelatine. However, these promising results encourage controlled studies in larger number of patients in order to delineate the therapeutic profile of agomelatine in patients with ADHD. For now, the authors recommend the use of agomelatine if ADHD therapy with methylphenidate or atomoxetine is not indicated, for example, because of side effects, and if an ADHD patient suffers from additional sleep disorders. Furthermore, patients with ADHD might respond better to a combination of agomelatine and methylphenidate than each drug separately (Niederhofer 2012b).

24.5.4 Agomelatine, Nociception, and Analgesia

It has been reported that perception of pain has a circadian rhythm both in rodents and humans. Patients suffer less pain during dark photoperiod and it is attributed to high melatonin levels occurring at night and their possible analgesic effects (Laikin et al. 1981; Srinivasan et al. 2012). Melatonin's antinociceptive effects have been demonstrated in animals like mice and rats by using a number of experimental models of nociception, and this has been brought out extensively in an earlier review (Srinivasan et al. 2010).

On the other hand, there is a common relationship between pain and depression, and it is suggested that a common physiopathology may perhaps underlie these disorders. Simultaneous effectiveness against pain and depression is present in conventional antidepressants. Furthermore, melatonin has shown analgesic activity both experimentally (Srinivasan et al. 2012) and in different clinical entities: fibromyalgia (Citera et al. 2000; Acuña-Castroviejo et al. 2006; Reiter et al. 2007; Hussain et al. 2011), irritable bowel syndrome (Lu et al. 2005; Song et al. 2005), and migraine (Claustrat et al. 1997). Moreover, many studies showed that using variable doses of melatonin (3–6 mg/day) in subjects affected from fibromyalgia had significantly been effective on pain, sleep, daytime fatigue, and depression (Bruno et al. 2013).

The mechanism of the analgesic effect of melatonin is complex, being involved cholinergic receptors and α_1-receptors and α_2-adrenergic receptors located on the spinal cord. Moreover, as has been demonstrated in a number of experimental studies on animals, melatonin shows a direct antinociceptive effect through the activation of MT_1 and MT_2 melatonin receptors. This effect is inhibited by luzindol, an antagonistic of MT_1 and MT_2 receptors. The localization of these receptors in the nociceptive control key points, as thalamus, hypothalamus, dorsal horn of the spinal cord, spinal trigeminal tract, and trigeminal nucleus, suggest that antinociceptive actions of melatonin are mediated through melatonergic receptors (Srinivasan et al. 2012).

As discussed, agomelatine has been shown to be effective in treating patients with MDD and has high affinity on MT_1 and MT_2 receptors, which suggests its potential analgesic efficacy. Use of agomelatine may be effective in the treatment of MDD associated with painful conditions and it may offer the potential for effective management and control of neuropathic pain (Srinivasan et al. 2012).

The efficacy of agomelatine on depression, anxiety, cognition, and pain in a sample of drug-free patients with fibromyalgia has been evaluated. Agomelatine was administered at the single daily dose of 25 mg/day to 15 fibromyalgia "drug-free" female subjects during 12 weeks. Treatment with agomelatine significantly improved depression, anxiety, and pain in patients with fibromyalgia and was well tolerated. Further research is needed to fully evaluate the role of agomelatine as a potential pharmacological strategy for the treatment of fibromyalgia (Bruno et al. 2013). Furthermore, the clinical efficacy of agomelatine (25 mg/day) as an adjunct in fibromyalgia has been observed in a patient not responding to treatment with amitriptyline and analgesics (Medina-Ortiz et al. 2013).

Lal (2011) shows the clinical experience of the use of agomelatine in the treatment of patients with migraine and fibromyalgia, comorbid with neurological diseases, insomnia, and depression. In this complex clinical entity, agomelatine demonstrated its efficacy and safety. In other experimental and pivotal clinical trials, agomelatine (25 mg/day/6 months) administered to 20 patients suffering from migraine attacks decreased both the frequency and duration of migraine attacks and thus reduced the intensity of pain in these patients. Moreover, it also reduced significantly the severity of depression and normalized sleep disturbances (Tabeeva et al. 2011). Recently, Guglielmo et al. (2013) report two cases of patients with migraine successfully treated with agomelatine; one patient presented with comorbid depression, whereas the other had no comorbidities. Despite the limited clinical experience with agomelatine in the treatment of migraine, given its specific mechanism of action and similarity with melatonin, preclinical and clinical data suggest its possible role in the treatment of this disease.

24.5.5 Agomelatine and Anticonvulsant Properties

The majority of data indicate anticonvulsant properties of melatonin when applied at pharmacological dose in both animal models (Banach et al. 2011) and clinical investigations (Peled et al. 2001). Furthermore, melatonin was reduced in patients with epilepsy at baseline compared with controls, and increased threefold following seizures (Bazil et al. 2000). However, a few experimental studies have shown direct or indirect proconvulsant effects of melatonin (Banach et al. 2011). Antiepileptic activity of melatonin has been associated with its antioxidant activity as a free radical scavenger (Peled et al. 2001).

Agomelatine show anticonvulsant properties in mouse models of pentylenetetrazole- and pilocarpine-induced convulsions. However, in the strychnine-, electroshock-, and picrotoxin-induced seizure models, agomelatine caused no significant alterations in latency to convulsions and in time until death when compared to controls (Aguiar et al. 2012). Agomelatine has antioxidant activity in pilocarpine-induced seizure models. However, neither melatonin nor agomelatine has shown antioxidant effects on other seizure models when compared to controls (Aguiar et al. 2013).

The clinical efficacy of agomelatine as anticonvulsant is anecdotal. One case of a 42-year-old female patient with treatment-resistant chronic posthypoxic myoclonus improved with administration of the drug agomelatine (González de la Aleja et al. 2012).

24.6 CONCLUSIONS

Agomelatine is an antidepressant agent with a highly original pharmacological profile. Agomelatine is a melatonergic receptor (MT_1/MT_2) agonist and 5-HT_{2C} receptor antagonist that has showed antidepressant efficacy in animal models and clinical trials. Agomelatine is the first regulatory approved agent to incorporate a nonmonoaminergic mechanism.

Several experimental studies indicate that its antidepressant activity across a broad range of experimental procedures in animal models and its distinctive therapeutic profile in humans probably reflect a synergistic interplay of its melatonergic (agonist) and 5-HT_{2C} (antagonist) properties. Furthermore, agomelatine controls biological rhythms, reverses the effects of stress, and enhances neuroplasticity mechanisms and adult neurogenesis. These data may explain its antidepressant, anxiolytic, and sleep-regulating properties but also may be the basis for other therapeutic effects as those discussed in this chapter. However, for these potential indications, clinical studies with higher scientific level are required.

REFERENCES

Acuña-Castroviejo, D., G. Escames, and R.J. Reiter. 2006. Melatonin therapy for fibromyalgia. *J Pineal Res* 40: 98–99.

Aguiar, C.C., A.B. Almeida, P.V. Araújo et al. 2012. Anticonvulsant effects of agomelatine in mice. *Epilepsy Behav* 24: 324–328.

Aguiar, C.C., A.B. Almeida, P.V. Araújo et al. 2013. Effects of agomelatine on oxidative stress in the brain of mice after chemically induced seizures. *Cell Mol Neurobiol* 33: 825–835.

AlAhmed, S. and J. Herbert. 2008. Strain differences in proliferation of progenitor cells in the dentate gyrus of the adult rat and the response to fluoxetine are dependent on corticosterone. *Neuroscience* 157: 677–682.

AlAhmed, S. and J. Herbert. 2010. Effect of agomelatine and its interaction with the daily corticosterone rhythm on progenitor cell proliferation in the dentate gyrus of the adult rat. *Neuropharmacology* 59: 375–379.

Álamo, C. and F. López-Muñoz. 2009. New antidepressant drugs: Beyond monoaminergic mechanisms. *Curr Pharm Des* 15: 1559–1562.

Álamo, C. and F. López-Muñoz. 2010. Optimizando el tratamiento de los pacientes deprimidos. Depresión y ritmos circadianos: Relación farmacológica. El papel de la agomelatina. *Rev Psiquiatr Salud Ment (Barc)* 3: 3–11.
Álamo, C., F. López-Muñoz, and M.J. Armada. 2008. Agomelatina: Un nuevo enfoque farmacológico en el tratamiento de la depresión con traducción clínica. *Psiquiatr Biol* 15: 125–139.
Arnsten, A.F. and S.R. Pliszka. 2011. Catecholamine influences on prefrontal cortical function: Relevance to treatment of attention deficit/hyperactivity disorder and related disorders. *Pharmacol Biochem Behav* 99: 211–216.
Arnsten, A.F.T. 2011. Catecholamine influences on dorsolateral prefrontal cortical networks. *Biol Psychiatry* 69: e89–e99.
Bagley, J. and B. Moghaddam. 1997. Temporal dynamics of glutamate efflux in the prefrontal cortex and in the hippocampus following repeated stress: Effects of pretreatment with saline or diazepam. *Neuroscience* 77: 65–73.
Banach, M., E. Gurdziel, M. Jędrych, and K.K. Borowicz. 2011. Melatonin in experimental seizures and epilepsy. *Pharmacol Rep* 63: 1–11.
Banasr, M., M. Hery, E. Mocaer, and A. Daszuta. 2006. Agomelatine, a new antidepressant drug, increases cell proliferation, maturation and survival of newly generated granule cells in adult hippocampus. *Biol Psychiatry* 59: 1087–1096.
Bazil, C.W., D. Short, D. Crispin, and W. Zheng. 2000. Patients with intractable epilepsy have low melatonin, which increases following seizures. *Neurology* 55: 1746–1748.
Bessa, J.M., D. Ferreira, I. Melo et al. 2009. The mood improving actions of antidepressants do not depend on neurogenesis but are associated with neuronal remodeling. *Mol Psychiatry* 14: 764–773.
Bodinat, C., B. Guardiola-Lemaitre, E. Mocaër et al. 2010. Agomelatine, the first melatonergic antidepressant: Discovery, characterization and development. *Nat Rev Drug Discov* 9: 628–642.
Bonakis, A., N.T. Economou, S.G. Papageorgiou et al. 2012. Agomelatine may improve REM sleep behavior disorder symptoms. *J Clin Psychopharmacol* 32: 732–734.
Bonanno, G., R. Giambelli, L. Raiteri et al. 2005. Chronic antidepressants reduce depolarization-evoked glutamate release and protein interactions favouring formation of SNARE complex in hippocampus. *J Neurosci* 25: 3270–3279.
Bourin, M., E. Mocaer, and R. Porsolt. 2004. Antidepressant-like activity of S 20098 (agomelatine) in the forced swimming test in rodents: Involvement of melatonin and serotonin receptors. *J Psychiatry Neurosci* 29: 126–133.
Bruin, V.M., L.R. Bittencourt, and S. Tufik. 2012. Sleep-wake disturbances in Parkinson's disease: Current evidence regarding diagnostic and therapeutic decisions. *Eur Neurol* 67: 257–267.
Bruno, A., U. Micò, S. Lorusso et al. 2013. Agomelatine in the treatment of fibromyalgia: A 12-week, open-label, uncontrolled preliminary study. *J Clin Psychopharmacol* 33: 507–511.
Calabrese, F., R. Molteni, C. Gabriel et al. 2011. Modulation of neuroplastic molecules in selected brain regions after chronic administration of the novel antidepressant agomelatine. *Psychopharmacology (Berl)* 215: 267–275.
Calabrese, J.R., J.D. Guelfi, C. Perdrizet-Chevallier, and Agomelatine Bipolar Study Group. 2007. Agomelatine adjunctive therapy for acute bipolar depression: Preliminary open data. *Bipolar Disord* 9: 628–635.
Cardona, X., A. Avila, M. Martín-Baranera et al. 2012. Agomelatine for the treatment of depression in Parkinson's disease patients. *8th International Congress in Mental Dysfunction & Other Non-Motor Features in Parkinson's Disease and Related Disorders*, Berlin, Germany, May 3–6, 2012.
Chagraoui, A., P. Protais, T. Filloux, and E. Mocaër. 2003. Agomelatine (S 20098) antagonizes the penile erections induced by the stimulation of 5-HT2C receptors in Winstar rats. *Psychopharmacology (Berl)* 170: 17–22.
Citera, G., M.A. Arias, J.A. Maldonado-Cocco et al. 2000. The effect of melatonin in patients with fibromyalgia: A pilot study. *Clin Rheumatol* 19: 9–13.
Claustrat, B., J. Brun, M. Geoffriau et al. 1997. Nocturnal plasma melatonin profile and melatonin kinetics during infusion in status migrainosus. *Cephalalgia* 17: 511–517.
Crippa, J.A., J.E. Hallak, A.W. Zuardi et al. 2010. Agomelatine in the treatment of social anxiety disorder. *Prog Neuropsychopharmacol Biol Psychiatry* 34: 1357–1358.
Dagyte, G., I. Crescente, F. Postema et al. 2011. Agomelatine reverses the decrease in hippocampal cell survival induced by chronic mild stress. *Behav Brain Res* 218: 121–128.
Dagyte, G., A. Trentani, F. Postema et al. 2010. The novel antidepressant agomelatine normalizes hippocampal neuronal activity and promotes neurogenesis in chronically stressed rats. *CNS Neurosci Ther* 16: 195–207.

Da Rocha, F.F. and H. Correa. 2011. Is circadian rhythm disruption important in obsessive-compulsive disorder (OCD)? A case of successful augmentation with agomelatine for the treatment of OCD. *Clin Neuropharmacol* 34: 139–140.

De Berardis, D., N. Serroni, S. Marini et al. 2012. Agomelatine for the treatment of posttraumatic stress disorder: A case report. *Ann Clin Psychiatry* 24: 241–242.

Dowling, G.A., J. Mastick, J. Colling et al. 2005. Melatonin for sleep disturbances in Parkinson's disease. *Sleep Med* 6: 459–466.

Duman, R.S., J. Malberg, and S. Nakagawa. 2001. Regulation of adult neurogenesis by psychotropic drugs and stress. *J Pharmacol Exp Ther* 299: 401–407.

Dubovsky, S.L. and C. Warren. 2009. Agomelatine, a melatonin agonist with antidepressant properties. *Expert Opin Investig Drugs* 18: 1533–1540.

Elizalde, N., A.L. García-García, S. Totterdell et al. 2010. Sustained stress-induced changes in mice as a model for chronic depression. *Psychopharmacology (Berl)* 210: 393–406.

Fornaro, M. 2011a. Switching from serotonin reuptake inhibitors to agomelatine in patients with refractory obsessive-compulsive disorder: A 3 month follow-up case series. *Ann Gen Psychiatry* 10: 5.

Fornaro, M. 2011b. Agomelatine in the treatment of panic disorder. *Prog Neuropsychopharmacol Biol Psychiatry* 35: 286–287.

Froböse, T., H. Slawik, R. Schreiner et al. 2012. Agomelatine improves sleep in a patient with fatal familial insomnia. *Pharmacopsychiatry* 45: 34–36.

Froger, N., E. Palazzo, C. Boni et al. 2004. Neurochemical and behavioral alterations in glucocorticoid receptor impaired transgenic mice after chronic mild stress. *J Neurosci* 24: 2787–2796.

Fuchs, E. and G. Flugge. 1998. Stress, glucocorticoids and structural plasticity of the hippocampus. *Neurosci Biobehav Rev* 23: 295–300.

Fuchs, E., G. Flugge, F. Ohl, P. Lucassen et al. 2001. Psychosocial stress, glucocorticoids, and structural alterations in the tree shrew hippocampus. *Physiol Behav* 73: 285–291.

Fuchs, E., M. Simon, and B. Schmelting. 2006. Pharmacology of a new antidepressant: Benefit of the implication of the melatonergic system. *Int Clin Psychopharmacol* 21: S17–S20.

González de la Aleja, J., R.A. Saiz-Díaz, and P. de la Peña. 2012. Relief of intractable posthypoxic myoclonus after administration of agomelatine. *Clin Neuropharmacol* 35: 258–259.

Goodwin, G.M., R. Emsley, S. Rembry, F. Rouillon, and Agomelatine Study Group. 2009. Agomelatine prevents relapse in patients with major depressive disorder without evidence of a discontinuation syndrome: A 24-week randomized, double-blind, placebo-controlled trial. *J Clin Psychiatry* 70: 1128–1137.

Gressens, P., L. Schwendimann, I. Husson et al. 2008. Agomelatine, a melatonin receptor agonist with 5-HT(2C) receptor antagonist properties, protects the developing murine white matter against excitoxicity. *Eur J Pharmacol* 588: 58–63.

Gronli, J., C. Bramham, R. Murison et al. 2006. Chronic mild stress inhibits BDNF protein expression and CREB activation in the dentate gyrus but not in the hippocampus proper. *Pharmacol Biochem Behav* 85: 842–849.

Guardiola-Lemaitre, B. 2005. Agonistes et antagonistes des récepteurs mélatoninergiques: Effets pharmacologiques et perspectives thérapeutiques. *Ann Pharm Fr* 63: 385–400.

Guglielmo, R., G. Martinotti, M. Di Giannantonio, and L. Janiri. 2013. A possible new option for migraine management: Agomelatine. *Clin Neuropharmacol* 36: 65–67.

Hale, A., R. Corral, C. Mencacci et al. 2010. Superior antidepressant efficacy results of agomelatine versus fluoxetine in severe MDD patients: A randomized, double-blind study. *Int Clin Psychopharmacol* 25: 305–314.

Harder, A., K. Jendroska, F. Kreuz et al. 1999. Novel twelve-generation kindred of fatal familial insomnia from Germany representing the entire spectrum of disease expression. *Am J Med Gen* 87: 311–316.

Hussain, S.A., H. Al-Khalifa, N.A. Jasim, and F.I. Gorial. 2011. Adjuvant use of melatonin for treatment of fibromyalgia. *J Pineal Res* 50: 267–271.

Kasper, S., G. Hajak, K. Wulff et al. 2010. Efficacy of the novel antidepressant agomelatine on the circadian rest-activity cycle and depressive and anxiety symptoms in patients with major depressive disorder: A randomized, double-blind comparison with sertraline. *J Clin Psychiatry* 71: 109–120.

Kennedy, S.H. and R.A. Emsley. 2006. Placebo controlled trial of agomelatine in the treatment of major depressive disorder. *Eur Neuropsychopharmacol* 16: 93–100.

Kennedy, S.H., S. Rizvi, K. Fulton, and J. Rasmussen. 2008. A double-blind comparison of sexual functioning, antidepressant efficacy, and tolerability between agomelatine and venlafaxine XR. *J Clin Psychopharmacol* 28: 329–333.

Koda, K., Y. Ago, Y. Cong, Y. Kita et al. 2010. Effects of acute and chronic administration of atomoxetine and methylphenidate on extracellular levels of noradrenaline, dopamine and serotonin in the prefrontal cortex and striatum of mice. *J Neurochem* 114: 259–270.

Kupfer, D.J. 2006. Depression and associated sleep disturbances: Patient benefits with agomelatine. *Eur Neuropsychopharmacol* 16: S639–S643.

Ladurelle, N., C. Gabriel, A. Viggiano et al. 2012. Agomelatine (S20098) modulates the expression of cytoskeletal microtubular proteins, synaptic markers and BDNF in the rat hippocampus, amygdala and PFC. *Psychopharmacology (Berl)* 221: 493–509.

Laikin, M.I., C.H. Miller, M.L. Stott, and W.D. Winters. 1981. Involvement of the pineal gland and melatonin in murine analgesia. *Life Sci* 29: 2543–2551.

Lal, L. 2011. Clinical experience in using agomelatine (valdoxan) in the neurological practice. *Zh Nevrol Psikhiatr Im S S Korsakova* 111(11 Pt 1): 25–28.

Lemoine, P., C. Guilleminault, and E. Alvarez. 2007. Improvement in subjective sleep in major depressive disorder with a novel antidepressant, agomelatine: Randomized, double-blind comparison with venlafaxine. *J Clin Psychiatry* 68: 1723–1732.

Leproult, R., A. Van Onderbergen, M. L'Hermite-Baleriaux et al. 2005. Phase-shifts of 24h rhythms of hormonal release and body temperature following early evening administration of the melatonin agonist agomelatine in healthy older men. *Clin Endocrinol (Oxf)* 63: 298–304.

Li, N., R.J. Liu, J.M. Dwyer, M. Banasr et al. 2011. Glutamate *N*-methyl-D-aspartate receptor antagonists rapidly reverse behavioral and synaptic deficits caused by chronic stress exposure. *Biol Psychiatry* 69: 754–761.

Loiseau, F., C. Le Bihan, M. Hamon et al. 2006. Effects of melatonin and agomelatine in anxiety-related procedures in rats: Interaction with diazepam. *Eur Neuropsychopharmacol* 16: 417–428.

Lôo, H., A. Hale, and H. D'Haenen. 2002. Determination of the dose of agomelatine, a melatoninergic agonist and selective 5-HT2C antagonist, in the treatment of major depressive disorder: A placebo controlled dose range study. *Int Clin Psychopharmacol* 17: 239–247.

López-Muñoz, F. and C. Alamo. 2009. Monoaminergic neurotransmission: The history of the discovery of antidepressants from 1950s until today. *Curr Pharm Des* 15: 1563–1586.

Lu, W.Z., K.A. Gwee, S. Moochhalla, and Y.Y. Ho. 2005. Melatonin improves bowel symptoms in female patients with irritable bowel syndrome: A double blind placebo controlled study. *Aliment Pharmacol Ther* 22: 927–934.

Magariños, A.M., C.J. Li, J. Gal Toth et al. 2011. Effect of brain-derived neurotrophic factor haploinsufficiency on stress-induced remodeling of hippocampal neurons. *Hippocampus* 21: 253–264.

Medeiros, C.A., P.F. Carvalhedo de Bruin, L.A. Lopes et al. 2007. Effect of exogenous melatonin on sleep and motor dysfunction in Parkinson's disease: A randomized, double blind, placebo-controlled study. *J Neurol* 254: 459–464.

Medina-Ortiz, O., G. Rico, L. Oliveros, and N. Sánchez-Mora. 2013. Uso de agomelatina como tratamiento coadyuvante en la fibromialgia. *Reumatol Clin* 9: 328–329.

Milanese, M., D. Tardito, L. Musazzi et al. 2013. Chronic treatment with agomelatine or venlafaxine reduces depolarization-evoked glutamate release from hippocampal synaptosomes. *BMC Neurosci* 14: 75.

Millan, M.J. 2005. Serotonin 5-HT2C receptors as a target for the treatment of depressive and anxious states: Focus on novel therapeutics strategies. *Therapie* 60: 441–460.

Millan, M.J., A. Gobert, F. Lejeune et al. 2003. The novel melatonin agonist agomelatine (S20098) is an antagonist at 5-hydroxy-tryptamine 2C receptors, blockade of which enhances the activity of frontocortical dopaminergic and adrenergic pathways. *J Pharmacol Exp Ther* 306: 954–964.

Millan, M.J., F. Lejeune, and A. Gobert. 2000. Reciprocal autoreceptor and heteroreceptor control of serotonergic dopaminergic and noradrenergic transmission in the frontal cortex: Relevance to the actions of antidepressant agents. *J Psychopharmacol* 14: 114–138.

Molteni, R., F. Calabrese, S. Pisoni et al. 2010. Synergistic mechanisms in the modulation of the neurotrophin BDNF in the rat prefrontal cortex following acute agomelatine administration. *World J Biol Psychiatry* 11: 148–153.

Morera-Fumero, A.L. and P. Abreu-González. 2010. Diazepam discontinuation through agomelatine in schizophrenia with insomnia and depression. *J Clin Psychopharmacol* 30: 739–741.

Müller, H., F. Seifert, J.M. Maler et al. 2012. Agomelatine reduces craving in benzodiazepine addicts: A follow-up examination of three patients. *Singapore Med J* 53: e228–e230.

Musazzi, L., M. Milanese, P. Farisello et al. 2010. Acute stress increases depolarization-evoked glutamate release in the rat prefrontal/frontal cortex: The dampening action of antidepressants. *PLoS One* 5: e8566.

Nestlerm E.J., R. Gould, H. Manji et al. 2002. Preclinical models: Status of basic research in depression. *Biol Psychiatry* 52: 503–528.

Niederhofer, H. 2012a. Agomelatine treatment with adolescents with ADHD. *J Atten Disord* 16: 530–532.

Niederhofer, H. 2012b. Treating ADHD with agomelatine. *J Atten Disord* 16: 346–348.

Nowacka, M. and E. Obuchowicz. 2013. BDNF and VEGF in the pathogenesis of stress-induced affective diseases: An insight from experimental studies. *Pharmacol Rep* 65: 535–546.

Olié, J.P. and S. Kasper. 2007. Efficacy of agomelatine, a MT1/MT2 receptor agonist with 5-HT2C antagonistic properties, in major depressive disorder. *Int J Neuropsychopharmacol* 10: 661–673.

Païzanis, E., T. Renoir, V. Lelievre et al. 2010. Behavioural and neuroplastic effects of the new-generation antidepressant agomelatine compared to fluoxetine in glucocorticoid receptor-impaired mice. *Int J Neuropsychopharmacol* 13: 759–774.

Peled, N., Z. Shorer, E. Peled, and G. Pillar. 2001. Melatonin effect on seizures in children with severe neurologic deficit disorders. *Epilepsia* 42: 1208–1210.

Pérez-Cruz, C., M. Simon, B. Czéh et al. 2009. Hemispheric differences in basilar dendrites and spines of pyramidal neurons in the rat prelyimbic cortex: Activity- and stress-induced changes. *Eur J Neurosci* 29: 738–747.

Pinnock, S.B., R. Balendra, M. Chan et al. 2006. Interactions between nitric oxide and corticosterone in the regulation of progenitor cell proliferation in the dentate gyrus of the adult rat. *Neuropsychopharmacology* 32: 493–504.

Popoli, M., Z. Yan, B.S. McEwen, and G. Sanacora. 2011. The stressed synapse: The impact of stress and glucocorticoids on glutamate transmission. *Nat Rev Neurosci* 13: 22–37.

Quera-Salva, M.A., P. Lemoine, and C. Guilleminault. 2010. Impact of the novel antidepressant agomelatine on disturbed sleep-wake cycles in depressed patients. *Hum Psychopharmacol* 25: 222–229.

Quintana, H., G.J. Butterbaugh, W. Purnell, and A.K. Layman. 2007. Fluoxetine monotherapy in attention-deficit/hyperactivity disorder and comorbid non-bipolar mood disorders in children and adolescents. *Child Psychiatr Hum Dev* 37: 241–253.

Ramar, K. and E.J. Olson. 2013. Management of common sleep disorders. *Am Fam Physician* 88: 231–238.

Reiter, R.J., D. Acuña-Castroviejo, and D.X. Tan. 2007. Melatonin therapy for fibromyalgia. *Curr Pain Headache Rep* 11: 339–342.

San, L. and B. Arranz. 2008. Agomelatine: A novel mechanism of antidepressant action involving the melatonergic and the serotonergic system. *Eur Psychiatry* 6: 396–402.

Schenck, C.H., J.Y. Montplaisir, B. Frauscher et al. 2013. Rapid eye movement sleep behavior disorder: Devising controlled active treatment studies for symptomatic and neuroprotective therapy—A consensus statement from the International Rapid Eye Movement Sleep Behavior Disorder Study Group. *Sleep Med* 14: 795–806.

Song, G.H., P.H. Leng, K.A. Gwee et al. 2005. Melatonin improves abdominal pain in irritable bowel syndrome patients who have sleep disturbances: A randomised, double blind, placebo controlled study. *Gut* 54: 1402–1407.

Srinivasan, V., D.P. Cardinali, U.S. Srinivasan et al. 2011. Therapeutic potential of melatonin and its analogs in Parkinson's disease: Focus on sleep and neuroprotection. *Ther Adv Neurol Disord* 4: 297–317.

Srinivasan, V., E.C. Lauterbach, K.Y. Ho et al. 2012. Melatonin in antinociception: Its therapeutic applications. *Curr Neuropharmacol* 10: 167–178.

Srinivasan, V., S.R. Pandi-Perumal, D.W. Spence et al. 2010. Potential use of melatonergic drugs in analgesia. *Brain Res Bull* 81: 362–371.

Soumier, A., S. Lortet, C. Gabriel et al. 2008. Cellular and molecular mechanisms underlying increased adult hippocampal neurogenesis induced by agomelatine. *Eur Neuropsychopharmacol* 18: S350.

Stahl, S.M. 2007. Novel mechanism of antidepressant action: Norepinephrine and dopamine disinhibition (NDDI) plus melatonergic agonism. *Int J Neuropsychopharmacol* 10: 575–578.

Stein, D.J., A.A. Ahokas, C. Albarran et al. 2012. Agomelatine prevents relapse in generalized anxiety disorder: A 6-month randomized, double-blind, placebo-controlled discontinuation study. *J Clin Psychiatry* 73: 1002–1008.

Stein, D.J., A.A. Ahokas, and C. de Bodinat. 2008. Efficacy of agomelatine in generalized anxiety disorder: A randomized, double-blind, placebo-controlled study. *J Clin Psychopharmacol* 28: 561–566.

Tabeeva, G.R., A.V. Sergeev, and S.A. Gromova. 2011. Possibilities of preventive treatment of migraine with MT1 and MT2 agonist and 5-HT2c receptor antagonist agomelatin (valdoxan). *Zh Neurol Psikhiatr Im S S Korsakova* 111(9): 32–36.

Tardito, D., M. Milanese, T. Bonifacino et al. 2010. Blockade of stress-induced increase of glutamate release in the rat prefrontal/frontal cortex by agomelatine involves synergy between melatonergic and 5-HT2C receptor-dependent pathways. *BMC Neurosci* 11: 68.

Ushijima, K., T. Morikawa, H. To et al. 2006. Chronobiological disturbances with hyperthermia and hypercortisolism induced by chronic mild stress in rats. *Behav Brain Res* 173: 326–330.
Weerkamp, N.J., G. Tissingh, P.J. Poels et al. 2013. Nonmotor symptoms in nursing home residents with Parkinson's disease: Prevalence and effect on quality of life. *J Am Geriatr Soc* 61: 1714–1721.
Wiley, J.L., M.E. Dance, and R.L. Balster. 1998. Preclinical evaluation of the reinforcing and discriminative stimulus effects of agomelatine (S-20098), a melatonin agonist. *Psychopharmacology (Berl)* 140: 503–509.
Wong, E.Y. and J. Herbert. 2006. Raised circulating corticosterone inhibits neuronal differentiation of progenitor cells in the adult hippocampus. *Neuroscience* 137: 83–92.
Yamamoto, B.K. and L.P. Reagan. 2006. The glutamatergic system in neuronal plasticity and vulnerability in mood disorders. *Neuropsychiatr Dis Treat* 2: 7–14.
Ying, S.W., B. Rusak, P. Delagrange et al. 1996. Melatonin analogues as agonists and antagonists in the circadian system and other brain areas. *Eur J Pharmacol* 296: 33–34.
Zajecka, J., A. Schatzberg, S. Stahl et al. 2010. Efficacy and safety of agomelatine in the treatment of major depressive disorder: A multicenter, randomized, double-blind, placebo-controlled trial. *J Clin Psychopharmacol* 30: 135–144.
Zupancic, M. and C. Guilleminault. 2006. Agomelatine: A preliminary review of a new antidepressant. *CNS Drugs* 20: 981–992.

25 Melatonergic Antidepressant Agomelatine in Depressive and Anxiety Disorders

Domenico De Berardis, Michele Fornaro, Venkataramanujam Srinivasan, Stefano Marini, Nicola Serroni, Daniela Campanella, Gabriella Rapini, Luigi Olivieri, Felice Iasevoli, Alessandro Valchera, Giampaolo Perna, Maria Antonia Quera-Salva, Monica Mazza, Marilde Cavuto, Ida Potena, Giovanni Martinotti, and Massimo Di Giannantonio

CONTENTS

25.1 INTRODUCTION

Major depression (MD) is one of the most disabling and common psychiatric disorders. Recent data estimate a lifetime prevalence of MDD at 16.6% and the 1-year prevalence at 6.7% (Kessler et al. 2005; Sadock and Sadock 2005; De Berardis et al. 2008a). MDD is a leading cause of premature death and ongoing disability (Lopez et al. 2006). Psychopharmacological treatments include a number of antidepressant drugs, but over 60% of treated patients respond unsatisfactorily, and almost 20% of patients become refractory to the treatments (Fava 2003; Little 2009;

Parker and Brotchie 2010). Patients who respond satisfactory to the treatments benefit from reduced suicide rates, increased participation in the workforce, reduced secondary alcohol or other substance misuse, and decreased risk of cardiovascular disease (Hall et al. 2003; Hickie 2007). In clinical studies, patients with a reduction of 50% or more on the Hamilton Depression rating scale (HAM-D) total score at endpoint are considered responders to treatment; remission, which represents complete or near-complete symptom resolution including resolution of functional impairment, is commonly defined as HAM-D total score of ≤7 (De Berardis et al. 2008b).

The relationships between the endogenous circadian pacemaker and the development of depressive symptoms are complex and intriguing (Courtet and Olié 2012). The worsening of diurnal mood variation (DMV) in the early morning is a typical symptom of melancholic features of MD and is one of the time-linked symptoms that has promoted conjectures about the role of the circadian system in its pathogenesis (Wirz-Justice 2008). MD seems to be related to a disturbance in the central circadian clock function and not to a modification in a specific rhythm and the kind of rhythm abnormality may be highly variable in depressed patients, including phase advance or phase delay of rhythms and increase or decrease in the rhythm amplitude (Dallaspezia and Benedetti 2011). There is substantial evidence that circadian rhythms are more attenuated in MD than euthymic states, with decreased circadian amplitudes in core body temperature, motor activity, thyroid-stimulating hormone, norepinephrine (NE), and cortisol found in several studies (Coogan and Thome 2011).

Anxiety disorders (ADs) are also widespread mental disorders, with lifetime prevalence rates ranging from 10.4% to 28.8% and 12-month prevalence rates of about 18%, that can adversely affect the patient's quality of life, education, employment, social functioning, health care, and physical well-being (De Berardis et al. 2008b). ADs are a cluster of psychiatric disorders whose key features include excessive anxiety, fear, worry, avoidance, and compulsive rituals. The most prevalent anxiety disorders listed in the *Diagnostic and Statistical Manual of Mental Disorders* (DSM) include panic disorder with and without agoraphobia, obsessive–compulsive and related disorders, social phobia, generalized anxiety disorder, specific phobia, and trauma- and stressor-related disorders (posttraumatic stress disorder). ADs are often comorbid with other psychiatric disorders, including MD, substance abuse, and bipolar disorder (BD) (Tempesta et al. 2013).

25.2 PHARMACOLOGICAL CHARACTERISTICS OF AGOMELATINE

Agomelatine (Valdoxan®/Thymanax®) (S20098, *N*-[2-(7-methoxynaphth-1-yl)ethyl]acetamide) was first reported in the literature in 1992, among a series of synthetic naphthalene melatonin analogs, and acts as an agonist for melatonin MT1 and MT2 receptors and as an antagonist on serotonin 5HT2C receptors (Barden et al. 2005). Various animal models of abrupt shifts and disorganization of the light–dark cycle, of free-running conditions as well as of delayed sleep-phase syndrome have shown that agomelatine accelerates the resynchronization of circadian rhythms of locomotor activity and relevant biological parameters (i.e., body temperature, secretions of hormones) (Barden et al. 2005). The capacity of agomelatine to synchronize rest–activity rhythms in free-running animals requires the integrity of the suprachiasmatic nucleus (SCN) (Leproult et al. 2005).

The accelerating effect of agomelatine was particularly notable if treatment was started 3 weeks prior to the induced phase shift (Barden et al. 2005; Leproult et al. 2005). Agomelatine treatment did not cause any major change in corticosterone or adrenocorticotropic hormone concentrations, vasopressin, corticotropin-releasing hormone, and mineralocorticoid receptor mRNAs levels, which suggests that the mechanism of agomelatine action is not related to hypothalamic–pituitary adrenocortical axis changes (Tardito et al. 2012). It has also been demonstrated that agomelatine, a potent melatonin receptor agonist drug that strongly binds to and stimulates the activity of melatonin MT1 and MT2 receptors, showed cognitive enhancing properties, at least in preclinical studies (Conboy et al. 2009; Bertaina-Anglade et al. 2011).

As specified earlier, agomelatine shows agonistic activity with high affinity for melatonin MT1 and MT2 receptors and an antagonist activity with moderate affinity for 5HT2C

(Audinot et al. 2003; Sharpley et al. 2011; Norman 2012). Serotonin 5HT1A and 5HT2B receptors are not thought to be responsible for agomelatine clinical effects due to the low affinity of the drug for such receptors (Millan et al. 2003). No significant affinity for any of the monoamine transporters or for adrenergic, noradrenergic, dopaminergic, muscarinic, histaminic, and benzodiazepine receptors has been reported (Dubocovich 2006). The binding affinity of agomelatine for MT1 and MT2 is similar to melatonin. The literature reported that antidepressant efficacy could be related to melatonin secretion through monoaminergic mechanisms (Palazidou et al. 1992; Mitchell and Weinshenker 2009; Pecenak and Novotny 2013), even if controversial data regarding blood melatonin concentrations in MDD were reported (Wetterberg 1979; Rubin et al. 1992; Shafii et al. 1996; Sekula et al. 1997; Kripke et al. 2003; Crasson et al. 2004; Carvalho et al. 2006).

Moreover, in experiments conducted on animals, agomelatine has demonstrated the ability to increase adult hippocampal and prefrontal cortex neurogenesis, to enhance expression of brain-derived neurotrophic factor (BDNF), and to trigger several cellular signals, that is, protein kinase B (Akt), extracellular signal-regulated kinase 1/2 (ERK1/2), and glycogen synthase kinase 3β (GSK3β) (Pompili et al. 2013). It has also been reported that agomelatine may have beneficial effects on hippocampal neurogenesis in the stress-compromised brain of rats (Dagytė et al. 2011). Tardito et al. (2012) suggested that the molecular–cellular effects of agomelatine and, therefore, its antidepressant activity may be the result of a synergistic action between its agonism at MT1/MT2 and antagonism at 5-HT(2C) receptors. The antidepressant properties of agomelatine related to its effect on neurogenesis, cell survival, BDNF, activity-regulated cytoskeleton associated protein (Arc), and stress-induced glutamate release are due to this synergistic action.

After oral administration, agomelatine is rapidly (Tmax ranging from 0.5 to 4 h) well absorbed (80%) (San and Arranz 2008), but its bioavailability is relatively low (<5% at the therapeutic oral dose) due to its high first-pass metabolism (European Medicines Agency 2009), which may be of concern especially in the elderly patients or in subjects with liver disorders. In humans, agomelatine has a moderate volume of distribution of approximately 35 L, a plasma proteins binding of 90%–94% (albumin and alpha 1-acid glycoprotein), and a short plasma half-life (1–2 h) (Servier Laboratories Ltd. 2009). At the therapeutic levels, agomelatine blood concentration increases proportionally with dose; at higher doses, a saturation of first-pass effect may occur. About 90% of agomelatine is metabolized by cythocrome P450 (CYP) 1A2 (hydroxilation) and about 10% by CYP 450 2C9 (demethylation) isoformes. At higher serum concentrations, also CYP 450 2C19 is involved in the metabolism. Metabolites are conjugated with glucuronic acid and then sulphonated. About 80% of the drug is eliminated through urinary excretion of the metabolites (61%–81% of dose in humans), whereas a small amount of the metabolites undergoes fecal excretion (Dolder et al. 2008).

25.3 AGOMELATINE TREATMENT OF MAJOR DEPRESSION

25.3.1 Acute Phase Trials with Agomelatine versus Placebo

There are 11 acute phase trial studies (8 published and 3 unpublished) comparing agomelatine versus placebo (see Table 25.1). From published studies, several trials showed that agomelatine was more effective than placebo on the total HAM-D score (Loo et al. 2002; Kennedy and Emsley 2006; Olie and Kasper 2007; Heun et al. 2013). From unpublished studies, one unpublished trial reported no significant differences in HAM-D and clinical global impression scale (CGI) scores in agomelatine versus placebo compared to fluoxetine versus placebo groups (CL3-022); two studies (CL3-023 and CL3-024) were failed trials.

25.3.2 Antidepressant Efficacy in Active Comparator Trials

Agomelatine treatment efficacy, based on HAM-D, CGI, and Montgomery–Asberg depression rating scale (MADRS), has been rated by several studies. Treatment with agomelatine systematically

TABLE 25.1
Published Placebo-Controlled and/or Active Comparator Studies of Agomelatine in the Treatment of Major Depression

Authors	Study Design	Comparator/ Active Control	Number of Patients	Duration	Agomelatine Dosage (mg/Day)	Results
Loo et al. (2002)	Placebo-controlled dose range study	Placebo/paroxetine 20 mg/day	711	8 weeks	1, 5, and 25	Agomelatine 25 mg was statistically more effective than placebo.
Montgomery et al. (2004)	Randomized, double-blind, placebo-controlled discontinuation study	Placebo/paroxetine 20 mg/day	335	12 weeks	25	Agomelatine was effective and had less potential to cause discontinuation symptoms than paroxetine.
Kennedy and Emsley (2006)	Randomized, double-blind, placebo-controlled study	Placebo	212	6 weeks	25 and 50	Agomelatine 25 mg was effective but 50 mg may be beneficial for some patients without reducing tolerability.
Lemoine et al. (2007)	Randomized, double-blind comparison with venlafaxine study	Venlafaxine 75–150 mg/day	334	6 weeks	25 and 50	Agomelatine showed similar antidepressant efficacy with earlier and greater efficacy in improving subjective sleep as compared to venlafaxine.
Olié and Kasper (2007)	Double-blind, flexible dose, parallel-group, placebo-controlled study	Placebo	238	6 weeks	25 (with dose adjustment at 2 weeks to 50 mg/day in patients with insufficient improvement)	Agomelatine was significantly more efficacious than placebo. Agomelatine had a safety profile similar to placebo.
Kennedy et al. (2008)	Randomized, double-blind comparison with venlafaxine study	Venlafaxine XR 150 mg/day	277	12 weeks	50	Agomelatine showed similar antidepressant efficacy with a superior sexual side effect profile than venlafaxine.
Goodwin et al. (2009)	Randomized, double-blind, placebo-controlled study	Placebo	339	24 weeks	25 and 50	Agomelatine was more effective than placebo and prevented relapses without evidence of discontinuation symptoms.
Hale et al. (2010)	Randomized, double-blind comparison with fluoxetine study on severely depressed patients (HAM-D ≥ 25)	Fluoxetine 20–40 mg/day	515	8 weeks	25 and 50	Agomelatine was statistically more effective than fluoxetine.

Kasper et al. (2010)	Randomized, double-blind comparison with sertraline study	Sertraline 50–100 mg/day	313	6 weeks	25 and 50	Agomelatine was more effective than sertraline. Agomelatine improved circadian rest–activity cycle more than sertraline.
Zajecka et al. (2010)	Multicenter, randomized, double-blind, placebo-controlled study	Placebo	511	8 weeks	25 and 50	Agomelatine 50 mg showed greater and rapid reduction in all core symptoms of depression compared with placebo.
Stahl et al. (2010)	Randomized, double-blind, placebo-controlled study	Placebo	503	8 weeks	25 and 50	Agomelatine 25 mg was more effective than placebo over the course of the study, whereas agomelatine 50 mg provided evidence for its antidepressant efficacy until week 6 but not at study end.
Quera-Salva et al. (2011)	Randomized, double-blind comparison with escitalopram study	Escitalopram 10–20 mg/day	138	24 weeks	25 and 50	Agomelatine was as effective as escitalopram. Treatment with agomelatine improved morning condition, and reduced daytime sleepiness compared with escitalopram.
Martinotti et al. (2012)	Open-label parallel-group, randomized comparison with venlafaxine study	Venlafaxine 75–150 mg/day	60	8 weeks	25 and 50	Agomelatine antidepressant efficacy proved to be similar to that of venlafaxine during an 8-week treatment period.
Karaiskos et al. (2013)	Observational open-label, randomized, comparison with sertraline study	Sertraline 50–100 mg/day	40 depressed patients with nonoptimally controlled type 2 diabetes mellitus (DM)	4 months	25 and 50	Agomelatine was effective in the treatment of depression and anxiety as well as in the improvement of health-related behaviors, in depressed patients with non-optimally controlled type 2 DM
Heun et al. (2013)	Randomized, double-blind, placebo-controlled study	Placebo	222 elderly patients (151 in the agomelatine group, 71 in the placebo group)	8 weeks	25 and 50	Agomelatine improved depressive symptoms and was well tolerated in elderly depressed patients older than 65 years.

showed, at least, comparable efficacy with other antidepressants. More numerous studies have been conducted comparing agomelatine and venlafaxine. It is interesting that although antidepressant efficacy on the HAM-D was similar, the CGI improvement was significantly higher and statistically significant for agomelatine than for venlafaxine (Lemoine et al. 2007; Martinotti et al. 2012). Based on MADRS scores at the endpoint for response and remission rates, antidepressant efficacy was similar in the two treatment groups (Kennedy et al. 2008). After 6 weeks of treatment, the HAM-D final score as well as CGI was significantly better for agomelatine than for sertraline (Kasper et al. 2010). Over 8 weeks, the mean decrease in HAM-D total score was significantly greater with agomelatine than fluoxetine (Hale et al. 2010). Based on HAM-D scores, agomelatine was reported to be statistically noninferior to escitalopram at 6 weeks (Quera-Salva et al. 2011).

One study compared the efficacy of agomelatine and sertraline in the treatment of depression and anxiety in depressed patients with type 2 diabetes mellitus (Karaiskos et al. 2013). Agomelatine was effective in the treatment of depression and anxiety, as well as in the improvement of health-related behaviors, in depressed patients with nonoptimally controlled type 2 DM.

25.3.3 Anhedonia in Major Depression and Response to Agomelatine

Anhedonia is defined as a loss of interest and lack of reactivity to pleasurable stimuli. It is considered a core symptom of MD, a predictor of poor outcome (Spijker et al. 2001), a frequent residual symptom after treatment (Taylor et al. 2010), and related to dysfunctions of the brain reward system (Keedwell et al. 2005; Di Giannantonio and Martinotti 2012). In the first study where agomelatine was reported to be effective in the management of anhedonia, Di Giannantonio et al. (2011) found a noteworthy improvement in Snaith Hamilton rating scale (SHAPS). Moreover, after 8 weeks of treatment, agomelatine showed a more significant decrease compared to venlafaxine in SHAPS scores (Martinotti et al. 2012).

25.3.4 Sleep in Major Depressive Disorder

Sleep and daytime functioning are important aspects of major depressive disorder. Agomelatine showed an important difference in getting to sleep and quality of sleep in comparison with venlafaxine and sertraline (Kasper et al. 2010; Martinotti et al. 2012; Srinivasan et al. 2012). Agomelatine was superior to venlafaxine, but no statistically significant difference was found when compared to sertraline, in ease of awakening and integrity of behavior after awakening (Kasper et al. 2010).

Two open-label studies evaluated agomelatine efficacy on sleep parameters in patients with MD (Lopes et al. 2007; Quera-Salva et al. 2007). No change in rapid eye movement (REM) latency, amount of REM, or REM density was observed. Agomelatine improved sleep continuity and quality, and it increased sleep efficiency, time awake after sleep onset and the total amount of slow-wave sleep (SWS) (Lopes et al. 2007; Quera-Salva et al. 2007). Agomelatine treatment improved very early NREM and REM sleep (Lopes et al. 2007; Quera-Salva et al. 2007).

Recently, agomelatine has been evaluated on nighttime sleep and daytime condition compared to escitalopram (Quera-Salva et al. 2011). Agomelatine reduced sleep latency, preserved the number of sleep cycles, and reduced daytime sleepiness. In a recent open-label study, 80% of patients with MD receiving a flexible dose of agomelatine showed significant improvements at all visits in LSEQ (Quera-Salva et al. 2007).

25.3.5 Sexual Function

An important and not negligible side effect of antidepressants is represented by sexual dysfunction. Kennedy et al. (2010) found that treatment-related sexual dysfunctions were significantly lower in the agomelatine group, whereas venlafaxine was associated with greater deterioration in the domains of desire and orgasm of the sex effects questionnaire (SEQ). In a randomized, placebo-controlled, 8-week study involving healthy male volunteers, agomelatine showed to have better sexual acceptability than paroxetine. In fact, the psychotropic related sexual dysfunction questionnaire (PRSEXDQ), reported better scores for agomelatine, similar to placebo, compared to paroxetine (Montejo et al. 2010).

25.3.6 Anxiety Symptoms within Depression

Anxiety symptoms are common in patients with MDD. Some trial studies evaluating agomelatine treatment efficacy in depressed patients reported Hamilton anxiety (HAMA) scale scores. Final HAMA scores were similar when agomelatine was compared to paroxetine, fluoxetine, and venlafaxine (Montgomery et al. 2004; Kennedy et al. 2008; Hale et al. 2010). Agomelatine was superior in reducing HAMA scores compared to sertraline (Kasper et al. 2010).

25.3.7 Discontinuation Symptoms

Discontinuation symptoms in MDD have been evaluated in only one randomized, double-blind, placebo-controlled study, with paroxetine as active comparator (Montgomery et al. 2004). No significant differences were present in discontinuation-emergent signs and symptoms scale (DESS) between patients who stopped or continued agomelatine treatment. On the other hand, DESS scores were higher in patients discontinuing paroxetine, who reported insomnia, dreaming, dizziness, muscle ache, nausea, diarrhea, rhinorrhea, and chills. These data suggest that agomelatine is not associated with discontinuation symptoms. In active comparator trial, discontinuation rates were fewer for agomelatine than venlafaxine, sertraline, and fluoxetine, but similar to paroxetine (Loo et al. 2002; Montgomery et al. 2004; Lemoine et al. 2007; Kasper et al. 2010; Martinotti et al. 2012).

25.3.8 Responders, Remitters, and Relapse Prevention

Remission is the final goal of antidepressant treatment. Six studies reported respond and remission rates (three after 6 months of treatment and three in acute phase). At 3 months, the efficacy of agomelatine was superior to venlafaxine in CGI scores, but no significant differences in the proportion of responders and remitters were found (Kennedy et al. 2008). Responders' proportion was superior for agomelatine compared to sertraline by HAM-D, but no differences were found in the proportion of responders by CGI or remitters by HAM-D or CGI (Kasper et al. 2010; Karaiskos et al. 2013). Compared to placebo, in acute phase trials, response rates were significantly higher for agomelatine (Loo et al. 2002; Kennedy and Emsley 2006; Olie and Kasper 2007). Sparshatt et al. (2013) conducted a multicenter naturalistic evaluation of the use of agomelatine over a 2-year period, in order to provide a picture of its clinical value in the treatment of depression. Agomelatine was largely used in difficult-to-treat or refractory patients. After 12 weeks of treatment, a substantial number of patients, improved by at least one point of the CGI (severity) scale.

Two trial studies (one unpublished and one published) investigated long-term antidepressant effect of agomelatine treatment compared to placebo, regarding relapse prevention. The incidence of relapse over 6 months was significantly lower with agomelatine (Goodwin et al. 2009). No significant differences in relapse rates were shown in the unpublished study (CL3-021).

25.3.9 Serum Transaminases

Servier Laboratories reported that agomelatine may cause a dose-related elevated liver function test (LFT), specifically serum transaminases >3 times the upper limit of normal (Servier Laboratories 2011). The European Medicines Agency requires the monitoring of liver function during treatment at all doses (European Medicines Agency 2008).

Two studies reported notable aminotransferase elevations in 2.4% and 4.5% of patients in treatment with agomelatine 50 mg, but not with agomelatine 25 mg or with placebo (Stahl et al. 2010; Zajecka et al. 2010). These LFTs increases were isolated, mainly within the first month of treatment, and no clinical signs of liver damage were found. A higher proportion of patients with LFTs elevations, had a history of cholecystitis, gallbladder disorder, or hepatic steatosis. For these reasons, agomelatine is contraindicated in patients with hepatic impairment. Consequently, it is a condition of treatment that LFTs should be performed for all patients at initiation of treatment and then periodically after around 6, 12, and 24 weeks, and thereafter when clinically indicated (European Medicines Agency 2008). If an increase in serum transaminases occurs, blood liver function analyses must be repeated within 48 h and it is necessary to discontinue the therapy if such increase is three times the upper limit of the normal range. Liver function test must be evaluated until serum transaminases return to normal range.

25.3.10 Limitations of Agomelatine Trials in Major Depressive Disorder

Despite the majority of positive study results regarding agomelatine in the treatment of MDD, limitations of the reviewed studies should be considered. For example, inclusion and exclusion criteria employed in the trials may have somewhat favored individuals who would respond to treatment. In fact, patients with a recent history of suicidality, electroconvulsive therapy, psychotic features, or recent substance abuse were excluded in almost all studies. Moreover, most of the studies with an active comparator arm employed relatively low dosages of venlafaxine, paroxetine, sertraline, and fluoxetine, which may have improved the relative efficacy of agomelatine (De Berardis et al. 2011). Several trials with an active comparator arm did not have a placebo group (Lemoine et al. 2007; Kennedy et al. 2008; Quera-Salva et al. 2011; Martinotti et al. 2012). These studies reported high rates of response but the lack of placebo groups makes it difficult to place the high rates of response in a proper context. In addition to high response rates, two studies reported that the differences in antidepressant efficacy were not statistically significant when comparing agomelatine with venlafaxine or paroxetine (Loo et al. 2002; Lemoine et al. 2007; Martinotti et al. 2012). Thus, the similar rates of antidepressant efficacy between agomelatine and the comparator agents may have been affected by the relatively low dose of venlafaxine (75–150 mg/day) and paroxetine (20 mg/day) and by an inadequate power when comparing antidepressant efficacy between these agents.

However, despite these shortcomings, the placebo-controlled trials reported improvements in depression rating scale scores (i.e., 2–3 points) that were similar to responses reported to the Food and Drug Administration involving a number of agents approved for the treatment of MDD. In 2006, agomelatine was denied marketing authorization in Europe due to a reported lack of efficacy. Since that time, additional studies demonstrating agomelatine's efficacy have been published and in November 2008, the committee for medicinal products for human use of European Medicines Agency provided marketing authorization for treating MDD episodes in adults with agomelatine (European Medicines Agency 2008).

25.4 AGOMELATINE TREATMENT OF ANXIETY DISORDERS

25.4.1 Generalized Anxiety Disorder

To date, there are two published randomized, placebo-controlled trials (RCTs) that evaluated agomelatine efficacy and tolerability in generalized anxiety disorder (GAD). Stein et al. (2008) evaluated 121 patients diagnosed with GAD but no comorbid disorders, randomized to agomelatine (25–50 mg/day) or placebo for 12 weeks. Only nine patients failed to complete the trial (92.6% completers), and there were no differences in rates of withdrawal between agomelatine and placebo. Study results demonstrated significant superiority of agomelatine 25–50 mg as compared with placebo, and the difference between groups was statistically significant in favor of agomelatine from week 6 onward. Moreover, secondary outcome measures, including improvement in associated disability, were consistent with the efficacy of agomelatine. In particular, improvement in sleep symptoms on the self-rated Leeds Sleep Evaluation Questionnaire was more marked on agomelatine than on placebo, including the items for sleep initiation, quality of sleep, and sleep awakening. In this trial, agomelatine was as well tolerated as placebo without development of discontinuation symptoms. The most common emergent adverse events reported more frequently in the agomelatine than in the placebo groups were dizziness (7.9% vs. 3.4%) and nausea (4.8% vs. 1.7%). However, it should be considered that this clinical trial was for short-term (12 weeks) acute treatment, and therefore, longer trials were necessary to determine the extended response and maintenance of effect from agomelatine treatment in GAD.

In fact, more recently, Stein et al. (2012) evaluated the efficacy and tolerability of agomelatine in the prevention of relapse in patients with GAD. Patients who responded to a 16-week course of agomelatine 25–50 mg/day treatment were randomly assigned to receive continuation treatment with agomelatine (n = 113) or placebo (n = 114) for 26 weeks and the main outcome measure was time to relapse during the maintenance period. These authors reported that, during the 6-month maintenance period, in the intention-to-treat population, the proportion of patients who relapsed during the double-blind period in the agomelatine group (22 patients, 19.5%) was lower than in the placebo group (35 patients, 30.7%). The risk of relapse over 6 months was significantly lower with agomelatine than with placebo, and the risk of relapse over time was reduced by 41.8% for agomelatine-treated patients. Moreover, agomelatine was well tolerated throughout the study, and there were no differences in discontinuation symptoms after withdrawal of agomelatine in comparison to maintenance on agomelatine. The most frequent emergent adverse events with agomelatine were similar to those reported during the double-blind treatment period and included headache (11.3%), nasopharyngitis (9.9%), dizziness (8%), nausea (6.5%), dry mouth (5.7%), somnolence (5.0%), and fatigue (4.4%). Fourteen patients treated with agomelatine had at least one emergent potentially clinically significant abnormal liver enzyme value, but were not discontinued from the study and monitored for liver enzymes.

However, despite these positive observations, it should be noted that patients with GAD frequently have comorbid psychiatric and medical illnesses and such trials excluded significant psychiatric and medical comorbidity: therefore, future studies should be extended to, for example, primary-care settings to substantiate generalizability of these result to the general patient population.

25.4.2 Other Anxiety Disorders

Taken together, these findings support the notion that agomelatine, on the basis of its capability of restoring circadian rhythms, may be useful in the treatment of OCD, but data present in literature are mainly case reports and case series (De Berardis et al. 2013a,b). However, as expected, encouraging evidence emerged when agomelatine was used in the treatment of this disorder. Fornaro et al. (2011) reported the outcome of six treatment-refractory OCD patients with or without comorbid mood and/or other anxiety disorders who were switched from SSRIs to agomelatine 50 mg/day and followed up for 12 weeks. Three out of six patients, in particular those with relevant circadian rhythm subjective impairment, showed a Yale-Brown obsessive–compulsive scale (Y-BOCS) score reduction of ≥35%, suggesting a potential role of agomelatine in some SSRI-refractory cases.

25.5 CONCLUSIONS

Agomelatine seems to be effective in the treatment of MD and GAD.

However, concerning MD, it should be noted that some unpublished studies reported no significant differences in HAM-D and CGI scores between agomelatine versus placebo compared to fluoxetine/paroxetine versus placebo groups, but also no significant differences in relapse rates were shown when agomelatine was compared to placebo. On the other hand, the majority of published data reported agomelatine efficacy, compared to placebo and based on HAM-D, in the treatment of MD. When compared to other antidepressants (venlafaxine, sertraline, fluoxetine, and escitalopram), agomelatine showed, at least, comparable efficacy. The efficacy of agomelatine on the dimension of anhedonia may be of particular importance in the treatment of MD with anhedonic features. Some studies reported that agomelatine was similar to sertraline and superior to venlafaxine and escitalopram in the improvement of sleep parameters in patients with MDD. Lower deterioration in the domains of desire and orgasm of the SEQ were reported when agomelatine was compared to venlafaxine. In healthy male volunteers, agomelatine was shown to have better sexual acceptability than paroxetine.

Discontinuation rates for any cause were fewer for agomelatine than venlafaxine, sertraline, and fluoxetine. Moreover, data suggest that agomelatine is not associated with discontinuation symptoms even if this potential side effect warrants further investigation, especially regarding long-term risks. About responder and remission rates, data are contrasting, even if agomelatine was largely used in difficult-to-treat or refractory patients. The incidence of relapse over 6 months was significantly lower with agomelatine.

REFERENCES

Audinot, V., Mailliet, F., Lahaye-Brasseur, C., Bonnaud, A., Le Gall, A., Amossé, C., Dromaint, S. et al. 2003. New selective ligands of human cloned melatonin MT1 and MT2 receptors. *Naunyn Schmiedebergs Arch Pharmacol* 6:553–561.

Barden, N., Shink, E., Labbé, M., Vacher, R., Rochford, J., Mocaër, E. 2005. Antidepressant action of agomelatine (S 20098) in a transgenic mouse model. *Prog Neuropsychopharmacol Biol Psychiatry* 29:908–916.

Bertaina-Anglade, V., Drieu-La-Rochelle, C., Mocaër, E., Seguin, L. 2011. Memory facilitating effects of agomelatine in the novel object recognition memory paradigm in the rat. *Pharmacol Biochem Behav* 98:511–517.

Carvalho, L.A., Gorenstein, C., Moreno, R.A., Markus, R.P. 2006. Melatonin levels in drug-free patients with major depression from the southern hemisphere. *Psychoneuroendocrinology* 31:761–768.

Conboy, L., Tanrikut, C., Zoladz, P.R., Campbell, A.M., Park, C.R., Gabriel, C., Mocaër, E., Sandi, C., Diamond, D.M. 2009. The antidepressant agomelatine blocks the adverse effects of stress on memory and enables spatial learning to rapidly increase neural cell adhesion molecule (NCAM) expression in the hippocampus of rats. *Int J Neuropsychopharmacol* 12:329–341.

Coogan, A.N., Thome, J. 2011. Chronotherapeutics and psychiatry: Setting the clock to relieve the symptoms. *World J Biol Psychiatry* 12(Suppl 1):40–43.

Courtet, P., Olié, E. 2012. Circadian dimension and severity of depression. *Eur Neuropsychopharmacol* 22(Suppl 3):476–481.

Crasson, M., Kjiri, S., Colin, A., Kjiri, K., L'Hermite-Baleriaux, M., Ansseau, M., Legros, J.J. 2004. Serum melatonin and urinary 6-sulfatoxymelatonin in major depression. *Psychoneuroendocrinology* 29:1–12.

Dagytė, G., Crescente, I., Postema, F., Seguin, L., Gabriel, C., Mocaër, E., Boer, J.A., Koolhaas, J.M. 2011. Agomelatine reverses the decrease in hippocampal cell survival induced by chronic mild stress. *Behav Brain Res* 218:121–128.

Dallaspezia, S., Benedetti, F. 2011. Chronobiological therapy for mood disorders. *Expert Rev Neurother* 11:961–970.

De Berardis, D., Campanella, D., Serroni, N., Sepede, G., Carano, A., Conti, C., Valchera, A., Cavuto, M., Salerno, R.M., Ferro, F.M. 2008b. The impact of alexithymia on anxiety disorders: A review of the literature. *Curr Psychiatry Rev* 4:80–86.

De Berardis, D., Conti, C.M., Marini, S., Ferri, F., Iasevoli, F., Valchera, A., Fornaro, M. et al. 2013a. Is there a role for agomelatine in the treatment of anxiety disorders? A review of published data. *Int J Immunopathol Pharmacol* 26:299–304.

De Berardis, D., Di Iorio, G., Acciavatti, T., Conti, C., Serroni, N., Olivieri, L., Cavuto, M. et al. 2011. The emerging role of melatonin agonists in the treatment of major depression: Focus on agomelatine. *CNS Neurol Disord Drug Targets* 10:119–132.

De Berardis, D., S. Marini, M. Fornaro, V. Srinivasan, F. Iasvoli, C. Tomasetti, A. Valchera et al. 2013b. The melatonergic system in mood and anxiety disorders and the role of agomelatine: Implications for clinical practice. *Int J Mol Sci* 14:12458–12483.

De Berardis, D., Serroni, N., Campanella, D., Carano, A., Gambi, F., Valchera, A., Conti, C. et al. 2008a. Alexithymia and its relationships with C-reactive protein and serum lipid levels among drug naïve adult outpatients with major depression. *Prog Neuropsychopharmacol Biol Psychiatry* 32:1982–1986.

Di Giannantonio, M., Di Iorio, G., Guglielmo, R., De Berardis, D., Conti, C.M., Acciavatti, T., Cornelio, M., Martinotti, G. 2011. Major depressive disorder, anhedonia and agomelatine: An open-label study. *J Biol Regul Homeost Agents* 25:109–114.

Di Giannantonio, M., Martinotti, G. 2012. Anhedonia and major depression: The role of agomelatine. *Eur Neuropsychopharmacol* 22:505–510.

Dolder, C.R., Nelson, M., Snider, M. 2008. Agomelatine treatment of major depressive disorder. *Ann Pharmacother* 42:1822–1831.

Dubocovich, M.L. 2006. Drug evaluation: Agomelatine targets a range of major depressive disorder symptoms. *Curr Opin Investig Drugs* 7:670–680.

European Medicines Agency. 2008. CHMP assessment report for Valdoxan. Document reference EMEA/655251/2008, London, U.K.

European Medicines Agency. 2009. Evaluation of medicines for human use CHMP assessment report for Valdoxan. http://www.ema.europa.eu/docs/en_GB/document_library/Summary_of_opinion/human/000916/WC500150138.pdf (Accessed November 20, 2013).

Fava, M. 2003. Diagnosis and definition of treatment-resistant depression. *Biol Psychiatry* 15:649–659.

Fornaro M. 2011. Switching from serotonin reuptake inhibitors to agomelatine in patients with refractory obsessive-compulsive disorder: A 3 month follow-up case series. *Ann Gen Psychiatry* 10:5.

Goodwin G, Emsley R, Rembry S, Rouillon, F.; Agomelatine Study Group. 2009. Agomelatine prevents relapse in patients with major depressive disorder, without evidence of a discontinuation syndrome. *J Clin Psychiatry* 70:1128–1237.

Hale, A., Corral, R.M., Mencacci, C., Ruiz, J.S., Severo, C.A., Gentili, V. 2010. Superior antidepressant efficacy results of agomelatine versus fluoxetine in severe MDD patients: A randomized, double-blind study. *Int Clin Psychopharmacol* 25:305–314.

Hall, W.D., Mant, A., Mitchell, P.B., Rendle, V.A., Hickie, I.B., McManus, P. 2003. Association between antidepressant prescribing and suicide in Australia, 1991–2000: Trend analysis. *BMJ* 326:1008.

Heun, R., Ahokas, A., Boyer, P., Giménez-Montesinos, N., Pontes-Soares, F., Olivier, V.; Agomelatine Study Group. 2013. The efficacy of agomelatine in elderly patients with recurrent major depressive disorder: A placebo-controlled study. *J Clin Psychiatry* 74:587–594.

Hickie, I. 2007. Is depression overdiagnosed? No. *BMJ* 335:329.

Karaiskos, D., Tzavellas, E., Ilias, I., Liappas, I., Paparrigopoulos, T. 2013. Agomelatine and sertraline for the treatment of depression in type 2 diabetes mellitus. *Int J Clin Pract* 67:257–260.

Kasper, S., Hajak, G., Wulff, K., Hoogendijk, W.J., Montejo, A.L., Smeraldi, E., Rybakowski, J.K. et al. 2010. Efficacy of the novel antidepressant agomelatine on the circadian rest-activity cycle and depressive and anxiety symptoms in patients with major depressive disorder: A randomized, double-blind comparison with sertraline. *J Clin Psychiatry* 71:109–120.

Keedwell, P.A., Andrew, C., Williams, S.C., Brammer, M.J., Phillips, M.L. 2005. The neural correlates of anhedonia in major depressive disorder. *Biol Psychiatry* 58:843–853.

Kennedy, S.H., Emsley, R. 2006. Placebo-controlled trial of agomelatine in the treatment of major depressive disorder. *Eur Neuropsychopharmacol* 16:93–100.

Kennedy, S.H., Rizvi, S., Fulton, K., Rasmussen, J. 2008. A double-blind comparison of sexual functioning, antidepressant efficacy, and tolerability between agomelatine and venlafaxine XR. *J Clin Psychopharmacol* 28:329–333.

Kessler, R.C., Berglund, P., Demler, O., Jin, R., Merikangas, K.R., Walters, E.E. 2005. Lifetime prevalence and age-of-onset distributions of DSM-IV disorders in the National Comorbidity Survey Replication. *Arch Gen Psychiatry* 62:593–602.

Kripke, D.F., Youngstedt, S.D., Rex, K.M., Klauber, M.R., Elliott, J.A. 2003. Melatonin excretion with affect disorders over age 60. *Psychiatry Res* 118:47–54.

Lemoine, P., Guilleminault, C., Alvarez, E. 2007. Improvement in subjective sleep in major depressive disorder with a novel antidepressant, agomelatine: Randomized, double-blind comparison with venlafaxine. *J Clin Psychiatry* 68:1723–1732.

Leproult, R., Van Onderbergen, A., L'hermite-Balériaux, M., Van Cauter, E., Copinschi, G. 2005. Phase-shifts of 24-h rhythms of hormonal release and body temperature following early evening administration of the melatonin agonist agomelatine in healthy older men. *Clin Endocrinol* 63:298–304.

Little, A. 2009. Treatment-resistant depression. *Am Fam Physician* 80:167–172.

Loo, H., Hale, A., D'Haenen, H. 2002. Determination of the dose of agomelatine, a melatoninergic agonist and selective 5-HT(2C) antagonist, in the treatment of major depressive disorder: A placebo-controlled dose range study. *Int Clin Psychopharmacol* 17:239–247.

Lopes, M.C., Quera-Salva, M.A., Guilleminault, C. 2007. Non-REM sleep instability in patients with major depressive disorder: Subjective improvement and improvement of non-REM sleep instability with treatment (agomelatine). *Sleep Med* 9:33–41.

Lopez, A.D., Mathers, C.D., Ezzati, M., Jamison, D.T., Murray, C.J. 2006. Global and regional burden of disease and risk factors, 2001: Systematic analysis of population health data. *Lancet* 367:1747–1757.

Martinotti, G., Sepede, G., Gambi, F., Di Iorio, G., De Berardis, D., Di Nicola, M., Onofrj, M., Janiri, L., Di Giannantonio, M. 2012. Agomelatine versus venlafaxine XR in the treatment of anhedonia in major depressive disorder: A pilot study. *J Clin Psychopharmacol* 32:487–491.

Millan, M.J., Gobert, A., Lejeune, F., Dekeyne, A., Newman-Tancredi, A., Pasteau, V., Rivet, J.M., Cussac, D. 2003. The novel melatonin agonist agomelatine (S20098) is an antagonist at 5-hydroxytryptamine2C receptors, blockade of which enhances the activity of frontocortical dopaminergic and adrenergic pathways. *J Pharmacol Exp Ther* 306:954–964.

Mitchell, H.A., Weinshenker, D. 2009. Good night and good luck: Norepinephrine in sleep pharmacology. *Biochem Pharmacol* 79:801–809.

Montejo, A., Prieto, N., Terleira, A., Matias, J., Alonso, S., Paniagua, G., Naval, S. et al. 2010. Better sexual acceptability of agomelatine (25 and 50 mg) compared with paroxetine (20 mg) in healthy male volunteers: An 8-week, placebo-controlled study using the PRSEXDQ-SALSEX scale. *J Psychopharmacol* 24:111–120.

Montgomery, S.A., Kennedy, S.H., Burrows, G.D., Lejoyeux, M., Hindmarch, I. 2004. Absence of discontinuation symptoms with agomelatine and occurrence of discontinuation symptoms with paroxetine: A randomized, double-blind, placebo-controlled discontinuation study. *Int Clin Psychopharmacol* 19:271–280.

Murray, C.J., Lopez AD. 1997. Alternative projections of mortality and disability by cause 1990–2020: Global Burden of Disease Study. *Lancet* 349:1498–1504.

Norman, T.R. 2012. The effect of agomelatine on 5HT(2C) receptors in humans: A clinically relevant mechanism? *Psychopharmacology* 221:177–178.

Olie, J.P., Kasper, S. 2007. Efficacy of agomelatine, a MT1/MT2 receptor agonist with 5-HT2C antagonistic properties, in major depressive disorder. *Int J Neuropsychopharmacol* 10:661–673.

Palazidou, E., Papadopoulos, A., Ratcliff, H., Dawling, S., Checkley, S.A. 1992. NE uptake inhibition increases melatonin secretion, a measure of noradrenergic neurotransmission, in depressed patients. *Psychol Med* 22:309–315.

Parker, G., Brotchie, H. 2010. Do the old psychostimulant drugs have a role in managing treatment-resistant depression? *Acta Psychiatr Scand* 121:308–314.

Pecenak, J., Novotny, V. 2013. Agomelatine as monotherapy for major depression: An outpatient, open-label study. *Neuropsychiatr Dis Treat* 9:1595–1604.

Pompili, M., Serafini, G., Innamorati, M., Venturini, P., Fusar-Poli, P., Sher, L., Amore, M., Girardi, P. 2013. Agomelatine, a novel intriguing antidepressant option enhancing neuroplasticity: A critical review. *World J Biol Psychiatry* 14:412–431.

Quera-Salva, M.A., Hajak, G., Philip, P., Montplaisir, J., Keufer-Le Gall, S., Laredo, J., Guilleminault, C. 2011. Comparison of agomelatine and escitalopram on nighttime sleep and daytime condition and efficacy in major depressive disorder patients. *Int Clin Psychopharmacol* 26:252–262.

Quera Salva, M.A., Vanier, B., Laredo, J., Hartley, S., Chapotot, F., Moulin, C., Lofaso, F., Guilleminault, C. 2007. Major depressive disorder, sleep EEG and agomelatine: An open-label study. *Int J Neuropsychopharmacol* 10:691–696.

Rubin, R.T., Heist, E.K., McGeoy, S.S., Hanada, K., Lesser, I.M. 1992. Neuroendocrine aspects of primary endogenous depression. XI. Serum melatonin measures in patients and matched control subjects. *Arch Gen Psychiatry* 49:558–567.

Sadock, B.J., Sadock, V.A. 2005. *Kaplan Sadock's Synopsis of Psychiatry: Behavioral Sciences*. Lippincott Williams & Wilkins, Philadelphia, PA.

San, L., Arranz, B. 2008. Agomelatine: A novel mechanism of antidepressant action involving the melatonergic and the serotonergic system. *Eur Psychiatry* 23:396–402.

Sekula, L.K., Lucke, J.F., Heist, E.K., Czambel, R.K., Rubin, R.T. 1997. Neuroendocrine aspects of primary endogenous depression. XV: Mathematical modeling of nocturnal melatonin secretion in major depressives and normal controls. *Psychiatry Res* 69:143–153.

Servier Laboratories Ltd. 2009. Valdoxan (agomelatine) summary of product characteristics. http://emc.medicines.org.uk/medicine/21830/SPC/Valdoxan/. Accessed January 29, 2010.

Servier Laboratories Ltd. 2011. Valdoxan. http://www.servier.com/sites/default/files/ValdoxanSPC.pdf (Accessed November 20, 2013).

Shafii, M., MacMillan, D.R., Key, M.P., Derrick, A.M., Kaufman, N., Nahinsky, I.D. 1996. Nocturnal serum melatonin profile in major depression in children and adolescents. *Arch Gen Psychiatry* 53:1009–1013.

Sharpley, A.L., Rawlings, N.B., Brain, S., McTavish, S.F., Cowen, P.J. 2011. Does agomelatine block 5-HT2C receptors in humans? *Psychopharmacology* 213;653–655.

Sparshatt, A., McAllister Williams, RH., Baldwin, D.S., Haddad, P.M., Bazire, S., Weston, E., Taylor, P., Taylor, D. 2013. A naturalistic evaluation and audit database of agomelatine: Clinical outcome at 12 weeks. *Acta Psychiatr Scand* 128:203–211.

Spijker, J., Bijl, R.V., de Graaf, R., Nolen, W.A. 2001. Determinants of poor 1-year outcome of DSM-III-R major depression in the general population: Results of the Netherlands Mental Health Survey and Incidence Study (NEMESIS). *Acta Psychiatr Scand* 103:122–130.

Srinivasan, V., De Berardis, D., Shillcutt, S.D., Brzezinski, A. 2012. Role of melatonin in mood disorders and the antidepressant effects of agomelatine. *Expert Opin Investig Drugs* 21;1503–1522.

Stahl, S.M., Fava, M., Trivedi, M.H., Caputo, A., Shah, A., Post, A. 2010. Agomelatine in the treatment of major depressive disorder: An 8-week, multicenter, randomized, placebo-controlled trial. *J Clin Psychiatry* 71:616–626.

Stein, D.J., Ahokas, A., Albarran, C., Olivier, V., Allgulander, C. 2012. Agomelatine prevents relapse in generalized anxiety disorder: A 6-month randomized, double-blind, placebo-controlled discontinuation study. *J Clin Psychiatry* 73:1002–1008.

Stein, D.J., Ahokas, A.A., de Bodinat, C. 2008. Efficacy of agomelatine in generalized anxiety disorder: A randomized, double-blind, placebo-controlled study. *J Clin Psychopharmacol* 28:561–566.

Tardito, D., Molteni, R., Popoli, M., Racagni, G. 2012. Synergistic mechanisms involved in the antidepressant effects of agomelatine. *Eur Neuropsychopharmacol* 22:482–486.

Taylor, D.J., Walters, H.M., Vittengl, J.R., Krebaum, S., Jarret, R.B. 2010. Which depressive symptoms remain after response to cognitive therapy of depression and predict relapse and recurrence? *J Affect Disord* 123:181–187.

Tempesta, D., Mazza, M., Serroni, N., Moschetta, F.S., Di Giannantonio, M., Ferrara, M., De Berardis, D. 2013. Neuropsychological functioning in young subjects with generalized anxiety disorder with and without pharmacotherapy. *Prog Neuropsychopharmacol Biol Psychiatry* 45:236–241.

Wetterberg, L. 1979. Clinical importance of melatonin. *Prog Brain Res* 52:539–547.

Wirz-Justice, A. 2008. Diurnal variation of depressive symptoms. *Dialogues Clin Neurosci* 10:337–343.

Zajecka, J., Schatzberg, A., Stahl, S., Shah, A., Caputo, A., Post, A. 2010. Efficacy and safety of agomelatine in the treatment of major depressive disorder: A multicenter, randomized, double-blind, placebo-controlled trial. *J Clin Psychopharmacol* 30:135–144.

26 Effects of Melatonin and Melatonergic Drugs in Transgenic Mouse Models of Alzheimer's Disease

Beatriz B. Otalora, M. Angeles Rol, and Juan A. Madrid

CONTENTS

26.1 PATHOGENESIS OF ALZHEIMER'S DISEASE

Alzheimer's disease (AD) is an age-associated neurodegenerative disorder. The major clinical hallmarks of AD are progressive cognitive impairments—primarily in learning and memory—with behavioral abnormalities. The neuropathological hallmarks of AD encompass extracellular senile plaques of β-amyloid (Aβ) deposits and intracellular neurofibrillary tangles (NFTs) from the accumulation of hyperphosphorylated microtubule-associated protein tau. This is followed by neuronal and synaptic loss (Duyckaerts et al., 2009).

Approximately 10% of AD cases are classified as early onset (occur at ages of <50 years). These cases are linked to autosomal dominant mutations in genes that encode the amyloid precursor protein (APP) or the presenilins (PS1 and PS2). However, in more than 90% of cases, the precise etiology of the disease is still unknown. Mitochondrial dysfunction, oxidative stress, and neuroinflammation are believed to play major roles in the pathogenesis of AD (Marques et al., 2010).

26.2 CIRCADIAN RHYTHM ALTERATIONS IN ALZHEIMER'S DISEASE

Patients with AD show a profound disruption in their circadian rhythms (Wu and Swaab, 2007). This includes severe disturbances in their sleep–wake cycle, resulting in increased nocturnal activity and daytime sleep (Lee et al., 2007; Merlino et al., 2010; van Someren et al., 1996). These sleep disturbances exacerbate memory and cognitive impairments (Gerstner and Yin, 2010). Indeed, recent evidence suggests that in "healthy" brains, sleep promotes Aβ clearance (Xie et al., 2013), thus protecting the brain against potentially neurotoxic waste products that accumulate during wakefulness. Improving the quality of sleep in AD patients, therefore, is of great importance and necessary in the treatment process.

In addition to sleep disturbances, AD patients display a condition known as "sundowning" (Bliwise, 1994; Vitiello et al., 1992). This is characterized by increased agitated behavior and arousal, confusion, and emotional disturbances in the late afternoon or early evening. These alterations in their sleep–wake cycle are associated with phase shifts in the core body temperature rhythm (Satlin et al., 1995; Volicer et al., 2001).

In mammals, circadian rhythms in physiology, behavior, and the sleep–wake cycle are generated by the central pacemaker in the hypothalamic suprachiasmatic nuclei (SCN). Here, the activity of core clock genes generates and communicates circadian rhythms to the rest of the brain and body (Welsh et al., 2010). The activity of the SCN is synchronized daily by light, communicated to this brain structure by the retina through the glutamatergic retino-hypothalamic tract (RHT). SCN neurones are neurochemically and functionally heterogeneous, and cells containing the neuropeptides vasointestinal polypeptide (VIP) and vasopressin (AVP) are important in circadian rhythm generation, synchronization, and maintenance (Dibner et al., 2010). Clock genes are also expressed in many regions of the brain and body, including the hippocampus and pineal gland (Cermakian et al., 2011). Activity of these extra-SCN clock genes provides local circadian timing in a tissue-specific manner, which together with the SCN, form the body's circadian system.

Indeed, neural degeneration and tangles and significant reduction in the number of VIP- and AVP-expressing neurones are reported in the SCN of AD patients, compared with healthy age-matched individuals (Stopa et al., 1999; Wu and Swaab, 2007). Activity of other key signaling neurochemicals that are intrinsic (such as neurotensin) or extrinsic (importantly melatonin) to the SCN also shows diminished levels in AD (Stopa et al., 1999; Wu and Swaab, 2007). Further, individuals with AD have degeneration of the optic nerves and retinal ganglion cells, which presumably leads to decreased transmission of photic information to the SCN (Katz and Rimmer, 1989). Daily clock gene expression in the SCN and in other areas of the brain and body, especially in the pineal gland, also shows a loss or dampened rhythm and/or desynchrony in AD patients (Cermakian et al., 2011; Wu et al., 2006). Consequently, in neurodegenerative diseases, such as AD, reduced signal magnitude is seen both in SCN inputs and outputs (Coogan et al., 2013).

26.3 MELATONIN

Melatonin (*N*-acetyl-5methoxytryptamine) is a hormone which in vertebrates is primarily synthesized by the pineal gland, and is a key factor in transmitting timekeeping signals of the SCN to the rest of the brain and body (Pevet and Challet, 2011). The precursor for melatonin synthesis is the amino acid L-tryptophan, which is converted to 5-hydroxytryptophan and then to serotonin. Next, serotonin is acetylated by arylalkylamine *N*-acetyltransferase (AA-NAT) resulting in *N*-acetyl serotonin, which finally is methylated by Hydroxyindole-*O*-methyltransferase (HI-OMT) to melatonin. AA-NAT is the key enzyme in the synthesis of melatonin, and shows circadian rhythm in its activity/levels that is indirectly controlled by the SCN (Pandi-Perumal et al., 2006). The SCN controls the synthesis of melatonin via the hypothalamic paraventricular nucleus (PVN), which projects to the intermediolateral column (IMC) of the spinal cord and then to the pineal gland via the superior cervical ganglion (SCG). The release of noradrenaline at night from postganglionic sympathetic fibers acts on β-adrenergic receptors in the pinealocytes membrane and triggers an intracellular cascade which promotes enzymatic activity of AA-NAT. The synthesized melatonin is then released into the bloodstream and cerebrospinal fluid (CSF) and conveys temporal signals of the SCN to the rest of the brain and body. During the light phase, increased SCN electrical activity inhibits neurones of the PVN, which in turn, decreased AA-NAT activity in the pinealocytes and melatonin production. Nocturnal decrease in SCN electrical activity promotes PVN-driven melatonin production.

Melatonin exerts numerous physiological functions and displays pleiotropic effects throughout the body. For example, melatonin regulates circadian rhythms by modulating SCN's electrical activity and by phase-adjusting the circadian clock (Arendt and Skene, 2005). These chronobiological effects of melatonin are mediated mainly by its interactions with the G protein-coupled melatonin membrane receptors MT1 and MT2 (Pandi-Perumal et al., 2008). In addition to its role as a chronobiotic, melatonin also has neuroprotective, immunomodulatory, anti-inflammatory, and antioxidant properties and enhances cellular mitochondrial activity (Pandi-Perumal et al., 2006, 2008). Melatonin's protective effects against oxidative stress may partially involve its interactions with the enzyme quinone reductase 2 (QR2, also found in the literature as MT3) and the RORα receptors (retinoic acid receptor-related orphan receptor α), as well as its capacity to act as a direct scavenger of free radicals (Galano et al., 2011). That is, this neurohormone can act in several ways to reduce oxidative stress in the brain. Other less known pathways through which melatonin protects the brain include its association with melatonin-binding intracellular proteins, such as calmodulin and calreticulin, and mitochondrial complex I (Hardeland, 2009). Although melatonin's signaling pathways as a chronobiotic are established, the cellular processes involved in its neuroprotective, immunomodulatory, anti-inflammatory, and antioxidant properties are relatively less defined. Specific agonists, such as ramelteon, which are highly selective to the MT1/MT2 receptors (Kato et al., 2005) but have no direct free radical scavenging actions (Mathes et al., 2008), are useful tools by which the multiple pathways underlying melatonin's actions can be discerned.

26.4 MELATONIN AND ITS RECEPTORS IN ALZHEIMER'S DISEASE

Strikingly, decreased plasma and CSF melatonin levels are found in AD patients when compared to age-matched control subjects (Wu and Swaab, 2007). These reduced CSF melatonin levels in AD patients are attributed to profound reduction in melatonin secretion. Degeneration of the SCN, the pineal gland, or the neuronal connections between them may account for this reduction in melatonin levels. In addition, alteration in SCN and hippocampal melatonin MT1/MT2 receptor expression is reported in AD patients (Savaskan et al., 2002, 2005; Wu et al., 2007; Wu and Swaab, 2005). Together, changes in systemic and central melatonin levels and reduced brain melatonin-receptor signaling may contribute to the circadian rhythm alterations and cognitive impairments seen in patients with AD. Indeed, the declines in blood and CSF melatonin levels in AD patients parallel the progression of AD neuropathology, as determined by Braak stages (Wu et al., 2003; Zhou et al., 2003). Remarkably, CSF melatonin level is low in preclinical AD individuals (Wu et al., 2003), suggesting that measurements of melatonin levels can be a useful and novel early biomarker of AD.

Together, these observations strongly suggest that exogenous melatonin can be therapeutically used for circadian rhythm disturbances, high oxidative stress levels, and increased inflammatory responses that typically occur in the pathogenesis of AD (Figure 26.1). As such, to fully exploit the beneficial effects of melatonin-receptor signaling pathways in AD, agonists such as ramelteon, which has higher affinity and specificity for the melatonin receptors (MT1/MT2) with significantly longer systemic half-life than melatonin, have been developed (Kato et al., 2005). To date, however, little is known of the therapeutic effects of ramelteon in AD.

26.5 MELATONIN AND RAMELTEON IN THE TRANSGENIC MOUSE MODELS OF AD

Over the last two decades, a wide variety of transgenic mouse models of AD has been developed (see Table 26.1 for an overview). These models aim to replicate the AD-like neuropathologies (Aβ deposits and neurofibrillary tangles), behavioral abnormalities, and cognitive impairments commonly observed in human patients with AD. These cohorts of transgenic mouse models overexpress

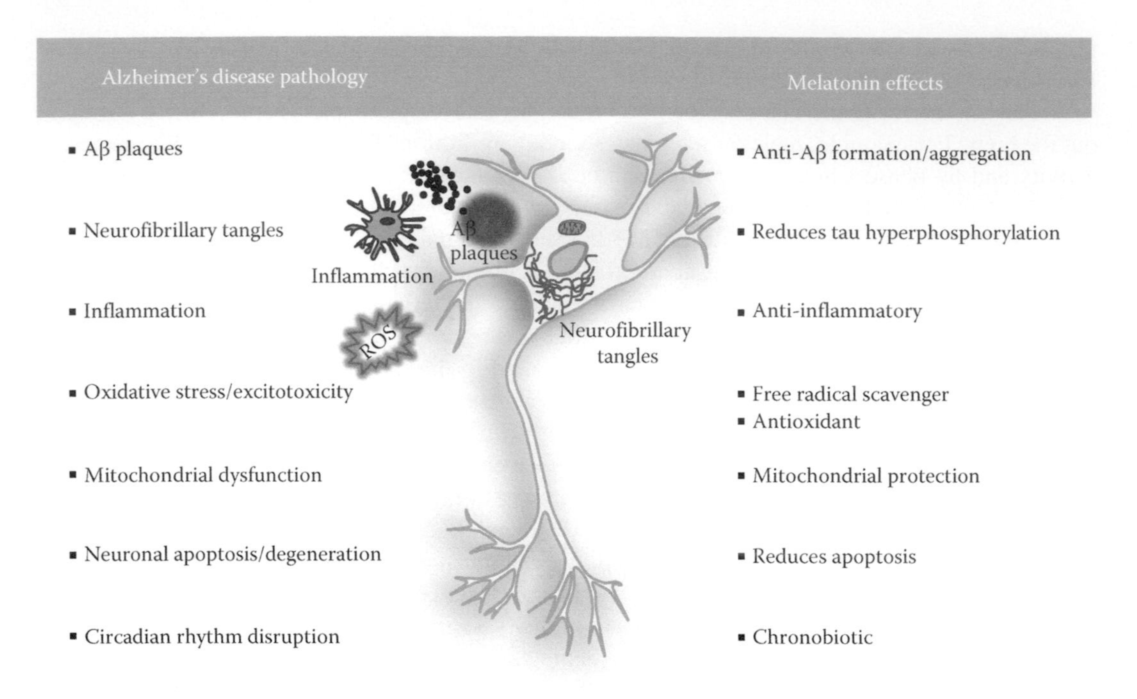

FIGURE 26.1 Hallmarks of Alzheimer's disease (AD) and melatonin as a therapeutic agent for treating AD.

three major genes linked with the early onset of AD—mutant forms of amyloid precursor protein (APP) and/or presenilins (PS1 and PS2) and/or tau. Overproduction and accumulation of the derivative proteins lead to the formation of senile plaques (SP), tau hyperphosphorylation, and the formation of NFT (Spires and Hyman, 2005).

Because of the relatively short life span (1–2 years) of the AD mouse models, longitudinal studies from these animals have revealed important insights into the physiological, neurological, and behavioral changes that occur at the various stages of AD—including a time before pathological state. However, these studies also highlight inherent problems when comparing and contrasting results from the different AD models generated thus far, and call for caution when interpreting the results. For example (see Table 26.1), the age at which the different models begin to develop plaques can significantly vary from model to model, and plaque development in the brain does not necessarily imply cognitive impairments or its progression at the behavioral level. Further, most of these transgenic AD models generated to date cannot fully recapitulate all of the etiological features of this disease as seen in humans, suggesting that AD is more complex. Nevertheless, these models are useful tools in which to study and better understand the pathogenesis of AD at the molecular, cellular, and behavioral levels. Further, these models can also be useful to test the therapeutic effects of potential drugs against the diverse pathologies of AD.

26.5.1 Effects of Melatonin and Ramelteon on Aβ Deposition and Oxidative Stress Levels

One of the mechanisms by which melatonin can protect against AD pathogenesis is through its anti-amyloidogenic and antioxidant properties. Table 26.1 gives a detailed summary of the various AD animal models used, duration of the treatments, animal's age, and effects of treatment. Pioneering biochemical studies (Pappolla et al., 1998; Poeggeler et al., 2001), and more recently ex vivo brain assays from APP/PS1 AD mice, intriguingly show that melatonin efficiently reduces Aβ aggregation (Olcese et al., 2009). This beneficial effect of melatonin on Aβ

TABLE 26.1
Experimental Studies Evaluating the Effects of Melatonin and Melatonergic Drugs in Transgenic Mouse Models of Alzheimer's Disease (AD)

Reference	Transgenic Model of AD	Age (Months)	Treatment Duration (Months)	Treatment Administration	Results
Matsubara et al. (2003)	Tg2576	4 ↓ up to 8, 9.5, 11, 15.5	4–11	Drinking water, melatonin: 0.5 mg/mL	Melatonin administration • Partially inhibits the expected time-dependent elevation of Aβ • Reduces abnormal nitration of proteins • Increases survival
Feng et al. (2004)	APP 695-V717I	4 ↓ up to 8	4	Drinking water, melatonin: 10 mg/kg	Melatonin supplementation • Alleviates learning and memory deficits • Increases choline acetyltransferase activity in the frontal cortex and hippocampus • Reduces the number of apoptotic neurones • Decreases Aβ deposits
Quinn et al. (2005)	Tg2576	14 ↓ up to 18	4	Drinking water, melatonin: 2.6 mg/kg	Melatonin treatment after the age of amyloid plaque deposition • Fails to modify brain levels of Aβ • Does not reduce brain lipid peroxidation • Fails to affect astrocytes and microglial cell activation surrounding amyloid plaques
Feng et al. (2006)	APP695-V717I	4 ↓ up to 8	4	Drinking water, melatonin: 10 mg/kg	Melatonin alleviates oxidative stress • Decreases brain lipid peroxidation • Increases brain glutathione content • Increases superoxide dismutase activity • Inhibits upregulation of apoptosis-related factors

(Continued)

TABLE 26.1 (*Continued*)
Experimental Studies Evaluating the Effects of Melatonin and Melatonergic Drugs in Transgenic Mouse Models of Alzheimer's Disease (AD)

Reference	Transgenic Model of AD	Age (Months)	Treatment Duration (Months)	Treatment Administration	Results
Olcese et al. (2009)	APP/PS1 ($APP^{K670N,M671L}$ $PS1^{M146V}$)	2–2.5 ↓ up to 7.5	5	Drinking water, melatonin: 0.5 mg/day	Melatonin treatment • Protects from cognitive impairment • Reduces Aβ deposition in hippocampus and entorhinal cortex • Facilitates removal of Aβ from the brain • Suppresses proinflammatory cytokine levels • Reduces levels of expression of antioxidant enzymes
Dragicevic et al. (2011)	APP/PS1 ($APP^{K670N,M671L}$ $PS1^{M146V}$)	18–20	1	Drinking water, melatonin: 0.5 mg/day	Melatonin treatment • Restores isolated hippocampal, cortical, striatal mitochondrial function
Garcia-Mesa et al. (2012)	3xTg-AD (PS1/M146V, APPswe, and tauP301L)	6 ↓ up to 12	6	Drinking water, melatonin: 10 mg/kg	Melatonin treatment • Decreases soluble amyloid oligomers • Decreases levels of tau with abnormal phosphorylation • Improves learning and memory • Reduces brain oxidative stress (reduction of lipoperoxidation, increased reduced GSH levels, increased activity of the antioxidant enzymes) • Protects against mitochondrial DNA reduction

McKenna et al. (2012)	B6C3-Tg (APPswe, PSEN1dE9)	3 ↓ up to 6 or 9	3 or 6	Drinking water, ramelteon: 3 mg/kg	Ramelteon treatment • Does not improve cognitive performance • Does not affect the quantities of Aβ plaques • Does not reduce apoptosis
Otalora et al. (2012)	B6C3-Tg (APPswe, PSEN1dE9)	3.5–5.5 ↓ up to 12	5.5	Drinking water, melatonin: 5 mg/kg	Melatonin treatment • Reduces hippocampal protein oxidation • Does not affect the amplitude of locomotor activity and body temperature rhythms
				Re-pelleted food, ramelteon: 2 mg/kg	Ramelteon treatment • Reduces hippocampal protein oxidation • Reduces the amplitude of the body temperature rhythm • Increases circadian rhythm fragmentation • Does not alter the free running period
Peng et al. (2013)	Tg2576	4 or 8 ↓ up to 8 or 12	4 or 8	Intraperitoneally, melatonin: 10 mg/kg	Melatonin treatment • Attenuates memory impairments and arrests the tau/Aβ pathologies only when it is provided at proper timing (8–12 months). It is suggested that targeting the activated Glycogen synthase kinase-3β plays a critical role in those effects

levels and deposits is also supported in vivo (Dragicevic et al., 2011; Feng et al., 2004; Garcia-Mesa et al., 2012; Matsubara et al., 2003; Olcese et al., 2009; Peng et al., 2013). In agreement with in vitro results, most of the experimental data in vivo using AD transgenic mice showed that the most beneficial effects of melatonin occur when treatment was started before plaque deposition, that is, at an early stage. Indeed, Matsubara et al. (2003) reported that in Tg2576 mice early melatonin treatment (beginning at 4 months of age, prior to the appearance of amyloid plaques) reduced Aβ peptides (Aβ1-40, Aβ1-42) and oxidative damage in brain tissues, leading to increased survival. In the brain, early melatonin treatment also inhibited extracellular Aβ deposition, increased antioxidant defense and decreased oxidation and apoptosis (Feng et al., 2004, 2006, see also Table 26.1). These neuroprotective effects of melatonin are also reported in the double APP/PS1 and triple 3xTg-AD mouse models (Garcia-Mesa et al., 2012; Olcese et al., 2009, see also Table 26.1). By contrast, melatonin seems not to provide any benefits in removing existing amyloid plaque deposition or preventing additional Aβ deposits, as well as to reduce oxidative damage in older animals (15 months old) (Quinn et al., 2005).

These data indicate that melatonin can regulate APP metabolism and prevent Aβ pathology, but fails to exert anti-amyloid effect when administered after Aβ deposition. Therefore, the benefit of melatonin can only be obtained if the therapy is given at an early stage of the disease process—underscoring the importance of identifying early biomarkers for AD. This has great relevance, since AD in humans is usually diagnosed relatively late in life, at a time when AD neurological damage is normally established. A recent study has added a further twist in melatonin's effect in AD, associating its therapeutic effects with the age-dependent temporal activity of glycogen synthase kinase-3β (GSK-3β) (Peng et al., 2013). This may provide yet further pharmaceutical targets to render the body and brain more susceptible to melatonin's actions.

To the best of our knowledge, only two studies have examined the effects of ramelteon, a selective MT1/MT2 receptor agonist, in AD mice models (McKenna et al., 2012; Otalora et al., 2012). Treatment with ramelteon seems to produce diverse effects on AD neuropathology. For example, long-term ramelteon treatment had no effects on Aβ plaque counts, and did not reduce apoptosis in the hippocampus and prefrontal cortex (McKenna et al., 2012)—albeit a dose response curve was not performed. On the other hand, ramelteon reduced protein oxidation in the hippocampus of transgenic mice (Otalora et al., 2012). Unlike melatonin, ramelteon has no known direct free-radical-scavenging properties (Mathes et al., 2008), suggesting that activation of the melatonin receptors (MT1/MT2) may be sufficient in providing protection against oxidative stress in some brain regions. This result is exciting and promising and underscores the complexity of the melatoninergic systems, thereby, warranting further work to understand the effects of ramelteon in the AD brain.

26.5.2 Effects of Melatonin on Neuroinflammation

Neuroinflammation plays a key role in the pathogenesis of the AD (Rojo et al., 2008; Rosales-Corral et al., 2012). These inflammatory responses are believed to be partly triggered by an Aβ-induced overexcitation of microglia cells in the brain. Overactivity of microglia causes further release of neurocytotoxins, such as proinflammatory cytokines and reactive oxygen species (ROS), which promote inflammatory responses, leading to neurodegeneration. Indeed, in the transgenic AD mouse models, proinflammatory cytokine expression is upregulated (Patel et al., 2005; Ruan et al., 2009).

Melatonin has been shown to reduce oxidative stress and proinflammatory cytokines in the rat hippocampus following Aβ peptide injection (Rosales-Corral et al., 2003). Compelling evidence also shows that melatonin treatment in APP/PS1 transgenic mice suppressed the levels of proinflammatory cytokines, including those of TNF-α in the hippocampus. Interestingly, the suppressive effects of melatonin on proinflammatory cytokines were not apparent in plasma or frontal cortex. This has led to the suggestion that melatonin may act specifically within the hippocampus with no widespread or global effects on the immune system (Olcese et al., 2009). Together, these results indicate that melatonin can act in a specific manner to reduce the levels of proinflammatory cytokines in AD brain.

26.5.3 Effects of Melatonin and Ramelteon on Spatial Memory Performance and Cognition

One of the most devastating effects of Aβ deposits and oxidative stress levels in AD brains is that they lead to brain inflammation, which consequently promotes excitotoxicity and neurodegeneration, particularly in the hippocampal areas. In humans, these effects translate into severe impairments in face and object recognition, diminished spatial learning, and memory performance and cognition, which are indeed amongst early indicators of AD (Blackwell et al., 2004; Laatu et al., 2003). However, mixed behavioral observations are reported in the transgenic mouse models of AD, which are related to the specific models used—that is, single versus double versus triple. For example, in the double APP/PS1 model, some studies have reported a link between AD pathological marker load and cognitive impairments (Arendash et al., 2001; Reiserer et al., 2007; Savonenko et al., 2005), while in others no such association has been found (Gruart et al., 2008; Park et al., 2010). This suggests that in some transgenic models although markers for AD are centrally present, object recognition and spatial memory processes are preserved (Otalora et al., 2012). Thus, in these animals learning and memory deficits are not related to the presence of amyloid plaques but may be associated with the ageing process.

Melatonin treatment in the APP695 V717I and 3xTg-AD transgenic models alleviates learning and memory impairments in passive avoidance task and the Morris water maze (Feng et al., 2004; Garcia-Mesa et al., 2012). Interestingly though, in the double APP/PS1 transgenic model, the ameliorating effects of melatonin were seen only in transgenic animals that showed significant cognitive impairments (Olcese et al., 2009), but not in models in which memory performance remained intact (Otalora et al., 2012).

Despite having enhanced activity at the melatonin receptors (Kato et al., 2005) and ameliorating effects on hippocampal protein oxidation in some transgenic mice (Otalora et al., 2012), the effects of ramelteon on behavior and cognitive performances in transgenic AD mice have been sparsely investigated. Nevertheless, thus far, results in the double APP/PS1 transgenic mice show that chronic ramelteon treatment at an early age (3–5 months) had no effect on spatial learning and memory (McKenna et al., 2012; Otalora et al., 2012). This suggests that the restoration of some AD pathologies may rely on more than MT1/MT2 receptor activation and that the global activity of melatonin is necessary for the amelioration of cognitive and memory deficits in the double transgenic AD mice.

26.5.4 Effects of Melatonin on Circadian System Functionality

In humans, one of the most profound behavioral and physiological alterations that occur prior to and following the onset of AD is the loss of coherent circadian system functionality. Indeed, in AD patients several SCN inputs and outputs are severely compromised. This includes neural degeneration and tangles in the SCN, loss or dampened rhythm amplitude of key clock gene expression in the pineal gland, and reduced brain and systemic melatonin levels (Stopa et al., 1999; Wu et al., 2006; Wu and Swaab, 2007). Consequently, patients with AD experience fragmented sleep–wake cycle, which exacerbates memory and other cognitive impairments (Gerstner and Yin, 2010). Therefore, improving the quality of sleep of AD patients is now an accepted part of the treatment process.

Until recently, most studies using transgenic AD animal models to investigate the neuropathology of this disease (e.g., Arendash et al., 2001; Gordon et al., 2002; Reiserer et al., 2007; Savonenko et al., 2005) have largely ignored the circadian system disruptions that can occur prior to and following the onset of AD pathology. As such, limited knowledge exists on the nature of the circadian rhythm disruptions in these models (Bedrosian et al., 2011; Gorman and Yellon, 2010; Otalora et al., 2012; Sterniczuk et al., 2010; Vloeberghs et al., 2004; Wisor et al., 2005) and how to improve circadian rhythm generation and maintenance in these animals—a useful tool to better understand the human sleep–wake cycle condition in AD.

In-depth knowledge of the circadian system functionality in these AD mice will provide a platform in which the effects of melatonin can be robustly tested. Studies performed in rodents show that melatonin treatment prevents circadian rhythm alterations caused by Aβ peptide 25–35 injection in the SCN (Furio et al., 2002). To the best of our knowledge, there is only one comprehensive study addressing the effects of melatonin and ramelteon on circadian system functionality in an APP/PS1 AD mouse model (Otalora et al., 2012, see also Table 26.1). Unfortunately, in this model of AD no major circadian system dysfunction was found. Therefore, when compared with controls, no net beneficial effects of melatonin or ramelteon treatment to the circadian system functionality were seen in these AD mice. Nevertheless, emerging results show that melatonin treatment can play a central role in ameliorating circadian rhythm disruptions in AD patients (Srinivasan et al., 2010). Therefore, future studies designed to investigate circadian rhythms alteration in AD mouse models should also assess the therapeutic effects of melatonin and ramelteon on circadian system functionality. This will provide better understanding of the role of the melatonergic systems in AD.

In spite of the usefulness of melatonin or melatonergic drugs as therapeutic treatments in AD, there are several points to consider. First, the effects of melatonin are dependent upon time-of-day of administration, duration of exposure, and melatonin receptors sensitivity. For example, exogenous melatonin treatment in the late subjective day (dusk) or in the early part of the night phase-advances the circadian clock. In contrast, delays of the circadian rhythms, or no responses, are seen when melatonin is given in the late subjective night or at dawn (Lewy et al., 1998). Second, melatonin receptors are widely expressed in numerous tissues of the body. Therefore, melatonin or melatonergic drugs treatments will exert pleiotropic effects, acting over multiple bodily targets which may lead to unwanted/unanticipated side effects.

26.6 CONCLUSION

Alzheimer's disease is complex and efforts to generate transgenic animal mouse models of AD that fully recapitulate all of the etiological features of this disease have encountered some limitations. Melatonin or melatonergic drugs have proven to be a promising tool in the treatment of AD and their circadian-related impairments, mostly by their versatility and multipronged actions. Melatonin acts both at the cellular and system levels to directly combat the neuropathology of AD and to ameliorate the circadian system's functionality, which misalignment during AD can accelerate the pathogenesis of this disease. Since some neuroanatomical structures that are strongly linked with AD, such as the hippocampus, show alteration in melatonin receptors, and the brain shows reduced levels of melatonin prior to and during AD, it is conceivable that the activity of this neurohormone can be used as a novel early biomarker for AD.

REFERENCES

Arendash GW, King DL, Gordon MN, Morgan D, Hatcher JM, Hope CE, Diamond DM. (2001). Progressive, age-related behavioral impairments in transgenic mice carrying both mutant amyloid precursor protein and presenilin-1 transgenes. *Brain Res* 891:42–53.

Arendt J, Skene DJ. (2005). Melatonin as a chronobiotic. *Sleep Med Rev* 9:25–39.

Bedrosian TA, Herring KL, Weil ZM, Nelson RJ. (2011). Altered temporal patterns of anxiety in aged and amyloid precursor protein (APP) transgenic mice. *Proc Natl Acad Sci USA* 108:11686–11691.

Blackwell AD, Sahakian BJ, Vesey R, Semple JM, Robbins TW, Hodges JR. (2004). Detecting dementia: Novel neuropsychological markers of preclinical Alzheimer's disease. *Dement Geriatr Cogn Disord* 17:42–48.

Bliwise DL. (1994). What is sundowning? *J Am Geriatr Soc* 42:1009–1011.

Cermakian N, Lamont EW, Boudreau P, Boivin DB. (2011). Circadian clock gene expression in brain regions of Alzheimer's disease patients and control subjects. *J Biol Rhythms* 26:160–170.

Coogan AN, Schutova B, Husung S, Furczyk K, Baune BT, Kropp P, Hassler F, Thome J. (2013). The circadian system in Alzheimer's disease: Disturbances, mechanisms, and opportunities. *Biol Psychiatry* 74:333–339.

Dibner C, Schibler U, Albrecht U. (2010). The mammalian circadian timing system: Organization and coordination of central and peripheral clocks. *Annu Rev Physiol* 72:517–549.

Dragicevic N, Copes N, O'Neal-Moffitt G, Jin J, Buzzeo R, Mamcarz M, Tan J et al. (2011). Melatonin treatment restores mitochondrial function in Alzheimer's mice: A mitochondrial protective role of melatonin membrane receptor signaling. *J Pineal Res* 51:75–86.

Duyckaerts C, Delatour B, Potier MC. (2009). Classification and basic pathology of Alzheimer disease. *Acta Neuropathol* 118:5–36.

Feng Z, Chang Y, Cheng Y, Zhang BL, Qu ZW, Qin C, Zhang JT. (2004). Melatonin alleviates behavioral deficits associated with apoptosis and cholinergic system dysfunction in the APP 695 transgenic mouse model of Alzheimer's disease. *J Pineal Res* 37:129–136.

Feng Z, Qin C, Chang Y, Zhang JT. (2006). Early melatonin supplementation alleviates oxidative stress in a transgenic mouse model of Alzheimer's disease. *Free Radic Biol Med* 40:101–109.

Furio AM, Cutrera RA, Castillo Thea V, Perez LS, Riccio P, Caccuri RL, Brusco LL, Cardinali DP. (2002). Effect of melatonin on changes in locomotor activity rhythm of Syrian hamsters injected with beta amyloid peptide 25–35 in the suprachiasmatic nuclei. *Cell Mol Neurobiol* 22:699–709.

Galano A, Tan DX, Reiter RJ. (2011). Melatonin as a natural ally against oxidative stress: A physicochemical examination. *J Pineal Res* 51:1–16.

Garcia-Mesa Y, Gimenez-Llort L, Lopez LC, Venegas C, Cristofol R, Escames G, Acuna-Castroviejo D, Sanfeliu C. (2012). Melatonin plus physical exercise are highly neuroprotective in the 3xTg-AD mouse. *Neurobiol Aging* 33:1124–1129.

Gerstner JR, Yin JC. (2010). Circadian rhythms and memory formation. *Nat Rev Neurosci* 11:577–588.

Gordon MN, Holcomb LA, Jantzen PT, DiCarlo G, Wilcock D, Boyett KW, Connor K, Melachrino J, O'Callaghan JP, Morgan D. (2002). Time course of the development of Alzheimer-like pathology in the doubly transgenic PS1+APP mouse. *Exp Neurol* 173:183–195.

Gorman MR, Yellon S. (2010). Lifespan daily locomotor activity rhythms in a mouse model of amyloid-induced neuropathology. *Chronobiol Int* 27:1159–1177.

Gruart A, Lopez-Ramos JC, Munoz MD, Delgado-Garcia JM. (2008). Aged wild-type and APP, PS1, and APP + PS1 mice present similar deficits in associative learning and synaptic plasticity independent of amyloid load. *Neurobiol Dis* 30:439–450.

Hardeland R. (2009). Melatonin: Signaling mechanisms of a pleiotropic agent. *Biofactors* 35:183–192.

Kato K, Hirai K, Nishiyama K, Uchikawa O, Fukatsu K, Ohkawa S, Kawamata Y, Hinuma S, Miyamoto M. (2005). Neurochemical properties of ramelteon (TAK-375), a selective MT1/MT2 receptor agonist. *Neuropharmacology* 48:301–310.

Katz B, Rimmer S. (1989). Ophthalmologic manifestations of Alzheimer's disease. *Surv Ophthalmol* 34:31–43.

Laatu S, Revonsuo A, Jaykka H, Portin R, Rinne JO. (2003). Visual object recognition in early Alzheimer's disease: Deficits in semantic processing. *Acta Neurol Scand* 108:82–89.

Lee JH, Bliwise DL, Ansari FP, Goldstein FC, Cellar JS, Lah JJ, Levey AI. (2007). Daytime sleepiness and functional impairment in Alzheimer disease. *Am J Geriatr Psychiatry* 15:620–626.

Lewy AJ, Bauer VK, Ahmed S, Thomas KH, Cutler NL, Singer CM, Moffit MT, Sack RL. (1998). The human phase response curve (PRC) to melatonin is about 12 hours out of phase with the PRC to light. *Chronobiol Int* 15:71–83.

Marques SC, Oliveira CR, Outeiro TF, Pereira CM. (2010). Alzheimer's disease: The quest to understand complexity. *J Alzheimers Dis* 21:373–383.

Mathes AM, Kubulus D, Waibel L, Weiler J, Heymann P, Wolf B, Rensing H. (2008). Selective activation of melatonin receptors with ramelteon improves liver function and hepatic perfusion after hemorrhagic shock in rat. *Crit Care Med* 36:2863–2870.

Matsubara E, Bryant-Thomas T, Pacheco Quinto J, Henry TL, Poeggeler B, Herbert D, Cruz-Sanchez F et al. (2003). Melatonin increases survival and inhibits oxidative and amyloid pathology in a transgenic model of Alzheimer's disease. *J Neurochem* 85:1101–1108.

McKenna JT, Christie MA, Jeffrey BA, McCoy JG, Lee E, Connolly NP, Ward CP, Strecker RE. (2012). Chronic ramelteon treatment in a mouse model of Alzheimer's disease. *Arch Ital Biol* 150:5–14.

Merlino G, Piani A, Gigli GL, Cancelli I, Rinaldi A, Baroselli A, Serafini A, Zanchettin B, Valente M. (2010). Daytime sleepiness is associated with dementia and cognitive decline in older Italian adults: A population-based study. *Sleep Med* 11:372–377.

Olcese JM, Cao C, Mori T, Mamcarz MB, Maxwell A, Runfeldt MJ, Wang L et al. (2009). Protection against cognitive deficits and markers of neurodegeneration by long-term oral administration of melatonin in a transgenic model of Alzheimer disease. *J Pineal Res* 47:82–96.

Otalora BB, Popovic N, Gambini J, Popovic M, Vina J, Bonet-Costa V, Reiter RJ, Camello PJ, Rol MA, Madrid JA. (2012). Circadian system functionality, hippocampal oxidative stress, and spatial memory in the APPswe/PS1dE9 transgenic model of Alzheimer disease: Effects of melatonin or ramelteon. *Chronobiol Int* 29:822–834.

Pandi-Perumal SR, Srinivasan V, Maestroni GJ, Cardinali DP, Poeggeler B, Hardeland R. (2006). Melatonin: Nature's most versatile biological signal? *FEBS J* 273:2813–2838.

Pandi-Perumal SR, Trakht I, Srinivasan V, Spence DW, Maestroni GJ, Zisapel N, Cardinali DP. (2008). Physiological effects of melatonin: Role of melatonin receptors and signal transduction pathways. *Prog Neurobiol* 85:335–353.

Pappolla M, Bozner P, Soto C, Shao H, Robakis NK, Zagorski M, Frangione B, Ghiso J. (1998). Inhibition of Alzheimer beta-fibrillogenesis by melatonin. *J Biol Chem* 273:7185–7188.

Park SW, Ko HG, Lee N, Lee HR, Rim YS, Kim H, Lee K, Kaang BK. (2010). Aged wild-type littermates and APPswe+PS1/dE9 mice present similar deficits in associative learning and spatial memory independent of amyloid load. *Genes Genomics* 32:63–70.

Patel NS, Paris D, Mathura V, Quadros AN, Crawford FC, Mullan MJ. (2005). Inflammatory cytokine levels correlate with amyloid load in transgenic mouse models of Alzheimer's disease. *J Neuroinflammation* 2:9.

Peng CX, Hu J, Liu D, Hong XP, Wu YY, Zhu LQ, Wang JZ. (2013). Disease-modified glycogen synthase kinase-3beta intervention by melatonin arrests the pathology and memory deficits in an Alzheimer's animal model. *Neurobiol Aging* 34:1555–1563.

Pevet P, Challet E. (2011). Melatonin: Both master clock output and internal time-giver in the circadian clocks network. *J Physiol Paris* 105:170–182.

Poeggeler B, Miravalle L, Zagorski MG, Wisniewski T, Chyan YJ, Zhang Y, Shao H et al. (2001). Melatonin reverses the profibrillogenic activity of apolipoprotein E4 on the Alzheimer amyloid Abeta peptide. *Biochemistry* 40:14995–15001.

Quinn J, Kulhanek D, Nowlin J, Jones R, Pratico D, Rokach J, Stackman R. (2005). Chronic melatonin therapy fails to alter amyloid burden or oxidative damage in old Tg2576 mice: Implications for clinical trials. *Brain Res* 1037:209–213.

Reiserer RS, Harrison FE, Syverud DC, McDonald MP. (2007). Impaired spatial learning in the APPSwe + PSEN1DeltaE9 bigenic mouse model of Alzheimer's disease. *Genes Brain Behav* 6:54–65.

Rojo LE, Fernandez JA, Maccioni AA, Jimenez JM, Maccioni RB. (2008). Neuroinflammation: Implications for the pathogenesis and molecular diagnosis of Alzheimer's disease. *Arch Med Res* 39:1–16.

Rosales-Corral S, Tan DX, Reiter RJ, Valdivia-Velazquez M, Martinez-Barboza G, Acosta-Martinez JP, Ortiz GG. (2003). Orally administered melatonin reduces oxidative stress and proinflammatory cytokines induced by amyloid-beta peptide in rat brain: A comparative, in vivo study versus vitamin C and E. *J Pineal Res* 35:80–84.

Rosales-Corral SA, Acuna-Castroviejo D, Coto-Montes A, Boga JA, Manchester LC, Fuentes-Broto L, Korkmaz A, Ma S, Tan DX, Reiter RJ. (2012). Alzheimer's disease: Pathological mechanisms and the beneficial role of melatonin. *J Pineal Res* 52:167–202.

Ruan L, Kang Z, Pei G, Le Y. (2009). Amyloid deposition and inflammation in APPswe/PS1dE9 mouse model of Alzheimer's disease. *Curr Alzheimer Res* 6:531–540.

Satlin A, Volicer L, Stopa EG, Harper D. (1995). Circadian locomotor activity and core-body temperature rhythms in Alzheimer's disease. *Neurobiol Aging* 16:765–771.

Savaskan E, Ayoub MA, Ravid R, Angeloni D, Fraschini F, Meier F, Eckert A, Muller-Spahn F, Jockers R. (2005). Reduced hippocampal MT2 melatonin receptor expression in Alzheimer's disease. *J Pineal Res* 38:10–16.

Savaskan E, Olivieri G, Meier F, Brydon L, Jockers R, Ravid R, Wirz-Justice A, Muller-Spahn F. (2002). Increased melatonin 1a-receptor immunoreactivity in the hippocampus of Alzheimer's disease patients. *J Pineal Res* 32:59–62.

Savonenko A, Xu GM, Melnikova T, Morton JL, Gonzales V, Wong MP, Price DL, Tang F, Markowska AL, Borchelt DR. (2005). Episodic-like memory deficits in the APPswe/PS1dE9 mouse model of Alzheimer's disease: Relationships to beta-amyloid deposition and neurotransmitter abnormalities. *Neurobiol Dis* 18:602–617.

Spires TL, Hyman BT. (2005). Transgenic models of Alzheimer's disease: Learning from animals. *NeuroRx* 2:423–437.

Srinivasan V, Kaur C, Pandi-Perumal S, Brown GM, Cardinali DP. (2010). Melatonin and its agonist ramelteon in Alzheimer's disease: Possible therapeutic value. *Int J Alzheimers Dis* 2011:741974.

Sterniczuk R, Dyck RH, Laferla FM, Antle MC. (2010). Characterization of the 3xTg-AD mouse model of Alzheimer's disease: Part 1. Circadian changes. *Brain Res* 1348:139–148.

Stopa EG, Volicer L, Kuo-Leblanc V, Harper D, Lathi D, Tate B, Satlin A. (1999). Pathologic evaluation of the human suprachiasmatic nucleus in severe dementia. *J Neuropathol Exp Neurol* 58:29–39.

van Someren EJ, Hagebeuk EE, Lijzenga C, Scheltens P, de Rooij SE, Jonker C, Pot AM, Mirmiran M, Swaab DF. (1996). Circadian rest-activity rhythm disturbances in Alzheimer's disease. *Biol Psychiatry* 40:259–270.

Vitiello MV, Bliwise DL, Prinz PN. (1992). Sleep in Alzheimer's disease and the sundown syndrome. *Neurology* 42:83–93.

Vloeberghs E, Van DD, Engelborghs S, Nagels G, Staufenbiel M, De Deyn PP. (2004). Altered circadian locomotor activity in APP23 mice: A model for BPSD disturbances. *Eur J Neurosci* 20:2757–2766.

Volicer L, Harper DG, Manning BC, Goldstein R, Satlin A. (2001). Sundowning and circadian rhythms in Alzheimer's disease. *Am J Psychiatry* 158:704–711.

Welsh DK, Takahashi JS, Kay SA. (2010). Suprachiasmatic nucleus: Cell autonomy and network properties. *Annu Rev Physiol* 72:551–577.

Wisor JP, Edgar DM, Yesavage J, Ryan HS, McCormick CM, Lapustea N, Murphy GM, Jr. (2005). Sleep and circadian abnormalities in a transgenic mouse model of Alzheimer's disease: A role for cholinergic transmission. *Neuroscience* 131:375–385.

Wu YH, Feenstra MG, Zhou JN, Liu RY, Torano JS, Van Kan HJ, Fischer DF, Ravid R, Swaab DF. (2003). Molecular changes underlying reduced pineal melatonin levels in Alzheimer disease: Alterations in preclinical and clinical stages. *J Clin Endocrinol Metab* 88:5898–5906.

Wu YH, Fischer DF, Kalsbeek A, Garidou-Boof ML, van der Vliet J, van Heijningen C, Liu RY, Zhou JN, Swaab DF. (2006). Pineal clock gene oscillation is disturbed in Alzheimer's disease, due to functional disconnection from the "master clock". *FASEB J* 20:1874–1876.

Wu YH, Swaab DF. (2005). The human pineal gland and melatonin in aging and Alzheimer's disease. *J Pineal Res* 38:145–152.

Wu YH, Swaab DF. (2007). Disturbance and strategies for reactivation of the circadian rhythm system in aging and Alzheimer's disease. *Sleep Med* 8:623–636.

Wu YH, Zhou JN, Van HJ, Jockers R, Swaab DF. (2007). Decreased MT1 melatonin receptor expression in the suprachiasmatic nucleus in aging and Alzheimer's disease. *Neurobiol Aging* 28:1239–1247.

Xie L, Kang H, Xu Q, Chen MJ, Liao Y, Thiyagarajan M, O'Donnell J et al. (2013). Sleep drives metabolite clearance from the adult brain. *Science* 342:373–377.

Zhou JN, Liu RY, Kamphorst W, Hofman MA, Swaab DF. (2003). Early neuropathological Alzheimer's changes in aged individuals are accompanied by decreased cerebrospinal fluid melatonin levels. *J Pineal Res* 35:125–130.

27 Melatonin's Mitochondrial Protective Role in Alzheimer's Mice

Role of Melatonin Receptors

Vedad Delic and Patrick C. Bradshaw

CONTENTS

27.1 ALZHEIMER'S DISEASE

Dementia affects 35 million people worldwide, and this total is predicted to increase to over 60 million people in 2030 and over 100 million people by 2050. Due to this sharp increase in prevalence and the high cost for healthcare for these individuals, treatments that prevent or even delay the onset of dementia are highly sought. In this regard, melatonin has been identified as a compound that may delay the onset of certain types of dementia. In this chapter, we will discuss the potential of melatonin as a treatment for Alzheimer's disease (AD), the most prevalent form of dementia, and the role that melatonin and its cellular receptors play in this process.

AD is characterized clinically by memory and functional deficits culminating in progressive cognitive decline. The vast majority of individuals with AD are 65 years of age or older. The likelihood of developing Alzheimer's doubles roughly every 5 years after the age of 65. The risk reaches nearly 50% after the age of 85. Another strong risk factor is family history, as both genetics and environmental factors play a role in disease development.

In the brain, AD mostly affects the hippocampal and cerebral cortical regions and is characterized by the presence of extracellular amyloid beta (Aβ) plaques, intracellular soluble Aβ, and also by the presence of intracellular neurofibrillary tangles (NFTs) caused by the hyperphosphorylated microtubule-associated tau protein. Familial or early onset forms of AD exist, make up around 5% of total AD cases, and are generally caused by mutations in the presenilin-1 (PS1) or presenilin-2 (PS2) genes located on chromosome 14 or the amyloid precursor protein (APP) gene located on chromosome 21. The most prevalent (~85%) of these familial forms are caused by mutations in PS1. About 95% of AD cases are sporadic, with age being the largest risk factor. The single largest genetic factor associated with late-onset Alzheimer's disease (LOAD) is the ε4 allele of the apolipoprotein E (APOE) gene (Corder et al. 1993; Strittmatter et al. 1993). It likely contributes to 20%–25% of LOAD cases. The *APOE* gene, coding for the major cholesterol-binding protein in the human brain, apoliprotein E, exists as three polymorphic alleles (ε2, ε3, and ε4) resulting in six different genotypes (ε2/ε2, ε2/ε3, ε2/ε4, ε3/ε3, ε3/ε4, and ε4/ε4) (Bu 2009; Mahley 1988). The presence of one ε4 allele increases the likelihood of a person developing AD by three to four times (Bertram and Tanzi 2008; Corder et al. 1993). Antecedent factors influencing AD progression are cardiovascular disease, smoking, hypertension, type II diabetes, obesity, and traumatic brain injury (TBI) (Mayeux and Stern 2012).

27.2 MITOCHONDRIAL DYSFUNCTION AS A CONTRIBUTING FACTOR IN AD

Mitochondrial dysfunction occurs with aging (Chomyn and Attardi 2003; Harman 1972; Kujoth et al. 2005; Wallace and Fan 2009) and also contributes to the pathophysiology of many diseases that occur as a normal part of aging. An association between mitochondrial dysfunction and AD is well established (Parker 1991; Perry et al. 1980; Sorbi et al. 1983). Redox imbalance and oxidative stress, which frequently develop as a result of mitochondrial dysfunction, have been observed in aging and neurodegenerative disorders such as AD. This oxidative stress may lead to damage of important mitochondrial components such as mitochondrial DNA, mitochondrial proteins such as the electron transport chain (ETC) components, and mitochondrial phospholipids, especially to cardiolipin, which contains a high percentage of unsaturated fatty acids. Damage to the ETC proteins or cardiolipin, required for ETC complex function, may result in decreased ETC oxygen consumption, ATP production, and mitochondrial membrane potential, and also increased ROS production. This damage to mitochondria has also been shown to result in dysregulation of mitochondrial dynamics (e.g., fission, fusion, transport, mitochondrial biogenesis, and mitophagy) and, if severe enough, results in mitochondrial-mediated apoptosis. The mitochondrial damage has been shown to be most severe in areas adjacent to amyloid plaques (Xie et al. 2013a).

Mitochondrial dysfunction can have varying effects depending on the degree. Small amounts of mitochondrial dysfunction, which typically slightly increase ROS production or decrease ATP levels, can activate retrograde signaling and be compensated for by increased mitochondrial biogenesis. Two signals for mitochondrial biogenesis, which occur as a result of mitochondrial dysfunction, include increased ROS production and AMP-activated protein kinase

(AMPK) activation. Both of these signals can lead to the activation of peroxisome proliferator-activated receptor gamma coactivator 1-alpha (PGC-1α), which binds nuclear respiratory factors-1 and 2, estrogen-related receptor-alpha, and other transcription factors that increase the expression of proteins targeted to mitochondria. When damage to mitochondria becomes too great to be compensated by the increased mitochondrial biogenesis and/or degradation of damaged mitochondria through mitophagy, mitochondrial dysfunction arises, which can lead to cell death.

Due to the fact that amyloid plaques are one of the two major phenotypic markers of AD, many strategies have been employed to pharmacologically decrease their formation, or once they have formed, to break them down and remove the plaque components from the brain. Unfortunately, anti-amyloid drugs have shown little success in AD treatment (Herrmann et al. 2011). Therefore, novel multipronged research strategies need to be employed, which attack other phenotypes of AD. With this in mind, much scientific evidence links mitochondrial dysfunction with AD, yet therapies targeting mitochondrial dysfunction in AD have yet to be methodically tested in human patients. Several research groups have discovered altered activities of the mitochondrial citric acid cycle enzymes and ETC complexes, predominately complex IV, in the brains from postmortem AD patients (Bubber et al. 2005) and in transgenic mouse models of familial AD (Hauptmann et al. 2008). Mitochondrial ATP synthase levels have also been found to be low in the AD brain (Schagger and Ohm 1995). Specifically, mitochondrial dysfunction is one of the earliest symptoms in double-mutant London and Swedish APP transgenic mice (Hauptmann et al. 2008). Mitochondrial dysfunction occurs in these mice at 3 months of age, at an age before extracellular deposits of Aβ peptide, and much earlier than memory impairment. Therefore, mitochondrial dysfunction may be an initiating event in AD, an idea which has been formalized in the mitochondrial cascade hypothesis of AD (Swerdlow et al. 2013; Swerdlow and Khan 2004).

An oligomeric form of Aβ, likely dimeric Aβ, has been described to inhibit cytochrome c oxidase of the mitochondrial ETC (Crouch et al. 2005). Through this interaction, as well as through an interaction with complex I (Munguia et al. 2006), Aβ stimulates increased reactive oxygen species production from the ETC, which may ultimately lead to cell death through opening of the mitochondrial permeability transition pore in the inner mitochondrial membrane. This event uncouples mitochondrial oxidative phosphorylation and releases mitochondrial factors such as cytochrome c and apoptosis-inducing factor (AIF) into the cytoplasm, where caspases execute the apoptotic program. Knockout of cyclophilin D, a mitochondrial matrix peptidylprolyl cis–trans isomerase and Aβ-binding protein (Du et al. 2008) that facilitates mitochondrial permeability transition pore opening, increases learning and memory and synaptic function in a mouse model of AD (Du et al. 2008). In this regard, APP_{swe} overexpression upregulates mitochondrial cyclophilin D (Manczak et al. 2010), which facilitates mitochondrial permeability transition pore opening and cell death. High Aβ levels, in addition to inhibiting the mitochondrial ETC, have been shown to decrease TCA cycle enzyme activities (Bubber et al. 2005), alter the rates of mitochondrial fission and fusion (Manczak et al. 2011), decrease the rate of mitophagy (Santos et al. 2010), and decrease axonal transport of mitochondria (Calkins et al. 2011).

But whether mitochondrial dysfunction plays an initiating role in human AD pathogenesis, specifically in the synaptic or neuronal loss associated with the late-onset form of AD is controversial. It may as yet prove that mitochondrial dysfunction is a result of other AD pathology. In this regard, some reports of normal mitochondrial function in AD models have been published. For example, mitochondrial function appeared mostly normal in presynaptic nerve terminals in mouse models of AD (Choi et al. 2012). But the ability of Aβ, a common component of AD plaques (Pereira et al. 1998); hyperphosphorylated tau, a component of NFTs (David et al. 2005); and apolipoprotein E4 (Chang et al. 2005), a major risk factor for late-onset AD, to each cause mitochondrial dysfunction lends strong support for a mitochondrial etiology for AD.

27.3 MELATONIN PROTECTS MITOCHONDRIAL, CELL, AND BRAIN FUNCTION IN AD MICE IN MANY WAYS

Melatonin has long been used to alleviate jet lag and to modify circadian rhythms. However, its potential use as an antioxidant in the treatment of disease is gaining momentum, especially since controlled-release formulas having a more favorable pharmacokinetic profile have been developed (Lemoine and Zisapel 2012). Melatonin can be used to increase the amount of total and REM sleep (Dijk and Cajochen 1997). Sleep itself can protect against AD as Aβ and other metabolic waste products are more quickly cleared from brains in the sleeping state than during wakefulness (Xie et al. 2013b). This has led some to suggest that disrupted circadian rhythms may be partly a cause instead of a result of AD (Bedrosian and Nelson 2012). Melatonin has an advantage over many other antioxidants such as Vitamin E in that it is freely permeable to the brain (Lahiri et al. 2004). Melatonin has been reported to extend the life span of APP transgenic mice (Matsubara et al. 2003) and prevent cognitive dysfunction in APP/PS1 mice (Olcese et al. 2009), although it failed to reverse cognitive dysfunction if melatonin treatment was initiated after the initiation of cognitive impairment (G. Arendash, unpublished data). Others also found little to no positive effect of melatonin on oxidative stress or amyloid burden when treatment was initiated late in life in a mouse AD model (Quinn et al. 2005). Melatonin was also shown to decrease immunoreactive Aβ levels (Olcese et al. 2009) and protein nitration (Matsubara et al. 2003) in the brain of AD mouse models. Melatonin is hypothesized to delay AD in mice through preventing Aβ oligomerization (Olcese et al. 2009; Pappolla et al. 1998), preventing reactive oxygen species-mediated damage (Ionov et al. 2011), and by stabilizing mitochondrial function in the presence of increased levels of Alzheimer's amyloid (Dragicevic et al. 2011) or phosphorylated tau (Peng et al. 2013). One study found that supplementation of Tg2576 mutant amyloid precursor protein-expressing Alzheimer's mice with melatonin from 8 to 12 months of the life span was critical, because treating the mice with melatonin from 4 to 8 months of life was ineffective (Peng et al. 2013).

27.4 MELATONIN RECEPTORS

In the plasma membrane, two receptors for melatonin, MT1 and MT2, have been identified. These melatonin receptors are seven-pass G-protein–coupled receptors (GPCRs) that can homo- or heterodimerize and can signal through multiple signaling pathways in different cells by using distinct G-protein alpha subunit isoforms. Melatonin has subnanomolar affinity for these receptors (Dubocovich and Markowska 2005), which allows the receptors to be activated when melatonin levels rise due to increased pineal gland secretion at night. The MT1 and MT2 genes, located on chromosomes 4q and 11q, respectively, show 55% identity and encode proteins of 350 and 365 amino acids in length. The intracellular portions of the receptors contain casein kinase 1 and 2, protein kinase A, and protein kinase C phosphorylation sites. MT1 and MT2 are especially found at high density in the suprachiasmatic nucleus in the brain (Liu et al. 1997) where they are bound by melatonin and influence the diurnal rhythm. mRNA levels of the MT1 receptor have also been shown to undergo diurnal variation (Guerrero et al. 1999). The classic action of melatonin receptor signaling is to dissociate heterotrimeric G proteins where the Gα and Gβγ subunits interact with various signaling pathways. In the suprachiasmatic nucleus following melatonin binding, the Gαi2 and Gαi3 isoforms have been described to inhibit adenylate cyclase activity (Brydon et al. 1999). This event decreases cAMP levels and decreases pituitary adenylate cyclase-activating protein (PACAP)-mediated CREB activation, affecting the circadian clock (Travnickova-Bendova et al. 2002).

In addition to inhibiting cAMP-mediated signaling, melatonin receptors can activate phospholipase C signaling through Gq-coupling (Chan et al. 2002). This activates Ca^{2+}-mediated and protein kinase C (PKC) phosphorylation cascades. These signals activate calmodulin kinase and mitogen-activated protein kinase (MAPK) signaling kinases including p38, JNK, and ERK. There is also evidence that melatonin receptor activation may lead to stimulation of the PI3 kinase/Akt pathway or opening of different ion channels such as voltage-gated calcium channels or large conductance calcium-activated potassium channels (Hardeland 2009b). In addition to these effects mediated by GPCRs, melatonin can also interact directly with quinone reductase 2, orphan nuclear receptors such as retinoid Z receptor (RZR) and retinoid acid receptor-related orphan receptor (ROR), calmodulin, calreticulin, and other proteins (Reiter et al. 2010). The melatonin receptor-related protein GPR50, an orphan GPCR that is 45% identical to melatonin receptors, can bind to MT1 and function as an antagonist (Levoye et al. 2006). From this collection of data, it is apparent that melatonin and melatonin receptors act through many different complex signaling mechanisms to affect cellular physiology.

The melatonin MT1 and MT2 receptors play complementary roles in cell function. MT1 receptors are found in similar, but not identical locations. MT1 is located in the suprachiasmatic nucleus, cerebellum, hippocampus, substantia nigra, and many other tissues of the body, while MT2 receptor expression is mostly restricted to the brain, including the suprachiasmatic nucleus, but expression has been found in a few other tissues as well, but not to the extent of MT1 (Dubocovich and Markowska 2005). In the suprachiasmatic nucleus, MT1 and MT2 receptors appear to play different roles. MT1 receptor activation leads to acute inhibition of neuronal activity, while MT2 activation leads to the phase shift of circadian rhythm (Hunt et al. 2001; Liu et al. 1997). Some of these studies using brain slices have also been confirmed in mice with specific knockout of the MT1 receptor or in mice administered 4P-PDOT, a specific inhibitor of MT2 (Dubocovich et al. 2005). The phenotypes of the MT1 and MT2 receptor knockouts are subtle (Jin et al. 2003) and phenotypes have mainly been found at the molecular level, with the exception of the lack of phase-shifting effect of melatonin in the MT1 receptor knockout mice and a deficit in learning and memory in the MT2 receptor knockout mice (Dubocovich et al. 2005, Larson et al. 2006).

27.5 MELATONIN RECEPTORS AND THE HIPPOCAMPUS

The hippocampus plays a key role in memory and is the main site of dysfunction in AD. Both MT1 and MT2 receptors are expressed in the hippocampus, specifically in the CA1 and CA3 regions, the subiculum, and the dentate gyrus (Musshoff et al. 2002). Administration of melatonin to hippocampal slices was shown to increase the neuronal firing rate, which was blocked by luzindole, which shows a slight selectivity for MT2 over MT1. Melatonin also showed an inhibition of long-term potentiation in hippocampal slices, which was inhibited by both luzindole and 4P-PDOT, a more specific inhibitor of MT2 (Wang et al. 2005). MT1 and MT2 knockout mice were used to confirm these observations (Dubocovich et al. 2005; Larson et al. 2006). To verify that MT2 is involved in learning and memory, MT2 knockout mice were tested in the elevated plus maze paradigm (Larson et al. 2006). A marked inhibition of learning was measured indicating a deficiency in long-term synaptic plasticity.

In human studies it was shown that the time of day that melatonin was administered drastically altered the effect of melatonin on learning behavior (Gorfine and Zisapel 2007). These studies implicated melatonin and the circadian clock in human memory processing and consolidation. In further neuroimaging studies, it was shown that melatonin administration was as effective as a 2 h nap in increasing performance in a verbal association test (Gorfine et al. 2007). Future studies should aim to confirm these initial human studies on the effects of melatonin on learning behavior and determine other paradigms where melatonin treatment may be beneficial.

27.6 MITOCHONDRIAL LOCALIZATION OF THE MT1 MELATONIN RECEPTOR IN MICE

Melatonin MT1 receptors have been localized to brain mitochondria in mice, whereas only a trace amount of MT2 receptors could be found in this subcellular localization (Wang et al. 2011). Interestingly, MT1 receptor levels declined in the R6/2 mouse model of Huntington's disease and this decline, including the decline in mitochondrial levels, was delayed by melatonin treatment, which also delayed disease pathology in these mice. Therefore, the decline in MT1 receptors induced by mutant Huntingtin protein may be a contributing factor to the disease. It is currently unclear what role, if any, that mitochondrial MT1 receptors play in brain physiology and if these declines occur in other neurodegenerative disorders such as AD, but identifying the mitochondrial role of MT1 receptors will likely prove instrumental in understanding the protective effect of melatonin treatment for neurodegenerative diseases.

27.7 AD PATIENTS HAVE DECREASED MELATONIN LEVELS

The involvement of melatonin in the pathogenesis of AD has been suggested from studies indicating that AD patients have decreased blood and cerebrospinal fluid (CSF) levels of melatonin (Maurizi 1997). In AD patients, ApoE4 allele status also plays a role in determining melatonin levels as APOE4 homozygotes have the lowest melatonin levels (Liu et al. 1999). Unexpectedly, in C6 glioma cells, expression of APOE4, the allele most associated with the development of AD, expression increased melatonin levels compared to expression of APOE3 or APOE2 (Liu et al. 2012b). Perhaps different results would be obtained in neuronal cells or in primary, nontransformed cells. In addition, lower CSF levels of melatonin have been strongly correlated with progression of AD neuropathology, and preclinical AD subjects already have decreased CSF melatonin levels. In addition to these findings suggesting that decreased melatonin levels are an early event in AD pathogenesis, epidemiologic studies have reported that melatonin treatment provides cognitive benefit to patients with mild cognitive impairment (MCI) as well as AD patients. The decreased melatonin levels in the AD brain may allow oxygen and nitrogen free radicals to damage sensitive neurons (Srinivasan et al. 2010).

27.8 CHANGES IN MELATONIN RECEPTOR LEVELS IN AGING, PARKINSON'S DISEASE, AND AD

MT1 and MT2 receptor levels have been shown to decline with aging in many tissues (with the exception of the thymus) in rats (Sanchez-Hidalgo et al. 2009) and mice. There was also a report of the MT1 receptor increasing with age in mice in the spleen (Bondy et al. 2010). The level of the MT2 receptor declined in the aged suprachiasmatic nucleus in mice (von Gall and Weaver 2008) and humans (Wu et al. 2007). But interestingly, the MT2 receptor increased with age in the hippocampus in gerbils (Lee et al. 2010). Therefore, the regulation of melatonin receptor levels is likely species-specific.

The MT1 receptor level was shown to increase in hippocampal sections CA-1-4 from AD patients (Savaskan et al. 2002), while the MT2 receptor levels declined in these and other regions such as the retina in AD patients (Savaskan et al. 2005, 2007). It is possible that the increased MT1 receptor expression is a compensatory mechanism in response to the decreased melatonin levels, while the decreased MT2 receptor levels and/or decrease in melatonin levels may contribute to the onset of AD. In addition to the decline in melatonin receptor levels with aging, both MT1 and MT2 receptors have been shown to decrease in the substantia nigra and the amygdala in Parkinson's disease (PD) patients (Adi et al. 2010). Melatonin has been shown to delay movement problems associated with PD in rodent models (Mayo et al. 2005). But the role that the MT1 and MT2 receptors play in this protection remains unclear.

27.9 MELATONIN TREATMENT HAS SHOWN BENEFICIAL EFFECTS IN AD PATIENTS

Preliminary data with human AD patients showed that melatonin supplementation decreased "sundowning," improved sleep, and slowed disease progression (Maurizi 2001). A slight improvement in cognitive function was observed when melatonin was given to AD patients (Asayama et al. 2003; Brusco et al. 1998, 2000). In addition, a retrospective study showed that melatonin treatment improved cognitive performance and sleep quality in patients with MCI (Furio et al. 2007).

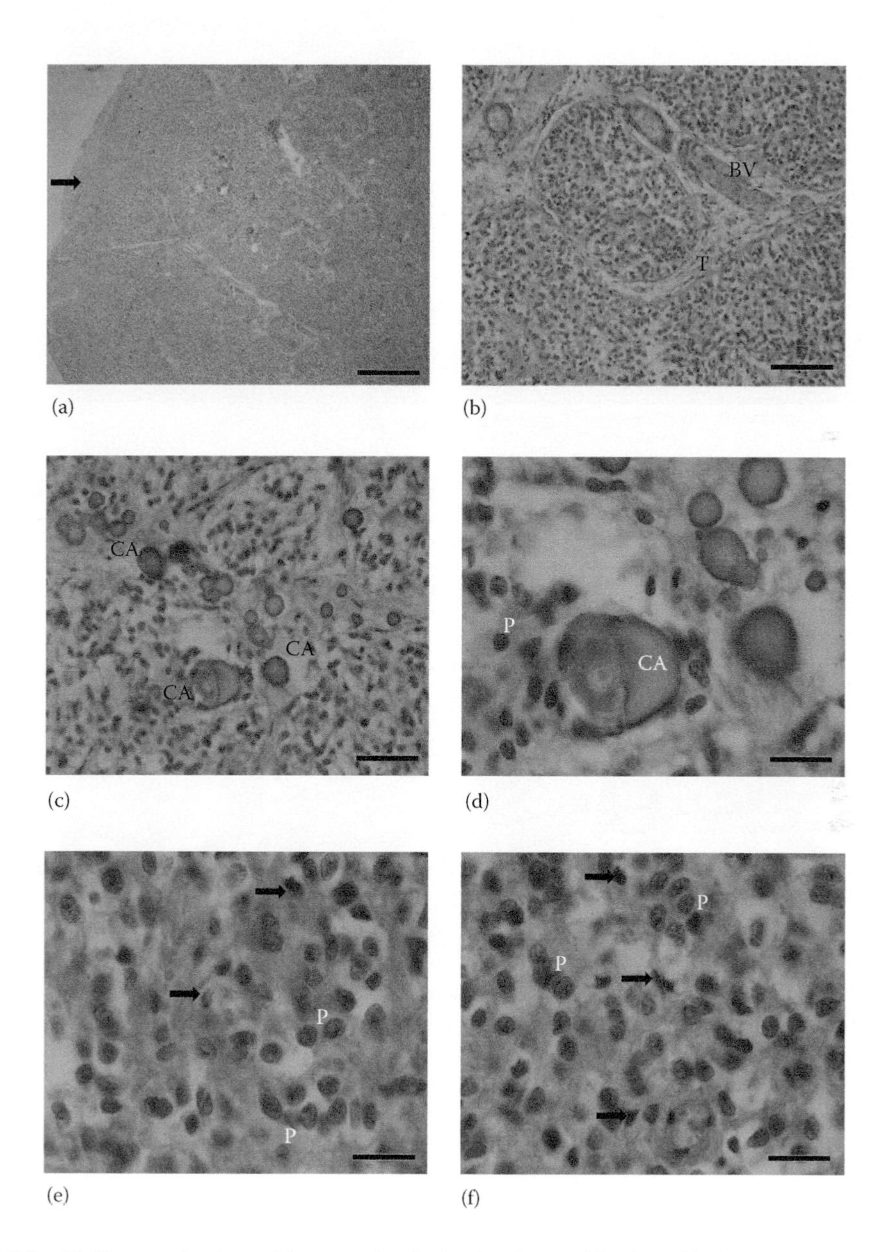

FIGURE 1.1 (a) Panoramic view of human pineal gland. ×4 magnification. Black arrow, capsule-irregular connective tissue. Hematoxylin & Eosin staining. Scale bar = 1250 μm. (b) Trabecular connective tissue separating cells. T, trabecula; BV, blood vessel. ×20 magnification. Hematoxylin & Eosin staining. Scale bar = 250 μm. (c) Panoramic view of human pineal gland. ×40 magnification. CA, corpora arenacea. Hematoxylin & Eosin staining. Scale bar = 125 μm. (d) Panoramic view of human pineal gland. ×40 magnification. CA, corpora arenacea; P, pinealocyte. Hematoxylin & Eosin staining. Scale bar = 50 μm. (e, f) Human pineal gland. ×100 magnification. P, pinealocyte; E, endothelium; Black arrow, interstitial cell. Scale bar = 50 μm.

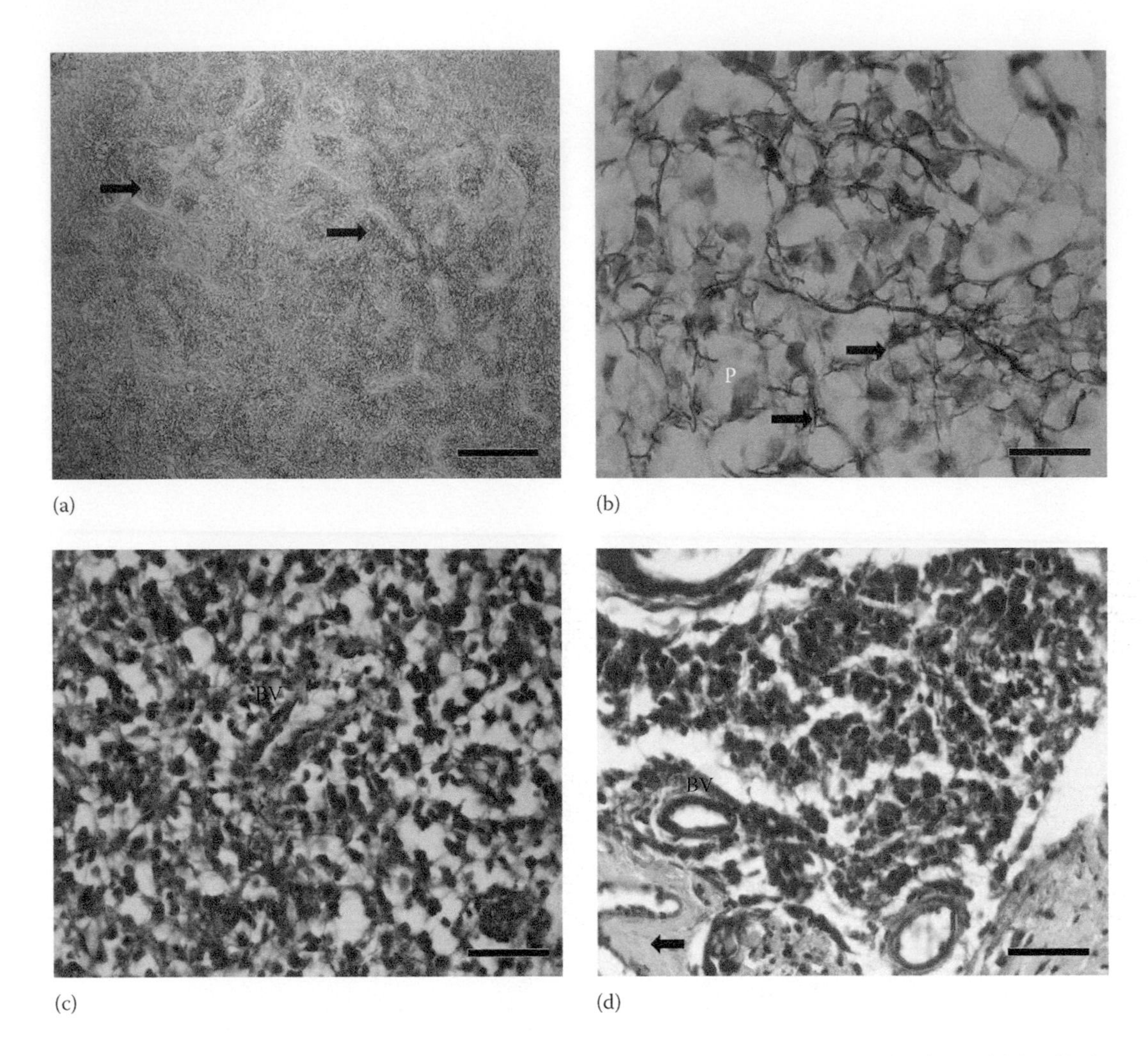

FIGURE 1.2 (a) General distribution of glial fibrillary acid protein (GFAP) positive interstitial cells in human pineal gland (dark brown areas). ×4 magnification. Anti-GFAP staining. Arrow shows dark brown areas. Scale bar = 1250 μm. (b) Interstitial cells and cell processes. Arrow shows cell process. ×100 magnification. Anti-GFAP staining. Scale bar = 50 μm. P, pinealocyte; black arrow, interstitial cell. (c) Parenchyma of the pineal gland. BV, blood vessel (erythrocytes are red in color), ×40 magnification. Gomori trichrome staining. Scale bar = 125 μm. (d) Parenchyma of the pineal gland, ×40 magnification. Gomori trichrome staining. Scale bar = 125 μm. Trabecular connective tissue separating cells. Arrow shows the green-stained connective septa. BV, blood vessel.

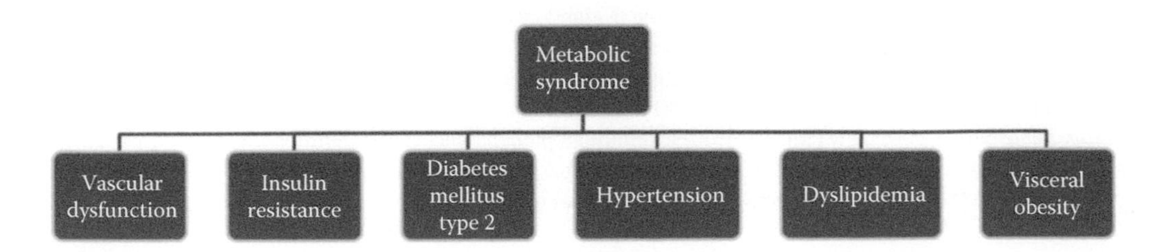

FIGURE 4.1 The metabolic abnormalities of metabolic syndrome.

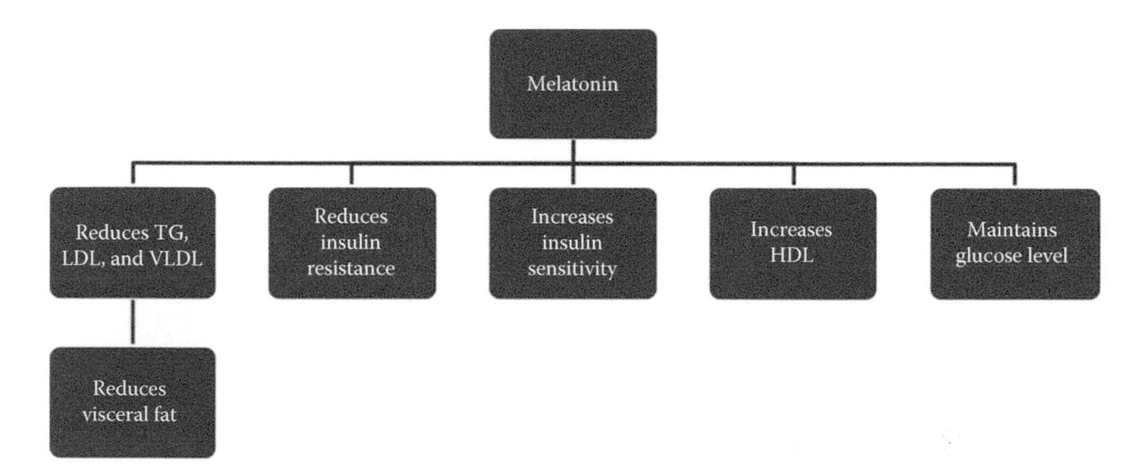

FIGURE 4.2 Melatonin's beneficial actions in ameliorating the signs and symptoms of metabolic syndrome.

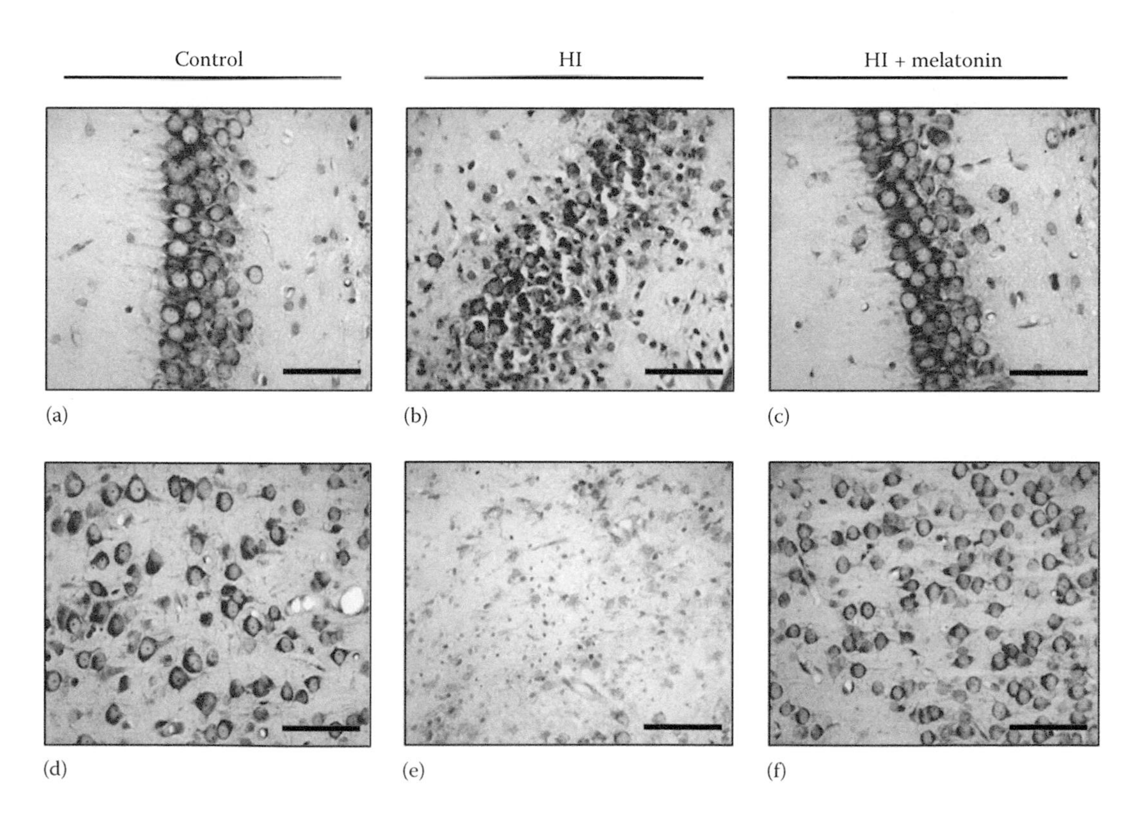

FIGURE 6.1 Nissl-stained brain sections corresponding to the surrounding areas of the CA1 region of the hippocampus (a–c) and cortex (d–f) from neonatal rats showing cell loss after hypoxia-ischemia (b and e) and recovery after melatonin administration (c and f). Bar: 100 μm.

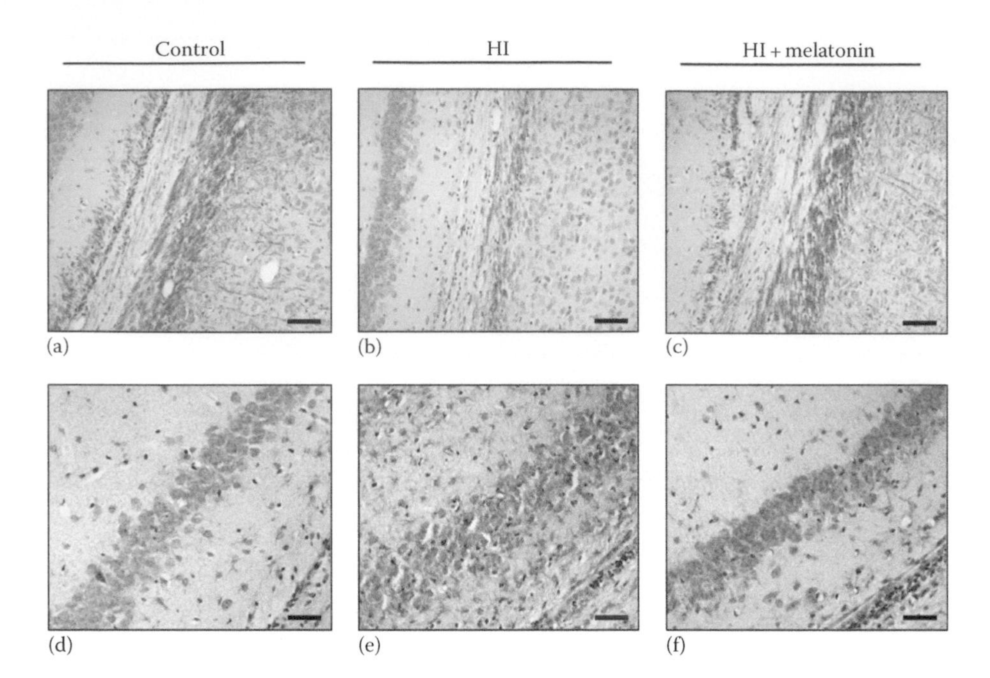

FIGURE 6.2 Myelin basic protein (a–c) and glial fibrillary acidic protein (d–f) immunolabeled brain sections corresponding to the surrounding areas of the CA1 region of the hippocampus (a–c) and the external capsule (d–f) from neonatal rats showing myelination deficit (b) and reactive gliosis (e) after hypoxia-ischemia and recovery after melatonin administration (c and f). Bar: 100 μm.

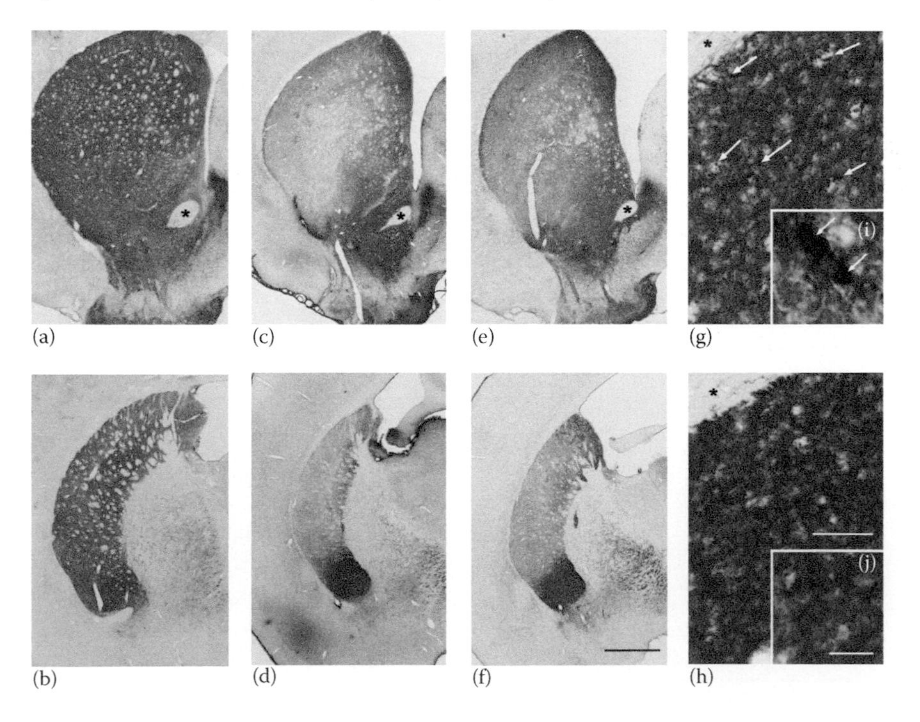

FIGURE 15.2 Low-power photomicrographs through the middle (a, c, e) and caudal (b, d, f) parts of the striatum in 1-month-old (a, b) and 10-month-old *zi/zi* rats without treatment (c, d) and the 10-month-old *zi/zi* rat with a 9-month exposure to drinking water containing melatonin (e, f). Sections were immunostained for TH. Scale bar = 1 mm. (g–j) High-power photomicrographs of TH-immunohistochemically stained sections of the NA of 10-month-old *zi/zi* rats without treatment (g, i) and 10-month-old *zi/zi* rats with a 9-month exposure to drinking water that contained melatonin (h, j). Arrows indicate swollen TH-immunoreactive fibers (g, i). ★ = anterior commissure. Scale bar = 1 mm (a–f), 50 μm (g, h), and 2 μm (i, j).

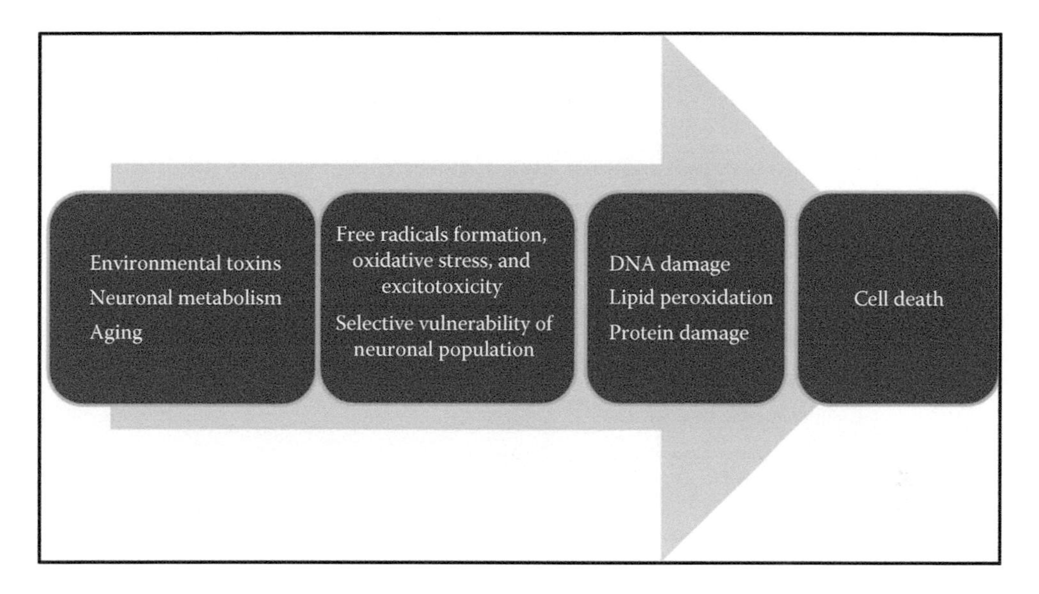

FIGURE 16.1 Various factors that trigger Parkinson's disease.

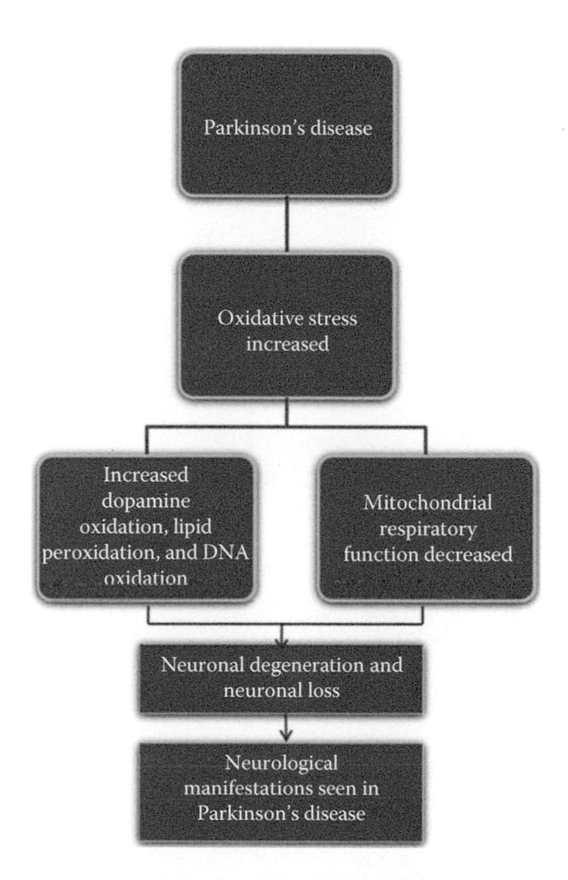

FIGURE 16.2 Role of oxidative stress in the pathogenesis of Parkinson's disease.

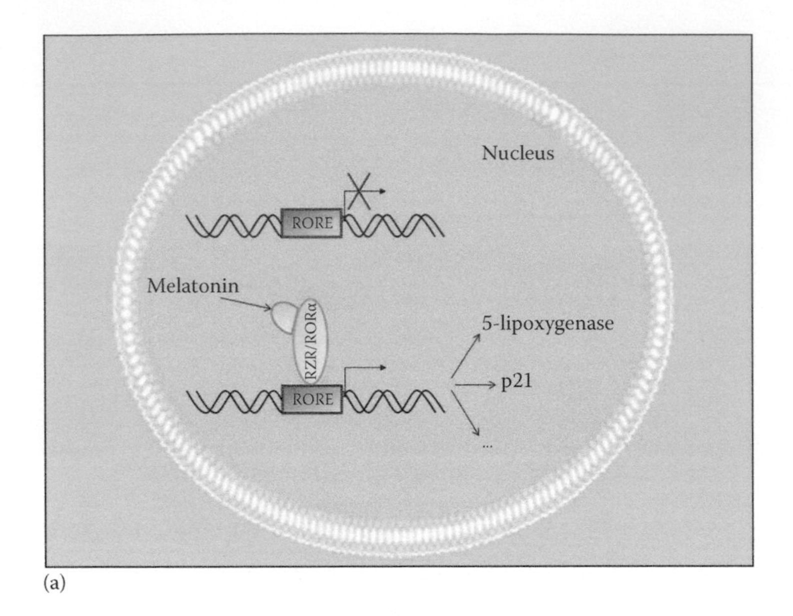

(a)

(b)

FIGURE 18.1 Melatonin receptors. (a) Upon melatonin binding, melatonin nuclear receptor RZR/RORα is recruited onto ROREs present in its target genes and drives their transcription. (b) Melatonin membrane receptors MT1 and MT2 activate G-proteins following melatonin binding. This results in inhibition of adenylate cyclise (AC) and consequent decrease in cAMP intracellular levels. This, in turn, leads to activation of p38MAPK (red arrows) and its downstream targets, such as p53.

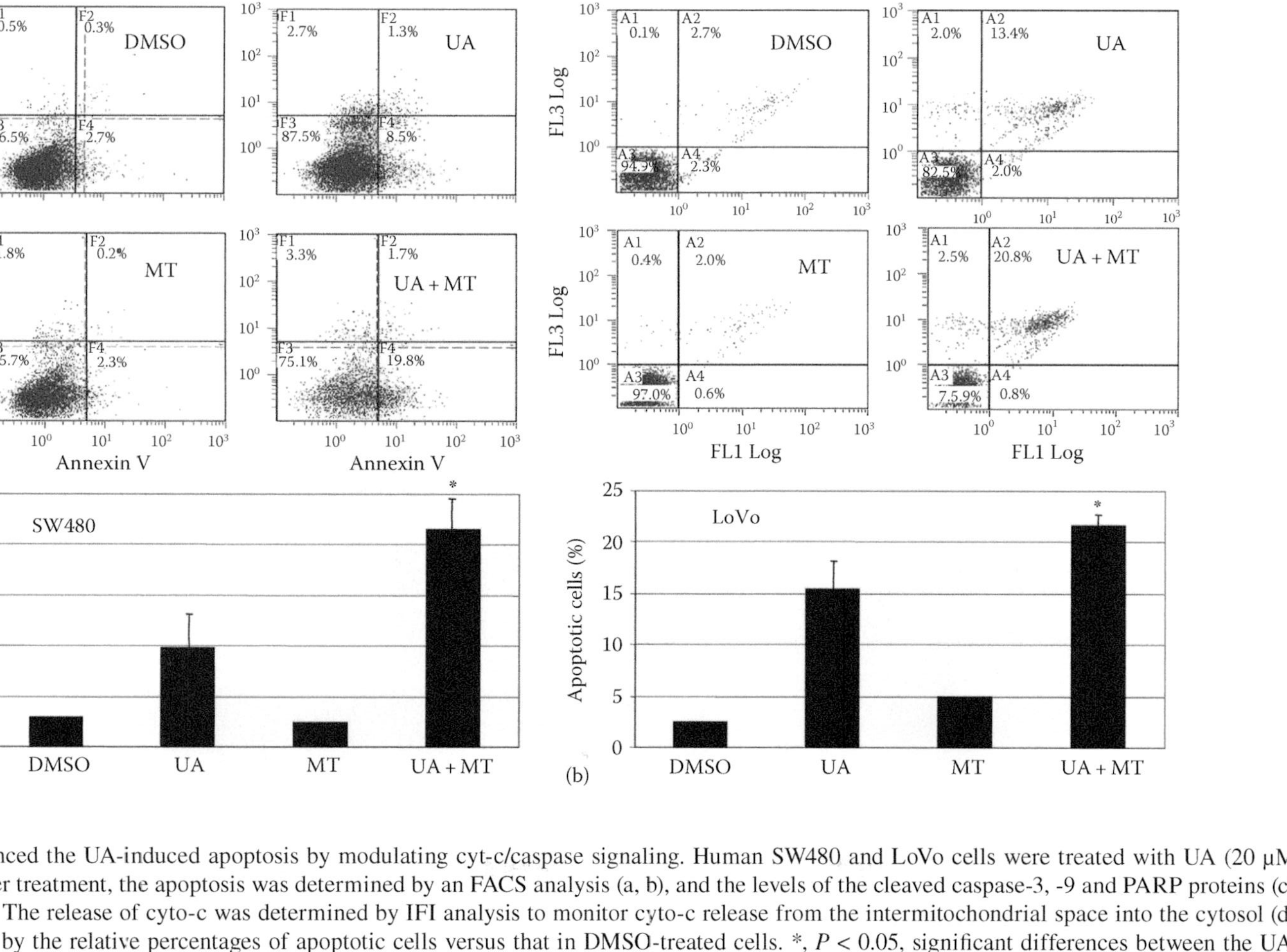

FIGURE 19.3 MT enhanced the UA-induced apoptosis by modulating cyt-c/caspase signaling. Human SW480 and LoVo cells were treated with UA (20 μM) and MT (1.0 mM). At 48 h after treatment, the apoptosis was determined by an FACS analysis (a, b), and the levels of the cleaved caspase-3, -9 and PARP proteins (c) were analyzed by Western blot. The release of cyto-c was determined by IFI analysis to monitor cyto-c release from the intermitochondrial space into the cytosol (d). The apoptoses are represented by the relative percentages of apoptotic cells versus that in DMSO-treated cells. *, $P < 0.05$, significant differences between the UA+MT-treated groups and the UA-treated groups. *(Continued)*

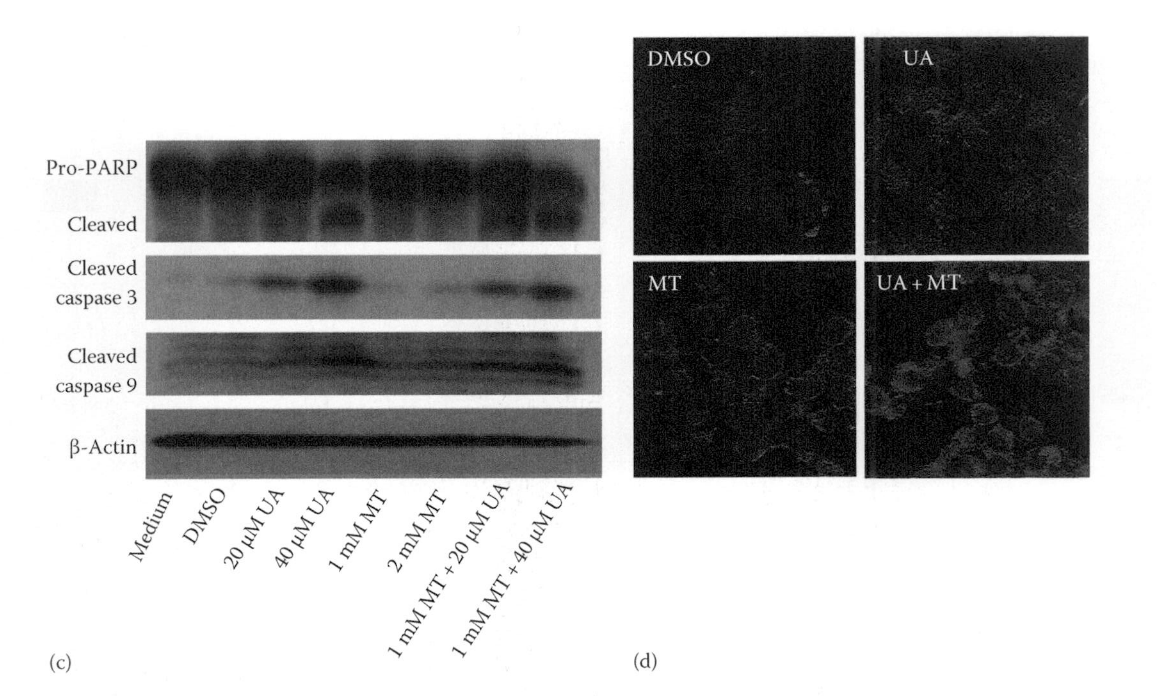

FIGURE 19.3 (*Continued*) MT enhanced the UA-induced apoptosis by modulating cyt-c/caspase signaling. Human SW480 and LoVo cells were treated with UA (20 μM) and MT (1.0 mM). At 48 h after treatment, the apoptosis was determined by an FACS analysis (a, b), and the levels of the cleaved caspase-3, -9 and PARP proteins (c) were analyzed by Western blot. The release of cyto-c was determined by IFI analysis to monitor cyto-c release from the intermitochondrial space into the cytosol (d). The apoptoses are represented by the relative percentages of apoptotic cells versus that in DMSO-treated cells. *, $P < 0.05$, significant differences between the UA+MT-treated groups and the UA-treated groups.

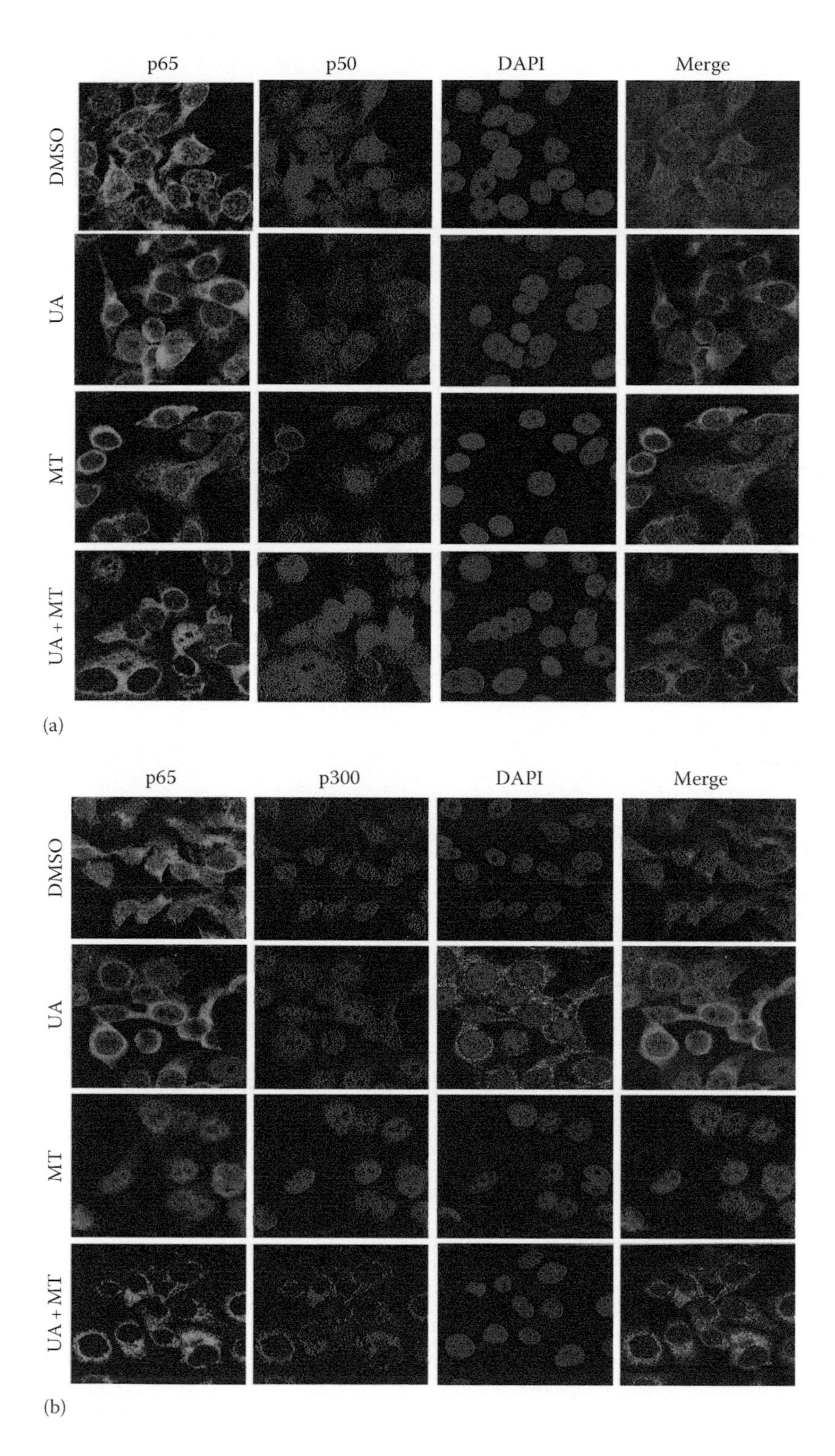

FIGURE 19.5 MT promoted the UA-induced translocation of p300 and NF-κB from nuclei to cytoplasm. Human LoVo cells grown on chamber slides were treated with UA (20 μM) and MT (1.0 mM). At 48 h after treatment, the subcellular localization of p50, p65, and p300 and the colocalization of p65 with p50 (a) or p300 (b) were examined by confocal microscopy analysis with a confocal microscope. More than 100 cells were inspected per experiment, and cells with typical morphology were presented.

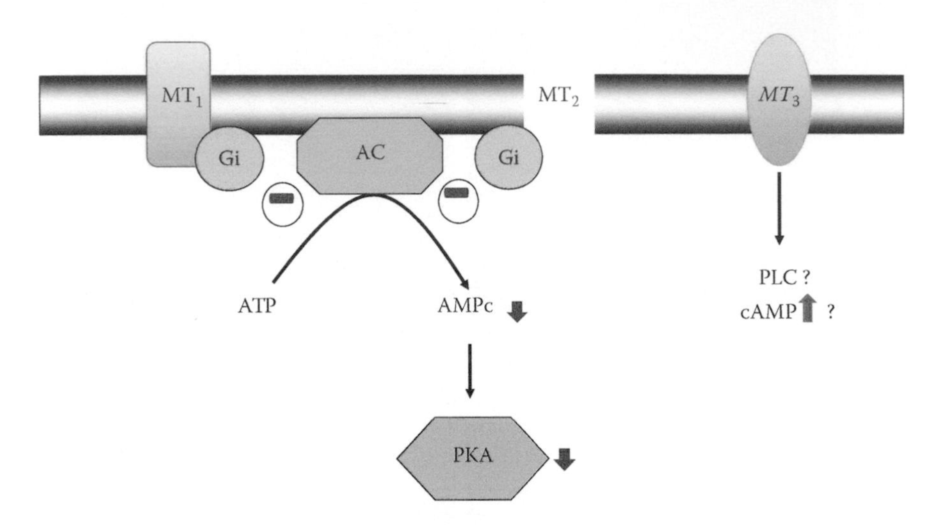

FIGURE 22.2 Melatonin receptors and second messenger cascades. Three melatonin receptors have been described. MT_1 and MT_2 are negatively coupled to adenylate cyclase, whereas the putative MT_3 could be coupled to PLC and/or positively to adenylate cyclase.

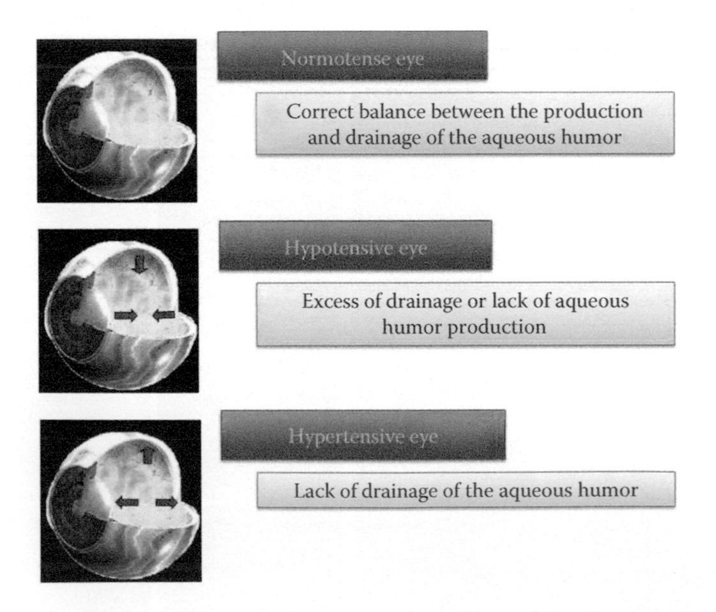

FIGURE 22.4 Ocular conditions related with IOP. Ocular shape and function depends on IOP. Under normal conditions, the production and drainage of the aqueous humor (responsible for IOP) is balanced. The imbalance can produce either an increase in the IOP and therefore a risk of glaucoma or a reduction that may collapse the ocular globe (see the arrows in the figure).

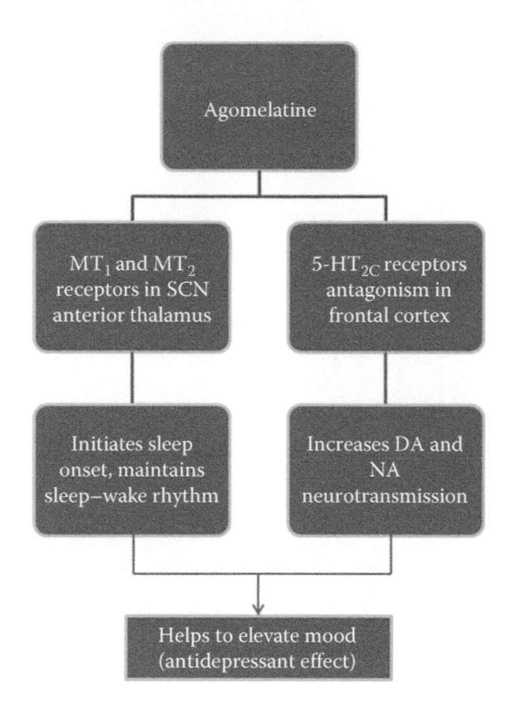

FIGURE 23.2 Mechanism of melatonergic antidepressant's actions in depressive disorders.

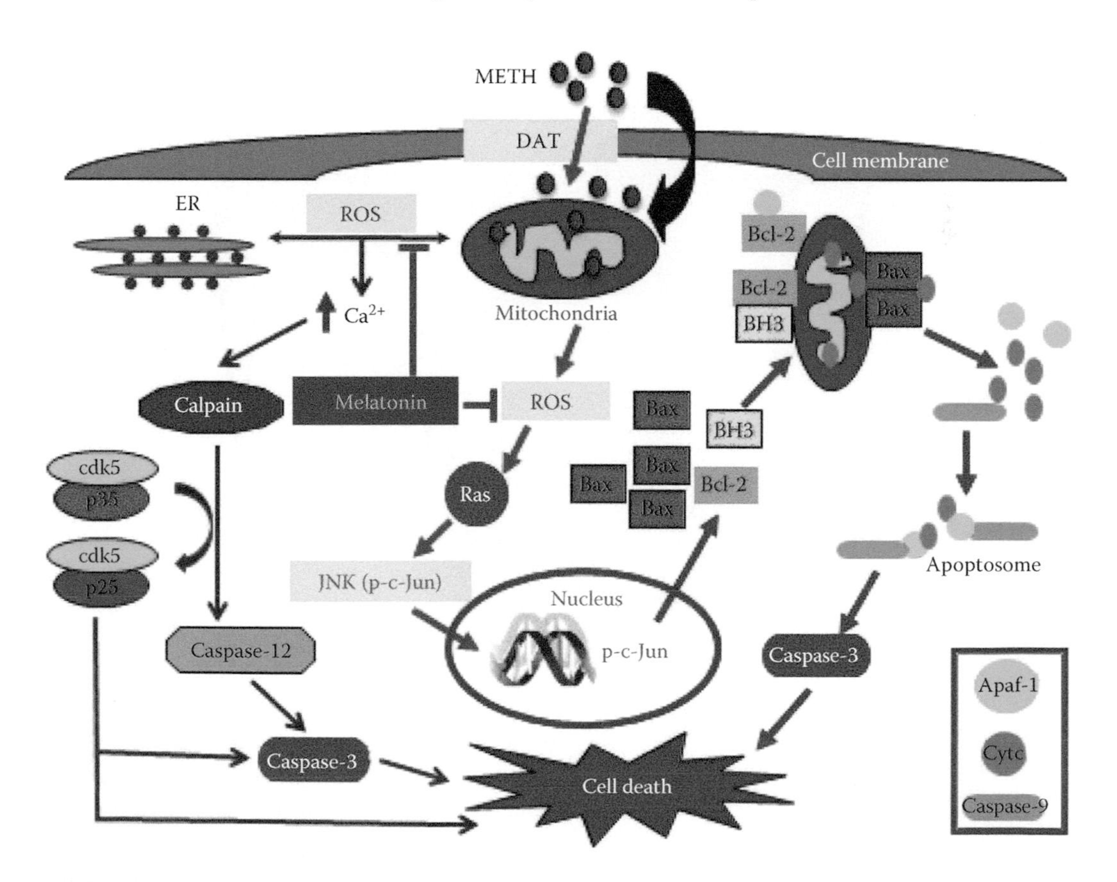

FIGURE 30.1 The proposed neuroprotective mechanisms of melatonin against methamphetamine (METH) toxicity–induced cell death in neuroblastoma SH-SY5Y cells. DAT, dopamine transporter; ROS, reactive oxygen species; JNK, c-Jun-N-terminal kinase; p-c-Jun, c-Jun phosphorylation; ER, endoplasmic reticulum; Cyt. c, cytochrome c; Apaf-1, apoptotic protease activating factor-1.

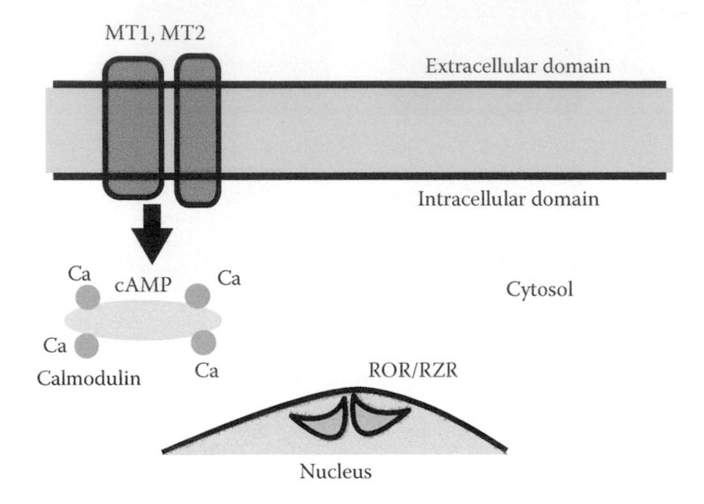

FIGURE 35.2 Schematic illustration of various melatonin receptors.

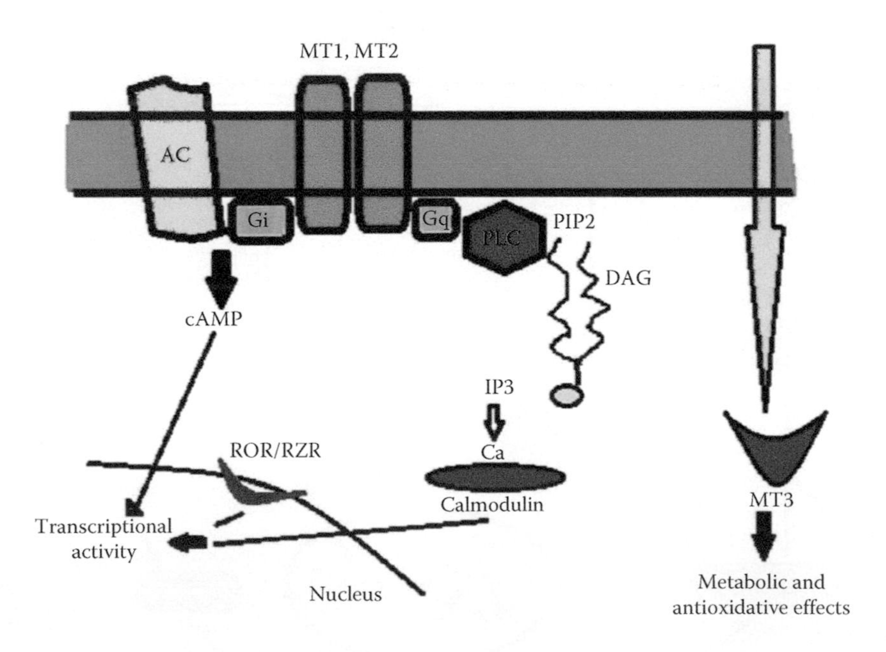

FIGURE 35.3 Schematic illustration of the antioxidative effects of Mel. DAG, diacylglycerol; IP3, inositol trisphosphate; cAMP, cyclic adenosine monophosphate; AC, adenylate cyclase; Gi, Gi protein; Gq, Gq protein; PIP2, phosphatidylinositol 4,5-bisphosphate; PLC, phospholipase C.

Several studies have also shown that melatonin increases sleep duration and quality in AD patients (Cardinali et al. 2011; Mahlberg et al. 2004; Mishima et al. 2000). These data suggest that melatonin may be a beneficial add-on drug for the treatment of AD. However, there has also been a large clinical study showing that melatonin had no effect on sleep or agitation in AD patients (Gehrman et al. 2009). Therefore, larger multicenter double-blind placebo-controlled studies are needed to clarify these results.

27.10 BENEFICIAL EFFECTS OF MELATONIN IN CELL AND MOUSE MODELS OF AD

Cell culture studies using AD model cells have also shown positive results following melatonin treatment. Most of the studies have been performed using mutant APP overexpression, which is a model of familial AD. A drawback of these studies is that these cells might not address the driving force behind the pathology in LOAD patients. The ApoE4 allele appears to be the strongest determinant of LOAD. ApoE4 can enhance or inhibit toxic fibril formation when bound to Aβ, depending upon the concentration (Naiki et al. 1997). Cell culture studies have shown that melatonin can bind to ApoE and inhibit toxic Aβ fibril formation to a much higher extent than melatonin can inhibit fibril formation by itself (Poeggeler et al. 2001). In culture, melatonin has also been shown to inhibit mitochondrial DNA damage and apoptosis induced by Aβ (Pappolla et al. 1997, 1999). Most of the culture and rodent studies attributed the beneficial effects of melatonin to a combination of direct antioxidant effects (Ionov et al. 2011) and the ability of melatonin to prevent toxic Aβ fibril formation (Olcese et al. 2009). However, neither of these explanations satisfactorily explains why melatonin was only able to prevent Aβ-mediated mitochondrial dysfunction in young isolated hippocampal neurons of low (~10) passage number, but not in senescent hippocampal neurons of high (~25) passage number (Dong et al. 2010). A possible explanation for these results is that melatonin receptor expression declined in the senescent neurons and that a portion of the protective effect is mediated by melatonin receptor signaling as well.

Melatonin has shown beneficial effects on cognitive function and plaque formation in many rodent studies of amyloid-β toxicity (Pandi-Perumal et al. 2013). There are too many to discuss in this chapter, so we will mention some of the most relevant concerning the ability of melatonin to positively impact mitochondrial function. A more detailed examination of this topic can be found in the following reviews (Cardinali et al. 2013; Cheng et al. 2006; Lin et al. 2013; Pandi-Perumal et al. 2013; Rosales-Corral et al. 2012b). First, a study showed that both exercise or melatonin treatment delayed many of the phenotypes in 3x-Tg AD mice. But only the combination of melatonin and exercise together was able to completely prevent the loss of mitochondrial ETC complex protein levels and increase levels of coenzyme Q9, the precursor to the ETC electron carrier CoQ10 (Garcia-Mesa et al. 2012). Another study showed that hippocampal injection of fibrillar Aβ into mice led to cellular and mitochondrial uptake that increased ROS production and led to a decrease in the respiratory control ratio that was slightly increased by the presence of melatonin in the drinking water. However, an Aβ-mediated inhibition of the mitochondrial F0F1-ATP synthase was not improved by melatonin administration (Rosales-Corral et al. 2012a). Lastly, we have shown that the ability of a 1-month melatonin treatment largely reverses mitochondrial dysfunction mediated by Aβ in Alzheimer's mice and this restorative effect was blunted in mice receiving melatonin and caffeine (Dragicevic et al. 2012). The mechanism through which caffeine partially blocks the effect of melatonin remains unknown, but it is likely through inhibition of melatonin receptor signaling.

27.10.1 Indole-3-Propionamide as an Alternative to Melatonin for the Treatment of Neurodegenerative Disease

Indole-3-propionic acid (IPA) (OXIGON™), a natural metabolite found in almost all organisms, and a related indole to melatonin, was reported to be a better antioxidant than melatonin and just like melatonin showed no pro-oxidant activity (Chyan et al. 1999). However, being an acid, it is hydrophilic and is slow to permeate the blood–brain barrier for the treatment of neurodegenerative diseases. Nonetheless, a phase II clinical trial using IPA to treat Friedreich's ataxia was initiated in 2012 (Gomes and Santos 2013). The same group of researchers who discovered the strong antioxidant activity of IPA subsequently published that indole-3-propionamide (IPAM) was more effective as an antioxidant than melatonin, but retained the hydrophobic nature of melatonin (Poeggeler et al. 2010). In addition, when i.p. injected into rats at 0.5 mg/kg, high IPAM levels could be measured in the brains for over 8 h, but melatonin and IPA levels were not even measurable after 1 h. IPAM also prevented Aβ aggregation in a thioflavin T assay just like melatonin and IPA. Therefore, IPAM may show the highest potential of any known indole antioxidant for the treatment of neurodegenerative diseases. IPAM was also shown to increase the life span of a rotifer species by 300% (Poeggeler et al. 2010). IPA treatment had no effect on the life span of a separate rotifer species (Snell et al. 2012). We published that IPA-induced restoration of mitochondrial function in N2a-APP_{swe} cells was partially blocked by luzindole (Dragicevic et al. 2011). Therefore, melatonin-related indoles such as IPA and IPAM may be utilizing melatonin receptors to protect the cells from the damage induced by Aβ. However, these observations should be verified in animal models. These results suggest that IPAM in addition to controlled-release melatonin may show beneficial effects in clinical trials for neurodegenerative diseases such as AD. Other melatonin-related compounds, such as the melatonin breakdown product AFMK, have also shown promise as a strong antioxidant in the protection of cells from oxidative stress induced by Aβ (Poeggeler et al. 2001).

27.10.2 Direct Effects of Melatonin and IPAM on Mitochondrial Function

Melatonin increases mitochondrial function through both direct and indirect mechanisms. Binding sites for melatonin have been found on mitochondrial membranes (Poon and Pang 1992; Yuan and Pang 1991). This may be partially explained by the recent localization of the melatonin MT1 receptor to mitochondria (Wang et al. 2011). But melatonin also binds mitochondrial complex I with an affinity of 150 pM (Hardeland 2009a). Mitochondrial ETC complex I and IV activities are stimulated by melatonin in the brain (Martin et al. 2000). Melatonin also preserves mitochondrial respiration in the aging mouse brain (Carretero et al. 2009), in mice given ruthenium red (Martin et al. 2000), or in mice given the ETC complex IV inhibitor cyanide (Yamamoto and Mohanan 2002). Like melatonin, but more potently, IPAM stimulates complex I and IV activity, and stabilizes mitochondrial function in the presence of mitochondrial toxins such as FCCP, doxorubicin, and antimycin A (Poeggeler et al. 2010).

27.11 ANTIOXIDANT SIGNALING THROUGH MELATONIN RECEPTORS

Melatonin treatment has been shown to increase the mRNA levels of many antioxidants in rat brain including Mn-SOD (SOD2) Cu/Zn-SOD (SOD1), and catalase (Garcia et al. 2010; Gunasingh et al. 2008; Kotler et al. 1998). Glutathione peroxidase and glutathione reductase have also been shown to be upregulated by melatonin in certain tissues (Carretero et al. 2009; Limon-Pacheco and Gonsebatt 2010; Pandi-Perumal et al. 2013). It is hypothesized that this upregulation of antioxidant gene expression occurs mainly through melatonin receptor signaling as luzindole, an inhibitor of melatonin receptors has been shown to partially or completely prevent upregulation of these antioxidant genes in various tissues and cell lines (Adamczyk-Sowa et al. 2013; Choi et al. 2011; Rezzani et al. 2006).

Superoxide dismutase converts superoxide to hydrogen peroxide and then either catalase in the peroxisomes or glutathione peroxidase in the mitochondria or cytoplasm converts the hydrogen peroxide to water. It has been shown that melatonin treatment of rats increased mitochondrial superoxide dismutase activity in old rat brain (Ozturk et al. 2012). It was also shown that aging upregulated mitochondrial glutathione peroxidase activity and melatonin prevented this response. Lastly, there was an aging-related decline in the superoxide dismutase to glutathione peroxidase ratio that was prevented by melatonin treatment. Therefore, melatonin treatment preserves the correct youthful ratio of these enzymes, preventing superoxide increases in the cell and oxidative damage.

Melatonin receptors are not always required for the antioxidant effect of melatonin on cells and tissues. In several cases, luzindole did not prevent the protective effects of melatonin indicating that melatonin's antioxidant effects can frequently be receptor-independent (Behan et al. 1999; Lahiri et al. 2009; Song et al. 2012). However, concerning the effect of melatonin on neural cells, melatonin was shown to prevent neural ischemic stroke injury, partially through a MT2-dependent mechanism as both luzindole and the more selective MT2 antagonist 4P-PDOT (4-phenyl-2-propionamidotetralin) partially blocked melatonin's protective effect (Chern et al. 2012). Similarly, luzindole blocked the protective effect of melatonin on the palmitic acid–induced increase in ROS levels and cell death in primary mouse astroglial cells (Wang et al. 2012a). Therefore, it will be important to determine if the protective effects of melatonin on mitochondrial dysfunction in aging and AD depends on the MT1 or MT2 receptors.

27.12 MELATONIN RECEPTOR AGONISTS FOR THE TREATMENT OF AD

There has been in vitro data implicating a role for melatonin receptor signaling in mediating a partial protection from Aβ-mediated mitochondrial dysfunction (Dragicevic et al. 2011), while two studies showed that AD model mice treated with the melatonin receptor agonist ramelteon showed little or no protection from AD pathology. In the first ramelteon treatment study, the AD model B6C3-Tg(APPswe,PSEN1dE9)85Dbo/J transgenic mouse strain (APP/PS1 mice) was used. The mice were treated with ~3 mg/kg/day of ramelteon and no change in cognitive behavior as judged by performance in a water maze was observed after 3 months of treatment (McKenna et al. 2012). Even 6 months of ramelteon treatment failed to yield reductions in amyloid plaque burden. The second study, also using an APP/PS1 mouse strain, determined a lack of effect of ramelteon on spatial memory performance, but did find that ramelteon slightly decreased hippocampal protein oxidation in the APP/PS1 mice (Bano Otalora et al. 2012). However, a study using the melatonin receptor agonist Neu-P11 (piromelatine) found that rats which had undergone intrahippocampal injection with Aβ performed better in novel object recognition and Y-maze tasks and showed less CA1 hippocampal cell loss when i.p. injected with 50 mg/kg Neu-P11. The Neu-P11 injected group even outperformed the melatonin-injected group (He et al. 2013). One further piece of data that supports the use of melatonin receptor agonists clinically for cognitive dysfunction is that ramelteon was shown to improve delirium in five patients after only 1 day of treatment (Furuya et al. 2012).

27.13 ROLE THAT MELATONIN RECEPTORS PLAY IN PROTECTING AD-ASSOCIATED MITOCHONDRIAL DYSFUNCTION

We have identified melatonin receptors, likely MT2 as being essential for the full mitochondrial protective effect of melatonin against Alzheimer's amyloid (Dragicevic et al. 2011). We also observed that a low concentration of caffeine or cAMP-dependent or cGMP-dependent phosphodiesterase inhibitors blocked melatonin from fully protecting against amyloid-mediated mitochondrial dysfunction. Therefore, melatonin receptor signaling may decrease adenylate cyclase activity to decrease cAMP levels to protect mitochondrial function in the hippocampus in AD and inhibiting cAMP-dependent phosphodiesterases may antagonize this response by restoring cAMP levels. Although the mechanism has not yet been fully elucidated, direct oxidant scavenging and direct inhibition of amyloid fibril formation are likely not the sole mechanisms of melatonin-mediated protection in AD. Melatonin receptor signaling, therefore, likely contributes to the melatonin-mediated prevention of cognitive dysfunction in AD mice. Discovering the signal transduction pathway between melatonin receptors and mitochondria is an essential next step in determining the suitability of melatonin, melatonin-related indoles, or melatonin receptor agonists for clinical trials.

The most straightforward hypothesis is that there is increased expression of antioxidant defense proteins as a result of MT2 receptor signaling that is protecting mitochondria from oxidative damage in AD model mice and cells.

27.14 ROLE THAT MELATONIN RECEPTORS PLAY IN PROTECTING FROM AGING-INDUCED LOSS OF CYTOCHROME c OXIDASE ACTIVITY IN MICE

In data unpublished at the time of print, we have performed experiments determining the role of melatonin receptors in preventing the decline in brain cytochrome c oxidase (COX) activity during aging. We used striatal tissue from knockout mice deficient in both the MT1 and MT2 melatonin receptors. We found that mice deficient in both MT1 and MT2 melatonin receptors showed 33% more loss of COX activity at 16 months of age as compared to the activity in young mice (67% loss in MT1/MT2 knockout mice compared to 50% loss in WT controls). In addition, melatonin treatment completely prevented the loss of COX activity in the WT mice, but the melatonin-treated MT1/MT2 receptor knockout mice lost 33% of COX activity by 16 months of age. These data indicate that roughly half of the effect of melatonin in preventing aging-related loss of COX activity is MT1 or MT2 receptor-dependent, while the other half of the protective effect is receptor-independent. Future studies will determine which of the melatonin receptors is required for protection from the aging-related loss of COX activity.

27.15 ROLE THAT MELATONIN RECEPTORS PLAY IN PROTECTING FROM AD-INDUCED ALTERATION OF COX ACTIVITY IN MICE

In further data unpublished at the time of print, we determined the role that melatonin receptors play in protection from $APP_{swe/PS1}$-induced alteration of COX activity in the striatum of mice. In contrast to what we were expecting, we did not find a decrease in COX activity caused by APP_{swe} expression as we and others have observed when using mice of different genetic backgrounds (Dragicevic et al. 2010; Manczak et al. 2006). In fact, we found a striking increase in COX activity from 13 to 16 months of age. Others have also found increased cytochrome c oxidase activity in APP-expressing mice, such as in the ventral striatum of APP23 mice partially backcrossed onto a C57B/6 background (Strazielle et al. 2003). This report showed that COX activity increased, but only in specific regions of the brain. COX activity was also shown to increase in Tg2576 mice at 5 months of age (Poirier et al. 2011) and in another report in these mice at 7 months of age (Cuadrado-Tejedor et al. 2013). Consistent with these data, mitochondrial electron transport genes are upregulated in Tg2576 mice (Reddy et al. 2004). However, many others have shown decreased COX activity in different brain regions of Tg2576 mice (Manczak et al. 2006; Valla et al. 2007; Varghese et al. 2011; Zhang et al. 2010). Decreased COX activity has also been shown in double and triple transgenic mouse models of AD combining mutant APP overexpression with overexpression of presenilin-1 and/or tau (Rhein et al. 2009; Wolf et al. 2012).

We used MT1/MT2 knockout mice crossed with $APP_{swe/PS1}$ transgenic mice and studied the effects on COX activity on striatal extracts from mice treated with or without melatonin. We found that melatonin treatment completely inhibited the $APP_{swe/PS1}$-mediated increase in COX activity under conditions of normal MT1 and MT2 expression and even led to a slight decrease in COX activity. However, in the MT1/MT2 knockout mice, melatonin treatment only led to a small non-statistically significant decrease in the $APP_{swe/PS1}$-mediated increase in COX activity (p = 0.3). Therefore, the melatonin receptors play an important role in the ability of melatonin to prevent AD-related changes in COX activity. It will be important to repeat these experiments using hippocampal or cerebral cortical tissue or a different genetic background of mice to determine if these trends hold under conditions where COX activity declines as a result of APP_{swe} expression.

27.16 MOLECULAR MECHANISMS THROUGH WHICH MELATONIN RECEPTOR SIGNALING PROTECTS MITOCHONDRIAL FUNCTION IN AGING AND DISEASE

The signaling pathway through which melatonin receptor signaling induces antioxidant gene expression and protects mitochondrial function has yet to be convincingly identified in the central nervous system (CNS). However, it is likely to signal through one of the following pathways described in the following. There are at least four examples of melatonin stimulating the Nrf2 pathway to protect against oxidative damage in the nervous system. Melatonin was shown to decrease okadaic acid-induced memory dysfunction in rats through upregulation of the Nrf2 signaling pathway and prevention of NF-κB activation, which leads to neuroinflammation (Mendes et al. 2013). Nrf2 is normally sequestered in the cytoplasm by Keap1. During oxidative and nitrosative stress, such as shown following ischemic brain injury, tyrosine 473 of Keap1 is nitrated, which prevents release of Nrf2 to the nucleus to induce transcription (Tao et al. 2013). Melatonin treatment prevents damage to Keap1, allowing Keap1 to release Nrf2 and Nrf2 to translocate to the nucleus, bind antioxidant response elements (ARE) in DNA, and transcribe protective genes as a response to the stress. Melatonin was also shown to activate Nrf2 to provide protection from early brain injury in a subarachnoid hemorrhage model in rats (Wang et al. 2012b), to protect against high-linear energy transfer (LET) carbon ion radiation in mouse brains (Liu et al. 2012a), and to protect the sciatic nerve from increased levels of pro-inflammatory cytokines and cell death in streptozotocin-induced diabetic neuropathy in rats (Negi et al. 2011).

Melatonin can also modulate expression of SIRT1, an nicotinamide adenine dinucleotide (NAD)-dependent protein deacetylase that deacetylates and activates many substrate proteins including PGC-1α, FoxO1, NF-κB, and p53. There have been several examples of melatonin downregulating SIRT1 in cancer cell lines (Cheng et al. 2013), while there have been at least three reports of melatonin upregulating or preventing the decline of SIRT1 expression in brain in response to a stress (Hardeland 2013). For example, melatonin prevented the decline in brain SIRT1 levels at 10 months of age in the SAMP8 mouse model of accelerated senescence (Gutierrez-Cuesta et al. 2008), in sleep-deprived rat hippocampus (Chang et al. 2009), and in isolated aged neurons from rats (Tajes et al. 2009).

Consistent with this activation of SIRT1 by melatonin, melatonin treatment has been shown to increase expression of the master mitochondrial transcriptional coactivator α in white adipocytes, turning a portion of them into brown adipocytes (Jimenez-Aranda et al. 2013). PGC-1α is known to induce gene expression of antioxidant genes such as SOD2 and glutathione peroxidase-1 (GPx1) in the brain (St-Pierre et al. 2006). However, the PGC-1α promoter has a CREB-binding site for transcriptional activation (Ashabi et al. 2012; Sheng et al. 2012). So it is possible that the ability of melatonin to decrease cAMP levels through adenylate cyclase inhibition may lead to decreased PGC-1α transcription. However, treating cardiac cells with catecholamines, which increase adenylate cyclase activity to increase cAMP levels, decreases PGC-1α activity (Arany et al. 2006). In addition, there was also an inverse correlation between CREB activation and PGC-1α activation in the heart of spontaneously hypertensive rats. Therefore, there is not always a direct correlation between the activities of CREB and PGC-1α.

There is diurnal variation in the expression of SIRT1 and PGC-1α in some tissues (Asher and Schibler 2011) similar to the cyclic variation in melatonin synthesis in the pineal gland. In liver and skeletal muscle it has been shown that this diurnal expression pattern of PGC-1α stimulates the expression of clock genes through the coactivation of the ROR family of orphan nuclear receptors. Expression of the *Bmal1* and *Rev-erb-α* genes were notably induced. This diurnal expression of PGC-1α is likely influenced by the similar cyclic expression pattern of the NAD-dependent SIRT1 deacetylase (Asher et al. 2008; Nakahata et al. 2008) and the diurnal variation in NAD levels and the NAD/NADH ratio caused by the circadian expression of nicotinamide phosphoribosyltransferase (NAMPT) (Nakahata et al. 2009; Ramsey et al. 2009). SIRT1 associates with the CLOCK-BMAL1 heterodimer and deacetylates BMAL1 and PER2, which destabilizes PER2 leading to its degradation. The rhythmic PGC-1α expression pattern is also likely influenced by the circadian oscillation of CREB activation as shown in both the suprachiasmatic nucleus (O'Neill et al. 2008) as well as in the peripheral tissues (Wang and Zhou 2010). The relation between the oscillations in these metabolic regulators, melatonin action, and AD remains relatively unexplored.

Since the enzymes that produce melatonin from tryptophan in the pineal gland are controlled by the circadian clock and because melatonin receptors are present in the suprachiasmatic nucleus in the hypothalamus, it has been speculated that melatonin binding to these receptors would influence the circadian clock machinery. Unexpectedly, only a limited number of studies have examined the role of melatonin or melatonin receptors on the circadian clock genes (Jung-Hynes et al. 2010). One study found that clock genes were downregulated in the adrenal cortex of a C57BL melatonin-deficient mouse strain compared to a C3H melatonin-proficient strain (Torres-Farfan et al. 2006). Another study found that melatonin, through binding to MT1 receptors, could decrease expression of PER1 and CLOCK, but had no effect on the levels of BMAL1 in primary striatal cultures from mice (Imbesi et al. 2009). In addition, a phase-dependent effect of rhythmic melatonin administration was found on circadian clock gene expression (*Per2* and *Bmal1*) in the heart of hypertensive rats (Zeman et al. 2009), but no effect was observed in the suprachiasmatic nucleus (Poirel et al. 2003). Specifically, the authors suggested that only melatonin applied during the dark phase of the 24 h cycle allowed a strong synchronization of circadian clock expression in the heart. It will be interesting to determine if expression changes or activation of PGC-1α, SIRT1, or CREB play a role in these effects of melatonin.

AMP kinase (AMPK) works upstream and in parallel to SIRT1 in neuroprotective pathways. The two pathways intersect to increase PGC-1α activity and increase mitochondrial biogenesis. AMPK directly phosphorylates PGC-1α at threonine-177 and serine-538 to increase PGC-1α-induced activation of the PGC-1α promoter (Jager et al. 2007). Phosphorylation of AMPK stimulates its kinase activity. Melatonin has been shown to have disparate effects on AMPK in different cancer cell lines (Hardeland 2013). In HT22 immortalized hippocampal cells, Aβ treatment increased AMPK phosphorylation, while melatonin prevented this activation. This AMPK activation was interpreted as being mediated by oxidative stress and melatonin presumably prevented the oxidative damage from occurring to prevent AMPK activation. In primary tissues such as in livers undergoing steatosis (Zaouali et al. 2013) and in muscle and livers from aged rats, especially when exercised (Mendes et al. 2013), melatonin treatment led to AMPK activation. The melatonin treatment led to increased physiological benefit from exercise in the aged rats. Studies should also be performed in aged and AD brain to determine if melatonin treatment leads to the activation of AMPK and if melatonin receptors play a role.

In aging and aging-related disease, there is a strong correlation between increased inflammation and decreased mitochondrial function. These two factors appear to be most centrally linked to the aging process. Since these two phenomena are so intricately linked, it is likely that one induces the other. Melatonin treatment can decrease chronic and acute inflammation by transcriptionally inhibiting iNOS and cyclooxygenase-2 transcriptional activation (Costantino et al. 1998; Cuzzocrea et al. 1997; Deng et al. 2006). Melatonin has also been shown to directly act on immune cells to decrease production of IL-6, IL-8, TNF-α, and adhesion molecules (Esposito and Cuzzocrea 2010). Astrocytes become activated in AD by proinflammatory cytokines and can upregulate iNOS to produce excess nitric oxide, which binds and inhibits cytochrome c oxidase of the ETC in both neurons and glia leading to energy decline and tissue dysfunction. Melatonin prevents iNOS upregulation by inhibiting the p38 MAPK signaling pathway activated by cytokine binding (Vilar et al. 2014). Including melatonin in the diet at 200 ppm for 8 weeks normalized the expression of many pro-inflammatory genes that were upregulated in the aged mouse brain (Sharman et al. 2004).

Melatonin has been shown to either inhibit or enhance autophagy dependent upon the tissue type and disease treatment (Coto-Montes et al. 2012). In many pathological conditions, reactive oxygen species, which may be required for autophagy induction, are increased. Under these conditions, melatonin treatment may decrease reactive oxygen species levels to decrease autophagic flux.

For example, autophagy increases in a rotenone-induced Parkinson's disease model and melatonin treatment decreased autophagy markers and autophagic cell death (Zhou et al. 2012). Similar observations were made in methamphetamine-induced autophagic cell death (Nopparat et al. 2010). However, in certain conditions melatonin treatment has also been shown to increase autophagy. One of the mechanisms through which melatonin receptor signaling may protect against AD is through modulation of the rate of autophagy. In this regard, autophagy is defective in AD brain due to defective lysosomal acidification causing an accumulation of autophagosomes in certain AD neurons (Wolfe et al. 2013). This effect was also observed in presenilin-1 (PS1) knockout and mutant mouse neurons and was identified to be caused by a requirement for WT PS1 in the maturation and sorting of a v-ATPase subunit to the lysosome. In addition, another group showed that the unfolded protein response increased autophagy in AD neurons (Scheper et al. 2011). Induction of autophagy relies on AMPK activation in many cell types (Meijer and Codogno 2007) and melatonin may activate AMPK under certain conditions as described earlier. Melatonin-induced autophagy has been shown to protect against neural cell death in early brain injury following a subarachnoid hemorrhage (Chen et al. 2013). Melatonin-induced autophagy has also been shown to provide neuroprotection from prion proteins (Jeong et al. 2012), and protect N2a cells from ischemia-reperfusion-induced cell death (Guo et al. 2010). In addition, melatonin prevented aging-related abnormalities in the autophagosomal–lysosomal system in the brain from the SAMP8 senescence-accelerated mouse model (Garcia et al. 2011). Unfortunately, little is yet known on the effect of melatonin on autophagy in the AD brain or AD model systems (Coto-Montes et al. 2012).

27.17 CONCLUSION

The very low toxicity of melatonin versus other potential AD therapeutics makes melatonin an obvious choice for human AD therapy if efficacy in slowing cognitive dysfunction can be convincingly demonstrated. Melatonin is more versatile than other antioxidants such as vitamin C or vitamin E because it can scavenge peroxynitrite as well as reactive oxygen species (Korkmaz et al. 2009). Melatonin is also a candidate therapy for many other neurodegenerative and aging-associated disorders. Novel studies on the molecular mechanisms of mitochondrial protection mediated by melatonin receptor signaling will lead to a better understanding of how melatonin can be used to hinder disease progression in AD.

ACKNOWLEDGMENTS

We would like to thank Stephen Bell and Krupa Curien for performing COX assays in the aged and AD mice. We would like to thank Dr. James Olcese and Dr. Gina O'Neil-Moffitt for providing the mouse brain samples for COX analysis. We would also like to thank Dr. Natasa Dragicevic, Dr. Gary Arendash, Dr. Chuanhai Cao, Neil Copes, and Clare Edwards for intellectual contribution to the melatonin research project in our laboratory. Lastly, we would also like to thank Ilknur, Tara, and Ela Bradshaw and Sandra Zivkovic for their support and understanding during the writing of this chapter.

REFERENCES

Adamczyk-Sowa M, Sowa P, Zwirska-Korczala K, Pierzchala K, Bartosz G, Sadowska-Bartosz I. 2013. Role of melatonin receptor MT(2) and quinone reductase II in the regulation of the redox status of 3T3-L1 preadipocytes in vitro. *Cell Biol Int*. 37:835–842.

Adi N, Mash DC, Ali Y, Singer C, Shehadeh L, Papapetropoulos S. 2010. Melatonin MT1 and MT2 receptor expression in Parkinson's disease. *Med Sci Monit*. 16:BR61–BR67.

Arany Z, Novikov M, Chin S, Ma Y, Rosenzweig A, Spiegelman BM. 2006. Transverse aortic constriction leads to accelerated heart failure in mice lacking PPAR-gamma coactivator 1alpha. *Proc Natl Acad Sci USA*. 103:10086–10091.

Asayama K, Yamadera H, Ito T, Suzuki H, Kudo Y, Endo S. 2003. Double blind study of melatonin effects on the sleep-wake rhythm, cognitive and non-cognitive functions in Alzheimer type dementia. *J Nippon Med Sch*. 70:334–341.

Ashabi G, Ramin M, Azizi P, Taslimi Z, Alamdary SZ, Haghparast A, Ansari N, Motamedi F, Khodagholi F. 2012. ERK and p38 inhibitors attenuate memory deficits and increase CREB phosphorylation and PGC-1alpha levels in Abeta-injected rats. *Behav Brain Res*. 232:165–173.

Asher G, Gatfield D, Stratmann M, Reinke H, Dibner C, Kreppel F, Mostoslavsky R, Alt FW, Schibler U. 2008. SIRT1 regulates circadian clock gene expression through PER2 deacetylation. *Cell*. 134:317–328.

Asher G, Schibler U. 2011. Crosstalk between components of circadian and metabolic cycles in mammals. *Cell Metab*. 13:125–137.

Bano Otalora B, Popovic N, Gambini J, Popovic M, Vina J, Bonet-Costa V, Reiter RJ, Camello PJ, Rol MA, Madrid JA. 2012. Circadian system functionality, hippocampal oxidative stress, and spatial memory in the APPswe/PS1dE9 transgenic model of Alzheimer disease: Effects of melatonin or ramelteon. *Chronobiol Int*. 29:822–834.

Bedrosian TA, Nelson RJ. 2012. Pro: Alzheimer's disease and circadian dysfunction: Chicken or egg? *Alzheimers Res Ther*. 4:25.

Behan WM, McDonald M, Darlington LG, Stone TW. 1999. Oxidative stress as a mechanism for quinolinic acid-induced hippocampal damage: Protection by melatonin and deprenyl. *Br J Pharmacol*. 128:1754–1760.

Bertram L, Tanzi RE. 2008. Thirty years of Alzheimer's disease genetics: The implications of systematic meta-analyses. *Nat Rev Neurosci*. 9:768–778.

Bondy SC, Li H, Zhou J, Wu M, Bailey JA, Lahiri DK. 2010. Melatonin alters age-related changes in transcription factors and kinase activation. *Neurochem Res*. 35:2035–2042.

Brusco LI, Marquez M, Cardinali DP. 1998. Monozygotic twins with Alzheimer's disease treated with melatonin: Case report. *J Pineal Res*. 25:260–263.

Brusco LI, Marquez M, Cardinali DP. 2000. Melatonin treatment stabilizes chronobiologic and cognitive symptoms in Alzheimer's disease. *Neuroendocrinol Lett*. 21:39–42.

Brydon L, Roka F, Petit L, de Coppet P, Tissot M, Barrett P, Morgan PJ, Nanoff C, Strosberg AD, Jockers R. 1999. Dual signaling of human Mel1a melatonin receptors via G(i2), G(i3), and G(q/11) proteins. *Mol Endocrinol*. 13:2025–2038.

Bu G. 2009. Apolipoprotein E and its receptors in Alzheimer's disease: Pathways, pathogenesis and therapy. *Nat Rev Neurosci*. 10:333–344.

Bubber P, Haroutunian V, Fisch G, Blass JP, Gibson GE. 2005. Mitochondrial abnormalities in Alzheimer brain: Mechanistic implications. *Ann Neurol*. 57:695–703.

Calkins MJ, Manczak M, Mao P, Shirendeb U, Reddy PH. 2011. Impaired mitochondrial biogenesis, defective axonal transport of mitochondria, abnormal mitochondrial dynamics and synaptic degeneration in a mouse model of Alzheimer's disease. *Hum Mol Genet*. 20:4515–4529.

Cardinali DP, Furio AM, Brusco LI. 2011. The use of chronobiotics in the resynchronization of the sleep/wake cycle. Therapeutical application in the early phases of Alzheimer's disease. *Recent Pat Endocr Metab Immune Drug Discov*. 5:80–90.

Cardinali DP, Pagano ES, Scacchi Bernasconi PA, Reynoso R, Scacchi P. 2013. Melatonin and mitochondrial dysfunction in the central nervous system. *Horm Behav*. 63:322–330.

Carretero M, Escames G, Lopez LC, Venegas C, Dayoub JC, Garcia L, Acuna-Castroviejo D. 2009. Long-term melatonin administration protects brain mitochondria from aging. *J Pineal Res*. 47:192–200.

Chan AS, Lai FP, Lo RK, Voyno-Yasenetskaya TA, Stanbridge EJ, Wong YH. 2002. Melatonin mt1 and MT2 receptors stimulate c-Jun N-terminal kinase via pertussis toxin-sensitive and -insensitive G proteins. *Cell Signal*. 14:249–257.

Chang HM, Wu UI, Lan CT. 2009. Melatonin preserves longevity protein (sirtuin 1) expression in the hippocampus of total sleep-deprived rats. *J Pineal Res*. 47:211–220.

Chang S, ran Ma T, Miranda RD, Balestra ME, Mahley RW, Huang Y. 2005. Lipid- and receptor-binding regions of apolipoprotein E4 fragments act in concert to cause mitochondrial dysfunction and neurotoxicity. *Proc Natl Acad Sci USA*. 102:18694–18699.

Chen J, Wang L, Wu C, Hu Q, Gu C, Yan F, Li J, Yan W, Chen G. 2013. Melatonin-enhanced autophagy protects against neural apoptosis via a mitochondrial pathway in early brain injury following a subarachnoid hemorrhage. *J Pineal Res*. 56(1):12–19. doi:10.1111/jpi.12086.

Cheng Y, Cai L, Jiang P, Wang J, Gao C, Feng H, Wang C, Pan H, Yang Y. 2013. SIRT1 inhibition by melatonin exerts antitumor activity in human osteosarcoma cells. *Eur J Pharmacol*. 715:219–229.

Cheng Y, Feng Z, Zhang QZ, Zhang JT. 2006. Beneficial effects of melatonin in experimental models of Alzheimer disease. *Acta Pharmacol Sin*. 27:129–139.

Chern CM, Liao JF, Wang YH, Shen YC. 2012. Melatonin ameliorates neural function by promoting endogenous neurogenesis through the MT2 melatonin receptor in ischemic-stroke mice. *Free Radic Biol Med*. 52:1634–1647.

Choi SI, Dadakhujaev S, Ryu H, Im Kim T, Kim EK. 2011. Melatonin protects against oxidative stress in granular corneal dystrophy type 2 corneal fibroblasts by mechanisms that involve membrane melatonin receptors. *J Pineal Res*. 51:94–103.

Choi SW, Gerencser AA, Ng R, Flynn JM, Melov S, Danielson SR, Gibson BW, Nicholls DG, Bredesen DE, Brand MD. 2012. No consistent bioenergetic defects in presynaptic nerve terminals isolated from mouse models of Alzheimer's disease. *J Neurosci*. 32:16775–16784.

Chomyn A, Attardi G. 2003. MtDNA mutations in aging and apoptosis. *Biochem Biophys Res Comm*. 304:519–529.

Chyan YJ, Poeggeler B, Omar RA, Chain DG, Frangione B, Ghiso J, Pappolla MA. 1999. Potent neuroprotective properties against the Alzheimer beta-amyloid by an endogenous melatonin-related indole structure, indole-3-propionic acid. *J Biol Chem*. 274:21937–21942.

Corder EH, Saunders AM, Strittmatter WJ, Schmechel DE, Gaskell PC, Small GW, Roses AD, Haines JL, Pericak-Vance MA. 1993. Gene dose of apolipoprotein E type 4 allele and the risk of Alzheimer's disease in late onset families. *Science*. 261:921–923.

Costantino G, Cuzzocrea S, Mazzon E, Caputi AP. 1998. Protective effects of melatonin in zymosan-activated plasma-induced paw inflammation. *Eur J Pharmacol*. 363:57–63.

Coto-Montes A, Boga JA, Rosales-Corral S, Fuentes-Broto L, Tan DX, Reiter RJ. 2012. Role of melatonin in the regulation of autophagy and mitophagy: A review. *Mol Cell Endocrinol*. 361:12–23.

Crouch PJ, Blake R, Duce JA, Ciccotosto GD, Li QX, Barnham KJ, Curtain CC et al. 2005. Copper-dependent inhibition of human cytochrome c oxidase by a dimeric conformer of amyloid-beta1–42. *J Neurosci*. 25:672–679.

Cuadrado-Tejedor M, Cabodevilla JF, Zamarbide M, Gomez-Isla T, Franco R, Perez-Mediavilla A. 2013. Age-related mitochondrial alterations without neuronal loss in the hippocampus of a transgenic model of Alzheimer's disease. *Curr Alzheimer Res*. 10:390–405.

Cuzzocrea S, Zingarelli B, Gilad E, Hake P, Salzman AL, Szabo C. 1997. Protective effect of melatonin in carrageenan-induced models of local inflammation: Relationship to its inhibitory effect on nitric oxide production and its peroxynitrite scavenging activity. *J Pineal Res*. 23:106–116.

David DC, Hauptmann S, Scherping I, Schuessel K, Keil U, Rizzu P, Ravid R, Drose S, Brandt U, Muller WE et al. 2005. Proteomic and functional analyses reveal a mitochondrial dysfunction in P301L tau transgenic mice. *J Biol Chem*. 280:23802–23814.

Deng WG, Tang ST, Tseng HP, Wu KK. 2006. Melatonin suppresses macrophage cyclooxygenase-2 and inducible nitric oxide synthase expression by inhibiting p52 acetylation and binding. *Blood* 108:518–524.

Dijk DJ, Cajochen C. 1997. Melatonin and the circadian regulation of sleep initiation, consolidation, structure, and the sleep EEG. *J Biol Rhythms*. 12:627–635.

Dong W, Huang F, Fan W, Cheng S, Chen Y, Zhang W, Shi H, He H. 2010. Differential effects of melatonin on amyloid-beta peptide 25–35-induced mitochondrial dysfunction in hippocampal neurons at different stages of culture. *J Pineal Res*. 48:117–125.

Dragicevic N, Copes N, O'Neal-Moffitt G, Jin J, Buzzeo R, Mamcarz M, Tan J et al. 2011. Melatonin treatment restores mitochondrial function in Alzheimer's mice: A mitochondrial protective role of melatonin membrane receptor signaling. *J Pineal Res*. 51:75–86.

Dragicevic N, Delic V, Cao C, Copes N, Lin X, Mamcarz M, Wang L, Arendash GW, Bradshaw PC. 2012. Caffeine increases mitochondrial function and blocks melatonin signaling to mitochondria in Alzheimer's mice and cells. *Neuropharmacology*. 63:1368–1379.

Dragicevic N, Mamcarz M, Zhu Y, Buzzeo R, Tan J, Arendash GW, Bradshaw PC. 2010. Mitochondrial amyloid-beta levels are associated with the extent of mitochondrial dysfunction in different brain regions and the degree of cognitive impairment in Alzheimer's transgenic mice. *J Alzheimers Dis*. 20(Suppl 2): S535–S550.

Du H, Guo L, Fang F, Chen D, Sosunov AA, McKhann GM, Yan Y et al. 2008. Cyclophilin D deficiency attenuates mitochondrial and neuronal perturbation and ameliorates learning and memory in Alzheimer's disease. *Nat Med*. 14:1097–1105.

Dubocovich ML, Hudson RL, Sumaya IC, Masana MI, Manna E. 2005. Effect of MT1 melatonin receptor deletion on melatonin-mediated phase shift of circadian rhythms in the C57BL/6 mouse. *J Pineal Res*. 39:113–120.

Dubocovich ML, Markowska M. 2005. Functional MT1 and MT2 melatonin receptors in mammals. *Endocrine*. 27:101–110.

Esposito E, Cuzzocrea S. 2010. Anti-inflammatory activity of melatonin in central nervous system. *Curr Neuropharmacol*. 8:228–242.

Furio AM, Brusco LI, Cardinali DP. 2007. Possible therapeutic value of melatonin in mild cognitive impairment: A retrospective study. *J Pineal Res*. 43:404–409.

Furuya M, Miyaoka T, Yasuda H, Yamashita S, Tanaka I, Otsuka S, Wake R, Horiguchi J. 2012. Marked improvement in delirium with ramelteon: Five case reports. *Psychogeriatrics*. 12:259–262.

Garcia JJ, Pinol-Ripoll G, Martinez-Ballarin E, Fuentes-Broto L, Miana-Mena FJ, Venegas C, Caballero B, Escames G, Coto-Montes A, Acuna-Castroviejo D. 2011. Melatonin reduces membrane rigidity and oxidative damage in the brain of SAMP8 mice. *Neurobiol Aging*. 32:2045–2054.

Garcia T, Esparza JL, Nogues MR, Romeu M, Domingo JL, Gomez M. 2010. Oxidative stress status and RNA expression in hippocampus of an animal model of Alzheimer's disease after chronic exposure to aluminum. *Hippocampus*. 20:218–225.

Garcia-Mesa Y, Gimenez-Llort L, Lopez LC, Venegas C, Cristofol R, Escames G, Acuna-Castroviejo D, Sanfeliu C. 2012. Melatonin plus physical exercise are highly neuroprotective in the 3xTg-AD mouse. *Neurobiol Aging*. 33:1124 e1113–e1129.

Gehrman PR, Connor DJ, Martin JL, Shochat T, Corey-Bloom J, Ancoli-Israel S. 2009. Melatonin fails to improve sleep or agitation in double-blind randomized placebo-controlled trial of institutionalized patients with Alzheimer disease. *Am J Geriatr Psychiatry*. 17:166–169.

Gomes CM, Santos R. 2013. Neurodegeneration in Friedreich's ataxia: From defective frataxin to oxidative stress. *Oxid Med Cell Longev*. 2013:487534.

Gorfine T, Yeshurun Y, Zisapel N. 2007. Nap and melatonin-induced changes in hippocampal activation and their role in verbal memory consolidation. *J Pineal Res*. 43:336–342.

Gorfine T, Zisapel N. 2007. Melatonin and the human hippocampus, a time dependent interplay. *J Pineal Res*. 43:80–86.

Guerrero HY, Gauer F, Pevet P, Masson-Pevet M. 1999. Daily and circadian expression patterns of mt1 melatonin receptor mRNA in the rat pars tuberalis. *Adv Exp Med Biol*. 460:175–179.

Gunasingh MJ, Philip JE, Ashok BS, Kirubagaran R, Jebaraj WC, Davis GD, Vignesh S, Dhandayuthapani S, Jayakumar R. 2008. Melatonin prevents amyloid protofibrillar induced oxidative imbalance and biogenic amine catabolism. *Life Sci*. 83:96–102.

Guo Y, Wang J, Wang Z, Yang Y, Wang X, Duan Q. 2010. Melatonin protects N2a against ischemia/reperfusion injury through autophagy enhancement. *J Huazhong Univ Sci Technol Med Sci*. 30:1–7.

Gutierrez-Cuesta J, Tajes M, Jimenez A, Coto-Montes A, Camins A, Pallas M. 2008. Evaluation of potential pro-survival pathways regulated by melatonin in a murine senescence model. *J Pineal Res*. 45:497–505.

Hardeland R. 2009a. Melatonin, mitochondrial electron flux and leakage: Recent findings and resolution of contradictory results. *Adv Stud Biol*. 1:207–230.

Hardeland R. 2009b. Melatonin: Signaling mechanisms of a pleiotropic agent. *Biofactors*. 35:183–192.

Hardeland R. 2013. Melatonin and the theories of aging: A critical appraisal of melatonin's role in antiaging mechanisms. *J Pineal Res*. 55:325–356.

Harman D. 1972. The biologic clock: The mitochondria? *J Am Geriatr Soc*. 20:145–147.

Hauptmann S, Scherping I, Drose S, Brandt U, Schulz KL, Jendrach M, Leuner K, Eckert A, Muller WE. 2008. Mitochondrial dysfunction: An early event in Alzheimer pathology accumulates with age in AD transgenic mice. *Neurobiol Aging*. 30(10):1574–1586.

He P, Ouyang X, Zhou S, Yin W, Tang C, Laudon M, Tian S. 2013. A novel melatonin agonist Neu-P11 facilitates memory performance and improves cognitive impairment in a rat model of Alzheimer' disease. *Horm Behav*. 64:1–7.

Herrmann N, Chau SA, Kircanski I, Lanctot KL. 2011. Current and emerging drug treatment options for Alzheimer's disease: A systematic review. *Drugs*. 71:2031–2065.

Hunt AE, Al-Ghoul WM, Gillette MU, Dubocovich ML. 2001. Activation of MT(2) melatonin receptors in rat suprachiasmatic nucleus phase advances the circadian clock. *Am J Physiol Cell Physiol*. 280:C110–C118.

Imbesi M, Arslan AD, Yildiz S, Sharma R, Gavin D, Tun N, Manev H, Uz T. 2009. The melatonin receptor MT1 is required for the differential regulatory actions of melatonin on neuronal 'clock' gene expression in striatal neurons in vitro. *J Pineal Res*. 46:87–94.

Ionov M, Burchell V, Klajnert B, Bryszewska M, Abramov AY. 2011. Mechanism of neuroprotection of melatonin against beta-amyloid neurotoxicity. *Neuroscience*. 180:229–237.

Jager S, Handschin C, St-Pierre J, Spiegelman BM. 2007. AMP-activated protein kinase (AMPK) action in skeletal muscle via direct phosphorylation of PGC-1alpha. *Proc Natl Acad Sci USA*. 104:12017–12022.

Jeong JK, Moon MH, Lee YJ, Seol JW, Park SY. 2012. Melatonin-induced autophagy protects against human prion protein-mediated neurotoxicity. *J Pineal Res*. 53:138–146.

Jimenez-Aranda A, Fernandez-Vazquez G, Campos D, Tassi M, Velasco-Perez L, Tan DX, Reiter RJ, Agil A. 2013. Melatonin induces browning of inguinal white adipose tissue in Zucker diabetic fatty rats. *J Pineal Res*. 55:416–423.

Jin X, von Gall C, Pieschl RL, Gribkoff VK, Stehle JH, Reppert SM, Weaver DR. 2003. Targeted disruption of the mouse Mel(1b) melatonin receptor. *Mol Cell Biol*. 23:1054–1060.

Jung-Hynes B, Reiter RJ, Ahmad N. 2010. Sirtuins, melatonin and circadian rhythms: Building a bridge between aging and cancer. *J Pineal Res*. 48:9–19.

Korkmaz A, Reiter RJ, Topal T, Manchester LC, Oter S, Tan DX. 2009. Melatonin: An established antioxidant worthy of use in clinical trials. *Mol Med*. 15:43–50.

Kotler M, Rodriguez C, Sainz RM, Antolin I, Menendez-Pelaez A. 1998. Melatonin increases gene expression for antioxidant enzymes in rat brain cortex. *J Pineal Res*. 24:83–89.

Kujoth GC, Hiona A, Pugh TD, Someya S, Panzer K, Wohlgemuth SE, Hofer T et al. 2005. Mitochondrial DNA mutations, oxidative stress, and apoptosis in mammalian aging. *Science*. 309:481–484.

Lahiri DK, Ge YW, Sharman EH, Bondy SC. 2004. Age-related changes in serum melatonin in mice: Higher levels of combined melatonin and 6-hydroxymelatonin sulfate in the cerebral cortex than serum, heart, liver and kidney tissues. *J Pineal Res*. 36:217–223.

Lahiri S, Singh P, Singh S, Rasheed N, Palit G, Pant KK. 2009. Melatonin protects against experimental reflux esophagitis. *J Pineal Res*. 46:207–213.

Larson J, Jessen RE, Uz T, Arslan AD, Kurtuncu M, Imbesi M, Manev H. 2006. Impaired hippocampal long-term potentiation in melatonin MT2 receptor-deficient mice. *Neurosci Lett*. 393:23–26.

Lee CH, Choi JH, Yoo KY, Park OK, Hwang IK, You SG, Lee BY, Kang IJ, Won MH. 2010. MT2 melatonin receptor immunoreactivity in neurons is very high in the aged hippocampal formation in gerbils. *Cell Mol Neurobiol*. 30:255–263.

Lemoine P, Zisapel N. 2012. Prolonged-release formulation of melatonin (Circadin) for the treatment of insomnia. *Expert Opin Pharmacother*. 13:895–905.

Levoye A, Dam J, Ayoub MA, Guillaume JL, Couturier C, Delagrange P, Jockers R. 2006. The orphan GPR50 receptor specifically inhibits MT1 melatonin receptor function through heterodimerization. *Embo J*. 25:3012–3023.

Limon-Pacheco JH, Gonsebatt ME. 2010. The glutathione system and its regulation by neurohormone melatonin in the central nervous system. *Cent Nerv Syst Agents Med Chem*. 10:287–297.

Lin L, Huang QX, Yang SS, Chu J, Wang JZ, Tian Q. 2013. Melatonin in Alzheimer's disease. *Int J Mol Sci*. 14:14575–14593.

Liu C, Weaver DR, Jin X, Shearman LP, Pieschl RL, Gribkoff VK, Reppert SM. 1997. Molecular dissection of two distinct actions of melatonin on the suprachiasmatic circadian clock. *Neuron*. 19:91–102.

Liu RY, Zhou JN, van Heerikhuize J, Hofman MA, Swaab DF. 1999. Decreased melatonin levels in postmortem cerebrospinal fluid in relation to aging, Alzheimer's disease, and apolipoprotein E-epsilon4/4 genotype. *J Clin Endocrinol Metab*. 84:323–327.

Liu Y, Zhang L, Zhang H, Liu B, Wu Z, Zhao W, Wang Z. 2012a. Exogenous melatonin modulates apoptosis in the mouse brain induced by high-LET carbon ion irradiation. *J Pineal Res*. 52:47–56.

Liu YJ, Meng FT, Wang LL, Zhang LF, Cheng XP, Zhou JN. 2012b. Apolipoprotein E influences melatonin biosynthesis by regulating NAT and MAOA expression in C6 cells. *J Pineal Res*. 52:397–402.

Mahlberg R, Kunz D, Sutej I, Kuhl KP, Hellweg R. 2004. Melatonin treatment of day-night rhythm disturbances and sundowning in Alzheimer disease: An open-label pilot study using actigraphy. *J Clin Psychopharmacol*. 24:456–459.

Mahley RW. 1988. Apolipoprotein E: Cholesterol transport protein with expanding role in cell biology. *Science*. 240:622–630.

Manczak M, Anekonda TS, Henson E, Park BS, Quinn J, Reddy PH. 2006. Mitochondria are a direct site of A beta accumulation in Alzheimer's disease neurons: Implications for free radical generation and oxidative damage in disease progression. *Hum Mol Genet*. 15:1437–1449.

Manczak M, Calkins MJ, Reddy PH. 2011. Impaired mitochondrial dynamics and abnormal interaction of amyloid beta with mitochondrial protein Drp1 in neurons from patients with Alzheimer's disease: Implications for neuronal damage. *Hum Mol Genet*. 20:2495–2509.

Manczak M, Mao P, Calkins MJ, Cornea A, Reddy AP, Murphy MP, Szeto HH, Park B, Reddy PH. 2010. Mitochondria-targeted antioxidants protect against amyloid-beta toxicity in Alzheimer's disease neurons. *J Alzheimers Dis*. 20(Suppl 2):S609–S631.

Martin M, Macias M, Escames G, Reiter RJ, Agapito MT, Ortiz GG, Acuna-Castroviejo D. 2000. Melatonin-induced increased activity of the respiratory chain complexes I and IV can prevent mitochondrial damage induced by ruthenium red in vivo. *J Pineal Res*. 28:242–248.

Matsubara E, Bryant-Thomas T, Pacheco Quinto J, Henry TL, Poeggeler B, Herbert D, Cruz-Sanchez F et al. 2003. Melatonin increases survival and inhibits oxidative and amyloid pathology in a transgenic model of Alzheimer's disease. *J Neurochem*. 85:1101–1108.

Maurizi CP. 1997. Loss of intraventricular fluid melatonin can explain the neuropathology of Alzheimer's disease. *Med Hypotheses*. 49:153–158.

Maurizi CP. 2001. Alzheimer's disease: Roles for mitochondrial damage, the hydroxyl radical, and cerebrospinal fluid deficiency of melatonin. *Med Hypotheses*. 57:156–160.

Mayeux R, Stern Y. 2012. Epidemiology of Alzheimer disease. *Cold Spring Harb Perspect Med*. 2:pii: a006239.

Mayo JC, Sainz RM, Tan DX, Antolin I, Rodriguez C, Reiter RJ. 2005. Melatonin and Parkinson's disease. *Endocrine*. 27:169–178.

McKenna JT, Christie MA, Jeffrey BA, McCoy JG, Lee E, Connolly NP, Ward CP, Strecker RE. 2012. Chronic ramelteon treatment in a mouse model of Alzheimer's disease. *Arch Ital Biol*. 150:5–14.

Meijer AJ, Codogno P. 2007. AMP-activated protein kinase and autophagy. *Autophagy*. 3:238–240.

Mendes C, Lopes AM, do Amaral FG, Peliciari-Garcia RA, Turati Ade O, Hirabara SM, Scialfa Falcao JH, Cipolla-Neto J. 2013. Adaptations of the aging animal to exercise: Role of daily supplementation with melatonin. *J Pineal Res*. 55:229–239.

Mishima K, Okawa M, Hozumi S, Hishikawa Y. 2000. Supplementary administration of artificial bright light and melatonin as potent treatment for disorganized circadian rest-activity and dysfunctional autonomic and neuroendocrine systems in institutionalized demented elderly persons. *Chronobiol Int*. 17:419–432.

Munguia ME, Govezensky T, Martinez R, Manoutcharian K, Gevorkian G. 2006. Identification of amyloid-beta 1–42 binding protein fragments by screening of a human brain cDNA library. *Neurosci Lett*. 397:79–82.

Musshoff U, Riewenherm D, Berger E, Fauteck JD, Speckmann EJ. 2002. Melatonin receptors in rat hippocampus: Molecular and functional investigations. *Hippocampus*. 12:165–173.

Naiki H, Gejyo F, Nakakuki K. 1997. Concentration-dependent inhibitory effects of apolipoprotein E on Alzheimer's beta-amyloid fibril formation in vitro. *Biochemistry*. 36:6243–6250.

Nakahata Y, Kaluzova M, Grimaldi B, Sahar S, Hirayama J, Chen D, Guarente LP, Sassone-Corsi P. 2008. The NAD^+-dependent deacetylase SIRT1 modulates CLOCK-mediated chromatin remodeling and circadian control. *Cell*. 134:329–340.

Nakahata Y, Sahar S, Astarita G, Kaluzova M, Sassone-Corsi P. 2009. Circadian control of the NAD^+ salvage pathway by CLOCK-SIRT1. *Science*. 324:654–657.

Negi G, Kumar A, Sharma SS. 2011. Melatonin modulates neuroinflammation and oxidative stress in experimental diabetic neuropathy: Effects on NF-kappaB and Nrf2 cascades. *J Pineal Res*. 50:124–131.

Nopparat C, Porter JE, Ebadi M, Govitrapong P. 2010. The mechanism for the neuroprotective effect of melatonin against methamphetamine-induced autophagy. *J Pineal Res*. 49:382–389.

O'Neill JS, Maywood ES, Chesham JE, Takahashi JS, Hastings MH. 2008. cAMP-dependent signaling as a core component of the mammalian circadian pacemaker. *Science*. 320:949–953.

Olcese JM, Cao C, Mori T, Mamcarz MB, Maxwell A, Runfeldt MJ, Wang L et al. 2009. Protection against cognitive deficits and markers of neurodegeneration by long-term oral administration of melatonin in a transgenic model of Alzheimer disease. *J Pineal Res*. 47:82–96.

Ozturk G, Akbulut KG, Guney S, Acuna-Castroviejo D. 2012. Age-related changes in the rat brain mitochondrial antioxidative enzyme ratios: Modulation by melatonin. *Exp Gerontol*. 47:706–711.

Pandi-Perumal SR, BaHammam AS, Brown GM, Spence DW, Bharti VK, Kaur C, Hardeland R, Cardinali DP. 2013. Melatonin antioxidative defense: Therapeutical implications for aging and neurodegenerative processes. *Neurotox Res*. 23:267–300.

Pappolla M, Bozner P, Soto C, Shao H, Robakis NK, Zagorski M, Frangione B, Ghiso J. 1998. Inhibition of Alzheimer beta-fibrillogenesis by melatonin. *J Biol Chem*. 273:7185–7188.

Pappolla MA, Chyan YJ, Poeggeler B, Bozner P, Ghiso J, LeDoux SP, Wilson GL. 1999. Alzheimer beta protein mediated oxidative damage of mitochondrial DNA: Prevention by melatonin. *J Pineal Res*. 27:226–229.

Pappolla MA, Sos M, Omar RA, Bick RJ, Hickson-Bick DL, Reiter RJ, Efthimiopoulos S, Robakis NK. 1997. Melatonin prevents death of neuroblastoma cells exposed to the Alzheimer amyloid peptide. *J Neurosci*. 17:1683–1690.

Parker WD, Jr. 1991. Cytochrome oxidase deficiency in Alzheimer's disease. *Ann N Y Acad Sci*. 640:59–64.

Peng CX, Hu J, Liu D, Hong XP, Wu YY, Zhu LQ, Wang JZ. 2013. Disease-modified glycogen synthase kinase-3beta intervention by melatonin arrests the pathology and memory deficits in an Alzheimer's animal model. *Neurobiol Aging*. 34:1555–1563.

Pereira C, Santos MS, Oliveira C. 1998. Mitochondrial function impairment induced by amyloid beta-peptide on PC12 cells. *Neuroreport*. 9:1749–1755.

Perry EK, Perry RH, Tomlinson BE, Blessed G, Gibson PH. 1980. Coenzyme A-acetylating enzymes in Alzheimer's disease: Possible cholinergic 'compartment' of pyruvate dehydrogenase. *Neurosci Lett.* 18:105–110.

Poeggeler B, Miravalle L, Zagorski MG, Wisniewski T, Chyan YJ, Zhang Y, Shao H et al. 2001. Melatonin reverses the profibrillogenic activity of apolipoprotein E4 on the Alzheimer amyloid Abeta peptide. *Biochemistry.* 40:14995–15001.

Poeggeler B, Sambamurti K, Siedlak SL, Perry G, Smith MA, Pappolla MA. 2010. A novel endogenous indole protects rodent mitochondria and extends rotifer lifespan. *PLoS One* 5:e10206.

Poirel VJ, Boggio V, Dardente H, Pevet P, Masson-Pevet M, Gauer F. 2003. Contrary to other non-photic cues, acute melatonin injection does not induce immediate changes of clock gene mRNA expression in the rat suprachiasmatic nuclei. *Neuroscience.* 120:745–755.

Poirier GL, Amin E, Good MA, Aggleton JP. 2011. Early-onset dysfunction of retrosplenial cortex precedes overt amyloid plaque formation in Tg2576 mice. *Neuroscience.* 174:71–83.

Poon AM, Pang SF. 1992. 2[125I]iodomelatonin binding sites in spleens of guinea pigs. *Life Sci.* 50:1719–1726.

Quinn J, Kulhanek D, Nowlin J, Jones R, Pratico D, Rokach J, Stackman R. 2005. Chronic melatonin therapy fails to alter amyloid burden or oxidative damage in old Tg2576 mice: Implications for clinical trials. *Brain Res.* 1037:209–213.

Ramsey KM, Yoshino J, Brace CS, Abrassart D, Kobayashi Y, Marcheva B, Hong HK et al. 2009. Circadian clock feedback cycle through NAMPT-mediated NAD^+ biosynthesis. *Science.* 324:651–654.

Reddy PH, McWeeney S, Park BS, Manczak M, Gutala RV, Partovi D, Jung Y et al. 2004. Gene expression profiles of transcripts in amyloid precursor protein transgenic mice: Up-regulation of mitochondrial metabolism and apoptotic genes is an early cellular change in Alzheimer's disease. *Hum Mol Genet.* 13:1225–1240.

Reiter RJ, Tan DX, Fuentes-Broto L. 2010. Melatonin: A multitasking molecule. *Prog Brain Res.* 181:127–151.

Rezzani R, Rodella LF, Bonomini F, Tengattini S, Bianchi R, Reiter RJ. 2006. Beneficial effects of melatonin in protecting against cyclosporine A-induced cardiotoxicity are receptor mediated. *J Pineal Res.* 41:288–295.

Rhein V, Song X, Wiesner A, Ittner LM, Baysang G, Meier F, Ozmen L et al. 2009. Amyloid-beta and tau synergistically impair the oxidative phosphorylation system in triple transgenic Alzheimer's disease mice. *Proc Natl Acad Sci USA.* 106:20057–20062.

Rosales-Corral S, Acuna-Castroviejo D, Tan DX, Lopez-Armas G, Cruz-Ramos J, Munoz R, Melnikov VG, Manchester LC, Reiter RJ. 2012a. Accumulation of exogenous amyloid-beta peptide in hippocampal mitochondria causes their dysfunction: A protective role for melatonin. *Oxid Med Cell Longev.* 2012:843649.

Rosales-Corral SA, Acuna-Castroviejo D, Coto-Montes A, Boga JA, Manchester LC, Fuentes-Broto L, Korkmaz A, Ma S, Tan DX, Reiter RJ. 2012b. Alzheimer's disease: Pathological mechanisms and the beneficial role of melatonin. *J Pineal Res.* 52:167–202.

Sanchez-Hidalgo M, Guerrero Montavez JM, Carrascosa-Salmoral Mdel P, Naranjo Gutierrez Mdel C, Lardone PJ, de la Lastra Romero CA. 2009. Decreased MT1 and MT2 melatonin receptor expression in extrapineal tissues of the rat during physiological aging. *J Pineal Res.* 46:29–35.

Santos RX, Correia SC, Wang X, Perry G, Smith MA, Moreira PI, Zhu X. 2010. A synergistic dysfunction of mitochondrial fission/fusion dynamics and mitophagy in Alzheimer's disease. *J Alzheimers Dis.* 20(Suppl 2):S401–S412.

Savaskan E, Ayoub MA, Ravid R, Angeloni D, Fraschini F, Meier F, Eckert A, Muller-Spahn F, Jockers R. 2005. Reduced hippocampal MT2 melatonin receptor expression in Alzheimer's disease. *J Pineal Res.* 38:10–16.

Savaskan E, Jockers R, Ayoub M, Angeloni D, Fraschini F, Flammer J, Eckert A, Muller-Spahn F, Meyer P. 2007. The MT2 melatonin receptor subtype is present in human retina and decreases in Alzheimer's disease. *Curr Alzheimer Res.* 4:47–51.

Savaskan E, Olivieri G, Meier F, Brydon L, Jockers R, Ravid R, Wirz-Justice A, Muller-Spahn F. 2002. Increased melatonin 1a-receptor immunoreactivity in the hippocampus of Alzheimer's disease patients. *J Pineal Res.* 32:59–62.

Schagger H, Ohm TG. 1995. Human diseases with defects in oxidative phosphorylation. 2. F1F0 ATP-synthase defects in Alzheimer disease revealed by blue native polyacrylamide gel electrophoresis. *Eur J Biochem.* 227:916–921.

Scheper W, Nijholt DA, Hoozemans JJ. 2011. The unfolded protein response and proteostasis in Alzheimer disease: Preferential activation of autophagy by endoplasmic reticulum stress. *Autophagy.* 7:910–911.

Sharman EH, Sharman KG, Ge YW, Lahiri DK, Bondy SC. 2004. Age-related changes in murine CNS mRNA gene expression are modulated by dietary melatonin. *J Pineal Res*. 36:165–170.

Sheng B, Wang X, Su B, Lee HG, Casadesus G, Perry G, Zhu X. 2012. Impaired mitochondrial biogenesis contributes to mitochondrial dysfunction in Alzheimer's disease. *J Neurochem*. 120:419–429.

Snell TW, Fields AM, Johnston RK. 2012. Antioxidants can extend lifespan of *Brachionus manjavacas* (Rotifera), but only in a few combinations. *Biogerontology*. 13:261–275.

Song N, Kim AJ, Kim HJ, Jee HJ, Kim M, Yoo YH, Yun J. 2012. Melatonin suppresses doxorubicin-induced premature senescence of A549 lung cancer cells by ameliorating mitochondrial dysfunction. *J Pineal Res*. 53:335–343.

Sorbi S, Bird ED, Blass JP. 1983. Decreased pyruvate dehydrogenase complex activity in Huntington and Alzheimer brain. *Ann Neurol*. 13:72–78.

Srinivasan V, Kaur C, Pandi-Perumal S, Brown GM, Cardinali DP. 2010. Melatonin and its agonist ramelteon in Alzheimer's disease: Possible therapeutic value. *Int J Alzheimers Dis*. 2011:741974.

St-Pierre J, Drori S, Uldry M, Silvaggi JM, Rhee J, Jager S, Handschin C et al. 2006. Suppression of reactive oxygen species and neurodegeneration by the PGC-1 transcriptional coactivators. *Cell*. 127:397–408.

Strazielle C, Sturchler-Pierrat C, Staufenbiel M, Lalonde R. 2003. Regional brain cytochrome oxidase activity in beta-amyloid precursor protein transgenic mice with the Swedish mutation. *Neuroscience*. 118:1151–1163.

Strittmatter WJ, Saunders AM, Schmechel D, Pericak-Vance M, Enghild J, Salvesen GS, Roses AD. 1993. Apolipoprotein E: High-avidity binding to beta-amyloid and increased frequency of type 4 allele in late-onset familial Alzheimer disease. *Proc Natl Acad Sci USA*. 90:1977–1981.

Swerdlow RH, Burns JM, Khan SM. 2013. The Alzheimer's disease mitochondrial cascade hypothesis: Progress and perspectives. *Biochim Biophys Acta*. doi:10.1016/j.bbadis.2013.09.010.

Swerdlow RH, Khan SM. 2004. A "mitochondrial cascade hypothesis" for sporadic Alzheimer's disease. *Med Hypotheses*. 63:8–20.

Tajes M, Gutierrez-Cuesta J, Ortuno-Sahagun D, Camins A, Pallas M. 2009. Anti-aging properties of melatonin in an in vitro murine senescence model: Involvement of the sirtuin 1 pathway. *J Pineal Res*. 47:228–237.

Tao RR, Huang JY, Shao XJ, Ye WF, Tian Y, Liao MH, Fukunaga K, Lou YJ, Han F, Lu YM. 2013. Ischemic injury promotes Keap1 nitration and disturbance of antioxidative responses in endothelial cells: A potential vasoprotective effect of melatonin. *J Pineal Res*. 54:271–281

Torres-Farfan C, Seron-Ferre M, Dinet V, Korf HW. 2006. Immunocytochemical demonstration of day/night changes of clock gene protein levels in the murine adrenal gland: Differences between melatonin-proficient (C3H) and melatonin-deficient (C57BL) mice. *J Pineal Res*. 40:64–70.

Travnickova-Bendova Z, Cermakian N, Reppert SM, Sassone-Corsi P. 2002. Bimodal regulation of mPeriod promoters by CREB-dependent signaling and CLOCK/BMAL1 activity. *Proc Natl Acad Sci USA*. 99:7728–7733.

Valla J, Schneider LE, Small AM, Gonzalez-Lima F. 2007. Quantitative cytochrome oxidase histochemistry: Applications in human Alzheimer's disease and animal models. *J Histotechnol*. 30:235–247.

Varghese M, Zhao W, Wang J, Cheng A, Qian X, Chaudhry A, Ho L, Pasinetti G. 2011. Mitochondrial bioenergetics is defective in presymptomatic Tg2576 AD mice. *Transl Neurosci*. 2:1–5.

Vilar A, de Lemos L, Patraca I, Martinez N, Folch J, Junyent F, Verdaguer E, Pallas M, Auladell C, Camins A. 2014. Melatonin suppresses nitric oxide production in glial cultures by pro-inflammatory cytokines through p38 MAPK inhibition. *Free Radic Res*. 48(2):119–128. doi:10.3109/10715762.2013.845295.

von Gall C, Weaver DR. 2008. Loss of responsiveness to melatonin in the aging mouse suprachiasmatic nucleus. *Neurobiol Aging*. 29:464–470.

Wallace DC, Fan WW. 2009. The pathophysiology of mitochondrial disease as modeled in the mouse. *Gene Dev*. 23:1714–1736.

Wang J, Zhou T. 2010. cAMP-regulated dynamics of the mammalian circadian clock. *Biosystems*. 101:136–143.

Wang LM, Suthana NA, Chaudhury D, Weaver DR, Colwell CS. 2005. Melatonin inhibits hippocampal long-term potentiation. *Eur J Neurosci*. 22:2231–2237.

Wang X, Sirianni A, Pei Z, Cormier K, Smith K, Jiang J, Zhou S et al. 2011. The melatonin MT1 receptor axis modulates mutant Huntingtin-mediated toxicity. *J Neurosci*. 31:14496–14507.

Wang Z, Liu D, Wang J, Liu S, Gao M, Ling EA, Hao A. 2012a. Cytoprotective effects of melatonin on astroglial cells subjected to palmitic acid treatment in vitro. *J Pineal Res*. 52:253–264.

Wang Z, Ma C, Meng CJ, Zhu GQ, Sun XB, Huo L, Zhang J et al. 2012b. Melatonin activates the Nrf2-ARE pathway when it protects against early brain injury in a subarachnoid hemorrhage model. *J Pineal Res*. 53:129–137.

Wolf AB, Braden BB, Bimonte-Nelson H, Kusne Y, Young N, Engler-Chiurazzi E, Garcia AN et al. 2012. Broad-based nutritional supplementation in 3xTg mice corrects mitochondrial function and indicates sex-specificity in response to Alzheimer's disease intervention. *J Alzheimers Dis*. 32:217–232.

Wolfe DM, Lee JH, Kumar A, Lee S, Orenstein SJ, Nixon RA. 2013. Autophagy failure in Alzheimer's disease and the role of defective lysosomal acidification. *Eur J Neurosci*. 37:1949–1961.

Wu YH, Zhou JN, Van Heerikhuize J, Jockers R, Swaab DF. 2007. Decreased MT1 melatonin receptor expression in the suprachiasmatic nucleus in aging and Alzheimer's disease. *Neurobiol Aging*. 28:1239–1247.

Xie H, Guan J, Borrelli LA, Xu J, Serrano-Pozo A, Bacskai BJ. 2013a. Mitochondrial alterations near amyloid plaques in an Alzheimer's disease mouse model. *J Neurosci*. 33:17042–17051.

Xie L, Kang H, Xu Q, Chen MJ, Liao Y, Thiyagarajan M, O'Donnell J et al. 2013b. Sleep drives metabolite clearance from the adult brain. *Science*. 342:373–377.

Yamamoto HA, Mohanan PV. 2002. Melatonin attenuates brain mitochondria DNA damage induced by potassium cyanide in vivo and in vitro. *Toxicology*. 179:29–36.

Yuan H, Pang SF. 1991. [125I]Iodomelatonin-binding sites in the pigeon brain: Binding characteristics, regional distribution and diurnal variation. *J Endocrinol*. 128:475–482.

Zaouali MA, Boncompagni E, Reiter RJ, Bejaoui M, Freitas I, Pantazi E, Folch-Puy E, Abdennebi HB, Garcia-Gil FA, Rosello-Catafau J. 2013. AMPK involvement in endoplasmic reticulum stress and autophagy modulation after fatty liver graft preservation: A role for melatonin and trimetazidine cocktail. *J Pineal Res*. 55(1):65–78. doi:10.1111/jpi.12051.

Zeman M, Szantoova K, Stebelova K, Mravec B, Herichova I. 2009. Effect of rhythmic melatonin administration on clock gene expression in the suprachiasmatic nucleus and the heart of hypertensive TGR(mRen2)27 rats. *J Hypertens Suppl*. 27:S21–S26.

Zhang XM, Xiong K, Cai Y, Cai H, Luo XG, Feng JC, Clough RW, Patrylo PR, Struble RG, Yan XX. 2010. Functional deprivation promotes amyloid plaque pathogenesis in Tg2576 mouse olfactory bulb and piriform cortex. *Eur J Neurosci*. 31:710–721.

Zhou H, Chen J, Lu X, Shen C, Zeng J, Chen L, Pei Z. 2012. Melatonin protects against rotenone-induced cell injury via inhibition of Omi and Bax-mediated autophagy in Hela cells. *J Pineal Res*. 52:120–127.

28 Melatonin and Its Therapeutic Implications in Injuries of the Developing Brain and Retina

Gurugirijha Rathnasamy, Eng-Ang Ling, and Charanjit Kaur

CONTENTS

28.1 INTRODUCTION

Melatonin (5-methoxy-*N*-acetyltryptamine), a derivative of the amino acid tryptophan, was first discovered by Lerner and his coworkers (Lerner et al., 1958) as an endogenous neurohormone secreted by the pineal gland. However, later findings revealed that melatonin is also synthesized by other tissues such as retina (Dubocovich, 1983), Harderian gland (Buzzell et al., 1990; Djeridane and Touitou, 2001), bone marrow (Conti et al., 2000), platelets (Champier et al., 1997), gastrointestinal tract (Bubenik, 2002), skin (Slominski et al., 2005), and lymphocytes (Carrillo-Vico et al., 2004). In addition to these extrapineal sources, a recent study demonstrated the synthesis of melatonin in the immature rat brains from fetal day 18 to the first week of postnatal life, without any influence from pineal gland (Jimenez-Jorge et al., 2007). Melatonin gained importance when it was identified to function as a chronobiotic substance regulating the circadian rhythms (Redman et al., 1983; Armstrong et al., 1986). Melatonin is known to participate in a wide range of physiological functions (Pandi-Perumal et al., 2006), to act as an antioxidant (Reiter et al., 2003) and an immune regulator (Carrillo-Vico et al., 2013) and has also been used in treating sleep disorders (Cummings, 2012; Ferracioli-Oda et al., 2013). Furthermore, it has been documented to have anticonvulsant action (Champney et al., 1996), antiaging property (Poeggeler, 2005), and oncostatic effect (Srinivasan et al., 2008).

Several experimental studies have suggested the neuroprotective action of melatonin in neuropathologies such as Parkinson's disease (Mayo et al., 2005a; Borah and Mohanakumar, 2009), Alzheimer's disease (Feng et al., 2004), ischemic brain injury (Pei et al., 2003; Pei and Cheung, 2004; Carloni et al., 2008), and neuropsychiatric disorders (Srinivasan et al., 2006b). In the earlier-mentioned conditions, oxidative stress, inflammation, and mitochondrial dysfunction were found to be the major cause of neuronal damage (Reiter, 1998). In conjunction with this, the authors reported a reduction in the amount of melatonin being produced (Pang et al., 1990; Fiorina et al., 1996;

Wu and Swaab, 2005). However, exogenous administration of melatonin counteracted the adverse effects due to oxidative stress and mitochondrial dysfunction resulting in a reduction of damage to the neurons (Sharma et al., 2006; Srinivasan et al., 2006a; Lin et al., 2008, 2013). A similar protective effect of melatonin was also observed in conditions such as age-related macular degeneration (Liang and Godley, 2003), glaucoma (Belforte et al., 2010), and ischemia reperfusion injury (Park et al., 2012) wherein the death of retinal ganglion cells (RGCs) and other neurons of the retina has been documented. At cellular level, the protective property of melatonin was attributed to its ability to increase the activity of antioxidants such as glutathione and superoxide dismutase, and its capability of being an electron donor and an acceptor.

However, when compared to the adult nervous system, the developing nervous system, owing to the lack of antioxidants, is highly susceptible to oxidative stress–mediated damage. Parallel to the findings in adult neural system, melatonin was demonstrated to render neuroprotection to the developing brain and retina (Watanabe et al., 2012; Kaur et al., 2013). In rodent models of hypoxic developing white matter damage, melatonin effectively reduced the amount of malondialdehyde (MDA) being formed and the apoptosis of oligodendrocytes (Kaur et al., 2010). In immature retina, the death of RGCs following hypoxic insult was attenuated with melatonin administration (Kaur et al., 2013). In addition, by reducing the expression of vasoactive factors, melatonin protected the integrity of blood–brain barrier (BBB) and blood–retinal barrier (Kaur et al., 2010, 2013; Yawno et al., 2012). Although the understanding of the broad spectrum of beneficial functions of melatonin during perinatal period is still lacking, this review summarizes the available knowledge on the protective role of melatonin in safeguarding the developing brain and retina.

28.2 ROLE IN FREE RADICAL SEQUESTRATION

Melatonin is a remarkable free radical scavenger and has the potential to induce antioxidant enzymes such as superoxide dismutase and glutathione peroxidase. It can neutralize most of the highly reactive oxygen and nitrogen molecules, which include hydroxyl radical (OH), hydrogen peroxide (H_2O_2), singlet oxygen, hypochlorous acid, peroxynitrite anion ($ONOO^-$), and/or peroxynitrous acid (Reiter et al., 2001, 2003). The presence of *O*-methyl and *N*-acetyl residues in melatonin is attributed to its free radical scavenging property (Poeggeler et al., 2002). Evidences gathered from several studies demonstrate the ability of melatonin to readily scavenge OH (Matuszak et al., 1997; Horstman et al., 2002). Melatonin's efficiency as an antioxidant has been paralleled to that of vitamin E (Scaiano, 1995). Furthermore, the metabolism of melatonin in response to oxidative stress is also of relative importance. In conditions such as ischemic stroke, melatonin was metabolized (Ritzenthaler et al., 2013), resulting in reduced excretion of melatonin. This catabolism is suggested to protect the brain from free radical–mediated damage as metabolites of melatonin were shown to possess antioxidant property (Acuña-Castroviejo et al., 2003; León et al., 2006; Schaefer and Hardeland, 2009). A single molecule of melatonin via its metabolites is reported to scavenge up to 10 free radical molecules (Tan et al., 2007). The attenuation of free radical chain, catalyzed by the breakdown of H_2O_2, by melatonin involved electron transfer. Melatonin by donating one electron to OH results in the formation of hydroxyl anion and melatonyl cation. This melatonyl cation reacts with superoxide radical to form the stable *N*1-acetyl-*N*2-formyl-5-methoxykynuramine (AFMK) (Poeggeler et al., 1994). Oxidation of melatonin could also be mediated by myeloperoxidase, cytochrome c (cyt c), hemoglobin, horseradish peroxidase, or reactive oxygen species (ROS) resulting in the generation of AFMK (Silva et al., 2000; Tesoriere et al., 2001; Semak et al., 2005; Ximenes et al., 2007). Pharmacological concentrations of AFMK were found to rescue the hippocampal neurons from oxidative stress and reduce lipid peroxidation (Tan et al., 2001). In addition, in the process of metabolic conversion from melatonin to AFMK, at least four free radicals are consumed (Tan et al., 2003). Furthermore, AFMK when degraded by arylamine formamidase results in the formation of a more potent radical scavenger, *N*1-acetyl-5-methoxykynuramine (AMK) (Hirata et al., 1974; Hardeland et al., 1993; Hardeland, 2005). The metabolite AMK not only interacts with reactive nitrogen species (Guenther et al., 2005) but also possesses an inhibitory property against cyclooxygenase2 (Mayo et al., 2005b), a key molecule primarily responsible for the production of proinflammatory molecules, thereby limiting the oxidative stress and inflammatory response (Hardeland, 2005).

Besides the earlier-mentioned characteristics, melatonin has been demonstrated to prevent the free radical–mediated injury by enhancing the antioxidants such as superoxide dismutase, peroxidase, and enzymes involved in glutathione synthesis (Hardeland, 2005) and by downregulating pro-oxidant systems. Pro-oxidants such as 5- and 12-lipo-oxygenases (Manev et al., 1998; Uz and Manev, 1998; Zhang et al., 1999) and nitric oxide synthase (NOS) (Pozo et al., 1994; Bettahi et al., 1996) were shown to be inhibited by melatonin. Melatonin was also demonstrated to prevent *N*-methyl-D-aspartate-induced excitotoxicity by inhibiting neuronal NOS (Escames et al., 2004) and subsequent nitric oxide (NO) production. Binding of melatonin to calmodulin, and making it less available for the enzymatic conversion of arginine to NO, was suggested to be the mechanism involved in the inhibition of NOS (Pozo et al., 1997; León et al., 2006). The inhibition of NO production is of major significance as it can hinder the generation of free radicals and peroxynitrite (Pandi-Perumal et al., 2006) in conditions such as ischemia (Warner et al., 2004) and multiple sclerosis (Cross et al., 1998).

Furthermore, melatonin was also shown to prevent cell death by mitigating the mitochondrial stress (León et al., 2005) and subsequent free radical production. In mitochondria, the energy transport chain (ETC) that yields adenosine triphosphate (ATP) due to electron leakage could also result in the generation of free radicals. Mitochondrial dysfunction and electron leakage from ETC have been reported in conditions such as hypoxia (Kim et al., 2006), ischemia (Petrosillo et al., 2006), hyperoxia (Freeman and Crapo, 1981), Parkinson's disease (Keane et al., 2011), and Alzheimer's disease (Benzi and Moretti, 1995). Increased electron leakage could result in increased generation of ROS in the mitochondria (Genova et al., 2003; Miwa and Brand, 2003), which could further result in apoptosis through the oxidization of glutathione and enhanced mitochondrial permeability transition pore opening (León et al., 2004). However, melatonin is reported to suppress mitochondrial damage and maintain the integrity of mitochondria (León et al., 2004). The remarkable ability of melatonin to donate single electron to the molecules of respiratory chain was attributed to the protective property exerted by melatonin on mitochondria (Hardeland, 2005). Accordingly, melatonin was suggested to interact with mitochondrial complexes and improve the electron flow (Martin et al., 2002) by donating and accepting electrons. In *in vitro* models of t-butyl hydroperoxide (t-BHP-)-induced mitochondrial oxidative stress, melatonin treatment was demonstrated to increase mitochondrial complex I and complex IV activity (Martin et al., 2000a) thereby improving the performance of ETC. Consistent with this, mitochondrial respiration and ATP synthesis were also found to be increased (Martin et al., 2000b, 2002). This increase in ATP synthesis was suggested to favor mitochondria in restoring the damage occurred to mitochondrial DNA (Reiter et al., 2003). Melatonin was also reported to be effective in inhibiting peroxynitrite-mediated suppression of mitochondrial respiration (Gilad et al., 1997). In animal models of ruthenium red–dependent mitochondrial damage, administration of melatonin was found to be beneficial as it increased mitochondrial respiration and restored glutathione peroxidase activity (Martin et al., 2000b). Furthermore, in skeletal muscle cells subjected to oxidative stress, melatonin was demonstrated to block the mitochondrial permeability transition pore opening (Hibaoui et al., 2009). Besides these, the metabolite of melatonin, AMK, was also suggested to undergo single-electron transfer reactions (Acuña-Castroviejo et al., 2003). In a manner similar to melatonin, AMK was reported to influence electron flux in the ETC and improved ATP synthesis (Acuña-Castroviejo et al., 2003).

28.3 MELATONIN AND DEVELOPING BRAIN

Injury to the developing brain leads to long-term neurological morbidity in infants that persists throughout the rest of their lives. Factors such as inflammation and oxidative stress have been majorly implicated in the pathogenesis of the immature brain damage (McAdams and Juul, 2012). Increased lipid peroxidation and DNA damage have been demonstrated in asphyxiated fetal brains indicating oxidative damage (Manoj et al., 2011). NO has been implicated in various pathophysiological processes, including hypoxic injury (Murugan et al., 2011) and hydrocephalus (Del Bigio et al., 2012). In addition, in the immature brains, in response to hypoxia, hypoxia–ischemia, or asphyxia, neuropathological events such as activation of microglial cells, astrocytosis, and apoptosis of oligodendrocytes and neurons were demonstrated (Van de Berg et al., 2002; Mallard et al., 2003; Kaur and Ling, 2009; Chen et al., 2012). In addition, in response to hypoxia or asphyxia, the initiation of inflammatory cascade through secretion of pro-inflammatory cytokines by various glial cells was implicated in causing damage to the developing white matter (Mallard et al., 2003; Deng et al., 2008). Activated microglia in the hypoxic white matter were found to express pro-inflammatory cytokines such as tumor necrosis factor-α (TNF-α) and interleukin-1β (IL-1β) (Deng et al., 2008). Besides these, several studies have demonstrated increased vascular permeability in the developing white matter following hypoxic injury (Kaur et al., 2006, 2008b). Factors such as vascular endothelial growth factor (VEGF) and NO were implicated in increasing the permeability of blood vessels (Kaur et al., 2006, 2008b). The involvement of various factors in the pathophysiology of perinatal brain damage has complicated the development of effective therapeutic possibilities. Though several therapeutic alternatives have been suggested to ameliorate the injury to the developing brain, the potential of these drugs in preventing the tissue damage is unknown. Antioxidant therapies have been suggested to prevent tissue damage due to hypoxia, asphyxia, etc. Recent evidences from both clinical and animal studies have reported that melatonin is effective in reducing the brain tissue damage due to various factors (Kaur et al., 2008b, 2010; Gitto et al., 2009; Buonocore et al., 2012; Robertson et al., 2013). For example, in a mouse model of transient focal cerebral ischemia, the damage to the gray and white matter was reduced following melatonin administration (Lee et al., 2005). Neurobehavioral outcome of these rodents was also found to be improved (Lee et al., 2005). In neonatal rodent models of white matter damage, administration of melatonin suppressed astrocytosis and microglial activation and improved myelination by promoting oligodendroglial maturation (Hutton et al., 2009; Hamada et al., 2010; Kaur et al., 2010). Furthermore, in compromised pregnancies, treatment with melatonin is documented to stimulate antioxidant expression in the placenta and increase umbilical blood flow, which could protect the developing fetus (Richter et al., 2009). Melatonin also exhibits anti-inflammatory properties (Welin et al., 2007; Esposito and Cuzzocrea, 2010).

28.4 MELATONIN AND OXIDATIVE STRESS IN THE DEVELOPING BRAIN

Oxidative stress has been well documented to contribute to the tissue damage in the developing brain caused by conditions such as hypoxia–ischemia (Vasiljevic et al., 2012), asphyxia (Fulia et al., 2001), hydrocephalus (Socci et al., 1999), and hyperoxia (Sifringer et al., 2010). In developing brains, affected by either hypoxia–ischemia or asphyxia or hydrocephalus, high concentrations of free radicals such as NO and ROS have been reported (Fulia et al., 2001). Lack of antioxidant system and high fatty acid content renders the developing brain extremely susceptible to free radical–mediated damage. In addition, in immature brains, the premyelinating oligodendrocytes are highly susceptible to oxidative stress in conditions such as hypoxia, due to the lack of antioxidants (Back et al., 2002), and undergo apoptosis through reduction of glutathione levels, increased lipid peroxidation, and increased caspase-3 expression (Kaur et al., 2010; Rathnasamy et al., 2011). However, administration of melatonin has been reported to neutralize the deleterious effects of free radicals and protect the developing brain. Clinically, in newborns affected by sepsis, melatonin reduced the serum MDA levels (Gitto et al., 2001). Similarly, in asphyxiated newborns, treatment with melatonin effectively reduced the MDA and nitrite/nitrate levels in the serum (Fulia et al., 2001). Besides these, melatonin, when administered prior to in utero asphyxia, prevented the formation of highly toxic OH in the immature brains of late gestation sheep fetus (Miller et al., 2005).

Maternal administration of melatonin was reported to protect the fetal brains from oxidative damage due to hypoxic–ischemic insults or asphyxia (Watanabe et al., 2012; Yawno et al., 2012). In rat models of hypoxic periventricular white matter damage, melatonin administration rendered protection to the white matter by significantly reducing the MDA level, which was elevated due to hypoxia (Kaur et al., 2010). In fetal sheep models of perinatal asphyxia, melatonin protected the white matter injury by reducing the serum concentrations of 8-isoprostanes (Welin et al., 2007). In addition, melatonin reduced the number of activated microglial cells in the white matter (Welin et al., 2007), which are well documented to generate free radicals (Dringen, 2005; Li et al., 2008; Kaur et al., 2009a; Rathnasamy et al., 2011). We have observed a reduction in ROS production in hypoxic microglial cultures when treated with melatonin (Kaur et al., 2008a). In microglial cells treated with amphetamine, melatonin suppressed the up-regulation of inducible NOS mRNA, which is a pro-oxidant (Tocharus et al., 2008). Recent studies have shown that melatonin decreases superoxide production in activated microglia by impairing nicotinamide adenine dinucleotide phosphate oxidase assembly (Zhou et al., 2008). Melatonin is further known to increase the antioxidant glutathione levels in hypoxic periventricular white matter and is also known to increase the activity of antioxidants (Barlow-Walden et al., 1995; Richter et al., 2009). Parallel to melatonin synthesis, the activity of antioxidants was also found to follow a day–night cycle (Díaz-Muñoz et al., 1985). The activity of glutathione peroxidase, which is considered as the major antioxidant responsible for abolishing free radicals in the brain, was correlated with the synthesis of melatonin (Barlow-Walden et al., 1995). Based on this, it was speculated that melatonin could abrogate free radical–induced damage to the brain by enhancing the activity of glutathione peroxidase (Barlow-Walden et al., 1995). Support to this comes from the study by Kotler et al. (1998), who demonstrated the ability of melatonin to induce the mRNA of antioxidants glutathione peroxidase, copper–zinc superoxide dismutase, and manganese superoxide dismutase in the rat brains. Consistent with these, Okatani et al. (2000) demonstrated that maternal administration of melatonin could enhance the activity of antioxidants, such as superoxide dismutase and glutathione peroxidase, in rat fetal brains. Taken together, it could be stated that melatonin, either by increasing the endogenous antioxidant activity or by reducing the number of activated microglia, could protect the immature brains against the harmful free radicals.

28.5 MELATONIN AND INFLAMMATION IN THE DEVELOPING BRAIN

Apart from oxidative stress, melatonin has the potential to terminate the harmful inflammatory reaction that develop in response to various pathological insults. Elevated cytokine levels have been implicated in perinatal brain injury (McAdams and Juul, 2012). Presence of cytokines such as TNF-α, IL-1β, IL-6, and IL-8 has been demonstrated in various models of developing brain injury (McAdams and Juul, 2012). In hypoxic or anoxic developing white matter, enhanced expression of cytokines such as TNF-α and IL-1β was reported (Kadhim et al., 2001; Deng et al., 2008). The cerebrospinal fluid of asphyxiated infants was found to contain elevated concentrations of IL-6 and IL-8 (Sävman et al., 1998). Moreover, activated glial cells, such as astrocytes and microglia, are reported to be involved in the production of cytokines in the hypoxic brain (Deng et al., 2008, 2010; Rathnasamy et al., 2011). Both clinical and experimental studies have shown the efficacy of melatonin in reducing inflammation in the developing brain (Gitto et al., 2001; Welin et al., 2007; Hutton et al., 2009). In asphyxiated neonatal rats, treatment with melatonin was demonstrated to reduce the number of activated microglia, which are implicated in inflammatory cascade (Welin et al., 2007). In neonatal animal models of ibotenate-induced white matter damage, melatonin suppressed the initial activation of microglia thereby preventing cytokine production (Husson et al., 2002). When administered prenatally, melatonin prevented inflammation in the immature brains of asphyxiated spiny mouse (Hutton et al., 2009). Melatonin was found to inhibit the inflammatory process by preventing cytokine production in activated microglia (Min et al., 2012). In lipopolysaccharide (LPS)-stimulated Raw264.7 cells, melatonin inhibited the excess production of TNF-α, IL-1β, IL-6, IL-8, and IL-10 (Xia et al., 2012). Melatonin was reported to impede NF-κB translocation

(Min et al., 2012), which is essential for cytokine production. In animal models of LPS-induced intrauterine fetal death and growth retardation, maternal administration of melatonin effectively suppressed LPS-evoked TNF-α concentration in fetal brains (Xu et al., 2007). Besides these, melatonin has been suggested to have a role in immune responses in the developing brain. Although melatonin has been demonstrated to be anti-inflammatory, it could also act as an immunomodulatory molecule enhancing immune functions. For example, administration of melatonin to postnatal rats increased the expression of major histocompatibility antigens class I and II and complement type 3 receptors in many regions of the brain (Kaur and Ling, 1999). In light of these points, it could be suggested that melatonin has the potential to influence the immune response depending on the systemic environment present.

28.6 OTHER NEUROPROTECTIVE PROPERTIES OF MELATONIN IN DEVELOPING BRAIN

Although the role of melatonin against oxidative stress and inflammation is well studied, there are few other neuroprotective properties of melatonin that could shield the developing brain from deleterious effects. In anesthesia-induced neurodegeneration models, melatonin was found to suppress the mitochondrial apoptotic pathway in developing brains by reducing the cytochrome c release and activation of caspase-3 (Yon et al., 2006). In conditions such as hypoxia or asphyxia, excitotoxicity due to excess glutamate is also considered to play a major role in causing damage to the developing brain (Levene, 1992; Murugan et al., 2011; Reddy et al., 2011). Excess glutamate has been demonstrated to cause apoptosis of neurons and oligodendrocytes, through excitotoxic mechanisms, in the developing brain (Portera-Cailliau et al., 1997; Follett et al., 2004; Murugan et al., 2011). Glutamate on binding to its receptors expressed by glial cells, such as microglia and astrocytes, could initiate a vicious cycle resulting in increased pro-inflammatory cytokine secretion (Chao and Hu, 1994; Sulkowski et al., 2013) leading to brain damage. Furthermore, excess glutamate was found to mask the neuroprotective effects of insulin-like growth factor (IGF)-1 (Sivakumar et al., 2010). A decrease in IGF-1 levels has been proposed to be a causative factor leading to cell death in the perinatal brain following hypoxic–ischemic insult (Wood et al., 2007; Lin et al., 2009). However, administration of melatonin was found to prevent excitotoxicity by reducing the concentrations of glutamate in the hypoxic white matter of neonatal rats (Sivakumar et al., 2010) by reducing the activity of glutaminase (Saenz et al., 2004). In addition, it markedly increased the levels of IGF-1 in the immature hypoxic rat brains (Sivakumar et al., 2010). Excess IGF-1 has been shown to ameliorate hypoxic damage to the gray and white matter (Guan et al., 2003). Furthermore, IGF-1 has been reported to protect the developing oligodendrocytes (Ness and Wood, 2002; Ness et al., 2004), which are highly susceptible to hypoxic–ischemic damage in the perinatal brains.

The oxidative stress and inflammation triggered in the developing brain in response to various stimuli could result in the BBB damage. Excess ROS and NO have been documented to disrupt the BBB (Lehner et al., 2011). Additionally, inflammatory cytokines such as TNF-α and IL-1β were documented to increase BBB permeability (Yang et al., 1999; Minagar and Alexander, 2003). As stated previously, in response to hypoxic insult, excess production of ROS, TNF-α, and IL-1β was documented in the developing brain (Kaur and Ling, 2009). Furthermore, following hypoxic insult, increased expression of VEGF and NO has been demonstrated in the hippocampus and white matter of neonatal rats (Kaur et al., 2008b, 2010). VEGF and NO have been reported to enhance BBB permeability (Schoch et al., 2002; Young et al., 2004). VEGF causes hyperpermeability in the brain microvasculature by altering the expression of zonula occludens-1 in the tight junctions of the brain endothelial cells (Fischer et al., 2002). Increased vascular permeability was also apparent with the leakage of rhodamine isothiocyanate (RhIC) in the developing white matter and hippocampus (Kaur et al., 2008b, 2010). The protective ability of melatonin was evident with the decreased expression of VEGF and NO and reduced RhIC leakage in hypoxic rats administered with melatonin (Kaur et al., 2008b, 2010).

28.7 MELATONIN AND THE DEVELOPING NEURAL RETINA

The developing neural retina, similar to the developing brain, has been reported to be affected by various insults such as hypoxia–ischemia, bright light exposure, hyperoxia, and asphyxia (Lawwill et al., 1977; Chemtob et al., 1993; Kaur et al., 2009b), which result in the death of RGCs and other retinal neurons. Although hardly any investigation is available on the potential of melatonin in ameliorating immature retinal damage, our studies on neonatal rat models showed that melatonin is remarkably protective against the adverse effects of hypoxia in the neonatal retina. Following hypoxic insult, in the developing retina, there was a widespread presence of apoptotic and necrotic cells, swollen Müller cell processes, and increased vascular leakage. RGCs are highly vulnerable to hypoxic insult, and hypoxia-mediated death of RGCs is influenced by several factors such as excitotoxicity, oxidative stress, and inflammation (Sivakumar et al., 2011; Kaur et al., 2012, 2013). In the developing retina, following a hypoxic insult, there was increased glutamate concentration and expression of glutamate receptors (Kaur et al., 2012; Sivakumar et al., 2013). The activation of glutamate receptors expressed on RGCs resulted in increased intracellular ROS production in response to hypoxic exposure (Sivakumar et al., 2013). Along with this, retinal microglial cells were also involved in the hypoxia-induced inflammatory response in immature retina (Sivakumar et al., 2011). Moreover, hypoxia-induced oxidative stress was evident with increased lipid peroxidation and reduced glutathione content (Kaur et al., 2013). The apoptosis of RGCs was evident with increased expression of cytosolic cytochrome-c and caspase-3 in them following hypoxic insult (Kaur et al., 2012, 2013). However, these changes were attenuated in retinas of hypoxic neonatal rats treated with melatonin (Kaur et al., 2009b). In hypoxic cultures of RGCs treated with melatonin, the expression of cytosolic cytochrome-c and caspase-3 was reduced when compared with hypoxic RGCs not treated by melatonin (Kaur et al., 2013). In retinal microglial cultures, hypoxia-induced cytokine production was abolished when treated with melatonin. Administration of melatonin to hypoxic neonatal rats resulted in reduced number of apoptotic and necrotic cells, and reduced the swelling of Müller cell processes (Kaur et al., 2009b). Concomitantly, melatonin increased the glutathione content and reduced the lipid peroxidation in hypoxic retina (Kaur et al., 2013). The increased vascular leakage in hypoxic retina, evidenced by increased expression of VEGF, and its receptors Flt-1 and FlK-1, was reduced with melatonin treatment (Kaur et al., 2009b, 2013). As stated previously, hypoxia-induced increased expression of VEGF and NO is implicated in causing increased vascular permeability. This was correlated with the increased vascular leakage of RhIC and horse–radish peroxidase (HRP) from the retinal and hyaloid vessels in the hypoxic retina (Kaur et al., 2009b). However, administration of melatonin was suggested to protect the integrity of blood–retinal barrier in the hypoxic retina by reducing the expression of VEGF and NO in the hypoxic developing retina. In addition, melatonin administration attenuated the leakage of RhIC and HRP from the blood vessels of retina (Kaur et al., 2009b). In light of these results, it could be stated that melatonin could be a promising therapeutic alternative for sight-threatening conditions such as retinopathy of prematurity in which hypoxia is implicated.

28.8 CONCLUSION

Melatonin is a highly potent free radical scavenger and a broad-spectrum antioxidant. It has been proven beneficial in ameliorating the oxidative stress observed in several neuropathologies such as Parkinson's disease, Alzheimer's disease, and stroke. Melatonin's ability to act as electron acceptor and donor makes it unique from other antioxidants, and this property is attributed to its role against free radicals and in maintaining mitochondrial homeostasis. Melatonin has been demonstrated to protect the developing brain and retina from injuries due to hypoxia–ischemia, asphyxia, or hyperoxia. Not only does melatonin protect the neural tissues in the postnatal infants, but it also has a remarkable ability to enter the fetal brains in utero when administered to pregnant mothers. Apart from this, melatonin also maintained the integrity of BBB and blood–retinal barrier in neonates.

In light of these outcomes, melatonin could be suggested as a potential therapeutic alternative in ameliorating damage to the developing brain and retina. However, extensive investigations on the beneficial role of melatonin and its metabolites in developing brain and retina are warranted.

REFERENCES

Acuña-Castroviejo D, Escames G, León J, Carazo A, Khaldy H (2003) Mitochondrial regulation by melatonin and its metabolites. *Adv Exp Med Biol* 527:549–557.

Armstrong SM, Cassone VM, Chesworth MJ, Redman JR, Short RV (1986) Synchronization of mammalian circadian rhythms by melatonin. *J Neural Transm Suppl* 21:375–394.

Back SA, Han BH, Luo NL, Chricton CA, Xanthoudakis S, Tam J, Arvin KL, Holtzman DM (2002) Selective vulnerability of late oligodendrocyte progenitors to hypoxia-ischemia. *J Neurosci* 22:455–463.

Barlow-Walden LR, Reiter RJ, Abe M, Pablos M, Menendez-Pelaez A, Chen LD, Poeggeler B (1995) Melatonin stimulates brain glutathione peroxidase activity. *Neurochem Int* 26:497–502.

Belforte NA, Moreno MC, de Zavalía N, Sande PH, Chianelli MS, Keller Sarmiento MI, Rosenstein RE (2010) Melatonin: A novel neuroprotectant for the treatment of glaucoma. *J Pineal Res* 48:353–364.

Benzi G, Moretti A (1995) Are reactive oxygen species involved in Alzheimer's disease? *Neurobiol Aging* 16:661–674.

Bettahi I, Pozo D, Osuna C, Reiter RJ, Acuna-Castroviejo D, Guerrero JM (1996) Melatonin reduces nitric oxide synthase activity in rat hypothalamus. *J Pineal Res* 20:205–210.

Borah A, Mohanakumar KP (2009) Melatonin inhibits 6-hydroxydopamine production in the brain to protect against experimental parkinsonism in rodents. *J Pineal Res* 47:293–300.

Bubenik GA (2002) Gastrointestinal melatonin: Localization, function, and clinical relevance. *Dig Dis Sci* 47:2336–2348.

Buonocore G, Perrone S, Turrisi G, Kramer BW, Balduini W (2012) New pharmacological approaches in infants with hypoxic-ischemic encephalopathy. *Curr Pharm Des* 18:3086–3100.

Buzzell GR, Menendez-Pelaez A, Troiani ME, McNeill ME, Reiter RJ (1990) Effects of short-day photoperiods and of *N*-(2,4-dinitrophenyl)-5-methoxytryptamine, a putative melatonin antagonist, on melatonin synthesis in the Harderian gland of the Syrian hamster, *Mesocricetus auratus*. *J Pineal Res* 8:229–235.

Carloni S, Perrone S, Buonocore G, Longini M, Proietti F, Balduini W (2008) Melatonin protects from the long-term consequences of a neonatal hypoxic-ischemic brain injury in rats. *J Pineal Res* 44:157–164.

Carrillo-Vico A, Calvo JR, Abreu P, Lardone PJ, García-Mauriño S, Reiter RJ, Guerrero JM (2004) Evidence of melatonin synthesis by human lymphocytes and its physiological significance: Possible role as intracrine, autocrine, and/or paracrine substance. *FASEB J* 18:537–539.

Carrillo-Vico A, Lardone PJ, Alvarez-Sanchez N, Rodriguez-Rodriguez A, Guerrero JM (2013) Melatonin: Buffering the immune system. *Int J Mol Sci* 14:8638–8683.

Champier J, Claustrat B, Besançon R, Eymin C, Killer C, Jouvet A, Chamba G, Fèvre-Montange M (1997) Evidence for tryptophan hydroxylase and hydroxy-indol-*O*-methyl-transferase mRNAs in human blood platelets. *Life Sci* 60:2191–2197.

Champney TH, Hanneman WH, Legare ME, Appel K (1996) Acute and chronic effects of melatonin as an anticonvulsant in male gerbils. *J Pineal Res* 20:79–83.

Chao CC, Hu S (1994) Tumor necrosis factor-alpha potentiates glutamate neurotoxicity in human fetal brain cell cultures. *Dev Neurosci* 16:172–179.

Chemtob S, Roy MS, Abran D, Fernandez H, Varma DR (1993) Prevention of postasphyxial increase in lipid peroxides and retinal function deterioration in the newborn pig by inhibition of cyclooxygenase activity and free radical generation. *Pediatr Res* 33:336–340.

Chen Y-C, Tain Y-L, Sheen J-M, Huang L-T (2012) Melatonin utility in neonates and children. *J Formos Med Assoc* 111:57–66.

Conti A, Conconi S, Hertens E, Skwarlo-Sonta K, Markowska M, Maestroni JM (2000) Evidence for melatonin synthesis in mouse and human bone marrow cells. *J Pineal Res* 28:193–202.

Cross AH, Manning PT, Keeling RM, Schmidt RE, Misko TP (1998) Peroxynitrite formation within the central nervous system in active multiple sclerosis. *J Neuroimmunol* 88:45–56.

Cummings C (2012) Melatonin for the management of sleep disorders in children and adolescents. *Paediatr Child Health* 17:331–336.

Del Bigio MR, Khan OH, da Silva Lopes L, Juliet PA (2012) Cerebral white matter oxidation and nitrosylation in young rodents with kaolin-induced hydrocephalus. *J Neuropathol Exp Neurol* 71:274–288.

Deng Y, Lu J, Sivakumar V, Ling EA, Kaur C (2008) Amoeboid microglia in the periventricular white matter induce oligodendrocyte damage through expression of proinflammatory cytokines via MAP kinase signaling pathway in hypoxic neonatal rats. *Brain Pathol* 18:387–400.

Deng YY, Lu J, Ling EA, Kaur C (2010) Microglia-derived macrophage colony stimulating factor promotes generation of proinflammatory cytokines by astrocytes in the periventricular white matter in the hypoxic neonatal brain. *Brain Pathol* 20:909–925.

Díaz-Muñoz M, Hernández-Muñoz R, Suárez J, Chagoya de Sánchez V (1985) Day–night cycle of lipid peroxidation in rat cerebral cortex and their relationship to the glutathione cycle and superoxide dismutase activity. *Neuroscience* 16:859–863.

Djeridane Y, Touitou Y (2001) Melatonin synthesis in the rat harderian gland: Age- and time-related effects. *Exp Eye Res* 72:487–492.

Dringen R (2005) Oxidative and antioxidative potential of brain microglial cells. *Antioxid Redox Signal* 7:1223–1233.

Dubocovich ML (1983) Melatonin is a potent modulator of dopamine release in the retina. *Nature* 306: 782–784.

Escames G, León J, Lopez LC, Acuna-Castroviejo D (2004) Mechanisms of *N*-methyl-D-aspartate receptor inhibition by melatonin in the rat striatum. *J Neuroendocrinol* 16:929–935.

Esposito E, Cuzzocrea S (2010) Antiinflammatory activity of melatonin in central nervous system. *Curr Neuropharmacol* 8:228–242.

Feng Z, Chang Y, Cheng Y, Zhang BL, Qu ZW, Qin C, Zhang JT (2004) Melatonin alleviates behavioral deficits associated with apoptosis and cholinergic system dysfunction in the APP 695 transgenic mouse model of Alzheimer's disease. *J Pineal Res* 37:129–136.

Ferracioli-Oda E, Qawasmi A, Bloch MH (2013) Meta-analysis: Melatonin for the treatment of primary sleep disorders. *PLoS One* 8:e63773.

Fiorina P, Lattuada G, Ponari O, Silvestrini C, DallAglio P (1996) Impaired nocturnal melatonin excretion and changes of immunological status in ischaemic stroke patients. *Lancet* 347:692–693.

Fischer S, Wobben M, Marti HH, Renz D, Schaper W (2002) Hypoxia-induced hyperpermeability in brain microvessel endothelial cells involves VEGF-mediated changes in the expression of zonula occludens-1. *Microvasc Res* 63:70–80.

Follett PL, Deng W, Dai W, Talos DM, Massillon LJ, Rosenberg PA, Volpe JJ, Jensen FE (2004) Glutamate receptor-mediated oligodendrocyte toxicity in periventricular leukomalacia: A protective role for topiramate. *J Neurosci* 24:4412–4420.

Freeman BA, Crapo JD (1981) Hyperoxia increases oxygen radical production in rat lungs and lung mitochondria. *J Biol Chem* 256:10986–10992.

Fulia F, Gitto E, Cuzzocrea S, Reiter RJ, Dugo L, Gitto P, Barberi S, Cordaro S, Barberi I (2001) Increased levels of malondialdehyde and nitrite/nitrate in the blood of asphyxiated newborns: Reduction by melatonin. *J Pineal Res* 31:343–349.

Genova ML, Pich MM, Biondi A, Bernacchia A, Falasca A, Bovina C, Formiggini G, Parenti Castelli G, Lenaz G (2003) Mitochondrial production of oxygen radical species and the role of Coenzyme Q as an antioxidant. *Exp Biol Med (Maywood)* 228:506–513.

Gilad E, Cuzzocrea S, Zingarelli B, Salzman AL, Szabo C (1997) Melatonin is a scavenger of peroxynitrite. *Life Sci* 60:PL169–PL174.

Gitto E, Karbownik M, Reiter RJ, Tan DX, Cuzzocrea S, Chiurazzi P, Cordaro S, Corona G, Trimarchi G, Barberi I (2001) Effects of melatonin treatment in septic newborns. *Pediatr Res* 50:756–760.

Gitto E, Pellegrino S, Gitto P, Barberi I, Reiter RJ (2009) Oxidative stress of the newborn in the pre- and postnatal period and the clinical utility of melatonin. *J Pineal Res* 46:128–139.

Guan J, Bennet L, Gluckman PD, Gunn AJ (2003) Insulin-like growth factor-1 and post-ischemic brain injury. *Prog Neurobiol* 70:443–462.

Guenther AL, Schmidt SI, Laatsch H, Fotso S, Ness H, Ressmeyer A-R, Poeggeler B, Hardeland R (2005) Reactions of the melatonin metabolite AMK (N1-acetyl-5-methoxykynuramine) with reactive nitrogen species: Formation of novel compounds, 3-acetamidomethyl-6-methoxycinnolinone and 3-nitro-AMK. *J Pineal Res* 39:251–260.

Hamada F, Watanabe K, Wakatsuki A, Nagai R, Shinohara K, Hayashi Y, Imamura R, Fukaya T (2010) Therapeutic effects of maternal melatonin administration on ischemia/reperfusion-induced oxidative cerebral damage in neonatal rats. *Neonatology* 98:33–40.

Hardeland R (2005) Antioxidative protection by melatonin: Multiplicity of mechanisms from radical detoxification to radical avoidance. *Endocrine* 27:119–130.

Hardeland R, Reiter RJ, Poeggeler B, Tan DX (1993) The significance of the metabolism of the neurohormone melatonin: Antioxidative protection and formation of bioactive substances. *Neurosci Biobehav Rev* 17:347–357.
Hibaoui Y, Roulet E, Ruegg UT (2009) Melatonin prevents oxidative stress-mediated mitochondrial permeability transition and death in skeletal muscle cells. *J Pineal Res* 47:238–252.
Hirata F, Hayaishi O, Tokuyama T, Seno S (1974) In vitro and in vivo formation of two new metabolites of melatonin. *J Biol Chem* 249:1311–1313.
Horstman JA, Wrona MZ, Dryhurst G (2002) Further insights into the reaction of melatonin with hydroxyl radical. *Bioorg Chem* 30:371–382.
Husson I, Mesplès B, Bac P, Vamecq J, Evrard P, Gressens P (2002) Melatoninergic neuroprotection of the murine periventricular white matter against neonatal excitotoxic challenge. *Ann Neurol* 51:82–92.
Hutton LC, Abbass M, Dickinson H, Ireland Z, Walker DW (2009) Neuroprotective properties of melatonin in a model of birth asphyxia in the spiny mouse (*Acomys cahirinus*). *Dev Neurosci* 31:437–451.
Jimenez-Jorge S, Guerrero JM, Jimenez-Caliani AJ, Naranjo MC, Lardone PJ, Carrillo-Vico A, Osuna C, Molinero P (2007) Evidence for melatonin synthesis in the rat brain during development. *J Pineal Res* 42:240–246.
Kadhim H, Tabarki B, Verellen G, De Prez C, Rona AM, Sebire G (2001) Inflammatory cytokines in the pathogenesis of periventricular leukomalacia. *Neurology* 56:1278–1284.
Kaur C, Ling EA (1999) Effects of melatonin on macrophages/microglia in postnatal rat brain. *J Pineal Res* 26:158–168.
Kaur C, Ling EA (2009) Periventricular white matter damage in the hypoxic neonatal brain: Role of microglial cells. *Prog Neurobiol* 87:264–280.
Kaur C, Sivakumar V, Ang LS, Sundaresan A (2006) Hypoxic damage to the periventricular white matter in neonatal brain: Role of vascular endothelial growth factor, nitric oxide and excitotoxicity. *J Neurochem* 98:1200–1216.
Kaur C, Sivakumar V, Foulds WS, Luu CD, Ling E-A (2009b) Cellular and vascular changes in the retina of neonatal rats after an acute exposure to hypoxia. *Invest Ophthalmol Vis Sci* 50:5364–5374.
Kaur C, Sivakumar V, Foulds WS, Luu CD, Ling E-A (2012) Hypoxia-induced activation of N-methyl-D-aspartate receptors causes retinal ganglion cell death in the neonatal retina. *J Neuropathol Exp Neurol* 71:330–347.
Kaur C, Sivakumar V, Ling EA (2008a) Melatonin and its therapeutic potential in neuroprotection. *Centr Nerv Syst Agents Med Chem* 8:260–266.
Kaur C, Sivakumar V, Ling EA (2010) Melatonin protects periventricular white matter from damage due to hypoxia. *J Pineal Res* 48:185–193.
Kaur C, Sivakumar V, Lu J, Tang FR, Ling EA (2008b) Melatonin attenuates hypoxia-induced ultrastructural changes and increased vascular permeability in the developing hippocampus. *Brain Pathol* 18: 533–547.
Kaur C, Sivakumar V, Robinson R, Foulds WS, Luu CD, Ling E-A (2013) Neuroprotective effect of melatonin against hypoxia-induced retinal ganglion cell death in neonatal rats. *J Pineal Res* 54:190–206.
Kaur C, Sivakumar V, Yip GW, Ling EA (2009a) Expression of syndecan-2 in the amoeboid microglial cells and its involvement in inflammation in the hypoxic developing brain. *Glia* 57:336–349.
Keane PC, Kurzawa M, Blain PG, Morris CM (2011) Mitochondrial dysfunction in Parkinson's disease. *Parkinsons Dis* 2011:716871.
Kim JW, Tchernyshyov I, Semenza GL, Dang CV (2006) HIF-1-mediated expression of pyruvate dehydrogenase kinase: A metabolic switch required for cellular adaptation to hypoxia. *Cell Metab* 3:177–185.
Kotler M, Rodriguez C, Sainz RM, Antolin I, Menendez-Pelaez A (1998) Melatonin increases gene expression for antioxidant enzymes in rat brain cortex. *J Pineal Res* 24:83–89.
Lawwill T, Crockett S, Currier G (1977) Retinal damage secondary to chronic light exposure. *Doc Ophthalmol* 44:379–402.
Lee EJ, Lee MY, Chen HY, Hsu YS, Wu TS, Chen ST, Chang GL (2005) Melatonin attenuates gray and white matter damage in a mouse model of transient focal cerebral ischemia. *J Pineal Res* 38:42–52.
Lehner C, Gehwolf R, Tempfer H, Krizbai I, Hennig B, Bauer H-C, Bauer H (2011) Oxidative stress and blood-brain barrier dysfunction under particular consideration of matrix metalloproteinases. *Antioxid Redox Signal* 15:1305–1323.
León J, Acuña-Castroviejo D, Escames G, Tan D-X, Reiter RJ (2005) Melatonin mitigates mitochondrial malfunction. *J Pineal Res* 38:1–9.
León J, Acuna-Castroviejo D, Sainz RM, Mayo JC, Tan DX, Reiter RJ (2004) Melatonin and mitochondrial function. *Life Sci* 75:765–790.

León J, Escames G, Rodríguez MI, López LC, Tapias V, Entrena A, Camacho E, et al. (2006) Inhibition of neuronal nitric oxide synthase activity by N1-acetyl-5-methoxykynuramine, a brain metabolite of melatonin. *J Neurochem* 98:2023–2033.

Lerner AB, Case JD, Takahashi Y, Lee TH, Mori W (1958) Isolation of melatonin, the pineal gland factor that lightens melanocytes. *J Am Chem Soc* 80:2587.

Levene M (1992) Role of excitatory amino acid antagonists in the management of birth asphyxia. *Biol Neonate* 62:248–251.

Li F, Lu J, Wu CY, Kaur C, Sivakumar V, Sun J, Li S, Ling EA (2008) Expression of Kv1.2 in microglia and its putative roles in modulating production of proinflammatory cytokines and reactive oxygen species. *J Neurochem* 106:2093–2105.

Liang FQ, Godley BF (2003) Oxidative stress-induced mitochondrial DNA damage in human retinal pigment epithelial cells: A possible mechanism for RPE aging and age-related macular degeneration. *Exp Eye Res* 76:397–403.

Lin CH, Huang JY, Ching CH, Chuang JI (2008) Melatonin reduces the neuronal loss, downregulation of dopamine transporter, and upregulation of D2 receptor in rotenone-induced parkinsonian rats. *J Pineal Res* 44:205–213.

Lin L, Huang Q-X, Yang S-S, Chu J, Wang J-Z, Tian Q (2013) Melatonin in Alzheimer's disease. *Int J Mol Sci* 14:14575–14593.

Lin S, Fan LW, Rhodes PG, Cai Z (2009) Intranasal administration of IGF-1 attenuates hypoxic-ischemic brain injury in neonatal rats. *Exp Neurol* 217:361–370.

Mallard C, Welin AK, Peebles D, Hagberg H, Kjellmer I (2003) White matter injury following systemic endotoxemia or asphyxia in the fetal sheep. *Neurochem Res* 28:215–223.

Manev H, Uz T, Qu T (1998) Early upregulation of hippocampal 5-lipoxygenase following systemic administration of kainate to rats. *Restor Neurol Neurosci* 12:81–85.

Manoj A, Ramachandra Rao K, Vishnu Bhat B, Venkatesh C, Bobby Z (2011) Oxidative stress induced DNA damage in perinatal asphyxia. *Curr Pediatr Res* 15:19–23.

Martin M, Macias M, Escames G, Leon J, Acuna-Castroviejo D (2000a) Melatonin but not vitamins C and E maintains glutathione homeostasis in t-butyl hydroperoxide-induced mitochondrial oxidative stress. *FASEB J* 14:1677–1679.

Martin M, Macias M, Escames G, Reiter RJ, Agapito MT, Ortiz GG, Acuna-Castroviejo D (2000b) Melatonin-induced increased activity of the respiratory chain complexes I and IV can prevent mitochondrial damage induced by ruthenium red in vivo. *J Pineal Res* 28:242–248.

Martin M, Macias M, Leon J, Escames G, Khaldy H, Acuna-Castroviejo D (2002) Melatonin increases the activity of the oxidative phosphorylation enzymes and the production of ATP in rat brain and liver mitochondria. *Int J Biochem Cell Biol* 34:348–357.

Matuszak Z, Reszka KJ, Chignell CF (1997) Reaction of melatonin and related indoles with hydroxyl radicals: EPR and spin trapping investigations. *Free Radic Biol Med* 23:367–372.

Mayo JC, Sainz RM, Tan DX, Antolin I, Rodriguez C, Reiter RJ (2005a) Melatonin and Parkinson's disease. *Endocrine* 27:169–178.

Mayo JC, Sainz RM, Tan DX, Hardeland R, Leon J, Rodriguez C, Reiter RJ (2005b) Anti-inflammatory actions of melatonin and its metabolites, N1-acetyl-N2-formyl-5-methoxykynuramine (AFMK) and N1-acetyl-5-methoxykynuramine (AMK), in macrophages. *J Neuroimmunol* 165:139–149.

McAdams RM, Juul SE (2012) The role of cytokines and inflammatory cells in perinatal brain injury. *Neurol Res Int* 2012:561494.

Miller SL, Yan EB, Castillo-Melendez M, Jenkin G, Walker DW (2005) Melatonin provides neuroprotection in the late-gestation fetal sheep brain in response to umbilical cord occlusion. *Dev Neurosci* 27:200–210.

Min KJ, Jang JH, Kwon TK (2012) Inhibitory effects of melatonin on the lipopolysaccharide-induced CC chemokine expression in BV2 murine microglial cells are mediated by suppression of Akt-induced NF-kappaB and STAT/GAS activity. *J Pineal Res* 52:296–304.

Minagar A, Alexander JS (2003) Blood-brain barrier disruption in multiple sclerosis. *Mult Scler* 9:540–549.

Miwa S, Brand MD (2003) Mitochondrial matrix reactive oxygen species production is very sensitive to mild uncoupling. *Biochem Soc Trans* 31:1300–1301.

Murugan M, Sivakumar V, Lu J, Ling EA, Kaur C (2011) Expression of N-methyl D-aspartate receptor subunits in amoeboid microglia mediates production of nitric oxide via NF-kappaB signaling pathway and oligodendrocyte cell death in hypoxic postnatal rats. *Glia* 59:521–539.

Ness JK, Scaduto RC, Jr., Wood TL (2004) IGF-I prevents glutamate-mediated bax translocation and cytochrome C release in O4+ oligodendrocyte progenitors. *Glia* 46:183–194.

Ness JK, Wood TL (2002) Insulin-like growth factor I, but not neurotrophin-3, sustains Akt activation and provides long-term protection of immature oligodendrocytes from glutamate-mediated apoptosis. *Mol Cell Neurosci* 20:476–488.

Okatani Y, Wakatsuki A, Kaneda C (2000) Melatonin increases activities of glutathione peroxidase and superoxide dismutase in fetal rat brain. *J Pineal Res* 28:89–96.

Pandi-Perumal SR, Srinivasan V, Maestroni GJM, Cardinali DP, Poeggeler B, Hardeland R (2006) Melatonin: Nature's most versatile biological signal? *FEBS J* 273:2813–2838.

Pang SF, Li Y, Jiang DH, Chang B, Xie BL (1990) Acute cerebral hemorrhage changes the nocturnal surge of plasma melatonin in humans. *J Pineal Res* 9:193–208.

Park SW, Lee HS, Sung MS, Kim SJ (2012) The effect of melatonin on retinal ganglion cell survival in ischemic retina. *Chonnam Med J* 48:116–122.

Pei Z, Cheung RT (2004) Pretreatment with melatonin exerts anti-inflammatory effects against ischemia/reperfusion injury in a rat middle cerebral artery occlusion stroke model. *J Pineal Res* 37:85–91.

Pei Z, Pang SF, Cheung RT (2003) Administration of melatonin after onset of ischemia reduces the volume of cerebral infarction in a rat middle cerebral artery occlusion stroke model. *Stroke* 34:770–775.

Petrosillo G, Di Venosa N, Pistolese M, Casanova G, Tiravanti E, Colantuono G, Federici A, Paradies G, Ruggiero FM (2006) Protective effect of melatonin against mitochondrial dysfunction associated with cardiac ischemia-reperfusion: Role of cardiolipin. *FASEB J* 20:269–276.

Poeggeler B (2005) Melatonin, aging, and age-related diseases: Perspectives for prevention, intervention, and therapy. *Endocrine* 27:201–212.

Poeggeler B, Saarela S, Reiter RJ, Tan DX, Chen LD, Manchester LC, Barlow-Walden LR (1994) Melatonin—A highly potent endogenous radical scavenger and electron donor: New aspects of the oxidation chemistry of this indole accessed in vitro. *Ann N Y Acad Sci* 738:419–420.

Poeggeler B, Thuermann S, Dose A, Schoenke M, Burkhardt S, Hardeland R (2002) Melatonin's unique radical scavenging properties—Roles of its functional substituents as revealed by a comparison with its structural analogs. *J Pineal Res* 33:20–30.

Portera-Cailliau C, Price DL, Martin LJ (1997) Excitotoxic neuronal death in the immature brain is an apoptosis-necrosis morphological continuum. *J Comp Neurol* 378:10–87.

Pozo D, Reiter RJ, Calvo JR, Guerrero JM (1994) Physiological concentrations of melatonin inhibit nitric oxide synthase in rat cerebellum. *Life Sci* 55:PL455–PL460.

Pozo D, Reiter RJ, Calvo JR, Guerrero JM (1997) Inhibition of cerebellar nitric oxide synthase and cyclic GMP production by melatonin via complex formation with calmodulin. *J Cell Biochem* 65:430–442.

Rathnasamy G, Ling EA, Kaur C (2011) Iron and iron regulatory proteins in amoeboid microglial cells are linked to oligodendrocyte death in hypoxic neonatal rat periventricular white matter through production of proinflammatory cytokines and reactive oxygen/nitrogen species. *J Neurosci* 31:17982–17995.

Reddy NR, Krishnamurthy S, Chourasia TK, Kumar A, Joy KP (2011) Glutamate antagonism fails to reverse mitochondrial dysfunction in late phase of experimental neonatal asphyxia in rats. *Neurochem Int* 58:582–590.

Redman J, Armstrong S, Ng KT (1983) Free-running activity rhythms in the rat: Entrainment by melatonin. *Science* 219:1089–1091.

Reiter RJ (1998) Oxidative damage in the central nervous system: Protection by melatonin. *Prog Neurobiol* 56:359–384.

Reiter RJ, Tan D-X, Manchester LC, Lopez-Burillo S, Sainz RM, Mayo JC (2003) Melatonin: Detoxification of oxygen and nitrogen-based toxic reactants. *Adv Exp Med Biol* 527:539–548.

Reiter RJ, Tan DX, Manchester LC, Qi W (2001) Biochemical reactivity of melatonin with reactive oxygen and nitrogen species: A review of the evidence. *Cell Biochem Biophys* 34:237–256.

Richter HG, Hansell JA, Raut S, Giussani DA (2009) Melatonin improves placental efficiency and birth weight and increases the placental expression of antioxidant enzymes in undernourished pregnancy. *J Pineal Res* 46:357–364.

Ritzenthaler T, Lhommeau I, Douillard S, Cho TH, Brun J, Patrice T, Nighoghossian N, Claustrat B (2013) Dynamics of oxidative stress and urinary excretion of melatonin and its metabolites during acute ischemic stroke. *Neurosci Lett* 544:1–4.

Robertson NJ, Faulkner S, Fleiss B, Bainbridge A, Andorka C, Price D, Powell E, et al. (2013) Melatonin augments hypothermic neuroprotection in a perinatal asphyxia model. *Brain* 136:90–105.

Saenz DA, Goldin AP, Minces L, Chianelli M, Sarmiento MI, Rosenstein RE (2004) Effect of melatonin on the retinal glutamate/glutamine cycle in the golden hamster retina. *FASEB J* 18:1912–1913.

Sävman K, Blennow M, Gustafson K, Tarkowski E, Hagberg H (1998) Cytokine response in cerebrospinal fluid after birth asphyxia. *Pediatr Res* 43:746–751.

Scaiano JC (1995) Exploratory laser flash photolysis study of free radical reactions and magnetic field effects in melatonin chemistry. *J Pineal Res* 19:189–195.

Schaefer M, Hardeland R (2009) The melatonin metabolite N-acetyl-5-methoxykynuramine is a potent singlet oxygen scavenger. *J Pineal Res* 46:49–52.

Schoch HJ, Fischer S, Marti HH (2002) Hypoxia-induced vascular endothelial growth factor expression causes vascular leakage in the brain. *Brain* 125:2549–2557.

Semak I, Naumova M, Korik E, Terekhovich V, Wortsman J, Slominski A (2005) A novel metabolic pathway of melatonin: Oxidation by cytochrome C. *Biochemistry* 44:9300–9307.

Sharma R, McMillan CR, Tenn CC, Niles LP (2006) Physiological neuroprotection by melatonin in a 6-hydroxydopamine model of Parkinson's disease. *Brain Res* 1068:230–236.

Sifringer M, Brait D, Weichelt U, Zimmerman G, Endesfelder S, Brehmer F, von Haefen C, et al. (2010) Erythropoietin attenuates hyperoxia-induced oxidative stress in the developing rat brain. *Brain Behav Immun* 24:792–799.

Silva SO, Ximenes VF, Catalani LH, Campa A (2000) Myeloperoxidase-catalyzed oxidation of melatonin by activated neutrophils. *Biochem Biophys Res Commun* 279:657–662.

Sivakumar V, Foulds WS, Luu CD, Ling EA, Kaur C (2011) Retinal ganglion cell death is induced by microglia derived pro-inflammatory cytokines in the hypoxic neonatal retina. *J Pathol* 224:245–260.

Sivakumar V, Foulds WS, Luu CD, Ling EA, Kaur C (2013) Hypoxia-induced retinal ganglion cell damage through activation of AMPA receptors and the neuroprotective effects of DNQX. *Exp Eye Res* 109: 83–97.

Sivakumar V, Ling EA, Liu J, Kaur C (2010) Role of glutamate and its receptors and insulin-like growth factors in hypoxia induced periventricular white matter injury. *Glia* 58:507–523.

Slominski A, Wortsman J, Tobin DJ (2005) The cutaneous serotoninergic/melatoninergic system: Securing a place under the sun. *FASEB J* 19:176–194.

Socci DJ, Bjugstad KB, Jones HC, Pattisapu JV, Arendash GW (1999) Evidence that oxidative stress is associated with the pathophysiology of inherited hydrocephalus in the H-Tx rat model. *Exp Neurol* 155:109–117.

Srinivasan V, Pandi-Perumal SR, Cardinali DP, Poeggeler B, Hardeland R (2006a) Melatonin in Alzheimer's disease and other neurodegenerative disorders. *Behav Brain Funct* 2:15.

Srinivasan V, Smits M, Spence W, Lowe AD, Kayumov L, Pandi-Perumal SR, Parry B, Cardinali DP (2006b) Melatonin in mood disorders. *World J Biol Psychiatry* 7:138–151.

Srinivasan V, Spence DW, Pandi-Perumal SR, Trakht I, Cardinali DP (2008) Therapeutic actions of melatonin in cancer: Possible mechanisms. *Integr Cancer Ther* 7:189–203.

Sulkowski G, Dabrowska-Bouta B, Chalimoniuk M, Struzynska L (2013) Effects of antagonists of glutamate receptors on pro-inflammatory cytokines in the brain cortex of rats subjected to experimental autoimmune encephalomyelitis. *J Neuroimmunol* 261:67–76.

Tan D-X, Hardeland R, Manchester LC, Poeggeler B, Lopez-Burillo S, Mayo JC, Sainz RM, Reiter RJ (2003) Mechanistic and comparative studies of melatonin and classic antioxidants in terms of their interactions with the ABTS cation radical. *J Pineal Res* 34:249–259.

Tan D-X, Manchester LC, Burkhardt S, Sainz RM, Mayo JC, Kohen R, Shohami E, Huo YS, Hardeland R, Reiter RJ (2001) N1-acetyl-N2-formyl-5-methoxykynuramine, a biogenic amine and melatonin metabolite, functions as a potent antioxidant. *FASEB J* 15:2294–2296.

Tan D-X, Manchester LC, Terron MP, Flores LJ, Reiter RJ (2007) One molecule, many derivatives: A never-ending interaction of melatonin with reactive oxygen and nitrogen species? *J Pineal Res* 42:28–42.

Tesoriere L, Avellone G, Ceraulo L, D'Arpa D, Allegra M, Livrea MA (2001) Oxidation of melatonin by oxo ferryl hemoglobin: A mechanistic study. *Free Radic Res* 35:633–642.

Tocharus J, Chongthammakun S, Govitrapong P (2008) Melatonin inhibits amphetamine-induced nitric oxide synthase mRNA overexpression in microglial cell lines. *Neurosci Lett* 439:134–137.

Uz T, Manev H (1998) Circadian expression of pineal 5-lipoxygenase mRNA. *Neuroreport* 9:783–786.

Van de Berg WDJ, Schmitz C, Steinbusch HWM, Blanco CE (2002) Perinatal asphyxia induced neuronal loss by apoptosis in the neonatal rat striatum: A combined TUNEL and stereological study. *Exp Neurol* 174:29–36.

Vasiljevic B, Maglajlic-Djukic S, Gojnic M, Stankovic S (2012) The role of oxidative stress in perinatal hypoxic-ischemic brain injury. *Srp Arh Celok Lek* 140:35–41.

Warner DS, Sheng H, Batinic-Haberle I (2004) Oxidants, antioxidants and the ischemic brain. *J Exp Biol* 207:3221–3231.

Watanabe K, Hamada F, Wakatsuki A, Nagai R, Shinohara K, Hayashi Y, Imamura R, Fukaya T (2012) Prophylactic administration of melatonin to the mother throughout pregnancy can protect against oxidative cerebral damage in neonatal rats. *J Matern Fetal Neonatal Med* 25:1254–1259.

Welin A-K, Svedin P, Lapatto R, Sultan B, Hagberg H, Gressens P, Kjellmer I, Mallard C (2007) Melatonin reduces inflammation and cell death in white matter in the mid-gestation fetal sheep following umbilical cord occlusion. *Pediatr Res* 61:153–158.

Wood TL, Loladze V, Altieri S, Gangoli N, Levison SW, Brywe KG, Mallard C, Hagberg H (2007) Delayed IGF-1 administration rescues oligodendrocyte progenitors from glutamate-induced cell death and hypoxic-ischemic brain damage. *Dev Neurosci* 29:302–310.

Wu YH, Swaab DF (2005) The human pineal gland and melatonin in aging and Alzheimer's disease. *J Pineal Res* 38:145–152.

Xia MZ, Liang YL, Wang H, Chen X, Huang YY, Zhang ZH, Chen YH, Zhang C, Zhao M, Xu DX, Song LH (2012) Melatonin modulates TLR4-mediated inflammatory genes through MyD88- and TRIF-dependent signaling pathways in lipopolysaccharide-stimulated RAW264.7 cells. *J Pineal Res* 53:325–334.

Ximenes VF, Fernandes JR, Bueno VB, Catalani LH, de Oliveira GH, Machado RG (2007) The effect of pH on horseradish peroxidase-catalyzed oxidation of melatonin: Production of N1-acetyl-N2-5-methoxykynuramine versus radical-mediated degradation. *J Pineal Res* 42:291–296.

Xu DX, Wang H, Ning H, Zhao L, Chen YH (2007) Maternally administered melatonin differentially regulates lipopolysaccharide-induced proinflammatory and anti-inflammatory cytokines in maternal serum, amniotic fluid, fetal liver, and fetal brain. *J Pineal Res* 43:74–79.

Yang GY, Gong C, Qin Z, Liu XH, Lorris Betz A (1999) Tumor necrosis factor alpha expression produces increased blood-brain barrier permeability following temporary focal cerebral ischemia in mice. *Brain Res Mol Brain Res* 69:135–143.

Yawno T, Castillo-Melendez M, Jenkin G, Wallace EM, Walker DW, Miller SL (2012) Mechanisms of melatonin-induced protection in the brain of late gestation fetal sheep in response to hypoxia. *Dev Neurosci* 34:543–551.

Yon JH, Carter LB, Reiter RJ, Jevtovic-Todorovic V (2006) Melatonin reduces the severity of anesthesia-induced apoptotic neurodegeneration in the developing rat brain. *Neurobiol Dis* 21:522–530.

Young PP, Fantz CR, Sands MS (2004) VEGF disrupts the neonatal blood-brain barrier and increases life span after non-ablative BMT in a murine model of congenital neurodegeneration caused by a lysosomal enzyme deficiency. *Exp Neurol* 188:104–114.

Zhang H, Akbar M, Kim HY (1999) Melatonin: An endogenous negative modulator of 12-lipoxygenation in the rat pineal gland. *Biochem J* 344(Pt 2):487–493.

Zhou J, Zhang S, Zhao X, Wei T (2008) Melatonin impairs NADPH oxidase assembly and decreases superoxide anion production in microglia exposed to amyloid-beta1–42. *J Pineal Res* 45:157–165.

29 Melatonin's Protection against Human Prion-Mediated Neurotoxicity

Jae-Kyo Jeong and Sang-Youel Park

CONTENTS

29.1 MELATONIN INHIBITS PRION-INDUCED APOPTOTIC SIGNALING BY REGULATING MITOCHONDRIAL APOPTOTIC PATHWAYS

Melatonin, also known as 5-methoxy-*N*-acetyltryptamine, is a natural hormone secreted by the pineal gland in the brain (Domínguez-Alonso et al. 2012; García et al. 2011; Luchetti et al. 2010; Paradies et al. 2010). It is stimulated by darkness and has been related to the circadian regulation of sleep (Crowley and Eastman 2013; García et al. 2011; Münch et al. 2005; Takaesu et al. 2012). Melatonin is also a potent scavenger of reactive oxygen species and is a regulator of mitochondrial homeostasis (Galano et al. 2011; García et al. 2011; Paradies et al. 2010). Some physiological effects of melatonin are related to neuroprotective effects associated with neurodegenerative diseases, including Alzheimer's, Parkinson's, and prion diseases (Figure 29.1) (Jeong et al. 2012; Singhal et al. 2011; Xiong et al. 2011).

Regulation of mitochondrial homeostasis influences the progression of neurodegenerative diseases, including Alzheimer's, Parkinson's, and prion diseases (Dragicevic et al. 2011; Jeong et al. 2012; Olcese et al. 2009; Paradies et al. 2010; Quintanilla et al. 2011; Singhal et al. 2011). In vivo experiments have shown that melatonin administration restores mitochondrial function by enhancing mitochondrial glutathione levels and ATP synthesis by stimulating NADH–coenzyme Q reductase (complex I) and cytochrome c oxidase (complex IV) activities in a transgenic mice model of Alzheimer's disease (Dabbeni-Sala et al. 2001). Melatonin also exerts a protective effect against the 1-methyl-4-phenylpyridine (*MPP*+)-induced neuronal cell death and mitochondrial dysfunction by the inhibition of p38 mitogen–activated protein kinase, p53, and GSK-3β phosphorylation (Absi et al. 2000). In addition, melatonin protects amyloid-beta peptide 25- to 35-induced mitochondrial dysfunction in hippocampal neurons (Dong et al. 2010). And also, it may protect against anesthesia-induced apoptotic neurodegeneration via improved mitochondrial homeostasis (Yon et al. 2006). In addition, a recent study suggested that melatonin administration protects from glutamate-induced oxytosis by an antioxidant effect that specifically targets mitochondria (Herrera et al. 2007). And also the typical pattern of neurotoxicity in prion disease is through mitochondrial damage.

Thus, the melatonin therapeutic study focused on the effect of melatonin on prion peptide (PrP)–mediated mitochondrial abnormalities. Recent report suggested that melatonin inhibited prion-mediated mitochondrial dysfunction (Figure 29.2) and neurotoxicity (Figure 29.1).

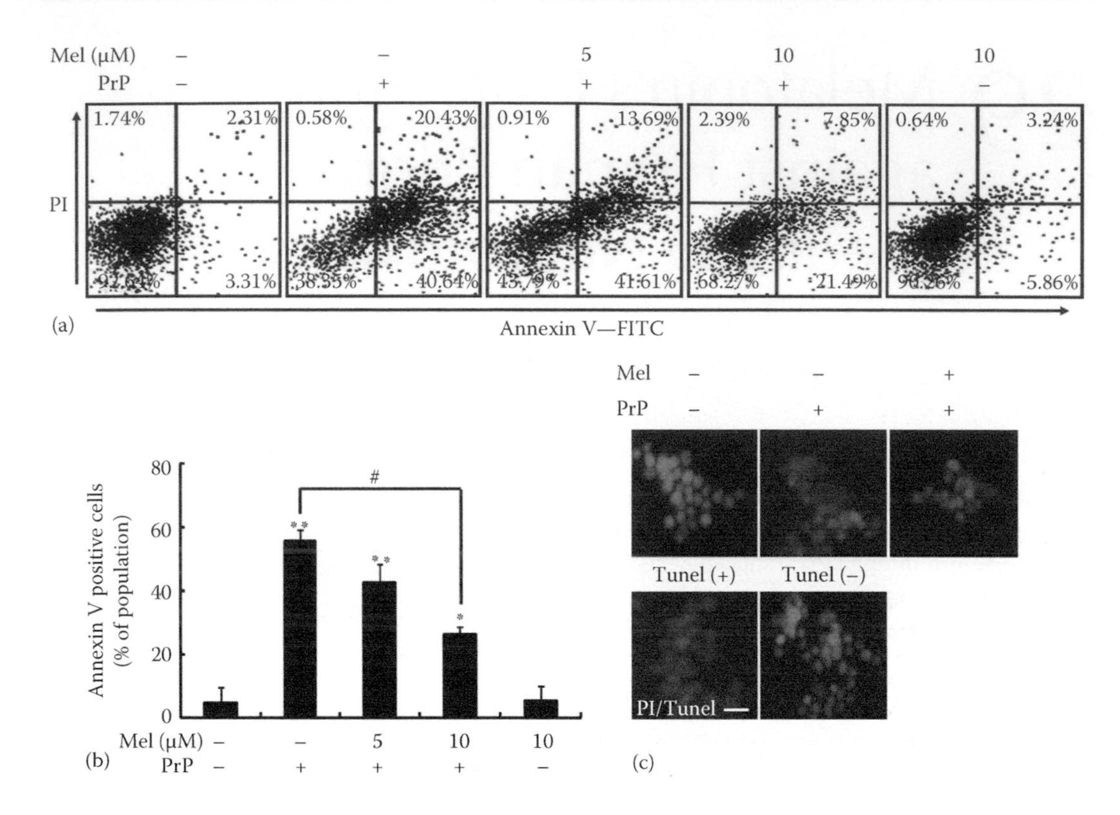

FIGURE 29.1 Administration of melatonin prevented neuronal cells from PrP (106–126)-induced cell death. (a) SH-SY5Y neuronal cells were pretreated with melatonin (12 h) in a dose-dependent manner and then exposed to 50 μM PrP (106–126) for 24 h. Cell viability was measured by annexin V assay. M1 represents the population of annexin V–positive cells. (b) Bar graph indicated the averages of annexin V–positive cells. $*P < 0.05$, $**P < 0.001$ significant differences between control and each treatment group and $^{\#}P < 0.01$ significantly different when compared with PrP (106–126)-treated group. (c) Representative immunofluorescence images of TUNEL-positive (gray) SH-SY5Y cells at 24 h after exposure to 50 μM PrP (106–126) in the absence or presence of melatonin (12 h). The cells were counterstained with PI (white) to show all cell nuclei. Magnification 400×, scale bar = 100 μm. (From Park, S.-Y.: Melatonin-induced autophagy protects against human prion protein-mediated neurotoxicity. *J. Pineal Res.* 2012. 53. 138–146. Copyright Wiley-VCH Verlag GmbH & Co. KGaA. Reprinted with permission.)

Cells treated with PrP (106–126) showed increased JC-1 monomers, indicating low MTP values, while melatonin treatment reduced PrP-induced JC-1 monomers, indicating high MTP values (Figure 29.2). Also, fluorescence image (Figure 29.2) showed that cells with green fluorescence (JC-1 monomer form) after PrP treatment indicated lower MTP, while the negative control cells and melatonin-treated cells had red fluorescence (JC-1 aggregates form), indicating high MTP values. Also, PrP treatment induced Bax translocation to the mitochondria and cytochrome c release to the cytosol in neuronal cells, while PrP-induced Bax translocation and cytochrome c release were inhibited with melatonin treatment (Figure 29.2). These observations support the idea that melatonin treatment prevents against PrP-mediated mitochondrial dysfunctions and neurotoxicity.

29.2 MELATONIN INHIBITS PRP-INDUCED MITOCHONDRIAL APOPTOTIC PATHWAY VIA ACTIVATING AUTOPHAGY

Melatonin reduces autophagy in some experimental systems, such as oxidative-injured acute pancreatitis (Eşrefoğlu et al. 2006) and colon cancer (Motilva et al. 2011). Many studies support an

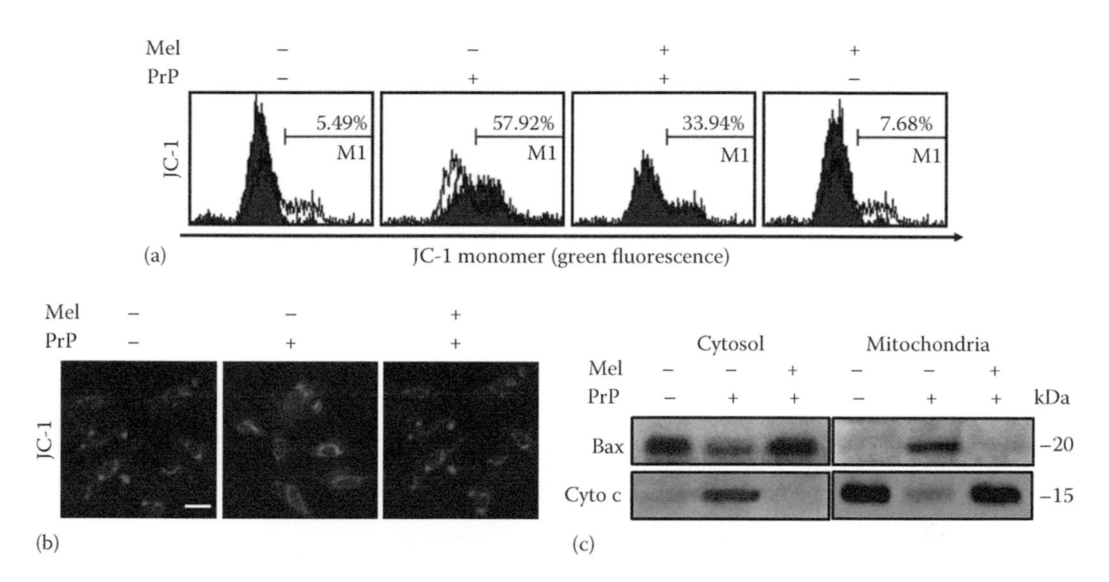

FIGURE 29.2 Protective effect of melatonin treatment on PrP (106–126)-mediated mitochondrial dysfunction. (a) SH-SY5Y cells were pretreated with 10 μM melatonin (12 h) and then exposed to 50 μM PrP (106–126) for 24 h. The treated cells were measuring JC-1 mono form (green) by flow cytometry. M1 represents the population of JC-1 monomeric cells. (b) Representative images of J-aggregate formation in cells treated as described in (a). The treated cells were measuring JC-1 aggregates form (red) and mono form (green) by confocal microscopy analysis. Scale bar = 50 μm. (c) Cells were homogenized in a mitochondrial buffer, seperated the cytosol and mitochondrial extracts, and analyzed by Western blotting using antibodies against cytochrome c and Bax protein. (From Park, S.-Y.: Melatonin-induced autophagy protects against human prion protein-mediated neurotoxicity. *J. Pineal Res.* 2012. 53. 138–146. Copyright Wiley-VCH Verlag GmbH & Co. KGaA. Reprinted with permission.)

active role of melatonin on autophagy (Caballero et al. 2009; Guo et al. 2010). Autophagy is a lysosomal process to degrade large structures such as cell organelles and protein aggregates (Glick et al. 2010). As such, autophagy can promote survival under various stress conditions, including immunity, cancer, and neuroprotection (Richard et al. 2011; Tili and Michaille 2011). Especially, modulation of the autophagy activation is thought to play an important role in neurodegenerative disorders and to be related to protection from mitochondrial damage (Graef and Nunnari 2011). Some reports showed that activation of autophagy prevents neurodegenerative disorders including Alzheimer's disease and Parkinson's disease by autophagy-mediated clearance of mitochondrial abnormalities (Filomeni et al. 2010; Heiseke et al. 2009; Vingtdeux et al. 2010; Wu et al. 2011). Also, a recent study showed that melatonin treatment protects neuronal cells against ischemia/reperfusion injury through autophagy enhancement (Guo et al. 2010). Thus, the study focused to the effect of melatonin-induced autophagy on PrP-mediated neurotoxicity and mitochondrial dysfunction. Treatment of SH-SY5Y neuronal cells with melatonin induced LC3-II, a late autophagosome marker, in a dose-dependent manner. In addition, the melatonin-induced LC3-II levels were inhibited by treatment with autophagy inhibitors (Figure 29.3). Also, melatonin treatment decreased PrP-induced mitochondrial apoptosis-related proteins, pGSK-3β, p-p38, and p53 (Figure 29.3). The protective effect of melatonin on PrP (106–126)-induced mitochondrial apoptotic pathway associated with the activation of p38 MAPK, p53, and GSK-3β was blocked by treatment with the autophagy inhibitor 3-MA and wortmannin (Figure 29.3).

Presently, the protective effect of melatonin was blocked by autophagy inhibitors during PrP-induced neuronal cell death (Figure 29.5) and mitochondrial damage (Figure 29.4).

Melatonin treatment prevented PrP-induced apoptosis, inhibited Bax translocation and cytochrome c release, and increased MTP values, whereas autophagy inhibitors, 3-MA, bafilomycin

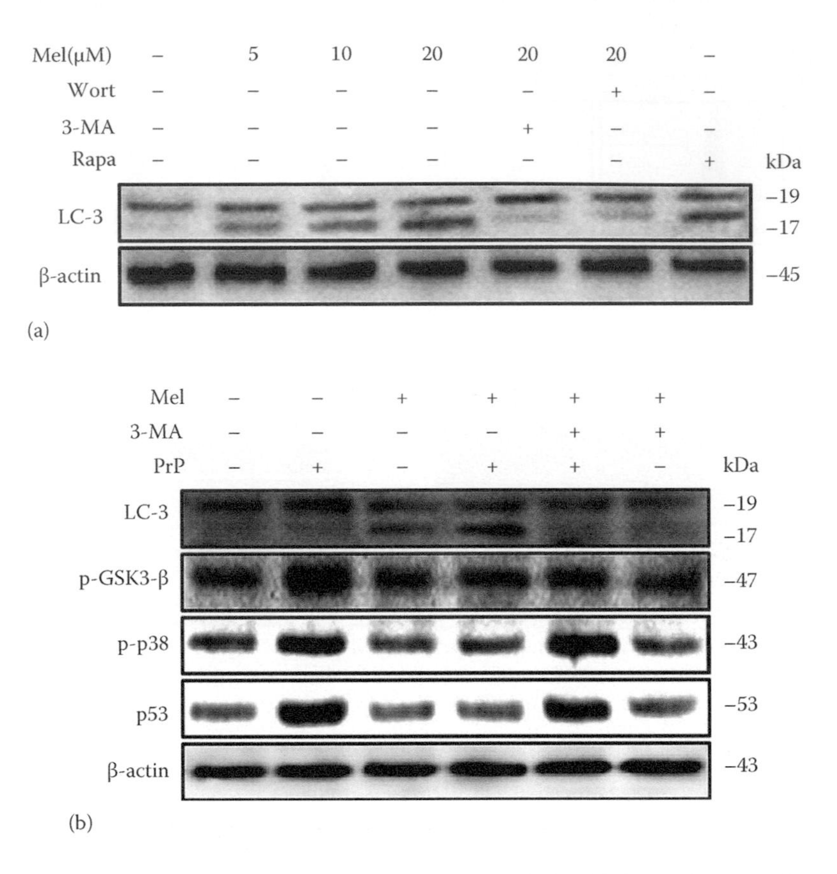

FIGURE 29.3 Administration of melatonin blocked PrP (106–126)-induced apoptotic pathway via regulation of autophagy. (a) SH-SY5Y cells were pretreated with melatonin (12 h) in a dose-dependent manner and then exposed to autophagy inhibitors including 3-MA (2 mM) and wortmannin (10 μM). Two hundred nanometer rapamycin was used as a positive control to induce autophagy. The treated cells were assessed for LC3 production by Western blot analysis. Results were normalized with β-actin. (b) Cells were treated with 10 μM melatonin (12 h) and then exposed to 50 μM PrP (106–126) with or without 2 mM 3-MA (autophagy inhibitor) for 24 h. The treated cells were assessed for p53, p-p38, p-GSK-3β, and LC-3 production by Western blot analysis. Results were normalized with β-actin. (From Park, S.-Y.: Melatonin-induced autophagy protects against human prion protein-mediated neurotoxicity. *J. Pineal Res.* 2012, 53, 138–146. Copyright Wiley-VCH Verlag GmbH & Co. KGaA.)

A1, and wortmannin, inhibited the protective effect of melatonin on prion-mediated neurotoxicity and mitochondrial dysfunction (Figure 29.4).

Investigation of the protective role of melatonin-induced autophagy in prion-mediated neurotoxicity showed that negative control siRNA-treated (NC-siRNA) cells were prevented PrP-induced neurotoxicity (Figure 29.5) and induced expression of LC3-II and ATG5 protein levels (Figure 29.5). In contrast, treatment with ATG-5 RNAi oligomer (ATG-5 siRNA) blocked the protective effect of melatonin on PrP-induced neurotoxicity (Figure 29.5) and inhibited the expression of LC3-II and ATG5 protein levels (Figure 29.5). Also, the protective effect of melatonin on PrP-induced reduction of MTP and an increase in p-p38 MAPK, p53, and p-GSK-3β protein levels were blocked by treatment with the ATG-5 siRNA.

These observations suggest that melatonin-induced autophagy plays a pivotal neuroprotective role in neurodegenerative diseases, including prion disease.

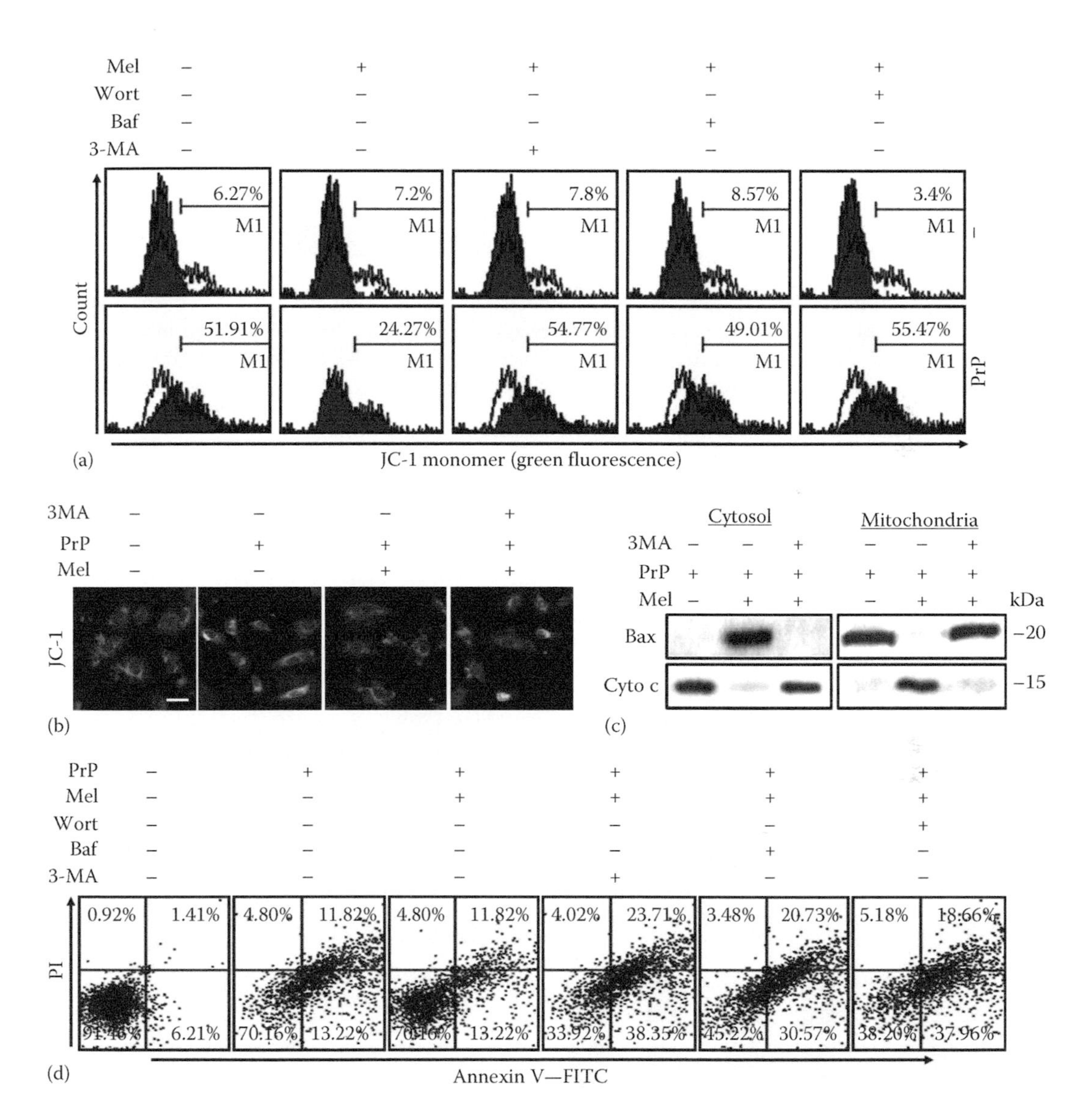

FIGURE 29.4 Melatonin-mediated autophagy protected PrP (106–126)-induced mitochondrial apoptotic pathway. (a) SH-SY5Y cells were treated with 10 μM melatonin (12 h) and then exposed to 50 μM PrP (106–126) with or without autophagy inhibitor (2 mM 3-MA, 100 nM bafilomycin A1, or 10 μM wortmannin) for 24 h. The treated cells were measuring JC-1 mono form (white) by flow cytometry. M1 represents the population of JC-1 monomeric cells. (b) Representative images of J-aggregate formation in SH-SY5Y cells treated as described (a). (c) Cells were homogenized in a mitochondrial buffer, seperated the cytosol and mitochondrial extracts, and analyzed by Western blotting using antibodies against cytochrome c and Bax protein. (d) SH-SY5Y cells were treated with 10 μM melatonin (12 h) and then exposed to 50 μM PrP (106–126) with or without autophagy inhibitor (2 mM 3-MA, 100 nM bafilomycin A1, or 10 μM wortmannin) for 24 h. Cell viability was measured by annexin V assay. *(Continued)*

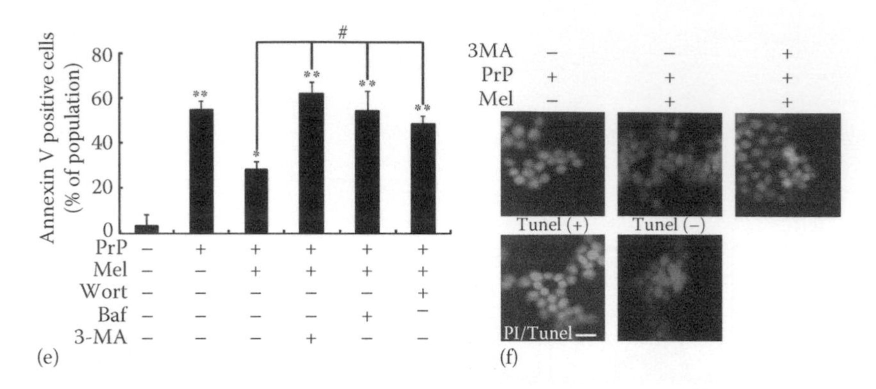

FIGURE 29.4 (*Continued*) Melatonin-mediated autophagy protected PrP (106–126)-induced mitochondrial apoptotic pathway. (e) Bar graph indicated the averages of annexin V positive cells. $*P < 0.05$, $**P < 0.001$ significant differences between control and each treatment group and $^{\#}P < 0.01$ significantly different when compared with PrP (106–126)- and melatonin-treated group. (f) Representative immunofluorescence images of TUNEL-positive (gray) SH-SY5Y cells at 24 h after exposure to 50 μM PrP (106–126) in the absence or presence of melatonin (12 h). The cells were counterstained with PI (white) to show all cell nuclei. Magnification 400×, *scale bar* = 100 μm. (From Park, S.-Y.: Melatonin-induced autophagy protects against human prion protein-mediated neurotoxicity. *J. Pineal Res.* 2012. 53. 138–146. Copyright Wiley-VCH Verlag GmbH & Co. KGaA. Reprinted with permission.)

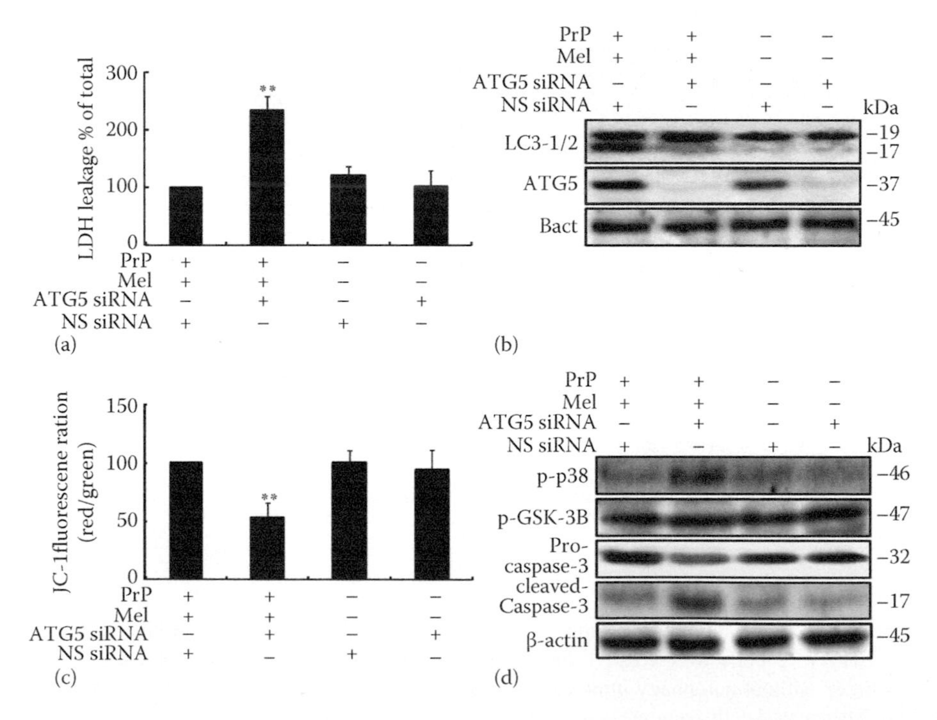

FIGURE 29.5 ATG5 knockdown affects the melatonin-mediated neuroprotective effect against PrP (106–126). (a) ATG-5 siRNA (ATG5 small interfering RNA)- or NS siRNA (nonspecific siRNA)-transfected SH-SY5Y cells were incubated with 50 μM PrP (106–126) for 24 h after exposure to 10 μM melatonin (12 h), and the release of lactate dehydrogenase into the cell culture supernatant from damaged cells was measured. $**P < 0.001$ significant differences between control and each treatment group. (b) Western blot analysis of LC3-1/2 and ATG5 conversion in the SH-SY5Y cells treated as described in (a). (c) Bar graph indicated the J-aggregate formation. $**P < 0.001$ significant differences between control and each treatment group. (d) Western blot analysis of p-p38, p-GSK-3β, caspase-3 conversion in the SH-SY5Y cells treated as described in (a). (From Park, S.-Y.: Melatonin-induced autophagy protects against human prion protein-mediated neurotoxicity. *J. Pineal Res.* 2012. 53. 138–146. Copyright Wiley-VCH Verlag GmbH & Co. KGaA. Reprinted with permission.)

Taken together, these reports demonstrate that melatonin treatment protects against PrP-induced neuronal cell death and PrP-mediated mitochondrial dysfunction by the regulation of autophagy activation. These results also suggest that autophagy inducers, including melatonin, may have clinical benefits when used as neuro-chemotherapy for prion diseases.

REFERENCES

Absi, E., A. Ayala, A. Machado, and J. Parrado. 2000. Protective effect of melatonin against the 1-methyl-4-phenylpyridinium-induced inhibition of complex I of the mitochondrial respiratory chain. *Journal of Pineal Research* 29(1):40–47.

Caballero, B., I. Vega-Naredo, V. Sierra, D. DeGonzalo-Calvo, P. Medrano-Campillo, J. M. Guerrero, D. Tolivia, M. J. Rodríguez-Colunga, and A. Coto-Montes. 2009. Autophagy upregulation and loss of NF-κB in oxidative stress-related immunodeficient SAMP8 mice. *Mechanisms of Ageing and Development* 130(11–12):722–730.

Crowley, S. J. and C. I. Eastman. 2013. Melatonin in the afternoons of a gradually advancing sleep schedule enhances the circadian rhythm phase advance. *Psychopharmacology* 225(4):825–837.

Dabbeni-Sala, F., S. Di Santo, D. Franceschini, S. D. Skaper, and P. Giusti. 2001. Melatonin protects against 6-OHDA-induced neurotoxicity in rats: A role for mitochondrial complex I activity. *The FASEB Journal* 15(1):164–170.

Domínguez-Alonso, A., G. Ramírez-Rodríguez, and G. Benítez-King. 2012. Melatonin increases dendritogenesis in the hilus of hippocampal organotypic cultures. *Journal of Pineal Research* 52(4):427–436.

Dong, W., F. Huang, W. Fan, S. Cheng, Y. Chen, W. Zhang, H. Shi, and H. He. 2010. Differential effects of melatonin on amyloid-β peptide 25–35-induced mitochondrial dysfunction in hippocampal neurons at different stages of culture. *Journal of Pineal Research* 48(2):117–125.

Dragicevic, N., N. Copes, G. O'Neal-Moffitt, J. Jin, R. Buzzco, M. Mamcarz, J. Tan et al. 2011. Melatonin treatment restores mitochondrial function in Alzheimer's mice: A mitochondrial protective role of melatonin membrane receptor signaling. *Journal of Pineal Research* 51(1):75–86.

Eşrefoğlu, M., M. Gül, B. Ateş, and M. A. Selimoğlu. 2006. Ultrastructural clues for the protective effect of melatonin against oxidative damage in cerulein-induced pancreatitis. *Journal of Pineal Research* 40(1):92–97.

Filomeni, G., I. Graziani, D. De Zio, L. Dini, D. Centonze, G. Rotilio, and M. R. Ciriolo. 2010. Neuroprotection of kaempferol by autophagy in models of rotenone-mediated acute toxicity: Possible implications for Parkinson's disease. *Neurobiology of Aging* 33(4):767–785.

Galano, A., D. X. Tan, and R. J. Reiter. 2011. Melatonin as a natural ally against oxidative stress: A physicochemical examination. *Journal of Pineal Research* 51(1):1–16.

García, J. J., G. Piñol-Ripoll, E. Martínez-Ballarín, L. Fuentes-Broto, F. J. Miana-Mena, C. Venegas, B. Caballero, G. Escames, A. Coto-Montes, and D. Acuña-Castroviejo. 2011. Melatonin reduces membrane rigidity and oxidative damage in the brain of SAMP8 mice. *Neurobiology of Aging* 32(11):2045–2054.

Glick, D., S. Barth, and K. F. Macleod. 2010. Autophagy: Cellular and molecular mechanisms. *The Journal of Pathology* 221(1):3–12.

Graef, M. and J. Nunnari. 2011. A role for mitochondria in autophagy regulation. *Autophagy* 7(10):1245–1246.

Guo, Y., J. Wang, Z. Wang, Y. Yang, X. Wang, and Q. Duan. 2010. Melatonin protects N2a against ischemia/reperfusion injury through autophagy enhancement. *Journal of Huazhong University of Science and Technology—Medical Sciences* 30(1):1–7.

Heiseke, A., Y. Aguib, C. Riemer, M. Baier, and H. M. Schatzl. 2009. Lithium induces clearance of protease resistant prion protein in prion-infected cells by induction of autophagy. *Journal of Neurochemistry* 109(1):25–34.

Herrera, F., V. Martin, G. García-Santos, J. Rodriguez-Blanco, I. Antolín, and C. Rodriguez. 2007. Melatonin prevents glutamate-induced oxytosis in the HT22 mouse hippocampal cell line through an antioxidant effect specifically targeting mitochondria. *Journal of Neurochemistry* 100(3):736–746.

Jeong, J.-K., M.-H. Moon, Y.-J. Lee, J.-W. Seol, and S.-Y. Park. 2012. Melatonin-induced autophagy protects against human prion protein-mediated neurotoxicity. *Journal of Pineal Research* 53(2):138–146.

Luchetti, F., B. Canonico, M. Betti, M. Arcangeletti, F. Pilolli, M. Piroddi, L. Canesi, S. Papa, and F. Galli. 2010. Melatonin signaling and cell protection function. *FASEB Journal* 24(10):3603–3624.

Münch, M., V. Knoblauch, K. Blatter, C. Schröder, C. Schnitzler, K. Kräuchi, A. Wirz-Justice, and C. Cajochen. 2005. Age-related attenuation of the evening circadian arousal signal in humans. *Neurobiology of Aging* 26(9):1307–1319.

Motilva, V., S. García-Mauriño, E. Talero, and M. Illanes. 2011. New paradigms in chronic intestinal inflammation and colon cancer: Role of melatonin. *Journal of Pineal Research* 51(1):44–60.
Olcese, J. M., C. Cao, T. Mori, M. B. Mamcarz, A. Maxwell, M. J. Runfeldt, L. Wang et al. 2009. Protection against cognitive deficits and markers of neurodegeneration by long-term oral administration of melatonin in a transgenic model of Alzheimer disease. *Journal of Pineal Research* 47(1):82–96.
Paradies, G., G. Petrosillo, V. Paradies, R. J. Reiter, and F. M. Ruggiero. 2010. Melatonin, cardiolipin and mitochondrial bioenergetics in health and disease. *Journal of Pineal Research* 48(4):297–310.
Quintanilla, R. A., P. J. Dolan, Y. N. Jin, and G. V. Johnson. 2011. Truncated tau and Abeta cooperatively impair mitochondria in primary neurons. *Neurobiology of Aging* 33(3):619.e25–619.e35.
Richard, T., A. D. Pawlus, M. L. Iglesias, E. Pedrot, P. Waffo-Teguo, J. M. Merillon, and J. P. Monti. 2011. Neuroprotective properties of resveratrol and derivatives. *Annals of the New York Academy of Sciences* 1215:103–108.
Singhal, N. K., G. Srivastava, D. K. Patel, S. K. Jain, and M. P. Singh. 2011. Melatonin or silymarin reduces maneb- and paraquat-induced Parkinson's disease phenotype in the mouse. *Journal of Pineal Research* 50(2):97–109.
Takaesu, Y., Y. Komada, and Y. Inoue. 2012. Melatonin profile and its relation to circadian rhythm sleep disorders in Angelman syndrome patients. *Sleep Medicine* 13(9):1164–1170.
Tili, E. and J. J. Michaille. 2011. Resveratrol, microRNAs, inflammation, and cancer. *Journal of Nucleic Acids* 2011:102431.
Vingtdeux, V., L. Giliberto, H. Zhao, P. Chandakkar, Q. Wu, J. E. Simon, E. M. Janle et al. 2010. AMP-activated protein kinase signaling activation by resveratrol modulates amyloid-beta peptide metabolism. *The Journal of Biological Chemistry* 285(12):9100–9113.
Wu, Y., X. Li, J. X. Zhu, W. Xie, W. Le, Z. Fan, J. Jankovic, and T. Pan. 2011. Resveratrol-activated AMPK/SIRT1/autophagy in cellular models of Parkinson's disease. *Neurosignals* 19(3):163–174.
Xiong, Y.-F., Q. Chen, J. Chen, J. Zhou, and H.-X. Wang. 2011. Melatonin reduces the impairment of axonal transport and axonopathy induced by calyculin A. *Journal of Pineal Research* 50(3):319–327.
Yon, J.-H., L. B. Carter, R. J. Reiter, and V. Jevtovic-Todorovic. 2006. Melatonin reduces the severity of anesthesia-induced apoptotic neurodegeneration in the developing rat brain. *Neurobiology of Disease* 21(3):522–530.

30 Melatonin's Protection against Methamphetamine Toxicity in Neuroblastoma Cells

Banthit Chetsawang

CONTENTS

Methamphetamine (METH) is a psychostimulant cationic lipophilic molecule with potent action on the central nervous system (CNS). The chemical structure of METH is similar to amphetamine and the neurotransmitter dopamine (DA). It is chemically related to amphetamine, but, at comparable doses, the effects of METH are much more potent, longer lasting, and more harmful to the CNS than amphetamine. METH is believed to exert euphoric effects by stimulating DA release resulting in feeling of pleasure and reward. METH is selectively taken up into dopaminergic terminals by dopamine transporter (DAT) and displacement of DA into the cytoplasm of the terminal and subsequent release of the DA into synaptic cleft, leading to much higher concentrations in the synapse, which can be toxic to nerve terminals (Fleckenstein et al., 2007). After the displacement to the cytoplasm by METH, DA rapidly auto-oxidizes to form potentially toxic substances including superoxide radicals, hydroxyl radicals, hydrogen peroxide, and DA quinones (Acikgoz et al., 1998; Larsen et al., 2002; Lazzeri et al., 2007). Increases in DA oxidation occurred only under conditions resulting in toxicity, suggesting that the oxidation of DA may contribute to the mechanism of METH-induced damage to DA terminals (LaVoie and Hastings, 1999). DA metabolism by monoamine oxidase (MAO) enzyme is also accompanied by increased production of hydrogen peroxide, which interacts with metal ions such as iron, whose level is elevated by METH treatment to form toxic hydroxyl radicals (Melega et al., 2007). The accumulated evidence indicates that METH can also cause oxidative stress by switching the balance between reactive oxygen species (ROS) production and the capacity of antioxidant enzyme systems to scavenge ROS (Jayanthi et al., 1998; Gluck et al., 2001). For example, METH administration causes decreases in the levels of copper and zinc superoxide dismutase (Cu/Zn-SOD), catalase, glutathione, and peroxiredoxins in the brain accompanied by elevated lipid peroxidation and substantial increases in the levels of protein carbonyls (Jayanthi et al., 1998; Chen et al., 2007).

Various studies have addressed the role of cellular and molecular events involved in METH-induced DA terminal degeneration and neuronal apoptosis within the striatum (Cadet et al., 2005; Staszewski and Yamamoto 2006; Krasnova and Cadet, 2009). Chronic METH abuse may create disturbances in dopaminergic systems of the brain that may predispose individuals to Parkinsonism. Parkinson's disease (PD) is the second most common neurodegenerative disease characterized by a progressive loss of dopaminergic neurons of the substantia nigra (SN) pars compacta, which send their axons to terminate at the striatum. The exact mechanism of this selective cell death is not understood. Generation of ROS caused by an oxidative stress in the SN is widely considered to lead to neuronal death (Ebadi et al., 1996).

30.1 NEUROTOXICITY OF METHAMPHETAMINE IN HUMAN NEUROBLASTOMA SH-SY5Y CELLS

It has been demonstrated that METH can cause an alteration in dopaminergic cells in the human brain (Krasnova and Cadet, 2009). DA has long been implicated as a key factor in METH neurotoxicity that involves imbalance between DA release and reuptake by the presynaptic nerve terminal. In the normal condition, DAT removes DA from the synapse and transports from cytoplasmic site into vesicles via vesicular monoamine transporter-2 for storage, release, and protection from oxidation and reactive consequences. The mechanisms by which METH causes neurotoxicity are not understood. SH-SY5Y human neuroblastoma cells were used to study the effect of METH toxicity on DA system in our study. Human neuroblastoma SH-SY5Y cells are often considered cathecholaminergic neuroblastoma cell lines, in principle that they express tyrosine hydroxylase (TH) activity (Lode et al., 1995) and have mechanisms of catecholamine uptake (Richards and Sadee, 1986). SH-SY5Y cells have been widely used as a model in PD (Wu et al., 2007).

30.1.1 Effect of Methamphetamine on Cell Viability in SH-SY5Y Cells

Our study demonstrated that METH induced dopaminergic SH-SY5Y cells death with dose- and time-dependent effect. The higher the METH concentrations or the longer METH exposure times, the greater the reduction in cell viability (Wisessmith et al., 2009; Suwanjang et al., 2010). METH-induced neurotoxicity has been demonstrated in animals (Cadet et al., 2007) and humans (McCann et al., 2008). In human brain, the reward circuit responds to METH by the activation of mesolimbic system (Vollm. et al., 2004) because DA and its metabolites in mesolimbic system are responsible for degenerative alteration. Several studies have shown that METH enters dopaminergic cells via DAT (Wisessmith et al., 2009) and passive diffusion (Krasnova and Cadet, 2009) and then induces DA release from synaptic vesicle into cytoplasm (Guillot et al., 2008). Many studies suggest that high level of DA auto-oxidize to generate ROS leading to oxidative stress, mitochondrial dysfunction and increase in free radical formation (Krasnova and Cadet 2009). It has been demonstrated that METH decreases mitochondrial membrane potential and increases the levels of ROS and the occurrence of apoptosis in SH-SY5Y cultured cells (Wu et al., 2007).

30.1.2 Effect of Methamphetamine-Induced Reduction in Tyrosine Hydroxylase Phosphorylation in SH-SY5Y Cells

The results of our study have shown that METH toxicity decreases phosphorylated TH at serine residue 40 (Ser40) levels in SH-SY5Y cells (Suwanjang et al., 2010). TH is the initial and rate-limiting enzyme in the biosynthesis of the catecholamine such as DA, noradrenaline, and adrenaline in the CNS. TH is a tetramer of identical 60 kDa subunits. The regulation of TH activity is mediated through phosphorylation of Ser19, Ser31, and Ser40. The residue of TH at Ser40 is phosphorylated by protein kinase A (PKA), protein kinase C (PKC), calcium/calmodulin-dependent protein kinase II (CaMKII), and mitogen-activated protein kinase (MAPK). It has been demonstrated that the phosphorylation of TH at Ser40 is associated with an increase in the activity of TH (Bevilaqua et al., 2001). TH has eventually been used as markers of the integrity of DA neurons and terminals after METH administration. For example, METH causes losses of phosphorylated TH in SK-N-SH cells (Klongpanichapak et al., 2008). Several studies suggest that METH can cause a decrease in TH levels in different brain regions of rat (Kaewsuk et al., 2009; Krasnova and Cadet, 2009) and mice (Thomas et al., 2008, 2009).

30.2 PROTECTIVE EFFECT OF MELATONIN ON METHAMPHETAMINE-INDUCED INDUCTION IN BAX AND CLEAVED CASPASE-3 IN SH-SY5Y CELLS

It has been reported that METH can cause differential changes in signaling cascades, which are involved in the mitochondrial cell death pathway (Davidson et al., 2001). One of Bcl-2 family members is Bax, a pro-apoptotic protein that affects the mitochondria outer membrane permeabilization and induced mitochondrial dysfunction (Goping et al., 1998; Kluck et al., 1999). Several studies point out that Bax expression correlates with cell death (Cheng et al., 2001; Jayanthi et al., 2001; Er et al., 2006). In our study, METH treatment resulted in extreme increase in Bax protein levels while slight increase in pro-apoptotic Bcl-2 protein levels in SH-SY5Y cells leading to marked increase in Bax/Bcl-2 ratio (Wisessmith et al., 2009). When Bcl-2 is overexpressed, on the other hand, it heterodimerizes with Bax, and cell death is repressed (Brenner et al., 2000; Tsujimoto and Shimizu, 2000). Under normal conditions, Bax exists as a monomer form either in the cytosol or loosely bound to the outer mitochondria membrane (Suzuki et al., 2000). When Bax is overexpressed in cells, the cell death signals are accelerated. Under pathological conditions, Bax undergoes a conformational change by oligomerization into large complexes to prepare to insert into the mitochondria outer membrane (Nechushtan et al., 1999). The Bax oligomerization translocates to mitochondria (Wolter et al., 1997) and induces their own channel by inserting into mitochondrial lipid bilayer membrane, which is the release site of cytochrome c from inside mitochondria into cytosol (Sharpe et al., 2004). Cytochrome c forms complexes with apoptotic protease activating factor-1 (Apaf-1), ATP, and procaspase-9, which trigger the caspase-dependent pathway by cleaving the inactive site of caspase-9 and changing it to the active form of caspase-9 (Zimmermann et al., 2001; Zhang et al., 2005). Active caspase-9 cleaves inactive part of procaspase-3, resulting in the activation of caspase-3, which then induced the biochemical execution of cell death (Zhang et al., 2005). METH-induced induction in cleaved caspase-3 in SH-SY5Y cells was demonstrated in our study (Wisessmith et al., 2009). Along the same line of our study, it has shown that amphetamine highly decreases the mitochondria membrane potential and induces the activation in caspase-2, -3, and -9 and the appearance of nuclear apoptotic morphology in rat cortical cell cultures (Cunha-Oliveira et al., 2006).

Melatonin and its metabolites have protective effects on a variety of oxidative stress–associated neuropathologies (Manda et al., 2007; Galano et al., 2011, 2013). Melatonin is a highly lipophilic and hydrophilic molecule, which easily crosses the blood–brain barrier and plasma membrane and enters several cellular compartments including mitochondria (Menendez-Pelaez and Reiter, 1993). Melatonin has the ability to maintain mitochondria homeostasis by increasing the activity of the complexes I and IV of the electron transport chain, promoting ATP synthesis under normal conditions and restoring its activities in pathological situations (Martin et al., 2002). Melatonin also directly scavenges ROS that is abundantly produced in mitochondria during normal metabolism, and additionally, it indirectly promotes the activities of antioxidant enzymes that help transform ROS to low-reactive products (Acuna-Castroviejo et al., 2001). Mitochondria play some functional roles related to antioxidative properties and cell survival such as the stimulation of antioxidant enzymes for the direct scavenging of toxic reagents and functional units of ATP synthesis that help to maintain cellular homeostasis, respectively (Leon et al., 2004). In our study, we examined the effective role of melatonin against METH-induced induction in death signaling proteins Bax and cleaved caspase-3 proteins using immunoblotting and immunocytochemistry. Treatment of

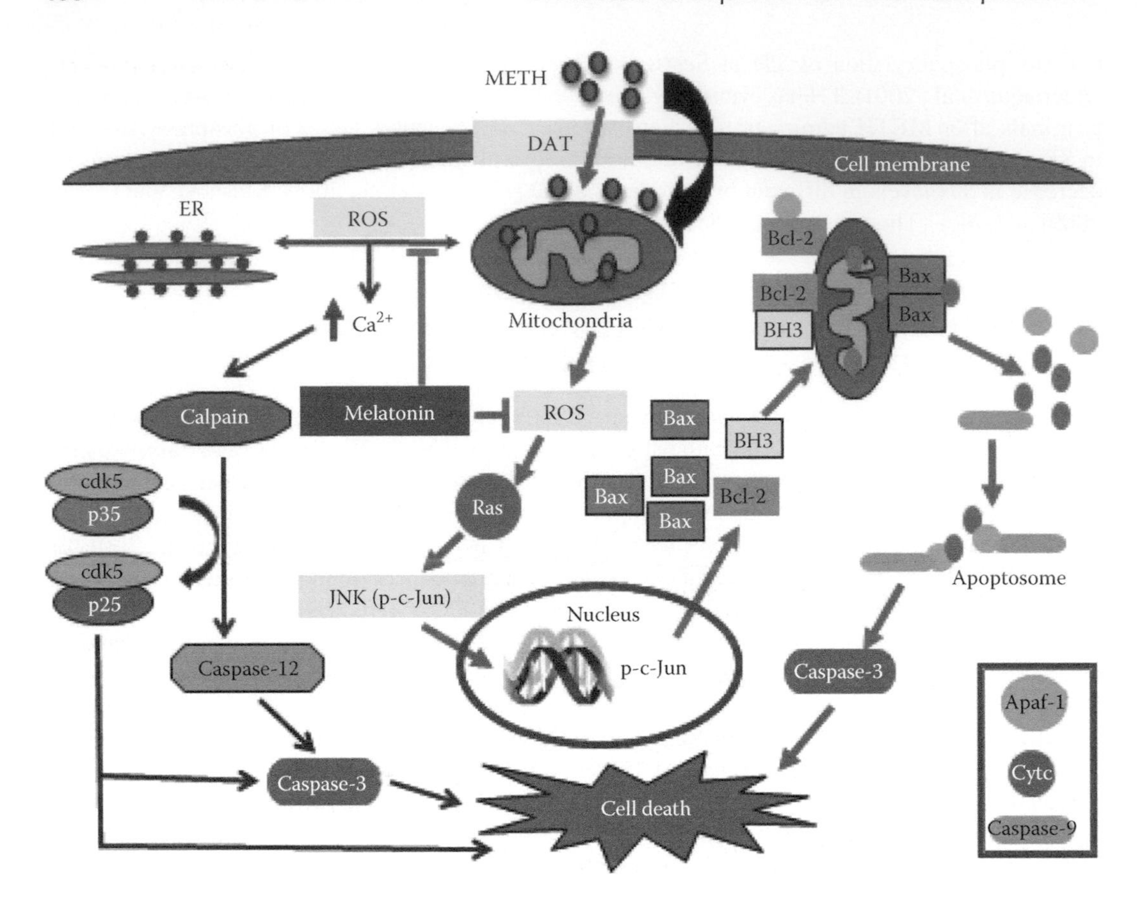

FIGURE 30.1 **(See color insert.)** The proposed neuroprotective mechanisms of melatonin against methamphetamine (METH) toxicity–induced cell death in neuroblastoma SH-SY5Y cells. DAT, dopamine transporter; ROS, reactive oxygen species; JNK, c-Jun-N-terminal kinase; p-c-Jun, c-Jun phosphorylation; ER, endoplasmic reticulum; Cyt. c, cytochrome c; Apaf-1, apoptotic protease activating factor-1.

SH-SY5Y cells with METH caused an increase in Bax and cleaved caspase-3. On the other hand, this increase was diminished with melatonin (Wisessmith et al., 2009). This result supports an evidence of induction in Bax expression that can lead to mitochondrial dysfunction and cell death.

A possible explanation for the neuroprotective effect of melatonin on reducing the toxic effect of METH in SH-SY5Y cells might be the attenuation of METH-induced increase in ROS formation that leads to shut down the downstream death signaling cascades (Figure 30.1). Usually, the ROS may activate downstream Ras-dependent and c-Jun-N-terminal kinase (JNK) signaling cascades, which subsequently increase the phosphorylation of transcription factor, c-Jun (Pirompul et al., 2013). Phosphorylated c-Jun may act as transcription factor, which activates the expression of Bax protein. High levels of Bax promote the opening of the mitochondrial permeability transition pore (MPTP) and the release of cytochrome c to cytosol, where it forms apoptosome complexes, which then cleave and activate the series of caspase enzyme, especially a common executioner caspase or caspase-3, and lead to cell death.

30.3 PROTECTIVE EFFECT OF MELATONIN ON METHAMPHETAMINE-INDUCED INDUCTION IN CALPAIN-DEPENDENT DEATH PATHWAY IN SH-SY5Y CELLS

Recently, it has been reported that METH can induce an increase in intracellular calcium (Ca^{2+}) concentrations both in in vivo and in vitro studies (Krasnova and Cadet, 2009). High levels of Ca^{2+} inside the cell activate several intracellular signaling cascades including calpain, which is a calcium-dependent cysteine protease. Calpain was localized in numerous cellular compartments such as mitochondria, cytosol, and nucleus. It is a member of death signaling pathways that is regulated by intracellular Ca^{2+} concentration. Moreover, calpain activity is also regulated by an endogenous calpain inhibitor named calpastatin (Sorimachi and Suzuki, 1998). Calpain has been implicated in certain pathological conditions in the CNS (Ray and Banik, 2003) including Alzheimer's disease (AD) and PD (Samantaray et al., 2007). Upregulation of calpain activity in microglia, astrocytes, and neurons have been observed in spinal cord of neurotoxin MPTP-treated mice (Ray et al., 1999; Chera et al., 2004). The unilateral intrastriatal infusion of neurotoxin, 6-OHDA, increases calpain activation in caudate-putamen and SN, and dopaminergic denervation in caudate-putamen of rat brain (Grant et al., 2009). In addition, it has been shown that the activation of calpain in rat striatal synaptosome by increasing Ca^{2+} concentrations leads to an increase in the cleavage of DAT, and this process is totally blocked by a calpain inhibitor (Veronika et al., 2008). In our study, we found that METH toxicity can induce reduction in the levels of calpastatin but induction in calpain activation in dopaminergic SH-SY5Y cells (Suwanjang et al., 2010). It has been reported that an overproduction of reactive oxygen radical stimulates Ca^{2+} influx, which, in turn, increases calpain expression and Bax/Bcl-2 ratio and then promotes cytochrome c release and apoptosis in C6 cell (Gao and Dou, 2000). It has been demonstrated that calpain has the capacity to mediate caspase-12 activation and Bcl-xl inactivation in ischemic-induced apoptotic glial cell cultures and amyloid beta peptide-induced cell death in cortical neuron cultures (Nakagawa and Yuan, 2000). Although the exact mechanisms of METH-induced calpain activation and cell death in neuronal cells are unknown, a possible explanation might be an increase in ROS formation, which, in turn, results in mitochondrial dysfunction. This dysfunction leads to a leakage of mitochondrial Ca^{2+} into cytosol. High levels of intracellular Ca^{2+} may contribute to the activation of calpain, and this activation eventually leads to a cleavage of cytoskeletal and myelin proteins and also an activation of pro-apoptotic factors such as Bax and caspase-3 (Choi et al., 2001; Ozaki et al., 2001).

Several studies have demonstrated that an increase in calpain activity is a hallmark in the pathophysiology of degenerative diseases (Nixon, 2000; Dufty et al., 2007). In addition, an increase in calpain/calpastatin ratio induces increase in the degradation of calpastatin in the brain of AD-like transgenic (Tg2576) mice, while a high level of calpastatin is able to diminish calpain activation in the cerebellum of this transgenic mice (Vaisid et al., 2007). Overexpression of calpastatin in PC12 cells can inhibit amyloid beta peptide (Aβ)-induced calpain activation, degradation of fodrin, protein kinase Cε, and β-catenin (Vaisid et al., 2008). Furthermore, an inhibition of calpain can prevent neuronal and behavioral deficits in MPTP-induced experimental parkinsonism (Crocker et al., 2003).

Either the overproduction of intracellular ROS or the inhibition of mitochondrial protein complexes disturbs the normal function of mitochondria and induces mitochondrial Ca^{2+} dysregulation, which leads to the activation of calpain and cell damage or death (Vosler et al., 2009). The restorative properties of melatonin on secretory functions, Ca^{2+} signals, and mitochondrial potential have been demonstrated in aged exocrine cells (Camello-Almaraz et al., 2008). Some of the actions of melatonin prevent a ruthenium red–induced reduction in the activity of complexes I and IV mitochondrial in rats (Martin et al., 2000). In our work, the protective effect of melatonin on METH-induced reduction in cell viability, calpastatin levels, mitochondrial function, phosphorylation of TH, and induction in calpain activity was determined in SH-SY5Y cells. The results of our study clearly demonstrated that melatonin has a potential effect on the protection and restoration of METH-induced calpain-dependent death processes in dopaminergic SH-SY5Y cells (Suwanjang et al., 2010). Our findings are consistent with the other studies that demonstrated that treatment with melatonin decreases inflammation and calpain expression in spinal cord injury of rats (Samantaray et al., 2008). Treatment with melatonin also reduces the neurotoxic effects of MPP^+ (toxic metabolite of MPTP) on the induction in calpain activity, activation in cleavage of cdk5/p35 to cdk5/p25, and increases in the apoptotic cell population in cerebellar granule neuron cultures (Alvira et al., 2006). Furthermore, melatonin can reduce calcium overload from mitochondria and can block MPTP-dependent cytochrome c release and caspase 3 activation (Jou et al., 2004; Yeung et al., 2008). In our study, we also observed the rescue effect of melatonin on reduction in calpastatin levels in METH-treated SH-SY5Y cells. This might be an indirect beneficial effect of melatonin regarding the prevention of cell death, as this death is brought about by calpain-dependent death processes. It has been demonstrated that treatment with NGF and cAMP decreases calpain activity and increases calpastatin levels in PC12 cells (Oshima et al., 1989). The overexpression of calpastatin protects loss of nigral DA neurons in MPTP-treated mouse (Crocker et al., 2003) and attenuates the release of apoptogenic factors from mitochondria in ischemic hippocampal and cortical neuron injury (Cao et al., 2007).

In conclusion, the results of our study point to the contribution of melatonin as a potential protective agent for calpain-dependent death processes. Finally, it is suggested that melatonin should be used as a calpain inhibitor in PD models before using it in therapies that implicate human patients.

30.4 PROTECTIVE EFFECT OF MELATONIN ON METHAMPHETAMINE TOXICITY–INDUCED DISTURBANCE IN MITOCHONDRIAL DYNAMICS IN SH-SY5Y CELLS

Mitochondria are intracellular double membrane organelles found in most eukaryotic cells. These organelles range from 0.5 to 10 micrometers (μm) in diameter. Mitochondria are described as the powerhouse of cell because they generate the cellular energy in the form of ATP. Mitochondria contain outer and inner membranes composed of phospholipid bilayers and proteins that separate four distinct compartments, outer membrane, inner membrane, intermembrane space (the space between outer and inner membranes), and the matrix (the center of organelle) (Rismanchi and Blackstone, 2007). The inner membrane is folded into cristae, which increase its ability to produce ATP (McBride et al., 2006). In addition to energy production, the importance of the remarkable ability of mitochondria to accumulate and buffer intracellular Ca^{2+} is becoming increasingly recognized. Calcium signals are crucial in the control of most physiological processes. Mitochondria can effectively buffer the cytoplasmic Ca^{2+} concentration via a set of specific transporters and pores. Several different mechanisms exist to achieve intracellular Ca^{2+} rise within cells for the purposes of communicating environmental signals into cells. Recently, mitochondria are known not only for supplying the energy used by the cell through oxidative phosphorylation, but also for their role in the induction of cell death through apoptosis (Kroemer et al., 1997).

The morphology of mitochondria is dynamic, often changing within a cell and from one cell type to the next. Recently, it has become clear that most mitochondria have more complex morphologies, ranging from long interconnected tubules to individual small spheres (Bereiter-Hahn and Voth, 1994). This highly dynamic phenomenon is regulated by a balance between two opposing processes: mitochondrial fission and mitochondrial fusion (Rismanchi and Blackstone, 2007). These dynamic processes are believed to ensure an appropriate distribution of mitochondria during cell proliferation and provide sufficient energy to a localized cytoplasmic region. However, little was known about molecular mechanisms regulating mitochondrial number and morphology. However, combined genetic, biochemical, and microscopic approaches have revealed the existence of a number of proteins participating in mitochondrial fusion and fission. Mediators of mitochondrial fission and fusion are best described in yeast, but many yeast mediators have mammalian homologues. In mammals, mitochondrial fission involves at least two proteins: dynamin-like protein 1 (DLP1, also referred to as Drp1) and a small molecule, Fis1. On the other hand, mitochondrial fusion is regulated by three large GTPase proteins: mitofusin 1 (Mfn1), mitofusin 2 (Mfn2), and optic atrophy protein 1 (OPA1). At steady state, the frequencies of fusion and fission events are balanced to maintain the overall morphology of the mitochondrial population. When this balance is perturbed, dramatic transitions in mitochondrial shape can occur (Nunnari et al., 1997). Unbalance between mitochondrial fission and fusion leads to abnormal mitochondrial morphology such as excessive mitochondrial fission and mitochondrial fragmentation and also perturbs the mitochondrial function. Normal mitochondrial function is necessary for normal respiratory function in cells (Karbowski and Youle, 2003). We studied the toxic effect of METH on mitochondrial dynamics in

SH-SY5Y cells. The results of our study demonstrated that METH toxicity decreased cell viability and increased the mitochondrial fission protein; Fis1 levels and oligomerization of Drp1 and these toxic effects of METH can be diminished by melatonin. Mitochondrial morphological study using transmission electron microscopy demonstrated that METH-treated cells exhibited a large proportion of small globular mitochondrial structures (Figure 30.2c, arrows), but control-untreated cells and METH-treated cells plus melatonin exhibited a large proportion of tubular networks of mitochondria (Figure 30.2a, arrows and Figure 30.2e, arrows, respectively) (Parameyong el al., 2013).

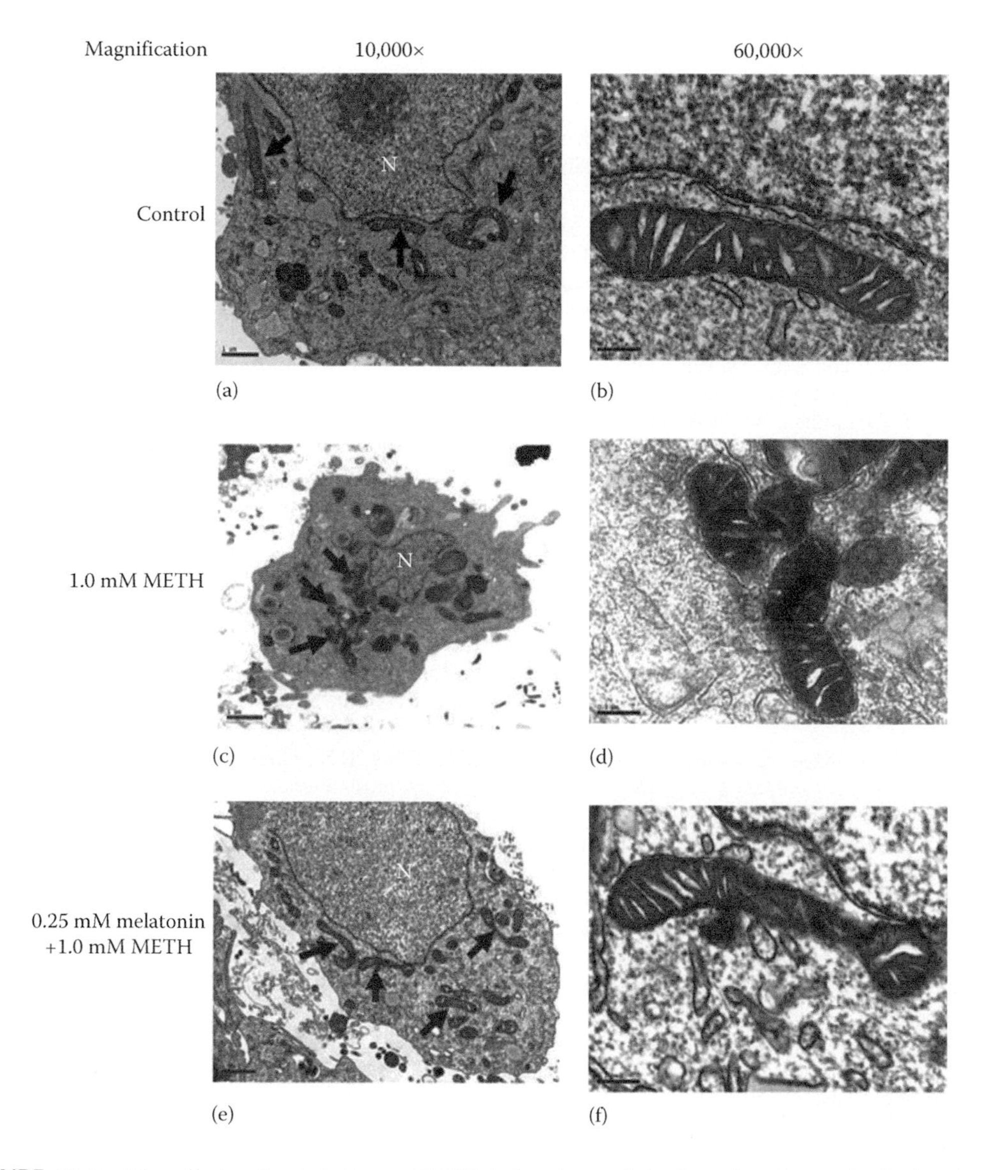

FIGURE 30.2 The effects of melatonin on METH-induced morphological alterations of mitochondria in SH-SY5Y cells. The control-untreated cells were incubated with culture medium for 24 h. Some cells were treated with 1.0 mM METH for 24 h with or without pretreatment with 0.25 mM melatonin for 1 h. The mitochondrial morphology was visualized under a transmission electron microscope. N, nucleus; Scale bar = 1.0 μm (10,000×) and 0.2 μm (60,000×). (From Parameyong, A., Charngkaew, K., Govitrapong, P., Chetsawang, B.: Melatonin attenuates methamphetamine-induced disturbances in mitochondrial dynamics and degeneration in neuroblastoma SH-SY5Y cells. *J. Pineal Res.* 2013. 55. 313. Copyright Wiley-VCH Verlag GmbH & Co. KGaA. Reproduced with permission.)

Our study has revealed that METH toxicity causes disturbances in mitochondrial dynamics in SH-SY5Y cells by induced mitochondrial fission pathways leading to mitochondrial fragmentation into small globular structures. In addition, melatonin exhibits the potential role to restore mitochondrial homeostasis mechanism in neurons affected by METH-induced toxicity.

ACKNOWLEDGMENTS

This work was supported by a research grant from the Thailand Research Fund and Mahidol University.

REFERENCES

Acikgoz, O., Gonenc, S., Kayatekin, B.M., Uysal, N., Pekcetin, C., Semin, I., and Gure, A. 1998. Methamphetamine causes lipid peroxidation and an increase in superoxide dismutase activity in the rat striatum. *Brain Res* 813:200–202.

Acuna-Castroviejo, D., Martin, M., Macias, M., Escames, G., Leon, J., Khaldy, H., and Reiter, R.J. 2001. Melatonin, mitochondria, and cellular bioenergetics. *J Pineal Res* 30:65–74.

Alvira, D., Tajes, M., Verdaguer, E., Acuna-Castroviejo, D., Folch, J., Camins, A., and Pallas, M. 2006. Inhibition of the cdk5/p25 fragment formation may explain the antiapoptotic effects of melatonin in an experimental model of Parkinson's disease. *J Pineal Res* 40:251–258.

Bereiter-Hahn, J. and Voth, M. 1994. Dynamics of mitochondria in living cells: Shape changes, dislocations, fusion, and fission of mitochondria. *Microsc Res Tech* 27:198–219.

Bevilaqua, L., Gramhan, M., and Dunkley, P. 2001. Phosphorylation of Ser19 alters the conformation of tyrosine hydroxylase to increase the rate of phosphorylation of Ser40. *J Bio Chem* 276:40411–40416.

Brenner, C., Cadiou, H., Vieira, H.L., Zamzami, N., Marzo, I., Xie, Z., Andrews, B.L., Duclohier, H., Reed, J.C., and Kroemer, G. 2000. Bcl-2 and Bax regulate the channel activity of the mitochondrial adenine nucleotide translocator. *Oncogene* 19:329–336.

Cadet, J.L., Jayanthi, S., and Deng, X. 2005. Methamphetamine-induced neuronal apoptosis involves the activation of multiple death pathways. Review. *Neurotox Res* 8:199–206.

Cadet, J.D., Krasnova, I.N., Jayanthi, S., and Lyles, J. 2007. Neurotoxicity of substituted amphetamines: Molecular and cellular mechanisms. *Neurotox Res* 11:183–202.

Camello-Almaraz, C., Gomez-Pinilla, P.J., Pozp, M.J., and Camello, P.J. 2008. Age-related alterations in Ca^{2+} signals and mitochondrial membrane potential in exocrine cells are prevented by melatonin. *J Pineal Res* 45:191–198.

Cao, G., Xing, J., Xiao, X., Liou, A.K., Gao, Y., Yin, X.M., Clark, R.S., Graham, S.H., and Chen, J. 2007. Critical role of calpain I in mitochondrial release of apoptosis-inducing factor in ischemic neuronal injury. *J Neurosci* 27:9278–9293.

Chen, H.M., Lee, Y.C., Huang, C.L., Liu, H.K., Liao, W.C., Lai, W.L., Lin, Y.R., and Huang, N.K. 2007. Methamphetamine downregulates peroxiredoxins in rat pheochromocytoma cells. *Biochem Biophys Res Commun* 354:96–101.

Cheng, E.H., Wei, M.C., Weiler, S., Flavell, R.A., Mak, T.W., Lindsten, T., and Korsmeyer, S.J. 2001. BCL-2, BCL-X(L) sequester BH3 domain-only molecules preventing BAX- and BAK-mediated mitochondrial apoptosis. *Mol Cell* 8:705–711.

Chera, B., Schaecher, K.E., Rocchini, A., Imam, S.Z., Sribnick, E.A., Ray, S.K., Ali, S.F., and Banik, N.L. 2004. Immunofluorescent labeling of increased calpain expression and neuronal death in the spinal cord of 1-methyl-4-phenyl-1,2,3,6-tetrahydropyridine-treated mice. *Brain Res* 1006:150–156.

Choi, W.S., Lee, E.H., Chung, C.W., Jung, Y.K., Jin, B.K., Kim, S.U., Oh, T.H., Saido, T.C., and Oh, Y.J. 2001. Cleavage of Bax is mediated by caspase-dependent or -independent calpain activation in dopaminergic neuronal cells: Protective role of Bcl-2. *J Neurochem* 77:1531–1541.

Crocker, S.J., Smith, P.D., Jackson-Lewis, V., Lamba, W.R., Hayley, S.P., Grimm, E., Callagham, S.M. et al. 2003. Inhibition of calpains prevents neuronal and behavioral deficits in an MPTP mouse model of Parkinson's disease. *J Neurosci* 23:4081–4091.

Cunha-Oliveira, T., Rego, A.C., Cardoso, S.M., Borges, F., Swerdlow, R.H., Macedo, T., and de Oliviera, C.R. 2006. Mitochondrial dysfunction and caspase activation in rat cortical neurons treated with cocaine or amphetamine. *Brain Res* 1089:44–54.

Davidson, C., Gow, A.J., Lee, T.H., and Ellinwood, E.H. 2001. Methamphetamine neurotoxicity: Necrotic and apoptotic mechanisms and relevance to human abuse and treatment. *Brain Res Rev* 36:1–22.

Dufty, B.M., Warner, L.R., Hou, S.T., Jiang, S.X., Gomez-Isla, T., Leenhouts, K.M., Oxford, J.T., Feany, M.B., Masliash, E., and Rohn, T. 2007. Calpain-cleavage of alpha-synuclein: Connecting proteolytic processing to disease-linked aggregation. *Am J Pathol* 170:1725–1738.

Ebadi, M., Srinivasan, S.K., and Baxi, M.D. 1996. Oxidative stress and antioxidant therapy in Parkinson's disease. *Prog Neurobiol* 48:1–19.

Er, E., Oliver, L., Cartron, P.F., Juin, P., Manon, S., and Vallette, F.M. 2006. Mitochondria as the target of the pro-apoptotic protein Bax. *Biochim Biophys Acta* 1757:1301–1311.

Fleckenstein, A.E., Volz, T.J., Riddle, E.L., Gibb, J.W., and Hanson, G.R. 2007. New insights into the mechanism of action of amphetamines. *Annu Rev Pharmacol Toxicol* 47:681–698.

Galano, A., Tan, D.X., and Reiter, R.J. 2011. Melatonin as a naturally against oxidative stress: A physicochemical examination. *J Pineal Res* 51:1–16.

Galano, A., Tan, D.X., and Reiter, R.J. 2013. On the free radical scavenging activities of melatonin's metabolites, AFMK and AMK. *J Pineal Res* 54:245–257.

Gao, G. and Dou, Q.P. 2000. N-terminal cleavage of Bax by calpain generates a potent proapoptotic 18-kDa fragment that promotes Bcl-2-independent cytochrome C release and apoptotic cell death. *J Cell Biochem* 80:53–72.

Gluck, M.R., Moy, L.Y., Jayatilleke, E., Hogan, K.A., Manzino, L., and Sonsalla, P.K. 2001. Parallel increases in lipid and protein oxidative markers in several mouse brain regions after methamphetamine treatment. *J Neurochem* 79:152–160.

Goping, I.S., Gross, A., Lavoie, J.N., Nguyen, M., Jemmerson, R., Roth, K., Korsmeyer, S.J., and Shore, G.C. 1998. Regulated targeting of BAX to mitochondria. *J Cell Biol* 143:207–215.

Grant, R.J., Sellings, L.H., Crocker, S.J., Melloni, E., Park, D.S., and Clarke, P.B. 2009. Effects of calpain inhibition on dopaminergic markers and motor function following intrastriatal 6-hydroxydopamine administration in rats. *Neuroscience* 158:558–569.

Guillot, T.S., Shepherd, K.R., Richardson, J.R., Wang, M.Z., Li, Y., Piers, C.E., and Gary, W.M. 2008. Reduced vesicular storage of dopamine exacerbates methamphetamine-induced neurodegeneration and astrogliosis. *J Neurochem* 106:2205–2217.

Jayanthi, S., Deng, X., Bordelon, M., McCoy, M.T., and Cadet, J.L. 2001. Methamphetamine causes differential regulation of pro-death and anti-death Bcl-2 genes in the mouse neocortex. *FASEB J* 15:1745–1752.

Jayanthi, S., Ladenheim, B., and Cadet, J.L. 1998. Methamphetamine-induced changes in antioxidant enzymes and lipid peroxidation in copper/zinc-superoxide dismutase transgenic mice. *Ann N Y Acad Sci* 844:92–102.

Jou, M.J., Peng, T.I., Reiter, R.J., Jou, S.B., Wu, H.Y., and Wen, S.T. 2004. Visualization of the antioxidative effects of melatonin at the mitochondrial level during oxidative stress-induced apoptosis of rat brain astrocytes. *J Pineal Res* 37:55–70.

Kaewsuk, S., Sae-Ung, K., Phansuwan-Pujito, P., and Govitrapong, P. 2009. Melatonin attenuates methamphetamine-induced reduction of tyrosine hydroxylase, synaptophysin and growth-associated protein-43 levels in the neonatal rat brain. *Neurochem Int* 55:397–405.

Karbowski, M. and Youle, R.J. 2003. Dynamics of mitochondrial morphology in healthy cells and during apoptosis. *Cell Death Differ* 10:870–880.

Klongpanichapak, S., Phansuwan-Pujito, P., Manuchair, E., and Govitrapong, P. 2008. Melatonin inhibits amphetamine-induced increase in alpha-synuclein and decrease in phosphorylated tyrosine hydroxylase in SK-N-SH cells. *Neurosci Lett* 436:309–313.

Kluck, R.M., Esposti, M.D., Perkings, G., Renken, C., Kuwana, T., Bossy-Wetzel, E., Goldberg, M. et al. 1999. The pro-apoptotic proteins, Bid and Bax, cause a limited permeabilization of the mitochondrial outer membrane that is enhanced by cytosol. *J Cell Biol* 147:809–822.

Krasnova, I.N. and Cadet, J.L. 2009. Methamphetamine toxicity and messengers of death. *Brain Res Rev* 60:379–407.

Kroemer, G., Zamzami, N., and Susin, S.A. 1997. Mitochondrial control of apoptosis. *Immunol Today* 18:44–51.

Larsen, K.E., Fon, E.A., Hastings, T.G., Edwards, R.H., and Sulzer, D. 2002. Methamphetamine-induced degeneration of dopaminergic neurons involves autophagy and upregulation of dopamine synthesis. *J Neurosci* 22:8951–8960.

LaVoie, M.J. and Hastings, T.G. 1999. Dopamine quinone formation and protein modification associated with the striatal neurotoxicity of methamphetamine: Evidence against a role for extracellular dopamine. *J Neurosci* 19:1484–1491.

Lazzeri, G., Lenzi, P., Busceti, C.L., Ferrucci, M., Falleni, A., Bruno, V., Paparelli, A., and Fornai, F. 2007. Mechanisms involved in the formation of dopamine-induced intracellular bodies within striatal neurons. *J Neurochem* 101:1414–1427.

Leon, J., Acuna-Castroviejo, D., Sainz, R.M., Mayo, J.C., Tan, D.X., and Reiter, R.J. 2004. Melatonin and mitochondrial function. *Life Sci* 75:765–790.

Lode, H., Bruchelt, G., Seitz, J., Gebhart, S., Gekeler, V., Niethammer, D., and Beck, J. 1995. Reverse transcriptase-polymerase chain reaction (RT-PCR) analysis of monoamine transporters in neuroblastoma cell lines: Correlations to meta-iodobenzylguanidine (MIBG) uptake and tyrosine hydroxylase gene expression. *Eur J Cancer* 31:586–590.

Manda, K., Ueno, M., and Anzai, K. 2007. AFMK, a melatonin metabolite, attenuates X-ray-induced oxidative damage to DNA, proteins and lipids in mice. *J Pineal Res* 42:386–393.

Martin, M., Macias, M., Escames, G., Riter, R.J., Agapito, M.T., Ortiz, G.G., and Acuna-Castroviejo, D. 2000. Melatonin-induced increased activity of the respiratory chain complexes I and IV can prevent mitochondrial damage induced by ruthenium red in vivo. *J Pineal Res* 28:242–248.

Martin, M., Macias, M., Leon, J., Escames, G., Khaldy, H., and Acuna-Castroviejo, D. 2002. Melatonin increases the activity of the oxidative phosphorylation enzymes and the production of ATP in rat brain and liver mitochondria. *Int J Biochem Cell Biol* 34:348–357.

McBride, H.M., Neuspiel, M., and Wasiak, S. 2006. Mitochondria: More than just a powerhouse. *Curr Biol* 16:R551–R560.

McCann, U.D., Kuwabara, H., and Kumar, A. 2008. Persistent cognitive and dopamine transporter deficits in abstinent methamphetamine users. *Synapse* 62:91–100.

Melega, W.P., Lacan, G., Harvey, D.C., and Way, B.M. 2007. Methamphetamine increases basal ganglia iron to levels observed in aging. *Neuroreport* 18:1741–1745.

Menendez-Pelaez, A. and Reiter, R.J. 1993. Distribution of melatonin in mammalian tissues: The relative importance of nuclear versus cytosolic localization. *J Pineal Res* 15:59–69.

Nakagawa, T. and Yuan, J. 2000. Cross-talk between the cysteine protease families activation of caspase-12 by calpain in apoptosis. *J Cell Biol* 150:887–894.

Nechushtan, A., Smith, C.L., Hsu, Y.-T., and Youle, R.J. 1999. Conformation of the Bax C-terminus regulates subcellular location and cell death. *EMBO J* 18:2330–2341.

Nixon, R.A. 2000. A "protease activation cascade" in the pathogenesis of Alzheimer's disease. *Ann N Y Acad Sci* 921:117–131.

Nunnari, J., Marshall, W.F., Straight, A., Murray, A., Sedat, J.W., and Walter, P. 1997. Mitochondrial transmission during mating in *Saccharomyces cerevisiae* is determined by mitochondrial fusion and fission and the intramitochondrial segregation of mitochondrial DNA. *Mol Biol Cell* 8:1233–1242.

Oshima, M., Koizumi, S., Fujita, K., and Guroff, G. 1989. Nerve growth factor-induced decrease in the calpain activity of PC12 cells. *J Biol Chem* 264:20811–20816.

Ozaki, Y., Blomgren, K., Ogasawara, M.S., Aoki, K., Furuno, T., Nakanishi, M., Sasaki, M., and Suzumori, K. 2001. Role of calpain in human sperm activated by progesterone for fertilization. *Biol Chem* 382:831–838.

Parameyong, A., Charngkaew, K., Govitrapong, P., and Chetsawang, B. 2013. Melatonin attenuates methamphetamine-induced disturbances in mitochondrial dynamics and degeneration in neuroblastoma SH-SY5Y cells. *J Pineal Res* 55:313–323.

Pirompul, N., Govitrapong, P., and Chetsawang, B. 2013. Farnesyltransferase inhibitor attenuates methamphetamine toxicity-induced Ras proteins activation and cell death in neuroblastoma SH-SY5Y cells. *Neurosci Lett* 545:138–143.

Ray, S.K. and Banik, N.L. 2003. Calpain and its involvement in the pathophysiology of CNS injuries and diseases: Therapeutic potential of calpain inhibitors for prevention of neurodegeneration. *Curr Drug Targets CNS Neurol Disord* 2:173–189.

Ray, S.K., Shields, D.C., Saido, T.C., Matzelle, D.C., Wilford, G.G., Hogan, E.L., and Banik, N.L. 1999. Calpain activity and translational expression increased in spinal cord injury. *Brain Res* 816:375–380.

Richards, M. and Sadee, W. 1986. Human neuroblastoma cell lines as models of catechol uptake. *Brain Res* 384:132–137.

Rismanchi, N. and Blackstone, C. 2007. Mitochondrial function and dysfunction in the nervous system. In *Molecular Neurology*, ed. S. Waxman, pp. 29–41. San Diego, CA: Elsevier.

Samantaray, S., Knaryan, V.H., and Banik, N.L. 2007. Parkinsonian neurotoxin rotenone activated calpain and caspase-3 leading to motoneuron degeneration in spinal cord of Lewis rats. *Neuroscience* 146:741–755.

Samantaray, S., Sribnick, E.A., Das, A., Knaryan, V.H., Matzelle, D.D., Yallapragada, A.V., Reiter, R.J., Ray, S.K., and Banik, N.L. 2008. Melatonin attenuates calpain upregulation, axonal damage and neuronal death in spinal cord injury in rats. *J Pineal Res* 44:348–357.

Sharpe, J.C., Arnoult, D., and Youle, R.J. 2004. Control of mitochondrial permeability by Bcl-2 family members. *Biochem Biophys Acta* 1644:107–113.

Sorimachi, H. and Suzuki, K. 1998. Calpain as a Ca^{2+} binding protein. *FEBS Lett* 43:1666–1674.

Staszewski, R.D. and Yamamoto, B.K. 2006. Methamphetamine-induced spectrin proteolysis in the rat striatum. *J Neurochem* 96:1267–1276.

Suwanjang, W., Phansuwan-Pujito, P., Govitrapong, P., and Chetsawang, B. 2010. The protective effect of melatonin on methamphetamine-induced calpain-dependent death pathway in human neuroblastoma SH-SY5Y cultured cells. *J Pineal Res* 48:94–101.

Suzuki, M., Youle, R.J., and Tjandra, N. 2000. Structure of Bax: Coregulation of dimer formation and intracellular localization. *Cell* 103:645–654.

Thomas, D.M., Francescutti-Verbeem, D.M., and Kuhn, D.M. 2009. Increases in cytoplasmic dopamine compromise the normal resistance of the nucleus accumbens to methamphetamine neurotoxicity. *J Neurochem* 109:1745–1755.

Thomas, S.G., Kennie, R.S., Jason, R.R., Min, Z.W., Yingjie, L., Piers, C.E., and Gary, W.M. 2008. Reduced vesicular storage of dopamine exacerbates methamphetamine-induced neurodegeneration and astrogliosis. *J Neurochem* 106:2205–2217.

Tsujimoto, Y. and Shimizu, S. 2000. Bcl-2 family: Life-or-death switch. *FEBS Lett* 466:6–10.

Vaisid, T., Barnoy, S., and Kosower, N.S. 2008. Calpastatin overexpression attenuates amyloid beta-peptide toxicity in differentiated PC12 cells. *Neuroscience* 156:921–931.

Vaisid, T., Kosower, N.S., Katzav, A., Chapman, J., and Barnoy, S. 2007. Calpastatin levels affect calpain activation and calpain proteolytic activity in APP transgenic mouse model of Alzheimer's disease. *Neurochem Int* 51:391–397.

Veronika, F., Martina, B., and Frantisek, J. 2008. Truncation of human dopamine transporter by protease calpain. *Neurochem Int* 52:1436–1441.

Vollm, B.A., de Araujo, I.E., Cowen, P.J., Rolls, E.T., Kringelbach, M.L., Smith, K.A., Jezzard, P., Heal, R.J., and Matthews, P.M. 2004. Methamphetamine activates reward circuitry in drug naive human subjects. *Neuropsychopharmacology* 29:1715–1722.

Vosler, P.S., Sun, D., Wang, S., Gao, Y., Kintner, D.B., Signore, A.P., Cao, G., and Chen, J. 2009. Calcium dysregulation induces apoptosis-inducing factor release: Cross-talk between PARP-1- and calpain-signaling pathways. *Exp Neurol* 218:213–220.

Wisessmith, W., Phansuwan-Pujito, P., Govitrapong, P., and Chetsawang, B. 2009. Melatonin reduces induction of Bax, caspase and cell death in methamphetamine-treated human neuroblastoma SH-SY5Y cultured cells. *J Pineal Res* 46:433–440.

Wolter, K.G., Hsu, Y.T., Smith, C.L., Nechushtan, A., Xi, X.G., and Youle, R.J. 1997. Movement of Bax from the cytosol to mitochondria during apoptosis. *J Cell Biol* 139:1281–1292.

Wu, C.W., Ping, Y.H., Yen, J.C., Chang, C.Y., Wang, S.F., Yeh, C.L., Chi, C.W., and Lee, H.C. 2007. Enhanced oxidative stress and aberrant mitochondrial biogenesis in human neuroblastoma SH-SY5Y cells during methamphetamine induced apoptosis. *Toxicol Appl Pharmacol* 220:243–251.

Yeung, H.M., Hung, M.W., and Fung, M.L. 2008. Melatonin ameliorates calcium homeostasis in myocardial and ischemia-reperfusion injury in chronically hypoxic rats. *J Pineal Res* 45:373–382.

Zhang, X., Chen, Y., Jenkins, L.W., Kochanek, P.M., and Clark, R.S. 2005. Bench-to-bedside review: Apoptosis/programmed cell death triggered by traumatic brain injury. *Crit Care* 9:66–75.

Zimmermann, K.C., Bonzon, C., and Green, D.R. 2001. The machinery of programmed cell death. *Pharmacol Ther* 92:57–70.

31 Melatonin's Beneficial Effects on Dopaminergic Neurons in the Rat Model of Parkinson's Disease

Manisha Gautam, Reshu Gupta, Ravinder Kumar Saran, and Bal Krishana

CONTENTS

31.1 INTRODUCTION

Melatonin is an endogenous neurohormone produced predominantly in the pineal gland. Administration of exogenous melatonin was found useful for abnormal sleep parameters in autism spectrum disorders (Rossignol and Frye, 2013). Melatonin plays an important role in the process of flap viability in the dorsal skin flap rat model (Kerem et al., 2013). It suppresses nitric oxide formation in glial cultures by proinflammatory cytokines through p38 mitogen–activated protein kinase inhibition (Vilar et al., 2013). Clinical trial of antenatal maternally administered melatonin decreases the level of oxidative stress in human pregnancies affected by preeclampsia (Hobson et al., 2013). Melatonin therapy is effective in protecting rats against the harmful cardiac inflammatory response that is characteristic of chronic *Trypanosoma cruzi* infection

(Oliveira et al., 2013). It was found to prevent cancer tumorigenesis and altered cancer correlates, such as sleep–wake and mood disturbances (Rondanelli et al., 2013). It attenuates dexamethasone-induced spatial memory impairment and dexamethasone-induced reduction of synaptic protein expressions in the mouse brain (Tongjaroenbuangam et al., 2013). It may be used to dampen interleukin-17-mediated inflammation that is enhanced by the increased levels of insulin and insulin-like growth factor 1 in obesity (Ge et al., 2013). Melatonin-enhanced autophagy protects against neural apoptosis via amitochondrial pathway in early brain injury following a subarachnoid hemorrhage in rats (Chen et al., 2013). It is a useful treatment to delay the cellular and behavioral alterations observed in PD (Gutierrez-Valdez et al., 2012). It was reported as a neuroprotective agent in the rodent models of PD (Singh et al., 2006; Singhal et al., 2012). It inhibits 6-hydroxydopamine (6-OHDA) production in the brain to protect against experimental parkinsonism in the mouse striatum (Borah and Mohanakumar, 2009). High treatment efficacy of melatonin was found useful in the treatment of sleep disorders in PD without dementia (Litvinenko et al., 2012).

The earlier-mentioned studies focused on the use of young animals to examine the in vivo effects of melatonin. However, since the degeneration of substantia nigra dopaminergic 9DA0 neurons that occurs in PD is more often than not confined to elderly individuals, it is of interest to determine whether the beneficial effects of melatonin against 6-OHDA in young adult rats can be extended to aged animals. Therefore, in the present study, beneficial effects of melatonin were investigated in the aged rat model of PD. Both behavioral and histochemical changes were examined after animals were pretreated with melatonin and subsequently administered the neurotoxin 6-OHDA into the rat striatum. No information available so far on the beneficial potential of melatonin in 6-OHDA-induced aged rat model of PD.

31.2 MATERIALS AND METHODS

31.2.1 Animals

Forty Sprague–Dawley rats, 3–4 months, young rats (weight 180 ± 50 g; (Margaret et al., 1998)) and 22–24 months, aged rats (weight 280 ± 50 g; Ghosh et al., 1984)) were housed in polyethylene cages with food and water available ad libitum in a temperature-controlled room (23 ± 3°C) with a 12:12 h light:dark cycle. The animal's procedure was in accordance with guidelines of the principles of Institutional Animal Ethical Committee for care and use of laboratory animals.

31.2.2 Animal Groups

The animals were randomly allocated to four groups of 10 rats each. Group 1 is the aged rat sham control group. Aged rats of group 2 received 8 μg 6-OHDA into the left striatum. Young rats of group 3 received first melatonin (500 μg/kg body weight/1 mL saline, i.p.) for seven consecutive days, and then 8 μg 6-OHDA was given into left striatum. Aged rats of group 4 received first melatonin (500 μg/kg body weight/1 mL saline, i.p.) for seven consecutive days, and then 8 μg 6-OHDA was injected into left striatum.

31.2.3 6-OHDA Lesions

Animals were anesthetized with ketamine–xylazine (Sigma, 50–100 mg/kg; 5–10 mg/kg, i.p.). Animals were placed into a David Kopf stereotaxic frame (INCO, Ambala). A lesion was made at the following coordinates: anterior–posterior, 0 mm; lateral, 3.5 mm; dorsoventral to the dura, 5.5 mm; from bregma. The rat tooth position on the tooth bar of the stereotaxis apparatus was fixed at 3.3 mm. After scalp incision, a burr hole was drilled over the injection sites, and a blunted 26-gauge cannula, connected to a 10 μL Hamilton syringe, was lowered to the injection site. The 6-OHDA solution was injected over a 4 min period, and the needle was left in place for an additional 5 min before retraction (Paxinos and Watson, 1986).

31.2.4 Quantitation of Rotational Behavior

Rats were tested for apomorphine-induced rotations in response to apomorphine (Sigma, 0.05 mg/kg, s.c.) at the base line (prelesion) and after 5 weeks of 6-OHDA-induced lesion (postlesion) in all groups for 30 min duration (Ungerstedt and Arbuthnott, 1970; Singh et al., 2006) by the help of rota count 8 (Columbus Instruments, USA).

31.2.5 Staircase Test

The staircase test was used to assess skilled forelimb use in the rat (Montoya et al., 1991). Six pellets (45 mg) were placed on both sides of every step, giving a total of 30 pallets available on each side. The rat was allowed to attempt to acquire as many food pellets as possible using each forelimb independently within a 15 min period. The number of pellets eaten by the rat was counted after each test period. Scores were assigned after 15 min intervals for both the left and right forelimbs. All the rats were tested for six consecutive days. In each test session, several measures were evaluated: the number of pellets eaten (successful reaches), the number of pellets taken (pellets eaten plus pellets grasped but dropped), and the success rate% (pellets eaten divided by pellets taken and multiplied by 100) were calculated separately.

31.2.6 Stepping Test

Each rat was held so that both hind limbs and one forelimb were raised off the surface of the table and held steady by the experimenter. The animal was moved across the surface of the wooden plank in such a way that the animal must bear its weight on the remaining forepaw. The number of adjusting steps were counted while the rat was slowly moved sideways along the wooden plank (0.9 m wooden plank at a consistent speed in 30 s), first in the forehand direction, that is, movement of the paw toward the torso to compensate for an outward lateral movement of the body, and then in the backhand direction, that is, movement of the paw away from the torso to compensate for the inward medial movement of the body (Schallert et al., 1979).

31.2.7 Initiation Time

Each rat was held with both hind limbs and one forelimb raised off the surface of the table and held steady by the experimenter. The rat was then moved across the surface of the table in such a way that the animal must bear its weight on the remaining forelimb. The time to actively initiate a forelimb movement was determined in these test sessions in the same way as for the stepping test. The time elapsed before the rat actively initiated movement with the unrestrained forelimb and started to step forward along the ramp (1.1 m) toward the home cage was recorded (Olsson et al., 1995).

31.2.8 Postural Balance Test

In this test, the rat was held in the same position as described in the stepping test, and the rat was titled by the experimenter toward the side of the paw touching the table. This resulted in loss of balance. The ability of the rat to regain balance by the tilting movement of the forelimb was monitored by a scoring system ranging from 0 to 3. Score 0 was recorded when the rat fell onto the side and there was no detectable muscle reaction in the forelimb. Score 1 represented a clear forelimb reaction, as seen by muscle concentration, but lack of success in recovering balance, for example, the rat still fell onto the side. Score 2 was given when the rat showed an incomplete recovery of balance, for example, the rat performed clear forelimb movement, but the placement of the paw, compared with the control rat, displayed digits not plainly split on the table but partially crossed over one another. Score 3 was given for a normal forelimb placement movement and total recovery of balance, similar to unlesioned control rats. The test was repeated six times a day on each side for consecutive days, giving a maximum score of 18 at each of the three tests based. Final results were expressed as the mean of the 3 days (Winkler et al., 1996).

31.2.9 Disengage Behavior Test

A blunt wooden probe touched the perioral region beneath the vibrissae of the rat repeatedly at 1 s intervals when the rat was engaged in eating a piece of milk chocolate. The latency of the orienting response, that is, turning of the head toward the stimulus, was recorded; an immediate response was scored as 1 s. Stimulation was discontinued if the rat did not respond within a period of 180 s. The rats were tested once daily on three consecutive days. Final results were expressed as the mean of the three tests (Schallert and Hall, 1988; Rozas and Labandeira García, 1997).

31.2.10 Cresyl Violet Staining

When the behavioral tests were complete, rats were sacrificed with an overdose of ketamine (200 mg/kg, i.p.) and perfused intracardially with heparin saline (0.1% heparin in 0.9% saline; 100 mL/rat) followed by paraformaldehyde (4% in phosphate buffer). Brains were removed and postfixed in 4% paraformaldehyde. Brain sections (5 μm cryostat coronal sections) were cut using a microtome. They were stained with cresyl violet and observed for neurons under microscope (Saji et al., 1988).

31.2.11 Tyrosine Hydroxylase Immunohistochemistry

Sections were treated for 10 min in 3% hydrogen peroxide, rinsed three times in 0.1 M PBS, and incubated in 2% normal goat serum with 0.1% Triton X-100 for 30 min prior to overnight incubation at 4°C with primary antibody (Sigma,1:500). The primary antibody utilized was rabbit anti-TH. After six washes in 0.1 M PBS (5 min each), sections were incubated in 0.1 M PB containing 1% normal goat serum and biotinylated goat anti-rabbit secondary antibody (1:200) for 60 min at 37°C. The sections were rinsed three times in PBS and incubated in avidin-biotinylated-peroxidase complex (Vectastain ABC-Elite kit, tertiary antibody) for 50 min at room temperature. Following thorough rinsing with PBS, staining was visualized by incubation in 3,3-diaminobenzidine solution with nickel enhancement. After immunostaining, sections were mounted and counterstained before dehydrating in ascending alcohol, cleared in xylene and coverslipped in DPX (Reum et al., 2002). To quantify the number of TH-positive neurons, the positive neurons were counted using a microscope (Olympus) connected to a camera (Olympus) at a magnification of 40×.

31.2.12 Glial Fibrillary Acidic Protein

Sections were deparaffinized in xylene (two or three changes) and were hydrated in alcohol (100%, 95%, and 70% to tap water) and then rinsed in distilled water. Sections were treated for 10 min in 3% hydrogen peroxide, rinsed three times in 0.1 M PBS, and incubated in 2% normal goat serum with 0.1% Triton X-100 for 30 min prior to overnight incubation at 4°C with primary antibody. The primary antibody utilized was rabbit antiglial fibrillary acidic protein (Sigma). After six washes in 0.1 M PBS (5 min each), sections were incubated in 0.1 M PB containing 1% normal goat serum and biotinylated goat antirabbit secondary antibody for 60 min at 37°C. The sections were rinsed three times in PBS and incubated in avidin-biotinylated-peroxidase complex (Vectastain ABC-Elite kit, tertiary antibody) for 50 min at room temperature. Following thorough rinsing with PBS, staining was visualized by incubation in 3,3-diaminobenzidine solution with nickel enhancement. This immunostaining allowed the determination of the extent of dopaminergic cell degeneration. After immunostaining, sections were mounted and counterstained before dehydrating in ascending alcohol, cleared in xylene and coverslipped in DPX (Segovia et al., 1998, Svingos et al., 1998).

31.2.13 Statistical Analysis

All values are presented as mean ± SEM. The significance of difference between prelesion and postlesion within the group was determined by paired Student's t test. ANOVA was used to find the significance of difference between the values of all the groups for prelesion and postlesion separately. A p-value of less than 0.05 was regarded as being statistically significant (Subbakrishna, 2000).

31.3 RESULTS

Rats subjected to melatonin and receiving stereotaxic injection of 6-OHDA had significant effects in various behavioral tests (Table 31.1). There were no statistical significant differences between prelesion values of group 1 (Student's t test, $p > 0.1$). Comparative analysis (Student's t test, $p < 0.001$) between prelesion values and postlesion values of group 2 was found to be highly significant in apomorphine-induced rotational behaviors, stepping tests, initiation times, staircase tests, disengage times, and postural balance tests. Statistically significant differences (Student's t test, $p < 0.001$) in various behavior tests were found between prelesion and postlesion values of group 3 in apomorphine-induced rotational behaviors, stepping tests, initiation times, disengage times, and $p < 0.05$ in staircase tests and postural balance tests. Significant differences (Student's t test, $p < 0.001$) between prelesion and postlesion values of group 4 were found in apomorphine-induced rotational behaviors, staircase test success rates, stepping tests, initiation times and disengage times, and $p < 0.05$ in postural balance tests. Statistical evaluations revealed significant differences between all the groups (ANOVA, $p < 0.001$) in apomorphine-induced rotational behaviors, staircase test success rates and disengage times, and (ANOVA, $p < 0.05$) in stepping tests, initiation times, and postural balance tests (Table 31.1).

Cresyl violet staining was performed in all the rat groups. The animals of group 1 exhibited normal dopaminergic neurons in striatal brain sections. Cresyl violet stained nissl bodies of the dopaminergic neurons. No neuronal loss was seen in these brain sections. Significant loss of striatal DA neurons was observed in group 2 animals. Injected 6-OHDA alone resulted in almost complete loss of DA neurons compared to group 1. Partially protected DA neurons were found in melatonin-treated animals of groups 3 and 4. The number of dopaminergic neurons in groups 3 and 4 was found significantly more than that in animals of group 2. However, there was a slight decrease in dopaminergic neurons in group 4 in comparison to that of group 3.

Tyrosine hydroxylase (TH) immunohistochemistry was also investigated in all groups of the rats. Microscopic examination of slides showed abundant dark brown–stained DA neurons in the brain sections of group 1. Dark brown–stained TH-immunoreactive neurons were greatly reduced in brain sections of group 2. It is due to the loss of normal dopaminergic neurons. The brain sections of group 3 and 4 showed abundant dark brown–stained neurons. The dark brown–colored neurons were found significantly more in groups 3 and 4 than that of group 2 (Figure 31.1). The aged rats showed less protection in neurons loss (40%) than that of young (60%) rats (Figure 31.2).

Glial fibrillary acidic protein (GFAP) immunohistochemistry was also performed in all groups. In aged control rats' brain sections, the resting form of astrocytes was found to be greatly decreased. After the 6-OHDA lesion, astrocytes are changed to activated forms. The density of these GFAP-ir astrocytes is also higher on the lesioned side compared with the nonlesioned side. Further, the number of GFAP-ir astrocytes is significantly more in aged rat model of PD than that of young rat (Figure 31.3).

31.4 DISCUSSION

Parkinson's disease is a progressive neurodegenerative disorder characterized by tremor, rigidity, bradykinesia, postural instability, and loss of DA neurons in substantia nigra pars compacta (Beal, 1998). It is one of the most common progressive neurodegenerative disorders, affecting over 1% of

TABLE 31.1
Effect of Melatonin on Behavior Tests in All Groups of Rats before and after 6-OHDA Lesion in Young (Group 3) and Aged Rats (Group 4)

Behavior Tests		Group 1		Group 2		Group 3		Group 4	
		Prelesion	Postlesion	Prelesion	Postlesion	Prelesion	Postlesion	Prelesion	Postlesion
Apomorphine-induced rotational		9±2	10±2	10±2	$256\pm18^{**}$	10±1	$134\pm10^{**}$	11±1	$156\pm8^{**\#\#}$
Staircase test	Taken	23±1	25±2	24±2	$11\pm1^{**}$	25±2	$20\pm3^{*}$	25±2	$17\pm3^{*\#}$
	Eaten	22±1	23±1	22±1	$5\pm1^{**}$	23±2	$16\pm2^{*}$	23±2	$14\pm2^{*\#}$
	Success rate	96±2	92±2	92±2	$33\pm3^{**}$	92±3	$80\pm4^{*}$	92±3	$72\pm4^{**\#\#}$
Stepping test	In forehand direction	11±1	12±1	12±2	$3\pm1^{**}$	12±1	$7\pm1^{**}$	11±2	$6\pm1^{**\#}$
	In backhand direction	14±1	15±1	16±1	$4\pm1^{**}$	13±1	$8\pm2^{**}$	13±1	$5\pm1^{**\#}$
Initiation time		4±1	5±1	3±1	$18\pm2^{**}$	5±1	$10\pm2^{**}$	6±1	$14\pm1^{**\#}$
Postural balance test score		14±1	16±1	13±1	$6\pm2^{**}$	14±3	$11\pm2^{*}$	14±3	$7\pm1^{*\#}$
Disengage behavior		5±1	4±1	5±1	$134\pm17^{**}$	6±1	$68\pm8^{**}$	9±1	$78\pm6^{**\#\#\#\#}$

* $p<0.05$ versus prelesion value of the same group. ** $p<0.001$ versus prelesion value of the same group.
\# $p<0.05$ versus postlesion value of group 1. ## $p<0.001$ versus postlesion value of group 1.

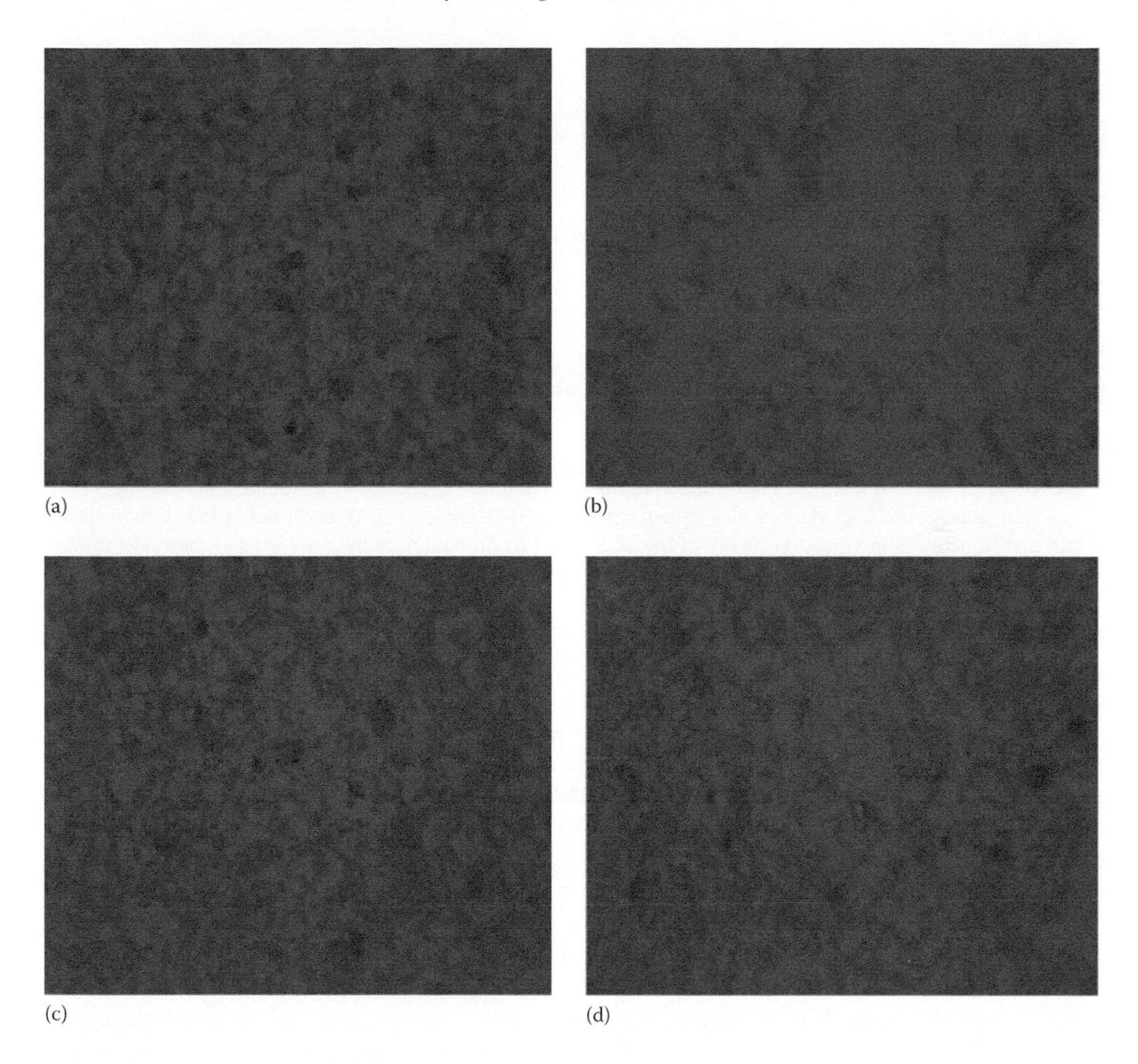

FIGURE 31.1 Photomicrographs of tyrosine hydroxylase immunohistochemistry in the striatal sections. Magnification 400×. (a) Abundant neurons are seen in control striatum of aged rats. (b) Neurons are almost absent in 6-OHDA-treated aged rats. (c) Protected neurons in melatonin-treated young rats prior to 6-OHDA injection. (d) Protected neurons in melatonin-treated aged rats prior to 6-OHDA injection.

the aged people in Western countries (Wirdefeldt et al., 2011). The etiology of DA neurons death is not known. However, reported data suggest oxidative stress as the probable candidate to mediate the original unknown cause. Similar changes were also produced by 6-OHDA-induced PD models (Schober, 2004).

The results of the current study showed that intraperitoneal administration of melatonin improved a hemi-Parkinson's condition in rats caused by intrastriatal application of the neurotoxin 6-OHDA. Various behavior and histological tests were used as an index of striatal dopaminergic function. Melatonin corrected these behaviors and histological tests in 6-OHDA-induced young and aged rat models of PD. Aged rats showed less neuroprotection in DA neuron loss than young rats.

These finding are consistent with the results of previous studies. In this regard, Singh et al. (2006) have reported that rats pretreated intraperitoneally with melatonin for a period of 7 days, and subsequently lesioned with 6-OHDA, showed decreases in the number of apomorphine-induced rotations, and improved postures, staircase tests, disengage times, stepping tests, and initiation times compared to 6-OHDA-lesioned animals treated with vehicle group. It was also reported that

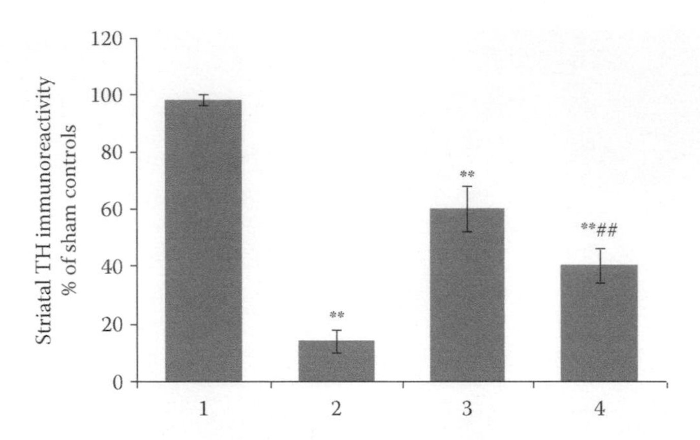

FIGURE 31.2 Neuroprotective effect of melatonin in young and aged rats. 1: histodiagram showing TH-ir neurons in sham control. 2: Marked loss of neurons in 6-OHDA-induced aged rat model of PD. 3: Significant protection of neurons in young rat model of PD. 4: Significant protection of neurons in aged rat model of PD. *Note*: ** P<0.001, highly significant, paired student t' test; ## P<0.001, highly significant, ANOVA.

6-OHDA-induced animals showed decreases in the number of steps in forehand and backhand directions, postural balance apomorphine-induced rotations, and initiation times. Hamdi (1998) reported that the striatum had a higher affinity of DA to D_2 receptors after oral administration of melatonin. It may be produced through conformational changes in the D_2 receptor–binding site. Therefore, melatonin may have a modulatory influence on the DA system (Aguiar et al., 2002). In other studies, administration of melatonin affected locomotor activity (Burton e al., 1991, Tenn and Niles, 1997), blocked L-dopa-induced movement disorder (Cotzias et al., 1971), and inhibited the apomorphine-induced rotation (Jenner et al., 1992; Pierrefiche and Laborut, 1995). It was also reported that melatonin was able to counteract the decrease in striatal TH immunoreactivity and the loss of complex I activity produced in rats after acute 6-OHDA administration (Joo et al., 1998; Dabbeni-Sala et al., 2001; Gutierrez-Valdez et al., 2012).

But there is controversy in earlier studies regarding the protective potential of melatonin in PD. Anton-Tay et al. reported improved motor activity in PD patients given high doses of melatonin (Anton-Tay et al., 1971). Later replication of studies employing melatonin treatment either failed to confirm such treatment effects (Shaw et al., 1973) or actually reported worsening by melatonin (Willis and Armstrong, 1999). In the original clinical study where melatonin was administered to PD patients (Anton-Tay et al., 1971), there were difficulties with experimental design and interpretation. Firstly, the definition of each patient's condition prior to treatment and their response after the treatment commenced and during drug withdrawal was not standardized (Guarduika-Lemaitre, 1997). Secondly, the doses of melatonin utilized were large (1.2 g/day) making a meaningful interpretation as to the treatment mechanism difficult to decipher (Guarduika-Lemaitre, 1997). Thirdly, the melatonin was mixed with 2% alcohol solution, a combination that will increase the sedative effect of melatonin (Willis and McLennan, 2001). Fourthly, a bolus of melatonin (5 mg/kg) given immediately after 6-OHDA stereotaxic injection failed to change apomorphine-induced contralateral rotation. This may be due to the short biological half-life (20 min) of melatonin (Dabbeni-Sala et al., 2001). These factors were improved in later studies. Melatonin was given 1 h after 6-OHDA lesion and thereafter daily for 7 days (Aguiar et al., 2002). Osmotic minipumps filled with a solution of melatonin were placed in the subcutaneous tissue between the scapulae. The delivery rate was constant at 1 ± 0.15 μL/h (50 ± 7.5 μg melatonin/h). This produced a plasma concentration of 1660 ± 240 pg melatonin/mL for at least 7 days (Dabbeni-Sala et al., 2001). In the present study, melatonin was injected for 7 days to increase melatonin bioavailability.

Inappropriate activation of apoptosis by dopamine and/or its oxidation products has been hypothesized to initiate nigral cell loss in PD (Offen et al., 1996). Increased reactive oxygen

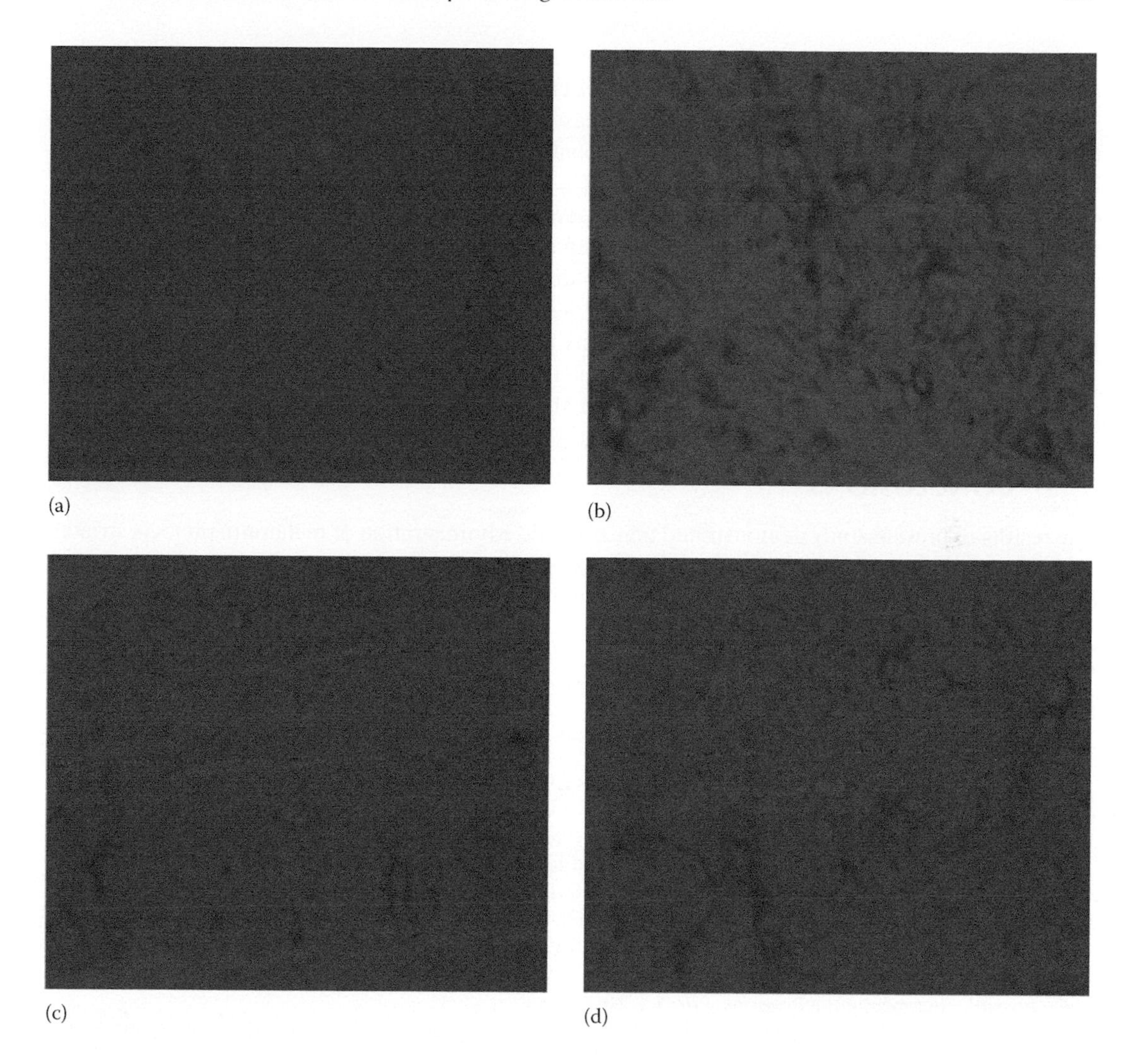

FIGURE 31.3 Photomicrographs of glial fibrillary acidic protein immunohistochemistry in the striatal sections. Magnification 400×. (a) Very less GFAP-ir astrocytes in control striatum of aged rats. (b) Numerous and activated GFAP-ir astrocytes in 6-OHDA-treated aged rats. (c) Numerous GFAP-ir astrocytes in melatonin-treated young rats prior to 6-OHDA injection. (d) GFAP-ir astrocytes in melatonin-treated aged rats prior to 6-OHDA injection.

species may participate in 6-OHDA neurotoxicity, as reported by decrease in brain glutathione (GSH) (e.g., −22% in striatum) and loss in superoxide dismutase (SOD) activity (−22% in striatum) (Perumal et al., 1992, Hodgson and Fridovich, 1975, Kumar et al., 1995). Furthermore, 6-OHDA appears to be more toxic to complex I than 1-methyl-4-phenylpyridinium ion (MPP+) (Glinka and Youdim, 1995). Inhibition of complex I stimulates mitochondrial formation of superoxide free radicals (Hasegawa et al., 1990), hydrogen peroxide, and hydroxyl radicals (Sewerynek et al., 1995) and may also trigger apoptotic mechanisms (Olanow and Tatton, 1999).

The neuroprotective effects of melatonin appear to be mediated by the antioxidant capacity. It stimulates antioxidant enzymes such as SOD, glutathione peroxidase, and glutathione reductase (Tomás-Zapico and Coto-Montes, 2005). Melatonin prevented an increase in lipid peroxidation and a decrease in TH immunoreactivity in the striatum after a single dose of MPTP, arguing that melatonin was able to prevent the damage caused by this drug in the striatal dopaminergic axons (Acuña-Castroviejo et al., 1996). It also prevented both cell damage and DNA fragmentation in MPTP-induced mouse brain (Ortiz et al., 2001). The 6-OHDA toxicity is based on the direct inhibition of complex I of the electron transport chain in the mitochondria (Glinka et al., 1997). This inhibition causes energy depletion and increases free radical concentration in the mitochondria (Tooyama et al., 1994). Melatonin was able to counteract the loss of complex I activity produced in rats after acute 6-OHDA administration (Joo et al., 1998; Dabbeni-Sala et al., 2001).

In the present study, the extent of protection against 6-OHDA-induced dopaminergic neurotoxicity is less in aged animals than that of young animals. It may be due to neurochemical and cellular changes in the nigrostriatal dopaminergic changes during aging. There are reductions in striatal levels of DA as well as DA receptors (Antolín et al., 2002). Neuronal loss in the substantia nigra reaches about 50% by the ninth decade of life in humans (Horvath and Davis, 1990). Reduction in high-affinity DA uptake sites (Allard and Marecusson, 1989), DA transporter messenger RNA (Bannon et al., 1992), and TH messenger RNA (Mattson et al., 1995) also become evident as individuals age.

31.5 CONCLUSIONS

The results of present study demonstrated that systemic administration of melatonin protects striatal dopaminergic neurons against 6-OHDA neurotoxicity in the aged rat. The effect was accompanied by a significant recovery in behavioral and histological tests. Further studies need to be done to find the melatonin's mechanism of neuroprotection.

REFERENCES

Acuña-Castroviejo, D., Coto-Montes, A., Monti M. G. et al. 1996. Melatonin is protective against MPTP-induced striatal and hippocampal lesions. *Life Sci.* 60(2):23–29.

Aguiar, L. M. V., Vasconcelos, S. M. M., Sousa, F. C. F. et al. 2002. Melatonin reverses neurochemical alterations induced by 6-OHDA in rat striatum. *Life Sci.* 70(9):1041–1051.

Allard, P. and Marecusson, J. O. 1989. Age-correlated loss of dopamine uptake sites labeled with [^{3}H] GBR 12935 in human putamen. *Neurobiol. Aging* 10:661–664.

Antolín, I., Mayo, J. C., Sainz, R. M. et al. 2002. Protective effect of melatonin in a chronic experimental model of Parkinson's disease. *Brain Res.* 943(2):163–173.

Anton-Tay, F., Diaz, J. L., and Fernandez-Guardiola, A. 1971. On the effect of melatonin upon human brain. Its possible therapeutic implications. *Life Sci.* 10:841–850.

Bannon, M. J., Poosch, M. S., Xia, Y. et al. 1992. Dopamine transporter mRNA content in human substantia nigra decreases precipitously with age. *Proc. Natl. Acad. Sci. USA* 89:7095–7099.

Beal, M. F. 1998. Mitochondrial dysfunction in neurodegenerative diseases. *Biochim. Biophys. Acta* 1366:211–223.

Borah, A. and Mohanakumar, K. P. 2009. Melatonin inhibits 6-hydroxydopamine production in the brain to protect against experimental parkinsonism in rodents. *J. Pineal Res.* 47(4):293–300.

Burton, S., Daya, S., and Potgeiter, B. 1991. Melatonin modulates apomorphine-induced rotational behaviour. *Experentia* 47:466–469.

Chen, J., Wang, L., Wu, C. et al. 2013. Melatonin-enhanced autophagy protects against neural apoptosis via amitochondrial pathway in early brain injury following a subarachnoid hemorrhage. *J. Pineal Res.* 56:12–19. doi: 10.1111/jpi.12086.

Cotzias, G. C., Tang, L. C., Miller, S. T. et al. 1971. Melatonin and abnormal movements induced by L-DOPA in mice. *Science* 173:450–452.

Dabbeni-Sala, F. S. D. S., Franceschini, D., Skaper, S. D. et al. 2001. Melatonin protects against 6-OHDA-induced neurotoxicity in rats: A role for mitochondrial complex I activity. *FASEB J.* 15(1):164–170.

Ge, D., Dauchy, R. T., Liu, S. et al. 2013. Insulin and IGF1 enhance IL-17-induced chemokine expression through a GSK3B-dependent mechanism: A new target for melatonin's anti-inflammatory action. *J. Pineal Res.* 55(4):377–387.

Ghosh, M. N. 1984. *Fundamentals of Experimental Pharmacology*. Calcutta, India: Scientific Book Agency.

Glinka, Y., Gassen, M., and Youdim, M. B. H. 1997. Mechanism of 6-hydroxydopamine neurotoxicity. *J. Neural Transm. Suppl.* 50:55–66.

Glinka, Y. Y. and Youdim, M. B. 1995. Inhibition of mitochondrial complexes I and IV by 6-hydroxydopamine. *Eur. J. Pharmacol.* 292:329–332.

Guarduika-Lemaitre, B. 1997. Toxicology of melatonin. *J. Biol. Rhythms* 12:697–706.

Gutierrez-Valdez, A. L., Anaya-Martínez, V., Ordoñez-Librado, J. L. et al. 2012. Effect of chronic L-DOPA or melatonin treatments after dopamine deafferentation in rats: Dyskinesia, motor performance, and cytological analysis. *ISRN Neurol.* 2012:1–15.

Hamdi, A. 1998. Melatonin administration increases the affinity of D2 dopamine receptors in the rat striatum. *Life Sci.* 63(23):2115–2120.

Hasegawa, E., Takeshige, K., Oishi, T. et al. 1990. 1-Methyl-4-phenylpyridinium (MPP+) induces NADH-dependent superoxide formation and enhances NADH-dependent lipid peroxidation in bovine heart submitochondrial particles. *Biochem. Biophys. Res. Commun.* 170:1049–1055.

Hobson, S. R., Lim, R., Gardiner, E. E. et al. 2013. Phase I pilot clinical trial of antenatal maternally administered melatonin to decrease the level of oxidative stress in human pregnancies affected by pre-eclampsia (PAMPR): Study protocol. *BMJ Open* 3(9):e003788.

Hodgson, E. K. and Fridovich, I. 1975. The interaction of bovine erythrocyte superoxide dismutase with hydrogen peroxide: Chemiluminescence and peroxidation. *Biochemistry* 14:5299–5303.

Horvath, T. B. and Davis, K. L. 1990. Central nervous system disorders in aging. In *Handbook of the Biology of Aging*, eds. E. L. Schneider and J. W. Rowe, p. 306. San Diego, CA: Academic Press.

Jenner, P., Schapira, A. H., and Marsden, C. D. 1992. New insights into the cause of Parkinson's disease. *Neurology* 42:2241–2250.

Joo, W. S., Jin, B. K., Park, C. W. et al. 1998. Melatonin increases striatal dopaminergic function in 6-OHDA-lesioned rats. *Neuroreport* 9(18):4123–4126.

Kerem, H., Akdemır, O., Ates, U. et al. 2014. The effect of melatonin on a dorsal skin flap model. *J. Invest. Surg.* April;27(2):57–64.

Kumar, R., Agarwal, A. K., and Seth, P. K. 1995. Free radical-generated neurotoxicity of 6-hydroxydopamine. *J. Neurochem.* 64:1703–1707. (Published erratum appears in *J. Neurochem.* 65, 1906, 1995)

Litvinenko, I. V., Krasakov, I. V., and Tikhomirova, O. V. 2012. Sleep disorders in Parkinson's disease without dementia: A comparative randomized controlled study of melatonin and clonazepam. *Zh. Nevrol. Psikhiatr. Im. S. S. Korsakova* 112(12):26–30.

Margaret, O. J., Zimeng, Y., Rachel, C. et al. 1998. Pharmacokinetics and metabolism of [^{14}C] dichloroacetate in male Sprague–Dawley rats. *Drug Metab. Dispos.* 26(11):1134–1143.

Mattson, M. P., Lovell, M. A., Furukawa, K. et al. 1995. Neurotrophic factors attenuate glutamate-induced accumulation of peroxides, elevation of [Ca^{2+}] and neurotoxicity, and increase antioxidant enzyme activities in hippocampal neurons. *J. Neurochem.* 65:1740–1751.

Montoya, C. P., Campbell-Hope, L. J., Pemberton, K. D. et al. 1991. The "Staircase test": A measure of independent forelimb reaching and grasping abilities in rats. *J. Neurosci. Methods* 36:219–228.

Offen, D., Ziv, I., Sternin, H. et al. 1996. Prevention of dopamine-induced cell death by thiol antioxidants: Possible implications for treatment of Parkinson's disease. *Exp. Neurol.* 141:32–39.

Olanow, C. W. and Tatton, W. G. 1999. Etiology and pathogenesis of Parkinson's disease. *Annu. Rev. Neurosci.* 22:123–144.

Oliveira, L. G., Kuehn, C. C., Santos, C. D. et al. 2013. Protective actions of melatonin against heart damage during chronic Chagas disease. *Acta Trop.* 128(3):652–658.

Olsson, M., Nikkhah, G. Bentlage, C. et al. 1995. Forelimb akinesia in the rat Parkinson model: Differential effects of dopamine agonists and nigral transplants as assessed by a new stepping test. *J. Neurosci.* 15:3863–3875.

Ortiz, G. G., Elena Crespo-López, M., Morán-Moguel, C. et al. 2001. Protective role of melatonin against MPTP-induced mouse brain cell DNA fragmentation and apoptosis in vivo. *Neuroendocrinol. Lett.* 22(2):101–108.

Paxinos, G. and Watson, C. 1986. *The Rat Brain in Stereotaxic Coordinates.* Sydney, New South Wales, Australia: Academic Press.

Perumal, A. S., Gopal, V. B., Tordzro, W. K. et al. 1992. Vitamin E attenuates the toxic effects of 6-hydroxydopamine on free radical scavenging systems in rat brain. *Brain Res. Bull.* 29:699–701.

Pierrefiche, G. and Laborit, H. 1995. Oxygen free radicals, melatonin, and aging. *Exp. Gerontol.* 30:213–227.

Reum, T., Olshausen, F., Mazel, T. et al. 2002. Diffusion parameters in the striatum of rats with 6-hydroxydopamine induced lesions and with fetal mesencephalic grafts. *J. Neurosci. Res.* 70(5):680–693.

Rondanelli, M., Faliva, M. A., Perna, S. et al. 2013. Update on the role of melatonin in the prevention of cancer tumorigenesis and in the management of cancer correlates, such as sleep–wake and mood disturbances: Review and remarks. *Aging Clin. Exp. Res.* 5:499–510.

Rossignol, D. A. and R. E. Frye. 2013. Melatonin in autism spectrum disorders. *Curr. Clin. Pharmacol.* [Epub ahead of print].

Rozas, G. and Labandeira García, J. L. 1997. Drug-free evaluation of rat models of parkinsonism and nigral grafts using a new automated rotarod test. *Brain Res.* 749(2):188–199.
Saji, M., Blau, A. D., and Volpe, B. T. 1988. Prevention of transneuronal degeneration in the substantia nigra reticulata by ablation of the subthalamic nucleus. *Exp. Neurol.* 141:120–129.
Schallert, T. and Hall, S. 1988. 'Disengage' sensorimotor deficit following apparent recovery from unilateral dopamine depletion. *Behav. Brain Res.* 30:15–24.
Schallert, T., Ryck, M. D., Whishaw, I. Q. et al. 1979. Excessive bracing reactions and their control by atropine and L-DOPA in an animal analog of parkinsonism. *Exp. Neurol.* 64:33–43.
Schober, A. 2004. Classic toxin-induced animal models of Parkinson's disease: 6-OHDA and MPTP. *Cell Tissue Res.* 318:215–224.
Segovia, J., Vergara, P., and Brenner, M. 1998. Astrocyte-specific expression of tyrosine hydroxylase after intracerebral gene transfer induces behavioral recovery in experimental parkinsonism. *Gene Ther.* 5(12):1650–1655.
Sewerynek, E., Melchiorri, D., Ortiz, G. G. et al. 1995. Melatonin reduces H_2O_2-induced lipid peroxidation in homogenates of different rat brain regions. *J. Pineal Res.* 1:51–56.
Shaw, K. M., Stern, G. M., and Sandler, M. 1973. Melatonin and parkinsonism. *Lancet* 1:271–279.
Singh, S., Ahmed, R., Krishana, B. et al. 2006. Neuroprotection of the nigrostriatal dopaminergic neurons by melatonin in hemiparkinsonium rat. *Indian J. Med. Res.* 124(4):419–426.
Singhal, N. K., Srivastava, G., Agrawal, S. et al. 2012. Melatonin as a neuroprotective agent in the rodent models of Parkinson's disease: Is it all set to irrefutable clinical translation? *Mol. Neurobiol.* 45(1):186–199.
Subbakrishna, D. K. 2000. Statistics for neuroscientists. *Ann. Indian Acad. Neurol.* 3(2):55–67.
Svingos, A. L., Clarke, C. L., and Pickel, V. M. 1998. Cellular sites for activation of ∂-opioid receptors in the rat nucleus accumbens shell: Relationship with met^5-enkephalin. *J. Neurosci.* 18(5):1923–1933.
Tenn, C. C. and Niles, L. P. 1997. Mechanisms underlying the antidopaminergic effect of clonazepam and melatonin in serotonin. *Neuropharmacology* 36:1659–1663.
Tomás-Zapico, C. and Coto-Montes, A. 2005. A proposed mechanism to explain the stimulatory effect of melatonin on antioxidative enzymes. *J. Pineal Res.* 39(2):99–104.
Tongjaroenbuangam, W., Ruksee, N., Mahanam, T. et al. 2013. Melatonin attenuates dexamethasone-induced spatial memory impairment and dexamethasone-induced reduction of synaptic protein expressions in the mouse brain. *Neurochem. Int.* 63(5):482–491.
Tooyama, I., McGeer, E. G., and Kawamata, T. 1994. Retention of basic fibroblast growth factor immunoreactivity in dopaminergic neurons of the substantia nigra during normal aging in humans contrasts with loss in Parkinson's disease. *Brain Res.* 656:165–168.
Ungerstedt, U. and Arbuthnott, G. W. 1970. Quantitative recording of rotational behavior in rats after 6-hydroxy-dopamine lesions of the nigrostriatal dopamine system. *Brain Res.* 24:485–493.
Vilar, A., deLemos, L., Patraca, I. et al. 2014. Melatonin suppresses nitric oxide production in glial cultures by pro-inflammatory cytokines through p38MAPK inhibition. *Free Radic. Res.* 48:119–128.
Willis, G. L. and Armstrong, S. M. 1999. A therapeutic role for melatonin antagonism in experimental models of Parkinson's disease. *Physiol. Behav.* 66:785–795.
Willis, G. L. and McLennan, C. A. 2001. Pinealectomy and dopamine replacement therapy in models of Parkinson's disease. *A Satellite Symposium of the 34th International Congress of Physiological Sciences on the Theme Melatonin and Biological Rhythms*, p. 10.
Winkler, C., Sauer, H., Lee, C. S. et al. 1996. Short-term GDNF treatment provides long-term rescue of lesioned nigral dopaminergic neurons in a rat model of Parkinson's disease. *J. Neurosci.* 16:7206–7215.
Wirdefeldt, K., Adami, H. O., Cole, P. et al. 2011. Epidemiology and etiology of Parkinson's disease: A review of the evidence. *Eur. J. Epidemiol.* 26:S1–S58.

32 Antiepileptic Effects of Melatonin

M. Hakan Seyithanoğlu, Mehmet Turgut, Meliha Gündağ Papaker, Mustafa Ayyıldız, Süleyman Kaplan, and Saffet Tüzgen

CONTENTS

32.1 INTRODUCTION

Epilepsy is a very important chronic neurological disorder, which is characterized by recurrent spontaneous seizure discharges (Dichter 1994). Experimental epilepsy models used for the development of new antiepileptic drugs have played all-important role, but there is no unique experimental model that could be useful for all types of epilepsy (Cakil et al. 2011). It is well known that active oxygen free radicals have a role in the mechanism of epileptic discharges (Mori et al. 1990, Yildirim et al. 2011). Melatonin (Mel) is one of the anticonvulsant substances that reduce the epileptiform activity (Copolla et al. 2004, Fariello et al. 1997, Gloor and Testa 1974, Golombek et al. 1992, Maurizi 1985, Moezi et al. 2011, Reiter 2000, Saracz and Rosdy 2004, Tan et al. 2003, Yahyavi-Firouz-Abadi et al. 2006). It has been suggested that Mel has anticonvulsant (Copolla et al. 2004, Fariello et al. 1997, Golombek et al. 1992, Maurizi et al. 1985, Peled et al. 2001, Reiter et al. 2000, Tan et al. 2003, Yahyavi-Firouz-Abadi et al. 2006) and proconvulsant properties (Elkhayat et al. 1995, Sandyk et al. 1992, Stewart and Leung 2005). Since the results of in vitro experiments are not enough to show the anticonvulsant effect of Mel, in vivo experiments are needed to be performed (Banach et al. 2011).

Mel, *N*-acetylated-5-methoxytryptamine, is a hormone largely being produced by pineal gland with reproductive changes, hypnotic effect, hypothermic response, immunomodulatory, anticancer, antiaging, and antioxidant effects (Brzezinski et al. 1997). It affects the plasma membrane receptors and controls electrical activity in neurons and regulates the central nervous system (CNS).

32.2 EFFECTS OF MELATONIN AT MOLECULAR LEVEL

Mel is metabolized primarily in the liver and then transmitted to excrete from the kidney. Circulating Mel is hydroxylated by the enzyme microsomal cytochrome P-450 monooxygenase-isoenzymes, CYP1A2, CYP1A1, and CYP1B1, and turns into 6-hydroxy-Mel and is excreted in the urine. A smaller amount of glucuronide creates conjugate. Mel remains only 1% unchanged (Claustrat et al. 2005). The Mel in the brain being metabolized is metabolized into kynurenic acid, which is an anticonvulsive substance (Munos Hayos et al. 1998). Besides kynurenic acid, Mel is also metabolized into *N*-acetyl-*N*-formyl-5-metoxy-kinuramine (AFMK) and *N*-acetyl-5-metoxy-kinuramine (AMK). These metabolites are strong antioxidants and are inhibitors of cyclooxygenase 2. Thus, they can be considered to be potent selective anti-inflammatory agents (Kabuto et al. 1998, Mayo et al. 2005).

Mel is a molecule of which the description could not yet exactly be made. Depending on where and how to have an effect, it can act as a tissue factor, antioxidant, paracoid, or autacoid (Yahyavi-Firouz-Abadi et al. 2006). Its main effect is observed upon the CNS, and it can have an effect with or independent from the receptor. Mel acts through two receptors: (1) MT1 (Mel 1a, ML1a); and (2) MT2 (Mel1b, ML1b). These receptors are located within the suprachiasmatic nucleus of the hypothalamus, the pars tuberalis of the pituitary, and cardiac blood vessels (MT1) and the retina and hippocampus (MT2) (Wurtman 2013). The effector systems involved in MT1 and MT2 receptor signaling through high-affinity G-protein-coupling include adenyl cyclase, phospholipase C, phospholipase A2, potassium channels, and possibly guanylyl cyclase and calcium channels (Pang and Brown 1983). Activations of these receptors separates into alpha and beta dimers interacting with different G-protein effector molecules. MT1 receptor is more prevalent than MT2 in the hippocampus (Dubocovich 1995, Morgan et al. 1994). MT3, less well known as a low-affinity membrane receptor, is located in kidney, brain, and various peripheral organs (Wurtman 2013). It reduces intraocular pressure by regulating the activity of calcium and calmaduline (Nosjean et al. 2000, Pintor et al. 2001).

Mel is a strong scavenger for reactive oxygen species (ROS) and reactive nitrogen species (RNS). It stimulates antioxidant enzyme activity that converts ROS into harmless molecules (Reiter 2000). Furthermore, Mel reduces the leakage of electron in the mitochondria and impairs the production of free radicals (Kabuto et al. 1998). All of these processes not only reduce the DNA damage but also decrease lipid and protein peroxidation (Mohanan and Yamamoto 2002). In certain circumstances observed in other antioxidants, such as ascorbate and tocopherol Mel does not have the feature to act as a pro-oxidant and easily passes the blood–brain barrier, and this plays an important role in the protection of the brain and reduction of neuron excitability (Reiter 2000, Reiter et al. 2007).

Excitatory neurotransmitters such as glutamate and aspartate relate to the expression and generation of epileptic seizures in mammalian brains (Sejima et al. 1997). After its interaction with the *N*-methyl-D-aspartate (NMDA)-glutamate receptor subtypes, glutamate stimulates the entry of Ca^{+2}, and accordingly, the neuronal nitric oxide synthase (nNOS) and nitrous oxide (NO) production increases. NO plays an important role in learning and long-term potentiation but can be toxic if being excessively released (Urushitani et al. 2001). Depending on the contents of the stimulator agent, NO concentrations can be proconvulsant or anticonvulsant (Prast and Philippu 2000). Falling levels of NO has an anticonvulsant effect, but with the increase in the level, NO reacts with superoxide anions and causes peroxynitrite formation (Lipton et al. 1993).

At the highest concentration of nocturnal Mel, it inhibits the entry of calcium into neurons and binds the calcium–calmodulin complex. Thus, it inhibits the nNOS activity and reduces NO production and accordingly reduces the NMDA excitatory effect (Leon et al. 2000, Munos Hayos et al. 1998). Nevertheless, the inhibitory neurotransmitters such as gamma-aminobutyric acid (GABA) and glycine reduce neuronal excitation. GABA interacting with GABA receptors and increased Cl^- entry, which causes hyperpolarization of the cell membrane, shows an antiepileptic effect (Meldreum and Garthwaite 1990). Indoleamine increases the brain GABA concentration and receptor affinity and potentiates the brain inhibitory transmission by GABAergic synapses (Rosenstein et al. 1990, Wan et al. 1999). Mel increases the levels of GABA by regulating GABA–benzodiazepine receptor complex in rat brains (Castroviejo et al. 1986). It was shown that indoline binds to calcium ions and Mel acts as an antagonist to L-type calcium channels (Acuna-Castroviejo et al. 1997). Another physiological effect of Mel is reducing striatal dopaminergic activity through D1 and D2 dopaminergic receptors (Stewart 2001, Sweis 2005).

Finally, Mel is not only metabolized to kynurenic acid, but also converted to AFMK and AMK. These metabolites can be considered as potential anti-inflammatory agents because they are anti-oxidant and cyclooxygenase 2 inhibitors (Kabuto et al. 1998, Meldreum and Garthwaite 1990).

32.3 ANIMAL STUDIES ON MELATONIN

The first studies to support a possible relationship between Mel and epilepsy were not only about observation of epileptogenic effects of Mel being intraventriculary injected to rats but also have shown severe convulsions inducing pinealectomized and parathyroidectomized rats (Fariello et al. 1997, Reiter et al. 1973). With the suppression of Mel in pinealectomized rats, an increase in brain damage was observed in animals being stimulated with kainic acid to form epilepsy, and thus the neuroprotective role of Mel was supported (Manev et al. 1996).

Penicillin-induced experimental epilepsy is similar to human myoclonic petit mal epilepsy (Guo and Yao 2009). Yildirim and Marangoz (2006) reported that Mel has an anticonvulsant effect on the penicillin-induced epileptiform activity in rats. They injected Mel, at doses of 20, 40, and 80 µg, intracerebroventricularly 10 min before the penicillin application (Yildirim and Marangoz 2006). They have found that Mel significantly decreased the frequency of epileptiform activity dose dependently compared with penicillin group and also increased the latency of onset of the spike activity (Yildirim and Marangoz 2006).

Most recently, Forcelli et al. (2013) have been examined the effect of Mel on the anticonvulsant action of phenobarbital in pentylenetetrazole (PTZ)-induced neonatal rats. They found that Mel, when used alone, did not affect seizure onset latency, seizure score against score 4 or 5 seizures (Forcelli et al. 2013). When they used combination of Mel and phenobarbital, Mel significantly potentiated the effect of phenobarbital by increasing the seizure onset latency and decreasing the seizure severity (Forcelli et al. 2013). Especially, combination of the doses of 80 mg/kg Mel and 20 mg/kg phenobarbital completely removed score 4 and 5 seizures (Forcelli et al. 2013). Based on their findings, they speculated based on previous studies that GABAergic and NO system may have a role in the Mel's phenobarbital-potentiating effect (Forcelli et al. 2013).

Aguiar et al. (2012) reported that agomelatine, a potent MT1 and MT2 Mel receptor agonist and a 5-HT2C serotonin receptor antagonist, has anticonvulsant activity on the PTZ- and pilocarpin-induced seizures but not on the strychnine-, electroshock-, and picrotoxin-induced seizures in mice. At the doses of 25 and 50 mg/kg, agomelatine significantly delayed the latency of convulsion, and at the doses of 50 and 75 mg/kg, agomelatine postponed the time of death in the PTZ-induced seizure (Aguiar et al. 2012). Only high dose of agomelatine (75 mg/kg) significantly delayed the latency of convulsion and the time of death in the pilocarpine-induced seizures (Aguiar et al. 2012). They proposed that these anticonvulsant effects may be related to GABAergic mechanism (Aguiar et al. 2012).

Moczi et al. (2011) explained that Mel and agmatine have an additive effect in diminishing PTZ-induced seizure threshold in mice, probably via $ML_{1/2}$ receptors. In this research, Mel (at the doses of 40 and 80 mg/kg) and agmatine (at the doses of 10 and 20 mg/kg) displayed anticonvulsant effect (Moezi et al. 2011). Agmatine (5 mg/kg) plus noneffective dose of Mel (20 mg/kg) and agmatine (5 mg/kg) plus effective dose of Mel (80 mg/kg) have also anticonvulsant effect. $ML_{1/2}$ receptor antagonist luzindole (2.5 mg/kg) inhibited anticonvulsant effects of Mel, agmatine, and interaction groups, but at a dose of 0.5 mg/kg prazosin, ML_3 receptor antagonist did not block anticonvulsant activity of all groups (Morgan et al. 1994).

In 1998, Mevissen and Ebert (1998) showed that in epileptic animals being dosed with 75 mg/kg and higher doses of Mel, the discharge threshold increased significantly, and in amygdala-kindled rats, generalized seizures have been suppressed. In another study, Mel has shown a weak anticonvulsant effect on pilocarpine-induced seizures (Cousta-lotufo et al. 2002). Bikjdaouene et al. (2003) suggested that giving 100 mg/kg sc Mel before the addition of PTZ reduces the duration of the first seizure, significantly increasing the latent period. The role of Mel in seizure activity has been shown in pinealectomized animals and has increased seizure activity in gerbils after pinealectomy (Rudeen et al. 1980).

In addition, the epileptogenic process became easier in pinealectomized animals in the model with pilocarpine stimulation (Turgut et al. 2006). Again, in a study where pinealectomy was performed and epilepsy was induced in pregnant rats, it was suggested that damage occurred in the hippocampal neurogenesis and neuronal maturation in the offspring of these animals and that this can be inhibited by applying Mel (Turgut et al. 2006). A research was made on the effect of taking pineal gland on the kindling model, and it was seen that there was a significant effect on the development of pinealectomy on the amygdala kindling, and the number of stimulations required to reach stage 5 was reduced (Janjoppi et al. 2006). Physical exercise programs turned back the acceleration in the process of kindling in pinealectomized animals (De Lacerda et al. 2007).

Importantly, it has been reported that Mel (50 mg/kg) significantly increases the electroconvulsive threshold (Manev et al. 1996). In addition, 25 mg/kg carbamazepine and phenobarbital in subprotective doses increases the activity of antielectroshock (Meldreum and Garthwaite 1990). This effect is reverted by subconvulsive doses of bicuculine, aminophyline, and picrotoxin (Bikjdaouene et al. 2003). This shows that antielectroshock activity of Mel is depending on the GABAergic and purinergic neurotransmission (Borowicz et al. 1999). Hippocampal sections in animal studies, Mel causes to increase in the GABA binding areas, show that the mechanism of action may result from positive regulations in GABAergic transmission (Rosenstein et al. 1990). Yahyavi-Firouz-Abadi et al. (2006) investigated the role of NO pathway in the anticonvulsant effect of Mel in PTZ-induced seizures in mice. In this study, Mel (40 and 80 mg/kg) increased the threshold of PTZ-induced seizure (Yahyavi-Firouz-Abadi et al. 2006). Combinations of noneffective dose of Mel (10 and 20 mg/kg) and L-arginine (30 and 60 mg/kg) have anticonvulsant activity (Yahyavi-Firouz-Abadi et al. 2006). This effect was inhibited by *N*(G)-nitro-L-arginine methyl ester (L-NAME, 30 mg/kg) (Yahyavi-Firouz-Abadi et al. 2006). This case shows that NO system play a role in the anticonvulsant effect of Mel (Yahyavi-Firouz-Abadi et al. 2006). Pretreatment of L-NAME and *N*(G)-nitro-L-arginine (nonspecific synthase [NOS] inhibitors) prevented the anticonvulsant effect of Mel completely (Yahyavi-Firouz-Abadi et al. 2006). Pretreatment of 7-nitroindazole, a preferential neuronal NOS inhibitor, also blocked the anticonvulsant effect of Mel (Yahyavi-Firouz-Abadi et al. 2006). Based on their results, they suggest that NO pathway have a role in the anticonvulsant effect of Mel (Yahyavi-Firouz-Abadi et al. 2006). Yahyavi-Firouz-Abadi et al. (2006) also reported that Mel increases anticonvulsant and proconvulsant properties of morphine by way of a mechanism that may involve the NO pathway.

32.4 HUMAN STUDIES ON MELATONIN

Clinical findings have shown that the peak point of daily seizures' profile coincides with night hours. Although the mechanism responsible for the frequency of seizures is still uncertain, the convulsive sensitivity depends on time-dependent biological signal produced by intrinsic neural pendulum. Interestingly, as well as the daily secretion of Mel in active epilepsy may be normal, in some patients not treated for epilepsy, the night level may be doubled at night. The seizure activity increases during menstruation and pregnancy and decreases during menopause. Thus, in relation to increase and decrease in Mel in the reproductive stage, the seizure rhythm changes in epileptic women (Reiter 1995).

In one of the first studies showing the potential benefits of Mel on epilepsy, Mel was given to children with refractory epilepsy who did not respond to conventional therapy as a single medicine with a 5–10 mg single evening dose, and it was seen that the seizure frequency decreased. In most of the studies, Mel was given as an addition to other antiepileptic drugs. In another study, a child with severe seizures was given 3 mg/day oral Mel in addition to the conventional treatment 30 min before the bedtime for a period of 3 months, and a significant improvement in seizure activity especially at night was observed (Munos Hayos et al. 1998, Peled et al. 2001). Other studies have shown the fluctuations in Mel rhythm for epilepsy with various origins, and compared to seizure-free children, Mel levels were found to be low in these children (Guo and Yao 2009, Paprocka et al. 2010).

Mel is able to control the convulsive crises by affecting GABA and glutamate receptors (Molina-Carballo et al. 2007, Munos-Hoyos et al. 1998). A 7-year-old child with intractable seizures was given 3.4 mcg/kg/h Mel intravenously, but an improvement was not seen in EEG (Brueske et al. 1981). In 1995, Champney (Champney et al. 1996, Champney and Peterson 1993) has given a ketogenic diet and Mel 60 mg orally to a 6-year-old child with intractable nocturnal epilepsy, and seizures have been taken under control. However, seizures emerged again when the Mel dose fell below 20 mg (Gupta et al. 2004). After 2 years, the current treatment was changed with other anticonvulsants, and the patient's seizures improved (Champney 1992).

Volunteers were given a single dose of 1.25 μg Mel administered iv; within 5 h, a progressive decrease was recorded in the amplitude of electrical activity, and the epileptiform discharges decreased (Anton-Tay 1974). In another study carried out by the same research team, patients were given rapid release Mel in oral 2 g divided doses for 30 days (Anton-Tay 1974). Together with a reduction in the frequency of seizures in all patients, there was also observed a decrease in epileptiform activity (Anton-Tay 1974).

Since Mel is protective against oxidative stress and neuronal damage associated with epilepsy, it was focused on studies in which Mel was used alone or together with other anticonvulsants in order to assess oxidative changes in status epilepticus patients (Gupta et al. 2004). In monotherapy epileptic patients being treated with 6–9 mg/day of Mel for 14 days, Mel decreased the oxidative stress caused by the effect of Mel on glutathione reductase and glutathione peroxidase and serves as a neuroprotector (Gupta et al. 2004).

In a study carried out on 25 patients with sleep disorders associated with seizure-induced anxiety and mental retardation, an improvement in sleep–wake disorders has been provided with doses of Mel started with a dose of 3 mg/day and increasing up to 9 mg/day (Copolla et al. 2004). Again, in a group of 23 patients with intractable seizures, Mel was given orally at bedtime, and an improvement in sleep problems associated with seizure severity was observed (Elkhayat et al. 2010).

According to the authors, seizure sensitivity does not depend on changes in endogenous Mel per day, because the highest seizure threshold is seen in the morning hours when the levels of endogenous Mel are the lowest (Mevissen and Ebert 1998). Other endogenous substances such as glucocorticoids may be responsible for an increase in the sensitivity of seizures at night (Weiss et al. 1993).

The fact that endogenous Mel in the brain is far below the amount needed to influence the sensitivity of seizure is an indicator that the protective effect of Mel is inadequate. The increase in endogenous Mel production in untreated patients with active epilepsy laid the foundations for the use of Mel in the treatment of epilepsy (Molina-carballo et al. 1997). Antiepileptic effect of the selective Mel receptor agonist ramelteon was demonstrated in rats, but the effectiveness in humans has not been studies (Mevissen and Ebert 1998).

32.5 CONCLUSION

According to the literature, Mel has anticonvulsant properties not only in animal models but also in human epilepsy (Munos-Hoyos et al. 1998). In fact, Mel available in the brain tissue and CSF is a natural anticonvulsant (Maurizi 1985). Due to lipophilic features, the levels of Mel in the brain are 3–10 times higher than serum, and the brain uptakes Mel selectively (Pang and Brown 1983). In the treatment of epilepsy, it is known that Mel has some effects on neuronal functions by depressing them. It functions by potentiating the GABAergic system and affecting the same receptors with benzodiazepines. Therefore, Mel and benzodiazepines show significant similarities, and there are similarities in the molecular structure of metabolites (Champney and Peterson 1993, Golombek et al. 1992). During the identification of anticonvulsant substances in vitro, despite the advances in cellular neurophysiology and biochemical structure, tests cannot take place in animal models; only through animal testing, anticonvulsant compounds can be defined, and the targeted areas in the brain can be reached (Mevissen and Ebert 1998). By creating electroshock chemical models, it is possible to evaluate the mechanism of possible anticonvulsant effect of GABA, glutamate, acetylcholine, glycine, and strychnine (Mares and Kubova 2006, Velisek 2006). When this hormone is associated with antiepileptic drugs, the incidence of tonic–clonic seizure character is reduced (Peled et al. 2001). Golombek et al. (1992) have suggested that the effect of Mel increases the Cl^- ion uptake by GABA-dependent chloride channels and is depending on the membrane ion permeability. This finding may explain the anticonvulsant property of Mel.

REFERENCES

Acuna-Castroviejo, D., G. Escames, and J. Leon. 1997. Interaction between calcium ionophore A-23187 and melatonin in the rat striatum. *J Physiol Biochem* 53: 119.

Aguiar, C.C., A.B. Almeida, P.V. Araújo, G.S. Vasconcelos, E.M. Chaves, O.C. do Vale, D.S. Macêdo, F.C. de Sousa, G.S. Viana, and S.M. Vasconcelos. 2012. Anticonvulsant effects of agomelatine in mice. *Epilepsy Behav* 24(3): 324–328. doi: 10.1016/j.yebeh.2012.04.134.

Anton-Tay, F. 1974. Melatonin: Effects on brain function. *Adv Biochem Psychopharmacol* 11(0): 315–324.

Banach, M., E. Gurdziel, M. Jedrych, and K.K. Borowicz. 2011. Melatonin in experimental seizures and epilepsy. *Pharmacol Rep* 63(1): 1–11.

Bikjdaouene, L., G. Escames, J. Leon, J.M. Ferrer, H. Khaldy, F. Vives, and D. Acuna-Castroviejo. 2003. Changes in brain amino acids and nitric oxide after melatonin administration in rats with pentylenetetrazole-induced seizures. *J Pineal Res* 35(1): 54–60.

Borowicz, K.K., R. Kaminski, M. Gasior, Z. Kleinrok, and S.J. Czuczwar. 1999. Influence of melatonin upon the protective action of conventional anti-epileptic drugs against maximal electroshock in mice. *Eur Neuropsychopharmacol Mar* 9(3): 185–190.

Brueske, V., J. Allen, T. Kepic, W. Meissner, R. Lee, G. Vaughan, and U. Weinburg. 1981. Melatonin inhibition of seizure activity in man. *Electroencephalogr Clin Neurophysiol* 51: 209.

Brzezinski, A. 1997. Melatonin in humans. *N Engl J Med* 336(3): 186–195.

Cakil, D., M. Yildirim, M. Ayyildiz, and E. Agar. 2011. The effect of co-administration of the NMDA blocker with agonist and antagonist of CB1-receptor on penicillin-induced epileptiform activity in rats. *Epilepsy Res* 93: 128–137.

Castroviejo, D.A., R.E. Rosenstein, H.E. Romeo, and D.P. Cardinali. 1986. Changes in gamma-aminobutyric acid high affinity binding to cerebral cortex membranes after pinealectomy or melatonin administration to rats. *Neuroendocrinology* 43: 24–31.

Champney, J.C. 1992. Novel anticonvulsant action of chronic melatonin in gerbils. *Neuroreports* 3: 1152–1154.

Champney, T.H., W.H. Hanneman, M.E. Legare, and K. Appel. 1996. Acute and chronic effects of melatonin as an anticonvulsant in male gerbils. *J Pineal Res* 20: 79–83.

Champney, T.H. and S.L. Peterson. 1993. Circadian, sensorial, pineal and melatonin influences on epilepsy. In: Yu, H.-S. and R.J. Reiter (Eds.), *Melatonin Biosynthesis, Physiological Effect and Clinical Applications*. London, U.K.: CRC Press, pp. 478–494.

Claustrat B., J. Brun, and G. Chazot. 2005. The basic physiology and pathophysiology of melatonin. *Sleep Med Rev* 9: 11–24.

Copolla, G., G. Iervolino, M. Mastrosimone, G. La Torre, F. Ruiu, and A. Pascotto. 2004. Melatonin in wake–sleep disorders in children, adolescents and young adults with mental retardation with or without epilepsy: A double-blind, cross-over, placebo-controlled trial. *Brain Dev* 26: 373–376.

Cousta-lotufo, L.V., M.M. Fontales, I.S. Lima, A.A. Oliveria, V.S. Nascimento, V.M. de Bruin, and G.S. Viana. 2002. Attenuating effect of melatonin on pilocarpine-induced seizures in rats. *Comp Biochem Physiol C Toxicol Pharmacol* 131: 521–529.

De Lacerda A.F.S., I. Janjoppi, E.A. Scorza, E. Lima, D. Amado, E.A. Cavalheiro, and R.M. Arıda. 2007. Physical exercise program reverts the effect of pinealectomy on the amygdala kindling development. *Brain Res Bull* 74: 216–220.

Dichter, M.A. 1994. The epilepsies and convulsive disorders. In: Isselbacher, K.J. (Ed.), *Harrison's Principles of Internal Medicine*. New York: McGraw-Hill, pp. 2223–2233.

Dubocovich, M.L. 1995. Melatonin receptors: Are there multiple subtypes? *Trends Pharmacol Sci* 16: 50–56.

Elkhayat, H.A., S.M. Hassanein, H.Y. Tomoum, I.A. Elhamid, T. Asaad, and A.S. Elwakkad. 2010. Melatonin and sleep-related problems in children with intractable epilepsy. *Pediatr Neurol* 42: 249–254.

Fariello, R.G., G.A. Bubenik, and G.M. Brown. 1997. Epileptogenic action of intraventriculary injected anti-melatonin antibody. *Neurology* 27: 267–270.

Forcelli, P.A., C. Soper, A. Duckles, K. Gale, and A. Kondratyev. 2013. Melatonin potentiates the anticonvulsant action of phenobarbital in neonatal rats. *Epilepsy Res* 107: 217–223. http://dx.doi.org/10.1016/j.eplepsyres.2013.09.013.

Gloor, P. and G. Testa. 1974. Generalized penicillin epilepsy in the cat: Effects of intracarotid and intravertebral pentylenetetrazol and amobarbital injections. *Electroencephalogr Clin Neurophysiol* 36: 499–515.

Golombek, D.A., D.F. Duque, M.G. De Brito Sanchez, L. Burin, and D.P. Cardinali. 1992a. Time dependent anticonvulsant activity of melatonin in hamsters. *Eur J Pharmacol* 210: 253–258.

Golombek, D.A., E. Escolar, L.J. Burin, M.G. De Brito Sanches, D.D. Fernandez, and D.P. Cardinalli. 1992b. Chronopharmacology of melatonin. Inhibition by benzodiazepine antagonism. *Chronobiol Int* 9: 124–131.

Guo, J.F. and B.Z. Yao. 2009. Serum melatonin levels in children with epilepsy or febrile seizures. *Zhongguo Dang Dai Er Ke Za Zhi* 11: 288–290.

Gupta, M., S. Aneja, and K. Kohli. 2004a. Add-on melatonin improves quality of life in epileptic children on valproate monotherapy: A randomized, double-blind, placebo-controlled trial. *Epilepsy Behav* 5: 316–321.

Gupta, M., Y.K. Gupta, S. Agarwal, S. Aneja, M. Kalaivani, and K. Kohli. 2004b. Effects of add-on melatonin administration on antioxidant enzymes in children with epilepsy taking carbamazepine monotherapy: A randomized, double-blind, placebo-controlled trial. *Epilepsia* 45: 1636–1639.

Gupta, M., Y.K. Gupta, S. Agarwal, S. Aneja, and K. Kohli. 2004c. A randomized, double-blind, placebo controlled trial of melatonin add on therapy in epileptic children on valproate monotherapy: Effect on glutathione peroxidase and glutathione reductase enzymes. *Br J Clin Pharmacol* 58: 542–547.

Janjoppi, L., A.F. Silva De Lacerda, F.A. Scorza, D. Amado, E.A. Cavalheiro, and R.M. Arida. 2006. Influence of pinealectomy on the amygdala kindling development in rats. *Neurosci Lett* 392: 150–153.

Kabuto, H., I. Yokoi, and N. Ogawa. 1998. Melatonin inhibits iron induced epileptic discharges in rats by suppressing peroxidation. *Epilepsia* 39: 237–243.

Leon, J., M. Macias, G. Escames, E. Camacho, H. Khaldy, M. Martin, A. Espinosa, M.A. Gallo, and D. Acuno-Castroviej. 2000. Structure-related inhibition of calmodulin-dependent neuronal nitric-oxide synthase activity by melatonin and synthetic kynurenines. *Mol Pharmacol* 58: 967–975.

Lipton, S.A., Y.B. Choi, and Z.H. Pan. 1993. Redox-based mechanism for the neuroprotective and neurodestructive effect of nitric oxide and related nitroso-compounds. *Nature* 364: 626–632.

Manev, H., U.Z. Tolga, A. Kharlamov, and J.Y. Joo. 1996. Increased brain damage after stroke or excitotoxic seizures in melatonin-deficient rats. *FASEB J* 10: 1546–1551.

Mares, P. and H. Kubova. 2006. Electrical stimulation-induced models of seizures. In: Pitkanen, A., P.A. Schwartzkroin, and S.L. Moshe (Eds.), *Models of Seizures and Epilepsy*. Burlington, MA: Elsevier, p. 159.

Maurizi, C.P. 1985. Could supplementary dietary tryptophan and taurine prevent epileptic seizures. *Neurosci Lett* 150: 112–116.

Mayo, J.C., R.M. Sainz, D.X. Tan, R. Hardeland, J. Leon, C. Rodriquez, and R.J. Reiter. 2005. Anti-inflammatory action of melatonin and its metabolites, *N*-acetyl-*N*-formyl-5-methoxykynuramine (AFMK) and *N*-acetyl-5-methoxykynuramine (AMK) in macrophages. *J Neuroimmunol* 165: 139–149.

Meldreum, B. and J. Garthwaite. 1990. Excitatory amino acid neurotoxicity and neurodegenerative disease. *Trends Pharmacol Sci* 11: 379–387.

Mevissen, M. and U. Ebert. 1998. Anticonvulsant effect of melatonin in amygdala-kindled rats. *Neurosci Lett* 257: 13–16.

Moezi, L., H. Shafaroodi, A. Hojati, and A.R. Dehpour. 2011. The interaction of melatonin and agmatine on pentylenetetrazole-induced seizure threshold in mice. *Epilepsy Behav* 22: 200–206.

Mohanan, P.V. and H. Yamamoto. 2002. Preventive effect of melatonin against brain mitochondrial DNA damage, lipid peroxidation and seizures induced by kainic acid. *Toxicol Lett* 129: 99–105.

Molina-Carballo, A., A. Munos-Hoyos, and R.J. Reiter. 1997. Utility of high doses of melatonin as adjunctive anticonvulsant therapy in a child with severe myoclonic experience: Two years experience. *J Pineal Res* 23: 97–105.

Molina-Carballo, A., A. Munos-Hoyos, M. Sanches-Forte, J. Uberos Fernandes, F. Moreno-Madrid, and D. Acuna-Castroviejo. 2007. Melatonin increases following convulsive seizures may be related to its anticonvulsant properties at physiological concentrations. *Neuropediatrics* 38: 122–125.

Morgan, P.J., P. Barret, H.E. Howell, and R. Helliwell. 1994. Melatonin receptors: Localization, molecular pharmacology and physiological significance. *Neurochem Int* 24: 101–146.
Mori, A., M. Hiramatsu, I. Yokoi, and R. Edamatsu. 1990. Biochemical pathogenesis of posttraumatic epilepsy. *Pav J Biol Sci* 25: 54–62.
Munos-Hayos, A., A. Molina-Carballo, M. Macias, T. Rodriques-Cabezas, E. Martin-Medina, E. Narbona Lopez, A. Valenzuella-Ruiz, and D. Acuna-Castroviejo. 1998. Comparison between tryptophan methoxyindole and kynurenine metabolic pathways in normal and preterm neonates and in neonates with acute fetal distress. *Eur J Endocrinol* 139: 89–95.
Munos-Hoyos, A., M. Sanches Forte, and A. Molina-Carballo. 1998. Melatonins role as an anticonvulsant and neuronal protector: Experimental and clinic evidence. *J Child Neurol* 13: 501–509.
Nosjean, O., M. Ferro, F. Coge, P. Beauverger, J.M. Henlin, F. Lefoulon, and J.L. Fauchere. 2000. Identification of the melatonin-binding site MT3 as the quinone reductase 2. *J Biol Chem* 275: 31311–31317.
Pang, S.F. and G.M. Brown. 1983. Regional concentration of melatonin in the rat brain in the light and dark period. *Life Sci* 33: 1199–1204.
Paprocka, J., R. Dee, and E. Jamroz. 2010. Melatonin and childhood refractory epilepsy—A pilot study. *Med Sci Monit* 16: 389–396.
Peled, N., Z. Shorer, and G. Pillar. 2001. Melatonin effect on seizures in children with severe neurologic deficit disorders. *Epilepsia* 42: 1208–1210.
Pintor, J., L. Martin, T.E. Pelaez, C.H. Hoyle, and A. Peral. 2001. Involvement of melatonin MT3 receptors in the regulation of intraocular pressure in rabbits. *Eur J Pharmacol* 416: 251–254.
Prast, H. and A. Philippu. 2000. Nitric oxide as modulator of neuronal function. *Prog Neurobiol* 64: 51–68.
Reiter, R.J. 1995. Functional pleiotropy of the neurohormone melatonin: Antioxidant protection and neuroendocrine regulation. *Front Neuroendocrinol* 16: 383–415.
Reiter, R.J. 2000. Melatonin: Lowering the high price of free radicals. *News Physiol Sci* 15: 246–250.
Reiter, R.J., D.E. Blask, J.A. Talbot, and M.B. Barner. 1973. Nature and the time course of seizures associated with surgical removal of the pineal gland from parathyroidectomized rats. *Exp Neurol* 38: 386–397.
Reiter, R.J., D.X. Tan, L.C. Mancester, and H. Tamura. 2007. Melatonin defeats neurally-derived free radicals and reduces the associated neuromorphological and neurobehavioral damage. *J Physiol Pharmacol* 58: 5–22.
Rosenstein, R.E., H.E. Chuluyan, M.C. Diaz, and D.P. Cardinalli. 1990. GABA as a presumptive paracrine signal in the pineal gland. Evidence on an intrapineal GABAergic system. *Brain Res Bull* 25: 339–344.
Rudeen, P.K., R.C. Philo, and S.K. Symmes. 1980. Antiepileptic effects of melatonin in the pinealectomized Mongolian gerbil. *Epilepsia* 21: 149–154.
Sandyk, R., N. Tsagas, and P.A. Anninos. 1992. Melatonin as a proconvulsive hormone in humans. *Int J Neurosci* 63: 125–135.
Saracz, J. and B. Rosdy. 2004. Effect of melatonin on intractable epilepsies. *Orv Hetil* 145: 2583–2587.
Sejima, H., M. Ito, and K. Kishi. 1997. Regional excitatory and inhibitory amino acid concentration in pentylenetetrazole kindling and kindled rat brain. *Brain Dev* 19: 171–175.
Stewart, L.S. 2001. Endogenous melatonin and epileptogenesis: Facts and hypothesis. *Intern J Neurosci* 107: 77–85.
Stewart, L.S. and L.S. Leung. 2005. Hippocampal melatonin receptors modulate seizure threshold. *Epilepsia* 46: 473–480.
Sweis, D. 2005. The uses of melatonin. *Arc Dis Child Educ Pract Ed* 90: 74–77.
Tan, D.X., L.C. Manchester, R. Hardeland, S. Lopez Burillo, J.C. Mayo, R.M. Sainz, and R.J. Reite. 2003. Melatonin a hormone, a tissue factor, an autocoid, a paracoid, and an antioxidant vitamin. *J Pineal Res* 34: 75–78.
Turgut, M., Y. Uyanıkgil, U. Ateş, M. Baka, and M. Yurtseve. 2006. Pinealectomy stimulates and exogenous melatonin inhibits harmful effects of epileptiform activity during pregnancy in the hippocampus of newborn rats: an immunohistochemical study. *Child Nerv Syst* 22: 481–488.
Urushitani, M., T. Nakamizo, and R. Inoue. 2001. *N*-Methyl-D-aspartate receptor-mediated mitochondrial Ca(2+) overload in acute excitotoxic motor neuron death: A mechanism distinct from chronic neurotoxicity after Ca(2+) influx. *J Neurosci Res* 63: 377–387.
Velisek, L. 2006. Models of chemically-induced acute seizures. In: Pitkanen, A., P.A. Schwartzkroin, and S.L. Moshe (Eds.), *Models of Seizures and Epilepsy*. Burlington, MA: Elsevier. pp. 127–152.
Wan, Q., H.Y. Man, F. Liu, J. Braunton, H.B. Niznik, S.F. Pang, G.M. Brown, and Y.T. Wang. 1999. Differential modulation of GABA(A) receptor function by Mel (1a) and Mel (1b) receptors. *Nat Neurosci* 2: 401–403.
Weiss, G., K. Lucero, and M. Fernandez. 1993. The effect of adrenalectomy on the circadian variation in the rate of kindled seizure development. *Brain Res* 612: 354–356.

Wurtman, R. 2013. Physiology and clinical use of melatonin. http://www.uptodate.com/contents/physiology-and-clinical-use-of melatonin?detectedLanguage=en&source=search_result&search=epilepsy+and+melatonin&selectedTitle=1~150&provider=noProvider.

Yahyavi-Firouz-Abadi, N., P. Tahsili-Fahadan, K. Riazi, M.H. Ghahremani, and A.R. Dehpour. 2006. Involvement of nitric oxide pathway in the acute anticonvulsant effect of melatonin in mice. *Epilepsy Res* 68: 103–113.

Yildirim, M., M. Ayyildiz, and E. Agar. 2010. Endothelial nitric oxide synthase activity involves in the protective effect of ascorbic acid against penicillin-induced epileptiform activity. *Seizure* 19: 102–108.

Yildirim, M. and C. Marangoz. 2006. Anticonvulsant effects of melatonin on penicillin-induced epileptiform activity in rats. *Brain Res* 1099: 183–188.

33 Analgesic Effects of Melatonin

Selçuk Yavuz, Mehmet Turgut,
Özlem Yalçınkaya Yavuz, and Teoman Aydın

CONTENTS

33.1 INTRODUCTION

Melatonin (Mel), *N*-acetyl-5-methoxytryptamine, is an endogenous chronobiotic and neurohormone produced mainly by the pineal gland, which may have an important role in the regulation of various neuroendocrine functions, including circadian rhythm, mood regulation, and sleep, in addition to its effects on antimitotic activity, immunological functioning, antioxidant effect, vasoregulation, and pain modulation (Figure 33.1) (Brzezinski et al. 2005, Dominguez-Rodriguez et al. 2010, Doolen et al. 1998, Lewy et al. 2006, Maestroni 2001, Pandi-Perumal et al. 2008, Reiter et al. 2009, Srinivasan et al. 2008). Analgesic effects of Mel have been shown in both experimental and clinical studies (Citera et al. 2000, Lu et al. 2005, Naguib et al. 2003b, Pang et al. 2001, Press et al. 1998, Tu et al. 2004, Ulugol et al. 2006, Yousaf et al. 2010).

33.2 BIOSYNTHESIS AND METABOLISM OF MELATONIN

Firstly, Lerner identified Mel neurohormone in 1958 (Lerner et al. 1958). L-tryptophan is the amino acid precursor of Mel. Biochemically, synthesis of serotonin from tryptophan is a two-step process. Importantly, serotonin is converted into *N*-acetylserotonin by the enzyme arylalkylamine *N*-acetyltransferase (AANAT), while *N*-acetylserotonin is catalyzed into Mel by the enzyme hydroxyindole-*O*-methyltransferase (HIOMT) (Figure 33.2) (Axelrod and Wurtman 1968). Afterward, Mel is released into the third ventricle directly while the remainder is secreted into circulation, and it is distributed to whole body (Cardinali and Pevet 1998, Tricoire et al. 2003).

In vertebrates, circulating Mel is produced mainly in the pineal gland (Claustrat et al. 2005). *N*-acetylation of serotonin is the rate-limiting reaction (Ackermann and Stehle 2006). This reaction's enzyme, HIOMT, is regulated by the circadian pacemaker in the suprachiasmatic nucleus (SCN) via the light–dark cycle. Mel secretion increases at night and decreases during the day (Figure 33.3) (Klein and Moore 1979).

Anatomically, the SCN controls pineal secretion through a multisynaptic neural pathway descending fiber tracts from the paraventricular nucleus of the hypothalamus project to the superior cervical ganglion. From this point, postganglionic sympathetic fibers reach the pineal gland and modulate Mel synthesis through presynaptic secretion of noradrenaline (NA) (Buijs et al. 1998). At night,

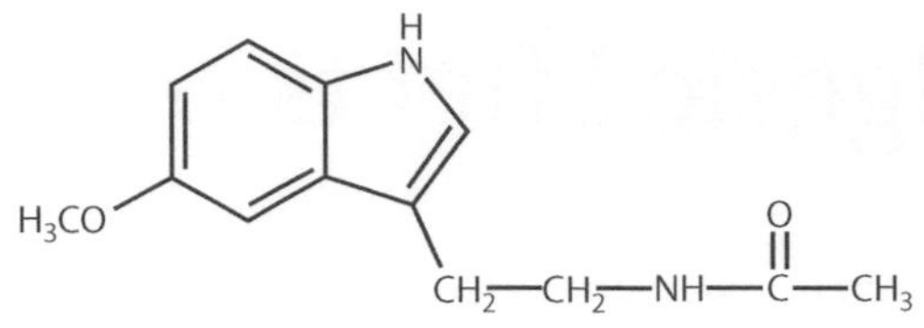

FIGURE 33.1 Chemical structure of melatonin.

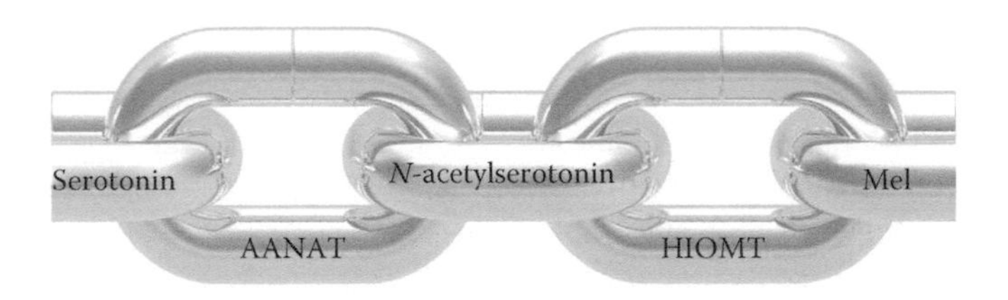

FIGURE 33.2 Schematic drawing of melatonin synthesis pathway. AANAT, arylalkylamine *N*-acetyltransferase; HIOMT, hydroxyindole-O-methyltransferase; Mel, melatonin.

FIGURE 33.3 Schematic representation of circadian pattern of melatonin. Mel, melatonin.

released NA stimulates postsynaptic α_1 and β adrenoceptors. This stimulation causes an increase in 3′,5′-cyclic adenosine monophosphate (cAMP) accumulation, which promotes subsequent activation of AANAT and biosynthesis of Mel (Klein and Moore 1979, Sugden 1989). Nocturnal exposure to light represses production of Mel immediately via degradation of pineal AANAT (Figure 33.4) (Gastel et al. 1998).

Cytochrome P-450 monooxygenase enzymes (CYPA2 and CYP1A) metabolize circulating Mel in the liver (Claustrat et al. 2005). In addition, Mel in tissues, especially in the nervous system, is metabolized into kynurenic acid via oxidative pyrrole ring cleavage (Hardeland et al. 2009, Hussain et al. 2011). Antioxidant and anti-inflammatory properties of Mel are associated with these metabolites, and lastly, a conjugated form of Mel is eliminated through the kidney (Skene et al. 2001).

33.3 MELATONIN RECEPTORS AND TRANSDUCTION SYSTEMS

Mel may diffuse through cell membranes; therefore, it executes its actions in all tissues of the body that are receptor dependent or receptor independent (Srinivasan et al. 2010). There are a total of four different Mel receptor subtypes: two are nuclear receptors, while the remaining two are membrane-associated (MT_1 and MT_2) receptors (Dubocovich and Markowska 2005). MT_1 and MT_2 are the members of the family of seven-transmembrane G-protein-coupled cell surface receptors

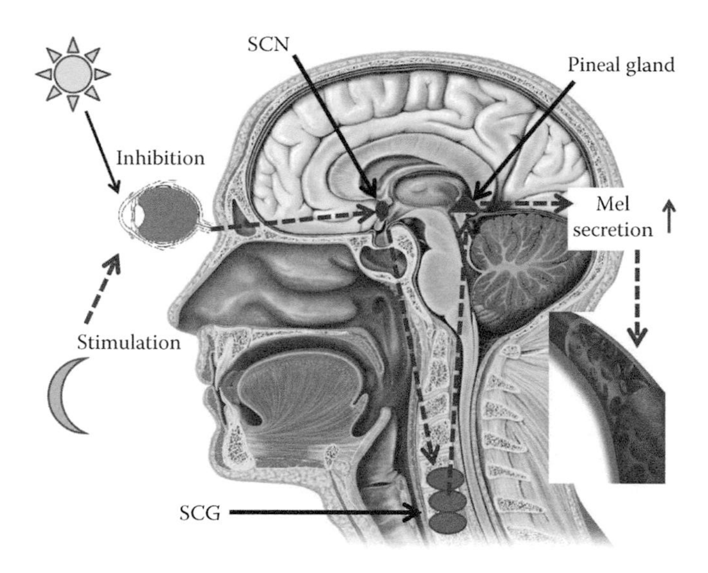

FIGURE 33.4 Anatomical structures related to melatonin secretion. Mel, melatonin; SCG, superior cervical ganglion; SCN, suprachiasmatic nucleus.

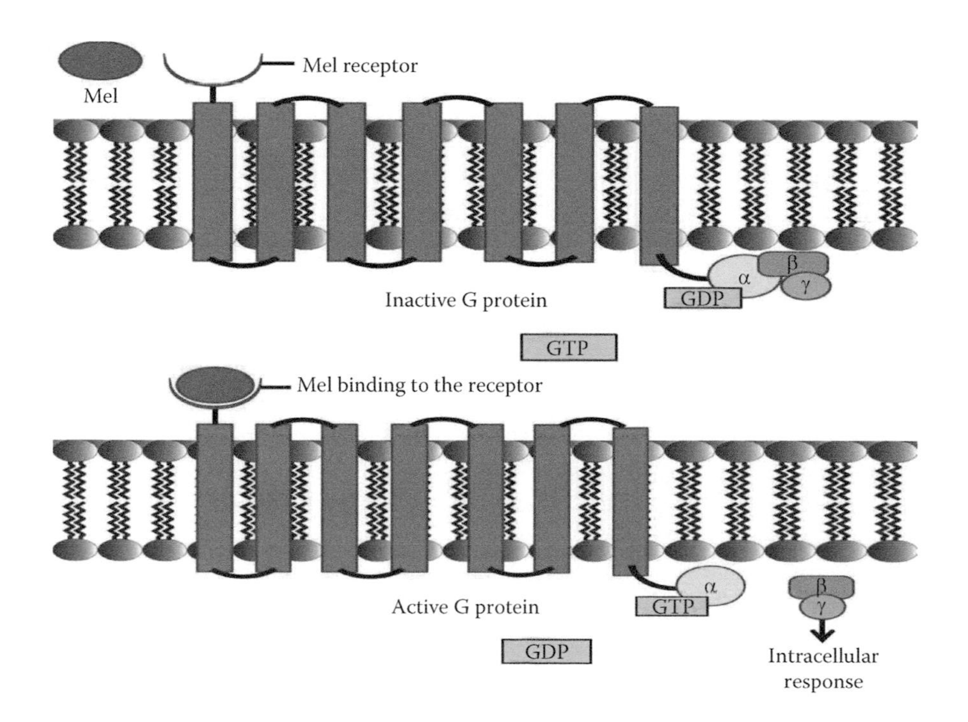

FIGURE 33.5 A total of seven-transmembrane domains known as G-protein-coupled receptors. Mel, melatonin.

(Figure 33.5) (Reppert 1997). In addition, several authors assert a Mel receptor subtype initially called MT_3, but recent studies provide evidence that MT_3 is more closely akin to that of a cytosolic quinine reductase 2 enzyme, not to a receptor (Mailliet et al. 2005).

Mel binds retinoid orphan nuclear hormone receptors (RZRα and RZRβ) as well (Becker-Andre et al. 1994). Park et al. (1997) investigated the distribution of RZRβ mRNA in the nervous system using in situ hybridization in rats. They showed that the RZRβ mRNA was expressed in the dorsal horn of the spinal cord, but not in the motor control region. These results clearly show that RZRβ plays a particular role as a transcription factor in the nervous system (Park et al. 1997). On the other hand, RZRα is involved in inflammatory reactions. Thus, Mel can repress 5-lipooxygenase (5-LOX) expression via RZRα receptors (Steinhilber et al. 1995). It is possible that some actions of Mel on pain regulation may be mediated via nuclear receptors, but details of the topic are still unknown (Ambriz-Tututi et al. 2009).

Activation of MT_1 and MT_2 receptors by Mel generally leads to a decrease in cAMP concentration. Both antinociceptive and anti-inflammatory effects of Mel occur via this pathway (Browning et al. 2000). Further, it is reported that Mel can block calmodulin (CaM) (Soto-Vega et al. 2004). This finding is crucial because CaM kinase II is found in the nervous system and spinal cord, which regulates calcium signaling. Some authors revealed an increased expression and phosphorylation of CaM kinase II after noxious stimulation (Fang et al. 2002).

Both MT_1 and MT_2 receptors are identified in neural tissues to take part in nociceptive transmission. Based on autoradiography studies, it has been shown that Mel receptors are expressed in the hypothalamus, thalamus, and anterior part of the pituitary gland, spinal trigeminal tract, trigeminal nucleus, and the dorsal horn of the spinal cord (Weaver et al 1989, Williams et al. 1995). Analgesic effects of Mel may be related to activation of Mel receptors, which exist in both spinal cord and different brain regions (Srinivasan et al. 2010). It has been demonstrated that Mel has various effects that are mostly inhibitory on spinal nociception (Laurido et al. 2002, Noseda et al. 2004).

33.4 EFFECTS OF MELATONIN ON NOCICEPTION

Mel produces a significant decrease in nociceptive response to painful stimuli. In animal studies of acute, inflammatory, or neuropathic pain, it is shown that Mel has strong and long-term antinociceptive effects (Bilici et al. 2002, Cuzzocrea et al. 1999, Mantovani et al. 2006, Padhy and Kumar 2005, Raghavendra et al. 2000, Ray et al. 2004). Effects of Mel on pain were investigated in 1969 firstly, suggesting mice are less susceptible to painful stimuli in the dark and more susceptible to morphine (Morris and Lutsch 1969). They proposed the result owing to high plasma Mel levels during darkness (Morris and Lutsch 1969). In another study, authors showed that exogenic Mel is able to generate analgesia in mice (Lakin et al. 1981). They also reported that the antinociception derived by Mel can be inhibited by naloxone, suggesting the involvement of opioid receptors in Mel actions (Lakin et al. 1981). Many experimental studies using rats also demonstrate that the analgesic effects of Mel are dose dependent (Naguib et al. 2003b, Noseda et al. 2004). In experimental studies, no significant analgesic effect was found when 0.1–20 mg/kg Mel is given (Pang et al. 2001). However, an important increase in analgesic effects was reported when 25 mg/kg or higher Mel was administered (Ray et al. 2004).

Many authors investigated endotoxin-induced hyperalgesia as a model to assess antinociceptive effects of Mel in inflammation using injection of either latex, lipopolysaccharide (LPS), capsaicin, or formalin (Mantovani et al. 2006, Padhy and Kumar 2005, Raghavendra et al. 2000, Ray et al. 2004). It has been reported that Mel has reduced the pain induced by inflammatory mediators (Padhy and Kumar 2005).

In other inflammatory models, Mel was shown to attenuate hyperalgesic and inflammatory responses, and Mel reduces both nociception and the inflammation in animal models (Raghavendra et al. 2000). Naltrexone cannot reverse the reduction of LPS-induced hyperalgesia by Mel. In vitro tests clearly demonstrate that Mel blocked TNF-α release from macrophages (Raghavendra et al. 2000). Moreover, Mel inhibits secretion of inflammatory mediators, in addition to the aggregation of polymorphonuclears at the inflammatory site (Cuzzocrea et al. 1999). In addition, Mel decreases the levels of nitric oxide and malondialdehyde, which are associated with inflammation (Bilici et al. 2002).

It is well known that any damage or disease that affects the nervous system can cause neuropathic pain (McCleane 2003). Mel can reduce neuropathic pain in animals. In an animal neuropathic pain model, the sciatic nerve was cut and ligated after 2–3 weeks; Mel significantly reduced thermal hyperalgesia and antiallodynic effects (Ulugol et al. 2006). Thus, it is obvious that Mel has antinociceptive and antiallodynic effects. These effects have been shown in diabetic rat animal model (Arreola-Espino et al. 2007). Moreover, oral administration of Mel substantially alleviated formalin-induced allodynia (Arreola-Espino et al. 2007).

In humans, analgesic effects of Mel on pain have not been studied, but it is a promising alternative treatment for patients with chronic conditions like fibromyalgia (FM). In a double-blind, placebo-controlled study, FM patients took fluoxetine alone or in combination with various doses of Mel (Hussain et al. 2011). Combination Mel and fluoxetine showed a highly important reduction in depressive symptoms and a significant reduction in anxiety score and fatigue (Hussain et al. 2011). They concluded that combination of Mel with fluoxetine decreases the complaints of the patients (Hussain et al. 2011).

In a systematic review, a total of eight studies were reviewed for the efficacy of Mel in irritable bowel syndrome (IBS), and it was concluded that exogenous Mel reduces pain and recovers IBS symptoms (Mozaffari et al. 2010). Saha et al. (2007) also demonstrated that Mel is effective in healing extracolonic complaints in IBS.

The relationship between migraine and secretion of Mel is examined in three studies (Brun et al. 1995, Claustrat et al. 1997, Masruha et al. 2008). In the first study, female patients urine specimens were collected, and authors found that urinary Mel level is significantly higher in controls compared to migraine patients (Brun et al. 1995). In a clinical study, a total of 146 migraine patients and 74 control subjects were included (Masruha et al. 2008). They showed that the urinary concentration of 6-sulphatoxymelatonin was significantly higher in patients without pain and control subjects than in migraine sufferers (Masruha et al. 2008). Based on their findings, they concluded that there may be a correlation between Mel secretion and pain, and Mel may be useful for the treatment of chronic pain conditions (Claustrat et al. 1997, Masruha et al. 2008, Peres et al. 2004).

In a randomized, double-blind, placebo-controlled trial, 32 females with myofascial temporomandibular disorder were included to investigate the effects of Mel on pain and sleep (Vidor et al. 2012). Based on their findings, they concluded that Mel provides a relief in pain compared with placebo and it improves sleep quality, while the effect of Mel on pain is not dependent of its effect on sleep quality (Vidor et al. 2012).

In a systematic review, Yousaf et al. (2010) investigated the use of Mel as an analgesic or anxiolytic in perioperative patients. They found 9 out of 10 studies statistically significant with respect to the reduction of preoperative anxiety with Mel premedication compared to placebo (Yousaf et al. 2010). Based on their findings, they concluded that Mel premedication is effective in alleviating anxiety in patients in preoperative period, but its analgesic effects are not clear in the perioperative period (Yousaf et al. 2010).

33.5 MECHANISMS OF ANALGESIC ACTION OF MELATONIN

MT_1 and MT_2 Mel receptors may be involved in the analgesic action of Mel. This possibility is investigated with the usage of luzindole, an unspecific MT_1/MT_2 receptor antagonist, or by the usage of specific MT_2 antagonists 4P-PDOT and K185 (Ambriz-Tututi and Granados-Soto 2007, Mantovani et al. 2006, Onal et al. 2004, Wu et al. 1998, Yoon et al. 2008).

In an experimental study using rats, it was demonstrated that intracerebroventricular injection of luzindole completely antagonized the antinociceptive effect of Mel (Laurido et al. 2002). Likewise, oral or intrathecal administration of highly selective MT_2 receptor antagonists 4P-PDOT and K185 decreased the antiallodynic effects of Mel (Ambriz-Tututi and Granados-Soto 2007). Subcutaneous or intrathecal treatment with naltrexone reduced the antiallodynic effects induced by oral or intrathecal Mel (Ambriz-Tututi and Granados-Soto 2007, Arreola-Espino et al. 2007). Based on these findings, we suggest that MT_2 receptors in spinal cord and opioid receptors play a role in the antiallodynic effect of Mel. Recently, some authors found that the antiallodynic effects of Mel were blocked by naltrexone and the MT_2 receptor antagonists, luzindole, and 4P-PDOT (Ambriz-Tututi and Granados-Soto 2007, Arreola-Espino et al. 2007). Thus, it is obvious that the antinociceptive effects of Mel are related to opioid and MT_2 receptor activation (Ambriz-Tututi et al. 2009).

In light of these studies, several Mel analogues have been developed, and some of them are more effective than Mel in binding to MT_1 and MT_2 Mel receptors (Garratt and Tsotinis 2007). In particular, 2-bromomelatonin binds to Mel receptors with a high affinity compared to Mel (Duranti et al. 1992). It has been reported that intraperitoneal injection of 2-bromomelatonin at 30 and 45 mg/kg doses results in substantial antinociception (Naguib et al. 2003a). It has been reported that intrathecal administration of 2-bromomelatonin is effective in decreasing mechanical nociception (Onal et al. 2004).

Evidence increasingly indicates that reactive oxygen species (ROS) are implicated in neuropathic and inflammatory pain (Close et al. 2005, Sung and Wong 2007). Recurrent painful stimulus causes an increase in oxidative stress and free radical production, and several studies confirm a relationship between hyperalgesia and increased free radical production (Kim et al. 2004, Rokyta et al. 2004, Schwartz et al. 2009, Wang et al. 2004). The analgesic activity of Mel may depend on ROS scavenging, because Mel and its metabolites are strong antioxidants (Tan et al. 2007).

33.6 CONCLUSION

Although the mechanism underlying Mel-induced analgesia is under debate, the potential role of Mel cannot be ignored. Experimentally, the potential role of Mel in the treatment of painful disorders has been demonstrated in different small animal models. As with all new drugs, additional well-designed randomized controlled trials are necessary to compare the effects of Mel on pain with other pharmacological interventions. Due to its high safety profile, it is possible that Mel has an important role as a new analgesic therapy in future.

REFERENCES

Ackermann, K. and J. H. Stehle. 2006. Melatonin synthesis in the human pineal gland: Advantages, implications, and difficulties. *Chronobiol Int* 23(1–2):369–379.

Ambriz-Tututi, M. and V. Granados-Soto. 2007. Oral and spinal melatonin reduces tactile allodynia in rats via activation of MT2 and opioid receptors. *Pain* 132(3):273.

Ambriz-Tututi, M., H. I. Rocha-González, S. L. Cruz, and V. Granados-Soto. 2009. Melatonin: A hormone that modulates pain. *Life Sci* 84(15):489–498.

Arreola-Espino, R., H. Urquiza-Marin, M. Ambriz-Tututi, C. I. Araiza-Saldana, N. L. Caram-Salas, H. I. Rocha-Gonzalez, T. Mixcoatl-Zecuatl, and V. Granados-Soto. 2007. Melatonin reduces formalin-induced nociception and tactile allodynia in diabetic rats. *Eur J Pharmacol* 577(1–3):203–210.

Axelrod, J. and R. J. Wurtman. 1968. Photic and neural control of indoleamine metabolism in the rat pineal gland. *Adv Pharmacol* 6(Pt A):157–166.

Becker-Andre, M., I. Wiesenberg, N. Schaeren-Wiemers, E. Andre, M. Missbach, J. H. Saurat, and C. Carlberg. 1994. Pineal gland hormone melatonin binds and activates an orphan of the nuclear receptor superfamily. *J Biol Chem* 269(46):28531–28534.

Bilici, D., E. Akpinar, and A. Kiziltunc. 2002. Protective effect of melatonin in carrageenan-induced acute local inflammation. *Pharmacol Res* 46(2):133–139.

Browning, C., I. Beresford, N. Fraser, and H. Giles. 2000. Pharmacological characterization of human recombinant melatonin MT(1) and MT(2) receptors. *Br J Pharmacol* 129(5):877–886.

Brun, J., B. Claustrat, P. Saddier, and G. Chazot. 1995. Nocturnal melatonin excretion is decreased in patients with migraine without aura attacks associated with menses. *Cephalalgia* 15(2):136–139.

Brzezinski, A., M. G. Vangel, R. J. Wurtman, G. Norrie, I. Zhdanova, A. Ben-Shushan, and I. Ford. 2005. Effects of exogenous melatonin on sleep: A meta-analysis. *Sleep Med Rev* 9(1):41–50.

Buijs, R. M., M. H. Hermes, and A. Kalsbeek. 1998. The suprachiasmatic nucleus-paraventricular nucleus interactions: A bridge to the neuroendocrine and autonomic nervous system. *Prog Brain Res* 119:365–382.

Cardinali, D. P. and P. Pevet. 1998. Basic aspects of melatonin action. *Sleep Med Rev* 2(3):175–190.

Citera, G., M. A. Arias, J. A. Maldonado-Cocco, M. A. Lazaro, M. G. Rosemffet, L. I. Brusco, E. J. Scheines, and D. P. Cardinalli. 2000. The effect of melatonin in patients with fibromyalgia: A pilot study. *Clin Rheumatol* 19(1):9–13.

Claustrat, B., J. Brun, and G. Chazot. 2005. The basic physiology and pathophysiology of melatonin. *Sleep Med Rev* 9(1):11–24.

Claustrat, B., J. Brun, M. Geoffriau, R. Zaidan, C. Mallo, and G. Chazot. 1997. Nocturnal plasma melatonin profile and melatonin kinetics during infusion in status migrainosus. *Cephalalgia* 17(4):511–517; discussion 487.

Close, G. L., T. Ashton, A. McArdle, and D. P. M. MacLaren. 2005. The emerging role of free radicals in delayed onset muscle soreness and contraction-induced muscle injury. *Comp Biochem Physiol A Mol Integr Physiol* 142(3):257–266.

Cuzzocrea, S., G. Costantino, E. Mazzon, and A. P. Caputi. 1999. Regulation of prostaglandin production in carrageenan-induced pleurisy by melatonin. *J Pineal Res* 27(1):9–14.

Dominguez-Rodriguez, A., P. Abreu-Gonzalez, J. J. Sanchez-Sanchez, J. C. Kaski, and R. J. Reiter. 2010. Melatonin and circadian biology in human cardiovascular disease. *J Pineal Res* 49(1):14–22.

Doolen, S., D. N. Krause, M. L. Dubocovich, and S. P. Duckles. 1998. Melatonin mediates two distinct responses in vascular smooth muscle. *Eur J Pharmacol* 345(1):67–69.

Dubocovich, M. L. and M. Markowska. 2005. Functional MT1 and MT2 melatonin receptors in mammals. *Endocrine* 27(2):101–110.

Duranti, E., B. Stankov, G. Spadoni, A. Duranti, V. Lucini, S. Capsoni, G. Biella, and F. Fraschini. 1992. 2-Bromomelatonin: Synthesis and characterization of a potent melatonin agonist. *Life Sci* 51(7):479–485.

Fang, L., J. Wu, Q. Lin, and W. D. Willis. 2002. Calcium-calmodulin-dependent protein kinase II contributes to spinal cord central sensitization. *J Neurosci* 22(10):4196–4204.

Garratt, P. J. and A. Tsotinis. 2007. Synthesis of compounds as melatonin agonists and antagonists. *Mini Rev Med Chem* 7(10):1075–1088.

Gastel, J. A., P. H. Roseboom, P. A. Rinaldi, J. L. Weller, and D. C. Klein. 1998. Melatonin production: Proteasomal proteolysis in serotonin *N*-acetyltransferase regulation. *Science* 279(5355):1358–1360.

Hardeland, R., D. X. Tan, and R. J. Reiter. 2009. Kynuramines, metabolites of melatonin and other indoles: The resurrection of an almost forgotten class of biogenic amines. *J Pineal Res* 47(2):109–126.

Hussain, S. A., I. I. Al-Khalifa, N. A. Jasim, and F. I. Gorial. 2011. Adjuvant use of melatonin for treatment of fibromyalgia. *J Pineal Res* 50(3):267–271.

Kim, H. K., S. K. Park, J. L. Zhou, G. Taglialatela, K. Chung, R. E. Coggeshall, and J. M. Chung. 2004. Reactive oxygen species (ROS) play an important role in a rat model of neuropathic pain. *Pain* 111(1):116–124.

Klein, D. C. and R. Y. Moore. 1979. Pineal *N*-acetyltransferase and hydroxyindole-*O*-methyltransferase: Control by the retinohypothalamic tract and the suprachiasmatic nucleus. *Brain Res* 174(2):245–262.

Lakin, M. L., C. H. Miller, M. L. Stott, and W. D. Winters. 1981. Involvement of the pineal gland and melatonin in murine analgesia. *Life Sci* 29(24):2543–2551.

Laurido, C., T. Pelissie, R. Soto-Moyano, L. Valladares, F. Flores, and A. Hernandez. 2002. Effect of melatonin on rat spinal cord nociceptive transmission. *Neuroreport* 13(1):89–91.

Lerner, A. B., J. D. Case, Y. Takahashi, T. H. Lee, and W. Mori. 1958. Isolation of melatonin, the pineal gland factor that lightens melanocytes. *J Am Chem Soc* 80(10):2587.

Lewy, A. J., J. Emens, A. Jackman, and K. Yuhas. 2006. Circadian uses of melatonin in humans. *Chronobiol Int* 23(1–2):403–412.

Lu, W. Z., K. A. Gwee, S. Moochhalla, and K. Y. Ho. 2005. Melatonin improves bowel symptoms in female patients with irritable bowel syndrome: A double-blind placebo-controlled study. *Aliment Pharmacol Ther* 22(10):927–934.

Maestroni, G. J. 2001. The immunotherapeutic potential of melatonin. *Expert Opin Investig Drugs* 10(3):467–476.

Mailliet, F., G. Ferry, F. Vella, S. Berger, F. Coge, P. Chomarat, C. Mallet et al. 2005. Characterization of the melatoninergic MT3 binding site on the NRH:quinone oxidoreductase 2 enzyme. *Biochem Pharmacol* 71(1–2):74–88.

Mantovani, M., M. P. Kaster, R. Pertile, J. B. Calixto, A. L. Rodrigues, and A. R. Santos. 2006. Mechanisms involved in the antinociception caused by melatonin in mice. *J Pineal Res* 41(4):382–389.

Masruha, M. R., D. S. de Souza Vieira, T. S. Minett, J. Cipolla-Neto, E. Zukerman, L. C. Vilanova, and M. F. Peres. 2008. Low urinary 6-sulphatoxymelatonin concentrations in acute migraine. *J Headache Pain* 9(4):221–224.

McCleane, G. 2003. Pharmacological management of neuropathic pain. *CNS Drugs* 17(14):1031–1043.

Morris, R. W. and E. F. Lutsch. 1969. Daily susceptibility rhythm to morphine analgesia. *J Pharm Sci* 58(3):374–376.

Mozaffari, S., R. Rahimi, and M. Abdollahi. 2010. Implications of melatonin therapy in irritable bowel syndrome: A systematic review. *Curr Pharm Des* 16(33):3646–3655.

Naguib, M., M. T. Baker, G. Spadoni, and M. Gregerson. 2003a. The hypnotic and analgesic effects of 2-bromomelatonin. *Anesth Analg* 97(3):763–768.

Naguib, M., D. L. Hammond, P. G. Schmid, M. T. Baker, J. Cutkomp, L. Queral, and T. Smith. 2003b. Pharmacological effects of intravenous melatonin: Comparative studies with thiopental and propofol. *Br J Anaesth* 90(4):504–507.

Noseda, R., A. Hernandez, L. Valladares, M. Mondaca, C. Laurido, and R. Soto-Moya. 2004. Melatonin-induced inhibition of spinal cord synaptic potentiation in rats is MT2 receptor-dependent. *Neurosci Lett* 360(1–2):41–44.

Onal, S. A., S. Inalkac, S. Kutlu, and H. Kelestimur. 2004. Intrathecal melatonin increases the mechanical nociceptive threshold in the rat. *Agri* 16(4):35–40.

Padhy, B. M. and V. L. Kumar. 2005. Inhibition of *Calotropis procera* latex-induced inflammatory hyperalgesia by oxytocin and melatonin. *Mediators Inflamm* 2005(6):360–365.

Pandi-Perumal, S. R., I. Trakht, G. M. Brown, and D. P. Cardinali. 2008. Melatonin, circadian dysregulation, and sleep in mental disorders. *Prim Psychiatry* 15(5):77.

Pang, C. S., S. F. Tsang, and J. C. Yang. 2001. Effects of melatonin, morphine and diazepam on formalin-induced nociception in mice. *Life Sci* 68(8):943–951.

Park, H. T., Y. J. Kim, S. Yoon, J. B. Kim, and J. J. Kim. 1997. Distributional characteristics of the mRNA for retinoid Z receptor beta (RZR beta), a putative nuclear melatonin receptor, in the rat brain and spinal cord. *Brain Res* 747(2):332–337.

Peres, M. F. P., E. Zukerman, F. da Cunha Tanuri, F. R. Moreira, and J. Cipolla-Neto. 2004. Melatonin, 3 mg, is effective for migraine prevention. *Neurology* 63(4):757.

Press, J., M. Phillip, L. Neumann, R. Barak, Y. Segev, M. Abu-Shakra, and D. Buskila. 1998. Normal melatonin levels in patients with fibromyalgia syndrome. *J Rheumatol* 25(3):551–555.

Raghavendra, V., J. N. Agrewala, and S. K. Kulkarni. 2000. Melatonin reversal of lipopolysaccharides-induced thermal and behavioral hyperalgesia in mice. *Eur J Pharmacol* 395(1):15–21.

Ray, M., P. K. Mediratta, P. Mahajan, and K. K. Sharma. 2004. Evaluation of the role of melatonin in formalin-induced pain response in mice. *Indian J Med Sci* 58(3):122–130.

Reiter, R. J., S. D. Paredes, L. C. Manchester, and D. X. Tan. 2009. Reducing oxidative/nitrosative stress: A newly-discovered genre for melatonin. *Crit Rev Biochem Mol Biol* 44(4):175–200.

Reppert, S. M. 1997. Melatonin receptors: Molecular biology of a new family of G protein-coupled receptors. *J Biol Rhythms* 12(6):528–531.

Rokyta, R., P. Stopka, V. Holecek, K. Krikava, and I. Pekarkova. 2004. Direct measurement of free radicals in the brain cortex and the blood serum after nociceptive stimulation in rats. *Neuro Endocrinol Lett* 25(4):252–258.

Saha, L., S. Malhotra, S. Rana, D. Bhasin, and P. Pandhi. 2007. A preliminary study of melatonin in irritable bowel syndrome. *J Clin Gastroenterol* 41(1):29–32.

Schwartz, E. S., H. Y. Kim, J. Wang, I. Lee, E. Klann, J. M. Chung, and K. Chung. 2009. Persistent pain is dependent on spinal mitochondrial antioxidant levels. *J Neurosci* 29(1):159–168.

Skene, D. J., E. Papagiannidou, E. Hashemi, J. Snelling, D. F. Lewis, M. Fernandez, and C. Ioannides. 2001. Contribution of CYP1A2 in the hepatic metabolism of melatonin: Studies with isolated microsomal preparations and liver slices. *J Pineal Res* 31(4):333–342.

Soto-Vega, E., I. Meza, G. Ramirez-Rodriguez, and G. Benitez-King. 2004. Melatonin stimulates calmodulin phosphorylation by protein kinase C. *J Pineal Res* 37(2):98–106.

Srinivasan, V., S. R. Pandi-Perumal, D. W. Spence, A. Moscovitch, I. Trakht, G. M. Brown, and D. P. Cardinali. 2010. Potential use of melatonergic drugs in analgesia: Mechanisms of action. *Brain Res Bull* 81(4):362–371.

Srinivasan, V., D. W. Spence, S. R. Pandi-Perumal, I. Trakht, A. I. Esquifino, D. P. Cardinali, and G. J. Maestroni. 2008. Melatonin, environmental light, and breast cancer. *Breast Cancer Res Treat* 108(3):339–350.

Steinhilber, D., M. Brungs, O. Werz, I. Wiesenberg, C. Danielsson, J. P. Kahlen, S. Nayeri, M. Schrader, and C. Carlberg. 1995. The nuclear receptor for melatonin represses 5-lipoxygenase gene expression in human B lymphocytes. *J Biol Chem* 270(13):7037–7040.

Sugden, D. 1989. Melatonin biosynthesis in the mammalian pineal gland. *Experientia* 45(10):922–932.

Sung, C. S. and C. S. Wong. 2007. Cellular mechanisms of neuroinflammatory pain: The role of interleukin-1beta. *Acta Anaesthesiol Taiwan* 45(2):103.

Tan, D. X., L. C. Manchester, M. P. Terron, L. J. Flores, and R. J. Reiter. 2007. One molecule, many derivatives: A never-ending interaction of melatonin with reactive oxygen and nitrogen species? *J Pineal Res* 42(1):28–42.

Tricoire, H., M. Moller, P. Chemineau, and B. Malpaux. 2003. Origin of cerebrospinal fluid melatonin and possible function in the integration of photoperiod. *Reprod Suppl* 61:311–321.

Tu, Y., R. Q. Sun, and W. D. Willis. 2004. Effects of intrathecal injections of melatonin analogs on capsaicin-induced secondary mechanical allodynia and hyperalgesia in rats. *Pain* 109(3):340–350.

Ulugol, A., D. Dokmeci, G. Guray, N. Sapolyo, F. Ozyigit, and M. Tamer. 2006. Antihyperalgesic, but not antiallodynic, effect of melatonin in nerve-injured neuropathic mice: Possible involvements of the l-arginine–NO pathway and opioid system. *Life Sci* 78(14):1592–1597.

Vidor, L. P., I. L. S. Torres, I. C. Custódio de Souza, F. Fregni, and W. Caumo. 2012. Analgesic and sedative effects of melatonin in temporomandibular disorders: A double-blind, randomized, parallel-group, placebo-controlled study. *J Pain Symptom Manage* 46(3):422–432.

Wang, Z., F. Porreca, S. Cuzzocrea, K. Galen, R. Lightfoot, E. Masini, C. Muscoli, V. Mollace, M. Ndengele, and H. Ischiropoulos. 2004. A newly identified role for superoxide in inflammatory pain. *J Pharmacol Exp Ther* 309(3):869–878.

Weaver, D. R., S. A. Rivkees, and S. M. Reppert. 1989. Localization and characterization of melatonin receptors in rodent brain by in vitro autoradiography. *J Neurosci* 9(7):2581–2590.

Williams, L. M., L. T. Hannah, M. H. Hastings, and E. S. Maywood. 1995. Melatonin receptors in the rat brain and pituitary. *J Pineal Res* 19(4):173–177.

Wu, J., Q. Lin, D. J. McAdoo, and W. D. Willis. 1998. Nitric oxide contributes to central sensitization following intradermal injection of capsaicin. *Neuroreport* 9(4):589–592.

Yoon, M. H., H. C. Park, W. M. Kim, H. G. Lee, Y. O. Kim, and L. J. Huang. 2008. Evaluation for the interaction between intrathecal melatonin and clonidine or neostigmine on formalin-induced nociception. *Life Sci* 83(25):845–850.

Yousaf, F., E. Seet, L. Venkatraghavan, A. Abrishami, and F. Chung. 2010. Efficacy and safety of melatonin as an anxiolytic and analgesic in the perioperative period: A qualitative systematic review of randomized trials. *Anesthesiology* 113(4):968–976.

34 Effects of Melatonin on the Proliferative and Differentiative Activity of Neural Stem Cells

Saime İrkören, Heval Selman Özkan, and Mehmet Turgut

CONTENTS

34.1 INTRODUCTION

Melatonin (Mel) is a major indole released from the pineal gland and shows various biological effects such as the regulation of circadian rhythms, seasonal reproduction, and body temperature via high-affinity receptors on the cell membrane (Figure 34.1) (Brydon et al. 1999, Niles et al. 2013, Sharma et al. 2006, Sotthibundhu et al. 2010, Witt-Enderby et al. 2000, 2006).

Today, it is widely accepted that Mel is crucial for various functions of the endocrine and immune systems like pubertal development, circadian cycle, and seasonal adaptation in mammals (Shochat et al. 1997, Smirnov 2001, Taguchi et al. 2004). It induces many physiological effects via two high-affinity G protein–coupled receptors, called MT1 and MT2, using various signaling pathways through Gq, Gi, and Gs G proteins (Bellenchi et al. 2013, Fu et al. 2011, Lee et al. 2004, Reiter 1980). In addition to decreased cyclic adenosine monophosphate (cAMP) signaling resulting with inhibition of adenylate cyclase (AC) activity, it has been suggested that Mel activates the mitogen-activated protein kinase (MAPK)-extracellular signal-regulated kinase (ERK) pathway (Chan et al. 2002, Reiter 1991, Roy and Belsham 2002, Sahar and Sassone-Corsi 2012, Salti et al. 2000, Shimazu et al. 2013, Turgut and Kaplan 2001). Furthermore, it has been reported that Mel is an antioxidant agent with protective capacity against ischemic cerebral proteins (Niles et al. 2013).

There is also evidence that Mel interacts with other cellular (Niles et al. 2013, Van Lint et al. 1996). It potentiates a neurotransmitter gamma-aminobutyric acid receptor (GABAAR)-mediated current in the hypothalamus by the MT1 receptor, through inhibiting this current in the hippocampus with the MT2 subtype (Niles et al. 2013). Meanwhile, both of the Mel receptor subtypes are linked to the inhibition of cAMP production; the mechanisms that generate the differences in GABAA receptor responses apparently involve other differing signaling pathways. Consistent with this, transfection studies with human Mel receptors in the human embryonic kidney (HEK) cells show that the MT2, but not the MT1, receptor is linked to inhibition of the cyclic guanosine monophosphate (cGMP)

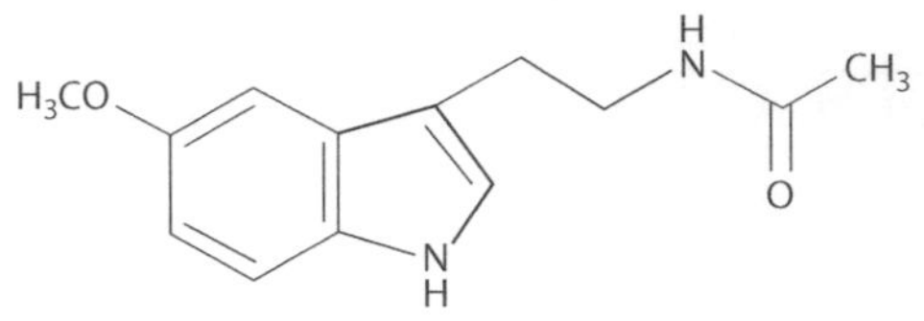

FIGURE 34.1 Chemical structure of melatonin.

pathway (Brydon et al. 1999, Niles et al. 2013, Sharma et al. 2006, Smirnov 2001, Taguchi et al. 2004, Witt-Enderby et al. 2000). Thus, Mel can act together with many cellular targets to produce its various effects (Bellenchi et al. 2013).

At present, there is wide agreement on stem cells playing a pivotal role in tissue repair and plasticity. Stem cells are expressed and can be harvested from embryonic tissue, from umbilical cord plasma, as well as from mature tissue such as peripheral blood, bone marrow, liver, and adipose tissue. They are not focused in any specific cell type as yet, though they have the ability to, on the one hand, reproduce themselves and also differentiate into any specialized cell type (de Munter and Wolters 2013). Neural stem cells (NSCs) are multipotent and can differentiate into glial cells, astrocytes, and oligodendrocytes. Harvesting of these cells is demanding though, and limited number of cells can be obtained (de Munter and Wolters 2013).

Multipotential functions and they were found in the developing and adult brain tissues, particularly in the anterior subventricular and subgranular regions of the hippocampus (Eriksson et al. 1998, Niles et al. 2013). In addition, the proliferative and differentiative functions of the NSCs are not constant; rather they are controlled by lots of humoral factors under physiological and pathophysiological situations (Willaime-Morawek and van der Kooy 2008). Thus, the identification of the molecules regulating NSCs activity may provide the understanding of the neural development and physiology and the generation of novel therapy approaches against diseases involving neural cells.

Various factors have an important effect upon embryonic and adult NSCs survival. Basically, it is divided into intrinsic and extrinsic factors. From these, intrinsic factors contain transcription factors and epigenetic status of chromosomes. Growth factors, morphogens, proteoglycans, cytokines, and hormones influence differentiation and proliferation of NSCs. There many studies in the literature concerning the effects of Mel over the proliferation and differentiation of NSCs (Niles et al. 2013, Sejima et al. 1997). This chapter focuses on the demanding dependence on a highly effective neuroprotectant, which is to be beneficial in the management of traumatic nerve traumas, which generally involves young people; that is why the maximum extent of functional healing is preferred.

34.2 PRODUCTION MECHANISMS OF MELATONIN IN THE PINEAL GLAND

Today, it is accepted that Mel, a kind of indoleamine derived from tryptophan, is found in microorganisms, eukaryotes, herbs, and muticellular organisms. Due to its origin, it is thought that Mel shows many functions derived throughout evolution (Escames et al. 2010, Fu et al. 2011, Niles et al. 2013). In circulation, Mel is produced in the pineal gland as well as in peripheral tissues, and secreted in a circadian manner (Niles et al. 2013). It can enter the central nervous system (CNS) by promptly crossing the blood–brain barrier (BBB), also by the pineal recess, and in damaged brain, directly from the circulation due to permeable BBB. Mel has various crucial roles, comprising control of diurnal patterns, along with visual, procreative, cerebrovascular, endocrine, in addition neuroimmunological actions (Niles et al. 2013). Moreover, Mel has a neuroprotective effect in many pathological conditions of the CNS, including Parkinson, Alzheimer, and also ischemic injury of brain (Barrett et al. 2003, Ladran et al. 2013, Reiter 1980, Shochat et al. 1997, Smirnov 2001). Furthermore, its antioxidant function has attracted much consideration (Bellenchi et al. 2013,

Chan et al. 2002, Lee et al. 2004). Currently, it has been shown that Mel effects cell growth and differentiation of the NSCs (Escames et al. 2010, Kong et al. 2008, Moriya et al. 2007, Niles et al. 2004). However, its vital roles under altered conditions have remained unidentified.

The biosynthesis of Mel has a total of four enzymatic stages. First, tryptophan hydroxylase catalyzes the transformation of tryptophan to 5-hydroxytryptophan, and later decarboxylated via aromatic amino acid decarboxylase produces serotonin. Then, the enzyme arylalkylamine *N*-acetyltransferase (AANAT) converts serotonin to *N*-acetylserotonin by hydroxyindole-*O*-methyltransferase (HIOMT), resulting in the production of Mel (Falcon et al. 2009, Fu et al. 2011). In the blood, Mel concentrations can be up to 0.5 nM (Falcon et al. 2009, Fu et al. 2011, Wan et al. 1999).

It is well known that Mel is also produced in many tissues and organs of the body and this extrapineal production of Mel is more than pineal production (Wan et al. 1999). The genes of the key enzymes for Mel synthesis, AANAT and HIOMT, exist in many tissues (Tan et al. 2003). Several high-affinity cell surface and nuclear receptors for Mel were determined (Butcher et al. 2011). The surface receptors affect intracellular Ca^{2+} levels, AC activity, and inositol phosphate turnover (Butcher et al. 2011). Additionally, nuclear Mel receptors belonging to the retinoid-related orphan receptor and the retinoid Z receptor family have the higher binding affinity for Mel in the nano or subnanomolar area (Butcher et al. 2011, Niles et al. 2013). Currently, the expression of MT1 receptor mRNA was found in C17.2 NSC series (Ramasamy et al. 2013). Moreover, Mel increases glial cell line–derived neurotrophic factor (GDNF) in glioma cell production and C17.2 NSC line (Kong et al. 2008).

34.3 NEURAL STEM/PROGENITOR CELLS

Neural stem/progenitor cells (NSPCs) exist at several locations and times during embryonic and adult development. NSPCs are responsible for the development of the growing brain throughout embryogenesis; besides, NSPCs play a role in learning and memory but do not typically contribute to regenerative repair in adults. Neurogenesis shows the entire process of neuronal progress in embryonic phases. Adult neurogenesis is sensitive to external and internal stimuli at nearly every stage (Fu et al. 2011). Previously, it was known as that there are no stem cells in adult brain according to statement by Ramon and Cajal (1952). Now the concept of adult neurogenesis is generally accepted due to the whelming information collected in the last 20 years (Fu et al. 2011, Moriya et al. 2007). NSCs are primarily located in the "neurogenic" areas, that is, the subgranular region (SGZ) in the hippocampus and the subventricular zone (SVZ) of the lateral ventricles where new neuroblasts are unendingly produced and migrate through the stump migratory stream (RMS) to the olfactory bulb (Butcher et al. 2011, Moriya et al. 2007).

Various subtypes of NSPCs can be reported by their expression of unique markers; however, the extracellular signals and intracellular factors in charge for the regulation of NSPC fate and specialization often overlap. Induced NSPCs (iNSPCs) can be produced from healthy and diseased individuals of somatic cells, regulation of NSPCs is progressively important, and iNSPCs have the potential to serve as a refreshing platform for cell-based replacement treatments.

34.4 EFFECTS OF MELATONIN UPON THE NEURAL STEM CELLS

Mel is an important agent in neurogenesis process as suggested in increasing evidence. Nevertheless, there is no clarified role of Mel in the regulation of neurogenesis. It increases neurogenesis in embryologic life, but there are no clear data on adult neural precursor cells from the SVZ (Taguchi et al. 2004). Furthermore, embryonic and adult NSCs have distinct belongings. For example, adult NSCs produce astrocytes, whereas embryonic stem cells are devoted to produces neurons (Niles et al. 2013, Taguchi et al. 2004). Proliferation and neuronal differentiation of embryonic NSCs can be regulated by Mel signaling throughout wide concentrations, including pharmacological doses (Willaime-Morawek and van der Kooy 2008). Moriya et al. (2007) and Niles et al. (2013)

demonstrated that Mel modulates the proliferative and differentiative capacity of the NSCs in mouse brain on embryonic day 15.5 in a concentration and exposure-timing-dependent way. Pharmacological concentrations of Mel decreased the EGF-induced proliferation during the proliferation period, but increased neural differentiation without impacting astroglial specialization. In contrast, these concentrations of Mel diminished 1% fetal bovine serum-induced neurogenesis from NSCs during the differentiation period (Niles et al. 2013). Then again, physiological densities of Mel failed to affect the proliferation and differentiation of the NSCs whether the exposure became in the differentiation and proliferation periods. These results show that pharmacological concentrations of Mel have possible modulatory actions on proliferation and neural differentiation of NSCs, but not physiological concentrations. In addition to these physiological functions, Mel also defends neurons from ischemia–reperfusion injury (Niles et al. 2004, 2013). These observations have reported that Mel has the ability to prevent the death of NSCs. The effects connected with Mel on neurogenesis from the NSCs in the CNS, even so, are not known. Some authors showed that this indole affects cell development as well as differentiation of the PC12 cells and neural blastoma (Niles et al. 2004, Sotthibundhu et al. 2010). Mel has been shown to be a potent free radical scavenger capable of defending oxidative stress in a count of biological systems (Castro et al. 2005, Falcon et al. 2009, Kong et al. 2008, Niles et al. 2004, 2013, Sharma et al. 2008, Taguchi et al. 2004, Tan et al. 2003, Turgut and Kaplan 2001). Due to Mel being lipid soluble, it readily passes across the plasma membrane layer into the cell and connects to a number of reactive products (Niles et al. 2013). Radical scavenging actions are closely related with the concentration of Mel (Butcher et al. 2011, Niles et al. 2013, Stefulj et al. 2001, Taguchi et al. 2004). The effects of 3-methyl-1-phenyl-2-pyrazolin-5(MCI-186), which is known to be an antioxidant (Niles et al. 2013), on the proliferation and the differentiation of the NSCs to confirm the idea that Mel has certain characteristics of the NSCs through radical scavenging activities are being investigated.

In a current research by Sotthibundhu et al. (2010) and Taguchi et al. (2004), it has been revealed that Mel modulates the particular proliferative capability of precursor cells coming from adult mouse SVZ in a concentration-based study (Chan et al. 2002, Taguchi et al. 2004). In addition to increasing the proliferation, their data additionally show that Mel helps the shift of precursor cells into neurons. Still, this specific stimulatory influence is refused at the maximum Mel concentration (10 mM). Likewise, previous studies showed that, when compared with lower doses, 10 mM Mel decreased this mitotic growth associated with astrocytes in both pineal secreted and exogenously implemented conditions (Taguchi et al. 2004).

Under various cell culture conditions, the NSCs can express neurotrophins. This potential was not afflicted with their morphological state, excluding the situation of GDNF mRNA expression that was reduced in cells going through differentiation in FBS-supplemented media. Present evidence shows that NSPCs express MT1 receptors increases the raising data that NSCs can interact with various modulators (Witt-Enderby et al. 2000), and also indicates an earlier function for Mel in the development of CNS. Furthermore, its potential with various neurotrophic components including GDNF or G protein–coupled receptors may have crucial significances for modifying treatment techniques in neurodegenerative diseases, because Mel induces GDNF expression in NSCs.

Recently, Kong et al. (2008) demonstrated that Mel improves the viability of NSCs, which can be according to the current document that Mel improves the growth of the NSPCs in the dentate gyrus of rat pups. Their data demonstrates increasing the viability of NCSs and MT facilitates differentiation of NSC into TH (+) neurons (Kong et al. 2008).

34.5 CONCLUSIONS

Based on current literature data, it is evident that Mel not only has neural protection properties in animal models but is also involved in the proliferation and differentiation of human nerve tissue (Barrett et al. 2003, Niles et al. 2013, Turgut and Kaplan 2001). The results of the studies given earlier suggest that the multifunctional molecule Mel with NSC might be a helpful therapeutic agent regarding treatment method of neural traumas. Based on the findings of these studies, we can say that Mel is secure, nontoxic, and obtainable in natural form for human use as a substance (Taguchi et al. 2004). The outcomes of various experimental reports provide essential data for the efficient layout and setup of clinical studies using Mel as a neuroprotective therapy for traumatic CNS lesions (Kong et al. 2008, Ladran et al. 2013, Niles et al. 2013). This particular finding can also clarify the anticonvulsant effects related with Mel.

REFERENCES

Barrett, P., S. Conway, and P.J. Morgan. 2003. Digging deep-structure-function relationships in the melatonin receptor family. *J Pineal Res* 35(4):221–230.

Bellenchi, G.C., F. Volpicelli, V. Piscopo, C. Perrone-Capano, and U. di Porzio. 2013. Adult neural stem cells: An endogenous tool to repair brain injury? *J Neurochem* 124(2):159–167.

Brydon, L., F. Roka, L. Petit, P. de Coppet, M. Tissot, P. Barrett, P.J. Morgan, C. Nanoff, A.D. Strosberg, and R. Jockers. 1999. Dual signaling of human Mella melatonin receptors via G(i2), G(i3), and G(q/ll) proteins. *Mol Endocrinol* 13(12):2025–2038.

Butcher, A.J., R. Prihandoko, K.C. Kong, A.J. Butcher, R. Prihandoko, K.C. Kong, P. McWilliams et al. 2011. Differential G-protein-coupled receptor phosphorylation provides evidence for a signaling barcode. *J Biol Chem* 286(13):11506–11518.

Castro, L.M., M. Gallant, and L.P. Niles. 2005. Novel targets for valproic acid: Up-regulation of melatonin receptors and neurotrophic factors in C6 glioma cells. *J Neurochem* 95(5):1227–1236.

Chan, A.S., F.P. Lai, R.K. Lo, T.A. Voyno-Yasenetskaya, E.J. Stanbridge, and Y.H. Wong. 2002. Melatonin MT1 and MT2 receptors stimulate c-Jun N-terminal kinase via pertussis toxin-sensitive and -insensitive G proteins. *Cell Signal* 14(3):249–257.

de Munter, J.P. and E.C. Wolters. 2013. Autologous stem cells in neurology: Is there a future? *J Neural Transm* 120(1):65–73.

Eriksson, P.S., E. Perfilieva, T. Bjork-Eriksson, A.M. Alborn, C. Nordborg, D.A. Peterson, and F.H. Gage. 1998. Neurogenesis in the adult human hippocampus. *Nat Med* 4(11):1313–1317.

Escames, G., A. López, J.A. García, L. García, D. Acuña-Castroviejo, J.J. García, and L.C. López. 2010. The role of mitochondria in brain aging and the effects of melatonin. *Curr Neuropharmacol* 8(3):182–193.

Falcon, J., L. Besseau, M. Fuentes, S. Sauzet, E. Magnanou, and G. Boeuf. 2009. Structural and functional evolution of the pineal melatonin system in vertebrates. *Ann N Y Acad Sci* 1163:101–111.

Fu, J., S.D. Zhao, H.J. Liu, Q.H. Yuan, S.M. Liu, Y.M. Zhang, E.A. Ling, and A.J. Hao. 2011. Melatonin promotes proliferation and differentiation of neural stem cells subjected to hypoxia in vitro. *J Pineal Res* 51(1):104–112.

Kong, X., X. Li, Z. Cai, N. Yang, Y. Liu, J. Shu, L. Pan, and P. Zuo. 2008. Melatonin regulates the viability and differentiation of rat midbrain neural stem cells. *Cell Mol Neurobiol* 28(4):569–579.

Ladran, I., N. Tran, A. Topol, and K.J. Brennand. 2013. Neural stem and progenitor cells in health and disease. *Wiley Interdiscip Rev Syst Biol Med* 5(6):701–715. doi: 10.1002/wsbm.1239.

Lee, E.J., T.S. Wu, M.Y. Lee, T.Y. Chen, Y.Y. Tsai, J.I. Chuang, and G.L. Chang. 2004. Delayed treatment with melatonin enhances electrophysiological recovery following transient focal cerebral ischemia in rats. *J Pineal Res* 36(1):33–42.

Moriya, T., N. Horie, M. Mitome, and K. Shinohara. 2007. Melatonin influences the proliferative and differentiative activity of neural stem cells. *J Pineal Res* 42(4):411–418.

Niles, L.P., K.J. Armstrong, L.M. Rincón Castro, C.V. Dao, R. Sharma, C.R. McMillan, L.C. Doering, and D.L. Kirkham. 2004. Neural stem cells express melatonin receptors and neurotrophic factors: Colocalization of the MT1 receptor with neuronal and glial markers. *BMC Neurosci* 5:41.

Niles, L.P., Y. Pan, S. Kang, and A. Lacoul. 2013. Melatonin induces histone hyperacetylation in the rat brain. *Neurosci Lett* 541:49–53.

Ramasamy, S., G. Narayanan, S. Sankaran, Y.H. Yu, and S. Ahmed. 2013. Neural stem cell survival factors. *Arch Biochem Biophys* 534(1–2):71–87.

Ramon, Y. and S. Cajal. 1952. Structure and connections of neurons. *Bull Los Angel Neuro Soc* 17(1–2):5–46.

Reiter, R.J. 1980. The pineal and its hormones in the control of reproduction in mammals. *Endocr Rev* 1(2):109–131.

Reiter, R.J. 1991. Melatonin: The chemical expression of darkness. *Mol Cell Endocrinol* 79(1–3):153–158.

Roy, D. and D.D. Belsham. 2002. Melatonin receptor activation regulates GnRH gene expression and secretion in GT1–7 GnRH neurons. Signal transduction mechanisms. *J Biol Chem* 277(1):251–258.

Sahar, S. and P. Sassone-Corsi. 2012. Circadian rhythms and memory formation: Regulation by chromatin remodeling. *Front Mol Neurosci* 5:37.

Salti, R., F. Galluzzi, G. Bindi, F. Perfetto, R. Tarquini, F. Halberg, and G. Cornélissen. 2000. Nocturnal melatonin patterns in children. *J Clin Endocrinol Metab* 85(6):2137–2144.

Sejima, H., M. Ito, K. Kishi, H. Tsuda, and H. Shiraishi. 1997. Regional excitatory and inhibitory amino acid concentration in pentylenetetrazole kindling and kindled rat brain. *Brain Dev* 19(3):171–175.

Sharma, R., C.R. McMillan, C.C. Tenn, and L.P. Niles. 2006. Physiological neuroprotection by melatonin in a 6-hydroxydopamine model of Parkinson's disease. *Brain Res* 1068(1):230–236.

Sharma, R., T. Ottenhof, P.A. Rzeczkowska, and L.P. Niles. 2008. Epigenetic targets for melatonin: Induction of histone H3 hyperacetylation and gene expression in C17.2 neural stem cells. *J Pineal Res* 45(3):277–284.

Shimazu, T., M.D. Hirschey, J. Newman, W. He, K. Shirakawa, N. Le Moan, C.A. Grueter et al. 2013. Suppression of oxidative stress by beta-hydroxybutyrate, an endogenous histone deacetylase inhibitor. *Science* 339(6116):211–214.

Shochat, T., R. Luboshitzky, and P. Lavie. 1997. Nocturnal melatonin onset is phase locked to the primary sleep gate. *Am J Physiol* 273(1 Pt 2):364–370.

Smirnov, A.N. 2001. Nuclear melatonin receptors. *Biochemistry (Mosc)* 66(1):19–26.

Sotthibundhu, A., P. Phansuwan-Pujito, and P. Govitrapong. 2010. Melatonin increases proliferation of cultured neural stem cells obtained from adult mouse subventricular zone. *J Pineal Res* 49(3):291–300.

Stefulj, J., M. Hortner, M. Ghosh, K. Schauenstein, I. Rinner, A. Wölfler, J. Semmler, and P.M. Liebmann. 2001. Gene expression of the key enzymes of melatonin synthesis in extrapineal tissues of the rat. *J Pineal Res* 30(4):243–247.

Taguchi, A., T. Soma, H. Tanaka, T. Kanda, H. Nishimura, H. Yoshikawa, Y. Tsukamoto et al. 2004. Administration of CD34+ cells after stroke enhances neurogenesis via angiogenesis in a mouse model. *J Clin Invest* 114(3):330–338.

Tan, D.X., L.C. Manchester, R. Hardeland, S. Lopez-Burillo, J.C. Mayo, R.M. Sainz, and R.J. Reiter. 2003. Melatonin: A hormone, a tissue factor, an autocoid, a paracoid, and an antioxidant vitamin. *J Pineal Res* 34(1):75–78.

Turgut, M. and S. Kaplan. 2001. Effects of melatonin on peripheral nerve regeneration. *Recent Pat Endocr Metab Immune Drug Discov* 5(2):100–108.

Van Lint, C., S. Emiliani, and E. Verdin. 1996. The expression of a small fraction of cellular genes is changed in response to histone hyperacetylation. *Gene Expr* 5(4–5):245–253.

Wan, Q., H.Y. Man, F. Liu, J. Braunton, H.B. Niznik, S.F. Pang, G.M. Brown, and Y.T. Wang. 1999. Differential modulation of GABAA receptor function by Mella and Mellb receptors. *Nat Neurosci* 2(5):401–403.

Willaime-Morawek, S. and D. van der Kooy. 2008. Cortex- and striatum-derived neural stem cells produce distinct progeny in the olfactory bulb and striatum. *Eur J Neurosci* 27(9):2354–2362.

Witt-Enderby, P.A., R.S. MacKenzie, R.M. McKeon, E.A. Carroll, S.L. Bordt, and M.A. Melan. 2000. Melatonin induction of filamentous structures in non-neuronal cells that is dependent on expression of the human mt1 melatonin receptor. *Cell Motil Cytoskeleton* 46(1):28–42.

Witt-Enderby, P.A., N.M. Radio, J.S. Doctor, and V.L. Davis. 2006. Therapeutic treatments potentially mediated by melatonin receptors: Potential clinical uses in the prevention of osteoporosis, cancer and as an adjuvant therapy. *J Pineal Res* 41(4):297–305.

35 Role of Melatonin in Collagen Synthesis

Özgür Taşpınar, Mehmet Turgut, Engin Taştaban, and Teoman Aydın

CONTENTS

35.1 INTRODUCTION

35.1.1 Biochemical Structure of Collagen

Classically, collagen is one of the most plentiful structural proteins present in humans. It is well known that collagen includes 1/3 of all protein and 3/4 of the waterless weight of skin. Additionally, it is the most common part of the extracellular matrix (ECM) in humans and has a triple helical structure (Brinckmann 2005). Biochemically, it is made up of restating uncommon amino acids, including glycine (35%), proline and hydroxyproline (21%), and alanine (11%) (Figure 35.1) (Brinckmann 2005).

To the best of our knowledge, a total of 28 different types of collagen comprising at least 46 distinct polypeptide chains have been acknowledged in vertebrates, and many other proteins comprise collagenous areas as well (Brinckmann 2005; Veit et al. 2006). The defining feature of collagen is a graceful structural motif with three polypeptide elements in a left-handed, polyproline II–type helical conformation coil around each other, with a one-residue stagger to form a right-handed triple helix (Brazel et al. 1987). It is widely accepted that it plays a significant role in building the fundamental framework in several tissues, such as bone, skin, tendon, and cartilage (Bachinger et al. 2010). Basically, the collagen family can be categorized into subtypes according to their structural and functional goods (Heino 2007). Structurally, the primary building of the collagen molecule has been identified as (glycine-X-Y)(n) and the collagen structure is generally characterized by the glycine-proline-hyptripeptide (Ramshaw et al. 1998). Glycine makes up 1/3 of all amino acid (Brodsky and Persikov 2005). The triple-helix structure is significant for specific cellular functions, including adhesion and ECM activation and for enzymatic functions such as hydroxylation of lysines and prolines on propeptide by various enzymes (Fields 1995). Although its catabolism by matrix metalloproteinases is not entirely understood to date, it is generally accepted that the triple-helix structure is required for the catabolism of both the collagen molecule and the cell superficial receptors of macrophages (Lauer-Fields et al. 2002).

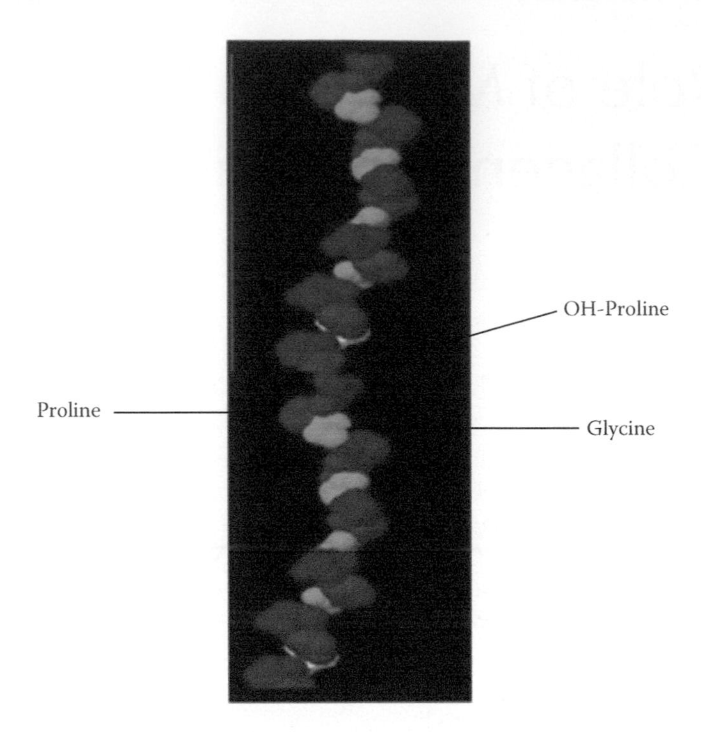

FIGURE 35.1 Biochemical structure of the collagen.

35.1.2 Pineal Gland, Melatonin and Its Receptors

The pineal gland is a little endocrine gland in the brain and the serotonin-derived Mel, *N*-acetyl-5-metyoxytryptamine, is the dominant hormone of the pineal gland. Now, it is accepted that Mel has neuroprotective, free radical scavenging, antioxidative, oncostatic, immunomodulator, and analgesic effects. Furthermore, it regulates wake/sleep patterns and seasonal functions (Macchi and Bruce 2004). Numerous studies revealed that Mel is a known forceful antioxidant and anti-inflammatory agent, by reducing oxidative stress leading to mitochondrial dysfunction (Acuna Castroviej et al. 2002). Mechanism of Mel action is mediated by the binding of the indoleamine to membrane receptors, MT1 and MT2, or retinoid-related orphan nuclear hormone receptor family (RZR/ROR) (Figure 35.2) (Cutando et al. 2011).

Recently, the presence of Mel in hard tissue like bone and teeth has received great attention for researchers. Then, many studies have investigated the effect of Mel on bone remodeling, osteoporosis, osseointegration of dental implants, and dentine formation (Jie et al. 2013). Moreover, there are studies demonstrating antifibrotic effects of Mel in controlling collagen contents in tissues in the current literature (Ogeturk et al. 2008). It is accepted that Mel is an endogenous antioxidant, regulates circadian rhythms, sleep, and immune system activity, behaves as a free radical scavenger, and eliminates oxygen free radicals and reactive intermediates (Allegra et al. 2003).

Moreover, Mel also has an unintended antioxidant influence by enhancing the levels of potential antioxidants, such as glutathione peroxidase (GPx), superoxide dismutase (SOD), and glutathione (GSH) (Rozov et al. 2003). Recent studies have revealed that Mel has a cytoprotective effect on different experimental models of acute liver injuries and it decreases fibroblast proliferation and collagen synthesis, indicating that Mel may have a therapeutic effect on acute and chronic liver injury (Figure 35.3) (Cruz et al. 2005).

On the other hand, Janus and Dabrowski (1994) reported that Mel hormone has various effects on the intact skin. In an experimental study, an increased collagen content in the abdominal cavities of pinealectomized rats has been reported, suggesting the possible role of Mel in the reduction of

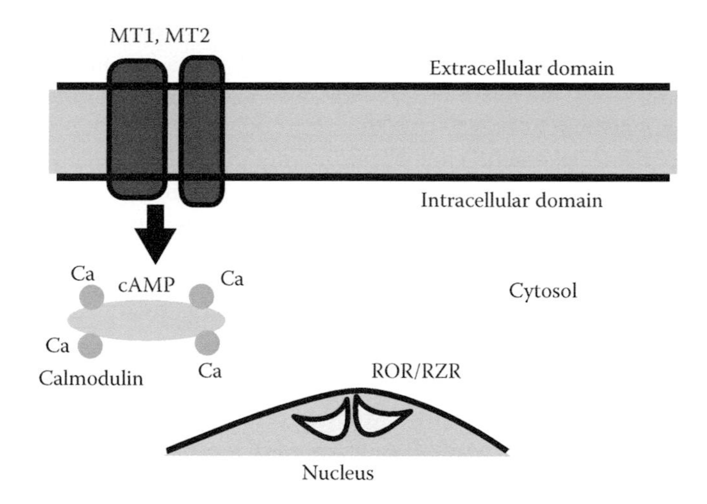

FIGURE 35.2 **(See color insert.)** Schematic illustration of various melatonin receptors.

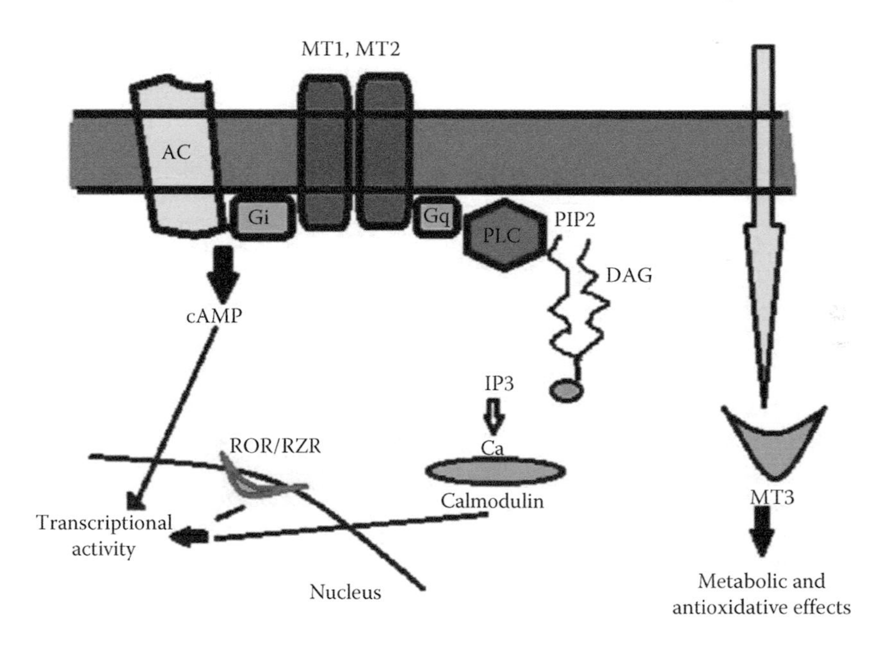

FIGURE 35.3 **(See color insert.)** Schematic illustration of the antioxidative effects of Mel. DAG, diacylglycerol; IP3, inositol trisphosphate; cAMP, cyclic adenosine monophosphate; AC, adenylate cyclase; Gi, Gi protein; Gq, Gq protein; PIP2, phosphatidylinositol 4,5-bisphosphate; PLC, phospholipase C.

collagen accumulation (Drobnik and Dabrowski 1996). In addition, Mel has been found to increase the synthesis of prostaglandin E1 (PGE1), which is an inhibitor of collagen production (Drobnik and Dabrowski 1996).

Collagen fibers provide a support for migrating and proliferating cells and they also determine the tensile strength of wounds (Binnebosel et al. 2009). The last step of the repair is the contraction of the wound and remodeling of the scar, and through contraction, the wound margins are brought closer to each other by myofibroblasts or fibroblasts (Gabbiani et al. 1971). Basically, scar remodeling includes two opposite processes: collagen synthesis and breakdown (Desmouliere et al. 1995). It is reported that remodeling leads to reorganization of collagen fibers, making them better organized and denser, and also to a decrease in the cellularity of the granulation tissue

(Desmouliere et al. 1995). Developments regarding the therapeutic strategies on the repair of both superficial wounds and various internal organs are based on a detailed understanding of different types of wound healing. Thus, the repair process remains under the controlling effects of local and endocrine factors (Drobnik and Dabrowski 2000).

In addition, neuroendocrine reaction to myocardial infarction (MI) is modified by Mel (Ciosek and Drobnik 2012). New data suggest that Mel that is secreted by the pineal gland is intricate in the regulation of collagen deposition in wounds. This effect is dependent on the target organ, the applied dose, and the time of pineal hormone submission (Pugazhenthi et al. 2008). Later, while Mel reduces collagen accumulation in the superficial wound, pinealectomy has the opposite effect. It has been found that it elevates the collagen saturated of wounds in rats with intact hearts, while Mel treatment of the pinealectomized rats reversed the influence of pineal gland removal and controlled collagen content in the superficial wounds (Macchi and Bruce 2004). In contrast to these findings, pineal hormone Mel was observed to elevate collagen levels in the infarcted heart scar (Drobnik et al. 2011). It is accepted that effects of Mel observed in vivo are dependent on the direct response of cells sequestered from the granulation tissue of the wounds (Drobnik and Dabrowski 2000). Both surgical and pharmacological pinealectomies concentrated the collagen level in the MI scar, and the substitution with Mel of the pinealectomized rats normalized collagen content in the infarcted heart scar tissue (Drobnik and Dabrowski 2000). These inconsistencies suggest that various types of wound repair are based on different regulatory mechanisms.

35.1.3 Effects of Melatonin on Collagen Content in the Wound

In a previous study, an increased fibrosis of the retroperitoneal space was detected in pinealectomized rats by Cunnane et al. (1979). Afterward, this finding was supported by the studies on methysergide (an antagonist of serotonin the substrate for synthesis of Mel)-induced fibrosis in humans (Elkind et al. 1968). Based on this finding, it is hypothesized that the patients with retroperitoneal fibrosis as well as heart and pulmonary fibrosis may be treated with inhibition of Mel production by using methysergide (Cunnane et al. 1979). Similarly, an inhibited construction of PGE1, an inhibitor of collagen production, has been found in pinealectomized rats (Cunnane et al. 1979, Reiter 1991). Altogether, the exceeding documents support the hypothesis by Cunnane et al. (1979) that the pineal gland is intricate in fibrosis instruction. Importantly, they found that sunset injections of Mel retarded the restoration of full-thickness circular skin wounds and improved the wound surface area associated to a control (Cunnane et al. 1979). Further, they found that all doses, ranging from 10 μg/100 g to 100 μg/100 g body weight (b.w.), improved the wound surface area Days 7–14 after the damage (Cunnane et al. 1979). On Day 14, although the healing of the control rats was nearly finished, the wound surface area was still large in the Mel-treated animals (Drobnik et al. 1995). Nevertheless, a pinealectomy procedure has the opposite influence to Mel and has been found to decrease the surface area of a wound, thereby accelerating healing process (Drobnik et al. 1995). Based on these findings, it may be concluded that pinealectomy procedure fundamentally reduces the serum Mel levels.

To study the various substances of the ECM in the wounded tissue, some authors implanted sponges subcutaneously in the lumbar region of the rats (Drobnik et al. 1995). It was found that the application of Mel (30 μg/100 g b.w.) reduced the level of collagen in the granulation tissue of the sponge, whereas a pinealectomy exerted the opposite influence and elevated the collagen content (Drobnik et al. 1995). Furthermore, administration of Mel at a 30 μg/100 g b.w. dose to the pinealectomized rats reversed the influence of the pinealectomy and regulated collagen content in the granulation material (Cunnane et al. 1979). Moreover, a reduction of the collagen level and an inhibition of full-thickness rounded wound healing were observed in rats kept in unbroken dark (Cunnane et al. 1979). They noted that in unbroken dark, the rhythm of Mel secretion persisted, and an increasing background level of Mel was still found (Cunnane et al. 1979).

Moreover, Mel has been found to have no influence on the unpolymerized collagen (salt soluble) level in granulation tissue (Drobnik and Dabrowski 1996). The latter effect of Mel is dependent on the time that the indoleamine is applied. Therefore, a morning injection of Mel proliferates the level of collagen in the wound, while a sunset administration of the pineal indoleamine lowers it (Drobnik and Dabrowski 2000). This experiment suggests that the Mel answer to tissue differs through the day (Drobnik and Dabrowski 2000). Bulbuller et al. (2005) found that sunset Mel administration (30 μg/100 g b.w.) reduced the hydroxyproline level in intestine anastomotic wounds in rats, owing to a reduced mechanical conflict of the intestinal wounds. Pinealectomy improved the level of hydroxyproline; nevertheless, it reversed the influence of pineal gland removal and lowered both the level of hydroxyproline in the wound and the bursting force when Mel was applied to the pinealectomized rats (Bulbuller et al. 2005). In the incisional skin wound of the similar experimental model, Mel reduced hydroxyproline satisfied in the scar and lowered the breaking strength of the wound (Bulbuller et al. 2005). In contrast, a pinealectomy elevated the hydroxyproline level in the incisional wound and this influence was supplemented by a rise of the breaking strength (Bulbuller et al. 2005). Nevertheless, administration of Mel (30 μg/100 g b.w.) to the pinealectomized rats lowered the hydroxyproline level and reduced the breaking strength of the incisional wounds (Bulbuller et al. 2005).

Nowadays, it has been demonstrated that transforming growth factor (TGF)-beta and basic fibroblast growth factor (bFGF) play a significant role in the collagen construction of fibroblasts and in Schwann cell activity (Turgut et al. 2006a). In a study by Turgut et al. (2006b), the rats were divided into a control group, a Mel-treated group, a surgical pinealectomy group, and a group treated with Mel following pinealectomy. They formerly underwent a surgical sciatic nerve transection and primary suture anastomosis (Turgut et al. 2006b). At 2 months after anastomosis, the animals were sacrificed, and unilateral sciatic nerve samplings, including the anastomotic section, were removed from two animals in each group and processed for immunohistochemical study (Turgut et al. 2006b). On behalf of both antibody, immunoreactivity was assessed using a semiquantitative scoring system; strong TGF-b1 and/or bFGF expression was observed in the epineurium of animals that underwent pinealectomy; nonetheless, not at all or weak expression was observed in animals in the control and Mel treatment groups (Turgut et al. 2006c). Founded on these data, they suggested that both TGF-b1 and bFGF play significant roles in the control of collagen accumulation and neuroma establishment at the anastomotic place and that the pineal neurohormone Mel has a useful influence on nerve regeneration (Turgut et al. 2006c).

In an ischemic wound model in rats, Mel (100 μg/100 g b.w.) did not vary the hydroxyproline content in the wounded set (Ozler et al. 2011). The hydroxyproline concentration in a dorsal burn injury followed by a full-thickness midline skin incision was not influenced by Mel at a dose of 10 mg/kg b.w. practical for 2 days (Basak et al. 2003). It has also been found that when Mel (at a dose of 100 mg/kg) declines the hydroxyproline level in the liver, a dimethylnitrosamine injury of the liver occurs, followed by fibrosis (Tahan et al. 2004). Furthermore, Mel has also been found to decrease fibrosis, as measured by the hydroxyproline content of the livers of rats with ethanol-induced injuries (Drobnik and Dabrowski 1999).

Interestingly, a close relationship has been observed between Mel and Zn or Mg levels in the serum samples of patients with intervertebral disc herniations (IVDs), and the levels of these elements might be affected by the presence of the degeneration process and serum Mel level or vice versa (Turgut et al. 2006d). The consequences achieved in a rodent model of sciatic nerve neuroma formation presented that there was a positive correlation between macroscopic and microscopic observations and that Mel enhanced axonal regeneration, presumably due to its inhibitory influence on neuroma establishment (Turgut et al. 2006d). Pugazhenthi et al. (2008) observed that at a dose of 1.2 mg/kg applied intradermally, Mel enhanced the healing process of full-thickness incisional wounds; an improvement of scarring from Days 14 to 21 post-injury. Histological analyses showed that the collagen fibers of Mel-treated wounds were more matured and resembled fibers in the whole skin, which are more random in construction (Pugazhenthi et al. 2008). The controlling effect of the pineal gland on collagen content was not limited to the injured tissue, nonetheless was also observed in the intact skin (Drobnik and Dabrowski 1999). Therefore, when Mel was applied in the sunset, it reduced the level of collagen in the skin, while a pinealectomy exerted an opposite influence (Drobnik and Dabrowski 1999). In addition, when Mel (30 μg/100 g b.w.) was applied to the pinealectomized animals, it normalized the collagen level in the whole skin (Drobnik and Dabrowski 1999). To explain whether the observed in vivo influence of Mel was an effect of the direct influence of the pineal indoleamine on the cells synthesizing collagen in the wound, in vitro experiments were carried out (Drobnik and Dabrowski 1999).

Electron microscopic studies have showed, primarily, fibroblasts in the culture, along with certain myofibroblasts (Carossino et al. 1996). At a concentration of 10–7 M, Mel was found to rise collagen growth in the culture, nonetheless lower concentrations were unsuccessful (Carossino et al. 1996). The inhibitory influence of Mel observed in vivo cannot be explained by a direct influence of the pineal indoleamine on fibroblasts in the wound. Based on these findings, it is hypothesized that the incidental things of Mel via controlling systems are accountable for the last influence observed in vivo. Minor concentrations of the pineal indoleamine were found to inspire the proliferation of fibroblasts, whereas Mel (100–400 μg/mL) was observed to have an inhibitory effect on the increase of fibroblasts isolated from normal and sclerodermic skin (Carossino et al. 1996). Some experiments have also suggested that the Mel membrane receptors are present in fibroblasts (Choi et al. 2011). Mel-induced increase of Cu/Zn-SOD and glutathione reductase expression in human corneal fibroblasts has been found to be inhibited by lusindol, the nonselective Mel receptor antagonist (Choi et al. 2011). Furthermore, lusindol has been found to decrease the stimulatory influence of Mel on lipid metabolism in the cells (Maestroni 1993). Steady manifestation of both MT1 and MT2 was shown in mouse embryonic fibroblasts (Ishii et al. 2008). Mel was also found to relate with nuclear and cytoskeletal structures of the cell, and is thought to modify its function (Finocchiaro and Glikin 1998).

In brief, the outcomes presented earlier obviously suggest that the pineal gland and its secretory product, Mel, inhibit the accumulation of collagen in healing tissue in different types of wounds, such as circular wounds, incisional wounds, intestinal anastomotic wounds, and ethanol-induced liver fibrosis, etc. (Cunnane et al. 1979). Therefore, the effect of Mel on the healing process is not dependent on the mechanism of healing and the type of the wound. In the case of incisional wounds healed, by first intention, or a sponge implanted subcutaneously, resembling healing by the second intention, Mel has been reported to have certain effects on the collagen content (Cunnane et al. 1979). Accordingly, a similar lowering effect of Mel on collagen accumulation has been observed in liver injuries when pharmacological doses of Mel are given (10 mg/100 g b.w.) (Cunnane et al. 1979). The dose of Mel that reverses the effects of a pinealectomy and normalizes the collagen content in the wounds of pinealectomized animals compared to controls. (Bulbuller et al. 2005). That dose of Mel (30 μg/100 g b.w.) could be seen as being equivalent to the total Mel that expresses a physiological effect. Furthermore, it has been documented that Mel has been found to increase collagen maturation in full-thickness skin wounds (Bulbuller et al. 2005).

Importantly, Turgut et al. (2006c) have suggested that Mel activates the recovery process in degenerated IVD tissue, possibly by stimulating TGF-beta 1 activity. Accordingly, many authors reported that pineal secretory Mel has been shown to stimulate the regeneration process (Turgut et al. 2005a,b, 2006c). To the best of our information, however, this is the first report investigating the involvement of the pineal hormone Mel in the repair of rat IVDs. According to their findings, Mel administration meaningfully inhibits collagen accumulation in the creation of neuroma in the suture repair site and thus increases nerve regeneration (Turgut et al. 2006d). Importantly, they found that the pineal indoleamine reversed the antiangiogenic effects of indometacin in both wound models (Ganguly et al. 2010). The Mel-induced promotion of angiogenesis is possibly linked to the upgraded levels of VEGF and matrix metalloproteinase-2 and decreased tissue inhibitor of metalloprotease-2 (Ganguly et al. 2010).

35.1.3.1 Effects of Melatonin on the Regulation of an Inflammatory Process in the Wound

It is known that Mel has an effect on inducible nitric oxide synthase (iNOS) activity during the wound healing. The iNOS activity and NO content in the wound are thought to induce angiogenesis and granulation tissue formation (Ozler et al. 2011). For these reasons, the inhibition of pineal indoleamine iNOS activity and decreased iNOS protein levels during the inflammatory phase have been recommended (Ozler et al. 2011). In another study, it has been reported that the effect of Mel upon wound healing is inhibited by the MT2 receptor blockade (Konturek et al. 2008). During last two decades, many studies revealed that Mel has antioxidative effects on the wound healing (Konturek et al. 2008). Some authors noted that the malonyldialdehyde level was distinctly reduced by Mel in ischemic wound tissue, but the activities of SOD and GPx were not inclined by the pineal indoleamine (Konturek et al. 2008; Ozler et al. 2011). Mel is thought to have an important role in cell membrane stabilization by lowering lipid oxidation and reducing the augmentation of sulfhydryl groups in proteins (Ganguly and Swarnakar 2009).

35.2 CONCLUSION

Mel is the dominant hormone of the pineal gland. Recent studies have shown that Mel exerts its cytoprotective things in several untried models of acute liver injury and reduces fibroblast proliferation and collagen production, indicating that Mel may have therapeutic things on acute and chronic liver injury through its antioxidant properties. Studies have reported antifibrotic effects of Mel, as well as its properties in controlling collagen levels in different tissues.

REFERENCES

Acuna, C.D., G. Escames, A. Carazo, J. León, H. Khaldy, and R.J. Reiter. 2002. Melatonin, mitochondrial homeostasis and mitochondrial-related diseases. *Curr Top Med Chem* 2, 133–151.

Allegra, M., R.J. Reiter, D.X. Tan, C. Gentile, L. Tesoriere, and M.A. Livrea. 2003. The chemistry of melatonin's interaction with reactive species. *J Pineal Res* 34, 1–10.

Bachinger, H.P., K. Mizuno, J. Vranka, and S. Boudko. 2010. Collagen formation and structure. In: Mander, L. and Liu, H.W. (eds.), *Comprehensive Natural Products II: Chemistry and Biology*. Elsevier Limited, Amsterdam, the Netherlands, pp. 469–530.

Basak, P.Y., F. Agalar, F. Gultekin, E. Eroglu, I. Altuntas, and C. Agalar. 2003. The effect of thermal injury and melatonin on incisional wound healing. *Ulus Travma Acil Cerrahi Derg* 9, 96–101.

Binnebosel, M., K. Junge, D.A. Kaemmer, C.J. Krones, S. Titkova, M. Anurov, V. Schumpelick, and U. Klinge. 2009. Intraperitoneally applied gentamicin increases collagen content and mechanical stability of colon anastomosis in rats. *Int J Colorectal Dis* 24, 433–440.

Brazel, D., I. Oberbaumer, H. Dieringer, W. Babel, R.W. Glanville, R. Deutzmann, and K. Kühn. 1987. Completion of the amino acid sequence of the α1 chain of human basement membrane collagen (type IV) reveals 21 non triplet interruptions located within the collagenous domain. *Eur J Biochem* 168, 529–536.

Brinckmann, J. 2005. Collagens at a glance. *Top Curr Chem* 247, 1–6.

Brodsky, B. and A.V. Persikov. 2005. Molecular structure of the collagen triple helix. *Adv Protein Chem* 70, 301–339.

Bulbuller, N., O. Dogru, H. Yekeler, Z. Cetinkaya, N. Ilhan, and C. Kirkil. 2005. Effect of melatonin on wound healing in normal and pinealectomized rats. *J Surg Res* 123, 3–7.

Carossino, A.M., A. Lombardi, M. Matucci-Cerinic, A. Pignone, and M. Cagnoni. 1996. Effect of melatonin on normal and sclerodermic skin fibroblasts proliferation. *Clin Exp Rheumatol* 14, 493–498.

Choi, S.L., S. Dadakhujaev, H. Ryu, T. Im Kim, and E.K. Kim. 2011. Melatonin protects against oxidative stress in granular corneal dystrophy type 2 corneal fibroblasts by mechanisms that involve membrane melatonin receptors. *J Pineal Res* 51, 94–103.

Ciosek, J. and J. Drobnik. 2012. The function of the hypothalamoneurohypophysial system in rats with myocardial infarction is modified by melatonin. *Pharmacol Rep* 64, 1442–1454.

Cruz, A., F.J. Padillo, E. Torres, C.M. Navarrete, J.R. Muñoz-Castaneda, F.J. Caballero, J. Briceño et al. 2005. Melatonin prevents experimental liver cirrhosis induced by thioacetamide in rats. *J Pineal Res* 39, 143–150.

Cunnane, S.C., M.S. Manku, and D.F. Horrobin. 1979. The pineal and regulation of fibrosis: Pinealectomy as a model of primary biliary cirrhosis: Roles of melatonin and prostaglandins in fibrosis and regulation of T lymphocytes. *Med Hypotheses* 5, 403–414.

Cutando, A., J. Aneiros-Fernandez, A. Lopez-Valverde, S. Arias-Santiago, J. Aneiros-Cachaza, and R.J. Reiter. 2011. A new perspective in oral health: Potential importance and actions of melatonin receptors MT1, MT2, MT3, and RZR/ROR in the oral cavity. *Arch Oral Biol* 56, 944–950.

Desmouliere, A., M. Redaed, I. Darbi, and G. Gabbiani. 1995. Apoptosis mediates the decrease in cellularity during the transition between granulation tissue and scar. *Am J Pathol* 146, 55–66.

Drobnik, J. and R. Dabrowski. 1996. Melatonin suppresses the pinealectomy induced elevation of collagen content in a wound. *Cytobios* 85, 51–58.

Drobnik, J. and R. Dabrowski. 1999. Pinealectomy-induced elevation of collagen content in intact skin is suppressed by melatonin application. *Cytobios* 100, 49–55.

Drobnik, J. and R. Dabrowski. 2000. The opposite effect of morning or afternoon application of melatonin on collagen accumulation in the sponge-induced granuloma. *Neuro Endocrinol Lett* 21(3), 209–212.

Drobnik, J., J. Janus, R. Dabrowski, and A. Szczepanowska. 1995. The healing process, collagen and chondroitin-4-sulphate accumulation in the wound are controlled by the pineal gland. *Pol J Endocrinol* 46, 411–418.

Drobnik, J., D. Slotwinska, S. Olczak, D. Tosik, A. Pieniazek, K. Matczak, A. Koceva-Chyla, and A. Szczepanowska. 2011. Pharmacological doses of melatonin reduce the glycosaminoglycan level within the infarcted heart scar. *J Physiol Pharmacol* 62, 29–35.

Elkind, A.H., A.P. Friedman, A. Bachman, S.S. Siegelman, and O.W. Sacs. 1968. Silent retroperitoneal fibrosis with methysergide therapy. *JAMA* 206, 1041–1044.

Fields, G.B. 1995. The collagen triple-helix: Correlation of conformation with biological activities. *Connect Tissue Res* 31, 235–243.

Finocchiaro, L.M. and G.C. Glikin. 1998. Intracellular melatonin distribution in cultured cell lines. *J Pineal Res* 24, 22–34.

Gabbiani, G., G.B. Ryan, and G. Majno. 1971. Presence of modified fibroblasts in granulation tissue and the possible role in wound contraction. *Experientia* 27, 549–550.

Ganguly, K., A.V. Sharma, R.J. Reiter, and S. Swarnakar. 2010. Melatonin promotes angiogenesis during protection and healing of indometacin-induced gastric ulcer: Role matrix metalloproteinase-2. *J Pineal Res* 49, 130–140.

Ganguly, K. and S. Swarnakar. 2009. Induction of matrix metalloproteinase-9 and -3 in nonsteroidal anti-inflammatory drug-induced acute gastric ulcers in mice: Regulation by melatonin. *J Pineal Res* 47, 43–55.

Heino, J. 2007. The collagen family members as cell adhesion proteins. *Bioessays* 29, 1001–1010.

Ishii, H., N. Tanaka, M. Kobayashi, M. Kato, and Y. Sakuma. 2008. Gene structures biochemical characterization and distribution of rat melatonin receptors. *J Physiol Sci* 59, 37–47.

Janus, D. and R. Dabrowski. 1994. The effect of pinealectomy and exogenous melatonin on some connective tissue elements in the skin (in Polish) XXX. *Congress of Polish Biochemical Society*, Szczecin, Poland, p. 309.

Jie, L., Z. Hong-yu, F. Wen-Guo, D. Wei-Guo, F. Shen-Li, H. Hong-Wen, and H. Fang. 2013. Melatonin influences proliferation and differentiation of rat dental papilla cells in vitro and dentine formation in vivo by altering mitochondrial activity. *J Pineal Res* 54, 170–178.

Konturek, P.C., S.J. Konturek, G. Burnat, T. Brzozowski, and R.J. Reiter. 2008. Dynamic physiological and molecular changes in gastric ulcer healing achieved by melatonin and its precursor L-tryptophan in rats. *J Pineal Res* 45, 180–190.

Lauer-Fields, J.L., D. Juska, and G.B. Fields. 2002. Matrix metalloproteinases and collagen catabolism. *Biopolymers* 66, 19–32.

Macchi, M. and J. Bruce. 2004. Human pineal physiology and functional significance of melatonin. *Front Neuroendocrinol* 25, 177–195.

Maldonado, M.D., A.W. Siu, M. Sanchez-Hidalgo, D. Acuna-Castroviejo, and G. Escames. 2006. Melatonin and lipid uptake by murine fibroblasts: Clinical implications. *Neuroendocrinol Lett* 27, 601–608.

Ogeturk, M., I. Kus, H. Pekmez, H. Yekeler, S. Sahin, and M. Sarsilmaz. 2008. Inhibition of carbon tetrachloride mediated apoptosis and oxidative stress by melatonin in experimental liver fibrosis. *Toxicol Ind Health* 24, 201–208.

Ozler, M., A. Korkmaz, B. Uysal, K. Simsek, C. Ozkan, T. Topal, and S. Oter. 2011. Effects of topical melatonin and vitamin E in rat ischemic wound model. *J Exp Integr Med* 1, 123–129.

Pugazhenthi, K., M. Kapoor, A.N. Clarkson, I. Hall, and I. Appleton. 2008. Melatonin accelerates the process of wound repair in full-thickness incisional wounds. *J Pineal Res* 44, 387–396.

Riquet, F.B., W.F. Lai, J.R. Birkhead, L.F. Suen, G. Karsenty, and M.B. Goldring. 2000. Suppression of type I collagen gene expression by prostaglandins in fibroblasts is mediated at the transcriptional level. *Mol Med* 6, 705–719.

Rozov, S.V., E.V. Filatova, A.A. Orlov, A.V. Volkova, A.R. Zhloba, E.L. Blashko, and N.V. Pozdeyev. 2003. *N*1-acetyl-*N*2-formyl-5-methoxykynuramine is a product of melatonin oxidation in rats. *J Pineal Res* 35, 245–250.

Tahan, V., R. Ozaras, B. Canbakan, H. Uzun, S. Aydin, B. Yildirim, H. Aytekin, G. Ozbay, A. Mert, and H. Senturk. 2004. Melatonin reduces dimethylnitrosamine-induced liver fibrosis in rats. *J Pineal Res* 37, 78–84.

Turgut, M., S. Kaplan, K. Metin, Y.B. Koca, E. Soylu, B. Sahin, Z.B. Ateşlier, and H.K. Başaloğlu. 2006a. Effects of constant lightness, darkness and parachlorophenylalanine treatment on tail regeneration in the lizard ophisops elegans macrodactylus: Macroscopic, biochemical and histological changes. *Anat Histol Embryol* 35, 155–161.

Turgut, M., G. Oktem, S. Uslu, M.E. Yurtseven, H. Aktuğ, and A. Uysal. 2006b. The effect of exogenous melatonin administration on trabecular width, ligament thickness and TGF-beta(1) expression in degenerated intervertebral disk tissue in the rat. *J Clin Neurosci* 13, 357–363.

Turgut, M., G. Oktem, A. Uysal, and M.E. Yurtseven. 2006c. Immunohistochemical profile of transforming growth factor-b1 and basic fibroblast growth factor in sciatic nerve anastomosis following pinealectomy and exogenous melatonin administration in rats. *J Clin Neurosci* 13, 753–758.

Turgut, M., Y. Uyanikgil, M. Baka, A.T. Tunç, A. Yavaşoğlu, M.E. Yurtseven, and S. Kaplan. 2005a. Pinealectomy exaggerates and melatonin treatment suppresses neuroma formation of transected sciatic nerve in rats: Gross morphological, histological and stereological analysis. *J Pineal Res* 38, 284–291.

Turgut, M., A. Uysal, M. Pehlivan, G. Oktem, and M.E. Yurtseven. 2005b. Assessment of effects of pinealectomy and exogenous melatonin administration on rat sciatic nerve suture repair: An electrophysiological, electron microscopic, and immunohistochemical study. *Acta Neurochir (Wien)* 147, 67–77.

Turgut, M., C. Yenisey, O. Akyüz, Y. Ozsunar, M. Erkus, and T. Biçakçi. 2006d. Correlation of serum trace elements and melatonin levels to radiological, biochemical, and histological assessment of degeneration in patients with intervertebral disc herniation. *Biol Trace Elem Res* 109, 123–134.

Veit, G., B. Kobbe, D.R. Keene, M. Paulsson, M. Koch, and R. Wagener. 2006. Collagen XXVIII, a novel von Willebrand factor A domain-containing protein with many imperfections in the collagenous domain. *J Biol Chem* 281, 3494–3504.

36 Melatonin Use in Peripheral Nerve Injury

Mehmet Emin Önger, Süleyman Kaplan, Ebru Elibol, Ömür Gülsüm Deniz, and Berrin Zühal Altunkaynak

CONTENTS

36.1 INTRODUCTION

Peripheral nerves consisting of clustered peripheral axons provide a common pathway to convey nerve impulses. Nerve injuries are generally classified depending on the degree of injury and symptoms as well (Turgut and Kaplan, 2011). Melatonin (Mel), the pineal neurohormone, has therapeutic effects on peripheral nerve regeneration in pathologic conditions or traumatic events (Odaci and Kaplan, 2009). In some current studies, a number of neuroscientists demonstrated that Mel has an effect on the morphologic features of the regenerating peripheral nerve tissue, which suggests its neuroprotective, free radical scavenging, antioxidative, and analgesic effects throughout that regeneration processes (Turgut and Kaplan, 2011).

36.2 HISTOLOGY OF PERIPHERAL NERVE

A peripheral nerve is a bundle of peripheral axons of neurons, which forms very complex arborization and widespread pathway to convey nerve impulses (Geuna et al., 2009; Turgut and Kaplan, 2011). Each nerve contains many axons, which are surrounded by a fine connective tissue, named as the "endoneurium." These axons form the nerve fascicles within a layer of connective tissue, named as the "perineurium." A dense connective tissue layer, called "epineurium," surrounds the entire peripheral nerve (Turgut and Kaplan, 2011) (Figures 36.1 and 36.2), which is derived from the central nervous system (CNS) and divided into two parts: the cranial and the spinal nerves. Whereas the morphology of the nerve trunks is relatively similar in all body districts, they can show some differences in terms of the presence and number of fascicles, the fiber-type composition and location (Geuna et al., 2009). Generally, peripheral nerves can be categorized into three main subsets according to fiber-type composition: sensory, motor, and mixed nerves. Motor nerves are derived from somatic and autonomic motor neurons in the CNS, whereas sensory nerves are derived from pseudounipolar neurons in the sensory ganglia (Williams, 1999).

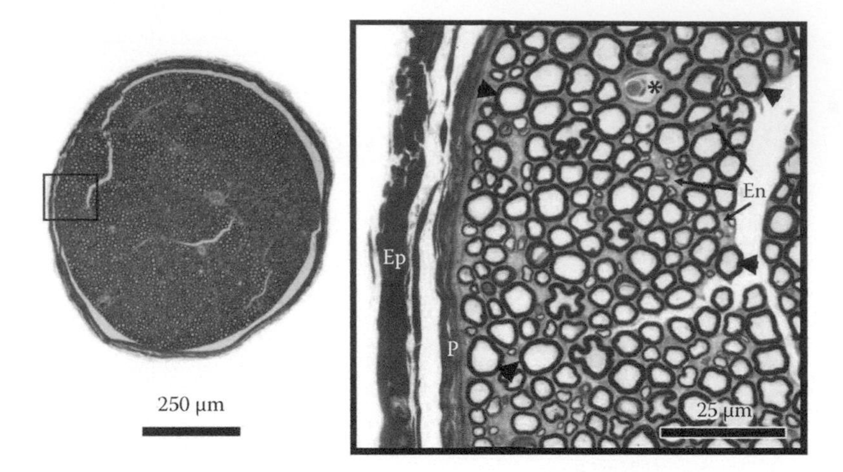

FIGURE 36.1 A transverse section of a rat sciatic nerve. As seen in the picture, rat sciatic nerve is consisting of one fascicule. It covers from outside to inside by epineurium and perineurium. Each nerve fiber is surrounded by endoneurium. The epineurium (Ep), perineurium (P), and endoneurium (En) are shown at higher magnification. Arrowheads: myelinated axons, *: blood vessel, toluidine blue staining.

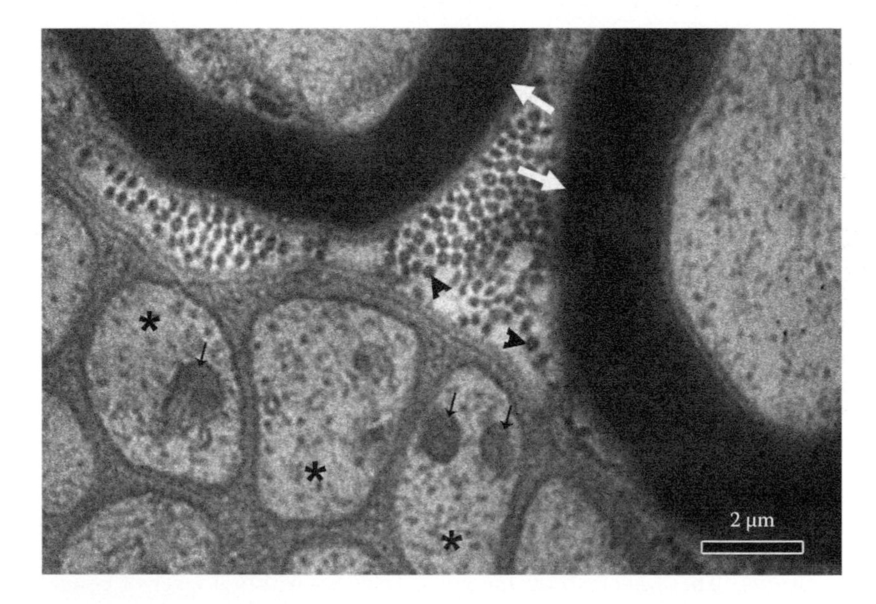

FIGURE 36.2 Electron micrograph of two myelinated axons (white arrows) and several unmyelinated axons can be seen. Black arrows indicate mitochondria seen in the axoplasm (*). Arrowheads show collagen fibers of endoneurium that are cut in transverse direction.

36.3 PERIPHERAL NERVE INJURY

Traumatic injuries of a peripheral nerve result in partial or total loss of both motor and sensory functions because of interruption between both tips of axons, proximal and distal. Nerve fibers degeneration begins from distal to the lesion and eventual death of axotomized neurons (Flores et al., 2000). In addition, the peripheral nerve injuries can result in not only significant functional loss but also substantially decreased quality of life. It is well known that after injuries, permanently impaired sensory and motor functions and secondary problems would be seen (Jaquet et al., 2001; Rosberg et al., 2005). Injury to peripheral nerve trunks may result in several aspects of nerve fiber injury. The axonal fate is essential in order to determine the extension and recovery after nerve injury. Once the traumatic injury of a peripheral nerve occurs, complex morphologic and metabolic changes start almost immediately at the injury site (Flores et al., 2000). These complex changes also occur in the perikaryon, in the segments both proximal and distal to the injury site. In the proximal segment, axons begin to degenerate a little bit far from the injury site. This degeneration length is based on the severity of the lesion and may extend over one or more segments (Geuna et al., 2009). Macrophage stimulates by means of interleukin releasing to production of growth factor, which is secreted by Schwann cells. After that, Schwann cells begin to proliferate within the tube and stimulate the outgrowth of axonal sprouting via molecular messengers, such as growth factors and adhesion molecules (Odaci and Kaplan, 2009). Therefore, Schwann cells play an important role in peripheral nerve regeneration at the injury site. These cells elaborate regeneration processes, which include physical conduits in order to guide axons to their target. Regeneration level directly depends on the extension of these Schwann cell processes rather than on axonal growth (Son and Thompson, 1995). Initially, newly regenerated axons will be deprived of myelin even if the parent axon was a myelinated fiber. Later, these newly regenerated axons will get myelin as time goes by (Flores et al., 2000).

The neuroprotective, antioxidative, and free radical scavenging properties of Mel has been recently growing among scientist. Mel is a known antioxidant and anti-inflammatory agent, and there is a lot of both experimental and clinical evidence for its beneficial effects against oxidative stress (Acuña-Castroviejo et al., 2007, 2011).

36.4 ANTIOXIDANT CAPACITY OF MELATONIN

Mel is one of the most significant molecules because of its free radical scavenging and antioxidant properties at both physiological and pharmacological concentrations in vivo (Hardeland and Pandi-Perumal, 2005; Reiter et al., 2005). Mel and its metabolic derivatives function as direct free radical scavengers and also stimulate several antioxidative enzymes (Reiter et al., 2010). Investigations of the protection properties of Mel show the ability of the hormone to efficiently scavenge those free radicals (Turgut et al., 2009; Yurt et al., 2013). Further scavenging evidence of Mel was elegantly demonstrated by the surgical removal of the pineal gland, which further increased nuclear DNA damage (Tan et al., 1994). This experiment indicates that either Mel is an important antioxidant when administered exogenously at pharmacological doses or physiological Mel concentrations are effective to diminish the destruction of DNA by free radicals (Tan et al., 1994). In addition, it was shown that Mel is significantly effective to protect neuroblastoma cell death induced by oxidative stress and glutamate excitotoxicity (Pappolla et al., 1997). Furthermore, appropriate doses of Mel enhance cell viability via the prevention of oxidative damage as a result of both free-radical formation and increased intracellular calcium levels related to glutamate excitotoxicity (Sewerynek et al., 1995; Antolín et al., 1996; Giusti et al., 1996; Matuszak et al., 1997; Shaikh et al., 1997).

In recent years, researchers have paid more attention to possible protective effects of Mel following traumatic peripheral nerve injuries, especially the sciatic nerve and its pathological conditions, since Mel administration could be therapeutic postoperatively. While some studies report toxic effects of Mel on peripheral nerve regeneration, most scientists indicate that Mel has protective effects on peripheral nerve pathologies (Odaci and Kaplan, 2009).

36.5 PERIPHERAL NERVE INJURY AND MELATONIN

The regeneration potential of Mel on peripheral nerve following the peripheral nerve transection or injury has been investigated on a vast scale. In one study, recoverable effect of Mel on both collagen content and neuroma formation was shown (Turgut et al., 2005a). According to this study, Mel suppressed collagen production and the development of neuroma formation after peripheral nerve transaction (Turgut et al., 2005a). In another study, the effects of Mel on both peripheral nerve repair and regeneration was also investigated (Turgut et al., 2005b). In this study, researchers suggest that exogenous Mel administration dramatically improves nerve regeneration in the rat sciatic nerve suture repair. Apparently, the beneficial effects of Mel on neuroma formation and nerve regeneration are an especially interesting option for the treatment of peripheral nerve injury clinically, especially in case of Mel deficiency (Turgut et al., 2005b). Another study was designed to investigate transforming growth factor (TGF)-β and basic fibroblast factor (bFGF) in the anastomotic region of the sciatic nerve immunohistochemically (Turgut et al., 2006). TGF-β and bFGF play an important role in Schwann cell activity and in collagen production by fibroblasts as well (Tatagiba et al., 2002). TGF-β and bFGF expression was investigated after pinealectomy and exogenous Mel administration (Turgut et al., 2006). It was shown that strong expression of TGF-β1 and bFGF was observed in the epineurium of nerves taken from pinealectomized animals, whereas no or weak staining was seen in control and Mel treatment groups. Therefore, scientists argue that Mel has a therapeutic effect on nerve regeneration (Turgut et al., 2006) (Figures 36.3 and 36.4). Sayan et al. (2004) investigated ischemia/reperfusion (I/R)-induced alterations in peripheral nerve. This study was conducted to evaluate the possible protective effect of Mel on rat sciatic nerve exposed to 2 h ischemia followed by 3 h of reperfusion. In I/R groups, significant axonal damage, common axonal shrinkage and swollen axons, and striking morphological changes of myelin sheath were observed. Degenerative changes were also seen in Schwann cells. In the Mel treatment groups, either degeneration of myelin sheath was decreased or remarkable improvement of axons was seen. Schwann cells were seemingly normal. Briefly, pretreatment with Mel healed I/R injury of the sciatic nerve (Sayan et al., 2004). Shokouhi et al. (2008) also investigated the dose-dependent protective effects of Mel on neural damage and lipid peroxidation after traumatic peripheral nerve injury. According to their results, 10 mg/kg Mel treatment reduced trauma-induced myelin breakdown and axonal changes in the peripheral nerve. Furthermore, 50 mg/kg Mel treatment almost entirely healed any ultrastructural changes. Therefore, researchers suggest that Mel has a strong neuroprotective effect at a dose of 50 mg/kg and it can protect peripheral nerve from any damage after blunt trauma (Shokouhi et al., 2008).

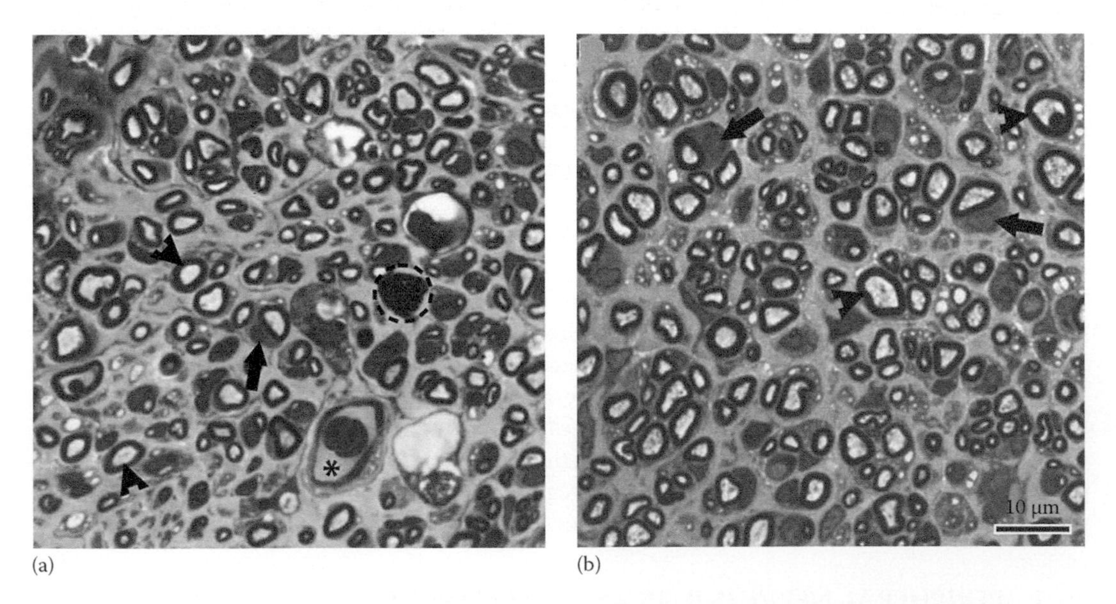

FIGURE 36.3 Light microscope images of regenerating peripheral nerve in the control (a) and the Mel treated (b) groups. In addition to increased myelinated axon number, remarkable myelination can also be seen in Mel treated group (b). That means Mel treatment has a positive effect on the nerve fibers regeneration in comparison to the control group. Arrows: Schwann cells; arrowheads: myelinated axons; *: blood vessel; dashed circle: Mast cell; toluidine blue staining.

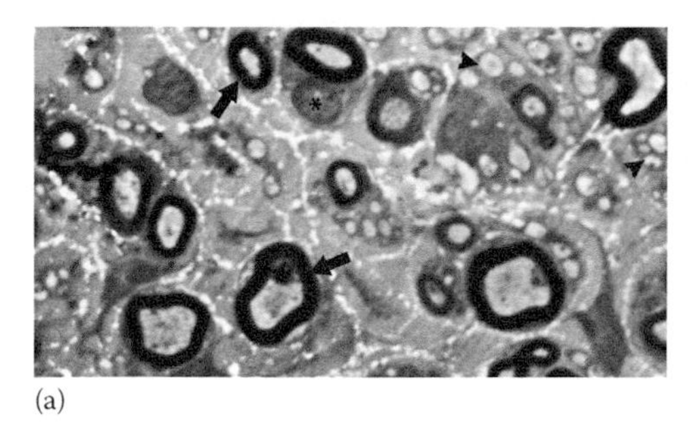
(a)

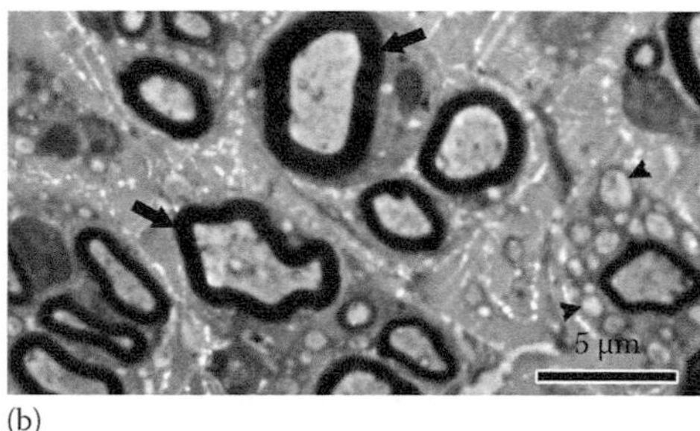

(b)

FIGURE 36.4 Myelinated and unmyelinated fibers of regenerating peripheral nerve in control (a) and Mel-treated (b) groups are shown at the electron microscopic level. It can be seen from these pictures that after transection of sciatic nerve, the regeneration of nerve in the Mel treated group is more prominent than the control group. Mel increases not only the diameter of nerve fibers but also has a positive effect on the thickness of myelin sheath. Arrows show myelinated axons; arrowheads: unmyelinated axons, *: Schwann cell.

36.6 CONCLUSION

Currently, some studies demonstrated that pineal neurohormone Mel has an effect on the histological and physiological features of the nerve tissue, suggesting its neuroprotective, free radical scavenging, antioxidative, and analgesic effects in degenerative diseases of peripheral nerves, and it is broadly accepted that Mel has a useful effect on axon length and sprouting after traumatic injury to peripheral nerves (Reiter et al., 2000; Sayan et al., 2004; Turgut and Kaplan, 2011). Although there are some studies that indicate toxic effect of Mel on peripheral nerve (Piezzi and Cavicchia, 1981; Lehman and Johnson, 1999), most of the studies clearly show positive effects of Mel on the number of axons, thickness of myelin sheath by inhibition of collagen accumulation and neuroma formation following traumatic events to peripheral nerves by means of various experimental injury models (Turgut et al., 2006, 2010; Turgut and Kaplan, 2011). Nevertheless, further experimental and randomized controlled clinical studies are vital to identify the clinical use of Mel as well as clearly refine protective or toxic effects of it.

This is an overview of recent patents and current literature in terms of the effects of Mel peripheral nerve injury referring to electrophysiological, biochemical, and microscopical findings, in addition to functional observations.

REFERENCES

Acuña-Castroviejo, D., Escames, G., Rodriguez, M. I., Lopez, L. C. 2007. Melatonin role in the mitochondrial function. *Front Biosci* 12:947–963.

Acuña Castroviejo, D., López, L. C., Escames, G., López, A., García, J. A., Reiter, R. J. 2011. Melatonin-mitochondria interplay in health and disease. *Curr Top Med Chem* 11:221–240.

Antolín, I., Rodríguez, C., Saínz, R. M., Mayo, J. C., Uría, H., Kotler, M. L., Rodríguez-Colunga, M. J., Tolivia, D., Menéndez-Peláez, A. 1996. Neurohormone melatonin prevents cell damage: Effect on gene expression for antioxidant enzymes. *FASEB J* 10:882–890.

Flores, A. J., Lavernia, C. J., Owens, P. W. 2000. Anatomy and physiology of peripheral nerve injury and repair. *Am J Orthop* 29:167–173.

Geuna, S., Raimondo, S., Ronchi, G., Di Scipo, F., Tos, P., Czaja, K., Fornaro, M. 2009. Histology of the peripheral nerve and changes occurring during nerve regeneration. *Int J Neurosci* 87:27–46.

Giusti, P., Lipartiti, M., Franceschini, D., Schiavo, N., Floreani, M., Manev, H. 1996. Neuroprotection by melatonin from kainite-induced excitotoxicity in rats. *FASEB J* 10:891–896.

Hardeland, R., Pandi-Perumal, S. R. 2005. Melatonin, a potent agent in antioxidative defense: Actions as a natural food constituent, gastrointestinal factor, drug and prodrug. *Nutr Metab* 2:22.

Jaquet, J. B., Luijsterburg, A. J., Kalmijn, S., Kuypers, P. D., Hofman, A., Hovius, S. E. 2001. Median, ulnar, and combined median–ulnar nerve injuries: Functional outcome and return to productivity. *J Trauma* 51:687–692.

Lehman, N. L., Johnson, L. N. 1999. Toxic optic neuropathy after concomitant use of melatonin, zoloft, and a high-protein diet. *J Neuroophthalmol* 19:232–234.

Matuszak, Z., Reszka, K. J., Chignell, C. F. 1997. Reaction of melatonin and related indoles with hydroxyl radicals: EPR and spin trapping investigations. *Free Radic Biol Med* 23:367–372.

Odaci, E., Kaplan, S. 2009. Chapter 16: Melatonin and nerve regeneration. *Int Rev Neurobiol* 87:317–335.

Pappolla, M. A., Sos, M., Omar, R. A., Bick, R. J., Hickson-Bick, D. L. M., Reiter, R. J., Efthimiopoulos, S., Robakis, N. K. 1997. Melatonin prevents death of neuroblastoma cells exposed to the Alzheimer amyloid peptide. *J Neurosci* 17:1683–1690.

Piezzi, R. S., Cavicchia, J. C. 1981. Effects of cold and melatonin on the microtubules of the toad sciatic nerve. *Anat Rec* 200:115–120.

Reiter, R. J., Manchester, L. C., Tan, D. X. 2010. Neurotoxins: Free radical mechanisms and melatonin protection. *Curr Neuropharmacol* 8:194–210.

Reiter, R. J., Tan, D. X., Maldonado, M. D. 2005. Melatonin as an antioxidant: Physiology versus pharmacology. *J Pineal Res* 39:215–216.

Reiter, R. J., Tan, D. X., Osuna, C., Gitto, E. 2000. Actions of melatonin in the reduction of oxidative stress. *J Biomed Sci* 7:444–458.

Rosberg, H. E., Carlsson, K. S., Dahlin, L. B. 2005. Prospective study of patients with injuries to the hand and forearm: Costs, function, and general health. *Scand J Plast Reconstr Surg Hand Surg* 39:360–369.

Sayan, H., Ozacmak, V. H., Ozen, O. A., Coskun, O., Arslan, S. O., Sezen, S. C., Aktas, R. G. 2004. Beneficial effects of melatonin on reperfusion injury in rat sciatic nerve. *J Pineal Res* 37:143–148.

Sewerynek, E., Poeggeler, B., Melchiorri, D., Reiter, R. J. 1995. H_2O_2-induced lipid peroxidation in rat brain homogenates is greatly reduced by melatonin. *Neurosci Lett* 195:203–205.

Shaikh, A. Y., Xu, J., Wu, Y., He, L., Hsu, C. Y. 1997. Melatonin prevents bovine cerebral endothelial cells from hyperoxia induced DNA damage and death. *Neurosci Lett* 229:193–197.

Shokouhi, G., Tubbs, R. S., Shoja, M. M., Hadidchi, S., Ghorbanihaghjo, A., Roshangar, L., Farahani, R. M., Mesgari, M., Oakes, W. J. 2008. Neuroprotective effects of high-dose vs low-dose melatonin after blunt sciatic nerve injury. *Childs Nerv Syst* 24:111–117.

Son, Y. J., Thompson, W. J. 1995. Schwann cell processes guide regeneration of peripheral axons. *Neuron* 14:125–132.

Tan, D. X., Reiter, R. J., Chen, L. D., Poeggeler, B., Manchester, L. C., Barlow-Walden, L. R. 1994. Both physiological and pharmacological levels of melatonin reduce DNA adduct formation induced by the carcinogen safrole. *J Carcinog* 15:215–218.

Tatagiba, M., Rosahl, S., Gharabaghi, A., Blömer, U., Brandis, A., Skerra, A., Sami, M., Schwab, M. E. 2002. Regeneration of auditory nerve following complete sectioning and intrathecal application of the IN-1 antibody. *Acta Neurochir (Wien.)* 144:181–187.

Turgut, M., Kaplan, S. 2011. Effects of melatonin on peripheral nerve regeneration. *Recent Pat Endocr Metab Immune Drug Discov* 5:100–108.

Turgut, M., Kaplan, S., Unal, Z. B., Altunkaynak, B. Z., Unal, D., Sahin, B., Bozkurt, M., Yurtseven, M. E. 2009. Effects of pinealectomy on morphological features of blood vessel in chicken: An electron microscopic and stereological study. *J Exp Clin Med* 26:112–118.

Turgut, M., Kaplan, S., Unal, B. Z., Bozkurt, M., Yürüker, S., Yenisey, C., Sahin, B., Uyanıkgil, Y., Baka, M. 2010. Stereological analysis of sciatic nerve in chickens following neonatal pinealectomy: An experimental study. *J Brachial Plex Peripher Nerve Inj* 5:10.

Turgut, M., Oktem, G., Uysal, A., Yurtseven, M. E. 2006. Immunohistochemical profile of transforming growth factor-beta1 and basic fibroblast growth factor in sciatic nerve anastomosis following pinealectomy and exogenous melatonin administration in rats. *J Clin Neurosci* 13:753–758.

Turgut, M., Uyanikgil, Y., Baka, M., Tunc, A. T., Yavaşoğlu, A., Yurtseven, M. E., Kaplan, S. 2005a. Pinealectomy exaggerates and melatonin treatment suppresses neuroma formation of transected sciatic nerve in rats: Gross morphological, histological and stereological analysis. *J Pineal Res* 38:284–291.

Turgut, M., Uysal, A., Pehlivan, M., Oktem, G., Yurtseven, M. E. 2005b. Assessment of effects of pinealectomy and exogenous melatonin administration on rat sciatic nerve suture repair: An electrophysiological, electron microscopic, and immunohistochemical study. *Acta Neurochir (Wien.)* 147:67–77.

Williams, P. L. 1999. *Gray's Anatomy.* Churchill Livingstone, London, U.K.

Yurt, K. K., Kayhan, E., Altunkaynak, B. Z., Tümentemur, G., Kaplan, S. 2013. Effects of the melatonin on the kidney of high fat diet fed obese rats: A stereological and histological approach. *J Exp Clin Med* 30:153–158.

37 Neuroprotective Role of Melatonin and Its Agonists

Neuroprotective Role of Melatonin against Dexamethasone-Induced Neurotoxicity

Aziza B. Shalby and Hanaa H. Ahmed

CONTENTS

37.1 INTRODUCTION

Melatonin (5-methoxy-*N*-acetyltryptamine) was first isolated and discovered by Lerner and coworkers in the late 1950s (Lerner et al., 1960). In vertebrates, the synthesis of melatonin takes place primarily in the pineal gland and the retina, as the synthesis and release of melatonin show a marked circadian variation, being at higher levels during the night and at lower levels during the day (Tosini and Menaker, 1998). While it is well established that melatonin is a major regulator of circadian rhythm, abundant evidence suggests that this hormone may be involved in the regulation of a variety of physiological processes, such as sleep, reproduction, immune and vascular response, etc. (Hardeland et al., 2011). In recent years, it has been demonstrated that melatonin may modulate the function of various types of neurons in the central nervous system (CNS) by modifying the activity of ligand- and voltage-gated ion channels (Yang et al., 2011a). These actions of melatonin on central neurons are mediated by distinct intracellular pathways via the activation of different subtypes of melatonin receptors (Huang et al., 2013).

37.2 SYNTHESIS AND DEGRADATION OF MELATONIN

The synthesis of melatonin in the pineal gland involves several steps. First, L-tryptophan, which is taken up from the cerebral vessels, is converted to serotonin. Serotonin is subsequently metabolized by the rate-limiting arylalkylamine *N*-acetyltransferase (AANAT) to *N*-acetyl-5-hydroxytryptamine. The final step of the synthesis pathway is the conversion of *N*-acetyl-5-HT to melatonin by hydroxyindole-*o*-methyltransferase. Melatonin is a lipophilic hormone, which is widely distributed throughout the human body (Reiter, 1991; Ekmekcioglu, 2006). The degradation of melatonin involves several steps of enzymatic reactions, and approximately 90% of melatonin is cleaned out through a single passage via the liver (Pardridge and Mietus, 1980; Huang et al., 2013). For many years, melatonin was thought to be almost exclusively catabolized by hepatic P450 monooxygenases, followed by conjugation of the resulting 6-hydroxymelatonin to give the main urinary metabolite 6-sulfatoxymelatonin. This may be largely true for the circulating hormone, but not necessarily for tissue melatonin. Especially, in the CNS, oxidative pyrrole-ring cleavage prevails and no 6-hydroxymelatonin was detected after melatonin injection into the cisterna magna (Hirata et al., 1974). This may be particularly important because much more melatonin is released via the pineal recess into the cerebrospinal fluid than into the circulation (Tricoire et al., 2002). The primary cleavage product is *N*1-acetyl-*N*2-formyl-5-methoxykynuramine (AFMK), which is deformylated, either by arylamine formamidase or hemoperoxidases to *N*1-acetyl-5-methoxykynuramine (AMK). Surprisingly, numerous—enzymatic (indoleamine 2,3-dioxygenase, myeloperoxidase), pseudoenzymatic (oxoferryl hemoglobin, hemin), photocatalytic or free-radical—reactions lead to the same product, AFMK (Hardeland, 2005). Recent estimations have revealed that pyrrole-ring cleavage contributes to about one-third of the total catabolism, but the percentage may be even higher in certain tissues. Other oxidative catabolites are cyclic 3-hydroxymelatonin (c3OHM), which can also be metabolized to AFMK, and a 2-hydroxylated analog, which does not cyclize, but turns into an indolinone (Hardeland, 2005). Additional hydroxylated or nitrosated metabolites have been detected, which appear to represent minor quantities only. AFMK and AMK also form metabolites by interactions with reactive oxygen and nitrogen species (Hardeland, 2005).

37.3 BIOLOGICAL FUNCTIONS

Melatonin is highly pleiotropic (Figure 37.1). Classical effects are attributed to Gi protein–coupled membrane receptors, MT1 and MT2, differing in ligand affinity (Jin et al., 2003). Both are involved in a circadian feedback to the suprachiasmatic nuclei (SCN). MT2 is required for efficient phase-shifting. MT1, having a higher affinity, causes acute suppressions of neuronal firing. These actions involve decreases in phsopho cAMP responsive element binding protein (pCREB) levels stimulated by pituitary adenylyl cyclase activating peptide (PACAP). Other effects may be related to nuclear receptors of lower ligand sensitivity, orphan receptor family (RZR/ROR α and RZR β) (Carlberg, 2000), but in these cases, functional significance and target genes are less clear. Although a lot of good and solid work has been carried out on the control of circadian and seasonal rhythms (Reiter, 1993), including receptor-mediated actions on the mammalian SCN (Jin et al., 2003), and we would like to focus here on selected nonclassical effects, according to most recent developments. Melatonin exhibits immunomodulatory properties, which are mediated via membrane and nuclear receptors (Guerrero and Reiter, 2002). Data were reported on the activation of T, B, NK cells and monocytes, thymocyte proliferation, release of cytokines (IL-1, IL-2, IL-6, IL-12, and IFNγ), met-enkephalin, other immunoopioids, and antiapoptotic effects, including glucocorticoid antagonism. Signaling mechanisms are only partially understood, and some findings are contradictory. In thymocytes and lymphocytes, cAMP is decreased via MT1or MT2 receptors. However, melatonin also potentiated VIP-induced rises of cAMP in lymphocytes. Anti-inflammatory actions of melatonin are related to the inhibition of prostaglandin E_2 (PGE2) effects, and in particular, cyclooxygenase-2 (COX-2) downregulation, which may be transmitted by its metabolite AMK (Mayo et al., 2005). Immunomodulation seems to be part of antitumor effects described for melatonin. Other oncostatic actions involve MT1/MT2-dependent suppression of linoleic acid uptake or estrogen receptor downregulation (Blask et al., 2002). A developing area is antioxidative protection. Even if unjustified claims based on suprapharmacological doses remain unconsidered, a remarkable body of evidence exists showing protection in numerous cell culture and in vivo systems (Srinivasan et al., 2005). A special but important aspect is melatonin's role in neuroprotection. Antioxidant actions are observed at different levels, including attenuation of radical formation by antiexcitatory and anti-inflammatory effects. This is not restricted to scavenging, although melatonin efficiently interacts

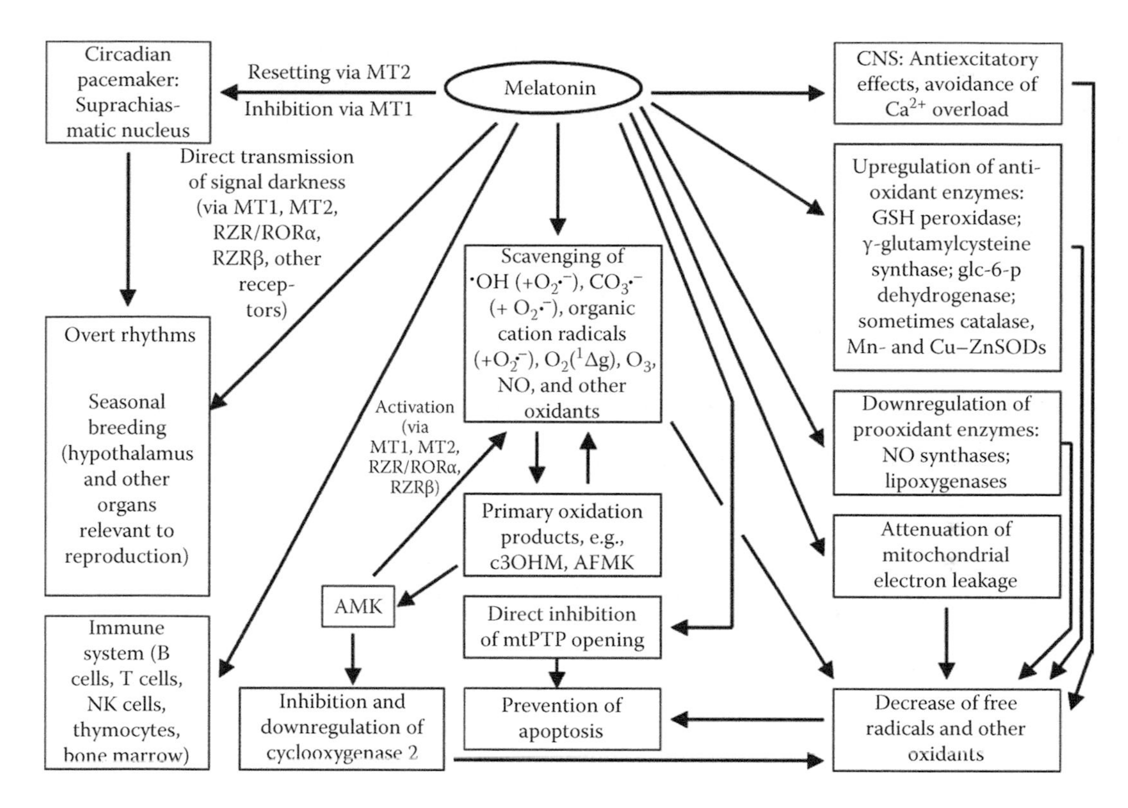

FIGURE 37.1 Overview of some major effects of melatonin on circadian and seasonal rhythms, immunomodulation, as an antiinflammatory, antioxidant, and antiapoptotic agent, including actions of its metabolites c3OHM, AFMK, and AMK. (From Hardeland, R. et al., *Int. J. Biochem. Cell Biol.*, 38(3), 313, 2006.)

with various reactive oxygen and nitrogen species as well as organic radicals, but includes upregulation of antioxidant enzymes (glutathione peroxidase, glutathione reductase, γ-glutamylcysteine synthase, glucose 6-phosphate dehydrogenase, sometimes Cu-, Zn- and Mn-superoxide dismutases and catalase) and downregulation of prooxidant enzymes (NO synthases and lipoxygenases; Hardeland, 2005). Mechanisms of the enzyme inductions have not been identified, whereas suppression of Ca^{2+}-dependent NOS may involve melatonin binding to calmodulin, an effect also playing a role in cytoskeletal rearrangements. Other studies related NOS downregulation to membrane receptors. Additional antioxidant effects may be mediated by binding to quinone reductase 2, which had previously been assumed to represent another melatonin receptor. Antioxidative protection is particularly evident in senescence-accelerated mice. Recently, mitochondrial effects of melatonin have come into the focus of interest, which comprise safeguarding of respiratory electron flux, reduction of oxidant formation by lowering electron leakage, effects shared by AMK, and inhibition of opening of the mitochondrial permeability transition pore (Hardeland et al., 2006).

37.4 DEXAMETHASONE-INDUCED NEUROTOXICITY

Glucocorticoids are adrenal steroids secreted during stress and their numerous actions are essential for the stress response (Lee et al., 2002). However, excess of glucocorticoids, through the activation of glucocorticoid receptors, has been implicated in the reduction of the hippocampal volume observed in depression (Sterner and Kalynchuk, 2010). While the primary function of glucocorticoids is the mobilization of energy to respond to the stressor, they can also favor the cells to undergo apoptosis (Joëls, 2008). Specifically, dexamethasone, a synthetic glucocorticoid, has been reported to induce cellular death in different kinds of cells, namely, dexamethasone induces apoptosis in striatal, hippocampal, cerebellar granule neurons, and glial cells (Budni et al., 2011). The possible mechanism may be one of the following mechanisms. One is that the inhibition of the PI3K/Akt signaling pathway enhances dexamethasone-induced cell death (Nuutinen et al., 2006). Another possibility is that glucocorticoid induces the increase of extracellular glutamate in the hippocampus, which can be prevented by blocking N-methyl-d-aspartate (NMDA) receptors. Moreover, excessive glucocorticoid reduces the expression and impairs brain-derived neurotrophic factor (BDNF) function, which damages the hippocampus and other brain areas (Kunugi et al., 2010). It is interesting to mention that glucocorticoids exert antiproliferative effects in many cell types (Hong et al., 2011) by the inhibition of cell cycle progression or induction of cell death (Budni et al., 2011).

37.5 NEUROPROTECTIVE EFFECTS OF MELATONIN AGAINST NEUROTOXICITY OF DEXAMETHASONE

Melatonin mechanisms are related to headache pathophysiology in many ways, including its anti-inflammatory effect, toxic free radical scavenging, reduction of proinflammatory cytokine upregulation, nitric oxide synthase activity and dopamine release inhibition, membrane stabilization, gamma-Aminobutyric acid (GABA) and opioid analgesia potentiation, glutamate neurotoxicity protection, neurovascular regulation, serotonin modulation, and the similarity of chemical structure to that of indomethacin. Melatonin can directly act as free radical scavenger and indirectly induce the expression of some genes linked to the antioxidant defense system (Garcia et al., 2010). Melatonin has a powerful scavenging activity for hydroxyl and peroxyl radicals over that of both glutathione and vitamin E. Melatonin also conserves the activities of the antioxidant enzymes via enhancing gene expression of these enzymes (Sharman et al., 2007). Melatonin has a phenol group that provides a proton to detoxify hydroxyl or lipid peroxy radicals and thus can reduce lipid peroxidation induced by dexamethasone (Allegra et al., 2003). Also, melatonin has the ability to serve as a metal chelator to reduce metal toxicity (Gulcin et al., 2003). Melatonin stabilizes membranes against free radicals and can resist the rigidity of the biological membranes caused by dexamethasone (Albendea et al., 2007). Moreover, melatonin effectively reduces lipid peroxidation induced by dexamethasone (Yang et al., 2011b). Melatonin is a potent antioxidant that protects DNA, lipids, and proteins from free radical damage (Mayo et al., 2003). The ability of melatonin to serve as a metal chelator for Fe^{3+} ions provides an evidence supporting its role as a neuroprotective agent (Garcia et al., 2010; Ahmed et al., 2013). Several melatonin metabolites that are generated when melatonin interacts with toxic reactants are themselves able to increase the efficiency of the electron transport chain in the inner mitochondrial membrane with a consequent impairment of free radical generation (Reiter et al., 2002). Exogenous melatonin increases the activities of GPx and Cu–ZnSOD/MnSOD in the brain (Rodriguez et al., 2004). Also, melatonin administration increases the levels of each of Cu–ZnSOD, MnSOD, GPx, and CAT in the hippocampus (Gómez et al., 2005).

Exogenous melatonin elevates both brain Bcl-2 and BDNF levels. Melatonin acts through the mitochondrial pathway and blocks the spill of cytochrome c to the cytosol and thus prevents activation of caspases, increasing cellular content of Bcl-2 in old rats, thus reduces apoptosis. Melatonin regulates the complex Bax/Bcl-2 and antagonizes apoptosis through the activation of Mitogen activated protein kinases/Extracellular signal-regulated kinases (MAPK/ERK pathway) and inhibition of the stress kinases JNK and p38 MAPK in neuronal cells (Luchetti et al., 2009). Melatonin and its activated receptors have been linked to the regulation of neurotrophic factors, including BDNF. Both G-proteins-mediated signaling and other pathways such as extracellular signal-regulated kinase (ERK) may contribute to melatonin action on BDNF (Imbesi et al., 2008; Ahmed et al., 2013).

Melatonin exerts beneficial effects on cholinergic neurotransmission in brain by increasing the activities of choline acetyl transferase (ChAT) enzyme in the frontal cortex and hippocampus

(Weinstock and Shoham, 2004; Ahmed et al., 2013). Melatonin possesses an electron-rich aromatic indole ring and functions as donar and directly detoxifies free radicals and thus can enhance brain Ach activity.

37.6 CONCLUSION

Melatonin has several neuroprotective effects against neurotoxicity of dexamethasone acting as an anti-inflammatory, antioxidant, and neuroprotective agent. So, melatonin represents a good therapeutic approach for intervention against progressive neurological damage associated with dexamethasone.

REFERENCES

Ahmed HH, Estefan SF, Mohamd EM, Farrag ARH, Salah RS. (2013). Does melatonin ameliorate neurological changes associated with Alzheimer's disease in ovariectomized rat model? *Ind J Clin Biochem* 28(4):381–389.

Albendea CD, Gómez-Trullén EM, Fuentes-Broto L, Miana-Mena FJ, Millán-Plano S, Reyes-Gonzales MC, Martínez-Ballarín E, García JJ. (2007). Melatonin reduces lipid and protein oxidative damage in synaptosomes due to aluminium. *J Trace Elem Med Biol* 21(4):261–268.

Allegra M, Reiter RJ, Tan DX, Gentile C, Tesoriere L, Livrea MA. (2003). The chemistry of melatonin's interaction with reactive species. *J Pineal Res* 34(1):1–10.

Blask DE, Sauer LA, Dauchy RT. (2002). Melatonin as a chronobiotic/anticancer agent: Cellular, biochemical, and molecular mechanisms of action and their implications for circadian-based cancer therapy. *Curr Top Med Chem* 2(2):113–132.

Budni J, Romero A, Molz S, Martín-de-Saavedra MD, Egea J, DelBarrio L, Tasca CI, Rodrigues AL, López MG. (2011). Neurotoxicity induced by dexamethasone in the human neuroblastoma SH-SY5Y cell line can be prevented by folic acid. *Neuroscience* 190:346–353.

Cardinali DP, Pévet P. (1998). Basic aspects of melatonin action. *Sleep Med Rev* 2(3):175–190.

Carlberg C. (2000). Gene regulation by melatonin. *Ann N Y Acad Sci* 917:387–396.

Ekmekcioglu C. (2006). Melatonin receptors in humans: Biological role and clinical relevance. *Biomed Pharmacother* 60(3):97–108.

Garcia T, Esparza JL, Nogués MR, Romeu M, Domingo JL, Gómez M. (2010). Oxidative stress status and RNA expression in hippocampus of an animal model of Alzheimer's disease after chronic exposure to aluminum. *Hippocampus* 20(1):218–225.

Gómez M, Esparza JL, Nogués MR, Giralt M, Cabré M, Domingo JL. (2005). Pro-oxidant activity of aluminum in the rat hippocampus: Gene expression of antioxidant enzymes after melatonin administration. *Free Radic Biol Med* 38(1):104–111.

Guerrero JM, Reiter RJ. (2002). Melatonin-immune system relationships. *Curr Top Med Chem* 2(2):167–179.

Gulcin I, Buyukokuroglu ME, Kufrevioglu OI. (2003). Metal chelating and hydrogen peroxide scavenging effects of melatonin. *J Pineal Res* 34(4):278–281.

Hardeland R. (2005). Antioxidative protection by melatonin: Multiplicity of mechanisms from radical detoxification to radical avoidance. *Endocrine* 27(2):119–130.

Hardeland R, Cardinali DP, Srinivasan V, Spence DW, Brown GM, Pandi-Perumal SR. (2011). Melatonin—A pleiotropic, orchestrating regulator molecule. *Prog Neurobiol* 93(3):350–384.

Hardeland R, Pandi-Perumal SR, Cardinali DP. (2006). Melatonin. *Int J Biochem Cell Biol* 38(3):313–316.

Hirata F, Hayaishi O, Tokuyama T, Seno S. (1974). In vitro and in vivo formation of two new metabolites of melatonin. *J Biol Chem* 249(4):1311–1313.

Hong D, Chen HX, Yu HQ, Wang C, Deng HT, Lian QQ, Ge RS. (2011). Quantitative proteomic analysis of dexamethasone-induced effects on osteoblast differentiation, proliferation, and apoptosis in MC3T3-E1 cells using SILAC. *Osteoporos Int* 22(7):2175–2186.

Huang H, Wang Z, Weng SJ, Sun XH, Yang XL. (2013). Neuromodulatory role of melatonin in retinal information processing. *Prog Retin Eye Res* 32:64–87.

Imbesi M, Uz T, Manev H. (2008). Melatonin receptor agonist ramelteon activates the extracellular signal-regulated kinase 1/2 in mouse cerebellar granule cells. *Neuroscience* 155(4):1160–1164.

Jin X, vonGall C, Pieschl RL, Gribkoff VK, Stehle JH, Reppert SM, Weaver DR. (2003). Targeted disruption of the mouse Mel(1b) melatonin receptor. *Mol Cell Biol* 23(3):1054–1060.

Joëls M. (2008). Functional actions of corticosteroids in the hippocampus. *Eur J Pharmacol* 583(2–3):312–321.
Kunugi H, Hori H, Adachi N, Numakawa T. (2010). Interface between hypothalamic-pituitary-adrenal axis and brain-derived neurotrophic factor in depression. *Psychiatry Clin Neurosci* 64(5):447–459.
Lee AL, Ogle WO, Sapolsky RM. (2002). Stress and depression: Possible links to neuron death in the hippocampus. *Bipolar Disord* 4(2):117–128.
Lerner AB, Case JD, Takahadhi Y. (1960). Isolation of melatonin and 5-methoxy indole-3-acetic acid from bovine pineal glands. *J Biol Chem* 235:1992–1997.
Luchetti F, Betti M, Canonico B, Arcangeletti M, Ferri P, Galli F, Papa S. (2009). ERK MAPK activation mediates the antiapoptotic signaling of melatonin in UVB-stressed U937 cells. *Free Radic Biol Med* 46(3):339–351.
Mayo JC, Sainz RM, Tan DX, Hardeland R, Leon J, Rodriguez C, Reiter RJ. (2005). Anti-inflammatory actions of melatonin and its metabolites, *N*1-acetyl-*N*2-formyl-5-methoxykynuramine (AFMK) and *N*1-acetyl-5-methoxykynuramine (AMK), in macrophages. *J Neuroimmunol* 165(1–2):139–149.
Mayo JC, Tan DX, Sainz RM, Lopez-Burillo S, Reiter RJ. (2003). Oxidative damage to catalase induced by peroxyl radicals: Functional protection by melatonin and other antioxidants. *Free Radic Res* 37(5):543–553.
Nuutinen U, Postila V, Mättö M, Eeva J, Ropponen A, Eray M, Riikonen P, Pelkonen J. (2006). Inhibition of PI3-kinase-Akt pathway enhances dexamethasone-induced apoptosis in a human follicular lymphoma cell line. *Exp Cell Res* 312(3):322–330.
Pardridge WM, Mietus LJ. (1980). Transport of thyroid and steroid hormones through the blood-brain barrier of the newborn rabbit: Primary role of protein-bound hormone. *Endocrinology* 107(6):1705–1710.
Reiter RJ. (1991). Pineal melatonin: Cell biology of its synthesis and of its physiological interactions. *Endocr Rev* 12(2):151–180.
Reiter RJ. (1993). The melatonin rhythm: Both a clock and a calendar. *Experientia* 49(8):654–664.
Reiter RJ, Tan DX, Burkhardt S. (2002). Reactive oxygen and nitrogen species and cellular and organismal decline: Amelioration with melatonin. *Mech Ageing Dev* 123(8):1007–1019.
Rodriguez C, Mayo JC, Sainz RM, Antolín I, Herrera F, Martín V, Reiter RJ. (2004). Regulation of antioxidant enzymes: A significant role for melatonin. *J Pineal Res* 36(1):1–9.
Sharman EH, Bondy SC, Sharman KG, Lahiri D, Cotman CW, Perreau VM. (2007). Effects of melatonin and age on gene expression in mouse CNS using microarray analysis. *Neurochem Int* 50(2):336–344.
Srinivasan V, Pandi-Perumal SR, Maestroni GJ, Esquifino AI, Hardeland R, Cardinali DP. (2005). Role of melatonin in neurodegenerative diseases. *Neurotox Res* 7(4):293–318.
Sterner EY, Kalynchuk LE. (2010). Behavioral and neurobiological consequences of prolonged glucocorticoid exposure in rats: Relevance to depression. *Prog Neuropsychopharmacol Biol Psychiatry* 34(5):777–790.
Tosini G, Menaker M. (1998). The clock in the mouse retina: Melatonin synthesis and photoreceptor degeneration. *Brain Res* 789(2):221–228.
Tricoire H, Locatelli A, Chemineau P, Malpaux B. (2002). Melatonin enters the cerebrospinal fluid through the pineal recess. *Endocrinology* 143(1):84–90.
Weinstock M, Shoham S. (2004). Rat models of dementia based on reductions in regional glucose metabolism, cerebral blood flow and cytochrome oxidase activity. *J Neural Transm* 111(3):347–366.
Yang X, Yang Y, Fu Z, Li Y, Feng J, Luo J, Zhang Q, Wang Q, Tian Q. (2011a). Melatonin ameliorates Alzheimer-like pathological changes and spatial memory retention impairment induced by calyculin A. *J Psychopharmacol* 25(8):1118–1125.
Yang XF, Miao Y, Ping Y, Wu HJ, Yang XL, Wang Z. (2011b). Melatonin inhibits tetraethylammonium-sensitive potassium channels of rod ON type bipolar cells via MT2 receptors in rat retina. *Neuroscience* 173:19–29.

38 Does Melatonin Have a Protective Role against Side Effects of Nonsteroidal Anti-Inflammatory Drugs?

Ebru Elibol, Berrin Zühal Altunkaynak, Ömür Gülsüm Deniz, Mehmet Emin Önger, and Süleyman Kaplan

CONTENTS

38.1 INTRODUCTION

Melatonin (Mel) is an important signal molecule that is known as the basic secretory product of the pineal gland. It is in a component of unicellular organisms, plants, and fungi and is additionally found in other organisms, such as animals and humans (Odaci and Kaplan, 2009; Aygun et al., 2012). Nonsteroidal anti-inflammatory drugs (NSAIDs) are widely used for the purpose of anti-inflammation, antipyretic, and analgesia (Aygun et al., 2012). Although they are common prescription, toxic side effects of NSAIDs have been widely investigated (Canan et al., 2008). This chapter summarizes the physiology of Mel and discusses the potential therapeutic uses of Mel on the possible toxic effects of NSAIDs.

38.2 WHAT IS MELATONIN?

Mel, which is also chemically known as *N*-acetyl-5-methoxytryptamin (Erlich and Apuzzo, 1985), is a hormone found in animals, plants, and some pathogens (Caniato et al., 2003; Paredes et al., 2009). It is produced in small amounts in extra-pineal sites, for example, the retina, gastrointestinal system, ovary, skin, immune system, and other cerebral structures except from the pineal gland (Siu et al., 2006; Reiter et al., 2007a,b). In mammals, circulating levels of Mel vary in a daily cycle, thereby allowing the entrainment of the circadian rhythms of several biological functions (Altun and Ugur-Altun, 2007). This hormone has many important roles in the maintenance of numerous biological and physiologic functions of the body (Tunc et al., 2006).

38.3 BIOSYNTHESIS AND METABOLISM

Mel is synthesized within the pinealocytes of the pineal gland and it constitutes the majority of parenchymal cells. An amino acid, tryptophan, is needed for the synthesis of Mel by the cells. Tryptophan, which is taken up from the bloodstream into pinealocyte, converted to serotonin, first is metabolized into 5-hydroxytryptophan by 5-hydroxylase and is then transformed into 5-hydroxytryptamine (serotonin) by L-aromatic amino acid decarboxylase (dopa decarboxylase). *N*-acetyltransferase (NAT) first acetylates serotonin into *N*-acetylserotonin, which is finally converted to Mel by hydroxyindole-*O*-methyltransferase (HIOMT) (Erlich and Apuzzo, 1985; Cagnacci, 1996).

When Mel reaches the sympathetic nerve fibers of the pineal gland, which are terminated between the parenchymal cells, norepinephrine is released from these nerve endings. The release of norepinephrine from these nerve endings is suppressed in light, but its secretion increases in dark. Norepinephrine, which is secreted by the pineal gland and in the dark, is connected to β-adrenergic receptors within the structure of pinealocyte membrane. First, adenylate cyclase is activated in the cell by stimulating these receptors and cAMP increases (Figure 38.1). Thereafter, NAT occurs and the synthesis of Mel is completed. ά-adrenergic receptors are also available in pinealocyte cell membranes and the regulation of pineal gland function has the task of enhancing β-stimulation (Erlich and Apuzzo, 1985; Cagnacci, 1996). Since Mel, which has both lipid- and water-soluble characteristics, has low molecular weight, it is rapidly excreted from the cell by passive diffusion after being produced by pinealocytes (Cagnacci, 1996).

The pineal gland is outside the blood–brain barrier (BBB) and therefore, Mel passes directly into the bloodstream. Later, it reaches all biological fluids and tissues of the organism (Turgut et al., 2007b). Mel is also produced in a number of other areas from pineal gland, for example the gastrointestinal tract (Altun and Ugur-Altun, 2007; Paredes et al., 2009). Also, the biochemical and tissue analysis were shown the presence of Mel in many body fluids and tissues such as cerebrospinal fluid, saliva, lymph, amniotic fluid, urine, semen, retina, and sciatic nerve. During pregnancy, this hormone passes through the placenta into fetus from mother and after the birth it passes with milk to newborns (Erlich and Apuzzo, 1985; Cagnacci, 1996). Thus, Mel is important for the development of all body systems, especially nervous system. Experimental pinealectomy models have investigated the role of Mel on the development of newborn animals. At this point, Tunc et al. (2006) observed that pinealectomy significantly reduces the Purkinje cell number in cerebellar cortex. Turgut et al. (2007a) investigated the role of pinealectomy/Mel on the cervical spinal cord development. Their results indicated that pinealectomy procedure significantly reduces neuron number in the gray matter and the volume of white matter of the C6 segment of cervical spinal cord in the chicken. Tunc et al. (2006) studied the effects of pinealectomy on the development of cerebellar granule cells. The results of the study suggest that the granule cell loss in the cerebellar cortex is due to developmental retardation in the early postnatal period due to absence of pineal gland and/or Mel deficiency.

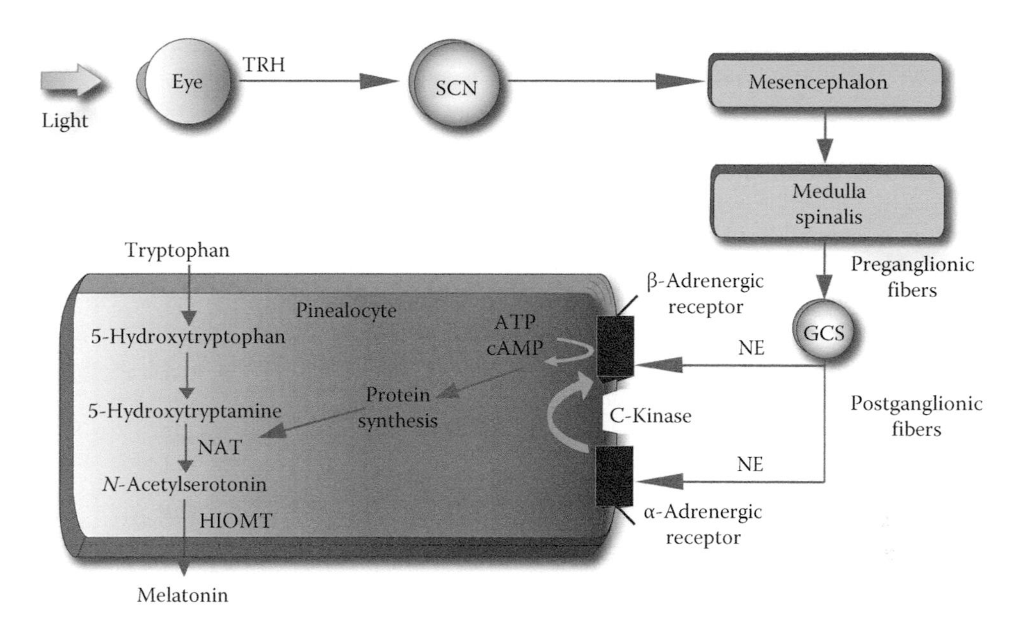

FIGURE 38.1 Mel biosynthesis under physiological conditions. TRH, tractus retinohypothalamicus; NSC, nucleus suprachiasmaticus; NPV, nucleus paraventricularis; GCS, ganglion cervicale superior; NE, norepinephrine; HIOMT, hydroxyindole-*O*-methyltransferase; NAT, *N*-acetyltransferase; ATP, adenosine-triphosphate; CAMP, cyclic adenosine monophosphate. (Modified from Konturck, S.J. et al., *J. Physiol. Pharmacol.*, 58, 23, 2007.)

38.4 REGULATION OF MELATONIN SECRETION

Circadian rhythm of Mel secretion is regulated by nucleus suprachiasmaticus in the hypothalamus. This nucleus inhibits the synthesis of Mel in the pineal gland, depending on the intensity of light (Cagnacci, 1996). Mel, which has in the blood half-life of 10–40 min, is metabolized in the liver and kidneys (Erlich and Apuzzo, 1985; Cagnacci, 1996). In all mammals including humans, the greater amount of Mel is secreted at night. Reported to be pineal, NAT and HIOMT enzymatic activities, which play a role in the synthesis of Mel, are very high in the night. In humans, the Mel hormone starts to secrete usually from 9:00 to 10:00 p.m. in the evening. Maximum concentration in the blood (50–70 pg/mL) is observed between 24:00 and 04:00 h. The Mel secretion significantly decreases from 07:00 to 9:00 a.m. in the morning (Erlich and Apuzzo, 1985; Cagnacci and Volpe, 1996). The level of Mel in the blood varies with age. In newborns, Mel concentration in the blood is low and increases by third month. After this month, the circadian rhythm of Mel is obvious. The concentration of Mel in the blood reaches the maximum level at around 8 years of age. During puberty, significantly decreasing Mel level shows a continuous decrease with age (Erlich and Apuzzo, 1985).

38.5 FUNCTIONS OF MELATONIN

38.5.1 Effects on Endocrine System and Metabolism

Mel not only has a general inhibitory effect on the hypothalamus, pituitary, and gonads (Erlich and Apuzzo, 1985, Cagnacci and Volpe, 1996; Diaz Lopez et al., 2005), but it also suppresses the release of gonadotropin-releasing hormone (GnRH from the hypothalamus and inhibits the secretion of luteinizing hormone (LH) from the front part of the pituitary gland. As a result, this effect on the hypothalamus and the anterior pituitary decreases the production of gonadal hormones (Guney et al., 2013). In addition, Mel hormone has the effect on the other endocrine organs such as thyroid and adrenal gland.

Showing an inhibitory effect on the functions of the thyroid, this hormone decreases the secretion of glucocorticoid and mineralocorticoid in the adrenal gland (Campino et al., 2011).

In the literature, the effects of Mel on the metabolism were well studied. Obesity is the accumulation of excessive fat in adipose tissue, and it is one of the most important metabolic problems in the world at the present time, which affects both genders and all age groups. Atılgan et al. (2013) reported that Mel injection reduced oxidative stress parameters in rat testes. According to the study of Srinivasan et al. (2013), Mel increased insulin sensitivity and glucose tolerance in animals fed with either high fat or high sucrose diet. Yurt et al. (2013) examined effects of fatty diet–induced obesity and Mel on kidney in female rats by histological and quantitative methods. Their results indicated that Mel treatment after obesity may contribute structural and functional healing from obesity-caused renal deformities. Also, Mel affects the bone metabolism as Turgut et al. (2005a) suggested that pineal gland/Mel might have an osteoinductive effect on bone formation.

38.5.2 Strengthening of the Immune System

Mel, by effecting the synthesis and secretion of thyrotropin-releasing hormone (TRH) from the hypothalamus, shows the activity of immunomodulators, and antistress. The decline of immune functions was observed during the blocking of Mel synthesis like pinealectomy and reported being induced again when exogenous Mel was quantitative (Oner et al., 2004). It reduces cholesterol levels in circulation, decreases the risk of hypertension and atherosclerosis (Cagnacci, 1996), and may also be important for the vascular development (Turgut et al., 2009).

38.5.3 Regulation of Sleep Circadian Rhythm and Body Temperature

If Mel is not secreted at a sufficient level during the night, sleep time is shortened. The case of least or nonregular secretion events results in sleeping difficulties and frequent awakenings. Previous experimental studies indicate that the external application of Mel results in increased sleep time (Erlich and Apuzzo, 1985). The preoptic area of the hypothalamus is the heat center and here is Mel hormone sensitive receptor in neurons. Hormone affects the region via these receptors and leads to a decrease in body temperature (Cagnacci and Volpe, 1996).

38.5.4 Antioxidant Effect of Melatonin

Some experimental studies indicated that Mel has antioxidant properties and prevents oxidative damage caused especially by lipid peroxidation (Sewerynek et al., 1995; Longoni et al., 1998; Stastica et al., 1998; Zang et al., 1998). It can easily reach all organelles, even the core of the cell, due to both hydrophilic and lipophilic components, and thus play a role in protecting the DNA construct against oxidative damage (Arendt, 1988). Mel is an antioxidant that can penetrate the mitochondria at the cellular level and protects the mitochondria from oxidative damage. This antioxidant effect of Mel on mitochondria is an important feature, since mitochondrial oxidation has an important role in the initiation of apoptotic cell death (Cardinali et al., 2013). In addition, because of the powerful antioxidant property of Mel, it stimulates the activity of antioxidant enzymes such as superoxide dismutase, glutathione peroxidase, and glutathione reductase (Reiter et al., 1997).

38.5.5 Effect of Melatonin on Hippocampus

The tissues of the central nervous system (CNS) are the most affected tissues by oxidative damage in the organism (Kus et al., 2002, 2004). Although the brain has a rather small percentage (2%) of body weight, it consumes a large part of the inhaled oxygen (20%). The products of oxygen are extremely toxic and nerve tissues are known to be more vulnerable to this toxicity. Antioxidant enzyme levels at the neural tissues of CNS are relatively lower than at other tissues.

On the other hand, we should state that the brain has high concentration of polyunsaturated fatty acids (PUFA), and therefore, oxidative process can easily progress (Kus et al., 2004). If an organism confronts molecular deterioration resulting from oxidative damage, it prevents premature aging and diseases caused by free radicals (Kus et al., 2002).

38.5.6 Protective Effects of Melatonin

The oxidative damage to proteins is reflected by a decrease in the levels of protein thiols, as a result of oxidation of protein thiol groups by free radicals and an increase in levels of advanced oxidation protein products, which are terminal products of protein exposure to free radicals (Cakatay et al., 2003). And also, the oxidation of proteins leads to the cleavage of the polypeptide chain and to the formation of cross-linked protein aggregates (Eskiocak et al., 2007).

Mel can protect against protein degradation caused by free radicals. It has been suggested by Abe et al. (1994) that if treated with Mel, glutathione -depleted newborn rats did not develop cataracts. Protective effect of Mel may be related to its antioxidant effect. Another study suggests that treatment of Mel partly prevented the increases in protein oxidation after hypoxia in the brain tissue (Eskiocak et al., 2007). Molecular oxidation within the nucleus may cause DNA fragmentation and mRNA mutagenic changes (Siu et al., 2006). Oxidative DNA damage has been contributed to the general decline in cellular functions that are associated with a variety of diseases such as Alzheimer disease, amyotrophic lateral sclerosis (ALS), Parkinson's disease, atherosclerosis, neuronal injuries, degenerative disease of the human temporomandibular joint, cataract, rheumatoid arthritis, multiple sclerosis (Barziali and Yamamoto, 2004).

Eskiocak et al. (2007) have suggested that Mel stimulates a number of antioxidant enzymes. Since Mel can pass through membranes and barriers, then accumulate in cell nuclei by means of lipophilic property, it protects DNA against oxidative stress (El-Azizi et al., 2005; Hussein et al., 2005). Free radicals are produced during lipid peroxidation. If generation of lipid peroxidation reacts with the PUFA in the membranes, free radicals may cause their own destruction (Reiter et al., 1997). It has a protective effect on ascorbate-Fe^{2+} lipid peroxidation of PUFA in rat brain microsomes (Leaden and Catalá, 2005). Tütüncüler et al. (2005) have showed that Mel prevents lipid peroxidation during hypoxia ischemia.

Free-radical damage is highly effective on the brain tissue because of its high utilization of oxygen; for this reason, Mel production is very important for brain during embryonic development (Halliwell and Cutteridge, 1985). Baydas et al. (2003) and Osuna et al. (2002) have shown that homocysteine-induced lipid peroxidation leads to neurotoxicity and Mel protects neural tissues against it.

It has been noted that the human brain has Mel receptors that are located in the regions such as the hippocampus (Savaskan et al., 2001, 2005). Mel's antioxidative and neuroprotective properties may be expressed as synaptic plasticity of pyramidal neurons in the hippocampus (El-Sherif et al., 2003), regulation of the expression of cell adhesion molecules (Baydas et al., 2002), and serotonin release (Monnet, 2002).

Effectiveness of the Mel on the peripheral nerve injury was widely investigated. Stereological analysis of Kaplan et al. (2011) showed that sciatic nerve of Mel-treated animals had significantly higher axon number than nontreated ones after injury. Likewise, Turgut et al. (2005b) reported that surgical pinealectomy caused a proliferation of connective tissue and large neuroma formation at the proximal end of the sciatic nerve. Statistically significant reduction in connective tissue content of the same region was observed in pinealectomized animals treated with Mel.

38.5.7 Neuroprotective Effects of Melatonin

Mel easily crosses BBB and is found in large amounts in the brain because it can be influenced by exogenous administration (Dubocovich et al., 2003). It has free radical scavenging activities and lipophilic–hydrophilic properties (Reiter, 1998), and so Mel is a neuroprotective agent in a wide affecting the CNS (Antolín et al., 2002; Srinivasan et al., 2006). It can prevent injury owing to its radical scavenging function (Tan et al., 1993; Reiter et al., 2000). Additionally, Mel can preserve the cell membrane, organelles, and core toward free radical damage (Aygun et al., 2012). It has been reported that Mel also has a strong antiapoptotic signaling. Additionally, it induces 1-methyl-4-phenyl-1,2,3,6-tetrahydropyridine (MPTP) and significantly reduces mitochondrial DNA damage in the substantia nigra (Pandi-Perumal et al., 2006).

38.6 NONSTEROIDAL ANTI-INFLAMMATORY DRUGS

NSAIDs have been used to relieve the symptoms such as pain, fever, and inflammation associated with rheumatoid arthritis, sports injuries, and temporary pain for many years. Although NSAIDs have side effects, they are still used by millions of patients in the world. All NSAIDs act through inhibiting cyclooxygenase (COX), which produces prostaglandins (PGs) from arachidonic acid. COX is a special enzyme. Because the basal production of PGs through COX-2 may participate in neuronal homeostasis, COX-2 in CNS may have an unstable functionality (Zarghi and Arfaei, 2011). Additionally, COX-2 has been associated with pro-inflammatory activities in the brain. However, two major aspects of COX should be considered. First, under normal conditions, in the CNS, COX-2 contributes to fundamental brain functions such as synaptic activity. Second, neuroinflammation is a much more controlled reaction than inflammation in the peripheral nervous system and mostly sustained by activation of microglia (Minghetti, 2004).

38.6.1 Neurotoxicity of Nonsteroidal Anti-Inflammatory Drugs

PGs stimulate astrocytic glutamate release into the synaptic cleft; thus they take roles normal and abnormal functions of organs and systems in human (Sanzgiri et al., 1999; Siu et al., 2000). COX is an enzyme that catalyses formation of PGs from arachidonic acid (Siu et al., 2000). COX-1 is an isoform of COX, which is found constitutively in many cell types but another isoform of COX named COX-2 is found selectively in neurons of the cerebral cortex, hippocampus, and amygdale (Andreasson et al., 2001). On the other hand, COX-2 upregulation in response to lipopolysaccharide mediates after fever induction and contributes to changes in the BBB. Therefore, it is believed that it contributes to CNS inflammatory processes (Li et al., 1999). Neuronal COX-2 increases due to increasing oxidative stress, and this may also be detrimental to neurons (Yermakova and O'Banion, 2001). Recent experimental and clinical studies on Parkinson's disease point out that selective COX-2 inhibition prevented microglial activation and cell loss of dopaminergic neurons in the substantia nigra by dopaminergic neurotoxin MPTP or 6-hydroxydopamine (6-OHDA) (Esposito et al., 2007). For these reasons, some COX-2 inhibitors have therapeutic efficacy against neurodegeneration related to inflammatory reaction (Aygun et al., 2012) (Figure 38.2).

In one study, it has been shown that NSAIDs can enhance the heat shock response as a reaction to hyperthermia and other toxic conditions by the induction of heat shock proteins (Hsps) (Batulan et al., 2005). The major inducible member of the Hsp family such as Hsp70 functions as a chaperone during protein synthesis, intracellular transport, and degradation of abnormally folded proteins (Jolly and Morimoto, 2000). Additionally, the heat shock cognate proteins and Hsps proteins inhibit apoptosis (Mosser and Morimoto, 2004). Lee et al. (2001) suggest that Hsp70 protects against hyperthermia, oxidative stress, glutamate excitotoxicity, and ischemia in the nervous system.

Several in vivo studies have reported that there are inconsistent effects of NSAIDs on the neurotoxicity of MPTP or MPP+ (1-methyl-4-phenylpyridinium) (Esposito et al., 2007). Some NSAIDs such as aspirin provide neuroprotection from MPTP at the striatal and nigral levels (Teismann and Ferger, 2001); contrary to this study, other NSAIDs (e.g., diclofenac sodium [DS]) have no protective effect on MPTP toxicity. Celecoxib is a selective COX-2 inhibitor and aggravated MPP+-induced striatal dopamine depletion in rats (Sairam et al., 2003). Moreover, ibuprofen may inactivate microglial proliferation by inhibiting cell cycle (Elsisi et al., 2005). Additionally, Chang et al. (2005) suggest

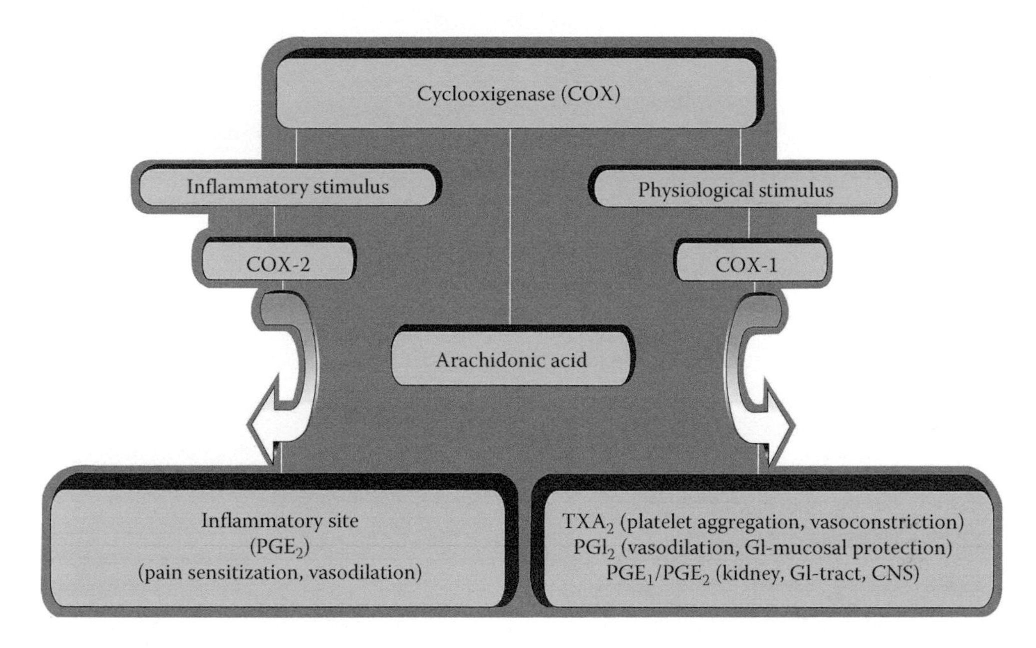

FIGURE 38.2 Effects of NSAIDs on inflammations and COX pathway. (Modified from Aygun, D. et al., *Histol. Histopathol.*, 27, 417, 2012.)

that NSAIDs significantly block the cell cycle at the G0/G1 phase, inducing cytotoxicity and cell death. It has also been observed that the administration of DS, which is another NSAI drug, was ineffective in the Parkinsonian model induced by other neurotoxins 1-methyl-4-phenylpyridinium or 6-hydroxydopamine (Asanuma and Miyazaki, 2007). Treatment of indomethacin, ibuprofen, ketoprofen, and diclofenac in PC12 cells significantly potentiate MPP-induced cell death (Morioka et al., 2004). Furthermore, it has been reported that NSAIDs induce apoptosis in different types of cells (Yamazaki et al., 2002). The NSAI agents such as tolmetin and sulindac inhibit liver tryptophan 2,3-dioxygenase activities and alter brain neurotransmitter levels (Dairam et al., 2006). Therefore, it may be concluded that NSAIDs may cause cell death by not only COX inhibition but also other mechanisms.

Mechanism of action of such drugs on cell death is still unclear. However, multifactorial mechanisms may be acceptable, including diminishing the effects of antioxidants such as Mel, apoptosis, ROS, activity of the caspase-dependent cascade, activation of the peroxisome proliferator-activated receptor (PPAR), arrested cell cycle, and increasing the intracellular accumulation of toxic agents by inhibiting the activities of multidrug resistance proteins (MRPs) (Kusuhara et al., 1998; Klampfer et al., 1999; Piqué et al., 2000).

38.6.2 Prenatal Exposure of Nonsteroidal Anti-Inflammatory Drugs

The toxic effect of NSAIDs during development has been widely investigated (Aygun et al., 2012; Ekici et al., 2012; Zengin et al., 2013). Guven et al. (2013) reported that prenatally subjected DS, which is one of the most commonly used NSAIDs, may lead to morphometric and pathological changes in uterine horns and ovaries. Also, it was detected that DS affects vascular development in terms of morphometric and structural aspects (Zengin et al., 2013). In addition to the development of cardiovascular and reproductive systems, prenatally exposed NSAIDs are a major problem for the development of the nervous system (Figures 38.3 through 38.5) (Tunc et al., 2006; Gokcimen et al., 2007; Ragbetli et al., 2007; Canan et al., 2008; Ozyurt et al., 2011; Ekici et al., 2012).

In the study of Gokcimen et al. (2007), the total neuron number in one side of the cornu ammonis and gyrus dentatus of the hippocampal formation in control and drug treated (DS) groups of male

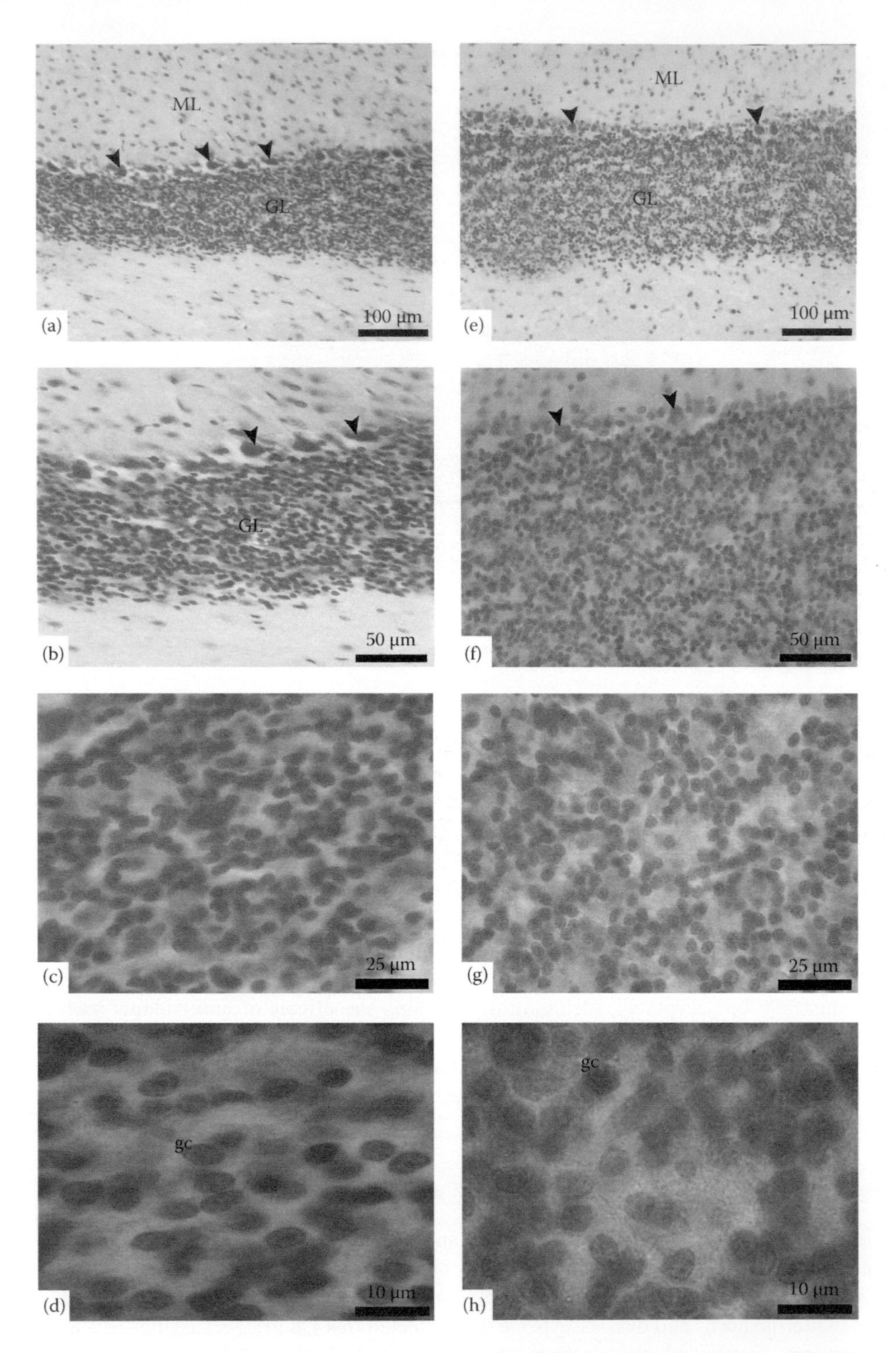

FIGURE 38.3 Prenatal exposure of diclofenac sodium substantially reduced the number of Purkinje cells as well as granular cells in the rat cerebellum. Pictures in (a–d) belong to the saline group and (e–h) are taken from the drug-exposed group. As seen from these pictures, the density of both type of cells, that is, Purkinje and granular cell, was lower than the DS-exposed group. This comment is based on the stereological analysis of the cells in the cerebellum, not qualitative analysis (Ragbetli et al., 2007). ML, molecular layer; GL, granular layer; arrowheads show the Purkinje cells. H&E staining.

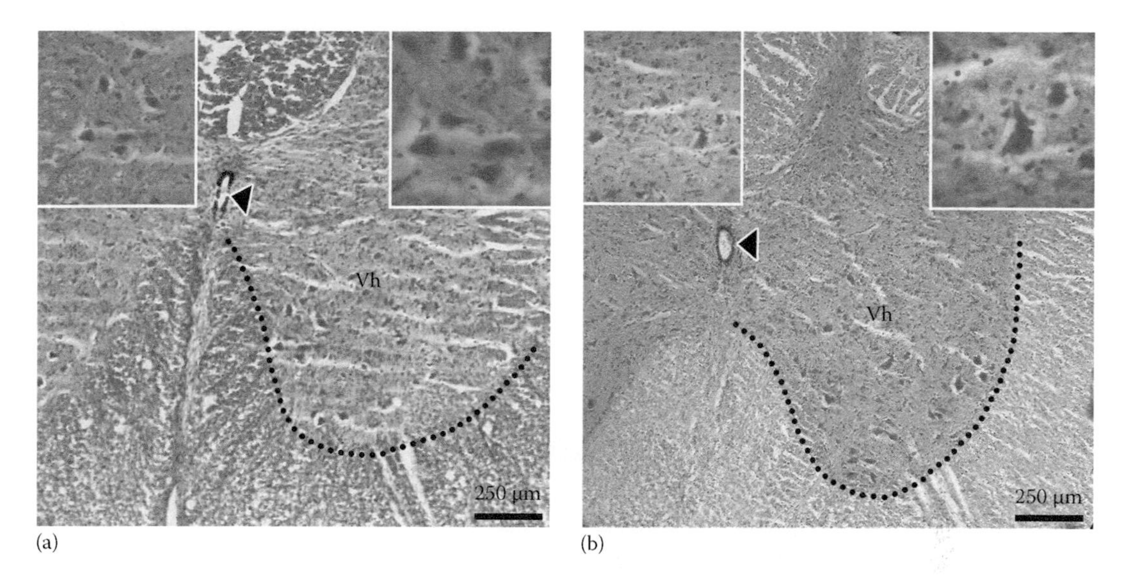

FIGURE 38.4 Effects of prenatal exposure of DS on the spinal cord of male rats were investigated in a previous study (Ozyurt et al., 2011). Two groups were used in this study: saline group and DS-exposed group. It could be seen that DS exposure substantially decreases the neuron number. (a) is saline group and (b) is DS-exposed group. Vh, ventral horn; arrowheads show central canal of spinal cord. H&E staining.

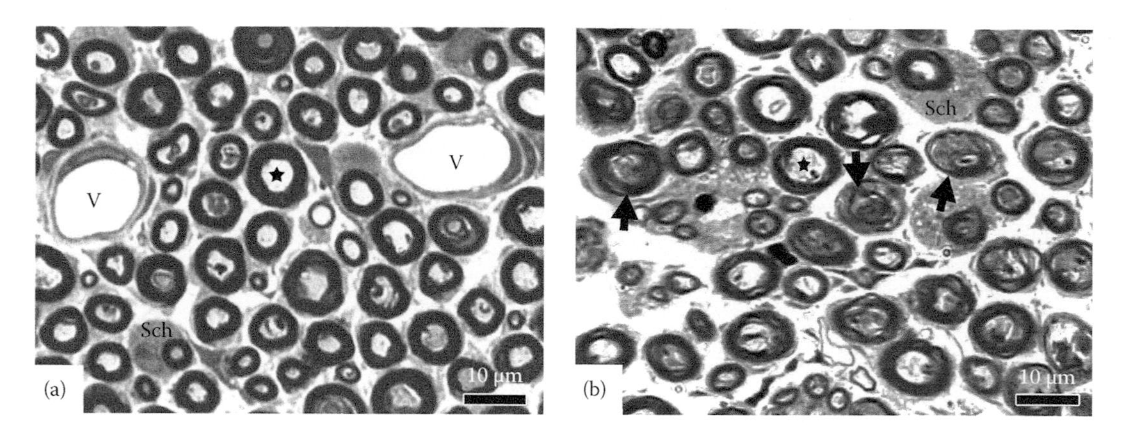

FIGURE 38.5 Prenatal exposure of saline and DS was investigated in rats. It was seen that DS exposure not only decreased the number of axons but also destroyed the morphology of nerve fibers (b) in comparison to saline group (a) (Canan et al., 2008). Arrows show the destroying of nerve fiber sheath. (*) Normal axon and its sheath; Sch, Schwann cell; V, vessel.

rats was estimated using the optical fractionator technique. Results of the study suggested that DS causes neuronal loss in the hippocampus. Ragbetli et al. (2007) also investigated the effects of DS on the development of the cerebellum. In this study, they estimated the total number of Purkinje cells of the cerebellum in a control and DS-treated groups using stereological methods. They found that the total number of Purkinje cells in the offspring of drug-treated rats was significantly lower than in the offspring of control animals. So, their results indicated that the Purkinje cells of a developing cerebellum were affected by the administration of DS at the prenatal period. Ozyurt et al. (2011) showed that the development of neurons and volume of cervical spinal cord were affected by the prenatal administration of DS (Figure 38.4).

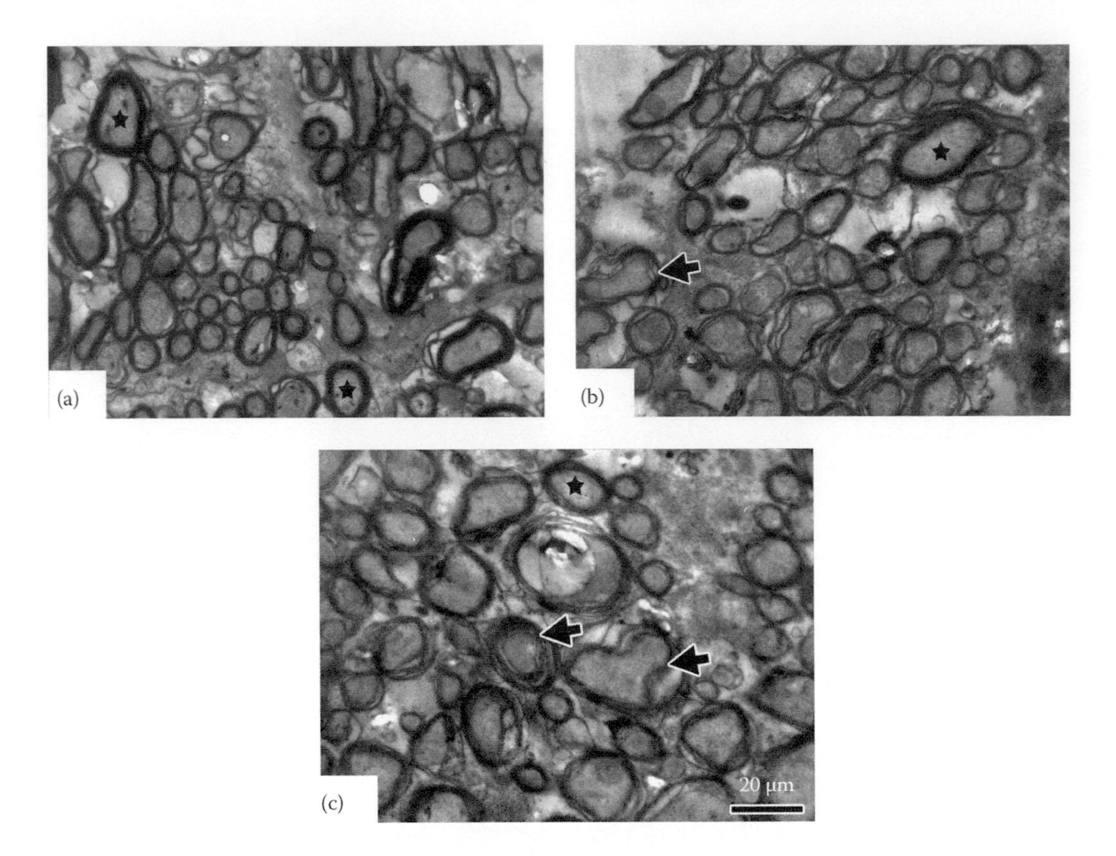

FIGURE 38.6 The effects of prenatal exposure of DS were investigated; in this experiment, three animal groups were used: control (a), saline (b), and DS (c). DS exposure did not change the total number of myelinated optic nerve fiber number, but the morphology of the fibers substantially impaired. It was demonstrated that both DS and saline exposure did some structural and morphometric changes in the rat optic nerve. Even DS exposure made some changes in the optic nerve of 4-week-old of male and female rats, but these changes could be repaired by the time as seen in 20-week-old rats (Colakoglu et al., 2013). (*) Normal axon, arrows: impaired axons.

It has been shown that NSAIDs not only impair the CNS development and damage the neural activity but also destroy the morphology of peripheral nerves (Figure 38.5) and optic nerves (Figure 38.6), (Colakoglu et al., 2013). Canan et al. (2008) examined the development of sciatic nerve in prenatally DS exposed and control rats. They evaluated the nerves in terms of axon number, cross-sectional area of axon and myelin sheet thickness, as well as the ultrastructure of nerve fibers. Their results showed that DS exposure made morphometric changes such as axonal loss and decrease in the nerve cross-section area as well as axon cross-section area and deterioration of myelin sheaths. On the other hand, Ayranci et al. (2013) found that prenatal exposure of DS did not affect axon number in rats, but it can alter the morphology of the male and female median nerve. This discrepancy may be attributed to type of nerve, since Canan et al. (2008) used sciatic nerve but Ayranci et al. (2013) used median nerve.

38.6.3 Neuroprotective Effects of Melatonin against Neurotoxicity of NSAIDs

The level of neurotransmitters, which act as chemical messengers in the brain, may change by NSAIDs (Sairam et al., 2003). Tolmetin and sulindac, which are NSAIDs, have a potential to induce adverse effects in patients suffering with neurological disorders such as Parkinson's disease. These agents inhibit tryptophan 2,3-dioxygenase with a concomitant increase in 5-HT levels in

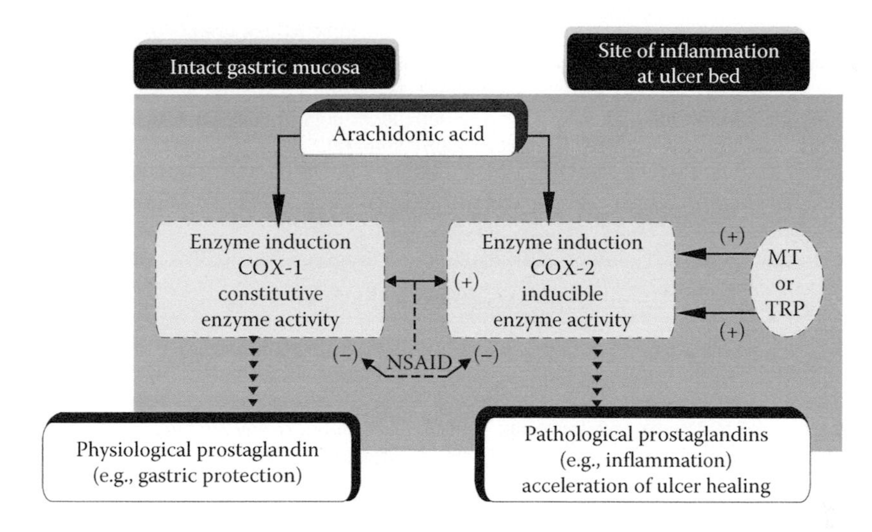

FIGURE 38.7 Mechanism of Mel on pathological conditions. (Modified from Aygun, D. et al., *Histol. Histopathol.*, 27, 417, 2012.)

the hippocampus and reduce dopamine levels in the striatum (Dairam et al., 2006). A physiological neuroprotective effect of Mel on neurodegeneration in the nigrostriatal system in Parkinson's disease has been shown (Sharma et al., 2006). Mel inhibits the adverse effect of NSAIDs. It has been reported that PGs markedly enhance Mel synthesis at night (Voisin et al., 1993). Nevertheless, NSAIDs inhibit PG synthesis by inhibiting COX, and they have the ability to reduce Mel synthesis (Vane, 1971). Its synthesis is reduced by NSAIDs such as aspirin and ibuprofen in humans (Murphy et al., 1996). For this reason, NSAIDs may usually cause neurotoxicity by decreasing the synthesis of Mel in the tissues and organs. Some NSAIDs such as DS cause drug toxicity by oxidative stress and apoptosis (Inoue et al., 2004). Mel and its metabolites are potent free radical scavengers and antioxidants, which protect cells from damage induced by a variety of oxidants (Catalá, 2007). It has also been suggested that Mel has a strong antiapoptotic signaling function (Pandi-Perumal et al., 2006). Figure 38.7 shows the mechanism of Mel on pathological conditions.

38.7 CONCLUSION

Mel is a hormone found naturally in the tissues of body, and it is used as medicine that is usually made synthetically in the laboratory. It is also used for treatment of the inability to fall asleep (insomnia); delayed sleep phase syndrome (DSPS); insomnia associated with attention deficit-hyperactivity disorder (ADHD); insomnia due to certain high blood pressure medications called beta-blockers; and sleep problems in children with developmental disorders including autism, cerebral palsy, and mental retardation (Andersen et al., 2008). The antioxidant role of Mel may be of potential use for conditions in which oxidative stress is involved in the pathophysiologic processes (Reiter et al., 2000). The multiplicity of actions and variety of biological effects of Mel suggest the potential for a range of clinical and wellness-enhancing uses. Protective roles of Mel in preventing toxic side effects of NSAIDs are not well known. New experimental studies are needed to explore the real way of Mel action on the DS toxicity.

REFERENCES

Abe, M., Reiter, R. J., Orhii, P. B., Hara, M., Poeggeler, B. 1994. Inhibitory effect of melatonin on cataract formation in newborn rats: Evidence for an antioxidative role for melatonin. *J Pineal Res* 17:94–100.

Altun, A., Ugur-Altun, B. 2007. Melatonin: Therapeutic and clinical utilization. *Int J Clin Pract* 61:835–845.

Andersen, I. M., Kaczmarska, J., McGrew, S. G., Malow, B. A. 2008. Melatonin for insomnia in children with autism spectrum disorders. *J Child Neurol* 23:432–485.

Andreasson, K. I., Savonenko, A., Vidensky, S., Goellner, J. J., Zhang, Y., Shaffer, A., Kaufmann, W. E., Worley, P. F., Isakson, P., Markowska, A. L. 2001. Age-dependent cognitive deficits and neuronal apoptosis in cyclooxygenase-2 transgenic mice. *J Neurosci* 21:8198–8209.

Antolín, I., Mayo, J. C., Sainz, R. M., del Brío Mde, L., Herrera, F., Martín, V., Rodríguez, C. 2002. Protective effect of melatonin in a chronic experimental model of Parkinson's disease. *Brain Res* 943:163–173.

Arendt, J. 1988. Melatonin. *Clin Endocrinol* 29:205–209.

Asanuma, M., Miyazaki, I. 2007. Common anti-inflammatory drugs are potentially therapeutic for Parkinson's disease? *Exp Neurol* 206:172–178.

Atilgan, D., Parlaktas, B. S., Uluocak, N., Erdemir, F., Kilic, S., Erkorkmaz, U., Ozyurt, H., Markoc, F. 2013. Weight loss and melatonin reduce obesity-induced oxidative damage in rat testis. *Adv Urol* 2013:836121. doi:10.1155/2013/836121.

Aygun, D., Kaplan, S., Odaci, E., Onger, M. E., Altunkaynak, M. E. 2012. Toxicity of non-steroidal anti-inflammatory drugs: A review of melatonin and diclofenac sodium association. *Histol Histopathol* 27:417–436.

Ayranci, E., Altunkaynak, B. Z., Aktas, A., Ragbetli, M. C., Kaplan, S. 2013. Prenatal exposure of diclofenac sodium affects morphology but not axon number of the median nerve of rats. *Folia Neuropathol* 51:76–86.

Barziali, A., Yamamoto, K. 2004. DNA damage responses to oxidative stress. *DNA Repair* 3:1109–1115.

Batulan, Z., Nalbantoglu, J., Durham, H. D. 2005. Nonsteroidal anti-inflammatory drugs differentially affect the heat shock response in cultured spinal cord cells. *Cell Stress Chaperones* 10:185–196.

Baydas, G., Kutlu, S., Naziroglu, M., Canpolat, S., Sandal, S., Ozcan, M., Kelestimur, H. 2003. Inhibitory effects of melatonin on neural lipid peroxidation induced by intracerebroventricularly administered homocysteine. *J Pineal Res* 34:36–39.

Baydas, G., Nedzvetsky, V. S., Nerush, P. A., Kirichenko, S. V., Demchenko, H. M., Reiter, R. J. 2002. A novel role for melatonin: Regulation of the expression of cell adhesion molecules in the rat hippocampus and cortex. *Neurosci Lett* 326:109–112.

Cagnacci, A. 1996. Melatonin in relation to physiology in adult humans. *J Pineal Res* 21:200–213.

Cagnacci, A., Volpe, A. 1996. Influence of melatonin and photoperiod on animal and human reproduction. *J Endocrinol Invest* 19:382–411.

Cakatay, U., Telci, A., Kayali, R., Tekeli, F., Akçay, T., Sivas, A. 2003. Relation of aging with oxidative protein damage parameters in the rat skeletal muscle. *Clin Biochem* 36:51–55.

Campino, C., Valenzuela, F. J., Torres-Farfan, C., Reynolds, H. E., Abarzua-Catalan, L., Arteaga, E., Trucco, C., Guzmán, S., Valenzuela, G. J., Seron-Ferre, M. 2011. Melatonin exerts direct inhibitory actions on ACTH responses in the human adrenal gland. *Horm Metab Res* 43:337–342.

Canan, S., Aktaş, A., Ulkay, M. B., Colakoglu, S., Ragbetli, M. C., Ayyildiz, M., Geuna, S., Kaplan, S. 2008. Prenatal exposure to a non-steroidal anti-inflammatory drug or saline solution impairs sciatic nerve morphology: A stereological and histological study. *Int J Dev Neurosci* 26:733–738.

Caniato, R., Filippini, R., Piovan, A., Puricelli, L., Borsarini, A., Cappelletti, E. M. 2003. Melatonin in plants. *Adv Exp Med Biol* 527:593–597.

Cardinali, D. P., Pagano, E. S., Bernasconi, P. A. S., Reynoso, R., Scacchi, P. 2013. Melatonin and mitochondrial dysfunction in the central nervous system. *Horm Behav* 63:322–330.

Catalá, A. 2007. The ability of melatonin to counteract lipid peroxidation in biological membranes. *Curr Mol Med* 7:638–649.

Chang, H. M., Tseng, C. Y., Wei, I. H., Lue, J. H., Wen, C. Y., Shieh, J. Y. 2005. Melatonin restores the cytochrome oxidase reactivity in the nodose ganglia of acute hypoxic rats. *J Pineal Res* 39:206–214.

Colakoglu, S., Aktaş, A., Raimondo, S., Turkmen, A. P., Altunkaynak, B. Z., Odaci, E., Geuna, S., Kaplan, S. 2013. Effects of prenatal exposure to diclofenac sodium and saline on the optic nerve of 4- and 20-week-old male rats: A stereological and histological study. *Biotech Histochem* 89(2):136–144. doi:10.3109/10520295.2013.827741.

Dairam, A., Antunes, E. M., Saravanan, K. S., Daya, S. 2006. Nonsteroidal anti-inflammatory agents, tolmetin and sulindac, inhibit liver tryptophan 2,3-dioxygenase activity and alter brain neurotransmitter levels. *Life Sci* 79:2269–2274.

Diaz Lopez, B., Diaz Rodriguez, E., Urquijo, C., Alvarez, C. 2005. Melatonin influences on the neuroendocrine-reproductive axis. *Ann N Y Acad Sci* 1057:337–354.

Dubocovich, M. L., Rivera-Bermudez, M. A., Gerdin, M. J., Masana, M. I. 2003. Molecular pharmacology, regulation and function of mammalian melatonin receptors. *Front Biosci* 8:d1093–d1108.

Ekici, F., Keskin, İ., Aslan, H., Erişgin, Z., Altunkaynak, B. Z., Gökçimen, A., Odaci, E., Kaplan, S. 2012. Does prenatal exposure to diclofenac sodium affect the total number of cerebellar granule cells in male juvenile and adult rats? *J Exp Clin Med* 29:52–57.

El-Aziz, M. A., Hassan, H. A., Mohamed, M. H., Meki, A. R., Abdel-Ghaffar, S. K., Hussein, M. R. 2005. The biochemical and morphological alterations following administration of melatonin, retinoic acid and *Nigella sativa* in mammary carcinoma: An animal model. *Int J Exp Pathol* 86:383–396.

El-Sherif, Y., Tesoriero, J., Hogan, M. V., Wieraszko, A. 2003. Melatonin regulates neuronal plasticity in the hippocampus. *J Neurosci Res* 72:454–460.

Elsisi, N. S., Darling-Reed, S., Lee, E. Y., Oriaku, E. T., Soliman, K. F. 2005. Ibuprofen and apigenin induce apoptosis and cell cycle arrest in activated microglia. *Neurosci Lett* 375:91–96.

Erlich, S. S., Apuzzo, M. L. J. 1985. The pineal gland: Anatomy, physiology and clinical significance. *J Neurosurg* 63:321–341.

Eskiocak, S., Tutunculer, F., Basaran, U. N., Taskiran, A., Cakir, E. 2007. The effect of melatonin on protein oxidation and nitric oxide in the brain tissue of hypoxic neonatal rats. *Brain Dev* 29:19–24.

Esposito, E., Di Matteo, V., Benigno, A., Pierucci, M., Crescimanno, G., Di Giovanni, G. 2007. Non-steroidal anti-inflammatory drugs in Parkinson's disease. *Exp Neurol* 205:295–312.

Gokcimen, A., Rağbetli, M. C., Baş, O., Tunc, A. T., Aslan, H., Yazici, A. C., Kaplan, S. 2007. Effect of prenatal exposure to an anti-inflammatory drug on neuron number in cornu ammonis and dentate gyrus of the rat hippocampus: A stereological study. *Brain Res* 1127:185–192.

Guney, M., Ayranci, E., Kaplan, S. 2013. Development and histology of the pineal gland in animals. Turgut M. (Ed.) In *Step by Step Experimental Pinealectomy in Animals for Researchers*, Chapter II, pp. 33–52. New York: Nova Science Publishers.

Guven, D., Altunkaynak, B. Z., Ayranci, E., Kaplan, S., Bildircin, F. D., Kesim, Y., Ragbetli, M. C. 2013. Stereological and histopathological evaluation of ovary and uterine horns of female rats prenatally exposed to diclofenac sodium. *J Obstet Gynaecol* 33:258–263.

Halliwell, B., Gutteridge, J. M. C. 1985. Oxygen radicals and the nervous system. *Trends Neurosci* 8:22–26.

Hussein, M. R., Abu-Dief, E., Abd El-Reheem, M. H., Abd-Elrahman, A. 2005. Ultrastructural evaluation of the radioprotective effects of melatonin against X-ray-induced skin damage in Albino rats. *Int J Exp Pathol* 86:45–55.

Inoue, A., Muranaka, S., Fujita, H., Kanno, T., Tamai, H., Utsumi, K. 2004. Molecular mechanism of diclofenac-induced apoptosis of promyelocytic leukemia: Dependency on reactive oxygen species, Akt, Bid, cytochrome and caspase pathway. *Free Radic Biol Med* 37:1290–1299.

Jolly, C., Morimoto, R. I. 2000. Role of the heat shock response and molecular chaperones in oncogenesis and cell death. *J Natl Cancer Inst* 92:1564–1572.

Kaplan, S., Piskin, A., Ayyildiz, M., Aktas, A., Koksal, B., Ulkay, M. B., Turkmen, A. P., Bakan, F., Geuna, S. 2011. The effect of melatonin and platelet gel on sciatic nerve repair: An electrophysiological and stereological study. *Microsurgery* 31:306–313.

Klampfer, L., Cammenga, J., Wisniewski, H. G., Nimer, S. D. 1999. Sodium salicylate activates caspases and induces apoptosis of myeloid leukemia cell lines. *Blood* 93:2386–2394.

Konturek, S. J., Konturek, P. C., Brzozowski, T., Bubenik, G. A. 2007. Role of melatonin in upper gastrointestinal tract. *J Physiol Pharmacol* 58:23–52.

Kus, I., One, H., Ozogul, C., Ayar, A., Ozen, O. A., Sarsilmaz, M., Kelestimur, H. 2002. Effects of estradiol benzoate on the ultrastructure of the pinealocyte in the ovariectomized rat. *Neuroendocrinol Lett* 23:405–410.

Kus, I., Sarsilmaz, M., Ozen, O. A., Turkoglu, A. O., Pekmez, H., Songur, A., Kelestimur, H. 2004. Light and electron microscopic examination of pineal gland in rats exposed to constant light and constant darkness. *Neuroendocrinol Lett* 25:102–108.

Kusuhara, H., Matsuyuki, H., Matsuura, M., Imayoshi, T., Okumoto, T., Matsui, H. 1998. Induction of apoptotic DNA fragmentation by nonsteroidal anti-inflammatory drugs in cultured rat gastric mucosal cells. *Eur J Pharmacol* 360:273–280.

Leaden, P. J., Catalá, A. 2005. Protective effect of melatonin on ascorbate-Fe^{2+} lipid peroxidation of polyunsaturated fatty acids in rat liver, kidney and brain microsomes: A chemiluminescence study. *J Pineal Res* 39:164–169.

Lee, J. E., Yenari, M. A., Sun, G. H., Xu, L., Emond, M. R., Cheng, D., Steinberg, G. K., Giffard, R. G. 2001. Differential neuroprotection from human heat shock protein 70 overexpression in in vitro and in vivo models of ischemia and ischemia-like conditions. *Exp Neurol* 170:129–139.

Li, S., Wang, Y., Matsumura, K., Ballou, L. R., Morham, S. G., Blatteis, C. M. 1999. The febrile response to lipopolysaccharide is blocked in cyclooxygenase-2(–/–), but not in cyclooxygenase-1(–/–) mice. *Brain Res* 825:86–94.

Longoni, B., Salgo, M. G., Pryor, W. A., Marchiafava, P. L. 1998. Effects of melatonin on lipid peroxidation induced by oxygen radicals. *Life Sci* 62:853–859.

Minghetti, L. 2004. Cyclooxygenase-2 (COX-2) in inflammatory and degenerative brain diseases. *J Neuropathol Exp Neurol* 63:901–910.

Monnet, F. P. 2002. Melatonin modulates [3H] serotonin release in the rat hippocampus: Effects of circadian rhythm. *J Neuroendocrinol* 14:194–199.

Morioka, N., Kumagai, K., Morita, K., Kitayama, S., Dohi, T. 2004. Nonsteroidal anti-inflammatory drugs potentiate 1-methyl-4- phenylpyridinium (MPP+)-induced cell death by promoting the intracellular accumulation of MPP+ in PC12 cells. *J Pharmacol Exp Ther* 310:800–807.

Mosser, D. D., Morimoto, R. I. 2004. Molecular chaperones and the stress of oncogenesis. *Oncogene* 23:2907–2918.

Murphy, P. J., Myers, B. L., Badia, P. 1996. Nonsteroidal anti-inflammatory drugs alter body temperature and suppress melatonin in humans. *Physiol Behav* 59:133–139.

Odaci, E., Kaplan, S. 2009. Melatonin and nerve regeneration. *Int Rev Neurobiol* 87:317–335.

Oner, H., Kus, I., Oner, J., Ogeturk, M., Ozan, E., Ayar, A. 2004. Possible effects of melatonin on thymus gland after pinealectomy in rats. *Neuroendocrinol Lett* 25:115–118.

Osuna, C., Reiter, R. J., García, J. J., Karbownik, M., Tan, D. X., Calvo, J. R., Manchester, L. C. 2002. Inhibitory effect of melatonin on homocysteine-induced lipid peroxidation in rat brain homogenates. *Pharmacol Toxicol* 90:32–37.

Ozyurt, B., Kesici, H., Alici, S. K., Yilmaz, S., Odaci, E., Aslan, H., Ragbetli, M. C., Kaplan, S. 2011. Prenatal exposure to diclofenac sodium changes the morphology of the male rat cervical spinal cord: A stereological and histopathological study. *Neurotoxicol Teratol* 33:282–287.

Pandi-Perumal, S. R., Srinivasan, V., Maestroni, G. J., Cardinali, D. P., Poeggeler, B., Hardeland, R. 2006. Melatonin: Nature's most versatile biological signal? *FEBS J* 273:2813–2838.

Paredes, S. D., Korkmaz, A., Manchester, L. C., Tan, D. X., Reiter, R. J. 2009. Phytomelatonin: A review. *J Exp Bot* 60:57–69.

Piqué, M., Barragán, M., Dalmau, M., Bellosillo, B., Pons, G., Gil, J. 2000. Aspirin induces apoptosis through mitochondrial cytochrome c release. *FEBS Lett* 480:193–196.

Ragbetli, M. C., Ozyurt, B., Aslan, H., Odaci, E., Gokcimen, A., Sahin, B., Kaplan, S. 2007. Effect of prenatal exposure to diclofenac sodium on Purkinje cell numbers in rat cerebellum: A stereological study. *Brain Res* 1174:130–135.

Reiter, R. J. 1998. Oxidative damage in the central nervous system: Protection by melatonin. *Prog Neurobiol* 56:359–384.

Reiter, R. J., Carneiro, R. C., Oh, C. S. 1997. Melatonin in relation to cellular antioxidative defense mechanisms. *Horm Metab Res* 29:363–372.

Reiter, R. J., Tan, D. X., Manchester, L. C., Simopoulos, A. P., Maldonado, M. D., Flores, L. J., Terron, M. P. 2007a. Melatonin in edible plants (phytomelatonin): Identification, concentrations, bioavailability and proposed functions. *World Rev Nutr Diet* 97:211–230.

Reiter, R. J., Tan, D. X., Osuna, C., Gitto, E. 2000. Actions of melatonin in the reduction of oxidative stress. A review. *J Biomed Sci* 7:444–458.

Reiter, R. J., Tan, D. X., Terron, M. P., Flores, L. J., Czarnocki, Z. 2007b. Melatonin and its metabolites: New findings regarding their production and their radical scavenging actions. *Acta Biochim Pol* 54:1–9.

Sairam, K., Saravanan, K. S., Banerjee, R., Mohanakumar, K. P. 2003. Non-steroidal anti-inflammatory drug sodium salicylate, but not diclofenac or celecoxib, protects against 1-methyl-4-phenyl pyridinium-induced dopaminergic neurotoxicity in rats. *Brain Res* 966:245–252.

Sanzgiri, R. P., Araque, A., Haydon, P. G. 1999. Prostaglandin E2 stimulates glutamate receptor-dependent astrocyte neuromodulation in cultured hippocampal cells. *J Neurobiol* 41:221–229.

Savaskan, E., Ayoub, M. A., Ravid, R., Angeloni, D., Fraschini, F., Meier, F., Eckert, A., Müller-Spahn, F., Jockers, R. 2005. Reduced hippocampal MT2 melatonin receptor expression in Alzheimer's disease. *J Pineal Res* 38:10–16.

Savaskan, E., Olivieri, G., Brydon, L., Jockers, R., Kräuchi, K., Wirz-Justice, A., Müller-Spahn, F. 2001. Cerebrovascular melatonin MT1-receptor alterations in patients with Alzheimer's disease. *Neurosci Lett* 308:9–12.

Sewerynek, E., Melchiorri, D. A., Chen, L., Reiter, R. J. 1995. Melatonin reduces both basal and bacterial lipopolysaccharide-induced lipid peroxidation in vitro. *Free Radic Biol Med* 19:903–909.

Sharma, R., McMillan, C. R., Tenn, C. C., Niles, L. P. 2006. Physiological neuroprotection by melatonin in a 6-hydroxydopamine model of Parkinson's disease. *Brain Res* 1068:230–236.

Siu, A. W., Maldonado, M., Sanchez-Hidalgo, M., Tan, D. X., Reiter, R. J. 2006. Protective effects of melatonin in experimental free radical related ocular diseases. *J Pineal Res* 40:101–109.

Siu, S. S., Yeung, J. H., Lau, T. K. 2000. A study on placental transfer of diclofenac in first trimester of human pregnancy. *Hum Reprod* 15:2423–2425.

Srinivasan, V., Ohta, Y., Espino, J., Pariente, J. A., Rodriguez, A. B., Mohamed, M., Zakaria, R. 2013. Metabolic syndrome, its pathophysiology and the role of melatonin. *Recent Pat Endocr Metab Immune Drug Discov* 7:11–25.

Srinivasan, V., Pandi-Perumal, S. R., Cardinali, D. P., Poeggeler, B., Hardeland, R. 2006. Melatonin in Alzheimer's disease and other neurodegenerative disorders. *Behav Brain Funct* 2:15.

Stastica, P., Ulanski, P., Rosiak, J. M. 1998. Melatonin as a hydroxyl radical scavenger. *J Pineal Res* 25:65–66.

Tan, D. X., Chen, L. D., Poeggeler, B., Manchester, L. C., Reiter, R. J. 1993. Melatonin: A potent, endogenous hydroxyl radical scavenger. *Endocr J* 1:57–60.

Teismann, P., Ferger, B. 2001. Inhibition of the cyclooxygenase isoenzymes COX-1 and COX-2 provide neuroprotection in the MPTP-mouse model of Parkinson's disease. *Synapse* 39:167–174.

Tunc, A. T., Turgut, M., Aslan, H., Sahin, B., Yurtseven, M. E., Kaplan, S. 2006. Neonatal pinealectomy induces Purkinje cell loss in the cerebellum of the chick: A stereological study. *Brain Res* 1067:95–102.

Turgut, M., Erdogan, S., Ergin, K., Serter, M. 2007b. Melatonin ameliorates blood-brain barrier permeability, glutathione, and nitric oxide levels in the choroid plexus of the infantile rats with kaolin-induced hydrocephalus. *Brain Res* 1175:117–125.

Turgut, M., Kaplan, S., Turgut, A. T., Aslan, H., Güvenç, T., Cullu, E., Erdogan, S. 2005b. Morphological, stereological and radiological changes in pinealectomized chicken cervical vertebrae. *J Pineal Res* 39:392–399.

Turgut, M., Kaplan, S., Unal, Z. B., Altunkaynak, Z., Unal, D., Sahin, B., Bozkurt, M., Yurtseven, M. E. 2009. Effects of pinealectomy on morphological features of blood vessel in chicken: An electron microscopic and stereological study. *J Exp Clin Med* 26:112–118.

Turgut, M., TurkkaniTunc, A., Aslan, H., Yazici, A. C., Kaplan, S. 2007a. Effect of pinealectomy on the morphology of the chick cervical spinal cord: A stereological and histopathological study. *Brain Res* 1129:166–173.

Turgut, M., Uyanikgil, Y., Baka, M., Tunc, A. T., Yavasoglu, A., Yurtseven, M. E., Kaplan, S. 2005a. Pinealectomy exaggerates and melatonin treatment suppresses neuroma formation of transected sciatic nerve in rats: Gross morphological, histological and stereological analysis. *J Pineal Res* 38:284–291.

Tütüncüler, F., Eskiocak, S., Baflaran, U. N., Ekuklu, G., Ayvaz, S., Vatansever, U. 2005. The protective role of melatonin in experimental hypoxic brain damage. *Pediatr Int* 47:434–439.

Vane, J. R. 1971. Inhibition of prostaglandin synthesis as a mechanism of action for aspirin-like drugs. *Nat New Biol* 231:232–235.

Voisin, P., Van Camp, G., Pontoire, C., Collin, J. P. 1993. Prostaglandins stimulate serotonin acetylation in chick pineal cells: Involvement of cyclic AMP-dependent and calcium/calmodulin-independent mechanisms. *J Neurochem* 60:666–670.

Yamazaki, R., Kusunoki, N., Matsuzaki, T., Hashimoto, S., Kawai, S. 2002. Nonsteroidal anti-inflammatory drugs induce apoptosis in association with activation of peroxisome proliferator-activated receptor gamma in rheumatoid synovial cells. *J Pharmacol Exp Ther* 302:18–25.

Yermakova, A. V., O'Banion, M. K. 2001. Downregulation of neuronal cyclooxygenase-2 expression in end stage Alzheimer's disease. *Neurobiol Aging* 22:823–836.

Yurt, K. K., Kayhan, E., Altunkaynak, Z., Tümentemur, G., Kaplan, S. 2013. Effects of the melatonin on kidney of high fat diet fed obese rats: A stereological and histological approach. *J Exp Clin Med* 30:153–158.

Zang, L. Y., Cosma, G., Gardner, H., Vallyathan, V. 1998. Scavenging of reactive oxygen species by melatonin. *Biochim Biophys Acta* 1425:469–477.

Zarghi, A., Arfaei, S. 2011. Selective COX-2 inhibitors: A review of their structure-activity relationships. *Iran J Pharm Res* 10:655–683.

Zengin, H., Kaplan, S., Tümkaya, L., Altunkaynak, B. Z., Rağbetli, M. Ç., Altunkaynak, M. E., Yilmaz, O. 2013. Effect of prenatal exposure to diclofenac sodium on the male rat arteries: A stereological and histopathological study. *Drug Chem Toxicol* 36:67–78.

39 Melatonin
Pharmacological Aspects and Clinical Trends

Emiliano Ricardo Vasconcelos Rios

CONTENTS

39.1 INTRODUCTION

Melatonin, *N*-acetyl-5-methoxytryptamine, is a tryptophan-derived molecule widely distributed in nature in both plant and animal sources, such as human milk, bananas, beets, cucumbers, and tomatoes. In humans, melatonin is synthesized primarily in the pineal gland but is also produced by other organs such as the retina, bone marrow, gastrointestinal tract, gonads, and immune system.

The classic pathway of melatonin formation involves four steps, starting with tryptophan 5-hydroxylase, followed by 5-hydroxytryptophan decarboxylation by aromatic amino acid decarboxylase, N-acetylation of serotonin by arylalkylamine-*N*-acetyltransferase (AANAT), and O-methylation of *N*-acetylserotonin by hydroxyindole-*O*-ethyltransferase (HIOMT). In extrapineal sites, other enzymes can be involved and regulation mechanisms can be different.

The synthesis is controlled by the suprachiasmatic nucleus (SCN), located in the hypothalamus, and GABA is the principal transmitter in SCN cells. The nervous fibers of the retina detect the brightness of the environment and transmit this information to the central nervous system (CNS). In addition, the SCN connects through the subparaventricular zone to the dorsomedial nucleus of the hypothalamus, which is crucial for producing circadian rhythms of sleep and waking, locomotive activity, feeding, and corticosteroid production.

During the day, the synthesis of melatonin, as well as the stream of the sympathetic activity, is limited. When it is dark, the liberation of norepinephrine occurs through the β-adrenergic receptors in the pinealocytes (cell type derived from nonrod and noncone photoreceptors). Activation of pineal beta-adrenergic receptors results in the elevation of cyclic AMP and an increase in the synthesis and activity of pineal AANAT. Beta-adrenergic blockers have been shown to suppress the nocturnal synthesis and secretion of melatonin in humans, suggesting a similar regulation of pineal melatonin production. Pandi-Perumal et al. (2005) have reported that melatonin declines with aging, but the mechanism has not been fully explained. A decrease in the number of beta-adrenergic receptors in the pinealocytes, decreased activity of AANAT, the key enzyme responsible for melatonin synthesis, and increased clearance of plasma melatonin, have all been documented with age.

The G_i protein–coupled metabotropic melatonin receptors MT1 and MT2 are the primary mediators of the physiological actions of melatonin, and binding to these receptors results in the inhibition of adenyl cyclase. A third melatonin binding site, MT3, was recently identified as the enzyme quinine oxydoreductase 2 (QR2). Another protein, GPR50 (also known as H9 and ML1X), is an orphan GPCR that is about 45% identical overall to human MT1 and MT2. Although structurally related to MT1 and MT2 receptors, GPR50 does not directly bind melatonin, but it influences the binding of melatonin to MT1.

No definitive guidelines have been formulated for the clinical evaluation of patients with low melatonin levels, mainly because a "melatonin deficiency syndrome" has not yet been defined as an independent entity. The secretion of melatonin is usually measured by analyzing serum or salivary concentrations. The salivary concentration is considered to be equivalent to the serum concentration, except in the elderly or in patients with dry mouth. However, more substantive clinical evidence is required before any precise recommendations can be made.

Besides its chronobiotic role, where melatonin is involved in the regulation of physiological and neuroendocrine functions, such as synchronization of seasonal reproductive rhythms, and regulation of circadian cycles, several pharmacological effects of melatonin have been reported in mammals, including sedative, antioxidant, anxiolytic, antidepressant, anticonvulsant, and analgesic activities.

39.2 EPIDEMIOLOGICAL ASPECTS AND CLINICAL USE OF MELATONIN IN ADULTS AND CHILDREN

It has been proposed that melatonin may have significant therapeutic effects. In some countries such as Argentina, China, Poland, United States of America and New Zealand, melatonin has become recently available as either an *over-the-counter* (OTC) drug or food supplement. In the United States, ramelteon, a melatonin receptor agonist, is among the 10 drugs listed by the FDA for treatment of sleep disorders. There are some widely accepted indications for the therapeutic use of melatonin and also perspectives for its broader use. Melatonin has been proven to be useful in circadian rhythm disorders such as sleep disturbances, jet lag, sleep–wake cycle disturbances in blind people, and shift workers. For other indications, there is no definitive evidence for the use of melatonin. As the toxicity is remarkably low with no serious adverse effects reported so far, the investigation of additional indications for melatonin seems justified.

In Argentina, melatonin is a registered medicine for the treatment of sleep disorders in the elderly, particularly for those whose endogenous melatonin levels are low and it is often used as adjunctive treatment in Alzheimer's disease (AD), sleep disorders, and memory disturbances.

Lerner and Case (1960) described the first use of melatonin in humans, where 200 mg of melatonin was injected intravenously into a volunteer, with resulting light sedation. Subsequent investigators tested several doses (0.5–1.25 mg/kg, p.o.) in healthy volunteers and people with Parkinson's disease (PD) and epilepsy (Antón-Tay et al. 1971). This study showed that induction of sleep occurred 15–20 min after melatonin administration, in epileptic and normal individuals, and people with Parkinson showed significant improvement in several tests of performance.

Cramer et al. (1974) carried out an important clinical study by administering melatonin to healthy volunteers (50 mg, i.v.) during distinct schedules (diurnal and nocturnal). Using questionnaires and sleepgraphic evaluation, they observed melatonin functioned as a powerful sleep inducer. Nocturnal injections of melatonin significantly reduced latency before the start of sleep, although total sleep was maintained unaltered either in the quality or in its architecture.

There are several studies of melatonin use for sleep disturbances in children. In 1991, the first record of melatonin administration (0.5 mg/day) was reported in a 9-year-old boy with amaurosis and mental delay secondary to congenital toxoplasmosis. This child presented with a deeply altered sleep–wake rhythm, which returned to normal after receiving melatonin. Jan et al. (1994) administered melatonin to 15 children with insomnia and visual deficit in doses of 2.5–10 mg/day for 1 week with promising results. In the following year, the same group reported the administration of melatonin to around 90 children with alterations of sleep–wake rhythm (insomnia, rhythm in free-course, syndrome of the delay of phase, etc.), also with good clinical outcomes. Zhdanova et al. (1996) described the experience of 16 children with Angelman syndrome that is associated with difficulty in initiating and maintaining sleep. Melatonin, in a dose of 0.3 mg/day, corrected sleep patterns and improved diurnal attention level.

It is important to point out that no side effects have been reported with melatonin following short- or long-term administration (Cavallo and Ritschel 1996). Chronic administration does not appear to alter the baseline secretion of GH, TSH, testosterone, LH, and prolactin in man. However, reduced secretion of LH has been reported in women (Mantovani et al. 2006). Despite the apparent absence of adverse reactions, the use of the melatonin in clinical practice, especially in children, still requires further research. An understanding of the functions and mechanisms of action of melatonin can offer insights for its potential clinical use.

39.3 PHARMACOLOGICAL ASPECTS INVOLVED IN ACTIONS OF MELATONIN

39.3.1 Role of Melatonin in Sleep–Wake Disorders

Increased prevalence of insomnia is noted among the elderly population. Age progression is associated with shortened sleep, loss of consolidated deep sleep, increased inability to maintain sleep, and impaired daytime functioning. The significance of good sleep to the overall quality of life in older age is significant. Restorative sleep must be considered as a neuroprotective strategy that can potentially improve the course and outcome of several brain disorders, and thus the quality of life of the affected individuals and their family members. Moreover, it can substantially reduce health care costs, in particular, the costs associated with lifetime treatment of some neurodegenerative disorders.

In studies of healthy older people, mostly males, the mean melatonin levels measured in plasma, or the urinary hMT6s excreted during the night, were not significantly different from those in younger adults. A meta-analysis of literature data in subjects of age ranging from 21 to 82 years concluded that peak plasma melatonin levels in the 50–65 and the 65–80 years age groups are significantly lower (by 36% and 43%, respectively) than that in the 20–35 years age group. Furthermore, the amounts of hMT6s excreted decreases by approximately 36% between the age of 21–33 years and 49–85 years. This observation is consistent with previous and later studies showing a similar decline in hMT6s in elderly subjects without sleep complaints, which is exacerbated in patients with insomnia aged 55 years and older. From the hMT6s values reported for subjects, it may be concluded that the production of 8 mg hMT6s/night should be considered normal for this age group (Pandi-Perumal et al. 2005).

Traveling across three or more time zones may lead to jet lag, which is a consequence of circadian misalignment that occurs after crossing time zones too rapidly for the circadian system to keep pace. Melatonin has hypnotic properties and it has been used to alleviate "jet lag" and milder forms of insomnia. As mentioned before, it is available commercially in the United States as a "nutritional supplement" and can be obtained in other countries via the Internet. The sleep promoting and circadian regulating effects of melatonin have been attributed to the two subtypes of human melatonin receptors. It is believed that in the CNS, the amplitude of central circadian rhythmicity involves the MT1 receptor, while MT2 receptors are involved in the entrainment of circadian rhythms.

When melatonin is taken on arrival at the destination, between 22:00 and midnight, it can correct sleep disturbances, mental inefficiency, and daytime fatigue (cumulatively known as "jet lag"). The biological rhythm disorganization caused by the rapid change of environment (and associated light/dark cues) apparently can be corrected by melatonin. The benefit is likely to be greater as more time zones are crossed, and less for westward flights. However, melatonin taken before travel can actually worsen symptoms as opposed to the benefit of melatonin initiated immediately upon arrival.

Melatonin has also been used to alter sleep architecture in narcolepsy, a disorder of disturbed circadian sleep–wake rhythm and rapid-eye-movement (REM) sleep deficit. Changes in REM sleep patterns similar to those of narcolepsy also occur in animals and humans after removal of the pineal gland. Varying doses of melatonin (2–20 mg/daily) can improve sleep quality, accelerate sleep initiation, and improve sleep maintenance without significantly altering memory, in contrast to benzodiazepines.

39.3.2 Antioxidant Role of Melatonin

The antioxidant effects of melatonin have been well described and include both direct as well as indirect effects. Melatonin, similar to other indoleaminic derivatives, reversed the pro-oxidant effects caused by glutamate, kainic, quinolenic, Alzheimer amyloid peptide, homocysteine, and bacterial lipopolysaccharide. Melatonin also reduced the severity of ischemia–reperfusion injury in brain. Melatonin is a very efficient ·OH scavenger, while, in an exhaustive study extended to a wide number of antioxidants, Melatonin had an elevated reaction rate with ·OH and ROO·. Melatonin's ability to reduce lipid peroxidation is assumed to be related, at least partially, to its direct scavenging activity. However, several other mechanisms may be involved in its protective effects against these neurotoxins, for example, by stabilizing cell membranes, allowing them to effectively resist free radical toxicity, and by stimulating antioxidative enzymes.

Melatonin's antioxidant activity exceeds that of glutathione (GSH) and vitamin E, which are well-known antioxidants. Melatonin also acts as an indirect antioxidant through the activation of the major antioxidant enzymes, including SOD, CAT, and GPx. The free radical scavenging ability of melatonin has implications for variety of diseases, including age-associated neurodegenerative diseases and cancer.

Thus, melatonin could exert its indirect antioxidant actions through several mechanisms, which in turn would lead to the maintenance of cell integrity and protection against oxidative stress.

In addition, it has been found to decrease the formation of malonildialdeide (MDA) in the Thiobarbituric acid reactive substances (TBARS) test in brain homogenates. Furthermore, melatonin reportedly suppressed lipid peroxidation induced by iron, lipopolysaccharides, nitric oxide, and kainate. There is a substantial body of evidence for the protective effect of melatonin on DNA, lipids, and proteins, which are exposed to a number of endogenous and exogenous free radical-generating processes. Moreover, melatonin has been shown to inhibit cell death in those exposed to MPTP, a neurotoxin that causes permanent symptoms of PD, and β-amyloid peptides.

The melatonin binding site has been purified and characterized as the enzyme quinone reductase 2 (QR2). The physiological role of this enzyme is unknown. Recent results obtained by different groups suggest that: (1) inhibition of QR2 may lead to "protective" effects and (2) overexpression of this enzyme may have deleterious effects. The inhibitory effect of melatonin on QR2 observed in vitro may explain the protective effects reported for melatonin in different animal models, such as cardiac or renal ischemia—effects that have been attributed to the controversial antioxidant properties of the hormone. The development of specific ligands for each of these melatonin binding sites is necessary to link physiological and/or therapeutic effects.

39.3.3 Melatonin as Anticonvulsant Drug

Melatonin, a neuromodulator, has been shown to have antiepileptic activity in animal studies using different seizure models as well as in cases of childhood epilepsy. Several mechanisms for the anticonvulsant activity of melatonin have been suggested. Melatonin exerts neuroprotection due to its antioxidant, antiexcitotoxic, and free radical scavenging properties within the CNS. It has also been demonstrated to be safe in humans even in high pharmacological doses.

Melatonin stabilizes the electrical activity of the CNS and causes rapid synchronization of the electroencephalogram. By contrast, pinealectomy predisposes animals to seizures. In mice, intracerebroventricular administration of melatonin protected against seizures induced by kainate, glutamate, and *N*-methyl-D-aspartate (NMDA). Melatonin depresses brain excitability through inhibition of the glutamate-mediated response of the striatum to motor cortex stimulation and inhibition of neuronal excitation produced by either NMDA or L-arginine. Similarly, melatonin antagonises the seizure-producing effects of cyanide and ferric chloride. The anticonvulsant effect of melatonin has also been demonstrated in amygdala-kindled rats (Malhotra et al. 2004).

Findings imply the involvement of the L-arginine/NO pathway in melatonin-induced modulation of seizure susceptibility in mice. Combination of noneffective doses of melatonin (10 and 20 μg/kg) and the nitric oxide synthase (NOS) substrate L-arginine (30, 60 mg/kg) showed a significant anticonvulsant activity. This effect was reversed by the NOS inhibitor *N*(G)-nitro-L-arginine methyl ester (L-NAME, 30 mg/kg), implying a NO-dependent mechanism for melatonin effect (Yahyavi-Firouz-Abadi et al. 2006).

The induction of seizures by intravenous infusion of pentylenetetrazole (PTZ) is a standard experimental model of clinical myoclonic seizures. This model is proven to be more sensitive than i.p. PTZ administration, and allows better detection of modulatory effects on convulsive tendency. PTZ increases activity in major epileptogenic centers of the forebrain like the amygdala and piriform cortex. Neurochemical evidence suggests that PTZ binds to the picrotoxin site of the GABA receptor complex and blocks the GABA-mediated inhibition. The GABAergic effect of melatonin possibly depends on the combined effect on membrane ion permeability with increasing chloride ion influx through the $GABA_A$-dependent chloride channel. Stimulation of the inhibitory neurotransmitter, GABA and/or activation of NMDA receptor appear to be factors involved in the initiation and generalization of the PTZ-induced seizures. Melatonin increases the PTZ-induced clonic seizure threshold in mice and this effect might be due to an increase in constitutive nitric oxide activity.

Some of these experimental data have been corroborated by clinical studies in patients with epilepsy. Elkhayat et al. (2010) conducted a study on 23 children with intractable epilepsy and 14 children with controlled seizures. Children with intractable epilepsy received oral melatonin before bedtime, and showed a significant reduction in seizure severity and improvement of sleep-related phenomena. However, because of the paucity of well-controlled studies, melatonin cannot, as yet, be recommended in any form of epilepsy, although it may have some role as an adjuvant therapy for children with intractable seizures.

Melatonin treatment suppresses epileptiform seizures in both adults and children, and chronic high-dose melatonin is used as adjunctive anticonvulsant therapy in intractable seizures. Also, changes are found in day–night melatonin levels during convulsions in healthy children and in children with febrile or epileptic convulsions.

39.3.4 Antidepressant Role

Nocturnal melatonin levels are low in subjects with Major Depressive Disorder and panic disorder compared with nonpsychiatric patients. This is particularly marked in subjects with abnormal pituitary–adrenal responses to exogenous corticoids (abnormal dexamethasone suppression) who also have disturbed corticoid-secretion patterns. Healthy individuals with a dysthymic disposition (mild or episodic depression) also had lower-than-normal nocturnal melatonin levels as did subjects with melancholic depression. By contrast, higher-than-normal melatonin levels have been observed in manic subjects during the manic phase.

In a genetic animal model of depression, acute administration of either a melatonin receptor agonist or antagonist was ineffective, whereas chronic administration of the agonist revealed an antidepressant effect. Similarly, studies have shown that melatonin reduces the duration of immobility in the forced swim test. In the tail suspension test, the antidepressant effects of melatonin have been shown to involve the L-arginine-nitric oxide pathway. This was confirmed by the observation that pretreatment with L-arginine antagonized these effects (Ergün et al. 2006).

Melatonin synthesis is stimulated by the action of norepinephrine on intact beta-receptors, and this provides further theories on the mechanism of the antidepressant effect of melatonin. Several of the tricyclic antidepressants dramatically increase melatonin synthesis in humans. Thus, it is possible that the action of norepinephrine in affective disorders is mediated in part by effects on melatonin synthesis. Tricyclics often exert sedative effects and for this reason are often administered at night, an appropriate time to enhance melatonin rhythm amplitude. Furthermore, beta-receptor blockers depress melatonin secretion and can cause neuropsychiatric problems, such as nightmares, insomnia, lassitude, dizziness, and depression.

Agomelatine is a melatonin derivative, and a member of a novel class of antidepressants. It acts as an agonist at melatonin MT1 and MT2 receptors, and as a specific antagonist at 5-HT_{2C} receptors. The drug has anxiolytic properties and a beneficial effect on sleep without affecting REM activity. In general, agomelatine is well tolerated with few adverse effects and is currently being investigated for other indications such as seasonal affective disorder, generalized anxiety disorder, and bipolar depression.

39.3.5 Antinociceptive Role

It has been clearly demonstrated that melatonin, administered by the intraperitoneal, intraplantar or intracerebroventricular route, elicits significant and dose-dependent antinociception in mice, when assessed in the behavioral model of nociception induced by an intraplantar injection of glutamate or capsaicin. It has been shown that melatonin exerts an antinociceptive effect in several chemical models, and influences mechanical allodynia, but not thermal hyperalogia, in mice and rats. Moreover, melatonin administration causes a synergistic analgesic effect with morphine and diazepam, suggesting its possible use as an adjunct medicine for patients with pain.

The effects of melatonin result from activation of MT1 and MT2 melatonin receptors, which leads to reduced cyclic AMP formation and reduced nociception. In addition, melatonin is able to activate opioid receptors indirectly, to open K^+ channels, and to inhibit expression of 5-lipoxygenase and cyclooxygenase 2. Melatonin also inhibits the production of pro-inflammatory cytokines, modulates GABAa receptor function, and acts as a free radical scavenger. In addition, melatonin receptors constitute potential targets for developing analgesic drugs, and their activation may prove to be a useful strategy to generate analgesics with a novel mechanism of action.

39.4 CURRENT TRENDS OF PHARMACOLOGICAL USE OF MELATONIN

39.4.1 Neurodegenerative Diseases and Melatonin and MT Receptors

CNS expression of melatonin receptors in the human brain appears to be affected by pathological conditions such as AD and PD (PD). For example, significant alterations in the intensity and distribution of MT1 and MT2 immunoreactivities were observed in postmortem brain from Alzheimer's patients; MT1 immunoreactivity was significantly increased in Alzheimer's hippocampus, whereas MT2 immunoreactivity in Alzheimer's hippocampus was decreased. Transgenic mice deficient in MT2 receptors demonstrated deficient hippocampal long-term potentiation (LTP). Furthermore, when wild-type mice were tested in an elevated plus maze on 2 consecutive days, they showed shorter transfer latencies to enter a closed arm on the second day. This experience-dependent behavior did not occur in MT2 knockouts. These results suggest that MT2 receptors may participate in hippocampal synaptic plasticity and in memory processes (Imbesi et al. 2006).

Studies have described a potent neuroprotective action against the toxicity of amyloid-β in AD. In patients with this disease, cerebrospinal fluid melatonin levels have been found to be significantly reduced. In fact, application of melatonin prevented the death of neuroblastoma cells exposed to amyloid-β peptide. Since then, the antifibrillogenic actions have been demonstrated in vitro, also in the presence of profibrillogenic apoE4 or apoE3, and in vivo in transgenic mouse models of AD. It seems feasible that the efficacy of melatonin to improve clinical conditions in mild cognitive impairment depends partly on the effective neuroprotective effect seen at an early phase of the disease. In a study of 14 patients at various stages of AD, melatonin supplementation for 22–35 months improved sleep and significantly reduced the incidence of "sundowning." Furthermore, patients experienced no cognitive or behavioral deterioration during the study period (Srinivasan et al. 2005).

There are also experimental data that suggest a role of melatonin in PD, another neurodegenerative disorder. This disease is characterized by the progressive deterioration of dopamine-containing neurons in the pars compacta of the substantia nigra in the brain stem due to the oxidation of dopamine. There is evidence that melatonin may reduce dopamine auto-oxidation under experimental conditions, although its administration did not slow progression of the PD.

Recently, there has been growing interest among researchers about the apparent protective effects of melatonin after traumatic events to the peripheral nerves, especially the sciatic nerve. The administration of melatonin may be beneficial after surgery. Although there are numerous studies that have indicated a protective effect of melatonin on diseases of peripheral nerves, there are also some authors reporting toxic effects of melatonin on peripheral nerves.

39.4.2 Melatonin and Neoplastic Disease

A link between the pineal gland, melatonin and neoplastic disease has been demonstrated in various experimentally induced animal tumors. Some studies have shown that melatonin can inhibit the development and/or growth of various experimental animal tumors and some human cell lines in vitro, but its role in human malignancy is unclear. However, depressed nocturnal melatonin concentrations, or nocturnal excretion of the main melatonin metabolite, 6-sulfatoxymelatonin, were found in various tumor types (breast cancer, prostate cancer, colorectal cancer, endometrial cancer, cervical cancer, lung cancer, and stomach cancer), whereas in other tumor types (Hodgkin's sarcoma, osteosarcoma, ovarian cancer, laryngeal cancer, and urinary bladder cancer), melatonin levels were not changed or showed great variations among individuals.

Some clinical studies performed mainly by Lissoni's group suggest that the administration of melatonin (in relatively high doses, either alone or in combination with IL-2) is able to favorably influence the course of advanced malignant disease in humans, and lead to an improvement in quality of life. However, these observations require to be verified by independent and controlled studies (Bartsch et al. 2002; Hrushesky 2001).

REFERENCES

Antón-Tay, F., Diaz, J.L., and Fernandez-Guardiola, A. 1971. On the effect of melatonin upon human brain. Its possible therapeutic implications. *Life Sci* 10:841–850.

Bartsch, C., Bartsch, H., and Karasek, M. 2002. Melatonin in clinical oncology. *Neuroendocrinol Lett* 23:30–38.

Cavallo, A. and Ritschel, W.A. 1996. Pharmacokinetics of melatonin in human sexual maturation. *J Clin Endocrinol Metab* 81:1882–1886.

Cramer, H., Rudolph, J., Consbruch, U., and Kendel, K. 1974. On the effects of melatonin on sleep and behavior in man. *Adv Biochem Psychopharmacol* 11:187–191.

Elkhayat, H.A., Hassanein, S.M., Tomoum, H.Y., Abd-Elhamid, I.A., Asaad, T., and Elwakkad, A.S. 2010. Melatonin and sleep-related problems in children with intractable epilepsy. *Pediatr Neurol* 42(4):249–254.

Ergün, Y., Ergün, U.G., Orhan, F.O., and Küçük, E. 2006. Co-administration of a nitric oxide synthase inhibitor and melatonin exerts an additive antidepressant-like effect in the mouse forced swim test. *Med Sci Monit* 12(9):307–312.

Hrushesky, W.J.M. 2001. Melatonin cancer therapy. In: Bartsch, C., Bartsch, H., Blask, D.E., Cardinali, D.P., Hrushesky, W.J.M., and Mecke, D. (eds.), *The Pineal Gland and Cancer*, pp. 476–508. Springer-Verlag, Berlin, Germany.

Imbesi, M., Uz, T., Yildiz, S., Arslan, A.D., and Manev, H. 2006. Drug- and region-specific effects of pretrated antidepressant and cocaine treatment on the content of melatonin MT(1) and MT(2) receptor mRNA in the mouse brain. *Int J Neuroprot Neuroregener* 2:185–189.

Jan, J.E., Espezel, H., and Appleton, R.E. 1994. The treatment of sleep disorders with melatonin. *Dev Med Child Neurol* 36:97–107.

Kneen, R. and Appleton, R.E. 2006. Alternative approaches to conventional antiepileptic drugs in the management of paediatric epilepsy. *Arch Dis Child* 91(11):936–941.

Lerner, A.B. and Case, J.D. 1960. Melatonin. *Fed Proc* 19:590–592.

Malhotra, S., Sawhney, G., and Pandhi, P. 2004. The therapeutic potential of melatonin: A review of the science. *Med Gen Med* 6(2):46.

Mantovani, M., Kaster, M.P., Pertile, R., Calixto, J.B., Rodrigues, A.L., and Santos, A.R. 2006. Mechanisms involved in the antinociception caused by melatonin in mice. *J Pineal Res* 41(4), 382–389.

Palm, L., Blennow, G., and Wetterberg, L. 1991. Correction of non-24 Hs sleep-wake cycle by melatonin in a blind retarded boy. *Ann Neurol* 29(3):336–339.

Pandi-Perumal, S.R., Zisapel, N., Srinivasan, V., and Cardinali, D.P. 2005. Melatonin and sleep in aging population. *Exp Gerontol* 40:911–925.

Ravindra, T., Lakshmi, N.K., and Ahuja, Y.R. 2006. Melatonin in pathogenesis and therapy of cancer. *Indian J Med Sci* 60(12):523–535.

Srinivasan, V., Pandi-Perumal, S.R., Maestroni, M.J.G., Esquifino, A., Harderland, R., and Cardinali, D.P. 2005. Role of melatonin in neurodegenerative diseases. *Neurotox Res* 7:293–318.

Terzolo, M., Piovesan, A., Puligheddu, B., Torta, M., Osella, G., Paccotti, P., and Angeli, A. 1990. Effects of long-term, low-dose, time-specified melatonin administration on endocrine and cardiovascular variables in adult men. *J Pineal Res* 9:113–124.

Yahyavi-Firouz-Abadi, N., Tahsili-Fahadan, P., Riazi, K., Ghahremani, M.H., and Dehpour, A.R. 2006. Involvement of nitric oxide pathway in the acute anticonvulsant effect of melatonin in mice. *Epilepsy Res* 68(2):103–113.

Zhdanova, I.V., Wagstaff, J., and Wurtman, R.J. 1996. Melatonin and sleep in Angelman syndrome children. *APSS 10th Annual Meeting*, Washington, DC, abstract 58.

40 Protective Effects of Melatonin in Radiation-Induced Nephrotoxicity

Eda Kucuktulu and Uzer Kucuktulu

CONTENT

Production of reactive oxygen species (ROS) is one of the early effects of ionizing radiation, which induces the cellular antioxidant defense enzymes such as superoxide dismutase and glutathione peroxidase [1]. ROS and free radicals react with cellular macromolecules (i.e., nucleic acids, lipids, and carbohydrates) and cause damage. Oxidative damage to living cells can be estimated with measurable major biomarkers of lipid peroxidation such as penthane, isoprostane, and aldehytic products measurable in tissue and body fluids; DNA-hydroxylation products and microscopic indices of damage such as chromosomal aberrations and micronuclei and protein hydroxylation products such as oxidized amino acids can also be detected [2].

˙OH radicals are believed to be responsible for an estimated 60%–70% of tissue damage induced by ionizing radiation [3]. Melatonin is a potent antioxidant, which shows its effect either directly or indirectly. Melatonin (*N*-acetyl-5-metoxytriptamine) is a known agent preventing the oxidative damage of the toxins and radiotherapy with its free radical scavenging capacity [4–6]. In in vitro experiments, it was shown that melatonin was 5- and 14-fold more potent than glutathione and mannitol, respectively, in scavenging hydroxyl radicals. The indoly (melatonyl) radicals are formed when the melatonin interacts with ˙OH radicals. This newly formed radical has very low toxicity. When melatonin scavenges ˙OH radicals, a less toxic reactant is generated by removing highly toxic OH radicals. After some molecular rearrangement, the indolyl radical scavenges a second ˙OH to form cyclic 3-hydroxymelatonin. In addition to antioxidant activity of the melatonin itself, the compounds generated also have antioxidant potential. For example, the product N^1-acetyl-N^2-formyl-5-methoxykynuramine (AFMK) may also be an effective scavenger. The antioxidant potential of melatonin is augmented by these consecutive reactions [7] (Figure 40.1).

Indirect effect: Melatonin decreases the activity of nitric oxide synthase, a pro-oxidative enzyme [8]. Additionally, melatonin increases the activity of some important antioxidant enzymes at a molecular level, including superoxide dismutase and glutathione peroxidase [9], and stimulates the synthesis of antioxidant enzymes. Furthermore, melatonin has a potential to protect antioxidative enzymes from oxidative damage. Melatonin increases protein levels of antioxidant enzymes through nuclear factor erythroid 2-related factor 2(Nrf2), by increasing mRNA upregulation of Nrf2 by melatonin, which resulted in an increased expression of antioxidant enzyme heme oxigenase-1 (HO-1) [10].

Any radioprotector, despite being an effective, convenient compound to protect humans from the damaging effects of ionizing radiation, should be nontoxic and should not protect tumor from radiation effect. Melatonin is a nontoxic endogenous substance, which is synthesized by pineal gland of human brain and has been reported to take part in the regulation of various physiological and

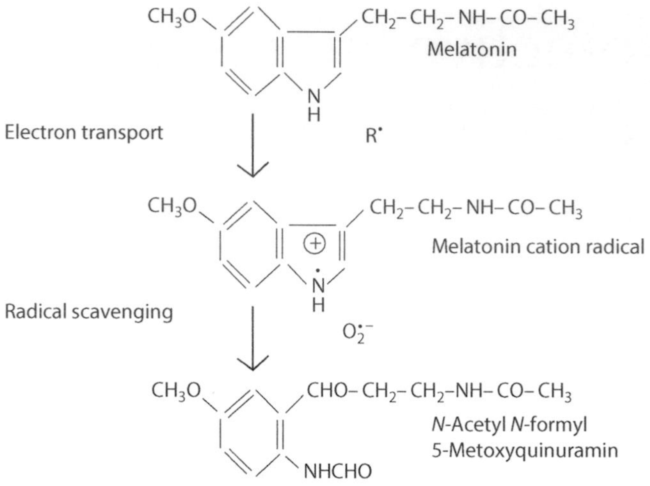

CH_3O $CH_2-CH_2-NH-CO-CH_3$
Melatonin
N
H
Electron transport R•

CH_3O $CH_2-CH_2-NH-CO-CH_3$
Melatonin cation radical
N
H
Radical scavenging $O_2^{\bullet -}$

CH_3O $CHO-CH_2-CH_2-NH-CO-CH_3$
N-Acetyl *N*-formyl
5-Metoxyquinuramin
NHCHO

FIGURE 40.1 The mechanism of free radical scavenging of melatonin.

pathological processes [11]. The concentration of melatonin peaks at night in the darkness, and it is at the lowest level during the daytime [3]. The half-life of melatonin ranges between 30 and 57 min [12]. The potential toxicity of melatonin has been investigated in different animal species in a wide range of doses, from physiological to pharmacological concentrations. Vijayalaxmi et al. observed in an in vitro study on peripheral lymphocytes that micronutrients ratio increases when they are exposed to 15.6 Gy of radiotherapy. They also observed that melatonin caused a drop of micronutrient count when added to the cell cultures in concentrations 0.5–2.0 mmol 20 min before the incubation [11]. They also compared the effects of melatonin given in doses of 5 and 10 mg/kg doses given 1 h before radiotherapy and observed that melatonin in 10 mg/kg dose had better results. In a randomized double-blind clinical trial conducted by Sebra et al., melatonin when administered orally to healthy adult males in the dose of 0.5–2.0 mmol for 28 days caused no toxicity [3]. In this study, melatonin 10 mg/kg intraperitoneally was used and provided sufficient protection. But in this experimental study, melatonin itself was found to be a cause of kidney function elevation when added to the radiotherapy protocol. In previous kidney studies with chemotherapeutic agents and melatonin, this effect of melatonin was not mentioned [13]. This condition may be the results of the low doses of melatonin (5 mg/kg) used in these studies. Several studies, including the earlier-mentioned studies, indicated that both the acute and chronic toxicity of melatonin is extremely rare. In our study, it was shown that despite BUN elevations caused by melatonin, it also had protective effects of kidney histopathology both in light and electron microscopic level. Additionally, melatonin rarely interacts with other medications, and it may even reduce the potential side effects of other substances with its free radical scavenging properties [14].

Radioprotectors should have certain properties to be used in clinical settings. Any compound protecting normal tissue has the risk to protect the tumor tissue as well. Their protective effect both in the normal and the tumor tissue should be known quantitatively and the therapeutic gain should be calculated. Radioprotectors have potential risks of protecting tumor tissue from the effects of radiation as protecting the normal tissue. The protective effects on both tumor and normal tissue of any radioprotector should be quantitatively known and a therapeutic gain can be calculated. Ideally, the dose–response effect of the compound should be evaluated both in the normal and the tumor tissue and ideal dose be determined [15].

Melatonin, as a member of regulatory factors that control cell proliferation and loss, is the only chronobiotic that is a hormonal neoplastic cell growth regulator. Melatonin, at physiological concentrations, has a cytostatic effect that inhibits cancer cell proliferation, whereas at pharmacological concentrations, it shows cytotoxic activity on cancer cells. At both physiological and pharmacological concentrations, melatonin acts as a differentiating agent in some cancer cells and lowers their invasive and metastatic status through alterations in adhesion molecules and maintenance of gap junctional intercellular communication. In other cancer cell types, melatonin, either alone or in combination with other agents, induces apoptotic cell death [16–18]. Melatonin reduces tumor growth in experimental models in vivo and proliferation and invasive properties of cancer cells in culture [19,20].

Mills et al. reported in a meta-analysis that melatonin provides substantial reduction in risk of death with low adverse effects and low cost. They suggested a great potential for melatonin in treating cancer in humans [21]. This property makes melatonin an outstanding radioprotector among other radioprotectors. Despite not being able to protect tumor cells from radiation, melatonin has been shown to have the potential of directly treating the tumors.

The first study on the antioxidative effects of melatonin was conducted using ultraviolet light (UV), which has many features in common with ionizing radiation [22]. In a report of Tan et al., melatonin revealed to scavenge ˙OH radicals generated in vitro from H_2O_2 exposed to UV light. In vivo studies have shown that the damage caused by UV light can be prevented by melatonin. The potency of melatonin in protecting against OH generated by ionizing radiation was confirmed under in vivo conditions using a similar model. There are a number of other studies in the literature reporting the protective efficacy of melatonin against ionizing radiation in both in vivo and in vitro settings [3,11].

Most of the studies concerning the capacity of melatonin to prevent kidney toxicity recruited chemotherapeutic agents. Hara et al., in their study on melatonin use in kidney toxicity, showed that melatonin sustained GSH/GSSG ratio, prevented lipid peroxidation caused by cisplatin, and normalized antioxidant substance glutathione peroxidase levels [13]. Kılıc et al. studied the effect of melatonin on cisplatin-induced nephrotoxicity and concluded that melatonin treatment decreased the cisplatin-induced tubular necrosis; they found melatonin shows its effect through increased expression of Nrf2 and HO-1 [23].

In antibiotic-induced nephrotoxicity animal models such as antracyclin antibiotics and gentamycin nephrotoxicity, melatonin as an antioxidant has been investigated [24]. In a rat model of acute renal damage, melatonin was shown to increase the expression of the antioxidant and detoxification enzyme HO-1 and through this activity improved makers of oxidative stress [23].

Several mechanisms including hypoxia, free radicals, inflammation, and apoptosis are thought to be involved in chemotherapeutic agents–induced nephrotoxicity. Excessive production of free radicals, such as superoxide anion, hydrogen peroxide, and hydroxyl radicals, and the occurrence of lipid peroxidation due to oxidative stress are associated with chemotherapeutic agents–induced renal dysfunction [13]. These properties have similarities with radiotherapy-induced renal damage mechanisms. It was shown that cisplatin caused structural damages in kidneys similar to radiotherapy-induced changes such as tubular necrosis [25].

It has been shown in many clinical and experimental studies that kidneys are highly sensitive to radiation injuries. In many clinical and experimental studies, kidneys are shown to be highly sensitive to radiation injuries. Radiation nephropathy develops mount even years after radiotherapy. Since clinical findings and time elapsed for the development of radiation nephropathy differs, it is more important for children and patients with longer life expectancy. The radiation injury develops earlier and increases in severity with higher radiation doses. Radiation nephropathy was well documented in a large case series published by Kunkler et al. over 50 years ago. These were men who had undergone therapeutic irradiation for seminomas [26]. Radiation nephropathy occurred in about 20% of sufficiently irradiated subjects and could take various clinical forms: acute radiation nephritis, chronic radiation nephritis, malignant hypertension, benign hypertension [25].

The pathogenesis of the radiation nephropathy remains controversial. The direct functional relationships between tubuli, glomeruli, and blood vessels will not enable conclusions on the damage developing in separate compartments. Jongejan et al. reported a simultaneous decline in GFR and urine osmolality after radiation but did not describe any difference in radiation sensitivity between tubules and glomeruli [27]. Glatstein et al. suggested the glomeruli as the site of initial pathologic changes [28]. In literature, many studies reported that radiation-induced glomerular changes appear diffuse and they precede tubular alterations [29–31]. However, chronic renal failure is observed primarily in those animals in which glomerular injury is combined with severe tubular injury and tubulointerstitial fibrosis [32]. In clinical chronic renal disease, the degree of renal dysfunction [33] does not correlate with glomerular changes [34]. In our study, in the radiotherapy-given group, the narrowing of proximal and distal tubules was observed with glomerular damage. Our study showed complex damage of glomerular, tubular, and interstitial cells in acute nephropathy period and supported the findings of Cohen and coworkers [35].

Our study is the first in the literature that shows that melatonin has radioprotective effects on kidneys, in terms of histopathology when administered concomitant with radiotherapy. If in further studies, it can be shown to be nephroprotective when administered orally and minimal BUN elevations can be evaluated pharmacologically, melatonin may become a drug of choice with its anticancer and antioxidant properties [36].

As a conclusion, melatonin, a known antioxidant agent, has a radioprotective effect on irradiated kidneys. After a 6-month follow-up period, light and electron microscopic findings showed that kidneys in the melatonin and radiotherapy group were healthier compared with the radiotherapy-only group.

REFERENCES

1. Zhang B, Su Y, Wang Y. (2005). Involvement of peroxiredoxin I in protecting cells from radiation-induced death. *J Radiat Res* 46: 305–312.
2. Shirazi A, Ghobadi G, Ghazi-Khansari M. (2007). A radiobiological review on melatonin: A novel radioprotector. *J Radiat Res* 48: 263–272.
3. Vijayalaxmi ML, Reiter RJ, Tan DX. (2004). Melatonin as a radioprotective agent: A review. *Int J Radiat Oncol Biol Phys* 59: 639–653.
4. Martinez-Cayuela M. (1955). Oxygen free radicals and human disease. *Biochimie* 77: 147–161.
5. Edwards JC, Chapman D, Cramp WA. (1984). The effects of ionizing radiation on biomembrane structure and function. *Prog Biophys Mol Biol* 43: 71–93.
6. Verma SP, Sonwalker N. (1991). Structural changes in plasma membranes prepared from irradiated Chinese hamster V79 cells as revealed by Raman spectroscopy. *Radiat Res* 126: 27–35.
7. Reiter RJ, Tan DX, Acuna-Castroviejo D. (2000). Melatonin: Mechanisms and actions as an antioxidant. *Curr Topics Biophys* 24: 171–183.
8. Majsterek I, Gloc E, Blasiak J. (2005). A comparison of the action of amifostine and melatonin on DNA damaging effects and apoptosis induced by idarubicin in normal and cancer cells. *J Pineal Res* 38: 254–263.
9. Rodriguez C, Mayo JC, Sainz RM. (2004). Regulation of antioxidant enzymes: A significant role for melatonin. *J Pineal Res* 36: 1–9.
10. Negi C, Kumar A, Sharma SS. (2011). Melatonin modulates neuroinflammation and oxidative stress in experimental diabetic neuropathy effects on NF-κB and Nrf2 cascades. *J Pineal Res* 50: 124–131.
11. Vijayalaxmi ML, Reiter RJ, Herman TS. (1999). Melatonin and protection from genetic damage in blood and bone marrow: Whole body irradiation studies in mice. *J Pineal Res* 27: 221–225.
12. Lane EA, Moss HB. (1985). Pharmacokinetics of melatonin in man: First pass hepatic metabolism. *J Clin Endocrinol Metab* 61: 1214–1216.
13. Hara M, Yoshida M, Nishijima H. (2001). Melatonin: A pineal secretory product with antioxidant properties protects against cisplatin-induced nephrotoxicity in rats. *J Pineal Res* 30: 129–138.
14. Reiter RJ, Tan DX, Sainz RM. (2002). Melatonin: Reducing the toxicity and increasing the efficacy of drugs. *J Pharm Pharmacol* 54: 1299–1321.
15. Andreassen CN, Grau C, Lindegoard JC. (2003). Chemical radioprotection: A critical review of amifostine as a cytoprotector in radiotherapy. *Semin Radiat Oncol* 13: 62–72.
16. Blask DE, Sauer LA, Dauchy RT. (2002). Melatonin as a chronobiotic/anticancer agent: Cellular, biochemical and molecular mechanisms of action and their implications for circadian-based cancer therapy. *Curr Top Med Chem* 2: 113–132.

17. Casado-Zapico S, Rodringuez-Blanco J, Garcia-Santos G. (2010). Synergistic antitumor effect of melatonin with several chemotherapeutic drugs on human Ewing sarcoma cancer cells: Potentiation of the extrinsic apoptotic pathway. *J Pineal Res* 48(1): 72–80.
18. Rodriguez-Garcia A, Mayo JC, Hevia D. (2013). Phenotypic changes caused by melatonin increased sensitivity of prostate cancer cells to cytokine-induced apoptosis. *J Pineal Res* 54(1): 33–45.
19. Cos S, Mediavilla MD, Fernandez R. (2002). Melatonin induce apoptosis in MCF-7 human breast cancer cells in vitro. *J Pineal Res* 32: 90–94.
20. Manda K, Bhatia AL. (2003). Melatonin-induced reduction in age-related accumulation of oxidative damage in mice. *Biogerontology* 4: 133–139.
21. Mills E, Wu P, Seely D. (2005). Melatonin in the treatment of cancer: A systematic review of randomized controlled trials and meta-analysis. *J Pineal Res* 31: 360–366.
22. Karbownik M, Reiter RJ. (2000). Antioxidative effects of melatonin in protection against cellular damage caused by ionizing radiation. *Proc Soc Exp Biol Med* 225: 9–22.
23. Kılıc U, Kılıc E, Tuzcu Z. (2013). Melatonin suppresses cisplatin induced nephrotoxicity via activation of Nrf-2/HO-1 pathway. *Nutr Metab* 10: 7.
24. Dziegiel P, Suder E, Surowiak P. (2002). Role of exogenous melatonin in reducing the nephrotoxic effect of daunorubicin and doxorubicin in the rat. *J Pineal Res* 33: 95–100.
25. Perez CA, Halperin EC. (2008). *Principles and Practice of Radiation Oncology*, 5th edn. Lippincott Williams & Wilkins, Philadelphia, PA.
26. Kunkler PB, Farr RF, Luxton RW. (1952). The limit of renal tolerance to X-rays. *Br J Radiol* 25: 190–201.
27. Jongejan HTM, Van Der Kogel AJ, Provoost AP. (1987). Radiation nephropathy in young and adult rats. *Int J Radiat Oncol Biol Phys* 13: 225–232.
28. Glatstein E, Fajanda LF, Brown JM. (1977). Radiation injury in the mouse kidney-I sequential light microscopic study. *Int J Radiat Oncol Biol Phys* 2: 933–943.
29. Robbins MEC, Wooldridge MJA, Jaenke RS. (1991). A morphological study of radiation nephropathy in the pig. *Radiat Res* 126: 317–327.
30. Stephens LC, Robbins MEC, Thames HD. (1995). Radiation nephropathy in the rhesus monkey: Morphometric analysis of glomerular and tubular alterations. *Int J Radiat Oncol Biol Phys* 31: 865–873.
31. Madrazo AA, Churg J. (1976). Radiation nephritis. Chronic changes following moderate doses of radiation. *Lab Invest* 34: 283–290.
32. Robbins MEC, Stephens LC, Thames HD. (1994). Radiation response of the monkey kidney following contralateral nephrectomy. *Int J Radiat Oncol Biol Phys* 30: 347–354.
33. Bohle A, Mackensen-Haen S, Gise H. (1990). The consequences of tubulo-interstitial changes for renal function in glomerulopathies. *Pathol Res Pract* 186: 135–144.
34. Ong ACM, Fine LG. (1994). Tubular-derived growth factors and cytokines in the pathogenesis of tubulointerstitial fibrosis: Implications for human renal disease progression. *Am J Kidney Dis* 23: 205–209.
35. Cohen EP, Robbins EC. (2003). Radiation nephropathy. *Semin Nephrol* 23: 486–499.
36. Kucuktulu E, Yavuz AA, Umit C. (2012). Protective effect of melatonin against radiation induced nephrotoxicity in rats. *Asian Pac J Cancer Prev* 13: 4101–4105.

41 Melatonin and Familial Mediterranean Fever

Engin Taştaban, Mehmet Turgut, and Teoman Aydın

CONTENTS

41.1 FAMILIAL MEDITERRANEAN FEVER

Hereditary recurrent fever are a group of inherited systemic disorders characterized by episodes of fever with a variety of localized inflammatory manifestations such as fever, diarrhea, abdominal pain, rash, arthralgia, or arthritis. Familial Mediterranean fever (FMF, MEFV) is the prototype of this family of diseases. The other hereditary recurrent fevers include the tumor necrosis factor (TNF) receptor-associated periodic syndrome, hyperimmunoglobulinemia D with periodic fever syndrome, familial cold autoinflammatory syndrome, Muckle-Wells syndrome, and neonatal onset multisystem inflammatory disease. It is accepted that FMF is a recessively inherited disorder in individuals of non-Ashkenazi Jewish, Armenian, Arab, and Turkish ancestry (Eisenstein et al. 2013). The gene responsible for FMF was first reported in 1997 (Eisenstein et al. 2013). It is characterized by episodes of fever with localized inflammation, often affecting serosal membranes, joints, and skin (Eisenstein et al. 2013). Between the fever episodes, patients with most of these syndromes generally feel healthy and function normally (Eisenstein et al. 2013).

41.2 PINEAL GLAND AND EFFECTS OF MELATONIN

It is known that circadian rhythm is a 24 h cycle in the physiological and behavioral processes that is endogenously generated and can be entrained by external factors, particularly the daylight. Melatonin (Mel) produced by the pineal gland is the major hormone regulating the circadian rhythm and produced with a maximal secretion at night. It regulates circadian and seasonal rhythms (Malpaux et al. 2001). Characteristically, optimum Mel production is only achieved in complete darkness, and the concentration in blood remains high during sleepiness. Mel is not only a pineal hormone, but also has additional functions as a local tissue factor and leukocyte-derived cell hormone with paracrine and autocrine actions (Tan et al. 2003).

Today, it is known that Mel has a lot of physiological functions related with seasonal changes, circadian rhythm regulation, sleep, reproduction, and cardiovascular function (Di Bella et al. 2013). It also modulates the functions of the immune and hemopoietic systems (Di Bella et al. 2013). Many authors documented a direct correlation between Mel production and the circadian and seasonal variations in the immune system (Nelson and Drazen 2000; Calvo et al. 2013). Importantly, Mel can stimulate the immune response and correct immune deficiencies secondary to acute stress, viral diseases, or drug treatment. It has been reported that binding of Mel to its specific receptors resulted in an upregulation of cytokine production and immune function (Maestroni 1998). Also, Mel may act on specific membrane receptors expressed on immunocompetent cells with MT_2 receptors (Drazen and Nelson 2001; Carrillo-Vico et al. 2003). Moreover, the role of Mel has been suggested in various rheumatic diseases, especially in rheumatoid arthritis (RA). Also, it has been shown that it altered functioning of the hypothalamo–pituitary–adrenal axis and of the pineal gland in patients with RA, important factors in the appearance of the clinical circadian symptoms of the disease (Cutolo et al. 2005). Sulli et al. (2002) revealed that Mel serum levels at 08:00 p.m. and 08:00 a.m. UTC were found to be higher in RA patients than in control subjects. They have suggested that clinical symptoms of RA, including morning gelling, stiffness, and swelling, which are more evident in the early morning, might be related to the neuroimmunomodulatory effects of Mel (Sulli et al. 2002).

41.3 EFFECTS OF MELATONIN ON FAMILIAL MEDITERRANEAN FEVER

In previous studies, overproduction of proinflammatory cytokines like TNF-α, interleukin (IL)-1 and IL-6 have been reported in FMF (Gang et al. 1999; Akcan et al. 2003). The gene of FMF encodes for a protein of 781-amino acids product, called pyrin or marenostrin and the expression of pyrin is induced by inflammatory mediators such as TNF-α and IL-1 and IL-6 (Centola et al. 2000). It has been speculated that a mutated pyrin leads to uncontrolled inflammation by the inhibition of apoptosis of leukocytes and the production of IL-1 (Eisenstein et al. 2013). Importantly, Ozen et al. (2001) found an increased apoptosis in the neutrophil and monocyte during FMF attacks. Then, they have concluded that the increased apoptosis may be the explanation of the self-limited nature of the FMF attacks (Ozen et al. 2001).

In the current literature, the pathogenesis of FMF and the immunoregulatory role of Mel in FMF are not well described. Nevertheless, Mel may function as a regulator of the inflammatory cell compartment with a potent antioxidant potential able to reduce the oxidative environment of chronic inflammation and to regulate leukocyte function and number, thus contributing to the control of inflammation in tissues acting as both an activator and inhibitor of the inflammatory and immune responses (Bonnefont-Rousselot and Collin 2010). It has been suggested that Mel is able to prevent or reduce the inflammatory-derived activation of a variety of enzymes, including phospholipase A2, lipoxygenase, and cyclooxygenases (Radogna et al. 2010).

Interestingly, Mel stimulates the production of natural killer cells, monocytes, and leukocytes, but it also increases the production of Interleukin IL-2–6–10–12 and Interferon-gamma (IFN-γ) by the mononucleate cells, promoting a T helper-1 (Th-1) lymphocyte response (Di Bella et al. 2013). Also, it has been reported that Th1/Th2 balance may exhibit diurnal rhythm in healthy humans, and Mel may regulate diurnal variation in IFNγ/IL-10 ratio (Petrovsky and Harrison 1997). Garcia-Maurino et al. (1997) found that Mel enhances the production of IL-2, IL-6, and IFN-γ in human circulating CD4 cells.

In a previous study, it has been reported that the peak plasma levels of Mel in healthy subjects coincide with a peak in the secretion of lipopolysaccharide-stimulated IFN-γ in whole blood, suggesting an association between IFN-γ production and Mel secretion (Petrovsky and Harrison 1997). Furthermore, Mel can also activate Th1 lymphocytes by increasing IL-12 production by antigen-presenting cell (Garcia-Maurino et al. 1999). Afterward, it has been suggested that these effects may be related with the reduction of matrix metalloproteinases (Ekmekcioglu et al. 2001), reduction of immunological injury by regulation of macrophage activity (Mei et al. 2002), reduction of oxidative stress (Hagar et al. 2007), inhibition of nitric oxide (NO) production (Mei et al. 2005), inhibition of nuclear factor-kappa beta (NF-kB) activity (Li et al. 2005), and reduction of proinflammatory cytokines (Mazzon et al. 2006).

Clinically, it has been demonstrated that Mel has antioxidant, immunomodulatory, and anti-inflammatory functions, owing to the inhibition of NO production, reducing activation of the transcription factor NF-kB, the expression of cyclooxygenase and prostaglandins, and the recruitment of polymorphonuclear cells to the site of inflammation (Ha et al. 2005; Maldonado et al. 2007; Reiter et al. 2007; Peyrot and Ducrocq 2008; Laste et al. 2012). In particular, the anti-inflammatory effect of Mel may be related to the inhibition of NOS (Pei et al. 2003) and to the prevention of translocation of the NF-kB with the subsequent reduction of proinflammatory genes, such as COX-2, and proinflammatory cytokines (Esposito and Cuzzocrea 2010). The effect of Mel on the suppression of proinflammatory cytokine production has been studied by Raghavendra et al. (2001). They have demonstrated that Mel suppresses the production of TNF-α (Raghavendra et al. 2001). Also, Wang et al. (2005) demonstrated that Mel decreases the production of proinflammatory cytokines, including TNF-α and IL-1β from Kupffer cells in fibrotic rats. Majewska et al. (2007) suggested that Mel suppresses cell-mediated immune responses partly through inhibiting the production of IL-12 in antigen-presenting cells. In another study, it has been shown that Mel protects against experimental reflux esophagitis by repressing the upregulation of TNF-α, IL-1β, and IL-6 (Lahiri et al. 2009). Moreover, Jung et al. (2009) demonstrated that intraperitoneal administration of Mel in rats suppressed the mRNA expression of TNF-α, IL-1β, IL-6, and inducible NOS. Veneroso et al. (2009) also found that Mel administration at a lower dose decreases the mRNA levels of proinflammatory cytokines and protein level of iNOS and COX-2 in rats induced with cardiac inflammatory injury by acute exercise.

Recently, Musabak et al. (2011) demonstrated that Mel levels are high in patients with FMF at night, 03:30 a.m. UTC, and day time, 10:00 a.m. UTC, in contrast to those of healthy controls. Based on their findings, they have concluded that high levels of Mel during nighttime and daytime in the FMF patients may contribute to disease pathogenesis by stimulating other immune cells (Musabak et al. 2011).

41.4 CONCLUSION

Consequently, it is widely accepted that Mel may play an important role in FMF, possibly due to its anti-inflammatory effects. There is no doubt that further research investigating the inflammatory markers and changes with Mel-containing drugs may be of value to better understand the role of anti-inflammatory effect of Mel in FMF.

REFERENCES

Akcan, Y., Y. Bayraktar, S. Arslan, D. H. Van Thiel, B. C. Zerrin, and O. Yildiz. 2003. The importance of serial measurements of cytokine levels for the evaluation of their role in pathogenesis in familial Mediterranean fever. *Eur. J. Med. Res.* 8(7): 304–306.

Bonnefont-Rousselot, D. and F. Collin. 2010. Melatonin: Action as antioxidant and potential applications in human disease and aging. *Toxicology* 278(1): 55–67.

Calvo, J. R., C. Gonzalez-Yanes, and M. D. Maldonado. 2013. The role of melatonin in the cells of the innate immunity: A review. *J. Pineal Res.* 55(2): 103–120.

Carrillo-Vico, A., A. Garcia-Perganeda, L. Naji, J. R. Calvo, M. P. Romero, and J. M. Guerrero. 2003. Expression of membrane and nuclear melatonin receptor mRNA and protein in the mouse immune system. *Cell Mol. Life Sci.* 60(10): 2272–2278.

Centola, M., G. Wood, D. M. Frucht, J. Galon, M. Aringer, C. Farrell, D. W. Kingma et al. 2000. The gene for familial Mediterranean fever, MEFV, is expressed in early leukocyte development and is regulated in response to inflammatory mediators. *Blood* 95(10): 3223–3231.

Cutolo, M., B. Villaggio, K. Otsa, O. Aakre, A. Sulli, and B. Seriolo. 2005. Altered circadian rhythms in rheumatoid arthritis patients play a role in the disease's symptoms. *Autoimmun. Rev.* 4(8): 497–502.

Di Bella, G., F. Mascia, L. Gualano, and L. Di Bella. 2013. Melatonin anticancer effects: Review. *Int. J. Mol. Sci.* 14(2): 2410–2430.

Drazen, D. L. and R. J. Nelson. 2001. Melatonin receptor subtype MT2 (Mel 1b) and not mt1 (Mel 1a) is associated with melatonin-induced enhancement of cell-mediated and humoral immunity. *Neuroendocrinology* 74(3): 178–184.

Eisenstein, E. M., Y. Berkun, and E. Ben-Chetrit. 2013. Familial Mediterranean fever: A critical digest of the 2012–2013 literature. *Clin. Exp. Rheumatol.* 31(3 Suppl 77): 103–107.

Ekmekcioglu, C., P. Haslmayer, C. Philipp, M. R. Mehrabi, H. D. Glogar, M. Grimm, V. J. Leibetseder, T. Thalhammer, and W. Marktl. 2001. Expression of the MT1 melatonin receptor subtype in human coronary arteries. *J. Recept. Signal. Transduct. Res.* 21(1): 85–91.

Esposito, E. and S. Cuzzocrea. 2010. Antiinflammatory activity of melatonin in central nervous system. *Curr. Neuropharmacol.* 8(3): 228–242.

Gang, N., J. P. Drenth, P. Langevitz, D. Zemer, N. Brezniak, M. Pras, J. W. van der Meer, and A. Livneh. 1999. Activation of the cytokine network in familial Mediterranean fever. *J. Rheumatol.* 26(4): 890–897.

Garcia-Maurino, S., M. G. Gonzalez-Haba, J. R. Calvo, M. Rafii-El-Idrissi, V. Sanchez-Margalet, R. Goberna, and J. M. Guerrero. 1997. Melatonin enhances IL-2, IL-6, and IFN-gamma production by human circulating CD4+ cells: A possible nuclear receptor-mediated mechanism involving T helper type 1 lymphocytes and monocytes. *J. Immunol.* 159(2): 574–581.

Garcia-Maurino, S., D. Pozo, A. Carrillo-Vico, J. R. Calvo, and J. M. Guerrero. 1999. Melatonin activates Th1 lymphocytes by increasing IL-12 production. *Life Sci.* 65(20): 2143–2150.

Ha, E., B. K. Choe, K. H. Jung, S. H. Yoon, H. J. Park, H. K. Park, S. V. Yim et al. 2005. Positive relationship between melatonin receptor type 1B polymorphism and rheumatoid factor in rheumatoid arthritis patients in the Korean population. *J. Pineal Res.* 39(2): 201–205.

Hagar, H. H., A. El-Medany, E. El-Eter, and M. Arafa. 2007. Ameliorative effect of pyrrolidinedithiocarbamate on acetic acid-induced colitis in rats. *Eur. J. Pharmacol.* 554(1): 69–77.

Jung, K. H., S. W. Hong, H. M. Zheng, D. H. Lee, and S. S. Hong. 2009. Melatonin downregulates nuclear erythroid 2-related factor 2 and nuclear factor-kappaB during prevention of oxidative liver injury in a dimethylnitrosamine model. *J. Pineal Res.* 47(2): 173–183.

Lahiri, S., P. Singh, S. Singh, N. Rasheed, G. Palit, and K. K. Pant. 2009. Melatonin protects against experimental reflux esophagitis. *J. Pineal Res.* 46(2): 207–213.

Laste, G., I. C. de Macedo, J. Ripoll Rozisky, F. Ribeiro da Silva, W. Caumo, and I. L. Torres. 2012. Melatonin administration reduces inflammatory pain in rats. *J. Pain Res.* 5: 359–362.

Li, J. H., J. P. Yu, H. G. Yu, X. M. Xu, L. L. Yu, J. Liu, and H. S. Luo. 2005. Melatonin reduces inflammatory injury through inhibiting NF-kappaB activation in rats with colitis. *Mediators Inflamm.* 2005(4): 185–193.

Maestroni, G. J. 1998. The photoperiod transducer melatonin and the immune-hematopoietic system. *J. Photochem. Photobiol. B* 43(3): 186–192.

Majewska, M., K. Zajac, M. Zemelka, and M. Szczepanik. 2007. Influence of melatonin and its precursor L-tryptophan on Th1 dependent contact hypersensitivity. *J. Physiol. Pharmacol.* 58(Suppl 6): 125–132.

Maldonado, M. D., F. Murillo-Cabezas, M. P. Terron, L. J. Flores, D. X. Tan, L. C. Manchester, and R. J. Reiter. 2007. The potential of melatonin in reducing morbidity-mortality after craniocerebral trauma. *J. Pineal Res.* 42(1): 1–11.

Malpaux, B., M. Migaud, H. Tricoire, and P. Chemineau. 2001. Biology of mammalian photoperiodism and the critical role of the pineal gland and melatonin. *J. Biol. Rhythms* 16(4): 336–347.

Mazzon, E., E. Esposito, C. Crisafulli, L. Riccardi, C. Muia, P. Di Bella, R. Meli, and S. Cuzzocrea. 2006. Melatonin modulates signal transduction pathways and apoptosis in experimental colitis. *J. Pineal Res.* 41(4): 363–373.

Mei, Q., J. M. Xu, L. Xiang, Y. M. Hu, X. P. Hu, and Z. W. Xu. 2005. Change of nitric oxide in experimental colitis and its inhibition by melatonin in vivo and in vitro. *Postgrad. Med. J.* 81(960): 667–672.

Mei, Q., J. P. Yu, J. M. Xu, W. Wei, L. Xiang, and L. Yue. 2002. Melatonin reduces colon immunological injury in rats by regulating activity of macrophages. *Acta Pharmacol. Sin.* 23(10): 882–886.

Musabak, U., G. Kilciler, A. Uygun, M. Kantarcioglu, Z. Polat, R. I. Sagkan, and S. Bagci. 2011. Melatonin and its day and night rhythm of alterations in familial Mediterranean fever: A brief research letter. *Open Rheumatol. J.* 5: 13–17.

Nelson, R. J. and D. L. Drazen. 2000. Melatonin mediates seasonal changes in immune function. *Ann. N. Y. Acad. Sci.* 917: 404–415.

Ozen, S., D. Uckan, E. Baskin, N. Besbas, H. Okur, U. Saatci, and A. Bakkaloglu. 2001. Increased neutrophil apoptosis during attacks of familial Mediterranean fever. *Clin. Exp. Rheumatol.* 19(5 Suppl 24): S68–S71.

Pei, Z., P. C. Fung, and R. T. Cheung. 2003. Melatonin reduces nitric oxide level during ischemia but not blood-brain barrier breakdown during reperfusion in a rat middle cerebral artery occlusion stroke model. *J. Pineal Res.* 34(2): 110–118.

Petrovsky, N. and L. C. Harrison. 1997. Diurnal rhythmicity of human cytokine production: A dynamic disequilibrium in T helper cell type 1/T helper cell type 2 balance? *J. Immunol.* 158(11): 5163–5168.

Peyrot, F. and C. Ducrocq. 2008. Potential role of tryptophan derivatives in stress responses characterized by the generation of reactive oxygen and nitrogen species. *J. Pineal Res.* 45(3): 235–246.

Radogna, F., M. Diederich, and L. Ghibelli. 2010. Melatonin: A pleiotropic molecule regulating inflammation. *Biochem. Pharmacol.* 80(12): 1844–1852.

Raghavendra, V., V. Singh, A. V. Shaji, H. Vohra, S. K. Kulkarni, and J. N. Agrewala. 2001. Melatonin provides signal 3 to unprimed CD4(+) T cells but failed to stimulate LPS primed B cells. *Clin. Exp. Immunol.* 124(3): 414–422.

Reiter, R. J., D. X. Tan, M. P. Terron, L. J. Flores, and Z. Czarnocki. 2007. Melatonin and its metabolites: New findings regarding their production and their radical scavenging actions. *Acta Biochim. Pol.* 54(1): 1–9.

Sulli, A., G. J. Maestroni, B. Villaggio, E. Hertens, C. Craviotto, C. Pizzorni, M. Briata, B. Seriolo, and M. Cutolo. 2002. Melatonin serum levels in rheumatoid arthritis. *Ann. N. Y. Acad. Sci.* 966: 276–283.

Tan, D. X., L. C. Manchester, R. Hardeland, S. Lopez-Burillo, J. C. Mayo, R. M. Sainz, and R. J. Reiter. 2003. Melatonin: A hormone, a tissue factor, an autocoid, a paracoid, and an antioxidant vitamin. *J. Pineal Res.* 34(1): 75–78.

Veneroso, C., M. J. Tunon, J. Gonzalez-Gallego, and P. S. Collado. 2009. Melatonin reduces cardiac inflammatory injury induced by acute exercise. *J. Pineal Res.* 47(2): 184–191.

Wang, H., W. Wei, N. P. Wang, S. Y. Gui, L. Wu, W. Y. Sun, and S. Y. Xu. 2005. Melatonin ameliorates carbon tetrachloride-induced hepatic fibrogenesis in rats via inhibition of oxidative stress. *Life Sci.* 77(15): 1902–1915.

42 Effects of Melatonin on Osteoblastic and Osteoclastic Activities of Bone Tissue

Meliha Gündağ Papaker, M. Hakan Seyithanoğlu, Mehmet Turgut, Süleyman Kaplan, and Saffet Tüzgen

CONTENTS

42.1 INTRODUCTION

Melatonin (Mel), known as *N*-acetyl-5-methoxytryptamine chemically, is secreted not only by pineal gland but also by retina, gastrointestinal tract, and bone marrow (Sánchez-Barceló et al. 2010) and plays in many physiological processes as a regulatory role as well as bone metabolism (Cardinali et al. 2003; Gitto et al. 2009). Bone remodeling is under the control of the function of osteoclasts and osteoblasts (Manolagas 2000). Parathyroid hormone (PTH), estradiol (E2), and growth hormone controlled the balance between the activities of these two types of cells (Manolagas 2000). Mel promotes bone formation and prevents bone degradation through the promotion of the osteoblast differentiation and activation and increases osteoprotegerin expression that prevents osteoclast differentiation and scavenging of the free radicals that are responsible for bone resorption (Sánchez-Barceló et al. 2010; Turgut et al. 2005) (Figure 42.1).

42.2 MELATONIN RECEPTORS

MT1 and MT2 receptors have been determined in mammals (Dubocovich and Markowska 2005). These receptors are membrane-associated G-protein-coupled and disperse not only in peripheral tissues but also in the central nervous system (Slominski et al. 2012). MT1 and MT2 locate on cells of the suprachiasmatic nucleus (SCN) at high densities (Von Gall et al. 2002). These two receptors demonstrate ligand-binding properties identically and mediate lots of signal pathways, such as inhibition of adenylyl cyclase (AC) and activation of phospholipase C (Von Gall et al. 2002). The MT1 receptor is coupled to the inhibition of AC and activation of phospholipase C (Von Gall et al. 2002).

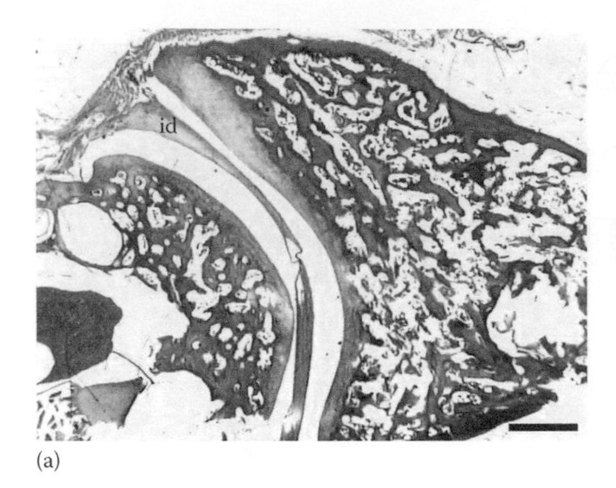

(a)

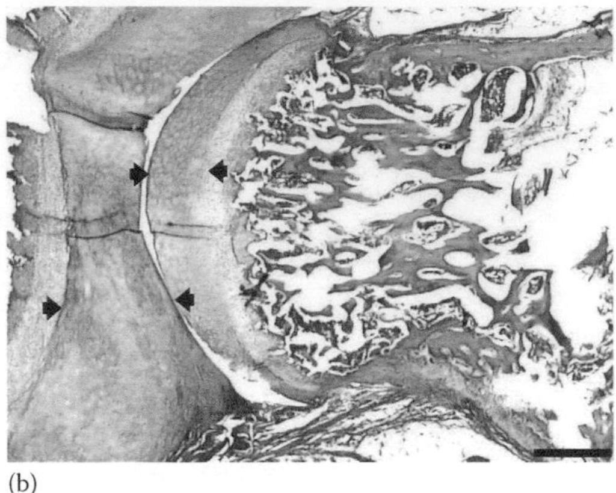

(b)

FIGURE 42.1 Pinealectomy seriously affects the development of bone formation and maturation. During these processes, osteoblasts do not mature as osteocytes; even many osteoblasts are produced, but most of them could not complete their maturation into osteocytes. Although normal maturation processes of bone are seen in the control vertebral body (a), the same development could not been observed in the corpus of vertebra in the pinealectomized animals (b). As a result of the absence or decreasing level of melatonin in the circulated blood, after pinealectomy, the maturation of osteoblasts into osteocytes is not completed as shown by arrowheads in (b).

Following inhibition of AC, this subsequently suppresses protein kinase A activity and phosphorylation of cyclicadenosinemonophosphate (cAMP) and cAMP-responsive element binding protein (Von Gall et al. 2002).

Recently, some studies have shown that osteoblasts and osteoclasts express MT1 and MT2 receptors (Slominski et al. 2012). In addition, osteoblasts express MT1 receptor, and the expression reduces with age (Satomura et al. 2007).

42.3 MELATONIN AND CIRCADIAN RHYTHM

Mel secretion is regulated by circadian rhythms and its secretion increases during at night (Cardinali et al. 2003). At daytime, Mel receptor density increases in SCN (Gauer et al. 1994). Furthermore, some animal studies have shown that an elevated binding site density at night to compare the day time (Vanecek et al. 1990). Circadian rhythms are related with age and peak concentration between the age of 1 and 3 years and declines progressively with increasing age (Waldhauser et al. 1988). There is a significant decrease in the Mel secretion around menopause and secretion is increased in response to activity, whereas decreased by immobility (Sack et al. 1986; Yocca and Friedman 1984). Some studies shows that bone cycle demonstrate a circadian rhythm with a rise in bone resorption during night (Ledger et al. 1995; Sairanen et al. 1994).

42.4 MELATONIN AND BONE PHYSIOLOGY

Mel may suppress the peroxisome proliferator-activated receptor gamma (PPAR-γ) thus, via hMSC, decreases adipogenesis (Zhang et al. 2010). Mel inhibits adipogenesis and simultaneously increases osteogenesis of hMSC in a dose-dependent manner, when it is added to the medium (Zhang et al. 2010). Furthermore, runt-related transcription factor 2 (Runx2) expression is enhanced by Mel, thereby it induces the formation of osteoblastic cells (Zhang et al. 2010).

Human bone cell studies have demonstrated that, in a dose-dependent manner, Mel increases bone proliferation and differentiation markers such as alkaline phosphatase, osteopontin, osteocalcin, or procollagen type 1 c-peptide (Nakade et al. 1999; Satomura et al. 2007; Sethi et al. 2010). Moreover, animal studies have also shown that, in a dose-dependent manner, Mel decreases the formation of mineralized matrix (Park et al. 2011; Roth et al. 1999; Satomura et al. 2007). Likewise, Mel has been shown to induce IGF-1 synthesis and, thus, increase the functions of osteoblast and prevent adiposis (Canalis 2009; Ostrowska et al. 2001).

Mel secretion is usually regulated by *N*-acetyltransferase (NAT) activation (Liu and Borjigin 2005). Studies suggest that estrogen inhibits NAT synthesis (Hayashi and Okatani 1999; Weiss and Crayton 1970). For this reason, in adolescent women, in response to an increased level of estrogen, NAT protein and the synthesis of Mel decrease (Liu and Borjigin 2005). In vivo studies, in which Mel was injected intraperitoneally into mice, have shown that there is new bone formation on the femoral cortex surface (Ladizesky et al. 2003).

Both the PTH and calcitonin control pineal secretion (Shoumura et al. 1992). Some authors suggest that the inhibition of Mel synthesis to white fluorescent light decreases the calcium concentration (Hakanson and Bergstrom 1981). Exogenous Mel administration suppresses this effect (Hakanson and Bergstrom 1981). Furthermore, β-adrenoceptor blockers' administration decreases serum calcium concentration in rats; this effect was prevented by the Mel administration (Hakanson et al. 1987). In brief, this study shows that hypocalcemia inhibits Mel, and Mel upregulates the blood calcium levels (Hakanson and Bergstrom 1981).

Some in vitro studies strongly support the stimulatory effects of Mel on both differentiation and the activity of osteoblasts (Sack et al. 1986). When compared with control cells Mel, the presence of Mel in preosteoblast culture produced early cell differentiation and expression of bone marker proteins (Sack et al. 1986). These effects were inhibited by the Mel receptor antagonist luzindole (Sack et al. 1986). These results clearly demonstrate that Mel increases the differentiation of human osteoblasts.

Activity of osteoclasts is under the control of paracrine factors, which are produced by osteoblasts (Cardinali et al. 2003). Osteoclast differentiating factor expresses from the marrow stromal cells and osteoblasts and is stimulated by PTH and 1.25-dihydroxycholecalcipherol (Cardinali et al. 2003). The receptor activator of nuclear factor-κ B ligand (RANKL) and the receptor activator of nuclear factor-κ B (RANK) have significant functions that differentiate precursor cells into osteoclast cells (Bell 2001). Both RANK and RANKL on the surface of osteoclasts activate bone resorption (Cardinali et al. 2003). Binding to RANK on osteoclastic cells, RANKL stimulates osteoclast differentiation (Cardinali et al. 2003). In mouse osteoblast study, micromolar doses of Mel reduce the expression of RANK mRNA and increase mRNA and osteoprotegerin levels (Koyama et al. 2002). Likewise, Mel promotes osteoblast differentiation and osteoprotegerin synthesis, which prevents the binding of RANKL to its receptor and inhibits osteoclast differentiation (Bell 2001).

Osteoclastic cells generate free radicals, which contribute to the process of bone degradation and resorption (Fraser et al. 1996). Mel directly neutralizes free radicals and stimulates antioxidative enzymes (Reiter et al. 2001, 2009a,b). When the free radical products exceed the capacity of the antioxidant systems, oxidative stress occurs, which increases bone resorption although the mechanisms are unknown; however, some studies show that oxidative stress may induce RANKL and so induce osteoclastic cell differentiation (Baek et al. 2010; Lee et al. 2005).

In animal studies, pinealectomy, which causes low levels of Mel, has been indicated to cause a decrease in antioxidants (Ozler et al. 2010; Sahna et al. 2004). Increased reactive oxygen species have been associated with decreased endogenous antioxidants and also osteoporosis (Baek et al. 2010). Some studies show that deficient intake of antioxidants increases fracture risk in smokers compared to smokers with sufficient intake of antioxidants (Melhus et al. 1999). Recently, Kaplan et al. (2011) showed that Mel has a positive effect on nerve regeneration, which is the sciatic nerve that innervates the bone, increasing the time of nerve regeneration after cutting, or injury might result in a delayed bone repair (Figure 42.2).

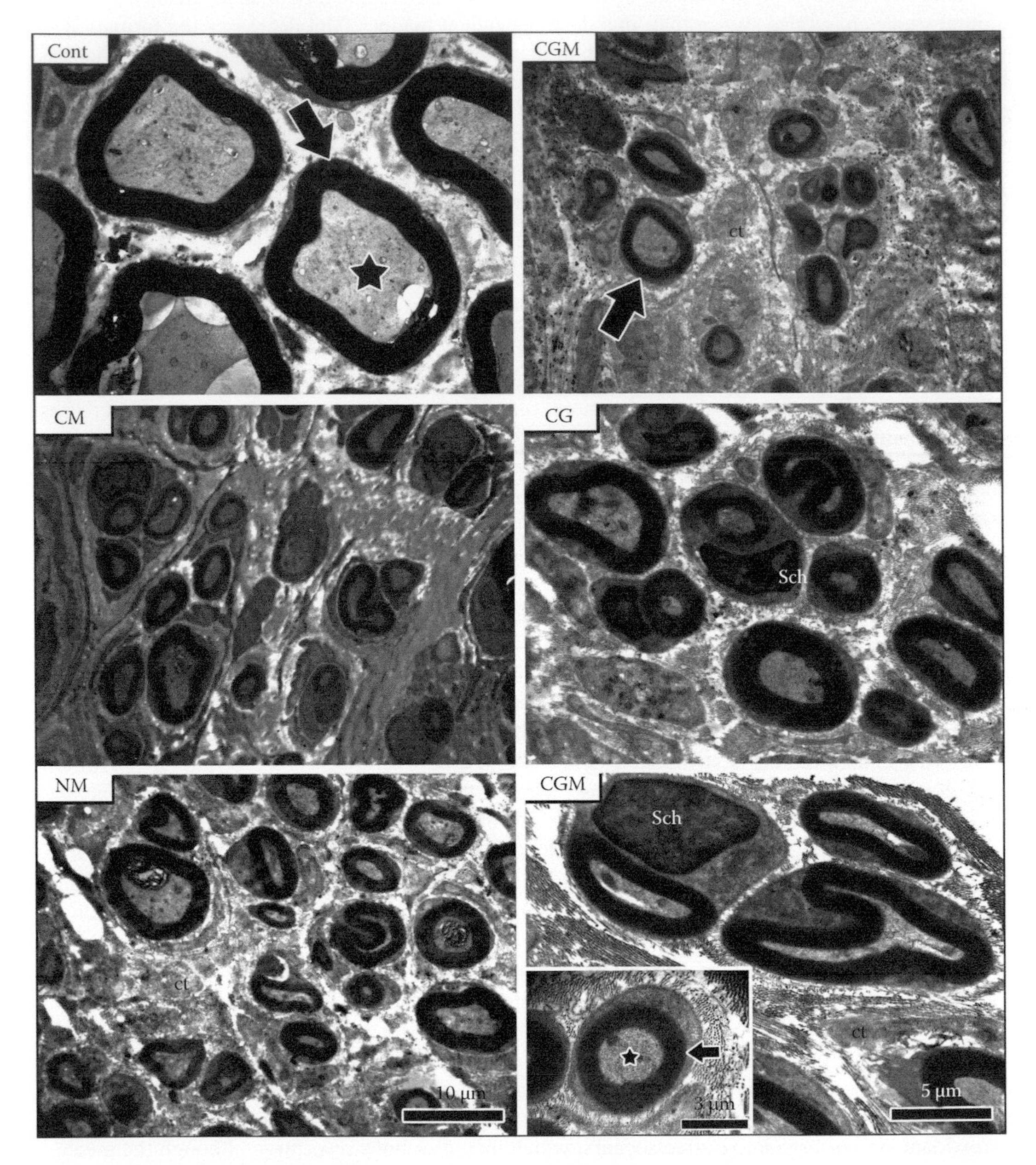

FIGURE 42.2 As seen in the control group (Cont), the myelinated nerve fibers are intact, and their diameters are larger than that in other groups, which made a cutting process to create a gap between the two ends of sciatic nerve. Mel has a positive effect on the regeneration of sciatic nerve after cutting. As seen in the figure, although axon number density was low in the collagen nerve conduit + platelet gel + Mel (CGM) and collagen nerve conduit + Mel (CM) groups in comparison with other groups, density of fibers in the autologous nerve graft + Mel (NM) group was higher than that of CGM and CM groups. These pictures give a general idea on each application, but the real change in the nerve could be seen after quantitative analysis such as stereological analysis of nerve fiber number and thicknesses of myelin sheath and axon cross-section area. These analyses were made and published by Kaplan et al. (2011). It was observed that not only Mel but also autologous platelet gel has a positive effect on the nerve regeneration (Kaplan et al. 2011). (*) Axoplasm; Sch, Schwann cell; arrows indicate myelinated nerve fibers as well as myelin sheath, ct, connective tissue.

42.5 EFFECTS OF MELATONIN ON HISTOLOGICAL FEATURES OF BONE TISSUE

Mel on bone metabolism consists of antiresorptive effects, and anabolic effects have been demonstrated by numerous in vitro studies (Zhang et al. 2010). Moreover, Mel affects the repair of bone defect through increasing angiogenesis, reducing inflammation, bone cell proliferation, and differentiation of osteogenic cells (Ramrez-Fernandez et al. 2013). In addition to these, in vitro studies show that Mel induces the synthesis of type 1 collagen fibers in human osteoblast in micromolar concentration (Nakade et al. 1999).

Bone defect due to neoplasm, trauma, surgery, or secondary effect through some bone disease is common, and bone repair may be categorized into three stages such as inflammatory, proliferative, and remodelling phases (Ganguly et al. 2010). Lots of studies indicated that Mel might play a significant role in the process of bone healing on account of inflammatory cell infiltration, regulation of bone cells, its antioxidant properties, promotion of angiogenesis, granulation tissue formation, and mineral matrix deposition (Calvo-Guirado et al. 2011; Halıcı et al. 2010; Ramírez-Fernández et al. 2013). Furthermore in this stage it was observed the clot formation, ischemia and reperfusion injury, and inflammatory cells infiltration (Halıcı et al. 2010). Neutrophils produce free oxygen radicals in this period that have a negative effect on bone defect healing (Halıcı et al. 2010). Mel is an important free radical scavenger and antioxidant at physiological and pharmacological doses (Halıcı et al. 2010). Thus, Mel may benefit in suppressing the effect of free oxygen radicals and accelerate healing process (Cetinus et al. 2005; Gitto et al. 2009).

It has been reported that angiogenesis, collagen deposition, differentiation of osteoblasts and fibroblasts, and granulation tissue formation occur in the proliferative phase (Ramírez-Fernández et al. 2013). Mel promotes angiogenesis in addition to osteoblast proliferation and differentiation (Ramírez-Fernández et al. 2013). Soybir et al. (2003) reported an increase in the number of blood vessels due to Mel application to wounds in rats. Yamada et al. (2008) suggest that the new bone regeneration was dependent on angiogenesis. Some studies illustrate that Mel reduces the number of apoptotic cells in nucleus pulposus of the disk tissue and cartilage end plate of the vertebra (Oktem et al. 2006). Also Mel inhibits the expression of inducible nitric oxide synthase (iNOS), which plays an important role in osteoporosis (Oktem et al. 2006).

42.6 CONTRASTING VIEWS ABOUT EFFECTS OF MELATONIN ON BONE FORMATION

Although there are some studies demonstrating the opposite effect of Mel, even most of studies show a role of Mel on the bone protection. Suzuki and Hattori (2002) investigated the effects of Mel upon osteoclasts and osteoblasts in a culture. They used specific markers such as tartrate-resistant acid phosphatase and alkaline phosphatase on osteoclasts and osteoblasts, respectively (Suzuki and Hattori 2002). They observed an inhibition of the both cell types by Mel incubation after 6 h, but only osteoblasts decreased after 18 h (Suzuki and Hattori 2002). Ostrowska et al. (2003) showed that pinealectomy correlated with increased levels of bone-forming markers, and high levels of Mel correlated with low levels of bone-forming markers.

42.7 MELATONIN AND OSTEOPOROSIS

The effects of Mel on bone metabolism were considered in some studies about osteoporosis. Feskanich et al. (2009) reported that more than 20 years of night shift work is associated with high risk of wrist and hip fracture over 8-year follow-up. Night shift work causes disruption in Mel secretion such as circadian rhythm disturbances (Reiter et al. 2009).

Uslu et al. (2007) suggest that Mel could prevent osteoporosis via inhibiting the iNOS, which generates nitric oxide. They showed on ovariectomized rat model that Mel decreased the expression of iNOS and epiphyseal cartilage of spinal column (Uslu et al. 2007). Uslu et al. (2007) used the same rat model and showed cortical thickness of femur and trabecular thickness of vertebra after treatment with Mel. Suzuki et al. (2008) produced bromomelatonin (1-benzyl-2,4,6-tribromomelatonin), which is a synthetic Mel derivate, and it increased the bone mineral density of ovariectomized rats, and therefore, it may be used for the treatment of osteoporosis.

Estrogens have a positive impact on bone formation, and some animal studies show a synergistic effect of Mel and E2 combination therapy (Ladizesky et al. 2003). However, Mel has also antiestrogenic actions under different circumstances (Sánchez-Barceló et al. 2010).

42.8 MELATONIN AND ADOLESCENT IDIOPATHIC SCOLIOSIS

Mel deficiency has been considered as etiologic factors in adolescent idiopathic scoliosis (AIS) although the condition is unknown (Cheng et al. 1999). Some studies have shown that iliac crest biopsy and scoliosis patients have disrupted function of osteoclastic and osteoblastic cells (Cheng et al. 1999). Experimentally, Turgut et al. (2005) showed that in pinealectomized chickens, total number of osteocytes in cervical vertebra was significantly lower than the control group, which has nonpinealectomized chickens. As a result, they suggested that Mel stimulates bone formation (Turgut et al. 2005). On the other hand, they showed that pineal gland transplantation in chicken after pinealectomy has no significant effect on the development of spinal deformation and levels of serum Mel in their another study (Turgut et al. 2003). Sobajima et al. (2003) have determined the Mel receptor mRNA in rabbit spinal cord. Nevertheless, they have not found any important difference between hereditary lordoscoliotic rabbit and control group (Sobajima et al. 2003).

Some authors have exhibited the correlation levels of serum Mel and vertebral curve progression in 40 AIS patients with moderate-to-severe levels (Machida et al. 2009). Of the 16 patients with low levels of Mel that were treated with Mel, a stable scoliosis was developed in 12, while 4 had progressive scoliosis (Machida et al. 2009). This study suggests that Mel supplementation may prevent the progression of scoliosis especially in mild cases (Machida et al. 2009). Evidently, further studies are needed to see the benefits of Mel on AIS.

42.9 CONCLUSION

A lot of in vitro as well as animal studies showed that Mel increases the differentiation of osteoblasts. Likewise, these studies suggested that Mel might be applied as a medical agent to promote bone regeneration especially in bone fracture, bone distraction, and some osteotomies. Today, it is widely accepted that Mel affects bone protection and prevents bone resorption, but only few studies show controversial results. However, further studies are warranted to strengthen on the evidence regarding the benefits of Mel on bone metabolism.

REFERENCES

Baek, K.H., K.W. Oh, W.Y. Lee, S.S. Lee, M.K. Kim, H.S. Kwon, E.J. Rhee et al. 2010. Association of oxidative stress with postmenopausal osteoporosis and the effects of hydrogen peroxide on osteoclast formation in human bone marrow cell cultures. *Calcif Tissue Int* 87(3):226–235.

Bell, N.H. 2001. Advances in the treatment of osteoporosis. *Curr Drug Targets Immune Endocr Metabol Disord* 1(1):93–102.

Calvo-Guirado, J.L., M.P. Ramírez-Fernández, G. Gómez-Moreno, J.E. Maté-Sánchez, R. Delgado-Ruiz, J. Guardia, L. López-Marí et al. 2011. Melatonin stimulates the growth of new bone around implants in the tibia of rabbits. *J Pineal Res* 49(4):356–363.

Canalis, E. 2009. Growth factor control of bone mass. *J Cell Biochem* 108(4):769–777.

Cardinali, D.P., M.G. Ladizesky, V. Boggio, R.A. Cutrera, and C. Mautalen. 2003. Melatonin effects on bone: Experimental facts and clinical perspectives. *J Pineal Res* 34(2):81–87.

Cetinus, E., M. Kilinç, M. Uzel, F. Inanç, E.B. Kurutaş, E. Bilgic, and A. Karaoguz. 2005. Does long-term ischemia affect the oxidant status during fracture healing? *Arch Orthop Trauma Surg* 125(6):376–380.

Cheng, J.C., X. Guo, and A.H. Sher. 1999. Persistent osteopenia in adolescent idiopathic scoliosis. A longitudinal follow up study. *Spine* 24(12):1218–1222.

Dubocovich, M.L. and M. Markowska. 2005. Functional MT1 and MT2 melatonin receptors in mammals. *Endocrine* 27(2):101–110.

Feskanich, D., S.E. Hankinson, and E.S. Schernhammer. 2009. Nightshift work and fracture risk. The nurses' health study. *Osteoporos Int* 20(4):537–542.

Fraser, J.H., M.H. Helfrich, H.M. Wallace, and S.H. Ralston. 1996. Hydrogen peroxide, but not superoxide, stimulates bone resorption in mouse calvariae. *Bone* 19(3):223–226.

Ganguly, K., A.V. Sharma, R.J. Reiter, and S. Swarnakar. 2010. Melatonin promotes angiogenesis during protection and healing of indomethacin induced gastric ulcer: Role of matrix metaloproteinase-2. *J Pineal Res* 49(2):130–140.

Gauer, F., M. Masson-Pevet, and P. Pevet. 1994. Daily variations in melatonin receptor density of rat pars tuberalis and suprachiasmatic nuclei are distinctly regulated. *Brain Res* 641(1):92–98.

Gitto, E., S. Pellegrino, P. Gitto, I. Barberi, and R.J. Reiter. 2009. Oxidative stress of the newborn in the pre- and postnatal period and the clinical utility of melatonin. *J Pineal Res* 46(2):128–139.

Hakanson, D.O. and W.H. Bergstrom. 1981. Phototherapy induced hypocalcemia in newborn rats: Prevention by melatonin. *Science* 214(4522):807–809.

Hakanson, D.O., R. Penny, and W.H. Bergstrom. 1987. Calcemic responses to photic and pharmacologic manipulation of serum melatonin. *Pediatr Res* 22(4):414–416.

Halıcı, M., M. Öner, A. Güney, Ö. Canöz, F. Narin, and C. Halıcı. 2010. Melatonin promotes fracture healing in the rat model. *Eklem Hastalik Cerrahisi* 21(3):172–177.

Hayashi, K. and Y. Okatani. 1999. Mechanisms underlying the effects of estrogen on nocturnal melatonin synthesis in peripubertal female rats: Relation to norepinephrine and adenylate cyclase. *J Pineal Res* 26(3):178–183.

Kaplan, S., A. Pişkin, M. Ayyildiz, A. Aktaş, B. Köksal, M.B. Ulkay, A.P. Türkmen, F. Bakan, and S. Geuna. 2011. The effect of melatonin and platelet gel on sciatic nerve repair: An electrophysiological and stereological study. *Microsurgery* 31(4):306–313.

Koyama, H., O. Nakade, Y. Takada, T. Kaku, and K.H. Lau. 2002. Melatonin at pharmacologic doses increases bone mass by suppressing resorption through down-regulation of the RANKL-mediated osteoclast formation and activation. *J Bone Miner Res* 17(7):1219–1229.

Ladizesky, M.G., V. Boggio, L.E. Albornoz, P.O. Castrillon, C. Mautalen, and D.P. Cardinali. 2003. Melatonin increases oestradiol-induced bone formation in ovariectomized rats. *J Pineal Res* 34(2):143–151.

Ledger, G.A., M.F. Burritt, P.C. Kao, W.M. O'Fallon, B.L. Riggs, and S. Khosla. 1995. Role of parathyroid hormone in mediating nocturnal and age-related increases in bone resorption. *J Clin Endocrinol Metab* 80(11):3304–3310.

Lee, N.K., Y.G. Choi, J.Y. Baik, S.Y. Han, D.W. Jeong, Y.S. Bae, N. Kim, and S.Y. Lee. 2005. A crucial role for reactive oxygen species in RANKL-induced osteoclast differentiation. *Blood* 106(3):852–859.

Liu, T. and J. Borjigin. 2005. *N*-acetyltransferase is not the rate-limiting enzyme of melatonin synthesis at night. *J Pineal Res* 39(1):91–96.

Machida, M., J. Dubousset, T. Yamada, and J. Kimura. 2009. Serum melatonin levels in adolescent idiopathic scoliosis prediction and prevention for curve progression—A prospective study. *J Pineal Res* 46(3):344–348.

Manolagas, S.C. 2000. Birth and death of bore cells: Basic regulatory mechanisms and implications for the pathogenesis and treatment of osteoporosis. *Endocr Rev* 21(2):115–137.

Melhus, H., K. Michaelsson, L. Holmberg, A. Wolk, and S. Ljunghall. 1999. Smoking, antioxidant vitamins, and the risk of hip fracture. *J Bone Miner Res* 14(1):129–135.

Nakade, O., H. Koyama, H. Ariji, A. Yajima, and Kaku T. 1999. Melatonin stimulates proliferation and type I collagen synthesis in human bone cells in vitro. *J Pineal Res* 27(2):106–110.

Oktem, G., S. Uslu, S.H. Vatansever, H. Aktug, M.E. Yurtseven, and A. Uysal. 2006. Evaluation of the relationship between inducible nitric oxide synthase (iNOS) activity and effects of melatonin in experimental osteoporosis in the rat. *Surg Radiol Anat* 28(2):157–162.

Ostrowska, Z., B. Kos-Kudla, M. Nowak, E. Swietochowska, B. Marek, J. Gorski, D. Kajdaniuk, and K. Wolkowska. 2003. The relationship between bone metabolism, melatonin and other hormones in sham-operated and pinealectomized rats. *Endocr Regul* 37(4):211–224.

Ostrowska, Z., B. Kos-Kudla, E. Swietochowska, B. Marek, D. Kajdaniuk, and N. Ciesielska-Kopacz. 2001. Influence of pinealectomy and long-term melatonin administration on GH-IGF-I axis function in male rats. *Neuro Endocrinol Lett* 22(4):255–262.

Ozler, M., K. Simsek, C. Ozkan, E.O. Akgul, T. Topal, S. Oter, and A. Korkmaz. 2010. Comparison of the effect of topical and systemic melatonin administration on delayed wound healing in rats that underwent pinealectomy. *Scand J Clin Lab Invest* 70(6):447–452.

Park, K.H., J.W. Kang, E.M. Lee, J.S. Kim, Y.H. Rhee, M. Kim, S.J. Jeong, Y.G. Park, and S.H. Kim. 2011. Melatonin promotes osteoblastic differentiation through the BMP/ERK/Wnt signaling pathways. *J Pineal Res* 51(2):187–194.

Ramírez-Fernández, M.P., J.L. Calvo-Guirado, J.E. de-Val, R.A. Delgado-Ruiz, B. Negri, G. Pardo-Zamora, D. Peñarrocha, C. Barona, J.M. Granero, and M. Alcaraz-Baños. 2013. Melatonin promotes angiogenesis during repair of bone defects: A radiological and histomorphometric study in rabbit tibiae. *Clin Oral Invest* 17(1):147–158.

Reiter, R.J., S.D. Paredes, L.C. Manchester, and D.X. Tan. 2009a. Reducing oxidative/nitrosative stress: A new discovered genre for melatonin. *Crit Rev Biochem Mol Biol* 44(4):175–200.

Reiter, R.J., D.X. Tan, T.C. Eren, L. Fuentes-Broto, and S.D. Paredes. 2009b. Light-mediated perturbations of circadian timing and cancer risk. *Integr Cancer Ther* 8(4):354–360.

Reiter, R.J., D.X. Tan, L.C. Manchester, and W. Qi. 2001. Biochemical reactivity of melatonin with reactive oxygen and nitrogen species: A review of the evidence. *Cell Biochem Biophys* 34(2):237–256.

Roth, J.A., B.G. Kim, W.L. Lin, and M.I. Cho. 1999. Melatonin promotes osteoblast differentiation and bone formation. *J Biol Chem* 274(31):22041–22047.

Sack, R.L., A.J. Lewy, D.L. Erb, W.M. Vollmer, and C.M. Singer. 1986. Human melatonin production decreases with age. *J Pineal Res* 3(4):379–388.

Sahna, E., H. Parlakpinar, N. Vardi, Y. Cigremis, and A. Acet. 2004. Efficacy of melatonin as protectant against oxidative stress and structural changes in liver tissue in pinealectomized rats. *Acta Histochem* 106(5):331–336.

Sairanen, S., R. Tahtela, K. Laitinen, S.L. Karonen, and M.J. Valimaki. 1994. Nocturnal rise in markers of bone resorption is not abolished by bedtime calcium or calcitonin. *Calcif Tissue Int* 55(5):349–352.

Sánchez-Barceló, E.J., M.D. Mediavilla, D.X. Tan, and R.J. Reiter. 2010. Scientific basis for the potential use of melatonin in bone diseases: Osteoporosis and adolescent idiopathic scoliosis. *J Osteoporos* 2010:830231.

Satomura, K., S. Tobiume, R. Tokuyama, Y. Yamasaki, K. Kudoh, E. Maeda, and M. Nagayama. 2007. Melatonin at pharmacological doses enhances human osteoblastic differentiation in vitro and promotes mouse cortical bone formation in vivo. *J Pineal Res* 42(3):231–239.

Sethi, S., N.M. Radio, M.P. Kotlarczyk, C.T. Chen, Y.H. Wei, R. Jockers, and P.A. Witt-Enderby. 2010. Determination of the minimal melatonin exposure required to induce osteoblast differentiation from human mesenchymal stem cells and these effects on downstream signaling pathways. *J Pineal Res* 49(3):222–238.

Shoumura, S., H. Chen, S. Emura, M. Utsumi, D. Hayakawa, T. Yamahira, K. Terasawa, A. Tamada, M. Arakawa, and H. Isono. 1992. An in vitro study on the effects of melatonin on the ultrastructure of the hamster parathyroid gland. *Histol Histopathol* 7(4):715–718.

Slominski, R.M., R.J. Reiter, N. Schlabritz-Loutsevitch, R.S. Ostrom, and A.T. Slominski. 2012. Melatonin membrane receptors in peripheral tissues: Distribution and functions. *Mol Cell Endocrinol* 351(2):152–166.

Sobajima, S., A. Kin, I. Baba, K. Kanbara, Y. Semoto, and M. Abe. 2003. Implication for melatonin and its receptor in the spinal deformities of hereditary lordoscoliotic rabbits. *Spine* 28(6):554–558.

Soybir, G., C. Topuzlu, O. Odabaş, K. Dolay, A. Bilir, and F. Köksoy. 2003. The effects of melatonin on angiogenesis and wound healing. *Surg Today* 33(12):896–901.

Suzuki, N. and A. Hattori. 2002. Melatonin suppresses osteoclastic and osteoblastic activities in the scales of goldfish. *J Pineal Res* 33(4):253–258.

Suzuki, N., M. Somei, K. Kitamura, R.J. Reiter, and A. Hattori. 2008. Novel bromomelatonin derivatives suppress osteoclastic activity and increase osteoblastic activity: Implications for the treatment of bone diseases. *J Pineal Res* 44(3):326–334.

Turgut, M., S. Kaplan, A.T. Turgut, H. Aslan, T. Güvenç, E. Cullu, and S. Erdogan. 2005. Morphological, stereological and radiological changes in pinealectomized chicken cervical vertebrae. *J Pineal Res* 39(4):392–399.

Turgut, M., C. Yenisey, A. Uysal, M. Bozkurt, and M.E. Yurtseven. 2003. The effects of pineal gland transplantation on the production of spinal deformity and serum melatonin level following pinealectomy in the chicken. *Eur Spine J* 12(5):487–494.

Uslu, S., A. Uysal, G. Oktem, M. Yurtseven, T. Tanyalçin, and G. Başdemir. 2007. Constructive effect of exogenous melatonin against osteoporosis after ovariectomy in rats. *Anal Quant Cytol Histol* 29(5):317–325.

Vanecek, J., E. Kosar, and J. Vorlicek. 1990. Daily changes in melatonin binding sites and the effect of castration. *Mol Cell Endocrinol* 73(2–3):165–170.

Von Gall, C., J.H. Stehle, and D.R. Weaver. 2002. Mammalian melatonin receptors: Molecular biology and signal transduction. *Cell Tissue Res* 309(1):151–162.

Waldhauser, F., G. Weiszenbacher, E. Tatzer, B. Gisinger, M. Waldhauser, M. Schemper, and H. Frisch. 1988. Alterations in nocturnal serum melatonin levels in humans with growth and aging. *J Clin Endocrinol Metab* 66(3):648–652.

Weiss, B. and J. Crayton. 1970. Gonadal hormones as regulators of pineal adenyl cyclase activity. *Endocrinology* 87(3):527–533.

Yamada, Y., T. Tamura, K. Hariu, Y. Asano, S. Sato, and K. Ito. 2008. Angiogenesis in newly augmented bone observed in rabbit calvarium using a titanium cap. *Clin Oral Implants Res* 19(10):1003–1009.

Yocca, F.D. and E. Friedman. 1984. Effect of immobilization stress on rat pineal beta-adrenergic receptor-mediated function. *J Neurochem* 42(5):1427–1432.

Zhang, L., P. Su, C. Xu, C. Chen, A. Liang, K. Du, Y. Peng, and D. Huang. 2010. Melatonin inhibits adipogenesis and enhances osteogenesis of human mesenchymal stem cells by suppressing PPAR gamma expression and enhancing Runx2 expression. *J Pineal Res* 49(4):364–372.

43 Phyto-Melatonin
A Novel Therapeutic Aspect of Melatonin in Nature's Way

Chandana Haldar, Somenath Ghosh, and Amaresh Kumar Singh

CONTENTS

43.1 INTRODUCTION

Around nineteenth century, some preliminary studies indicated that plants may possess melatonin (Mel; Kolár et al., 1995), but clear-cut indications that the plants are having Mel were supported by scientists later during 1995 (Dubbels et al., 1995). However, the existence of phyto-melatonin was reported by scientists in micro- and macro-alga (Balzer et al., 1998), particularly in red alga (Rhodophyta, Lorenz and Lüning, 1998), metazoans (Hardeland, 1999), and other photoautotrophic microorganisms (Hardeland and Poeggeler, 2003). Among the other plant groups, the existence of Mel is still controversial and needs further proof. Phyto-melatonin is now accepted to be present only in angiosperms, in their different parts like fruits, seeds, etc., including some medicinal herbs (Reiter et al., 2007). The reason behind this fact is mainly the lack of specific molecular and biochemical approaches to detect the same in plants other than angiosperms.

However, in recent years, new methodologies for the extraction of Mel from plants have been developed (Mercolini et al., 2008) along with the effective protocol designed by scientists to detect the presence of Mel in plants (Cao et al., 2006). Studies with radioisotope tracer techniques have provided clues that Mel can be produced in plants using the tryptophan precursor, which is the common ontogenic molecule for serotonin and indole-3 acetic acid (IAA, Murch et al., 2000). However, some reports are there to suggest that plants can absorb Mel even from soil (Tan et al., 2007). The first speculation regarding phyto-melatonin was reported in 1997 (Murch et al., 1997), and evidence regarding the synthesis of phyto-melatonin was also reported (Tan et al., 2007) with the help of radioisotope tracer technique. The study suggested that radiolabeled tryptophan was incorporated into serotonin and was further converted to Mel. The recovering concentration of radiolabeled serotonin in the plant was higher than Mel under low light condition. But the ratio metric concentration reversed under high light concentration suggesting a strong evidence for Mel synthesis by higher plants. Many of the microorganisms like fungi and bacteria also contain Mel (Hardeland and Poeggeler, 2003). Upon their degradation, it may be possible that Mel may directly come to soil and can be directly absorbed by a number of plants ranging from dianoflagellates (Mueller and Hardeland, 1999) to angiosperms where phyto-melatonin may be directly absorbed by roots and can be directly transported to fruits (Manchester et al., 1995). Notable conclusion may be that plants can either synthesize Mel by some biochemical mechanisms that are yet to be explored or can procure Mel from different sources. Thus, like in animals, Mel is capable enough to modulate a number of physiological functions in plants.

43.2 DISTRIBUTION AND AMOUNT OF PHYTO-MELATONIN IN PLANTS

The occurrences of Mel have been studied in different plants that are either of economic importance or can be consumed directly by humans or other herbivores (Reiter et al., 2007). The ubiquitous distribution of Mel is reported in leaves, roots, fruits, and seeds of different flowering plants. Mel is present in banana having a concentration of 0.655 ng/g when detected by GC-MS (Badria, 2002), but the level was significantly high (1 ng/g of plant tissue) when detected by HPLC-MS (Dubbels et al., 1995). On the other hand, most of the authors have regarded the radioimmunoassay (RIA) as the best possible technique for the detection of any biomolecule, and the amount Mel present in most of plant tissues has now been detected by this technique. RIA is sensitive enough to detect the trace amount of Mel in dry parts of plants like seed (Manchester et al., 2000). Thus, using this technique, the highest value was measured in seeds of white mustard (189 ng/g of plant tissue) and black mustard (123 ng/g of plant tissue; Manchester et al., 2000). The corns (*Zea mays*), which are another economically important edible plant tissue, contain Mel (~1.878 ng/g of plant tissue; Badria, 2002). Thus, it is clear that corns possess phyto-melatonin. However, no one has reported that the amount of Mel in plants is habitat or niche dependent.

43.3 ROLE OF MELATONIN IN PLANTS

43.3.1 Circadian Time Management

The differential synthesis of Mel in plants is dependent upon the duration of scotophase of light–dark cycle that suggests that Mel is playing a crucial role in providing circadian timing in plants likewise in animals (Kolár and Machácková, 2001).

43.3.2 Free Radical Scavenging Property

Phyto-melatonin level increases in different parts of the plant just after UV irradiation (Tettamanti et al., 2000) suggesting that phyto-melatonin is helpful in plants against free radical as an antioxidant similarly in animals (Tan et al., 2007). Thus, it can be suggested that Mel is positively acting as a free radical scavenger as well as a natural antioxidant in plants particularly in the tissues that are not capable enough to detoxify the free radicals enzymatically like dry seeds (Hardeland et al., 2007).

43.3.3 Miscellaneous Functions

In plants, they protect against harsh environments (Posmyk et al., 2008), promote vegetative growth (Hernández-Ruiz et al., 2004), and attenuate apoptosis (Lei et al., 2004). In most of the cases, phyto-melatonin can perform these physiological functions either by modulating the mainstream molecular mechanisms or by being a part of subcellular mechanisms that are modulated by a number of other plant hormones or factors (Tan et al., 2007).

43.4 ROLE OF MELATONIN IN ANIMALS

Mel is secreted during the dark phase of day–night cycle (Claustrat et al., 2005) and thus regarded as *chemical expression of darkness* (Reiter, 1991). Mel is regarded as clock and calendar (Reiter, 1993) depending upon its magnitude and duration of secretion during different times of year. There are handful of literatures available depicting the role of Mel in animals. Further, many authors have reported about the role of Mel in the regulation of various physiological functions in animals like the regulation of circadian rhythm (Cajochen et al., 2003), immune modulation (Carrillo-Vico et al., 2013), reproduction (Haldar and Thapliyal, 1977a), seasonality (Wehr, 1997), stress management (Gupta and Haldar, 2013), and apoptosis (Sainz et al., 2003) in different animal models ranging from lizard (Haldar and Thapliyal, 1977b) to ruminants (Kaushalendra and Haldar, 2012). In human trials under different neuronal disorders and related clinical symptoms (like Alzheimer's syndrome), Mel is reported to intervene the same and is reported to exert some beneficial results (Cardinali et al., 2010). Most importantly, the regulation of circadian rhythm in a daily manner (Dubocovich, 2007) and seasonality in annual manner (Butler et al., 2010) has become particularly an important and unique aspect among the multifactorial functions of Mel.

43.5 MELATONIN AND SEASONALITY

43.5.1 Seasonality in Reproduction

The manifestation of reproductive seasonality can only be observed in wild or semi-wild animals exposed to the challenges of nature during different phases of their life cycle. They use the environmental cues and decide the best time fit for the reproduction or perpetuation of species in the most energy-efficient manner (Nelson and Demas, 1997). Melatonin orchestrates the function of different gonadal steroids, which provide sufficient intimation to animals when to reproduce and when to improve the defense mechanism of body (Nelson and Drazen, 2000). Thus, depending upon the time of reproduction, seasonally breeding animals are differentiated as *long-day breeders* and *short-day breeders* (Hansen, 1985). In terms of the duration of Mel secretion (the calendar), in long-day breeders, the yearly peak of Mel reverses the gonadal activity and thus regarded as *anti-gonadotropic hormone* (as in case of golden hamster, Reiter et al., 2009). But, in case of short-day breeders, the yearly peak of Mel coincides with gonadal hormone and reproductive activity and is designated as *progonadotropic hormone* (e.g., in sheep and goat).

43.5.2 Seasonality in Immune Functions

Immunity, in general, is defined as the counterstrategy of the homeostatic mechanism of the body being provoked by any pathogenic invasion. Immune system is now accepted as an open circuit and is being regulated by different cells (e.g., T- and B-cells), cytokines, chemokines, lymphokines, antibodies, and nevertheless by a number of hormones and factors. Seasonality in immune function is not affected by domestication and is still prevalent in domesticated or semidomesticated animals (Martin et al., 2008) even in humans (Nelson and Demas, 1996). Thus, modulation of immune functions by Mel has become an important aspect of study, and handful of literatures are available emphasizing on differential functions of Mel in the immune regulation of submammalians (Yadav and Haldar, 2013), and in most cases, Mel has been proved to be an *immune-enhancing* neurohormone (Gupta and Haldar, 2013).

43.6 EFFECT OF PHYTO-MELATONIN IN ANIMALS

The role of phyto-melatonin in animals was not well documented. Only few literatures are available regarding the phyto-melatonin implications in the management of animal health. A trial study on human females in Japan reported an increase in the urinary profile of 6-sulfatoxymelatonin (Oba et al., 2008) upon the consumption of the six varieties of phyto-melatonin-rich vegetables. This study has only speculated about the role of phyto-melatonin-rich vegetables in the prevention of cardiovascular and cancer diseases.

43.7 GOATS, PHYTO-MELATONIN, AND BEYOND

The goats, apart from an emerging model for veterinary research and a ruminant short-day breeder, are making their economical importance in today's worldwide scenario due to their delicious meat, nutritious milk, and different byproducts produced from meat and milk. But the most concerning issue is the management of goat health due to their high mortality rate in tropical and subtropical countries like India (Abubakar et al., 2008).

In most parts of world, goats are fed with corn (*Zea mays*) for improving their milk quality. The corn is a naturally available nutritious food for goats and rich in phyto-melatonin. But unlike phytoestrogen, the role(s) of phyto-melatonin in the regulation of different physiological functions in animals is totally lacking. Thus, practical studies and implementations are needed to explore out the role of phyto-melatonin in goat health management.

We have identified this lacuna and conducted experiments on the role of phyto-melatonin in the modulation of immunity and general health of goat in terms of both physiology and energetic aspect. In the subsequent part of the chapter, we will be discussing only the effect of phyto-melatonin treatment in general health management and immune modulation of goats during winter (as it is the most challenging season in terms of cold stress for both the sexes and in terms of gestational stress for females in particular) and will try to provide an idea that how phyto-melatonin (as a dietary supplement) can be used as a novel therapeutic aspect of Mel in goat health management in herbal manner.

43.8 PHYTO-MELATONIN IN GOAT HEALTH MANAGEMENT

The general indications of good health can be manifested by the significant outrages in immunological, hormonal, and metabolic parameters. These help in the maintenance of body homeostasis in a proper and systematic manner. As a part of our preliminary experiment, we will be expressing the result only (unpublished data).

43.8.1 Effect of Phyto-Melatonin as Dietary Supplement on Body Homeostasis

Corn is not the natural food of goats in India, and hence, we used the maize as a dietary supplement with the normal food for goats. Hence, it was necessary to check if there was any side effect(s) of maize and/its metabolites on the digestive system as well as excretory system of goats as these two are the major systems that maintain the body homeostasis. Our results have suggested that there was no significant variation in the AST and ALT for liver function test and urea, BUN, and creatinine levels for renal function test in experimental groups. Further, there were no sex dependent variations suggesting that neither the corns (as a source of phyto-melatonin) nor their metabolites are affecting the body homeostasis in a detrimental manner or having any negative impact on goat health.

43.8.2 Effect on Body Mass

Upon supplementation, we found that there was a significant sex-dependent increase in body mass in the phyto-melatonin-treated groups, being higher in females as a reason that it provoked the basal metabolism (anabolism) of the body, and as a result of the same, body mass has increased. To confirm the same, we further studied the different circulatory metabolic parameters.

43.8.3 Effect on Metabolic Parameters

An increase in body mass in phyto-melatonin supplementation groups has provided us a clue that anabolism may be higher in animals upon treatment. We checked the circulatory level of glucose (a readymade source of energy) that was significantly high in treated groups particularly in females. Cholesterol, the sustainable source of energy, is found to be significantly higher in females of supplemented groups than males; however, the circulatory level of protein in supplemented group was only higher in females than males. Thus, we may conclude that there may be an increase in body metabolic processes upon treatment, and for that, a higher requirement of energy was needed. To balance, the same circulatory level of glucose was increased. However, protein, which is being used as a source of energy only under severe pathological or starving condition, was found to be unaffected. Simultaneously and most interestingly, the cholesterol level was also high in the phyto-melatonin-treated groups, and this higher level of cholesterol may be deposited in the body and thus increasing the body mass.

Higher cholesterol in circulation proved to be physiologically beneficial in goats. The higher circulatory level of glucose has suggested that body metabolism is high. An increase in peripheral cholesterol has suggested that this may be used as a source of energy in near or far future. The higher circulatory level of glucose suggests that body metabolism is high while an increase in peripheral cholesterol suggests that this may be used as a source of energy in near or far future. The next issue is to find out which energy-demanding process of goat is modulated using this energy, as during winter, two most energy-demanding mega events (reproduction and immune modulation) are occurring simultaneously. To explore the this issue, we studied the hematological, immunological, and hormonal parameters.

43.8.4 Effect on Hematological Parameters

The hematological parameters including total RBC count and %Hb content were found to be significantly high in phyto-melatonin-supplemented groups, being higher in females. Thus, higher level of metabolic parameters may be providing higher fitness level to females than males during stressful months of winter.

43.8.5 Effect on Immune Parameters

In the cell-mediated immune parameters, the total leukocyte count, %lymphocyte count, and %stimulation ratio of peripheral blood mononuclear cells were found to be significantly high in the supplemented groups. These results may be explained that, phyto-melatonin supplementation might have increased the peripheral Mel level and that increased Mel might have increased the peripheral cell–mediated immune parameters being in agreement with other reports (Carrillo-Vico et al., 2013).

43.8.6 Effect on Hormonal Parameters

We noted a significant increase in circulatory Mel level in both the sexes in phyto-melatonin-supplemented groups, being higher in females. Upon supplementation, testosterone level was unaffected, but estrogen level in females was significantly higher. Thus, we may conclude that in males, the higher level of Mel may have increased the immune parameters, and to cope up with this higher energy demand, metabolic parameters were also increased. But, for females, Mel itself, and with elevated estrogen level, may increase inflammatory factors or cytokines in body that are in agreement with previous reports in other animals that estrogen may upregulate inflammatory cytokines (Calippe et al., 2008). Thus, under the influence of both the hormones, the higher immune status of the females was maintained, and they showed higher levels in metabolic and hematological parameters than males. In males, Mel alone was sufficient enough to well manage the immune functions as testosterone is an immune suppressor.

43.8.7 Effect on Free Radical Parameters

The free radicals can be estimated by estimating the activities of their scavenger enzymes. The main free radical scavenging enzymes in the system are superoxide dismutase (SOD), catalase (CAT), and glutathione peroxidases (GPx). The good enough marker for lipid peroxidation is malonaldehyde (MDA) estimation. MDA level was lower in both the sexes of phyto-melatonin supplementation, being significantly lower in females. We noted the SOD, CAT, and GPx in the blood of goats upon phyto-melatonin supplementation and found there was a significant increase in SOD, CAT, and GPx activities in phyto-melatonin treatment than control group; however, sex-dependent variation was statistically nonsignificant. Thus, in both the sexes of phyto-melatonin supplementation group, increased Mel level has maintained its parallelism with increased levels of free radical scavenging enzymes as metabolic activity has increased. But lipid peroxidation, which generally depicts the level of cellular disintegrate (as MDA level is a universal marker of lipid peroxidation caused due to cell membrane disruption; Wong-ekkabut et al., 2007), was found to be low, thus speculating that free radical generation was only a causal effect of increased metabolism.

43.9 CONCLUSION

For the first time, the role of phyto-melatonin as a protective molecule with improved effect on the health and immunity of Indian goat *Capra hircus* is being proposed in the most beneficial manner. The effect of phyto-melatonin supplementation can be brought back to normal, and this dietary supplement might be utilizing the similar pathway as commercial Mel. There are so many less expensive and readily available sources of phyto-melatonin that requires the proper knowledge of exploitation of these sources for the extreme benefit of animals as well as of the human beings in near or far future.

43.10 FUTURE PROSPECTIVE

Depending upon our preliminary work, we may propose that phyto-melatonin is having the full potential to perform the same function in the regulation of apoptosis, circadian rhythm, seasonality, and stress management in animals also. However, more detailed molecular approaches are needed to explore out the exact mechanism of action of this phyto-compound in animals.

REFERENCES

Abubakar, M., Ali, Q., Khan, H.A. 2008. Prevalence and mortality rate of peste des petitis ruminant (PPR): Possible association with abortion in goat. *Trop. Anim. Health Prod.* 40: 317–321.

Badria, F.A. 2002. Melatonin, serotonin, and tryptamine in some Egyptian food and medicinal plants. *J. Med. Food* 5: 153–157.

Balzer, I., Bartolomaeus, B., Höcker, B. 1998. Circadian rhythm of melatonin content in Chlorophyceae. *Proceedings of the Conference News from the Plant Chronobiology Research.* Markgrafenheide, Germany, pp. 55–56.

Butler, M.P., Turner, K.W., Park, J.H., Schoomer, E.E., Zucker, I., Gorman, M.R. 2010. Seasonal regulation of reproduction: Altered role of melatonin under naturalistic conditions in hamsters. *Proc. Biol. Sci.* 277: 2867–2874.

Cajochen, C., Kräuchi, K., Wirz-Justice, A. 2003. Role of melatonin in the regulation of human circadian rhythms and sleep. *J. Neuroendocrinol.* 15: 432–437.

Calippe, B., Douin-Echinard, V., Laffargue, M., Laurell, H., Rana-Poussine, V., Pipy, B. et al. 2008. Chronic estradiol administration in vivo promotes the proinflammatory response of macrophages to TLR4 activation: Involvement of the phosphatidylinositol 3-kinase pathway. *J. Immunol.* 180: 7980–7988.

Cao, J., Murch, S.J., O'Brien, R., Saxena, P.K. 2006. Rapid method for accurate analysis of melatonin, serotonin and auxin in plant samples using liquid chromatography-tandem mass spectrometry. *J. Chromatogr. A* 1134: 333–337.

Cardinali, D.P., Furio, A.M., Brusco, L.I. 2010. Clinical aspects of melatonin intervention in Alzheimer's disease progression. *Curr. Neuropharmacol.* 8: 218–227.

Carrillo-Vico, A., Lardone, P.J., Alvarez-Sánchez, N., Rodríguez-Rodríguez, A., Guerrero, J.M. 2013. Melatonin: Buffering the immune system. *Int. J. Mol. Sci.*14: 8638–8683.

Claustrat, B., Brun, J., Chazot, G. 2005. The basic physiology and pathophysiology of melatonin. *Sleep Med. Rev.* 9: 11–24.

Dubbels, R., Reiter, R.J., Klenke, E., Goebel, A., Schnakenberg, E., Ehlers, C. et al. 1995. Melatonin in edible plants identified by radioimmunoassay and by high performance liquid chromatography-mass spectrometry. *J. Pineal Res.* 18: 28–31.

Dubocovich, M.L. 2007. Melatonin receptors: Role on sleep and circadian rhythm regulation. *Sleep Med.* 8: 34–42.

Gupta, S., Haldar, C. 2013. Physiological crosstalk between melatonin and glucocorticoid receptor modulates T-cell mediated immune responses in a wild tropical rodent, *Funambulus pennanti. J. Steroid Biochem. Mol. Biol.* 134: 23–36.

Haldar, C., Thapliyal, J.P. 1977a. Effect of pinealectomy on the annual testicular cycle of *Calotes versicolor. Gen. Comp. Endocrinol.* 32: 395–399.

Haldar, C., Thapliyal J.P. 1977b. Development of pineal complex in *Calotes versicolor. Arch. Anat. Histol. Embryol.* 60: 139–146.

Hansen, P.J. 1985. Photoperiodic regulation of reproduction in mammals breeding during long days versus mammals breeding during short days. *Anim. Reprod. Sci.* 9: 301–315.

Hardeland, R. 1999. Melatonin and 5-methoxytryptamine in non-metazoans. *Reprod. Nutr. Dev.* 39: 399–408.

Hardeland, R., Pandi-Perumal, S.R., Poeggeler, B. 2007. Melatonin in plants: Focus on a vertebrate night hormone with cytoprotective properties. *Funct. Plant Sci. Biotech.* 1: 32–45.

Hardeland, R., Poeggeler, B. 2003. Non-vertebrate melatonin. *J. Pineal Res.* 34: 233–241.

Hernández-Ruiz, J., Cano, A., Arnao, M.B. 2004. Melatonin: A growth-stimulating compound present in lupin tissues. *Planta* 220: 140–144.

Kaushalendra and Haldar, C. 2012. Correlation between peripheral melatonin and general immune status of domestic goat, *Capra hircus*: A seasonal and sex dependent variation. *Small Rumin. Res.* 107: 147–156.

Kolár, J., Machácková, I. 2001. Occurrence and possible function of melatonin in plants. A review. *Endocytobio. Cell Res.* 14: 75–84.

Kolár, J., Machácková, I., Illnerová, H. Prinsen, E., Van Dongen, W., Van Onckelen, H.A. 1995. Melatonin in higher plants determined by radioimmunoassay and liquid chromatography-mass spectometry. *Biol. Rhythm Res.* 26: 406.

Lei, X.Y., Zhu, R.Y., Zhang, G.Y., Dai, Y.R. 2004. Attenuation of cold induced apoptosis by exogenous melatonin in carrot suspension cells: The possible involvement of polyamines. *J. Pineal Res.* 36: 126–131.

Lorenz, M., Lüning, K. 1998. Detection of endogenous melatonin in the marine red macroalgae *Porphyra umbilicalis* and *Palmaria palmata* by enzyme-linked immunoassay (ELISA) and effects of melatonin administration on algal growth. *Proceedings of the Conference News from the Plant Chronobiology Research.* Markgrafenheide, Germany, pp. 42–43.

Manchester, L.C., Poeggeler, B., Alvares, F.L., Ogden, G.B., Reiter, R.J. 1995. Melatonin immunoreactivity in the photosynthetic prokaryote *Rhodospirillum rubrum*: Implications for an ancient antioxidant system. *Cell. Mol. Biol. Res.* 41: 391–395.

Manchester, L.C., Tan, D.X., Reiter, R.J., Park, W., Monis, K., Qi, W. 2000. High levels of melatonin in the seeds of edible plants. Possible function in germ tissue protection. *Life Sci.* 67: 3023–3029.

Martin, L.B., Weil, Z.M., Nelson, R.J. 2008. Seasonal changes in vertebrate immune activity: Mediation by physiological trade-offs. *Philos. Trans. R. Soc. Lond. B Biol. Sci.* 363: 321–339.
Mercolini, L., Addolorata Saracino, M., Bugamelli, F., Ferranti, A., Malaguti, M., Hrelia, S. et al. 2008. HPLC-F analysis of melatonin and resveratrol isomers in wine using an SPE procedure. *J. Sep. Sci.* 31: 1007–1114.
Mueller, U., Hardeland, R. 1999. Transient accumulations of exogenous melatonin indicate binding sites in the dinoflagellate *Gonyaulax polyedra*. In: *Studies on Antioxidants and Their Metabolites*, R. Hardeland (ed.), pp. 140–147. Goettingen, Germany: Cuvilier.
Murch, S.J., Krishna Raj, S., Saxena, P.K. 2000. Tryptophan is a precursor for melatonin and serotonin biosynthesis in in vitro regenerated St John's wort (*Hypericum perforatum* L. cv. Anthos) plants. *Plant Cell Rep.* 19: 698–704.
Murch, S.J., Simmons, C.B., Saxena, P.K. 1997. Melatonin in feverfew and other medicinal plants. *Lancet* 350: 1598–1599.
Nelson, R.J., Demas, G.E. 1996. Seasonal changes in immune function. *Q. Rev. Biol.* 71: 511–548.
Nelson, R.J., Demas, G.E. 1997. Role of melatonin in mediating seasonal energetic and immunologic adaptations. *Brain Res. Bull.* 44: 423–430.
Nelson, R.J., Drazen, D.L. 2000. Melatonin mediates seasonal changes in immune function. *Ann. N. Y. Acad. Sci.* 917: 404–415.
Oba, S., Nakamura, K., Sahashi, Y., Hattori, A., Nagata, C. 2008. Consumption of vegetables alters morning urinary 6-sulfatoxymelatonin concentration. *J. Pineal Res.* 45: 17–23.
Posmyk, M.M., Kuran, H., Marciniak, K., Janas, K.M. 2008. Pre-sowing seed treatment with melatonin protects red cabbage seedlings against toxic copper ion concentrations. *J. Pineal Res.* 45: 24–31.
Reiter, R.J. 1991. Melatonin: The chemical expression of darkness. *Mol. Cell. Endocrinol.* 79: 153–158.
Reiter, R.J. 1993. The melatonin rhythm: Both a clock and a calendar. *Experientia* 49: 654–664.
Reiter, R.J., Tan, D.X., Manchester, L.C., Paredes, S.D., Mayo, J.C., Sainz, R.M. 2009. Melatonin and reproduction revisited. *Biol. Reprod.* 81: 445–456.
Reiter, R.J., Tan, D.X., Manchester, L.C., Simopoulos, A.P., Maldonado, M.D., Flores, L.J. et al. 2007. Melatonin in edible plants (phytomelatonin): Identification, concentrations, bioavailability and proposed functions. *World Rev. Nutr. Diet.* 97: 211–230.
Sainz, R.M., Mayo, J.C., Rodriguez, C., Tan, D.X., Lopez-Burillo, S., Reiter, R.J. 2003. Melatonin and cell death: Differential actions on apoptosis in normal and cancer cells. *Cell Mol. Life Sci.* 60: 1407–1426.
Tan, D.X., Manchester, L.C., Di Mascio, P., Martinez, G.R., Prado, F.M., Reiter, R.J. 2007. Novel rhythms of *N*1-acetyl-*N*2-formyl-5-methoxykynuramine and its precursor melatonin in water hyacinth: Importance for phyto-remediation. *FASEB J.* 21: 1724–1729.
Tettamanti, C., Cerabolini, B., Gerola, P., Conti, A. 2000. Melatonin identification in medicinal plants. *Acta Phytother.* 3: 137–144.
Wehr, T.A. 1997. Melatonin and seasonal rhythms. *J. Biol. Rhythms* 12: 518–527.
Wong-ekkabut, J., Xu, Z., Triampo, W., Tang, I-M., Tieleman, D. P., Monticelli, L. 2007. Effect of lipid peroxidation on the properties of lipid bilayers: A molecular dynamics study. *Biophys. J.* 93: 4225–4236.
Yadav, S.K., Haldar, C. 2013. Reciprocal interaction between melatonin receptors (Mel(1a), Mel(1b), and Mel(1c)) and androgen receptor (AR) expression in immune regulation of a seasonally breeding bird, *Perdicula asiatica*: Role of photoperiod. *J. Photochem. Photobiol. B* 122: 52–60.

Index

A

D

E

F

G

H

I

K

L

M

Q

R

S

T

U

V

W

X

Y

Z